18th International Workshop on Advanced Computing and Analysis Techniques in Physics Research (ACAT 2017)

AI Reloaded

Journal of Physics: Conference Series
Volume 1085

Seattle, Washington, USA
21 – 25 August 2017

ISBN: 978-1-5108-7371-1
ISSN: 1742-6588

Printed from e-media with permission by:

Curran Associates, Inc.
57 Morehouse Lane
Red Hook, NY 12571

Some format issues inherent in the e-media version may also appear in this print version.

Printed by Curran Associates, Inc. (2018)

For permission requests, please contact the Institute of Physics
at the address below.

Institute of Physics
Dirac House, Temple Back
Bristol BS1 6BE UK

Phone: 44 1 17 929 7481
Fax: 44 1 17 920 0979

techtracking@iop.org

Additional copies of this publication are available from:

Curran Associates, Inc.
57 Morehouse Lane
Red Hook, NY 12571 USA
Phone: 845-758-0400
Fax: 845-758-2633
Email: curran@proceedings.com
Web: www.proceedings.com

TABLE OF CONTENTS

PLENARY

TRACK 1: COMPUTING TECHNOLOGY FOR PHYSICS RESEARCH, SEPTEMBER 2018

TRACK 2: DATA ANALYSIS – ALGORITHMS AND TOOLS

TRACK 3: COMPUTATIONS IN THEORETICAL PHYSICS: TECHNIQUES AND METHODS

IOP Conf. Series: Journal of Physics: Conf. Series **1085** (2018) 011001 doi:10.1088/1742-6596/1085/1/011001

Abstract

This volume of the IOP Conference Series is dedicated to scientific contributions presented at the 18th International Workshop on Advanced Computing and Analysis Techniques in Physics Research (ACAT 2017). Each year ACAT has a motto, and this year the motto was "AI Reloaded". The conference took place on the week of August 21-25, 2017, at University of Washington in Seattle, Washington, USA.

The 18th edition of ACAT brought people together from the many different fields of the physics and computing research. Topics ranging from automated loop-integral calculations, the cutting edge of machine learning in particle physics and applications of machine learning to infrastructure, to novel tools used to aid in these accomplishments were presented. Side meetings and collaborations were also hosted looking forward to computing in particle physics into the late 2020's.

This edition of the workshop brought together more than 190 participants from all areas of the world. There were 23 invited speakers covering important topics on advanced computing and analysis techniques. There were two panel discussions, one on machine learning and the other on diversity. Finally, there was also a viewing of the 2017 solar eclipse. There were 95 parallel session presentations and more than 60 posters. The posters and parallel sessions were presented in three tracks: Computing Technology for Physics Research, Data Analysis - Algorithms and Tools, and Computations in Theoretical Physics: Techniques and Methods. ACAT 2017 thanks nVidia and the University of Washington for its support.

Special thanks go to the track liaisons: Niko Neufeld, Sergei Gleyzer, and Ayres Freitas, along with the track co-conveners: Graeme Stewart, Mira Girone, Shih-Chieh Hsu, Gregory Golovanov, Toby Burnett, Stephen Jones, and Fukuko Yuasa. Their work on the scientific program and the publication preparation was invaluable. The local organizers would also like to thank ACAT's International Advisory Committee (IAC). The ACAT IAC would like to thank all referees for their work on making sure the contributions are published in the proceedings. Many thanks for Federico Carminati for leading the Scientific Committee, and the conference chair Denis Perret-Gallix (LAPP/IN2P3/CNRS-KEK), who started the ACAT series in 1990. Just prior to submission of these proceedings Denis passed away (1949-2018), and the ACAT community will sorely miss him. The next edition of ACAT will be dedicated in his memory.

Full information for ACAT 2017 can be found at https://indico.cern.ch/event/567550.

IOP Conf. Series: Journal of Physics: Conf. Series **1085** (2018) 011002 doi:10.1088/1742-6596/1085/1/011002

Peer review statement

All papers published in this volume of *Journal of Physics: Conference Series* have been peer reviewed through processes administered by the proceedings Editors. Reviews were conducted by expert referees to the professional and scientific standards expected of a proceedings journal published by IOP Publishing.

IOP Conf. Series: Journal of Physics: Conf. Series **1085** (2018) 022001 doi:10.1088/1742-6596/1085/2/022001

New developments in FORM

Takahiro Ueda

Nikhef Theory Group, Science Park 105, 1098 XG Amsterdam, The Netherlands

E-mail: `tueda@nikhef.nl`

Abstract. Symbolic computation is an indispensable tool for theoretical particle physics, especially in the context of perturbative quantum field theory. In this work, I will review FORM, one of the computer algebra systems widely used in higher-order calculations, its design principles and advantages. The newly released version 4.2 will also be discussed.

1. Introduction

Recent rapid progress in higher-order calculations in theoretical particle physics would not have been possible without automatization based on symbolic computation software. Indeed, some computer algebra systems arose out of real needs in cumbersome theoretical calculations in perturbative quantum field theory; for example, `SCHOONSHIP` [1, 2], `REDUCE` [3, 4], FORM [5, 6, 7] and `GiNaC` [8].

FORM is often referred as the (portable) successor to legendary[1] `SCHOONSHIP`, which was programmed in assembly language and super fast compared to other programs at that time. Many large-scaled cutting-edge perturbative computations have been performed with FORM since its birth in the late '80s, because of its high performance in processing gigantic expressions, required in this field; nowadays in extreme cases the intermediate expression size can reach a few terabytes.

This work aims to review FORM and give an overview of its design principles. The recently released new version FORM 4.2 will also be briefly discussed.

2. Design principles in FORM

Some computer algebra systems store a mathematical expression in a tree data structure. For example, an expression $(a+b)(c+d)$ can be represented as

$$(a+b)(c+d) \quad \Longrightarrow$$

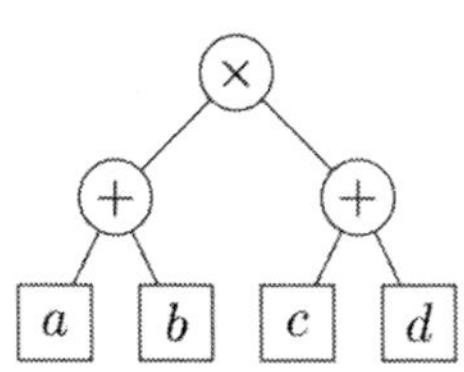

,

where the operators '$\times$' and '$+$' are expressed by nodes in the tree and have operands as their children. This is a natural design decision to represent an expression, which has an origin from

[1] Personally I never have had and perhaps never will have the chance to run `SCHOONSHIP` or access to any legacy computer on which `SCHOONSHIP` could run.

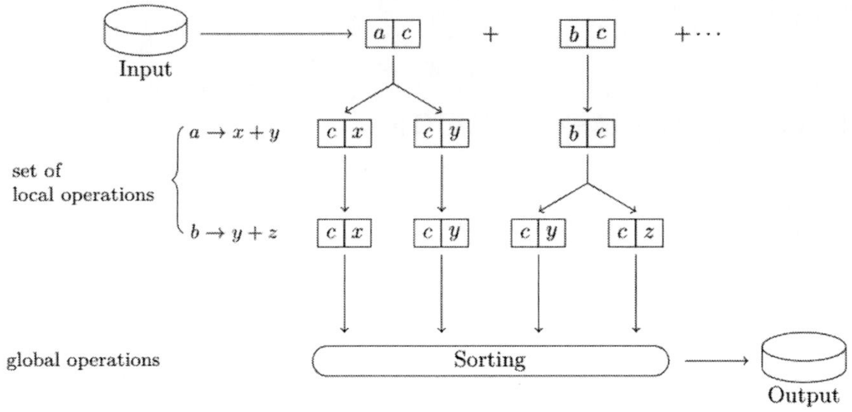

Figure 1. Processing a mathematical expression in FORM.

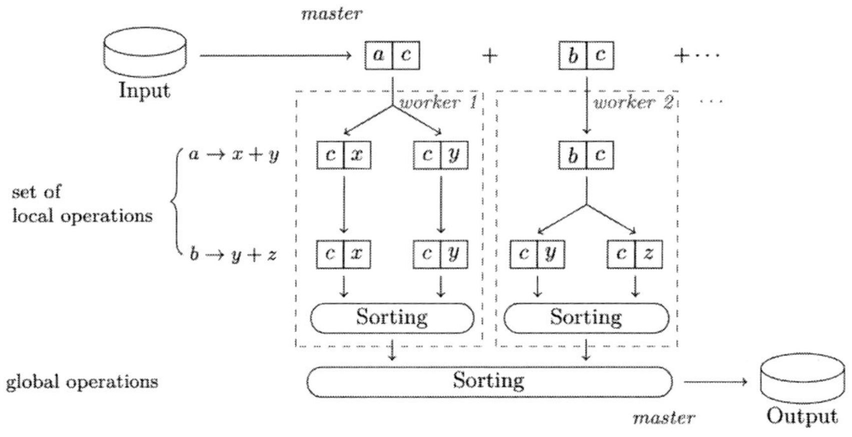

Figure 2. Parallelization of FORM.

LISP-influenced systems. Such a tree structure can keep any input expression in the exact form that the user intended. Symbolic manipulations like simplification may utilize the recursive tree structure and would be elegantly implemented in computer code. However, tree structures without any fixed size commonly require many dynamic memory allocations from the heap. When an expression tree becomes extremely large and does not fit in physical memory, the operating system typically assigns virtual memory on a disk, leading to intolerably slow random access to the disk in every symbolic manipulation.

In FORM, any input expressions are always expanded unless the user intervenes, and then an expression is stored as a sequence of sorted terms:

$$(a+b)(c+d) = ac + ad + bc + bd \quad \Longrightarrow \quad \boxed{a\,|\,c}\;\boxed{a\,|\,d}\;\boxed{b\,|\,c}\;\boxed{b\,|\,d}\,.$$

The expansion may generate many terms but each term tends to be rather small, expected to easily fit in physical memory or even in CPU cache. Indeed, FORM restricts the size of each term after the expansion to be smaller than some fixed size given by a setup parameter, which enables use of fast memory pools and eliminates many dynamic memory allocations at run-time.

Most of the operations on expressions in FORM are restricted to be 'local' in such a way that they act on each individual term and do not need any reference to other parts of the expression.

Listing 1. Differentiating polynomials.

```
1 Symbol x,n;
2 Local F1 = 1 + x;
3 Local F2 = 2 - 3*x + 5*x^2;

4 id x^n? = n * x^(n-1);

5 Print;
6 .end
```

FORM applies a set of local operations to terms in an expression, term by term, which may generates many terms in general. After that, non-local (or 'global') operations like sorting are performed. This is illustrated in Figure 1. The sorting is based on the merge sort algorithm and only requires sequential access to disks, instead of random access. Furthermore, after the sorting, terms with similar subterms tend to come together; therefore, the incremental encoding works well to compress the total data size, which speeds up the reading and writing. Such an efficient design makes FORM be a suitable choice for handling gigantic expressions exceeding physical memory, in comparison with many of the other computer algebra systems.

Another benefit of the restriction on local operations is that it is relatively straightforward to parallelize processing terms in the master-worker model. The master distributes terms over workers, each worker processes the distributed terms and partially sorts the result, and then the master collects all the results and performs the final sorting (Figure 2). This idea was implemented as ParFORM [9] using MPI and TFORM [10] based on POSIX Threads.

3. Programming in FORM

A good toolkit must provide a good user interface that is suitable and reasonable for the user's purpose. In the FORM programming language, one of the basic operations is term rewriting. The user writes a set of substitutions that will be applied on every term in an imperative programming style, by which one can build algorithms for, e.g., differentiations, integrations and function transformations.

Listing 1 shows a complete example. If FORM is installed on a computer, one can run this example by saving it as a text file, e.g., `example.frm` and executing `form example.frm` from a terminal window. The FORM language is strongly-typed, thus one has to declare symbols or other types of objects before their use. Line 1 declares symbols `x` and `n`. All statements in FORM end with a semicolon '`;`'. Although user-defined objects are case-sensitive, all built-in objects and command names are case-insensitive. Line 2 and 3 define (local) expressions `F1` and `F2`, which will be manipulated in the following code. Line 4 is an executable statement to apply a replacement, which will be acted on every term in the current active expressions, `F1` and `F2` in this case. The command `id` is an abbreviation of `identify`. The question mark '`?`' indicates `n?` is a wildcard, and this statement replaces x^n by nx^{n-1} for any $n \in \mathbb{Z}$, implementing differentiation of polynomials with respect to x. Line 5 instructs FORM to print the expressions at the end of the current *module* (see below). Line 6 is a special instruction to indicate the end of the program.

Note that an `id` statement is allowed to generate more than one term, like `id x = y + z;`, but the left-hand side of an `id` statement cannot have more than one term. For example, the statement `id y + z = x;` is forbidden because it is not a local operation and FORM just prints a compilation error. When an operation generates more than one term, the next operation is applied to each term in a depth-first manner.

IOP Conf. Series: Journal of Physics: Conf. Series **1085** (2018) 022001 doi:10.1088/1742-6596/1085/2/022001

Listing 2. Defining a procedure for differentiating polynomials and its call.

```
#procedure derivative(x,m)
  #do i=1,`m'
    id `x'^n? = n * `x'^(n-1);
  #enddo
#endprocedure

#call derivative(y,2)
```

Listing 3. A FORM program consisting of two modules.

```
Symbol a,b,x;
Local F = a*x + x^2;

id x = a + b;

Print;
.sort

if (count(b,1) == 1);   * select terms with b^1
  multiply a/b;
endif;

Print;
.end
```

A remarkable feature of the FORM language is that it embeds a very powerful preprocessor, which can modify the input for the compiler. A preprocessor directive line in FORM starts with a hash mark '#', like C-preprocessor directives. Indeed, complicated FORM programs virtually cannot be constructed without using the preprocessor directives. For example, the `#procedure` and `#call` directives are used for procedural programming, in which one makes use of reusable subroutines to construct programs or even bigger subroutines.

Listing 2 is an example of a procedure for differentiating polynomials.[2] The defined procedure can be *called* as `#call derivative(y,2)`, which leads to two sequential `id` statements that give the second derivative with respect to y. Two parameters are passed as preprocessor variables, `x` and `m`, and expanded by backquote/quote pairs inside the procedure. One should realize that this call is done as a text substitution by the preprocessor, hence at compile-time, not at run-time.

A FORM program consists of one or more basic executing blocks that are referred to as modules. A module ends with a special instruction starting with a period '`.`' such as `.sort` and `.end`. When the FORM compiler reaches such an instruction and recognizes the termination of a module, compiled code in the module is executed. All active expressions are processed and then the results are sorted. For example, the program in Listing 3 consists of two modules. First, FORM compiles the first module and executes the code in it. The result for expression `F` is sorted and printed. After that, FORM compiles the second module and executes the code. Note that in this example removing the `.sort` instruction does not change the final result. Inserting a

[2] This procedure is not perfect. It could be optimized such that multiple `id` statements would be combined into one at compile-time, and actually there is a buggy case; `#call derivative(n,1)` gives a result that most likely the user does not want.

Listing 4. A code snippet with compile-time optimization based on run-time information obtained in a previous module.

```
#define HaveX "0"

* Check whether any expressions have x.
if (count(x,1));
  redefine HaveX "1";
endif;

.sort

#if `HaveX'
* This part is skipped at compile-time
* if there is no x in all expressions.
#endif
```

`.sort` instruction may give a speed-up of the program if the sorting gives a much less number of terms by merging or cancellation among the terms, while sorting itself has some overheads. The user has full control of whether or not expressions are sorted at some point by inserting or not inserting a `.sort` instruction.

The fact that a module is compiled after the execution of previous modules means that one can optimize code by the preprocessor at compile-time with run-time information from the previous modules. Listing 4 is an example of compile-time optimization, where a code block is conditionally skipped depending on run-time information obtained in the previous module. Such compile-time optimization techniques make sense when one has to manipulate million of terms at run-time. Tricky and sometimes mystifying metaprogramming with the preprocessor is after all justified by productivity and improvement of the performance of the code.

4. Form 4.2

Currently, the latest stable version of Form is the version 4.2.0 and was released in July 2017 [11]. In comparison with the previous version 4.1, this release contains more than 50 bug fixes and more than 20 new features, including

- generating all matches by `id all`,
- automatic expansion of rational polynomials,
- dictionaries for textual manipulation of output,
- temporary storage for spectator terms,
- stochastic local search for polynomial optimization,
- zero-dimensional sparse tables.

Many of the introductions of new features and improvements in Form have been inspired by use in actual research projects. For this version, the Forcer [12] program and implementing R*-operation on tensor integrals [13] were the main driving force of the development.

Here I would like to focus on a problem that was made solvable in Form 4.2: graph isomorphism by pattern matching on the fly. The standard way of perturbative calculations in quantum field theory begins with generating Feynman diagrams. Feynman diagram generators on the market do the job, but their momentum assignment to diagrams may be different from what the succeeding programs expect. Therefore, one needs to recognize the topology of a

IOP Conf. Series: Journal of Physics: Conf. Series **1085** (2018) 022001 doi:10.1088/1742-6596/1085/2/022001

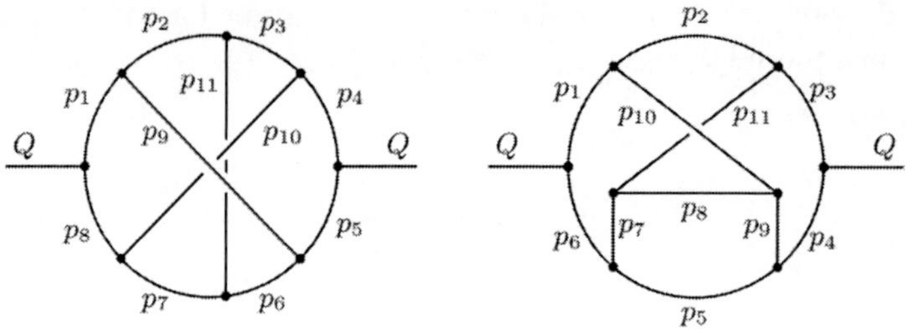

Figure 3. Two isomorphic graphs.

Listing 5. Finding isomorphic mappings.

```
1  CFunction map;              * for isomorphic mapping
2  CFunction vx(symmetric);    * for vertices
3  Vector Q,p1,...,p11,q1,...,q11;

4  Local F1 = vx(Q,p1,p8)*vx(p1,p2,p9)*vx(p2,p3,p11)*
5             vx(p3,p4,p10)*vx(Q,p4,p5)*vx(p5,p6,p9)*
6             vx(p6,p7,p11)*vx(p7,p8,p10);

7  Local F2 = vx(Q,p1,p6)*vx(p1,p2,p10)*vx(p2,p3,p11)*
8             vx(Q,p3,p4)*vx(p4,p5,p9)*vx(p5,p6,p7)*
9             vx(p7,p8,p11)*vx(p8,p9,p10);

10 id vx(Q,q1?,q8?)*vx(q1?,q2?,q9?)*vx(q2?,q3?,q11?)*
11    vx(q3?,q4?,q10?)*vx(Q,q4?,q5?)*vx(q5?,q6?,q9?)*
12    vx(q6?,q7?,q11?)*vx(q7?,q8?,q10?)
13    = map(q1,...,q11);

14 Print;
15 .end
```

generated diagram and reassign the momenta as in the notation of the succeeding programs. More problems related to Feynman diagrams and their solutions are found in [14].

For simplicity, we consider a graph isomorphism problem of *undirected* graphs.[3] The question is stated as follows: Are two graphs in Figure 3 isomorphic after relabeling the edges? If so, what is the isomorphic mapping to obtain one of them from the other?

This problem can be solved in FORM as in Listing 5. Two topologies are expressed as products of a symmetric function, `vx`, whose arguments are vectors representing edges connected to each vertex, and stored in expressions `F1` and `F2`. Lines 10 to 13 perform a pattern matching for the isomorphic test. The pattern is constructed from expression `F1` by replacing the internal edges `p1`, ..., `p11` with wildcards `q1?`, ..., `q11?`. If the pattern matches to `F2`, then the two graphs are isomorphic and we find the isomorphic mapping as a `map` function in the result.[4] One can also obtain all automorphic mappings via generating all matches by `id all` instead of `id`.

[3] The way described here can be extended to the case of *directed* graphs, namely, products of `vx` with *minus vectors* representing momentum directions like `vx(Q,p1,-p8)`. However, the cost of pattern matching with minus vectors is rather high and in practice it is much faster to decompose the problem into pattern matching without directions and then fixing the directions.

[4] In general, the pattern match also occurs when `F2` contains `F1` as its subgraph. Such a case can be easily excluded by checking graph invariants like the number of vertices.

IOP Conf. Series: Journal of Physics: Conf. Series **1085** (2018) 022001 doi:10.1088/1742-6596/1085/2/022001

Listing 6. Generation of a pattern at compile-time.

```
1 #$graph1 = F1;
2 #opendictionary wild
3   #do i=1,11
4     #add p`i': "q`i'?"
5   #enddo
6 #closedictionary
7 #usedictionary wild($)
8   id `$graph1' = map(q1,...,q11);
9 #closedictionary
```

Listing 7. Generation of a pattern at run-time.

```
$graph1 = F1;
id $graph1 * replace_(<p1,q1?>,...,<p11,q11?>)
  = map(q1,...,q11);
```

Listing 5 works for this example with the specific topology, but is inflexible from the point of view that the explicit pattern is directly written down in the code and it does not work with arbitrary topologies. A more preferable way is to generate a pattern from one of the given expressions on the fly. Two possible solutions are shown in the following. The both uses a *$-variable*, which stores a small expression in physical memory and is accessible both at compile-time and run-time.

The first solution uses dictionaries, a new feature introduced in the version 4.2. Dictionaries store textual mappings that are applied to the output, and originally intended for translation of the output into, for example, the LaTeX format like `mu1` to `\mu_1`. They can be also applied to text expansion of $-variables in the preprocessor, and hence lead to the code in Listing 6. Line 1 stores expression `F1` into a $-variable and line 8 expands it in the preprocessor, but with text replacements defined at line 4. As the result, the same pattern to that in Listing 5 is obtained at compile-time.

Another solution, shown in Listing 7, constructs the pattern at run-time. It makes use of the fact that the left-hand side of an `id` statement can be manipulated at run-time to a certain extent. Of course this causes some overheads at run-time, thus the compile-time solution of Listing 6 is preferable if possible, for example, in the case that `F1` is given in the previous module and one needs to perform the isomorphic test for many graphs with `F1`.

5. Summary

An overview of FORM was presented from viewpoints of its design principles and how one can write his or her program in the FORM language. The latest version 4.2 and its application to isomorphic test problems were also discussed. Hopefully, users can write more creative and even smarter programs with the newly added features. The source code of FORM is hosted on GitHub and available from `https://github.com/vermaseren/form`.

Acknowledgments

I am grateful to my collaborators in the FORM project and its physics applications. This work is supported by the ERC Advanced Grant no. 320651, "HEPGAME".

IOP Conf. Series: Journal of Physics: Conf. Series **1085** (2018) 022001 doi:10.1088/1742-6596/1085/2/022001

References

[1] Veltman M 1967 *CERN Preprint*
[2] Strubbe H 1974 *Comput. Phys. Commun.* **8** 1–30
[3] Hearn A C 1968 *Proceedings for the ACM Symposium on Interactive Systems for Experimental Applied Mathematics* pp 79–90
[4] `http://reduce-algebra.sourceforge.net/`
[5] Vermaseren J A M 1991 *Symbolic Manipulation with FORM, Tutorial and Reference Manual* (CAN)
[6] Vermaseren J A M 2000 *Preprint* math-ph/0010025
[7] Kuipers J, Ueda T, Vermaseren J A M and Vollinga J 2013 *Comput. Phys. Commun.* **184** 1453–67 (*Preprint* 1203.6543)
[8] Bauer C W, Frink A and Kreckel R 2000 *J. Symb. Comput.* **33** 1 (*Preprint* cs/0004015)
[9] Tentyukov M, Fliegner D, Frank M, Onischenko A, Retey A, Staudenmaier H M and Vermaseren J A M 2004 *Proceedings for 7th International Workshop on Computer Algebra in Scientific Computing (CASC 2004) St. Petersburg, Russia, July 12-19, 2004* (*Preprint* cs/0407066)
[10] Tentyukov M and Vermaseren J A M 2010 *Comput. Phys. Commun.* **181** 1419–27 (*Preprint* hep-ph/0702279)
[11] Ruijl B, Ueda T and Vermaseren J 2017 *Preprint* 1707.06453
[12] Ruijl B, Ueda T and Vermaseren J A M 2017 *Preprint* 1704.06650
[13] Herzog F and Ruijl B 2017 *JHEP* **05** 037 (*Preprint* 1703.03776)
[14] Herzog F, Ruijl B, Ueda T, Vermaseren J A M and Vogt A 2016 *PoS* **LL2016** 073 (*Preprint* 1608.01834)

IOP Conf. Series: Journal of Physics: Conf. Series **1085** (2018) 022002 doi:10.1088/1742-6596/1085/2/022002

Deep Learning usage by Large Experiments

Benjamin Nachman

Physics Division, Lawrence Berkeley National Laboratory, Berkeley, CA 94704, USA

E-mail: bpnachman@lbl.gov

Abstract. The modern machine learning revolution holds great promise for high energy physics. Enormous detectors have been designed to search for rare signals in order to expose the smallest distance scales in nature. Modern tools like deep neural networks will be required to fully explore the sub-nuclear universe. Using the radiation pattern inside high energy jets as an example, we show how neural networks are being used in new ways to produce a class of techniques that were not previously possible.

1. Introduction

Advances in machine learning have led to a "deep learning revolution" that has resulted in many tools and techniques for quickly and reliably training sophisticated neural networks with many layers. Complex neural networks can learn directly from low-level features that reside in the high-dimensional datasets common in large experiments. New ideas for deep neural networks (DNN) are being developed at a fast pace, but Ref. [1–3] are rather comprehensive introductions/reviews.

In many applications to large experiments like ATLAS and CMS at the Large Hadron Collider (LHC), DNNs are 'simply' replacing shallower learning algorithms. For example, the ATLAS experiment has recently used deep neural networks to combine a small number of high-level features for highly Lorentz boosted boson, top quark, and bottom quark tagging [5–7]. Unsurprisingly, performance is essentially the same as training shallower classifiers with the same inputs. Deep neural networks are most useful when the relevant dimensionality of the inputs is very large, $\sim \mathcal{O}(100)$ or bigger.

Two areas from large experiments in the high energy physics community making use of high dimensional data with deep neural networks are neutrino physics and jet physics. Neutrino physics detectors require a large volume to compensate for the small interaction rate. The data from liquid argon time projection chambers (e.g. NOνA, DUNE [8], MicroBooNE [9]) or vats of liquid scintillator surrounded by photomultiplier tubes (e.g. JUNO [10]) can be represented as a series of images with translationally invariant features. This makes convolutional neural networks (CNNs) a natural choice for this setting and there are now many proposed and applied techniques of CNN's to neutrino physics. Jets are collimates sprays of particles resulting from high energy quark and gluon production. The radiation pattern inside jets contains a wealth of information about the particle that initiated them. There are many proposals for using deep learning with this radiation pattern; see Ref. [4] for a recent review. The remainder of this talk focuses on representing jets as an image [20][1] to illustrate the application of deep neural networks for classification [21–26], regression [28], and generation [29–31].

2. The Jet Image and Special Relativity

Jet images are a two-dimensional fixed representation of the radiation pattern inside a jet. Most applications with jet images use a gray-scale digital image where the pixel intensity corresponds to the momentum deposited in a particular region of the detector. Figure 1 illustrates how the energy pattern in the cylindrical coordinates of a typical collider physics detector are unrolled to form a two-dimensional jet image. Unlike natural images of scenery or faces, jet images do not have any clear sharp features, do not have smooth pixel intensities, and are mostly empty (low occupancy). This is a challenge for applying standard techniques out-of-the-box.

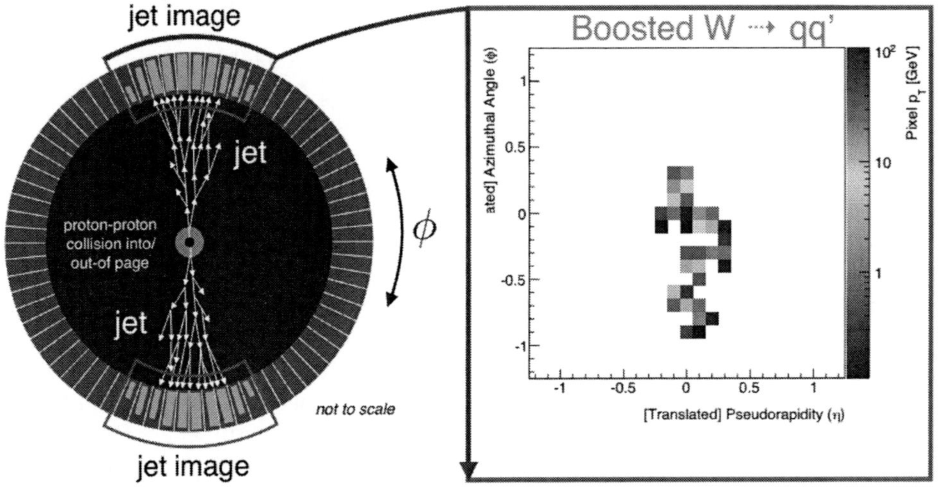

Figure 1. Left: a transverse view of a particle collision where the incoming particles are into/out of the page. Jets are collimated sprays of particles that are measured as localized high energy deposits. The cylindrical geometry is unrolled to form a jet image; Right: a typical jet image from a Lorentz boosted W boson decay. The pixel intensity is the momentum transverse to the incoming particles (p_T). Image reproduced from Ref. [30].

Despite the differences with natural images, jet images are a powerful way to represent the radiation pattern inside a jet because one can directly visualize the physical information that is available for learning. For example, Fig. 2 shows two jet images where a clear physical story can describe their differences and builds a foundation for applying sophisticated techniques for exploiting the differences between the images.

The first step in any image-processing analysis is image pre-processing. The symmetries of spacetime are known symmetries of jet images. Removing them during pre-processing can help machines to learn faster and smarter, but must be used with caution. For example, Fig. 3 shows how jet image rotations (which are not proper rotations in space) can distort the information content of a jet image. Other standard pre-processing steps like image normalization are known to distort the information inside a jet image [21]. Some pre-processing can nonetheless be useful and is an opportunity to inject domain knowledge.

[1] There are also many proposals for non-image-based techniques. See e.g. [11–19].

IOP Conf. Series: Journal of Physics: Conf. Series **1085** (2018) 022002 doi:10.1088/1742-6596/1085/2/022002

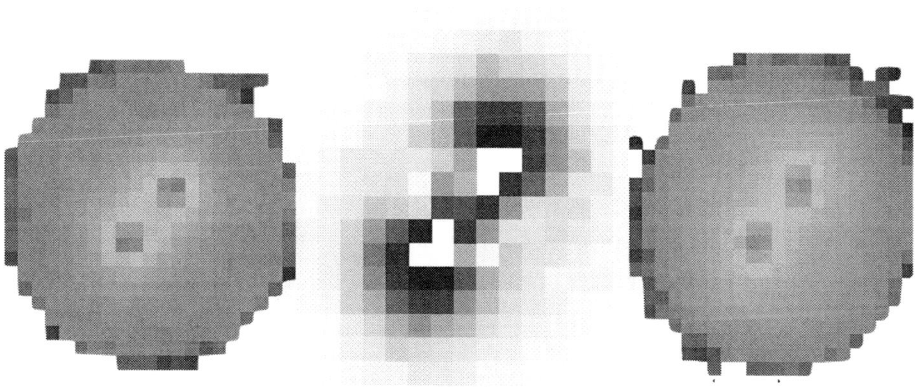

Figure 2. The average over many jet images from a boosted Higgs boson decay into two bottom quarks (left), a gluon [3]splitting into two bottom quarks (right), and the difference between the left and right images (center). The two energy lobes in the left and right images represent the 'subjets' formed from nearby quarks. The Higgs boson does not participate in the strong force and therefore the quarks it produces have equal and opposite strong-force-charge; they thus radiate like a dipole. In contrast, the gluon has net strong force charge and therefore radiates much more away from the two energy lobes.

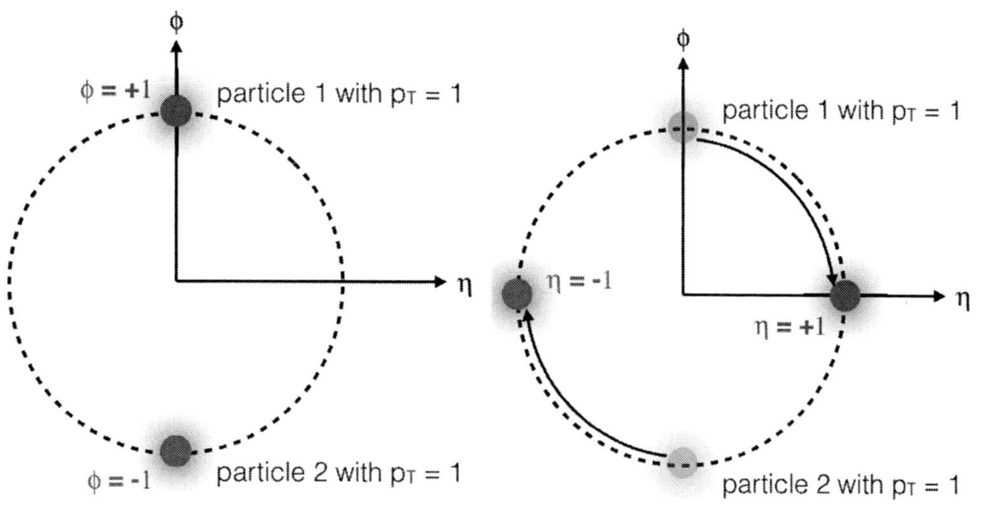

Figure 3. The jet mass of an image is derived from the four-vector sum of the pixel intensities, each treated as massless. The jet mass is a powerful feature for distinguishing interesting jets from generic jets. The images shown here have a simplified 'jet' with only two hit pixels. Rotating the image by $\pi/4$ does not preserve the image mass if the pixel intensities are left unchanged. This introduces a rotation-dependent smearing to the jet mass that can reduce the discrimination potential of a classifier.

[3] Technically, this is implemented as a Higgs boson decaying into two bottom quarks where the color strings are 'by-hand' distorted to make this Higgs act like a gluon in its radiation pattern. In this way, the Higgs kinematics are the same between the two samples. Due to color-connections between the gluon-like Higgs and the beam remnants, there is a dependence on the rest of the event for the right plot but not the left plot.

IOP Conf. Series: Journal of Physics: Conf. Series **1085** (2018) 022002 doi:10.1088/1742-6596/1085/2/022002

3. Classification

The most basic machine learning task is differentiating between a small number of categories (classification). Machine learning algorithms can be trained to identify the particle that originated the jet by analyzing the radiation pattern in the jet image. Figure 4 shows a Receiver Operating Characteristic (ROC) for the binary classification task of distinguishing jets from Lorentz boosted W bosons and generic quarks/gluons. The classifier is trained directly on the pixel intensities without any other guidance (aside from pre-processing). A reoccurring theme is that the machine learning algorithm can match or out-perform existing techniques. Understanding what the machine has learned beyond existing algorithms is a very important and active subject of research.

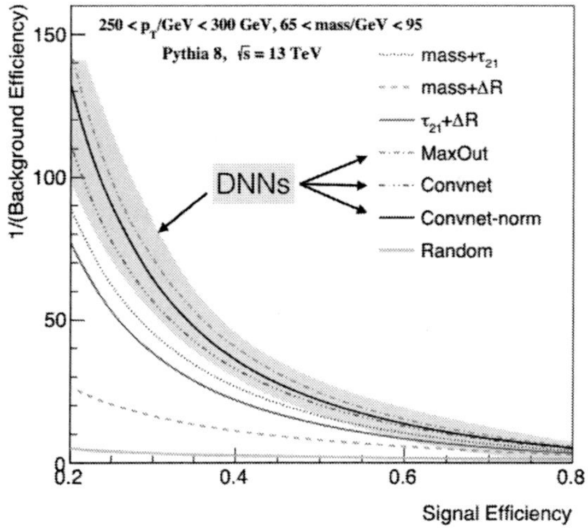

Figure 4. An ROC curve that shows the tradeoff between the probability to correctly identify a Lorentz boosted W boson jet and the probability to mis-classify a generic quark/gluon jet. Various deep neural network architectures are shown in the shaded region. The other observables are known (simple) features used in existing techniques for jet classification. Reproduced from Ref. [21].

One opportunity that is also a key challenge of deep neural networks is their ability to utilize subtle features within a jet. Traditional classification techniques in high energy physics use fully supervised learning where labeled examples are produced from simulation. However, subtle features are difficult to accurately model and therefore the classifier trained on simulation may be sub-optimal when applied on (unlabeled) data. There are now multiple proposals for training with data directly [32–34]. One proposal based on weak supervision is shown in Fig. 5.

4. Regression

A natural generalization of classification is to cases when there are (possibly infinitely) many classes. Jet image-based regression techniques have recently been applied for image de-noising.

IOP Conf. Series: Journal of Physics: Conf. Series **1085** (2018) 022002 doi:10.1088/1742-6596/1085/2/022002

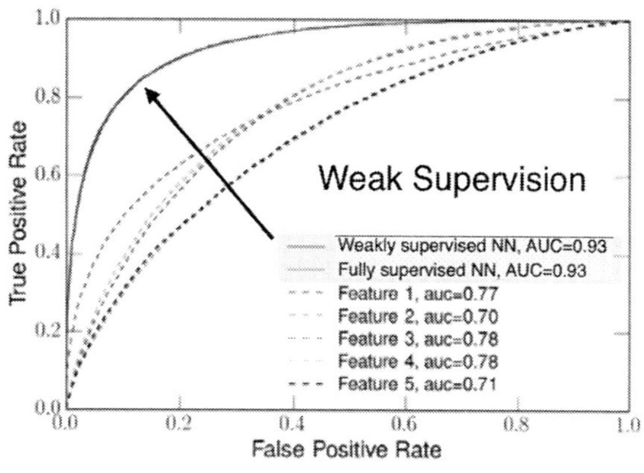

Figure 5. An ROC curve that shows how the performance of a traditional fully supervised learner with per-example labels can be matched with a technique that only knows the class proportions. Reproduced from Ref. [32].

At the LHC, many proton-proton collisions happen simultaneously, but only (at most) one is interesting. Using information from different detectors, a noise image can be predicted from the various measurements. The noise jet image can be subtracted from the original noisy image, after which traditional or DNN-based classification can proceed. This is illustrated in Fig. 6. Extensions, modifications, and variations of this idea are currently under study by the ATLAS and CMS co nsity and not just the nois

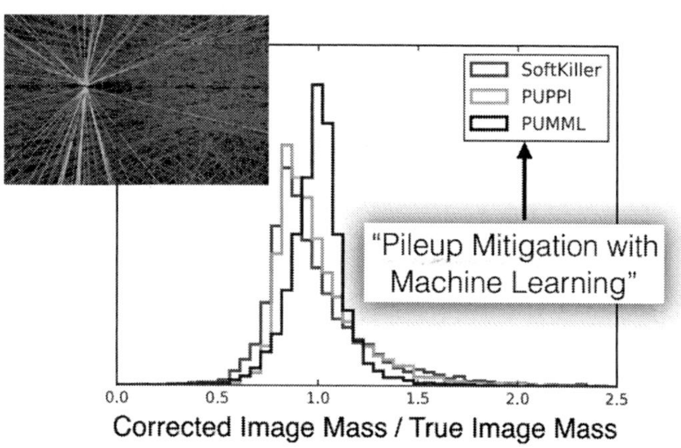

Figure 6. Top left: At the LHC, many collisions occur at the same time. Only one is of interest and the particles produced by the other collisions are a source of noise. Applying a jet-image-based CNN for de-noising ('PUMML') is very effective at correcting the jet mass back to the value without noise. In comparison, existing techniques ('Softkiller' and 'PUPPI') introduce a bias and have a worse resolution. Reproduced from Ref. [28].

IOP Conf. Series: Journal of Physics: Conf. Series **1085** (2018) 022002 doi:10.1088/1742-6596/1085/2/022002

5. Generation

Rounding out the applications of neural networks is generation, i.e. producing new images which can be used for subsequent classification or inference. One technique that has gained attention in high energy physics is the adversarial neural network (GAN) [35]. Neural networks are simply (compositions of) functions, so a neural network generator is a mapping between noise and structured images. In the GAN setup, this function from noise to structured images is complemented by a classifier network that tries to distinguish generated images from real images. The two networks are iteratively optimized and when the classifier is maximally confused, the generator will be effective at producing realistic looking images. High energy physics analysis relies heavily on physics-based simulation, but these simulators can be very slow (up to $\mathcal{O}$(min/event)). An exciting possibility is to use GANs to replace or augment physics-based simulations in order to speed them up. Figure 7 shows how the jet mass distribution can be learned by a GAN even though it was only shown examples of raw pixel intensities during training. One can naturally extend the jet image technology to multiple causally connected layers as is typical of a longitudinally segmented calorimeter. This extended geometry is shown in Fig. 9. Calorimeter simulation is a bottleneck for simulation at the LHC an speeding it up with a high-fidelity neural network simulator has the potential to save immense resources and empower a broad set of physics analyses.

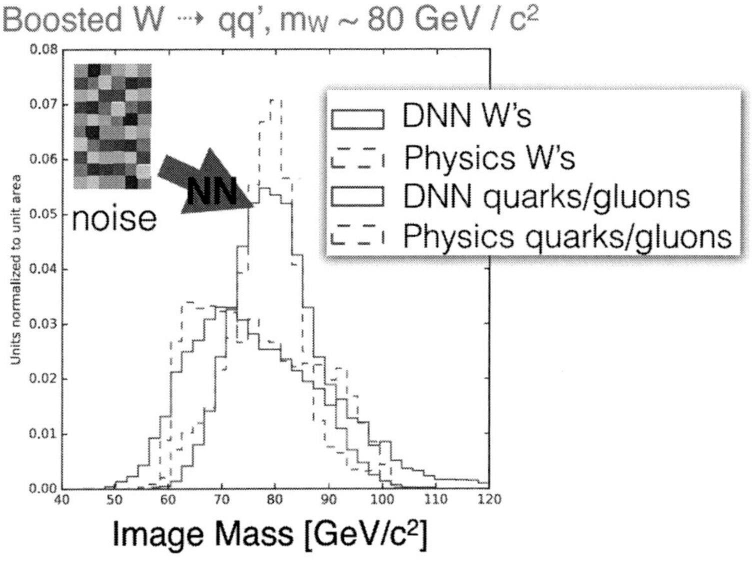

Figure 7. The jet mass distribution for signal (Lorentz boosted W bosons) and background (generic quarks and gluons) for a physics-based-simulator as well as a GAN trained on simulation images. Reproduced from Ref. [30].

6. Conclusions and Outlook

Modern machine learning tools have attracted widespread interest in the high energy physics community. DNN classification, regression, and generation are powerful tools to fully exploit the

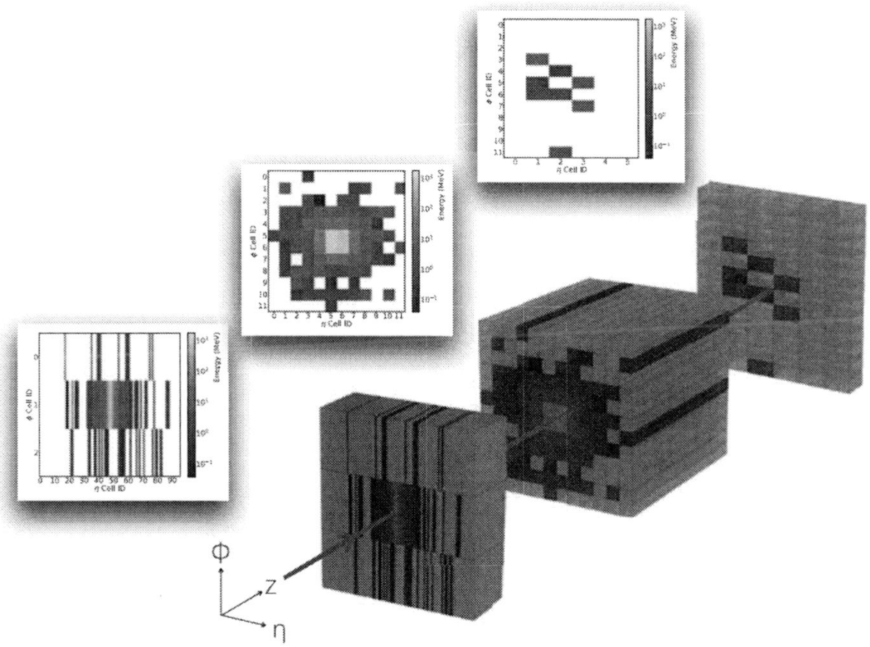

Figure 8. The setup of a three-layer calorimeter. The energy deposition in each layer resembles a grayscale jet image. Reproduced from Ref. [29].

physics program at the LHC and other large experiments with high-dimensional data. Some of these algorithms have already been studied using data or full detector simulations at the LHC. Figure 9 shows a CMS collaboration comparison of various techniques, including one developed by CMS in another context [18, 19]. For the tasks studied so far, it appears that any modern tool with access to the full radiation pattern has the same (excellent) performance. Taking these techniques to data and understanding exactly what the machines are learning will add robustness to the approaches and may even help us to learn something new about nature.

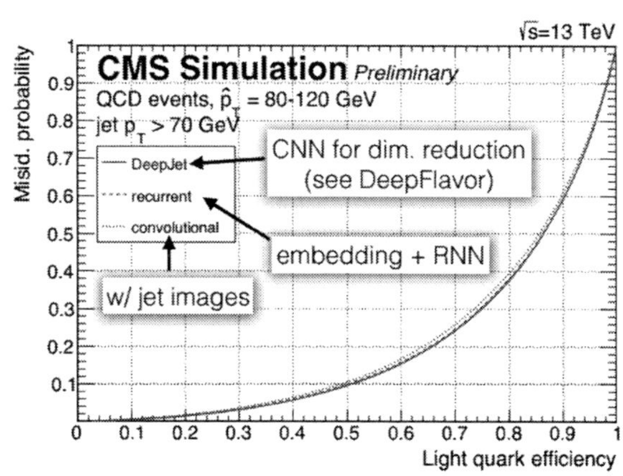

Figure 9. An ROC curve trading off the probability to correctly label a quark jet as such with the probability to mis-label a gluon-initiated jet as a quark jet. Three different DNN architecture types are compared. Reproduced from Ref. [27].

7. Acknowledgments
This work was supported by the U.S. Department of Energy, Office of Science under contract DE-AC02-05CH11231.

References
[1] I. Goodfellow, Y. Bengio and A. Courville,'Deep Learning," (2016) MIT Press.
[2] J. Schmidhuber,'Deep Learning in Neural Networks: An Overview," Neural Networks 61 (2015) 85.
[3] Y. LeCun, Y. Bengio and G. Hinton,'Deep Learning," Nature 521 (2015) 436.
[4] A. J. Larkoski, I. Moult and B. Nachman,"Jet Substructure at the Large Hadron Collider: A Review of Recent Advances in Theory and Machine Learning," [arXiv:1709.04464 [hep-ph]].
[5] ATLAS Collaboration,"Machine Learning Techniques for the Identification of Hadronically Decaying W Bosons and Top Quarks in ATLAS," [ATL-PHYS-PUB-2017-004].
[6] ATLAS Collaboration,"Performance of Top Quark and W Boson Tagging in Run 2 with ATLAS," [ATLAS-CONF-2017-064].
[7] ATLAS Collaboration,"Optimisation and performance studies of the ATLAS *b*-tagging algorithms for the 2017-18 LHC run ," [ATL-PHYS-PUB-2017-013].
[8] R. Acciarri *et al.* [MicroBooNE Collaboration], "Convolutional Neural Networks Applied to Neutrino Events in a Liquid Argon Time Projection Chamber," JINST **12**, no. 03, P03011 (2017). [arXiv:1611.05531 [physics.ins-det]].
[9] A. Aurisano *et al.*, "A Convolutional Neural Network Neutrino Event Classifier," JINST **11**, no. 09, P09001 (2016). [arXiv:1604.01444 [hep-ex]]. see also F. Psihas et al., these proceedings.
[10] L. Wang et al., "Muon reconstruction of JUNO experiment with convolutional neural networks", these proceedings.
[11] J. Pearkes, W. Fedorko, A. Lister and C. Gay, "Jet Constituents for Deep Neural Network Based Top Quark Tagging," [arXiv:1704.02124 [hep-ex]].
[12] G. Louppe, K. Cho, C. Becot and K. Cranmer, "QCD-Aware Recursive Neural Networks for Jet Physics," [arXiv:1702.00748 [hep-ph]].
[13] A. Butter, G. Kasieczka, T. Plehn and M. Russell, "Deep-learned Top Tagging using Lorentz Invariance and Nothing Else," [arXiv:1707.08966 [hep-ph]].
[14] K. Datta and A. J. Larkoski, "Novel Jet Observables from Machine Learning," [arXiv:1710.01305 [hep-ph]].
[15] K. Datta and A. Larkoski, "How Much Information is in a Jet?," JHEP **1706**, 073 (2017).
[16] J. A. Aguilar-Saavedra, J. H. Collins and R. K. Mishra, "A generic anti-QCD jet tagger," [arXiv:1709.01087 [hep-ph]].
[17] ATLAS Collaboration,"Identification of Jets Containing b-Hadrons with Recurrent Neural Networks at the ATLAS Experiment," [ATL-PHYS-PUB-2017-003]. see also M. Paganini et al., these proceedings.
[18] CMS Collaboration,"Heavy flavor identification at CMS with deep neural networks," [CMS-DP-2017-005].
[19] CMS Collaboration,"CMS Phase 1 heavy flavour identification performance and developments," [CMS-DP-2017-013].
[20] J. Cogan, M. Kagan, E. Strauss and A. Schwarztman, "Jet-Images: Computer Vision Inspired Techniques for Jet Tagging," JHEP **1502**, 118 (2015). [arXiv:1407.5675 [hep-ph]].
[21] L. de Oliveira, M. Kagan, L. Mackey, B. Nachman and A. Schwartzman, "Jet-images - deep learning edition," JHEP **1607**, 069 (2016). [arXiv:1511.05190 [hep-ph]].
[22] P. Baldi, K. Bauer, C. Eng, P. Sadowski and D. Whiteson, "Jet Substructure Classification in High-Energy Physics with Deep Neural Networks," Phys. Rev. D **93**, no. 9, 094034 (2016). [arXiv:1603.09349 [hep-ex]].
[23] G. Kasieczka, T. Plehn, M. Russell and T. Schell, "Deep-learning Top Taggers or The End of QCD?," JHEP **1705**, 006 (2017). [arXiv:1701.08784 [hep-ph]].
[24] J. Barnard, E. N. Dawe, M. J. Dolan and N. Rajcic, "Parton Shower Uncertainties in Jet Substructure Analyses with Deep Neural Networks," Phys. Rev. D **95**, no. 1, 014018 (2017). [arXiv:1609.00607 [hep-ph]].
[25] P. T. Komiske, E. M. Metodiev and M. D. Schwartz, "Deep learning in color: towards automated quark/gluon jet discrimination," JHEP **1701**, 110 (2017). [arXiv:1612.01551 [hep-ph]].
[26] ATLAS Collaboration,"Quark versus Gluon Jet Tagging Using Jet Images with the ATLAS Detector," [ATL-PHYS-PUB-2017-017].
[27] CMS Collaboration,"New Developments for Jet Substructure Reconstruction in CMS," [CMS-DP-2017-027]. see also M. Stoye et al., these proceedings.
[28] P. T. Komiske, E. M. Metodiev, B. Nachman and M. D. Schwartz, "Pileup Mitigation with Machine Learning (PUMML)," [arXiv:1707.08600 [hep-ph]]. See also E. M. Metodiev et al., these proceedings.
[29] M. Paganini, L. de Oliveira and B. Nachman,"CaloGAN: Simulating 3D High Energy Particle Showers in

Multi-Layer Electromagnetic Calorimeters with Generative Adversarial Networks," [arXiv:1705.02355 [hep-ex]]. see also L. de Oliveira et al., these proceedings.
[30] L. de Oliveira, M. Paganini and B. Nachman, "Learning Particle Physics by Example: Location-Aware Generative Adversarial Networks for Physics Synthesis," [arXiv:1701.05927 [stat.ML]].
[31] S. Vallecorsa, GeantV Collaboration, "A Machine Learning tool for fast simulation in GeantV," these proceedings.
[32] L. M. Dery, B. Nachman, F. Rubbo and A. Schwartzman, "Weakly Supervised Classification in High Energy Physics," JHEP **1705**, 145 (2017). [arXiv:1702.00414 [hep-ph]].
[33] E. M. Metodiev, B. Nachman and J. Thaler, "Classification without labels: Learning from mixed samples in high energy physics," [arXiv:1708.02949 [hep-ph]].
[34] G. Louppe, M. Kagan and K. Cranmer, "Learning to Pivot with Adversarial Networks," [arXiv:1611.01046 [stat.ME]].
[35] I. Goodfellow, et al. "Generative Adversarial Networks," [arXiv:1406.2661 [stat.ML]].

IOP Conf. Series: Journal of Physics: Conf. Series **1085** (2018) 022003 doi:10.1088/1742-6596/1085/2/022003

Machine learning challenges in theoretical HEP

Stefano Carrazza
Theoretical Physics Department, CERN, Geneva, Switzerland
E-mail: stefano.carrazza@cern.ch[1]

Abstract. In these proceedings we perform a brief review of machine learning (ML) applications in theoretical High Energy Physics (HEP-TH). We start the discussion by defining and then classifying machine learning tasks in theoretical HEP. We then discuss some of the most popular and recent published approaches with focus on a relevant case study topic: the determination of parton distribution functions (PDFs) and related tools. Finally, we provide an outlook about future applications and developments due to the synergy between ML and HEP-TH.

1. Introduction

Over the past several years machine learning (ML) has become one of the most popular and powerful sets of techniques and tools used for multidisciplinary scientific research. Such popularity has been continuously growing in the past years thanks to the increasing number of methodological developments, the availability of faster hardware with strong computational capabilities such as modern GPUs and coprocessors, and finally, the great interest and investment from the private sector.

The recent enthusiasm has led to a new underlying code development strategy where several tools based on ML are available as open source projects, some examples are TensorFlow [1], scikit-learn [2], Keras [3], Theano [4] among others. Easy access to these tools has simplified the integration of new modern techniques in many research fields, in particular for those where a large amount of data is available.

The dissemination of the innumerable applications and developments based on ML has given way to conferences such as ICML[2], NIPS[3] and ACAT, but also to the composition of specialized working groups, e.g the IML LHC working group at CERN[4] which promotes the integration of new techniques for specific requirements in experimental analysis.

In the next sections we focus the discussion on ML in theoretical High Energy Physics (HEP-TH). We start by defining the categories of applications observed in that field. Then we take as a case study the recent development achieved for the determination of parton distribution functions (PDFs), where several techniques from ML are employed. Finally, we conclude by listing the most plausible development directions for the integration of ML in HEP-TH.

[1] Preprint: CERN-TH-2017-212
[2] http://icml.cc
[3] http://nips.cc
[4] https://iml.web.cern.ch/

IOP Conf. Series: Journal of Physics: Conf. Series **1085** (2018) 022003 doi:10.1088/1742-6596/1085/2/022003

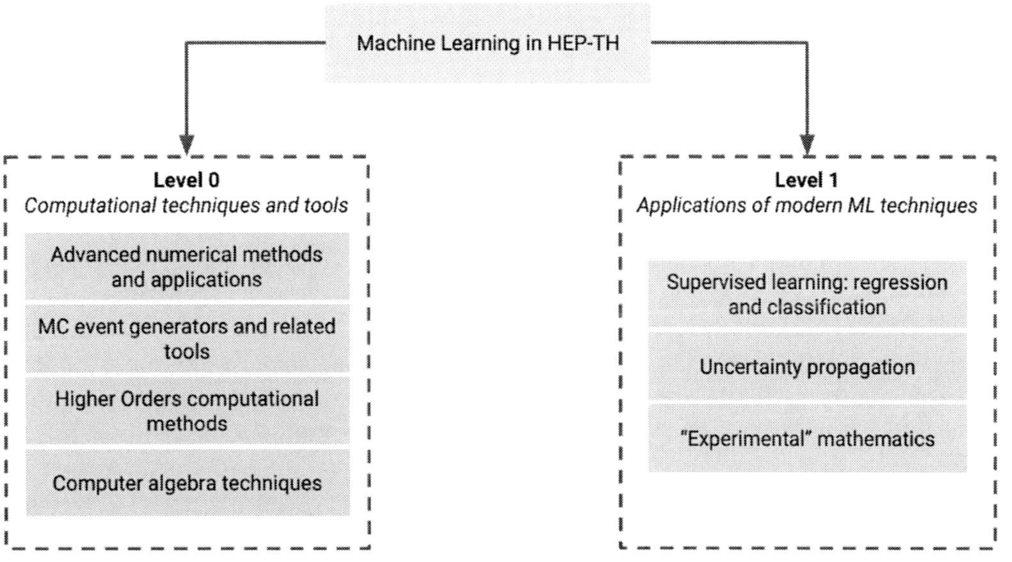

Figure 1. One possible graphical schematic representation of how machine learning is identified in HEP-TH applications.

2. Identifying machine learning applications in HEP-TH

While classifying ML applications in experimental physics may seem simple because the experimental analysis usually requires model regression, classification and noise filtering to extract useful information from large datasets. It becomes challenging when trying to determine applications from the point of view of theoretical physics.

When talking about ML techniques in theoretical High Energy Physics (HEP-TH) we may naively imagine that the usage of such tools is not ideal. In fact, we can argue that the theoretical physicists are trained to decode nature by constructing conjectures and building theoretical frameworks which remap the complexity of the measured observable in conceptual model rules. Therefore, this approach is in contrast with a model determination through a ML black box model, of which a physical interpretation is very difficult or nearly impossible to achieve.

However, nowadays such misleading interpretation is disappearing and machine learning is becoming a set of useful tools that helps achieving practical results in situations where the theory has a strong computational requirement or when it requires the determination of free parameters. So, before continuing our discussion, we think it is important to define and classify how ML is translated in HEP-TH. In Figure 1 we summarize one plausible graphical representation of the two branches of applications that we will discuss now in turn: *Level-0* and *Level-1*.

2.1. Level-0: computational techniques and tools

The first level contains machine learning in terms of computational techniques and tools. In this category we find the great majority of topics presented in the Track 3 section of the ACAT conference. These applications contain the most robust implications of computing in theoretical physics. Due to the specific nature of the problems addressed in HEP-TH these tools should be considered as part of ML applications because their development has contributed in exhaustive manner to the development of new techniques and methods in this field.

A non-exhaustive list of topics involved in *Level-0* are summarized in the points below:

- Advanced numerical methods and applications: some examples are algorithms for Monte Carlo and Quasi Monte Carlo integration, techniques for subtraction schemes and regularization of Feynman integrals, e.g. [5], resummation techniques [6].
- Monte Carlo event generators: in this category we have the methodological developments established by MC codes such as POWHEG [7], MadGraph_aMG5 [8], Pythia [9], Herwig [10], MCFM [11] and several others. The possibility to reuse events independently from the parton distribution functions thanks to reweighting and weight storage techniques such as APPLgrid [12] and FastNLO [13].
- Higher order computational methods: numerical techniques for N-loop integrals as OneLoop [14], QCDLoop [15], LoopTools [16]; parton level generators at NNLO, e.g. DYNNLO [17] and the more recent N3LO [18].
- Computer algebra techniques: some examples of algebra systems developed for the HEP-TH community as FORM [19] and QGRAF [20] for the translation of Feynman diagram rules into compact analytic expressions for its numerical evaluation.

2.2. Level-1: applications of ML modern techniques

This second level contains the ML applications in "sensu stricto" i.e. using ML modern techniques used in data sciences. This kind of application usually requires hybrid projects where experimental data and theory are included together. Nevertheless, during the last few years there has been a strong development of very successful tools based on these techniques.

Some examples for this category are:

- Supervised learning, such as regression and classification: parton distribution [21] and fragmentation functions [22] determination, Monte Carlo tunes [23], reweighting techniques, jet discrimination through deep convolutional neural networks [24].
- Techniques for uncertainty estimation and combination: in this category we find several tools from the PDF4LHC15 recommendation [25], and some recent methods to provide a reliable MC uncertainty for simulations, e.g. by modeling jet predictions [26].
- "Experimental" mathematics: applications based on machine learning optimization algorithms which may lead to the determination of multivariate densities [27, 28] and sampling for integral evaluation [29].

In the next section we describe the recent innovative achievements obtained by the parton distribution function community in HEP-TH using ML methods.

3. Case study: the proton structure determination

The Quantum Chromodynamics (QCD) theory describes the proton structure in terms of partons, e.g. quarks and gluons, but due to the non-perturbative regime of confinement we are unable to evaluate from QCD the momentum fraction of the proton carried by each parton. However, in order to avoid our lack of knowledge we introduce the concept of parton distribution functions (PDFs). These PDFs are then inferred from data of all relevant processes, together with the theoretical knowledge on how they affect the cross section which can be calculated approximately in perturbation theory.

3.1. The NNPDF approach

The NNPDF collaboration uses ML techniques to obtain a PDF determination. In contrast to other problems in ML, where finding an accurate and fast algorithm is enough, we are not only interested in the best PDF fit, but also in obtaining an uncertainty estimate of the PDF determination. The procedure we employ is described in full details in [30] and can be summarized as:

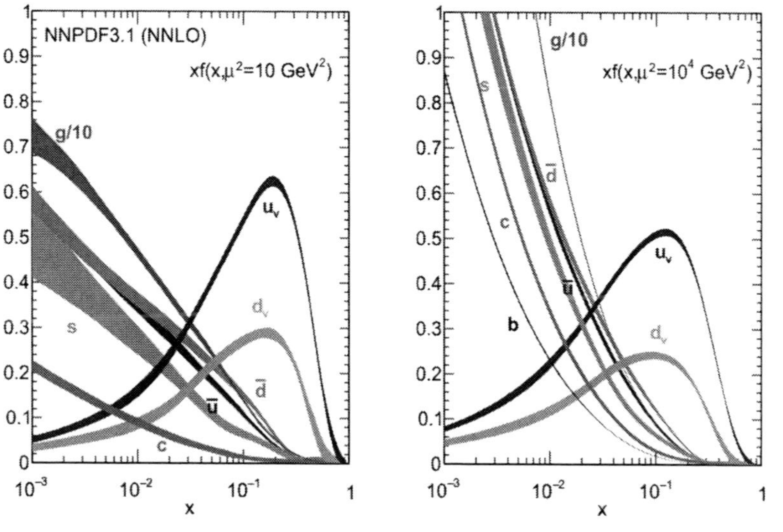

Figure 2. The NNPDF3.1 NNLO PDFs, evaluated at $\mu^2 = 10$ GeV^2 (left) and $\mu^2 = 10^4$ GeV^2 (right) from [21].

- Monte Carlo generation of artificial data. Experimental data, with central values, errors and their correlations are used to generate further artificial data, consistent with the covariance matrix provided by the experiment.
- Neural network fit to artificial data. A genetic algorithm is used to fit each artificial data set to a neural network representing the PDF (more details in [31]).
- Predictions are later obtained by computing statistical estimators (such as means, quantiles and standard deviations) over the set of neural networks. Figure 2 illustrates the current state of the art PDFs, NNPDF3.1 [21], obtained with the NNPDF framework.

The points above summarize the main difficulties in PDF fits. The first consists in the inclusion of multiple experimental data which introduces indirect constraints on the PDFs. The data included in such fits are based on several datasets measured during the past decades and based on different physical processes: deep-inelastic scattering, fixed target Drell-Yan and hardronic data. This data is obtained through diverse experimental techniques and statistical analyses, therefore yielding possible inconsistencies and tensions among themselves. In order to limit these effects the NNPDF methodology propagates the uncertainty of the experimental data on the PDFs by performing the Monte Carlo artificial replica generation based on the covariance matrix provided by each experiment.

The second most relevant problem consists in the choice of an unbiased functional form able to adapt and allow the propagation of data uncertainties into PDF errors while keeping under control physical requirements such as momentum sum rule conservation and positivity constraints. The NNPDF collaboration has successfully employed neural networks based on feed-forward multilayer perceptron architecture to model each single PDF flavor entering the fit procedure. The large number of parameters from these PDFs are then trained through a genetic optimization algorithm.

NNPDF has released several PDF sets of global unpolarized determinations in the last years[5],

[5] `http://nnpdf.mi.infn.it/`

IOP Conf. Series: Journal of Physics: Conf. Series **1085** (2018) 022003 doi:10.1088/1742-6596/1085/2/022003

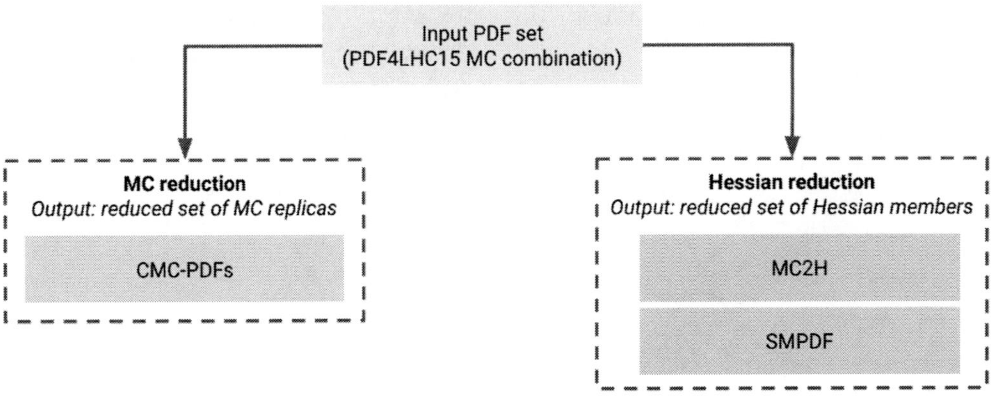

Figure 3. Graphical representation of the reduction algorithms used in the PDF4LHC15 recommendation [25].

together with the more recent polarized PDFs [32] and Fragmentation Functions [22]. One of the current most important tasks of the NNPDF collaboration is to improve the performance and quality of the optimization having in mind that in the next years the number of the new measurements from LHC will increase.

3.2. PDF4LHC15 tools for LHC Run II

Another issue concerning PDF determination consists in deciding which set of PDF should be used for the calculation of a PDF-dependent quantity. The question addressed here is how to obtain the best combined PDF uncertainty from individual PDF sets. This kind of issue is studied by the PDF4LHC working group which releases a recommendation document usually every three years. The latest one published in 2015 [25] aims to release combined PDF sets based on methods with clear statistical interpretation.

The PDF4LHC15 prescription starts by constructing a prior Monte Carlo combined set of PDFs derived from global determinations, namely MMHT14 [33], CT14 [34] and NNPDF3.0 [30]. These sets satisfy requirements such as: similar dataset, DGLAP solution and α_s. The prior construction is obtained by converting the Hessian sets to MC replicas following the procedure described in [35]. After that we apply algorithms to reduce the redundant information stored in the prior and deliver sets of PDFs with the same properties of the prior but smaller number of replicas.

Compression algorithm

The CMC-PDFs [36] implement the compression algorithm of MC replicas. This algorithm is designed to extract a subset of replicas which preserves as much as possible the underlying statistical distribution of the prior MC PDF set. This algorithm pre-computes for the input PDF the moments (central value, variance, skewness and kurtosis), the statistical distance (Kolmogorov distance) and the correlations for each flavor in a grid of x points. These estimators are then compared to subsets of replicas selected by a genetic algorithm. The procedure terminates when a tolerance value for the error function comparing these estimators is flat.

From a practical point of view in the context of the PDF4LHC15 the initial $N_{\rm rep} = 900$ replicas of the prior set have been reduced to 100 replicas. Further technical investigations based on clustering algorithms have also been performed in [37].

MC2H
The MC2H algorithm [38] was first introduced to convert MC PDF sets into a Hessian representation. This algorithm performs the principal component analysis (PCA) of the PDF covariance matrix in a predefined grid of x nodes for all flavors at a given initial scale. The eigenvectors obtained from the PCA are the basis of the Hessian representation. This representation consists in simple linear combinations of the input MC replicas. The MC2H procedure also allows the reduction of the number of required replicas of the input PDF set because we can reject the eigenvectors associated to small eigenvalues therefore obtaining a reduced set of Hessian replicas.

SMPDF
Starting from the MC2H algorithm we have developed the Specialized Minimal PDFs (SMPDF) [39] which consists in obtaining the smallest set of Hessian replicas for a given physical process. The SMPDF algorithm performs an interactive PCA reduction on top of the PDF covariance matrix computed in a grid of x points determined as the region where the PDF-process correlation is maximal. We have also provided a public web-interface for the construction of SMPDFs available and documented in [40].

4. Outlook

The advantages and results obtained thanks to ML have been essential for several developments of this research field from PDF determination to Monte Carlo event generation. In the next months and years new applications will be achieved. It is already possible to summarize the two main directions for these developments.

The first consist in the development of tools for the estimation and propagation of uncertainties. New methods to deal with uncertainty will possibly improve the determination of PDFs among other applications. This point also includes the new ideas about reweighting techniques in the context of higher order calculations.

The second development branch consists in the construction of new modern neural network architectures for problem specific applications together with efficient new gradient based methods. Such tools will open the possibility to obtain better methods for multidimensional density estimation. Consequently, we will obtain better sampling algorithms relevant for improved and faster integration algorithms in Monte Carlo event generators and similar tools.

In these proceedings we provided a short overview of successful applications of ML in HEP-TH but this is just the prelude of a new era where ML and HEP-TH are in synergy producing innovative and unique results.

Acknowledgements

S. C. is supported by the HICCUP ERC Consolidator grant (614577) and by the European Research Council under the European Union's Horizon 2020 research and innovation programme (grant agreement n° 740006).

References

[1] M. Abadi, et al. TensorFlow: Large-scale machine learning on heterogeneous systems, 2015. Software available from tensorflow.org.
[2] Pedregosa, F. *et al.* Journal of Machine Learning Research, volume 12, 2825–2830, 2011
[3] F. Chollet. keras. https://github.com/fchollet/keras, 2015.

[4] Theano Development Team [arXiv:1605.02688], May 2016,
[5] R. Boughezal, X. Liu and F. Petriello, JHEP **1703**, 160 (2017) doi:10.1007/JHEP03(2017)160 [arXiv:1612.02911 [hep-ph]].
[6] G. Altarelli, R. D. Ball and S. Forte, Nucl. Phys. B **799** (2008) 199 doi:10.1016/j.nuclphysb.2008.03.003 [arXiv:0802.0032 [hep-ph]].
[7] P. Nason, JHEP **0411** (2004) 040 doi:10.1088/1126-6708/2004/11/040 [hep-ph/0409146].
[8] J. Alwall *et al.*, JHEP **1407** (2014) 079 doi:10.1007/JHEP07(2014)079 [arXiv:1405.0301 [hep-ph]].
[9] T. Sjostrand, S. Mrenna and P. Z. Skands, JHEP **0605** (2006) 026 doi:10.1088/1126-6708/2006/05/026 [hep-ph/0603175].
[10] M. Bahr *et al.*, Eur. Phys. J. C **58** (2008) 639 doi:10.1140/epjc/s10052-008-0798-9 [arXiv:0803.0883 [hep-ph]].
[11] R. Boughezal, J. M. Campbell, R. K. Ellis, C. Focke, W. Giele, X. Liu, F. Petriello and C. Williams, Eur. Phys. J. C **77** (2017) no.1, 7 doi:10.1140/epjc/s10052-016-4558-y [arXiv:1605.08011 [hep-ph]].
[12] T. Carli, D. Clements, A. Cooper-Sarkar, C. Gwenlan, G. P. Salam, F. Siegert, P. Starovoitov and M. Sutton, Eur. Phys. J. C **66** (2010) 503 doi:10.1140/epjc/s10052-010-1255-0 [arXiv:0911.2985 [hep-ph]].
[13] T. Kluge, K. Rabbertz and M. Wobisch, doi:10.1142/9789812706706_0110 hep-ph/0609285.
[14] A. van Hameren, Comput. Phys. Commun. **182** (2011) 2427 doi:10.1016/j.cpc.2011.06.011 [arXiv:1007.4716 [hep-ph]].
[15] S. Carrazza, R. K. Ellis and G. Zanderighi, Comput. Phys. Commun. **209** (2016) 134 doi:10.1016/j.cpc.2016.07.033 [arXiv:1605.03181 [hep-ph]].
[16] T. Hahn and M. Perez-Victoria, Comput. Phys. Commun. **118** (1999) 153 doi:10.1016/S0010-4655(98)00173-8 [hep-ph/9807565].
[17] S. Catani, L. Cieri, G. Ferrera, D. de Florian and M. Grazzini, Phys. Rev. Lett. **103** (2009) 082001 doi:10.1103/PhysRevLett.103.082001 [arXiv:0903.2120 [hep-ph]].
[18] C. Anastasiou, C. Duhr, F. Dulat, F. Herzog and B. Mistlberger, Phys. Rev. Lett. **114** (2015) 212001 doi:10.1103/PhysRevLett.114.212001 [arXiv:1503.06056 [hep-ph]].
[19] B. Ruijl, T. Ueda and J. Vermaseren, arXiv:1707.06453 [hep-ph].
[20] P. Nogueira, J. Comput. Phys. **105** (1993) 279. doi:10.1006/jcph.1993.1074
[21] R. D. Ball *et al.* [NNPDF Collaboration], Eur. Phys. J. C **77** (2017) no.10, 663 doi:10.1140/epjc/s10052-017-5199-5 [arXiv:1706.00428 [hep-ph]].
[22] V. Bertone *et al.* [NNPDF Collaboration], Eur. Phys. J. C **77** (2017) no.8, 516 doi:10.1140/epjc/s10052-017-5088-y [arXiv:1706.07049 [hep-ph]].
[23] P. Skands, S. Carrazza and J. Rojo, Eur. Phys. J. C **74** (2014) no.8, 3024 doi:10.1140/epjc/s10052-014-3024-y [arXiv:1404.5630 [hep-ph]].
[24] P. T. Komiske, E. M. Metodiev and M. D. Schwartz, JHEP **1701**, 110 (2017) doi:10.1007/JHEP01(2017)110 [arXiv:1612.01551 [hep-ph]].
[25] J. Butterworth *et al.*, J. Phys. G **43** (2016) 023001 doi:10.1088/0954-3899/43/2/023001 [arXiv:1510.03865 [hep-ph]].
[26] S. Carrazza, Acta Phys. Polon. B **48**, 947 (2017) doi:10.5506/APhysPolB.48.947 [arXiv:1704.00471 [hep-ph]].
[27] Likas, Aristidis Computer physics communications, Vol. 135, 2, 167–175, 2001
[28] D. Krefl, S. Carrazza, B. Haghighat and J. Kahlen, arXiv:1712.07581 [stat.ML].
[29] J. Bendavid, arXiv:1707.00028 [hep-ph].
[30] R. D. Ball *et al.* [NNPDF Collaboration], JHEP **1504** (2015) 040 doi:10.1007/JHEP04(2015)040 [arXiv:1410.8849 [hep-ph]].
[31] S. Carrazza and N. P. Hartland, arXiv:1711.09991 [hep-ph].
[32] E. R. Nocera *et al.* [NNPDF Collaboration], Nucl. Phys. B **887** (2014) 276 doi:10.1016/j.nuclphysb.2014.08.008 [arXiv:1406.5539 [hep-ph]].
[33] L. A. Harland-Lang, A. D. Martin, P. Motylinski and R. S. Thorne, Eur. Phys. J. C **75** (2015) no.5, 204 doi:10.1140/epjc/s10052-015-3397-6 [arXiv:1412.3989 [hep-ph]].
[34] S. Dulat *et al.*, Phys. Rev. D **93** (2016) no.3, 033006 doi:10.1103/PhysRevD.93.033006 [arXiv:1506.07443 [hep-ph]].
[35] G. Watt and R. S. Thorne, JHEP **1208** (2012) 052 doi:10.1007/JHEP08(2012)052 [arXiv:1205.4024 [hep-ph]].
[36] S. Carrazza, J. I. Latorre, J. Rojo and G. Watt, Eur. Phys. J. C **75** (2015) 474 doi:10.1140/epjc/s10052-015-3703-3 [arXiv:1504.06469 [hep-ph]].
[37] S. Carrazza and J. I. Latorre, arXiv:1605.04345 [hep-ph].
[38] S. Carrazza, S. Forte, Z. Kassabov, J. I. Latorre and J. Rojo, Eur. Phys. J. C **75** (2015) no.8, 369 doi:10.1140/epjc/s10052-015-3590-7 [arXiv:1505.06736 [hep-ph]].
[39] S. Carrazza, S. Forte, Z. Kassabov and J. Rojo, Eur. Phys. J. C **76** (2016) no.4, 205 doi:10.1140/epjc/s10052-016-4042-8 [arXiv:1602.00005 [hep-ph]].
[40] S. Carrazza and Z. Kassabov, PoS PP **@LHC2016** (2016) 020 [arXiv:1606.09248 [hep-ph]].

IOP Conf. Series: Journal of Physics: Conf. Series **1085** (2018) 022004 doi:10.1088/1742-6596/1085/2/022004

New results on g-2 calculation[1]

Stefano Laporta

Dipartimento di Fisica e Astronomia, Università di Padova,
Istituto Nazionale Fisica Nucleare, Sezione di Padova,
Via Marzolo 8,
I-35131 Padova, Italy[2]

E-mail: `stefano.laporta@pd.infn.it`

Abstract. In this paper I summarize the current situation of the theoretical calculation of the electron g-2, including the result from the recent calculation of the QED mass-independent 4-loop contribution with a precision of 1100 digits.

A particle of mass m and spin s possesses a magnetic moment μ

$$\mu = g \frac{eh}{4\pi mc} s \, , \tag{1}$$

where g is the gyromagnetic ratio. According to the Dirac's theory [1], an electron has $g = 2$. This agreed with the experimental measurements until Kusch and Foley [2] measured a value of the *anomaly* a_e slightly different from zero:

$$a_e = \frac{g-2}{2} = 0.001\ 15(4) \, . \tag{2}$$

The deviation is due to the interaction of the electron with photons; using Q.E.D. Schwinger [3,4] was able to calculate at the first order that

$$a_e = \frac{\alpha}{2\pi} = 0.001\ 161\ldots \, , \tag{3}$$

where α is the fine structure constant

$$\alpha = \frac{e^2}{\hbar c} \approx \frac{1}{137} \, . \tag{4}$$

The current measurements of a_e are based on the Penning trap method, developed by the group at the University of Washington. This trap uses an axial magnetic field and a quadrupole electric field; the anomaly is expressed as the ratio of two frequencies, which can measured to a very high precision. For the development of this technique, the Nobel prize in Physics 1989 was awarded to H. Dehmelt. Their final results were [5]:

$$a_{e^-}^{exp} = 1\ 159\ 652\ 188.4(4.3) \times 10^{-12} \quad (4.3\ \text{ppb}) \, , \tag{5}$$
$$a_{e^+}^{exp} = 1\ 159\ 652\ 187.9(4.3) \times 10^{-12} \quad (4.3\ \text{ppb}) \, . \tag{6}$$

[1] Invited talk to ACAT 2017, University of Washington, Seattle, 21-25 August 2017
[2] since 1 September 2017

IOP Conf. Series: Journal of Physics: Conf. Series **1085** (2018) 022004 doi:10.1088/1742-6596/1085/2/022004

$A_1^{(4)}$	$A_1^{(6)}$	$A_1^{(8)}$	$A_1^{(10)}$
−2.973... [12]	1.49(20) [31]	−1.434(138) [24]	7.795(336) [29]
−0.328 478 965 579 ... [8,9]	1.195(26) [32]	−1.5098(384) [35]	6.599(223) [29]
	1.17611(42) [33]	−1.7283(35) [25]	
	1.181259(40) [34]	−1.9144(35) [26]	
	1.181241456587200006... [10]	−1.9106(20) [28]	
		−1.91298(84) [29]	
		−1.912245764926445... [11]	

Table 1. Numerical results of the evaluations of $A_1^{(4)}$, $A_1^{(6)}$, $A_1^{(8)}$ and $A_1^{(10)}$.

The most recent value obtained with the same technique by the Harvard group is [6, 7]:

$$a_{e^-}^{exp} = 1\ 159\ 652\ 180.73(.28) \times 10^{-12} \quad (0.24 \text{ ppb}) \ . \tag{7}$$

In the standard model

$$a_e^{SM} = a_e^{QED} + a_e^{\text{weak}} + a_e^{\text{hadr}} \ . \tag{8}$$

The QED contribution can be split up in mass-independent and mass-dependent parts:

$$a_e^{QED} = A_1 + A_2\left(\frac{m_e}{m_\mu}\right) + A_2\left(\frac{m_e}{m_\tau}\right) + A_3\left(\frac{m_e}{m_\mu}, \frac{m_e}{m_\tau}\right) \ . \tag{9}$$

The functions $A^{(i)}$ can be expanded in power series

$$A_i = A_i^{(2)}\left(\frac{\alpha}{\pi}\right) + A_i^{(4)}\left(\frac{\alpha}{\pi}\right)^2 + A_i^{(6)}\left(\frac{\alpha}{\pi}\right)^3 + A_i^{(8)}\left(\frac{\alpha}{\pi}\right)^4 + A_i^{(10)}\left(\frac{\alpha}{\pi}\right)^5 + \ldots \ . \tag{10}$$

The mass-independent coefficients at 1, 2 and 3 loop are known in analytical form [3, 4, 8–10]:

$$A_1^{(2)} = \frac{1}{2} \ , \tag{11}$$

$$A_1^{(4)} = \frac{197}{144} + \frac{1}{12}\pi^2 - \frac{1}{2}\pi^2 \ln 2 + \frac{3}{4}\zeta(3) = -0.328\ 478\ 965\ 579\ \ldots \ , \tag{12}$$

$$\begin{aligned} A_1^{(6)} &= \frac{83}{72}\pi^2\zeta(3) - \frac{215}{24}\zeta(5) + \frac{100}{3}\left[\left(a_4 + \frac{1}{24}\ln^4 2\right) - \frac{1}{24}\pi^2\ln^2 2\right] \\ &- \frac{239}{2160}\pi^4 + \frac{139}{18}\zeta(3) - \frac{298}{9}\pi^2\ln 2 + \frac{17101}{810}\pi^2 + \frac{28259}{5184} = 1.181\ 241\ 456\ \ldots \ , \end{aligned}$$

where $\zeta(n) = \sum_{i=0}^{\infty} i^{-n}$, $a_n = \sum_{i=0}^{\infty} 2^{-i} i^{-n}$. In table 1 we list some older theoretical evaluations of the two, three and four loop coefficients. In Ref. [11] I have evaluated up to 1100 digits of precision the 4-loop contribution $A_1^{(8)}$, finalizing a twenty-year effort [13–19] begun after the completion of the calculation of $A_1^{(6)}$ [10]. The first digits of the result are

$$A_1^{(8)} = -1.912245764926445574152647167439830054060873390658725345171329848\ldots \ . \tag{13}$$

The full-precision result is shown in table 3. The result (13) is in excellent agreement (0.9σ) with the numerical value

$$A_1^{(8)}(\text{Ref. [29]}) = -1.91298(84) \ , \tag{14}$$

latest result of a really impressive pluridecennial effort [20–29], and with the independent value

$$A_1^{(8)}(\text{Ref. [30]}) = -1.87(12) \ . \tag{15}$$

IOP Conf. Series: Journal of Physics: Conf. Series **1085** (2018) 022004 doi:10.1088/1742-6596/1085/2/022004

contribution	value in units of 10^{-12}
$A_1^{(2)}(\alpha/\pi)$	1 161 409 733.631(720)
$A_1^{(4)}(\alpha/\pi)^2$	− 1 772 305.065(3)
$A_1^{(6)}(\alpha/\pi)^3$	14 804.203
$A_1^{(8)}(\alpha/\pi)^4$	− 55.667
$A_1^{(10)}(\alpha/\pi)^5$	0.446(15)
$A_2^{(4)}(m_e/m_\mu)(\alpha/\pi)^2$	2.804
$A_2^{(6)}(m_e/m_\mu)(\alpha/\pi)^3$	− 0.092
$A_2^{(8)}(m_e/m_\mu)(\alpha/\pi)^4$	0.026
$A_2^{(10)}(m_e/m_\mu)(\alpha/\pi)^5$	− .0002
$A_2^{(4)}(m_e/m_\tau)(\alpha/\pi)^2$	0.010
$A_2^{(6)}(m_e/m_\tau)(\alpha/\pi)^3$	− 0.0008
a_e(hadronic v.p.)	1.866(11)
a_e(hadronic v.p.,NLO)	− 0.223(1)
a_e(hadronic v.p.,NNLO)	0.028(1)
a_e(hadronic l-l)	0.035(10)
a_e(weak)	0.0297(5)

Table 2. Contributions to a_e.

At 5-loop level there is only the numerical evaluation by the Kinoshita's group

$$A_1^{(10)}(\text{Ref. [29]}) = 6.599(223)\ . \tag{16}$$

Concerning the mass-dependent part $A_2(r)$, $A_2^{(4)}(r)$ is known in analytical form [36], as well as $A_2^{(6)}(r)$ [37–41]; the first terms of the expansion for small r of the 4-loop coefficient $A_2^{(8)}(r)$ are known analytically [42, 43] . $A_2^{(10)}(m_e/m_\mu)$ and $A_2^{(10)}(m_e/m_\tau)$ have been calculated numerically [29]; the first terms of the expansion for small mass rations of $A_3^{(6)}(m_e/m_\mu, m_e/m_\tau)$ and $A_3^{(8)}(m_e/m_\mu, m_e/m_\tau)$ are known analytically [43]. The hadronic and weak contribution are

$$\begin{aligned}
a_e(\text{hadronic v.p.}) &= \ \ 1.866(11) \times 10^{-12} \quad \text{(see Ref. [44])}\ , && (17)\\
a_e(\text{hadronic v.p.,NLO}) &= -0.223(1) \times 10^{-12} \quad \text{(see Ref. [45])}\ , && (18)\\
a_e(\text{hadronic v.p.,NNLO}) &= \ \ 0.028(1) \times 10^{-12} \quad \text{(see Ref. [45])}\ , && (19)\\
a_e(\text{hadronic l-l }) &= \ \ 0.035(10) \times 10^{-12} \quad \text{(see Ref. [46])}\ , && (20)\\
a_e(\text{weak}) &= \ \ 0.0297(5) \times 10^{-12} \quad \text{(see Ref. [47])}\ . && (21)
\end{aligned}$$

Inserting Eq.s(11-16,17-21), the known $A_2^{(j)}$ and the measurement of the fine structure constant [48, 49]

$$\alpha^{-1} = 137.035\ 998\ 996(85)\ (0.62\ \text{ppb})\ ,$$

into Eq.s(8-10) one finds

$$a_e^{\text{th}} = 1\ 159\ 652\ 182.031(15)(15)(720) \times 10^{-12}\ , \tag{22}$$

where the first error comes from $A_1^{(10)}$, the second one from the hadronic and electroweak corrections, the last one from α, respectively. The values of the single contributions to a_e are listed in table 2. Conversely, assuming the validity of the theory and using the experimental measurement (7) one finds

$$\alpha^{-1}(a_e) = 137.035\ 999\ 1500(18)(18)(330)(0.25\ \text{ppb})\ ,$$

where the errors come from $A_1^{(10)}$, hadronic and electroweak corrections, and a_e, respectively.

```
-1.91224576492644557415264716743983005406087339065872534517132984800603844398065170614276089270000363158375584153314732700563785149128545391
90280432705027382230434557895704556272930994129669976027778221157847203390641519081665270979708674381150121551479722743221642734319279759586
07405005783738496070187432831402483802519224946074229855893046350614049225266343109442400023563568812806206454940132249775943004292888367617
48899236915180878086989705263578533753776964117024536196013497574494361268486175162606832387186747303831505962741878015305514879400536977798
36946427868432691843117588958115974356695043304834907361342658649953116387811743475385423488364085584441882237217456706871041823307430517443
05573945961171550858961148995261266061246994073118403927472340023464969531735482584817998224097373710773657404645135211230912425281111372153
02154453721014811121159848970884223279879720484201445122828451516585236561786594592600991733031721302865467212345340500349104700728924487200
61604426132544906900043191519823004748818149431103849537829940629675867875385249781946989793132162197975750676701142904897962085050785592...
```

Table 3. First 1100 digits of $A_1^{(8)}$.

Methods of calculation of $A_1^{(8)}$

I sketch here the methods of calculation used in the literature, only for comparison. For further information on the technical aspects of my calculation of the 4-loop coefficient $A_1^{(8)}$, see Ref. [51] in these proceedings.

In QED the contributions to g-2 at n loops can be expressed as combinations of n-loop 4-dimensional Feynman integrals, belonging to a variety of Feynman diagrams.

- In Ref. [29], the n-loop 4-dimensional integrals are transformed in $(3n-2)$-dimensional integrals of (huge) rational functions of Feynman parameters. The integrals are computed using the MonteCarlo adaptative routine VEGAS [50]; an enormous amount of computing time is needed to sample adequately the integrands; for more information, see Ref. [20–29].
- My method, used in [11], consists in:
  - (i) reduction of contributions from each Feynman diagram to a small number (334 for $A_1^{(8)}$) of n-loop D-dimensional master integrals by using a suitable algorithm [10, 13];
  - (ii) determination of systems of difference or differential equations satisfied by the master integrals [13];
  - (iii) high precision calculation of these integrals by solving these systems of equations by means of rapidly convergent series expansions [13, 14];

  This method allowed to obtain 1100 digits of $A_1^{(8)}$ (and up to 9800 digits for some selected important integrals). See Ref. [51] for further details.
- In Ref. [30] the contributions of the various diagrams are reduced to combinations of a small number of master integrals. Most of these master integrals are computed with MonteCarlo methods.
- I note the alternative approach recently introduced by S.Volkov in Ref. [52, 53]. It is also based on MonteCarlo integration. It seems promising at 5-loop level.

References

[1] Dirac P A M 1928 *Proc. Roy. Soc. Lond.* A **117** 610
[2] Foley H M and Kusch P 1948 *Phys. Rev.* **73** 412
[3] Schwinger J 1948 *Phys. Rev.* **73** 416
[4] Schwinger J 1949 *Phys. Rev.* **75** 898
[5] Van Dyck R S, Schwinberg P B and Dehmelt H G 1987 *Phys. Rev. Lett.* **59** 26
[6] Hanneke D, Fogwell S and Gabrielse G 2008 *Phys. Rev. Lett.* **100** 120801
[7] Hanneke D, Hoogerheide S F and Gabrielse G 2011 *Phys. Rev.* A **83** 052122
[8] Petermann A 1957 *Helv. Phys. Acta* **30** 407
[9] Sommerfield C M 1957 *Phys. Rev.* **107** 328
[10] Laporta S and Remiddi E 1996 *Phys. Lett.* B **379** 283
[11] Laporta S 2017 *Phys. Lett.* B **772** 232
[12] Karplus R and Kroll N M 1950 *Phys. Rev.* **77** 536
[13] Laporta S 2000 *Int. J. Mod. Phys.* A **15** 5087
[14] Laporta S 2001 *Phys. Lett.* B **504** 188
[15] Laporta S 2001 *Phys. Lett.* B **523** 95

[16] Laporta S 2003 *Acta Phys. Polon.* B **34** 5323
[17] Laporta S, Mastrolia P and Remiddi E 2004 *Nucl. Phys.* B **688** 165
[18] Laporta S 2008 *Int. J. Mod. Phys.* A **23** 5007
[19] Laporta S 2009 *Subnucl. Ser.* **45** 409
[20] Kinoshita T and Lindquist W B 1983 *Phys. Rev.* D **27** 867
[21] Kinoshita T and Lindquist W B 1983 *Phys. Rev.* D **27** 877
[22] Kinoshita T and Lindquist W B 1983 *Phys. Rev.* D **27** 886
[23] Kinoshita T and Lindquist W B 1989 *Phys. Rev.* D **39** 2407
[24] Kinoshita T and Lindquist W B 1990 *Phys. Rev.* D **42** 636
[25] Kinoshita T and Nio M 2006 *Phys. Rev.* D **73** 013003
[26] Aoyama T, Hayakawa M, Kinoshita T and Nio M 2007 *Phys. Rev. Lett.* **99** 110406
[27] Aoyama T, Hayakawa M, Kinoshita T and Nio M 2008 *Phys. Rev.* D **77** 053012
[28] Aoyama T, Hayakawa M, Kinoshita T and Nio M 2012 *Phys. Rev. Lett.* **109** 111807
[29] Aoyama T, Hayakawa M, Kinoshita T and Nio M 2015 *Phys. Rev.* D **91** no.3 033006; Erratum: 2017 *Phys. Rev.* D **96** no.1 019901
[30] Marquard P, Smirnov A V, Smirnov V A , Steinhauser M and Wellmann D 2017 *preprint* arXiv:1708.07138 [hep-ph]
[31] Levine M J and Wright J 1971 *Phys. Rev. Lett.* **26** 1351
[32] Cvitanovic P and Kinoshita T 1974 *Phys. Rev.* D **10** 4007
[33] Kinoshita T 1990 *Quantum Electrodynamics* ed Kinoshita T (Singapore: World Scientific) pp 218-321
[34] Kinoshita T 1995 *Phys. Rev. Lett.* **75** 4728
[35] Hughes V W and Kinoshita T 1999 *Rev. Mod. Phys.* **71** S133
[36] Elend H H 1966 *Phys. Lett.* **20** 682
[37] Samuel M A and Li G W 1991 *Phys. Rev.* D **44** 3935; Erratum: 1993 *Phys. Rev.* D **48** 1879
[38] Li G W, Mendel R and Samuel M A 1993 *Phys. Rev.* D **47** 1723
[39] Laporta S and Remiddi E 1993 *Phys. Lett.* B **301** 440
[40] Laporta S 1993 *Nuovo Cim.* A **106** 675
[41] Passera M 2007 *Phys. Rev.* D **75** 013002
[42] Kataev A L 2012 *Phys. Rev.* D **86** 013010
[43] Kurz A, Liu T, Marquard P and Steinhauser M 2014 *Nucl. Phys.* B **879** 1
[44] Nomura D and Teubner T 2013 *Nucl. Phys.* B **867** 236
[45] Kurz A, Liu T, Marquard P and Steinhauser M 2014 *Phys. Lett.* B **734** 144
[46] Prades J, de Rafael E and Vainshtein A 2009 *Lepton Dipole Moments* ed Roberts B L and Marciano W J (Singapore: World Scientific) pp 303-319
[47] Mohr P J , Taylor B N and Newell D B 2012 *Rev. Mod. Phys.* **84** 1527
[48] Bouchendira R, Clade P, Guellati-Khelifa S, Nez F and Biraben F 2011 *Phys. Rev. Lett.* **106** 080801
[49] Mohr P J, Newell D B and Taylor B N 2016 *Rev. Mod. Phys.* **88** no.3 035009
[50] Lepage G P 1978 *J. Comput. Phys.* **27** 192
[51] Laporta S "High-precision calculation of the 4-loop contribution to the electron g-2 in QED" in *these proceedings*
[52] Volkov S A 2016 *J. Exp. Theor. Phys.* **122** 1008
[53] Volkov S A 2016 *Zh. Eksp. Teor. Fiz.* **149** 1164

IOP Conf. Series: Journal of Physics: Conf. Series **1085** (2018) 022005 doi:10.1088/1742-6596/1085/2/022005

Generative models for fast simulation

S.Vallecorsa[1,2]

[1] CERN, Geneva, Switzerland

[2] Gangneung-Wonju National University, Gangneung, South Korea

E-mail: sofia.vallecorsa@cern.ch

Abstract. Machine Learning techniques have been used in different applications by the HEP community: in this talk, we discuss the case of detector simulation. The need for simulated events, expected in the future for LHC experiments and their High Luminosity upgrades, is increasing dramatically and requires new fast simulation solutions. We will present results of several studies on the application of computer vision techniques to the simulation of detectors, such as calorimeters. We will also describe a new R&D activity, within the *GeantV* project, aimed at providing a configurable tool capable of training a neural network to reproduce the detector response and replace standard Monte Carlo simulation. This represents a generic approach in the sense that such a network could be designed and trained to simulate any kind of detector and, eventually, the whole data processing chain in order to get, directly in one step, the final reconstructed quantities, in just a small fraction of time. We will present the first three-dimensional images of energy showers in a high granularity calorimeter, obtained using Generative Adversarial Networks.

1. Introduction

Particle transport simulation is a building block of any physics experiment from detector design and R&D to the final steps of analysis and comparison to theoretical models. The traditional approach is based on Monte Carlo methods, which is a time consuming procedure requiring a tremendous amount of computation. For the past few years Monte Carlo simulations have represented more than 50% of the WLCG (LHC Computing Grid) workload and the simulation demands expected for the High Luminosity LHC runs in 2025 will increase significantly [1]. Unfortunately, WLCG is expected to hit the power consumption limits in several of the current centres but, at the same time, it is foreseeable that its budget will stay flat at best, or even decrease. All considered, the LHC experiments expectation is that the exploitation of the High Luminosity LHC will require a factor of 100 improvement in simulation throughput with the same computing resources available today [1].

Several initiatives have been trying to approach this problem from different perspectives. On one hand, techniques that try to reduce the quantity of data that needs to be simulated are being studied by the experiment. One example is the overlay technique that attempts to reduce the amount of simulated events needed to correctly describe physics quantities in high pile-up environment while improving data-simulation comparison. The idea is to mix data and simulation in order to reduce CPU time and memory usage [2].

On the other hand, different level of optimisations are being studied, in order to speed-up the existing simulation software and make it better suited to leverage on the latest hardware advancement. Multi-threading and a task based approach are being introduced in different frameworks, for example in GaudiMP [3]. Event level parallelism and multi-threading are being implemented in Geant4 [4]) to improve throughput.

A new prototype for particle transport simulation, *GeantV*, is being developed to improve physics accuracy and performance, in particular on modern architectures, such as the Intel Xeon Phi and (GP)GPUs [5]. It is aimed at introducing fine grained parallelism to achieve a factor 5 speedup with respect to *Geant4* by exploiting vectorisation, concurrency and locality. The project has raised large interest within different LHC experiments: improved geometry algorithms such as the VecGeom library [6] and a new SIMD library (VecCore [7]) have been developed within *GeantV* and are now used by a much larger community.

It is clear, however that, in order to increase the performance of simulation applications by two orders of magnitude, new methods and algorithms are to be studied as a complement to the basic work of parallelizing and optimizing the detailed simulation.

Fast simulation approaches are already used by the HEP community to reduce computation time, typically for cases in which it is possible to trade-off accuracy for speed (e.g. searches, upgrade studies, ...). Both the ATLAS and CMS experiments, for example, use pre-simulated EM-showers libraries to replace the detailed simulation of their forward calorimeters [9] [10]. GFlash [11] simulates electromagnetic and hadronic showers using parametrizations for the longitudinal and lateral profile. Another example is Delphes [8]: a C++-based framework for fast simulation of general purpose experiments, capable of simulating tracking systems (in magnetic field), calorimeters and muon systems. Each of these solutions can reach different performance in terms of speed improvements (x10 - x1000) and different levels of accuracy (generally around 10% with respect to full simulation). See, for example the figure 1 (left), comparing energy shower shapes in the FCC electromagnetic calorimeter, simulated using *Geant4* and GFlash.

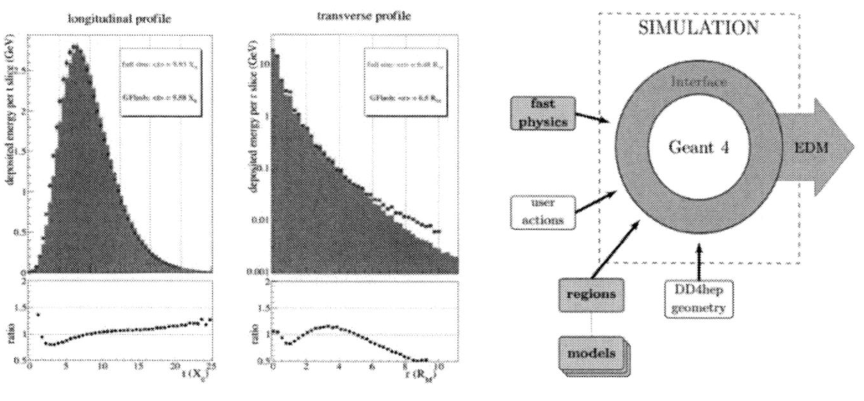

Figure 1. (left) FCC calorimeter showers GFlash results and comparison to full simulation. (right) Gaudi framework for integrating fast and full simulation [11]

The shortcoming of most of the fast simulation approaches used so far is the fact that they remain very specific to individual experiments. Instead, a generic coherent simulation framework, capable of combining different simulation types, is essential to facilitate and speedup data analysis and the comparison of the data. The idea behind such a simulation framework would be to use a single input for the detector geometries needed by the different simulation types and an identical analysis event data model [13]. Figure 1 shows, on the right, an example of a generic simulation framework that is being developed for the FCC communities [13], using the Gaudi architecture [12]. The ATLAS Integrated Simulation Framework is another example of this approach [16]. *Geant4* also has a mechanism sending particles through different simulations defined in certain detector regions.

IOP Conf. Series: Journal of Physics: Conf. Series **1085** (2018) 022005 doi:10.1088/1742-6596/1085/2/022005

A new R&D activity has recently started within *GeantV* to develop a generic, fully customizable, fast simulation framework: the structure of *GeantV* simulation steering allows to attach different simulation techniques to specific particle types, or energy ranges or single sub-detectors. This results in the possibility of seamlessly switching from detailed simulation to fast simulation in order to balance accuracy and speed. *GeantV* fast simulation framework is intended to make a varied palette of tools available to the user. Among those tools, a new deep learning based simulation is being developed. The recent developments in deep learning, coupled to the advancements of computing hardware, provide, in fact, the opportunity to replace complex algorithms with deep neural networks capable of reproducing the same results at a much higher speed.

2. Machine Learning in High Energy Physics

The use of machine learning methodologies in High Energy Physics is not a new approach [17, 18] and a lot of work has been done in that regard. The highly stochastic and non deterministic nature of High Energy Physics combined with complex and intricate nature of interactions makes the traditional approaches very time consuming [19]. One of the pioneering work in the quest for the elusive Higgs boson through machine learning was undertaken at Fermilab [20]. Further work proved that deep learning replaced the tedious feature engineering procedures and provided similar performance using low level features [21] as well as discovering novel high level features providing higher classification accuracy [22]. Image recognition networks employing convolutional and fully connected layers to classify between jets from single hadronic particles and overlapping jets from pairs of collimated hadronic particles resulted in modest performance improvement [23]. Moreover, the work of machine learning for HEP is not just limited to discrimination but also regression [24] and triggering [25]. A recent work even discussed analogies between quantum wave function and deep convolutional arithmetic circuits (convolutional layers with linear activation and pooling layers) [26]. A new frontier is that of physics simulation: only recently machine learning has been used to replace simulation [24], but the interest in these techniques is rapidly spreading through the experimental community.

3. Generative Models and Generative Adversarial Networks

Generative models like Generative Stochastic Networks [27], Variational AutoEncoders [28] and Generative Adversarial Networks [29] seem particularly suited to replace Monte Carlo simulation. The production of realistic samples is straightforward, as these techniques can model complicated probability function and deal with multi-modal output. Their ability to perform interpolation and to recover missing data has been proven several times [30, 31], two features that are particularly useful when dealing with particle physics simulations.

Introduced by I. Goodfellow in 2014, Generative Adversarial Networks consist of two networks, a generator and a discriminator, competing against each other [29]. Following the original implementation, a mathematical formulation of the problem can be expressed as follows. The generator is designed to learn some data distribution, starting from a prior on input noise variables $p_z(z)$: it can be represented as a differentiable function mapping z to the data space as $G(z;\theta_g)$ with parameters θ_g. The discriminator network is represented by a second differentiable function $D(x;\theta_d)$ that outputs a single scalar: it represents the probability that x came from the data rather than p_z. The value function, V, is defined as:

$$\min_G \max_D V(D,G) = \mathbb{E}_{x\sim p_g(x)}[logD(x)] + \mathbb{E}_{z\sim p_z(z)}[log(1-D(G(z)))] \quad (1)$$

The first term on the right hand side in the above equation denotes the loss for real data while the second term is the loss for generated data. The discriminator tries to maximize V, while generator tries to minimize it and the problem translates in finding the equilibrium saddle point.

Initially GANs were introduced as a Multi Layer Perceptron (MLP) networks but these networks suffered from high rate of non-convergence [29]. GAN shortcomings were further investigated and improvements were suggested to meet them. LAPGAN presented the idea of employing convolutional neural networks with the adversarial training [32]. Another version of GAN, called Deep Convolutional GAN or DCGAN, was also presented exploring unsupervised learning with Convolutional Neural Networks for discriminator and generator [33]. DCGAN model was explored in detail and architectural guidelines, like using strided convolutions instead of upsampling, or deconvolution for avoiding pooling layers, were presented. Batch normalization was suggested and fully connected layers were completely removed from the networks except the first layer in Generator and the last in the Discriminator. They also advocated Rectified Linear Units as most suitable activation functions for Generators and leaky Rectified Linear Units for Discriminators. Image generation was also conditioned to a label in a number of ways. An MLP implementation incorporated class labels as one hot vectors mapped to additional hidden dense layer before being combined with hidden layer generated by latent noise [34]. Another application of GAN employed Convolutional Neural Networks while both the Generator and Discriminator were conditioned on the class label implemented as embedding layer [35]. SGAN demonstrated semi-supervised approach by training the discriminator to predict labels for images and then the Generator utilizing these labels for generation [37]. Image generation and classification through GAN was further improved through techniques like feature matching, minibatch discrimination, historical averaging, label smoothing and batch normalization [36]. ACGAN adopts the semi-supervised approach and demonstrates that introduction of a label results in faster convergence and stable performance [38].

4. The LAGAN and CaloGAN applications

In High Energy Physics, some detector outputs can be also be interpreted in the form of images, thus the same techniques that are used for image recognition can be employed for detector output analysis. The LAGAN or Location Aware GAN [41] is one of the first applications of this approach, successfully reconstructing two-dimensional representations of jet images in calorimeters. To account for the specificity of jet images with respect to typical image reconstruction problems, namely their sparsity and highly non-linearly location-dependent data, the LAGAN networks make use of two-dimensional convolutional layers, locally connected layers and leaky rectified linear units, following work previously done on jet classification [42]. CaloGAN networks [43] extend the LAGAN work to generate images of energy deposited by particles travelling through a simplified detector geometry resembling the ATLAS LAr calorimeter: three instrumented layers in the radial (z) direction, with different thicknesses and different transverse segmentations. As an example, figure 4 represent the three transverse energy showers, one for each of the layers, created by a 10 GeV e^+ incident perpendicular to the center of the detector. The three images corresponds to the three calorimeter layers.

The CaloGAN architecture implements a LAGAN unit per each calorimeter layer plus a trainable transfer unit to preserve layer correlations. The result is a concatenation of two-dimensional images that reproduce the full three-dimensional picture. Figure 3 (left) shows a comparison of different energy shower shape variables and other event level variables as simulated by *Geant4* and CaloGAN. Different levels of agreement are reached depending on the variables and particle types and, although not perfect, the results prove that the CaloGAN network can reproduce successfully the main shower features.

The images are also conditioned on energy: the discriminator calculates the reconstructed energy per layer as well as the total energy. These quantities are used to build a specific loss components introduced to penalize the absolute deviation between the nominal energy and the reconstructed energy. Figure 3 (right), shows the energy response corresponding to different primary energy values. It is interesting to note that the system was trained over the $[0, 100]$

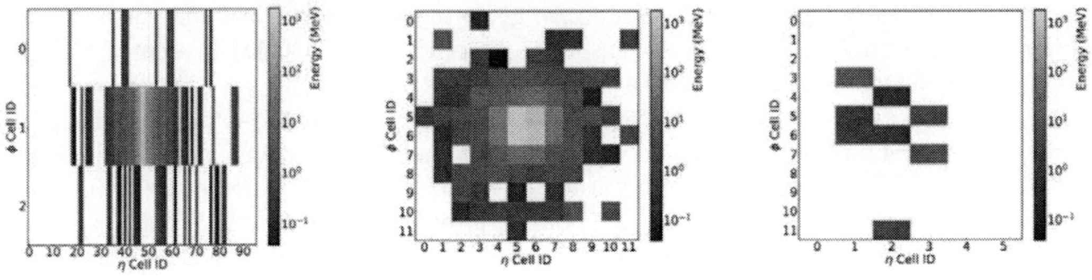

Figure 2. The three two-dimensional images, one for each calorimeter layer, represent together a 10 GeV e^+ incident perpendicular to the center of the detector.

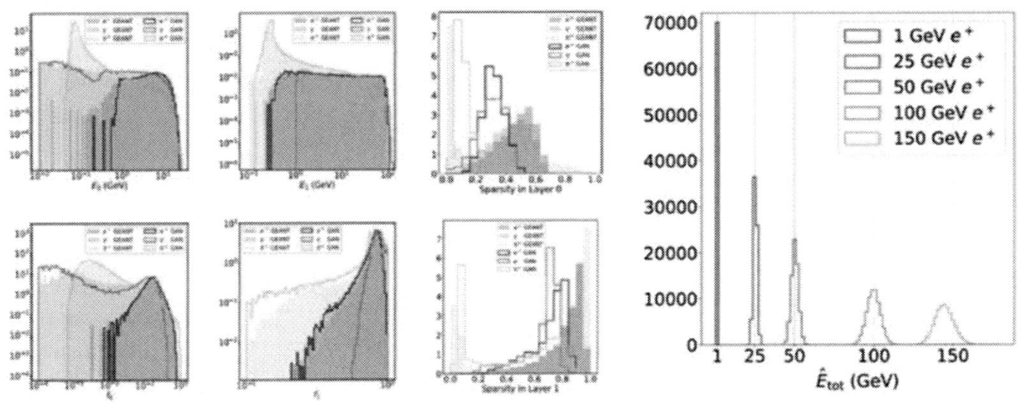

Figure 3. (left) Comparison of typical energy shower shape variables for the Geant and CaloGAN datasets for electrons, photons and pions: total deposited energy, fraction of measured energy and sparsity in the first and second layer. (right) CaloGAN energy response for primary particle energies of 1 GeV, 25 GeV, 50 GeV, 100 GeV, and 150 GeV.

GeV energy range, but it was still capable of generating images with energies lying outside the training range (150 GeV), although with broader width and shifted towards the training domain. This fact represent a first hint of the network extrapolation capability although the authors do no investigate the level at the which the extrapolated samples show the expected shower distributions. At the time of this writing, additional work is needed to fully validate the quality of the generated images with respect to detailed simulation or existing fast simulation methods. Nevertheless, CaloGAN represents a milestone in the development of deep learning application to HEP simulation.

5. The three-dimensional GAN

The *GeantV* three-dimensional GAN application steps further simulating three-dimensional calorimeter showers as a whole. It is intended as a first proof of concept to understand the level of accuracy achievable and to what extend the approach can be generalised to different detectors. It represent, in fact, the first step towards a fully integrated fast simulation tool configurable and trainable according to the different user needs in terms of physics and the nature of the detector.

IOP Conf. Series: Journal of Physics: Conf. Series **1085** (2018) 022005 doi:10.1088/1742-6596/1085/2/022005

The CLIC electromagnetic calorimeter design (ECAL) is chosen as an example of future high granularity detector [44]. The data is simulated using the DD4hep software framework [45]: it consists of energy showers produced by incoming particles travelling through the detector. They are generated with *Geant4*: each calorimeter cell is characterized by the energy recorded in it and three indices (iX, iY, iZ), identifying the position of the cell. The ECAL is made of 25 instrumented layers and for each of them, a 25x25 array of cells is considered. The energy of the incoming particle is sampled from a uniform [0-500 GeV] spectrum, a much larger energy range than the one tested by CaloGAN. Fig. 4 shows, on the right, an example energy shower, produced by a 100 GeV electron inside ECAL.

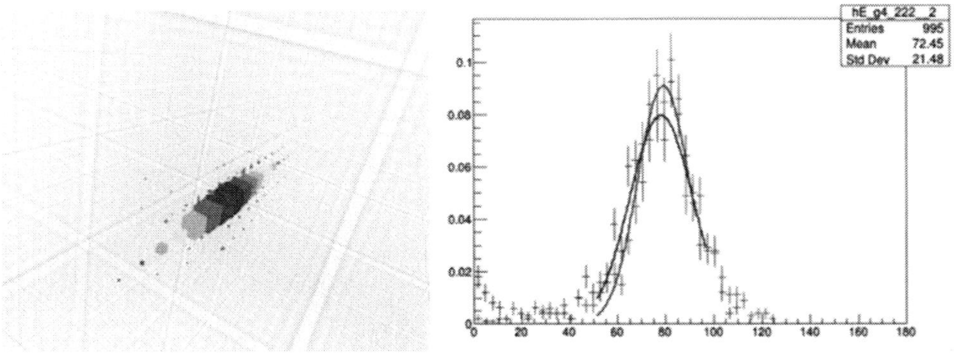

Figure 4. Typical energy showers produced by a 100 GeV electron inside the electromagnetic calorimeter. The grey lines represent identify the calorimeter cell. The color and hit dimensions correspond to the amount of energy deposited in the cell.

The *GeantV* GAN performs three-dimensional image reconstruction: the discriminator as well as generator exploit three-dimensional convolutional layers: the convolution parameters are adjusted to match the energy shower shapes. Usage of three-dimensional layers leads to a large number of training parameters, and therefore we keep the network as simple as possible. The model has been developed using Keras [47] and Tensorflow [48]. There are four convolutional layers in both the generator and discriminator with leaky rectified linear units activation functions. Batch normalization layers are added after the convolutional layers for improved performance [49]. The input image is mapped to two categorical outputs obtained by the final flattened layers. The first output denotes whether images are real or generated while the second output is an auxiliary classifier. Using this basic architecture two different experiments were conducted. In the first implementation the auxiliary classifier is the particle type (e.g. electron or photon) and a sigmoidal activation function generated the discriminator output. In the second version the auxiliary output is the energy of the incoming particle (a rectified linear unit constitutes the discriminator output). The generator uses a latent noise vector initialized to a gaussian probability distribution as well as the desired class of the image to generate the primary particle energy. To assess the performance of the networks a detailed study of the generated calorimeter response is performed. Figure 4 for example, shows the typical single cell energy response: the mean and width of the cell energy deposition are compared to *Geant4* prediction. The result is generally better for cells receiving large amount of energy. The average transverse and longitudinal shower shapes are also compared against the detailed simulation (see figure 5.) Figure 6 shows instead the discriminator energy response corresponding to 100, 150 and 300 GeV input energies: the network energy prediction is good both in terms of central value and the width of the distribution, with average offset and spread well below 10%. Even more promising, from the point of view of designing a reliable simulation engine,

IOP Conf. Series: Journal of Physics: Conf. Series **1085** (2018) 022005 doi:10.1088/1742-6596/1085/2/022005

is the fact that the network is capable of correctly describing how the energy shower shape distributions change depending on the primary particle energy: an example is shown in figure 7: the longitudinal shower shapes generated by GAN are compared to the detailed simulation predictions for different primary particle energies. Further work is currently ongoing to compare the three-dimensional GAN results to different fast simulation tools and to study the possibility to generalise the tool to different detectors via algorithm meta-optimisation and hyper-parameter scans.

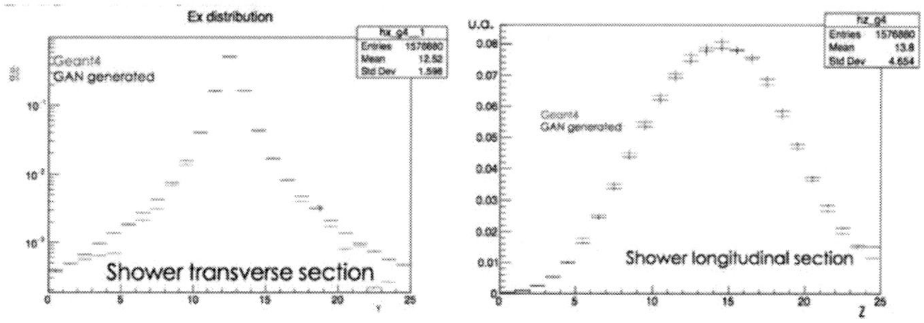

Figure 5. The transverse and longitudinal shower shapes generated for 100 GeV electrons.

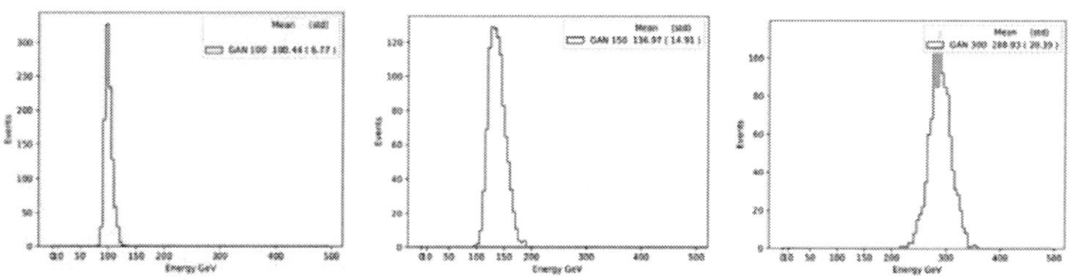

Figure 6. The discriminator energy prediction corresponding to 100, 150 and 300 GeV primary energies.

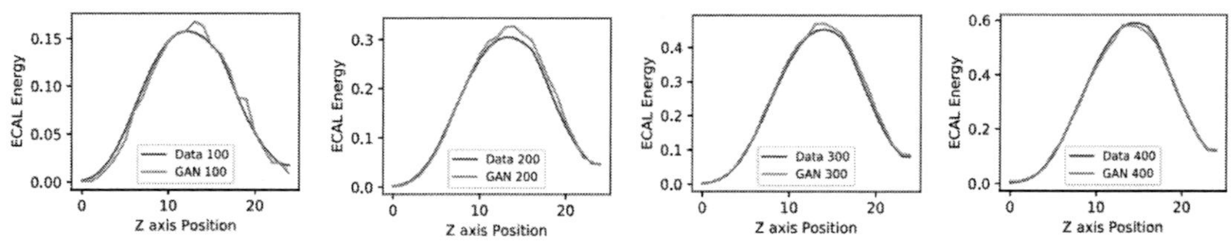

Figure 7. The longitudinal shower shapes generated by three-dimensional GAN for different primary particle energies (100, 200, 300, 400) GeV (orange). The GAN shape is compared to the corresponding detailed simulation results (blue).

6. Conclusion

Monte Carlo production has been so far a major fraction of WLCG computing workload and the High Luminosity LHC experiments needs will scale up orders of magnitude. A generic framework with common fast simulation algorithm and strategies for mixing detailed and fast simulation could bring great benefit to HEP and also to those smaller communities that could not afford to develop their own simulation framework. In this context generative models, relying on the possibility to interpret detector response as images, seem natural candidates to speedup simulation. Generative Adversarial Networks, in particular, require relatively small amount of data to train and are the subject of many ongoing studies. Their performance as imaging tools for calorimeter simulation is very promising, and from a computing resources perspective, the gain in time needed to generate a shower is huge: for example, using the three-dimensional GAN takes only $O(10^{-3})$ ms compared to the typical one minute time needed to generate it using the detailed approach. This corresponds to a speedup of more than 6 orders of magnitude. The accent should therefore be on optimising the computing resources needed to train the networks, studying parallelisation on clusters and cross-platform development.

7. Acknowledgement

The authors wish to acknowledge the contribution of Intel to the *GeantV* project through the Intel Performance Computing Centre (IPCC) program. We also want to acknowledge the very useful technical contributions of the CERN openlab.

References

[1] Ian Bird. Workshop Introduction, Context of the Workshop: Half-way through Run2; Preparing for Run3, Run4. (2016). https://indico.cern.ch/event/ 555063/contributions/2236372/ WLCG Workshop.
[2] The ATLAS collaboration. arXiv:1603.02934 (2016).
[3] . M. Clemencic at al. Journal of Physics: Conference Series, Volume 608, conference 1
[4] GEANT4 Collaboration, Nuclear Instruments and Methods in Physics Research A 506, 250 (2003).
[5] G Amadio et al 2016 J. Phys.: Conf. Ser. 762 012019.
[6] J Apostolakis et al 2015 J. Phys.: Conf. Ser. 608 012023
[7] https://github.com/root-project/veccore
[8] The DELPHES 3 collaboration, de Favereau, J., Delaere, C. et al. J. High Energ. Phys. (2014) 2014: 57.
[9] The ATLAS collaboration, ATL-SOFT-PROC-2017-005
[10] Iammanco, Andrea. (2014). Journal of Physics: Conference Series. 513. 022012. 10.1088/1742-6596/513/2/022012.
[11] G. Grindhammer and M. Rudowicz and S. Peters, Nuclear Instruments and Methods in Physics Research Section A: Accelerators, Spectrometers, Detectors and Associated Equipment, 290-2, 1990, 469-488.
[12] Gaudi webpage, http://cern.ch/gaudi
[13] J. Hrdinka, A. Zaborowska, A. Salzburger, B. Hegner. PoS(EPS-HEP2015)248
[14] G. Aad et al. (ATLAS), Phys. Lett. B716, 1 (2012), arXiv:1207.7214 [hep-ex].
[15] S. Chatrchyan et al. (CMS), Phys. Lett. B716, 30 (2012), arXiv:1207.7235 [hep-ex].
[16] E. Ritsch, ATLAS Detector Simulation in the Integrated Simulation Framework applied to the W Boson Mass Measurement, Universitaet Innsbruck, 2014, http://physik.uibk.ac.at/hephy/theses/diss_er.pdf
[17] P. Vannerem et al. 1999, hep-ex/9905027.
[18] R.K. Bock et al. Nuclear Instruments and Methods in Physics Research Section A: Accelerators, Spectrometers, Detectors and Associated Equipment, 516,2511-528,2004
[19] Bhat, Pushpalatha C, FERMILAB-PUB-10-425-PPD.
[20] Engineering Applications of Artificial Intelligence, 22, 8, 1203 - 1217, 2009
[21] P. Baldi, P. Sadowski and D. Whiteson, arXiv: 1402.4735
[22] Sadowski, Peter and Collado, Julian and Whiteson, Daniel and Baldi, Pierre, Proceedings of the 2014 International Conference on High-Energy Physics and Machine Learning - Volume 42, 2014
[23] D. Whiteson et al. arXiv:1603.09349
[24] A. Vitek at al. 2013 5th International Conference on Intelligent Networking and Collaborative Systems. 121-126
[25] V V Gligorov and M Williams, Journal of Instrumentation, 8, 02, 2013
[26] Yoav Levine et al., arxiv.org:1704.01552, 2017
[27] Yoshua Bengio et al., arxiv:1306.1091, 2013
[28] D. Kingma et al., arxv:1312.6114, 2013.
[29] I.J.Goodfellow et al. 2014. arXiv:stat.ML/1406.2661.
[30] , Scott E. Reed et al., arxiv:1605.05396, 2016.

IOP Conf. Series: Journal of Physics: Conf. Series **1085** (2018) 022005 doi:10.1088/1742-6596/1085/2/022005

[31] Ranzato, Susskind, Mnih, Hinton, IEEE CVPR 2011.
[32] EmilyL.Denton,SoumithChintala,ArthurSzlam,andRobertFergus.2015.Deep Generative Image Models using a Laplacian Pyramid of Adversarial Networks. CoRR abs/1506.05751 (2015). http://arxiv.org/abs/1506.05751.
[33] A. Radford et al. 2015.Arxiv e-prints: 1511.06434
[34] M. Mirza and S. Osindero. 2014. Conditional Generative Adversarial Nets. Arxiv e-prints: 1411.1784
[35] J. Gauthier. 2014. Class Project for Stanford CS231N: Convolutional Neural Networks for Visual Recognition, Winter semester 2014, 5 (2014), 2
[36] T. Salimans et al. 2016. Improved techniques for training gans. In Advances in Neural Information Processing Systems. 2234?2242.
[37] A.Odena.2016.Semi-SupervisedLearningwithGenerativeAdversarialNetworks. ArXiv e-prints (June 2016). arXiv:stat.ML/1606.01583
[38] A. Odena, C. Olah, and J. Shlens. 2016. Conditional Image Synthesis With Auxiliary Classi er GANs. ArXiv e-prints (Oct. 2016). arXiv:stat.ML/1610.09585
[39] JiajunWu,ChengkaiZhang,TianfanXue,BillFreeman,andJoshTenenbaum. 2016. Learning a Probabilistic Latent Space of Object Shapes via 3D Generative- Adversarial Modeling. In Advances In Neural Information Processing Systems. 82?90.
[40] M. Pierini. 2016. The LCD dataset. (2016). https://indico.hep.caltech.edu/ indico/contributionDisplay.py?sessionId=9&contribId=16&confId=102 DSHEP at the Simons Foundation.
[41] M. Paganini, L. de Oliveira, B. Nachman.arXiv preprint. arXiv:1701.05927.(2017).
[42] J. Barnard, E. N. Dawe, M. J. Dolan and N. Rajcic, arXiv preprint. arXiv:1609.00607
[43] M. Paganini, L. de Oliveira, and B. Nachman. arXiv preprint. arXiv:1705.02355 (2017).
[44] The CLIC collaboration. http://cds.cern.ch/record/2254048
[45] M Frank et al 2014 J. Phys.: Conf. Ser. 513 022010
[46] M. Pierini. DS@HEP at the Simons Foundation (2016).
[47] F. Chollet et al. 2015. Keras. https://github.com/fchollet/keras. (2015).
[48] M. Abad et al., TensorFlow: Large-scale machine learning on heterogeneous systems, 2015. Software available from tensorflow.org
[49] TimSalimans,IanGoodfellow,WojciechZaremba,VickiCheung,AlecRadford, and Xi Chen. 2016. Improved techniques for training gans. In Advances in Neural Information Processing Systems. 2234?2242.

IOP Conf. Series: Journal of Physics: Conf. Series **1085** (2018) 022006 doi:10.1088/1742-6596/1085/2/022006

HEP Software Development in the Next Decade; the Views of the HSF Community

E Sexton-Kennedy[1]
[1]Fermilab, P.O.Box 500, Batavia, IL 60510-5011, USA

E-mail: sexton@fnal.gov

Abstract. We have known for a while now that projections of computing needs for the experiments running in 10 years from now are unaffordable. Over the past year the HSF has convened a series of workshops aiming to find consensus on the needs, and produce proposals for research and development to address this challenge. At this time many of the software related drafts are far enough along to give a clear picture of what will result from this process. This talk will synthesize and report on some of the key elements to come out of this community work.

1. Motivation

When considering the big picture for HEP in the 10 year timescale the community must realize that we are driven by: 1. the physics objectives of our experiments and the timeline for taking that data; 2. the evolution of technologies we use including hardware, operating systems, compilers, and facility provisioning systems for Grid/Cloud and HPC use; 3. Resource limitations or limitations imposed by resource providers. Ultimately we need to guarantee that we can do the desired physics at the scale and for the cost that funding agencies can pay.

2. Cost Projections

Both CMS and ATLAS have made projections of the cost of their HL-LHC program by scaling up their current computing models. These have to be seen as an upper limit, or starting point to bring down, with research and development efforts over the coming decade. For example, CMS used the time per event and event sizes per tier of its current simulation software being used for HL-LHC detector TDR studies. Clearly the software will evolve to take better advantage of detector handles and algorithmic improvements. The point is to highlight the need for this work. The projections are shown in figures 1 and 2 below.

In order to arrive at a cost, it is also necessary to track the evolution of CPU and storage technologies. In this paper, I will use plots shown at the workshop opening the community process at the beginning of 2017[1]. CERN tracks the cost of its CPU server purchases over the year and uses what it knows about market trends and product roadmaps to project costs out to 2026. Comparing the plots in figures 3 and 4 below to the results in 2015 there has been a reduction in uncertainty of almost a factor of 2. Being aggressive and assuming a 20% increase/year for both CPU and disk, these plots project a cost of 2CHF/HS06 and 0.02CHF/GB respectively. Putting it together CMS would have to request 1.6 petaHS06-s over the course of the year (31.5million seconds) which if procured on owned machines would be 50 million HS06 costing about 100 million CHF. It would need 5 exabytes of disk costing another 100 million CHF. ATLAS and CMS projections agree to within the large factor of two margin of error, so combined the experiments would need 400 million CHF. That is comparable to the run 4 detector upgrade costs. This unaffordability is the challenge that drove the need for community building, and a desire for common solutions to mitigate costs as much as possible.

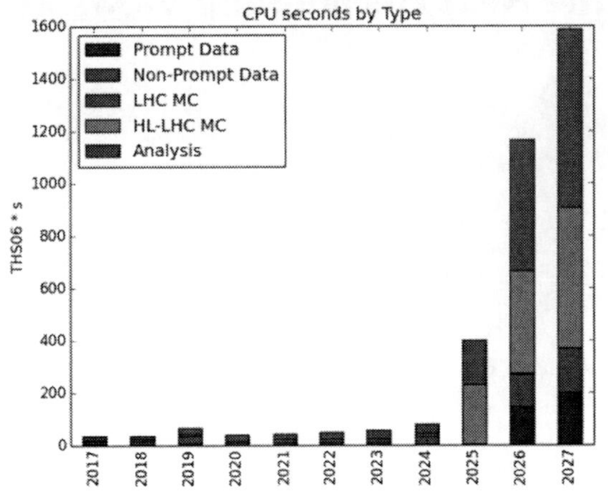

Figure 1. Required CPU second by Type as a function of year.

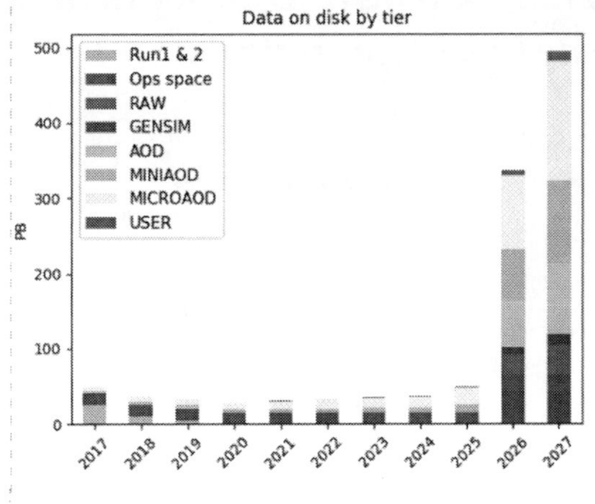

Figure 2. Required data on disk as a function of year.

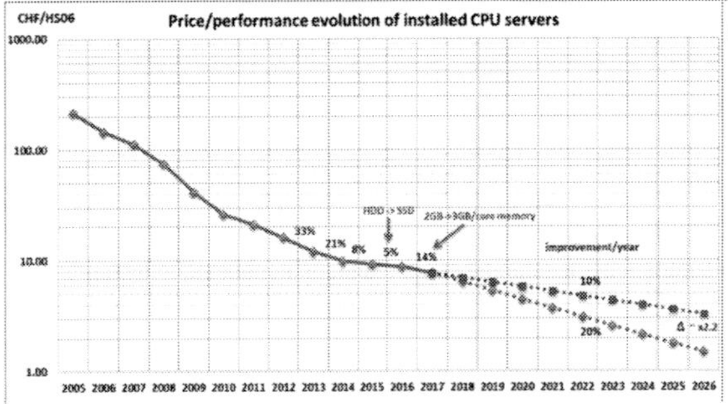

Figure 3. Price per performance evolution of installed CPU as a function of year

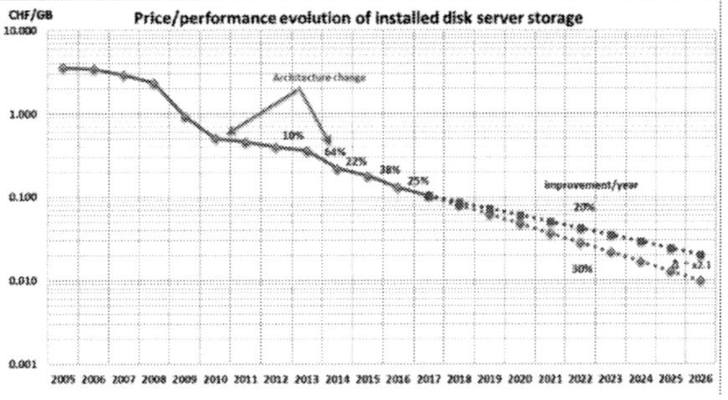

Figure 4. Price per performance evolution of disk server storage as a function of year

3.The Community White Paper, CWP Process

A coordinated response to the above technical and budgetary challenges is clearly needed. In the fall of 2016 the WLCG leadership charged the HEP Software Foundation, HSF to write a community white paper presenting an overall strategy and roadmap for carrying out a "software upgrade" analogous to the detector upgrades for HL-LHC [2]. The process consisted of a series of workshops held throughout the first half of 2017 starting with a kickoff meeting in January, and finishing up with a workshop at the end of June. Participation was open, and all were invited, but special care was taken to ensure representation across international funding agencies and HEP sub-disciplines.

4. Domains covered by the CWP

During the first workshop about a dozen working groups were formed with an initial list of participants who worked on creating charges that defined their area's scope. There was not enough time in the conference to cover all of the work, by all of these groups so I chose to summarize the work of the groups that drive CPU costs such as simulation, generators, reconstruction, frameworks and those that drive the data movement and storage costs such as analysis.

4.1. Simulation

In the domain of simulation there is already a dominant and common simulation engine and related tools like geometry modelers. That is the Geant [3] family of projects. Geant modernization projects have delivered a multi-threaded application (Geant4) capable of running in experiment frameworks on the GRID well before Run 2 giving the LHC experiments enough time to physics validate it. Continuing on that success there is also an R&D project focused on development of a fully vectorized version (GeantV). Adoption of this next step in modernization will be even more difficult then the last multi-threaded step. This CWP working group makes the case that the best strategy for moving forward is to continue this multi-pronged approach. Firstly Geant4 will continue to deliver new or refined functionalities both in physics coverage and accuracy, whilst introducing software performance improvements whenever possible. At the same time, continue the R&D program; investigating new approaches that aim to benefit from modern computing architectures. Its main feature is track-level parallelization, bundling particles with similar properties from different events to process them in a single thread. This approach combined with SIMD vectorization coding techniques and use of data locality is expected to yield large speed-ups, which are being measured in a realistic prototype under development. In addition, the work on Fast Simulation is accelerating with a view to producing a flexible framework that permits Full and Fast simulation to be combined for different particles in the same event. The overriding requirement is to ensure the support of experiments through continuous maintenance, support and improvement of the Geant simulation toolkit with minimal API changes visible to these experiments at least until production versions of potentially alternative engines, such as those resulting from ongoing R&D work, become available, integrated, and validated by experiments. The agreed ongoing strategy to meet this goal is to ensure that new developments resulting from the GeantV R&D program can be deployed immediately in Geant4. GeantV will need to make intrusive data marshaling API changes required to get the dramatic speed improvement that can take advantage of the wide vectors available on modern chips. This in turn will have to wait for the experiment frameworks to catch up and support toolkits like GeantV before any wide adoption in LHC experiments can proceed.

Of course there is more that goes into making simulation samples for experiments: 1. The outputs of Geant, hits in the detectors, have to be made into digitized detector signals. A further complication of this step is that the hard scatter signals (dependent on the generator process being modeled) has to be combined with a large number of minimum bias (low pT QCD events) interactions according to a model of the luminosity profile expected or observed. The issues with simulation overlays is not unique to the LHC. Unless the detector is deep underground it will have model cosmic overlay events. 2. Monte Carlo truth information hast bee handled and passed to analysis. 3. Digitized signals must then be compressed into the RAW data format of the experiment, so that the same reconstruction is run as you would run on the collected detector data.

The following program of work was agreed upon by the simulation community:

1. Continued work on VecGeom, facilitating experiment adoptions and back porting to Geant4 in order to make the modeling of complex detectors computationally more efficient, modular, and transparent.

2. Promote common solutions for detector geometry descriptions that service the needs of simulation and experiment reconstruction, DD4hep.

3. Review the physics models assumptions, approximations and limitations of the simulation engine in order to achieve higher precision, and to extend the validity of models up to FCC energies.

4. Improve the technical performance on emerging computing architecture, e.g., Single Instruction Multiple Data (SIMD) vectorization, Non-uniform Memory Access (NUMA) hierarchies, offloading to Many Integrated Core (MIC), Graphic Processing Unit (GPU), and Tensor Processing Unit (TPU) architectures;

5. Explore fast simulation options, including common frameworks for fast tuning and validation using Machine Learning (ML);

6. Develop techniques for background modeling, including contributions of multiple hard interactions overlapping the event of interest (data overlay, ML)
7. Revisit digitization algorithms to improve performance by means of vectorization and sub-system parallelization techniques. Exploring opportunities for code sharing among experiments when sharing readout electronics.

4.2. *Physics Generators*

Generator software is another domain in which the community shares tools. Historically generator packages come from small groups of theorist authors that have not needed to worry about computational performance. During the Tevatron and first run of the LHC, the event generation phase of the simulation took a very small fraction of the total time needed to simulate and reconstruct monte carlo samples. This has changed in Run 2 as the have there is a need to invest in higher precision calculations. The fact that theoretical approaches as well as implementations can vary offers an opportunity to validate and cross-check the predictions in a similar way that ATLAS and CMS provide systematic checks of experimental measurements. How much precision for which analysis samples is worth the cost, is still an open question. There is a lots of diversity in generator options and a wide difference in cost among the options. ATLAS reports that in 2016 they spent 20% of their GRID capacity on the generation phase of their application chain. Their average CPU time per event went from 1.6sec./event to 91.4sec/event. In CMS there was an increase as well but not quite as dramatic. It should be noted that the computational challenges involved with modeling QCD at higher orders lend themselves to super computer exploitation. Two areas of future work, with the goal of increasing the precision of generators needed for analyses of the HL-LHC while aiming to reduce our computing requirements, have been identified; 1. the development of a new scalable adaptive Monte-Carlo integrator that can take advantage of highly parallel computing architectures and can be used by existing frameworks; 2. work on re-weighting methods which attach to a given partonic event a new weight corresponding to a new theory. Those new weights allow to predict accurately all the differential distributions at parton-level and to perform a single detector simulation for all the models under consideration. The application of such re-weighting will allow a large reduction in the time spent on the generation of beyond standard model samples.

4.3. *Reconstruction*

Whether it is reconstructing the simulated data or detector data, reconstruction is the most expensive processing step for many experiments operating 10 years from now. The reconstruction working group has outlined the needs and opportunities to achieve affordability by working on the software tools and methods. They have identified the following broad areas of R&D for algorithms and approaches:

1. Enhanced vectorization programming techniques
2. Algorithms and data structures to efficiently exploit many-core architectures
3. Algorithms and data structures for non-x86 computing architectures (e.g., GPUs, FGAs)
4. Enhanced QA/QC approaches for reconstruction techniques
5. Real-time analysis
6. Precision physics-object reconstruction, identification and measurement techniques
7. Fast software trigger and reconstruction algorithms for high-density environments

Another avenue of attack is to design detectors with reconstruction costs in mind. ATLAS made a factor of 2 difference in cost between their LOI detector and their TDR. CMS collaborators have studied how variations on the tracker layout can significantly change the cost of seeding and track finding [4]. In fact detector handles, which add information to the event that enable new algorithms such as 4D pattern recognition, offer the most hope for reducing costs.

4.4. *Analysis*

In the analysis domain it is very difficult to separate thoughts about the software from the facility. Analysis is IO and compression limited so more capable facilities, the so called big data facilities, like Spark clusters, can enable new methods of doing analysis that were not available previously. As outlined in section 2, data movement and storage is a significant cost driver and since analysis drives the use of storage any innovations in analysis can make analysis and time to incite more affordable. We learned at the Analysis Ecosystem workshop[5] that while the different experiments draw the line between central and distributed analysis differently, we all do the same things. Namely filter, slice and flatten the experiment's centrally produced analysis samples, until they can fit on a user's personal storage space. Partially this is the nature of the problem; users need rapid turn around when creating plots for publication. Partially this is the result of using the same tool set, ROOT[6]. Historically analysis users are our least savvy software and computing colleagues. As the scale of data needing access increases by an order of magnitude these users will need more help. The Analysis working group outlines many promising avenues of investigation for doing this:

1. Establish PyROOT support and identify other needed measures to promote python as a first class language.
2. Modularization and plugin support plan and roadmap for ROOT
3. Inventory the needed bridges, connectors and converters, identify which are missing or need further work, develop workplans for them
4. Interfacing external ML tools: inventory the needs and develop workplans
5. ROOT I/O performance analysis and tuning
6. Assess the relevance of AaaS and notebook for the analysis community
7. R&D on how new hardware technologies could affect the analysis tool chain, Spark clusters and similar...

4.5. Frameworks

Frameworks are not cost drivers themselves, however they enable novel approaches, as already mentioned in the section discussing vectorized Geant. Without support in the experiment frameworks for innovative techniques, those approaches can not be integrated and deployed for the wides spread use in the production workflows of the experiments. It is recognized that frameworks have a central role in supporting the porting of applications onto advanced architectures and HPCs.

The group of people who have architected and designed previous experiment frameworks is not large, and the CWP effort has brought together most of the people involved in this over the last 2 decades. The biggest challenge in this area is the social challenge of dealing with large sets of legacy software, even though it is recognized that much of that software will have to be modernized. As a starting point this group has attempted to define common terms that help us understand that the different experiments do not have widely different needs. There is disagreement about the feasibility of a common solution even though it is recognized that this is needed for modernization and sustainability. The group proposes to investigate the possibilities of increasing the sharing of existing frameworks and framework components. In an attempt to answer the question, "Is it possible to reduce the number of existing frameworks?" Event without the answer to that question, there are some obvious areas in which common work can be done. Those include:

1. Investigate alternatives to TBB which include support for heterogeneous resources and HPC facilities.
2. Improve the understanding of the interplay between workload management and framework, and the tradeoffs between multi-threaded application, and heavy weight cooperative process frameworks.
3. Devise ways of accommodating multi-event tools like GeantV in the framework.

4.6. Crosscutting Topics and Solved Problems

There are many other topics addressed by CWP working groups that I did not have time to cover even in this summarized form but which do deserve mentioning. The area of Software Development, Deployment and Validation/Verification is a cross cutting area that is absolutely vital to the advancement of all of the domains covered above. These tools allow us to build on our long legacy through detailed regression testing. The software captures our domain knowledge, from matrix element calculations, to the detector understanding built into the reconstruction, and without the tools to prove that new techniques do not destroy that understanding, we can not adopt and deploy them.

The area of machine learning is another cross cutting technique that is applicable to analysis, reconstruction and simulation. It is a promising tool that offers exciting possibilities to solve these challenges.

Attention to Careers, Staffing and Training is a necessary investment to successful carry out this ambitious program of work.

While the areas of conditions databases and visualization are solved problems, the target of opportunity in these areas is to increase commonality among experiments and thereby decrease manpower. The Visualization working group has also concluded that there is a need to modernize and adopt industry tools, thereby getting much more bang for the same about of effort.

5. Conclusions

There is a field wide, large challenge ahead of us on the timescale of 10+ years. Limitations of power, cooling, and funding will not allow us to keep our current software paradigm, nor our current mode of computing resource provisioning. In the face of so much uncertainty it is vital that we continue to talk to each other. We need to attack this challenge from many angles; by maximizing the scientific computing infrastructures available to do science 10 years from now; by developing innovative and efficient software solutions for the domains we know today; by designing detectors that give us handles to minimize computational costs, throughout the data flow, from detector to insight. I have attempted to summarize the work done in 2017 to define a software roadmap for R&D targeting their challenge. As of this writing, the first draft of the CWP itself, it is under review and should be delivered by the end of Oct. For the most up to date information about the status of the paper check the HSF CWP web page [7].

6. References

[1] Helge Meinhard, and Bernd-Panzer Steindel, "Computing Evolution: Technology and Markets" HSF workshop San Diego, 2017 https://indico.cern.ch/event/570249/contributions/2404412/

[2] Charge for Producing a HSF Community White Paper, July 2016, http://hepsoftwarefoundation.org/assets/CWP-Charge-HSF.pdf

[3] Geant4 project, "Geant" [software], version 10.3.2, 2017. Available from https://github.com/ Geant4/geant4/releases/tag/v10.3.2 [accessed 2017-08-17]

[4] V. Krutelyov et. al. "Impact of Tracker Layout and Algorithmic Choices of Cost on Computing at High Pileup", 38th International Conference on High Energy Physics, August 2016 https:// pos.sissa.it/282/194/pdf

[5] The HSF Analysis Workshop, May 2017 https://indico.cern.ch/event/613842/

[6] Rene Brun and Fons Rademakers, ROOT - An Object Oriented Data Analysis Framework, Proceedings AIHENP'96 Workshop, Lausanne, Sep. 1996, Nucl. Inst. & Meth. in Phys. Res. A 389 (1997) 81-86. See also "ROOT" [software], https://doi.org/10.5281/zenodo.848819

[7] The HEP Software Foundation, "A Roadmap for HEP Software and Computing R&D for the 2020s" http://hepsoftwarefoundation.org/activities/cwp.html

Acknowledgments

IOP Conf. Series: Journal of Physics: Conf. Series **1085** (2018) 022006 doi:10.1088/1742-6596/1085/2/022006

This work was supported by Fermilab; Operated by Fermi Research Alliance, LLC under Contract No. DE-AC02-07CH11359 with the United States Department of Energy.
The HSF Community White Paper workshops have been supported and organized by many funding agencies and institutions including the University of California at San Diego, the San Diego Supercomputer Center (SDSC), CERN, Nikhef, the Laboratoire d'Annecy de Physique des Particules (LAPP) and the S2I2-HEP project supported by National Science Foundation grants ACI-1558216, ACI-1558219, and ACI-1558233.

IOP Conf. Series: Journal of Physics: Conf. Series **1085** (2018) 022007 doi:10.1088/1742-6596/1085/2/022007

Games and loop integrals

Ben Ruijl

ETH Zurich, Wolfgang-Pauli-Str. 27, 8093 Zürich, Switzerland

E-mail: benruyl@gmail.com

Abstract. In order to numerically integrate complicated cross sections, large expressions have to be sampled many times, which is a time consuming operation. Inspired by the use in games, artificial intelligence methods such as Monte Carlo Tree Search and Local Stochastic Search are used to simplify the expressions. As a result, large expressions can be compressed by more than a factor 20.

1. Introduction

Monte Carlo integration is the preferred way to compute complicated cross sections of scattering processes [1, 2, 3, 4]. The disadvantage of this method is that it converges slowly and thus requires a large number of samples. Sampling large multivariate polynomials is very time consuming, which makes Monte Carlo methods slow. The expressions that arise from Quantum Chromodynamics (QCD) are polynomials in many variables. The number of variables could range from a few to several hundreds. Analogously, the number of terms range from ten thousand terms to millions of terms [5]. In some extreme cases, the executable code that performs the evaluation of the expression could take up a few gigabytes. If we are able to reduce the number of operations required to evaluate these expressions, sampling becomes faster. As a result, Monte Carlo integrators can obtain precise results much faster.

We describe two methods to reduce the number of operations. The first is Horner's rule for multivariate polynomials, which is extracting variables outside brackets [6] (sec. 2.1). For multivariate expressions the order of these variables is called a *Horner scheme*. The second is called Common Subexpression Elimination (CSEE) [7], which is performed after the Horner scheme has been applied (sec. 2.2). We will investigate two methods of finding a Horner scheme that yields a near-minimal number of operations after the Horner scheme and CSEE have been applied. The first is the tree search algorithm Monte Carlo Tree Search (MCTS), which is also used for finding the best moves in games. For example, MCTS is an important component of AlphaGo [8]. The second method is a local search, which has been successful in solving the Travelling Salesman problem [9].

Our final local search algorithm is able to reduce the computation time of numerical integration from weeks to days or even hours. The methods have been implemented in FORM [10] and are used by at least one other research group.

The remainder of this work is structured as follows. First, we discuss methods for expression simplification in section 2. In section 3 we discuss the experimental setup. Next, we discuss MCTS in section 4. In section 5 we examine local search methods. We discuss performance and results in section 6. Finally, we present the conclusion in section 7.

IOP Conf. Series: Journal of Physics: Conf. Series **1085** (2018) 022007 doi:10.1088/1742-6596/1085/2/022007

2. Horner schemes and common subexpression elimination

Expression simplification is a widely studied problem. Some examples are Horner schemes [7], common subexpression elimination (CSEE) [11], partial syntactic factorisation [12] and Breuer's growth algorithm [13]. Much research is put into simplifications using more algebraic properties, such as factorisation, especially because of its interest for cryptographic research [14, 15]. Simplification methods that depend on factorisation have the major problem of being notoriously slow. Horner schemes and CSEE do not require sophisticated mathematics: only the commutative and associative properties of the operators are used. The expressions we are considering often have more than twenty variables and more than a hundred thousand terms [5]. In this regime, computationally expensive methods are infeasible. Therefore, we consider using basic methods such as Horner schemes and CSEE.

2.1. Horner Schemes

Horner's rule reduces the number of multiplications in an expression by lifting variables outside brackets [6, 7, 16]. For multivariate expressions Horner's rule can be applied sequentially, once for each variable. The order of this sequence is called the *Horner scheme*. Take for example:

$$x^2z + x^3y + x^3yz \to x^2(z + x(y(1 + z))) \,. \tag{1}$$

Here, first the variable x is extracted (i.e., x^2 and x) and second, y. The number of multiplications is now reduced from 9 to 4. However, the order x, y is chosen arbitrarily. One could also try the order y, x:

$$x^2z + x^3y + x^3yz \to x^2z + y(x^3(1 + z)) \,, \tag{2}$$

for which the number of multiplications is 6. Evidently, this is a sub-optimal Horner scheme. There are $n!$ orders of extracting variables, where n is the number of variables, and it turns out that the problem of selecting an optimal ordering is NP-hard [16].

2.2. Common subexpression elimination

The number of operations can be reduced further by applying common subexpression elimination (CSEE). This method is well known from the fields of compiler construction [11] and computer chess [17], where it is applied to much smaller expressions or subtrees than what we are considering here. Figure 1 shows an example of a common subexpression in a tree representation of an expression. The shaded expression $b(a+e)$ appears twice, and its removal means removing one superfluous addition and one multiplication.

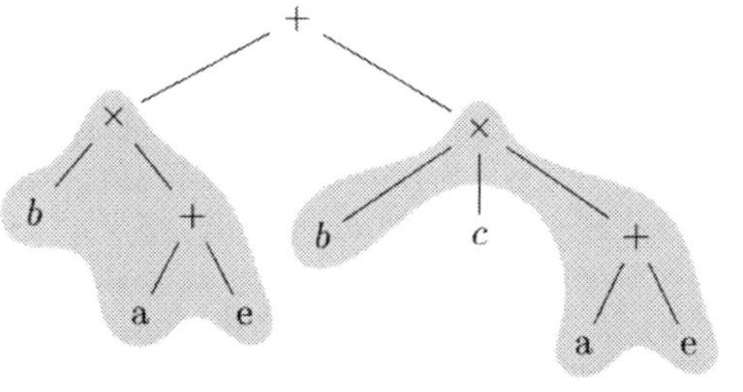

Figure 1: A common subexpression (shaded) in an associative and commutative tree representation.

CSEE is able to reduce both the number of multiplications and the number of additions, whereas Horner schemes are only able to reduce the number of multiplications.

	variables	terms	operations	eval. time (s)
res(7,4)	13	2561	29 163	0.001
res(7,5)	14	11 379	142 711	0.03
res(7,6)	15	43 165	587 880	0.13
res(9,8)	19	4 793 296	83 778 591	25.0
HEP(σ)	15	5716	47 424	0.008
HEP(F_{13})	24	105 058	1 068 153	0.4
HEP(F_{24})	31	836 009	7 722 027	3.0
HEP(b)	107	193 767	1 817 520	2.0

Table 1: The number of variables, terms, operations, and the evaluation time of applying a single Horner scheme and CSEE in seconds, for our eight (unoptimised) benchmark expressions. The time measurement is performed on a 2.4 GHz Xeon computer. All expressions fit in memory (192 GB).

3. Experimental setup

We use eight large benchmark expressions, four from mathematics and four from real-world High Energy Physics (HEP) calculations. In table 1 statistics for the expressions are displayed. We show the number of variables, terms, operations, and the evaluation time of applying a Horner scheme and CSEE.

The expressions called res(7,4), res(7,5), res(7,6), and res(9,8) are resolvents and are defined by $\text{res}(m,n) = \text{res}_x(\sum_{i=0}^{m} a_i x^i, \sum_{i=0}^{n} b_i x^i)$, as described in [12]. The number of variables is $m+n+2$. The polynomial res(9,8) is the largest polynomial we have tested and has been included to test the boundaries of our hardware.

The High Energy Physics expressions represent scattering processes for the future International Linear Collider, a likely successor to the Large Hadron Collider [5]. A standard method of calculating the probability of certain collision events is by using perturbation theory. As a result, for each order of perturbations, additional expressions are calculated as corrections to previous orders of precision. The HEP polynomials of table 1 are second-order corrections to various processes.

HEP(σ) describes parts of the process $e^+e^- \to \mu^+\mu^-\gamma$. HEP($F_{13}$), HEP($F_{24}$), and HEP($b$) are obtained from the process $e^+e^- \to \mu^+\mu^- u\bar{u}$. The results can be used to obtain next-generation precision measurements for electron-positron scattering [5]. These four HEP polynomials represent classes of polynomials with approximately the same behaviour.

4. Tree search methods

Recently, Monte Carlo Tree Search has been shown to yield good quality Horner schemes [18]. We will describe its characteristics, so that we can see if we can improve its performance.

Monte Carlo Tree Search (MCTS) is a tree search method that has been successful in games such as Go, Hex, and other applications with a large state space [19, 20]. It works by selectively building a tree, expanding only branches it deems worthwhile to explore. MCTS consists of four steps, which are displayed in figure 2. The first step (2a) is the selection step, where a leaf or a not fully expanded node is selected according to some criterion (see below). Our choice is node z. In the expansion step (2b), a random unexplored child of the selected node is added to the tree (node y). In the simulation step (2(c)), the rest of the path to a final node is completed using random child selection. Finally a score Δ is obtained that signifies the score of the chosen path through the state space. In the backpropagation step (2d), this value is propagated back through the tree, which affects the average score (winrate) of a node (see below). The tree is built iteratively by repeating the four steps.

IOP Conf. Series: Journal of Physics: Conf. Series **1085** (2018) 022007 doi:10.1088/1742-6596/1085/2/022007

In the game of Go, each node represents a player move and in the expansion phase the game is played out, in basic implementations, by random moves. In the best performing implementations heuristics and pattern knowledge are used to complement a random playout [19]. The final score is 1 if the game is won, and 0 if the game is lost. The entire tree is built, ultimately, to select the best first move.

For our purposes, we need to build a complete Horner scheme, variable by variable. As such, each node will represent a variable and the depth of a node in the tree represents the position in the Horner scheme. Thus, in figure 2(c) the partial Horner scheme is x,z,y and the rest of the scheme is filled in randomly with unused variables. The score of a path in our case, is the improvement of the path on the number of operations: the original number of operations divided by the number of operations after the Horner scheme and CSEE have been applied. We note that for our purposes the *entire* Horner scheme is important and not just the first variable.

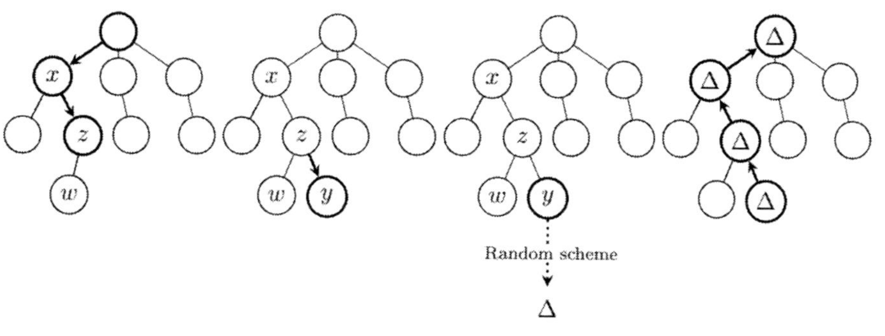

Figure 2: An overview of the four phases of MCTS: selection (a), expansion (b), simulation (c), and backpropagation (d). The selection of a not fully expanded node is done using the best child criterion (in our case Upper Confidence bounds applied to Trees (UCT)). Δ is the number of operations left in the final expression, after the Horner scheme and CSEE have been applied. See also [21].

In many MCTS implementations UCT (formula 3) is chosen as the selection criterion [21, 22]:

$$\underset{\text{children } c \text{ of } s}{\text{argmax}} \ \bar{x}(c) + 2C_p\sqrt{\frac{2\ln n(s)}{n(c)}}, \tag{3}$$

where c is a child node of node s, $\bar{x}(c)$ the average score of node c, $n(c)$ the number of times the node c has been visited, C_p the exploration-exploitation constant, and argmax the function that selects the child with the maximum value. This formula balances exploitation, i.e., picking terms with a high average score, and exploration, i.e., selecting nodes where the child has not been visited often compared to the parent. The C_p constant determines how strong the exploration term is: for high C_p the focus will be on exploration, and for low C_p the focus will be on exploitation.

We consider the role of the exploration-exploitation constant C_p in the UCT formula. We notice that in the first iterations of the simulation there is as much exploration as there is in the final iterations, since C_p remains constant throughout the search. For example, the final 100 iterations of a 1000 iterations MCTS run are used to explore new branches even though we know in advance that there is likely not enough time to reach the final nodes. Thus we would like to modify the C_p to change during the simulation to emphasise exploration early in the search and emphasise exploitation towards the end.

We introduce a new, dynamic exploration-exploitation parameter T that decreases linearly with the iteration number [23]:

$$T(i) = C_p \frac{N-i}{N} \,, \tag{4}$$

where i is the current iteration number, N the preset maximum number of iterations, and C_p the initial exploration-exploitation constant at $i = 0$.

We modify the UCT formula to become:

$$\underset{\text{children } c \text{ of } s}{\text{argmax}} \; \bar{x}(c) + 2T(i)\sqrt{\frac{2 \ln n(s)}{n(c)}} \,, \tag{5}$$

where c is a child of node s, $\bar{x}(c)$ is the average score of child c, $n(c)$ the number of visits at node c, and $T(i)$ the dynamic exploration-exploitation parameter of formula (4).

The role of T is similar to the role of the temperature in Simulated Annealing [24]: in the beginning of the simulation there is much emphasis on exploration, the analogue of allowing transitions to energetically unfavourable states. During the simulation the focus gradually shifts to exploitation, analogous to annealing. Hence, we call our new UCT formula "Simulated Annealing UCT (Simulated Annealing - UCT (SA-UCT))".

In the past related changes have been proposed. For example, Discounted UCB [25] and Accelerated UCT [26] both modify the average score of a node to discount old wins over new ones. The difference between our method and past work is that the previous modifications alter the importance of exploring based on the history and do not guarantee that the focus shifts from exploration to exploitation. In contrast, this work focuses on the exploration-exploitation constant C_p and on the role of exploration during the simulation.

We implemented four improvements over UCT. (1) The final iterations are used effectively. (2) There is more exploration in the middle and at the bottom of the tree. This is due to more nodes being expanded at lower levels, because the T is lowered. As a consequence, we see that (3) more branches reach the end states. As a result, (4) there is exploration near the bottom, where there was none for the random playouts.

In order to analyse the effect of SA-UCT on the fine-tuning of C_p (the initial temperature), we perform a sensitivity analysis on C_p and N [23]. We find that (1) SA-UCT increases the region of C_p that gives good results by and order of magnitude, and (2) produces Horner schemes of the same quality. Thus, we may conclude that SA-UCT reduces the fine-tuning problem without overhead.

5. Stochastic Local Search

There are some intrinsic shortcomings to using a tree representation, especially if the depth of the search tree becomes (too) large. We noticed that many branches do not reach the bottom when there are more than 20 variables (we remind the reader that the problem depth is equivalent to the number of variables) as is the case with many of our expressions. MCTS determines the scores of a branch by performing a random play-out. If the branch is not constructed all the way to the bottom, the final nodes are therefore random (no optimisation). For Horner schemes, the *entire* scheme is important, so sub-optimal selection of variables at the end of the scheme can have a significant impact.

The issue of poor optimisation at the bottom of the tree motivated us to look for a method that is symmetric in its optimisation: both the beginning and the end should be optimised equally well. In this section we (re)consider which class of algorithms is best suited for the Horner scheme problem. The Horner scheme problem belongs to the class of permutation problems. Many algorithms for optimising permutation problems have been suggested in the literature, such as Stochastic Hill Climbing (SHC) [27], Simulated Annealing (SA) [24], Tabu Search [28],

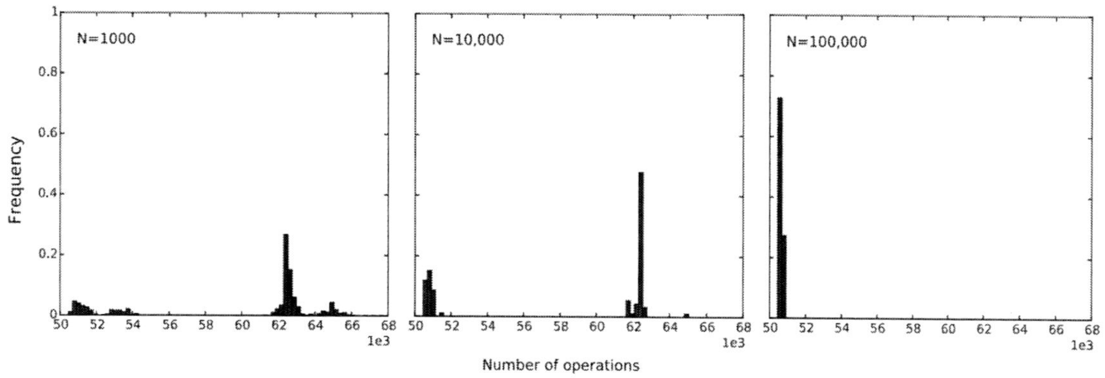

Figure 3: The distribution of the number of operations for the HEP(F_{13}) expression with 1 swaps for 1000 iterations (left), 10 000 iterations (middle) and 100 000 iterations (right) at $T = 0$. There appears to be a local minimum around 62 000. However, as the number of iterations is increased, the local minimum becomes smaller relative to the global minimum region (middle) and completely disappears (right). We may conclude that the apparent local minimum is not a local minimum, but a saddle point, since SHC is able to escape from the 'minimum'.

Ant Colony Optimisation [29], and Evolutionary Algorithms [30]. Since measurements take weeks per algorithm, we limit ourselves to two. In [31] we provide qualitative motivations for focusing on SHC and SA. In summary, the absence of heuristics and the high cost of sampling to tune parameters make the other options less interesting.

A Stochastic Hill Climbing procedure, also known as iterative improvement local search, has two parameters: (1) the number of iterations N, and (2) the neighbourhood structure, which defines the transition function [9]. Fewer tunable parameters is an advantage in our use case, since the evaluation time is often too long for extensive tuning. We find that running half of the simulations with a neighbourhood structure that makes minor changes to the state (i.e., a single shift of a variable), and running the other half with a neighbourhood structure that involves larger changes (i.e., the mirroring of a random sublist) is a good strategy for all of our benchmark expressions [31]. Consequently, only the computation time remains as an actual parameter. A Stochastic Hill Climbing procedure only moves to a neighbour if the evaluation score (number of operations) is improved. As a consequence, SHC could get stuck in local minima.

To obtain an idea on the number of local minima, we measure how often the simulation gets stuck: if there are many local minima, we expect the simulation to get stuck often. In figure 3, we show the distribution of HEP(F_{13}) for 1000, 10 000, and 100 000 SHC runs respectively. For 1000 and 10 000 runs we see two peaks: one at the global minimum near 51 000 and one at an apparent local minimum near 62 000. As the number of iterations is increased, the weight shifts from the apparent local minimum to the global minimum: at 1000 iterations, there is a probability of 27.5% of arriving in the region of the global minimum, whereas this is 36.3% at 10 000 iterations. Apparently the local minimum is 'leaking': given sufficient time, the search is able to escape. The figure on the right with 100 000 iterations confirms the escaping possibility: the apparent local minimum has completely disappeared. Thus, the local minimum is in reality a saddle point, since for a true local minimum there is no path with a lower score leading away from the minimum. Since SHC requires many iterations to escape from the saddle point, only a few transitions reduce the number of operations.

We observe that apparent local minima disappear for our other benchmark expressions as well. SHC runs with 100 000 iterations approach the global minimum for all of our benchmark expressions. For example, for HEP(F_{13}) mentioned above, the result is 50636 ± 57 and for

IOP Conf. Series: Journal of Physics: Conf. Series **1085** (2018) 022007 doi:10.1088/1742-6596/1085/2/022007

	vars	original	MCTS 1k	MCTS 10k	SHC 1k	SHC 10k	$E_{min,4}$ 1k
res(7,4)	13	29 163	$(3.86 \pm 0.1) \cdot 10^3$	$(3.84 \pm 0.01) \cdot 10^3$	$(3.92 \pm 0.28) \cdot 10^3$	3834 ± 26	3819 ± 9
res(7,5)	14	142 711	$(1.39 \pm 0.01) \cdot 10^4$	13768 ± 28	13841 ± 441	13767 ± 21	13770 ± 5
res(7,6)	15	587 880	$(4.58 \pm 0.05) \cdot 10^4$	$(4.54 \pm 0.01) \cdot 10^4$	46642 ± 3852	$(4.61 \pm 0.25) \cdot 10^4$	$(4.55 \pm 0.16) \cdot 10^4$
res(9,8)	19	83 778 591	$(5.27 \pm 0.25) \cdot 10^6$	$(4.33 \pm 0.31) \cdot 10^6$	$(4.13 \pm 0.34) \cdot 10^6$	$(4.03 \pm 0.17) \cdot 10^6$	$(3.97 \pm 0.18) \cdot 10^6$
HEP(σ)	15	47 424	4114 ± 14	4087 ± 5	4226 ± 257	4082 ± 58	4075 ± 25
HEP(F_{13})	24	1 068 153	$(6.6 \pm 0.2) \cdot 10^4$	$(6.47 \pm 0.08) \cdot 10^4$	$(5.99 \pm 0.51) \cdot 10^4$	$(5.80 \pm 0.55) \cdot 10^4$	$(5.37 \pm 0.40) \cdot 10^4$
HEP(F_{24})	31	7 722 027	$(3.80 \pm 0.06) \cdot 10^5$	$(3.19 \pm 0.04) \cdot 10^5$	$(3.16 \pm 0.23) \cdot 10^5$	$(3.06 \pm 0.23) \cdot 10^5$	$(2.98 \pm 0.09) \cdot 10^5$
HEP(b)	107	1 817 520	$(1.81 \pm 0.04) \cdot 10^5$	$(1.65 \pm 0.08) \cdot 10^5$	$(1.50 \pm 0.08) \cdot 10^5$	$(1.40 \pm 0.06) \cdot 10^5$	$(1.44 \pm 0.04) \cdot 10^5$

Table 2: SHC compared to MCTS. The MCTS results for all expressions except res(9,8) and HEP(b) are from [18]. All the values are statistical averages over at least 100 runs. SHC results have a larger standard deviation, and thus the expected value of the minimum is often lower than these values (see last column).

HEP(σ) the result is 4078 ± 9. The small standard deviations indicate that no runs get stuck in local minima (at least not in local minima significantly higher than the standard deviation).

From these results we may conclude that true local minima, from which a Stochastic Hill Climbing cannot escape, are rare for Horner schemes.

6. Performance of SHC vs. MCTS

Below, we compare the results of Stochastic Hill Climbing to the previous best results from MCTS, for our eight benchmark expressions res(7,4), res(7,5), res(7,6), res(9,8), HEP(σ), HEP(F_{13}), HEP(F_{24}) and HEP(b). The results of all the MCTS runs except for res(9,8), and HEP(b) are taken from [18].[1] The results are displayed in table 2.

The results for MCTS with 1000 and 10 000 iterations are obtained after considerable tuning of C_p and after selecting whether the scheme should be constructed forward or in reverse (i.e., the scheme is applied backwards [23]).

For smaller problems, we observe that the averages of SHC are on a par with MCTS. However, we see that the standard deviations of SHC are higher than MCTS. This is because for MCTS the first nodes are fixed rather fast, which limits the variety. Consequently, we expect SHC to outperform MCTS if several runs are performed in parallel. Indeed, this is what we see in the last column of table 2. The standard deviations of MCTS are often an order of magnitude smaller than those of SHC, so the benefits of running MCTS in parallel are smaller. We may conclude that SHC has a better minimal behaviour if run in parallel.

For our largest expressions, HEP(F_{13}), HEP(F_{24}) and HEP(b), we observe that SHC with 1000 iterations yields better results than MCTS with 10 000 iterations. For HEP(F_{24}), the average of SHC with 1000 iterations is about 20% better than the average for MCTS with 1000 iterations. In fact, the results are slightly better than MCTS with 10 000 iterations. If we take the $E_{min,4}$ into account, the expected value for HEP(F_{24}) is an additional 7% less.

The fact that SHC outperforms MCTS when the number of variables is larger than 23, may be due to the fact that there are not sufficient iterations for the branches to reach the bottom, making the choice of the last variables essentially random (see section 5 and [23]). This may also be the reason why for MCTS it is important whether the scheme is constructed forward or in reverse: if most of the performance can be gained by carefully selecting the last variables, building the scheme in reverse will yield better performance.

SHC is 10 times faster (in clock time) than MCTS, since most of the time is spent in the evaluation function. It is able to make reductions up to a factor of 26 for our largest expression.

[1] We only consider optimisations by Horner schemes and CSEE. Additional optimisations that are mentioned in [18], such as 'greedy' optimisations, can just as well be applied to the results of SHC.

IOP Conf. Series: Journal of Physics: Conf. Series **1085** (2018) 022007 doi:10.1088/1742-6596/1085/2/022007

7. Conclusions

Monte Carlo Tree Search (MCTS) with UCT has proven to be a good candidate to simplify large expression [18]. A downside of this method is that the constant C_p that governs exploration versus exploitation has to be tuned. The quality of the final scheme largely depends on this constant. We have modified the UCT algorithm so that C_p decreases linearly with the iteration number [23]. As a result, the final iterations are spent on *optimising* the end of the tree, instead of *exploring*. We show that using this modified UCT, the sensitivity to C_p is decreased by at least a factor 10 [23, 32]. Thus, the tuning is simplified.

Tree search methods, even with SA-UCT, have the problem that the beginning of the tree is optimised more than the end. For Horner schemes this does not lead to optimal solutions. Therefore we considered other algorithms that optimise uniformly. Since sampling is slow for our use case, tuning many parameters is infeasible. For this reason, we preferred straightforward algorithms over sophisticated ones. In the case of Horner schemes, we have found that one of the most basic algorithms, Stochastic Hill Climbing, yields the best results.

Stochastic Hill Climbing provides a search method with two parameters: (1) the number of iterations (computation time) and (2) the neighbourhood structure, which is a tunable parameter. We found that running half of the simulations with a neighbourhood structure that makes minor changes to the state (i.e., a single shift of a variable), and running the other half with a neighbourhood structure that involves larger changes (i.e., the mirroring of a random sublist) is a good strategy for all of our benchmark expressions. Consequently, only the computation time remains as an actual parameter. From our experimental results we arrive at three subconclusions: (1) SHC obtains similar results to MCTS for expressions with around 15 variables, (2) SHC outperforms MCTS for expressions with 24 or more variables, and (3) SHC requires ten times fewer samples than MCTS to obtain similar results. Therefore we may conclude that SHC is more than 10 times faster [31].

The result that a basic algorithm such as SHC performs well is surprising, since Horner schemes have at least two properties that make the search hard: (1) there are no known local heuristics, and (2) evaluations could take several seconds. In the previous sections we have shown that the performance of SHC is so good, because the state space of Horner schemes is flat and has few local minima.

The number of operations is linearly related to the time it takes to perform numerical evaluations. The difference between the number of operations for the unoptimised and the optimised expression is more than a factor 24 compared. As a consequence, we are able to perform numerical integration (via repeated numerical evaluations) at least 24 times faster.

For High Energy Physics, the contribution is immediate: numerical integration of processes that are currently experimentally verified at CERN can be done significantly faster. Our algorithms are implemented in the open source symbolic manipulation system FORM 4.2 [10] and are used by multiple research groups.

Acknowledgements

This work is supported by the ERC Advanced Grant no. 320651, "HEPGAME".

References

[1] Berends F A, Kuijf H, Tausk B and Giele W T 1991 *Nucl. Phys.* **B357** 32–64

[2] Stelzer T and Long W F 1994 *Comput. Phys. Commun.* **81** 357–371 (*Preprint* hep-ph/9401258)

[3] Campbell J M and Ellis R K 1999 *Phys. Rev.* **D60** 113006 (*Preprint* hep-ph/9905386)

[4] Boughezal R, Campbell J M, Ellis R K, Focke C, Giele W, Liu X, Petriello F and Williams C 2017 *Eur. Phys. J.* **C77** 7 (*Preprint* 1605.08011)

[5] Fujimoto J, Ishikawa T, Kato K, Kaneko T, Nakazawa N, Shimizu Y, Vermaseren J and Yasui Y 2006 *Nucl. Phys. Proc. Suppl.* **160** 150–154 [,150(2006)]

[6] Horner W 1819 *A New Method of Solving Numerical Equations of All Orders by Continuous Approximation* (W. Bulmer & Co) dover reprint, 2 vols 1959 URL `http://books.google.nl/books?id=2fLpGwAACAAJ`

[7] Knuth D E 1997 *The Art of Computer Programming, Volume 2 (3rd Ed.): Seminumerical Algorithms* (Boston, MA, USA: Addison-Wesley Longman Publishing Co., Inc.) ISBN 0-201-89684-2

[8] Silver D, Schrittwieser J, Simonyan K, Antonoglou I, Huang A, Guez A, Hubert T, Baker L, Lai M, Bolton A, Chen Y, Lillicrap T, Hui F, Sifre L, Driessche G v d, Graepel T and Hassabis D 2017 *Nature* **550** 354–359 ISSN 0028-0836 URL `http:https://doi.org/10.1038/nature24270`

[9] Aarts E and Lenstra J K (eds) 1997 *Local Search in Combinatorial Optimization* 1st ed (New York, NY, USA: John Wiley & Sons, Inc.) ISBN 0471948225

[10] Ruijl B, Ueda T and Vermaseren J 2017 (*Preprint* `1707.06453`)

[11] Aho A V, Sethi R and Ullman J D 1988 *Compilers: Principles, Techniques and Tools* (Addison-Wesley)

[12] Leiserson C E, Li L, Maza M M and Xie Y 2010 Efficient Evaluation of Large Polynomials *In Proc. International Congress of Mathematical Software - ICMS 2010* (Springer)

[13] Breuer M A 1969 *Commun. ACM* **12** 333–340 ISSN 0001-0782 URL `http://doi.acm.org/10.1145/363011.363153`

[14] Wang P S 1978 *Mathematics of Computation* **32** 1215–1231 ISSN 00255718, 10886842 URL `http://www.jstor.org/stable/2006346`

[15] Wu W, Chen J and Feng Y 2014 *Science China Mathematics* **57** 2123–2142 ISSN 1869-1862 URL `http://dx.doi.org/10.1007/s11425-014-4850-y`

[16] Ceberio M and Kreinovich V 2004 *SIGSAM Bull.* **38** 8–15 ISSN 0163-5824 URL `http://doi.acm.org/10.1145/980175.980179`

[17] Adelson-Velsky G M, Arlazarov V L and Donskoy M V 1988 *Algorithms for Games* (New York, NY, USA: Springer-Verlag New York, Inc.) ISBN 0-387-96629-3

[18] Kuipers J, Plaat A, Vermaseren J and van den Herik J 2013 *Computer Physics Communications* URL `http://www.sciencedirect.com/science/article/pii/S0010465513001689`

[19] Lee C S, Wang M H, Chaslot G, Hoock J B, Rimmel A, Teytaud O, Tsai S R, Hsu S C and Hong T P 2009 *IEEE Trans. Comput. Intellig. and AI in Games* **1** 73–89

[20] Hayward R 2009 *International Computer Games Association (ICGA)* **32** 114–116

[21] Browne C, Powley E, Whitehouse D, Lucas S, Cowling P, Rohlfshagen P, Tavener S, Perez D, Samothrakis S and Colton S 2012 *Computational Intelligence and AI in Games, IEEE Transactions on* **4** 1–43 ISSN 1943-068X

[22] Kocsis L and Szepesvri C 2006 Bandit based Monte-Carlo Planning *In: ECML-06. LNCS 4212* (Springer) pp 282–293

[23] Ruijl B, Vermaseren J, Plaat A and van den Herik H J 2014 *Proceedings of ICAART Conference 2014* **1** 724–731 (*Preprint* `1312.0841`) URL `http://arxiv.org/abs/1312.0841`

[24] Kirkpatrick S, Gelatt C D and Vecchi M P 1983 *Science* **220** 671–680

[25] Kocsis L and Szepesvri C 2006 Discounted UCB. Video Lecture in the lectures of PASCAL Second Challenges Workshop 2006, Slides: `http://www.lri.fr/sebag/Slides/Venice/Kocsis.pdf`.

[26] Hashimoto J, Kishimoto A, Yoshizoe K and Ikeda K 2012 *Lecture Notes in Computer Science* **7168** 1 – 12 URL `http://id.nii.ac.jp/0023/00008786`

[27] Hoos H and Stützle T 2004 *Stochastic Local Search: Foundations & Applications* (San Francisco, CA, USA: Morgan Kaufmann Publishers Inc.) ISBN 1558608729

[28] Glover F 1986 *Computers & Operations Research* **13** 533 – 549 ISSN 0305-0548 applications of Integer Programming URL `http://www.sciencedirect.com/science/article/pii/0305054886900481`

[29] Dorigo M and Gambardella L M 1997 *Trans. Evol. Comp* **1** 53–66 ISSN 1089-778X URL `http://dx.doi.org/10.1109/4235.585892`

[30] Goldberg D E 1989 *Genetic Algorithms in Search, Optimization and Machine Learning* 1st ed (Boston, MA, USA: Addison-Wesley Longman Publishing Co., Inc.) ISBN 0201157675

[31] Ruijl B, Vermaseren J, Plaat A and van den Herik H J 2014 Why Local Search Excels in Expression Simplification (*Preprint* `1409.5223`) URL `http://arxiv.org/abs/1409.5223`

[32] Ruijl B, Vermaseren J, Plaat A and van den Herik H J 2014 *Proceedings of the 11th International Workshop on Boolean Problems* URL `http://arxiv.org/abs/1405.6369`

IOP Conf. Series: Journal of Physics: Conf. Series **1085** (2018) 022008 doi:10.1088/1742-6596/1085/2/022008

Machine Learning in High Energy Physics Community White Paper

Abstract. Machine learning is an important applied research area in particle physics, beginning with applications to high-level physics analysis in the 1990s and 2000s, followed by an explosion of applications in particle and event identification and reconstruction in the 2010s. In this document we discuss promising future research and development areas in machine learning in particle physics with a roadmap for their implementation, software and hardware resource requirements, collaborative initiatives with the data science community, academia and industry, and training the particle physics community in data science. The main objective of the document is to connect and motivate these areas of research and development with the physics drivers of the High-Luminosity Large Hadron Collider and future neutrino experiments and identify the resource needs for their implementation. Additionally we identify areas where collaboration with external communities will be of great benefit.

E-mail: Sergei.Gleyzer@cern.ch

Authors: Kim Albertsson[1], Piero Altoe[2], Dustin Anderson[3], Michael Andrews[4], Juan Pedro Araque Espinosa[5], Adam Aurisano[6], Laurent Basara[7], Adrian Bevan[8], Wahid Bhimji[9], Daniele Bonacorsi[10], Paolo Calafiura[9], Mario Campanelli[8], Louis Capps[2], Federico Carminati[11], Stefano Carrazza[11], Taylor Childers[12], Elias Coniavitis[13], Kyle Cranmer[14], Claire David[15], Douglas Davis[16], Javier Duarte[17], Martin Erdmann[18], Jonas Eschle[19], Amir Farbin[20], Matthew Feickert[21], Nuno Filipe Castro[5], Conor Fitzpatrick[22], Michele Floris[11], Alessandra Forti[23], Jordi Garra-Tico[24], Jochen Gemmler[25], Maria Girone[11], Paul Glaysher[15], Sergei Gleyzer[26], Vladimir Gligorov[27], Tobias Golling[28], Jonas Graw[2], Lindsey Gray[17], Dick Greenwood[29], Thomas Hacker[30], John Harvey[11], Benedikt Hegner[11], Lukas Heinrich[14], Ben Hooberman[31], Johannes Junggeburth[32], Michael Kagan[33], Meghan Kane[34], Konstantin Kanishchev[7], Przemysław Karpiński[11], Zahari Kassabov[35], Gaurav Kaul[36], Dorian Kcira[3], Thomas Keck[25], Alexei Klimentov[37], Jim Kowalkowski[17], Luke Kreczko[38], Alexander Kurepin[39], Rob Kutschke[17], Valentin Kuznetsov[40], Nicolas Köhler[32], Igor Lakomov[11], Kevin Lannon[41], Mario Lassnig[11], Antonio Limosani[42], Gilles Louppe[14], Aashrita Mangu[43], Pere Mato[11], Helge Meinhard[11], Dario Menasce[44], Lorenzo Moneta[11], Seth Moortgat[45], Meenakshi Narain[46], Mark Neubauer[31], Harvey Newman[3], Hans Pabst[36], Michela Paganini[47], Manfred Paulini[4], Gabriel Perdue[17], Uzziel Perez[48], Attilio Picazio[49], Jim Pivarski[50], Harrison Prosper[51], Fernanda Psihas[52], Alexander Radovic[53], Ryan Reece[54], Aurelius Rinkevicius[40], Eduardo Rodrigues[6], Jamal Rorie[55], David Rousseau[56], Aaron Sauers[17], Steven Schramm[28], Ariel Schwartzman[33], Horst Severini[57], Paul Seyfert[11], Filip Siroky[58], Konstantin Skazytkin[39], Mike Sokoloff[6], Graeme Stewart[59], Bob Stienen[60], Ian Stockdale[61], Giles Strong[5], Savannah Thais[47], Karen Tomko[62], Eli Upfal[46], Emanuele Usai[46], Andrey Ustyuzhanin[63], Martin Vala[64], Sofia Vallecorsa[65], Justin Vasel[66], Mauro Verzetti[67], Xavier Vilasís-Cardona[68], Jean-Roch Vlimant[3], Ilija Vukotic[69], Sean-Jiun Wang[26], Gordon Watts[70], Michael Williams[71], Wenjing Wu[72], Stefan Wunsch[25], Omar Zapata[73]

[1] Lulea University of Technology
[2] NVidia
[3] California Institute of Technology
[4] Carnegie Mellon University
[5] LIP Lisboa
[6] University of Cincinnati
[7] Universita e INFN, Padova
[8] University of London

IOP Conf. Series: Journal of Physics: Conf. Series **1085** (2018) 022008 doi:10.1088/1742-6596/1085/2/022008

[9] Lawrence Berkeley National Laboratory
[10] Universita e INFN, Bologna
[11] CERN
[12] Argonne National Laboratory
[13] Universitaet Freiburg
[14] New York University
[15] Deutsches Elektronen-Synchrotron
[16] Duke University
[17] Fermi National Accelerator Laboratory
[18] RWTH Aachen University
[19] Universität Zürich
[20] University of Texas at Arlington
[21] Southern Methodist University
[22] Ecole Polytechnique Federale de Lausanne
[23] University of Manchester
[24] University of Cambridge
[25] Karlsruher Institut für Technologie
[26] University of Florida
[27] Centre National de la Recherche Scientifique
[28] Université de Genève
[29] Louisiana Tech University
[30] Purdue University
[31] University of Illinois at Urbana-Champaign
[32] Max Planck Institut für Physik
[33] SLAC National Accelerator Laboratory
[34] SoundCloud
[35] University of Milan
[36] Intel
[37] Brookhaven National Laboratory
[38] University of Bristol
[39] Russian Academy of Sciences
[40] Cornell University
[41] University of Notre Dame
[42] University of Melbourne
[43] University of California Berkeley
[44] Universita & INFN, Milano Bicocca
[45] Vrije Universiteit Brussel
[46] Brown University
[47] Yale University
[48] University of Alabama
[49] University of Massachusetts
[50] Princeton University
[51] Florida State University
[52] Indiana University
[53] College of William and Mary
[54] University of California, Santa Cruz
[55] Rice University
[56] Universite de Paris Sud 11
[57] University of Oklahoma
[58] Masaryk University
[59] University of Glasgow
[60] Radboud Universiteit Nijmegen
[61] Altair Engineering
[62] Ohio Supercomputer Center
[63] Yandex School of Data Analysis
[64] Technical University of Kosice

IOP Conf. Series: Journal of Physics: Conf. Series **1085** (2018) 022008 doi:10.1088/1742-6596/1085/2/022008

[65] Gangneung-Wonju National University
[66] Indiana University
[67] University of Rochester
[68] University of Barcelona
[69] University of Chicago
[70] University of Washington
[71] Massachusetts Institute of Technology
[72] Chinese Academy of Sciences
[73] OProject and University of Antioquia

IOP Conf. Series: Journal of Physics: Conf. Series **1085** (2018) 022008 doi:10.1088/1742-6596/1085/2/022008

1. Introduction

The main objectives of particle physics in the post-Higgs boson discovery era is to exploit the full physics potential of both the Large Hadron Collider (LHC) and its upgrade, the high luminosity LHC (HL-LHC), in addition to present and future neutrino experiments. The HL-LHC will deliver data from 100 times the luminosity compared to the LHC, bringing quantitatively and qualitatively new challenges due to event size, data volume, and complexity. The physics reach of the experiments will be limited by the physics performance of algorithms and computational resources. Machine learning (ML) applied to particle physics promises to provide improvements in both of these areas.

Incorporating machine learning in particle physics workflows will require significant research and development over the next five years. Areas where significant improvements are needed include:

- **Physics performance** of reconstruction and analysis algorithms;
- **Execution time** of computationally expensive parts of event simulation, pattern recognition, and calibration;
- **Realtime implementation** of machine learning algorithms;
- **Reduction of the data footprint** with data compression, placement and access.

1.1. Motivation

The experimental high-energy physics (HEP) program revolves around two main objectives: probing the Standard Model (SM) with increasing precision and searching for new physics. Both tasks require the identification of rare signals in immense backgrounds. Substantially increased levels of pile-up at the HL-LHC will make this a significant challenge.

Machine learning algorithms are already the state-of-the-art in event and particle identification, energy estimation and pile-up suppression applications in HEP. Despite their present advantage, machine-learning algorithms still have significant room for improvement in their exploitation of the full potential of the dataset.

1.2. Brief Overview of Machine Learning Algorithms in HEP

This section provides a brief introduction to the most important machine learning algorithms in HEP, introducing key vocabulary (in *italic*).

Machine learning methods are designed to exploit large datasets in order to reduce complexity and find new features in data. The current most frequently used machine learning algorithms in HEP are Boosted Decision Trees (BDTs) and Neural Networks (NN).

Typically, variables relevant to the physics problem are selected and a machine learning *model* is *trained* for *classification* or *regression* using signal and background events (or *instances*). Training the model is the most human- and CPU-time consuming step, while the application, the so called *inference* stage, is relatively inexpensive. BDTs and NNs are typically used to classify particles and events. They are also used for regression, where a continuous function is learned, for example to obtain the best estimate of a particle's energy based on the measurements from multiple detectors.

Neural Networks have been used in HEP for some time; however, improvements in training algorithms and computing power have in the last decade led to the so-called Deep Learning revolution, which has had a significant impact on HEP. Deep Learning is particularly promising when there is a large amount of data and features, as well as symmetries and complex non-linear dependencies between inputs and outputs.

There are different types of deep neural networks used in HEP: fully-connected (FCN), convolutional (CNN) and recurrent (RNN). Additionally, neural networks are used in the context of Generative Models, when a Neural Network is trained to mimic multidimensional distributions to generate any number of new instances. Variational AutoEncoders (VAE) and more recent Generative Adversarial Networks (GAN) are two examples of such generative models used in HEP.

A large set of Machine Learning algorithms are devoted to time series analysis and prediction. They are in general not relevant for HEP where events are independent from each other. However, there is more and more interest in these algorithms for Data Quality and Computing Infrastructure monitoring, as well as those physics processes and event reconstruction tasks where time is an important dimension.

1.3. Structure of the Document

Applications of machine learning algorithms motivated by HEP drivers are detailed in Section 2, while Section 3 focuses on outreach and collaboration with the machine learning community. Section 4 focuses on the machine learning software in HEP and discusses the interplay between internally and externally developed machine learning tools. Recent progress in machine learning was made possible in part by emergence of suitable hardware for training complex models, thus in Section 5 the resource requirements of training and applying machine learning algorithms in HEP are discussed. Section 6 discusses ways of training the HEP community in machine learning. Finally, Section 7 presents the roadmap for the near future.

2. Machine Learning Applications and R&D

This chapter describes the science drivers and high-energy physics challenges where machine learning can play a significant role in advancing the current state of the art. These challenges are selected because of their relevance and potential and also due to similarity with challenges faced outside the field. Despite similarities, major R&D work will go in adapting and evolving such methods to match the particular HEP requirements.

2.1. Detector Simulation

Particle discovery relies on the ability to accurately compare the observed detector response data with expectations based on the hypotheses of the Standard Model or models of new physics. While the processes of subatomic particle interactions with matter are known, it is intractable to compute the detector response analytically. As a result, Monte Carlo simulation tools, such as GEANT [1], have been developed to simulate the propagation of particles in detectors to compare with the data.

For the HL-LHC, on the order of trillions of simulated collisions are needed in order to achieve the required statistical accuracy of the simulations to perform precision hypothesis testing. However, such simulations are highly computationally expensive. For example, simulating the detector response of a single LHC proton-proton collision event takes on the order of several minutes. A particularly time consuming step is the simulation of particles incident on the dense material of a calorimeter, the detector used to measure the energy deposited by the particles. Radiative and nuclear interactions result in the production of a multitude of secondary particles, collectively referred to as a shower. The high interaction probability and resulting high multiplicity of particles passing through the dense material make the simulation of such processes very expensive. This problem is further compounded when particle showers overlap, as is frequently the case in the core of a jet of particles produced by high energy quarks and gluons.

Fast simulation is the process of replacing the slowest components of the simulation chain with computationally efficient approximations. Often such approximations have been done by using simplified parameterizations or particle shower look-up tables. These are computationally fast but often suffer from insufficient accuracy for high precision physics measurements and searches.

Recent progress in high fidelity fast generative models, such as GANs and VAEs, which are able to sample high dimensional feature distributions by learning from existing data samples, offer a promising alternative for simulation. A simplified first attempt at using such techniques saw orders of magnitude increase in simulation speed over existing fast simulation techniques [2], but such generative models

IOP Conf. Series: Journal of Physics: Conf. Series **1085** (2018) 022008 doi:10.1088/1742-6596/1085/2/022008

have not yet reached the required accuracy. Developing these techniques for realistic detector models and understanding how to reach the required accuracy is still needed. The fast advancement in the ML community of such techniques makes this a highly promising avenue to pursue.

Although fast simulation is a necessity, some data analyses will require the highest fidelity simulations from GEANT. There are a large number of parameters that can be used to tune various aspects of the simulation properties. Performing such tuning over many-dimensional parameter space is highly non-trivial. Again, machine learning may offer a solution. Modern optimization techniques, such as Bayesian Optimization, allow for global optimization of the simulator without the detailed knowledge of its internal details [3]. Applying such techniques to simulation tuning may further improve the output of the simulations.

2.2. Real Time Analysis and Triggering

The traditional approach to data analysis in particle physics assumes that the interesting events recorded by a detector can be selected in real-time (a process known as *triggering*) with a reasonable efficiency, and that once selected, these events can be affordably stored and distributed for further selection and analysis at a later point in time. However, the enormous production cross-section and luminosity of the LHC mean that these assumptions break down.[1] In particular there are whole classes of events, for example beauty and charm hadrons or low-mass dark matter signatures, which are so abundant that it is not affordable to store all of the events for later analysis. In order to fully exploit the physics reach of the LHC, it will increasingly be necessary to perform more of the data analysis in real-time [4].

This topic is discussed in some detail in the Reconstruction and Software Triggering chapter, but it is also an important driver of machine learning applications in HEP. Machine learning methods offer the possibility to offset some of the cost of applying reconstruction algorithms, and may be the only hope of performing the real-time reconstruction that enables real-time analysis in the first place. For example, the CMS experiment uses boosted decision trees in the Level 1 trigger to approximate muon momenta. One of the challenges is the trade-off in algorithm complexity and performance under strict inference time constraints. In another example, called the HEP.TrkX project, deep neural networks are trained on large resource platforms and subsequently perform fast inference in online systems.

Real-time analysis poses specific challenges to machine learning algorithm design, in particular how to maintain insensitivity to detector performance which may vary over time. For example, the LHCb experiment uses neural networks for fast fake-track and clone rejection and already employs a fast boosted decision tree for a large part of the event selection in the trigger [5]. It will be important that these approaches maintain performance for higher detector occupancy for the full range of tracks used in physics analyses. Another related application is speeding up the reconstruction of beauty, charm, and other lower mass hadrons, where traditional track combinatorics and vertexing techniques may become too computationally expensive.

In addition, the increasing event complexity particularly in the HL-LHC era will mean that Machine Learning techniques may also become more important to maintaining or improving the efficiency of traditional triggers. Examples of where ML approaches can be useful are the triggering of electroweak events with low-energy objects; improving jet calibration at a very early stage of reconstruction allowing jet triggers thresholds to be lowered; or supernovae and proton decay triggering at neutrino experiments.

2.3. Object Reconstruction, Identification, and Calibration

The physical processes of interest in high energy physics experiments occur on time scales too short to be observed directly by particle detectors. For instance, a Higgs boson produced at the LHC will decay within approximately 10^{-22} seconds and thus decays essentially at the point of production. However, the decay products of the initial particle, which are observed in the detector, can be used

[1] They may well also break down in other areas of high-energy physics in due course.

to infer its properties. Better knowledge of the properties (e.g. type, energy, direction) of the decay products permits more accurate reconstruction of the initial physical process.

Particles are observed in a detector through the energy they deposit when traversing material, which is subsequently digitized. Reconstruction is the process of converting the raw digital signals in the detector into the physical properties of particles. Particle Physics detectors are usually composed of several sub-detectors, each taking advantage of specific interaction mechanisms to detect the passage of a specific type of particle and measure its properties. There are a variety of sub-detector technologies, but most belong to one of three categories:

- Tracking Detectors: These detectors measure the trajectory of electrically-charged particles by spatially locating ionization. Usually trackers are placed in a magnetic field, such that the particle momentum can be inferred from the curvature of the trajectory. Very precise tracking detectors, such as those that employ silicon, provide sufficient spatial resolution to enable locating the particle creation and/or decay point. The ionization also allows for the identification of the particle type.
- Calorimeters: These detectors measure the energy of incident particles by causing them to interact and lose their energy in material and counting the secondary particles produced in such interactions. Highly segmented calorimeters measure the profile of the energy deposition and can also identify the particle type.
- Particle Identification: These detectors are aimed at determining a specific particle type using a variety of techniques.

Algorithmic reconstruction typically involves several steps that turn the data from the detector electronics (*raw* measurements) into higher level data objects, corresponding to the physical particles that were detected (*features*):

- Feature Extraction: The signal from the passage of particles through a detector element, e.g. a calorimeter cell, is observed above noise in the raw electronic output associated with the element. This signal is then characterized.
- Pattern Recognition: The pattern of signals in geometrically adjacent detector elements is associated with the passage of a signal or group of particles. In calorimeters, this step is commonly referred to as clustering.
- Object Characterization: Properties of the objects are measured. In tracking detectors, this step means fitting a pattern of "hits" to a helix. In calorimeters, this step extracts the energy, location, and other properties of the cluster that for example characterize the shape of the cluster.
- Combined reconstruction: Objects in different detectors are associated together to create a refined particle candidate.

Machine learning can in principle be applied at any of these steps. For example, experiments have trained ML algorithms on the features from combined reconstruction algorithms to perform particle identification for decades. In the past decade BDTs have been one of the most popular techniques in this domain. More recently, experiments have been able to extract better performance with deep neural networks.

An active area of research is the application of DNNs to the output of feature extraction in order to perform particle identification and extracting particle properties. This is particularly true for calorimeters or time projection chambers (TPCs), where the data can be represented as a 2D or 3D image and the problems can be cast as computer vision tasks, in which neural networks are used to reconstruct images from pixel intensities. These neural networks are adapted for particle physics applications by optimizing network architectures for complex, 3-dimensional detector geometries and training them on suitable signal and background samples derived from data control regions. Applications include identification and measurements of electrons and photons from electromagnetic showers, jet properties including substructure and b-tagging, taus and missing energy. Promising deep learning architectures for these tasks include convolutional, recurrent and adversarial neural networks. A particularly important application is to Liquid Argon TPCs (LArTPCs), which are the chosen detection technology for the flagship neutrino program.

IOP Conf. Series: Journal of Physics: Conf. Series **1085** (2018) 022008 doi:10.1088/1742-6596/1085/2/022008

For tracking detectors, pattern recognition is the most computationally challenging step. In particular, it becomes computationally intractable for the HL-LHC. The hope is that machine learning will provide a solution that scales linearly with LHC collision density. A current effort called HEP.TrkX investigates deep learning algorithms such as long short-term memory (LSTM) networks for track pattern recognition on many-core processors.

2.4. End-To-End Deep Learning

The vast majority of analyses at the LHC use high-level features constructed from particle four-momenta, even when the analyses make use of machine learning. A high-profile example of such variables are the seven, so-called MELA variables, used in the analysis of the final states $\mathrm{H} \to ZZ \to 4\ell$. While a few analyses, first at the Tevatron, and later at the LHC, have used the four-momenta directly, the latter are still high-level relative to the raw data. Approaches based on the four-momenta are closely related to the Matrix Element Method, which is described in the next section.

Given recent spectacular advances in image recognition based on the use of raw information, we are led to consider whether there is something to be gained by moving closer to using raw data in LHC analyses. This so-called *end-to-end deep learning* approach uses low level data from a detector together with deep learning algorithms. One obvious challenge is that low level data, for example, detector hits, tend to be both high-dimensional and sparse. Therefore, there is interest in also exploring automatic ways to compress raw data in a controlled way that does not necessarily rely on domain knowledge.

2.5. Sustainable Matrix Element Method

The Matrix Element (ME) Method [6–9] is a powerful technique which can be utilized for measurements of physical model parameters and direct searches for new phenomena. It has been used extensively by collider experiments at the Tevatron for standard model (SM) measurements and Higgs boson searches [10–15] and at the LHC for measurements in the Higgs and top quark sectors of the SM [16–22]. The ME method is based on *ab initio* calculations of the probability density function $\mathcal{P}$ of an event with observed final-state particle momenta $\mathbf{x}$ to be due to a physics process ξ with theory parameters $\boldsymbol{\alpha}$. One can compute $\mathcal{P}_\xi(\mathbf{x}|\boldsymbol{\alpha})$ by means of the factorization theorem from the corresponding partonic cross-sections of the hard-scattering process involving parton momenta $\mathbf{y}$ and is given by

$$\mathcal{P}_\xi(\mathbf{x}|\boldsymbol{\alpha}) = \frac{1}{\sigma_\xi^{\text{fiducial}}(\boldsymbol{\alpha})} \int d\Phi(\mathbf{y}_{\text{final}})\, dx_1\, dx_2\, \frac{f(x_1)f(x_2)}{2sx_1x_2}\, |\mathcal{M}_\xi(\mathbf{y}|\boldsymbol{\alpha})|^2\, \delta^4(\mathbf{y}_{\text{initial}} - \mathbf{y}_{\text{final}})\, W(\mathbf{x},\mathbf{y}) \quad (1)$$

where and x_i and $\mathbf{y}_{\text{initial}}$ are related by $y_{\text{initial},i} \equiv \frac{\sqrt{s}}{2}(x_i, 0, 0, \pm x_i)$, $f(x_i)$ are the parton distribution functions, $\sqrt{s}$ is the collider center-of-mass energy, $\sigma_\xi^{\text{fiducial}}(\boldsymbol{\alpha})$ is the total cross section for the process ξ (with $\boldsymbol{\alpha}$) times the detector acceptance, $d\Phi(\mathbf{y})$ is the phase space density factor, $\mathcal{M}_\xi(\mathbf{y}|\boldsymbol{\alpha})$ is the matrix element (typically at leading-order (LO)), and $W(\mathbf{x},\mathbf{y})$ is the probability density (aka "transfer function") that a selected event $\mathbf{y}$ ends up as a measured event $\mathbf{x}$. One can use calculations of Eqn. 1 in a number of ways (e.g. likelihood functions) to search for new phenomena at particle colliders.

The ME method brings in several unique and desirable features, most notably it (1) does not require training data being an *ab initio* calculation of event probabilities, (2) incorporates all available kinematic information of a hypothesized process, including all correlations, and (3) has a clear physical meaning in terms of the transition probabilities within the framework of quantum field theory.

One drawback to the ME Method is that it has traditionally relied on LO matrix elements, although nothing limits the ME method to LO calculations. Techniques that accommodate initial-state QCD radiation within the LO ME framework using transverse boosting and dedicated transfer functions to integrate over the transverse momentum of initial-state partons have been developed [23]. Another challenge is development of the transfer functions which rely on tediously hand-crafted fits to full simulated Monte-Carlo events.

The most serious difficulty in the ME method that has limited its applicability to searches for beyond-the-SM physics and precision measurements is that it is very *computationally intensive*. If this limitation is overcome, it would enable more widespread use of ME methods for analysis of LHC data. This could be particularly important for extending the new physics reach of the HL-LHC which will be dominated by increases in integrated luminosity rather than center-of-mass collision energy.

Accurate evaluation of Eqn. 1 is computationally challenging for two reasons: (1) it involves high-dimensional integration over a large number of events, signal and background hypotheses, and systematic variations and (2) it involves sharply-peaked integrands[2] over a large domain in phase space. In reference to point (1), the matrix element $\mathcal{M}_\xi(\mathbf{y}|\boldsymbol{\alpha})$ in the method involves all partons in the $n \to m$ process, so when the 4-momentum of particles are not completely measured experimentally (e.g. neutrinos), one must integrate over the missing information which increases the dimensionality of the integration. In reference to point (2), a clever technique to re-map the phase space in order to reduce the sharpness of integrate in that space in an automated way (MADWEIGHT [24]) is often used in conjunction with a matrix element calculation package (MADGRAPH_aMCNLO [25]). In practice, evaluation of definite integrals by the ME approach invokes techniques such as importance sampling (see VEGAS [26, 27] and FOAM [28]) or recursive stratified sampling (see MISER [29]) Monte Carlo integration. Acceleration of some of these techniques on modern computing architectures has been achieved, for example concurrent phase space sampling in VEGAS on GPUs.

Despite the attractive features of the ME method and promise of further optimization and parallelization of the evaluation of Eqn. 1, the computational burden of the ME technique will continue to limit is range of applicability for practical data analysis without new and innovative approaches. The primary idea put forward in this section is to utilize modern *machine learning techniques to dramatically speed up the numerical evaluation of Eqn. 1* and therefore broaden the applicability of the ME method to the benefit of HL-LHC physics.

Applying neural networks to numerical integration problems is plausible but not new (see [30–32], for example). The technical challenge is to design a network which is sufficiently rich to encode the complexity of the ME calculation for a given process over the phase space relevant to the signal process. Deep Neural Networks (DNNs) are strong candidates for networks with sufficient complexity to achieve good approximation of Eqn. 1, possibly in conjunction with smart phase-space mapping such as described in [24]. Promising demonstration of the power of Boosted Decision Trees [33, 34] and Generative Adversarial Networks [35] for improved Monte Carlo integration can be found in [36]. Once a set of DNNs representing definite integrals of the form of Eqn. 1 is generated to good approximation, evaluation of the ME method calculations via the DNNs will be very fast. These DNNs can be thought of as preserving the essence of ME calculations in a way that allows for fast forward execution. They can enable the ME method to be both *nimble* and *sustainable*, neither of which is true today.

The overall strategy is to do the expensive full ME calculations as infrequently as possible, ideally once for DNN training and once more for a final pass before publication, with the DNNs utilized as a good approximation in between. A future analysis flow using the ME method with DNNs might look something like the following: One performs a large number of ME calculations using a traditional numerical integration technique like VEGAS or FOAM on a large CPU resource, ideally exploiting acceleration on many-core devices. The DNN training data is generated from the phase space sampling in performing the full integration in this initial pass, and DNNs are trained either *in situ* or *a posteriori*. The accuracy of the DDN-based ME calculation can be assessed through this procedure. As the analysis develops and progresses through selection and/or sample changes, systematic treatment, etc., the DNN-based ME calculations are used in place of the time-consuming, full ME calculations to make the analysis nimble and to preserve the ME calculations. Before a result using the ME method is published, a final pass using full ME calculation would likely be performed both to maximize the numerical precision or sensitivity of the results and to validate the analysis evolution via the DNN-based approximations.

[2] a consequence of imposing energy/momentum conservation in the processes

IOP Conf. Series: Journal of Physics: Conf. Series **1085** (2018) 022008 doi:10.1088/1742-6596/1085/2/022008

There are several activities which are proposed to further develop the idea of a Sustainable Matrix Element Method. The first is to establish a cross-experiment group interested in developing the ideas presented in this section, along with a common software project for ME calculations in the spirit of [37]. This area is very well-suited for impactful collaboration with computer scientists and those working in machine learning. Using a few test cases (e.g. $t\bar{t}$ or $t\bar{t}h$ production), evaluation of DDN choices and configurations, developing methods for DNN training from full ME calculations and direct comparisons of the integration accuracy between Monte Carlo and DNN-based calculations should be undertaken. More effort should also be placed in developing compelling applications of the ME method for HL-LHC physics. In the longer term, the possibility of Sustainable-Matrix-Element-Method-as-a-Service (SMEMaaS), where shared software and infrastructure could be used through a common API, is proposed.

2.6. *Matrix Element Machine Learning Method*

The matrix element method is based in the fact that the physics of particle collisions is encoded in the distribution of the particles' four-momenta and with their flavors. As noted in the previous section, the fundamental task is to approximate the left-hand side of Eq. (1) for all (exclusive) final states of interest. In the matrix element method, one proceeds by approximating the right-hand side of Eq. (1). But, since the goal is to compute $\mathcal{P}_\xi(\mathbf{x}|\boldsymbol{\alpha})$, and given that billions of fully simulated events will be available, and that the simulations use exactly the same inputs as in the matrix element method, namely, the matrix elements, parton distribution functions, and transfer (or response) functions, one can ask whether a more direct machine learning approach can be developed to approximate $\mathcal{P}_\xi(\mathbf{x}|\boldsymbol{\alpha})$ without the need to execute the calculation of the right-hand side of Eq. (1) explicitly. We believe the answer is yes, provided that a key advantage of the matrix element method can be replicated, namely, the fact that the method provides a function that depends explicitly on the model parameters $\boldsymbol{\alpha}$. Simulated events are typically simulated at fixed values of the model parameters. In order to replicate the advantage of the matrix element method, it would seem necessary to simulate events over an ensemble of model parameter points. But, that is a huge computational burden. The question is: is there a way to sidestep? Perhaps.

The matrix element method is technically feasible because there are now many codes that provide access to the square of the matrix elements as a function of $\boldsymbol{\alpha}$. Consequently, it would be possible to build a parameterized set of simulated events by reweighting each simulated event using the weighting function

$$w(\boldsymbol{\alpha}, \boldsymbol{\alpha}_0) = \frac{|\mathcal{M}_\xi(\mathbf{y}|\boldsymbol{\alpha})|^2}{|\mathcal{M}_\xi(\mathbf{y}|\boldsymbol{\alpha}_0)|^2}, \tag{2}$$

where $\boldsymbol{\alpha}_0$ denotes the values of parameters at which the events were simulated. In practice, in order to keep the dynamic range of the weights within reasonable bounds, one presumably would simulate sets of events at a few reasonable choices of the parameters and use the weights to interpolate between these fixed parameter points.

What could one do with these parameterized simulated events? One could take advantage of the mathematical fact (shown more than quarter century ago) that, given a sufficiently expressive parameterized function $f(x, \omega)$, with parameters ω, fitted by minimizing either the quadratic loss or cross entropy using data comprising two classes of objects of equal size, for example, signals with density $s(x)$ assigned a target of unity and backgrounds with density $b(x)$ assigned a target of zero, then asymptotically — that is, for arbitrarily large training samples,

$$f(x, \omega) = \frac{s(x)}{s(x) + b(x)}. \tag{3}$$

This result was derived in the context of neural networks. However, it is in fact entirely independent of the nature of the function $f(x, \omega)$.

We can exploit this result to approximate $\mathcal{P}_\xi(\mathbf{x}|\boldsymbol{\alpha})$ directly using, for example, DNNs. For each simulated event, in the training set, one could sample $\boldsymbol{\alpha}$ from a known distribution $q(\boldsymbol{\alpha})$, thereby yielding an ensemble of triplets $\{(\mathbf{x}, \boldsymbol{\alpha}, w(\boldsymbol{\alpha}, \boldsymbol{\alpha}_0))\}$. Call this ensemble the "signal". Sample $\mathbf{x}$ from another known distribution $p(\mathbf{x})$ and for each $\mathbf{x}$ sample $\boldsymbol{\alpha}$ from $q(\boldsymbol{\alpha})$. Call the ensemble of pairs $\{(\mathbf{x}, \boldsymbol{\alpha})\}$ the "background". From Eq. (3), we have

$$f(x, \omega) = \frac{s(\mathbf{x}, \boldsymbol{\alpha})}{s(\mathbf{x}, \boldsymbol{\alpha}) + p(\mathbf{x})q(\boldsymbol{\alpha})}, \tag{4}$$

from which we find

$$\mathcal{P}_\xi(\mathbf{x}|\boldsymbol{\alpha}) = \frac{s(\mathbf{x}, \boldsymbol{\alpha})}{q(\boldsymbol{\alpha})} = p(\mathbf{x}) \left(\frac{f}{1-f} \right). \tag{5}$$

If this could be made to work, it would be a direct machine learning alternative to the matrix element method, which incorporates the advantages of the former with the added advantage that the transfer function is automatically incorporated without the need to model it explicitly and the calculation of $\mathcal{P}_\xi(\mathbf{x}|\boldsymbol{\alpha})$ would be fast because it would be given directly in terms of the DNN, $f(x, \omega)$.

2.7. *Learning the Standard Model*

New physics may manifest itself as unusual or rare events. One approach is to accurately identify the Standard Model processes and search for anomalies. Classifying the Standard Model events is a challenging task, as it consists of many complicated physics processes. Multi-class machine learning algorithms are well-suited for this classification problem. Once an event is classified as likely to be from a known physics process, it can be filtered out, and remaining events can be further analyzed for hints of new physics. Additionally, unsupervised machine learning techniques can be applied to remaining events to cluster them together. This approach would also be useful in identifying detector problems.

2.8. *Theory Applications*

The theoretical physics community has a number of challenges where machine learning can make an impact. These include areas of theoretical model optimization with hundreds of parameters, searches for new models, understanding and estimation of the parton distribution functions and possibly quantum machine learning. The following details one such application: learning of the parton distribution functions with machine learning.

Making progress towards the objectives of the HL-LHC program (see Section 1) requires not only obtaining experimental measurements of the physical processes but also reliable theory inputs to compare to. This becomes increasingly challenging as the experimental data gets more precise. There are numerous examples of phenomenologically relevant processes where the experimental uncertainty is comparable to the estimate of the theoretical uncertainty of the corresponding calculation.

Furthermore, the theory does not predict the value of all the inputs required for the computations (for example the value of the strong coupling constant α_S evaluated at the Z mass), and there are situations where the equations resulting from theory cannot be solved to describe the physics adequately, and the corresponding theory inputs must be obtained from data instead. A more complex example is the determination of Parton Distribution Functions (PDFs): Quantum Chromodynamics (QCD) describes the proton collisions at high energy in terms of *partons* (e.g. quarks and gluons), but it is not possible to calculate directly from QCD the momentum carried by each quark or gluon within a proton since QCD is not solvable in its confined regime. Our lack of theoretical knowledge about the characterization of partons within a proton is embedded into a suitable definition of the Parton Distribution Functions (approximately the momentum densities of each of the partons). The PDFs then need to be determined from experimental data. The NNPDF collaboration uses Machine Learning techniques to obtain a PDF determination that is accurate enough to be suitable for high-precision collider data comparison. The NNPDF fitting procedure is described in full details in [38].

The idea is to combine data from all relevant physical processes and fit a neural network representing each PDF. The difficulty of the procedure steams from the fact that multiple experimental inputs need to be combined to obtain a PDF fit. Each of these inputs adjoins only indirect constraints on the PDFs, leaving some regions of the PDF completely unconstrained by data. NNPDF fit includes around 50 datasets from different physical processes, and results that are not always consistent among themselves. Therefore it is crucial to propagate the uncertainty of the experimental inputs into uncertainty on the PDFs.

While the dataset is small, each experimental point has an indirect relation to the PDFs, as it is the result of the convolution of one or two PDFs with the corresponding partonic cross section. Code has been developed to compute these convolutions `APFELgrid` [39]. Future research directions include the possibility of using standard ML frameworks to efficiently express the PDF fitting problem. The uncertainties of the theory calculations need to be taken into account as well in the fits. A fully systematic treatment of theory errors in PDFs is a topic of research where Machine Learning could play an important role. The dominant uncertainties in the data are no longer statistical and instead arise from correlated systematics. Determining those systematics accurately is non trivial on the side of the experimental analyses and can have a major impact on the resulting PDFs. The problem grows more complex when ML techniques for which there is no simple recipe to estimate the uncertainty are used extensively in the experimental analysis. Taking full advantage of these advanced methods requires interdisciplinary research and communication on topics such as developing regularization schemes for experimental covariance matrices.

In conclusion, it is not only important to obtain the best fit PDF, but also a reliable estimation of the uncertainty, which in turn requires controlling the uncertainty of the experimental and theoretical inputs.

2.9. Uncertainty Assignment

Until fairly recently, little attention has been paid to the problem of assigning a measure of uncertainty to the output of machine learning algorithms. However, there is a growing recognition that the lack of such measures is a serious deficiency that needs to be addressed. This is particularly problematic in particle physics where accurate uncertainty assignments are crucial for assessing the quality of quantitative results. Uncertainty assignment is especially urgent if machine learning methods are to be used for regression.

Standard methods exist in the statistics literature to quantify the uncertainty when an explicit likelihood function is available. Unfortunately, however, the overlap between the machine learning and statistics communities has been minimal at best. The problem of uncertainty offers many opportunities for fruitful collaborations between physicists, statisticians, computer scientists, and machine learning practitioners, and is something that ought to be vigorously encouraged. One area ripe for collaboration is in the use of Bayesian methods to quantify uncertainty using, for example, Hamiltonian Monte Carlo (HMC) sampling of posterior densities. One serious issue is that HMC is computationally prohibitive for DNNs. Therefore, new and more effective ways to perform large scale Bayesian calculations that go beyond HMC need to be developed, and it would be particularly useful to develop methods that can take advantage of multicore machines as discussed in Section 5.

2.10. Monitoring of Detectors, Hardware Anomalies and Preemptive Maintenance

Data-taking of current complex HEP detectors is continuously monitored by physicists taking shifts to check the quality of the incoming data. Typically, hundreds of histograms have been defined by experts and shifters are alerted when an unexpected deviation with respect to a reference occurs. It is a common occurrence for a new type of problem to remain unseen for a non-negligible period of time because such a situation had not been foreseen by the expert.

A whole class of ML algorithms called anomaly detection can be useful in such situations. They are able to learn from data and produce an alert when a deviation is seen. By monitoring many variables at the same, time such algorithms are sensitive to subtle signs forewarning of imminent failure, so that preemptive maintenance can be scheduled. Such techniques are already used in industry applications.

One challenge is that normal drifts in environmental conditions can induce drifts in the data. Beyond just reporting a problem, the natural next step is to connect an anomaly detection algorithm to the appropriate action: restart an online computer, contact an on-call expert, or similar. In the long term, the hardware and data structures of future detectors should be designed to facilitate the operation of anomaly detection algorithms.

2.11. Computing Resource Optimization and Control of Networks and Production Workflows

Data operations is one of the significant challenges for the upcoming HL-LHC. In the current infrastructure, LHC experiments rely on in-house solutions for managing the data. While these approaches work reasonably well today, machine learning can help automate and improve the overall system throughput and reduce operational costs.

Machine Learning can be applied in many areas of computing infrastructure, workflow and data management. For example, dataset placement optimization and reduction of transfer latency can lead to a better usage of site resources and an increased throughput of analysis jobs. One of the current examples is predicting the "popularity" of a dataset from dataset usage, which helps reduce disk resource utilization and improve physics analysis time turn-over.

Data volume in data transfers is one of the challenges facing the current computing systems as thousand of users need to access thousands of datasets across the Grid. There is an enormous amount of metadata collected by application components, such as information about failures, file accesses, etc. Resource utilization optimization based on this data, including Grid components and software stack layers, can improve overall operations. Understanding the data transfer latencies and network congestion may improve operational costs of hardware resources.

Networks are going to play a crucial role in data exchange and data delivery to scientific applications in HL-LHC era. The network-aware application layer and configurations may significantly affect experiment's daily operations. ML can be applied to network security in order to identify anomalies in network traffic; predict network congestion; detect bugs via analysis of self-learning networks, and optimize WAN paths based on user access patterns.

3. Collaborating with other communities

3.1. Introduction

Discovery science provides a challenge that attracts brilliant minds eager to push the boundaries of scientific understanding of nature. Particle physics has a rich problem domain that offers avenues for intellectual reward. The goal is to achieve a vibrant collaboration between the data science and high-energy physics communities by finding a common language and working together to further science. Both communities can benefit from such collaboration. The HEP community can explore new research directions and applications of machine learning, novel algorithms, and direct collaboration on HEP challenges. The ML community can benefit from a diverse set particle physics problems with unique challenges in scale and complexity, and a large community of researchers that can expand the machine learning horizon by contributing to solving problems relevant to both communities. For example, the treatment of systematic uncertainties is an important topic for the HEP and ML communities. By working together on common challenges, the two fields can further progress in solving such problems.

There are a number of existing examples of collaboration between HEP and ML that have produced fruitful results through mostly local connections (e.g. [2, 40]). The HEP community should continue such collaborations and look for additional collaborations with the ML community.

Domain knowledge can present a barrier to collaboration. The HEP community needs to define its challenges in a language that the ML community can understand. This may involve stripping the domain knowledge entirely, or retaining necessary information with clear and concise explanations as to its relevance. Machine learning likewise has a significant amount of domain knowledge. Ideas and solutions provided by both communities should be presented in an understandable way for scientists without in-depth knowledge.

3.2. Academic Outreach and Engagement

Direct collaboration between HEP researchers and computer scientists working in the area of machine learning is an important possible driver of innovation in the area of machine learning applications described in Section 2. It is important for high-energy physics to engage and collaborate with the academic community focused on new machine learning algorithms and applications as they will naturally be interested in applying new ideas to interesting and complex data provided by HEP.

Conferences and workshops are a core aspect of the academic ML community, and organizing or contributing to key conferences is a means of gaining interest. Organizing sessions or mini-workshops within major ML conferences, such as NIPS, would increase the familiarity of HEP within the ML community and jump-start future collaborations. This has been explored in single cases [41] but is not an established, regular workshop series. At the same time inviting ML experts to HEP workshops, as done at [42] and the DS@HEP series [43–45], can foster greater long-term collaboration. There should be coordinated efforts to:

- Organize workshops and conferences open to external collaborators to discuss applications, algorithms and tools
- Organize thematic workshops around topics relevant to HEP

HEP should reach out to other scientific communities with similar challenges, for example astrophysics/cosmology, medium energy nuclear physics and computational biology. This can lead to more active partnerships to better collaborate on ideas, techniques, and algorithms.

3.3. Machine Learning Challenges

To engage the wider ML community, challenges such as the Higgs Boson Challenge (2014) or the Flavor Physics Challenge [41, 42] have been organized on Kaggle. These types of challenges draw considerable attention from the Machine Learning community and additional similar challenges should be organized in the future.

Organization of a challenge requires a well documented dataset, a starting-kit and an evaluation metric to rank the solutions. This forces the organizers to simplify the problem as much as possible, while retaining its intrinsic complexity. The drawback of challenges is that once they are launched, participants priority is winning the challenge and not eventual collaboration with HEP. It is important to foresee upfront a way to integrate incoming solutions, for example via forums and post-challenge workshops (like [42]) where a diversity of competitive algorithms can be presented. The challenge dataset and evaluation metric should be released publicly so that further developments can continue.

3.4. Collaborative Benchmark Datasets

There is a strong incentive for HEP to develop public benchmark datasets, beyond just challenges. Access to a dataset makes the discussion much more concrete and productive. Within the HEP community, common datasets enable comparisons of algorithms with much better accuracy, which is very useful for research and development. The same benchmark datasets can also be used for teaching, tutorials and training.

These benchmark datasets could be built based on public simulation engines or released by experiments within the bounds of their data access policy. Even a small subset of an experiment's simulated data can be the base of a very valuable benchmark dataset. For example, the CMS experiment has released a significant amount of its simulated and collected data via the CERN Open Data Portal [46].

To be maximally useful, the subsequent guidelines should be followed when designing a benchmark dataset:

- Simplify the dataset as much as possible
- Document the dataset to make it understandable by a non-HEP expert
- Create methodologies and metrics for the evaluation of proposed solutions, and document them
- Prepare an integration plan for incoming ideas and solutions
- Feedback results of successful applications

The HEP community should organize and curate a variety of such benchmark datasets covering its current physics drivers and make them publicly available. To improve the reproducibility of results and algorithm comparisons, some of the data used for evaluation of the solutions should be kept private. Additionally, after investing heavily into producing highly-detailed and realistic simulations, the HEP community can provide the machine learning community with labeled datasets with high statistical power to test algorithms and develop novel ideas.

3.5. Industry Engagement

Industry has been focused on the development and adoption of machine learning techniques. In addition to algorithm and software development, one of the promising areas is the adoption of dedicated specialized hardware and high performance co-processors. GPUs, FPGAs, and high core count co-processors all have the potential to dramatically increase performance of machine learning applications relevant to the HEP community. One of the challenges is gaining the human expertise for development and implementation. Industry interactions bring specific technology opportunities and access to specialized expertise that can be difficult to hire and support internally.

There are specific areas of development where industry has expressed interest in collaborating with HEP. Automated resource provisioning, data placement, and scheduling are similar to industrial applications to improve efficiency. Applications such as data quality monitoring, detector health monitoring and preventative maintenance can be automated using techniques developed for other industrial quality control applications. There are two more forward looking areas that coincide with HEP physics drivers, namely computer vision techniques for object identification and real-time event classification. These present a challenge to industry due to its complexity and benefit outside of HEP.

CERN OpenLab is a public-private partnership that accelerates the development of cutting-edge solutions for the LHC community and wider scientific research. CERN OpenLab has established the infrastructure to maintain non-disclosure agreements, to arrange ownership of intellectual property, and to provide an interface between CERN and industry. As part of its upcoming phases, OpenLab plans to explore machine learning applications for the benefit of LHC experiments computing and the HL-LHC. Such initiatives and industry partnerships should be supported in the future.

3.6. Machine Learning Community-at-large Outreach

Another form of engagement is using the communications mediums to broadcast our challenges and attract interested collaborators. There are a variety of channels which can be leveraged to increase the visibility of our problems and research opportunities in the ML community. These can be popular forums such as reddit, personal or official blogs, social media, or direct contact with influential personalities.

Podcasts have shown to be a great vehicle for reaching a large audience. Listeners are keen to consume material that is outside of their immediate problem domain in a way that is easy to digest. There is an abundance of machine learning podcasts with a large base of listeners that can be targeted for outreach:

- Linear Digressions (co-hosted by former ATLAS Ph.D. Katie Malone)
- Partially Derivative
- Talking Machines
- Data Skeptic
- Becoming a Data Scientist Podcast

- Not So Standard Deviations
- This Week in ML & AI

Another form of engagement is through outreach-style blog posts to explain HEP challenges in a way that is easy to understand by the public.

Another outreach opportunity is to make HEP related presentations at Machine Learning Meetups across the world to generate awareness, engage community, foster cross pollination of ideas between HEP and industry. Some popular ML meetups are:

- NYC: `https://www.meetup.com/NYC-Machine-Learning/`
- Berlin: `https://www.meetup.com/Advanced-Machine-Learning-Study-Group/`
- SF: `https://www.meetup.com/SF-Bayarea-Machine-Learning/`

In conclusion, existing outreach efforts should be expanded to attract greater collaboration between the HEP and ML communities. By understanding and speaking the same language, the two communities can better collaborate and find solutions to present and future challenges.

4. Machine Learning Software and Tools

Machine learning does not exist without software. There are a large variety of algorithms written in different programming languages and general software frameworks that combine many classes of methods into one package. The following sections focus on specific topics and challenges related to machine learning software design in HEP.

4.1. Software Methodology

Presently, there are two machine learning software methodologies in high-energy physics. The first approach focuses on HEP-developed ML toolkits, such as the Toolkit for Multivariate Analysis (TMVA) in ROOT, while the second approach relies on externally developed software, of which there are many examples. Historically, a variety of approaches and competition among them has led to important breakthroughs in the field. On the other hand, having too many choices increases repetition and leads community segmentation and possible issues with reproducibility.

4.2. I/O and Programming Languages

The sheer amount of data accumulated by HEP experiments requires a close look at data access optimization. To train and apply ML techniques on these data, efficient I/O becomes critical, especially for training. I/O performance is very dependent on data formats. Moreover, support for reading data in different formats is required for certain use-cases.

Exploration of new file systems and methods to improve I/O limitations are important and the following R&D studies should take place:

- Explore new file systems to assess I/O limitations;
- Use alternative industry approaches such as Google BigQuery to explore various data access patterns;
- Explore parallel data processing platforms such as Apache Spark for ML training.

Although particle physics has been reliant on C++ over the past decade, the machine learning community has explored other programming languages, in particular the python-based ecosystem.

4.3. Software Interfaces to Acceleration Hardware

Modern machine learning software significantly benefits from the use of hardware accelerators such as GPUs. At the same time, ML users should not be forced to write platform-dependent code. Various interfaces to different hardware architectures are needed in order to make efficient use of the available computing resources. The emergence of the Open Computing Language (OpenCL) allows programming of high-level interfaces that can run on various hardware platforms.

Machine learning tools often provide different sets of APIs to develop and train the models in one language, and various bindings to use trained models in other programming languages. This is a convenient model for many HEP applications, such as the trigger system, where application latency puts stringent requirements on the software and hardware used.

4.4. Parallelization and Interactivity

Training ML algorithms takes a significant amount of time and parallelization at various levels is desired, such as the parallelization of the computations within a single model. Another type of parallelism is data parallelism that targets the processing phase of the training with data partitioning and model training using distributed workers. Frameworks like Apache Spark and ideas such as batch training offer promise in this area.

Often one needs to produce many different machine learning models, for example while tuning hyper-parameters or performing k-fold cross-validation, and distribution of these algorithms is key to the reduction of the overall training time. ML algorithm inference significantly benefits from parallelization as well. For example, in particle physics trigger systems, the stringent latency requirements impose constraints on the type of algorithms that can be easily parallelized in the hardware.

The availability of interactive frameworks, for example Jupyter notebooks, allows for rapid prototype development and testing of ML tools. Such frameworks also ease the connection between the description of models and the data, providing straightforward means of visualizing models and data. HEP has started to exploring interactive frameworks, such as the Service for Web Based Analysis (SWAN). One of the challenges is availability of adequate hardware resources for these systems.

4.5. Internal and External ML tools

Internally developed ML software, such as the Toolkit for Multivariate Analysis (TMVA, [47]), has been developed to apply a variety of machine learning algorithms to HEP challenges. Currently, most published HEP analyses with machine learning have made use of TMVA. There is also software developed in HEP, such as NeuroBayes [48, 49] and RuleFit, that have gained popularity outside of HEP.

At the same time, the ML landscape has evolved and many different ML tools have emerged and gained popularity. There are a growing number of published results based on externally developed tools. The latter, often developed directly by industry for specific applications, are constantly undergoing development, incorporating the latest algorithms from academia. Currently, both internal and external tools are used by the HEP community. TMVA has also undergone significant development in recent years. In addition, there are smaller tools developed in the HEP community, extending either internal or external ML tools for specific use cases and applications within HEP experiments, such as `hep_ml` [50] or `tmva-branch-adder` [51].

This begs the question: what aspects of ML development and use should the HEP community focus on in the next 5-10 years? There are several aspects to consider including data formats, community size, and interfaces.

4.5.1. Machine Learning Data Formats HEP and ML communities currently make use of different data formats. HEP heavily relies on the ROOT software framework for data storage, data processing, and data analysis. The machine learning community uses a large variety of formats, as shown in Figure 1. This figure also shows the relationship of machine learning data formats with ROOT: the ROOT file format is very flexible, though it requires a significant investment to properly use. Table 1 summarizes the current machine learning toolkits and file formats they support.

IOP Conf. Series: Journal of Physics: Conf. Series **1085** (2018) 022008 doi:10.1088/1742-6596/1085/2/022008

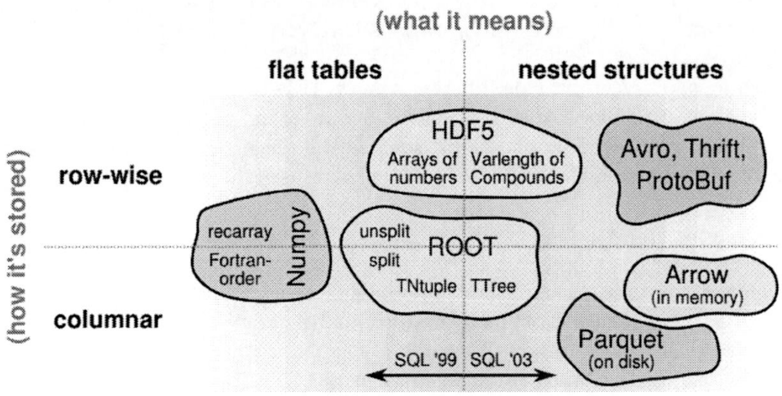

Figure 1. Existing data-formats used by ML communities.

Table 1. This table lists various data formats (rows) and ML tools (columns). The ✓ indicates that there is a native solution, while × means that the conversion from one data-format to another is straightforward. The following notation has been used to denote the data-formats: **T** Trees, **F** flat tables, **M** sparse matrices, **R** row-wise arrays, **C** column-wise arrays **S** static data structures

	TMVA	TensorFlow	Theano	Scikit Learn	R	Spark ML	VW	libFM	RGF	Torch
ROOT [**T, C**]	✓									
CSV [**F**]		✓	✓	✓	✓	✓	×	×	×	✓
libSVM [**M**]							×	✓	×	
VW [**M**]							✓			
RGF [**M**]									✓	
NumPy [**R**]		✓	✓	✓	✓	✓	×	×	×	✓
Avro [**S, R**]					✓	✓				
Parquet [**S, C**]					✓	✓				
HDF5 [**S**]		×	×	×						✓
R df [**R**]					✓					

4.5.2. Desirable HEP-ML Software and Data Format Attributes A desirable data format should have the following attributes: high read-write speed for efficient training, sparse readability without loading the entire dataset into RAM, compression, and common use by the machine learning community.

HEP machine learning applications require highly performant and flexible algorithms to address the variety of use cases. Some applications, such as triggering, also have to work under tight latency constraints of the order of a few microseconds and below. The data sets are extremely large, which comes with I/O challenges as described in Section 4.2. This is expected to become even more challenging, as the LHC continues to ramp-up and deliver increasingly large amounts of data.

As discussed in Section 4.3, Machine Learning tools use a number of languages. To use these tools it will therefore be important to offer adequate support. C++ converters or similar tools are also needed to make sure the training result can be efficiently evaluated.

Advantages of using the **external tools** are the size of the community that uses and supports them, being able to easily keep up with progress in the industry and profit from the forefront of the ML research. It should also be noted that some of the recent industrial efforts to develop and maintain ML-tools rely on resources far beyond that of basic research. The deep learning tools of the previous and current generations constitute a demonstration of corresponding quality.

Disadvantages of using external tools are that there are too many choices, they are not guaranteed to be supported over the lifetime of particle physics experiments, and it can be difficult to adapt them to HEP specific requirements which may not be among the priorities of the ML community.

Advantages of using **internal tools** are that decisions about long-term support remain in the community, and the tools can be adapted to the specific needs of HEP. Disadvantages include the challenges in incorporating new algorithms and ideas on a timely basis and a possible lack of resources for long-term maintenance.

4.5.3. Interfaces and Middleware One approach to bridge the gap between internal and external tools is by building interfaces. Some researchers prefer to convert their data to the file formats used by external tools and to work exclusively with external tools. This has the advantage of working as close as possible with the ML community's tools and its documentation, and is as close to what is used by ML researchers as possible. At the same time, interfaces have been built between TMVA and external machine learning tools, allowing for their use and direct comparison between their performance. Currently, interfaces to R, scikit-learn, keras and tensorflow have been developed. Those have the advantage of providing a homogeneous interface and require little training overhead for those already knowledgable in using TMVA.

A more general approach to file format conversion is to build middleware solutions that export HEP-specific formats like ROOT to formats used by external machine learning tools. Existing middleware solutions are shown in Table 2.

Approaches to bridge the different languages and data formats inside and outside of HEP include providing interfaces or building middleware solutions that translate HEP-specific data formats to external ML tools. It is a topic of current research to determine the most efficient solution.

Table 2. Middleware solutions translating the ROOT data format to other formats.

PyROOT	Python module that allows the user to interact with ROOT. [52]
root_numpy	Interface between ROOT and NumPy. [53]
root_pandas	Interface between ROOT and Pandas dataframes. [54]
uproot	A high throughput I/O interface between ROOT and NumPy. [55]
c2numpy	Pure C-based code to convert ROOT data into Numpy arrays which can be used in C/C++ frameworks. [56]
root4j	The hep.io.root package contains a simple Java interface for reading ROOT files. This tool has been developed based on freehep-rootio. [57]
root2npy	The `go-hep` package contains a reading ROOT files. This tool has been developed based on freehep-rootio. [57]
root2hdf5	Converts ROOT files containing TTrees into HDF5 files. [58]

5. Computing and Hardware Resources

A typical high-energy physics data model consists of a hierarchy of increasingly refined data stores. Each store provides a refined view of a list of "events", the self contained records that capture the state of the detector at the time when a particle interaction occurs. At the bottom of the hierarchy is the raw data, a byte-stream of the readout from detector electronics. At the top of the hierarchy are the "high-level" physics objects, such as electrons or jets, providing descriptive information about the quality and topology of physics events. The data stores are typically processed by independent copies of identical code processed in batch computing queues. The result of this processing is filtered data and extracted physics parameters.

At present, training of machine learning algorithms is done using dedicated or private resources. These vary in configuration and processing power, depending on the size of the data and complexity of the algorithm. For a given event, the evaluation of algorithms is performed on a single core producing a single discriminator or regressor output. In order to progress to evaluation of complex machine learning algorithms, more computing power is needed in both the training and evaluation stages, as larger amounts of data are needed to feed models with tens or hundreds of thousands of parameters. This implies the expansion of the current computing model to include architectures that are well suited to machine learning tasks, such as many integrated core (MIC), graphical processing units (GPUs) and tensor processing units (TPUs). This is a fundamental departure from the single-core or few-core jobs. These architectures provide a significant computational speed improvement for both training and evaluation of ML algorithms, but require dedicated hardware, drivers, and software configuration.

Similarly, the locality and bandwidth of large data stores will need to be optimized in order to avoid bottlenecks in training and evaluation for analysis. Data placement and the need to use dedicated hardware indicate that a transition to HPC, or HPC-like, architectures may be needed to achieve the desired performance. Due to significant synergy with the direction of industry in this respect, use of commercially available resources should be considered for future high-energy physics computing models.

In the following subsections we will discuss the resource needs for the physics drivers mentioned earlier: fast simulation, real-time analysis, object and event reconstruction and particle identification. The limitations of the current computing model are discussed as well as how those physics driver needs can be met in the future.

5.1. Resource Requirements

The popularity of deep learning methods is to a large extent due to the possibility of training these models in a reasonable amount of time with large scale parallelism. In particular, the training stage requires repeated simultaneous access to many data elements and specialized hardware has been developed for training deep learning models.

In contrast, inference can be an operation applied to a single data element at a time and needs to be performed only once. Inference has less demands on I/O and is limited only by the computing power and model complexity. Because inference has real-time applications in high-energy physics, latency and throughput constraints are the main challenges.

A typical HEP application can require up to 1 GPU-week to train a single model. To obtain the best results and understand the performance of the model, an average of 100 hyper-parameter points optimization is typically performed. A single project could therefore easily require up to a full GPU-year for training. The speed up of the training process can be obtained by means of faster and more capable hardware, parallelization of single training and by splitting training over multiple nodes. Resources, such as GPUs, TPUs and MIC need to be evaluated in the context of realistic benchmark particle physics applications.

If different ML techniques can achieve equivalent physics performance but require different processing power, it is important to quantify what is driving the performance gains and what level of performance-processing power trade-off is maintainable to achieve the required physics goals.

5.2. Graphical Processing Units

GPUs have been extremely useful in speeding up the training of complex deep learning models. Many researchers in HEP are currently relying on private or university GPU clusters to perform machine learning training. Unfortunately, there is no centralized GPU resource available for general HEP usage. Availability of greater and more modern GPU resources would significantly reduce the training time and assist many on-going *R&D* efforts in the community.

5.3. High Performance Computing

Resource-rich many-core processors such as MIC, GPUs, and TPUs are vital to the optimization of the training time of most modern machine learning algorithms, including deep neural networks, generative adversarial networks, autoencoders, etc. Availability of High Performance Computing (HPC) resources equipped with many-core processors and high-performance network storage are essential to distributed running of large-scale machine learning algorithms. Current efforts to bring and expand the availability of HPC resources in high-energy physics computing will be vital to the successful progress of the application of machine learning techniques to current and future experiments.

5.4. Field Programmable Gate Arrays

Field Programmable Gate Arrays (FPGAs) provide an efficient and low-latency application of machine learning algorithms directly at the level of hardware, as desirable for HEP trigger systems. The following ML algorithms are more suitable for FPGAs due to their simpler parallelization: boosted decision trees, random forests and decision rule ensembles. For example the CMS experiment currently uses boosted decision trees in FPGAs in the trigger system to estimate muon momenta. Further research and development is needed in this area to apply more advanced machine learning techniques like deep learning directly in the hardware. One of the challenges is the limited availability of floating point operations gates and the precision needed to maintain the best performance. The possibility of coupling the FPGAs with a CPU with significant random-access memory (RAM) allows the shift of some of these operations to RAM.

5.5. Opportunistic Resources

The current HEP computing model is based on a tiered structure where computing resources are mostly large data centers providing CPU resources for collaboration. Although existing resources are gradually moving towards supporting GPUs, it is unlikely to reach all HEP computing centers in the near future. Therefore opportunistic resources are a possible option for training machine learning applications.

Currently, cloud solutions provided by the industry run ML workflows on dedicated hardware and offer interfaces for training machine learning models. The scientific community should work closely with cloud providers to harmonize our analysis computing needs and data access patterns with their business models. Costs of the cloud resources should be compared with the costs of procuring these resources independently.

In order to make the best use of resources available to the community, all resources should ideally be made available through a unique work queue. That implies some uniformization of the software stack, and several specific requirements in the resource management system, especially in terms of data movement.

5.6. Data Storage and Availability

Data storage limitations will have a major impact on machine learning applications. Presently, to train machine learning algorithms, it has been possible to take advantage of increases in statistics of Monte-Carlo simulated events needed for other use cases. Further machine learning progress may require more simulated data than what is available today. How to produce and store these additional large amounts of data is a challenge that needs to be overcome.

Availability of data at PByte/EByte scale represents another challenge for the ML community. A good solution must provide access to a large data volume for hundred or thousand of users simultaneously. Additionally, data movement might need to be automatized to make the training data available transparently at high speed local storage with use of an automatic caching mechanism.

The success of Apache Spark and Google BigQuery platforms may serve as a model. In addition to the regular HEP workflows, data streaming, transformation and readout in mini-batches may be required to train models over large data sets.

5.7. Software Distribution and Deployment

To efficiently use the resources described in previous subsections, machine learning software needs to be available on the computing resources. Platforms, such as the CERNVM File System (cvmfs), are very useful for distributing software instead of requiring local installation. Additional tools like docker containers for application shipping can be useful in providing homogeneous software environments across the different systems. Another challenge is that the software layer needs to be agnostic to the hardware back-end.

5.8. Machine Learning As a Service

Current cloud providers rely on the machine learning as a service model, allowing for the efficient use of common resources, and make use of interactive machine learning tools. Machine Learning As a Service is not yet widely used in HEP, but examples of successful publications which used Machine Learning As a Service exist, e.g. [59]. Specialized HEP services for interactive analysis, such as CERN's Service for Web-based Analysis (SWAN) may play an important role in adoption of machine learning tools in HEP workflows. In order to use these tools more efficiently, sufficient and appropriately tailored hardware resources described in this chapter are needed.

6. Training the community

In order to address the communication barrier and to speak the same language, the HEP community should be trained in ML concepts and terminology as part of a standard curriculum. The training should focus on well-maintained and well-documented software packages. It should provide lectures on general ML concepts and hands-on tutorials on specific tools based on concrete examples.

Being able to apply machine learning to practical HEP problems requires the understanding of basic ML concepts and algorithms. For this, regular data science lecture series and seminars, like [60], are very useful. At the university level, courses dedicated to machine learning applications in physics research are an excellent way to train undergraduate and graduate students. For example, the "Deep Learning in Physics Research" course with 60 participants consisting of 12 lectures and exercises which are performed on 20 GPUs of the VISPA internet platform [61].

Experiments currently have training activities for newcomers that focus on analysis software and introduction to domain knowledge [62]. Machine learning should next be incorporated into the incoming collaborators training efforts of the experiments. As discussed in the Community White Paper report on Careers and Training [63], ensuring the development and availability of resources for knowledge transfer is likewise essential to ML.

7. Roadmap

7.1. Introduction

In this chapter we discuss the roadmap towards implementation of the research and development areas described in Section 2.

7.2. Timeline

The incorporation of machine learning into particle physics experiments must respect two primary time lines: the schedules of HL-LHC and funding agencies, and the experiments' need for extensive validation of the algorithms.

The current LHC schedule has Run 3 starting in 2021 and the HL-LHC starting in 2026. As software processes and algorithms are re-imagined, their implementation must fit into these time frames if they are to maximize their benefit to the particle physics community. To fit this schedule, a newly proposed implementation would need to have a demonstration in 2018 to prove viability. Two years later, in 2020, the proposed idea needs to attain a level of maturity to be included in the HL-LHC Technical Design Report. The project should then be further refined towards a large scale test around the middle of Run 3, about 2022. Run 3 is scheduled to end in late 2023, after which point the project must then be adapted to the HL-LHC software and physics analysis environment as it will be relied on by the experiment.

7.3. Steps to Deployment

The path of taking a machine learning idea from conception to community-wide acceptance and deployment will entail several stages, as appropriate. There are ample opportunities to make the process more efficient. For example, in many steps having common data sets, as discussed in Section 3.4, will likely accelerate the progress.

(i) **Problem formulation and data set preparation**: Problem formulation is the first step in building a machine learning algorithm. The inputs and desired output need to be established. The training and validation data sets must be identified and simulated. In many cases, these data sets are large, and resources must be identified to possibly create and store the data. In most cases, the data needs to be processed into a form suitable for input into the algorithm. Since these steps are often lengthy, common data sets with well-defined problems will be very helpful.

(ii) **Feasibility and demonstration**: Given a dataset, appropriate machine learning algorithms need to be investigated and evaluated for their ability to solve the problem. In some cases, such studies can be preformed on simplified data sets.

(iii) **First application**: An application of the solution to one or a few specific physics analysis examples where the ML technique significantly improves the physics result. The incorporation of the technique into the computing work-flow will likely be very specific to the application and require significant manual intervention.

(iv) **Scaling and optimization**: Evolving from a demonstration to a general solution requires the use of realistic data sets with full detector simulation, noise, etc. Furthermore, the solution will also require optimization to achieve nominal physics and computing performance. A good practice would be to apply the solution to a specific physics analysis. This stage will likely require significant computing resources to scale solutions to the full detector and data sets.

(v) **Integration and Validation**: The solution needs to be incorporated into the experimental software and work-flow and must be validated.

As an example, consider the simulation physics driver. An effort has recently started to build generative models that can significantly accelerate the simulation of particle showers in calorimeters. These early efforts are based on simplified data sets specifically created for this problem, without the complications of realistic data and limited to a small section of the calorimeter. The first papers [2] use generative adversarial networks to generate calorimetric data which are reasonably faithful but still require tuning. The next step involves exploration of neural network architectures and systematic hyper-parameter scans on HPCs to achieve the required performance. The technique can be applied to several possible searches at LHC that involve boosted objects, where simulation samples require full GEANT-based simulation and are therefore limited in statistics due to resource limitations.

The process of employing the new technique in a publication will elicit scrutiny by the full experiment, effectively validating the technique. Once the technique is accepted, it can be generalized beyond this first application and then incorporated into the experiment's software for use by others. Finally, as the technique is applied to an increasing number of physics analyses, the technique will be incorporated into the experiment's production work-flows.

A similar deployment and integration model can be applied to all of the major research and development ideas described in Chapter 2.

IOP Conf. Series: Journal of Physics: Conf. Series **1085** (2018) 022008 doi:10.1088/1742-6596/1085/2/022008

8. Conclusions

As particle physics moves into the post-Higgs boson discovery era, the physics drivers of the High-Luminosity Large Hadron Collider and future neutrino experiments will require increasingly more powerful identification and reconstruction algorithms to extract rare signals from copious and challenging backgrounds.

Machine learning algorithms are already state of the art in many areas of particle physics and will likely be called on to take on a greater role in solving upcoming data analysis and event reconstruction challenges. In this document we have outlined the promising areas of research and development applications of machine learning in particle physics and focused on addressing the most important science drivers with these initiatives. Additionally, we identified the need for greater collaboration with external communities in machine learning and a need to train the particle physics community in machine learning. We also provided a roadmap for acceptance and implementation of described research and development initiatives into the workflows of particle physics experiments.

References

[1] S. Agostinelli et al. "GEANT4: A Simulation toolkit." *Nucl. Instrum. Meth.* A506 (2003), pp. 250–303. DOI: 10.1016/S0168-9002(03)01368-8.

[2] M. Paganini, L. de Oliveira, and B. Nachman. "CaloGAN: Simulating 3D High Energy Particle Showers in Multi-Layer Electromagnetic Calorimeters with Generative Adversarial Networks" (2017). arXiv: 1705.02355 [hep-ex].

[3] P. Ilten, M. Williams, and Y. Yang. "Event generator tuning using Bayesian optimization." *JINST* 12.04 (2017), P04028. DOI: 10.1088/1748-0221/12/04/P04028. arXiv: 1610.08328 [physics.data-an].

[4] S. Benson et al. "The LHCb Turbo Stream." *Journal of Physics: Conference Series* 664.8 (2015), p. 082004. URL: http://stacks.iop.org/1742-6596/664/i=8/a=082004.

[5] V. V. Gligorov and M. Williams. "Efficient, reliable and fast high-level triggering using a bonsai boosted decision tree." *JINST* 8 (2013), P02013. DOI: 10.1088/1748-0221/8/02/P02013. arXiv: 1210.6861 [physics.ins-det].

[6] K. Kondo. "Dynamical Likelihood Method for Reconstruction of Events With Missing Momentum. 1: Method and Toy Models." *J. Phys. Soc. Jap.* 57 (1988), pp. 4126–4140. DOI: 10.1143/JPSJ.57.4126.

[7] F. Fiedler et al. "The Matrix Element Method and its Application in Measurements of the Top Quark Mass." *Nucl. Instrum. Meth.* A624 (2010), pp. 203–218. DOI: 10.1016/j.nima.2010.09.024. arXiv: 1003.1316 [hep-ex].

[8] I. Volobouev. "Matrix Element Method in HEP: Transfer Functions, Efficiencies, and Likelihood Normalization." *ArXiv e-prints* (Jan. 2011). arXiv: 1101.2259 [physics.data-an].

[9] F. Elahi and A. Martin. "Using the modified matrix element method to constrain $L_\mu - L_\tau$ interactions." *Phys. Rev.* D96.1 (2017), p. 015021. DOI: 10.1103/PhysRevD.96.015021. arXiv: 1705.02563 [hep-ph].

[10] V. M. Abazov et al. "A precision measurement of the mass of the top quark." *Nature* 429 (2004), pp. 638–642. DOI: 10.1038/nature02589. arXiv: hep-ex/0406031 [hep-ex].

[11] A. Abulencia et al. "Precision measurement of the top quark mass from dilepton events at CDF II." *Phys. Rev.* D75 (2007), p. 031105. DOI: 10.1103/PhysRevD.75.031105. arXiv: hep-ex/0612060 [hep-ex].

[12] T. Aaltonen et al. "First Measurement of ZZ Production in panti-p Collisions at $\sqrt{s} = 1.96$-TeV." *Phys. Rev. Lett.* 100 (2008), p. 201801. DOI: 10.1103/PhysRevLett.100.201801. arXiv: 0801.4806 [hep-ex].

[13] T. Aaltonen et al. "Inclusive Search for Standard Model Higgs Boson Production in the WW Decay Channel using the CDF II Detector." *Phys. Rev. Lett.* 104 (2010), p. 061803. DOI: 10.1103/PhysRevLett.104.061803. arXiv: 1001.4468 [hep-ex].

[14] V. M. Abazov et al. "Observation of Single Top Quark Production." *Phys. Rev. Lett.* 103 (2009), p. 092001. DOI: 10.1103/PhysRevLett.103.092001. arXiv: 0903.0850 [hep-ex].

[15] T. Aaltonen et al. "First Observation of Electroweak Single Top Quark Production." *Phys. Rev. Lett.* 103 (2009), p. 092002. DOI: 10.1103/PhysRevLett.103.092002. arXiv: 0903.0885 [hep-ex].

[16] S. Chatrchyan et al. "Observation of a new boson at a mass of 125 GeV with the CMS experiment at the LHC." *Phys. Lett.* B716 (2012), pp. 30–61. DOI: 10.1016/j.physletb.2012.08.021. arXiv: 1207.7235 [hep-ex].

[17] S. Chatrchyan et al. "Measurement of the properties of a Higgs boson in the four-lepton final state." *Phys. Rev.* D89.9 (2014), p. 092007. DOI: 10.1103/PhysRevD.89.092007. arXiv: 1312.5353 [hep-ex].

[18] G. Aad et al. "Measurements of Higgs boson production and couplings in the four-lepton channel in pp collisions at center-of-mass energies of 7 and 8 TeV with the ATLAS detector." *Phys. Rev.* D91.1 (2015), p. 012006. DOI: 10.1103/PhysRevD.91.012006. arXiv: 1408.5191 [hep-ex].

[19] V. Khachatryan et al. "Measurement of spin correlations in $t\bar{t}$ production using the matrix element method in the muon+jets final state in pp collisions at $\sqrt{s} = 8$ TeV." *Phys. Lett.* B758 (2016), pp. 321–346. DOI: 10.1016/j.physletb.2016.05.005. arXiv: 1511.06170 [hep-ex].

[20] V. Khachatryan et al. "Search for a Standard Model Higgs Boson Produced in Association with a Top-Quark Pair and Decaying to Bottom Quarks Using a Matrix Element Method." *Eur. Phys. J.* C75.6 (2015), p. 251. DOI: 10.1140/epjc/s10052-015-3454-1. arXiv: 1502.02485 [hep-ex].

[21] G. Aad et al. "Search for the Standard Model Higgs boson produced in association with top quarks and decaying into $b\bar{b}$ in pp collisions at $\sqrt{s} = 8$ TeV with the ATLAS detector." *Eur. Phys. J.* C75.7 (2015), p. 349. DOI: 10.1140/epjc/s10052-015-3543-1. arXiv: 1503.05066 [hep-ex].

[22] G. Aad et al. "Evidence for single top-quark production in the *s*-channel in proton-proton collisions at $\sqrt{s}$ =8 TeV with the ATLAS detector using the Matrix Element Method." *Phys. Lett.* B756 (2016), pp. 228–246. DOI: 10.1016/j.physletb.2016.03.017. arXiv: 1511.05980 [hep-ex].

[23] J. Alwall, A. Freitas, and O. Mattelaer. "The Matrix Element Method and QCD Radiation." *Phys. Rev.* D83 (2011), p. 074010. DOI: 10.1103/PhysRevD.83.074010. arXiv: 1010.2263 [hep-ph].

[24] P. Artoisenet et al. "Automation of the matrix element reweighting method." *JHEP* 12 (2010), p. 068. DOI: 10.1007/JHEP12(2010)068. arXiv: 1007.3300 [hep-ph].

[25] J. Alwall et al. "The automated computation of tree-level and next-to-leading order differential cross sections, and their matching to parton shower simulations." *JHEP* 07 (2014), p. 079. DOI: 10.1007/JHEP07(2014)079. arXiv: 1405.0301 [hep-ph].

[26] G. P. Lepage. "A new algorithm for adaptive multidimensional integration." *Journal of Computational Physics* 27.2 (1978), pp. 192–203. ISSN: 0021-9991. DOI: http://dx.doi.org/10.1016/0021-9991(78)90004-9. URL: http://www.sciencedirect.com/science/article/pii/0021999178900049.

[27] T. Ohl. "Vegas revisited: Adaptive Monte Carlo integration beyond factorization." *Comput. Phys. Commun.* 120 (1999), pp. 13–19. DOI: 10.1016/S0010-4655(99)00209-X. arXiv: hep-ph/9806432 [hep-ph].

[28] S. Jadach. "Foam: A general-purpose cellular Monte Carlo event generator." *Computer Physics Communications* 152.1 (2003), pp. 55–100. ISSN: 0010-4655. DOI: http://dx.doi.org/10.1016/S0010-4655(02)00755-5. URL: http://www.sciencedirect.com/science/article/pii/S0010465502007555.

[29] W. H. Press and G. R. Farrar. "RECURSIVE STRATIFIED SAMPLING FOR MULTIDIMENSIONAL MONTE CARLO INTEGRATION." *Submitted to: Comp.in Phys.* (1989).

[30] Z. Zhe-Zhao, W. Yao-Nan, and W. Hui. "Numerical integration based on a neural network algorithm." 8 (Aug. 2006), pp. 42–48.

[31] L. y. Xu and L. j. Li. "The New Numerical Integration Algorithm Based on Neural Network." *Third International Conference on Natural Computation (ICNC 2007)*. Vol. 1. Aug. 2007, pp. 325–328. DOI: 10.1109/ICNC.2007.730.

[32] L. Yan, J. Di, and K. Wang. "Spline Basis Neural Network Algorithm for Numerical Integration." *International Journal of Mathematical, Computational, Physical, Electrical and Computer Engineering* 7.3 (2013), pp. 458–461. ISSN: PISSN:2010-376X, EISSN:2010-3778. URL: http://waset.org/Publications?p=75.

[33] J. Friedman, T. Hastie, and R. Tibshirani. "Additive logistic regression: a statistical view of boosting (With discussion and a rejoinder by the authors)." *Ann. Statist.* 28.2 (Apr. 2000), pp. 337–407. DOI: 10.1214/aos/1016218223.

[34] J. H. Friedman. "Greedy function approximation: A gradient boosting machine." *Ann. Statist.* 29.5 (Oct. 2001), pp. 1189–1232. DOI: 10.1214/aos/1013203451.

[35] I. J. Goodfellow et al. "Generative Adversarial Networks." *ArXiv e-prints* (June 2014). arXiv: 1406.2661 [stat.ML].

[36] J. Bendavid. "Efficient Monte Carlo Integration Using Boosted Decision Trees and Generative Deep Neural Networks" (2017). arXiv: 1707.00028 [hep-ph].

[37] *MeMEMta: Modular Matrix Element Implementation.* URL: https://github.com/MoMEMta.

[38] R. D. Ball et al. "Parton distributions for the LHC Run II." *JHEP* 04 (2015), p. 040. DOI: 10.1007/JHEP04(2015)040. arXiv: 1410.8849 [hep-ph].

[39] V. Bertone, S. Carrazza, and N. P. Hartland. "APFELgrid: a high performance tool for parton density determinations." *Comput. Phys. Commun.* 212 (2017), pp. 205–209. DOI: 10.1016/j.cpc.2016.10.006. arXiv: 1605.02070 [hep-ph].

[40] T. Likhomanenko, D. Derkach, and A. Rogozhnikov. "Inclusive Flavour Tagging Algorithm." *J. Phys. Conf. Ser.* 762.1 (2016), p. 012045. DOI: 10.1088/1742-6596/762/1/012045. arXiv: 1705.08707 [hep-ex].

[41] *ALEPH Workshop @ NIPS 2015 - Applying (machine) Learning to Experimental Physics (ALEPH) and «Flavours of Physics» challenge.* Neural Information Processing Systems (NIPS). 2015. URL: http://yandexdataschool.github.io/aleph2015/.

[42] *Heavy Flavour Data Mining Workshop.* Zurich, 2016. URL: https://indico.cern.ch/event/433556/.

[43] *Data Science @ LHC 2015 Workshop.* 2015. URL: https://indico.cern.ch/event/395374/.

[44] *DS@HEP at the Simons Foundation.* 2016. URL: https://indico.hep.caltech.edu/indico/conferenceDisplay.py?confId=102.

[45] *DS@HEP.* 2017. URL: https://indico.fnal.gov/conferenceDisplay.py?confId=13497.

[46] URL: opendata.cern.ch.

[47] A. Hoecker et al. "TMVA: Toolkit for Multivariate Data Analysis." *PoS* ACAT (2007), p. 040. arXiv: physics/0703039.

[48] M. Feindt. "A Neural Bayesian Estimator for Conditional Probability Densities" (2004). arXiv: physics/0402093.

[49] M. Feindt and U. Kerzel. "The NeuroBayes neural network package." *Nucl.Instrum.Meth.* A559 (2006), pp. 190–194. DOI: 10.1016/j.nima.2005.11.166.

[50] A. Rogozhnikov. ***hep_ml** python package.* URL: https://github.com/arogozhnikov/hep_ml.

[51] P. Seyfert. ***pseyfert/tmva-branch-adder**: Version 0.5.0.* DOI: 10.5281/zenodo.1149690. URL: https://doi.org/10.5281/zenodo.1149690.

[52] *PyROOT.* URL: https://root.cern.ch/pyroot.

[53] N. Dawe et al. ***scikit-hep/root_numpy**: 4.7.3.* DOI: 10.5281/zenodo.842249. URL: https://doi.org/10.5281/zenodo.842249.

[54] I. Babuschkin et al. ***scikit-hep/root_pandas**: 0.3.2.* DOI: 10.5281/zenodo.1188928. URL: https://doi.org/10.5281/zenodo.1188928.

[55] J. Pivarski et al. ***scikit-hep/uproot**: 2.8.17.* DOI: 10.5281/zenodo.1219218. URL: https://doi.org/10.5281/zenodo.1219218.

[56] *c2numpy.* URL: https://github.com/diana-hep/c2numpy.

[57] *root4j.* URL: https://github.com/diana-hep/root4j.

[58] *root2hdf5.* URL: http://www.rootpy.org/commands/root2hdf5.html.

[59] R. Aaij et al. "Search for the lepton flavour violating decay $\tau^- \to \mu^-\mu^+\mu^-$." *JHEP* 02 (2015), p. 121. DOI: 10.1007/JHEP02(2015)121. arXiv: 1409.8548 [hep-ex].

[60] *Third Machine Learning in High Energy Physics Summer School.* 2017. URL: https://indico.cern.ch/event/613571/.

[61] URL: https://vispa.physik.rwth-aachen.de.

[62] L. Bel. "The LHCb Starterkit." *Proceedings of the 38th International Conference on High Energy Physics (ICHEP2016). 3-10 August 2016. Chicago, USA.* 2016, p. 334. URL: http://pos.sissa.it/cgi-bin/reader/conf.cgi?confid=282.

[63] *HEP Software Foundation Community White Paper Working Group — Training, Staffing and Careers.* Report Number: HSF-CWP-2017-02 (In preparation). URL: https://hepsoftwarefoundation.org/organization/cwp.html.

IOP Conf. Series: Journal of Physics: Conf. Series **1085** (2018) 022009 doi:10.1088/1742-6596/1085/2/022009

AI Reloaded: ACAT2017 Conference Summary

Pushpalatha C. Bhat

Fermi National Accelerator Laboratory (Fermilab), Batavia, IL 60510

pushpa@fnal.gov

Abstract. The first edition of the Advanced Computing and Analysis Techniques in Physics Research (ACAT) workshop series was held at Fermilab in the year 2000, which was in fact the 7th meeting of its parent the original AIHENP workshop series that started in 1990 in Lyon, France. ACAT 2017 was the 18th workshop in the series, which is growing stronger and increasingly successful. It covered the areas of computing in physics research, computations in theoretical physics, and data analysis algorithms and tools. In this summary article, I will give some historical perspective on the workshop series and provide highlights of this workshop. Since Artificial Intelligence (AI) was the featured theme, I will briefly discuss machine learning (ML) and comment on the latest developments in ML applications in physics research covered at this workshop.

1. Introduction

The 18th edition of the ACAT (formerly AIHENP) workshop returned to its roots and featured Artificial Intelligence with the theme "AI reloaded." Coinciding with the first session of the conference, a spectacular celestial phenomenon, a total solar eclipse that came to be called "The Great American Eclipse", swept across the United States from Northwest to the Southeast, bringing beautiful sights of the eclipse and an exciting atmosphere to the start of the conference.

2. History of the Workshop Series

The workshop series started as the *Artificial Intelligence in High Energy and Nuclear Physics* (AIHENP) workshop [1] held in March 1990 in Lyon, France. The series was renamed as *Advanced Computing and Analysis Techniques* (ACAT) [2] when it was held at Fermilab in October 2000. ACAT 2017 would be the 18th workshop in the series, growing ever stronger and successful. The meeting is held every 18 months, alternating between spring and late summer/autumn and rotating through various regions of the world – Asia, Europe and Americas. Among the pioneers of the early series were Denis Peret-Gallix, Rene Brun, Fred James, Bruce Denby, Slava Ilyin, Andrei Kataev, Christian Kiesling, Jos Vermaseren and Monique Werlen.

The "Raison d'être" of the conference series has been to bring together theory, experimental physics and computing communities to "bridge" the disciplines and communities. The general strategy for the program has been to have three parallel "tracks" of sessions at each meeting: Track 1 dealing with Computing Technology for Physics Research, Track 2 on topics of Data Analysis Algorithms and

Tools, and Track 3 on Computations in Theoretical Physics, Techniques and Methods. Occasionally, a fourth track has been added to the meeting, depending on the venue and relevance to the ongoing events. The Fermilab workshop, for example, had a fourth track on Very Large Scale Computing. That meeting brought many trailblazers and pioneers in the advanced computing field: Bjarne Stroustroup (father of C++), Ian Foster (father of Grid Computing), Stephen Wolfram (creator of Mathematica), Alex Szalay and Robert Ryne (Large Scale computing).

The ACAT workshops have also highlighted a major theme for each of the meetings. Just to note some examples: World Wide Web was the theme for the 1992 meeting where Tim Berners-Lee gave a plenary talk, Parallel Distributed Computing was the theme for the 1996 meeting, Large Scale Simulations in 2000, Quantum Computing in 2007, Cloud Computing in 2011. The meetings organizers have always tried to sense major paradigm shifts that are coming our way and highlighted them. The theme for ACAT 2017 has been, very aptly, "AI Reloaded".

3. This Meeting

This meeting amply met and surpassed the expectations on many fronts. As mentioned earlier, the major theme for the meeting was chosen to be "AI Reloaded", given the recognition by the high energy physics (HEP) community of the enormous positive impact the machine learning (ML) methods have already had in our field's major discoveries and studies, and the enormous potential of the ML/AI methods in the future [3,4].

The meeting drew quite a diverse gathering of about 200 participants. As has been the practice, the meeting was organized as a combination of common plenary sessions and three parallel sessions each following one of the three tracks of the meeting. Twenty-four invited plenary talks and eighty-six parallel session talks were given. About seventy posters were presented to a very keen audience in two successful poster sessions.

Another important feature at each ACAT meeting has been a number of round table discussions that are organized on topics of high relevance for the field and the times. There were four round table discussions on: Machine Learning, Using Heterogeneous Resources for HEP Computing, Analytical vs. Numerical Methods for NNLO+ Calculations for the Large Hadron Collider (LHC), and Diversity.

4. Highlights from the Three Parallel Tracks

4.1. Computing Technology for Physics Research

The main topic areas covered in talks in this track included (a) Grids and Cloud-federation - the continuing transition to virtual, standardized, commodity-based computing infrastructure (b) Heterogeneous Architectures, particularly on the use of Graphics Processing Units (GPUs) in HEP experiment triggers and promising applications of Field-Programmable Gate Arrays (FPGAs) in lattice QCD computing; (c) Online Systems and Triggers; (d) Modernizing tools for Software builds; (e) Containers, and their rising importance for software deployment and issues of their portability and scalability on supercomputers; (f) Machine Learning Tools for fast simulation; (g) Challenges in HEP Visualization in events at the LHC and HL-LHC (High-Luminosity LHC) and (h) Data Preservation Projects at the Tevatron, LHC and INFN.

4.2. Computations in Theoretical Physics

Several new developments in methods for symbolic calculations and loop integrals calculations were presented. Developments in event generators and their optimizations, and higher order radiative

IOP Conf. Series: Journal of Physics: Conf. Series **1085** (2018) 022009 doi:10.1088/1742-6596/1085/2/022009

corrections were discussed. Specific tools such as Loopedia, a new database for loop integrals, and Go_HEP, a new language for concurrent programming, open source codes such as pySecDec which provides tools for numerical evaluation of multiscale integrals, were discussed. It was clear that Machine Learning has been making its way into theoretical physics from the presentations on ML usage in PDFs to MC tools [5], Active Learning, and discussion on possible venues to use ML to accelerate lattice QCD.

4.3. Data Analysis: Algorithms and Tools

Several applications of ML algorithms such as Neural Networks (NN) and Boosted Decision Trees (BDT) were presented. In fact, many flavors of Deep Neural Networks (DNN) are being explored and exploited in HEP applications, especially by large experiments [6]. The use of ML is now ubiquitous in triggers and every aspect of event reconstruction and physics analysis – tracking, object ID, energy corrections, signal/background discrimination, simulations and so on. There were presentations on implementing them in end-to-end event reconstruction. Some examples of ML applications presented were (a) NN and BDT algorithms used in triggers in Belle-II, LHC, (b) event pile-up mitigation at the LHC, (c) event reconstruction in neutrino experiments, (d) use of DNN for online and offline tracking, (e) ML in theoretical physics as alluded to in the previous section. There were also talks on use of expert systems and cellular automata.

5. Machine Learning

Since this meeting's featured theme was AI, and that Machine Learning is becoming ubiquitous in HEP and elsewhere, I think it is worthwhile to muse over what it is, how it is still a burgeoning field and what to expect in terms of future applications and meetings.

In the context of physics data analysis, Machine Learning is the paradigm for automated learning of input-output mapping (or stimuli-response mapping) or "functional fitting" from data, using computer algorithms (or in hardware), requiring little *a priori* information about the mapping or function to be learned. A method that can approximate a continuous non-linear function to arbitrary accuracy is called a *universal approximator* and neural networks are an example of such a universal approximator. In the past nearly three decades since the AIHENP/ACAT workshop series started in 1990, when some of us who were the proponents and early adopters struggled to gain acceptance in the mainstream, these methods have gained gradual acceptance in HEP and other sciences, and have now become the "state of the art" in HEP data analyses. Some of the most important physics results in HEP, in the past over two decades, have come from the use of multivariate and/or machine learning methods. While in the 1990s, at the beginning of my talks, I had to proclaim that "we are riding the wave of the future", that future is here and the use of MVA/ML methods is expected to only grow in relevance and importance.

5.1. Deep Learning and all that Jazz

Recent explosion in interest and applications of ML come from the availability of lots of computing power including the GPUs as well as algorithmic advances. Multi-scale feature learning with multiple hidden layers as in deep neural networks, where each higher-level layer of neurons learns increasingly higher-level features in the data, has revolutionized image recognition capabilities. Deep learning makes it possible to use raw data as inputs without having to derive "intelligent" discriminating variables to input to the classifier algorithm. One could use both raw inputs as well as derived variables in a DNN to achieve the best possible learning. When raw variables are inputs, it has been shown that the DNN performs better than a shallow NN. The so-called "Drop-out" algorithm has been shown to help avoid over-fitting of the NNs. Unsupervised pre-training with stacked auto-encoders to build up the representation of the input space and then using supervised learning of the input-output mapping as a fine-tuning of optimization is shown to outperform supervised learning in a DNN where all hidden layers at once start out with randomized initial weights. A lot of processing power and lots of training data are needed indeed, and DNNs are generally implemented in GPUs and large amounts of training data can be generated with small perturbations in the initial training datasets.

Many flavors of deep learning are possible and have been implemented and used. Many of these were presented at this meeting. Apart from the deep feed-forward fully connected (FF FC) neural networks, Convolutional Neural Networks (CNN), Recurrent Neural Networks (RNN) and Generative Adversarial Networks (GAN) are some of the popular architectures being explored and applied in particle physics research. The CNNs have found novel applications in event classification in neutrino experiments. The applications of deep learning at the LHC have ranged from simple object-ID and signal/background discrimination and regression to fast simulation of events [7], pileup mitigation (in CMS and ATLAS) [8], and end-to-end reconstruction [9]. In some of the simpler applications of the first category, it is not clear that the resources applied to using the complicated architecture is justified. However, the current explorations will provide us with the experience and higher level of understanding leading to better heuristics.

Now that the Standard Model (SM) of particle physics is complete, in the sense that all the expected matter and force particles and the Higgs boson which provides for electroweak symmetry breaking have been discovered, new physics beyond the SM is vigorously being pursued by the HEP experimental community. Also very interesting for data analysis and theoretical interpretations are the ongoing efforts of the community to extract ML internal parameters to glean new insights on the process under study. We have the sophisticated, very powerful tools of ML in our hands. But the stakes are very high with regards to any claims of discovery of new physics. It is critical to establish robust methods of validation as well as proper handling of systematic uncertainties with the use of ML. As the vanguard of these powerful methods and tools, the ML community also has serious responsibilities in this regard.

6. Thoughts for Future Meetings

The ACAT2017 meeting covered an amazing array of diverse topics and applications and succeeded in bridging the theory, experiment and computing communities in high energy physics and their persuasions. Expansion into related areas for future meetings could include attracting contributions from astrophysics where advanced computing and ML applications have been in vogue for some time, and accelerator physics where ML is being used in controls applications and simulations. It is needless to say that it would be very fruitful to interact with statisticians and build strong connections. Since AI has been not only transforming scientific research but also our society and our daily lives, it would also be of great value to have invited plenary talks on other major advances in computing and AI/ML in fields outside the realm of physics. These issues as well as the interplay of High Performance Computing and Machine Learning will be on the agenda at the next ACAT (March 11-15, 2019) in Saas-Fee (Switzerland) [10].

7. Acknowledgments

I would like to thank the local organizing committee chaired by Gordon Watts and co-chair Shih-Chieh Hsu, the Program Committee chaired by Federico Carminati, and track coordinators N. Neufeld, M. Girone, G. Stewart, S. Gleyzer, G. Golovanov, A. Haas, T. Burnett, A. Freitas, S. Jones, F. Yuasa, and the International Advisory Committee. I thank D. Perret-Gallix and S. Gleyzer for reading this

IOP Conf. Series: Journal of Physics: Conf. Series **1085** (2018) 022009 doi:10.1088/1742-6596/1085/2/022009

paper and making useful suggestions. I also thank all the excellent speakers, presenters and participants for a very delightful and productive meeting.

My research is supported by Fermilab which is managed by Fermi Research Alliance, LLC (FRA), under the contract number DE-AC02-07CH11359 with the U.S. Department of Energy.

References

[1] New Computing Techniques in Physics Research, Proc. 1st International Workshop on Software Engineering, Artificial Intelligence and Expert Systems in High Energy and Nuclear Physics, Lyon (March 19-24, 1990), ed. By D. Perret-Gallix and W.Wojcik, CNRS, Paris 1990.

[2] Advanced Computing and Analysis Techniques in Physics Research, Proc. VII International Workshop, Fermilab, Batavia, IL, (Oct. 16-20, 2000), ed. By Pushpalatha C. Bhat and Matthias Kasemann, AIP Conference Proceedings Vol. 583, NY, USA 2001.

[3] P.C. Bhat, "Multivariate Analysis Methods in Particle Physics," Annu. Rev. Nucl. Part. Sci., Vol. 61 (2011) 281-309.

[4] http://lpc.fnal.gov/computing/HEPMachineLearningCommunityWhitePaper.pdf

[5] S. Carrazza, "Machine Learning Challenges in Theoretical HEP," these proceedings.

[6] B. Nachtman, "Deep Learning usage by Large Experiments," and references therein, these proceedings

[7] S. Vallecorsa, "Generative Models for fast simulation," these proceedings

[8] P.T. Komiske, E.M. Metodiev, B. Nachman and M.D. Schwartz, "Learning to remove Pileup at the LHC with Jet Images," these proceedings.

[9] M. Andrews, M. Paulini, S. Gleyzer, and B. Poczos, "End-to end Event Classification of High-Energy Physics Data," these proceedings.

[10] https://indico.cern.ch/event/708041/

IOP Conf. Series: Journal of Physics: Conf. Series **1085** (2018) 022010 doi:10.1088/1742-6596/1085/2/022010

Software Citations and the ACAT Community

Daniel S. Katz

Assistant Director for Scientific Software and Applications, National Center for Supercomputing Applications (NCSA); Research Associate Professor, Computer Science (CS); Research Associate Professor, Electrical and Computer Engineering (ECE); Research Associate Professor, School of Information Sciences (iSchool); University of Illinois Urbana-Champaign, Urbana, Illinois, 61801, USA

E-mail: d.katz@ieee.org

Abstract. Software is essential for the bulk of research today. It appears in the research cycle as infrastructure (both inputs and outputs, software obtained from others before the research is performed and software provided to others after the research is complete), as well as being part of the research itself (e.g., new software development). To measure and give credit for software contributions, the simplest path appears to be to overload the current paper citation system so that it also can support citations of software. A multidisciplinary working group built a set of principles for software citation in late 2016. Now, in ACAT 2017 and its proceedings, we want to experimentally encourage those principles to be followed, both to provide credit to the software developers and maintainers in the ACAT community and to try out the process, potentially finding flaws and places where it needs to be improved.

1. Introduction

The purpose of this paper is to explain an experiment being performed by the 18th International Workshop on Advanced Computing and Analysis Techniques in Physics Research (ACAT 2017). We believe that software is essential in most research, and in particular, ACAT is focused on work that is commonly implemented in software. In late 2016, a set of software citation principles were published that are intended to promote and guide the use of citations of software, similar to citations of papers. In this paper, we bring together software citations and ACAT, explain how software can be cited in ACAT-related work, and urge that it is cited in the ACAT proceedings and in the future ACATs and related meetings, so that the contributions of software developers will be recognized and rewarded, and the amount and quality of software that is shared will increase.

2. The role of software in research

Two types of evidence demonstrate the increased role and importance of software in today's research. While neither of these is specific to the ACAT community, there is no reason to think that these general trends are not also true in physics research in general, and there is reason to believe that they are even more true for the ACAT community, which includes a strong computational component as shown in its name, "Advanced Computing and Analysis Techniques in Physics Research."

The first type is evidence from surveys, where researchers are asked how important software is to them. Two recent surveys, one of UK academics at Russell Group Universities [1, 2], and one of members of (US) National Postdoctoral Research Association [3, 4] found that 67% / 63% of respondents said, "my research would not be possible without software," where the two results correspond to the two surveys. 21% / 31% said, "my research would be possible but harder," while just 10% / 6% said, "it would make no difference."

The second type is evidence from papers, where the discussions in any form of software in papers are counted, either manually or via natural language processing and machine learning. An informal scan of 6 months of *Science* in mid-2013 found that about half the papers were software-intensive projects, and most of the other papers also relied on some software. A formal study of 90 randomly selected papers in the biology literature in 2015 found that 80% mentioned software, and that those articles mentioned an average of 4.85 software packages [5]. A more recent study of *Nature* in Jan–Mar 2017 found software mentioned in 32 of 40 research articles, with an average of 6.5 software packages mentioned per article [6].

The software that is used in research can be thought of as part of the research process. Some software can be considered inputs to the process, where it is known in advance of the research, and the research plan relies on it. Some software is developed or improved as part of the research itself. And some software that is produced during a research project is intended to be shared with others who can then use it for their own research. Even if the software developed during research is not produced with the intent of sharing it, some limited sharing might be required by community and publisher and funder efforts toward reproducibility and open science. In this paper, where we discuss software citation, we focus on the software that is shared between researchers, either as inputs or outputs. We call this software-as-infrastructure.

3. Software citation principles

In 2015 and 2016, a FORCE11[1] Software Citation working group developed a set of software citation principles [7]. The group started July 2015, co-led by Arfon Smith and Daniel S. Katz. In September 2015, a working group in the WSSSPE3 workshop that focused on credit and citation, co-led by Kyle E. Niemeyer and Daniel S. Katz decided to join forces with the FORCE11 group, with Niemeyer joining as a co-lead of that group. The group grew to about 60 members, including researchers, developers, publishers, repository developer and maintainers, and librarians.

This group started by recognizing that current citation system was created for papers and books, and to develop a system for software citation, we either need to overload the current system to add software or to completely rework the citation system from scratch. The group decided to focus on the overloading path, as the reworking path seems too difficult. The challenge the group tried to address was not just how to identify software in a paper, but how to identify software used within the research process.

Work was done on GitHub (`https://github.com/force11/force11-scwg`) and on the FORCE11 web site (`https://www.force11.org/group/software-citation-working-group`). The group reviewed existing community practices and developed a set of use cases for software citation. It then drafted a software citation principles document, which started with previously published data citation principles [8], updated them based on software use cases and related work, and further updated them based on working group discussions. This led to a draft that was subject to community feedback and review, as well as feedback at a workshop at FORCE2016 in April 2016. Discussion was generally via GitHub issues, and changes in the document were tracked. In late 2016, the paper and its reviews were published [7]. The paper includes a set of six principles (general statements), use cases (where the principles should apply), and discussion (suggestions on how to apply the principles).

[1] FORCE11 (`https://www.force11.org`) is a community of scholars, librarians, archivists, publishers and research funders that has arisen organically to help facilitate the change toward improved knowledge creation and sharing.

The software citation principles are:

(i) Importance. Software should be considered a legitimate and citable product of research. Software citations should be accorded the same importance in the scholarly record as citations of other research products, such as publications and data; they should be included in the metadata of the citing work, for example in the reference list of a journal article, and should not be omitted or separated. Software should be cited on the same basis as any other research product such as a paper or a book, that is, authors should cite the appropriate set of software products just as they cite the appropriate set of papers.

(ii) Credit and Attribution. Software citations should facilitate giving scholarly credit and normative, legal attribution to all contributors to the software, recognizing that a single style or mechanism of attribution may not be applicable to all software.

(iii) Unique Identification. A software citation should include a method for identification that is machine actionable, globally unique, interoperable, and recognized by at least a community of the corresponding domain experts, and preferably by general public researchers.

(iv) Persistence. Unique identifiers and metadata describing the software and its disposition should persist – even beyond the lifespan of the software they describe.

(v) Accessibility. Software citations should facilitate access to the software itself and to its associated metadata, documentation, data, and other materials necessary for both humans and machines to make informed use of the referenced software.

(vi) Specificity. Software citations should facilitate identification of, and access to, the specific version of software that was used. Software identification should be as specific as necessary, such as using version numbers, revision numbers, or variants such as platforms.

In May 2017, the FORCE11 Software Citation Working Group ended, and a new Software Citation Implementation Working Group started, co-chaired by Neil Chue Hong, Martin Fenner, and Daniel S. Katz. This group has the goal of moving the software citation principles to implementation, with this conference experiment as an example of progress. Those interested in following the new group can join it at `https://www.force11.org/group/software-citation-implementation-working-group`.

4. Applying the principles

In general, the principles work to add a step into the software workflow: publishing a version of software to make it citable. As such, implementing the principles in practice for ACAT can be seen as having two parts:

4.1. Making your software citable

If a developer has software they want people to cite, the first step is to publish that software. If the software is on GitHub, which is common today, the developer can follow the steps in GitHub's guide (`https://guides.github.com/activities/citable-code/`). Otherwise, the developer can submit the software to an archive such as Zenodo or figshare. In either case, the submitter needs to supply the metadata for the publication, including the authors, the title, the version, and possibly any citations that software itself needs to make, such as to software that it uses. Once this is done, the service that accepts the software will provide a DOI. The developer can then create a CITATION file with a suggested citation that includes this DOI, update the README file similarly, or otherwise tell people how to cite the software. The developer can also write a software paper and ask people to cite that, but this is secondary, just since our current system doesn't work well for software.

IOP Conf. Series: Journal of Physics: Conf. Series **1085** (2018) 022010 doi:10.1088/1742-6596/1085/2/022010

4.2. Citing someone else's software

The author who wants to cite someone else's software should first check for a CITATION file or a README file. If either file says how to cite the software itself, the author should do that, and if not, the author should do their best to follow the principles. Specifically, they should try to include all contributors to the software, and if this is not clear, they can just name the project. They should try to include a method for identification that is machine actionable, globally unique, interoperable, perhaps as a URL to a release or a company product number, if no DOI is available. If there is a landing page that includes metadata, they should point to that, not directly to the software (e.g., the GitHub repo URL.) They should include specific version/release information. And, if there's a software paper, they can cite this too, but not in place of citing the software itself.

5. Examples

This proceedings uses a citation style that is somewhat unhelpful for software, and that seems focus on compactness at the cost of clarity, which may not be fully appropriate in a time where most papers are distributed electronically, and having DOIs or URLs where a work can be found are probably more important to the reader than saving a few photons. Thus, work is needed to fit software into this style, and the examples that follow should be considered a starting point that others can adapt to better meet the needs of the community. In addition, the `iopart-num` BibTeX package produced by Mark A. Caprio and provided to the proceedings' authors ideally should be updated to include a software type.

Some examples of citations for unpublished software are:

- Geant4 Project 2017 Geant [software] version 10.3.2 Available from `https://github.com/Geant4/geant4/releases/tag/v10.3.2` [accessed 2017-08-17]
- Eigen Project 2017 Eigen [software] version 3.3.4 Available from `https://bitbucket.org/eigen/eigen/` [accessed 2017-08-17]
- Python Project 2017 Python [software] version 3.6.2 Available from `https://www.python.org/downloads/release/python-362/` [accessed 2017-08-17]
- LLVM Project 2017 LLVM Core [software] version 4.0.1 Available from `http://releases.llvm.org/download.html#4.0.1` [accessed 2017-08-17]
- R Project 2017 R [software] version 3.4.1 Available from `https://cran.r-project.org/src/base/R-3/` [accessed 2017-08-17]
- TensorFlow 2017 Project TensorFlow [software] version 1.3.0 Available from `https://github.com/tensorflow/tensorflow/releases/tag/v1.3.0` [accessed 2017-08-17]
- Collobert R, Farabet C, Kavukcuoglu K, Chintala S, Leonard N, Tompson J, Zagoruyko J, Massa F, Dundar A, Jin J, et al. 2017 Torch [software] commit a0bf77ff070ca27eb2de31c6465f8ffa4e399be2 available from `https://github.com/torch/torch7` [accessed 2017-08-17]

And here are some examples of citations of published software:

- Pfenninger S and Pickering B 2017 calliope-project/calliope [software] Release v0.5.2 *Zenodo* `https://doi.org/10.5281/zenodo.810012`
- Heinrich L and Cranmer K 2017 diana-hep/packtivity [software] Initial Zenodo Release *Zenodo* `https://doi.org/10.5281/zenodo.309302`
- Stasto A, Xiao B and Zaslavsky D 2014 SOLO [software] version 1 *figshare* `https://doi.org/10.6084/m9.figshare.1033996.v1`

- Dawe EN, Ongmongkolkul P and Stark G 2017 root_numpy: The interface between ROOT and NumPy *Journal of Open Source Software* **2.16** `https://doi.org/10.21105/joss.00307`
- Rademakers F, Canal P, Naumann A, Couet O, Moneta L, GANIS G, Vassilev V, Piparo D, Bellenot B, wverkerke, et al. 2017 root-project/root: Release v6-11/02 *Zenodo* `https://doi.org/10.5281/zenodo.1003159`

6. ACAT software citation experiment

The intent of this paper is to follow up on a plenary talk given at ACAT 2017 about an experiment in which the conference organizers encourage conference speakers to cite the software they talk about in their proceedings papers. In addition to this talk, the proceedings instructions also made this request. Of course, the goal of this experiment is to move the ACAT community towards citing software, in order to give the authors of the software credit when it is used.

7. Conclusions

Software is essential for the bulk of research today as researchers tell us directly and via their papers. However, it is not cited in papers nearly as often as it is used, leading to developers not getting sufficient credit for their work. In order to change this, the FORCE11 Software Citation working group developed a set of software citation principles that permit software to be published and then cited in papers as other research is. This paper has described the principles, and has provided some examples of how physics software can be cited. This is intended to encourage ACAT authors to cite the software they use in the papers they write in these proceedings. In the future, we will be able to examine these proceedings and see what has worked and what needs to be improved, as well as to examine proceedings from future years to track changes.

References

[1] Hettrick S 2014 It's impossible to conduct research without software, say 7 out of 10 UK researchers URL `http://bit.ly/2B8y6Iz`

[2] Hettrick S, Antonioletti M, Carr L, Chue Hong N, Crouch S, De Roure D, Emsley I, Goble C, Hay A, Inupakutika D, Jackson M, Nenadic A, Parkinson T, Parsons M I, Pawlik A, Peru G, Proeme A, Robinson J and Sufi S 2014 UK research software survey 2014 URL `https://doi.org/10.5281/zenodo.14809`

[3] Nangia U and Katz D S 2017 *figshare* URL `https://doi.org/10.6084/m9.figshare.5328442.v3`

[4] Nangia U and Katz D S 2017 Survey of National Postdoctoral Association - Dataset URL `https://doi.org/10.5281/zenodo.843607`

[5] Howison J and Bullard J 2016 *Journal of the Association for Information Science and Technology* **67** 2137–2155 ISSN 2330-1643 URL `https://doi.org/10.1002/asi.23538`

[6] Nangia U and Katz D S 2017 Understanding Software in Research: Initial Results from Examining Nature and a Call for Collaboration *Proceedings of the 13th IEEE International Conference on eScience (eScience 2017)* URL `https://doi.org/10.1109/eScience.2017.78`

[7] Smith A M, Katz D S, Niemeyer K E and FORCE11 Software Citation Working Group 2016 *PeerJ Computer Science* **2** e86 URL `https://doi.org/10.7717/peerj-cs.86`

[8] Data Citation Synthesis Group 2014 Joint Declaration of Data Citation Principles Martone, M. (ed), FORCE11, San Diego, CA URL `https://doi.org/10.25490/a97f-egyk`

The ALICE O^2 common driver for the C-RORC and CRU read-out cards

Boeschoten P and Costa F for the ALICE collaboration

E-mail: `pascal.boeschoten@cern.ch`, `filippo.costa@cern.ch`

Abstract. ALICE (A Large Ion Collider Experiment) is the heavy-ion detector designed to study the strongly interacting state of matter realized in relativistic heavy-ion collisions at the CERN Large Hadron Collider (LHC). A major upgrade of the experiment is planned during the 2019-2020 long shutdown. In order to cope with a data rate 100 times higher than during LHC Run 1 and with the continuous read-out of the Time Projection Chamber (TPC), it is necessary to upgrade the Online and Offline Computing to a new common system called O^2. The O^2 read-out chain will use commodity x86 Linux servers equipped with custom PCIe FPGA-based read-out cards. This paper discusses the driver architecture for the cards that will be used in O^2: the PCIe v2 x8, Xilinx Virtex 6 based C-RORC (Common Readout Receiver Card) and the PCIe v3 x16, Intel Arria 10 based CRU (Common Readout Unit). Access to the PCIe cards is provided via three layers of software. Firstly, the low-level PCIe (PCI Express) layer responsible for the userspace interface for low-level operations such as memory mapping the PCIe BAR (Base Address Registers) and creating scatter-gather lists, which is provided by the PDA (Portable Driver Architecture) library developed by the Frankfurt Institute for Advanced Studies (FIAS). Above that sits our userspace driver which implements synchronization, controls the read-out card – e.g. resetting and configuring the card, providing it with bus addresses to transfer data to and checking for data arrival – and presents a uniform, high-level C++ interface that abstracts over the differences between the C-RORC and CRU. This interface – of which direct usage is principally intended for high-performance read-out processes – allows users to configure and use the various aspects of the read-out cards, such as configuration, DMA transfers and commands to the front-end. The top layer consists of a Python wrapper and command-line utilities that are provided to facilitate scripting and executing tasks from a shell, such as card resetting; performing benchmarks; reading or writing registers; and running test suites. Additionally, the paper presents the results of benchmarks in various test environments. Finally, we present our plans for future development, testing and integration.

1. Introduction

ALICE[1] (A Large Ion Collider Experiment) is an experiment at the CERN LHC[2] (Large Hadron Collider) studying the Quark-Gluon Plasma – a state of matter which existed shortly after the Big Bang – exploiting heavy-ion collisions. In 2019-2020, the LHC will undergo its second long shutdown (LS2) to prepare for Run 3, after which it will run at significantly higher luminosity. During the shutdown ALICE will upgrade its detectors and software systems to handle the higher data rate of over 3.4 TB/s. The new computing software suite called O^2[3] will do both offline and online processing. Part of this suite is the ReadoutCard[4] library containing the driver for the two read-out cards used during Run 3. In this paper, we will discuss the design and capabilities of the ReadoutCard library, the supported cards, and present benchmark results.

IOP Conf. Series: Journal of Physics: Conf. Series **1085** (2018) 032001 doi:10.1088/1742-6596/1085/3/032001

Figure 1: C-RORC

Figure 2: CRU

2. Cards

The cards supported by the ReadoutCard library are the C-RORC[5] (Common Readout and Receiver Card) seen in figure 1, and the CRU (Common Readout Unit) shown in figure 2. Both are custom-built FPGA-based PCIe cards. The C-RORC, is based on a Xilinx Virtex 6 FPGA. It supports up to PCIe v2x8, with the current firmware implementing v2x4 only (as the additional bandwidth was unnecessary). The card can take up to 12 optical links and distribute the data over 6 DMA channels to the host PC at a total maximum throughput of 1.6 GB/s. It is currently in use in ALICE for Run 2 and 25 units will be reused for some of the subdetectors during Run 3.

The CRU is based on the Intel Arria 10 FPGA, using PCIe v3x16, supporting 48 optical links, and 2 DMA channels with a total throughput of 27 GB/s measured using the current firmware and software. It will be the main read-out card for ALICE during Run 3 with 461 units to be used.

3. Software

The software stack is subdivided into three layers:

- Low-level PCIe system handling, provided by the PDA[6] library
- A userspace driver with C++ interface for high-performance read-out, providing the core functionality
- A Python interface and command-line utilities that use the C++ interface

The last two are referred to as the ReadoutCard library, which is part of the O^2 software suite.

3.1. PDA

PDA – the result of extensive work described in D. Eschweiler's thesis[7] – consists of a UIO (Userspace IO) kernel module that handles the creation of scatter-gather lists for DMA buffers, and a userspace C library that interfaces with the kernel module and provides additional utilities such as:

- PCI device enumeration.
- Registering DMA memory targets with the IOMMU (Input Output Memory Mapping Unit). This component can take multiple separate memory regions and present them as a contiguous virtual memory space to the PCI device. This is analogous to how the regular MMU (Memory Mapping Unit) works, presenting scattered physical memory regions as a single virtual space to processes running on the CPU. Both the IOMMU and MMU also provide protection against invalid memory accesses.
- Generation of scatter-gather lists, which combines regions of memory into a single list for the card's DMA engine. This is essential if an IOMMU is not enabled or not present.

IOP Conf. Series: Journal of Physics: Conf. Series **1085** (2018) 032001 doi:10.1088/1742-6596/1085/3/032001

- Memory mapping of the device's BAR (Base Address Register), a range of device registers memory-mapped to a virtual address space so they can accessed with regular memory reads and writes.

3.2. Userspace driver
The driver wraps the PDA functions in C++ classes, and builds upon the tools they provide to configure and control the cards. For each card, the ReadoutCard's C++ interface exposes to the user a single instance – system-wide – of a DmaChannelInterface implementation. The class instantiated is specific to the type of the card, and it operates the card's machinery. Read-out processes use it for high-performance data taking. It provides a queue-style interface, to push DMA target addresses and pop them when the transfer is complete. To instantiate this class, the user must provide a DMA buffer, which will be registered internally with PDA to create the scatter-gather list. The buffer can be any type of shared memory if the IOMMU is enabled, since the IOMMU can present the memory as a single contiguous region to the card. However, if the IOMMU is not available, it is essential to use hugepage-backed shared memory to ensure the buffer meets the required physical contiguousness for the cards' DMA engines. This is enforced by the driver through a lookup in the process's memory mappings.

In addition, the interface can provide multiple instances of BarInterface classes per card. These are essentially wrappers around the BAR, providing simple read and write access with bounds checking. They may also prohibit reads and/or writes to certain address ranges under certain conditions to prevent malfunctions.

The library uses "Boost Exception" for its error handling. When these exceptions are raised, arbitrary data can be added to them while they are caught and rethrown, bubbling up to the final catch site with information about possible causes and potential fixes for the error. This is useful for issues where the program itself cannot easily deduce the cause of the error. In these cases, the rich diagnostic messages generated from the exception allow users to potentially fix the issues themselves quickly, without referring to in-depth documentation or contacting developers.

3.3. Python interface
The Python interface is a wrapper around a subset of the C++ interface: the BarInterface. It aims to provide easy access for scripting purposes, and other situations where flexibility and ease of use is required. For example, for development of the card's firmware or detector. We expect it will be mainly used by card firmware developers, and detector teams testing configuration which are passed through the card to the detector front-ends.

3.4. Command-line utilities
The module contains several utility programs to assist with development, debugging and administration. Every program will display detailed information and a usage example when ran with the –help option. Most programs will also provide more detailed output when given the –verbose option. The available programs are:

roc-bench-dma DMA throughput and stress-testing benchmarks.

roc-flash Flashes firmware from a file onto the card.

roc-flash-read Reads from the card's flash memory.

roc-list-cards Lists the read-out cards present on the system, along with their type, PCI address, vendor ID, device ID, serial number, and firmware version.

roc-reg-read Reads from registers on a card's BAR.

roc-reg-write Writes to registers on a card's BAR.

roc-reg-reset Resets a card channel.

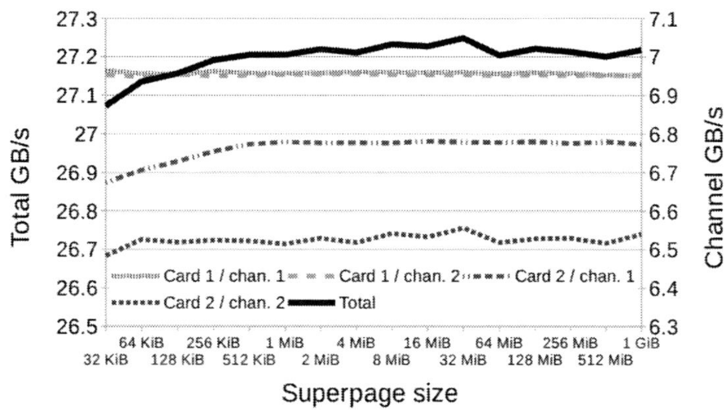

Figure 3: CRU benchmark results

4. Buffer layout & data format

The buffer of a DMA channel is referred to as the channel buffer. In real-world running conditions, these will be large – in the order of tens of gigabytes – to accommodate latency in the processing chain. It is allocated by the read-out task and registered with the driver before starting a data taking run. The read-out is then responsible for subdividing the channel buffer into "superpages". These are the logical units of transfer for the driver, and will be filled with multiple data pages by the cards. They are passed to the driver as a memory region within the DMA buffer, ranging in size from 32 kilobytes to gigabytes. The C-RORC and CRU are cards with very different interfaces, so the driver will then take care of the transfer in a card-specific way. For the older C-RORC, superpages are filled up with multiple transfers. The CRU, whose firmware was developed in close cooperation with the driver, natively supports superpages: it is given a pointer and size, and simply fills the entire superpage with data.

The in-memory data format consists of 8 kilobyte DMA pages. They each contain a header, the payload, a trailer, and padding in case the data does not fill up the DMA page. This rigid structure aids recovery or inspection of data in case of transfer errors. In addition, the driver will scan the superpage and generate a table of indices corresponding to DMA pages that start at "heartbeats", special DMA pages that indicate synchronization intervals or conditions. This lookup table speeds up handling and processing along the read-out chain.

5. Benchmark

Throughput was measured on a Dell R730 server (2x Intel E5-2630 v3 CPUs, 2133 MHz DDR4 RAM) fitted with two CRU cards, using various superpage sizes, recording individual DMA channel throughputs and summing them to arrive at a total throughput for both cards. The operating system used was CERN CentOS 7. The required throughput specified for the O^2 system is 5.625 GB/s. Additionally, we set a development target for 6.9 GB/s to allow for headroom.

Data was gathered using a script[8] that simultaneously starts instances of the roc-bench-dma program for each DMA channel. The cards were configured to use an internal data generator during the minute-long runs. To optimize the throughput on this multi-socket server, it was necessary to bind the DMA processes' threads and memory allocation to the CPU whose PCIe lanes are directly connected to the corresponding card. To achieve this, the script uses the "numactl" utility to launch the processes.

Figure 3 shows the results of the benchmark, with total and channel throughputs displayed as a function of superpage size. Little variance in total throughput is shown, even between the smallest supported superpage sizes and very large ones. For all channels, the minimum required throughput was exceeded. Both channels of card 1 were consistently performing similarly, exceeding the target throughput. Both of card 2's channels did not reach the target throughput. Card 2 channel 1 showed a steady increase up to 1 MiB superpages, then remained steady. Card 2 channel 2 performed inconsistently and is responsible for what jitter there is in the total throughput. Note that the DMA engines on all of these cards and channels are identical. More work is being done to understand the throughput differences between channels.

6. Conclusions

In this paper, we have described a userspace driver for high-performance read-out built upon PDA, its supporting utilities, the cards and data structures involved, and presented benchmark results demonstrating the driver's capabilities. Our next steps are:

- Testing of new server models coming from different companies to identify the best candidate for the O^2 farm.
- Investigate throughput differences between individual channels when running with multiple cards.
- Adapt the driver to CRU firmware changes, following experiment requirements.
- Finalize documentation and software interface, so detector teams can use the O^2 software and connect their own modules to the O^2 framework.

References

[1] ALICE Collaboration 2008 "The ALICE experiment at the CERN LHC" *JINST* **3** S08002
[2] Evans L and Bryant P 2008, "LHC Machine" *JINST* **3** S08001
[3] The ALICE Collaboration 2015, "Technical Design Report for the Upgrade of the Online-Offline Computing System" *CERN-LHCC-2015-006, ALICE-TDR-019*
[4] Boeschoten P, "ReadoutCard" [software], version 0ff4a1e, 2017, Available from https://github.com/AliceO2Group/ReadoutCard [accessed 2017-09-22]
[5] ALICE and ATLAS Collaborations 2015, "The C-RORC PCIe card and its application in the ALICE and ATLAS experiments" *JINST* **10** no.02 C02022
[6] Eschweiler D and Lindenstruth V 2014 "The portable driver architecture", Proceedings of the 16th Real-Time Linux Workshop, Open Source Automation Development Lab (OSADL)
[7] Eschweiler D 2015 "Efficient device drivers for supercomputers", dblp computer science bibliography, *Goethe University Frankfurt, Frankfurt am Main, Germany*
[8] Boeschoten P, "readoutcard-multi-bench" [software], version v1.0.0, 2017. Available from https://github.com/PascalBoeschoten/readoutcard-multi-bench [accessed 2017-09-22]

IOP Conf. Series: Journal of Physics: Conf. Series **1085** (2018) 032002 doi:10.1088/1742-6596/1085/3/032002

Design and Execution of make-like Distributed Analyses

R F von Cube, M Erdmann, B Fischer, R Fischer, M Rieger
III. Physics Institute A, RWTH Aachen University, Germany
E-mail: rfischer@physik.rwth-aachen.de

Abstract. In particle physics, workflow management systems are primarily used as tailored solutions in dedicated areas such as Monte Carlo production. However, physicists performing data analyses are usually required to steer their individual workflows manually, which is time-consuming and often leads to undocumented relations between particular workloads. We present a generic analysis design pattern that copes with the sophisticated demands of end-to-end HEP analyses. The approach presents a paradigm shift from executing parts of the analysis to defining the analysis. The clear interface and dependencies between individual workloads then enables a make-like execution.

Our tools allow to specify arbitrary workloads and dependencies between them in a lightweight and scalable structure. Further features are multi-user support, automated dependency resolution and error handling, central scheduling, and status visualization. The WLCG infrastructure is supported including CREAM-CE, DCAP, SRM and GSIFTP. Due to the open structure, additional computing resources, such as local computing clusters or Dropbox storage, can be easily added and supported. Computing jobs execute their payload, which may be any executable or script, in a dedicated software environment. Software packages are installed as required, and input data is retrieved on demand.

The management system is explored by a team performing ttbb and ttH cross section measurements.

1. Introduction

Modern high-energy physics data analyses are growing in complexity and scale, compare figure 1. Adding to the complexity are for instance the number of advanced analysis techniques like multivariate classifiers, sometimes requiring graphics processing units, the number of simultaneous analyses, e.g. for various final state channels or event categories. The scale increases as more data get recorded, and more simulated events are needed to describe the data with proper statistics. Issues of scale are typically addressed via high throughput computing concepts involving a large number of computing jobs and many computing sites, e.g via the Worldwide LHC Computing Grid (WLCG)[1]. The user is presented with the task to execute the analysis under these conditions. This involves a lot of bookkeeping to make sure that each required file is at the right place at the right time. Due to the complexity, often only the physicist knows the exact order in which certain scripts need to be executed, which files need to be copied, or where to perform the analysis. New team members or students are often challenged to learn many technical details on how to run the analysis. Finally, if people leave a group, they often take key knowledge with them. The focus of analysis documentation in publications or conference talks is mostly on what was performed, not on how it was done on a technical level. These factors contribute to making it difficult to extend or replicate an existing analysis.

In this paper, we explore the application of workflow management systems for user analyses. If the user defines what is the input data, what does the input depend on, and how it should be processed, the system should be able to execute the task. The goal is to run the analysis `make`-like [2] in a distributed manner.

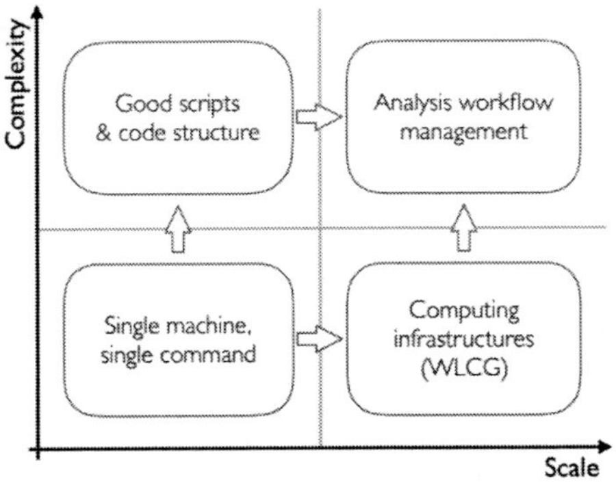

Figure 1. The figure shows different coping mechanisms for user analysis challenges with respect to their scale and their complexity.

2. Workflow Systems in High-Energy Physics

A data analysis comprises certain steps. Typical examples of analysis steps in high-energy physics are event selection, training of classifiers, statistical inference, plotting, or auxiliary estimations of scale factors or other relevant information. Often one part of the analysis consists of a number of steps. An example could be a data driven background estimation that first needs to create data tuples with relevant information, then performs a fit in a sideband, and subsequently creates control plots.

In more general terms, this can be described as a workflow, i.e. one or multiple related data analyses, that comprise smaller workloads. The workflows are related via a clear interface. For example, all dependencies have to be run first and the output of prior workloads needs to be accessible by subsequent workloads. In order to execute a workload, computing resources are needed, in particular, computing and storage resources. This perspective on data analyses shows how analogous it is to workflow management systems.

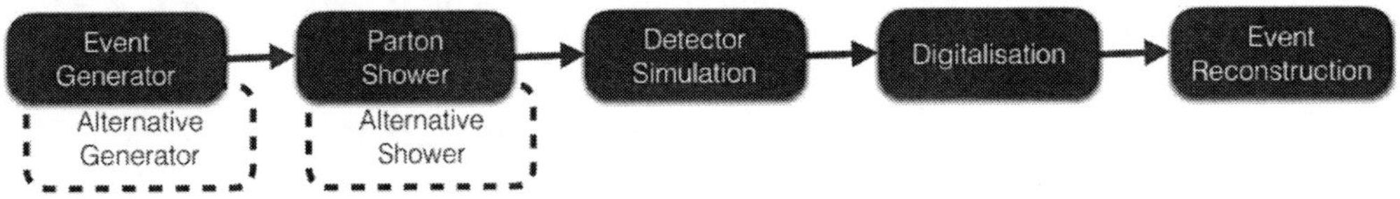

Figure 2. Typical Monte Carlo event simulation workflow.

Therefore, it is natural to look for existing workflow management systems related to high-energy physics. The typical example that can be found is illustrated in figure 2. It shows a chain from event generation to the event reconstruction that is typically used to created data sets of simulated collision events. A couple of observations can be made from these workflows. The type of depicted simulation workflow is one-dimensional, static, and recurring. Except for variations in certain steps of the chain, i.e. a alternative event generators, the software requirements are homogeneous. Because simulation is so important to analyse recorded data and because the whole generation requires enormous computing and storage resources, special infrastructures are built for it. Among these are databases, storage systems, workload management systems.

User analyses on the other hand are not one-dimensional. One workload might need input from several others and the output might be needed by multiple subsequent workloads. They are not static since they need to adapt to newly discovered challenges during the analysis, to new recipes developed within large collaborations, and to requirements made during a review process. The scale of analyses is much smaller than for event simulation, while the requirements are more diverse. In summary, existing workflow management systems are not applicable to user data analyses.

3. Workflow Management for User Analyses

In this section two approaches to create workflow management systems for user analyses are presented. The first one is the report-based approach. It was tailored from scratch for high-energy physics. The main idea is that each time any payload is executed a report with metadata, i.e. regarding created output files, is created. This system should perform well if the storage system is slow especially with respect to file lookups.

The seconds approach is target based. Instead of creating a reports with metadata, the system checks the existence of necessary input files prior to scheduling a workload. This system should perform well for quick access to storage. In addition, this approach was chosen to test whether existing workflow management systems can be adapted to the high-energy physics environment. It is based on the Luigi pipelining tool[3].

Both systems are written in Python and will be presented in the following sections.

3.1. Report-based Workflow Management

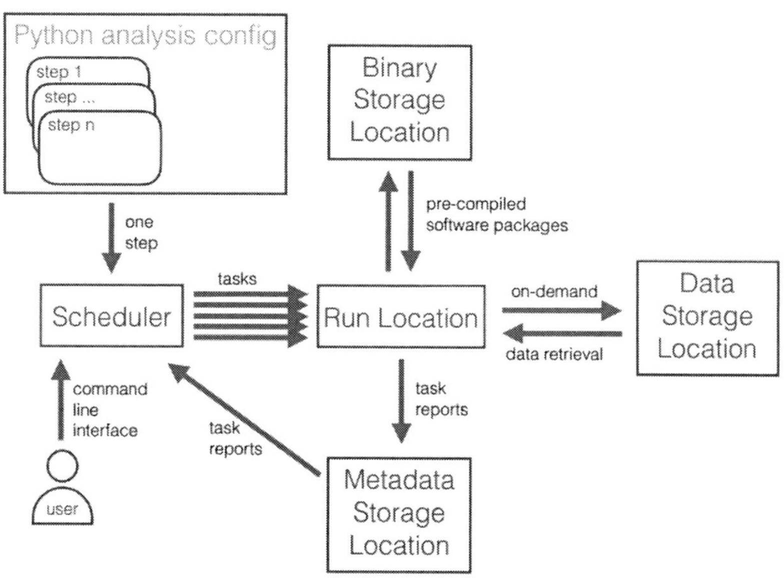

Figure 3. Scheme depicting the interplay between various components of the report-based approach in order to execute an analysis step.

The execution scheme is shown in Figure 3. Each step is executed in one or multiple tasks. The Scheduler is the component that determines which tasks are unfinished. Unfinished tasks are sent to the Run Location configured for this step. Multiple tasks can be combined into one computing job if the user reckons the execution duration for one task is too short for one job. Once the job starts at the Run Location, the software environment is created using setup scripts and the Software Manager. After the environment is set up, the required input data is downloaded. If all required software and files are available the payload is executed. After a task is finished, the step stores metadata and provenance information in JSON [4] task reports on a Storage Location for meta data.

The next time the Scheduler identifies unfinished tasks it first downloads newly created task reports. Reports contain information on whether a task was performed successfully and if not what was the cause of the error. This allows to resubmit unsuccessful tasks already while the computing job is still running on other tasks. They also contain information on output Data Collections that were created including information on Storage Locations of output files. Subsequent steps can use these collections as their input Data Collection such that the Scheduler can schedule the execution of appropriate tasks for the subsequent step.

Because steps can spawn tens of thousands of tasks, reports are summarized into one file. Thereby, each task report only has to be read once. An example for useful metadata that can be evaluated by subsequent steps is the number of originally generated events for samples containing simulated events. This number can be used later to calculate event weights and to scale the samples appropriately.

The user can specify whether dependent steps need to be finished entirely or whether subsequent steps can already work with parts of the output data. For example, not all tasks of an event selection step have to be done for a following reconstruction step to start running on parts of the data. A step performing a statistical inference on the other hand might need all available reconstructed data to produce proper results.

3.2. Target-based Workflow Management

The target-based approach is depicted in figure 4. It shares many similarities with the report-based system. The main difference is that it does not create metadata reports. This reduces the complexity of the system because it does not need to manage reports or summarize them. The reduction in complexity is traded off against lookups whether required input files exist. The number of files that need to be checked can be of the order of tens of thousands.

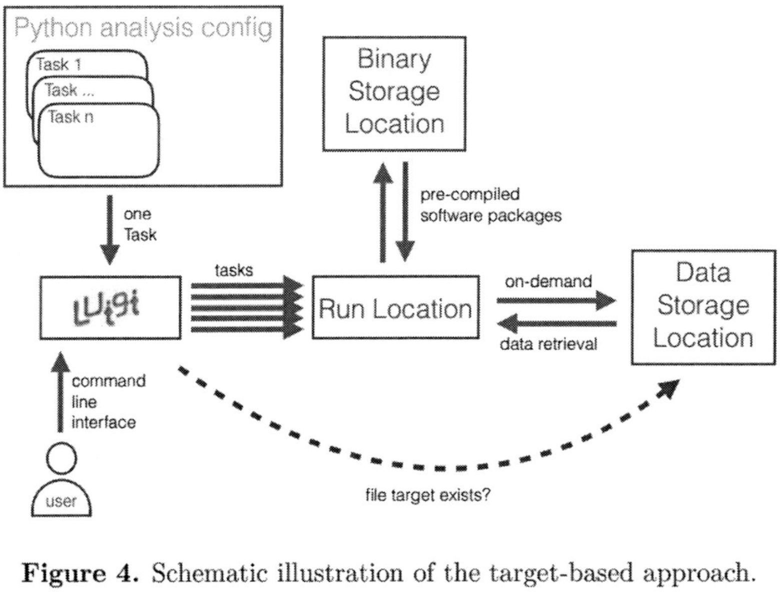

Figure 4. Schematic illustration of the target-based approach.

IOP Conf. Series: Journal of Physics: Conf. Series **1085** (2018) 032002 doi:10.1088/1742-6596/1085/3/032002

4. Conclusion

Both Workflow Management system presented here were tested alongside cross section measurements at the LHC for the top quark pair associated bottom quark pair production and the top quark pair associated Higgs production, respectively. Despite different requirements regarding central storage for data and metadata, both approaches worked well in conjunction with dCache system[5]. Many small reports files were handled equally well as the large number of file lookups.

Both systems provide building blocks for user analyses without limitations on the payload software that is used. Rather than providing an analysis framework, they can be understood as a design pattern for user analyses. In both cases all parameters needed for analyses are transparently encoded and thereby documented within the workflow descriptions. Thus, bookkeeping tasks and undocumented states of files and intermediate steps are prevented. This enables to reproduce the exact same results if the workflow is re-executed. From the point of view of the user, the systems allow to focus on physics topics rather than on technicalities of the execution. In summary, workflow management systems for user analysis present a paradigm shift from executing the analysis to defining it.

Acknowledgments

This work is supported by the Ministerium für Wissenschaft und Forschung, Nordrhein-Westfalen, the Bundesministerium für Bildung und Forschung (BMBF) and the Helmholz Alliance Physics at the Terascale.

References

[1] Bird I 2011 *Annual Review of Nuclear and Particle Science* **61** 99–118
[2] Free Software Foundation, Inc 2016 GNU Make URL `https://www.gnu.org/software/make`
[3] Spotify Luigi URL `https://github.com/spotify/luigi`
[4] 2013 The JSON Data Interchange Format Tech. Rep. Standard ECMA-404 1st Edition / October 2013 ECMA URL `http://www.ecma-international.org/publications/files/ECMA-ST/ECMA-404.pdf`
[5] Fuhrmann P, Behrmann G, Bernardt C *et al.* 2012 dCache, agile adoption of storage technology (Computing in High Energy and Nuclear Physics, New York City (USA), 21 May 2012 - 25 May 2012) URL `https://bib-pubdb2.desy.de/record/140491`

IOP Conf. Series: Journal of Physics: Conf. Series **1085** (2018) 032003 doi:10.1088/1742-6596/1085/3/032003

Supercomputers, Clouds and Grids powered by BigPanDA for Brain studies

A Beche[1], K De[2], F Delalondre[1], F Schuermann[1], A Klimentov[3] and R Mashinistov[2,a]

[1]Swiss Federal Institute of Technology in Lausanne, Route Cantonale, 1015 Lausanne, Switzerland
[2]University of Texas at Arlington, 701 South Nedderman Drive, Arlington, TX 76019, US
[3]Brookhaven National Laboratory, P.O. Box 5000, Upton, NY 11973-5000, US

[a]rmashinistov@gmail.com

Abstract. The PanDA WMS - Production and Distributed Analysis Workload Management System - has been developed and used by the ATLAS experiment at the LHC (Large Hadron Collider) for all data processing and analysis challenges. BigPanDA is an extension of the PanDA WMS to run ATLAS and non-ATLAS applications on Leadership Class Facilities and supercomputers, as well as traditional grid and cloud resources. The success of the BigPanDA project has drawn attention from other compute intensive sciences such as biology. In 2017, a pilot project was started between BigPanDA and the Blue Brain Project (BBP) of the Ecole Polytechnique Federal de Lausanne (EPFL) located in Lausanne, Switzerland. This proof of concept project is aimed at demonstrating the efficient application of the BigPanDA system to support the complex scientific workflow of the BBP, which relies on using a mix of desktop, cluster and supercomputers to reconstruct and simulate accurate models of brain tissue.

1. Introduction

The Production and Distributed Analysis (PanDA) system [1] was designed to meet ATLAS [2] requirements for a data-driven workload management system for production and distributed analysis processing capable of operating at the LHC data processing scale. It has been used in the ATLAS experiment since 2005 and is now expanding into a BigPanDA [3] project to extend PanDA as a meta application, providing location transparency of processing and data management, for High Energy Physics (HEP) and other data-intensive sciences, and a wider exascale community. The success of the projects to use BigPanDA beyond HEP and traditional HTC resources has drawn attention from other compute intensive sciences such as neuroscience.

In 2017, a pilot project was started between BigPanDA and the Blue Brain Project (BBP) [4] of the Ecole Polytechnique Federal de Lausanne (EPFL) located in Lausanne, Switzerland. This proof of concept project is aimed to demonstrate efficient application of the PanDA Workload Management System, initially developed for HEP applications, for the supercomputer-based reconstructions and simulations offering a radically new approach for understanding the multilevel structure and function of the brain.

In the first phase, the goal of this joint project is to support the execution of BBP software on a variety of distributed computing systems powered by BigPanDA. The targeted systems for demonstration are: x86 BBP clusters located in Geneva and Lugano (Switzerland), the Titan Supercomputer [5] operated by Oak Ridge Leadership Computing Facility (OLCF) and Cloud based resources (Amazon).

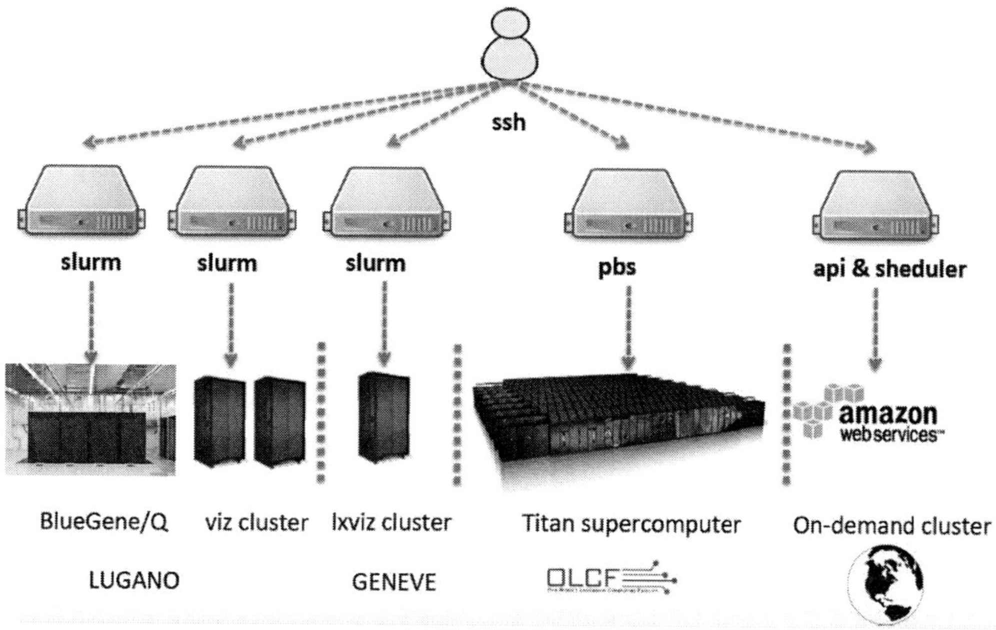

Figure 1. Variety of BBP distributed computing systems considered in the project.

2. Evolution of the computing model of the Blue Brain Project

Brain simulation software applications are highly parallelized using MPI, thus, they should run on HPC clusters with high throughput and low latency interconnect (InfiniBand). The main challenge in the HPC world is that the software stack (scheduler, MPI libraries etc.) is deeply system dependent and portability very difficult to achieve. In order to accelerate science, BBP workflow is no longer exclusively running on dedicated resources but also on world-class supercomputers in Leadership Computing Facilities such as Oak-Ridge National Laboratory (ORNL) - Titan or Argonne National Laboratory (ANL) - MIRA [6]. This heterogeneity in the available systems imposed to streamline the workflow and dataflow in order to lower the entry barrier to various system for scientists. This simplification has began with the software deployment using the NIX[7] package manager allowing an immutable system-agnostic deployment of the whole BBP software stack. The second stone towards simplification is workload scheduling. The current workflow requires every scientist to connect to a remote login node and schedule jobs from there using the local system resource manager (Slurm on BBP systems, Moab/PBS on Titan, Cobalt on MIRA etc.) complexifying the learning curve to run the exact same software. BigPanDA is approaching the problem by confining the scheduling specificities into the pilot job hiding it completely from the scientists and allowing them to run exact same jobs independently of the desired HPC system. Finally the last step will be to standardized on data movement, more precisely to abstract the stage-in stage-out of the data to the job phases to the scientists. While there is strong signal in favor of using Globus GridFTP [8] for achieving it, this will deserve a non-negligible part of the BigPanDA/BBP proof of concept to validate it and perhaps move it towards production one day. To summarize, BBP computing model is moving from system-dependent implementation to system-agnostic workflow with well-defined building blocks to achieve simplicity: NIX for software deployment, BigPanDA for Job submission and investigating GridFTP for data transfers.

3. BigPanDA based portal for BBP applications

3.1. Accelerating science impact with PanDA WMS

PanDA is a Workload Management System (WMS) designed to support the execution of distributed workloads and workflows via pilots [9]. Pilot-capable WMS enable high throughput of jobs execution via multi-level scheduling while supporting interoperability across multiple sites. PanDA as a basis technology delivers transparency of data and its processing in a distributed computing environment to the scientists [10]. It provides execution environments for a wide range of experimental applications, automates centralized data processing, enables data analytics for dozens of research groups, supports custom workflow of individual scientists, provides a unified view of distributed worldwide resources, presents status and history of workflow through an integrated monitoring system, archives and curates all workflow, manages distribution of data as needed for processing or scientists access, and provides other features.

The rich menu of features provided, coupled with support for heterogeneous computing environments, makes PanDA ideally suited for scientific data processing. HEP and astro-particle experiments COMPASS and AMS has chosen PanDA as WMS for data processing and analysis. Experiments nEDM with LSST will evaluate PanDA and ALICE is interested in PanDA evaluation for OLCF. Joint Institute for Nuclear Researches (Russia) is considering PanDA as main WMS for NICA collider. Also another several PanDA instances deployed beyond ATLAS at OLCF, Taiwan, Amazon EC2, Russia.

3.2. Software components of the BigPanDA Portal

Two most important components of the PanDA WMS are server and pilots. Jobs are submitted to the PanDA server via simple python-based client application. The PanDA server is the main component, which provides a task queue managing all job information centrally. The PanDA server receives jobs into the task queue, upon which a brokerage module operates to prioritize and assign work on the basis of job type, priority, software availability, input data and its locality, and available computing resources. Pilots launched on the supercomputer or cluster retrieve jobs from the PanDA server in order to run them via the native resource manager like pbs, slurm or others. Pilots use resources efficiently; they exit immediately if no job is available and the submission rate is regulated according to workload. Each pilot executes a job, detects zombie processes, reports job status to the PanDA server, and recovers failed jobs.

To hide execution complexity and simplify manual tasks by end-users, we an interface providing the reduced intuitive feature set like computing tasks definition and submission for execution on the chosen resource, monitoring the states of the system, jobs and resources.

Authentication/authorization module of the Portal integrated with BBP SSO and LDAP groups aware. Unified web form to define new custom tasks allows users to select the applications from the predefined list, provide the input/output files names together with additional parameters for execution. Then defined task can be submitted for execution on specific resource linked to the predefined queue at the PanDA server. After that the status of the tasks can be monitored with the built-in monitoring web interface. Also the portal provides the REST API streamlining the bulk tasks submission.

We have created the prototype of the data management system (DMS) that allows to connect to the storages of different types. The system supports the file replication mechanism and ensures consistency of the replicas. All files involved in the computing, their replicas and access rules are described in a special file catalog. Together distributed data management system and file catalog allows quick integration of third-party storage systems. The DMS system being a part of the portal not yet integrated with the BBP environment and workflow. Wide variety of the used computing resources requires the core data movement and storage technologies to be chosen. This strategic choice will be strictly followed later while development of the DMS plugins and integration with storage technologies. Currently the Globus - a secure, unified interface to research the data - is considered as underlying technology for DMS. Globus was built by the computer scientists at the University of Chicago and Argonne National Laboratory to meet the needs and requirements of the research community. It presents a secure, unified interface to identities and storage across Globus-connected sites, within the visibility and access control limits set by each site. The portal scheme is shown on figure 2.

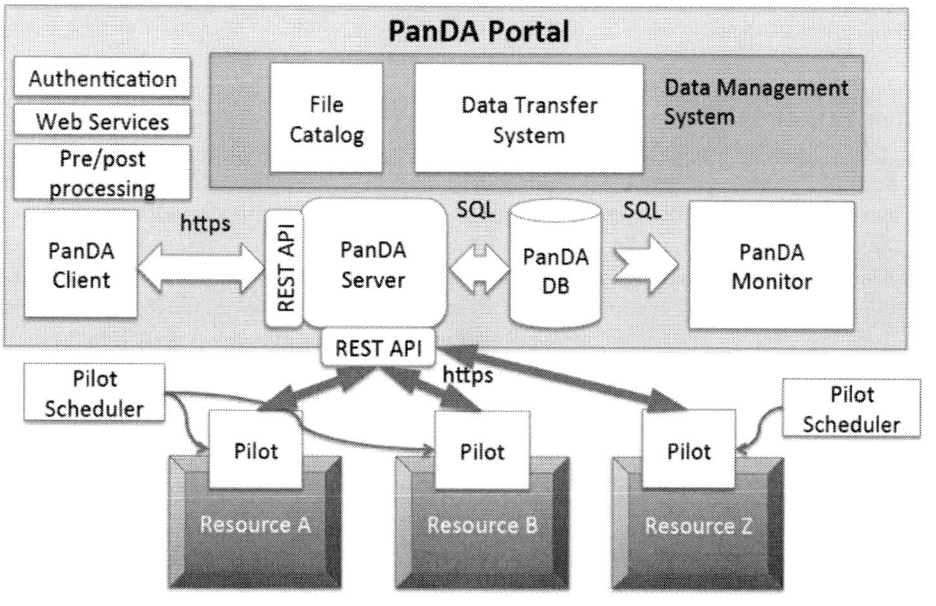

Figure 2. Software components of the PanDA Portal

The main components are:

- GUI/Web services - graphical user interface and web service that provides authentication and simple web interface to make a jobs definition. Also support of the API provided at this level.
- Pre/post processing tool - software component that support the pipelines - several sequenced computational steps. The tool translates the user task definition to the sets of standard PanDA jobs corresponding the pipeline steps and submit them to the PanDA server via the client. Jobs belonging to the same step can run in parallel while jobs of different steps are running by defined order.
- PanDA Server - local Panda server installed and configured on the BBP VM. The server is the heart of the system factorized as a general WMS service. Server provides mapping of all jobs in the system to all available resources. Server provides an internal API to interact with PanDA Client and Monitor.
- PanDA Monitor provides the overall information about the status of the system, submitted jobs and resources.
- The pilot launched on resource retrieves jobs from the server and handles payload execution. Resource specific schedulers are responsible to launch the pilots and maintain appropriate number of running pilots.
- Pilot Schedulers - pilot schedulers manages the pilots submission to available resources defining how many pilots to run on each resource.

- Data Management System - lightweight experiment data management tool. It consists of general file catalog to store metadata and distributed file transfer system to move data between heterogeneous data storages. At the moment an integration of the DMS is in progress.

The following resources were integrated for the Portal at BBP: Intel x86-NVIDIA GPU based BBP clusters located in Geneva (47 TFlops) and Lugano (81 TFlops); BBP IBM BlueGene/Q supercomputer (0.78 PFLops and 65 TB of DRAM memory) [10] located in Lugano; the Titan Supercomputer with peak theoretical performance 27 PFlops operated by the Oak Ridge Leadership Computing Facility (OLCF); Cloud based resources such as Amazon Cloud.

4. Adaptation of the PanDA pilots

Significant role within supported computing resources for this project is played by the Titan supercomputer. This is one of the largest supercomputers in the world. Titan is a hybrid-architecture Cray XK7 system with a theoretical peak performance exceeding 27 petaflops. Titan features 18,688 compute nodes, (each with one 16-core AMD Opteron CPU and 1 NVIDIA Kepler K20X GPU), 299,008 x86 cores, a total system memory of 710 terabytes, and a high-performance proprietary network. The combination of these technologies allows Titan to achieve up to 10 times the speed of its predecessor, the Jaguar supercomputer, while consuming the same average power load and occupying the same physical footprint. The BigPanDA project provided the first important demonstration of the capabilities that a workload management system (WMS) can have on improving the uptake and utilization of LCF from both application and systems points of view. This demonstration was done for ATLAS and now same approach is being applied with the variety of diverse scientific applications even outside of HEP area.

On Titan the pilots run on the front-end nodes, which allows them to communicate with the PanDA server, since front end nodes have connectivity to the Internet. Worker nodes use Cray's Gemini interconnect for inter-node MPI messaging but have no network connection to the outside world. The interactive front-end machines and the worker nodes use a shared file system which makes it possible for the pilots to stage-in input files that are required by the payload and stage-out the produced output files at the end of the job. The pilots submit payloads to the worker nodes using the local batch system (PBS) via the SAGA (Simple API for Grid Applications) interface [11]. It also uses the SAGA facilities for monitoring and management of PanDA jobs running on Titan's worker nodes.

All of the target systems for demonstration except Titan supercomputer use SLURM as local resource management system. In this case we use a wrapper script that will launch pilot jobs in adopted environment at clusters and supercomputers.

A payload also can be executed in cloud environment, which will need special pilot adaptation. In this case pilots launching system must be aware of cloud environment and have permissions and other means to request new virtual machines (VMs) and destroy unnecessarily. Some extra configuration of local resource management system will also be required to deal with VMs. At the moment the adaptation of the pilots for this case is in progress.

For specific cases users also need to strictly separate working environment of their workflows and usage of their own quota at clusters. That also imply slight modification of pilot jobs. Simple yet powerful approach can use wrapper that will switch user context using SUID bit in Linux operation systems. That wrapper can also do other cluster specific actions with user environment or administrative tasks. We have successfully conducted initial tests for this case.

5. Conclusion

The first phase of the pilot project lasted 6 months with the goal to support the execution of BBP software on a variety of distributed computing systems powered by BigPanDA. The targeted systems chosen for demonstration for demonstration shown on figure 3 and included: Intel x86-NVIDIA GPU based BBP clusters located in Geneva and Lugano, BBP IBM BlueGene/Q supercomputer [12] located in Lugano, the Titan Supercomputer operated by the Oak Ridge Leadership Computing Facility (OLCF), and Amazon Cloud.

The project demonstrated that the software tools and methods for processing large volumes of experimental data, which have been developed initially for experiments at the LHC accelerator, can be successfully applied to other scientific fields.

Next "preproduction" phase of the project is under investigation currently. The next goal is to implement the Data Management System for considered scope of the systems and seamlessly integrate it into the BigPanDA portal. Globus is considered as the basic technology. The main target is to establish PanDA portal as a service with 8/5 support.

Acknowledgement
This work was funded in part by the U.S. Department of Energy, Office of Science, High Energy Physics and Advanced Scientific Computing Research under Contracts DE-SC0008635, DE-SC0016280; Russian Ministry of Science and Education under Contract no 14.Z50.31.0024 and by Blue Brain Project. We would like to acknowledge that this research used resources of the Oak Ridge Leadership Computing Facility at the Oak Ridge National Laboratory, which is supported by the Office of Science of the U.S. Department of Energy under Contract no. DE-AC05-00OR22725.

References
[1] Maeno T et al 2011 *J. Phys.: Conf. Ser.* vol 331 072024 (Bristol: IOP Publishing) p 6
[2] ATLAS Collaboration, Aad J et al 2008 *J. Inst.* vol 3 S08003 (Bristol: IOP Publishing) p 407
[3] Buncic P et al 2015 *J. Phys.: Conf. Ser.* vol 608 012040 (Bristol: IOP Publishing) p 8
[4] Markram H 2006 *Nat. Rev. Neurosci.* vol 7 (London: Nature Publishing Group) pp 153-160
[5] Titan supercomputer: https://www.olcf.ornl.gov/titan
[6] MIRA supercomputer: https://www.alcf.anl.gov/mira
[7] NIX – The Purely Functional Package Manager: https://nixos.org/nix/
[8] Globus GridFTP: http://toolkit.globus.org/toolkit/docs/latest-stable/gridftp/
[9] Nilsson P et al 2011 *J. Phys.: Conf. Ser.* vol 331 062040 (Bristol: IOP Publishing)
[10] De K et al 2016 *EPJ Web of Conferences* vol 108 01003 (Les Ulis: EDP Sciences) p 10
[11] SAGA progect, "SAGA" [software], version 0.47.2, 2018. Available from https://github.com/radical-cybertools/saga-python [accessed 2018-02-28]
[12] BlueGene/Q SC: https://www-03.ibm.com/systems/technicalcomputing/solutions/bluegene/

IOP Conf. Series: Journal of Physics: Conf. Series **1085** (2018) 032004 doi:10.1088/1742-6596/1085/3/032004

Testing the limits of an LVS - GridFTP cluster as a replacement for BeSTMan

E Fajardo[1], C Pottberg[1], B Bockelman[2], G Attebury[2], T Martin[1] and F Würthwein[1]

[1] University of California San Diego, La Jolla, CA, USA

[2] University of Nebraska, Lincoln, NE, USA

E-mail: emfajard@ucsd.edu

Abstract. The Worldwide LHC Computing Grid (WLCG) is the largest grid computing infrastructure in the world pooling the resources of 170 computing centers (sites). One advantage of grid computing is that multiple copies of data can be distributed across different sites, allowing user access that is independent of geographic location or software. Each site is able to communicate using software stacks collectively referred to as "middleware". One key middleware piece is the storage element (SE), which provides remote POSIX-like access to a site's storage.

The middleware stack managed by the Open Science Grid (OSG) used a storage resource manager (SRM) protocol implementation that, among other things, allowed sites to load-balance servers providing the Grid File Transfer Protocol (GridFTP) interface. OSG is eliminating the use of an SRM entirely and is transitioning to a solution based solely on GridFTP load-balanced at the network level with Linux Virtual Server (LVS). LVS is a core component of the Linux kernel, so this change increases both maintainability and reduces complexity of the site. In this document, we outline our methodologies and results from the large scale testing of an LVS+GridFTP cluster for data reads. Additionally, we discuss potential optimizations to the cluster to maximize total throughput.

1. Introduction

The most common model for interacting with storage on the Worldwide LHC Computing Grid (WLCG) is the *storage element* (SE). In this model, the site provides a POSIX-like storage system and exposes it via a variety of remote access protocols. Clients interact directly with the storage element to transfer files in and out, and data management layers work to catalog the SEs contents and orchestrate inter-SE transfers.

The storage element model provides rich functionality (at the cost of complexity compared to, for example a cache-based model) and a common implementation involves providing a service that speaks the *storage resource management* (SRM) protocol. SRM is used to provide metadata services, manage storage allocations, and load-balance transfer protocols such as GridFTP. Due to a variety of reasons not explored here, the former use cases have fallen out of use, leaving only load-balancing.

A niche protocol such as SRM is not needed to load-balance network services; various techniques are well-established. In this paper, we explore the use of one system, Linux Virtual Server (LVS), to load balance data services provided by a number of sites in the US.

IOP Conf. Series: Journal of Physics: Conf. Series **1085** (2018) 032004 doi:10.1088/1742-6596/1085/3/032004

2. Current storage architecture

2.1. GridFTP

A common need for a computing facility on the grid is to expose their storage system to external grid clients. The Globus implementation [1] of the Grid File Transfer Protocol (GridFTP) over two parties solves this problem and it also provides authentication and authorization. A single GridFTP server transfer throughput can scale up to the limitation of its own hardware network capabilities. Hence for sites that have more bandwidth available, more than one server its needed to avoid bottlenecking.

2.2. BestMan

The Worldwide LHC Computing Grid (WLCG)[2] adopted the user of Storage Resource Manager (SRM) specification for data transfers between sites and also for storing the output of computing jobs at remote sites. One of the implementations of the SRM is the Berkeley Storage Manager(BeStMan) [3] developed at LBNL (Lawrence Berkeley Nationa Laboratory) and the suggested SRM solution to sites by the Open Science Grid (OSG) [4].

BeStMan also provided the functionality of load balancing (several GridFTP servers), authentication and authorization. Hence it allowed a site to fully use its available bandwidth (by using several GridFTP servers)while only advertising a single BeStMan server.

On 2012 LBNL dropped the support for BeStMan on the OSG, leaving OSG to maintaining its codebase and providing security and performance updates. While OSG has focused on simplifying and removing features, this software product is still over 150,000 lines of code - providing motivation for finding a simpler setup.

2.3. Hadoop File System

The Hadoop File System (HDFS) [5] is a distributed file system that works in commodity, heterogeneous hardware and can store tens of PetaBytes. HDFS features automated data replication (usually in a factor of 3) and recovery of lost or corrupted files. Since 2009 several sites in the OSG use HDFS as their file storage solution [6].

The architecture of HDFS is composed of a name node (NN) and several data nodes (DN). The name node has all the meta information about files in the system, while data nodes are where the files actually reside. The interface between the GridFTP servers and the name node can happen via a FUSE mount in the GridFTP server that offer a POSIX interface to the HDFS namespace. Or it can happen through the the GridFTP HDFS plug-in which converts GridFTP requests into HDFS requests using the HDFS libraries. The latter is the preferred and best performance solution.

3. LVS

LVS is a Linux Kernel based load balancing system project. The project combines the IP Virtual Server (IPVS) code that has been present in the Linux kernel since 2.4 with a variety of user level daemons and tools eg. Keepalived (http://www.keepalived.org).

3.1. IPVS

IPVS uses Layer 4, or transport layer, switching to provide load and reliability balancing for a cluster of Real Servers while providing a single IPv4(v6) Virtual Server address for access.

3.2. Keepalived

IPVS uses features of the kernel in combination with the daemon Keepalived. Keepalived uses the VRRP protocol and series of checks to dynamically manage the pool of real servers based on their health. It then directs IPVS to perform the required routing to have the Real Servers respond.

3.3. Direct Routing

At UCSD we use the LVS in the direct routing mode. This mode is used when the both the real servers, and the Virtual Servers have real world routable IPv4(v6) network address and can be directly accessed over the world wide network. Other models are available in LVS eg. Nat (but they were not tested for this work).

3.4. Linux Networking Customizations

In LVS the Director, which has the virtual server address, forwards all TCP/IP requests it gets to the correct real server after performing various book keeping duties (eg. tracking connections). The mechanisms on how this works depends if the underlaying network is IPv4 or IPv6.

3.4.1. IPv4 For IPv4 there are two key mechanisms to make this work. The interface on the Virtual Server and all real servers are given the same IPv4 network address. This creates a problem where multiple servers are responding to the same IPv4 address via the AARP protocol. Arptables is then used to filter these responses so that only the Virtual Server receives the corresponding ARP responses from the real Servers.

3.4.2. IPv6 In addition to the ARP filtering with IPv4 the link local interface assigned to the Virtual Server is the same as the one of the Real Servers. This allows the real server to only respond to requests on the local network for those requests unless the director has decided otherwise.

3.5. GridFTP Protocol and IPVS

The GridFTP protocol is similar to the FTP protocol in that it uses both a control and a data channel. The control channel operates to create the initial connection and to establish the data channels. As is most often the case the data channel operates in passive mode. This means that the GridFTP server starts up a process that listens in a port in a set range. The client then connects to that port and starts the data transfer. However this creates a problem since the data channel port is random and assigned after the Director has forwarded the control channel request.

For GridFTP the the protocol allows a workaround for this problem. The GridFTP server can be configured to tell the client to connect to a specific hostname for the data channel. By setting this configuration variable (data_interface) to the Real Servers hostname the client can be directed to directly connect to the data port on the GridFTP server selected by the LVS Keepalived.

4. LVS and GridFTP scalability

Scalability tests were performed to confirm that LVS was a good enough replacement for BeStMan . In this section we describe the procedures and the results of the scaling exercises that were performed.

4.1. Test procedure

The objective of the test was to generate a load of a given number of clients requesting from the LVS server and measure the total throughput. GlideinWMS (Glide-in based Workload Management System)[7] was used to submit several jobs (clients) at the same time to requests files through the LVS. Each job consisted of a cycle of submissions requesting a file at random, sleeping for a given time and then requesting another one.

In order to obtain the most number of distributed cores available and to do so without impacting production operations in the cluster a sleeper pool was used. A sleeper pool consist of a batch system pool in which compute nodes are configured to advertise more cores than what they have, under the premise that those extra slots will consume a negligible amount of resources(CPU, memory and disk).

All the jobs in these tests used the *gfal-copy* command line utility to request the files from the LVS server and dumped them into */dev/null*. To prevent HDFS from being a bottleneck and isolate the performance of the LVS setup, the set of files for the test were guaranteed to have a replication factor of ten in Hadoop.

4.2. LVS Directors and GridFTP Server Hardware Setup
The two LVS redirectors, and 6 GridFTP servers at UCSD run on identical single CPU blade servers.

(i) Intel(R) Xeon(R) CPU E5-2609 v4 @ 1.70GHz
(ii) 64GB RAM
(iii) Intel Corporation 82599 10 Gigabit Network Interface
(iv) Two 80GB SSD Drives

4.3. Results
We measured the total throughput of the LVS system at different concurrencies (parallel running jobs) and found out that the throughput was correlated with the number of GridFTP clients, see Figures 1 and 2, which is consistent with the results from [6]. The average and maximum total throughput are summarized in Table 1. It is worth noticing than although done with different hardware (but same network capacity) the maximum throughput achieved is more than twice than from the previous result in [6]. Finally we fit in Figure 3 the average throughput at each level of concurrency and can see that although the throughput scales with the number of clients as previously noted in [6], it is only true up to one point when the the total performance starts to decrease.

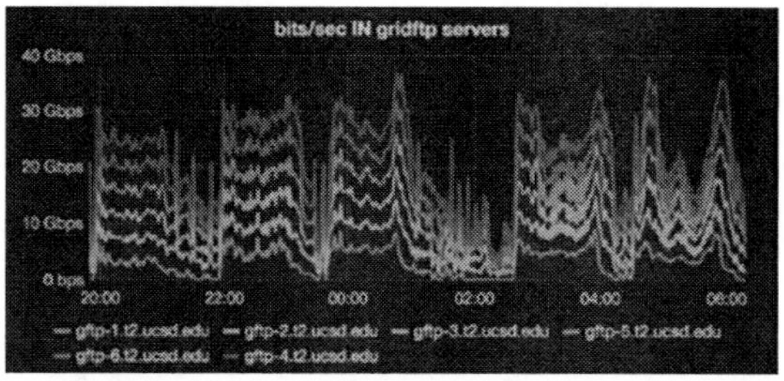

Figure 1. Combined throughput with different number of GridFTP clients

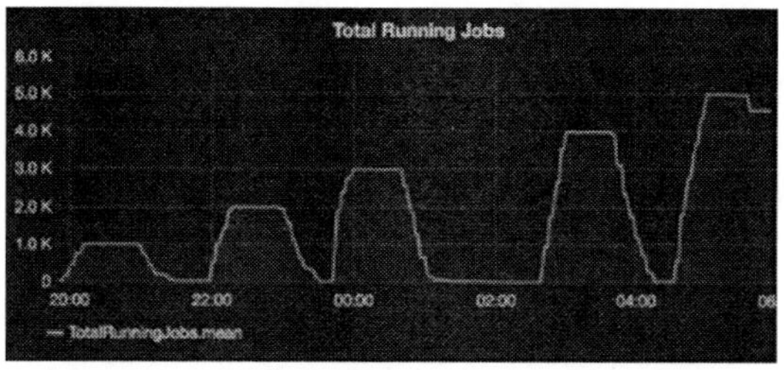

Figure 2. Number of running parallel jobs.

5. Conclusions
In this paper we show how the architecture of the Linux Virtual System in combination for the Globus Grid File transfer Protocol meets the needs of replacing BeStMan as a SE solution for OSG grid sites: load balancing a set of GridFTP servers while providing a single grid entry point to a site's storage. Moreover its bandwidth throughput is equal or higher than the one provided by BeStMan , while at the same time offering high availability, IPV6 compatibility and a single advertising point. There are still some problems in the future as more sites adopt this since installation and configuration of LVS requires more expertise than the BeStMan one.

Parallel clients	Average Throughput $Gbit/s$	Maximum Throughput $Gbit/s$
1000	22.945	32.934
2000	26.853	34.211
3000	28.384	37.190
4000	26.263	36.328
5000	23.334	37.076

Table 1. Average and maximum throughput at different concurrency levels

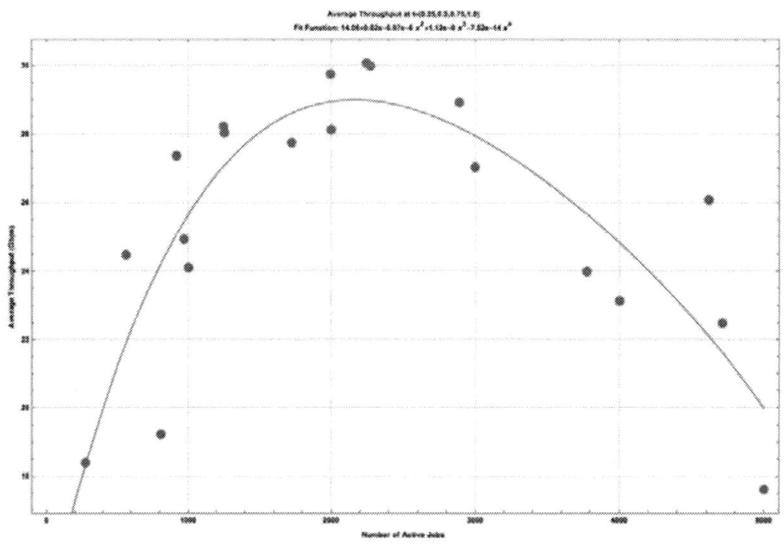

Figure 3. Fit for the average throughput at different concurrency levels

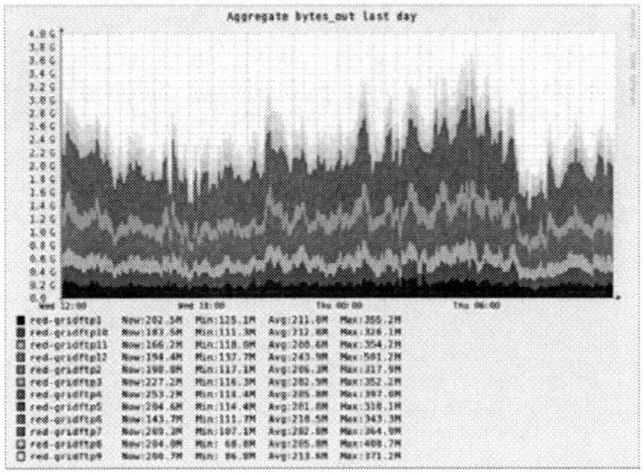

Figure 4. Tipical throughout of gridftp servers at UNL

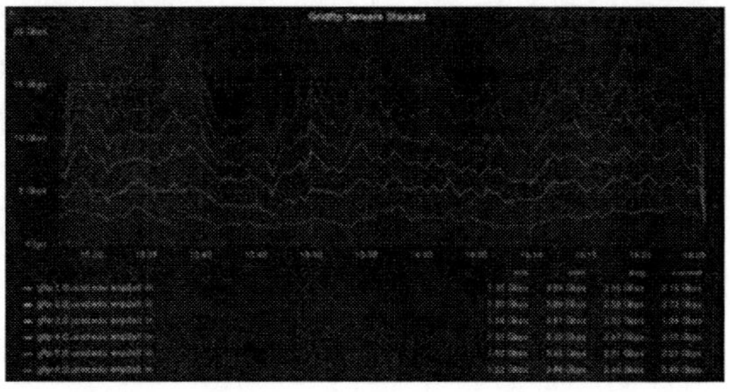

Figure 5. Tipical throughout of GridFTP servers at UCSD

However for sites like Nebraska and UCSD which have migrated, they have seen good results, see figures 4 and 5.

Looking into the future, similar tests should be performed once Hadoop 3 is available to see if there is an actual difference in performance. Finally, although reasonable maximum throughputs were achieved in this test, they did not achieve the maximum physical installed bandwidth ($60Gbit/s$) of the site (six GridFTP servers with $10Gbit/sec$ network interfaces) . We believe this is mostly caused because of the way sleeper pool works, where to many jobs can land in a single computing host and are throttled by the $1Gbit/sec$ network interfaces.

Acknowledgments

This work is supported in part by the National Science Foundation through awards PHY-1148698 and ACI-1321762.

References

[1] Allcock W, Bresnahan J, Kettimuthu R, Link M, Dumitrescu C, Raicu I and Foster I 2005 The globus striped gridftp framework and server *Proceedings of the 2005 ACM/IEEE Conference on Supercomputing* SC '05 (Washington, DC, USA: IEEE Computer Society) pp 54– ISBN 1-59593-061-2 URL `https://doi.org/10.1109/SC.2005.72`

[2] Bird I 2011 Computing for the large hadron collider vol 61 (Annual Reviews) pp 99–118 URL `http://www.annualreviews.org/doi/abs/10.1146/annurev-nucl-102010-130059`

[3] Berkeley storage manager (bestman) URL `https://sdm.lbl.gov/bestman/`

[4] Pordes R, Petravick D, Kramer B, Olson D, Livny M, Roy A, Avery P, Blackburn K, Wenaus T, Wrthwein F, Foster I, Gardner R, Wilde M, Blatecky A, McGee J and Quick R 2007 *Journal of Physics: Conference Series* **78** 012057 URL `http://stacks.iop.org/1742-6596/78/i=1/a=012057`

[5] Attebury G, Baranovski A, Bloom K, Bockelman B, Kcira D, Letts J, Levshina T, Lundestedt C, Martin T, Maier W, Pi H, Rana A, Sfiligoi I, Sim A, Thomas M and Wuerthwein F 2009 Hadoop distributed file system for the grid *2009 IEEE Nuclear Science Symposium Conference Record (NSS/MIC)* pp 1056–1061 ISSN 1082-3654

[6] Amin A, Bockelman B, Letts J, Levshina T, Martin T, Pi H, Sfiligoi I, Thomas M and Wüerthwein F 2011 *Journal of Physics: Conference Series* **331** 052016 URL `http://stacks.iop.org/1742-6596/331/i=5/a=052016`

[7] Sfiligoi I, Bradley D C, Holzman B, Mhashilkar P, Padhi S and Wurthwein F 2009 The pilot way to grid resources using glideinwms *2009 WRI World Congress on Computer Science and Information Engineering* vol 2 pp 428–432

IOP Conf. Series: Journal of Physics: Conf. Series **1085** (2018) 032005 doi:10.1088/1742-6596/1085/3/032005

Provisioning of data locality for HEP analysis workflows

C Heidecker, M J Schnepf, E Kuehn, M Fischer, M Giffels and G Quast

KIT - Karlsruhe Institute of Technology, Germany

E-mail: {christoph.heidecker, matthias.schnepf, eileen.kuehn, max.fischer, manuel.giffels, guenter.quast}@kit.edu

Abstract. The heavily increasing amount of data produced by current experiments in high energy particle physics challenge both end users and providers of computing resources. The boosted data rates and the complexity of analyses require huge datasets being processed in short turnaround cycles. Usually, data storages and computing farms are deployed by different providers, which leads to data delocalization and a strong influence of the interconnection transfer rates. The CMS collaboration at KIT has developed a prototype enabling data locality for HEP analysis processing via two concepts. A coordinated and distributed caching approach that reduce the limiting factor of data transfers by joining local high performance devices with large background storages were tested. Thereby, a throughput optimization was reached by selecting and allocating critical data within user work-flows. A highly performant setup using these caching solutions enables fast processing of throughput dependent analysis workflows.

1. Introduction

Performance of data analysis applications is primarily bounded by how fast data is read and processed [1]. While processing speed depends on local resources, read speed may depend on a series of connections to a remote data storage. A close proximity of processing and storage resources can improve bandwidth by avoiding bottlenecks and congestion of shared connections. The term *data locality* refers to this proximity of processing applications and data sources.

For High Energy Physics (HEP), the granularity of the Worldwide LHC Computing Grid (WLCG) [2] allows data locality only at the scale of computing centres. In contrast, distributed storage frameworks [3, 4] achieve data locality at the scale of racks and hosts. However, this approach requires high-performance storage on processing nodes with sufficient volume to hold all data. Given large data volumes and only partial data access at any time, as is common in HEP, the utilisation of such a system is low.

In the past, we developed on-demand data locality at host granularity [1, 5, 6, 7, 8] for HEP analyses: individual, high-performance caches in a batch system are coordinated to provide popular data from background storage. This significantly improves processing speed of recurring analyses, which also frees resources for unique analyses.

However, our research suggests that host granularity for data locality is not always desirable: optimal locality provides excess read speed compared to processing speed, and considerably complicates scheduling. Instead, data locality is best approached as a spectrum (see Figure 1), where performance limits and infrastructure complexity are balanced.

IOP Conf. Series: Journal of Physics: Conf. Series **1085** (2018) 032005 doi:10.1088/1742-6596/1085/3/032005

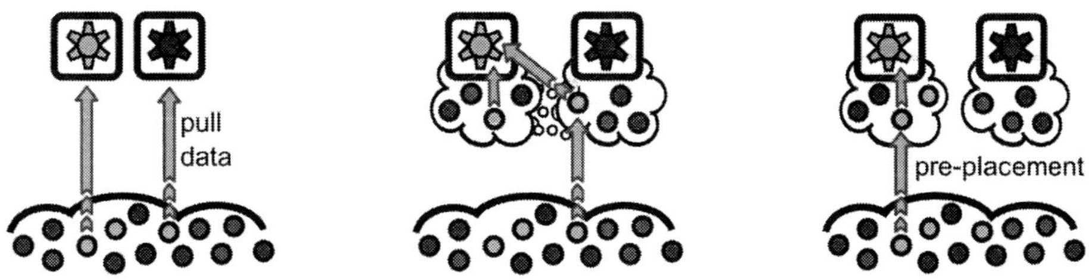

(a) Global access (b) Adjacent access (c) Local access

Figure 1: Spectrum of data locality: Treating data locality as a binary decisions, data consumers access data either globally or locally. However, this also defines the extremes of a spectrum, where data is accessed with a given adjacency, that is a *level of data locality*.

2. Caching for on-demand data locality

Data locality via caching can be split into two separate challenges: First, selecting which input data to provide in order to improve performance. Second, directing consumers to the location of their input data. This translates to two interleaved scheduling problems, both suitable to optimise data processing.

We use a generalised model for data processing, derived from HEP batch processing but not strictly bound to it: Data analysis is organised as *workflows*, each applying a specific analysis application to a *dataset*, a subset of the entire data volume. This is realised by executing multiple *jobs* concurrently, instances of the same application processing non-overlapping portions of the dataset. The execution environment is distributed, with distinct execution hosts sharing a single global namespace to separate data storage. Notably, we do not restrict us to a specific level of detail at which data is read or processed.

2.1. Locality preselection

The processing infrastructure we target is already suitable to efficiently process a fraction of workflows. This is already satisfied if some workflows perform simulation, which require practically no input bandwidth. From our experience, many data processing workflows are also not optimized enough to exhaust bandwidth. In effect, there is a fraction of workflows which does not benefit from caching at all.

Any data accessed only by such workflows can be culled from caching. This reduces the effective data volume, in turn increasing the relative size of caches. Thus, even with small cache sizes, acceptable cache hit rates and low cache trashing are achievable. While many workflow have no direct benefit of caching, they still benefit from processing resources being freed faster.

2.2. Concurrency optimisation

While caches work with individual jobs, our goal is the optimisation of total workflow throughput. This goal must be reflected in the strategy to select and distribute data for caching: the concurrent data access by many jobs of a workflow is the optimisation scenario[1]. Regardless of infrastructure details, this can be effectively achieved by reducing congestion on each data source.

For this goal, individual caches are beneficial as additional, not necessarily faster, data sources. Workflow dataset portions should be spread over as many caches as possible with

[1] Notably, an isolated job has little benefit from caching in our use case. SSD and HDD caches offer bandwidths of 1 Gbit/s to 5 Gbit/s compared to network bandwidths of 1 Gbit/s to 10 Gbit/s.

IOP Conf. Series: Journal of Physics: Conf. Series **1085** (2018) 032005 doi:10.1088/1742-6596/1085/3/032005

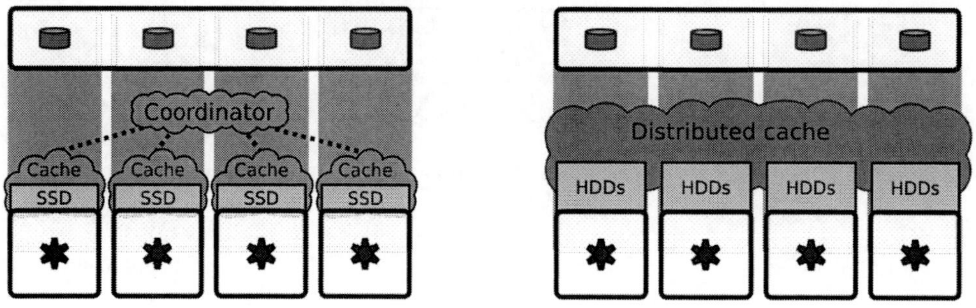

(a) Coordinated caching (b) Distributed caching

Figure 2: Caching Scenarios: Coordinated caching uses isolated caches, which form a single service via coordination. Distributed caching forms a single cache, which is delocalised across multiple hosts.

the splitting by jobs. Notably, global storage is also a viable data source; a fraction of each dataset should be excluded from caching to exploit this capacity.

3. Prototypes for HEP batch systems caching

To study the spectrum of on-demand data locality, we have implemented two caching scenarios (see Figure 2): First, caches local to processing hosts are coordinated to implement a global caching strategy. Second, a single cache distributed across all processing hosts of the cluster. For comparability, the prototypes for each scenario are deployed on the same processing hosts.

The prototypes are deployed in the KIT CMS Tier 3 analysis cluster, alongside other extensions [10]. Jobs are scheduled and deployed with the HTCondor batch system [9]: On the one hand, this allows us to inject locality information into job scheduling. On the other hand, services are deployed on a recent operating system, with job dependencies supplied via docker [11]. We operate at a resolution of individual files, since this is well-defined before a job is actually run. For technical details of our middleware, see previous publications [1, 5, 6, 7, 8].

3.1. Coordinated caching

To achieve ideal data locality, each cache is accessible only from its associated processing host. Jobs read data either from the local cache, or the global storage. Only metadata is accessible across hosts. To form a single service working at the scale of entire workflows, an additional coordination layer is required. Data selection is performed at this global level, and portions of the selected data assigned to specific caches.

Coordination allows the highest control, and thus optimisation, of locality. Both data selection and data locality can be dynamically adjusted to match the currently running and queued workflows. Since components exchange only metadata, the system is highly scalable both horizontally and vertically. In addition, processing nodes can be operated outside of dedicated clusters, as long as a local cache is available.

3.2. Distributed caching

To allow easier data locality, each cache is accessible also from adjacent processing hosts. We achieve this with a single, distributed cache pool spanning across all nearby hosts. Jobs read data from an arbitrary host of the pool, or the global storage. The entire cache pool acts as one service, being able to service entire workflows. While there is no distinction between local and adjacent access, data and jobs do not require assignment to specific hosts.

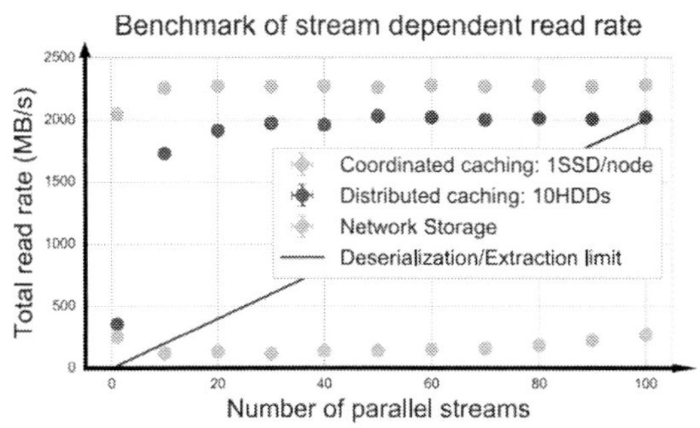

Figure 3: Performance of data sources: Throughput of synthetic, concurrent data streaming. All streams are equally distributed across 4 processing hosts. SSD and HDD devices are local to hosts, while network storage is shared. The Deserialisation/Extraction limit is extrapolated from analysis workflows.

Distribution simplifies job and data management, eliminating the risk of cache misses due to placement. A global strategy can be directly applied to the entire cache pool. As the entire cache pool is used at once, concurrency of data access is automatically smoothed out between hosts; cache devices on each node do not have to deliver a broad performance range. Finally, capacity of resources between hosts can be used, such as fast inter-rack network.

4. Experience and evaluation

We have used processing hosts in a single rack for controlled evaluation of the two caching scenarios. All data accesses use POSIX, with an overlay file system [12] transparently merging cache and background storage. While coordinated caching directly uses SSD devices, distributed caching uses CEPH [13] to integrate HDDs on all hosts. Notably, the HDD data capacity is 20 times the total SSD data capacity, but we use 2.5 times as many HDD devices. For controlled throughput tests, data has been manually pinned to caches.

4.1. Concurrent throughput

We have deployed both synthetic performance tests (see Figure 3) and orchestrated analysis workflows for testing. In general, caches add significant additional bandwidth of 5 Gbit/s per host. SSD caches offer superior response for few concurrent accesses - even with fewer devices, they offer superior performance for any number of streams. However, the HDD cache still provides comparable overall throughput.

For the synthetic test, throughput for both cache setups saturates at 25% cluster utilisation. However, HEP analysis workflows cannot process this throughput, even under ideal conditions - the ROOT [14] dataformat adds overhead for extracting and deseralising data. Even at full cluster utilisation, both cache scenarios offer sufficient throughput for analyses to be compute bound. As a result, caching a fraction of input data is sufficient for optimal performance.

4.2. Applicability and use-cases

We have found both caching scenarios to significantly improve data analysis throughput. Since both scenarios overprovision throughput, they are equally sufficient for our environment. However, their differences in complexity and scalability make them ideal for different use cases in the future.

Coordinated caching benefits from processing hosts having no additional dependencies on infrastructure. Due to the shared-nothing architecture, caches can be volatile, and can be added and removed from a cluster at any time. This allows dynamically deploying isolated processing hosts outside of dedicated computing centers [15].

Distributed caching offers significantly lower complexity for configuration. Given the lower placement constraint to achieve locality, scheduling requires less tuning. Overall, we feel that distributed caching is the correct choice for dedicated computing centres.

5. Summary and outlook

Data locality is a requirement to exploit available processing resources of modern hardware. Caching offers an effective way to provide data locality on-demand. We have developed and evaluated two similar but distinct caching scenarios: Coordinated caching shares only the most basic information, making it ideal for dynamic processing resources. Distributed caching enhances many processing hosts at once, allowing for a simple adoption in an existing cluster.

Still, the approach offers potential for future improvements and applications. From a conceptual point of view, the two scenarios can be combined - by coordinating cache pools, efficient data analysis can be enabled across clusters. Furthermore, extending support to grid protocols such as xrootd, caching and thus data analysis could be extended to opportunistic resources as well.

Acknowledgment

The authors wish to thank the German Helmholtz Association, and the Karlsruhe School of Elementary Particle and Astroparticle Physics (KSETA) for supporting and funding the work. The authors acknowledge also support by the state of Baden-Württemberg through bwHPC and the German Research Foundation (DFG).

References

[1] Fischer M, Metzlaff C, Giffels M, Jung C, Kuehn E, Hauth T and Quast G 2015 High performance data analysis via coordinated caches, *CJ. Phys.: Conf. Ser.* **664** 092008
[2] Eck C et al., 2005, LHC computing Grid : Technical Design Report, *Technical Design Report LCG*, https://cds.cern.ch/record/840543
[3] The Hadoop project homepage, **URL** http://hadoop.apache.org/
[4] Dean J and Ghemawat S 2004 MapReduce: Simplified Data Processing on Large Clusters, *OSDI'04: Sixth Symposium on Operating System Design and Implementation*
[5] Fischer M, Giffels M, Jung C, Kuehn E, and Quast G 2015 Tier 3 batch system data locality via managed caches, *J. Phys.: Conf. Ser.* **608** 012018
[6] Fischer M and Kuehn E 2016 Data Locality via Coordinated Caching for Distributed Processing, *CLOUD COMPUTING 2016*
[7] Fischer M, 2016, Coordinated caching for high performance calibration using $Z \to \mu\mu$ events of the CMS experiment *Karlsruher Institut für Technologie (KIT)* http://dx.doi.org/10.5445/IR/1000067354
[8] Heidecker C 2017 C Absolute jet energy scale determination at the CMS detector for LHC run 2 at $\sqrt{s} = 13$ TeV using an optimized processing setup, URL: https://ekp-invenio.physik.uni-karlsruhe.de/record/48877 [Accessed 2017-10-12]
[9] Thain D, Tannenbaum T and Livny M 2005 Distributed computing in practice: the Condor experience, *Concurrency Computat.: Pract. Exper.* **17** 32356 (doi: 10.1002/cpe.938)
[10] Fischer F et al. 2017 On-demand provisioning of HEP compute resources on cloud sites and shared HPC centers, *Journal of Physics: Conference Series*
[11] Schnepf M 2017 Calculation of cross-section limits for the production of single top quarks in association with a Higgs boson using container technologies, URL: https://ekp-invenio.physik.uni-karlsruhe.de/record/48876 [Accessed 2017-10-12]
[12] Junjiro Okajima, "aufs" [software], version 4.13.5, Available from http://aufs.sourceforge.net/ [accessed 2017-10-16]
[13] Sage Weil, Scott A. Brandt, Ethan L. Miller, Darrell D. E. Long, Carlos Maltzahn, 2006, "Ceph: A Scalable, High-Performance Distributed File System", *USENIX Association* Proceedings of the 7th Conference on Operating Systems Design and Implementation (OSDI 06), http://ceph.com/wp-content/uploads/2016/08/weil-ceph-osdi06.pdf
[14] CERN, "ROOT - Data Analysis Framework" [software], version 5.34, Available from https://root.cern.ch/ [accessed 2017-10-16]
[15] Schnepf MJ et al. 2017 Mastering Opportunistic Computing Resources for HEP, *Journal of Physics: Conference Series* Proceedings of 18th International Workshop on Advanced Computing and Analysis Techniques (to be published)

IOP Conf. Series: Journal of Physics: Conf. Series **1085** (2018) 032006 doi:10.1088/1742-6596/1085/3/032006

The management of heterogeneous resources in Belle II

Malachi Schram, Vikas Bansal, Antonio Ledesma

Pacific Northwest National Laboratory 902 Battelle Boulevard, 99352 - Richland, WA, USA

E-mail: `Malachi.Schram@pnnl.gov`

Abstract. The Belle II experiment at the SuperKEKB collider in Tsukuba, Japan, will start taking physics data in early 2018. The proposed plan is to accumulate 50 ab^{-1} which is approximately 50 times more data than the Belle experiment. The collaboration expects to manage and process approximately 190 PB of data samples. Computing at this scale requires efficient and coordinated use of the compute resources in North America, Asia and Europe. We present the general Belle II computing model, the distributed data management system, and the results of recent network data transfer tests. Additionally, we present how U.S. Belle II is using virtualization techniques to augment computing resources by leveraging Leadership Class Facilities (LCFs).

1. Introduction

The Belle II experiment is the successor of the Belle experiment [1] at the KEK laboratory in Tsukuba, Japan. The Belle experiment measured charge-parity (CP) violation in the B^0 system predicted by the theory of Kobayashi and Maskawa [2]. The successful confirmation of the prediction led to the Nobel Prize to both theorists.

The Standard Model of particle physics is an incomplete description of the fundemental forces of nature. The Belle II experiment is planned to collect 50 ab^{-1} of data. With this data physicists will be able to performed precision measurements that will provide stringent tests of the SM and discover or contrain new physics.

Precision flavor physics measurements to be performed by Belle II are complementary to the direct search for new particles at the LHC. If new physics is found at the LHC, flavor physics measurements are essential to identify the kind of new physics.

2. Belle II Computing Model

The Belle II collaboration was officially founded in December 2008. Today, it has more than 725 members from over 104 institutes in 24 different countries. With collaborators located in North America, Asia, Europe, and Australia it is distributed around the world.

Beam collisions are expected to start in Spring of 2018. A data sample of about 50 times the size collected by the Belle experiment is expected to be recorded by Belle II by 2025. Its data rate is predicted to be on equivalent to the LHC [3] Run I. Belle II will record events at a

	2017	2018	2019	2020	2021
CPU [kHepSPEC06]	20.11	27.56	58.90	69.71	82.97
Storage [PB]	0.31	0.81	5.04	6.50	9.28
Networking In/Out [Gbps]	0.30/0.30	0.49/0.36	1.06/0.26	1.56/0.31	1.89/0.83
CPU @ PNNL [kHEPSpecs]	0.41	6.35	40.52	29.01	41.58
CPU @ LCF [kHEPSpecs]	19.70	21.21	18.38	40.70	41.39

Table 1. U.S. Belle II computing resources requirements

rate of 6 kHz, corresponding to 11 PB of data per year from 2022 onwards. It will collect 100 PB raw data volume by 2024 with total data volume approaching 190 PB.

Figure 1 illustrates the Belle II computing model for the first three years of operations. KEK and PNNL will host a complete replica of the raw data and will process and distribute the produced data to the regional data centers. An estimation of the U.S. Belle II computing requirements

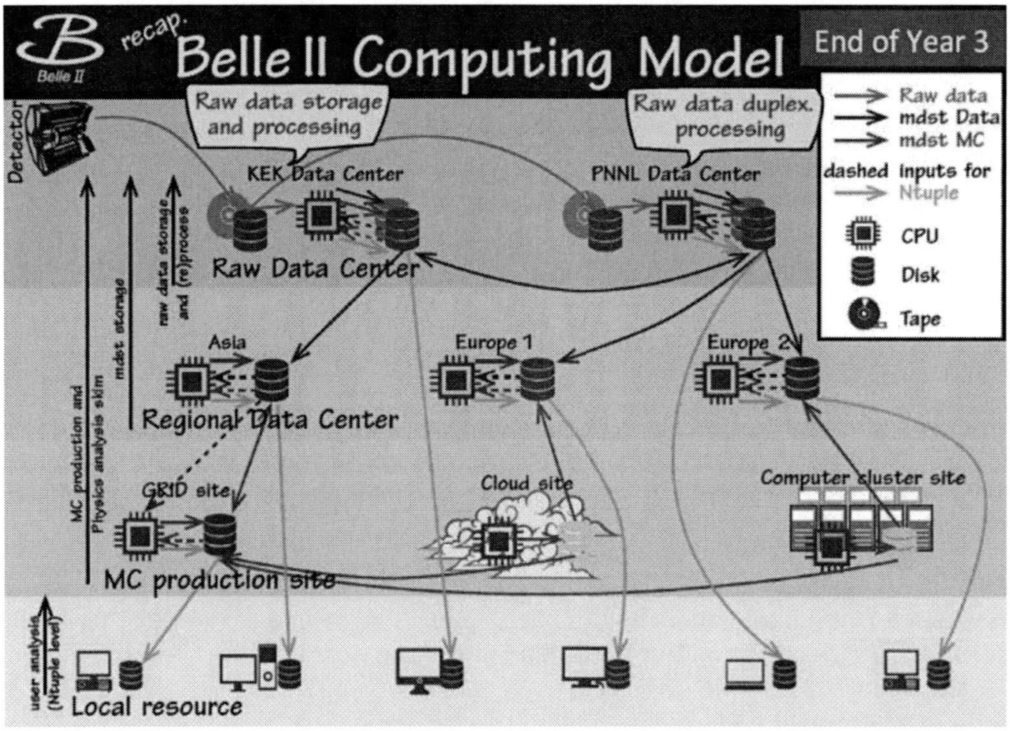

Figure 1. Belle II Computing Model for the first three years of operations.

through 2021 is provided in Table 1. U.S. Belle II also takes advantage of Leadership Class Facilities (LCFs) for MC samples production as show in Figure 2. To ensure that appropriate network bandwidths are available on network routes between major sites in North America, Asia, Europe, and Australia, they are routinely tested for network traffic throughput rates as part of data challenges. Virtual LAN (VLAN) setups were established between sites to perform network data challenges both trans-Atlantic and trans-Pacific using FTS3 [4] service. A site-to-site matrix was deployed to monitor and capture network information for latency and network

IOP Conf. Series: Journal of Physics: Conf. Series **1085** (2018) 032006 doi:10.1088/1742-6596/1085/3/032006

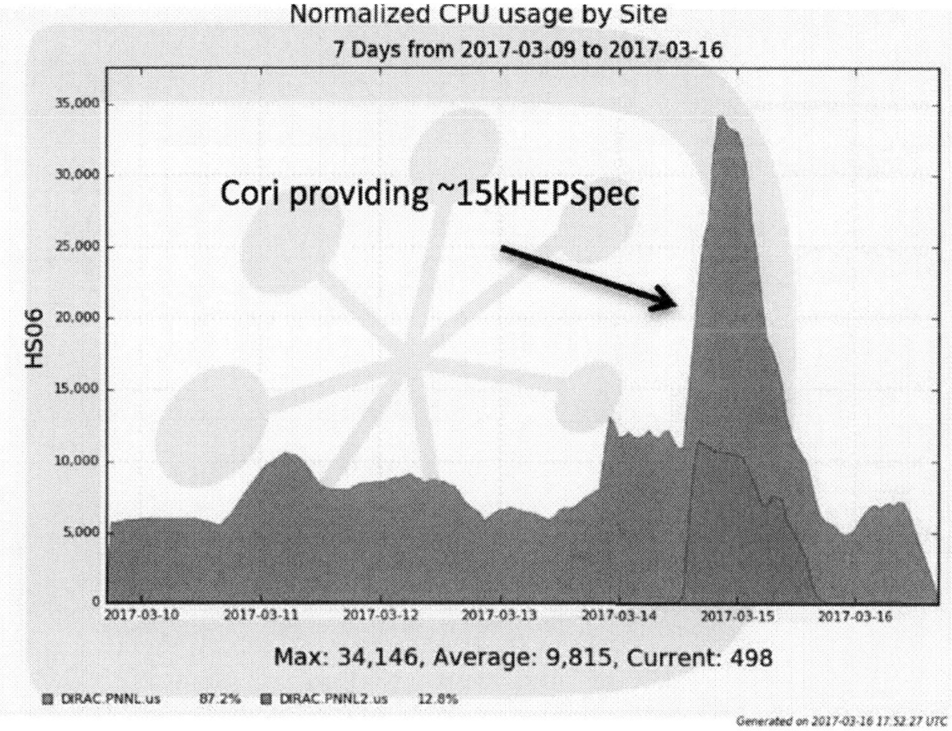

Figure 2. NERSC LCF - Cori providing 15 kHEPSpec to Belle II MC production

packet drop rates using MaDDash [5] and perfSONAR [6] services. Network information between sites will be inserted in a distributed database that will guide optimal data routing path between any two endpoints.

Belle II was recently included in the LHCONE [7] Acceptable Use Policy (AUP). This will enable Belle II member sites to interconnect via the well managed infrastructure of LHC network and hence provide maximal throughput possible. A key component of the LHCONE is that all the sites are "trusted" which alleviates the need for firewalls (as it slows down data transfer rates). Also, bandwidths via public Internet is limited and not feasible to support high data transfer rates. Matrix in Fig. 3 shows recent network data challenge results in terms of achieved bandwidth between some of the major site pairs. The numbers satisfy the bandwidth requirements assuming a 50% safety margin and continuous operations over 10 months period in a year.

Belle II has chosen DIRAC [8] to provide key functionality for their distributed computing model. DIRAC will orchestrate all its sub-components that can as well be distributed at many sites to achieve Belle II computing operations. It will also manage computing resources across all sites.

Belle II collaboration is developing various components and extensions to DIRAC, including a fabrication system that manages compute tasks over all sites, a distributed data management system, web portals for monitoring purposes, and dedicated wrappers around the Belle II reconstruction framework. Figure 4 illustrates an overview of DIRAC grid middle-ware with its major components integrating resources and other grid services to clients.

Distributed Data Management System (DDMS) is the DIRAC component responsible for managing all of Belle II data over grid. It is implemented using RPC service/client based model that is powered by MySQL database server. Many remote agents are designed and deployed

Source→ Destination↓	KEK (Gbps)	PNNL (Gbps)	DESY (Gbps)	KIT (Gbps)	CNAF (Gbps)	NAPOLI (Gbps)	SiGNET (Gbps)
KEK		6.2	11.0	5.0	9.2	15.0	3.0
PNNL	16.0		10.0	6.0	14.0	10.0	-
DESY	6.0	6.6		8.0	8.0	8.0	3.0
KIT	5.6	4.8	8.0		8.0	6.0	3.0
CNAF	18.0	14.0	10.0	6.0		8.0	3.0
NAPOLI	16.0	6.0	3.0	3.0	3.0		3.0
SiGNET	1.6	0.6	5.0	5.0	5.0	2.0	

Figure 3. Measured bandwidth in Gbps between some of the Belle II grid sites

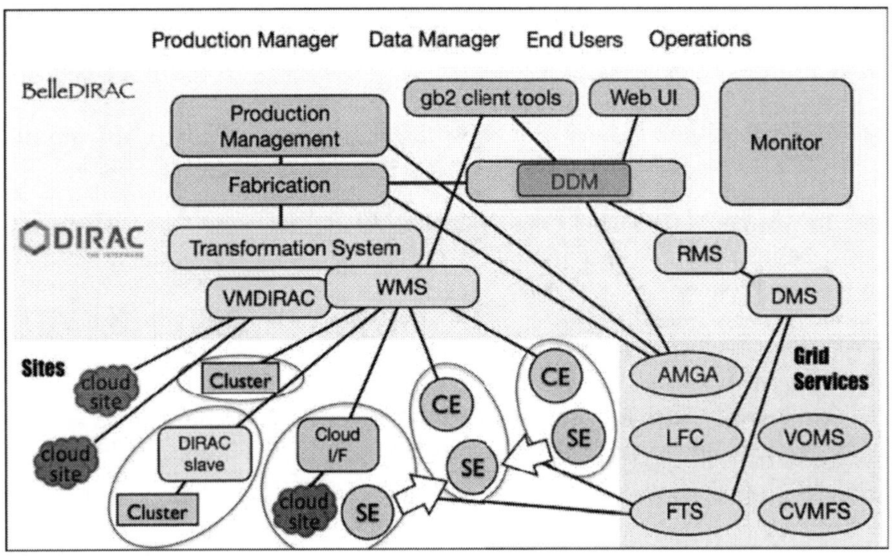

Figure 4. DIRAC, the "interware", between clients at top and grid and computing resources at bottom.

to perform the automation of choosing data destination, performing data transfer and data verification, assessing data popularity and finally cleaning of the data. One of the challenge is to resolve any race condition for a given file with multiple operations such as replicate and delete. DDMS prioritize operations as per Belle II policy and take respective actions. DDMS in on going Belle II MC production activity has shown to achieve 20k/hour file replication rates and 10k/hour file deletion rates. One of the reason deletion rates are lower is because of sanity checks we perform on each file before deleting it. We also plan to boast the rate by using a more popular HTTP over SRM protocol.

The simulation samples for the Belle II experiment have been produced in a globally distributed manner, in accordance with the distributed computing model. Figure 5 illustrates the output of the DIRAC-based production system for seven MC production campaigns between 2013 and 2016. About 27 billion events have been simulated so far with concurrent jobs running on heterogeneous grid environment reaching 25K (> 200 kHEPSpec). All produced data is again distributed to storage resources at various sites.

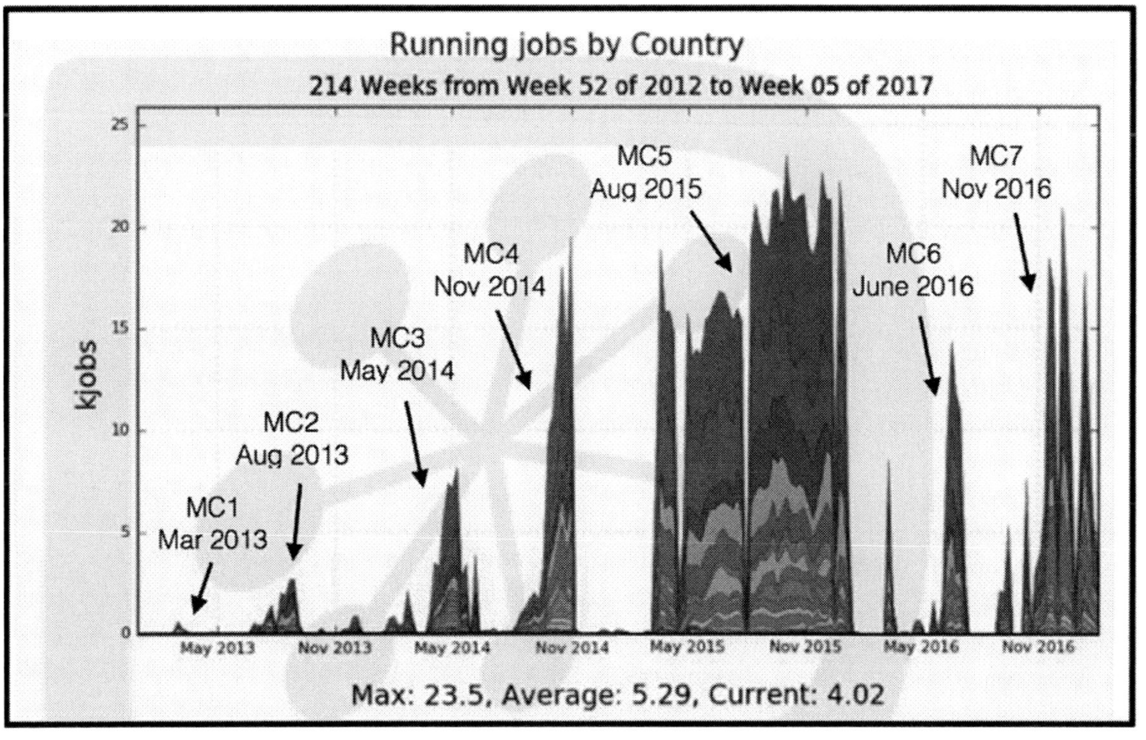

Figure 5. MC Jobs at Belle II MC production campaigns between 2013 and 2016. Y-axis shows number of production jobs and X-axis spans time.

3. Conclusions

The large increase in sensitivity of Belle II over Belle comes with a significantly larger data sample. The SuperKEKB accelerator is designed to provide 50 times more data until the year 2025. This data volume is on par with the LHC Run I and is a challenge for the new computing system. The strategy involves PNNL to host a complete replica of raw data for the first three years in operation and distribute processed data to Europe. Belle II grid brings together heterogeneous computing resources including LCFs. To ensure optimal network bandwidths, we routinely perform data challenges over the Belle II network which is now part of LHCONE AUP. Recent network data challenges achieved Belle II networking requirements. Our computing model is facilitated by DIRAC and we have successfully tested the prototype over MC production campaigns. DDMS, responsible for Belle II data management, is performing as expected.

More details about the computing at Belle II can be found in Chapter 14 of the Belle II Technical Design Report [9].

References

[1] A. Abashian *et al.*, Nucl. Instrum. Meth. A **479**, (2002) 117.
[2] M. Kobayashi and T. Maskawa, Prog. Theor. Phys., **49**, (1973) 652.

IOP Conf. Series: Journal of Physics: Conf. Series **1085** (2018) 032006 doi:10.1088/1742-6596/1085/3/032006

[3] http://jinst.sissa.it/LHC/
[4] http://fts3-service.web.cern.ch/
[5] http://software.es.net/maddash
[6] http://www.perfsonar.net/
[7] https://twiki.cern.ch/twiki/bin/view/LHCONE/LhcOneAup
[8] http://diracgrid.org/
[9] T. Abe *et al.*, arXiv:1011.0352 [physics.ins-det].

IOP Conf. Series: Journal of Physics: Conf. Series **1085** (2018) 032007 doi:10.1088/1742-6596/1085/3/032007

CODE-RADE a user centric system for delivery of science software.

S H T Murray[1] **and B Becker**[2]

[1] Center for High Performance Computing, CSIR, Rosebank, Cape Town, South Africa
[2] Meraka Cyberinfrastructure, CSIR, Lynwood, Pretoria, , South Africa

E-mail: [1]`smurray@csir.co.za`

E-mail: [2]`bbecker@csir.co.za`

Abstract. Scientific computing can be distilled to the execution of sets of applications that are combined into complex workflows. Due to both the complexity and number of scientific packages and computing platforms, delivering these applications to end users has always been a significant challenge through the grid era, and remains so in the cloud era. In this contribution we describe a platform for user-driven, continuous integration and delivery of research applications in a distributed environment - project CODE-RADE. Starting with 6 hypotheses describing the problem at hand, we put forward technical and social solutions to these. Combining widely-used and thoroughly-tested tools, we show how it is possible to manage the dependencies and configurations of a wide range of scientific applications, in an almost fully-automated way. Finally, we will give some insight into how this platform could be extended to address issues of reproducibility and collaboration in scientific research in Africa.

1. Introduction

Since the first computing program exceeded the capabilities of its original execution host, and was migrated to another platform, scientists have squandered countless hours attempting to get programs to run as opposed to actually doing the science they intended to do.

1.1. The paradox of plentiful computing

It is a central theme of scientific computing that there are never enough resources available to any particular researcher. Congruent to this, there has been massive and near-pervasive investment in HPC and distributed computing resources. There has been a coordinated national effort in South Africa to build a well-connected distributed computing platform[1], fully integrated and connected to the national research network[2]. This effort has even been extended on a trans-continental scale, via the signature of a memorandum of understanding between African and European infrastructure providers[3]. With the creation of the Africa-Arabia Regional Operations Centre[1], several sites across Africa are available to researchers to execute their scientific workflows, and support several different virtual organisations, or communities of practice. Despite this effort in interoperability and integration, many of these resources are under-utilised at any given time. This seems to point to a paradox of scientific computing, where there is an over-provision of resources, yet scientists are still impeded from working.

[1] See "`http://www.africa-grid.org`" for details

1.2. Barriers to Entry
This situation exists due to various barriers to entry. One of the main benefits of a distributed computing platform is also one of the main barriers to entry : the complexity of the environment. Researchers seldom need to take into account the configuration of their application or workflow for different platforms, and it is practically unfeasible to insist on homogeneous hardware and configurations across the sites. This results in significantly increased overhead on the part of the researcher to ensure that their application will properly execute at any of the given resources, which thus leads to oversubscription at certain sites and underutilisation at others[2].

CODE-RADE aims to reduce or remove the barrier to entry for new applications, thus eliminating the paradox of plentiful computing.

2. CODE-RADE Design and Philosophy
The porting and integration of new applications is ordinarily done by site administrators who attempt to compile and install the researcher's applications according to the local procedures and policies. This usually leads to a significant lead-time and depends critically on the ability of the site administrator to interpret the needs of the researcher. There can also be much ambiguity regarding the resolution of dependencies, optimisation and versions. We then come up against a tension between domain expertise and administrative privileges. The *researcher*, not the site administrator is usually the one that best understands how to compile the application. The site administrator has the knowledge of the local setup which would integrate the application appropriately and efficiently. This overhead is multiplied by the number of distinct application stacks which are in use in scientific domains, and exacerbated by the multitude of dependencies and optimisation configurations. For the physical sciences domain, this can be appreciated by considering the wide array of applications registered in the EGI Applications Database[4]. [3]

2.1. CODE-RADE Design Hypotheses
CODE-RADE provides a means to transparently integrate a new application into an arbitrary site configuration, that is both user-driven and satisfies the constraints imposed by site administrators. We reconsidered the way applications are built and deployed in scientific research and computational infrastructure. This was done as far as possible from first principles, by constructing seven hypotheses of bringing applications to distributed infrastructure. In what follows, we discuss the considerations of these design hypotheses and how they have influenced the development and results of the project. **1: Every researcher is an application.** The core aspect of initial communication between a research infrastructure provider and a researcher is the *application*. **2: No software is an island.** Every scientific application has a dependency list. No definition of an application can be complete without an explicit expression of the list of its dependencies, which should extend as far as possible down userspace. **3: All applications need an environment.** As with any organisms, biological or digital, it requires an environment in which to express its functionality. **4: There is more than one environment.** By restricting ourselves to a single environment, we lose main advantages of distributed systems, and erect unncessary barriers to entry for researchers. **5: Solutions Decay.** A "solution" in this case refers to a software expression of the method for creating an executable version of the application. The exploration of these solutions, including the attempts which failed, are important components of the system knowledge. **6: Humans need not apply.** If we assume that the previous hypotheses are true, it stands to reason that most of the work to actuall

[2] Another effect is the reduction in the capacity of the researcher and experts at sites to effectively collaborate with each other, due to geographical and technical divergence.

[3] For physics alone, there is a complexity and scale which is prohibitive to any particular site administrator. Considers the applications maintained in the Homebrew or EasyBuild repositories - about 600 and 1000 respectively - the situation seems untenebale.

compile and integrate applications can be automated. **7: This is not hard.** While it may seem that we are proposing a radical shift from common practice, which may seem discouraging, the fact is that continuous integration[5] and continuous delivery[6] has been common practice in the software development world for several years.

Along with these hypotheses, CODE-RADE attempts to solve problems which impede fluid collaboration and exploitation of computational resources. First of all, the system aims to reduce or remove entire barriers to entry represented by user applications, through the use of continuous integration and delivery, automated builds and version-controlled source code repositories. This allows all involved to know exactly where the application is in its lifecycle, and indeed which build of the application is currently being used.

Automated continuous delivery ensures that once applications are integrated, they are immediately available at sites which wish to host them, further reducing the workload on site administrators.

3. CODE-RADE in detail

3.1. Actors

The CODE-RADE platform identifies four kinds of actors which have distinct roles to play : Researchers, Research Software Engineers (RSE), Infrastructure Operators (Ops), Automated Agents (Bots).

Researchers are the final users of the application, they are not necessarily the authors of the software. Researchers may not know the details necessary to compile and optimise applications for specific environments, they are adept in interpreting the results of these applications and are the end consumers of them. The research software engineer, which in some cases is the researcher themselves, thus provides a means to express the application on the various targets in an executable state. This proves to the resource providers that applicatons can indeed execute properly on target infrastructures is needed before they will agree to actually host and excute the applications on their sites. Indeed, the characteristics of these target infrastructures are defined by the infrastructure operations team, which know them best. The role of the automtaed agents is to execute the code provided by the research software engineers and see whether it passes the tests set by the infrastructure operations team, at every commit pushed to the repository. This provides a direct link between change in the software and expression of that software (source code version linked to executable version), and since the criteria of the tests are predefined, once they pass applications can be integrated without human intervention.

The entire workflow can be driven by researchers themselves, once setup.

3.2. Components

The implementation of CODE-RADE consists essentially of three generic components : Continuous integration tool,Version controlled software repository,Automated software delivery tool. There are several choices which one could use to build a similar platform to CODE-RADE by choosing a particular tool to fulfill each component. In our case, we have selected Jenkins[7], GitHub[8] and CERN Virtual Machine Filesystem[9]. The reasoning and justification for these specific choices are given in some more detail below in section 4. The components have specific roles to play in each phase of the CODE-RADE workflow, which we describe next.

3.3. CODE-RADE workflow

Although the main aim of CODE-RADE is to ensure that applicaitons are available for execution on arbitrary remote sites, there are collateral issues which have importance to a community. We identify several goals in CODE-RADE which have their own flow of actions and we thus define the generic workflow for application **integration, delivery, review** and **publication**.

The most common workflow is 3 steps : Build, Test, Deliver. The first step (build) refers to the synchronisation of the source code with relevant updates. It is initiated by the Research Software Engineer: a commit is made to a repository, detailing a change, which the RSE authors. This triggers an event on Jenkins, which pulls the latest commit and executes the scripts in the repository which will build and test the code. The second step (Test) is conditionally executed on the success of the previous step. This step checks the viability of the built application and executes whatever tests are specified, then installs the application into the integration environment. Finally the third step (Deliver) cleans the built application and installs it into the deploy environment, then ships the executable code to the repository for subsequent replication.

Although there are several tools for building widely-used applications (see Section 4 below), we have chosen a convention within the project to have three scripts which perform the compilation, functional testing and integration into the delivery environment respectively.

Each job, as defined on Jenkins, consists of at least these three tasks. If each passes successfully, a job to update the CVMFS repository is triggered, which puts the repository into transaction, pulls in the newly built applications and module files, then publishes a new version of the repository. This is automatically and transparently available at sites after a few seconds, once the CVMFS client detects the change in the repository hash. In order to make it easier for end users or client applications to check which version of the repository they are using[4], an integer increment is added to a file in the repository which is linked to the job which resulted in the latest version.

In short, a commit to a repository by a researcher or RSE will result in the application being automatically delivered to arbitrary endpoints, as long as quality and functional tests set by Ops and researchers respectively are passed.

4. Discussion

Our particular combination of existing tools (Github, Jenkins and CVMFS) has resulted in a powerful platform with which to exploit any existing computational infrastructure, in an automated user-driven way. Some aspects which together make CODE-RADE a novel and powerful platform are highlighted.

4.1. CODE-RADE is cross-platform

CODE-RADE takes a software description of a scientific tool and builds a digital expression which can be executed on arbitrary architectures. This makes CODE-RADE different from more traditional package managers, since these only build for the architecture on which the user is currently. By simulating the execution environment which may in principle be available at geographically distributed sites.

We currently define a target according to a matrix of characteristics, which are encoded as variables in the build and deploy environments : **ARCH**: The CPU architecture of i the execution environment, **OS**: The operating system of the execution environment, **SITE**: A meta-variable encompassing aspects of specialised hardware (GPU, Myrinet, GPFS, etc) and other aspects (software licenses, available middleware etc.)

For each combination of these axes, specific optimisations can be applied, ensuring that the final result (compiled application) is both properly compiled, and will execute efficiently on the target site. By building on as many environments as possible, for as many combinations as possible, CODE-RADE promotes diversity in computing platforms, without exacerbating the paradox of plentiful computing.

[4] This can be determined more accurately using the cvmfs client tools, but this may not be intuitive to users.

4.2. CODE-RADE is atomic
Successfully built git hashes are linked to a cvmfs hash, permitting reproducibility across sites and allowing one to link research outputs to the underlying versions and builds. This is then citable via Zenodo digital object identifiers (doi) from the cvmfs hash, permitting the work done on the infrastructure to be citable and credit given, and critically be reproducible. The researcher has fine-grained control over dependencies, version, and targets that are built and built against.

4.3. CODE-RADE is community-based
There is no restriction on the applications that can be integrated, so long as the software is accesible online in someway. Anyone can contribute applications, resources, code review, testing, or where ever their skill sets lie. The entire community can see the build status and reinitiate failed builds via new pull requests. After a user updates their git repository, a git pull request is done, a build is executed. The results of the build are publicly viewable (`https://ci.sagrid.ac.za`), errors are viewable by the user doing the pull request (and everybody else), via a full build transcript. This can be fixed by merely redoing a pull request with what ever changes are required to fix the errors. This is completely indepdent of staff at sites.

4.4. CODE-RADE is Automated
There is a heavy relilance on automation in CODE-RADE, via automated agents to reduce bias, lead time and is user-driven. As an example, failures on jenkins cause issues on github, and the user is then able to fix these at their discretion. Successful subsequent builds will also close these issues. Notifications are sent to relevant slack channels [5] and bots within slack permit actions to be taken, for instance postponing issues, reassigning, or closing.

5. Summary, Conclusion and Further work
Making the best use of computational infrastructure comes down to running applications. Maintaining and porting them is a hard and often thankless task. We think we have come up with a solution to solve many parts of this and making the work citable. We've built an automated porting system, which will deliver functional, tested, relevant software to your site that is reproducible. The work done is then citable to give credit where credit is due for work that often goes uncredited. It eases the communication blocks in collaborative work, often highly distributed. Everyone is welcome to come on in and help us build it.

Acknowledgements
The authors would like to thank contributors to the project for ideas, discussion and code The original idea co-developed by one of the authors (BB) and Fanie Riekert (University of the Free State). Input and critique was provided by Dane Kennedy and Sakhile Masoka (Centre for High Performance Computing, Cape Town) and Peter van Heusden (South African National Bioinformatics Institute). The authors recognise the technical support of the EGI CVMFS Task Force support, in particular Catalin Condurache (European Grid Initiative and Science and Technology Facilities Council, UK). Several discussions on CODE-RADE design and extension were had with Timothy Carr (University of Cape Town e-Research Centre).

CODE-RADE is supported by the Sci-GaIA project under grant 654237 of the European Commission's Horizon 2020 programme

[5] See "`https://www.slack.com`"

IOP Conf. Series: Journal of Physics: Conf. Series **1085** (2018) 032007 doi:10.1088/1742-6596/1085/3/032007

References

[1] B Becker et al. "The South African National Compute Grid" *IST-Africa 2009 Conference Proceedings*, ed P Cunningham, M Cunningham.

[2] "SANReN South African National Research Network" [online] Available at: `http://www.sanren.ac.za/`. [Accessed 25 Feb 2018].

[3] S Andreozzi "Mou between EGI.eu and CSIR Meraka (African-Arabian Region)", Mou between EGI.eu and CSIR Meraka (African-Arabian Region), Apr 2015. [online] Available at: `https://documents.egi.eu/public/ShowDocument?docid=2407`. [Accessed: 25 Feb 2018].

[4] "EGI Applications Database", EGI Applications Database. [online] Available at: `https://appdb.egi.eu`. [Accessed: 25 Feb 2018].

[5] P Duvall *Continuous Integration*. 2007. Pearson Educational Publishers.

[6] J Humble, D Farley *Continuous delivery: reliable software releases through build, test, and deployment automation* 2010 Addison-Wesley Longman.

[7] J Smart *Jenkins: The Definitive Guide* 2011 OReilly.

[8] See https://developer.github.com/v3

[9] J Blomer, P Buncic, and T Fuhrmann "CernVM-FS: Delivering Scientific Software to Globally Distributed Computing Resources," *Proceedings of the First International Workshop on Network-aware Data Management, New York, NY, USA*, 2011, pp. 4956.

IOP Conf. Series: Journal of Physics: Conf. Series **1085** (2018) 032008 doi:10.1088/1742-6596/1085/3/032008

LEAF: A data cache and access system across remote sites

Y D Cheng, Q Xu, C Wang

Institute of High Energy Physics, CAS, Beijing, China

E-mail: chyd@ihep.ac.cn

Abstract. Nowadays, distributing computing technologies has been widely used in high energy physics field. Computing job is usually scheduled to the sites where the input data was pre-staged in. This model will lead to some problems including low CPU utilization, inflexibility, and difficulty in highly dynamic cloud environment. The paper proposed a data cache and access system, which presents one same file system view at local and the remote sites, supporting directly data access on demand. Then the computing job can run everywhere no need to know where data is located. The system was implemented based on streaming and caching mechanism. The test results showed the performance was better than traditional file system on high-latency WAN.

1. Introduction

Nowadays, distributing computing technologies has been widely used in high energy physics field. The Worldwide LHC Computing Grid (WLCG) [1] is one of the largest distributed computing system in the world. More than 500 thousand CPU cores and 980 PB storage from 167 sites are involved in WLCG collaboration until April 2017. To process the BES-III experiment's data the distributed computing infrastructure [2] based on DIRAC middleware [3] has been setup and became operational since 2013. Current BES-III distributed infrastructure includes 18 resource centers from 7 countries, providing access to about 3000 CPU cores and 0.5 PB of disk space. But most of sites are lack of dedicated administrator. Furthermore, public clouds such as Amazon AWS, Microsoft Azure, and Alibaba Aliyun are gradually popular today in High Energy Physics (HEP) area.

Computing in high energy physics is typical data-intensive application, and it needs a large amount of input data or output data in one task. And these data is usually organized in file system. Each file contains thousands of event data. Currently local computing model is typically cluster composed of storage system and computing nodes. The job runs on the computing nodes, fetching input data from storage system or putting output data into storage system. Grid is one kind of distributed computing system, which can be considered as global cluster of local cluster. Cloud in HEP is something like cluster or grid on virtual machines. The typical architecture of HEP computing system is depicted in figure 1.

When a job is submitted to a distributed computing system such as grid or cloud, it will be finally scheduled to one site containing a local cluster. Then the input data should be moved into the same cluster too. For example, WLCG deploys File Transfer Service (FTS) [3] which is responsible for distributing the majority of LHC data across the WLCG infrastructure. The data is pre-staged into one site, then the job can be scheduled to the site where the input data is. In such a distributed computing model, there are some problems as follows:

1) If one site does not have enough storage space, it cannot keep enough input data, and then there is no enough jobs running on the site. So it will potentially lead to low CPU utilization.

IOP Conf. Series: Journal of Physics: Conf. Series **1085** (2018) 032008 doi:10.1088/1742-6596/1085/3/032008

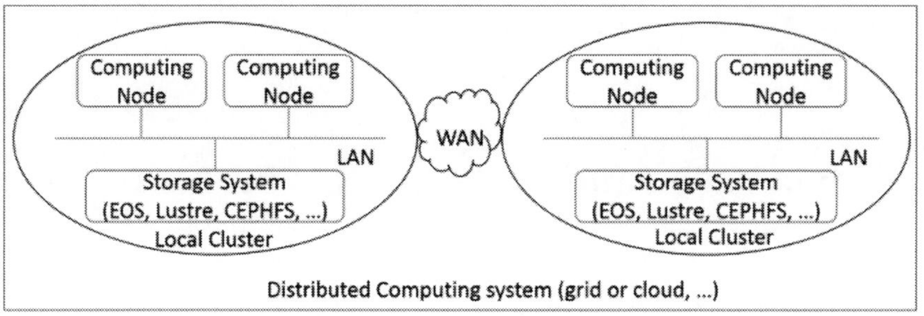

Figure 1. The typical architecture of HEP computing system

2) It is usually up to site administrators to decide which data should be transferred. If the data distribution is imbalance, it is difficult to schedule jobs fairly across different sites.

3) Virtual computing resources can be created in public cloud on demand, but it is not possible to transfer a large amount of data to remote cloud immediately, and some user's jobs cannot run without input data.

4) The whole file is transferred to remote site, but user's job only is usually interested in a few of events in the file, maybe only 1/1000, event more less [4]. So too much data is transferred based on the granularity of file.

The feasible solution to solve the above problems is to mount local file system directly on remote sites, which presents the same file system view at local and remote sites, then applications can run everywhere. However, traditional file systems such as Lustre, gluster, EOS cannot work well on high latency wide area network. We designed and implemented a data cache and access system – LEAF based on streaming and caching mechanism. This paper will describe the architecture, implementation and test results of LEAF.

2. The design and implementation

2.1. Design conception

LEAF aims at providing a same file system view at local and the remote sites, and good access speed over WAN as well. Client requests can be served as soon as one small fraction of file is available before the whole file is fully downloaded. It also should be portable, compatible, scalable, secure and reliable.

LEAF has two kinds of instances including one main site where all data is stored and serval remote sites where user can access data from main site on demand. To implement the system, we adopted the following design conceptions.

1) File system view consistency between main site and remote site. It is achieved by periodical metadata synchronization from main site and real-time update during file access.

2) File system syntax conversion between POSIX and file transfer. Data is transferred using multi-stream, chunk, non-block and other technologies to improve the performance across WAN.

3) Easy to deploy in internal network. Some small sites or computing nodes on public cloud do not have public IP, or network sockets are restricted by firewall. However, it is usually allowed to use HTTP protocol to access external resources. LEAF services running on remote sites can be easily deployed in internal network by exploiting HTTP protocol.

4) Easy to be compatible with existing HEP software. The existing HEP applications such as BES Offline Software System (BOSS) [5] should use the system to access remote data without modification. LEAF is implemented as an Xrootd [6] client pluglin, and can be transparently called by ROOT framework, or use XrootdFS (a Posix Filesytem for XrootD) to mount as a local file system.

2.2. Architecture

The architecture of LEAF is depicted in Figure 2. There are three components including TransferD, CacheD and Xrootd Client Plugin. TransferD is deployed in main site. CacheD is deployed in remote sites. Xrootd Client Plugin is a client module.

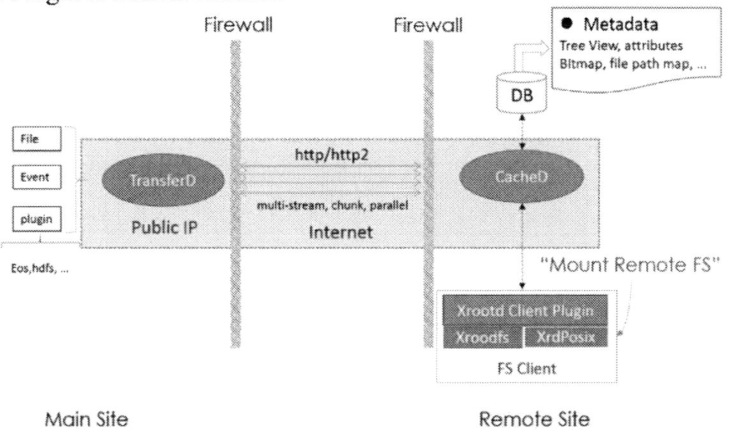

Figure 2. The architecture of LEAF composed of three components, TransferD, CacheD and Xrootd client plugin

2.3. Implementation

2.3.1. Data transfer service

Data transfer service is used to move data between main site and remote site, which has two components including TransferD and client library. TransferD is a daemon running at main site, which is configured with public IP address. Client library is deployed at remote sites, which is called by CacheD. Client does not need public IP. TransferD supports different data sources including local file system, storage system like Hadoop, and database such as Hbase.

Data transfer service is developed based on Tornado web framework [7]. Tornado is a Python web framework and asynchronous networking library. By using non-blocking network I/O, Tornado can scale to tens of thousands of open connections, making it ideal for long polling, WebSockets, and other applications that require a long-lived connection to each user.

Data transfer service can download data from main site or upload data into main site. If data transfer service receives a request, it will download or upload data using multi-streams in parallel to improve the performance. Data transfer service client routines are similar to file system POSIX calls including stat, readdir, read, write, link and so on. The parameters of these routines are also like POSIX, such as file path, mode, offset, buffer, etc.

2.3.2. Disk cache service

Disk cache service is used to cache data in remote site, which has three components including CacheD, DB and Client tool. CacheD is a daemon running in remote site, which receives requests from client, and then checks the information in the DB and decides to send data from cache or get data from remote site. DB stores file metadata of CacheD and bitmap which indicates which blocks of one file has been cached. Currently DB supports MySQL and ramcloud system [8].

If a file is opened for read, CacheD firstly checks if it exists in the cache. If it exists, CacheD will open the cached file and return fd. If it does not exist in the cache, CacheD will create an empty file with the same size as the original file in main site. If one application read a file, CacheD will check if the requested block exists in the cache. If the block is in the cache, CacheD will return block data to the application. If the request block does not exist in the cache, CacheD will get fix-sized block from

IOP Conf. Series: Journal of Physics: Conf. Series **1085** (2018) 032008 doi:10.1088/1742-6596/1085/3/032008

main site and then modify bitmap related to the file metadata, then return the block data. The workflow of LEAF read operation is depicted in figure 3.

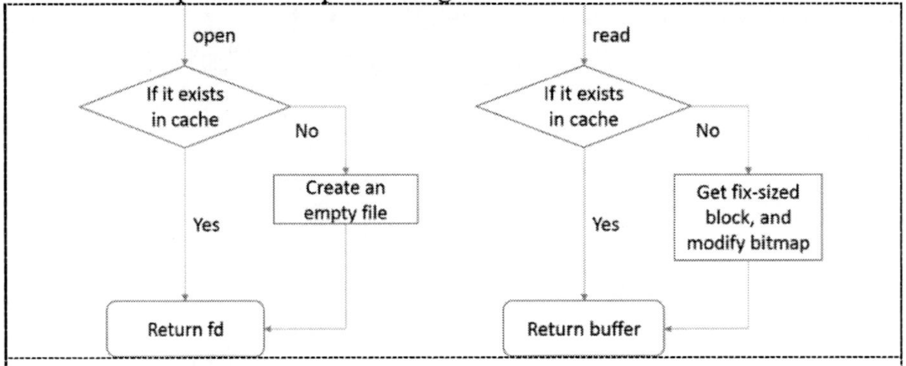

Figure 3. The workflow of LEAF read operations

If a file is opened for write or create, CacheD firstly checks if it exists in the cache. If it exists, CacheD will return an error because it is allowed to modify existing file in remote site. If it does not exist in the cache, CacheD will create an entry in main site, then create an empty file in remote site. CacheD then opens the empty file and return fd. CacheD then continue to receive data from application and write data into the file. After the write operation finishes, CacheD uploads the file to the main site, and update file metadata in main site as well. The workflow of LEAF write operation depicted in figure 4.

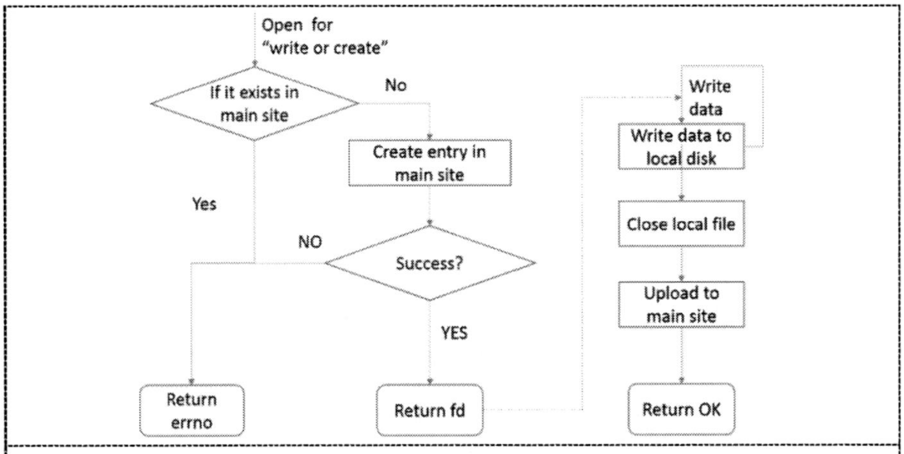

Figure 4. The workflow of LEAF write operations

2.3.3. Xrootd Client Plugin

LEAF client is implemented as an Xrootd Client Plugin to ensure HEP software access data transparently. Usually HEP software support ROOT framework [9] which can access file by using xrootd API, for example XrdPosix. LEAF can be accessed by most HEP software. If some software do not support ROOT framework, LEAF can be mounted as local file system through XrootdFS. XrootdFS is a Posix filesystem for an Xrootd storage cluster. It is based on FUSE (Filesystem in Userspace) and runs in user space. LEAF presents a local file system, and any software can access it. The implementation of Xrootd Client is depicted in figure 5.

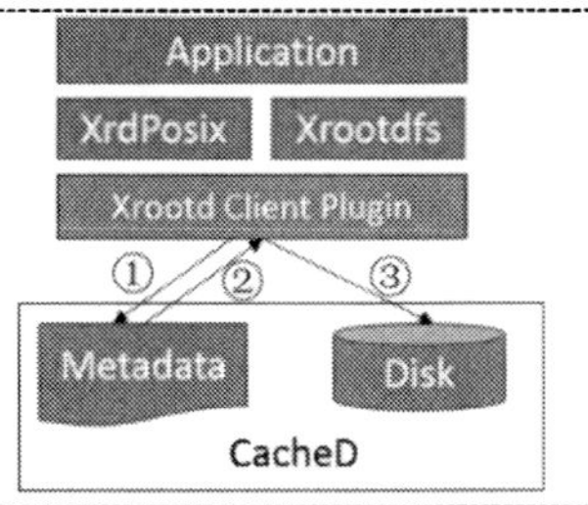

Figure 5. The implantation of Xrootd Client Plugin in LEAF

Xrootd Client Plugin can get data either from XrdPosix or XrootdFS, then it firstly connect CacheD to check if the block is in cache. If yes, CacheD will tell the physical path of the file. If not, it will connect CacheD to get the block from main site, then CacheD returns the physical path of the file. Finally, Xrootd Client Plugin gets real data from disk using xrootd API.

3. Test bed and results

We have setup a test bed including two sites between Beijing and Chengdu to evaluate the performance of the system. The distance of the two sites is about 2000 KM. The network latency is about 35 ms, bandwidth is 1 Gpbs, and iperf benchmark tool reported about 80 MB/s. The test results showed LEAF data transfer performance can achieve nearly 80 MB/s, and the performance is getting better with the increasing of stream number as depicted in figure 6.

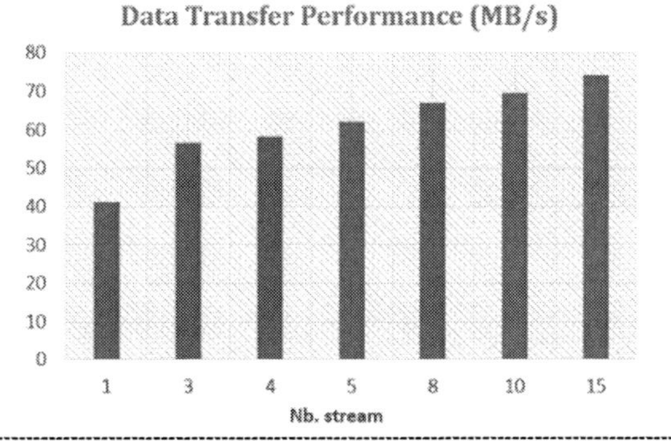

Figure 6. The performance of LEAF data transfer system between Beijing and Chengdu with 1Gpbs network link

As we know, TCP throughput is not only decided by bandwidth, and it is directly impacted by latency and packet loss. So traditional network file system such as NFS, EOS and Luster cannot work well in wide area network. The paper [10] showed NFS performance has degraded drastically while network latency was increased from 10 ms to 90 ms. We used tc tool to simulate network latency and found that the performance of EOS and lustre was decreased by 97.8% and 99.6% in the case of 100 ms latency. Then we tested the performance of LEAF with 1Gbps link and 100 ms latency. The test results showed the performance was decreased by 31%, much better than EOS and Lustre. The results was given in figure 7.

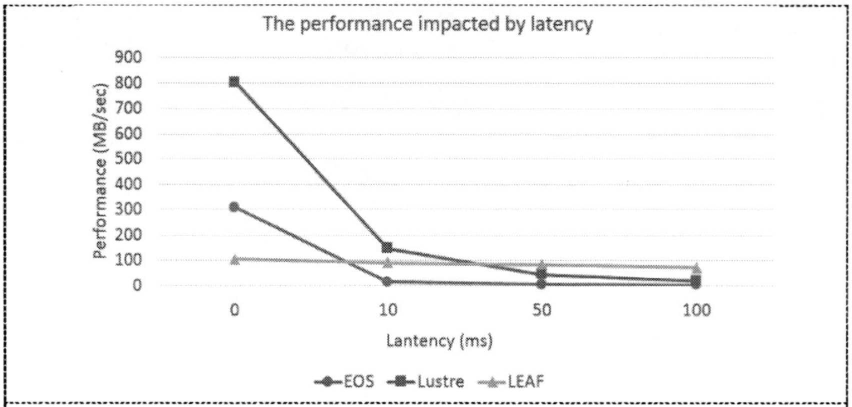

Figure 7. The performance impacted by latency. It showed LEAF is not impacted by network latency greatly

4. Summary

Distributed computing such as grid and cloud is widely used in high energy physics field. It is more flexible to access data directly from remote sites on demand. But traditional file systems won't work well over high-latency network. LEAF is designed as an extension of main storage system, which aims at improving data access performance over WAN. The test results showed LEAF was not impacted by network latency greatly and it had good performance across remote sites.

5. Acknowledgements

This presentation is supported by National Natural Science Foundation of China (No. 11575223) and National key Research Program of China "Scientific Big Data Management System" (No.2016YFB1000600).

References

[1] Shiers J. The worldwide LHC computing grid (worldwide LCG)[J]. Computer physics communications, 2007, 177(1): 219-223.
[2] Belov S, Suo B, Deng Z Y, et al. Design and operation of the BES-III distributed computing system[J]. Procedia Computer Science, 2015, 66: 619-624.
[3] Ayllon A A, Salichos M, Simon M K, et al. FTS3: New Data Movement Service For WLCG[J]. J.phys. Conf.ser, 2014, 513(3):032081.
[4] Liu B. High performance computing activities in hadron spectroscopy at BESIII[J]// J.phys. Conf.ser, 2014, 523(1):012008.
[5] Wei-Dong Li, Ya-Jun Mao, Yi-Fang Wang. Chapter 2 The BES-III Detector and Offline Software[J]. International Journal of Modern Physics A, 2009, 24(supp01):9-21.
[6] Dorigo A, Elmer P, Furano F, et al. XROOTD - A highly scalable architecture for data access[J]. Wseas Transactions on Computers, 2005, 4(4):348-353.
[7] Dory M, Parrish A, Berg B. Introduction to Tornado: Modern Web Applications with Python[M]. " O'Reilly Media, Inc.", 2012.
[8] Ousterhout J, Gopalan A, Gupta A, et al. The RAMCloud storage system[J]. ACM Transactions on Computer Systems (TOCS), 2015, 33(3): 7.
[9] Antcheva I, Ballintijn M, Bellenot B, et al. ROOT — A C++ framework for petabyte data storage, statistical analysis and visualization[J]. Computer Physics Communications, 2009, 182(6):1384-1385.
[10] Radkov P, Yin L, Goyal P, et al. A Performance Comparison of NFS and iSCSI for IP-Networked Storage[C]// Usenix Conference on File & Storage Technologies Usenix Association. 2004:101--114.

IOP Conf. Series: Journal of Physics: Conf. Series **1085** (2018) 032009 doi:10.1088/1742-6596/1085/3/032009

Parallel computing of SNiPER based on Intel TBB

J H Zou[1], T Lin[1], W D Li[1], X T Huang[2], T Li[2], Z Y Deng[1], G F Cao[1], Z Y You[3]

[1] Institute of High Energy Physics, Chinese Academy of Sciences, Beijing, China
[2] Shandong University, Jinan, China
[3] Sun Yat-sen University, Guangzhou, China

E-mail: zoujh@ihep.ac.cn, lintao@ihep.ac.cn

Abstract. SNiPER is a general purpose offline software framework for high energy physics experiments. During the development, we pay more attention to the requirements of neutrino and cosmic ray experiments. And now it has been adopted by many experiments. It is necessary for us to implement parallel computing to improve the data processing efficiency. Intel Threading Building Blocks (TBB), as a powerful high level library, emancipates us from trivial and complex details of raw threads. In this proceeding, we implemented an event level parallel computing wrapper for the original serial SNiPER based on Intel TBB. We have taken parallel computing into account at the beginning of SNiPER's design, it is possible to achieve multithreading in a non-invasive way. The SNiPER kernel module and Intel TBB are absolutely de-coupled. Features of multithreading are transparent to most users, except a few conventions such as global variables. This approach will significantly reduce the costs of migration from serial to parallel computing. However, it is more complicated for critical resources handling, such as disk I/O and memory management. There are also some attempts for these challenges.

1. Introduction

Offline software plays an important role in current high energy physics experiments. It helps physicists to achieve results accurately and efficiently. Under the fast development of particle detection technology and information technology, the volume of data collected in HEP becomes larger and larger, the requirement for offline data processing becomes more and more complicated and challenging. However, there are always some generic patterns in the offline software of this domain. Software framework can be independently implemented to present these patterns. It acts as the skeleton of a software system to provide a unified environment for programming and execution. Accordingly, users are able to concentrate on their specific tasks without concern for technical details. There are already many widely used frameworks, such as Gaudi [1], Art [2] and so on.

Now several Non-collider particle physics experiments are under construction in China, such as Jiangmen Underground Neutrino Observatory (JUNO) [3] and Large High Altitude Air Shower Observatory (LHAASO) [4]. All of them are comprised of very large scale detectors, which will produce huge amount of data in the near future. Meanwhile, there are some specific requirements for these experiments. Based on the experience of previous experiments, we implemented an offline software framework from scratch, which is named as Software for Non-collider Physics Experiments (SNiPER) [5]. Although the specific requirements of JUNO and

IOP Conf. Series: Journal of Physics: Conf. Series **1085** (2018) 032009 doi:10.1088/1742-6596/1085/3/032009

LHAASO are taken into account, SNiPER is implemented for general purpose that can be used by other experiments.

As a general purpose framework, SNiPER is customizable and extensible for different experiments, while its kernel module and main functionalities are independent of any experiments. For each specific experiment, physicists can implement their own data processing modules with the interfaces provided by SNiPER, as shown in Figure 1. These modules, such as simulation and reconstruction algorithms, can be dynamically loaded and configured during execution. Therefore we can perform various data processing jobs in a unified style.

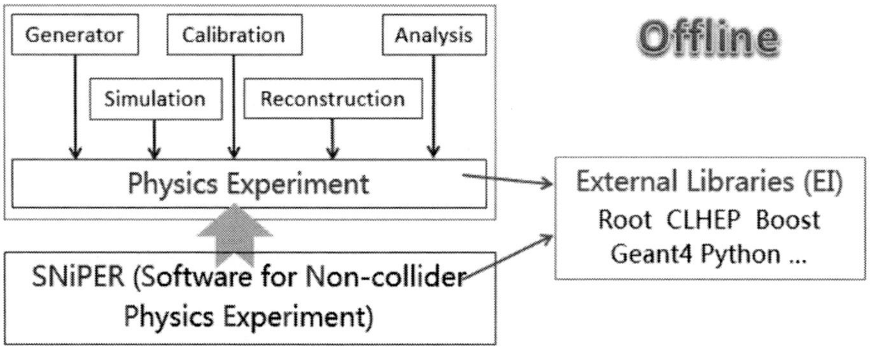

Figure 1. A typical offline software system based on SNiPER

At present, SNiPER has been officially adopted by JUNO, and successfully used in JUNO detector design studies [6, 7]. The practice proves the functionalities of SNiPER. It achieves our original goal. There is a lot of room for improvement, including parallel computing development.

2. Parallel computing schema of SNiPER

The data amounts of future experiments will be very huge, the processing of which will take a great number of computing resources. Then we encounter this urgent challenge, how to improve the execution efficiency of SNiPER. Due to technical limits, it is hard to increase CPU clock speed rapidly in recent years. Alternatively increasing cores number became the trend in the CPU industry. Accordingly, it is natural to apply parallel computing from the view of software developers. This became a hot spot in high energy physics field. Many software have supported parallel computing with varying degrees, including the framework GaudiHive [8].

In our long term plan, we will accomplish a mixture of multithreading and MPI programming for SNiPER. Multithreading is used on each multi-core node, meanwhile MPI is used for cross nodes parallelism. The two methods can be used separately or jointly. This paper focuses on our first step of multithreading programming.

There are a lot of existing modules developed by users. We want to find a way to minimize the costs of migration from serial to parallel computing. Our data consist of a large number of events. Every event is independent of each other, or correlated to nearby events only. And the processing procedure is similar for all events. We can use this characteristic to simplify the parallel computing schema of SNiPER. It is the underlying events parallelism from the view of framework. In this schema, there can be more than one parallel flows of execution, but each event is handled in a flow of execution uninterruptedly. As a result, the procedure of each event is almost the same as the serial version. Hence existing modules can be migrated easily. Moreover, this schema is also able to support the MPI implementation of SNiPER in the future.

IOP Conf. Series: Journal of Physics: Conf. Series **1085** (2018) 032009 doi:10.1088/1742-6596/1085/3/032009

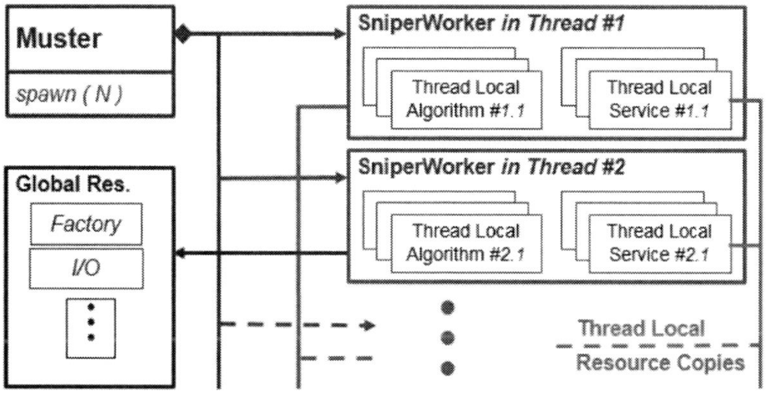

Figure 2. Parallel computing schema of multi-threaded SNiPER

Figure 2 shows the schema of multi-threaded SNiPER. There is a Worker in each thread, which holds a local copy of a SNiPER Task component and its belongings (algorithms and services). All Workers are managed by the global Muster (Multiple SNiPER Task Scheduler). However, sharing resources, such as physical memory and I/O streams, have to be treated carefully.

3. Implementation of SNiPER Muster

Intel TBB [9] is a widely used C++ template library for task parallelism. It provides rich features for general purpose parallelism. Developers are able to think parallelism at the higher level avoiding dealing with the details of raw threads. It is easier for us to reach a better dynamic workload balancing with Intel TBB tasks and scheduler. According to the advantages, we chose to implement SNiPER Muster based on Intel TBB.

It's a coincidence that a class in SNiPER is also named Task (but start with an upper case T). The Task component performs like a lightweight application manager. Data processing procedures are configured and managed in the Task components. In a serial SNiPER application, it handles the event loop continuously. However, we can handle a single event by one invoking each time. There is another important feature in SNiPER, more than one independent Task instances can exist simultaneously.

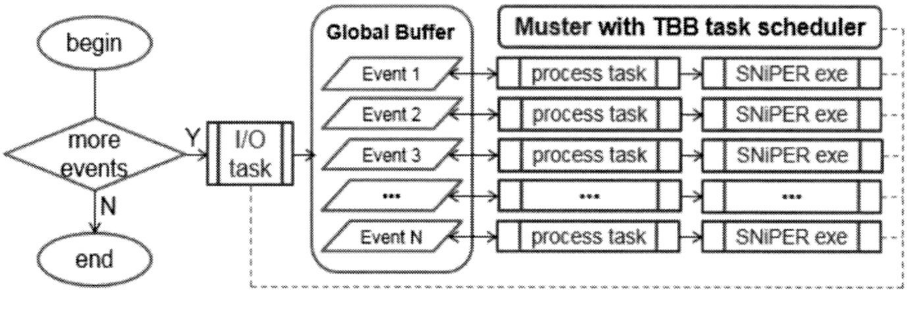

Figure 3. Task scheduling of the SNiPER Muster

In the multi-threaded SNiPER, the traditional event loop is broken up by the Muster. As shown in Figure 3, different events are dispatched to different Workers. The Muster is in charge

IOP Conf. Series: Journal of Physics: Conf. Series **1085** (2018) 032009 doi:10.1088/1742-6596/1085/3/032009

of the creation and scheduling of Workers, which is mainly based on Intel TBB's scheduler. We can simply map a SNiPER Task component to an Intel TBB task in each Worker. Then Workers can be executed in threads concurrently.

When a Worker is invoked, it grabs and locks an event from the memory until it completes. During the execution of a Worker, the event is handled by a thread local SNiPER Task component. This feature ensures that a Worker looks the same as a serial SNiPER application. Existing algorithms and services can be migrated almost transparently. The only restrictions are global variables and sharing resources. Once a module is migrated, it can be used in both serial and multi-threaded context.

There is only serial computing in the original SNiPER. But we have taken into account the requirements of SNiPER in a Worker at the beginning. So it's possible for us to implement Muster as a non-invasive wrapper of the SNiPER kernel module. By this way, users are able to switch between serial and parallel computing seamlessly.

4. I/O and memory management

In Figure 3, there is also an underlying I/O task Worker and a global buffer. The I/O task Worker represents the integration of I/O streams and Intel TBB. It is invoked according to the utilization ratio of the global buffer. The global buffer represents the memory management for all events. Besides that, there is a buffer locally managed by the SNiPER Task component in each Worker. We name it as local buffer to distinguish from the global buffer.

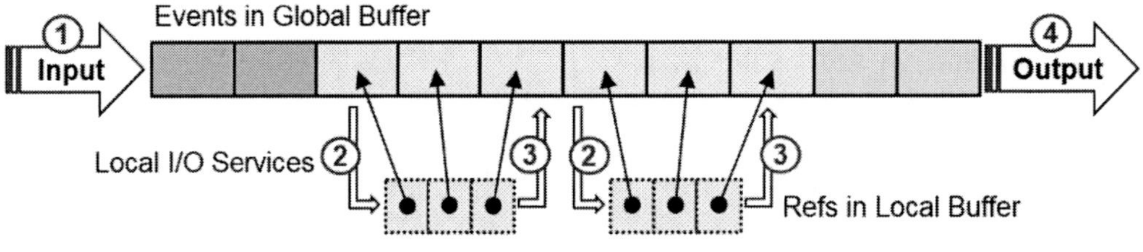

Figure 4. The data flow of multi-threaded SNiPER

A status code is assigned to each event. The event status can be READY, OCCUPIED or COMPLETE. As shown in Figure 4, the event data flow and status transition are expressed in 4 steps:

(1) When the utilization ratio of global buffer is low, the I/O task Worker is invoked for reading. New events are read from input stream, filled into the front of global buffer, and then marked as READY.

(2) A Worker requests a READY event from global buffer. A reference to the event is filled in the local buffer. Then the event is marked as OCCUPIED. In fact, a Worker can request a number of events each time for optimization.

(3) When an event is processed by the Worker, the reference is removed from the local buffer. And the event is marked as COMPLETE.

(4) When the utilization ratio of global buffer is high, the I/O task Worker is invoked for writing. COMPLETE events at the end of global buffer are sent to output stream and then removed.

From the previous description, we can notice that step 2 and 3 can be executed concurrently by Workers. Meanwhile, there are some advantages for us to simplify the programming:

IOP Conf. Series: Journal of Physics: Conf. Series **1085** (2018) 032009 doi:10.1088/1742-6596/1085/3/032009

- The I/O streams are associated to the global buffer, but not Workers. So that the I/O streams are invoked serially. We can avoid the complexity of stream synchronization.
- The SNiPER Task component in each Worker is executed serially, except that it has a customized I/O service. Instead of operating streams, it grubs event by reference from the global buffer. This operation is fast, and the lock can be very thin.
- The right order of events is kept in the global buffer. There is no need for sorting when we write events out.

The JUNO I/O service is still using ROOT 5. Since we applied a mechanism of data lazy loading, we have to add very fat locks in the I/O stream module. Otherwise there are potential conflicts when Workers access events data simultaneously. In order to make this prototype execute properly for JUNO, there are still a lot of things to do. ROOT 6 has made great progress for multithreading [10]. Now we are working on the migration from ROOT 5 to ROOT 6, which should be helpful to solve that problem.

5. Performance measurement

Due to the event I/O issue, this performance measurement uses a dummy algorithm without I/O operations. Following are the details of the testing node:

- Intel(R) Core(TM) i7-6700HQ CPU, 4 cores, Hyper-Threading is off
- 16 GiB Memory
- Ubuntu 16.04 with GCC 5.4
- Intel(R) TBB 4.4

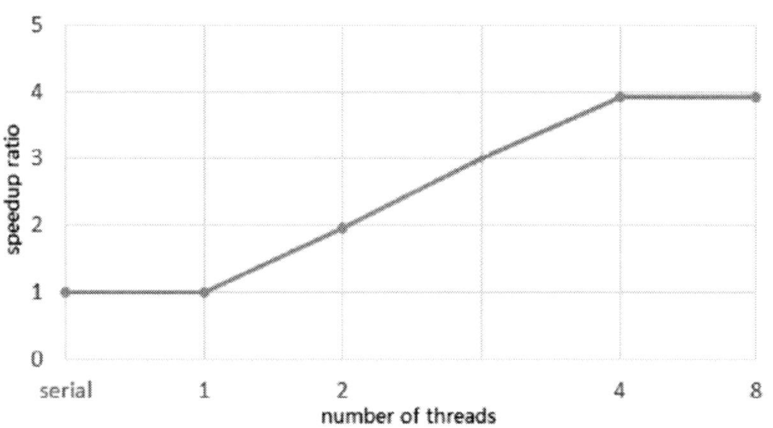

Figure 5. Speedup ratio vs. number of threads

20,000 events are executed three times for each case. The average time is used to calculate the speedup ratio. The result is shown in Figure 5. Since there is no effect from I/O operation, the speedup is nearly linear until all of the 4 cores are busy, just as we expected.

6. Conclusions

We have developed the general purpose framework, SNiPER. It has been adopted by JUNO, LHAASO and several other experiments. Its performance in serial data processing is well demonstrated in practice. Its parallel computing functionalities are under development and being optimized. In this paper, we present the result of multithreading programming based on Intel TBB. A non-invasive wrapper, Muster, is implemented as a SNiPER Task scheduler. We can handle events concurrently with Muster and Workers in threads. Muster is almost transparent to ordinary users. This will minimize the costs of migration from serial to parallel computing. We also present a prototype for I/O and memory management. However, critical resource handling is still a challenge. There is still a lot of room for improvement.

Acknowledgements

This work is supported by Joint Large-Scale Scientific Facility Funds of the NSFC and CAS (Grant No. U1532258), the Youth Innovation Promotion Association of Chinese Academy of Sciences (Grant No. 2017021), the National Natural Science Foundation of China (Grant No. 11605221) and the Strategic Priority Research Program of the Chinese Academy of Sciences (Grant No. XDA10010900).

References

[1] Barrand G *et al.* 2001 GAUDI - A software architecture and framework for building HEP data processing applications *Comput. Phys. Commun.* **140** 45
[2] C Green *et al.* 2012 *J. Phys.: Conf. Ser.* **396** 022020
[3] An F *et al.* (JUNO) 2016 *J. Phys.* **G43** 030401
[4] Cao Zhen *et al.* 2010 A future project at tibet: the large high altitude air shower observatory (LHAASO) *Chinese Phys. C* **34** 249
[5] Zou J H, Huang X T, Li W D, Lin T, Li T, Zhang K, Deng Z Y and Cao G F 2015 *J. Phys.: Conf. Ser.* **664** 072053. See also "SNiPER" [software], version 1.0, Available from `https://github.com/SNiPER-Framework/sniper/releases/tag/v1.0` [accessed 2018-03-20]
[6] Lin T, Zou J, Li W, Deng Z, Fang X, Cao G, Huang X and You Z (JUNO) 2017 *22nd International Conference on Computing in High Energy and Nuclear Physics (CHEP 2016) San Francisco, CA, October 14-16, 2016* arXiv:1702.05275 [physics.ins-det]
[7] Huang X T, Li T, Zou J H, Lin T, Li W D, Deng Z Y and Cao G F 2016 *PoS (ICHEP2016)* 1051 URL `https://pos.sissa.it/282/1051`
[8] M. Clemencic *et al.* 2014 *J. Phys.: Conf. Ser.* **513** 022013
[9] James Reinders. *Intel Threading Building Blocks.* O'Reilly Media, Sebastopol, CA, 2007.
[10] B Bellenot *et al.* 2015 *J. Phys.: Conf. Ser.* **664** 062006

Improving science yield for NASA *Swift* with automated planning technologies

A Tohuvavohu

Swift Mission Operations Center
Department of Astronomy and Astrophysics, The Pennsylvania State University
University Park, PA 16802

E-mail: aaronb@swift.psu.edu

Abstract. The Neil Gehrels *Swift* Observatory is a uniquely capable mission, with three on-board instruments and rapid slewing capabilities. It serves as a fast-response satellite observatory for everything from gravitational-wave counterpart searches to cometary science. *Swift* averages 125 different observations per day, and is consistently over-subscribed, responding to about one-hundred Target of Oportunity (ToO) requests per month from the general astrophysics community, as well as co-pointing and follow-up agreements with many other observatories. Since launch in 2004, the demands put on the spacecraft have grown consistently in terms of number and type of targets as well as schedule complexity. To facilitate this growth, various scheduling tools and helper technologies have been built by the *Swift* team to continue improving the scientific yield of the *Swift* mission. However, these tools have been used only to assist humans in exploring the local pareto surface and for fixing constraint violations. Because of the computational complexity of the scheduling task, no automation tool has been able to produce a plan of equal or higher quality than that produced by a well-trained human, given the necessary time constraints. In this proceeding we formalize the *Swift* Scheduling Problem as a dynamic fuzzy Constraint Satisfaction Problem (DF-CSP) and explore the global solution space. We detail here several approaches towards achieving the goal of surpassing human quality schedules using classical optimization and algorithmic techniques, as well as machine learning and recurrent neural network (RNN) methods. We then briefly discuss the increased scientific yield and benefit to the wider astrophysics community that would result from the further development and adoption of these technologies.

1. Introduction

The scheduling of astronomical observations has been widely studied, for both space and ground based observatories. While there now exist a large number of robotic ground based telescopes, *Swift* was the first semi-autonomous satellite observatory. Fast slewing and autonomous slew path determination and constraint checking were key design features of *Swift* to allow it to achieve its mission goal of detecting, localizing, and rapidly settling on short Gamma-Ray-Bursts (GRBs) before they fade [1]. Due to these capabilities, *Swift* was able to detect the first short GRB afterglow, and *Swift* science was the major contributor to the contemporary consensus that the progenitor of short GRBs are compact object mergers (NS-NS or NS-BH). *Swift* is a panchromatic mission with three instruments: the Ultraviolet/Optical Telescope, X-ray Telescope, and Burst Alert Telescope (UVOT, XRT, BAT) that are sensitive in the ultraviolet/optical, soft/hard X-ray, and gamma-ray regimes, with an energy range (non-

IOP Conf. Series: Journal of Physics: Conf. Series **1085** (2018) 032010 doi:10.1088/1742-6596/1085/3/032010

continuous) spanning from 2 eV up to 500 KeV. Due to its multi-wavelength capabilities, fast slewing, and rapid response time *Swift* has become an enormously popular observatory for many different types of astronomical targets. While *Swift* was originally launched as a GRB mission, over 85% of observing time is now spent on non-GRB science targets. These include everything from tracking and observing comets in the solar system, to exoplanet science, to high-energy astrophysical sources at cosmological distances. Along with normal targets, *Swift* also engages in large scale tiling campaigns, made possible by its rapid slewing and low re-pointing overhead. *Swift* currently performs weekly tiling of the Small Magellanic Cloud [2] and Galactic Bulge [3], as well as rapid-response tiling in response to astrophysical neutrino detections from ANTARES [4] and IceCube [5] and gravitational wave triggers from the LIGO/VIRGO Collaboration [6].

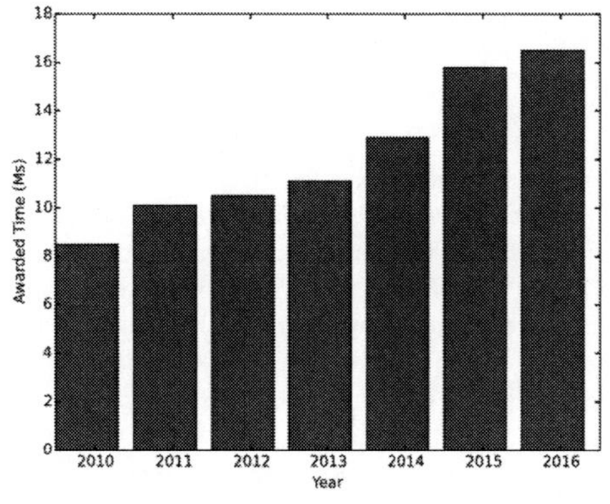

Figure 1: Amount of time awarded, in megaseconds, through the *Swift* Target of Opportunity (TOO) program, from 2010-2016.

Year	Avg. # Obs./day
2005	74
2010	85
2016	112

Table 1: Average number of observations-per-day over the lifetime of the mission. In 2017 there were some days with > 400 individual observations.

The unique capabilities of the spacecraft, and the ever evolving nature of *Swift* operations, have put unique demands on the operations team. Along with, and because of, *Swift*'s abilities, it also has the most complex, dense, and dynamic observing schedules of any satellite observatory. The complex scheduling goals (science targets) are accompanied by a host of very particular constraints, both astronautical and scientific, that these schedules must meet (see [7] for an example). In this proceeding we give a very brief overview of recent work performed with the goal of automating the creation of these sets of observing commands called Pre-Planned Science Timelines (PPSTs). Until now PPSTs have been built by (well trained) human Science Planners who were found to have a better understanding and command of the myriad constraints, and built markedly safer, more efficient, and higher quality PPSTs (with the assistance of a set of visualization tools) than any automation software available. However, building a PPST for one 24-hour period takes a human between 5-7 hours on average. In order to take maximal advantage of *Swift*'s fast-response capabilities, the goal is to build an automated system that can build PPSTs at greater than human-level quality in under 30 minutes by using state-of-the-art optimization and modeling techniques.

This will allow the *Swift* team to significantly increase the efficiency of the observatory and take maximal advantage of its fast-response capabilities when responding to ToO requests, while preserving as much pre-planned science as possible and maintaining spacecraft health and safety. This work will have a particularly large impact on *Swift*'s search for electromagnetic (EM) counterparts to gravitational-wave triggers as in [6] and other rapid follow-up projects.

In section 2 we provide a precise formulation of the *Swift* Scheduling Problem (SSP), the solution to which is a PPST. In section 3 we motivate our specific approach through the analysis of large sets of simulated plans. In Section 4 we briefly motivate and outline some recent work on constraint modeling. In Section 5 we conclude with next steps and the impact of implementation of the proposed system. Detailed analysis and results regarding constraint modeling/satisfaction will be reported elsewhere.

2. Formulation

We state the SSP as a dynamic fuzzy Constraint Satisfaction Problem (DF-CSP). We adopt this approach as opposed to regular CSP because the SSP is over-constrained. That is, any complete assignment of the CSP violates some constraint. The fuzzy CSP scheme allows us to extend the problem to constraints that are not *crisp*; constraints are assigned satisfaction degrees on the unit interval. We must also allow for new high priority ToO requests or transient detections (GRBs, Galactic transients, etc.). In this case, computing a completely new solution to the F-CSP is possible but has the distinct disadvantages of inefficiency and instability of the successive solutions, as described in [8]. So we adopt the full DF-CSP framework as a method of reusing the old solution when constraints are added or updated, by making local changes to the relevant portion of the schedule.

In general, a CSP problem is defined by a set of variables $\mathcal{X} = \{X_i\}_{i=1}^n$ that take values on their associated finite domains $\mathcal{D} = \{D_i\}_{i=1}^n$ under a set of constraints $\mathcal{C} = \{C_i\}_{i=1}^r$. We make it *fuzzy*, following the formalization in [9], by constructing $\mathcal{C}$ as a set of fuzzy constraints R_i with the membership function μ_{R_i} indicating to what extent some label ν satisfies the R_i:

$$\mathcal{C} = \left\{ R|\mu_{R_i} : \prod_{x_i \in var(R_i)} D_i \to [0,1] \right\}. \tag{1}$$

So $\mu_{R_i}(\nu)$ gives the degree of satisfaction of the relevant constraint. The ability to treat crisp constraints is preserved by the extreme cases of $\mu_R(\nu) = 0$ or 1. However, μ_{R_i} gives only the *local* degree of satisfaction; the degree to which the label ν satisfies the particular constraint. To judge whether a given ν is a solution to the full DF-CSP we need a measure of *global* constraint satisfaction:

$$\alpha(\nu) = \oplus \left\{ \mu_R(\nu_{var(R)}) | R \in \mathcal{C} \right\} \tag{2}$$

where $\oplus$ is an aggregation operator on the unit interval. With $\alpha(\nu)$ one can now define a solution to the DF-CSP as a compound label ν_X for which $\alpha(\nu_X) \geq \alpha_0$, where α_0 is called the *solution threshold*, which we determine based on need for optimality and run-time.

We also attach priorities p to each of the constraints, which allows some constraints to be treated as more or less important than others. This allows one to be confident that, while some fuzzy constraints can be violated, no label ν_X can be found as a solution which *completely* violates the most important constraints. This property matches the real-world need to preserve certain constraints that are critical to spacecraft health and safety. Adding priorities changes the specific definitions given in Eq. 1 and Eq. 2 slightly, a full formulation of the prioritized F-CSP can be found in [9]. Either way, we have a formalization of our problem, and a clear solution to aim for: a label ν_X whose global constraint satisfaction $\alpha(\nu_X)$ passes the defined threshold α_0.

All of the constraints, and associated cost function modeling by which we motivate the choice of priorities, can not be explicitly defined here, but we list most of the relevant hard and soft constraints below, with the most difficult to manage in red:

Hard Constraints

- Slew resources capped.
- Slew path constraints.
- Passive maintenance of XRT temperature between -55 C and -65 C.
- Reaction wheel momentum limit.
- Roll constrained by solar-panel pointing and star tracker catalog.

Soft Constraints

- Maximize observing/slew time ratio.
- Maximize number of science targets observed.
- Minimize # of UVOT filter wheel rotations.
- Maximize time spent observing targets > 8 hours from the sun.

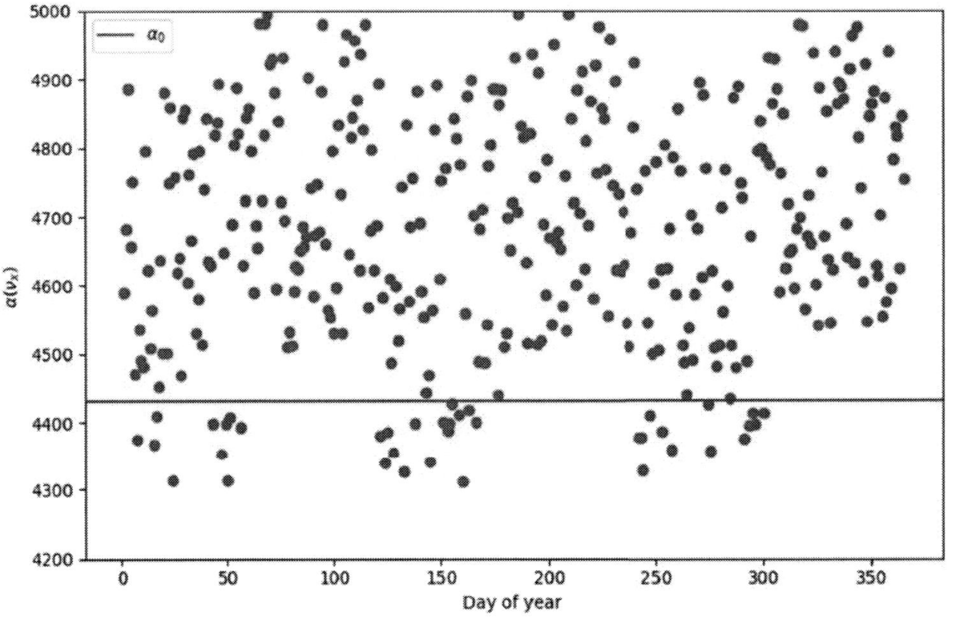

Figure 2: Distribution of $\alpha(\nu_x)$ for human constructed plans over a one-year period. The α_0 chosen for the solver is shown in red. Clear seasonal and otherwise periodic variations in the performance are due to ephemeride variability. Periodic structure can be found on scales ranging from O(seasons) to O(weeks).

3. Approach

Of course, one desires a strictly optimal solution to the scheduling problem, but combinatorial runtime is unacceptable: optimal scheduling for *Swift* is at least NP-hard. As the spacecraft has no consumables, a sub-optimal (but still highly optimized) plan that satisfies the various constraints and science priorities is acceptable. Early attempts set the solution threshold α_0 quite high. However, it was found that these extremely optimized plans were too fragile. That is, the dynamic constraints were balanced so precisely that any disturbance (by ToO upload or GRB detection) would often break a hard constraint. Based on an analysis of simulation results, the author has good reason to believe that the effect of adding a constructed metric for 'robustness' of a PPST to the list of soft constraints will be similar to that of a 'good-enough' plan found by a heuristic scheduler. To wit, we motivate here the use of a multi-strategy scheduling system that is comprised of a heuristics-informed seed scheduler, guided by a generational genetic algorithm [10] which performs a greedy search, a gap-filler using a heuristic-biased stochastic sampling

approach as in [11], and a final-pass constraint fixer based on the contention measure [12].
Large sets of planning simulations were performed for various planning periods over many seasons to allow for diversity in the set of ephemerides. The results of these simulations confirmed the experiential wisdom of the Planners, that there is "no one way to build a plan." That is, it was rigorously demonstrated that for any set of science targets that reasonably resembled the historical distribution of priorities for a given planning period, there exist multiple unique tours which fulfill the science priorities for the period, are safe for the spacecraft to perform (do not violate any hard constraints), and are reasonably similar in optimization score (soft constraints). Further it was shown that this holds for all relevant ephemerides, with a few marginal cases that can generally be alleviated with the assistance of a heuristic long-range planner.
While multiple unique acceptable solution classes exist, they were found to share general properties, with regard to target placement, that distinguish them from the majority of unacceptable tours. These properties are highly dependent on the details of the ephemeris for the given planning period. From the large simulation database, a set of meta-heuristics were derived for various classes of ephemerides. These meta-heuristics can be used to determine the appropriate scheduling strategy and partial search algorithm (local heuristic) for a given planning period. Tests of this approach have shown that once localized in a 'good' regime of the solution space by the meta-heuristic informed seed schedule, it is almost always possible to quickly converge to an acceptable, and sometimes locally optimal, plan.
In the analysis of the large class of simulated tours, it was found that the success of a given tour is significantly more dependent on the uncertainty in certain constraints than others. It was further demonstrated that an increased understanding and better model of those constraints can have outsize impacts on the success and optimization of a given PPST. In particular, the rate of convergence to an optimized plan can be significantly hampered by getting 'stuck' in unacceptable regions of the solution space because of an incomplete model of the momentum constraint. As such, a renewed effort at better understanding some of the constraints is valuable. A brief description of these efforts is given in Section 4.

4. Constraint modeling

In the course of investigating the solution space to the SSP, the critical constraint of preventing the saturation of the gyroscopic momentum wheels was found to have both the highest uncertainty in its modeling and the highest computational cost associated with that uncertainty. This motivates the construction of a better model. Constraint modeling for *Swift* is almost uniquely suited to a machine learning approach, as the spacecraft has performed $> 350{,}000$ individual observations so far in its mission. There is plenty of data to train on. In the investigation of the momentum constraint, previously unknown secular phenomena were discovered that make model building more difficult. However, over training periods of ~ 10 days a recurrent neural network (RNN) with 4 hidden layers and 50 nodes has shown promising results. This work will continue and results will appear in a future publication.

5. Next steps and impact

In this proceeding we formalized the SSP and motivated a particular approach to its solution; what remains is to finish building these planning algorithms into the *Swift* planning infrastructure. The ability to quickly construct high quality PPSTs in response to new ToOs or transient events will allow *Swift* to observe these targets in an optimal manner while minimizing the impact on pre-planned science priorities. Further, it will reduce inefficiency, continuous staffing needs, and pressure on members of the *Swift* Science Operations Team. Finally, these operational improvements will significantly benefit multi-messenger and localization projects, and ensure that *Swift* is able to maximally contribute to the new age of gravitational wave and multi-messenger astronomy.

IOP Conf. Series: Journal of Physics: Conf. Series **1085** (2018) 032010 doi:10.1088/1742-6596/1085/3/032010

Acknowledgments
This work was performed at the *Swift* Mission Operations Center at Penn State under support from the NASA *Swift* contract NAS5-00136. The author thanks his colleagues on the *Swift* FOT for their help in better understanding the spacecraft health & safety constraints, Jeffrey Gropp for ongoing work relevant to constraint modeling and automated plan creation, and Alex Ledger for useful discussion pertinent to algorithmic optimization.

References
[1] Gehrels N *et al.* 2004 *Astrophys. J.* **611** 1005–1020
[2] Kennea J A *et al.* 2018, in prep
[3] Shaw A W *et al.* 2017 (*AAS/High Energy Astrophysics Division* vol 16) p 400.01
[4] Adrian-Martinez S *et al.* 2016 *JCAP* **2** 062 (*Preprint* `astro-ph/1508.01180`)
[5] Evans P A *et al.* 2015 *MNRAS* **448** 2210–2223 (*Preprint* `astro-ph/1501.04435`)
[6] Evans P A *et al.* 2017 *Science* (*Preprint* `astro-ph/1710.05437`)
[7] Kennea J A *et al.* 2005 *UV, X-Ray, and Gamma-Ray Space Instrumentation for Astronomy XIV* (*Proceedings of SPIE* vol 5898) pp 341–351
[8] Verfaillie G and Schiex T 1994 *Proceedings of the Twelfth National Conference on Artificial Intelligence (Vol. 1)* AAAI '94 pp 307–312
[9] Luo X, man Lee J H, fung Leung H and Jennings N R 2003 *Fuzzy Sets and Systems* **136** 151 – 188
[10] Ortiz-Bayliss J C, Moreno-Scott J and Terashima-Marn H 2014 **512** 315–327
[11] Bresina J L 1996 *Proceedings of the Thirteenth National Conference on Artificial Intelligence - Volume 1* AAAI'96 pp 271–278
[12] Fraser S 2012 *Adaptive optimal telescope scheduling* Ph.D. thesis Liverpool John Moores University

A loosely coupled scalable cloud infrastructure

T Cui, Y D Cheng, Z J Cheng, X W Jiang

IHEP, 19B Yuquan Road, Beijing, 100049 China

Cuit@ihep.ac.cn

Abstract. Large-scale virtual computing system requires a loosely coupled virtual resource management solution. The solution should be able to flexibly increase or decrease virtual resources and to smooth upgrade the cloud platform. OpenStack provides large-scale virtualization solution such as Cells v2.0 and Tricircle/Trio2o. But it's so complex and difficult to be deployed and maintain for small cloud computing teams. We discuss a loosely coupled cloud infrastructure. It is based on a central database and plugin to achieve information collection, virtual machine scheduling and controlling and uses a simplified method to achieve unified management of network and images. It can flexibly expand or reduce cluster members, deploy different versions of OpenStack and even a heterogeneous cloud platform. The test bed works well at IHEP. We will show and analyse the infrastructure.

1. Introduction

IHEPCloud has been in production since 2014. The platform is dedicated to achieve virtualization of physical computing resources and to provide virtualized computing resources for the HPC computing environment. The IHEPCloud is built based on the OpenStack platform and uses GlusterFS [1] storage system to be the backend storage. The virtualized computing system consists of HTCondor [2] scheduler and VPManager middleware, CVMFS\AFS\EOS [3] storage system and OpenStack and so on. VPManager is a self-developed middleware and dynamically provides HTCondor with virtual resources based on the status of job accumulation in the queue. At the end of July 2017, Physics resources increased to 2K CPU cores. The size of EOS storage system is about 1.34PB. It serves more than 100 users and 4 scientific projects (BESIII, JUNO, LHAASO and CEPC).

As the development of local and cross-regional science projects at IHEP, local computing resources and remote computing resources have grown rapidly. Remote resources had been connected to IHEP through limited network bandwidth. We urgently need an expansion solution for cloud infrastructure. The solution should solve the problems of large-scale resources management, and cross-regional resources integration which are connected through limited network bandwidth. The solution should have a hierarchical and loosely coupled architecture. It also should be easy operation and maintenance.

Cells v2.0 [4] and Tricircle/Trio2o [5] [6] are large-scale virtualization solutions provided by OpenStack. Cells are better than Tricircle/Trio2o because of successful deployment experience at CERN. But it's also difficult to be deployed. First, Cells is complex and a tightly coupled solution. The cloud architecture isn't flexible. It is difficult to be deployed and maintain by a small cloud computing team. Second, RabbitMQ clusters are very sensitive to latency. It's unsuitable for network of limited bandwidth and unstable latency. Cells would be unstable for cross-regional cloud. Third, the performance of OpenStack goes down as computing resources increasing and frequent scheduling. A single OpenStack causes some problems, such as the risk of increasing, reducing or addition physical nodes, ceilometer component performance degradation, single point failures, etc. These problems may affect the operation of Cells.

IOP Conf. Series: Journal of Physics: Conf. Series **1085** (2018) 032011 doi:10.1088/1742-6596/1085/3/032011

We discuss a hierarchical architecture of loosely coupled OpenStack cluster. The core control layer consists of a central database and plugin. The plugin installed on the OpenStack sites weakly connect to central database through REST API interface. The architecture could simply and conveniently extend or shrink cloud platform by adding or reducing OpenStack unit. We will discuss the detail in this paper.

2. The VPManager middleware

The VPManager middleware [7] consists of five components, VCondor, VQuota, VMCtrl, NETDB and Image Mgmt. The VCondor and VQuota are responsible for virtual machine scheduling and quota controlling. The VMCtrl and NETDB are responsible for virtual machines controlling and environment configuration. Image Mgmt provides images and is responsible for image updating. The architecture is like Figure 1.

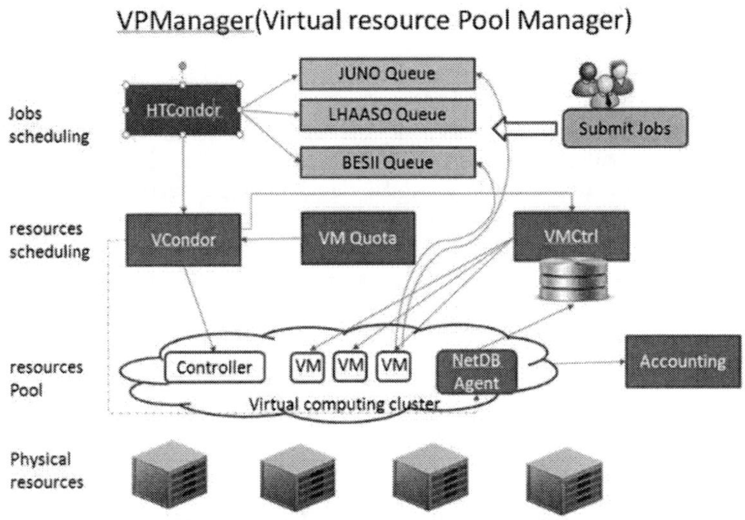

Figure 1. The VPManager Architecture

The running experience of the VPManager has shown us the feasibility to build an OpenStack cluster which controlled by central database. The central database just focuses on global data such as physical resources and virtual resources in this architecture. It doesn't focus on the OpenStack itself. The NETDB component is a bridge which connects central database and OpenStack and at the same time isolating them. If the NETDB plugin can collect virtual machine's information and execute operation such as launching or destroying virtual machine, it will be the only connection between central database and OpenStack. This is a loosely coupled relationship. We can build a Hierarchical, loosely coupled system architecture based on a central database and the NETDB plugin. Figure 2 shows the model.

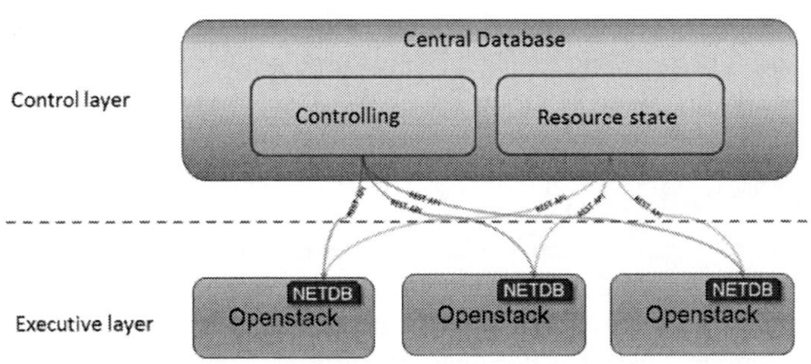

Figure 2. A Hierarchical, Loosely Coupled Model

In this model, A timestamp will be set to check each OpenStack site is alive or not. The NETDB plugin must refresh the timestamp to show an activity of the site. To start the NETDB plugin means an OpenStack site add to the cloud cluster. To terminate the NETDB means that the corresponding OpenStack site exits the cloud cluster. When the site exits cloud cluster, virtual machine on the site still running until the job is end. This design can prevent the problem that a site exits the cluster caused by link quality problems. Thus the controlling layer only focuses on centralized data and the global view of resources, the executing layer is responsible for resource controlling. It's as same as Cells v2.0 in terms of centralized data and resource controlling, but our model is more flexible and convenient and suitable for small team doing operation and maintenance.

The model we designed solves the problem of resources scheduling. To build a loosely coupled cloud infrastructure, a complete solution includes a core controlling component, network, OpenStack automatic configuration, image management and security etc. It always includes shared storage supported, VCondor/VQuota virtual scheduler, VMCtrl and so on. Figure 3 shows the infrastructure. We will discuss the infrastructure in detail below.

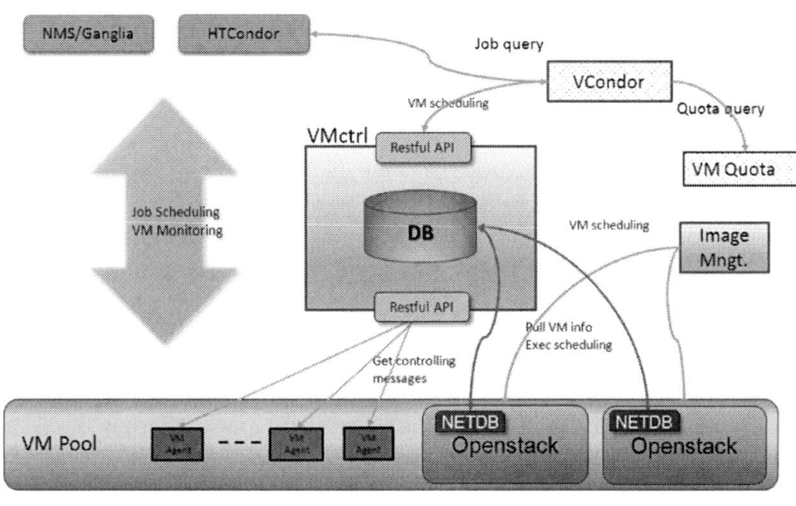

Figure 3. A Loosely coupled Cloud computing Architecture

3. A loosely coupled OpenStack cluster

3.1. Core controlling component

The core controlling component is designed according to the loosely coupled model and it includes a central database, a REST API interface and a plugin named NETDB. There are four types of data in this database. The first is the data of OpenStack environment such as flavor, network and image. This information will help the system sync OpenStack configuration when a new OpenStack is deployment. The second is resources data which include physical nodes and all virtual machines. It's collects by NETDB and provides service for scheduler. The third is scheduling data which added by resources scheduler (VCondor) and executed by NETDB to create or destroy virtual machine. The last is security data for security control and account management. The core controlling component provides a REST API interface to achieve data uploading or downloading. The web portal is based on HTTPS service and PHP script.

NETDB component is developed by C language. It is multi-versioned to accommodate the corresponding version of OpenStack. It's deployed on the OpenStack controller node and responsible for refreshing timestamp, collecting data such as number and state of virtual machines, physical and virtualized resources amount and free resources amount, and sending all information to central database. It also executes virtual resource operations such as launching or destroying virtual machine when the system gets task from central database.

3.2. Network design

A simplified network topology is deployed in IHEPCloud. 802.1Q protocol and linuxbridge mode are configured on neutron component and physical network switch. All rules on security-group are opened and subnets directly connect to physical network. The subnet gateway is configured in physical switch. These operations ensure smallest loss of network transmission efficiency. Virtual machine's IP address is valid local address. VMs can be accessed as same as physics hosts. Job-scheduler (HTCondor) can directly access all of the virtual machines. Cloud node interconnects each other through only one NIC. We continue to use this network design in multiple OpenStack cluster environment. IP address utilization should be fully planed and configured to central database. Each OpenStack gets theirs own subnets configuration from central database and maintain theirs DHCP service and broadcast area. New infrastructure does not focus on private network of tenants because it services HPC system. Virtual machines (UI and WN) are a part of the HPC system. But testing virtual machines should be isolated from HPC environment for security.

3.3. OpenStack automatic configuration

We design automatic configuration mechanism in OpenStack cluster environment to avoid manually configured errors. It is based on the central database and applied by the NETDB component. When an OpenStack is installed, NETDB will be running to register this site, download configuration information from central database and finished the configuration. Any modification will be sync on schedule. The configurations include flavor, network and subnet, images information and account information.

3.4. Image management

Automatic configuration and maintenance are based on the Puppet system in our HPC environment. The computing nodes need to update or put some patches for security and running problems almost every week. This is difficult for virtualized system because virtual machine is dynamical scheduled. Puppet can't be supported in virtual system because of the sync's efficiency problem. So we design a web images release system to service the OpenStack cluster environment and other users who build a private cloud. The image models are synced by puppet system as same as physical computing work node on schedule. NETDB synchronizes images to individual OpenStack unit when images are updated. Some urgent updating or patching will be done through the VMCtrl component during two image updating. Another problem of image management is instance's baking files. This is a very big row image located directory /var/lib/nova/instance/_base/. Normally, one image has multiple copies of backing files in the OpenStack cluster environment based on shared storage. The only difference of copies is name. This is great waste for unified Shared storage. To specify same image ID in the OpenStack cluster environment can solve this problem. Kilo and above version of OpenStack support this operation.

3.5. Security and Accounting

In the loosely coupled infrastructure, security protection includes three aspects. The first aspect is central database and REST API interface access control. The second aspect is OpenStack security protection. The third aspect is account authentication. There are four measures to build security protection in the loosely coupled infrastructure which is shown in Table 1.

measures	description
Iptables	Isolating services such as MySQL, SSHD etc. Open access for each OpenStack controller.
REST API interface	Hypertext Transfer Protocol Secure and tokens will be used for access security. The REST API interface is open only to the authorized source IP address.
Central database	Local IP address access limitation.
Account authentication	To support unified authentication system.

Table 1. Security measures

4. Summary

The loosely coupled cloud infrastructure aims to solve extension problem on the cloud computing system through a simple, weakly connected system architecture and design. It provides a simple mechanism to control resources and integrate OpenStack clusters. It is different from Cells v2.0 or Tricircle/ Trio2o. It's more simple and convenient. The loose coupling connection makes the system more convenience to extension or shrink. The hierarchical architecture ensures the separation of control and execution, each layer just focus on itself. Centralized virtual resource view makes global resource-based scheduling possible. This may be something like Cell v2.0. But in this infrastructure, the scheduling algorithm just focuses on the resources themselves but not OpenStack sites. The feature makes cloud system more flexible and reasonable. Now the system is still in the phase of test-bed and the running of test-bed proves that it can work well. In future, we'd like to do more research on the scheduling algorithm such as green energy-saving scheduling, regional scheduling, etc. We have a lot of work to do to further improve the design.

References

[1] GlusterFS Documentation, https://docs.gluster.org/en/latest/
[2] UW-Madison, HTConder, https://research.cs.wisc.edu/htcondor/manual/
[3] CERN IT data storage group, EOS, http://eos-docs.web.cern.ch/eos-docs/intro.html
[4] OpenStack Nova Cells v2, http://docs.openstack.org/developer/nova/devref/cells.html
[5] OpenStack Tricircle, https://wiki.openstack.org/wiki/Tricircle
[6] OpenStack Trio2o, https://wiki.openstack.org/wiki/Trio2o
[7] T Cui, Y D Cheng, The advances in IHEP Cloud facility, https://indico.cern.ch/event/531810/contributions/2331773/, Oct 2016

IOP Conf. Series: Journal of Physics: Conf. Series **1085** (2018) 032012 doi:10.1088/1742-6596/1085/3/032012

Optimizing ROOT IO For Analysis

B Bockelman[1], Z Zhang[1] and J Pivarski[2]

[1] Holland Computer Center, University Nebraska – Lincoln, Lincoln, NE 68588, USA
[2] Physics Department, Princeton University, Princeton, NJ 08544, USA

E-mail: `bbockelman@cse.unl.edu`

Abstract. The ROOT I/O (RIO) subsystem is foundational to most HEP experiments - it provides a file format, a set of APIs/semantics, and a reference implementation in C++. It is often found at the base of an experiment's framework and is used to serialize the experiment's data; in the case of an LHC experiment, this may be hundreds of petabytes of files! Individual physicists will further use RIO to perform their end-stage analysis, reading from intermediate files they generate from experiment data.

RIO is thus incredibly flexible: it must serve as a file format for archival (optimized for space) and for working data (optimized for read speed). To date, most of the technical work has focused on improving the former use case. We present work designed to help improve RIO for analysis. We analyze the real-world impact of a compression algorithm LZ4 to decrease decompression times (and the corresponding cost in disk space). We introduce new APIs that read RIO data in bulk, removing the per-event overhead of a C++ function call. We compare the performance with the existing RIO APIs for simple structure data and show how this can be complimentary with efforts to improve the parallelism of the RIO stack.

1. Introduction

The field of High Energy Physics is unique in how uniformly the community has standardized on a foundational framework, ROOT. ROOT [1] is an object-oriented C++ software framework originally developed at CERN. While the framework integrates most tools a physicist may need - from mathematical functions, to an XML parser, to a GUI - one of the most frequently-used features is the C++ object serialization capability. This capability is used as the file format of hundreds of petabytes of HEP experiment data. As nearly all experiments store their archival data in the format, it's also widely adopted by scientists producing their own derived data. This produces a delicate balancing act: the ROOT IO format must simultaneously meet the needs of HEP experiments and their users:

- Experiments need features like schema evolution and the ability to serialize complex, arbitrary C++ objects. Minimizing the storage footprint is a driving motivation (as they must manage tens of petabytes of data), and speed is low priority.
- Analysts only need to serialize simple objects; speed is paramount and there is less sensitivity to overall file size.

As experimental storage costs are easier to quantify than time spent waiting for results, the priority has historically been skewed toward meeting the experimental needs. In this paper, we explore new techniques to provide some specialization for analysis use cases in the ROOT framework. We argue specialization for analysis is critical as users may iterate across their private datasets many times and science can't proceed until the IO has finished.

We introduce these specializations through revisiting some tradeoffs. First, we explore a compression algorithm (LZ4 [2]) that has lower compression ratios but markedly faster decompression rates. We feel this is justified as the input to a typical analysis can often fit on a single hard drive, so a modest increase in dataset size may be acceptable. Second, we introduce the "bulk IO" APIs, which provide access to many events per ROOT framework library call. Here, the tradeoff is that the bulk IO drastically restricts the supported object types. We view this as acceptable as analysis events are often drastically simplified when compared to full experiment frameworks. Finally, we utilize some simple parallelism techniques to decompress file data in parallel even when the user may be only utilizing a single thread of control.

2. Background

The vast majority of the data generated by a HEP experiment is *event data*, which typically corresponds to a physical or simulated occurrence of the particle collision. Each event is decomposed into multiple C++ objects in ROOT that describe the understanding of the event. Within ROOT, an ordered list of event data is represented by the `TTree`. The `TTree` is further partitioned into branches called `TBranch`; each branch collects a set of similar objects from events (typically, objects of the same C++ type). For a more thorough discussion of the file organization, see [1].

When a C++ object is serialized, the resulting byte stream is stored in a memory buffer. Each branch contains one memory buffer. When the buffer is full, ROOT compresses the data and a `TBasket` is created. `TBasket` is derived from `TKey` and contains additional information specific to the `TTree` navigation logic. Hence, the serialized event data in a `TTree` consist of a sequence of compressed `TBaskets` and their metadata information.

As each event can be variable size, each `TBasket` for a branch may correspond to a different event range. In the worst case scenario, the contents of an event may be scattered across hundreds of megabytes of file data: this potentially-poor locality is addressed by creating *event clusters* – ranges at which all buffers are flushed to disk, even if the corresponding memory buffers are not full.

3. LZ4 Compression

By default, ROOT uses ZLIB [3] as the default compression algorithm; it provides a balance between compression ratios and decompression speed. The alternate algorithm, LZMA [4] is skewed toward a higher compression ratio and significantly slower decompression (detailed performance comparison is provided in [5]). LZMA is ideal for the largest datasets (or those infrequently read and kept on tape) but, for data analysis, total dataset size may only be a few terabytes and reading speed is the critical metric. We explore another compression algorithm - LZ4 - which sacrifices compression ratio for improved read performance.

Figure 1 shows compression ratios and decompression speed for various of compression levels of ZLIB and LZ4. We normalize performance to ZLIB-6 (higher levels result in higher compression ratios). For LZ4's compression levels, **hc** represents a high-compression variant of the algorithm. In general, LZ4 has lower compression ratios and faster decompression.

Decompression is not the only CPU cost when reading; for example, the uncompressed bytes must be converted to valid C++ objects. In Figure 2, we measure the decompression time separately from other CPU costs. We create ROOT files with an aggregate of 400MB of event data, but varying individual event size from 40 bytes to 4MB. At one extreme, we create a ROOT file with 100 events, each containing 1,000,000 floating point (FPs) numbers. On the other END, we create 10,000,000 events and each event contains 10 FPs. As seen in the figure, while decompression time is roughly consistent across event size, other CPU costs come to

IOP Conf. Series: Journal of Physics: Conf. Series **1085** (2018) 032012 doi:10.1088/1742-6596/1085/3/032012

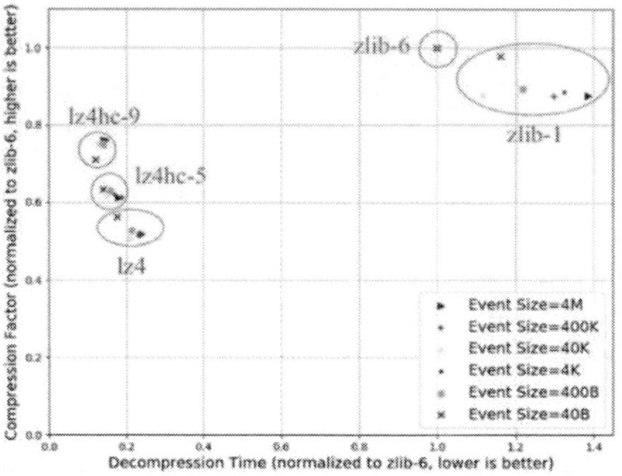

Figure 1. Comparison of Different Compression Algorithms

dominate as the event size decreases. Reading 10 small branches is not unreasonable, suggesting CPU overheads in ROOT IO for analysis (motivating the bulk IO work in Section 4).

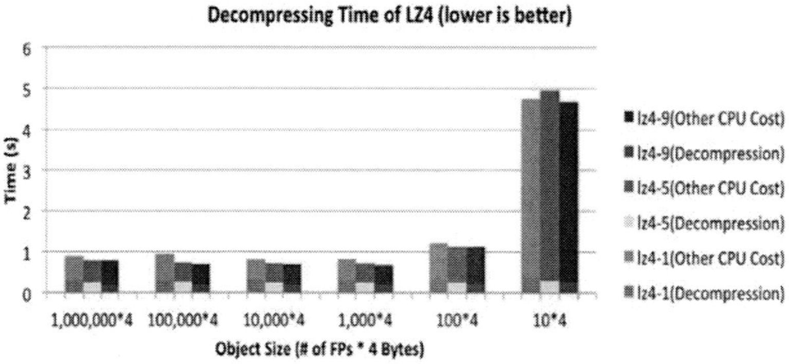

Figure 2. CPU costs for reading LZ4-compressed files

4. Bulk IO

The typical mechanism for iterating through data in a `TTree` is a handwritten for-loop: for each event in the tree, fetch the data for that event from ROOT into some proxy object(s) and execute a user-written code.

When the user code is computationally expensive, the cost of the library call to ROOT is amortized into effectively nothing. However, as Figure 2 shows, as the (uncompressed) bytes read per event decreases below one kilobyte, the ROOT overheads first become measurable, then eventually dominates for smaller event sizes. Accordingly, we introduce a new interface for ROOT to copy all events in an on-disk `TBasket` directly to a user-provided memory buffer. For the simplest of cases - primitives and C-style arrays of primitives - where the serialization can be done without a separate buffer or “fixing up” pointer contents, the user can request the serialized data be delivered to the buffer or deserialized data. By requesting the serialized data directly and deserializing directly in the event loop, the user can avoid an expensive scan from main memory.

Pragmatically, the user will not implement code for deserializing data themselves: rather, we have provided a header-only C++ facade around the data, allowing the user to work with a proxy object. This allows the compiler to inline the deserialization code in the correct place.

As is shown in Figure 3, this technique provides some drastic performance improvements. The bulk IO APIs - even if accessed from within various Python-based mechanisms - are up to 10x faster in events per second compared to the traditional C++ mechanisms ROOT users are familiar with.

The file for this study is a flat ntuple of px, py, pz, and mass of dimuons. The baskets of the mass branch, however, are not aligned with the baskets of px, py, and pz, so total momentum can be calculated without copying any basket data while total energy must copy basket data to create aligned arrays. The distinction between "viewing" and "copying" is another layer of copying, this time defensively, since a view into unowned data is unsafe. The Numpy calculation produces intermediate arrays and performs one operation per pass over the data, while Numba computes the whole expression in one pass without unnecessary allocation. We see that this difference between Numpy and Numba is a 20% effect, but that is dwarfed by differences with respect to `SetBranchAddress/GetEntry` and methods based on it, such as root_numpy, `TTreeReader`, and `TTree::Draw`. Also, the differences are almost entirely washed out by deflate (ZLIB) decompression— the value of bulk IO is only exposed by uncompressed and LZ4-compressed files.

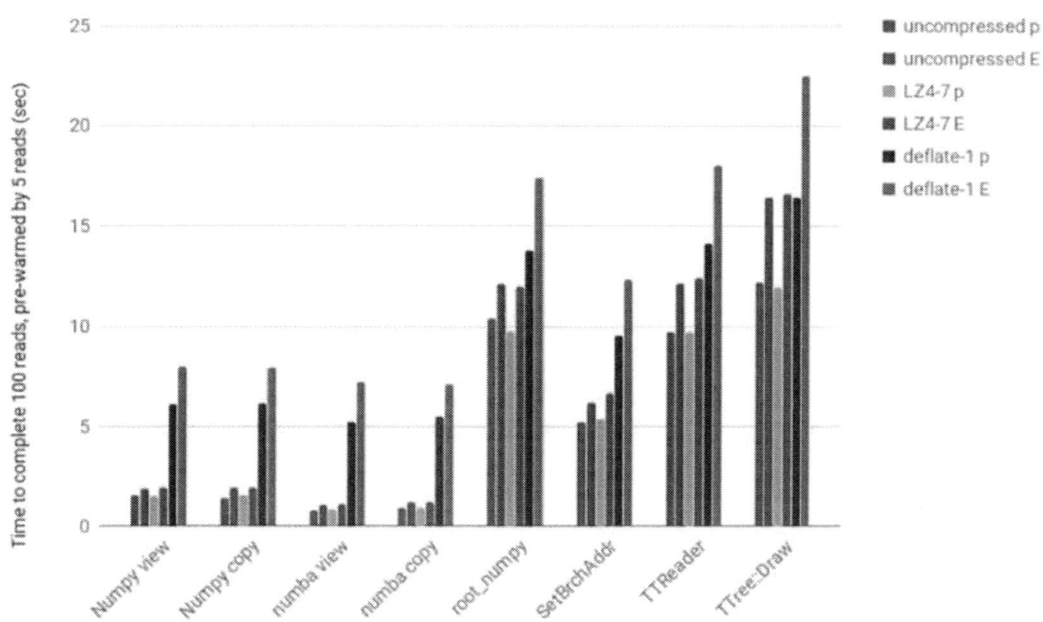

Figure 3. Performance of the bulk IO. Each access method was tested 6 times, using three different compression settings (uncompressed, LZ4, or ZLIB/deflate) and two different user calculations (momentum, p and energy, E). The energy calculation utilizes baskets that are poorly aligned with event clusters; this measurement should illustrate the cost of the code for this special case

5. Asynchronous Parallel Unzipping

To accelerate reading for analysis, we leverage Intel Thread Building Blocks (TBB) [6] to decompress multiple baskets in parallel. TBB allows users to break down high-level jobs into tasks that can run in parallel, allowing the library to manage and schedule threads for task execution. When ROOT encounters a new event cluster, we create one task per approximately

100KB of compressed baskets, returning control to the calling thread as soon as possible. The calling thread only blocks if the user requests event data whose decompression is not yet complete.

In Figure 4, we test decompression speed on synthetic event benchmarks. We create ROOT files with 500 - 50,000 events. The performance is measured by sequentially reading all events. The figure shows the average runtime per event, running on a desktop-class machine with Intel 4-core CPU and Ubuntu 14.04.

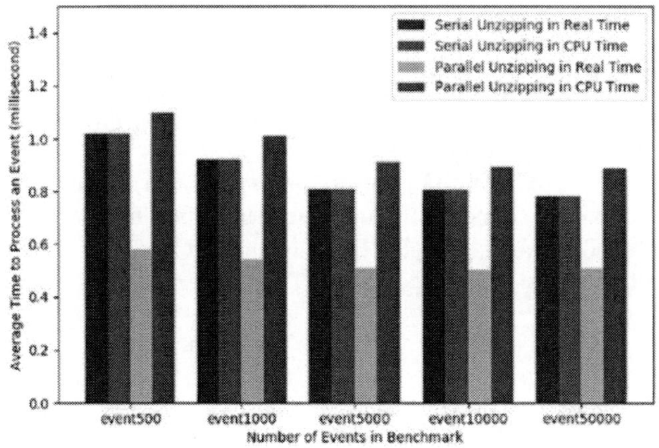

Figure 4. Performance between Serial and Parallel Unzipping on Event Benchmark

As background threads can decompress data prior to the main thread accessing it, As shown in Figure 4, parallel unzipping improves read performance with more noticeable improvements at higher event counts. Parallel unzipping takes 52% - 58% processing time of serial unzipping. These synchronization techniques require CPU cycles; this technique takes 8% - 13% more CPU cycles.

6. Conclusions

Analysisof ROOT data is a distinct use case from long-term archival - or data reconstruction - requiring fast IO with less sensitivity to compression ratio. In this paper, we discussed various ways to improve the read performance. For simple data types, bulk IO allows deserialization without multiple library calls. The LZ4 algorithm provides significantly faster decompression than the existing ZLIB. Finally, the asynchronous parallel unzipping baskets can further double the read performance comparing to the serial baseline.

Acknowledgments

This work was supported by the National Science Foundation under Grants ACI-1450323, ACI-1450377 and PHY-1624356. This research was done using resources provided by the Holland Computing Center of the University of Nebraska.

References

[1] Brun R and Rademakers F 1997 ROOT - An object oriented data analysis framework, *Nucl. Instr. Meth. Phys. Res.* **389** 81-86
[2] URL http://lz4.github.io/lz4/
[3] Deutsch P and Gailly J L 1996 *ZLIB Compressed Data Format Specification Version 3.3*
[4] URL http://www.7-zip.org/sdk.html
[5] Zhang Z and Bockelman B 2017 Exploring compression techniques for ROOT IO, *CoRR* **abs/1704.06976**
[6] Pheatt C 2008 Intel® Threading Building Blocks *J. Comput. Sci. Coll.* **23** 298

Data Knowledge Base for HENP Scientific Collaborations

V A Aulov[1, a], M V Golosova[1], M A Grigorieva[1,2,b], A A Klimentov[3], S Padolski[3], T Wenaus[3]

[1]Kurchatov complex of NBIC-technologies, National Research Centre Kurchatov Institute, 1 Akademika Kurchatova pl., Moscow 123182, Russia
[2]Tomsk Polytechnic University, 4a Usov Street, Building 19, Tomsk, Russia
[3]Brookhaven National Laboratory, PO Box 5000 Upton, NY 11973-5000, USA

[a]vasiliyaulov@gmail.com
[b]maria.grigorieva@cern.ch

Abstract. Contemporary scientific experiments produce significant amount of data as well as scientific publications based on this data. Since volumes of both are constantly increasing, it becomes more and more problematic to establish a connection between a given paper and the underlying data. However, such an association is one of the crucial pieces of information for performing various tasks, such as validating the scientific results presented in paper, comparing different approaches to deal with a problem or even simply understanding the situation in some area of science. Authors of this paper are working under the Data Knowledge Base (DKB) R&D project, initiated in 2016 to solve this issue for the ATLAS experiment at CERN. This project is aimed at developing of the software environment, providing the storage and a coherent representation of the basic information objects. In this paper authors present a metadata model developed for the ATLAS experiment, the architecture of the DKB system and its main components. Special attention is paid to the Kafka-based ETL subsystem implementation and mechanism for extraction of meta-information from the texts of ATLAS publications

1. Introduction

The life cycle of a scientific experiment has changed considerably during the last decades. Data and metadata related to the experiment must be stored and available for reprocessing for years or even decades, as scientists may need to consult it in order to plan further research, compare results to the ones achieved with different approaches, initiate reprocessing of some experimental data to maintain reproducibility. And while storing data is a problematic issue itself, but finding the required parts of it among the petabytes of scientific knowledge and then processing it in a proper way is even more complex problem. The Data Knowledge Base (DKB) project [1], dealing with the metadata of high energy physics (HEP) experiments (with ATLAS as an example), is one of the attempts to address this problem.

DKB project was started in 2016 as a joint effort by ATLAS experiment at CERN, Kurchatov Institute and Tomsk Polytechnic University. The initial idea of the project was to provide a fast and user-friendly access to relevant scientific information regarding the physical analysis -- data used in the research, hardware/software configuration, publications and so on.

IOP Conf. Series: Journal of Physics: Conf. Series **1085** (2018) 032013 doi:10.1088/1742-6596/1085/3/032013

2. Data Knowledge Base Architecture

One of the key points of DKB architecture (figure 1) is the possibility to interact with structured and unstructured (documentary) data sources. For example, plain text or PDF documents are considered as one of the forms of data representation and can be used as any other processing object. A special tool for metadata extraction from PDF documents, including papers and internal notes (drafts) -- PDFAnalyzer -- is one of the specific parts of DKB system.

Information, extracted from heterogeneous data sources is integrated into a number of internal DKB storages. Each of these storages provides optimization for specific types of search or aggregation requests. Continuous data integration is performed by Apache Kafka-based [2] ETL subsystem, discussed in section 4 of this paper.

Internal storages of DKB include:

- ***Ontological storage*** OpenLink Virtuoso [3]. It allows to store the meta-information in form of Semantic Web (representing objects with their properties and connections between them). Virtuoso internal storage contains metadata, integrated from CERN Document Server (CDS) [4] and search engine for the ATLAS Collaboration (GLANCE) [5], and information, extracted from PDF documents by PDFAnalyzer.
- ***ElasticSearch storage*** [6] provides indexing of metadata from Production System [7] and ATLAS Metadata Interface (AMI) [8], ensuring fast search and aggregation capabilities.
- ***Transitional storage***, based on Hadoop, is used to store the interim data processing results. For example, temporary storage of PDF documents from CDS.

DKB interfaces, providing access to these internal storages:

- JavaScript-based library Ontodia is used as the interface to Virtuoso, allowing to visualize, navigate and explore the data in the form of an interactive graph based on underlying data sources.
- Kibana runs on the top of the ElasticSearch storage, providing possibility to generate aggregated reports, perform data filtration and basic search requests.
- NodeJS interface provides full-text search within ElasticSearch by a set of keywords.

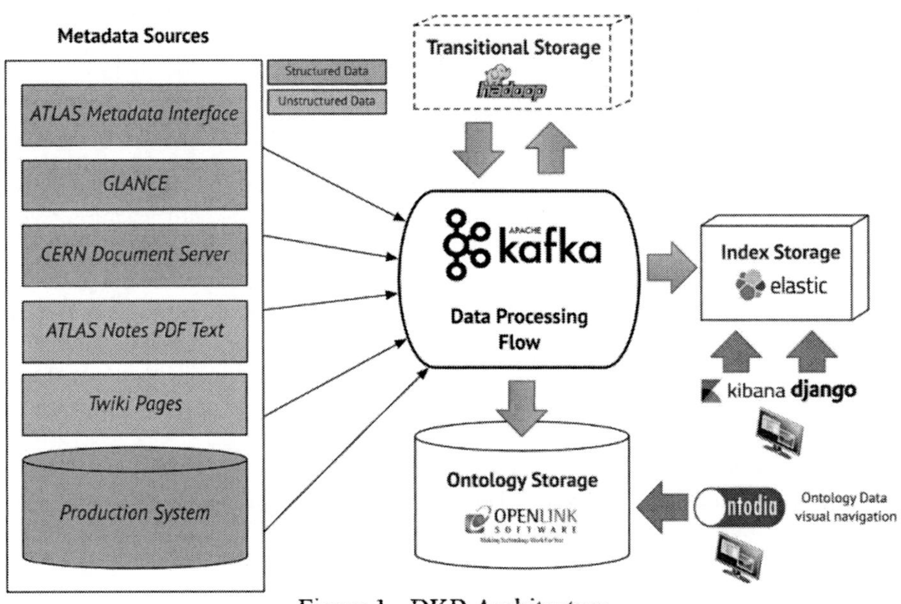

Figure 1 - DKB Architecture

3. ATLAS Ontological Data Model

To build a complete and coherent view of an experiment lifecycle various metadata must be integrated and interlinked. These metadata are strongly semantically connected and have a complex structure, but can be stored in loosely connected information subsystems of the experiment. With constantly growing volume of these metadata it is hardly possible to organize a relational storage, that would efficiently execute complex search requests, utilizing multiple interconnections (JOINs) between metadata items. The widely used technologies to solve the problem of strongly connected data are graph-based or ontology data model. In this study the ontological model of ATLAS experiment metadata was developed. It formalizes the description of scientific experiment phases, parameters, processes and other entities.

The main entities of this model are "Document", "DataSample" and "PhysicsExperiment". "Document" can be of several types: publication, internal document, journal issue, etc. All documents have attributes, corresponding to the ones in CDS meta-description, and obtained from PDF texts of these documents - experiment attributes and data samples. "Document" and "DataSample" can be associated to the "PhysicsExperiment" by a set of parameters: data samples, data processing project, Monte-Carlo simulation campaigns, keywords, physics group, the center-of-mass energy, integrated luminosity, data taking year.

"DataSample" is connected with "Task", which corresponds to the real computational task in distributed data processing system (Production System). "Task" is the most informative object, containing the meta-information about software versions, detector architecture and calibration metrics, the number of processed events, task name, status and timestamps, input/output data samples, physics process metadata, hashtags and others. Connections between "Task", "DataSample" and "Document" allow to gather complete information about physics analysis process.

4. ETL subsystem

The ETL subsystem of DKB, based on Apache Kafka, is organized as a set of independently developed modules, joined into a data flow topology.

Each of these modules implements a single ETL action: data extraction from an original source, transformation of the extracted data or load data into one of the internal storages.

Data *extraction modules* are run on a configured schedule by supervising Kafka Source Connector, which publishes extracted data to the corresponding Kafka topic(s) (a transitional storages between ETL stages).

Transformation process may consist of a number of steps, implemented as *transformation modules*, each responsible for a single logical operation: removing of excessive data, filtration, generation of surrogate keys or derived values, aggregations, retrieving additional information from other external sources (those that are not used as initial sources), and, finally, format conversion -- to prepare data for upload to the DKB integrated storages. These modules are organized into pipelines (or a more complicated topologies) by means of Kafka Streams library: every Kafka Processor is operating as a module supervisor, that makes sure that the module is running and ready to process input data as soon as it appear, taking care of restarting it in case of any failure and providing reliable communication mechanism between modules in standalone or distributed mode. These transformation topologies can be subscribed to one or more topics, and when new data are published by a Source Connector, transformation process begins with minimal delay. When a record, consumed from a source topic, passes through all the transformation steps, it is ready for the upload to some of DKB storages and goes to the topology sink (published to an output topic).

Load modules, final stages of ETL processes, responsible for data load to DKB internal storages, are subscribed to one or more of these output topics (now via supervising Kafka Sink Connectors). As soon as new data appears, load module start transferring it to the DKB storage -- and making available for user-oriented DKB services.

This ETL subsystem is oriented on a diverse development teams, as it allows to organize any ETL process as a combination of successive steps -- executable modules, sharing only input/output data format and obeying common set of rules -- each of which can be developed independently, implemented by means of any program language, used as a standalone program or reused in a number of different ETL processes. The mentioned set of rules is required to unify chaining of the modules, and is simple enough to follow: standard input/output streams for input/output data, conventional message delimiter and simple flow control specification (sending control symbol at the end of message processing).

This subsystem also allows to minimize the delay between consuming information from original sources and placing transformed data to the internal storages.

Apache Kafka is used to address general tasks (such as organization of transitional storage between separate processing stage, operating in a distributed mode, restarting modules without massive data reprocessing, stateful processing, e.g. (such as data filtration or aggregation, etc) and to organize independent transformation modules into a topology of successive processing steps. Variety of Kafka configuration parameters allows to tune thoroughly the ETL subsystem as a whole and every ETL process individually, making it very flexible and adaptable for different use cases.

5. Metadata Extraction from PDF Documents

Scientific data analysis requires usage of the correct datasets (sets of data files) for the computational tasks and providing relevant information in scientific papers. There are ATLAS tools, such as AMI/GLANCE interface, intended for finding correlations between papers and datasets -- however, currently it can not be considered as a universal tool covering all documents and datasets. Therefore, alternative options were considered. One possibility for solving the task is provided by the ATLAS Internal Notes (the documents describing physics analysis in details) which are referenced by published papers. Though papers themselves can refer to the datasets being used for physics analysis, it is usually done in form of human-oriented descriptions -- unlike the Internal Notes, where datasets are specified explicitly.

Unfortunately, there is no general agreement to define the data samples in the Internal Notes. This makes the datasets automated extraction rather complicated. To deal with this obstacle, evaluation of existing tools for PDF processing was carried out. As a result, the tool called PDFMiner [9] was chosen to perform the basic functions such as converting PDF to text, while further analysis and processing were implemented by a module developed from scratch - PDFAnalyzer.

PDFAnalyzer can extract two categories of metadata from the texts -- experiment attributes and datasets. Experiment attributes -- data taking year, energy of the center of mass, integrated luminosity and Monte-Carlo production campaigns -- are found by means of regular expressions. Extracting datasets is a more complex procedure since they can be specified in several ways, most prevalent of which are text lists and tables.

— *Dataset names in text lists* are the easiest to find – since all the names are expected to comply with the ATLAS Dataset Nomenclature [10], regular expressions are usually a sufficient option to find them. However, nomenclature cannot be fully relied on due to reasons such as PDF to TXT conversion problems.

— *Dataset tables* usually contain only parts of dataset names (for example, dataset identificators). An algorithm for table reconstruction based on XML representation of the page layout, which explicitly states the position, size and font of each symbol, was developed.

The corpus of 133 ATLAS Internal Notes published between 2016 and 2017 was selected to test the developed program. Each document was analyzed, as well as checked manually whether it contains datasets or not. Out of 133 documents:

— 95 documents where datasets were found and extracted by the program;

— 12 documents contained datasets, but the program failed to detect them correctly;

— 26 contained no datasets and were correctly identified as such.

Therefore, roughly 89% of the documents with datasets in them were analyzed correctly. 12 remaining documents either have dataset names in text lists, but these names are specified only partially; or require better methods of table handling, which are currently being developed.

IOP Conf. Series: Journal of Physics: Conf. Series **1085** (2018) 032013 doi:10.1088/1742-6596/1085/3/032013

Conclusion

A prototype of the described system provides continuous integration of metadata from heterogeneous sources, including extraction of information from unstructured (in terms of automated text analysis) PDF documents. This information is being gathered in the internal storages of DKB, and is accessible via number of interfaces, providing visualization of the metadata field in form of a connected graph (and navigation through it), google-like search and numerous aggregation tools. Further DKB development involves expansion of the ontological data model, addition of new metadata sources, development of services and data workflows, and advancing mechanisms of PDF documents analysis. The proposed scientific knowledge base will form a unified information field of any scientific experiment, and its architecture can be adapted to a variety of subject areas.

Acknowledgements

The work was supported by the Russian Ministry of Science and Education under contract No.14.Z50.31.0024 and by the Russian Science Foundation under contract №16-11-10280.

References

[1] M Grigorieva, V Aulov, A Klimentov and M Gubin 2016, Knowledge base of scientific experiment, *Open Systems. DBMS*, No. 4 of 2016, page 42
[2] Apache Kafka [Online]. Available: https://kafka.apache.org/ [accessed on: 27.02.2018]
[3] Erling O., Mikhailov I. RDF Support in the Virtuoso DBMS.. CEUR Proceedings, vol. 301. Proc. of the 1st Conference on Social Semantic Web, Leipzig, Germany, Sep 26- 28, 2007
[4] CERN Document Server [Online]. Available: https://cds.cern.ch [accessed on: 27.02.2018]
[5] ATLAS GLANCE [Online]. Available: https://atglance.web.cern.ch/atglance/ [accessed on: 27.02.2018]
[6] ElasticSearch [Online]. Available: https://www.elastic.co/products/elasticsearch [accessed on: 27.02.2018]
[7] K De, D Golubkov, A Klimentov, M Potekhin and A Vaniachine on behalf of the ATLAS Collaboration. Task Management in the New ATLAS Production System // Journal of Physics: Conference Series 513 (2014) 032078. doi:10.1088/1742-6596/513/3/032078. [Online]. Available: http://iopscience.iop.org/article/10.1088/1742-6596/513/3/032078/pdf [accesses on: 27.02.2018]
[8] ATLAS Metadata Interface [Online]. Available: http://ami.in2p3.fr/index.php/en/ [accessed on: 27.02.2018]
[9] PDFMiner [Online]. Available: http://euske.github.io/pdfminer/index.html [accessed on: 27.02.2018]
[10] ATLAS Dataset Nomenclature [Online]. Available: https://dune.bnl.gov/w/images/9/9e/Gen-int-2007-001_%28NOMENCLATURE%29.pdf [accesses on: 27.02.2018]

Increasing Parallelism in the ROOT I/O Subsystem

G Amadio[1], B P Bockelman[2], P Canal[3], D Piparo[1], E Tejedor[1], Z Zhang[2]
[1]CERN, [2]University of Nebraska-Lincoln, [3]FNAL
E-mail: amadio@cern.ch

Abstract. When processing large amounts of data, the rate at which reading and writing can take place is a critical factor. High energy physics data processing relying on ROOT is no exception. The recent parallelisation of LHC experiments' software frameworks and the analysis of the ever increasing amount of collision data collected by experiments further emphasised this issue underlying the need of increasing the implicit parallelism expressed within the ROOT I/O.

In this contribution we highlight the improvements of the ROOT I/O subsystem which targeted a satisfactory scaling behaviour in a multithreaded context. The effect of parallelism on the individual steps which are chained by ROOT to read and write data, namely (de)compression, (de)serialisation, access to storage backend, are discussed. Performance measurements are discussed through real life examples coming from CMS production workflows on traditional server platforms and highly parallel architectures such as Intel Xeon Phi.

1. Introduction

Particles collide at the Large Hadron Collider (LHC) approximately half a billion times per second. Each collision creates particles that decay into a complex cascade of many other particles that are recorded by the detectors as a series of electronic signals that are then digitally reconstructed and stored for further analysis. At the current rate, physicists have about 30 petabytes of new data per year to analyse [2], and this rate is expected to increase ten-fold in the next decade. Most of this data is stored and processed in ROOT's file format [1], hence optimizing ROOT to read and write all this data as fast as possible is of critical importance to the future of high energy physics.

1.1. The Meaning of ROOT I/O

One of the key innovations of ROOT is its columnar data format. With ROOT, if only a small subset of the data is needed for a given analysis, only those parts of the data file will be read, instead of the whole file. Another important aspect of ROOT I/O is its ability to perform serialisation and deserialisation of generic C++ objects via automatically generated object streamers. Moreover, ROOT can compress data before writing it to disk and decompress it before processing. Hence, ROOT I/O is much more than simply reading and writing from a regular file; it involves also (de)serialisation and (de)compression of data. Naturally, each of these processing phases can be parallelised independently. Since this form of parallelism is usually performed in a way that is invisible to the user, it is referred to as implicit parallelism, or implicit multi-threading (IMT) in ROOT. Support for implicit parallelism has been introduced

IOP Conf. Series: Journal of Physics: Conf. Series **1085** (2018) 032014 doi:10.1088/1742-6596/1085/3/032014

in ROOT 6.06. Since then, many improvements on this front have been added to newer versions, and this is what we discuss in the following sections.

2. Reading ROOT Data in Parallel

Each data set created in the ROOT file format may be read many times and by different users in their analyses, hence data reading performance is arguably the most important aspect to be considered when parallelising ROOT I/O and choosing what compression algorithm to use to store data intended for physicists to use.

2.1. Reading Data Columns

ROOT uses a columnar data format, and experiments' data files usually contain many different columns that contain the type, energy, momentum, and other properties of particles digitally reconstructed from collision events. One way in which it is possible to parallelise data reading is decompressing and deserialising each column in parallel. This implicit multi-threading feature has been introduced in ROOT 6.08. As a demonstration of the potential performance gains, two benchmarks have been run to read, decompress and deserialise data sets from CMS (GenSim) and ATLAS (xAOD), with approximately 70 and 200 different columns, respectively. Figure 1 shows that a speedup of about 3.5 can be obtained on a quad core laptop, a great improvement over running the code sequentially.

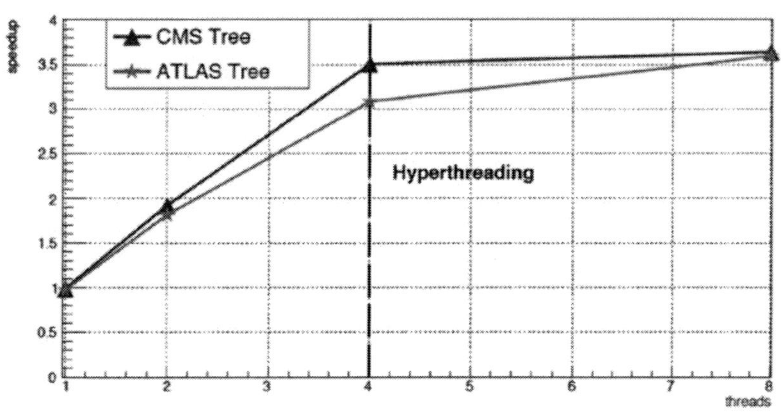

Figure 1. Parallel reading of multiple data columns.

2.2. Data Decompression

ROOT files usually store data in several compressed blocks, also referred to as baskets. When processing large files, many blocks can be decompressed and deserialised in parallel. Since decompression is compute-intensive, it is also possible to interleave parallel decompression with processing of decompressed data. This feature, currently under development, is planned for inclusion in ROOT 6.14. Figure 2 shows performance gains for the current implementation on a quad-core laptop (4 threads + hyperthreading).

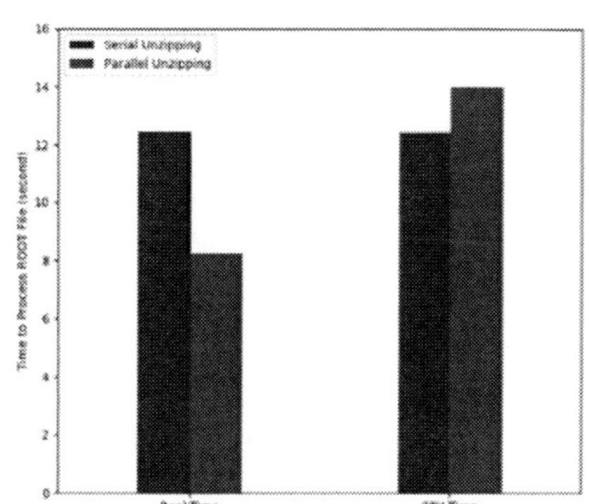

Figure 2. Parallel basket decompression.

IOP Conf. Series: Journal of Physics: Conf. Series **1085** (2018) 032014 doi:10.1088/1742-6596/1085/3/032014

3. Writing ROOT Data in Parallel

Even if reading ROOT data is much more common than writing, that is not to say that writing performance is not important for ROOT. Event reconstruction jobs process vast amounts of data from the detectors, so if the speed of writing out reconstructed data becomes a bottleneck, it may reduce the overall percentage of events that can be saved for offline analysis. In addition to allowing users to use one `TFile` per thread in their programs, ROOT has added new classes and features to support parallel writing, which are discussed next.

3.1. Writing Data Columns

In a similar fashion as what has been done for reading data columns from a file in parallel, ROOT can also serialise and compress multiple data columns in parallel when implicit multi-threading is enabled. This feature has become available in ROOT 6.08. The performance advantages of this parallelism have been demonstrated with a benchmark within CMSSW [3] that writes out two kinds of data sets: an I/O intensive reconstruction data set (RECO), and a data set with lighther I/O requirements, intended for analysis (AOD). Figure 3 shows that on the Intel Xeon Phi codenamed Knights Landing (KNL) [4], with IMT disabled, the throughput begins to degrade for the RECO data set after about 20 CMSSW streams (30 threads). When IMT is enabled, however, the throughput can be more than doubled and scales better to a larger number of threads. The number of streams is the number of framework tasks running in parallel; the number of threads is fixed to 1.5 times the number of streams. For AOD—a slimmer kind of data set—it is also possible to see an improvement, but since it is a less I/O intensive workload the gains are limited, as the performance for this data set is already close to that of not writing out any data at the end (red line in the figure).

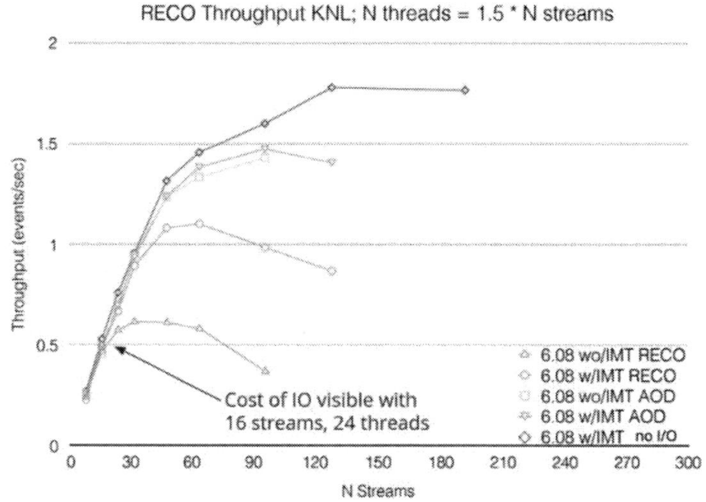

Figure 3. Parallel writing of multiple data columns.

3.2. Writing to an Output File from Multiple Threads

Since ROOT 6.06, it has been possible to write out data in parallel to multiple ROOT files, one per each thread of the user application. However, this creates the inconvenience of having many files with partial data to analyse at the end, or having an extra step to merge all data into a

IOP Conf. Series: Journal of Physics: Conf. Series **1085** (2018) 032014 doi:10.1088/1742-6596/1085/3/032014

single file. For the user, it is much more convenient to write out data in parallel from multiple threads into a single output file by performing the merging steps automatically in the background. This feature has been introduced in ROOT 6.10 with the addition of the `TBufferMerger` [5] class. A schematic of how it works is shown below in figure 4.

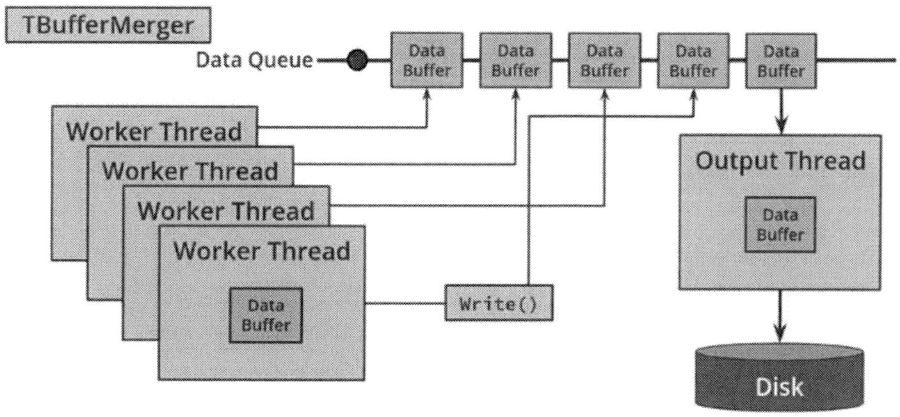

Figure 4. `TBufferMerger` schematic diagram. Worker threads place data in a queue, while the output thread merges one by one each buffer from the queue into the output file.

Users create a `TBufferMerger` [5] instead of a `TFile` [6], and get from it a `TMemFile`-like buffer to write into by calling `TBufferMerger::GetFile()`. Then, each worker thread can write to that `TMemFile`-like buffer, and the effect is that data will be pushed into a processing queue for merging into the output file. Once all data has been pushed into the queue and merged, the output thread closes the output file and exits. The code samples below show a simple example of filling a `TTree` in serial mode, and how that changes when using the `TBufferMerger`.

Sequential usage of `TFile`

```
void Fill(TTree &tree, int init, int count)
{
   int n = 0;

   tree->Branch("n", &n, "n/I");

   for (int i = 0; i < count; ++i) {
      n = init + i;
      tree.Fill();
   }
}

int WriteTree(size_t nEntries)
{
   TFile f("myfile.root");
   TTree t("mytree","mytree");

   Fill(&t, 0, nEntries);

   t.Write();

   return 0;
}
```

Parallel usage of `TFile` with `TBufferMerger`

```
void Fill(TTree *t, int init, int count);

int WriteTree(size_t nEntries, size_t nWorkers)
{
   size_t nEntriesPerWorker = nEntries/nWorkers;

   ROOT::EnableThreadSafety();
   ROOT::Experimental::TBufferMerger merger("myfile.root");

   std::vector<std::thread> workers;

   auto workItem = [&](int i) {
         auto f = merger.GetFile();
         TTree t("mytree", "mytree");
         Fill(t, i * nEntriesPerWorker, nEntriesPerWorker);
         f->Write(); // Send remaining content over the wire
     };

   for (size_t i = 0; i < nWorkers; ++i)
      workers.emplace_back(workItem,i);

   for (auto&& worker : workers) worker.join();

   return 0;
}
```

Figure 5. `TBufferMerger` code example.

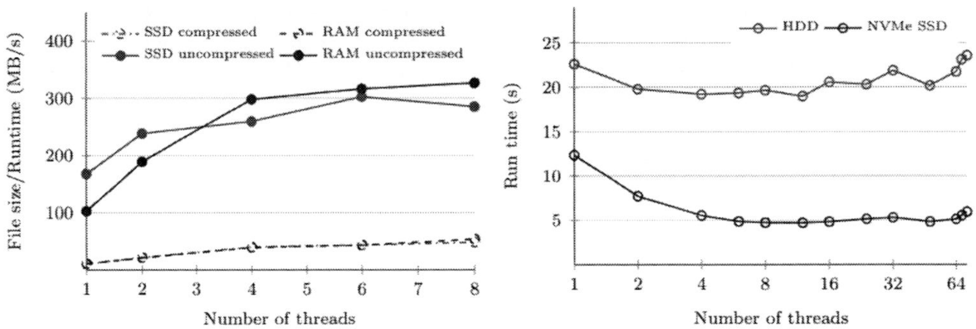

Figure 6. `TBufferMerger` writing performance to a hard-disk drive (HDD), a solid state drive (SSD), and a non-volatile memory solid state drive (NVMe SSD).

Write performance is shown in figure 6 for different output scenarios. The benchmarks consist in generating 1GB of pseudo-random numbers, and writing them out as a single column data file. On the left, red lines represent writing to a solid state disk (SSD), while blue lines represent writing data directly into system memory via tmpfs. On the right, two extremes in terms of hardware performance are compared: a standard hard drive (HDD) and a fast non-volatile memory solid state disk (NVMe SSD). Writing speed reaches over 320 MB/s on the SSD, which is near the hardware limit for uncompressed data. Speedup compared with sequential code is over three-fold on a quad core machine. When writing out compressed data, the CPU becomes the bottleneck due to the cost of compression. Hence, the speedup is similar, but writing speed decreases to about 50 MB/s. In this case, it is possible to scale to a larger number of threads until the limit of the disk is reached. The speedup depends on the workload and on the performance of the output device. On the right of figure 6, where only compressed data is written out, the fast NVMe SSD yields four times faster performance than a standard HDD.

3.3. Concurrency Improvements

Other concurrency improvements within ROOT were made possible by the development of the features just presented, such as reducing the usage of mutex locks in ROOT's interpreter where it is possible. After moving to a recent version of ROOT with these optimizations and using the `TBufferMerger` class, visible performance improvements have been achieved in CMSSW's output module [7].

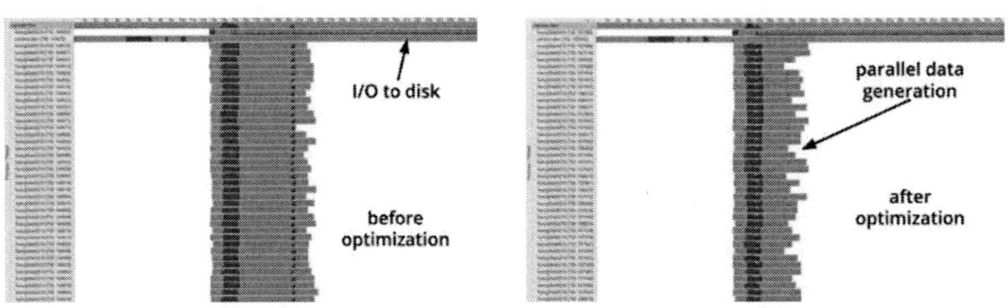

Figure 7. Concurrency optimisations in ROOT I/O.

In order to illustrate the effect of these optimizations, figure 7 shows VTune [8] screenshots of two runs of the benchmark on the right of figure 6, which were run on a 36-core Intel Xeon server, one before the optimisations (left), and another after (right). Each stripe represents one thread of execution (36 threads in total for this benchmark). Brown regions denote that the CPU is doing useful work and green regions denote that a thread is currently running, but not doing useful work. The single-threaded region at the beginning represents ROOT's startup, the short threads represent data generation, and the long thread at the end is the output thread writing to disk. Data generation threads are shorter after the optimisations due to the reduction in the usage of mutex locks. However, from figure 7 it is clear that what affects total running time the most is the single-threaded startup phase and the speed of the disk.

3.4. Parallel Merging of Files with `hadd` *Tool*
Finally, ROOT has incorporated several improvements to perform I/O in parallel, but users may still have old data in multiple files that could be processed and merged into a single file for further analysis. Since version 6.10, ROOT provides an option to the `hadd` tool for merging files in parallel.

4. Conclusion
Implicit parallelism has been introduced in ROOT beginning with version 6.06. Since then, many improvements have already been made to make ROOT I/O more performant. However, much work still lies ahead of us, if we want to cope with the expected ten-fold increase in luminosity at the LHC in the near future—which inevitably means similar increases in data production rate. For instance, in ROOT 7 old interfaces will be reformed, and in ROOT 6 new features are being added, such as more compression algorithms. Further optimizations are also planned in ROOT's interpreter and libraries. Nevertheless, current benchmarks show that ROOT is already not a bad performer when I/O is considered. Other work [9] has demonstrated that ROOT fares quite well if not even leads in performance when compared with alternatives like HDF5, protobuf, Parquet, etc, all that while offering many features not yet available in these alternative formats. ROOT is hence in the right path to continue to support the largest physics experiments in the next decade and beyond.

References
[1] Brun R and Rademakers F ROOT Object Oriented Data Analysis Framework *New computing techniques in physics research V. Proceedings, 5th International Workshop, AIHENP '96, Lausanne, Switzerland, September 2-6, 1996*
[2] CERN Computing Page `https://home.cern/about/computing`
[3] CMSSW Framework `https://github.com/cms-sw/cmssw`
[4] Sodani A Knights landing (KNL): 2nd Generation Intel® Xeon Phi™ processor *2015 IEEE Hot Chips 27 Symposium (HCS)* pp 1–24 URL `http://ieeexplore.ieee.org/document/7477467`
[5] `TBufferMerger` `https://root.cern.ch/doc/master/classROOT_1_1Experimental_1_1TBufferMerger.html`
[6] `TFile` `https://root.cern.ch/doc/master/classTFile.html`
[7] Riley D CMS Update *ROOT I/O Workshop, Oct 2017* URL `https://indico.fnal.gov/event/15154`
[8] Intel® VTune™ Amplifier XE `https://software.intel.com/en-us/intel-vtune-amplifier-xe`
[9] Blomer J A quantitative review of data formats for HEP analyses *18th International Workshop on Advanced Computing and Analysis Techniques in Physics Research (ACAT 2017)* URL `https://indico.cern.ch/event/567550/contributions/2628878`

IOP Conf. Series: Journal of Physics: Conf. Series **1085** (2018) 032015 doi:10.1088/1742-6596/1085/3/032015

Enabling High Availability Service with oVirt Virtualization and CephFS

M D Poat[1], J Lauret[1]
[1] Brookhaven National Laboratory, P.O Box 5000, Upton, New York 11973-5000, USA

Abstract. The online computing environment at STAR has generated demand for high availability of services (HAS) and a resilient uptime guarantee. Such services include databases, webservers, and storage systems that user and sub-systems tend to rely on for their critical workflows. Standard deployment of services on bare metal creates a problem if the fundamental hardware fails or loses connectivity. Additionally, the configuration of redundant failover nodes (a secondary Web service for example) requires constant syncing of configuration and content and sometimes manual interaction for switching hardware (DNS name comes to mind). For this project, we will focus on two tools: oVirt and Ceph. oVirt is an OpenSource virtualization management application that enables central management of hardware nodes, storage, and network resources used to deploy and monitor virtual machines. oVirt supports the deployment of a virtual environment for your data center leveraging automatic provisioning, live migration, and the ability to easily scale the number of hypervisors. oVirt enables the use of multiple storage technologies where you can store virtual machines, images, and templates within one or multiple storage systems. STAR's recent efforts focused on deploying a CephFS POSIX compliant distributed storage system, we would enable the ability to couple our Ceph storage with the oVirt virtualization management system. When designing an intricate system for hosting critical services, it is a requirement to circumvent single points of failure. This work will involve the testing and viability of such an approach along with test cases for high availability, live migration, and service on demand.

1. Introduction

The Relativistic Heavy Ion Collider (RHIC) is the only heavy ion collider in the USA. The Solenoidal Tracker at RHIC (STAR) experiment is the main running detector in operations since year 2000. During the RHIC running period, STAR operations have generated strong demand for high availability of services (HAS), guaranteed uptime, and overall strong reliability of our compute infrastructure. Currently many of our critical services such as our Webserver (serving all monitoring for the STAR experiment), Identity Provider (IDP) used for Single Sign On, and our database servers are all run on bare metal nodes with sometimes, a failover node providing a backup service. This setup creates a problem if the fundamental hardware does go down as the failover mechanisms typically require manual intervention and downtime usually occurs. Virtualization is a well-known established technology that enables portability to any working enclave. By moving towards a virtualized environment, STAR would gain the ability to easily take back-ups, clones, snapshots, and templates of our production virtualized systems. Virtual machines can be deployed on thin provisioned virtual disks and can be stored in a common shared storage such as CephFS.

2. CephFS

Ceph is a distributed storage system based on RADOS (Reliable Autonomic Distributed Object Store) [1]. Within STAR, we have deployed a 30 node, 240 TB raw storage CephFS cluster offering our users 80 TB of redundant safe storage (replication 3). In 2015, we published our first Ceph related research which was based on OpenStack Swift Object Storage and Ceph Object storage studying the architecture and IO performance of the two, the research was then followed by the deployment and study of CephFS which is the POSIX compliant layer on top of Ceph Object Storage, this work was presented at the CHEP 2015 conference [2]. The work shown at CHEP 2015 was then followed with a contribution for ACAT 2016 [3] where we studied the individual data placement techniques built into Ceph such as Primary Affinity, Journals on SSDs, and Cache Tiering. With this work we found that each technique had its own draw back and that Ceph would only perform as fast as its slowest component. Lastly, at the most recent CHEP conference (CHEP 2016) we published an additional research where we deployed SSD's in Ceph at a lower level in the software stack using disk caching techniques such as bcache and dm-cache [4]. We found that the performance of the SSDs laid on top of HDDs to act as a cache would outperform the bare HDDs when tested in a standalone environment. However, SSDs over HDDs tested as OSDs in Ceph the performance of the bcache/dm-cache devices performed below the plain HDD OSDs. We found that, in Ceph, drives that have Power Loss Protection (PLP) support will perform better in Ceph that drives without PLP support. The SSD drives that we used did not have PLP support but the HDDs did have PLP support. In the end, we concluded that the best configuration for our Ceph cluster was to use our 2 TB PLP supported HDDs as we gain maximum storage capacity with very fast write speed to Ceph.

3. oVirt

oVirt is a virtualization management application based on Red Hat Enterprise Virtualization (RHEV) [5]. oVirt has high availability features and a scalable architecture that may enable STAR to fulfil our niche requirements for our online enclave. Additionally, oVirt has support for multiple storage technologies such as our famed CephFS along with GlusterFS, NFS, iSCSI, and more. The oVirt framework consists of two main components, the oVirt engine and oVirt hypervisors. The oVirt engine provides the management interface and is responsible for managing hardware nodes, network resources, and is responsible for deploying and monitoring virtual machines that run on the hypervisor nodes. The hypervisor nodes are essentially "worker nodes" as they follow suite of requests from the oVirt engine.

Our initial plan for oVirt is to configure a resilient cluster with no single point of failure. In order to achieve this, we would reuse some of our over provisioned nodes to deploy our highly available virtual machines. We would enable the proper power management requirements on our nodes along with UPS backup power (IPMI or iDRAC) [6]. Our CephFS distributed storage system would be used as our main storage backend to store all virtual machines and content. Lastly, we would be deploying the self-hosted engine which will be discussed later in this paper.

4. Standard oVirt Deployment

A standard oVirt deployment typically requires 3 nodes at minimum in order to ensure some level of high availability. First you choose one node as the oVirt engine; you then install a required set of

IOP Conf. Series: Journal of Physics: Conf. Series **1085** (2018) 032015 doi:10.1088/1742-6596/1085/3/032015

packages from the oVirt repository [5], followed by running the engine-setup initialization tool that is required for preparing the node. Once your oVirt engine is all setup you can then deploy your hypervisor nodes (in our case we call them ovirt1.star and ovirt2.star). The oVirt engine web interface enables you to deploy your hypervisor nodes right from it, this feature is intuitive especially if you have a large set of nodes to deploy. Once all of the physical nodes are setup, we can then add CephFS in as the main Data and Image storage domains. Lastly, we can then create virtual machines that will be managed by the oVirt engine and can run on either ovirt1.star or ovirt2.star.

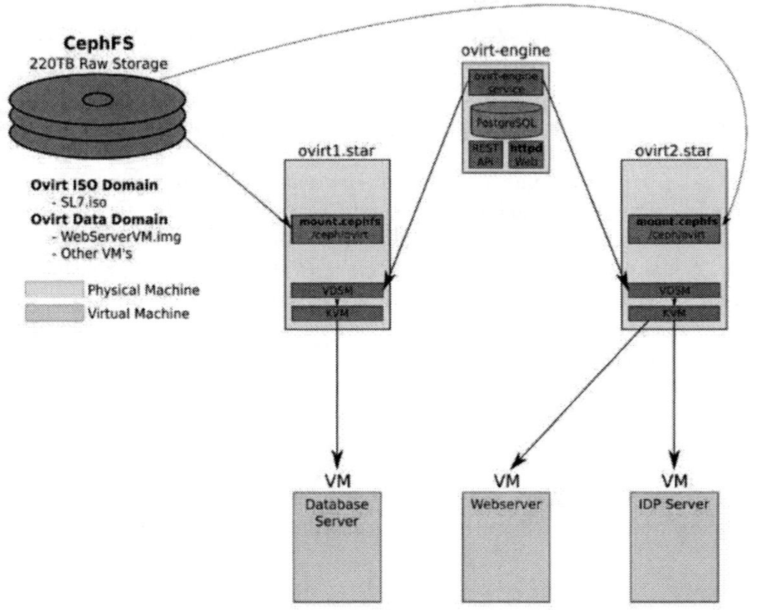

Figure 1. The diagram above is showing a Standard oVirt Deployment that is widely used within the oVirt community. The oVirt-engine is shown as the master node and the ovirt-engine service is shown managing the Virtual Desktop and Server Manager (VDSM) on the hypervisor nodes. CephFS is the main storage domain which is required to be mounted on all hypervisor nodes.

As shown in Figure 1, a Standard oVirt Deployment has been visually described and it can be seen how each service is implemented into the oVirt cluster. Within this deployment, if one of the hypervisor nodes (ovirt1.star or ovirt2.star) were to suffer a downtime, the running virtual machines on the failed hypervisor would failover to the other running hypervisor. The virtual machine may suffer a 1-2 minutes downtime, but it would soon recover. While a Standard oVirt Deployment may seem intuitive at first glance, there is a flaw in the setup. If, for example, the oVirt-engine machines were to suffer a failure and go down, the running virtual machines would become headless and there would be no control (i.e. no failover, no migration, no monitoring, etc.) over the hosts or virtual machines. If the oVirt-engine machine suffered a full catastrophic failure, recovering the oVirt environment would require downtime of the entire system and would not be pleasant.

5. Self-Hosted Engine Deployment

5.1 Deployment

The oVirt engine is the control center of the oVirt environment. The service talks directly to VDSM on the hosts to deploy, start, stop, migrate and monitor VM's [7]. Considering how imperative the oVirt engine is, it seems counterintuitive to run this service on a bare metal node as well. In oVirt, there is a feature called the Self-Hosted Engine. A self-hosted engine is a virtualized environment in which the oVirt Engine software runs on a virtual machine on the hosts managed by that engine [8]. Initially this may sound convoluted, but it works quite well. Essentially you have a physical host, that runs a virtual machine, the oVirt-engine software is installed on the virtual machine, the virtualized oVirt-engine then comes back to manage the initial host to run more highly available virtual machines. While the Self-Hosted engine may seem overly complex and unnecessary, the result in the end is worth it. The Self-Hosted engine needs at least two physical nodes that will both run software instances called 'ovirt-ha-agent' and 'ovirt-ha-broker'. The two nodes will negotiate using the ovirt-ha-agent to determine which host should run the self-hosted engine VM based on an internal scoring system, the host with the highest score will run the VM. The ovirt-ha-broker is responsible for monitoring the resources on the physical hosts themselves (CPU, load, memory, etc.).

5.2 Gluster

At the time of writing this paper the latest recommended version oVirt is version 4.1. oVirt 4.1 supports the Self-Hosted Engine feature, however there is no support for CephFS to act as the storage domain to store the Self-Hosted Engine virtual machine. When running through the deployment of the Self-Hosted Engine, the supported storage domains are GlusterFS, ISCSI, FC, NFSv3 or NFSv4. While CephFS can still be used as the storage domain for all other virtual machines, for our Self-Hosted oVirt deployment the next best thing would be a replicated GlusterFS storage system. GlusterFS is a scalable network filesystem suitable for data-intensive tasks such as cloud storage and media streaming [9]. GlusterFS is free open source software that can run on commodity hardware. Considering high availability as our main goal for our oVirt deployment, we decided to use our 3 physical hosts that will run our oVirt environment to also run a replication 3 GlusterFS storage system. We added 3 disks per machine, creating a 9 disks GlusterFS with replication 3. One out of three disks per machine is set as an arbiter volume; arbiter volumes only store file/directory names and metadata and do not store any actual real data. Enabling an arbiter volume will prevent a common issue in Gluster called "split-brains" [10]; a file is in split-brain when there is an inconsistency in either data or metadata and Gluster does not have enough information to ensure the integrity of the file (i.e. latest copy, correct metadata, etc.) . Split-brains can only really occur when there is a failure event or a simple reboot of one of the storage nodes without any arbiter volumes [11]. With proper configuration, we enabled our GlusterFS to seamlessly suffer a single node downtime with zero interruption of access to the file system. If we suffer a two nodes failure or downtime, the GlusterFS becomes read-only. After recovery of any downed hosts, GlusterFS will repair itself immediately to full replication 3.

IOP Conf. Series: Journal of Physics: Conf. Series **1085** (2018) 032015 doi:10.1088/1742-6596/1085/3/032015

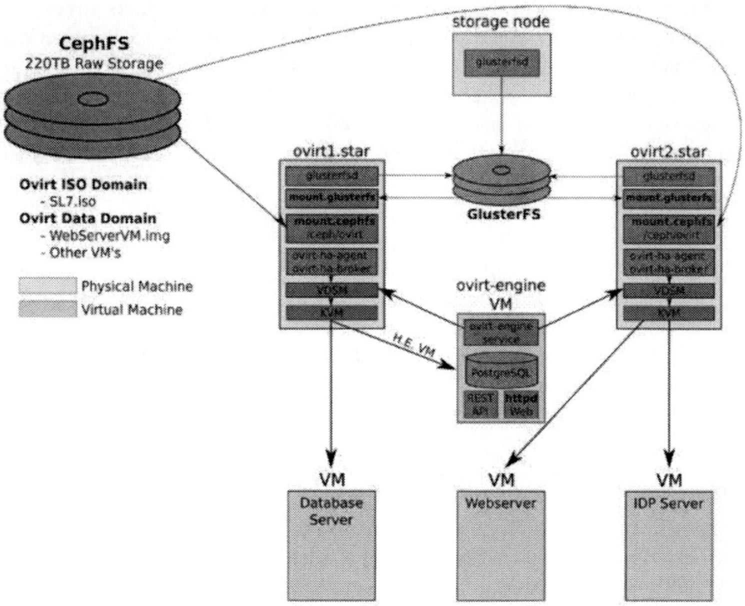

Figure 2. The diagram above is showing a Self-Hosted Engine oVirt Deployment where the oVirt engine is a highly available virtual machine itself. The GlusterFS filesystem is shown setup on ovirt1.star, ovirt2.star and a 3rd host titled storage node. The oVirt engine VM is stored in GlusterFS, the remaining VM's are all stored in CephFS.

As seen in Figure 2, we are showing a diagram of a Self-Hosted oVirt deployment. This setup first entails creating our 3 hosts replication 3 GlusterFS filesystem, followed by setting up our hypervisor nodes as GlusterFS clients. Once your initial storage is setup, we can move to the Self-Hosted engine deployment. First, we install ovirt-hosted-engine-setup and ovirt-engine-appliance packages from the oVirt repositories on our hypervisor nodes, once installed we then follow the simple deployment steps from running 'hosted-engine --deploy' which will take you through the setup to create the virtual machine and oVirt engine. The deployment setup also ensures the ovirt-ha-agent and ovirt-ha-broker packages are properly installed. Once the Self-Hosted oVirt engine is setup, the next steps are the same as any standard oVirt deployment. We simply deploy additional packages and configuration to any additional hypervisors nodes via the oVirt engine web interface, we re-add CephFS as our main storage domain, and from here we can create and deploy our highly available virtual machines for STAR production use case.

The Self-Hosted oVirt engine is a much more resilient setup compared to a standard oVirt engine deployment on bare metal. The reasons for this statement are as follows. When referring back to Figure 2, if we were to suffer a failure from our 'Storage Node' (which acts as one of the three GlusterFS hosts), then GlusterFS would continue working with replication 2 between ovirt1.star and ovirt2.star. In a different scenario, if we were to suffer a failure from ovirt2.star (which is running the two VM's titled Webserver and IDP server), then the oVirt-engine would immediately recognize the failed VM's and host, oVirt-engine would ensure ovirt2.star is actually down followed by restarting the downed VM's on ovirt1.star as a fail-over. Finally, in our last scenario if we were to suffer a failure from ovirt1.star (which is running the oVirt-engine VM itself and our Database VM), the ovirt-ha-agent and ovirt-ha-broker services running on ovirt2.star would recognize the oVirt-engine VM is down and would restart the VM on ovirt2.star. Once the oVirt-engine VM and services have recovered, any other downed VM's (Database VM) would then be restarted on ovirt2.star as well. With multiple passes of live testing of this configuration and successful fail-over mechanisms, it has been proven that a Self-Hosted oVirt engine setup truly is a highly available system.

6. Live Migration

Live migration is a feature of the KVM hypervisor that enables you to transfer live running virtual machines from one host to another with practically 0 down time [12]. With live migration, the virtual machine remains powered on and all user applications continue to run after the relocation/migration to the new host. The virtual machine RAM is actually copied from the source host to the destination host. In oVirt, live migration is a critical feature as it enables you to migrate production VM's to from one hypervisor to another hypervisor with no interruption. If a host needs to be taken out for maintenance or repair, it can be done without interrupting users or services.

To test live migration and to ensure that all acclaimed features work properly, we created a small test case demonstrating continuous functionality of a Web service. The test was to create a multi-stage CGI (a first stage passes to a second stage hidden parameters), making sure that between the two stages we a live migration and verify the information passed is preserved. In the CGI we made an entry in the first page, then migrated the virtual machine from one host to another, once migrated we would make the final entry and check the results. The result was that the entries from both the first page and second page are carried over properly. This test shows that virtual machine's RAM is properly copied during a live migration.

7. High Availability

High Availability is a feature in oVirt to ensure your virtual machines are always up [13]. To properly enable high availability there are few requirements that need to be met. First, your physical hypervisor nodes must have a power fencing interface such as an IPMI interface, iDRAC or the like [6]. An IMPI/iDRAC interface enables access to the baseboard management controller to manage the hosts low level controls such as the BIOS, hardware controls, and most importantly power management. The oVirt-engine requires access to this IPMI/iDRAC interface in order to power fence the host if needed, typically in cases of the host not responding or if the host power status needs to be verified. Furthermore, to enable high availability you must have equal available resources on your destination host for failover events. For example, if your highly available virtual machine has 4 CPU cores and 8 GB of memory, then there must be 4 CPU cores and 8 GB memory free on any destination host if you want to the virtual machine to successfully failover. Lastly, your hypervisors must be running the same CPU architecture (x86_64, etc.) and CPU family (Haswell, Sandy Bridge, etc.).

In STAR's oVirt environment we tested high availability in a live running oVirt cluster. The test was to simply pull the power chord for ~5 seconds on a hypervisor node running a highly available virtual machine to see how oVirt would react. Immediately after the power chord was removed, the oVirt-engine recognizes the downed VM and tries to restart it on another hypervisor. Additionally, the oVirt-engine checks the status of the downed hypervisor hosts via the IPMI/iDRAC interface, if the status of the machine returns that power is available but the host is not running, oVirt-engine will power fence the host to restart it. Meanwhile, while the hypervisor host is being power fenced, the virtual machine should be back up and running within a minute from the initial power chord removal. Lastly, within 3-5 minutes, the downed hypervisor host should be back up and running and re-added to oVirt as an available hypervisor host.

8. Conclusion

oVirt virtualization, a community free project branching from RHEV, will enable STAR to create a true highly available system helping to support our critical services. When deploying oVirt with the self-hosted engine feature, all single points of failure are eliminated. While the self-hosted engine does not support CephFS as its storage domain at this time, we have looked into this lack of support and have opened up a feature request with Red Hat to add CephFS as a self-hosted engine storage domain. However, apart from the self-hosted engine, all other virtual machines and content consequently can be stored in CephFS. For high availability to be enabled on our virtual machines in oVirt, there must be available resources on standby (2x CPU & Memory) in order for the VM to properly failover. This is a fair trade-off to ensure our critical nodes and services are always up. Virtualization will enable STAR to better use our over provisioned nodes to deploy a diverse set of virtual machines to meet our operation needs.

Acknowledgements
This work was supported by the Office of Nuclear Physics within the U.S. Department of Energy's Office of Science.

References

[1] Ceph http://www.ceph.com
[2] Poat M D, Lauret J and Betts W 2015 *POSIX and Object Distributed Storage systems – Performance Comparison Studies With Real-Life Scenarios in an Experimental Data Taking Context Leveraging OpenStack Swift & Ceph* J. Phys.: Conf. Ser. 664 042031
[3] Poat M D and Lauret J 2016 *Performance and Advanced Data Placement Techniques with Ceph's Distributed Storage System* J. Phys.: Conf. Ser. 762 012025
[4] CHEP 2016
[5] oVirt https://www.ovirt.org
[6] Integrated Dell Remote Access Controller 8 (iDRAC8) http://www.dell.com/support/manuals/us/en/19/idrac8-with-lc-v2.05.05.05/idrac8_2.05.05.05_ug-v1
[7] oVirt Engine Architecture https://www.ovirt.org/documentation/architecture/architecture/#engine
[8] oVirt Self-hosted engine deployment https://www.ovirt.org/documentation/self-hosted/chap-Introduction/
[9] Gluster https://www.gluster.org/
[10] Split Brain https://gluster.readthedocs.io/en/latest/Administrator%20Guide/Split%20brain%20and%20ways%20to%20deal%20with%20it/#split-brain
[11] Arbiter volumes and quorum options http://docs.gluster.org/en/latest/Administrator%20Guide/arbiter-volumes-and-quorum/
[12] Live Migration https://www.ovirt.org/documentation/vmm-guide/chap-Administrative_Tasks/#migrating-virtual-machines-between-hosts
[13] High Availability https://www.ovirt.org/develop/release-management/features/sla/ha-vm-reservation/

IOP Conf. Series: Journal of Physics: Conf. Series **1085** (2018) 032016 doi:10.1088/1742-6596/1085/3/032016

Three dimensional Generative Adversarial Networks for fast simulation

F Carminati[1], A Gheata[1], G Khattak[1], P Mendez Lorenzo[1], S Sharan[1] and S Vallecorsa[1,2]

[1] CERN, Geneva, Switzerland

[2] Gangneung-Wonju National University, Gangneung, South Korea

E-mail: sofia.vallecorsa@cern.ch

Abstract. We present the first application of three-dimensional convolutional Generative Adversarial Network to High Energy Physics simulation. We generate three-dimensional images of particles depositing energy in high granularity calorimeters. This is the first time such an approach is taken in HEP where most of data is three-dimensional in nature but it is customary to convert it into two-dimensional slices. The present work proves the success of using three dimensional convolutional GAN. Energy showers are well reproduced in all dimensions and show a good agreement with standard techniques (*Geant4* detailed simulation). We also demonstrate the ability to condition training on several parameters such as particle type and energy. This work aims at proving that deep learning techniques represent a valid fast alternative to standard Monte Carlo approaches. It is part of the *GeantV* project.

1. Introduction

High Energy Physics (HEP) software and simulation software, in particular, are going through an important phase of restructuring and optimisation for new computing architectures, in order to cope with the expected High Luminosity LHC computing needs [1]. The traditional simulation strategy is based on Monte Carlo methods. This is a time-consuming procedure requiring a large amount of computation. In order to increase the performance of simulation applications by one order of magnitude, it is important to study new methods and algorithms as a complement to the basic work of parallelizing and optimizing the existing code. Fast simulation approaches are already used by the particle physics community to reduce computation time. They are typically parameterized on physics quantities previously calculated or measured. This approach suffers from being specific to an individual experiment and detector and although faster than full simulation, it too would benefit from a performance improvement. Our proposal is to leverage state of the art deep learning algorithms to design a new, generic, fast simulation tool. The recent developments in deep learning provide a very promising avenue to replace complex algorithms with a suitably complex-structure deep neural network that is able to reproduce the results of the former at a much higher speed. The use of machine learning methodologies in HEP is not a new approach [2, 3] and a lot of work has been done in this field. Machine learning can provide also a fast alternative to simulation, where a reduced accuracy can be acceptable.

The idea is to treat simulation as a black-box and replace detailed Monte Carlo simulation with a machine learning-based tool that can be trained on a range of particle types, momenta and positions and detector configurations. Inputs are provided by elementary particles emerging from a collision or coming directly from a particle beam. Each particle is characterized by type, energy, direction of flight and the position it enters a detector, leading to 8-9 parameters per particle. More complex input types can also be considered such as *jets*, sprays of particles moving along a cone, or even whole *events*, collections of all particles emerging from the same primary collision. On the output side, there is also a large range of possibilities to consider. The most basic is the energy deposition in various parts of the detector. More complex outputs, such as detector response signals from data acquisition electronics can also be studied, as well as the final reconstructed particle quantities. Depending on the choice of input and output, the number of parameters may turn out to be very large and different techniques such as feature reduction will need to be applied to reduce dimensionality (and reduce learning time) while preserving correlations. Generative models, like Generative Adversarial Networks [4], seem particularly suited to replace Monte Carlo simulation. The production of realistic samples is straightforward and these techniques can model complicated probability function and deal with multi-modal outputs. Moreover, their ability to perform interpolation and to recover missing data has been already documented [5, 6]. These two features are particularly useful when dealing with particle physics simulations. This paper is structured as follow: after a short introduction about related work in section 2, the next section 3 introduces the calorimeter data set followed by a detailed description of the designed model in section 4. The adversarial training methodology adopted for the project is described in section 5. Preliminary results are presented in section 6. Section 7 concludes the discussion.

2. Related Work

Generative Adversarial Networks (GAN) consist of two networks, a generator and a discriminator, competing against each other [4]. The discriminator is constantly trying to catch the fake output produced by the generator, while the generator tries to reproduce images that look as realistic as possible. The discriminator should be updated in such a manner so as to be able to differentiate between the true data and the fake output produced by the generator. *Goodfellow et al.* [4] has shown that there is a unique solution to the problem with the generator covering the entire data distribution and discriminator getting totally confused [4]. Initially GANs were introduced as Multi Layer Perceptron networks [4]. Since then a lively research and many different applications have developed on the same subject, in particuaar in the field of image processing. A three-dimensional approach is a recent development [7] proving that adversarial training can successfully generate complex three dimensional structures. In HEP, the output of some detectors, e.g the calorimeters, can be also be interpreted in the form of images, thus the same techniques that are used for image recognition can be employed for detector output analysis. Similarly, image generation techniques can be used for detector simulation. The LAGAN or (Location Aware GAN) [8] and CaloGAN [9] generate the ATLAS Liquid Argon calorimeter output as a sequence of three two-dimensional images. Locally connected layers were used as well as propagation of images from previous layer to the next layer [8, 9]. The present work is intended to move further along this direction, simulating three-dimensional calorimeter showers as a whole.

3. Input data set

The data set that we use for our studies, was produced in the context of the CLIC detector design [10] and it is available under the DD4hep software framework [11]. It consists of energy showers generated with the *Geant4* software [13] inside the CLIC electromagnetic and hadronic calorimeters detector prototype [10]. This is an example of next generation highly granular calorimeters and therefore it represents a demanding use case simulation will have to face during High Luminosity LHC runs. This data set was produced in an effort to provide the HEP community with a common benchmark to perform tests, development and optimization of different machine learning techniques. Details of this work are available [12]. Here, we just want to highlight its most important features. The electromagnetic calorimeter (ECAL) is a sampling calorimeter consisting of 25 layers of Tungsten absorber with silicon sensors in between them. The first 17 absorber layers have a thickness of 2.4 mm, while the other 8 layers are twice as thick. Each plane of silicon sensor is 0.5 mm thick and it is segmented in (5.1 mm x 5.1 mm) cells. The inner radius of the calorimeter is 1.5 m. The hadron calorimeter (HCAL) is located behind the ECAL. Its barrel section consists of 60 layers of Steel and polystyrene scintillator sensors. Each sensor layer is segmented in (3.0 cm x 3.0 cm) cells, i.e. each HCAL cell corresponds approximately to a layer of 6x6 ECAL cells. Our study focuses on the electromagnetic calorimeter data, as a first step. Individual particles (electrons, photons and pions) travel through the tracking device and hit the inner surface of the electromagnetic calorimeter. An individual entry in the data set represents the energy deposit produced by the shower of particles generated by the collision of the incoming particle inside the detector. Each cell is characterized by the energy recorded in it and three indices (iX, iY, iZ), identifying the position of the cell. For each of the 25 ECAL sensor planes, a 25x25 array of cell is defined. The information stored for the ECAL shower is then E(iX, iY, iZ) for each of the 25x25x25 cells around the barycentre. The events are stored in a set of compressed HDF5 files. A 25x25x25 numpy array labeled ECAL includes the 25x25x25 energy deposits around the shower barycentre on each of the 25 planes. The three indices correspond to (iX, iY , iZ). The event also includes the information on the nature of the incoming particle and its energy. About 1M showers have been generated for each particle kind (electrons, photons and neutral pions). The energy of the incoming particle is sampled from a uniform [0-500 GeV] spectrum.

4. Models architecture

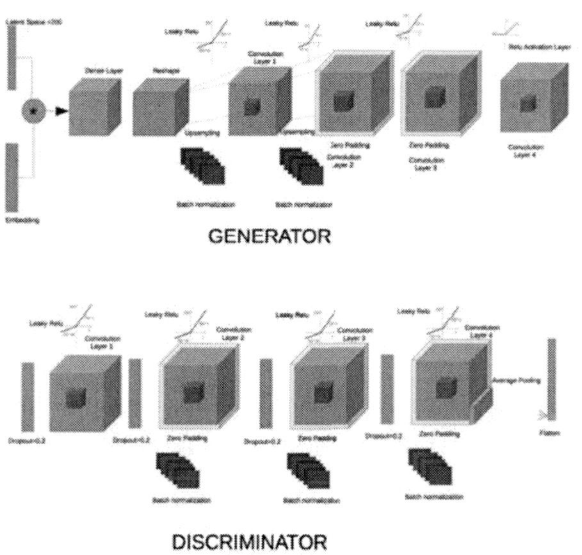

Figure 1. The three dimensional GAN model

Our model is implemented using keras [14] and tensorlow [15]. In the initial implementation, a three-dimensional reconstruction is implemented, incorporating image class labels (GAN images can be conditioned on some auxiliary input like particle type) [16]. The volumetric space for the image is 25x25x25 corresponding to the ECAL cells in the data set. A three dimensional, volumetric approach preserves the correct spatial relationships for the shower energy depositions, but it has the disadvantage of a much larger number of parameters. This difficulty is addressed by designing networks as simple as possible and by keeping the training times within acceptable limits. The generator, as well as the discriminator, consist of four convolutional layers as shown in figure 1. Leaky rectified linear unit activation functions are used, except for the last layer in the generator, where a rectified unit is used instead. Batch normalization layers are also included to improve performance. The generator uses a latent noise vector from a uniform probability distribution as well as the chosen class of the image to generate. The image class vector is embedded into the latent noise vector, thus implementing class conditioning on the generated images. An initial dense layer and two upsampling layers are also added to get the desired image dimensions of 25x25x25. The discriminator input is a batch of images to be discriminated. There are two categorial outputs activated by sigmoidal activation functions on the final flattened layer. The first output is the binary real/generated flag. The second output denotes the image class (particle type). In the present effort two classes electron and photon are considered making the second output also binary.

5. Adversarial training strategy

In the adversarial framework, the discriminator and the generator are trained against each other [4]. The training strategy follows a rather standard approach already described in [9].

Initially the generator is used to generate images from random class labels. The first batch of images provided to the discriminator are actual images from the data set and the discriminator is trained to make the correct prediction. It is then trained a second time on fake images. In order to keep the system balanced, the generator is also trained twice. The generator loss is defined as the discriminator loss when all generated images are real, effectively training the generator to reduce this loss. The networks are trained for 30 epochs on 100000 electrons of 100 GeV energy. This configuration is used to perform also a second experiment in which the primary particle energy is used to condition the training. The generator input noise vector is multiplied by the primary particle energy and the discriminator outputs are changed. The first output tells us whether the image is real or fake and the second output provides a floating-point value, predicting the input energy of the particle. A combined loss function is built as the sum of three terms: a binary cross entropy term accounting for the binary real/fake output, two mean relative error terms measuring how consistent is the discriminator energy prediction with respect to the primary particle energy and the sum of energy depositions generated in the calorimeter cells. It is important to note that, in this case, we are adding an energy regression test to the discriminator network. While not being the primary scope of this work, the regression test is an important cross check for our method, as it proves that we are capable of generating particle showers corresponding to specific energy values. In this second experiment, we train the GAN models on 200000 electrons generated with a uniform energy between 0 and 500 GeV. Training on a single NVIDIA GeForce GTX 1080 for 30 epochs requires slightly less than a day.

6. Preliminary results

The results are in good agreement with detailed simulation data. Energy showers are well reproduced in all dimensions and show a reasonable agreement with standard Monte Carlo techniques (*Geant4*). A detailed study to assess the quality of the GAN images, using typical high level calorimeter reconstruction variables is ongoing. In the meantime, we have studied the single cell energy response fitting it to a Gaussian: we obtain that the energy mean value and the standard deviation agree well within the statistical errors with the detailed Monte Carlo simulation results. In particular, the standard deviations, that measure the cell energy resolution, agree within 6% for most of the highly-populated cells (traditional accuracy of fast simulation models does not go below 10% [17]). Figure 2, on the left, represents an example of a three-dimensional image generated by the GAN for a 100 GeV electron. On the right, the shower average longitudinal section obtained from 1000 electrons is shown. The GAN generated showers are compared to the results of Geant4 full simulation. Table 1 summarises the preliminary

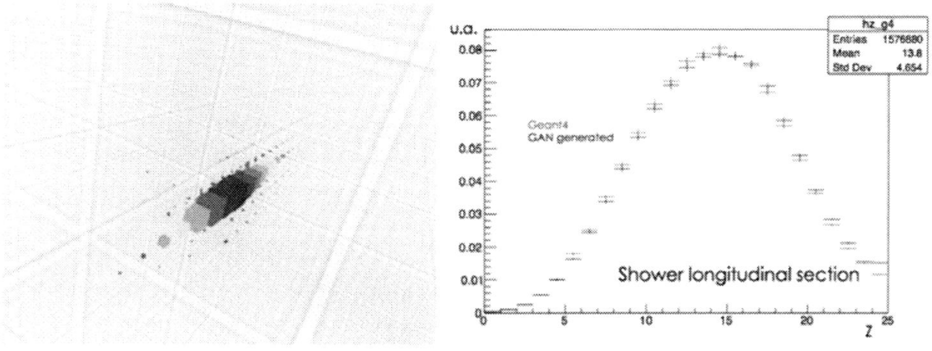

Figure 2. (left) The three-dimensional representation of an energy shower created by a 100 GeV electron as generated by the GAN, using particle type as conditioning information. (right) Longitudinal shower shapes for 100 GeV electrons: GAN result is compared to full *Geant4* simulation. The Z coordinate indicates the bin index in the longitudinal direction.

results obtained from the discriminator energy regression test. The performance of the network is remarkably good considered that no effort was made to optimise the network architecture for different energy values. The convolutional layers parameters were, in fact optimised to reproduce at best 100 GeV particles and, consistently, the best performance is achieved for showers generated at 100 GeV. This is expected since convolutional layers are sensitive to the spatial shape of the energy showers, which changes according to the primary particle energy.

Table 1. Preliminary discriminator energy regression test results. The discriminator associates an energy value to every image generated by the generator. For each energy point we generate 1000 particles and calculate mean value and spread for the discriminator response.

Primary Energy (GeV)	Discriminator Mean Energy (GeV)	Standard Deviation (GeV)
50	57	14
100	99	2
200	190	10
300	290	5
400	378	6
500	465	7

From the computing performance perspective, we run a very simple test comparing the time needed to generate a single shower using *Geant4* and the trained 3d GAN network: *Geant4* full simulation on an Intel Xeon E5-2683 processor takes approximately one minute, 3d GAN on the NVIDIA GeForce GTX 1080 just 0.04 milliseconds.

7. Conclusion

This study shows that Generative Adversarial Networks are a good candidate for fast simulation of high granularity detectors, typically forseen for the next generation accelerators. We have successfully generated images of shower energy deposition in three dimensions, including energy related information. At the time of this submission, the plots and results are still preliminary. We plan to improve the performance of the network by implementing large hyper-parameter scans and optimise network architecture according to the particle related quantities and detector geometry. This study is a first proof of concept for a much larger plan intended to provide a generic Deep Learning tool for fast simulation, to be integrated in existing simulation software, such as *Geant4*, or future projects, such as *GeantV* [18]. This work is part of the *GeantV* project.

8. Acknowledgement

The authors wish to acknowledge the contribution of Intel to the *GeantV* project through the Intel Performance Computing Centre (IPCC) program. We also want to acknowledge the very useful technical contributions of the CERN openlab.

References

[1] Bird I 2016 *WLCG wokshop*
[2] Aad G et al. 2012, *Phys. Lett.* B**716** 1 (*Preprint* hep-ex/1207.7214)
[3] Chatrchyan S et al. 2012, *Phys. Lett.* B**716** 30 (*Preprint* hep-ex/1207.7235)
[4] Goodfellow et al. 2014 (*Preprints* stat-ml/1406.2661)
[5] Odena A 2016 (*Preprints* stat-ml/1606.01583)
[6] Odena A, Olah C, Shlens J 2016 (*Preprints* stat-ml/1610.09585)
[7] Wu J et al 2016. *NIPS conf proc.*
[8] Paganini M et al 2017 (*Preprint* hep-ex/1701.05927)
[9] Paganini M et al 2017 (*Preprint* hep-ex/1705.02355)
[10] Alipour Tehrani, N et al 2017 (*Preprint* CERN/CLICdp-Note-2017-001)
[11] Frank M et al 2014 *J. Phys. Conf. Ser.***513** 022010
[12] M. Pierini 2016 *DS@HEP at the Simons Foundation*
[13] Geant4 project, "Geant4" [software], version 10.3.2, 2017. Available from https://github.com/Geant4/geant4/releases/tag/v10.3.2 [accessed 2017-08-17]
[14] Keras project, "Keras"[software]. version 2.12.2. Available from https://github.com/keras-team/keras/tree/2.1.2 [accessed on 2017-12-12]).
[15] TensorFlow Project, "TensorFlow" [software], version 1.4.0, Available from https://github.com/tensorflow/tensorflow/releases/tag/v1.3.0 [accessed 2017-05-17]
[16] Mirza M, Osindero S. 2014 (*Preprints* abs/1411.1784)
[17] M. Selvaggi 2014 *J. Phys.: Conf. Ser.* **523** 012033
[18] G Amadio et al 2016 *J. Phys.: Conf. Ser.* **762** 012019.

IOP Conf. Series: Journal of Physics: Conf. Series **1085** (2018) 032017 doi:10.1088/1742-6596/1085/3/032017

Pixel readout chip software emulators for the YARR DAQ system upgrade

N Whallon[1], T Heim[2], M Garcia-Sciveres[2], K Potamianos[3], H Oide[4] and S-C Hsu[1]

[1] Department of Physics, University of Washington, 3910 15th Ave. NE, Seattle, WA 98195-1560, USA, [2] LBNL, 1 Cyclotron Rd., Berkeley, CA 94720, USA, [3] DESY, Notkestraße 85, 22607 Hamburg, Germany, [4] INFN, 4 Via Monti Vincenzo, Genova, GE 16156, Italy

E-mail: alokin@uw.edu, timon.heim@cern.ch, mgarcia-sciveres@lbl.gov, karolos.potamianos@cern.ch, Hideyuki.Oide@cern.ch, schsu@uw.edu

Abstract. The Yet Another Rapid Readout (YARR) system is a DAQ system designed for pixel readout chips including the current generation ATLAS FE-I4 chip and the next generation RD53A chip, which is part of the development of new Pixel detector technology to be implemented in High-Luminosity Large Hadron Collider experiments. YARR utilises a PCI-e FPGA card which acts as a simple gateway to pipe all data from pixel readout chips via the high speed PCI-e connection into the host system's memory. All data processing is done on a software level in the host CPU(s), utilising a data-driven, multi-threaded, parallel processing paradigm. YARR has recently been upgraded to interface with software emulators of pixel readout chips. These emulators offer many benefits: quick development of DAQ software; expansion of the developer base to users without access to readout hardware; preparation of DAQ software for upcoming readout chips, such as RD53A; implementation of Continuous Integration and unit tests to ensure code quality and maintainability. The design and capabilities of the FE-I4 and RD53A software emulators will be presented.

1. Introduction

Beyond Run 3, the Large Hadron Collider (LHC) will undergo a major upgrade to produce high instantaneous luminosity. This High-Luminosity LHC (HL-LHC) will be able to deliver 3000 fb^{-1} at $\sqrt{s} = 14$ TeV within 10 years after it begins operations.

The increased luminosity will also result in an increased number of pileup interactions and increased radiation doses to detectors. These factors motivate a complete replacement of the ATLAS [1] inner tracker. The new inner tracker, ITk [2], will require a finer granularity than before in order to resolve the increased number of pileup vertices. As such, the readout bandwith required by the new tracker will increase.

RD53A [3] is a readout chip which meets the criteria of HL-LHC pixel detectors. The YARR [4] data acquisition (DAQ) system is being developed to provide robust data taking and calibration software for RD53A. Before RD53A chips are produced, it is possible to develop the YARR DAQ for RD53A using software emulators. The design and capabilities of pixel readout chip software emulators are described in the following sections.

2. Yet Another Rapid Readout

The Yet Another Rapid Readout (YARR) system is a robust data acquisition (DAQ) system targetting R&D of next generation pixel readout chips, such as RD53A. It is also capable of interfacing with the current generation readout chip FE-I4 [5]. YARR DAQ software runs on a host computer which interfaces with the readout chips via a commercially available PCI-e FPGA card. The custom FPGA firmware implements only a basic buffer and communication blocks. The host computer handles running scan loops, sending commands to the readout chip, and data analysis on the data read back from the readout chip. Since all data processing is handled in software, it is possible to interface YARR also directly with software emulators of readout chips. These emulators need to be able to interpret commands sent by YARR, and produce and send back sensible hit data.

Software emulators of the FE-I4 and RD53A chips are in development primarily as a tool to aid in the development of YARR software. This is especially the case for RD53A, whose specifications are available but which has not yet been produced. The emulators also allow for users without physical readout chips to join the development efforts. Finally, the emulators enable the use of Continuous Integration (CI) for the YARR DAQ software, something which is simply not feasible if hardware is required, and which will ensure code quality and maintainability.

The basic structure of YARR is illustrated in Figure 1.

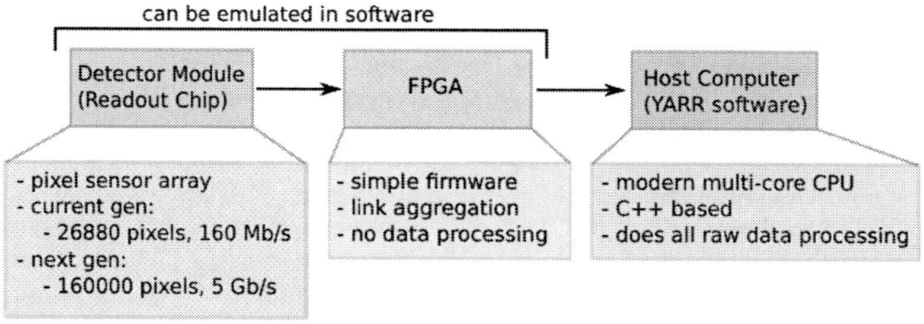

Figure 1. The basic struction of the YARR DAQ system.

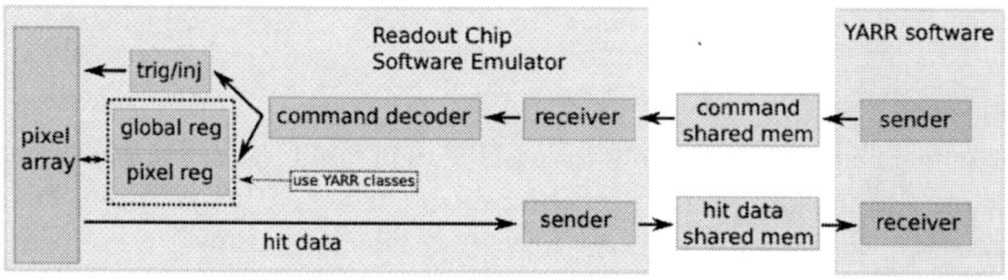

Figure 2. The basic struction of pixel readout chip software emulators.

3. Readout Chip Software Emulators

Software emulators of readout chips must be able to interface directly with YARR, correctly interpret the readout chip-specific commands sent by YARR, and return hit data to YARR when trigger commands are received.

The behavior of readout chips depends on values stored in Global Registers (GR) and Pixel Registers (PR). The GRs contain chip-level settings, such as global threshold voltages or pixel masking information, while the PRs contain pixel-level settings, such as local threshold voltage or local pixel enable. As such, the software emulators instantiate readout chip objects with virtual GRs and PRs in order to store the configuration of the readout chip. These registers are then configured when the emulators receive the appropriate commands from YARR.

When YARR sends trigger commands, the emulators loop over a virtual pixel array, model hit data, and send this data back to YARR in a format identical to what the real readout chips would send. YARR is then able to do the histogramming, analysis, and plotting with this data.

The basic structure of readout chip software emulators is illustrated in Figure 2. The specifics of decoding commands from YARR, configuring virtual GRs and PRs, and modelling hit data for the FE-I4 and RD53A software emulations are presented in the following subsections.

3.1. FE-I4 Command Decoding and Register Configuration
Configuring FE-I4 GRs is straightforward: the FE-I4 emulator receives a `WrRegister` command from YARR, which contains the address of the GR to write to, accompanied by the value to write to that GR. PRs, on the other hand, are configured via a Shift Register (SR), which requires a few extra commands and also relies on values stored in the GR. All PR and SR manipulation is done in double column (DC) loops in order to configure multiple pixels at the same time. How the pixel double columns are looped over is dictated by certain values in the GR. This command decoding and register configuration must all be done to the precise FE-I4 chip specifications by the emulator. A block diagram outlining the decoding is shown in Figure 3.

3.2. RD53A Command Decoding and Register Configuration
The scheme for configuring registers in RD53A is similar to FE-I4, with the exception of the SR. Instead of using a SR to configure pixels, RD53A has a virtual register which points to the address of specific pairs of pixels. If an RD53A `Write_Register` command specifies this virtual register as the register to write to, the paired value will be written directly to the PRs of the pixel pair pointed to by the virtual register, otherwise the paired value will be written directly to that GR. The command decoding and register configuration scheme used by the RD53A emulator is illustrated in Figure 4.

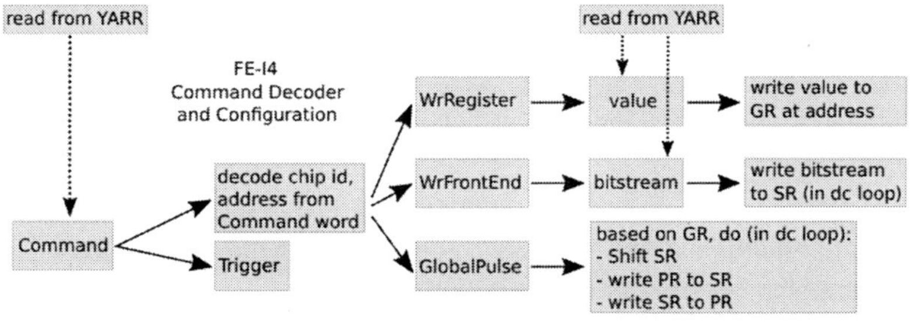

Figure 3. The FE-I4 emulator command decoder.

3.3. FE-I4 Hit Modelling
After chip configuration, scans typically send a series of commands to inject charge into the digital or analog front-end circuitry of pixels and triggers to read back their response. A simplified pixel analog front-end circuit is shown in Figure 5. A threshold voltage and the voltage resulting from the injected charge are routed into a comparator and the pixel will register a hit if the injected charge is above the threshold.

For the FE-I4 chip, the threshold voltage is determined by a local, per-pixel, threshold value combined with a global, per-chip, threshold value, stored in the PRs and GR, respectively. The

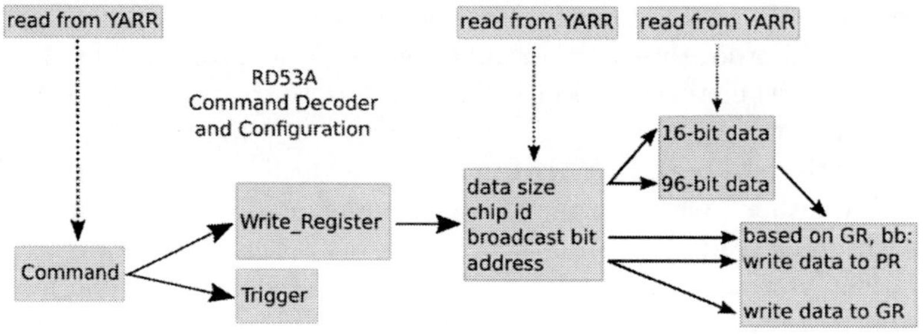

Figure 4. The RD53A emulator command decoder.

main hit metric is the time-over-threshold (ToT). Figure 6 shows an illustration of this process. The FE-I4 per-pixel feedback current adjustment is not modelled in the emulator and a linear charge to ToT conversion is assumed.

The FE-I4 emulator models hit data by comparing a fixed injection charge to a charge derived from the global and local threshold values stored in the GR and PRs, respectively. Per-pixel Gaussian smearing is applied to the local threshold value and Gaussian noise is applied to the injection charge, with the mean and standard deviation of the noise configurable per-pixel.

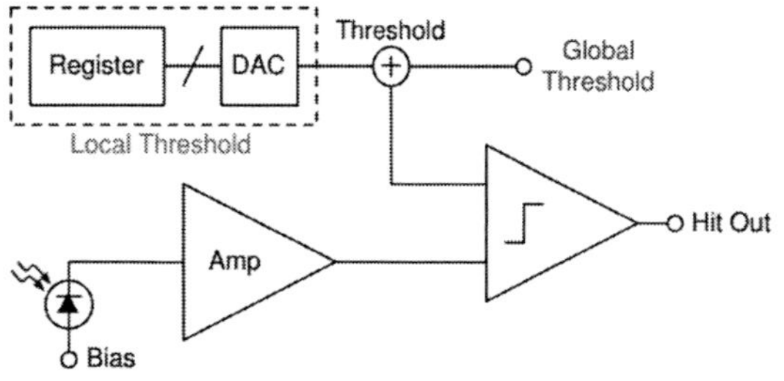

Figure 5. Simplified schematic of FE-I4 analog front-end.

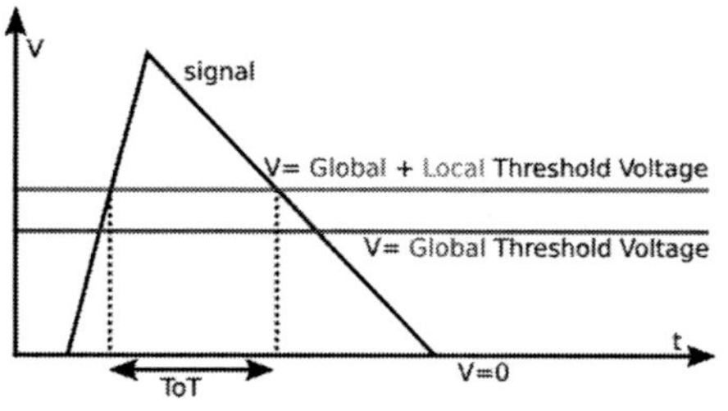

Figure 6. Illustration of FE-I4 ToT calculation.

3.4. RD53A Hit Modelling

Hit modelling for RD53A is very similar to FE-I4 with the exception of which threshold values are compared to the injection charge. RD53A implements 3 different analog front-end designs, and the RD53A emulator can currently models 2 of the designs, the linear and differential front-end design. The linear front-end design and modelling is identical to that of the FE-I4. A simplified schematic of the differential front-end design is shown in Figure 7.

ToT for the differential front-end design is calculated by shifting the signal from the injection charge and comparing the resulting signal to the original signal inverted and shifted by a different, larger, offset. This is illustrated in Figure 8.

The RD53A emulator applies per-pixel Gaussian smearing to the 2 local threshold values and per-pixel Gaussian noise is applied to the injection charge.

3.5. Software Emulator Scan Results

Interpretation of all commands necessary to run many scans has been implemented into the FE-I4 emulator. Some of these scans include digital scans, analog scans, noise scans, threshold scans, and threshold tunes. As a demonstration of the emulator's capabilities, threshold scans before and after threshold tunes are shown in Figure 9.

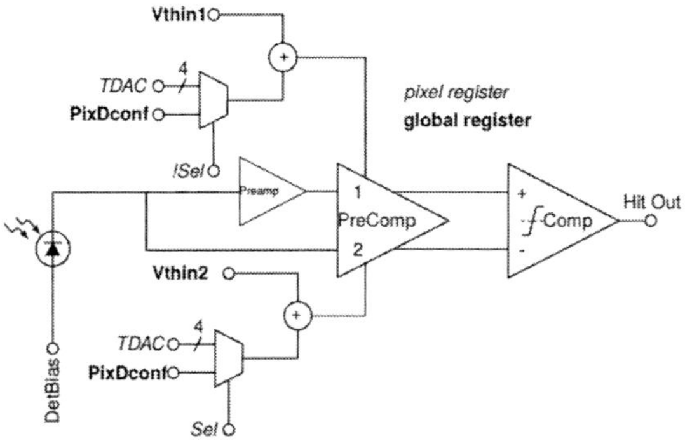

Figure 7. Simplified schematic of RD53A differential analog front-end.

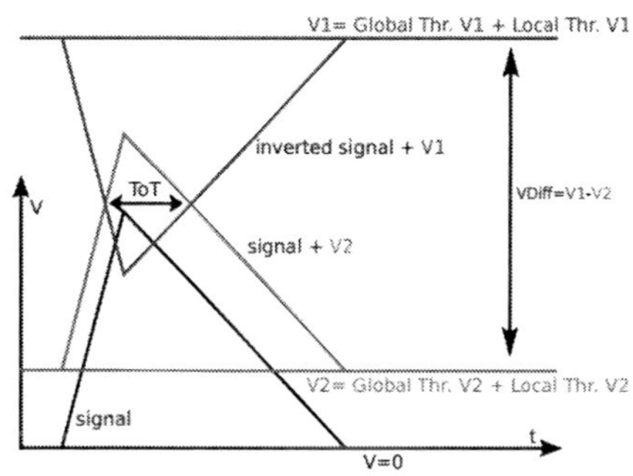

Figure 8. RD53A differential front-end ToT illustration.

The Gaussian smearing applied to the per-pixel local threshold values and noise is evident in the Gaussian shape of the pixel threshold distributions. After tuning the local and global threshold values, a much more narrow Gaussian threshold distribution can be acheived, centered around a user-determined value. This mirrors the behavior of real FE-I4 chips.

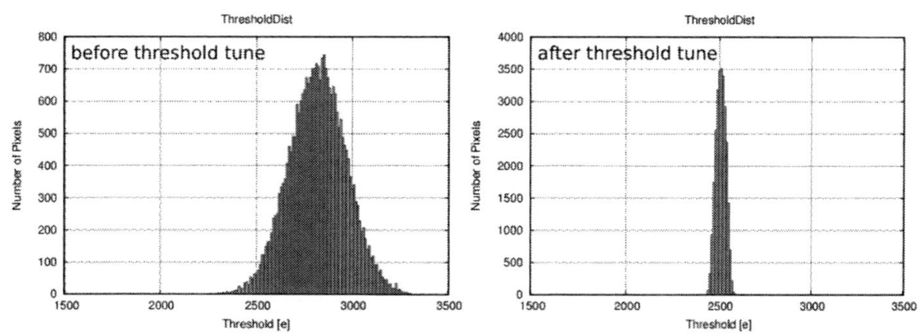

Figure 9. Threshold scans before (left) and after (right) applying a per-pixel threshold tune, using the FE-I4 software emulator.

4. Continuous Integration

One major benefit of software emulators is the ability to implement Continuous Integration (CI) for the YARR DAQ software. The CI workflow is illustrated in Figure 10.

When a developer pushes a commit of the YARR DAQ code, a GitLab [6] Runner will automatically check out the YARR repository, compile the YARR software, run a series of test scans using the software emulators, and check the resulting data. If there was an error in compilation, running the scans, or the resulting data does not match what it expected, CI will mark the commit as having failed, and notify the user. The user is then able to see which step failed, correct for the failure, and push a new commit. This system is aimed at increasing code quality and maintainability and would not be feasible if YARR had to interface with physical hardware readout chips to perform the tests.

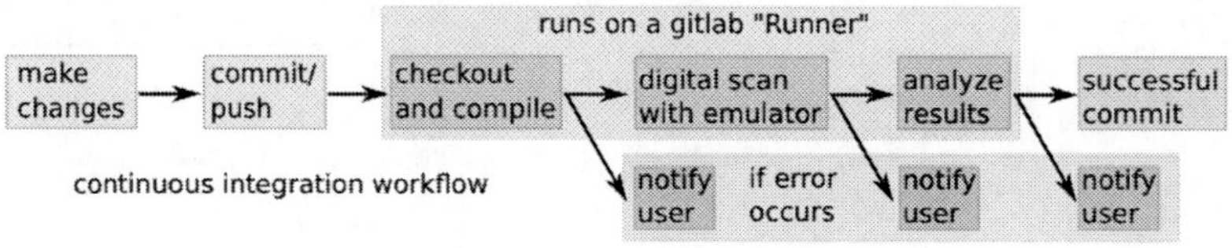

Figure 10. Diagram showing the continuous integration workflow.

5. Conclusion

A software emulator which interfaces with YARR has been successfully implemented for FE-I4 and one for RD53A is under development. The emulators are being used to enhance the DAQ software, to prepare YARR to interface physical RD53A chips when produced, and to ensure code quality and maintainability via the implementation of Continuous Integration.

Acknowledgments

The authors would like to acknowledge the support of the Department of Energy Office of Science Graduate Student Research Program which made this work possible.

References

[1] ATLAS Collaboration *et al* 2008 The atlas experiment at the cern large hadron collider *JINST* **3** S08003
[2] ATLAS Collaboration *et al* 2017 Technical design report for the atlas inner tracker strip detector *CERN-LHCC-2017-005 ATLAS-TDR-025*
[3] Garcia-Sciveres M 2015 Rd53a integrated circuit specifications *CERN-RD53-PUB-15-001*
[4] Heim T 2016 Performance of the insertable b-layer for the atlas pixel detector during quality assurance and a novel pixel detector readout concept based on pcie *CERN-THESIS-2016-085*
[5] Zivkovic V *et al* 2012 The fe-i4 pixel readout system-on-chip resubmission for the insertable b-layer project *JINST* **7** C02050
[6] GitLab Continuous Integration (GitLab CI) `https://docs.gitlab.com/ce/ci/` Accessed October 9, 2017

Upgrading the Fast Calorimeter Simulation in ATLAS

J Schaarschmidt, on behalf of the ATLAS collaboration
The University of Washington in Seattle, Washington 98105, USA
E-mail: jana.schaarschmidt@cern.ch

Abstract. The tremendous need for simulated samples now and even more so in the future, encourages the development of fast simulation techniques. The Fast Calorimeter Simulation is a faster though less accurate alternative to the full calorimeter simulation with GEANT4. It is based on parametrizing the longitudunal and lateral energy deposits of single particles in the ATLAS calorimeter. Principal component analysis and machine learning techniques are used to improve the performance and decrease the memory need compared to the current version of the ATLAS Fast Calorimeter Simulation. The parametrizations are expanded to cover very high energies and very forward detector regions, to increase the applicability of the tool. A prototype of this upgraded Fast Calorimeter Simulation has been developed and first validations with single particles show substantial improvements over the previous version.

1. Introduction

The success of the physics program of the ATLAS experiment [1] at the Large Hadron Collider (LHC) [2] strongly depends on the availability of simulated Monte Carlo (MC) samples with huge statistics, to accurately predict the detector response of any physics process. The increase of integrated and instantaneous luminosity resulting in a larger number of multiple-interactions (pile-up), as well as limited budgets and thus computing resources, motivate the development of fast simulation techniques. In ATLAS, about 90% of the simulation time with the GEANT4 [3] program is spent in the calorimeters. This is due to the complex shower development of an incoming primary particle, during which many additional particles are created. The ATLAS Fast Calorimeter Simulation (FastCaloSim) [4] was developed to provide a parameterized calorimeter response while decreasing the simulation time in the calorimeters by about one order of magnitude. An improved version of FastCaloSim is in development, and its main components and first validations are described in this proceedings contribution.
A short outline of the ATLAS simulation infrastructure is given in Section 2. Current fast simulation tools are described in Section 3. The improvements of the upgraded FastCaloSim are discussed in Section 4, and Section 5 is a summary.

2. The ATLAS Simulation Infrastructure

The ATLAS simulation infrastructure [5] is used to produce and validate the samples to be used in physics and performance studies. The production chain consists of the following steps:

- Event generation: Various MC generators are available to produce hard scatter or pile-up events and decay most of the emerging unstable particles. The event record is based on

IOP Conf. Series: Journal of Physics: Conf. Series **1085** (2018) 032018 doi:10.1088/1742-6596/1085/3/032018

the common HepMC format.

- Detector simulation: The GEANT4 toolkit is used to step through the various detector volumes and simulate the interactions of particles with the material according to the numerical models provided by the physics lists. It is also referred to as full simulation. Fast simulation can replace GEANT4 in defined detector parts. The output is stored in the G4HIT[1] format. Pile-up events typically undergo the full simulation and are then overlayed with the hard scatter event at the hit level.
- Digitization: In this step, the hits are converted into digits, which is the same format as that of the real detector. Typically, a digit is produced when a voltage or current on a readout-channel exceeds a given threshold during a time window. Some detector parts also include the signal shape as a function of time. Detector and electronics noise and cross talk effects are modelled as well. The output is raw data objects (RDO).
- Reconstruction: At the final step the digits are converted to real physics objects such as jets, muons, electrons, photons, etc., using various algorithms, such as jet finders, tracking or clustering software.

All the steps of the production chain undergo validation to ensure their correct functionality, performance and robustness.

3. Fast Simulation

While various algorithms for fast simulation exist (see Ref. [5]), only the most relevant ones are described here. In general, fast simulation is faster, less precise and easier to tune to data than the full simulation.

Frozen-showers [8] is a technique that replaces low-energetic particles from electromagnetic (EM) shower produced by GEANT4 by pre-simulated showers stored in a library. For the MC production for Run-2, it is used by default in the forward calorimeters (FCAL) in the full simulation.

Fatras [9] is a fast ATLAS track simulation, that employs the simplified reconstruction geometry rather than the simulation geometry, and uses an extrapolation engine to transport the track through the detector.

FastCaloSim [4] is a parameterized calorimeter response, based on the GEANT4 simulation of single particles, namely electrons and photons (representing EM showers) and charged pions (representing hadronic showers). The single particles are generated in a fine grid of energies and pseudorapidities[2]. Then, in the simulation step, a fitting parametrization is loaded into memory and the response (such as energy or energy density around the shower axis) is randomly sampled from that. The energy range covers values from 50 MeV to 1 TeV. In the upgraded version, this range is going to be expanded to up to 4 TeV. Each parametrization covers a range in η of size 0.05. A dedicated parametrization of the FCAL is not part of FastCaloSim, but this is planned for the new version.

FastCaloSim is successfully used in ATLAS, mostly for signal samples and systematic variations of backgrounds. During the MC production campaign of the first year of Run-2, about three billion MC events were processed or re-processed with ATLFAST2, while about 14 billion events were (re-)processed with the GEANT4 simulation. ATLFAST2 combines the full simulation in the inner detector and the muon spectrometer with the FastCaloSim. ATLFAST2F utilises Fatras and FastCaloSim, but it uses the full simulation in the muon system. FastCaloSim

[1] A hit object is a record of the deposited energy with time and position.

[2] ATLAS uses a right-handed coordinate system with its origin at the nominal interaction point (IP) in the centre of the detector and the z-axis along the beam pipe. The x-axis points from the IP to the centre of the LHC ring, and the y-axis points upwards. Cylindrical coordinates (r, ϕ) are used in the transverse plane, ϕ being the azimuthal angle around the z-axis. The pseudorapidity is defined in terms of the polar angle θ as $\eta = -\ln\tan(\theta/2)$.

reduces the simulation time by about a factor 10, and the additional use of Fatras reduces the simulation time by another factor 10, i.e., a factor 100 compared to full simulation.

Finally, there are also tools for fast digitization and fast reconstruction. If the entire production chain is replaced by fast tools, which is called the FastChain [6], then a speed-up of about a factor 30 compared to using only FastCaloSim can be achieved, however, only in the absence of pile-up. The ATLAS integrated simulation framework (ISF) [7] was developed to provide a flexible configuration of detailed or fast tools at every step of the chain. It also allows to use different simulators for different particles of the same event. This approach can be tailored to the specific needs of the various physics analyses.

4. The new FastCaloSim

The performance of the previous version of FastCaloSim is discussed for example in Ref. [10]. The new version aims to improve the performance, but also the speed and memory needs. The main features are briefly described here.

4.1. Energy parametrization

Along the longitudinal shower direction, energy is deposited in the various layers of the calorimeter that are hit by the shower. The energy distributions are obtained from the GEANT4 simulation, and they are converted into cumulative distributions in order to get a well-defined assignment of cumulative value versus the deposited energy, expressed as the fraction of the total shower energy.

The energy deposits per layer are correlated with each other. If, for example a large amount of energy is lost in the first layers, less will be deposited in the layers behind. The details will vary from shower to shower. In order to decorrelate those energy deposits, and to classify the showers by their longitudinal shower development, a principal component analysis (PCA) is used. A PCA is a transformation of a set of variables into a set of orthogonal and uncorrelated, so-called principal components. The first component has the largest variance. In order to achieve a better decorrelation, the events are divided into bins of the first (and/or second) component. The bins have approximately the same number of events, and typically a number of bins between 5 and 10 is chosen. A second PCA transformation is applied to the energy deposits per layer for the events in each of those bins, and hence the components of that second transformation are now largely decorrelated.

The information that needs to be stored, in each bin defined from the 1st PCA transformation, are the cumulative energy distribution and the PCA matrix of the 2nd PCA transformation. The information have to be loaded into memory for the simulation later on; it is therefore desired to keep the storage as small as possible. For that, a neural-network based regression is used. The training and evaluation of the regression is done with a multi-layer perceptron (MLP), as implemented in the TMVA [11] package. An iterative MLP training is employed, starting from two neurons up to ten neurons, using one hidden layer. After each training, the regression is evaluated and the agreement with the original distribution (from GEANT4) is tested. If the largest relative deviation between input and output is less than 5%, the iteration stops and the MLP weights are stored. If the regression fails, the histogram holding the cumulative energy distribution is rebinned iteratively and the bin contents and bin borders are stored instead.

Finally, in the simulation step, for each PCA bin a set of uncorrelated random numbers is drawn (one per calorimeter layer), the inverse PCA transformation is applied, and the regression output is calculated from the MLP weights. By construction, each PCA bin has the same probability and thus in the simulation this bin is determined randomly for each call of the energy simulation. A distribution of simulated energies is then a linear superposition of the deposited energy in each of the 5-10 PCA bins.

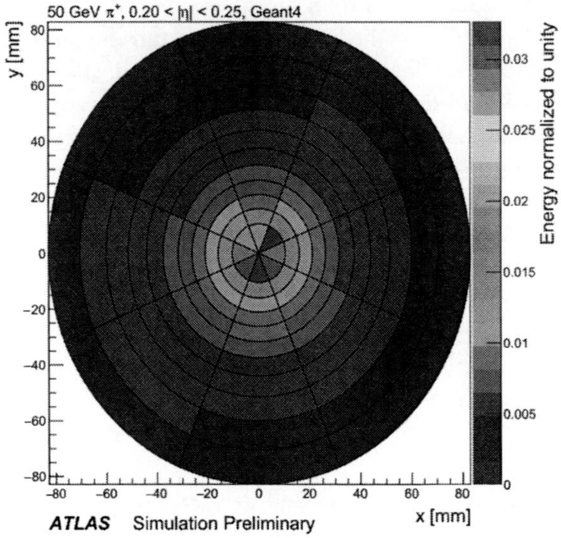

Figure 1: Example of an average shower shape histogram. This is the second layer of the EM barrel for central pions with 50 GeV total energy. Displayed is the averaged hit energy normalized to unity, in bins of α and r, presented in the x-y plane.

4.2. Shower Shape Parametrization

The lateral shower shape is characterised by the energy deposited around the shower axis, in a given calorimeter layer and PCA bin. The variables in which the shape is described are an angle α (around the axis) and a radius r (distance from the axis). The energy is taken from simulation hits, which allows a finer binning of the information than it would be possible with energy deposits on the calorimeter cell level. The binning is optimized such to get approximately same number of hits per bin, but a finer binning is used towards small radii where most of the energy is concentrated. An average 2D shower shape histogram is built from a sample of single particles simulated with GEANT4, for a given energy value and a narrow bin in η. An example of an average shower shape is displayed in Figure 1. A large number of these histograms is obtained and then stored. In a future step, these 2D histograms will be replaced by MLP weights from a 2D regression (analogous to the energy parametrisation), to achieve a much more compressed memory usage.

In the simulation step, these histograms are loaded into memory. They serve as probability density functions, and HIT positions are tossed randomly. Each hit is assigned the same amount of energy, i.e. the energy per layer (obtained from the energy simulation) divided by the number of hits.

The number of hits tossed from the 2D histograms is a critical parameter. If too many hits are tossed, the resulting shower shape will resemble the average shower shape, and then the energies don't fluctuate enough. If too few hits are tossed, the energy fluctuations could be too large and results will not be accurate. This parameter is still subject to further studies. The approach implemented to-date is to calculate the number of hits such that their stastistical precision equals the sampling term of the energy resolution in the calorimeter:

$$\frac{\sqrt{N}}{N} = \frac{\alpha}{\sqrt{E}}. \tag{1}$$

N is the number of hits, E the energy per layer, and the parameter α depends on the layer, ranging from 10% in the EM barrel to 100% in the FCAL.

4.3. Hit-to-Cell Assignment

Once the energy and the shower shape have been modelled, and a hit distribution is simulated, the hits need to be assigned to the calorimeter cells. The cell container is then eventually passed to the digitization step.

The simplified geometry tool is used to assign the hit positions to cells. This tool neglects the accordion shape of the electrodes in the Liquid Argon calorimeter (which assures a crack-less symmetry in phi), and assumes that cells are cuboids. In GEANT4 this accordion structure is included. In FastcaloSim, the accordion shape is emulated by applying a hit replacement probability function, also called *wiggle function*. This function slightly displaces the hit in phi direction before the geometry tool is called. Only if the hit is in the center of given cell, the probability that the hit is assigned to that cell is one. If the hit is positioned toward the edge of a cell, there is a non-zero probability that the hit is assigned to the neighbouring cell. The parameters of this hit replacement depend on the calorimeter layer, and they are tuned to GEANT4.

4.4. Prototype and First Validations

A first prototype of the upgraded FastCaloSim is available. It is implemented in the ISF, as part of the ATLAS software framework Athena [12]. Currently, the functionality is limited by the availability of the various parametrizations, but all main functionalities are included. To assess the performance of the new tool, first validations are performed on single particles. The goal is to resemble the full simulation as closely as possible. Comparisons are also made to the previous version of FastCaloSim. Figure 2 displays two examples of the validations for single particles in the central region of the calorimeter, i.e. $0.2 < |\eta| < 0.25$. These comparisons are obtained after the full chain, i.e. the events undergo simulation, digitization and finally reconstruction. A better modelling of the inner structure of hadronic showers is achieved, and also a notably better modelling of the longitudinal energy deposited by EM showers is seen. These are just a small subset of the validations that have been performed already, and more extensive studies using complex physics processes will follow.

5. Summary

FastCaloSim is a parameterized calorimeter response, based on the inputs from the detailed simulation with GEANT4, but approximately ten times faster. The improved FastCaloSim employs machine learning techniques to reduce memory need and to improve the modelling of the shower shape. PCA transformations are used to decorrelate the energy deposits per layer, and to define a binning that is used for both the energy and the shower shape parametrization. The accordion structure of some of the calorimeter parts is emulated with a dedicated hit replacement. The prototype is now available and first validations with central single particles show substantial improvements of the new FastCaloSim version compared to the previous version.

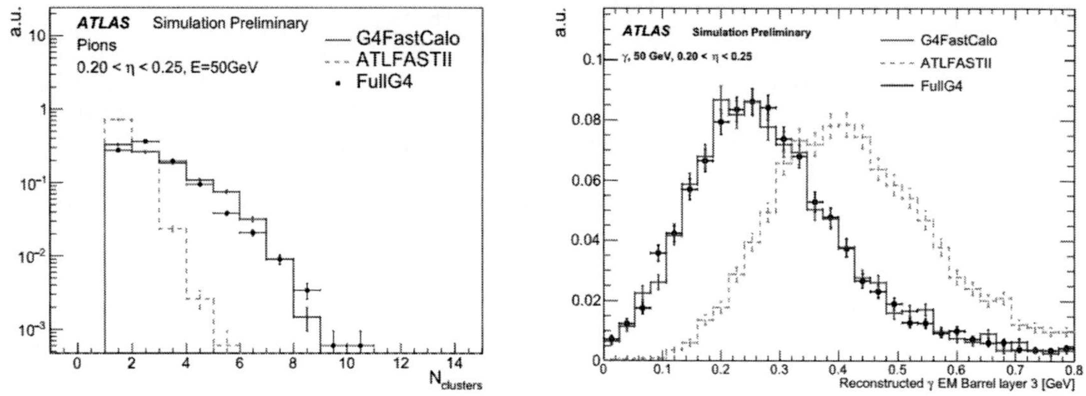

(a) (b)

Figure 2: Validation of the upgraded FastCaloSim with single particles. (a) shows the number of reconstruced clusters for a sample of 50 GeV pions. (b) displays the energy deposited in the third layer of the calorimeter for 50 GeV photons. A very good agremeent between the upgraded FastCaloSim (labelled as G4FastCalo) and the full simulation (labelled as FullG4) is achieved. The previous version of FastCaloSim (labelled as ATLFastII) is not able to reproduce these variables.

References

[1] ATLAS collaboration: The ATLAS experiment at the CERN Large Hadron Collider, *JINST* **3** S08003, 2008. http://dx.doi.org/10.1088/1748-0221/3/08/S08003

[2] Evans, Lyndon and Bryant, Philip: LHC Machine, *JINST* **3** S08001, 2008. http://dx.doi.org/10.1088/1748-0221/3/08/S08001

[3] Agostinelli, S. et al.: GEANT4 - A Simulation Toolkit, *Nucl. Instr. and Meth.* **A** 506, 2003. https://doi.org/10.1016/S0168-9002(03)01368-8

[4] ATLAS collaboration: The simulation principle and performance of the ATLAS fast calorimeter simulation FastCaloSim, ATL-PHYS-PUB-2010-013, 2010.

[5] ATLAS collaboration: The ATLAS Simulation Infrastructure, *Eur. Phys. J. C* **70** 823-874, 2010. http://dx.doi.org/10.1140/epjc/s10052-010-1429-9

[6] Basalaev, A. on behalf of the ATLAS collaboration: The Fast Simulation Chain for ATLAS, ATL-SOFT-SLIDE-2016-617, 2016.

[7] Ritsch, E. on behalf of the ATLAS collaboration: The ATLAS Integrated Simulation Framework, ATL-SOFT-SLIDE-2013-114, 2013.

[8] Gasnikova, K. on behalf of the ATLAS collaboration: Frozen-shower simulation of electromagnetic showers in the ATLAS forward calorimeter, ATL-SOFT-SLIDE-2016-735, 2016.

[9] Edmonds, K. et al.: The Fast ATLAS Track Simulation (FATRAS), ATL-SOFT-PUB-2008-001, 2008.

[10] ATLAS collaboration: Performance of the Fast ATLAS Tracking Simulation (FATRAS) and the ATLAS Fast Calorimeter Simulation (FastCaloSim) with single particles, ATL-SOFT-PUB-2014-01, 2014.

[11] Hoecker, A. et al.: TMVA - Toolkit for multivariate data analysis, *Proc. of Science* PoS ACAT 040, 2007. https://doi.org/10.1063/1.4771869

[12] ATLAS collaboration: ATLAS Computing Technical Design Report, ATLAS-TDR-017, CERN-LHCC-2005-022, 2005.

IOP Conf. Series: Journal of Physics: Conf. Series **1085** (2018) 032019 doi:10.1088/1742-6596/1085/3/032019

Making containers lazy with Docker and CernVM-FS

N Hardi, J Blomer, G Ganis and R Popescu
CERN
E-mail: nhardi@cern.ch

Abstract. Containerization technology is becoming more and more popular because it provides an efficient way to improve deployment flexibility by packaging up code into software micro-environments. Yet, containerization has limitations and one of the main ones is the fact that entire container images need to be transferred before they can be used. Container images can be seen as software stacks and High Energy Physics has long solved the distribution problem for large software stacks with CernVM-FS. CernVM-FS provides a global, shared software area, where clients only load the small subset of binaries that are accessed for any given compute job.

In this paper, we propose a solution to the problem of efficient image distribution using CernVM-FS for storage and transport of container images. We chose to implement our solution for the Docker platform, due to its popularity and widespread use. To minimize the impact on existing workflows our implementation comes as a Docker plugin, meaning that users will continue to pull, run, modify, and store Docker images using standard Docker tools.

We introduce the concept of a thin image, whose contents are served on demand from CernVM-FS repositories. Such behavior closely reassembles the lazy evaluation strategy in programming language theory. Our measurements confirm that the time before a task starts executing depends only on the size of the files actually used, minimizing the cold start-up time in all cases.

1. Introduction

Linux process isolation with containers became very popular with the advent of convenient container management tools such as Docker [1], rkt [2] or Singularity [3].

The Docker system, in particular, has had a large influence in the Linux container community because it introduced the de facto container image format, layered images and git-like workflow appreciated by container developers. Compared to full VM virtualization, containers benefit from close-to-zero performance overhead for both compute intensive and I/O intensive workloads.

However, despite the fact of being composed of shared read-only layers to save space and bandwidth, the price for obtaining images can be very high. This is because image handling is coarse grained, and the full image has to be transferred in order to be used. For example, to start a cluster of 1000 nodes using an image of 1 GB the network has to transfer 1 TB until all containers start executing the payload.

In this paper, we present a solution to the problem of efficient image distribution based on CernVM-FS [4] for storage and transport of container images.

The solution exploits the fact that, on average, only 6% of the image is required [5]. We developed a Docker plugin [6] to refine data reuse granularity from layers to files. Additionally, the download is delayed until the file is accessed for the first time. Using this approach only a fraction of the image is initially transferred and containers can start almost instantly.

IOP Conf. Series: Journal of Physics: Conf. Series **1085** (2018) 032019 doi:10.1088/1742-6596/1085/3/032019

We implemented our solution for the Docker platform, but the similar approach is applicable to other containerization technologies as well.

2. Docker system

The main components of a Docker ecosystem are the Docker daemon and Docker command line utility for controlling the daemon on every node, together with a central Docker registry which serves a repository for distributing and sharing the Docker images.

2.1. The Docker images

Docker images are described by image manifests [7] and they are composed of one or more read-only layers. The image manifest and layers are immutable. A layer is a tarball containing a part of the image's file system. For instance, the base layer could contain an Ubuntu system, the second layer adds additional software such as OwnCloud and third layer contains user's configuration and tweaks. Manifest and layers are content addressable, i.e., identified by a cryptographic hash that is calculated from their contents. Such an organization is convenient for data reuse. Changes in containers can be committed and saved but this just results in a new read-only layer and a new image manifest. It is not possible to physically remove files from a Docker image but only to hide them. Removing of a file or directory is realized by masking the original file system entry through a so called whiteout file in the scratch area.

2.2. The Docker workflow

The Docker command line tool provides git style subcommands for creating, running, managing and sharing containers. A common usage pattern emerged which involves obtaining a Docker image, creating and running a container, modifying and sharing it.

Conventionally the whole Docker image needs to be obtained before containers based on it can start. Multiple containers can reuse the same image and get a dedicated ephemeral scratch area. When a container stops, the corresponding scratch area is destroyed. In order to permanently keep changes, a new image needs to be created by "burning in" the scratch area as the topmost layer onto the stack of the read-only layers. Newly created Docker images can be uploaded to a Docker registry to be shared. Additionally, new images can be created with Dockerfiles [8] by the build command. It is possible to specify an existing image as a base image.

2.3. The Docker graphdriver

The graphdriver [9] is the component of the Docker daemon that stores new layers and combines them together to images using a union file system (e.g., AUFS [10], OverlayFS [11]). Graphdrivers operate with random layer IDs instead of hash values of layers, which is the very reason why we later introduce thin images. Graphdrivers implement a very slim interface and have limited access to other information about the rest of the system.

2.4. Docker plugin system

Docker v2 plugin interface is substantially reengineered and is available since Docker v1.12. As of Docker v1.13 (January 2017), there is support for graphdriver plugins in addition to the already existing plugin interfaces. Docker plugins are only loosely coupled to the daemon and communicate with it through a TCP or Unix socket using HTTP. The v2 plugins are also called "managed plugins". It is possible to obtain them from a Docker registry, and install and configure them using the Docker command line tool.

3. Using CernVM-FS for the image transport

Image handling in the Docker system can be improved in such a way that only a small fragment of the images is transferred before containers can start. On average only 6% of the image contents is actually read during the lifetime of a container. This comes partially from the way that container images are created. Usually users pick a base image that comes with a set of standard libraries and utilities. For instance, the Ubuntu base image brings almost 150 MB but other software layers that are added on top do not use all the software that comes with such a base image. This presents an opportunity for improvements that we addressed with our solution. It is also important to notice that this effect has much less impact with smaller base images such as Atomic Host [12] and Alpine Linux [13] or even micro CernVM [14]. The overhead coming from transferring whole images that are not carefully built becomes much more important with huge images that can go even up to several tens of gigabytes.

This effect becomes even stronger in the cluster environment where each cluster node tries to obtain such a big image from the central repository. When all nodes download the image at once, the network links can get saturated quickly. With modern software development techniques and continuous integration where software is updated very often, images are rebuilt on daily basis or more often.

In CernVM-FS the unit of transfer and reuse is a file. Big files are chunked into smaller pieces. File and chunk level deduplication is provided transparently. Files are transferred on demand but only after they are requested for the first time. The goal is to store extracted image layers into a CernVM-FS repository and later download only the files that are required for executing the given workload. This approach implicitly brings all the other well known benefits of using CernVM-FS for distributing software, including leveraging HTTP proxy caches, content addressable storage, limited cache size and others.

4. Integration of CernVM-FS and Docker

Our implementation consists of 3 main components.

(i) The Docker plugin based on the existing Docker graphdrivers, using either AUFS or OverlayFS as a union file system.
(ii) A CernVM-FS server and repository with additional "satellite services" to receive new layers from clients and publish them.
(iii) The utility docker2cvmfs for converting images from a Docker registry to thin images stored on CernVM-FS.

Our plugin is an extension of the built-in graphdrivers, so conventional Docker images are supported seamlessly. For CernVM-FS provided images, we introduce the concept of "thin images".

4.1. Running thin images

A "thin image" has all properties of a standard Docker image but instead of the actual data, it carries just the "thin image descriptor". The thin image descriptor is a JSON file with a list of layers that it replaces and the location (i.e., the name of the CernVM-FS repository) where these layers are available. The change in behavior happens once the graphdriver plugin recognizes that the imported Docker image is carrying the thin image descriptor instead of the actual layer data. Upon starting a container from a thin image, the information from the thin image descriptor is used to mount the corresponding CernVM-FS repositories, which are then used as part of the union mount. The graphdriver plugin comes with CernVM-FS client built into the plugin container image so no additional software except the plugin container itself needs to be set up to use thin images.

IOP Conf. Series: Journal of Physics: Conf. Series **1085** (2018) 032019 doi:10.1088/1742-6596/1085/3/032019

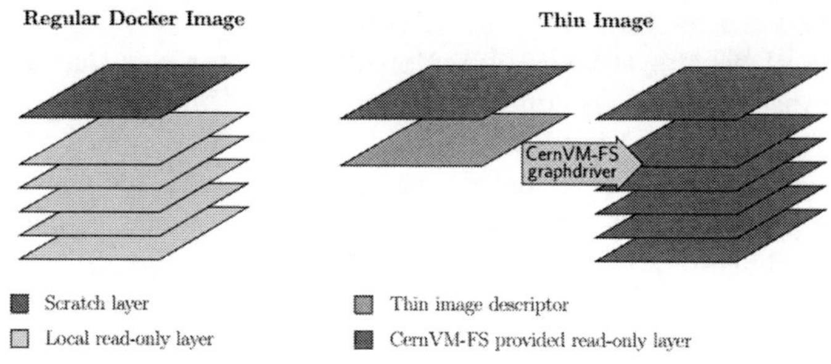

Figure 1. Comparing internal organization of normal and thin images. The original image layers are moved to a CernVM-FS repository and replaced by a "thin image descriptor".

4.2. Updating thin images
The Docker pull and run commands are straightforward, but there are some important details about Docker commit and build commands. The difficulty comes from the fact that CernVM-FS is a read-only file system from the client's viewpoint. Publishing new contents in a CernVM-FS repository takes place on a central server where a transaction is started, changes are made and then the transaction is committed or published. After changes are made and committed on the server, they will eventually propagate to clients. Publishing new contents from client to server is a new scenario for CernVM-FS that requires additional services. For the transport of the tarballs we choose Amazon S3 API [15], a widely used industry standard, in its open source implementation Minio [16]. The concept of active storage has recently become popular and usually is available alongside S3 storage (e.g., Amazon Lambda [17]). The purpose of active storage is to manipulate stored data on certain events (e.g., data is removed or uploaded). Minio supports active storage by calling a webhook. We introduce a "publishing service" which implements a webhook handler for Minio event notifications together with a status API. When new content is uploaded to Minio, the webhook is triggered and the publisher service is notified. Processing of notifications consists of the serialization of requests into a single queue, starting a CernVM-FS transaction, extracting the uploaded layer tarball into the repository and committing the transaction. On error, the transaction gets aborted and the next notification is processed. The status API allows for following the publication process from the client side. Clients remount the CernVM-FS repository once the new content is published.

4.3. Configuring the plugin
Configuration for the CernVM-FS and Minio clients can be supplied by bind mounting directories from a host file system to plugin's container file system. The Docker command line tool provides a subcommand for setting which directories should be bind mounted. The configuration format for the CernVM-FS client reuses the layout of the /etc/cvmfs directory, that usually provides repository public keys and repository configuration. The external configuration will overwrite the internal configuration. It is possible to supply default configuration together with the plugin.

It is not necessary to configure and use the Minio S3 client for running thin images. However, each worker needs to have the credentials and address of the publish status API in order to modify the thin images. The Minio client configuration is a JSON file that is created in the bind mounted directory.

4.4. Creating new thin images using docker2cvmfs
We developed a utility, docker2cvmfs, to fetch the layer tarballs from a Docker registry and create the thin image descriptor. This tool reads an image manifest and downloads each layer

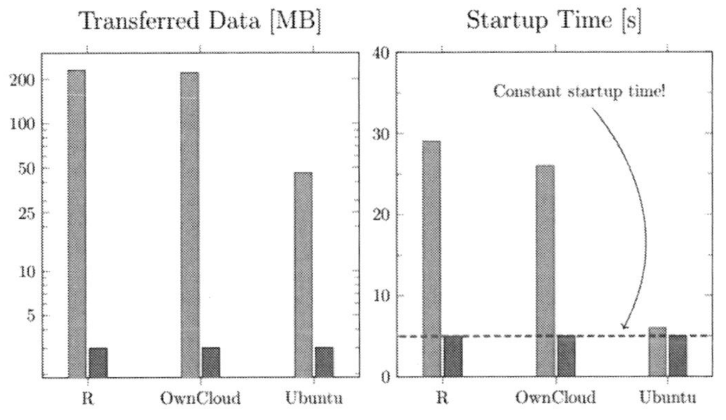

Figure 2. Transfer volume and startup time with the normal and corresponding thin images.

tarball which later needs to be published in a CernVM-FS repository. We use this tool also for debugging because it provides a convenient way to access image manifests.

4.5. Improvements in CernVM-FS

CernVM-FS is a special purpose file system that is already optimized for serving software and operating system binaries (e.g., the root file system of the CernVM virtual appliance). In particular, CernVM-FS supports POSIX operations needed by the software use case, which is often overlooked in other distributed file systems. Only a few tweaks were needed to support hosting container images, namely an improved handling of special files (device nodes) to support serving Docker image layers.

5. Evaluating the results

Figure 2 shows that the cold startup time with a thin image requires only a fraction of the image to be actually transferred. It is important to notice that no matter how large the original image is, the startup time and the amount of data transferred is constant and the same for all images in these tests.

For instance, the R-base Docker image of 200 MB compressed will result in only 6 MB transferred when converted to the thin image format.

6. Summary

In this contribution, we have shown how to integrate the Docker container engine with the CernVM File System for container image distribution. As a result, we removed the Docker Registry as a bottleneck for using Docker containers at the scale of a data center. As Docker image contents are distributed by CernVM-FS on demand and with file-level granularity, typical container start-up times are reduced by one to two orders of magnitude.

The CernVM-FS integration is implemented by means of a Docker graphdriver plugin. The integration is seamless in so far as regular Docker images can be used side-by-side with images from CernVM-FS and all Docker commands and workflows remain independent of the image source. To our knowledge, our Docker graphdriver implementing CernVM-FS access is the first 3rd party Docker graphdriver plugin.

Ongoing work is focused on a first exploitation phase under production conditions in the CERN data center. The CernVM-FS server side "satellite services" used to publish new Docker image layers is currently being generalized in order to support remote upload of contents to a CernVM-FS repository.

References

[1] Docker Inc Docker Engine URL `https://www.docker.com/`
[2] CoreOS Inc rkt URL `https://coreos.com/rkt/`
[3] Kurtzer G M, Sochat V and Bauer M W 2017 *PloS one* **12** e0177459
[4] Blomer J, Aguado-Sanchez C, Buncic P and Harutyunyan A 2011 *Journal of Physics: Conference Series* **331**
[5] Harter T, Salmon B, Liu R, Arpaci-Dusseau A C and Arpaci-Dusseau R H 2016 *Proc. 14th USENIX Conference on File and Storage Technologies (FAST'16)*
[6] Docker Inc Docker engine managed plugin system URL `https://docs.docker.com/engine/extend/`
[7] Docker Inc Docker image manifest specification URL `https://docs.docker.com/registry/spec/manifest-v2-2/`
[8] Docker Inc Dockerfile reference URL `https://docs.docker.com/engine/reference/builder/`
[9] Docker Inc About docker graphdrivers URL `https://docs.docker.com/storage/storagedriver/`
[10] Okajima J AUFS URL `https://aufs.sourceforge.net/`
[11] Brown N OverlayFS doc. URL `https://www.kernel.org/doc/Documentation/filesystems/overlayfs.txt`
[12] Project Atomic URL `https://projectatomic.io/`
[13] Alpine Linux URL `https://alpinelinux.org/`
[14] Blomer J, Berzano D, Buncic P, Charalampidis I, Ganis G, Lestaris G, Meusel R and Nicolaou V 2014 *Journal of Physics: Conference Series* vol 513 (IOP Publishing) p 032009
[15] Amazon Amazon S3 URL `https://aws.amazon.com/s3/`
[16] Minio, Inc Minio URL `https://minio.io/`
[17] Amazon Amazon Lambda URL `https://aws.amazon.com/lambda/`

IOP Conf. Series: Journal of Physics: Conf. Series **1085** (2018) 032020 doi:10.1088/1742-6596/1085/3/032020

A quantitative review of data formats for HEP analyses

J Blomer
CERN, Geneva, Switzerland
E-mail: jblomer@cern.ch

Abstract. The analysis of High Energy Physics (HEP) data sets often takes place outside the realm of experiment frameworks and central computing workflows, using carefully selected "n-tuples" or Analysis Object Data (AOD) as a data source. Such n-tuples or AODs may comprise data from tens of millions of events and grow to hundred gigabytes or a few terabytes in size. They are typically small enough to be processed by an institute's cluster or even by a single workstation. N-tuples and AODs are often stored in the ROOT file format, in an array of serialized C++ objects in columnar storage layout. In recent years, several new data formats emerged from the data analytics industry. We provide a quantitative comparison of ROOT and other popular data formats, such as Apache Parquet, Apache Avro, Google Protobuf, and HDF5. We compare speed, read patterns, and usage aspects for the use case of a typical LHC end-user n-tuple analysis. The performance characteristics of the relatively simple n-tuple data layout also provides a basis for understanding performance of more complex and nested data layouts. From the benchmarks, we derive performance tuning suggestions both for the use of the data formats and for the ROOT (de-)serialization code.

1. Introduction

The analysis of high energy physics data sets is typically an iterative and explorative process. From centrally curated experiment data and software frameworks, derived data sets are created for particular physics working groups or analyses. These data sets are often stored in the form of ROOT "n-tuples" [1], i. e., tabular data, or modestly nested tabular data (such as a vectors of jets for every event).

In this contribution, we presume that these "final data sets" are small enough to be processed by a single workstation. We further presume that such analysis data sets are independent from the software framework used to create them, so that they can potentially be processed using a variety of tools and data formats.

We evaluate several popular such libraries and data formats from industry and academia. We compare I/O performance, usability, and reliability against storage media failures. We are looking at the use case of I/O dominated, repeated, partial reading of the data set. It should be noted that for reading only a subset of the available columns of the data set, columnar data formats that store columns consequtively (instead of rows) have a natural advantage. As a benchmark, we use a typical LHC Run 1 analysis as described by the LHCb OpenData sample [2].

IOP Conf. Series: Journal of Physics: Conf. Series **1085** (2018) 032020 doi:10.1088/1742-6596/1085/3/032020

2. Data Formats and Libraries

The following list provides an overview of the data formats and access libraries that are used in this comparison.

ROOT ROOT [3] stores data in "trees", a collection of serialized, possibly nested C++ objects in columnar layout.

Protobuf Google protobuf [4] is not a data format per se but rather a serialization library for records of simple data types. A simple data format can be constructed, however, by concatenating serialized blobs with a header indicating each record's size.

SQLite SQLite [5] stores data as relational tables and provides data access routines through the SQL language.

HDF5 HDF5 [6] stores multi-dimensional arrays of simple data types. By changing the arrangement of the arrays, the user can effectively store data either column-wise or row-wise. HDF5 is popular in the High Performance Computing community, where it is particularly beneficial through its tight integration with the MPI I/O protocol.

Avro Apache Avro [7] provides row-wise serialization. It is mainly used in the Hadoop ecosystem. In contrast to Google Protobuf, Avro provides a full data format for collections of records out of the box. Its primary access library is written in Java, although a C port exists, too.

Parquet Apache Parquet [8] is a column-wise serialization format mainly used in the Hadoop ecosystem. Access libraries exist for Java and C++.

3. Sample Analysis

The LHCb OpenData sample comprises 8.5 million LHC Run 1 events describing $B^{\pm} \to K^{\pm}K^{+}K^{-}$ decays. Depending on the file format, the data set size is 1.1 GB to 1.5 GB. For every event, 26 values ("branches" in ROOT terminology) of floating point or integer type are stored. In order to approximate the B mass from the recorded Kaon candidates, 21 out of the 26 provided values are required. Furthermore, 2.4 million events can be skipped because one of the Kaon candidates is flagged as a Muon (we apply a cut). As an emulation of *reading the data set*, we calculate the sum of all 21 values of the 6.1 million non-cut events. As an emulation of *plotting from the data set*, we calculate the sum of only 2 values of all events.

Compared to ATLAS xAODs [9] or CMS MiniAODs [10], this is a simple data schema. It helps us understanding the performance base case.

4. Evaluation

With each library and in each format, the data set can be written and read with not more than a few hundred lines of code. No differences occurred in the floating point results. ROOT provides the most seamless integration by directly storing C++ classes. For the other libraries the data schema has to be explicitly specified. Only ROOT and SQLite provide an obvious method for schema evolution, i. e., adding a column to an already existing file. SQLite, on the other hand, provides no support for data compression. While HDF5 does in principal support data compression, it is omitted in these tests because compression requires significantly more code by the user who has to manually break-up the data tables in blocks.

4.1. Resilience against silent data corruption

Data on unmanaged storage is subject to a small, but non-neglible chance of silent data corruption. Silent data corruption is caused by media failures, which remain undetected during normal operation. It can also occur due to transmission errors when copying the data. In order to test the data formats' ability to detect silent corruption, we artificially introduce random bit

Table 1. Number of outcomes of 100 read trials with random bit flips. "No effect" means that the program run produced the correct result. "Crash" means an abnormal termination of the program run. "Err. Msg" means that the word "error" was emitted to stdout, while the program continued. The remaining cases are labeled as "undetected", that is an ordninary program run producing an incorrect result. The undetected errors for ROOT/LZ4 were meanwhile addressed by the ROOT developers.

Format	No effect	Crash	Err. Msg.	Undetected
ROOT (uncompressed)	68	-	-	32
ROOT (zlib)	27	41	32	-
ROOT (lz4)	60	2	4	34
ROOT (lzma)	27	37	36	-
Protobuf (uncompressed)	58	8	-	34
Protobuf (gzip)	-	100	-	-
SQlite	56	12	-	32
HDF5 (row-wise)	63	-	-	37
HDF5 (column-wise)	61	-	-	39
Parquet (uncompressed)	63	4	-	33
Parquet (zlib)	28	72	-	-
Avro-Java (uncompressed)	63	-	-	37
Avro-Java (zlib)	51	27	-	22

flips in the following way. For a reduced data set consisting of the first 500 000 events, we read the corresponding data file one hundred times. For each of these 100 reads, we flip one randomly selected bit in the file beforehand. So effectively we read 100 slightly different files for every data format. Possible results of reading such damaged files range from no change at all, e. g., when the bit flip happens to be at a position that is skipped during reading, to a crash of the program. A malicous result is a difference in the physics result, where no indication of a failure is given during the reading.

Table 1 summarizes outcome of the tests. For all of the data formats, except Avro, bit flip protection is a side effect of the checksums of the compression algorithm. Without compression, all data formats are subject to an occasional, undetected change of the physics result in case of silent corruption. Avro uses the zlib "raw mode" which explicitly omits the checksum even for compressed data. In ROOT, some of the bit flips trigger only error messages but no crash, which might be overlooked by a not carefully programmed application.

4.2. Performance

All performance measurements are performed on a machine with a 4 core Intel i7-6820HQ with hyper-threading, 2×16 GB RAM, a 1 TB Toshiba XG3 PCIe SSD and Gigabit Ethernet, which is supposed to resemble the performance of a typical workstation. The system runs a Linux 4.12 kernel, glibc 2.25, gcc 7.1.1 and the EOS 4.1.2 client [11]. The source code of the benchmarks is available on github[1].

Figure 1 shows the encoding effeciency of the different file formats. For this data set, the efficiency of compression is of the order of 20 % to 35 % while the difference between compression algorithms is of the order of 10 % to 15 %. Small file sizes are particularly performance-relevant for slow storage devices such as spinning hard drives or Gigabit attached storage.

[1] `https://github.com/jblomer/iotools/tree/acat17`

IOP Conf. Series: Journal of Physics: Conf. Series **1085** (2018) 032020 doi:10.1088/1742-6596/1085/3/032020

Figure 2 shows differences by a factor 10 and more between the file formats in reading throughput from fast, memory-class storage. For Avro, the Java library is used instead of the C library due to a tenfold performance penalty that is observed by the current C implementation. On fast storage, uncompressing data dominates the read spead, with the notable exception of the LZ4 compression algorithm.

The HDF5 column-wise performance penalty can be explained by a large number of seek calls due to the use of unchunked tables (cf. Figure 5). This is in contrast to the ROOT and Parquet columnar layouts, where the columnar layout is not applied globally but on blocks of a certain number of consequitve rows. These row groups are read into memory in a single go to avoid seeking on the storage medium.

Figures 3 and Figure 4 show the effect of a very sparse reading of columns. As expected, there is a performance boost for the columnar data formats while the performance of the row-wise formats remains more or less constant. The performance improvement of Parquet, although columnar, is surprisingly small. An analysis of the read pattern in Figure 5 reveals that the entire Parquet file is being read even though only a subset of columns is requested. This behavior is not an inherent problem of the data format but it can be traced back to the use of memory mapped I/O in the Parquet library.

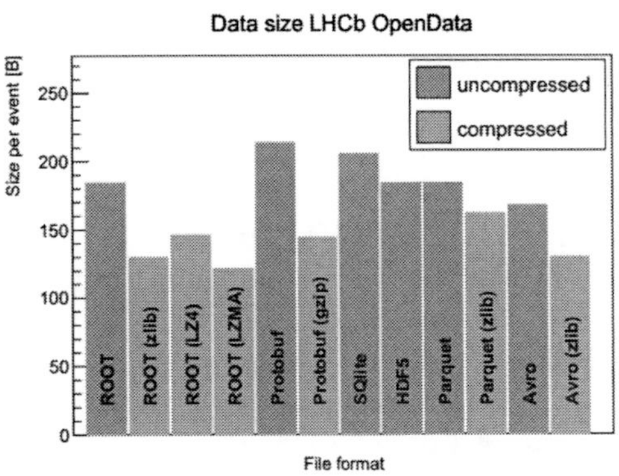

Figure 1. Encoding efficiency of the file formats.

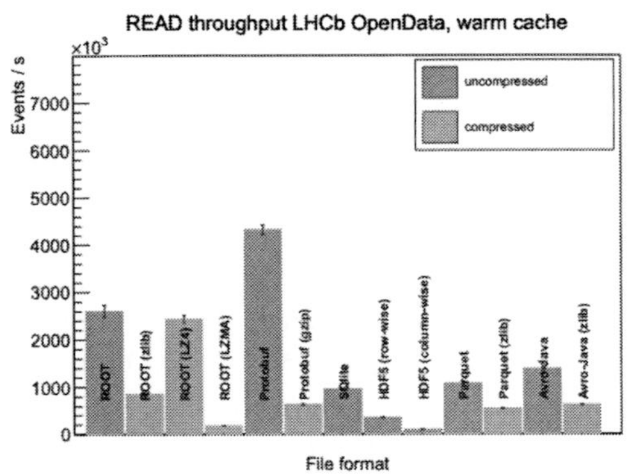

Figure 2. Throughput of reading data from from fast, memory-class storage.

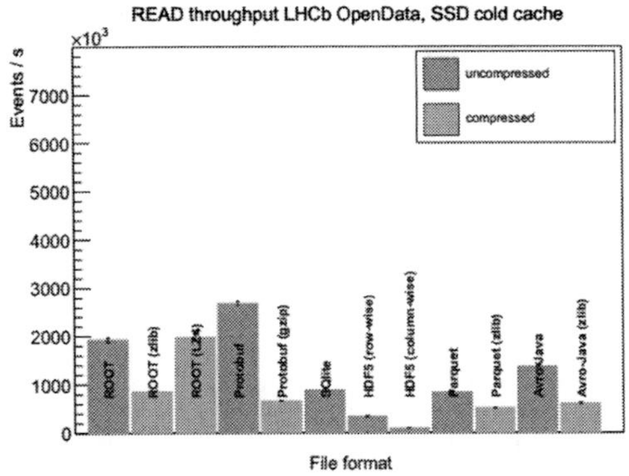

Figure 3. Throughput of reading full event data from the SSD with a cold hard disk cache.

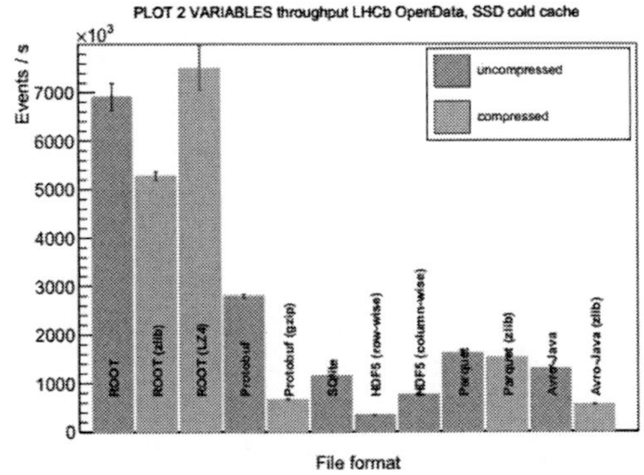

Figure 4. Throughput of sparsely reading data, typically done for plotting data.

IOP Conf. Series: Journal of Physics: Conf. Series **1085** (2018) 032020 doi:10.1088/1742-6596/1085/3/032020

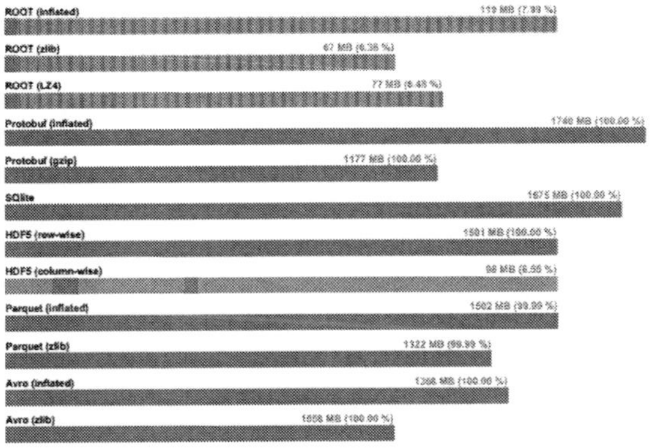

Figure 5. Visualization of read parts (red blocks) and unread parts (blue blocks) of the data file for the benchmark of plotting two columns. Data is collected by a recording fuse [12] file system.

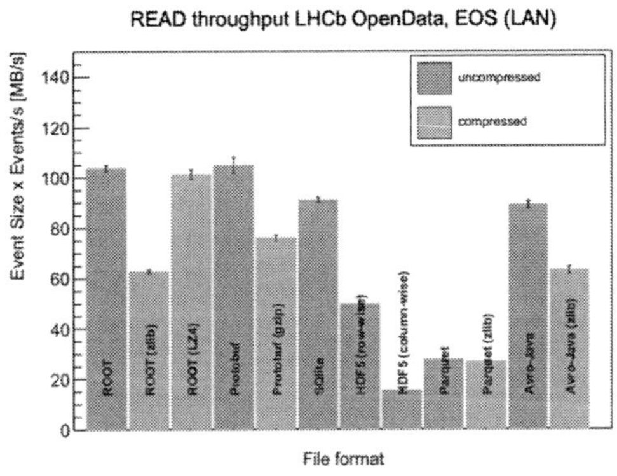

Figure 6. Throughput of reading data from an EOS mountpoint through a 1 GbE link with 20 ms round-trip time. Note that the throughput is shown in MB/s pointing out the network interface as a bottelneck.

Figure 6 shows that for the fastest file formats, ROOT and Google Protobuf, the 1 GbE network interface card becomes a bottleneck. Further tests with network latency increased by traffic shaping show that for high-latency (and thus low-throughput) links, the number of read bytes quickly becomes the dominating factor for performance.

5. Comparison of ROOT file format options and access modes

In this section, we look at the performance impact of different ways of using ROOT. All the following tests are performed on data in warm file system buffers in order to resemble fast storage devices and to emphasize performance differences.

Figure 7 shows the impact of column management to the ROOT serialization speed. When it is known beforehand that all of the columns of a data set need to be read, ROOT can be instructed to drop the columnar storage layout by setting the data record's "split level" to zero. In this test, ROOT has a significant overhead due to handling of potentially self-referencing data records. The overhead can be avoided if, for instance, the data record contains only simple, non-pointer data members. As indicated by the "fixed" data point, in this case ROOT has serialization performance comparable to Protobuf. Note that a proper patch has not yet been sent to the ROOT project though.

Figure 8 shows the impact of different event loop abstractions. Data values can be directly copied into memory areas ("SetBranchAddress") or, through the TTreeReader interface, bound to C++ variables in a type-safe manner. An unnecessary overhead in the type-safety checking of TTreeReader has been identified by the ROOT team and is being followed up. The TDataFrame abstraction, built on top of the TTreeReader interface, provides an interface similar to Python pandas. In this test, it shows some threefold speed-up on 4 cores for its automatic concurrent iteration through the data set. This is particularly beneficial to speed-up decompression.

Figure 9 shows the impact of ROOT splitting options and data record complexity. Two possible C++ representations of the LHCb OpenData data set, FlatEvent and DeepEvent, are sketched one the right hand side of the figure. We presume that particularly for large data records, users make use of ROOT's capability to automatically create columns from C++ class members. This "splitting" does not significantly change that storage layout compared to manual creation of branches. Some of the performance overhead of the DeepEvent class is due to larger

IOP Conf. Series: Journal of Physics: Conf. Series **1085** (2018) 032020 doi:10.1088/1742-6596/1085/3/032020

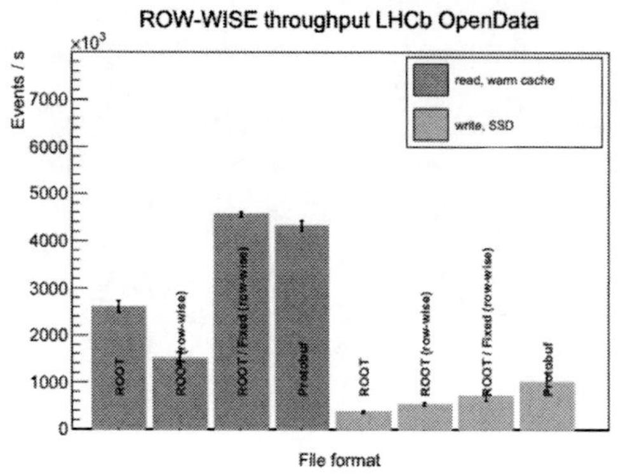

Figure 7. Read and write throughput of pure serialization of data with split level 0 in ROOT compared to Protobuf. "Fixed" refers to a patch avoiding additional code paths related to data records with pointers.

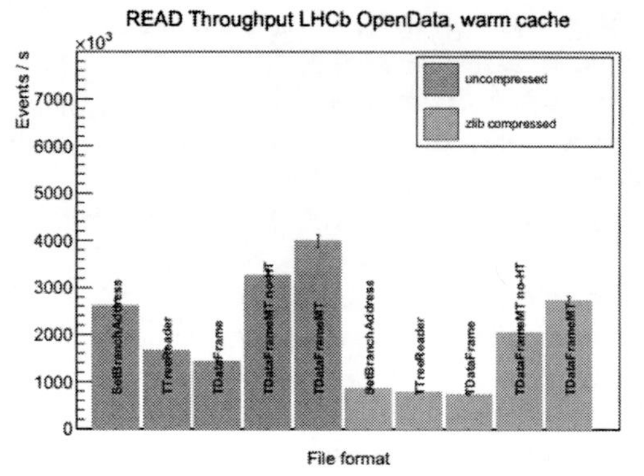

Figure 8. Comparison of event loop abstractions. The "MT" suffix indicates multi-threaded TDataFrame use, the "no-HT" suffix indicates that hyper-threading is turned off.

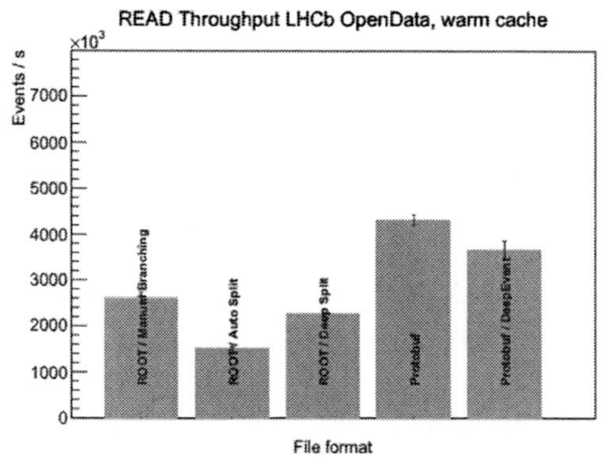


```
struct FlatEvent {
  double h1_px;
  double h2_px;
  double h3_px;
  double h1_py;
  ...
};
```

```
struct Kaon {
  double h_px;
  double h_py;
  ...
};
struct DeepEvent {
  std::vector<Kaon>
    kaons;
};
```

Figure 9. Impact of ROOT splitting options and event complexity. "Manual branching" refers to explicit creation of columns corresponding to the members of FlatEvent. "Auto Split" refers to FlatEvent data records that is parsed and atomatically transformed into columns by ROOT. "Deep Split" refers to DeepEvent data records automatically serialized by ROOT.

file sizes, as every data record also needs to store the size of the Kaon vector. This is reflected in both ROOT and Protobuf. The larger overhead of splitting the simpler data records of type FlatEvent compared to the DeepEvent data records is still under investigation.

6. Conclusion

The benchmarks in this paper show that the turn-around time of physics analyses can be very sensitive to the data format and access library. Overall, ROOT outperforms all other libraries except for few corner cases. In particular, pure serialization is faster with Protobuf, which benefits from not having to deal with column management.

Several smaller issues in ROOT's I/O code were identified in the course of these benchmarks and are currently followed up. The new TDataFrame abstraction for event loops is not only interesting for allowing for more more concise user code but also for its ability to automatically parallelize event loops. Currently, this is a unique ROOT feature. The new LZ4 compression algorithm provides a very interesting trade-off between storage efficiency and processing speed. In particular, for the evolving close-to memory-class storage such as 3D X-Point, LZ4 occurs as the best choice for analysis data sets.

7. Acknowledgements
I want to especially thank Philippe Canal and Axel Naumann for all their help in analyzing ROOT's behavior. I want to thank Jim Pivarski for pointing out the Avro Java library to me and for following up on Parquet's memory mapped I/O behavior with the developers. I want to thank Hervé Rousseau from CERN IT for providing access to their EOS test instance.

References
[1] Brun R and Rademakers F 1997 *Nuclear Instruments and Methods in Physics Research Section A: Accelerators, Spectrometers, Detectors and Associated Equipment A* **389** 81–86
[2] Rogozhnikov A, Ustyuzhanin A, Parkes C, Derkach D, Litwinski M, Gersabeck M, Amerio S, Dallmeier-Tiessen S, Head T and Gilliver G 2016 Particle physics analysis with the cern open data portal talk at the 22nd Int. Conf. on Computing in High Energy Physics (CHEP'16)
[3] Rademakers F, Brun R *et al.* 2017 ROOT - An Object-Oriented Data Analysis Framework. root-project/root: v6.10/04 URL `https://doi.org/10.5281/zenodo.848819`
[4] The Google Protobuf Project, 2017 "protobuf" [software], version 3.3.2 URL `https://github.com/google/protobuf/tree/v3.3.2`
[5] The SQLite project, 2017 "sqlite" [software], version 3.19.3 URL `https://www.sqlite.org/src/info/0ee482a1e0eae22e`
[6] The HDF Group, 2017 "hdf5" [software], version 1.10.0 patch 1 URL `https://support.hdfgroup.org/ftp/HDF5/releases/hdf5-1.10/hdf5-1.10.0-patch1/`
[7] The Apache Avro project, 2017 "avro" [software], version 1.8.2 URL `https://avro.apache.org/releases.html`
[8] The Apache Parquet project, 2017 "parquet-cpp" [software], version 1.2.0 URL `https://github.com/apache/parquet-cpp/tree/apache-parquet-cpp-1.2.0`
[9] Buckley A, Eifert T, Elsing M, Gillberg D, Koeneke K, Krasznahorkay A, Moyse E, Nowak M, Snyder S and van Gemmeren P 2015 *Journal of Physics: Conference Series* **664**
[10] Petrucciani G, Rizzi A and Vuosalo C 2015 *Journal of Physics: Conference Series* **664**
[11] The EOS project, 2017 "eos" [software], version 4.1.26 URL `https://github.com/cern-eos/eos`
[12] Henk C and Szeredi M Filesystem in Userspace (FUSE) `https://github.com/libfuse/libfuse` URL `https://github.com/libfuse/libfuse`

IOP Conf. Series: Journal of Physics: Conf. Series **1085** (2018) 032021 doi:10.1088/1742-6596/1085/3/032021

DAQExpert - An expert system to increase CMS data-taking efficiency

J-M Andre[5], U Behrens[1], J Branson[4], O Chaze[2], S Cittolin[4], C Contescu[5], G-L Darlea[6], C Deldicque[2], Z Demiragli[6], M Dobson[2], N Doualot[5], S Erhan[3], J R Fulcher[2], D Gigi[2], M Gladki[2], F Glege[2], G Gomez-Ceballos[6], J Hegeman[2], A Holzner[4], M Janulis[2,a], M Lettrich[2], F Meijers[2], E Meschi[2], R K Mommsen[5], S Morovic[2], V O'Dell[5], S J Orn[2], L Orsini[2], I Papakrivopoulos[7], C Paus[6], P Petrova[2], A Petrucci[8], M Pieri[4], D Rabady[2], A Racz[2], T Reis[2], H Sakulin[2], C Schwick[2], D Simelevicius[2,a], M Vougioukas[2] and P Zejdl[5,2]

[1] DESY, Hamburg, Germany
[2] CERN, Geneva, Switzerland
[3] University of California, Los Angeles, Los Angeles, California, USA
[4] University of California, San Diego, San Diego, California, USA
[5] FNAL, Chicago, Illinois, USA
[6] Massachusetts Institute of Technology, Cambridge, Massachusetts, USA
[7] Technical University of Athens, Athens, Greece
[8] Rice University, Houston, Texas, USA
[a] Also at Vilnius University, Vilnius, Lithuania

E-mail: `maciej.gladki@cern.ch`

Abstract. The efficiency of the Data Acquisition (DAQ) of the Compact Muon Solenoid (CMS) experiment for LHC Run 2 is constantly being improved. A significant factor affecting the data taking efficiency is the experience of the DAQ operator. One of the main responsibilities of the DAQ operator is to carry out the proper recovery procedure in case of failure of data-taking. At the start of Run 2, understanding the problem and finding the right remedy could take a considerable amount of time (up to many minutes). Operators heavily relied on the support of on-call experts, also outside working hours. Wrong decisions due to time pressure sometimes lead to an additional overhead in recovery time. To increase the efficiency of CMS data-taking we developed a new expert system, the DAQExpert, which provides shifters with optimal recovery suggestions instantly when a failure occurs. DAQExpert is a web application analyzing frequently updating monitoring data from all DAQ components and identifying problems based on expert knowledge expressed in small, independent logic-modules written in Java. Its results are presented in real-time in the control room via a web-based GUI and a sound-system in a form of short description of the current failure, and steps to recover.

1. Introduction

The data-taking efficiency of the Compact Muon Solenoid (CMS) [1] experiment at CERN in 2016 during proton-proton physics, understood as the ratio of recorded over delivered luminosity, was 92.7%, according to Web Based Monitoring (WBM). One of the major roles in operation is played by the Data Acquisition (DAQ) system, which controls the detector readout, event-building and operations of the event filtering farm [2]. The smooth operation of the experiment and its resulting high efficiency are ensured through constant supervision by a crew of operators and experts.

Being able to act within seconds during operations of LHC is one of the challenges of the operator. In order to simplify the operation of the complex DAQ system several tools exist to automate certain operations such as starting and stopping a run, configuring detector hardware and ensuring proper synchronization etc.[3]. These tools facilitate the actions initiated by operators. However, it is not uncommon that the system goes into erroneous states where deeper understanding of the system is required to solve the problem in an optimal way. Experts of all subsystems prepare recovery instructions needed to resume data taking in an optimal way and as quickly as possible. Operators need to effectively select and execute them to make sure CMS operates smoothly.

Over the course of the last few years the list of recovery instructions became larger and more complex. Despite its straightforward form, it takes considerable time for the operator to find the correct remedy when needed. This part of the intervention, the reaction time, defined as the time between the problem occurrence and the start of the corresponding recovery action, is one of two concerns. Wrong or suboptimal decisions were not uncommon. These led to increased recovery time, defined as the time elapsed from the beginning of the recovery action to the resuming of data taking. The sum of the reaction and recovery time are referred to as the intervention time.

Reducing the reaction time and selecting the optimal recovery action are the two areas for improvement that will reduce the overall intervention time and directly affect the efficiency of CMS. We also aim to facilitate the post mortem analysis for human DAQ experts, to aggregate and to report a statistical analysis of the performance of the system and to reduce the need for external help especially outside of working hours without a deterioration of CMS efficiency. In order to achieve these goals we have developed a new expert system, the DAQExpert. It completely replaces the Perl based DAQ Doctor [4] which addressed these issues in the Run 1 DAQ system, but which only had rudimentary functionalities for the Run 2 DAQ system.

1.1. Selecting the optimal recovery action

The time spent recovering is the main contributor to the overall intervention time in CMS and amounts to 82%. Therefore, it is essential to define and select the quickest set of actions needed to bring the system back to operation. Many different recovery procedures may be distinguished, e.g.: stopping and starting a run, (re)configuring and (re)initializing a specific subsystem, sending resynchronization or hard reset commands through the Timing Trigger and Control (TTC) system. They differ significantly in terms of impact they have on the system and how much time they take to complete. Some of them take a few seconds whereas others last for minutes. A common operator mistake is to issue the most general recovery procedure which solves most types of problems but takes longest time instead of using a faster problem specific action. DAQExpert was designed to help the operator make optimal decisions.

1.2. Reaction time

The second area to improve is the time between occurrence of the problem and the first recovery action. There are sound notifications in the control room to draw the attention of the shifter when data flow stops. Identification of the problem and selecting the recovery action takes from a few seconds up to some minutes. We define the *comparison period* as the period of six months of data-taking (August 2016 - July 2017, excluding the Year End Technical Stop that lasted from December 2016 to April 2016, the month of May of 2017 was excluded due to special conditions of data-taking). Figure 1 shows the histogram of a reaction time for *comparison period*. Most of the times the operator reacted within 20-30 seconds. The aim of the project is to reduce the reaction time.

1.3. Facilitating post mortem analysis

Numerous monitoring systems enable access to data necessary to analyze the problem, albeit with some caveats. Some data are available after the problem has been resolved, some only while the problem is ongoing. In many cases some insight can be gained from the monitoring data on how to avoid similar problems in the future. The aim of the project is also to streamline the post mortem analysis by making both historic and real-time data easily available in the same way at any moment.

2. Expert system and architecture

The system consists of three services. Firstly, the *Snapshot Service* aggregates all monitoring information into a snapshot and persists it. Secondly, the *Reasoning Service* applies the knowledge of the experts on the snapshots. Finally, the *Notification Manager* manages and delivers notifications, including e-mail and live suggestions and sound notifications in the control room. The *DAQExpert* brings all parts together in the form of a web application enabling browsing of the whole history of monitoring data to view the analysis and to manage notifications. The design of the *Reasoning Services* and technologies used were driven by external forces.

2.1. External forces

The project's lifetime is expected to be up to 2024 and different developers will maintain it. Thus, one of the main constraints was to avoid short-lived and highly specific technologies which would make it difficult for new team members to quickly become productive. It needs to be easy to understand and modifiable, requiring little effort to learn. Rule-based engines have beed evaluated in the past by the CMS DAQ group and were considered unsuitable due to their steep learning curve. Introducing an additional programming language was not an option. The solution needed to be based on the current skill set of the team members and as simple as possible.

Moreover, the nature of the project foresees frequent logic-related updates, possibly introduced by many authors. Whenever a new recovery solution is advised by sub-system experts the corresponding knowledge needs to be introduced into the system. Finally, high availability is essential, as downtime of the system implies lack of support for the operators and possibly extended downtime of CMS.

As a result, the new expert framework has been designed from the ground up. It allows to define the knowledge in imperative language popular in the group - Java. The knowledge is organized in small independent Logic Modules that are building blocks of the system.

2.2. Monitoring, aggregation and data flow

Multiple steps are needed before information from data sources is made available to the users. The monitoring data sources are queried to get information about data taking health. Monitoring data is collected from multiple heterogenous sources. They differ in terms of resolution, latency and time to retrieve. Diagnosing the problems of the data flow in real time requires quick access to key information. Post mortem analysis requires access to all relevant data at any time. To enable both, a so-called snapshot of the system is taken periodically, approximately every 2 seconds. It contains all necessary information to identify and understand the problem. Thus, the first step was to aggregate all the information in snapshots that bring together both structural data and instantaneous states of all components. The scope of the snapshot has been

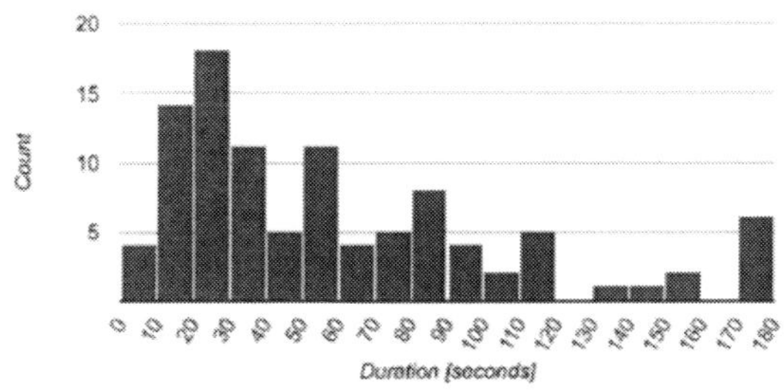

Figure 1. Reaction time histogram.

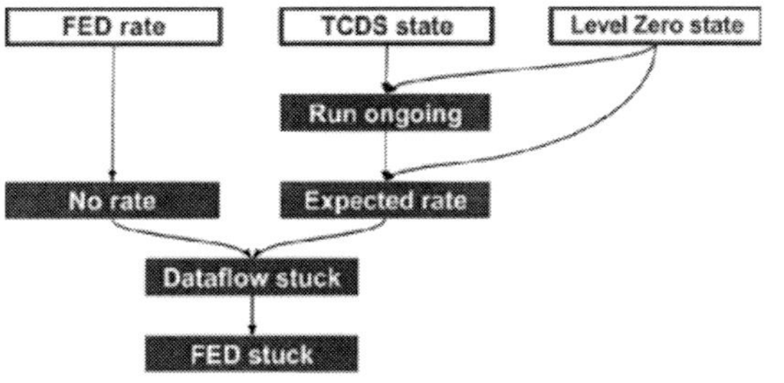

Figure 2. Subset of Logic Module hierarchy (white - observed parameters, blue - LM).

designed to give the full picture of the system's state at a given moment. Both real-time and post-mortem analysis are based on the same data from the snapshots.

While the snapshot is persisted for possible post-mortem analysis, it is simultaneously sent to the *Reasoning Service* for real time analysis. Notifications may result from the analysis and will be dispatched by the *Notification Service* and delivered to the clients.

2.3. Implementation of expert knowledge

It is common in the industry that the expert knowledge is defined in a declarative language. For instance, in a form of if-then-else rules expressed in a high-level rule-based language. As a result of external forces, this way of defining knowledge has not been used. A new custom made framework based on *Logic Modules* has been developed. A *Logic Module* (LM) is a building block of *DAQExpert* logic. Each LM focuses on one specific aspect of the behaviour of the DAQ system:

- each LM defines one condition,
- the definition of the condition is placed in the `satisfy` method,
- this method returns true if condition is satisfied and false otherwise,
- one LM can use results of other LMs.

The knowledge defined in the LMs is used to find the optimal solution when a problem occurs in the data-flow. LMs operate on the snapshot and verify that key indicators and metrics are in expected states and within their specified ranges. As LMs may use outputs of other LMs, there is a predefined order of firing them. Figure 2 presents a subset of LMs responsible for identifying one of many problems that may occur during data-taking. In the example there are five LMs and three parameters being monitored. The LMs are fired in top-down order allowing LMs to rely on others as indicated by arrows.

A LM may play two roles: it might be a sub-step of the analysis (*Expected rate*, *No rate*, *Dataflow stuck*) or final step of the analysis (*Front End Driver (FED) stuck*) delivering a description of the condition and a recovery suggestion. In the example shown in Figure 2, the following chain of LM was activated::

- *Run ongoing* - data taking was active according to the *TCDS state* (TCDS - Timing and Control Distribution System) and *Level Zero state* (Level Zero - Top level node in the run control system),
- *No rate* - the average *FED rate* was 0, there was no data flowing
- *Expected rate* - the *TCDS state* and *Level Zero state* indicate that no recovery action is being issued at the moment, a run is ongoing according to output of *Run ongoing* LM and the readout rate is expected to be non-zero.

- *Dataflow stuck* - this summarizes outputs of two LMs: *No rate* and *Expected rate*, and returns true if both are asserted, meaning that the data-flow is stuck,
- *FED stuck* - starts the analysis when the *Dataflow stuck* condition is found. It performs specific checks to find the FED(s) which caused the data flow to stop. This LM identifies the problem precisely thus it contains instructions for the operator, including a detailed description of the problem and a recovery suggestion.

3. Results

The current version of the service improves all of the outlined areas. Operators of the DAQ system at CMS now have a tool delivering recovery action suggestions in real-time - the Dashboard (see figure 3) at their disposal. Whenever a problem occurs the tool shows the description of the problem and the optimal steps to recover. It helps shifters to take correct decisions, reducing both reaction time and need of external human expert help.

Measuring the success of the project is quite a challenging task. The ultimate goal is to increase the data taking efficiency at the CMS. However, the overall efficiency is not an adequate figure of merit as it depends on many factors in addition to the quality of the DAQExpert suggestions. The detector is constantly being improved, there were major upgrades in both hardware and software which affected the final data taking efficiency of CMS. Feedback of operators and human experts is important as they are the target user groups of the system. As much as they appreciate the new system, the subjective verbal feedback cannot be used as a measure of success as it is not quantitive. Three parameters have been identified to adequately represent the success of the project: accuracy of the recovery action selection, reaction time of the operators and requests for external help. These are practically independent from other factors, which makes them a reliable means of comparison.

The DAQExpert was introduced in mid 2016, but operators were only instructed to use it in early 2017. All comparisons are based on *comparison period* (see section 1.2), corresponding to four months in 2016 without *DAQExpert* support and two months from 2017 where support was available.

3.1. Intervention time

Improvement was mostly observed in the number of optimal recovery decisions taken by operators in case of data-flow interruptions. As stated before, the recovery time is the majority of the total intervention time (82%), thus it is essential to minimize it by choosing the most appropriate and time-efficient recovery actions (optimal recovery actions).

There are two main categories of recovery actions: heavy - lasting usually for 3-4 minutes, initiated by e.g. stopping the run; and light - lasting 7-10 seconds implemented with commands through the TTC system. Suboptimal decisions mean unnecessary time spent on recovery or irrelevant actions that will not solve the problem leading to similar problematic condition after its completion. On average this means that each suboptimal decision adds between 93 and 125 seconds of overhead to the recovery time.

In 2016, during data-taking, operators had no automatic guidance and out of 36 data-taking failures that were diagnosed by the DAQExpert, 32 times the optimal decision was taken. In 2017 there were 29 data-taking failures and all of the decisions were optimal.

The reaction time is a secondary metric to improve, as it is a minor (18%) part of the total intervention time. There is a large variation depending on individual operator experience. The reaction time has slightly improved from 67 seconds in 2016 to 60 seconds in 2017. This is an area where the human factor plays an important role and bypassing the operators will certainly yield better results.

IOP Conf. Series: Journal of Physics: Conf. Series **1085** (2018) 032021 doi:10.1088/1742-6596/1085/3/032021

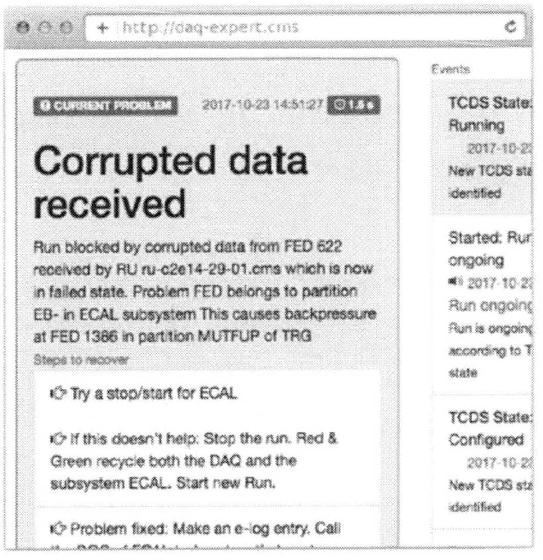

Figure 3. Dashboard - tool to show suggestions in real time.

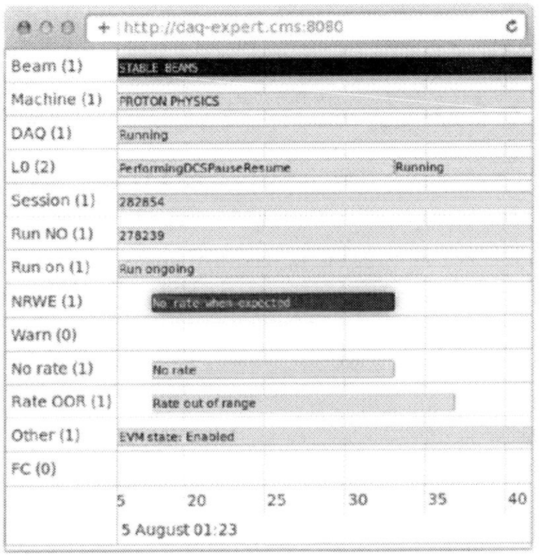

Figure 4. Browser - tool to browse time evolution and to enable post mortem analysis.

3.2. Non-quantitative results

An improvement in the post-mortem analysis has also been made. In the case of more demanding data-flow interruptions, where further investigation is required, the *DAQExpert* streamlines the chores of human experts. First, the go-back in time functions have been introduced that enable human experts to access the historic data easily with the timeline view (*Browser*), shown in figure 4. Secondly, there are more data available to support post-mortem analysis than ever before. All monitored quantities of the system are kept for some period of time for debugging purposes.

3.3. Summary

Both quantitative and qualitative results on the launching of the new system were presented in this report. The recovery action selection accuracy was improved by introducing the expert system. There were no non-optimal recovery decisions observed after deployment of *DAQExpert*, whereas, before they occurred 11% of the time. Additionally the average reaction time shrunk from 67 to 60 seconds, which means that the total intervention time, including reaction time and recovery procedure time, has been reduced by at least 10 seconds. One major improvement to be implemented in the future is to bypass the operator whenever possible. This will shorten the reaction time to a few seconds. External help demand has not yet been quantified. This area is of great interest for on-call human experts as it has the potential to reduce their workload. The number of phone calls (normalized by the number of problems) to the on-call experts before and after *DAQExpert* introduction may reveal trends in this area.

References

[1] CMS Collaboration 2008 *The CMS experiment at the CERN LHC* JINST 3 S08004

[2] Andre J M *et al* 2016 Performance of the new DAQ system of the CMS experiment for run-2 *Proc., 20th IEEE-NPSS Real Time Conf. (RT2016), Padua, Itally June 5-10, 2016*

[3] Andre J M *et al* 2017 New operator assistance features in the CMS Run Control System *22nd Int. Conf. on Computing in High Energy and Nuclear Physics, CHEP 2016, San Francisco, Usa, 10-14 Oct 2016*

[4] Bauer G *et al* 2014 Automating the CMS DAQ *J. Phys.: Conf. Ser.* 513 012031

IOP Conf. Series: Journal of Physics: Conf. Series **1085** (2018) 032022 doi:10.1088/1742-6596/1085/3/032022

Real-time heterogeneous stream processing with NaNet in the NA62 experiment

R Ammendola[2], M Barbanera[4], A Biagioni[1], P Cretaro[1], O Frezza[1], G Lamanna[3,4], F Lo Cicero[1], A Lonardo[1], M Martinelli[1], E Pastorelli[1], P S Paolucci[1], R Piandani[4], L Pontisso[1], D Rossetti[5], F Simula[1], M Sozzi[3,4], P Valente[1] and P Vicini[1]

[1] INFN Roma,P.le A.Moro,2 - 00185 Roma - Italy
[2] INFN Roma Tor Vergata, Via della Ricerca Scientifica,1 - 00133 Roma - Italy
[3] Pisa University, Largo B.Pontecorvo,3 - 56127 Pisa - Italy
[4] INFN Pisa, largo B.Pontecorvo,3 - 56127 Pisa - Italy
[5] nVIDIA corp., 2701 San Tomas Expressway Santa Clara, CA 95050 - USA

E-mail: `luca.pontisso@roma1.infn.it`

Abstract. The use of GPUs to implement general purpose computational tasks, known as GPGPU since fifteen years ago, has reached maturity. Applications take advantage of the parallel architectures of these devices in many different domains.

Over the last few years several works have demonstrated the effectiveness of the integration of GPU-based systems in the high level trigger of various HEP experiments. On the other hand, the use of GPUs in the DAQ and low level trigger systems, characterized by stringent real-time constraints, poses several challenges.

In order to achieve such a goal we devised NaNet, a FPGA-based PCI-Express Network Interface Card design capable of direct (zero-copy) data transferring with CPU and GPU (GPUDirect) while online processing incoming and outgoing data streams. The board provides as well support for multiple link technologies (1/10/40GbE and custom ones).

The validity of our approach has been tested in the context of the NA62 CERN experiment, harvesting the computing power of last generation NVIDIA Pascal GPUs and of the FPGA hosted by NaNet to build in real-time refined physics-related primitives for the RICH detector (i.e. the Cerenkov rings parameters) that enable the building of more stringent conditions for data selection in the low level trigger.

1. Introduction

The introduction of heterogeneity into the computing landscape reflects the effort in focusing the diversity of our computation abilities on achieving a global better performance in solving specific tasks. Being the architectures of CPU-GPU-FPGA vastly different, achieving such a collaborative computing requires reckoning with the varied characteristics of all devices.

We devised a heterogeneous approach with the aim of studying the usage of Graphic Processing Units (GPUs) in the Low Level trigger systems of High Energy Physics experiments. Such systems are real-time environments, requiring low and (almost) deterministic latency; their standard implementation is mostly on dedicated hardware (ASICs or FPGA only board).

In order to reduce and control the latency due to data transfer from network to GPU memory, a dedicated FPGA-based Network Interface Card (NIC) has been designed and developed within the INFN-funded project NaNet.

Our integrated system — a GPU-based trigger — consists of a server equipped with an x86-based CPU (managing the minimal software stack for setup of the configuration), hosting the GPU to perform the computing, and the NaNet board, which manages the data transfers and their online processing.

This configuration allows the exploitation of parallel algorithms tailored specifically on the architecture of GPUs to improve the performance in events selection in the Low Level Trigger.

In the following we give details of the system and report on results obtained with such system along with the implemented algorithm.

2. NaNet architecture

The purpose of bridging front-end electronics and the software trigger computing nodes of HEP Experiments requires a low-latency, high-throughput data transport mechanism.

NaNet features a modular design based on a low-latency PCIe RDMA NIC supporting different network link technologies. This layout is functional for a straightforward deployment in multiple scenarios: standard GbE (1000BASE-T) and 10GbE (10GBase-KR) plus a custom 34 Gbps APElink [1] and 2.5 Gbps deterministic latency KM3link [2].

Its key characteristics are i) the management of custom and standard network protocols in hardware, in order to avoid OS jitter effects and guarantee a deterministic behaviour of communication latency while achieving maximum capability of the adopted channel; ii) a processing stage which is able to reorganize data coming from detectors on the fly, in order to improve the efficiency of applications running on computing nodes; iii) a DMA engine performing zero-copy data transfers to or from application memory in order to avoid bounce buffers.

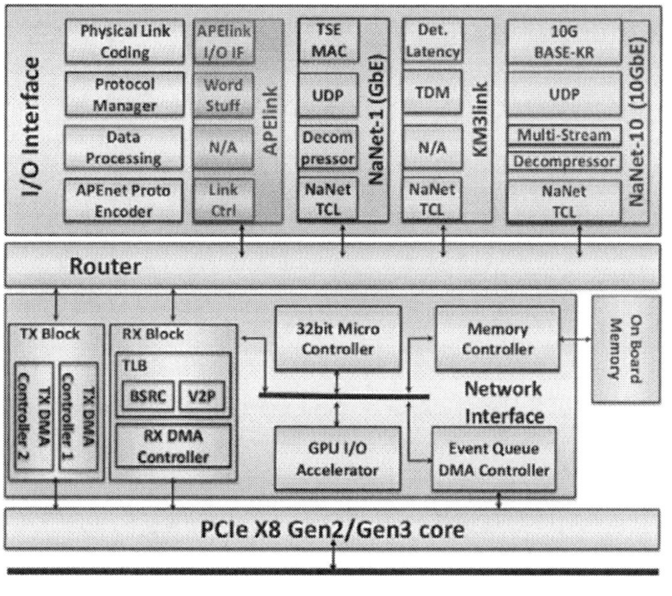

Figure 1: NaNet architecture schematic.

NaNet GPUDirect RDMA capability, inherited from the APEnet+ 3D torus NIC dedicated to HPC systems, extends its application range into the world of general-purpose computing on graphics processing units (GPGPU).

The I/O Interface module performs a 4-stages processing on the data stream: according to the OSI Model, the Physical Link Coding stage implements the channel physical layer (e.g. 10GBase-KR) while the Protocol Manager stage handles data/network/transport layers

IOP Conf. Series: Journal of Physics: Conf. Series **1085** (2018) 032022 doi:10.1088/1742-6596/1085/3/032022

(e.g. UDP); the Data Processing stage implements application-dependent manipulation on data streams (e.g. compression/decompression) while the APEnet Protocol Encoder performs protocol adaptation. This is carried out encapsulating inbound payload data into the APElink packet protocol (used in the inner NaNet logic) and decapsulating outbound APElink packets before re-encapsulating their payload into the output channel transport protocol.

2.1. Software stack

The NaNet board is capable of writing data coming from a network source directly into the GPU memory through the PCIe bus, using GPUDirect RDMA (see figure 2, left).

Following such write, a "receiving done" event is also DMA-written in a memory region called "event queue" and trapped in kernel-space by a device driver which sends a signal to the user application. This latter launches the consuming application on GPU.

Once the processing is completed, the results can eventually be sent via NaNet board to a remote device. Data are DMA-read directly from GPU memory; the kernel device driver — invoked by the user application on the host machine — instructs the NIC by filling a "descriptor" into a dedicated DMA-accessible memory region called "TX ring".

The presence of a new descriptor is notified over PCIe to NaNet by writing on a doorbell register so that the board can issue a "TX done" completion event in the "event queue".

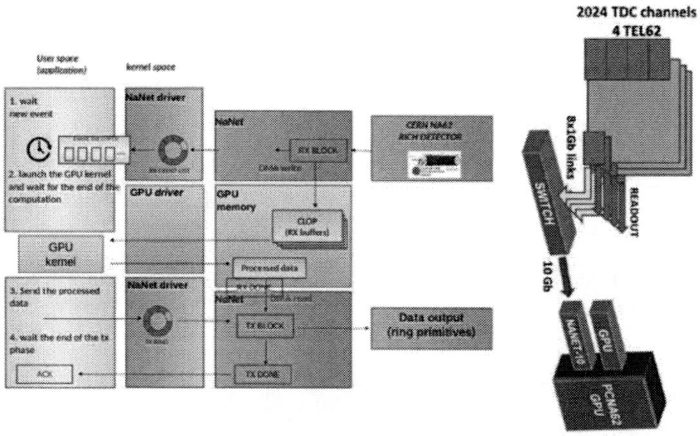

Figure 2: NaNet software stack (left) and GPU-RICH readout system (right).

2.2. NaNet-10

NaNet-10 is a PCIe Gen2/3 x8 network adapter implemented on the Terasic DE5-net board equipped with an Altera Stratix V FPGA featuring four SFP+ cages.

The board features a zero-latency full UDP/IP network stack on the 10GbE channels, as well as a high throughput PCIe core towards the host with GPUDirect capabilities [3].

3. Physics case: heterogeneous processing in the NA62 Low Level Trigger

The NA62 experiment at CERN is designed to measure the branching ratio of the ultra-rare decay of the charged kaon into a pion plus a neutrino-antineutrino pair.

The sub-detectors are located along the trajectory of the K^+ beam with the purpose of identifying decay particles and to measure their momentum and energy.

Due to an expected ~10 MHz rate of these decays, a cascade of three trigger levels is needed to reduce the rate by three orders of magnitude to manage the stream of raw data [4].

The low-level trigger (L0) performs the early selection of events with a time budget of ~1 ms. It is a synchronous real-time system implemented in hardware through FPGA boards.

The GPU-based processing stage is inset between the RICH readout and the L0 Trigger Processor (L0TP) with the task of computing in real-time physics-related primitives (i.e. centers and radii of Čerenkov ring patterns on the photomultipliers arrays), in order to improve the low level trigger discrimination capability.

Data from the detector photomultipliers (PMTs) are collected by four readout boards (TEL62) sending primitives to GPU as UDP datagrams on a dedicated GbE link (see figure 2, right) connected to a multiport GbE/10GbE switch. Packets are then received over the NaNet-10 10GbE channel. The onboard FPGA performs zero-copy of the raw data onto the GPU after decompressing and coalescing the event fragments scattered among the four streams.

Data are gathered and arranged in the GPU memory as a Circular List Of Persistent buffers (CLOP), according to a configurable time window which must be shorter than the total processing time to avert the overwriting of the buffers before they are consumed.

A system such as this, displaying improved processing capabilities and flexibility and based on different computing units, can be regarded as a heterogeneous "smart detector".

3.1. Overview of the histogram-based algorithm

In order to identify the ring patterns in the PMT hit map, we developed a specific pattern recognition histogram-based algorithm to leverage the massively parallel architecture of the GPU. Obviously, working in a real-time environment, this must be fast enough to cope with the input events rate while remaining within the L0TP 1 ms time budget.

Another constraint on the design is exerted by the need to be seedless, as data coming from other detectors are not available at this stage.

Preliminary to the ring reconstruction stage there is the indexing of events gathered and merged by NaNet into a CLOP. Many threads launched by a single Event Finder kernel look for event delimiter patterns in the buffer and the resulting vector of memory addresses is then passed to the processing kernel.

On a general level, the computing is arranged so that each event is processed by a whole block of threads, leveraging the GPU per-block shared memory faster than the global one.

Taking into account that $\sim 95\%$ of K^+ decays have only one charged particle in the final state, resulting in only one ring in the RICH detector, the overall processing of events is sped up using the histogram-based algorithm only for multi-rings events while a faster single-ring fit algorithm like Crawford [5] (a least square method, since iterative ones are not easily implemented in a parallel way) is employed for the most frequent cases.

The decision about which is the right one to use is based on a χ^2 test involving distances between the hits.

At a block level (i.e. for each event), every thread holds the coordinates of a single PMT hit. A χ^2 test between center and radius estimated with the Crawford algorithm using hits and their center of gravity is performed. If the revenue is less than a threshold value, the hypothesis of a single ring is accepted. If the test fails, the event is passed to the histogram-based method (details in figure 3).

A grid of 8×8 squares is established, associating each element to a thread for a total of 64 of them. For each square, the distances between its center and every hit are computed. These values are inserted into a histogram. The bins with the higher number of occurrences (above a threshold) are selected.

A comparison between the values of neighbouring grid elements makes us choose the local maxima.

These are the centers for the candidate rings. For each one we select the hits whose distances fall within an annular area (defined by maximum and minimum physics-related radii). For each

IOP Conf. Series: Journal of Physics: Conf. Series **1085** (2018) 032022 doi:10.1088/1742-6596/1085/3/032022

```
Step 1 Making a decision about single- or multi-ring approach
Require: device wide parallelization event→block
Require: block wide parallelization hit→thread
  function Crawford_fit() → (X_est, Y_est, R_est)
  evaluate χ² with distances between hits and (X_est, Y_est, R_est)
  if χ² < threshold then
    one ring found
  else
    passing to the histogram-based method
  end if
```

Figure 3: Illustrative code dealing with different paths in processing single or multi-ring events.

set of hits (associated to a candidate ring) we apply again the Crawford algorithm to evaluate center and radius with a better accuracy — see figure 4.

```
Step 2 Histogram
Require: device wide parallelization event→block
  parallelization element of grid→thread
  function histogram_selection() → (ring_cand, X_cand, Y_cand, R_cand)
  parallelization hit→thread
  for every ring_cand do
    function select_hits_in_annular_region() → (x_sel, y_sel)
  end for
  parallelization ring→thread
  function Crawford_fit() → (X_est, Y_est, R_est) using (x_sel, y_sel)
```

Figure 4: Schematic about parallel implementation of the histogram algorithm.

4. Experimental results

The whole GPU-RICH system equipped with the NaNet-10 board is currently installed at the NA62 site (working in a completely parasitic mode with respect to standard trigger), collecting data during the 2017 run.

The testbed is made up of a HP2920 switch, a NaNet-10 PCIe board plugged into a server made of X9DRG-QF dual-socket motherboard populated with Intel Xeon E5-2620 @2.00 GHz CPUs (i.e. Ivy Bridge architecture), 32 GB of DDR3 RAM, and a Kepler-class NVIDIA K20c GPU (until October, a NVIDIA Pascal P100 was employed thereafter).

Latencies of the stages corresponding to GPU processing using the K20c (event indexing and ring reconstruction) and sending UDP packets with the results to the L0TP are shown in figure 5. Data were taken during single bursts with beam intensity 18×10^{11}pps and a gathering time for NaNet-10 of 350 μs (dashed line on plots). The red area represents the latency of event indexing in the CLOP buffer (almost constant), the cyan area shows the ring reconstruction kernel latency and the blue is the sending stage.

On the left, data acquisition is performed processing only every eighth event (i.e. downscaling factor of 8) and the overall latency is below the time limit posed by gathering time. The plot on the right displays instead results with a downscaling factor of 4. The ring reconstruction time increases with the number of processed events and in this case the total latency exceeds constantly the time budget. This is an unwanted situation, leading to loss or corruption of data.

Having identified the ring-fitter kernel as the culprit for this condition, we investigated the possibility of increasing the computing speed with the availability of the new Pascal NVIDIA architecture. Our tests show (figure 6) a speedup factor of 5 on processing buffers with 2000 events: from ~500 ns/evt on K20c to ~100 ns on P100. This suggests that with a new GPU the system downscaling can comply with the time budget request also in a no downscaling mode.

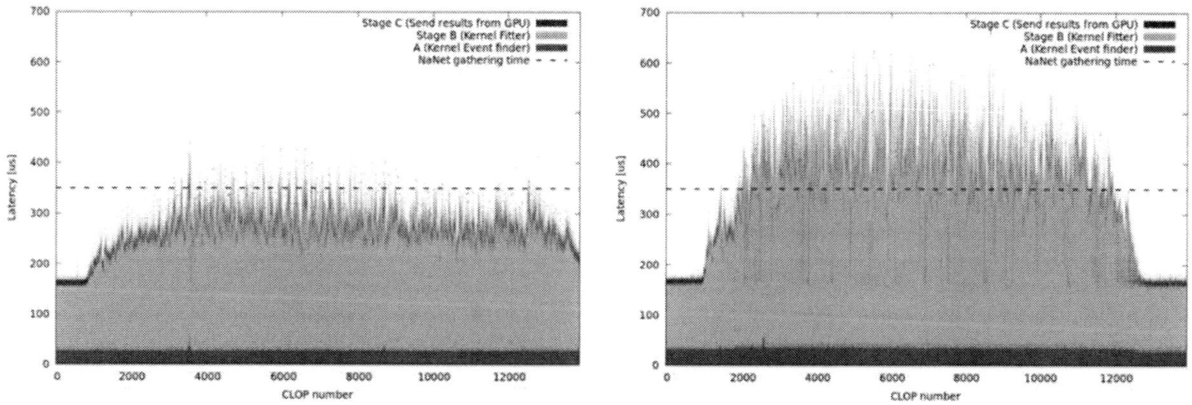

Figure 5: GPU processing latency.

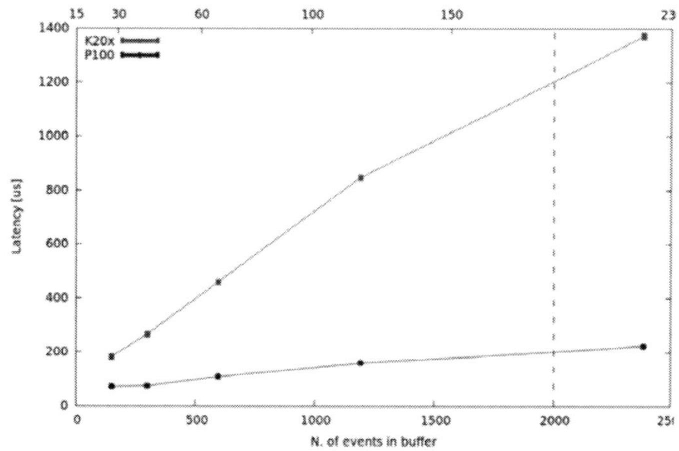

Figure 6: Ring reconstruction latency: NVIDIA Pascal (P100) vs K20c.

5. Conclusions

The widespread availability of heterogeneous architectures led us to assess their potential in the challenging real-time environment of low level trigger systems for High Energy Physics experiments. The whole NaNet-based framework turns out to be able to cope with the strict requirements for the NA62 RICH Low Level Trigger with respect to events rate and time budget.

The hurdle of the latency in the ring reconstruction might be overcome, according to our tests, with a GPU upgrade. We plan to install the new NVIDIA P100 in order to keep the process latency below the threshold without any downscaling factor and to prove the ability of the NaNet-based system to sustain the full detector throughput.

References

[1] R. Ammendola, A. Biagioni, O. Frezza, A. Lonardo, F. Lo Cicero, P. Paolucci, D. Rossetti, F. Simula, L. Tosoratto, and P. Vicini, "APEnet+ 34 Gbps Data Transmission System and Custom Transmission Logic," *Journal of Instrumentation*, vol. 8, no. 12, p. C12022, 2013.

[2] A. Aloisio, F. Ameli, A. D'Amico, R. Giordano, V. Izzo, and F. Simeone, "The NEMO experiment data acquisition and timing distribution systems," in *Nuclear Science Symposium and Medical Imaging Conference (NSS/MIC), 2011 IEEE*, pp. 147–152, Oct 2011.

[3] R. Ammendola et al., "Nanet-10: a 10GbE network interface card for the GPU-based low-level trigger of the NA62 RICH detector.," *Journal of Instrumentation*, vol. 11, no. 03, p. C03030, 2016.

[4] B. Angelucci et al., "The fpga based trigger and data acquisition system for the cern na62 experiment," *Journal of Instrumentation*, vol. 9, no. 01, p. C01055, 2014.

[5] J. Crawford, "A non-iterative method for fitting circular arcs to measured points," *Nuclear Instruments and Methods in Physics Research*, vol. 211, no. 1, pp. 223 – 225, 1983.

Calibration and Alignment Framework for the Belle II detector

T Bilka and J Kandra et al. on behalf of the Belle II Collaboration
Charles University, Ovocny trh 3, Prague 1, 116 36, Czech Republic
E-mail: jakub.kandra@karlov.mff.cuni.cz, bilka@ipnp.troja.mff.cuni.cz

Abstract. The Belle II experiment [1] is approaching its first physics run in 2018. Its full capability to operate at the precision frontier will need not only excellent performance of the SuperKEKB accelerator and the detector, but also advanced calibration methods combined with data quality monitoring. To deliver data in a form suitable for analysis as soon as possible, an automated Calibration Framework (CAF) has been developed. The CAF integrates various calibration algorithms and their input collection methods for event-level data. It allows execution of the calibration workflow using different backends from local machines to a computing cluster, resolution of dependencies among algorithms, management of the produced calibration constants, and database access across possible iterations. One of the main algorithms fully integrated in the framework uses Millepede II [2] to solve a large minimization problem emerging in the track-based alignment and calibration of the pixel and strip detector, the central drift chamber, and the muon system. Advanced fitting tools are used to properly describe the detector material and magnetic field and include measurements of different sub-detectors into a single global fit performed by Millepede. This talk will present the design of the calibration framework, the integration of the Millepede calibration, and its current performance.

1. Belle II detector

Belle II is an experiment under construction at the High Energy Accelerator Research Organization in Tsukuba, Ibaraki prefecture, Japan. As an upgrade of the Belle experiment, it is expected to accumulate a 50 times larger data sample ($\sim$ 50 ab^{-1}). Belle II will continue and extend the physics program of Belle, marking the beginning of the precision era in CP violation measurements in B-meson decays, and search for new physics. The SuperKEKB accelerator is a next generation high luminosity B-Factory designed to reach an instantaneous luminosity of $8 \times 10^{35} cm^{-2}s^{-1}$. The collider can provide a clean environment for production of B$\bar{\text{B}}$ pairs, $\tau^+\tau^-$ pairs and $c\bar{c}$ states. The Belle II detector utilizes a tracking system, particle identification detectors (PID), an electromagnetic calorimeter (ECL) and a K_L-muon (KLM) system (Fig. 1).

2. Calibration and alignment framework

The calibration and alignment framework (CAF) [3] provides automated calibration procedures of subdetectors of Belle II and is part of the Belle II analysis and simulation framework (basf2)[4]. The basf2 framework uses a modular architecture, C++ libraries and modules for event processing, together with extensive python bindings which allow for easy access to all framework features from python, the steering language of basf2. The basf2 framework controls all aspects of the detector simulation, reconstrution, analysis, and also data aquistion and data

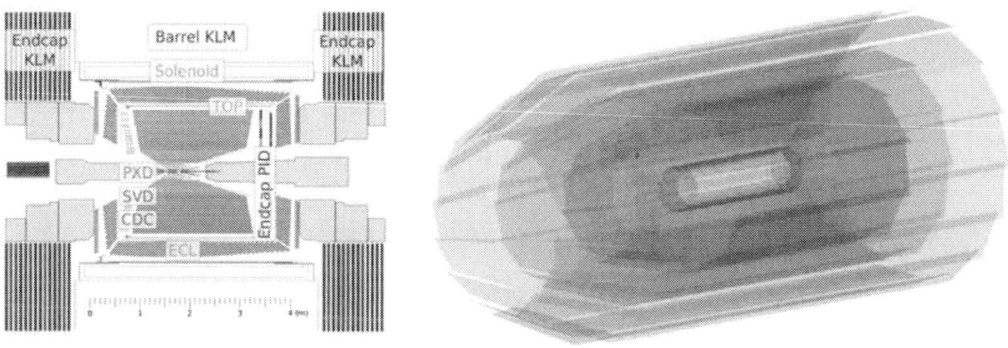

Figure 1. *Schematic cross-section of the Belle II detector (left):* The PXD (light green), the SVD (orange) and the CDC (blue) compose the tracking system. The TOP (pink) and the ARICH (violet) detector are particle identification detectors. The crystals of the ECL (yellow) surround the PID detectors. The outermost detectors are the barrel and endcap KLM (dark green).
The PXD and SVD in detail (right): Pixel sensors form first (gray) and second (blue) layer. Third (green), fourth (red), fifth (orange) and sixth (pink) layers compose strip detector.

processing. The CAF is a natural extension to the framework profiting from the mature state of basf2, extending its capabilities to calibration of the detectors using collected data, with the goal to deliver data with a quality suitable for physics analysis. The CAF standardizes and simplifies calibration development by exploiting the python side of basf2, where development is much more flexible to define data processing workflows, while the CPU-intensive calculations use compiled C++. The CAF should motivate developers to implement often existing algorithms within basf2 by providing an easy-to-use interface. This interface is similar to a well established access to data objects in basf2 and should thus be familiar to most basf2 developers, usually physicists.

The calibration procedures can be processed collectively in a predefined sequence. The procedures might run on input data with different run number and take into account variable time dependence of the calibration constants which are stored in a database. To reduce wall-clock time, independent procedures can be called in parallel. The calibration algorithm dependencies are resolved by CAF to determine the order and distribution of individual jobs, running data collection whose output is then passed to the specified algorithm. It is possible to manage and merge intermediate data objects (histograms, trees, or custom mergeable objects). The CAF should monitor the data quality to automate the workflow of the calibration process as much as possible. Data processing and final calibration constants should be monitored as well. The validity range of the constants is usually determined by CAF from input data, but can be adjusted by individual algorithms. The input data will be organized by run and experiment number, but the granularity can be arbitrary, even event-dependent.

The input data for calibration and alignment procedures comprise either simulated events offline or raw data, which could also be directly selected using the high level software trigger in the online configuration.

3. Design of CAF

The calibration and alignment framework is composed from two virtual C++ classes providing interfaces to calibration developers: *collector module* and *calibration algorithm*. A Python framework, with which the calibration user interacts via a Python interface [3], controls the

IOP Conf. Series: Journal of Physics: Conf. Series **1085** (2018) 032023 doi:10.1088/1742-6596/1085/3/032023

execution of collectors and algorithms.

1) **Collector module:**
Collector module builds data objects for the calibration algorithms. It has two parts: preparation and collection. In the first part, a user defines and initializes objects (trees, histograms, etc.), which are filled in the second part of collector module.
For example, *MillepedeCollector* is used to align the Belle II tracking detectors (PXD, SVD, KLM, CDC). This module re-fits the selected reconstructed tracks with General Broken Lines in the collection stage and outputs track data including all measurements with their uncertainities and the full track covariance matrix, later needed by the Millepede II algorithm (for explanation see **Alignment tools**).

2) **Calibration algorithm:**
Calibration algorithms is a base C++ class which developers inherit from to implement their own algorithm. It analyses collected data and determines calibration and/or alignment constants. The output status is one of:
   a) *Success*: The algorithm was successfully finished. Calibration or alignment constants are saved to current local database.
   b) *Iterate*: The algorithm was successfully finished, but iteration of collector and calibration is requested. The calibration or alignment constants are saved in current local database.
   c) *Not enough data*: The algorithm did not have enough data to successfully finish. No constants are saved.
   d) *Failure*: The algorithm failed for any other reason. No constants are saved.

3) **Python framework and user interface:**
To provide the user with workflow flexibility, the Python interface makes it possible to separate the collection and calibration stages and to include or exclude other related standard modules (simulation, reconstruction, etc.) and user-written modules. The Python interface permits parallel processing of data from multiple runs. Fast turn-around can be achieved by using a computing cluster as processing backend. During calibration development, it might be more convenient to run separated algorithms locally for testing and debugging. The backend, splitting jobs locally among processor cores, is easily changeable for a job submission system allowing for a very large number of parallel processing jobs.

4. Alignment tools

One important function of CAF in the future Belle II experiment will be the position alignment of the detector sensors. As the pixel detector is mounted directly to the beam pipe, it is mechanically independent from the rest of the detector. It is expected that the relative position of both PXD and SVD will need to be continually monitored and aligned to deliver high quality data. This might gain additional importance with the data reduction schemes for the pixel detector. Another task is monitoring of the primary beam spot, whose position is a crucial input of the time dependent CPV analysis. To deal with possible correlations among the sub detectors, we use a global alignment method using the Millepede II (MP-II) tool [2], integrated in basf2 and CAF. MP-II is used for track-based alignment of the vertex detector (PXD + SVD = VXD), the beam spot position and the K_L and muon system (KLM). An interface to the CDC already exists and is being tested with first cosmic data. The integration of several completely different sub detectors profits from the generic approach of the MP-II, General Broken Lines refit[5] and GENFIT toolkit[6], described below.

i) **Millepede II algorithm**
The Millepede II algorithm is a tool for determination of a large number of (for example)

alignment constants based on simultaneous (linearised) minimisation of measurement residuals with respect to all track and alignment parameters. As no approximations are used, except linearization, all correlations are kept in the solution. For nonlinearities, the procedure can be iterated.

ii) **General Broken Lines**
The General Broken Lines (GBL) method provides fast track refit with proper treatment of multiple scattering. The method introduces additional fit parameters: kink angles at predefined scattering points. A track seed is propagated in detector material to parametrize multiple scattering and populated with non-measurement scattering points. The constructed global covariance matrix of the track, containing correlations among detector layers induced by the multiple scattering, computed by GBL, is a crucial input of MP-II. In CAF it can be stored in binary files for MP-II or in an intermediate state in a ROOT tree.

ii) **GENFIT**
The GENFIT toolkit for generic track fitting in high energy physics experiments was heavily extended for the needs of the Belle II experiment, including integration of the General Broken Lines fitting method. GENFIT allows to seamlessly integrate measurements from different types of detectors and include them in the fit. It also contains all tools needed for extrapolation in the detector and handling its material distribution, important input to the General Broken Lines.

The alignment tools use initial positions of sensors, wires or plates and reconstructed track information. The calculated corrections to alignment constants of those elements are stored in a database and used for subsequent data collection in CAF. Due to the complexity of the experiment, it is desirable to check the validity of the used calibration procedures by ad-hoc studies or by developing an alternative method. An example is the drift chamber (CDC) where code inspired by the existing approach at Belle was used for the first acquired cosmic data. The MP-II technique might here be independently compared to a different procedure, for which CAF seems to be an ideal playground.

5. Types of calibration and alignment

The same or dedicated procedures might be employed at different levels of the calibration workflow. The introduced aproach can for example be used for a differential alignment of PXD, where the position and orientation of individual sensors are kept fixed, while the PXD is allowed to move as a whole or e.g. by layer. For such cases, a few hundred tracks are sufficient to determine accurate alignment parameters. On the other hand, for the finest calibration at sensor or wire level, many large track samples with different topologies, including magnet-off cosmics, are needed. In this sense we divide the calibration to several types, listed below.

I) **Online** calibration uses a small set of quickly reconstructed data. The online reconstruction uses a tailored set of quick algorithms for partial event reconstruction. The calibration of beam spot position is an example of a continuously monitored quantity important also for the accelerator, for which dedicated fast turn-around algorithms might be needed.

II) **Fast** calibration and alignment will determine constants required to produce analysis quality data. It will trade accuracy for speed. Calibration and alignment will be done during data conversion. Turn-around time should be less than one day from collection of the input raw data.

III) **Offline or reprocessing** calibration and alignment will run as many calibration algorithms as required to produce best-quality constants. Large datasets will be used including cosmic ray and field-off data.

IOP Conf. Series: Journal of Physics: Conf. Series **1085** (2018) 032023 doi:10.1088/1742-6596/1085/3/032023

6. Alignment and calibration examples

A) Vertex detector (VXD, PXD and SVD together)

- In simulated data (with only gaussian hit smearing to account for sensor resolution and only MC track finding), sensors are randomly misplaced by up to ~200μm (u, v, w) and misaligned by up to ~3.5 mrad (alpha, beta, gamma).
- After alignment, which considers all sensors to be rigid bodies with 6 parameters, residuals are ± 4.6 μm (du, dv, dw) and ± 0.1 mrad (dalpha, dbeta dgamma).
- Residual misalignment is almost independent on the degree of random misalignment. Dedicated studies are being performed for alignment of a systematically deformed detector.

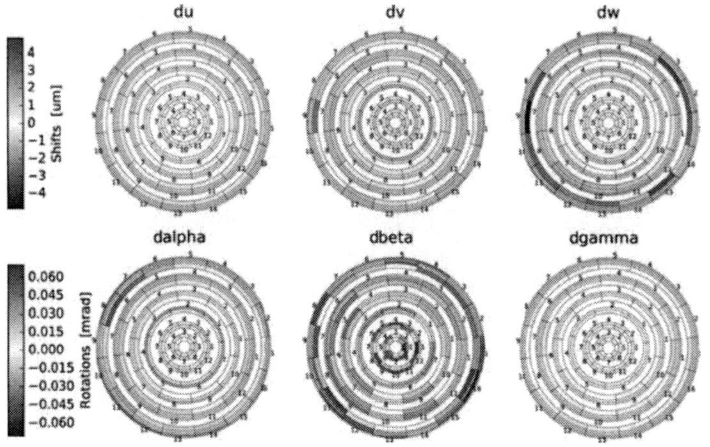

Figure 2. *Alignment of mis-aligned VXD in simulation:* The largest alignment corrections can be seen in wedge SVD sensors (innermost in last 3 SVD layers). The beta angle shows the largest residual misalignments as it is sensitive to the shorter size of the silicon sensors, making resolution in this angle worst.

B) Central drift chamber (CDC)

The CDC employs its own well established procedures for calibration of many different aspects of the drift chamber, including wire time offsets, x-t relation and resolution parametrization. It also has its own alignment procedure which was applied at layer level, allowing for layer shifts and rotations to be determined using the initial cosmics data. Example results showing the improvement in helix parameter estimation for cosmic tracks can be seen in fig 3.

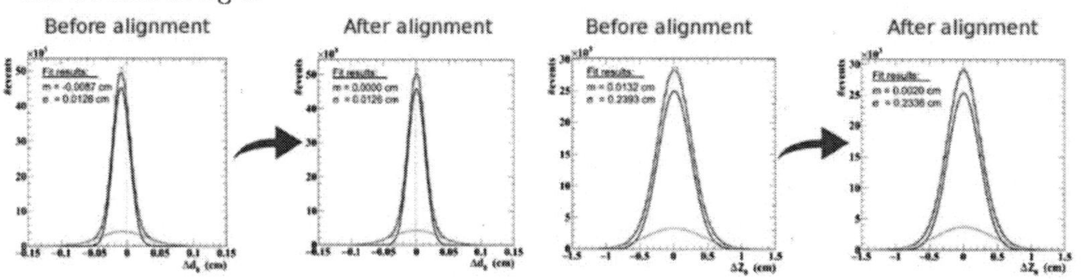

Figure 3. *Calibration of CDC using real data*: Distributions for Δd_0 (left double plots) and ΔZ_0 (right double plots) before and after using alignment procedure.

C) K_L and muon detector

KLM, divided into barrel (BKLM) and endcap (EKLM) parts employs a slightly different alignment approach. However, for both detectors the approach is very similar to the silicon detector, where the individual segments or modules of the detectors are considered as rigid bodies. In fig. 4, an example from a study of weak alignment modes using different simulated track samples can be seen for BKLM.

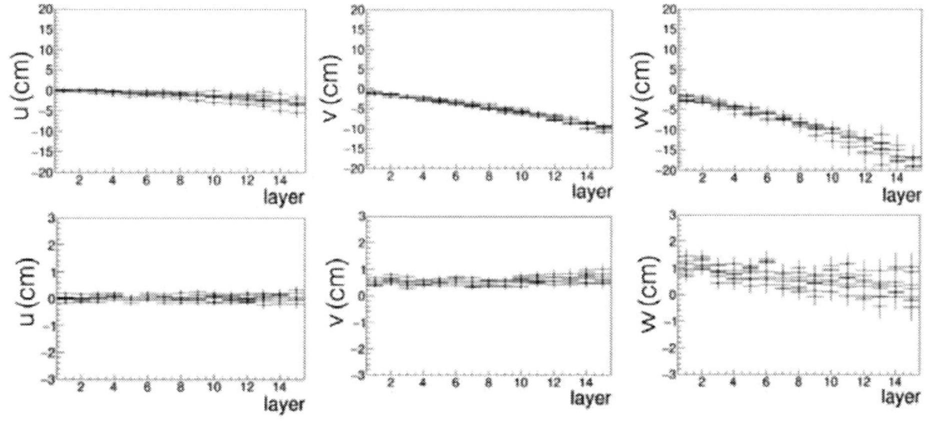

Figure 4. *BKLM alignment using simulation*: Comparison results of alignment using single muons (top line) and vertex-and-beam-constrained pairs of muons (bottom line). With the second sample, the vertex constraint and usage of the whole muon pair as a single object significantly reduces some weak modes in the alignment constants.

7. Summary

Calibration and alignment of the Belle II detector is a complex task which requires significant effort from many experienced physicists. The CAF framework presented here aims to allow experts to concentrate their effort on algorithm development and minimize time spent on coding and ensuring scalability. Large scale application of the Millepede II algorithm for several Belle II sub-detectors is one example, which will profit from CAF features. With the initial cosmic ray data, and with first physics data in 2018 without the full vertex detector, the algorithms are expected to evolve and in many cases be fully integrated within CAF. In 2018 they need to reach the performance expected for full detector operation and this paper presented some of the continuous work required to reach that goal.

Acknowledgments

This project was supported by Charles University, project GA UK no. 404316.

References

[1] *Z. Dolezal, S. Uno et al.:* Technical Design Report, arxiv: 1011.0352v1
[2] *V. Blobel, C. Kleinwort:* A new method for the high-precision alignment of track detectors, arXiv:hep-ex/0208021
[3] *D. Dossett:* Status of Alignment and Calibration Framework at Belle II Experiment, goo.gl/SEYfRB [5th August 2017]
[4] *D. Y. Kim* Software and Physics Simulation at Belle II arXiv:1510.09008 [physics.ins-det]
[5] *C. Kleinwort:* General Broken Lines as advanced track fitting method, arXiv:1201.4320 [physics.ins-det]
[6] *J. Rauch, T. Schluter:* GENFIT a Generic Track-Fitting Toolkit, J. Phys.: Conf. Ser. 608 012042 (2015)

IOP Conf. Series: Journal of Physics: Conf. Series **1085** (2018) 032024 doi:10.1088/1742-6596/1085/3/032024

Parallelization and vectorization of ROOT fitting classes

X Valls Pla[1,2], L Moneta[1]
[1]CERN, [2]Universitat Jaume I
E-mail: root-dev@cern.ch

Abstract.
We introduce parallelization and vectorization in ROOT mathematical and statistical libraries in order to take full advantage of new computer architectures and to maximize the CPU usage with an increasing amount of data to analyze.

As part of this effort, we define in ROOT new generic classes supporting a task-based parallelization mode, which can be used for a wide range of computational tasks in the field of High Energy Physics. We also include support for different SIMD libraries, by including in ROOT a new generic library for vectorization.

These different parallelization tools are applied together when parallelizing the minimization process for solving fitting problems. We report on the improvements obtained by adding the support for SIMD vectorization and multithreaded parallelization when fitting ROOT histograms and datasets represented by ROOT trees.

1. Introduction

Many operations in High Energy Physics (HEP) data processing can be classified in generic patterns or algorithms reproduced over several analyses, adapted to the specific problems at hand.

Our goal is to identify these patterns in the HEP analysis processes, and integrate in the ROOT [1] software framework a set of tools to parallelize them across multiple levels, starting from the most common patterns, such as MapReduce, or the ones that provide more potential, such as the integration of tools for vectorization and the adoption of SIMD types.

In this paper, we describe the generic tools developed in ROOT and their contribution to the parallelization of fitting in ROOT.

2. Tools for parallelism

The strategy that ROOT adapts to speed up computations consists in exploiting both data-level and task-level parallelism. For this purpose, ROOT provides several generic classes for the expression of parallelism at different levels.

The tool used to implement task-level parallelism, ThreadExecutor [2], provides the MapReduce [3] framework to partition the evaluation into tasks and distribute the computational workload among the system threads.

In order to accomodate data-level parallelism, we rely on a dedicated library—VecCore [4]—to abstract SIMD operations and data-types from the code, making the solution portable between architectures with different SIMD instructions sets.

2.1. Task-level parallelism: TThreadExecutor

TThreadExecutor is a task-oriented, multithreaded MapReduce tool for ROOT, which provides a simple programming model for parallel MapReduce operations on loops operating on independent data. Its interface includes operations such as *Map*, *Reduce*, *Foreach* and even clustered mapping with partial reduction. TThreadExecutor is currently implemented as a façade to Intel's Threading Building Blocks (TBB) [5], specifically the function to parallelize loops (`tbb::parallel_for`), exploiting TBB's scheduling and work stealing mechanisms.

Listing 1 exposes a basic example of the usage of TThreadExecutor. First we define our map function, which implements the task to be computed in parallel. Then we propose our reduction function accumulating the intermediate result from the map and we initialize TThreadExecutor and the pool of threads. Finally, by calling `TThreadExecutor::MapReduce`, we perform the operation in parallel.

```
auto mapFunc = [](const UInt_t &i){
    return i+1;
};

auto reduceFunc = [](const std::vector<UInt_t> &mapV){
    return std::accumulate(mapV.begin(), mapV.end());
};

ROOT::TThreadExecutor pool;
pool.MapReduce( mapFunction, ROOT::TSeq<int>(100), reductionFunction);
```

Listing 1: Use example for TThreadExecutor: parallel sum of the first 100 nonnegative integers.

Currently, TThreadExecutor is used in ROOT fitting to compute the maximum likelihood and the least squares functions in parallel. In TMVA [6], it is utilized to evaluate Boosted Decision Trees and to process Deep Neural Networks. It is also applied for performing implicitly multithreaded operations in ROOT I/O (reading, parallel writing, deserialization and decompression of tree branches in parallel) and in the parallel execution of functional chains in TDataFrame [7].

2.2. Data-level parallelism: VecCore

VecCore provides efficient vectorization on a variety of platforms by offering an abstraction layer on top of the libraries Vc [8], UME::SIMD [9] and the language extension CUDA, which are supported as optional backends as well as a fallback scalar one in case SIMD operations are not available. VecCore allows developing architecture-oblivious code which will map to the appropriate backend specific types, methods and instructions.

ROOT integrates VecCore by defining new SIMD array types, for example `ROOT::Double_v` or `ROOT::Float_v`, representing double and single precision vectors respectively.

3. Fitting parallelization

The computation process used for fitting is parallelized in two independent steps. We parallelize at task-level the fitting objective function, e.g. the maximum likelihood, and at data-level the model function which is provided by the user, e.g. a Gaussian function when fitting a histogram.

Figure 1 illustrates the fitting parallelization process both at task-level and data-level. We start by dividing the data into clusters of points of equal size (*Chunking*). This step is performed by TThreadExecutor and the user can specify the number of desired chunks. TThreadExecutor receives the clustered data and distributes the tasks among threads for their evaluation in the multithreaded Map stage (*Distribution*). In this stage, the model function is evaluated on the data points and, when the user provides a vectorized function, the evaluation is done on several points simultaneously (*Evaluation*). In both cases, the map function accumulates the results of the objective function evaluation into one partial result per thread.

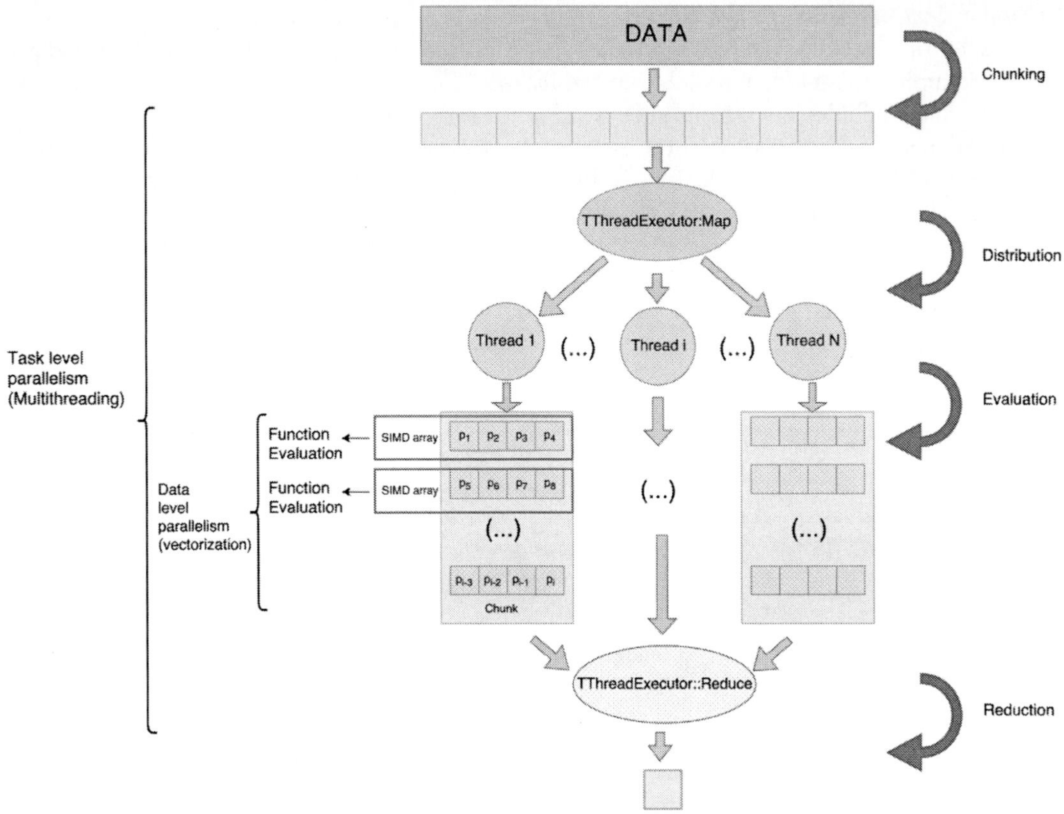

Figure 1. Depiction of the parallelization of the fitting. The parallelization at task level involves chunking, multithreaded mapping (distribution and evaluation) and reduction to a single result. At data level, we vectorize the model function evaluation.

The partial results are then returned to the main thread, which combines them into a final result representing the objective function of the fit in a reduction step applied by `TThreadExecutor::Reduce` (*Reduction*).

3.1. Case example

As a case example for benchmarking, we choose a fit of the diphoton invariant mass distribution resulting from a Higgs boson. Listings 2 and 3 offer a comparison of the scalar code for this benchmark fit against the code needed for a fully parallelized implementation of the same fit.

Note that in Listing 3 we provide parallelization and vectorization, while maintaining the previous programming model. This is an important achievement: users can benefit from multiple levels of parallelization without modifying their original code. Starting from line 9 in Listing 2 and line 13 in listing 3 the code related to the fitting operations is exactly the same. The only changes required for fitting the data in the vectorized and parallelized cases—Listing 3—are additions on top of the current workflow.

First, in order to profit from implicit task parallelization, the user only needs to call `EnableImplicitMT` (line 10 of Listing 3). This will activate all the mechanisms in ROOT that provide thread safety and parallelize all the operations that allow implicit multithreading in ROOT, such as the fitting. For vectorization, the user needs to provide a vectorized function,

IOP Conf. Series: Journal of Physics: Conf. Series **1085** (2018) 032024 doi:10.1088/1742-6596/1085/3/032024

```
//Example Fit: Implementation of the scalar function
double func(const double *data, const double *params)
{
   return params[0] * exp(-(*data + (-130.)) * (*data + (-130.)) / 2) +
          params[1] * exp(-(params[2] * (*data * (0.01)) - params[3] *
          ((*data) * (0.01)) * ((*data) * (0.01))));
}

auto f = TF1("fScalar", func, 100, 200, 4);
f.SetParameters(1, 1000, 7.5, 1.5);
TH1D h1f("h1f", "Test random numbers", 12800, 100, 200);
h1f.FillRandom("fScalar", 1000000);
h1f.Fit(&f);
```

Listing 2: Traditional scalar implementation where the model function is composed of a Gaussian signal component plus an exponential background function,

```
1  //Example Fit: Implementation of the vectorized function
2  ROOT::Double_v func(const ROOT::Double_v *data, const double *params)
3  {
4     return params[0] * exp(-(*data + (-130.)) * (*data + (-130.)) / 2) +
5            params[1] * exp(-(params[2] * (*data * (0.01)) - params[3] *
6            ((*data) * (0.01)) * ((*data) * (0.01))));
7  }
8
9  // Enable implicit parallelization
10 ROOT::EnableImplicitMT();
11
12 //This code is totally backwards compatible
13 auto f = TF1("fvCore", func, 100, 200, 4);
14 f.SetParameters(1, 1000, 7.5, 1.5);
15 TH1D h1f("h1f", "Test random numbers", 12800, 100, 200);
16 h1f.FillRandom("fvCore", 1000000);
17 h1f.Fit(&f);
```

Listing 3: Vectorized plus parallelized implementation.

which in this case only means changing the data parameter type and the return type —to the new `ROOT::Double_v` type, representing a SIMD array of double—as shown in line 2 of the Listing 3.

3.2. Results

We perform three different types of fits, χ^2 and Poisson likelihood in the binned case and Maximum Likelihood in the unbinned case, to the invariant mass distribution studying the speedup in each one of this cases. We measured the fitting times normalized to the number of objective function calls made by the fitter, as the nature of minimization problems makes the number of calls fluctuate between examples, due to numerical precision, and influence the measured times.

Figure 2(a) reports the speed up attained by the parallel fitting on a desktop server, a 4-core Haswell machine with 8 GB of RAM and distinct sets of vectorization instructions and different evaluation functions.

In Figure 2(b), we show the scaling of multithreaded operations on a 14-core Xeon processor with AVX2 and 64 GB of RAM. While it scales in the multithreaded scalar case, vectorization limits the total speed up.

It is also interesting to evaluate how different compilers behave. The results between compilers are similar (Figures 3(a) and 3(b)), with ICC slightly outperformed by Clang and GCC.

4. Conclusions

With the appropriate tools, we can exploit parallelism across multiple levels when fitting models in ROOT while preserving the user interfaces. We have shown how the users can vectorize and parallelize their original code and profit from a significant speedup in a convenient fashion.

Results differ between different fits, but in all those three implemented we obtain good speed

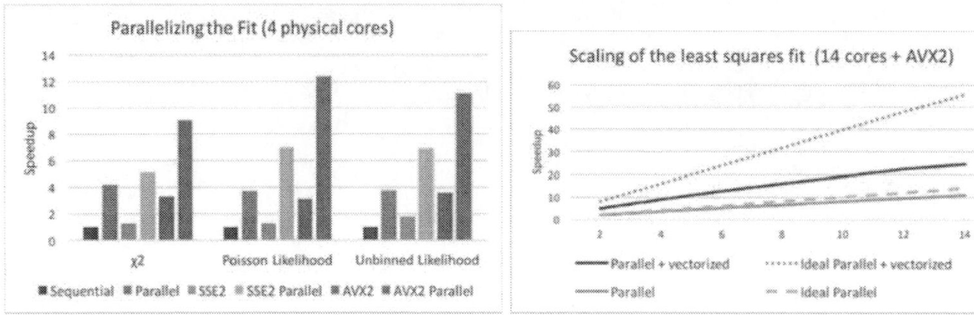

(a) Performance of the fit (4-cores Broadwell, 8GB RAM). Evaluated over 120k bins in the binned fits and 120k points in the unbinned fit.

(b) Scaling of the fit with an increasing number of cores.

Figure 2. Performance measures. Operations carried out in double precision.

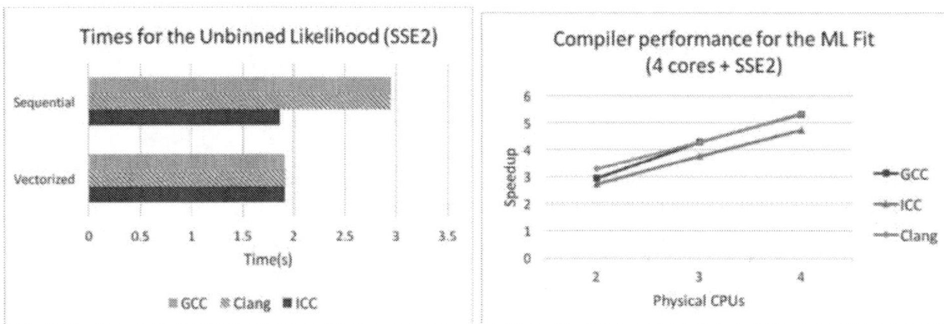

(a) Autovectorization turned off.

(b) Compiler performance.

Figure 3. Compiler comparison: gcc 6.2, clang 3.8, icc17. Carried out in double precision.

ups when vectorizing and remarkable ones when parallelizing, thanks to TThreadExecutor. The combination of both task-level and data-level parallelism produces the best results, and both speed ups and execution times are consistent across compilers.

5. References

[1] Brun R and Rademakers F 1997 Root - an object oriented data analysis framework *Proc. AIHENP'96 Wksh.* vol A389 ed in Phys N I M pp 81–86
[2] Piparo D, Tejedor E, Guiraud E, Ganis G, Mato P, Moneta L, Pla X V and Canal P 2017 Expressing parallelism with root *Journal of Physics: Conference Series* vol 898 (IOP Publishing) p 072022
[3] Dean J and Ghemawat S 2008 *Communications of the ACM* **51** 107–113
[4] Amadio G, Canal P and Wenzel S Veccore [software]: release v0.4.2, available from https://github.com/root-project/veccore/ [accessed 2017-10-28]
[5] Reinders J 2007 *Intel Threading Building Blocks: Outfitting C++ for Multi-core Processor Parallelism* (O'Reilly Media, Inc.)
[6] Hoecker A, Speckmayer P, Stelzer J, Therhaag J, von Toerne E, Voss H, Backes M, Carli T, Cohen O, Christov A *et al.* 2007 *arXiv preprint physics/0703039*
[7] Amadio G, Blomer J, Canal P, Ganis G, Guiraud E, Mato Vila P, Moneta L, Piparo D, Tejedor E and Valls Pla X 2018 Novel functional and distributed approaches to data analysis available in root *18th Int. Wksh. on Advanced Computing and Analysis Techniques in Physics Research* (IOP Publishing)
[8] Kretz M and Lindenstruth V 2012 *Software: Practice and Experience* **42** 1409–1430
[9] Karpinski P and McDonald J 2017 A high-performance portable abstract interface for explicit simd vectorization *PMAM@ PPoPP* pp 21–28

IOP Conf. Series: Journal of Physics: Conf. Series **1085** (2018) 032025 doi:10.1088/1742-6596/1085/3/032025

A federated Xrootd cache

E Fajardo[1], A Tadel[1], M Tadel[1], B Steer[2], T Martin[1], F Würthwein[1]

[1] University of California San Diego, La Jolla, CA, USA

[2] Monash University, Wellington Rd, Clayton VIC 3800, Australia

E-mail: emfajard@ucsd.edu

Abstract. With the shift in the LHC experiments from the computing tiered model where data was prefetched and stored at the computing site towards a bring data on the fly, model came an opportunity. Since data is now distributed to computing jobs using XrootD federation of data, a clear opportunity for caching arose.

In this document, we present the experience of installing and using a Federated Xrootd Cache (A Xrootd Cache consistent of several independent nodes). There is some fine tuning towards and scaling tests performed to make it fit for the CMS Analysis case.

Finally, we show how this federated cache can be expanded into a federation of caches in which the caches can be distributed among computing centers.

1. Introduction and Physics motivation

The LHC physics program, especially the large general purpose collaborations ATLAS and CMS, depend on a distributed compute and storage infrastructure comprising roughly 150 Petabytes of disk space and 100-200,000 hyperthreads of Intel X86 compute power. For CMS, roughly 1/3 of this compute power, and nearly all of the disk space is deployed to support physics data analysis. Typically, a thousand scientists are active in a given month. More than 1000 peer reviewed science papers have been produced this way by the two collaborations.

In CMS, two data formats are used for data analysis, AOD and MINIAOD, that differ in size per event by x10. While MINIAOD is used by more than 90% of the analysis activity, AOD occupies most of the disk space. Most of the analysis activity is focused on a single coherent set of datasets produced with a consistent set of releases. CMS uses a data popularity mechanism to place replicas of datasets based on their popularity [1]. This mechanism works on timescales of one year or more before it deletes the datasets replicas from disk.

The present paper describes initial exploratory work towards a much more agile data replication mechanism based on caches inside the CMS global data federation. Given that the experiment knows in advance that the MINIAOD format will be heavily used in analysis. We set aside a relatively small fraction of the total disk space for caching a complete version of it with the objective of making that format maximally available at sites dynamically with minimal explicit maintenance beyond specifying the namespace cached. Hence the cache allows for a more efficient use of the sites disk space in which the data being cached is the one on demand right now as opposed to the data that was popular several months ago.

By allowing distributed caches across site boundaries among sites that are "close" in terms of network latency, we can further economize on the amount of disk space used for serving data analysis.

The present work describes the basic ideas, and initial benchmarking towards such a system.

2. Xrootd and Data Federations

As demonstrated by the Anydata, Anytime, Anywhere project (AAA) [2] XRootd [3] was the missing piece of the puzzle allowing the HEP community to step up from the concept of distributed, multi-tiered storage to the level of global data federations with a single data-access entry point and a common data-access protocol. Hierarchical deployment of XRootd redirectors allows for a real-time discovery of sites where required data is available now, bypassing entirely the file catalogues where information might be stale or the site listed there might be temporarily inaccessible. While XRootd supported multi-storage deployments before, the novel idea of AAA was to extend and improve this XRootd functionality to seamlessly function in a global, multi-site environment. On the storage side, XRootd frontends or data-access plugins had to be developed for all storage solutions used by CMS (hdfs, dCache, and DPM).

With AAA extending from the United States based pilot to a global CMS data federation in 2014 [4], the usage of AAA quickly grew from the initial use-cases (fallback to remote data access, overflow of jobs from oversubscribed sites[5], direct remote access for data exploration and visualization) to the point of profoundly affecting the Computing model of CMS and also using remote data access for production jobs as well as for data access from jobs running on opportunistic resources. With all relevant CMS data being accessible through AAA, the data placement and data movement strategies started to change as well, becoming based on data popularity [1] or actual data access (by using XRootd caching proxy – just in time data placement).

3. XRootd File-based Caching Proxy

XRootd file-based caching proxy (known as Proxy File Cache (PFC) within XRootd code and documentation) acts as an intermediary between local XRootd clients and a remote XRootd service, either a single data-server or a complete data federation. As a proxy cache it serves data available on local disk directly and forwards requests for missing data to the remote service, serving the data to the requesting client and writing it into the cache. Caching proxy will only open a remote connection if data is actually not available locally or when a segment of a file not already in the cache is accessed by a client. This guarantees that data in the cache remains usable even when the data source is not available. Data can be automatically prefetched when it is known that most of the file will be read once it is open. As all caches, the available space is automatically managed and least recently used files get purged once disk usage reaches a predetermined level.

A set of caching proxy servers can be connected into a Caching Proxy Cluster by using a standard XRootd manager as an entry point into the distributed cache. Further, using the same mechanism, several caching clusters can be connected into a federation of proxy clusters. This makes sense when caching clusters in the federation are significantly closer to each other than to the source data federation.

3.1. Local Xrootd Cache hardware setup

For the pilot XrootD cache system at UCSD we tried to maximize the number of spindles while minimizing the amount of XrootD redirections a job needs to read its input data. Older worker nodes were chosen that had the capacity for 12 3.5" disks drives of modest size. The systems use two Intel Xeon X5650(6C/12HT) processors, 48GB of RAM, and 12 2TB HGST Hard drives each. We have 11 systems in our Xcache for a total disk capacity of 264TB. In order to make best use of the disks we retrofitted the systems with Mellanox MT27520 10Gbps network interfaces and connected them directly to our Arista 7150S switch using 10Gbps Mellanox AOC Cables.

3.2. Local tricks to increase use of the cache
The main use case for the cache is for CMS. We wanted CMS jobs to run at the UCSD Tier-2 site (and use the cache) for jobs that run over the datasets that were in the cache but not necessarily advertised as such via the global CMS PhEDEx data management system [6] (i.e jobs that would not originally be routed to the UCSD Tier-2 cluster). We decided to serve all MINIAOD(SIM) data from CMS for the 2016 data taking period, and corresponding simulations. This namespace was configured to be cached in the XRootd cache. Configuration changes were then made to the CMS overflow mechanism [5] to steer jobs to UCSD that pend for more than 6 hours in the global HTCondor pool of CMS[7], and declare files in the cached namespace as input. In addition there were some modifications needed for the site local configuration, so jobs requiring these datasets would not try to access the local storage system but the Xrootd cache cluster.

4. Scale tests and estimated throughput
Given the potential for the cache (two sites, and hundreds of terabytes of data) ensuring that it would scale with the current and future availability of compute nodes was a must. In this section we present the methodology and the scale results obtained while stressing the proxy cache cluster infrastructure. The base case for the test was the fact that CMS jobs read at $1MByte/s$ per core (client) [2]. All the measurements were done using clients that read at least twice this rate.

The scale tests for this project were divided into two different sections. Section involved holding the number of concurrent data requests fixed, while varying the rate at which individual processes requested data. Section4.2 investigated holding the total requested data rate fixed while varying the number of concurrent requests and individual data request rates.

Multiple large file requests were made from host machines at UCSD prior to the experiments to ensure that the cache was hot (desired files already present in the cache) when experiments were performed. This guaranteed a high cache hit/rate since requests were made for files from this same dataset.

GlideinWMS[8] was used to submit several jobs (clients) at the same time to request files from the cache, i.e create the artificial load. To gather the highest amount of distributed cores while reducing the impact on production operations in the cluster a sleeper pool was used. A sleeper pool consist of a batch system pool in which compute nodes are configured to advertise more cores than they have, under the premise that those extra slots will consume a negligable amount of resources(CPU, memory and disk). The sleeper pool is thus "overlayed" on top of the UCSD Tier-2 production system.

4.1. Fixed number of concurrent slots
For the this part, the number of concurrent jobs is fixed to 1000 and the data request rate is varied. The rate being specified in units of MB/10s reflects the nature of the script submitted in each job, which sleeps for 10 seconds between making remote file requests. The expected data-rate per job (see green data in Figure 1) is simply the specified data request rate divided by 10, changing to units MB/s.

4.2. Fixed Total Data Rate
In this section the total data request-rate is fixed at $5GB/s$ and the other parameters are varied. For example, this can be achieved with 1000 concurrent jobs making average requests of $5MB/s$, or equally with 5000 concurrent jobs making average requests of $1MB/s$. We would expect the system throughput to decrease when the number of clients increases because of the degradation of the service when disks in the cache cannot handle enough read requests. This is exactly what we see in Figure 2. Notice that in the worst case scenario at $1MB/sec$ (the expected data

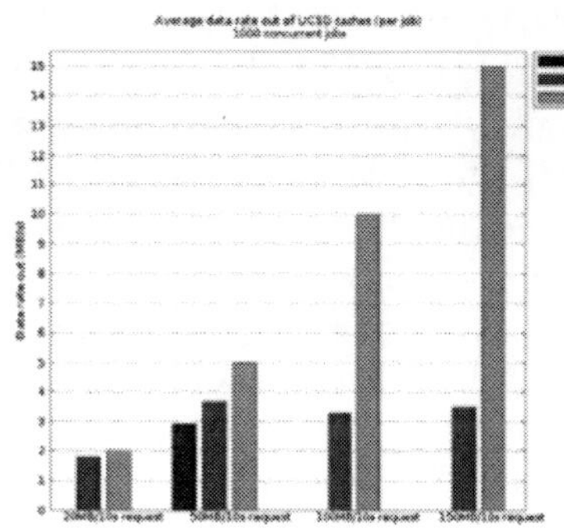

Figure 1. Average data-rate throughput (per job) for 1000 jobs

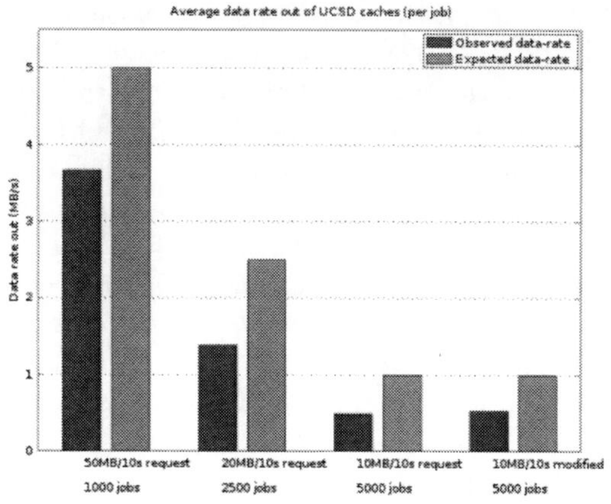

Figure 2. Average data-rate (per job) for a fixed total data request-rate.

rate) and usage of the entire cluster (5000 clients) the cache can only deliver up to 60% of the expected data-rate.

5. Conclusions and Future Work

We see in the future more sites setting up similar cache clusters like this one. In particular Caltech setting one up in which jobs running at either site can access seamlessly files from each other cache. This will require doing some these tests again across the Wide Area Network.

More work and tests will be needed to understand and mitigate the performance degradation when running at extreme concurrency levels if the data rates per core of CMS were to greatly increase in the future. However, given the rates seen today on the production Tier-2s we are confident based on these results that the existing cache will operate smoothly and efficiently. We are thus since running this cache as part of the regular production Tier-2 system for CMS.

Acknowledgments

This work is supported in part by the National Science Foundation through awards PHY-1148698 and PHY-1624356.

References

[1] Kuznetsov V, Li T, Giommi L, Bonacorsi D and Wildish T 2016 *Journal of Physics: Conference Series* **762** 012048 URL `http://stacks.iop.org/1742-6596/762/i=1/a=012048`

[2] Bauerdick L, Benjamin D, Bloom K, Bockelman B, Bradley D, Dasu S, Ernst M, Gardner R, Hanushevsky A, Ito H, Lesny D, McGuigan P, McKee S, Rind O, Severini H, Sfiligoi I, Tadel M, Vukotic I, Williams S, Wrthwein F, Yagil A and Yang W 2012 *Journal of Physics: Conference Series* **396** 042009 URL `http://stacks.iop.org/1742-6596/396/i=4/a=042009`

[3] Dorigo A, Elmer P, Furano F and Hanushevsky A 2005 Xrootd/txnetfile: A highly scalable architecture for data access in the root environment *Proceedings of the 4th WSEAS International Conference on Telecommunications and Informatics* TELE-INFO'05 (Stevens Point, Wisconsin, USA: World Scientific and Engineering Academy and Society (WSEAS)) pp 46:1–46:6 ISBN 960-8457-11-4 URL `http://dl.acm.org/citation.cfm?id=1391157.1391203`

[4] Bloom K and the Cms Collaboration 2014 *Journal of Physics: Conference Series* **513** 042005 URL `http://stacks.iop.org/1742-6596/513/i=4/a=042005`

[5] Sfiligoi I, Wuerthwein F, Bockelman B, Bradley D C, Tadel M, Bloom K, Letts J and Tadel A M 2012 *Journal of Physics: Conference Series* **396** 032102 URL `http://stacks.iop.org/1742-6596/396/i=3/a=032102`

[6] Rehn J 2006 Phedex high-throughput data transfer management system

[7] Balcas J, Belforte S, Bockelman B, Colling D, Gutsche O, Hufnagel D, Khan F, Larson K, Letts J, Mascheroni M, Mason D, McCrea A, Piperov S, Saiz-Santos M, Sfiligoi I, Tanasijczuk A and Wissing C 2015 *Journal of Physics: Conference Series* **664** 062031 URL `http://stacks.iop.org/1742-6596/664/i=6/a=062031`

[8] Sfiligoi I, Bradley D C, Holzman B, Mhashilkar P, Padhi S and Wurthwein F 2009 The pilot way to grid resources using glideinwms *2009 WRI World Congress on Computer Science and Information Engineering* vol 2 pp 428–432

IOP Conf. Series: Journal of Physics: Conf. Series **1085** (2018) 032026 doi:10.1088/1742-6596/1085/3/032026

A container model for resource provision at a WLCG Tier-2

G Roy[1], G Stewart[1], D Crooks[1], S C Skipsey[1] and D Britton[1]

[1] School of Physics and Astronomy, University of Glasgow, Kelvin Building, University Avenue, G12 8QQ, United Kingdom

E-mail: gareth.roy@glasgow.ac.uk

Abstract. Containers are more and more becoming prevalent in Industry as the standard method of software deployment. They have many benefits for shipping software by encapsulating dependencies and turning complex software deployments into single portable units. Similar to Virtual Machines, but with a lower overall resource requirement, greater flexibility and more transparency they are a compelling choice for software deployment. The use of containers is becoming attractive to WLCG experiments as a means to encapsulate their payloads, ensure that userland environments are consistent and to segregate running jobs from one another to improve isolation. Technologies such as Docker and Singularity are already being used and tested by larger WLCG experiments along with CERN IT.

Our purpose in this paper is to explore the use of containers at a medium to large WLCG Tier-2 as a method of reducing the manpower required to run such a site. By examining the requirements of WLCG payloads (such as the availability of CVMFS, Trust Anchors or VOMS information) a model of a contained compute platform is developed and presented. Along with providing compute it standardised monitoring solutions can be bundled to provide a complete toolbox for local System Administrators to provide resources quickly and securely.

1. Introduction

The use of linux containers has become prevalent throughout Industry with a large an varied ecosystem of tools developing and maturing. Containers use namespaces, a native Linux kernel feature, to isolate running processes from one another in their own wrapped up environment. The attraction of containers derives from the ability to wrap complex software packages, along with their many and varied dependancies, into a single unit which can be deployed and configured via simple tools.

The ability to wrap up complex software projects into a single easily deployable unit is a compelling model for running LHC experiment application payloads. Due to a reduced funding many of the Worldwide LHC Compute Grid (WLCG) sites are operating with limited manpower but are required to maintain large and complex stacks of middleware. There is much ongoing effort to investigate novel ways to simplify these systems and still support the computational requirements of the LHC experiments, containers may be one solution to this problem.

UKI-SCOTGRID-GLASGOW [2] (ScotGrid) has deployed some of these novel solutions in an attempt to simplify it's operations. ScotGrid it is a medium sized Tier-2 facility consisting of approximately 64000 HEPSPEC and 3.8 PB of Storage. It primarily supports ATLAS and LHCb, while acting as a CMS Tier-3 along with supporting other smaller experiments and local

IOP Conf. Series: Journal of Physics: Conf. Series **1085** (2018) 032026 doi:10.1088/1742-6596/1085/3/032026

user groups. At present ScotGrid currently provides 25% of it's computational resources as VAC. VAC is a VM lifecycle manager that implements the vacuum model on a group of autonomous worker nodes [5].

While VAC is simple to deploy, lightweight and auto-updates without the interaction of a system administrator the use of VMs leads to issues for the local site. As the VMs contain a complete running version of Linux they can be slow to start, and use a large amount of resources over and above that required by the running payload. They are also opaque to the local site monitoring making it difficult to get an accurate representation of what is happening on site resources.

Our goal is to take the idea of autonomous worker node from VAC and combine that with a standard site configuration (in the case of ScotGrid an HTCondor batch system and ARC CEs) through the use of containers. To accomplish this we use standard software tools (in this case Docker [1]) to develop a container that is capable of running WLCG application payloads, focusing in this paper on ATLAS multi-core payloads. Along with creating the container a set of monitoring tools and methods of deploying the container into production will be assessed.

2. Requirements

Before we can create a container image to run WLCG workloads it is essential to identify the requirements that such payloads would need. At present most workloads running on the WLCG infrastructure assume a Redhat based operating system such as Scientific Linux or Centos.

Along with this most CERN experimental software, or experiment workload management systems would expect a Grid site to provide:

- Access to experiment software and other WLCG packages usually via the Cern Virtual Machine Filesystem (CVMFS)[3].
- Access to EGI and Grid toolkits, libraries and software to allow interaction with Grid systems for file transfer, authentication etc.
- Availability of a common set of libraries and a known good set of packages.
- Up to date trust anchors and VO information to allow authentication and authorisation.

Additionally, to properly manage containers in production and reduce the overall workload of site administrators (which is especially important with the reduction in manpower available to many Grid sites due to flat-cash funding scenarios) the following is required:

- Lightweight resource provision, which is simple to deploy and makes best use of available hardware resources.
- Leverage existing infrastructure, and allow a staged migration to more novel methods of deployment.
- A complete set of monitoring tools that help to quickly identify problems and easily drill down between layers of information.
- Use standard and supported tools which allows access to a community of expertise to round out administrators knowledge.

3. Container Components

3.1. CVMFS

As mentioned in our statement of requirements gaining access to experiment software in most case requires the use of CVMFS. CVMFS is a fuse mounted caching filesystem that allows lazy access to software via a series of hierarchical squid proxies. In order to access the files stored within CVMFS from within a container we can either a) bind mount the CVMFS directory from

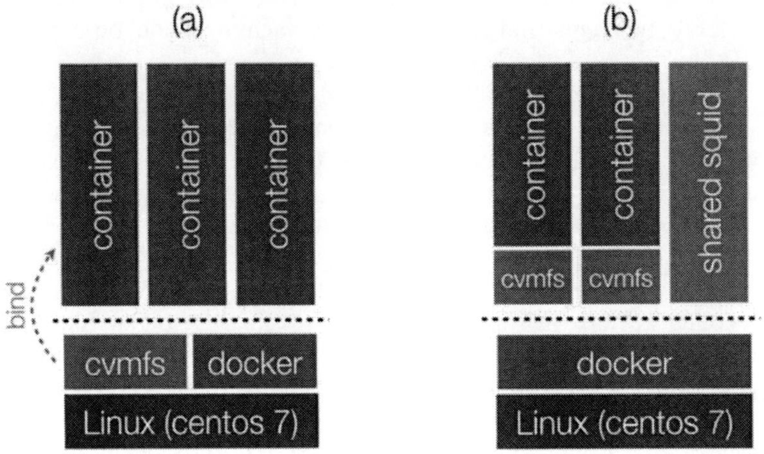

Figure 1: Methods of accessing CVMFS from with a container, either a) bind mounting from the host or b) via fuse within the container itself.

the container host to a directory within the container or b) use fuse from within the container itself to mount CVMFS. This is illustrated in Figure 1.

Each method of accessing CVMFS comes with trade-offs. Bind mounting CVMFS from the host requires that all required repositories are statically mounted via entries in the hosts `/etc/fstab`. This can in some cases cause systems to have difficulty in booting if the CVMFS directories are not available at startup. Bind mounting does allow a shared cache between all containers which improves performance and comes with a simple set of semantics in how volumes are added to the container.

Fuse mounting CVMFS within the container also has it's trade-offs, it requires `autofs` to be installed within the container and running as a service. As each container runs it's own instance of CVMFS there is no shared cache which reduces efficiency (this can be improved by including a shared squid on each container host as shown in Figure 1). Finally, the container requires escalated privileges specifically adding the MKNOD and SYS_ADMIN capabilities as shown below:

```
--cap-add MKNOD --cap-add SYS_ADMIN --device /dev/fuse.
```

The benefits of running CVMFS within the container itself are that repositories are no longer statically mounted and do not need to be known before the container is started, CVMFS is no longer required on the host so read-only linux distributions (such as coreos, rancheros) could be used to increase security.

In the remainder of this paper we use the method of bind mounting CVMFS from the container host as all repositories for running ATLAS multi-core production payloads are known before instantiation.

3.2. Middleware

For WLCG payloads to be able to perform authentication checks, stage data and communicate with Grid workload management servers it is essential to have the correct middleware available. In most cases these packages are installed from the EGI provided Unified Middleware Distribution (UMD) which contains meta-packages for worker node installations. However, to reduce the container size, these packages are not installed, instead a method similar to the one employed by ATLAS VAC virtual machines is used.

The grid environment is loaded from a CVMFS mounted "tarball" distribution available in the `grid.cern.ch` repository. In practice the grid environment is loaded via HTCondor at job runtime with a custom user-job-wrapper script that sets up the appropriate environment variables. This means that the Grid middleware packages are decoupled from the container image, reducing it's size. A second benefit of this method is it grants access to all of the required CA certificates and CRLs, again via CVMFS. These are updated regularly meaning they do not need to be installed within the container, nor does a CRL fetching service need to be included within the container image.

3.3. HEP Libraries
Each of the LHC experiments has a requirements for additional software packages to be available on a Grid worker node to allow proper execution of application payloads. All of the packages required that are not found in a bare installation of Scientific Linux or Centos are described by the meta-package, `HEP_OSlibs`. The contents of this meta-package are decided by representatives of the four LHC experiments via the Librarian and Integrators Meeting (LIM), the Architects Forum (AF) and WLCG Operations and Coordination meetings. To ensure that WLCG payloads correctly operate with our container environment we install the HEP_OSlibs within the container. Unfortunately this increases the size of the container by approximately 600MB, it is hoped that this dependancy can be removed with the move to Singularity [6] as an encapsulation mechanism of WLCG payloads.

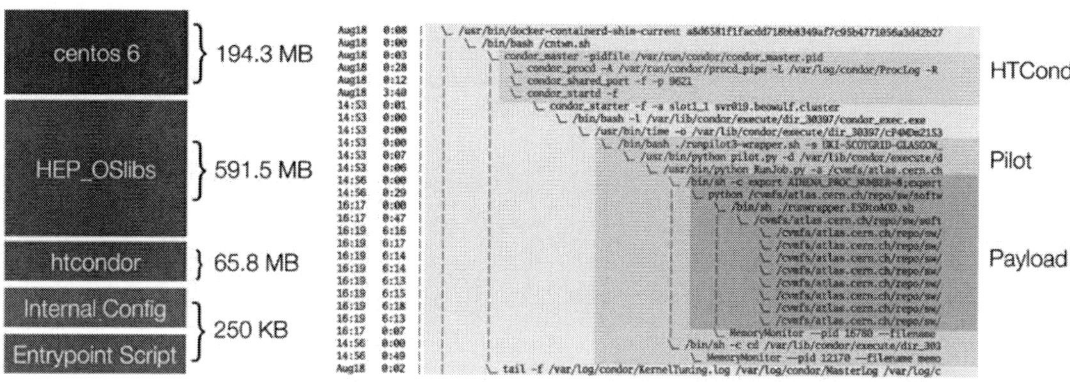

(a) Layers and size of final container image

(b) Process tree of container running an ATLAS multi-core payload

Figure 2: Details of final container layers, and example of running container.

3.4. Final Container
The different layers found in our final container is shown in Figure 2(a). The total size of this container is 851.9MB and is based initially on Centos 6 for backwards compatibility. As mentioned in our previous sections on requirements it includes `HEP_OSlibs` as a common baseline for WLCG experiment code.

Along with local site configuration, it contains HTCondor client packages to connect to ScotGrid's batch farm. The container is configured with standard pool accounts and has CGROUPS disabled as all resource utilisation is restricted at the container level. HTCondor

IOP Conf. Series: Journal of Physics: Conf. Series **1085** (2018) 032026 doi:10.1088/1742-6596/1085/3/032026

is also configured to use the Condor Connection Broker (CCB) to allow communication to key condor processes as container network access is restricted to the internal Docker network. The container itself is distributed to the hosts via a private Docker registry running on our head node, with each node in our test pool spawning one container per eight CPU cores.

An example ATLAS multi-core payload is shown in Figure 2(b), the process tree contains the HTCondor `startd` process, the ATLAS pilot and the multi-core payload running eight `athena.py` processes. Unlike a VM it is possible to observe these processes using `ps` from the container host making monitoring of payloads simple, this aids observability and traceability of any errant action.

4. Monitoring and Logging

Local monitoring is essential to enable sites to identify issues and optimise workloads, in this section we identify standard tools which can be deployed as containers that allow the gathering, aggregation and alerting on those metrics as well as capturing logging information for each running container.

4.1. Metrics

Gathering instantaneous metric for each running container is important to identify bottlenecks or issues with job performance. For metric generation we use the Google sponsored cAdvisor [7]. cAdvisor can be run as a container along with our containerised worker nodes. It provides process lists, cpu utilisation per core, memory utilisation and network usage for each running container. Figure 3 illustrates some of the information obtained from a running instance of an ATLAS multi-core production payload. The cAdvisor library is also integrated within Kubernetes which mean when moving to this as a deployment strategy low level metric monitoring remains the same.

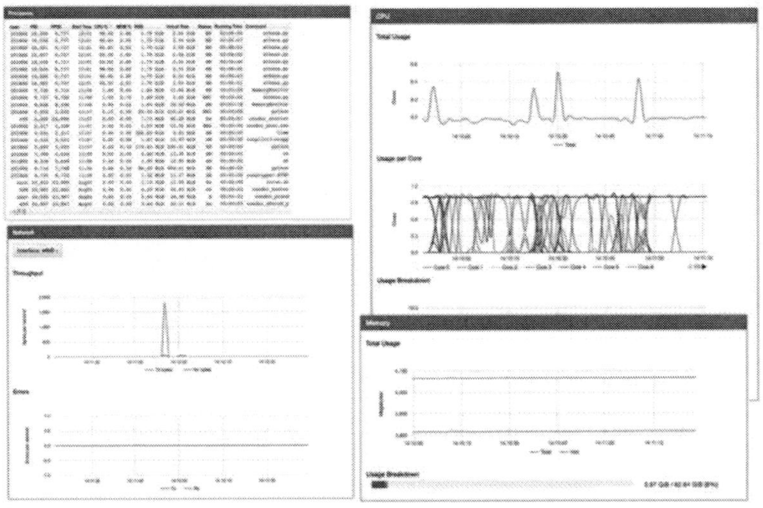

Figure 3: Graphical display of metrics gathered and presented by cAdvisor.

4.2. Aggregation

As well as interactively and graphically presenting system usage cAdvisor also exports it's metrics in a format that can be read by another software package, Prometheus [8]. Prometheus acts as an aggregation platform to collect metrics from multiple sources, store and aggregate them. It allows querying, plotting and alerting based on the information it finds. Prometheus uses a pull model for gathering metrics accessing http endpoints located at `/metrics` which has a number of benefits. The pull model allows Prometheus to scale by adding additional collectors to any size required and it gives a heartbeat for each container host to ensure the host is alive. An example of the metrics and graphing available via Prometheus is shown in Figure 4 where the CPU load is shown for four containers each running ATLAS multi-core production pilots.

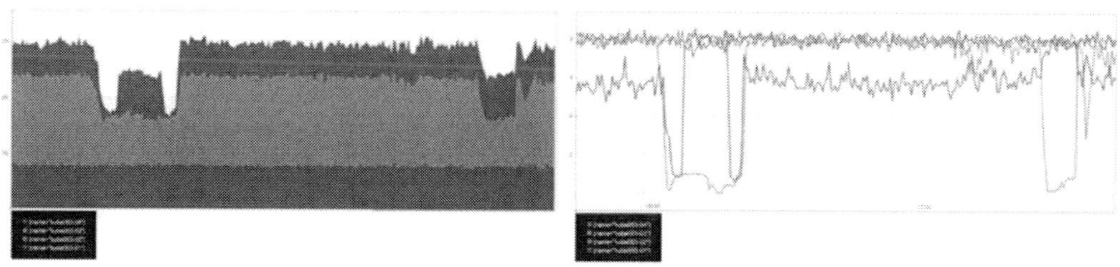

(a) CPU load stacked graph (b) CPU load

Figure 4: Plots showing the CPU load (a) Stacked and (b) Overlayed of one container host running four ATLAS mult-core containers.

4.3. Logging

As well as gathering metrics it is also important to gather logging information to diagnose problems or detect issues. As with cAdvisor a solution that can be deployed as a container would allow management in a similar fashion to the containerised worker nodes.

Using a combination of Logspout [9] and Oklog [10] we can export logging information from each running container, aggregate and search them. The configuration is shown in Figure 5. Logspout acts to collect logs and ships them to an Oklog instance which can be used to search via simple regex. Logspout is capable of sending logs to a number of different endpoints so as the number of logs scale Oklog could be replaced with an Elasticsearch service for better querying and alerting.

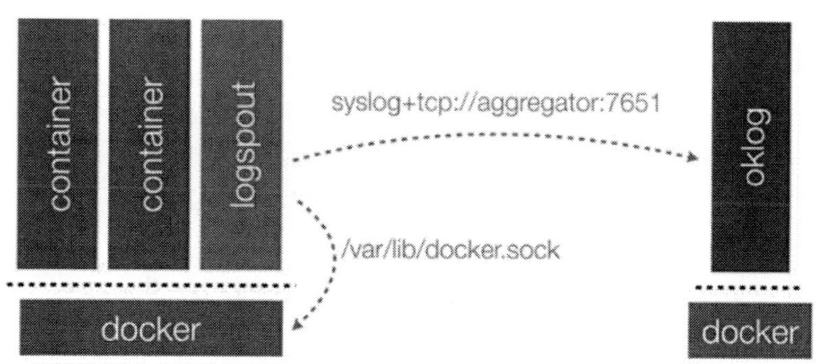

Figure 5: Logging configuration using container deployed logspout for collecting loging information and oklog for log aggregation.

IOP Conf. Series: Journal of Physics: Conf. Series **1085** (2018) 032026 doi:10.1088/1742-6596/1085/3/032026

5. Conclusions

In this paper we have examined the requirements of container to run WLCG application payloads. A nominal container model has been created and proven to work by running ATLAS muti-core payloads. A set of tools has been selected to allow sites to integrate monitoring and logging via containers as with the worker nodes.

In future work it is hoped to extend the worker node container to auto-expire and couple with a CI/CD pipeline to automate updates to containers as well as Integrate with Kubernetes (or other mechanism) to get auto-container redeploys (via services).

References

[1] *Docker.com*, "Docker" [software], version 1.12.6, 2017. Available from https://www.docker.com [accessed 18 Oct 2017]

[2] *ScotGrid* [online] Available at http://www.scotgrid.ac.uk/ [accessed 14 May 2015]

[3] *J Blomer et al.*; 2011 J. Phys.: Conf. Ser. 331 042003, "Distributing LHC application software and conditions databases using the CernVM file system"

[4] *docker-volume-cvmfs*, "docker-volume-cvmfs" [software], version v0.3.1, 2017. Available from https://gitlab.cern.ch/cloud-infrastructure/docker-volume-cvmfs/ [accessed 18 Oct 2017]

[5] *A McNab*; "Running jobs in the vacuum" (A McNab et al 2014 J. Phys.: Conf. Ser. 513 032065)

[6] *Singularity*, "singularity" [software], version 2.3.2-dist, 2017. Available from http://singularity.lbl.gov/ [accessed 18 Oct 2017]

[7] *cAdvisor*, "cadvisor" [software], version 0.26.3, 2017. Available from https://github.com/google/cadvisor [accessed 18 Oct 2017]

[8] *Prometheus*, "prometheus" [software], version 1.7.1, 2017. Available from https://prometheus.io/ [accessed 18 Oct 2017]

[9] *Logspout*, "logspout" [software], version 3.2.3, 2017. Available from https://github.com/gliderlabs/logspout [accessed 18 Oct 2017]

[10] *oklog*, "oklog" [software], version 0.2.2, 2017. Available from https://github.com/oklog/oklog [accessed 18 Oct 2017]

IOP Conf. Series: Journal of Physics: Conf. Series **1085** (2018) 032027 doi:10.1088/1742-6596/1085/3/032027

Federating distributed storage for clouds in ATLAS

F Berghaus[1], K Casteels[1], A Di Girolamo[2], C Driemel[1], M Ebert[1], F Furano[2], F Galindo[3], M Lassnig[2], C Leavett-Brown[1], M Paterson[1], C Serfon[2], R Seuster[1], R Sobie[1], R Tafirout[3] and R P Taylor[1]

[1] Department of Physics and Astronomy, University of Victoria, Finnerty Road, Victoria V8P 5C2, Canada
[2] CERN, Geneva 1211, Switzerland
[3] TRIUMF, Wesbrook Mall, Vancouver V6T 2A3 Canada

E-mail: `frank.berghaus@cern.ch`

Abstract. Input data for applications that run in cloud computing centres can be stored at distant repositories, often with multiple copies of the popular data stored at many sites. Locating and retrieving the remote data can be challenging, and we believe that federating the storage can address this problem. A federation would locate the closest copy of the data on the basis of GeoIP information. Currently we are using the dynamic data federation Dynafed, a software solution developed by CERN IT. Dynafed supports several industry standards for connection protocols like Amazon's S3, Microsoft's Azure, as well as WebDAV and HTTP. Dynafed functions as an abstraction layer under which protocol-dependent authentication details are hidden from the user, requiring the user to only provide an X509 certificate. We have setup an instance of Dynafed and integrated it into the ATLAS data distribution management system. We report on the challenges faced during the installation and integration. We have tested ATLAS analysis jobs submitted by the PanDA production system and we report on our first experiences with its operation.

1. Introduction

We aim to run data-intensive applications on globally distributed opportunistic resources that have no local grid storage. The ATLAS [1] experiment leverages a globally distributed system of infrastructure as a service (IaaS) clouds as part of its distributed computing system. These resources are integrated into the ATLAS distributed computing system using the Cloud Scheduler [2] technology developed at the University of Victoria. These IaaS resources are used opportunistically, and do not support any local grid infrastructure.

The workflows executed by high energy physics experiments often demand large volumes of input data or produce a significant volume of output data. We aim to use a data federation, such as Dynafed [3], to redirect the applications running on opportunistic resources to the optimal storage endpoint to retrieve input or deposit output data. We also aim to integrate storage solutions offered by cloud providers into the ATLAS distributed data management system using Dynafed.

In this paper we explain a system leveraging Cloud Scheduler and Dynafed which successfully executed functional test jobs as part of the ATLAS distributed computing system on the CERN OpenStack [4] cloud resource that read their input from and wrote their output to an object store implemented using Ceph [5] and exposing an S3 compatible gateway.

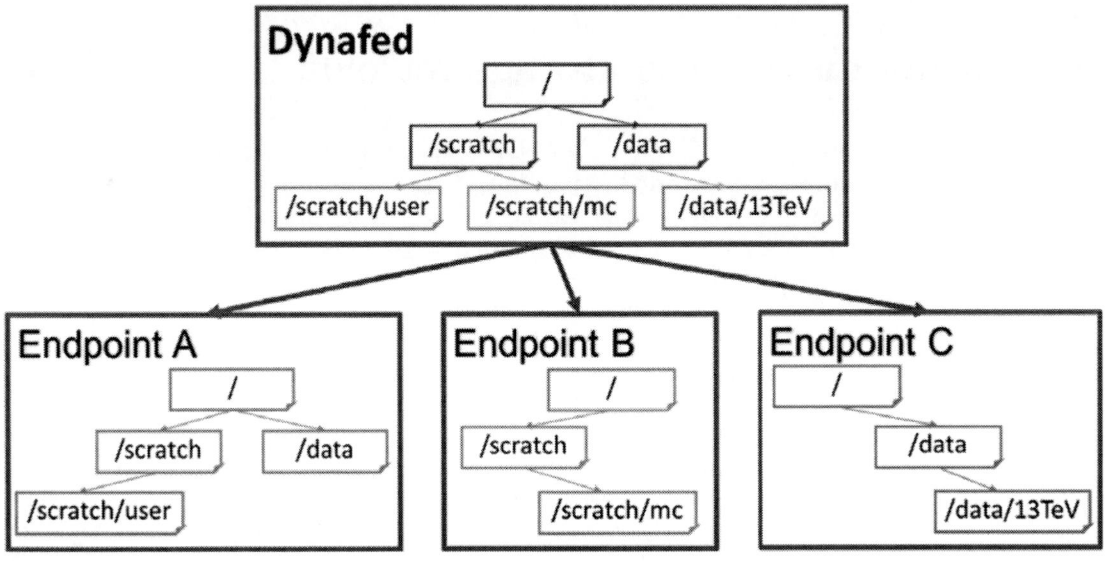

Figure 1. The dynamic federation is connected to multiple endpoints. Each endpoint may be a file system or an object store accessible using a protocol which allows redirection, such as HTTP. The dynamic federation provides a namespace that is a union of all the namespaces of the endpoints. That namespace is presented as a regular directory structure by the Dynafed. The content of a displayed directory is calculated when accessed.

2. Conceptual Design

The ATLAS experiment leverages the resources of the Worldwide LHC Computing Grid, WLCG [6]. The computing centres that are part of the WLCG and support ATLAS provide a global storage infrastructure for the experiment data and simulated events. They may be accessed using standard protocols, such as HTTP with WebDAV extensions. Dynafed supports storage backends that offer HTTP and WebDAV access and promises sufficient scalability to create the appearance of a single virtual namespace for the entire ATLAS data catalogue. Figure 1 shows how Dynafed could unify the namespaces of attached storage elements into a single namespace.

Dynafed allows the usage of cloud storage systems such as S3, SWIFT, and Azure. On the user-facing side Dynafed still presents an HTTP interface implementing authentication and authorization through an X509 public key infrastructure. Dynafed supports grid security infrastructure extensions of X509 with VOMS attributes [7]. Credentials may be presented as certificate and key or as a proxy, which allows the additional use of VOMS attributes. When Dynafed forwards clients to cloud storage systems, it translates their X509 credentials to pre-signed URL that permit, for a limited time, access to the cloud storage system.

3. Data Access

The dynamic federation used for this work was configured to use three endpoints: one at CERN, one at TRIUMF, and one at the University of Victoria. Each endpoint was a CephS3 object store. Table 1 illustrates the task division within the dynamic federation to handle client requests.

To access data through Dynafed a client makes a request for a file using HTTP optionally with WebDAV extensions. We will focus on WebDAV from here on. In our configuration the Dynafed only allows access to members of the ATLAS Virtual Organization. The client must provide X509 credentials with the request. The credential must be signed by a trusted certificate

IOP Conf. Series: Journal of Physics: Conf. Series **1085** (2018) 032027 doi:10.1088/1742-6596/1085/3/032027

Table 1. The dynamic web federation is an Apache server running the LCGDM implementation of WebDAV. The namespace usually managed by LCGDM has been replaced by the uniform general redirector (UGR) which translates the requests to the web file system to the connected endpoints. The endpoint modules handle the communication with the configured endpoints. All requests are cached in memory on the server as well as in a second-level cache which may be shared across multiple load-balanced servers.

Component		Purpose
Apache		Load the `lcgdm_dav` module and start up a WebDAV server
`lcgdm_dav` dmlite		Configure dmlite and load the uniform general redirector as namespace plugin
UGR		Configure authentication and endpoints
Plugins	dmlite WebDAV/HTTP S3 Azure	Communicate with endpoints on request
Memcached		Cache recent redirects to distributed object caching system

authority. If the credential contains VOMS extensions certifying the user to be a member of the ATLAS collaboration, access is granted. Without VOMS extensions the Dynafed checks the credential against all current members of the ATLAS collaboration and grants access if a match is found. Administrators and data management services have privileged accounts. Authorized clients are redirected to a signed URL on the closest CephS3 endpoint. Authorization is granted explicitly for reading, writing, listing, and/or deleting operations.

When a file is requested, the dynamic federation checks whether the locations of the file are already in its cache. If so, the cached entry is used, otherwise each endpoint is queried for the file after name translation to that endpoint. Dynafed waits for responses[1] to collect and cache. The resulting endpoints are evaluated for proximity to the client and the client is redirected to the closest copy. The Dynafed regularly polls all connected endpoints to determine if they are reachable. Should an endpoint be unresponsive, requests will not be forwarded to it.

4. Application Workflow

In order to integrate the dynamic federation into the ATLAS distributed computing and data management system, it was defined as a storage element associated with the `CERN-EXTENSION`[2] ATLAS site. It was configured to be accessible using WebDAV and flagged as special in the ATLAS grid information system to allow Rucio [8] to select a copy tool implementation which does not move or rename files.

The input datasets for analysis and production functional tests were transferred to the dynamic federation using the file transfer service at CERN. Once the transfers completed successfully the data was registered manually in the Rucio data catalogue.

With the input data registered in the data catalogue it was possible to run grid jobs against the data in the dynamic federation. The jobs were executed on virtual machines hosted on the CERN OpenStack using the Cloud Scheduler technology as illustrated in Figure 2. The resulting data and logs were uploaded to the CephS3 storage via Dynafed upon job completion.

Some additional development is required for full integration of cloud storage into the ATLAS

[1] Up to a given timeout set to 3 seconds here

[2] `CERN-EXTENSION` is an ATLAS site defined as a part of the `CERN-PROD` WLCG and GOCDB site.

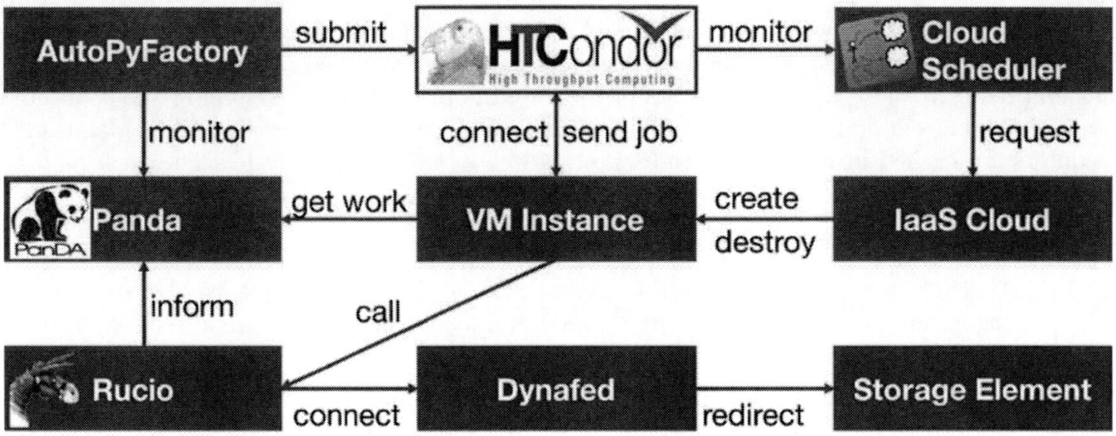

Figure 2. A client (AutoPyFactory [11] or Harvester [10]) submits pilot wrapper scripts to an HTCondor [9] queue. The queue is monitored by a Cloud Scheduler. The Cloud Scheduler makes requests to connected cloud interfaces in response to queued jobs. The cloud infrastructures create virtual machine instances and provide user-data to cloud-init running in the virtual machines for configuration. The VM instances are configured to connect to HTCondor and start consuming jobs from the queue. The jobs are pilot wrapper scripts which download the pilot. The pilot gets tasks from a PanDA [12] queue and uses Rucio to download input data and upload results. Rucio is configured to contact the dynamic federation using the HTTP protocol with WebDAV extensions. The federation forwards the Rucio client to the closest available storage element.

distributed data management: bulk transfers negotiated between storage endpoints using the HTTP protocol must be fully supported, and the data management system must be able to parse the checksums of files on cloud storage.

5. Summary

It was shown that ATLAS jobs can retrieve and deposit their data on a cloud storage system accessed via a dynamic federation using the HTTP protocol with WebDAV extensions. The jobs ran on virtual machine instances in a cloud and could be scheduled anywhere in the distributed cloud system currently running as part of the ATLAS production system. Further development is necessary to allow the execution of production or analysis jobs against the dynamic federation. Work is ongoing to integrate the dynamic federation with the Belle II [13] experiment and the DIRAC [14] workload management system. While this development is being pursued against opportunistic cloud resources it should also be useful in the context of volunteer resources [15].

References

[1] The ATLAS Collaboration 2008 The ATLAS Experiment at the CERN Large Hadron Collider *Journal of Instrumentation* **3** S08003
[2] Cloudscheduler project, "Cloudscheduler" [software], version 1.12, 2017, Available from `http://cloudscheduler.org/` [accessed 2017-10-20]
[3] Dynafed project, "Dynafed" [software], version 1.3.1, 2017, Available from `http://cern.ch/lcgdm/dynafed-dynamic-federation-project` [accessed 2017-10-20]
[4] Openstack project, "OpenStack" [software], various versions, Available from `https://www.openstack.org/` [accessed 2017-10-20]
[5] Ceph project, "Ceph" [software], various versions, Available from `http://ceph.com` [accessed 2017-10-20]
[6] Bird I 2011 Computing for the Large Hadron Collider *Ann. Rev. Nucl. Part. Sci.* **61** 99

[7] Foster I, Kesselman C and Tuecke S 2001 The Anatomy of the Grid: Enabling Scalable Virtual Organizations *IJHPCA* **15** 3 200 - 222
[8] Rucio project, "Rucio" [software], version 1.2.5, Available from `http://rucio.cern.ch` [accessed 2017-10-21]
[9] HTCondor project, "HTCondor" [software], version 8.4.3, Available from `http://research.cs.wisc.edu/htcondor/` [accessed 2018-02-06]
[10] PanDA project, "PanDA Harvester" [software], version 0.0.11, Available from `https://github.com/PanDAWMS/panda-harvester` [accessed 2018-02-06]
[11] J Caballero *et al* 2012 *J. Phys.: Conf. Ser.* **396** 032016
[12] Maeno T 2008 PanDA : Distributed Production and Distributed Analysis System for ATLAS *Journal for Physics* **119** 6
[13] Abe T *et al* 2010 *Belle II Technical Design Report* (KEK: KEK Report) vol 1
[14] DIRAC project, "DIRAC" [software], version BelleDIRAC, Available from `http://diracgrid.org` [accessed 2018-02-06]
[15] LHC@home project, "ATLAS@home" [software], version BOINC 7.8, Available from `http://lhcathome.web.cern.ch` [accessed 2017-10-21]

Leveraging the checkpoint-restart technique for optimizing CPU efficiency of ATLAS production applications on opportunistic platforms

D Cameron[1], J Elmsheuser[2], L Heinrich[3], W Lavrijsen[4], P Nilsson[2], V Tsulaia[4] and M Vogel[5] on behalf of the ATLAS Collaboration

[1]University of Oslo, P.b. 1048 Blindern, 0316 Oslo, Norway
[2]Brookhaven National Laboratory, PO Box 5000, Upton, NY 11973, USA
[3]New York University, 70 Washington Square South, New York, NY 10012, USA
[4]Lawrence Berkeley National Laboratory, 1 Cyclotron Road, Berkeley, CA 94720, USA
[5]University of Wuppertal, Gausstrasse 20, 42119 Wuppertal, Germany

E-mail: `VTsulaia@lbl.gov`

Abstract. Data processing applications of the ATLAS experiment, such as event simulation and reconstruction, spend considerable amount of time in the initialization phase. This phase includes loading a large number of shared libraries, reading detector geometry and condition data from external databases, building a transient representation of the detector geometry and initializing various algorithms and services. In some cases the initialization step can take as long as 10-15 minutes. Such slow initialization has a significant negative impact on overall CPU efficiency of the production job, especially when the job is executed on opportunistic, often short-lived, resources such as commercial clouds or volunteer computing. In order to improve this situation, we can take advantage of the fact that ATLAS runs large numbers of production jobs with similar configuration parameters (e.g. jobs within the same production task). This allows us to checkpoint one job at the end of its configuration step and then use the generated checkpoint image for rapid startup of thousands of production jobs. By applying this technique we can bring the initialization time of a job from tens of minutes down to just a few seconds. In addition to that we can leverage container technology for restarting checkpointed applications on the variety of computing platforms, in particular of platforms different from the one on which the checkpoint image was created.

We will describe the mechanism of creating checkpoint images of Geant4 simulation jobs with AthenaMP (the multi-process version of the ATLAS data simulation, reconstruction and analysis framework Athena) and the usage of these images for running ATLAS Simulation production jobs on volunteer computing resources (ATLAS@Home) and on Supercomputers.

1. Introduction

The ATLAS Experiment [1] processes its data at over 140 computing centers around the world using more than 4 million CPU-hours/day. Even at such a massive scale, the data processing of the experiment is resource-limited. The ATLAS physics program will strongly benefit from applying more compute resources to Monte Carlo simulation, and over the next decade the situation will become even more critical with the LHC [2] and ATLAS upgrade programs bringing an order of magnitude increase in computing requirements. In view of the steady demand for new computing resources, it becomes very important for the experiment not only to efficiently use all CPU power available to it, but also to pro-actively leverage computing resources it does not own, such as commercial clouds, supercomputers and volunteer computing.

In order to effectively use all available CPU-cores on a compute node, ATLAS is running the majority of its production workflows using AthenaMP [3] - a process-parallel version of the ATLAS data processing framework Athena. An AthenaMP job starts as a serial process (the master), which first goes through the application initialization phase, initializes and allocates as much memory as possible and then forks several event processors (workers). The workers run in parallel and process all events assigned to the given job. They share a substantial fraction of process memory and this memory sharing is provided solely by the Linux kernel copy-on-write (COW) mechanism. The serial part of an AthenaMP job (initialization in the master process) includes loading a large number of shared libraries, reading detector geometry and conditions data from external databases, building a transient representation of the detector geometry and initializing various components of the framework. In some cases, e.g. a slow connection to an external database, such an initialization step can take as long as 10-15 minutes. This may have a visible negative effect on CPU efficiency of the entire AthenaMP job, especially when the job is executed on an opportunistic resource with a maximum allowed wall time of 1-2 hours.

For speeding up the initialization step of AthenaMP jobs, we can benefit from the fact that in ATLAS production workflows a very large number of jobs can have almost identical job configuration. For example, all jobs within the same simulation production task have the same geometry and physics process configuration and only very few configuration parameters (e.g. input file name) change from job to job. This allows us to run one sample job from such a task, checkpoint it just after the initialization step and then use the generated checkpoint image for fast startup of all jobs within the same task, with only minimal modifications of the configuration parameters after the restart.

There are a number of use-cases where HEP software in general and ATLAS software in particular can benefit from leveraging the process checkpoint-restart technique [4]. In our studies so far we investigated how we can apply this technique to avoid the time-consuming initialization step of ATLAS Geant4 Simulation [5, 6] production jobs. In this paper we describe preliminary results obtained on two platforms: Volunteer Computing (ATLAS@Home [7]) and the Intel Knights Landing (KNL) supercomputer Cori at the National Energy Research Scientific Computing Center (NERSC), Berkeley, USA.

2. DMTCP - a tool for process checkpoint-restart

DMTCP (Distributed MultiThreaded Checkpointing) [8] is a free, open source tool which implements a process checkpoint-restart on the library level. It supports single-threaded, multi-threaded and multi-process applications running either on a single host or in a distributed environment. One of the main advantages of DMTCP is that it implements transparent checkpointing in user space with no modifications required either to user code or to the OS.

DMTCP supports two strategies for creating checkpoint images: a checkpoint can be triggered by an external process (dmtcp coordinator) or by the application itself. In the latter case the application decides when it is a good time to checkpoint itself and does so by calling the DMTCP API functions.

By default, when a checkpoint image is generated, DMTCP uses gzip for compressing the image file. If necessary, users can disable this feature via the DMTCP command-line interface. The compressed images use considerably less space on disk (a factor of 3 reduction was observed for ATLAS simulation images), but take a somewhat longer time to restart.

For our studies of AthenaMP checkpoint-restart, we used DMTCP version 2.4.5, which has been integrated into the ATLAS offline software release as an external package.

IOP Conf. Series: Journal of Physics: Conf. Series **1085** (2018) 032028 doi:10.1088/1742-6596/1085/3/032028

3. Checkpoint-Restart of AthenaMP

Athena/AthenaMP Monte Carlo production jobs are managed by the ATLAS workload management system Panda [9]. These jobs use specialized python wrappers, so called job transformations. One of the main purposes of a job transformation is to collect various configuration parameters via command-line options and to pass them over to the underlying Athena job. In order to support the process checkpoint-restart mechanism we introduced two new command-line options to the job transformation code: `--checkpoint` and `--restart`.

When the `--checkpoint` option is used, AthenaMP checkpoints itself at the end of the initialization step by calling the corresponding DMTCP API functions and then exits immediately. This operation generates a checkpoint image file as well as some auxiliary scripts, which are needed by DMTCP for restarting from this image later on. All these files are stored into an archive file for later usage.

In order to restart AthenaMP from a checkpoint image, the location of the checkpoint archive file is passed to the job transformation as a value of the `--restart` command-line option. In this case, instead of starting AthenaMP the normal way, the transformation first unpacks the checkpoint archive into the job's run directory and then sources the auxiliary DMTCP script, which restarts AthenaMP from the checkpoint image. The first thing AthenaMP does after the restart is to update some of the configuration parameters, e.g. the input file name and the number of events to process. AthenaMP reads this information from a special file created by the job transformation, which contains the current job configuration. After that AthenaMP proceeds with processing events in the usual way.

4. Portability of checkpoint images

Today ATLAS is running its production jobs in a heterogeneous computing environment. Jobs from the same production task can be executed at different production sites with potentially different computing platforms. Hence, from the perspective of the utilization of process checkpoint-restart, it is desirable to generate one checkpoint image once per task and to be able to use this image for launching jobs on heterogeneous platforms.

This is a challenging task, since DMTCP expects the same or very similar environment on the restart host compared to the checkpoint host. In order to work around this limitation we can leverage either virtual machine (VM) or container technologies. In particular, by using the same VM/container image for checkpoints and restarts the above requirement can be met.

Another issue, which we encountered during our tests and which can affect the portability of the checkpoint images, is the access to external resources (e.g. databases). For example, if a checkpointed job was configured to read data from a database which is accessible only for machines within a local network, then an attempt to restart such application at a remote site will lead to database connection errors. In order to deal with these kind of problems, the restarted application must be able to properly modify its configuration parameters (e.g. database connection string) required for accessing external resources.

5. Testing with production workloads

5.1. Volunteer Computing (ATLAS@Home)

ATLAS@Home is a research project which uses volunteer computing to execute detector simulation jobs with Geant4. On one hand it helps the experiment in finding free CPU cycles in a time of limited funding increases, and on the other hand it is an outreach tool to get the public involved in ATLAS. Like the majority of volunteer computing projects, ATLAS@Home is based on the BOINC [10] platform, which has built-in support for virtualization using VirtualBox [11]. Virtualization is required mainly because the majority of volunteers run the Windows operating system and Athena requires a Linux environment. ATLAS@Home is fully integrated into the

ATLAS workflow management system, such that from outside it looks like a regular grid site, and over the past 3 years ATLAS has simulated almost 200 million events using ATLAS@Home.

ATLAS@Home makes a good choice for a prototype platform to test the process checkpoint-restart technique for speeding up AthenaMP initialization time. Since jobs run inside a virtual machine, the environment can be fully controlled, i.e. the same VM image can be used for making checkpoint images and for restarting from them. There are several reasons which make fast initialization on ATLAS@Home rather important. Jobs in ATLAS@Home usually run for 1-2 hours, which increases the fraction of the initialization step in the overall job wall time with respect to "standard" production jobs running on the Grid. In addition volunteers may have a poor Internet connection which can lead to a very long initialization phase while data is read from external servers.

For testing the checkpoint-restart mechanism, we ran an interactive job inside ATLAS@Home VM, created a tarball with compressed checkpoint image and saved the tarball in the VM image. Preliminary results showed that the performance gains coming from restarting from a checkpoint image on ATLAS@Home strongly depend on external database access speed. In cases of fast database access ATLAS Geant4 simulation job initializes in ~4 minutes, while for nodes with slow database access this number can go up to ~20 minutes. By restarting from a checkpoint image we eliminate all need for database access during initialization and the restart usually takes 15-20 seconds, which is a rather significant improvement even for jobs with a fast database connection.

5.2. Intel KNL supercomputer

Cori is a Cray XC40 supercomputer at NERSC, Berkeley, USA. It is a unique system among supercomputers of its size with two different kinds of nodes: 2.4k Intel Xeon "Haswell" processor nodes (Cori Phase 1) and 9.7k Intel Xeon Phi "Knight's Landing" nodes (Cori Phase 2). ATLAS has been running Geant4 simulation production workflows on Cori KNL nodes since July 2017. In August 2017 the experiment simulated 135 million events on NERSC supercomputers, most of them on Cori Phase 2.

ATLAS simulation production jobs on Cori do not need to access external databases for reading detector geometry and condition data. Instead they use frozen snapshots of these databases stored in SQLite files on the Cori shared file system. Nevertheless, individual CPU cores on Intel KNL processors have considerably less processing power than, for example, CPU cores on Intel Haswell processors (1.4 GHz vs 2.3 GHz) and, as a result, on KNL we observe a significant slowdown of the serial initialization step of AthenaMP jobs.

To study performance gains from the usage of the process checkpoint-restart technique on Intel KNL, we created two types of checkpoint images: compressed and uncompressed. The image archive files were unpacked into the job run directories prior to submitting test jobs to the Cori batch queue, such that the jobs did not have to use additional time on unpacking archives. Using this approach we tested compressed and uncompressed checkpoint images and compared the results to startup times of conventional AthenaMP jobs. For each configuration we ran 300 single-node jobs. The results of our studies are presented in Table 1. The job startup time was measured between launching AthenaMP and the moment it forked worker processes. The results show that even with a compressed checkpoint image we can achieve a 13-fold speedup of job startup time with respect to conventional AthenaMP, while for uncompressed images the performance gains are much larger.

6. Summary and Outlook

We have successfully tested the DMTCP tool for checkpoint-restart of Geant4 simulation jobs in the multi-process framework AthenaMP. First tests with production ATLAS jobs on

Table 1. AthenaMP Simulation Startup Times on Cori KNL Nodes.

	Image size	Startup time (sec)	Startup speedup
`Conventional AthenaMP`	N/A	663.1 ± 22.8	1
`Compressed image`	550MB	50 ± 9.7	13.3x
`Uncompressed image`	1.8GB	20.8 ± 9.1	31.5x

ATLAS@Home and Intel KNL platforms have demonstrated that by restarting from checkpoint images we can considerably speedup job startup times.

There are however several improvements to be implemented before deploying on a larger scale. We plan to run more tests in order to demonstrate that by restarting from a checkpoint image we do not introduce a visible penalty into the event processing throughput. Also the process of creating and distributing checkpoint images must be automated and the results of the jobs restarted from checkpoint images must be validated for physics correctness.

7. Acknowledgments

The authors would like to thank our volunteer testers in ATLAS@Home for running the jobs for us and providing useful feedback.

The performance results on Cori KNL platform presented in this paper have been obtained by using resources of the National Energy Research Scientific Computing Center, a DOE Office of Science User Facility supported by the Office of Science of the U.S. Department of Energy under Contract No. DE-AC02-05CH11231.

References

[1] ATLAS Collaboration *JINST* **3** (2008) S08003
[2] Evans L and Bryant P "LHC Machine", *JINST* **3** (2008) S08001
[3] Calafiura P et al. on behalf of the ATLAS Collaboration "Running ATLAS workloads within massively parallel distributed applications using Athena Multi-Process framework (AthenaMP)", *J. Phys.: Conf. Ser.* **664** (2015) 072050 doi:10.1088/1742-6596/664/7/072050
[4] Arya K, Cooperman G, Dotti A and Elmer P "Use of checkpoint-restart for complex HEP software on traditional architectures and Intel MIC", *J. Phys.: Conf. Ser.* **523** (2014) 012015 doi:10.1088/1742-6596/523/1/012015
[5] Agostinelli S et al. GEANT4 Collaboration "Geant4 a simulation toolkit", *Nucl. Instrum. Meth. A* **506** (2003) 250 doi:10.1016/S0168-9002(03)01368-8
[6] ATLAS Collaboration (2010) "ATLAS Simulation Infrastructure", *Eur. Phys. J* **C70** (2003) 823
[7] Adam-Bourdarios C et al. on behalf of the ATLAS Collaboration "ATLAS@Home: Harnessing Volunteer Computing for HEP", *J. Phys.: Conf. Ser.* **664** (2015) 022009 doi:10.1088/1742-6596/664/2/022009
[8] DMTCP project, "DMTCP" [software], version 2.4.5, 2016. Available from https://github.com/dmtcp/dmtcp/releases/tag/2.4.5 [accessed 2016-07-19]
[9] Maeno T for the ATLAS Collaboration "PanDA: Distributed production and distributed analysis sys tem for ATLAS", *J. Phys.: Conf. Series* **119** (2008) 062036 doi:10.1088/1742-6596/119/6/062036
[10] Anderson D *Proc. 5th IEEE/ACM Int. Workshop on Grid Computing, GRID 04* (2004) pp 410
[11] VirtualBox project, "VirtualBox" [software], version 5.1.22, 2017. Available from https://www.virtualbox.org

IOP Conf. Series: Journal of Physics: Conf. Series **1085** (2018) 032029 doi:10.1088/1742-6596/1085/3/032029

An ATLAS distributed computing architecture for HL-LHC

S Campana[1] and T Wenaus[2] for the ATLAS Collaboration

[1] CERN, Geneva, Switzerland
[2] Brookhaven National Laboratory, Upton, NY 11973-5000, USA

Abstract. The ATLAS collaboration started a process to understand the computing needs for the High Luminosity LHC era. Based on our best understanding of the computing model input parameters for the HL-LHC data taking conditions, results indicate the need for a larger amount of computational and storage resources with respect to the projection of constant yearly budget for computing in 2026. Filling the gap between the projection and the needs will be one of the challenges in preparation for HL-LHC. While the gains from improvements in offline software will play a crucial role in this process, a different model for data processing, management, access and bookkeeping should also be envisaged to optimise resource usage. In this contribution we will describe a straw man of this model, founded on basic principles such as single event level granularity for data processing and virtual data. We will explain how the current architecture will evolve adiabatically into the future distributed computing system, through the prototyping of building blocks that would be integrated in the production infrastructure as early as possible, so that specific use cases can be covered much earlier with respect to the HL-LHC time scale. We will also discuss how such system would adapt to and drive the evolution of the WLCG infrastructure in terms of facilities and services.

1. Introduction

The High Luminosity data-taking period of the LHC (HL-LHC) scheduled starting in 2026 will present unprecedented challenges in terms of computing. The ATLAS experiment [1], after a major hardware upgrade, will collect and retain 10,000 proton-proton collision events per second, compared to the 1,000 collisions in the current LHC run. In addition, the complexity of the events will increase considerably, as up to 200 particle interactions will overlap (pile-up). Reconstructing one event is expected therefore to consume more CPU resources as the growth in reconstruction time exponentially increases with the pile-up. The event size is expected to increase linearly with the pile-up as it strongly correlates with the number of particles in the event. The LHC experiment computing infrastructures, founded on the Worldwide LHC Computing Grid (WLCG), organise computing resources distributed over 200 centers. The budget for computing hardware in WLCG is not expected to increase in the next years and at best will remain constant. Considering the needs to replace old hardware and the technology improvements we can expect roughly a 20% growth of processing capacity (CPUs) and a 15% growth of storage (disk and tape) every year for the same constant funding. In 2016 ATLAS presented a study of the computing challenges in HL-LHC, comparing the resource needs with the expected available resources assuming constant funding evolution. In this contribution we present an update of the study, based on a more refined model, a better understanding of the input parameters, the new ATLAS detector layout and recent improvements of the reconstruction algorithms. We show how, despite the recent improvements, the required level of resources is still considerably larger than

IOP Conf. Series: Journal of Physics: Conf. Series **1085** (2018) 032029 doi:10.1088/1742-6596/1085/3/032029

what we will realistically be able to obtain. We focus on the storage needs and present the architecture of a new computing model and infrastructure capable to at least partially address the likely shortage of resources.

2. The HL-LHC computing challenge

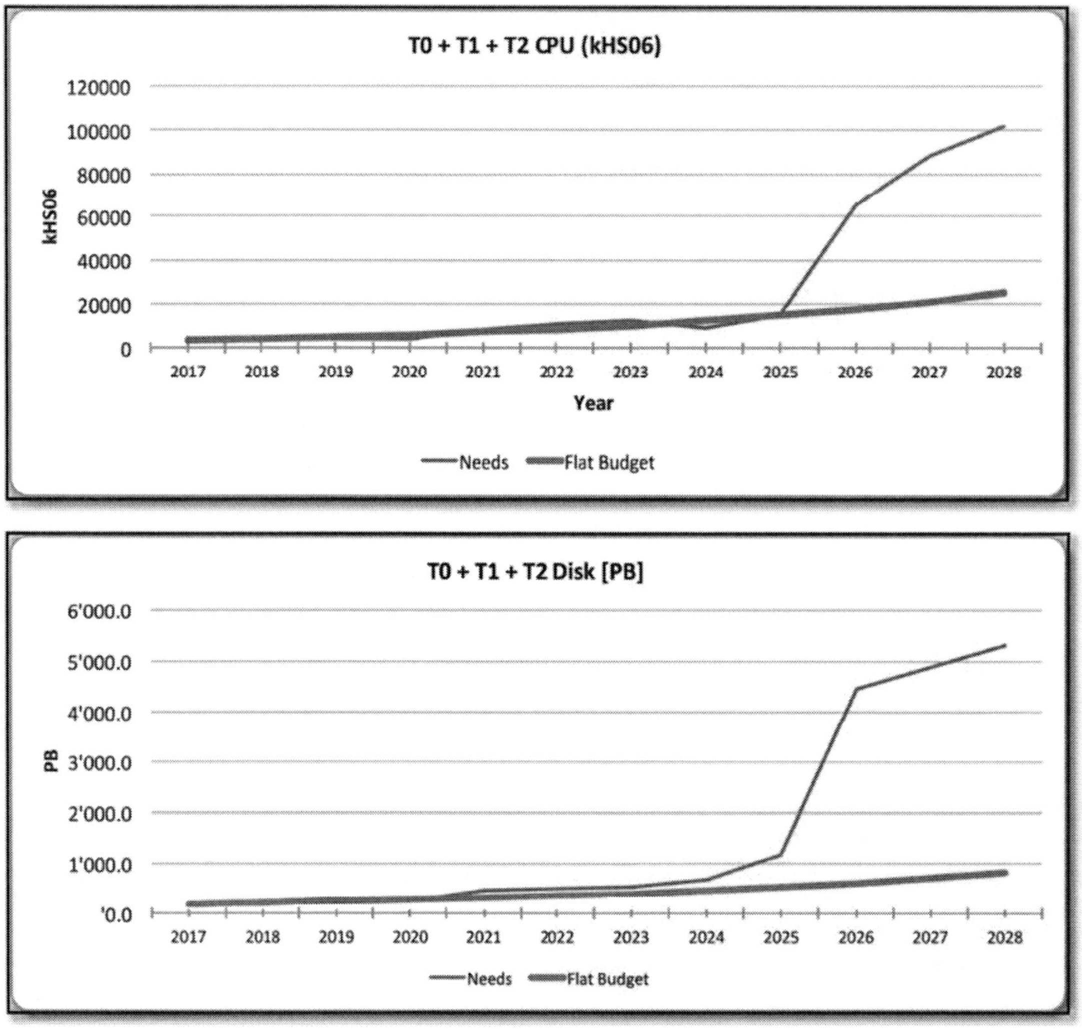

Figure 1. Estimates of the ATLAS CPU and storage (disk) needs in the next 10 years, in the top and bottom panel respectively. The needs (blue line) are compared with the projected available resources based on a constant funding model (red line)

The most recent study of the HL-LHC computing challenge for the four experiments was presented in Ref. [2]. In Ref. [3] ATLAS produced a more detailed analysis showing how different input parameters impact the expected resource needs. Such analysis estimated a gap of a factor 9 for CPUs and a factor 7 for storage between the needed resource and what likely will be available assuming constant funding. We recently updated the study and in particular:

1. The expected number of data-taking seconds is increased from 5.5 million to 7.3 million based on the current LHC running experience.
2. The contribution of the CERN computing facility is included in the model.
3. Data retained from previous data taking years is now properly accounted for.
4. The reconstruction time per event was lowered from 2880 HS06 seconds to 1300 HS06 seconds. The new number reflects the improvement after introducing the new layout of the inner detector and tuning the software accordingly.
5. The Monte Carlo needs are now estimated as 1.5 times more events than data, based on the current run experience. One third of such events can be produced with Fast Simulation while the rest should be simulated through Geant4 [4].

Figure 1 shows the evolution of the ATLAS computing needs in the next 10 years compared with the expected available resources obtained spending a constant budget every year. With the new parameters and model the gap in CPU needs decrease from a factor 9 to a factor 3.7. In fact, more improvements can be expected in the next years, as for the moment only the inner detector tracking was optimised, while more gains can be expected once the similar work is performed for calorimeters and the muon spectrometer.

There is no considerable improvement for storage however. The cost of disk space today exceeds by 4 times the cost of the same volume on tape, therefore we focus on disk. The vast majority of the disk space is needed to store analysis formats such as AODs and their derived ones (DAODs) for both data and Monte Carlo. The Computing Models evolved during Run-2 implementing data reduction based on popularity and implementing a lifetime model removing data from disk and tape once the datasets lifetime expires and the data is not recently accessed. Therefore, at the moment there is little contingency in further reducing the number of disk replicas.

3. The Run-2 analysis model

The DAODs were introduced in the ATLAS analysis model in 2014, during the LHC shutdown between Run-1 and Run-2 (2013-2015). The outputs of the reconstruction (AODs) are centrally processed through organised analysis campaigns called trains. Each train processes events in AOD formats, possibly updates some information, filters the ones suitable for a given analysis, removes the variables not relevant for that analysis and stores the derived format (DAODs) on disk. A train can run multiple selections (carriages) on the same events and produce multiple DAODs. The DAODs can be further processed through the ATLAS offline software (Athena) or the ROOT [5] analysis package, to produce NTUP files for final analyses. Today ATLAS runs approximately 20 trains for a total of 100 carriages, and therefore produces 100 DAOD types with the average size of 1% of the AODs each. The total volume of a full DAOD set is therefore similar to the one of the AOD it has been produced from. DAOD datasets are frequently accessed by analysers: to ensure data integrity and a rapid turnaround of the analyses, two copies of the most recent version of the DAODs are kept on disk. Eliminating the needs to store DAODs would reduce by more than a factor 2 the disk needs.

A possible data processing architecture for HL-LHC

A diagram of the proposed data processing architecture for HL-LHC is shown in Figure 2.

3.1 *The Event Whiteboard*

A considerable reduction of the storage needs can be achieved by reducing or eliminating the need to duplicate physical information between AODs and DAODs. Central to the new data processing architecture is a Data Knowledge Base (DKB) service, hosting the meta-information associated to all ATLAS events. Such an "Event Whiteboard" would contain all the relevant meta-data obtained from distributed computing services such as data management and workload management. It would in

IOP Conf. Series: Journal of Physics: Conf. Series **1085** (2018) 032029 doi:10.1088/1742-6596/1085/3/032029

addition allow to annotate at the event level user defined meta information. The derivation framework could at this point avoid creating DAOD files but simply annotate the events with a tag indicating that the event passes some selection criteria. The Whiteboard would then be able to provide event lists for a given tag together with the necessary information for the Offline I/O libraries to extract single events or event ranges. The workload management system would be able to define data processing units (jobs) accessing a defined list of events.

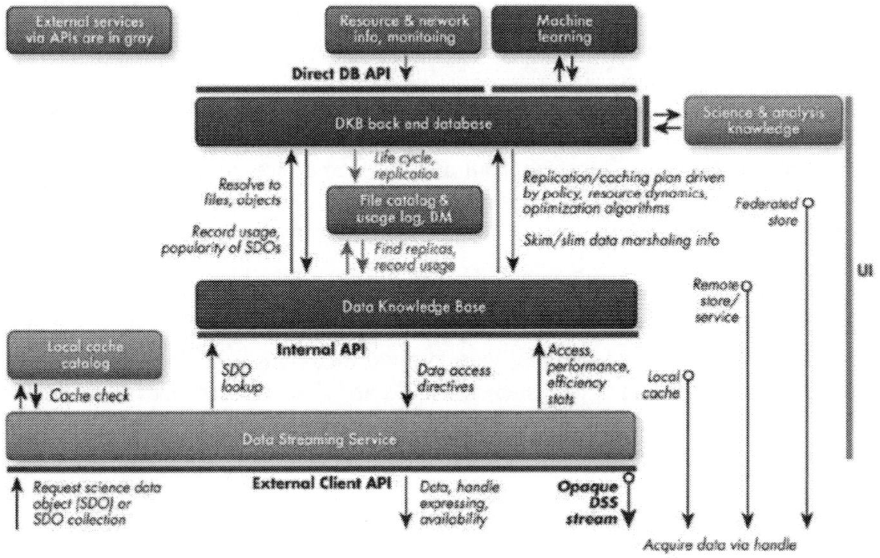

Figure 2. The Data Knowledge Base service and the interaction with external components, among which is the Event Streaming Service

3.2 *Asynchronous pre-fetching and Event Streaming Service*

Currently the ATLAS workload management system and particularly the jobs running at the worker nodes use two different data access techniques, depending on the workflow and storage I/O capabilities: the full copy of the input file containing the events to process into the worker node and the direct access using streaming protocols such as XRootD or HTTP from the storage endpoint. None of the two methods would be efficient in the scenario where the DAOD datasets are simply a logical collection of events. The full copy of the input file would end up downloading a large number of events not to be processed from the AOD, particularly in case the DAOD collection was the result of heavy skimming. Direct streaming would result in I/OWait while data are obtained from the storage locations, leaving idle CPU cycles. An Event Streaming Service would provide a solution. In its first implementation, the client on the worker node would pre-fetch event parcels (collections of events) asynchronously from the actual processing and store them on the local disk, so that CPUs would be able to open and process data locally. In a more advanced implementation, the pre-fetch would be directed at a service deployed in the network neighbourhood of the storage location. This service would prepare event parcels on disk caches under the direction of the workload management system, which would activate payloads accessing such parcels only when the parcels are available and possibly replicated in a convenient location. The system would be more robust and effective if the source data would be consolidated in a small number of storage locations interconnected with fast networks, while the event streaming service caches would need to be distributed as close as possible to the processing units and as broadly as possible. Today ATLAS processes 1 Exabyte of data per year. The event streaming service would need therefore to prepare that volume of DAODs, resulting in approximately 30 GB/s. There is however a considerable reuse of the DAODs (more than 10 times per sample).

Therefore a system delivering 3GB/s of cached data would suffice for Run-2 and Run-3 while one would need 10 times more for Run-4.
The potential gains are considerable: in principle we could eliminate all physical copies of DAODs for both data and Monte Carlo, consisting of 45% of the total disk needs. In reality many caveats apply and such estimates could be only the theoretical limit. The derivation framework does not only filter events but also retains only the relevant quantities based on the physics analysis that such derivation will support. Such reduction would be lost in the proposed model. Therefore we would likely continue to produce physical DAOD files but in a much smaller variety, say one per physics group (10 in total to be compared with 100 formats today). Additionally, today in producing DAOD files we order the event information to optimise data compression and we need to understand how that would happen in a model where no physical DAOD file is produced.

3.3 *Virtual Data*

A second application of the DKB consists in capturing all the necessary information to reproduce derived data. As of today, reproducing derived data is a tedious and error prone operation, requiring manual work. To ensure data integrity we store multiple replicas of the vast majority of the datasets and to avoid analysis reproducibility issues we tend to retain such copies for a long time. Only a very small fraction of such data is accessed after one year since it was produced. Capturing all necessary information to reproduce data in the Knowledge Base system and integrating the workload and data management system with it would allow treating part of the samples as *virtual data*. Knowing the cost to reproduce data, the cost to store replicas on different media and estimating the likelihood for that data to be accessed after a given period of time (or that data to be lost) the system could decide to retain physical copies or re-generate the data in case of need. The system could also decide to create more copies to facilitate the access or avoid the loss. The information sources could be multiple and extensible, including machine learning algorithms for anomaly detection and data popularity forecasts. The automation of the system is the key element in this model. The DKB would provide also the backbone for analysis and data preservation models being developed in ATLAS.

4. Conclusions

We presented an update of the computing challenges for HL-LHC in term of hardware resources. The computing and analysis model we use in Run-2 will not be able to accommodate the HL-LHC hardware needs in a constant funding profile. We propose a different analysis model based on event level processing granularity, virtual data and data reproducibility. While such a model presents a radical change with respect to the current system, single components can be developed and deployed early, complementing the data processing and storage tools we have available today.

References

[1] The ATLAS Experiment at the CERN Large Hadron Collider, **JINST 3** (2008) S08003

[2] S. Campana, "Software and Computing requirements and possible architecture", ECFA High Luminosity LHC Experiments Workshop, Aix-Les-Bains, France, (2016)

[3] S. Campana, "The ATLAS Computing Challenge for HL-LHC", Computing in High Energy and Nuclear Physics, San Francisco, United States, (2016)

[4] Agostinelli, S., Allison, J., Amako, K., et al., Geant4 – a simulation toolkit, Nuclear Instruments and Methods in Physics Research, A 506 (2003) 250-303

[5] Rene Brun and Fons Rademakers, ROOT - An Object Oriented Data Analysis Framework, Proceedings AIHENP'96 Workshop, Lausanne, Sep. 1996, Nuclear Instruments and Methods in Physics Research A 389 (1997) 81-86

IOP Conf. Series: Journal of Physics: Conf. Series **1085** (2018) 032030 doi:10.1088/1742-6596/1085/3/032030

The ATLAS Data Management System Rucio: Supporting LHC Run-2 and beyond

M Barisits[1], T Beermann[2], V Garonne[3], T Javurek[4], M Lassnig[2] and C Serfon[3] on behalf of the ATLAS collaboration

[1] University of Innsbruck, Austria
[2] CERN, Geneva, Switzerland
[3] University of Oslo, Norway
[4] University of Freiburg, Germany

E-mail: martin.barisits@cern.ch

Abstract. With this contribution we present some recent developments made to Rucio, the data management system of the High-Energy Physics Experiment ATLAS. Already managing 300 Petabytes of both official and user data, Rucio has seen incremental improvements throughout LHC Run-2, and is currently laying the groundwork for HEP computing in the HL-LHC era. The focus of this contribution are (a) the automations that have been put in place such as data rebalancing or dynamic replication of user data, as well as their supporting infrastructures such as real-time networking metrics or transfer time predictions; (b) the flexible approach towards inclusion of heterogeneous storage systems, including object stores, while unifying the potential access paths using generally available tools and protocols; (c) machine learning approaches to help with transfer throughput estimation; and (d) the adoption of Rucio for two other experiments, AMS and Xenon1t. We conclude by presenting operational numbers and figures to quantify these improvements, and extrapolate the necessary changes and developments for future LHC runs.

1. Introduction

Rucio [1] is the Distributed Data Management (DDM) system in charge of managing all ATLAS [2] data on the grid. The main purpose of the system is to help the collaboration to store, manage and process LHC data in a heterogeneous distributed environment. Typical tasks are: Transfer data to/from sites, delete data from sites, enforce the experiment's computing model. The system manages 300 Petabytes of physics data across more than 130 data centers globally, with more than 830 million files. Over the last year multiple improvements have been introduced to support the collaboration's data management needs for LHC Run-2 and beyond. These changes focused on the integration of new technologies, the automation of the system for optimization and reduction of manual work, and machine learning studies to better understand the usage of the system.

The paper is organized as follows: In Section 2 we present the integration of object store technologies into Rucio. In Section 3 we discuss two new approaches in system automation: Dynamic creation of additional replicas and automatic rebalancing of data. Section 4 continues with machine learning approaches for data management. In Section 5 we give a short overview of Rucio beyond ATLAS. Finally we conclude the article in Section 6.

IOP Conf. Series: Journal of Physics: Conf. Series **1085** (2018) 032030 doi:10.1088/1742-6596/1085/3/032030

2. Object stores

Within the ATLAS DDM system, the integration of a new storage type [3] is a constant need. The vast majority of cloud storage available in the market leverages an object-based storage architecture. It differs in many points from traditional file system storage and offers a highly scalable, simple and most common storage solution for the cloud. Object-based storage supports other authentication mechanisms than X509, usually based on (secret) access keys. The main idea was to integrate object-based storage as regular DDM end-point and to focus on the Amazon S3 (Simple Storage Service) protocol. Rucio allows the integration of non-posix namespace and has a flexible mapping protocol-endpoint. Two protocols have been implemented: s3:// and s3+rucio://. The protocol s3:// has been implemented with the boto python library and requires access keys. It is the prefered way on central machines we have access to in a secure way, e.g., central deletion. The protocol s3+rucio:// has been implemented with pure http python library and relies on pre-signed URLs, which guarantee (temporary read/write) access to the object. Currently, several institutes host object-based storage for ATLAS: BNL, Lancaster, RAL, CERN, MWT2. These storages have a Ceph implementation. Two use cases are supported in production: i) upload and download of log files and ii) deletion of objects generated by the ATLAS Event Service (AES) on object-based storage.

3. System automations

A significant part of the day to day operation is to optimize the usage of space. Rucio is currently using two daemons that automate the handling of replicas on the grid, either by automatically creating new replicas of datasets that are considered popular and therefore could benefit from extra replicas on the grid or by moving replicas around to balance the used and free space on storage.

3.1. Dynamic creation of additional replicas

This daemon [4] continuously monitors incoming user analysis jobs and the decides based on a multitude of parameters if and where to make extra copies of the input datasets. The system works as a cache for popular data. These extra replicas are created with a short lifetime, which guarantees that they can be deleted again quickly if not used anymore and space is needed. Because of the way the deletion works in Rucio if the replica continues to be used it will also stay on storage as long as enough space is available.

Every time a new job is defined a decision algorithm runs that uses the following information to make a decision:

- The past popularity of the input dataset. Rucio uses a tracer system that receives a trace every time a file is either used in grid jobs or downloaded. This information is then aggregated to get number of accesses for this dataset in the last 7 days.
- The existing replicas. If already more than 4 replicas exist on the grid the daemon will not create another one.
- The free space on potential destination sites.
- Network speed between potential sites hosting an existing replica and potential destination site.

The first two metrics are used to filter potential datasets that could benefit from extra replicas the latter two are combined to rank potential source and destination storage, where the ones with the most free space and a good connection are preferred. If a storage element has been picked to get a new replica it will get a penalty, so that for the next dataset it will be less likely to be picked again. It can prevent a single site with a lot of free space to be flooded with a lot of transfers. Figure 1 shows the kibana dashboard of this daemon.

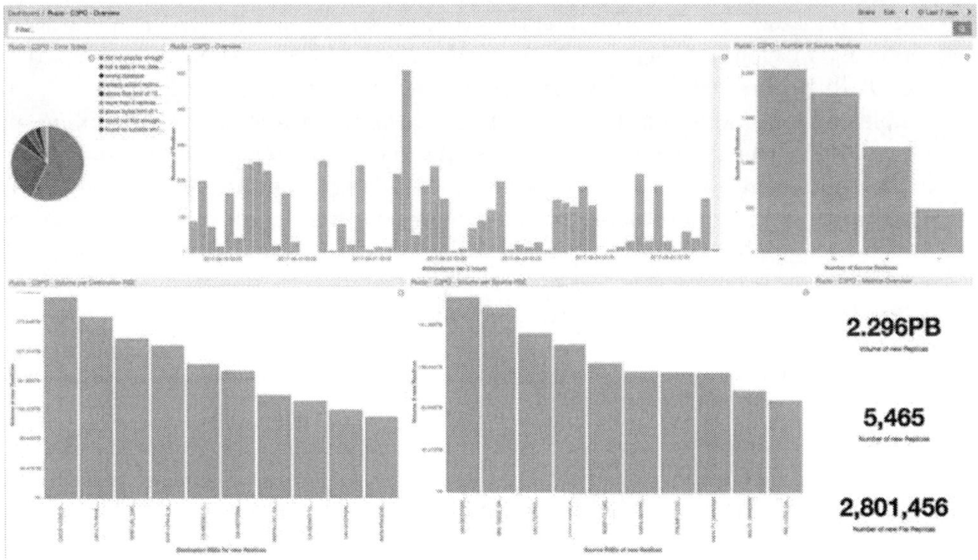

Figure 1. Kibana overview dashboard to monitor the decision made by the dynamic creation of additional replicas daemon

3.2. Rebalancing of data

The need of automatic rebalancing of data [5] arises due to the fact that some storage systems are at their capacity limit, but do not have any secondary data to delete. These are the storage resources which need data rebalanced to other storage systems in order to be functioning again. There are three reasons why these situations arise:

- Imbalances between computation resources and storage resources at a site. Thus, due to the higher computing capacity at a site more data is transferred and generated at the site than the storage element can hold. Mostly Tier-1 sites are affected by this phenomenon.
- Non optimal placement of data. This can either happen due to policies or users requesting replication. The data distribution policies are rather static, thus they do not make the most optimal decision at every point in time.
- In some cases computation jobs also request too large transfers or the job output is too large for a site to handle. This brings a storage element also into a situation of crisis.

The general idea of data rebalancing in ATLAS distributed data management is to move data away from the problematic storage endpoint while still adhering to the original data replication policy as much as possible. Data integrity is a key requirement in this process, thus, the original data is not deleted until the newly created data has been transferred and validated. Currently three different modes are supported for data rebalancing: Manual (emergency) rebalancing, automatic (emergency) rebalancing and automatic background rebalancing. The *manual rebalancing of data* mechanism gives DDM operators a tool to easily rebalance a given amount of data away from a storage element. This can be either done in emergency situations, such as a storage element being full without anything to delete, or also due to other data distribution needs. A rebalancing operation can be issued via the command line client interface by just specifying the storage from which to rebalance data, and the amount of bytes to rebalance. The *automatic emergency rebelancing* mode relieves DDM operators by autonomously detecting storage systems in crisis and rebalancing data away from them to make room for new data. Figure 2 shows a problematic data utilization workflow. The primary data occupancy (shown

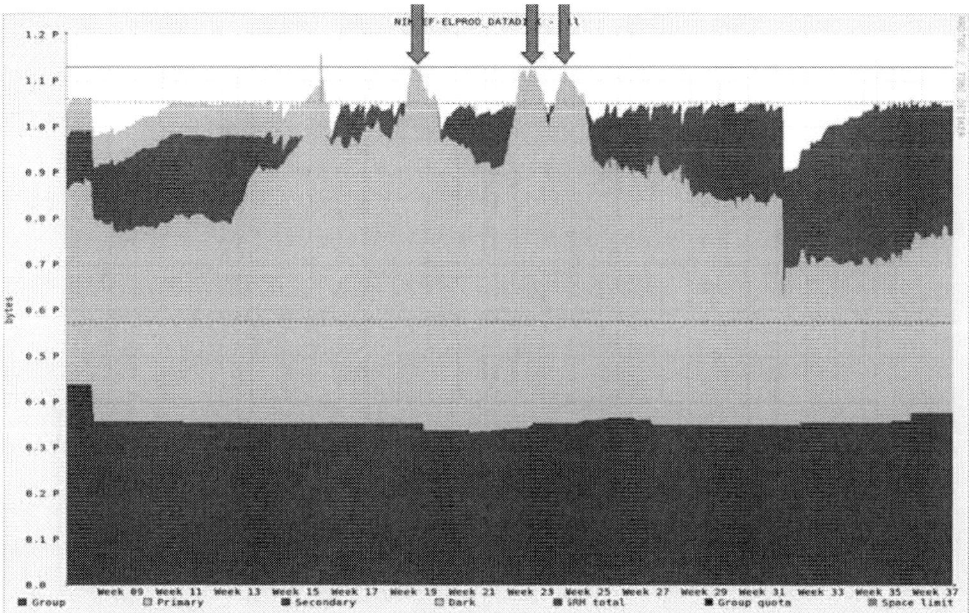

Figure 2. Problematic data utilization of a storage system.

as yellow and blue in the plot) rises, eventually replacing all secondary (shown as green) data and reaching the volume limit of the storage element. The concept of the automatic emergency rebalancing mode is to look for storages which are over the watermark and which do not have any secondary data to delete. In these cases, the mechanisms rebalance a fixed amount of data away from the storage. The idea of the *automatic background rebalancing* is to prevent emergency situations like this from even occurring. To this effect the automatic background rebalancing daemon trys to balance the amount of primary to secondary replicas at a set of storages, such as all Tier-1 storage elements. The idea behind this is, if the ratio is in balance, all storages have relatively the same amount of secondary replicas available for deletion.

4. Machine learning for data management

To quantify any improvements in data placement we need an estimator of potential transfer times. A global minimum solution for all expected transfers yields, in theory, an optimum placement strategy when assuming equivalence of computational resources. Geographical distance or network bandwidth are insufficient quantifiers though. The dynamic nature of both the involved applications and the networks themselves, as well as their physical deployment, require a more sophisticated method. Currently there is no sensible model available to estimate the duration of a particular collection of file transfers between data centres.

Long Short-Term Memory (LSTM) networks seem particularly suited to annotated time series data with recurring but non-periodic bursts [6]. We selected Keras as the software framework for the implementation since it supports LSTMs and opens the possibility for deep learning at later steps. We created the LSTM with one hidden layer at 512 neurons and one dense layer with one output, activated by a rectified linear unit (ReLU). The training input variates are *source*, *destination*, *activity*, *bytes*, *start timestamp*, and *end timestamp* of one transfer, using *end timestamp* as the label. The result of the model is shown in Figure 3. The absolute prediction error is less than 5 minutes across the majority of activities and links. There exists some multimodality, however it is constrained within an acceptable error range [7].

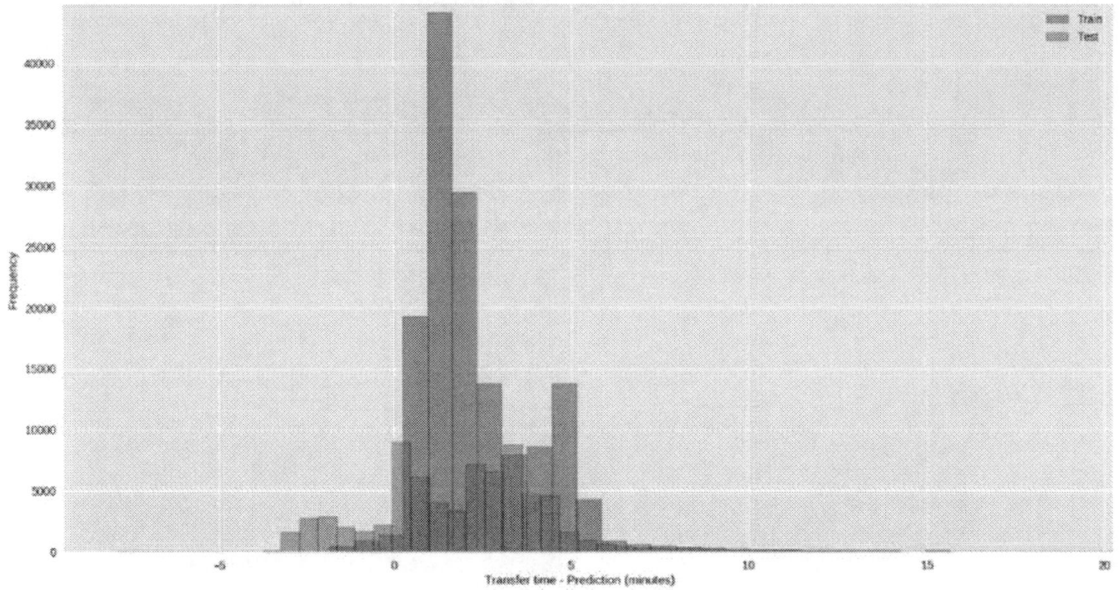

Figure 3. Single layer LSTM model absolute error across all activities and labels.

5. Beyond ATLAS

In addition to ATLAS, two other experiments are already running Rucio in production. The Alpha Magnetic Spectrometer (AMS) and Xenon1t experiments have expressed interested in both the namespace and transfer capabilities of Rucio, and subsequently installed a separate Rucio instance each. The most striking difference to ATLAS is the usage of a different database backend, namely MariaDB instead of Oracle, demonstrating the versatility of the Rucio persistence layer. Shared installations of the underlying File Transfer Service (FTS) instances at CERN and BNL are used by all experiments. Current developments in Rucio target an even easier deployment scheme for additional experiments using docker containers, new development workflows using *github.com* to facilitate non-ATLAS contributions, as well as monitoring improvements for experiments who cannot benefit from CERN-provided monitoring.

6. Conclusion

Rucio has demonstrated excellent performance since the beginning of LHC Run 2. In this paper we presented some of the recent developments that were done in order to reduce the operational burden as well as to exploit new storage technologies. Research activities based on machine learning will help us to further improve the performance and scalability of the system. Developed initially for the ATLAS experiment, Rucio is now used by two other non-LHC experiment and is being evaluated by others.

References

[1] Garonne V, Vigne R, Stewart G, Barisits M, Beermann T, Lassnig M, Serfon C, Goossens L, Nairz A, *Rucio - The next generation of large scale distributed system for ATLAS Data Management.* Journal of Physics: Conference Series 2014
[2] ATLAS Collaboration, *ATLAS Collaboration.* JINST **3** (2008) S08003
[3] Garonne V, Guan W, *Object-based storage integration within the ATLAS DDM system.* to appear in Journal of Physics Conference Series 2017
[4] Beermann T, Lassnig M, Barisits M, Garonne V, Serfon C, *C3PO - A Dynamic Data Placement Agent for ATLAS Distributed Data Management.* to appear in Journal of Physics Conference Series 2017

IOP Conf. Series: Journal of Physics: Conf. Series **1085** (2018) 032030 doi:10.1088/1742-6596/1085/3/032030

[5] Barisits M, Serfon C, Garonne V, Lassnig M, Beermann T, Javurek T, *Automatic rebalancing of data in ATLAS distributed data management.* to appear in Journal of Physics Conference Series 2017
[6] Hochreiter and Schmidhuber, J 1997 Neural Computation 9(8):1735-1780
[7] Lassnig M, Toler W, Vamosi R, Bogado J, *Predictive analytics for ATLAS Data Management using Machine Learning methods.* to appear in Journal of Physics Conference Series 2017

IOP Conf. Series: Journal of Physics: Conf. Series **1085** (2018) 032031 doi:10.1088/1742-6596/1085/3/032031

Global heterogeneous resource harvesting: the next-generation PanDA Pilot for ATLAS

A. Anisenkov[1,2], D. Drizhuk[3], W. Guan[4], M. Lassnig[5], P. Nilsson[6,*], D. Oleynik[7,8] on behalf of the ATLAS Collaboration

[1] Budker Institute of Nuclear Physics, Siberian Branch of Russian Academy of Sciences, 11, Akademika Lavrentieva prospect, Novosibirsk, 630090, Russia
[2] Novosibirsk State University, 2, Pirogova street, Novosibirsk, 630090, Russia
[3] National Research Centre Kurchatov Institute, 1, pl. Akademika Kurchatova, Moscow, 123182, Russia
[4] University of Wisconsin-Madison, Department of Physics, 4425 Chamberlin Hall Madison, WI 53706, USA
[5] CERN, European Laboratory for Particle Physics, CH-1211 Geneva 23
[6] Brookhaven National Laboratory, PO Box 5000, Upton, NY 11973-5000, USA
[7] University of Texas at Arlington, 502 Yates Street, Arlington, TX 76019-0059, USA
[8] Joint Institute for Nuclear Research, Joliot-Curie 6, Dubna, 141980, Russia

[*]Corresponding author: Paul.Nilsson@cern.ch

Abstract. The Production and Distributed Analysis system (PanDA), used for workload management in the ATLAS Experiment at the LHC for over a decade, has in recent years expanded its reach to diverse new resource types such as HPCs, and innovative new workflows such as the Event Service. PanDA meets the heterogeneous resources it harvests in the PanDA Pilot, which has embarked on a next-generation reengineering to efficiently integrate and exploit the new platforms and workflows. The new modular architecture is the product of a year of design and prototyping in conjunction with the design of a completely new component, Harvester, that will mediate a richer flow of control and information between Pilot and PanDA. Harvester will enable more intelligent and dynamic matching between processing tasks and resources, with an initial focus on HPCs, simplifying the operator and user view of a PanDA site but internally leveraging deep information gathering on the resource to accrue detailed knowledge of a site's capabilities and dynamic state to inform the matchmaking. This paper will give an overview of the new Pilot architecture, how it will be used in and beyond ATLAS, its relation to Harvester, and the work ahead.

1. Introduction

The PanDA [1] Pilot is responsible for executing payloads on the worker nodes on local resources, on grids or clouds, on (non-restrictive) High Performance Computers (HPCs) and on volunteer computers. It has been used in the ATLAS Experiment [2] at the LHC for well over a decade, and is in constant development. Over time, however, it has become more and more challenging to maintain the now aging code base and to add new features to it, which are often highly complex. It is a well-known problem in software engineering that eventually a solid code base will grow old and contain outdated and deprecated functions, at which point one has to make the decision whether to refactor the code significantly or to rewrite it. In our case, we had already spent years of partial refactoring when we decided to rewrite the original PanDA Pilot from scratch and create a Pilot 2 to benefit from a more modern design approach and get rid of all old code constructs in one go. Pilot 2 is a fresh start that provides more functionalities than the original Pilot, such as APIs for external usage, although some of its traditional responsibilities can be moved to a new member of the PanDA family, Harvester [3]. This new component is a service between the PanDA server and the Pilot especially designed for use on HPCs, and provides resource provisioning and workload shaping. It enables more intelligent and dynamic task matching with the resources, simplifying the operator and user view of a PanDA site but internally utilizing various kinds

IOP Conf. Series: Journal of Physics: Conf. Series **1085** (2018) 032031 doi:10.1088/1742-6596/1085/3/032031

of information about the resource and a site's capabilities to boost the performance of tasks as well as sites.

There are several functional overlaps between the Pilot and Harvester, such as payload setup and copy tools, that are implemented in the Pilot and imported and used by Harvester to avoid code duplication.

The Pilot 2 architecture follows a component-based approach which in comparison to the previous Pilot implementation allows to exploit an enhanced system flexibility, enables a clear workflow control, evolves the system according to modern functional use-cases and covers upcoming feature requests from new PanDA users.

2. Macroscopic system structure

The macroscopic system structure (Figure 1) shows the relations between the various Pilot components, how they communicate with each other and the flow of information. The main tasks or functions are sorted into several controller components, such as Job Control, Payload Control and Data Control. There is also a set of components with auxiliary functionalities, e.g. Setup, Information Service and Monitoring. There are two types of monitoring, one for internal use which monitors threads and one that is tied to the job and checks parameters that are relevant for the payload, e.g. size checks and that the payload is still producing output at a regular basis.

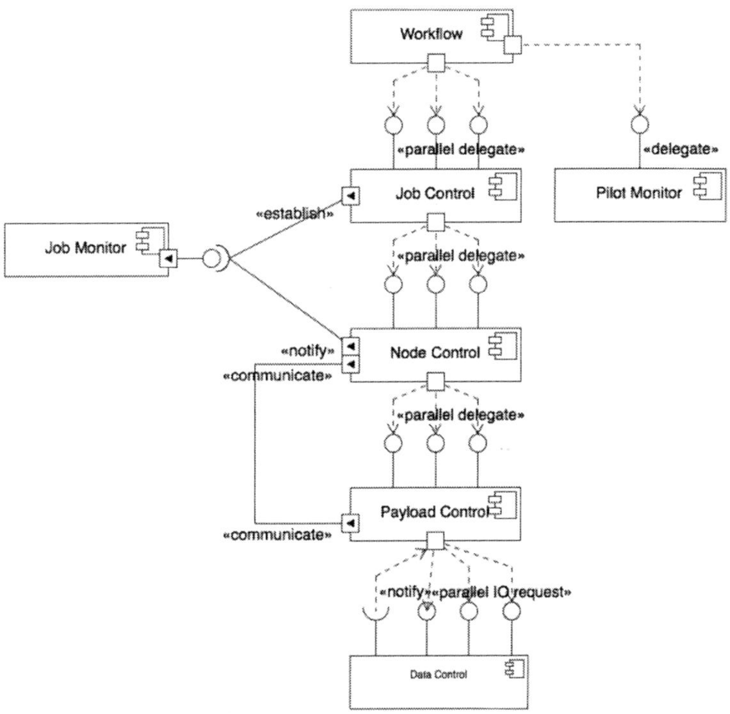

Figure 1. Macroscopic system structure.

Code that is specific to a PanDA Pilot user is handled by plug-in mechanisms and is stored in its own directories to keep it separated from the core Pilot code. New workflows can be defined in case the generic Pilot workflow is not sufficient, while still benefitting from core Pilot functionalities.

3. Parallel streaming data flow

Figure 2 shows the generic flow of the job objects and how they move through the internal queues in the different Pilot components. A job object is an entity that is containing all necessary information about the payload (e.g. software release version, parameters for payload setup, transfer type of input files, etc).

IOP Conf. Series: Journal of Physics: Conf. Series **1085** (2018) 032031 doi:10.1088/1742-6596/1085/3/032031

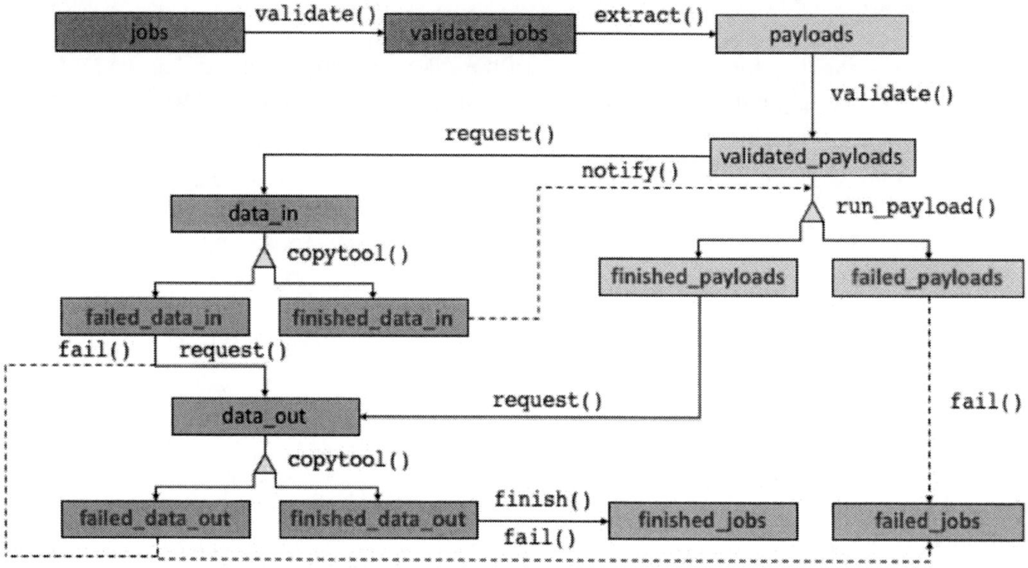

Figure 2. The generic flow of the job objects.

When a job has been verified, i.e. ready for execution, it is moved from the job queue in the Job Control to a validated job queue. The Payload Control takes it from there and executes it. When the payload has finished, any output is transferred by the Data Control to a relevant storage element (SE) along with the job log. As a last step, the Pilot informs the server about the final state of the job.

An Event Service [4] job is handled internally in a similar way to a 'normal' job but the workflow differs somewhat from the generic workflow (see Figure 3). While both job types have similar job definitions, there are differences especially with the stage-out process. In an Event Service job, input data is read directly from storage (also possible for normal jobs) and accessed via event ranges (unlike normal jobs). The output from a processed event range is immediately and asynchronously transferred to an object store, while a normal job only transfers the output at the end of the job to a normal SE.

An Event Service job may also use an additional service application, called the Prefetcher. This tool is an application that is launched by the Pilot. Its task is to asynchronously prefetch needed data from a remote location and store it locally, and is a key component of the Event Streaming Service (ESS) [4].

While this section has described workflows used by ATLAS, it should be stressed that Pilot 2 allows for other workflows to be implemented while still benefitting from core Pilot functionalities such as monitoring, copy tools, containerization of executable commands, etc.

4. Pilot APIs

Decoupling system logic into well-defined components with a corresponding common API helps to effectively consolidate system modules and in particular share code functionality, integrate and communicate with external applications like Harvester, as well as for dedicated testing.

Selected Pilot functionality is exposed to external users via the following APIs:

1. Data API: functions for performing stage-in and stage-out.
2. Communicator API: communication tools for external services.
3. Environment API: interface to the job execution environment on HPCs.
4. Services API: functions related to external service tools such as benchmarking and memory monitoring.

IOP Conf. Series: Journal of Physics: Conf. Series **1085** (2018) 032031 doi:10.1088/1742-6596/1085/3/032031

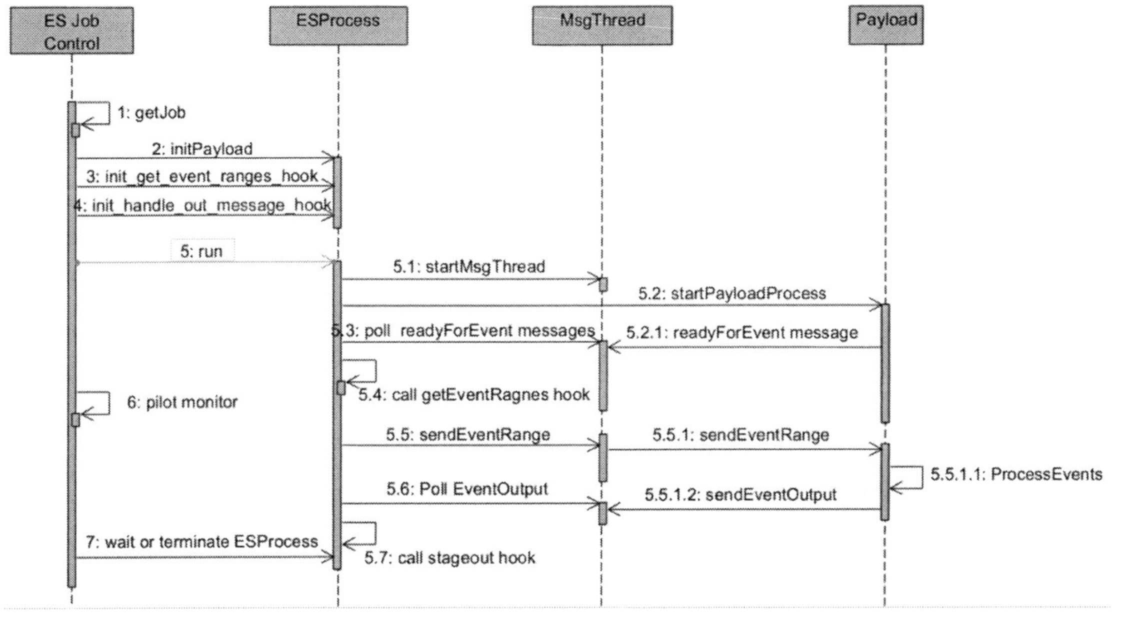

Figure 3. Sequence diagram for the Event Service workflow as used in ATLAS.

Server communications such as the getJob and updateJob commands can be issued using the Communicator API and may be used by both Pilot and Harvester. Furthermore, this API provides the interface to the Information Service component which has a set of low-level information providers to aggregate, prioritize, hide dependency to external storages and expose details for queues, sites, and storages, in a unified and structured way to all Pilot modules.

On HPCs, the Pilot is normally not running on the worker nodes but Harvester and the Yoda [5] tool can use the Pilot Environment API to determine how the payload should be set up. The Services API exposes benchmarking functions (currently an interface to the CERN Cloud Benchmark Suite [6]) that may be used by Harvester to add a benchmark step to the MPI rank 0 which would execute it before the actual payload. It could also be used on the resources where the Pilot is normally not used (such as on the ATLAS Tier-0) and also contains functions to interpret and forward the output of the benchmark tool to the relevant service. On normal grid sites, the current PanDA Pilot executes the benchmark suite during stage-in once per a hundred jobs and the results are gathered and studied on Elasticsearch [7].

The Services API also exposes functions that can be used for memory monitoring. Normally, the Pilot runs a memory monitoring tool in parallel with the payload to make sure the payload does not consume more resources than it is allowed to. The server is informed on each job update with current memory measurements during running which allows for real-time memory monitoring. On HPCs, Yoda can use the memory monitoring functions from the Services API to add a memory monitoring step to the MPI rank 0 which can execute the tool in parallel with the payload. The API contains functions for uploading the output to the relevant service.

5. Harvester-Pilot integration

The idea of the Harvester component is to facilitate the usage of PanDA on HPCs by moving away responsibilities from the Pilot, while still using some of the Pilot functionalities via APIs. Details about HPC queues and special workflows required for the HPC in question should not be handled directly by the Pilot. Instead, plug-ins to Harvester are being developed for each HPC. These plug-ins, as well as Harvester itself, will use relevant Pilot components to avoid code duplications; e.g. file transfers, job

IOP Conf. Series: Journal of Physics: Conf. Series **1085** (2018) 032031 doi:10.1088/1742-6596/1085/3/032031

definition downloads and functions for communicating with the server. In case Harvester is not used, the Pilot communicates directly with the server.

5.1. The road ahead

The Pilot 2 project is now in its second year of existence and is currently in the implementation stage. Basic functionality (file transfers, payload execution and monitoring) is in place and the first of the Pilot APIs, the Data API, was delivered in early 2017 for external use by Harvester. Supplementary components, such as the Information Service and Payload components, are being developed and the general code base is expanding rapidly. Additional APIs are in late planning and will be implemented next, including the Services API.

6. Summary

The PanDA Pilot has been in use in ATLAS for well over a decade. It was decided in 2016 to replace it with an entirely new product, Pilot 2, to more efficiently meet the demands of ATLAS and beyond on grids, clouds, volunteer computers and HPCs. It is being developed in parallel with another PanDA product, Harvester, located between the PanDA server and the Pilot, and will primarily be used on HPCs. Both Pilot 2 and Harvester are in rapid development and are expected to be used in production in 2018.

Acknowledgements

This material is based upon work supported by the U.S. Department of Energy, Office of Science, Office of High Energy Physics, under contract number DE-SC0012704, and was funded in part by the Russian Ministry of Education and Science under contract No 14.Z50.31.0024.

References

[1] ATLAS Collaboration, 2008 JINST 3 S08003

[2] F. H. Barreiro Megino et al., "PanDA for ATLAS distributed computing in the next decade", Proceedings of 22nd International Conference on Computing in High Energy and Nuclear Physics, J. Phys. Conf. Ser. 898 (2017)

[3] Harvester project, "Harvester" [software], version 0.0.13. Available from https://github.com/PanDAWMS/panda-harvester/releases/tag/v0.0.13 [accessed 2018-03-20]

[4] P. Calafiura et al, "The ATLAS Event Service: A New Approach to Event Processing", Proceedings of 21st International Conference on Computing in High Energy and Nuclear Physics, J. Phys. Conf. Ser. 664 (2015)

[5] P. Calafiura et al, "Fine grained event processing on HPCs with the ATLAS Yoda system", Proceedings of 21st International Conference on Computing in High Energy and Nuclear Physics, J. Phys. Conf. Ser. 664 (2015)

[6] Manfred Alef et al, "Benchmarking cloud resources for HEP", Proceedings of 22nd International Conference on Computing in High Energy and Nuclear Physics, J. Phys. Conf. Ser. 898 (2017)

[7] Elasticsearch project, "Elasticsearch" [software]. Available from https://www.elastic.co/downloads/elasticsearch [accessed 2018-03-20]

IOP Conf. Series: Journal of Physics: Conf. Series **1085** (2018) 032032 doi:10.1088/1742-6596/1085/3/032032

Belle II Conditions Database

M Ritter[1], L Wood[2], T Kuhr[1], M Bracko[3], T Elsethagen[2], K Fox[2], J Hall[2], C Pulvermacher[4] B Raju[2], M Schram[2], E Stephan[2],

[1] Ludwig-Maximilians University Munich, Excellence Cluster Universe, Boltzmannstr. 2, 85748 Garching, Germany
[2] Pacific Northwest National Laboratory, 902 Battelle Boulevard, Richland, WA, USA
[3]Jožef Stefan Institute, Jamova 39, 1000 Ljubljana, Slovenia
[4]High Energy Accelerator Research Organization (KEK), 1-1 Oho, Tsukuba 305-0801, Ibaraki, Japan

E-mail: Martin.Ritter@lmu.de, Lynn.Wood@pnnl.gov

Abstract. The Belle II experiment at KEK is preparing for taking first collision data in early 2018. For the success of the experiment it is essential to have information about varying conditions available to systems worldwide in a fast and efficient manner that is straightforward for both the user and maintainer. The Belle II Conditions Database was designed to make maintenance as easy as possible. To this end, a HTTP REST service was developed with industry-standard tools such as Swagger for the API interface development, Payara for the Java EE application server, and the Hazelcast in-memory data grid for support of scalable caching as well as transparent distribution of the service across multiple sites.

On the client side, the online and offline software has to be able to obtain conditions data from the Belle II Conditions Database in a robust and reliable way under very different situations. As such the client side interface to the Belle II Conditions Database has been designed with a variety of access mechanisms which allow the software to be used with and without an internet connection. Different methods to access the payload information are implemented to allow for a high level of customization per site and to simplify testing of new payloads locally. Changes to the conditions data are usually handled transparently but users can actively check whether an object has changed or register callback functions to be called whenever a conditions data object is updated. In addition a command line user interface has been developed to simplify inspection and modification of the database contents.

1. Introduction

The Belle II detector [1] and the SuperKEKB accelerator are currently under construction at the KEK laboratory in Tsukuba, Japan. The aim of this next generation B factory experiment is to collect 50 times more data than its predecessor Belle [2] and to use this data to search for new physics in a variety of B meson, charm hadron, or τ lepton decays with unprecedented precision. This requires detailed information on varying calibration and detector conditions to be available when analyzing the data.

The Belle II software Framework (basf2) [3] is a C++/Python framework to process the events recorded by the Belle II detector. Events are grouped in so-called runs which mark a data-taking period with stable operating conditions. Recorded events are processed one by one using a sequential set of algorithms called "modules." As all events are independent from each other we can benefit from current architectures with multiple cores by processing different events in parallel.

The Belle II Conditions Database is designed as a representational state transfer (REST) service. Communication is performed by standard HTTP using XML and JSON data. The database manages conditions data on run granularity. Binary objects called payloads which are defined for a certain interval of validity (IoV) are grouped into so-called global tags. The conditions database is by design agnostic to the contents of the payloads and only identifies them by name and revision number. The integrity of all payloads is verified using a checksum of the full content.

One requirement for the conditions client interface is that it needs to also work inside the data acquisition network which does not have a network connection to the outside. To facilitate this the client can also use a local, file-based backend which reads the list of payloads from an index file and the payload contents from files in a given directory. Command line tools written in Python allow easy downloading of all payloads and IoVs defined in a global tag to be used locally or uploading a locally prepared folder to be available for all users.

2. Server side

2.1. Implementation

The current configuration of the Belle II conditions database back-end is shown in Figure 1. The left and right sides represent two servers, one of which supports the REST API and database, while the other provides access to the payload data files. Each component in the figure above is implemented using Docker [4], a software framework that supports virtualization of user-space instances as opposed to full virtual machines, which reduces resource requirements, provides auto-restart functionality for each component separately, and provides a basis for scalability by instantiating multiple identical containers.

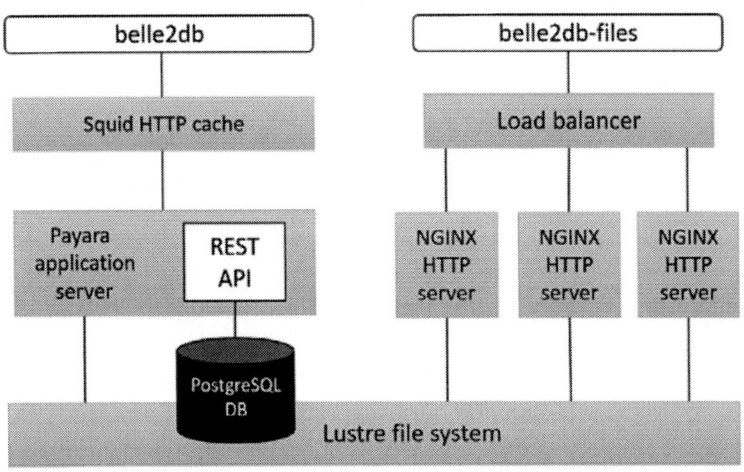

Figure 1. Current implementation of the conditions database back-end. Database and payload file access are handled separately by separate services.

Communication with the database uses standard HTTP using XML or JSON data. The choice of a standardized REST API makes the client coding independent of the actual database implementation details, and allows for easy caching support using a Squid [5] HTTP cache proxy server to reduce load on the API and database. The database itself is PostgreSQL, but the schema and procedures have been kept generic so other database applications could be used if necessary. To keep the database small, the payloads only consist of references to files on the separate payload file server. The payload files are currently assumed to be ROOT objects by the client, although there is nothing in the database design that limits the data type. The REST API interface is built from two commercially-available applications. Payara Micro [6] is a Java EE application server optimized for production operations in a containerized infrastructure such as Docker. Hazelcast [7] is a Java-based in-memory data grid. Data can be distributed evenly among separate nodes of a computer cluster, which provides horizontal scaling in both available storage space and processing power. This provides multiple benefits, including caching of frequently-used data in memory, transparent scalability, and load-balancing to reduce the query load on the database.

The file server currently consists of three NGINX [8] HTTP servers handled by a load balancer to distribute traffic evenly across the servers. The back-end file system is based on Lustre [9], an open-source parallel file system for high-performance computing environments. The Lustre file system is shared between all back-end instances.

2.2. Performance

The back-end has been evaluated with both direct access by Grid-based Belle II Monte Carlo campaigns and directed testing by the Belle II database group. The directed testing was done using Gatling [10], an open-source load stress testing tool for HTTP servers which allows custom test design through scripting. Testing of the database back-end was implemented by monitoring the usage patterns during Monte Carlo operations and then writing separate tests for the REST API and file server. The use of scripting allowed for much higher stress loads than was readily available from Grid-based testing. Recent performance results are shown in Figure 2. The top graph shows the database server responding successfully to 80 requests per second, while the bottom graph shows the payload file server responding to 180 file requests per second, with up to 10,000 open connections being handled simultaneously by the load-balanced HTTP servers. These tests correspond to nearly half of the payload bandwidth for the expected full-scale Belle II grid-based analysis cloud of 100,000 active nodes.

2.3. Future Improvements

Several improvements to the Belle II Conditions Database are in progress: Hazelcast supports cache clustering between remote sites, and will be evaluated as a means of supporting more localized database access worldwide. Placing portions of the cache at three or more sites will provide faster response times as well as backup capability for site or network outages. The PostgreSQL database is still a single-site instance and not currently scalable. Several options of supporting a distributed database are being investigated, including OpenStack Trove [11] and CockroachDB [12]. The replicated databases would be sited in tandem with the Hazelcast cache cluster sites. Authentication is currently not implemented, but is planned for services that would modify the database. The possibility of leveraging the X.509 authentication already present in the Belle II Grid computing interface is being investigated.

3. Client side

The choice of a standardized REST API makes the client implementation independent of the actual database implementation details and allows for a simple and flexible implementation of clients in different programming languages.

IOP Conf. Series: Journal of Physics: Conf. Series **1085** (2018) 032032 doi:10.1088/1742-6596/1085/3/032032

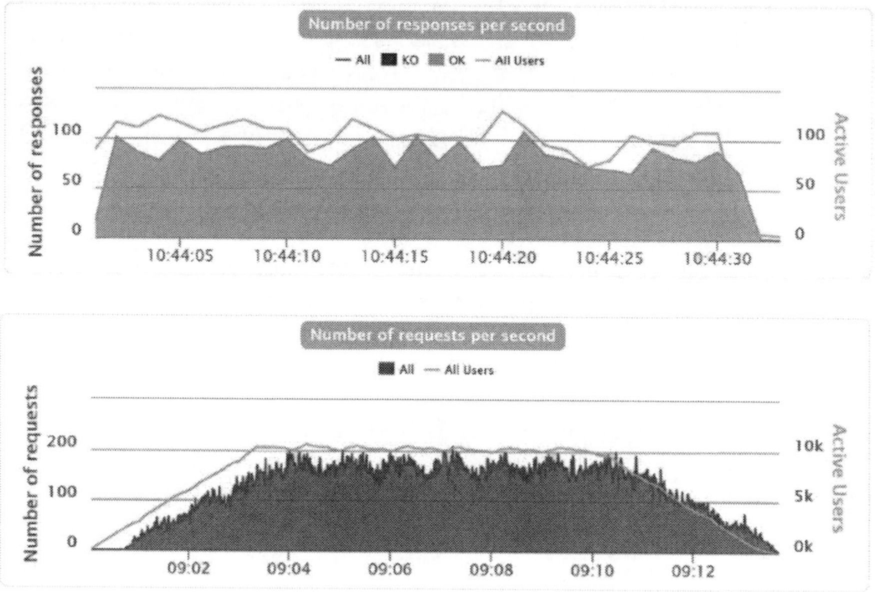

Figure 2. Performance results. Top graph shows Gatling stress testing of the database server; bottom graph show testing of the payload file server.

3.1. Access of Conditions Objects

The software framework assumes that payload contents are serialized ROOT [13] objects. User access to conditions objects is provided using two interface classes, one for single objects called `DBObjPtr` and one class for arrays of objects called `DBArray`. These class references payload objects, so called `DBEntry` objects in a global store, the `DBStore`. Multiple instances of the interface class all point to the same object. Access to the conditions objects is available in C++ and in Python where the class interface has been designed to be as close as possible to the already existing interface for event level data. Users familiar with the event level storage should have few problems accessing conditions data.

The interface classes always point to correct payload object for the current run, updates are transparent to the user. If the user needs to be aware when the object changes they can either manually check for changes or register a callback function to be notified on change. Figure 3 visualizes the relations between all the entities.

The objects in the global store are updated at the beginning of each run if needed. To allow for conditions which change inside of one run the client supports a so-called intra run granularity: Payload objects which inherit from an abstract base class `IntraRunDependency` can contain multiple objects for different parts of the run. The correct one will be selected transparently depending on certain criteria like event number or timestamp.

3.2. Creation of Conditions Data

To create payload objects we also provide two interface classes to simplify the procedure of preparing the objects, serializing them and committing them to the database. Users can just instantiate one of the creation classes, add objects to them and commit them to the configured database with a user supplied IoV. This includes support for intra run dependency. The possibility for a local file based database allows for easy preparation and validation of payloads as is needed during the calibration of the detector.

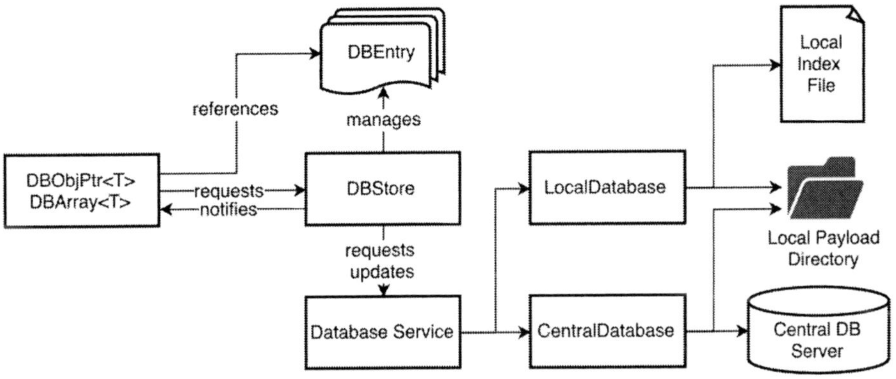

Figure 3. Relationships for the Conditions Database Interface. The user only interacts with the `DBObjPtr` and `DBArray` classes, everything else is handled transparently and can be configured independently.

Once all payloads are verified they can be uploaded to the central database using a standalone command line client developed in python. In addition this command line client allows inspection, modification and download of parts of the database contents.

3.3. Storage backends

Payloads in the central database are represented by logical filename, revision number and content checksum. While the central server provides a way to download all payloads directly via HTTP this is not the only possible distribution system. As the payloads are usually a large number of small files it might be beneficial to investigate alternative ways to distribute them. The database client can easily accommodate different storage implementations for the payloads by defining a cascade of storage backends to look for the actual payload content. After downloading the metadata on all needed payloads it will check all configured backends one by one for the needed files. If no storage backend contains the required payload it will automatically try to directly download the payload from the central server.

Currently only file based storage, for example shared filesystems or Cern-VM FS [14] are implemented. However it would be trivial to extend this to other technologies, for example key-value based storage systems like MICA [15]. Also hybrid systems where multiple payload objects are consolidated into larger archives similar to git packfiles [16] could be considered as well. This allows for a highly flexible distribution of the needed payload files which can be adapted as the content of the conditions database evolves.

The fallback solution to simply download missing payloads over HTTP makes it trivial to use the software without the need for explicit payload storage configuration.

4. Summary

Access to conditions data is a crucial part of the data processing. We have implemented a high level interface to the Belle II Conditions Database using a REST API. Access for the users is kept as simple as possible by requiring them just to instantiate one interface class to gain access to the correct data for any given event.

Distribution of the actual payload contents is highly flexible and can be expanded or configured easily. Any missing payloads will automatically be downloaded via HTTP so no special setup is required.

IOP Conf. Series: Journal of Physics: Conf. Series **1085** (2018) 032032 doi:10.1088/1742-6596/1085/3/032032

Acknowledgements
Portions of this work were carried out for the U.S. Department of Energy under Contract DE AC05 76RL01830 PNNL-SA-121968.

References
[1] Abe T *et al.* 2010 Belle II Technical Design Report *Preprint* `arXiv:1011.0352`
[2] Abashian A *et al.* 2000 The Belle Detector *Nucl. Instrum. Meth. A* **479** 117
[3] A Moll 2011 The Software Framework of the Belle II Experiment *J. Phys.: Conf. Series* **331** 032024
[4] Docker Platform, "Docker" [software], available from `http://www.docker.com/` [accessed 2017-10-26]
[5] Squid Caching Proxy, "Squid" [software], available from `http://www.squid-cache.org/` [accessed 2017-10-26]
[6] Payara Micro Server, "Payara Micro" [software], available from `http://www.payara.fish/payara_micro` [accessed 2017-10-26]
[7] Hazelcast In-Memory Data Grid, "Hazelcast" [software], available from `http://www.hazelcast.com` [accessed 2017-10-26]
[8] NGINX HTTP Server, "NGINX" [software], available from `http://www.nginx.com` [accessed 2017-10-26]
[9] Lustre parallel HPC file system, "Lustre" [software], available from `http://www.lustre.org` [accessed 2017-10-26]
[10] Gatling Load and Performance Test Tool, "Gatling" [software], available from `http://www.gatling.io` [accessed 2017-10-26]
[11] OpenStack Trove Database as a Service, "OpenStack Trove" [software], available from `http://www.openstack.org/software/releases/ocata/components/trove` [accessed 2017-10-26]
[12] CockroachDB Cloud-Based Database, "CockroachDB" [software], available from `http://www.cockroachlabs.com` [accessed 2017-10-26]
[13] Brun R and Rademakers F 1997 ROOT - An Object Oriented Data Analysis Framework, *Nucl. Instrum. Meth. A* **389** 81. See also "ROOT [software], Release v6.08.06, doi:10.5281/zenodo.848819
[14] P Buncic, C Aguado Sanchez, J Blomer, L Franco, A Harutyunian, P Mato and Y Yao 2010 CernVM a virtual software appliance for LHC applications *J. Phys.: Conf. Series* **219** 042003
[15] H Lim, D Han, D Andersen and M Kaminsky 2014 *11th USENIX Symposium on Networked Systems Design and Implementation (NSDI 14)* (Seattle: USENIX Association) pp 429–444
[16] Git Project, "Git" [software], available from `https://www.git-scm.com` [accessed 2017-10-23]

IOP Conf. Series: Journal of Physics: Conf. Series **1085** (2018) 032033 doi:10.1088/1742-6596/1085/3/032033

Modernising ATLAS Software Build Infrastructure

E Ritsch[1], G Gaycken[2], C Gumpert[1], A Krasznahorkay[1], W Lampl[3], E Moyse[4], E Obreshkov[5], K Potamianos[6], G A Stewart[7], A Undrus[8], F Winklmeier[9] *on behalf of the ATLAS Collaboration*

[1] European Laboratory for Particle Physics, CERN, [2] University of Bonn, [3] University of Arizona, [4] University of Massachusetts, Amherst, [5] The University of Texas at Arlington, [6] DESY, Hamburg and Zeuthen, [7] University of Glasgow, SUPA - School of Physics and Astronomy, [8] Brookhaven National Laboratory (BNL), [9] University of Oregon

E-mail: elmar.ritsch@cern.ch

Abstract. In the last year ATLAS has radically updated its software development infrastructure hugely reducing the complexity of building releases and greatly improving build speed, flexibility and code testing. The first step in this transition was the adoption of CMake as the software build system over the older CMT. This required the development of an automated translation from the old system to the new, followed by extensive testing and improvements. This resulted in a far more standard build process that was married to the method of building ATLAS software as a series of 12 separate projects from Subversion.

We then proceeded with a migration of the code base from Subversion to Git. As the Subversion repository had been structured to manage each package more or less independently there was no simple mapping that could be used to manage the migration into Git. Instead a specialist set of scripts that captured the software changes across official software releases was developed. With some clean up of the repository and the policy of only migrating packages in production releases, we managed to reduce the repository size from 62 GiB to 220 MiB.

After moving to Git we took the opportunity to introduce continuous integration so that now each code change from developers is built and tested before being approved.

With both CMake and Git in place we also dramatically simplified the build management of ATLAS software. Many heavyweight homegrown tools were dropped and the build procedure was reduced to a single bootstrap of some external packages, followed by a full build of the rest of the stack. This has reduced the time for a build by a factor of 2. It is now easy to build ATLAS software, freeing developers to test compile intrusive changes or new platform ports with ease. We have also developed a system to build lightweight ATLAS releases, for simulation, analysis or physics derivations which can be built from the same branch.

1. Introduction

In the first quarter of 2017 the ATLAS [1] collaboration completed a year-long process of modernising the build infrastructure of the Athena software stack. Athena is the workhorse of ATLAS' extensive software infrastructure. Athena production releases are deployed to the LHC Computing Grid [2] for Monte Carlo event generation, detector simulation and digitization, event reconstruction as well as generating derived data formats for physics analysis.

For this modernisation, new modern development workflows were established and custom made tools were replaced by community developed third-party tools. Approximately 4 million lines [3] of source code were migrated from the Subversion [4] version control system to Git [5].

IOP Conf. Series: Journal of Physics: Conf. Series **1085** (2018) 032033 doi:10.1088/1742-6596/1085/3/032033

Continuous integration tests and code reviews were established, which are both an essential component of the new workflow.

2. Structure of the ATLAS Athena repository

The ATLAS Athena repository is divided into individual packages and packages are structured in directories (see Figure 1). Each package contains a set of unique source code files, which typically get built into one dynamically linked library per package. There are, however, some packages that are statically linked against other packages. Packages may depend on each other at build- and at run-time. This implies that the packages must be built in a certain order so that all build-time dependencies are available upon building each individual package. As of summer 2017, the ATLAS Athena repository contains more than 2000 packages.

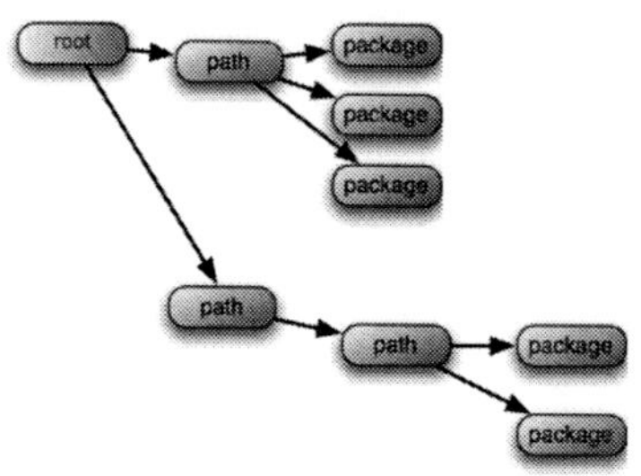

Figure 1. Structure of the ATLAS Athena repository. The repository is divided into packages. Packages are structured into (sub)directories.

3. Software development workflow

In any software project, the consistent application of a well structured development workflow plays a crucial role in the success of the project.

3.1. Past workflow

For many years, the ATLAS experiment applied a custom development workflow, which consisted of the following basic steps:

Code change: A developer changes an aspect of the software on their private working directory.

Tag package: The developer commits their private changes to the central `atlasoff` Subversion repository and creates a new version (called 'tag') of the package they modified.

Tag collection: The new package tag is added to all relevant nightly builds. This is done through a custom tool named the ATLAS Tag Collector [6].

Nightly builds and tests: On a daily basis all package tags that are collected for each respective nightly are built. After the completion of the build process (which takes several hours), a series of tests are executed. These tests include typical unit tests, integration tests, sanity checks as well as regression tests.

In the years 2015/2016 an internal report reviewing the ATLAS software infrastructure at the time identified the following problems, based on the experience of having applied the above workflow for a number of years:

- A number of internal tools were developed over time to facilitate the workflow applied by the ATLAS collaboration. Due to their internal nature, maintenance of these tools needed to be done by the collaboration itself. Moreover, developers joining the collaboration needed to first learn how to use these tools efficiently – and once a developer left the collaboration, this knowledge was no longer relevant to them even within the field of high energy physics. Some of these tools would become obsolete if a more modern and standard workflow were adopted.

IOP Conf. Series: Journal of Physics: Conf. Series **1085** (2018) 032033 doi:10.1088/1742-6596/1085/3/032033

- Dependencies of one package on other packages are documented in the source code – as it is required to successfully build each package. However, dependencies of packages on particular *versions* of other packages are *not* necessarily documented since the collaboration chose to not use this feature of the build configuration tool. This information may be documented elsewhere (for example on wiki pages) or, more commonly, is known only to the developer at the time the change was made. This often leads to manual trial and error approaches to identify compatible versions of dependencies of a given package when doing local development or updating nightly builds.
- Updates that require changes to multiple packages are tedious. It requires that all changed packages are tagged (i.e. versioned) separately, but the resulting package tags then need to be collected as one 'bundle' into the nightly builds. This 'bundle creation' is done manually by the developer when they request their package tags to be included in the next nightly build. The information about this bundle is not recorded in the source repository itself, rather in logs and emails that the ATLAS Tag Collector emits.
- Managing one access control list per package creates administrative overhead. Until the year 2012, individual developers were added manually (by the ATLAS Software librarian) to the access control list of a package if they requested write access to it. Later, ATLAS developed a custom tool (ATLAS SVN Access Manager) which allowed all developers with write access to a given package to grant write access to other developers. Effectively, developers still had to request write access to every package the first time they needed to modify it.

3.2. New workflow – ATLAS Flow

Among the most crucial changes in the recently updated ATLAS software infrastructure was the adoption of a new development workflow. The new workflow is heavily based on the GitLab Flow [7], which is sometimes referred to as *ATLAS Flow* (see Figures 2 and 3).

The new workflow is made possible due to the migration of the ATLAS Athena codebase from Subversion to Git (see section 4). Its main features are:

Release branches: A new Git branch is opened in the main code repository with every new major ATLAS Athena release that is made.

Adding bugfixes and features: New features are introduced in the `master` branch and in release branches, bugfixes are typically made directly in a release branch. The 'upstream first' policy (which is recommended in GitLab Flow) is not enforced in the ATLAS Flow. The reason is that bugs are mostly reported for released versions of the code, and those bugs are often difficult to reproduce in a branch different from the respective release branch (for example due to significant changes in the code between release branches and the `master` branch). Therefore, to reduce the 'time to production', the ATLAS Flow allows for bugfixes to be made in release branches directly.

Automated 'sweeps': Changes made to any release branch are automatically `cherry-picked` into the `master` branch on a daily basis. This is to ensure that all bugfixes and features are also included in future releases. These sweeps are done via our own cutom made script which run regularly every night. If there are conflicts they need to be handled manually by the developer or release coordinator.

4. Version control – From Subversion to Git

The most crucial change done by the ATLAS collaboration in its effort to modernise the software infrastructure was to move the source control system from Subversion to Git. Subversion had been used by the ATLAS collaboration for about 9 years (since the year 2008). Due to the custom structure of the ATLAS Athena repository stored in Subversion (named `atlasoff`, see Figure

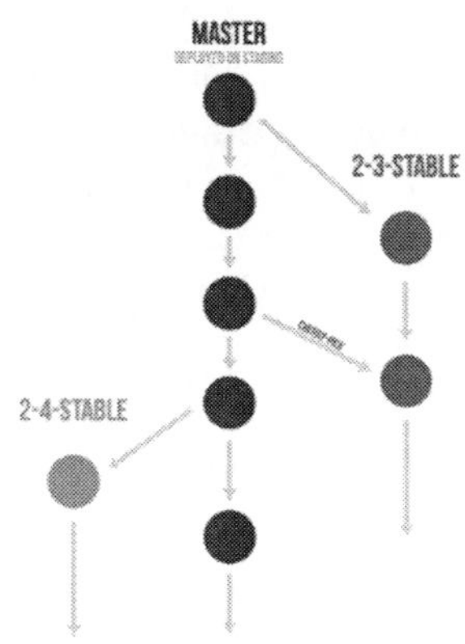

Figure 2. The original GitLab Flow [7] with release branches. Every time a new major release is made a corresponding Git branch is created. The 'upstream first' policy is recommended, which means that any updates (i.e. bugfixes) are done in the `master` branch first and then Git `cherry-pick`ed into relevant release branches if necessary.

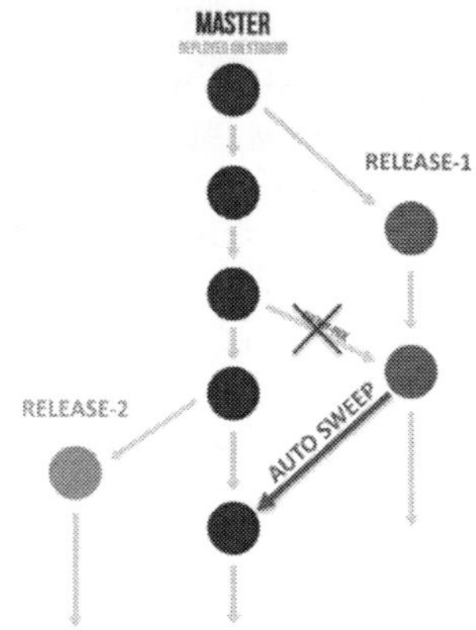

Figure 3. The new software development workflow employed by the ATLAS collaboration as of 2017. The workflow is derived from GitLab Flow. The difference is that ATLAS does not enforce the 'upstream first' policy, but allows for changes to be made directly in release branches. Automated daily 'sweeps' copy those changes into the `master` branch through Git cherry-picking.

4) a custom migration script was developed to copy relevant versions of individual packages from `atlasoff` into a new Git repository (`atlas/athena`). The `atlas/athena` Git repository is hosted on CERN's instance of GitLab Enterprise Edition [8].

Whereas packages were versioned independently in the `atlasoff` Subversion repository, in the `atlas/athena` Git repository the state of the entire set of packages is versioned at once. This means that a given version of the `atlas/athena` repository represents a given version of the entire ATLAS Athena stack. This eliminates the need for tools like the ATLAS Tag Collector that was used to define which package versions (tags) constitute a given ATLAS Athena build or release.

The following steps were carried out during the migration. Most of these steps were done in an automated fashion by the custom migration script, however, a small amount of manual intervention was required nevertheless:

- In order to facilitate debugging and 'code archaeology' tasks in the newly created `atlas/athena` Git repository, parts of the commit history were copied from the Subversion repository. The state of the code of any release from the last (approximately) 2 years was copied into respective *archive* branches in the Git repository.
- Files larger than 100 KiB were not copied from Subversion to Git. It was discovered that the majority of the files affected were plain text data files that had no good justification to reside inside the source code repository.
- A copyright statement was put at the head of each source file.

After applying the above steps, the initial `atlas/athena` Git repository had a size of approximately 220 MiB – the `atlasoff` Subversion repository had a size of approximately 62 GiB at the time of the migration. It is important to note that the size required can not be compared directly as only a partial commit history was migrated from Subversion to Git, and moreover, many obsolete packages and files were not migrated at all.

IOP Conf. Series: Journal of Physics: Conf. Series **1085** (2018) 032033 doi:10.1088/1742-6596/1085/3/032033

A triangular Git workflow [9] was established to keep the main Git repository (`atlas/athena`) clean and well structured (see Figure 5). Therefore, developers cannot directly modify this repository, but are required to create personal forks on CERN's GitLab instance which they can modify as required. Modifications that developers want to include in the official ATLAS Athena codebase need to be requested into `atlas/athena` via GitLab Merge Requests. The GitLab Merge Request interface is used to facilitate code reviews and to communicate automated build and test results. Every Merge Request that is created (whether automated or manually by a developer) triggers the following actions:

Multi-stage code review: ATLAS code reviewers are tasked to identify potential problems and weaknesses in the modifications before they are merged into the main Git repository. The code review is divided into three levels (thus 'multi-stage'): level 1, level 2 and expert level. Level 1 and 2 are handled by shifters. Level 1 reviewers inspect the code for logic errors, common programming errors, consistency with the rest of the codebase, conformity to style guides [15] and similar issues. Some code style guidelines are in process of being automated. Level 2 reviewers inspect the general structure of the code change and its integration into the ATLAS Athena software stack. If a modification is deemed critical and deep domain knowledge is required to assess its validity, a domain expert is included in the code review process. The time to review the code varies greatly depending on the number of changes done, their complexity, expereince of the developer and the experience of the reviewers.

Continuous Integration build and tests: To ensure that the code in the `atlas/athena` Git repository remains in a working and useable state at any time, any requested code change gets automatically compiled and a defined set of unit and integration tests are run on it. The continuous integration tool Jenkins [10] is used to schedule and distribute the build and test jobs on a set of dedicated build machines. As of summer 2017, 14 dedicated build machines form the backbone of the continuous integration system.

When a Merge Request passes the review stage as well as the corresponding build and tests, the release coordinator for the given target release branch is responsible for accepting the Merge Request at a suitable time.

5. Building binaries and releases – From CMT to CMake

The ATLAS collaboration moved from the Configuration Management Tool (CMT [11]) to CMake [12] to handle the setup of the software environment as well as the build process.

A converter was implemented to automatically translate CMT's `requirements` file format into CMake's `CMakeLists.txt` files for each individual package. In the majority of cases, the automatic conversion worked without issue. In a few cases, manual action was required to convert specific parts of the CMT syntax into the corresponding CMake syntax. The old CMT `requirements` files were removed from the code repository once the entire codebase could be built successfully with CMake.

Nightly builds of all release branches and the `master` branch are made on a daily basis. As of summer 2017, 7 different branches are included in the nightly builds. RPM [13] packages containing the build results are generated for every nightly build. They are installed on the CernVM File System [14] to make the build accessible globally to all developers. Builds are done in different 'flavors', where each flavor contains a different sub-set of all packages. Therefore they build faster and their release sizes are smaller compared to the full build. Currently, 8 different flavors of the Athena software are in use by the ATLAS collaboration. Among them are lightweight ATLAS releases for simulation, analysis or physics derivations.

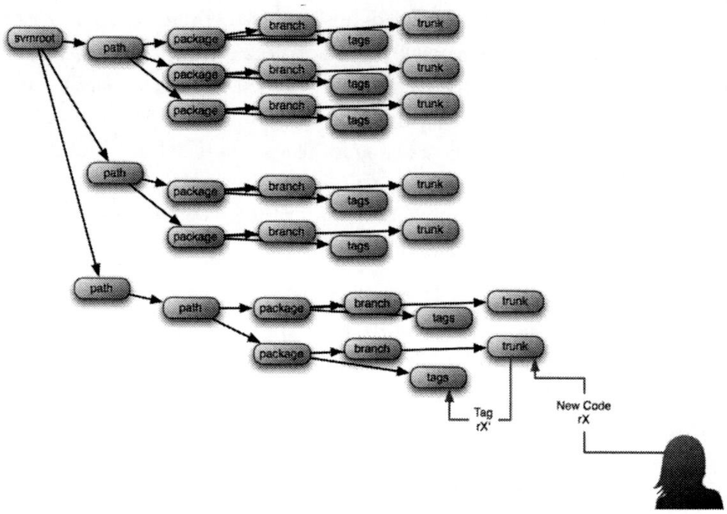

Figure 4. The structure of the ATLAS Athena repository as it is stored in Subversion (named **atlasoff**).

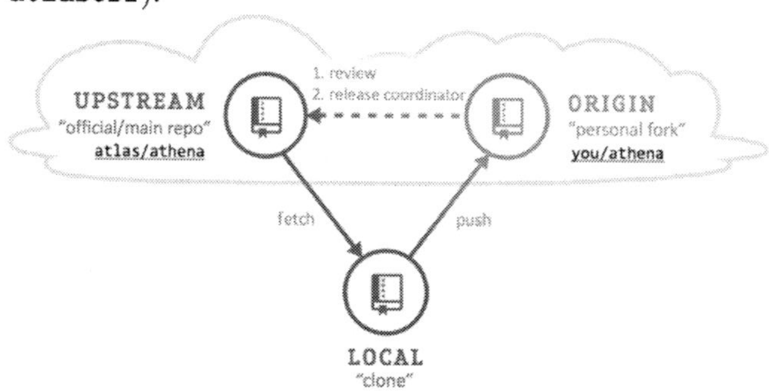

Figure 5. The ATLAS collaboration uses a triangular Git workflow to manage the interaction of developers with the Athena repository. (Image based on [9])

6. Tutorials and public documentation

In order to prepare the members of the ATLAS collaboration for the change of tools and workflow, a number of interactive tutorials were held before the changes were carried out.

A public website [16] was put in place to make the documentation on the new workflow and tools accessible and searchable for everyone. The website hosts step-by-step walkthroughs to help developers getting started with the new ATLAS software development environment.

7. Conclusions

After years of experience using custom software tools and workflows it became clear that these solutions were less than optimal when compared to well-established third-party tools and workflows. This not only needed the custom tools to be maintained by the ATLAS collaboration, but the custom workflows necessitated a learning phase for developers joining the collaboration – even for highly experienced HEP developers.

Moving the software infrastructure to well-established workflows and third-party tools alleviates much of the overhead that the custom infrastructure caused to the ATLAS

IOP Conf. Series: Journal of Physics: Conf. Series **1085** (2018) 032033 doi:10.1088/1742-6596/1085/3/032033

collaboration in the past. In particular the move to Git – as the de facto industry standard for source control management – was well received by the ATLAS developer community. Continuous integration builds and tests speed up development significantly as a faster feedback to the developer is provided. Code reviews improve the quality and consistency of the code, and they aid in spreading the knowledge of the codebase among the entire ATLAS developer community.

Though most of changes were disruptive in nature and the collaboration had to adopt the new workflows quickly, the current infrastructure is clearly an advantage for the ATLAS collaboration.

References

[1] ATLAS Collaboration, *The ATLAS Experiment at the CERN Large Hadron Collider*, JINST **3** (2008) S08003.
[2] WLCG Collaboration, *WLCG - Worldwide LHC Computing Grid*, http://wlcg-public.web.cern.ch/.
[3] G. Stewart, W. Lampl, The ATLAS Collaboration, *How to review 4 million lines of ATLAS code*, ATL-SOFT-PROC-2017-047, 2017, https://cds.cern.ch/record/2248494.
[4] The APACHE Software Foundation, *Subversion*, https://subversion.apache.org/.
[5] Git Project, *Git*, https://git-scm.com/.
[6] S. Albrand, J. Fulachier, J. Collot, F. Lambert, *The TAG Collector: A Tool for ATLAS Code Release Management*, 2005, https://cds.cern.ch/record/865644.
[7] S. Sijbrandij, GitLab Inc., *GitLab Flow*, 2014, https://about.gitlab.com/2014/09/29/gitlab-flow/.
[8] GitLab Inc., *GitLab*, https://about.gitlab.com/.
[9] M. Haggerty, GitHub Inc., *Git 2.5 including multiple worktrees and triangular workflows*, 2015, https://github.com/blog/2042-git-2-5-including-multiple-worktrees-and-triangular-workflows.
[10] Jenkins project, *Jenkins*, https://jenkins.io/.
[11] C. Arnault, G. Rybkin, *CMT*, http://www.cmtsite.net/.
[12] Kitware Inc., *CMake*, https://cmake.org/.
[13] Red Hat Inc. and Community, *RPM Package Manager*, http://rpm.org/.
[14] CERN, *CernVM File System (CernVM-FS)*, https://cernvm.cern.ch/.
[15] The ATLAS Collaboration, *ATLAS C++ coding guidelines, version 0.6* , http://atlas-computing.web.cern.ch/atlas-computing/projects/qa/draft_guidelines.html.
[16] The ATLAS Collaboration, *ATLAS Software Documentation*, https://atlassoftwaredocs.web.cern.ch/.

Speeding up software with VecCore

G Amadio[1], P Canal[2], D Piparo[1], S Wenzel[1]
[1]CERN, [2]FNAL

E-mail: amadio@cern.ch

Abstract. Portable and efficient vectorization is a significant challenge in large software projects such as GeantV, ROOT, and experiments' frameworks. Nevertheless, fully exploiting SIMD parallelism will be a required step in order to bridge the widening gap between the needs and availability of computing resouces for data analysis and processing in particle physics. Although there are SIMD libraries that wrap compiler intrinsics into a convenient interface, they do not always support all available architectures, or they only perform well in some of them. The VecCore library was created to address some of these performance and portability issues by providing a unified abstraction layer on top of existing libraries, such as Vc or UME::SIMD. In this article, we discuss VecCore's programming model for SIMD code and some use cases in HEP software packages such as VecGeom and GeantV.

1. Introduction

Experiments at CERN have an ever increasing demand for computational resources, be it for simulated collisions or data analysis. In the last few decades, this demand has been met by hardware upgrades of the Worldwide LHC Computing Grid. However, with the next runs of the LHC approaching, the expected increases in beam luminosity will push this demand far beyond the limits of what further hardware upgrades can reach. Therefore, in order to bridge the widening gap between the needs of the HEP community and the existing computing resources, HEP software will need to be optimized to be able to fully exploit SIMD and multithreading parallelism available in modern hardware.

One of the key areas through which performance can be substantially improved in HEP software is SIMD vectorization. Even so, writing efficient SIMD vectorized code is a significant challenge in many large software projects such as GeantV[1, 2], ROOT[3], and the experiments' frameworks. On the one hand, compilers cannot reliably auto-vectorize simulation and analysis software. One the other hand, while there are SIMD libraries that wrap the cumbersome compiler intrinsics functions into a more convenient interface, these libraries often target specific architectures, which means they may not work on ARM, for example. Therefore, it makes sense to be able to switch between libraries depending on the target architecture, to be able to use the best option on each platform.

2. The VecCore Library

The VecCore library was created in order to solve the lack of portability and unreliable performance problems usually associated with SIMD code. VecCore provides a simple API for users to express their SIMD-enabled algorithms that can be dispatched to different backend implementations, such as the SIMD libraries Vc [4] and UME::SIMD [5], or even CUDA, if the code has the proper

IOP Conf. Series: Journal of Physics: Conf. Series **1085** (2018) 032034 doi:10.1088/1742-6596/1085/3/032034

annotations. Vc in general has very good performance on systems that support the AVX2 SIMD instruction set, but its latest release, Vc 1.3.3, has no support for the more recent AVX512. Conversely, UME::SIMD focuses on scalar emulation for multiple architectures, while optimizing newer architectures first, which makes it perform quite well on AVX512, but the performance double precision types on AVX2 may not be as good as Vc[6]. In this paper we use the latest release, UME::SIMD 0.8.1.

Using VecCore, developers can write generic computational kernels using abstract types that map to the different concrete types in each backend. The API for vectorization provided by VecCore is architecture-agnostic, making it easy to fallback to scalar types on platforms that do not have SIMD instructions. Moreover, users do not have to change their code to switch between backends or target architecture. The API covers the essential parts of the SIMD instruction set that allows one to write many of the numerical algorithms needed by HEP software. The API functions are shown in Listing 1 and their operations are illustrated in Figure 2.

```
namespace vecCore {

template <typename T> struct TypeTraits;

template <typename T> using Mask   = typename TypeTraits<T>::MaskType;
template <typename T> using Index  = typename TypeTraits<T>::IndexType;
template <typename T> using Scalar = typename TypeTraits<T>::ScalarType;

// Vector Size
template <typename T> constexpr size_t VectorSize();

// Get/Set
template <typename T> Scalar<T> Get(const T &v, size_t i);
template <typename T> void Set(T &v, size_t i, Scalar<T> const val);

// Load/Store
template <typename T> void Load(T &v, Scalar<T> const *ptr);
template <typename T> void Store(T const &v, Scalar<T> *ptr);

// Gather/Scatter
template <typename T, typename S = Scalar<T>>
T Gather(S const *ptr, Index<T> const &idx);

template <typename T, typename S = Scalar<T>>
void Scatter(T const &v, S *ptr, Index<T> const &idx);

// Masking/Blending
template <typename M> bool MaskFull(M const &mask);
template <typename M> bool MaskEmpty(M const &mask);

template <typename T>
void MaskedAssign(T &dst, const Mask<T> &mask, const T &src);

template <typename T>
T Blend(const Mask<T> &mask, const T &src1, const T &src2);

} // namespace vecCore
```

Listing 1: VecCore API

3. Mandelbrot Set with VecCore

A simple example that can be used to demonstrate how an algorithm changes when implemented using the VecCore API is the Mandelbrot set, shown in Figure 1 as calculated by our code. The Mandelbrot set can be obtained by repeatedly computing the function $f(z) = z^2 + c$ starting with $z = 0$ for various values of c in the complex plane. The final color depends on how many iterations of $f(z)$ are necessary for the point to leave the circle $|z| < 2$. Listings 2 and 3 show the scalar and vectorized implementations, respectively. The vectorized version works on several points at a time. Both algorithms are single-threaded for simplicity.

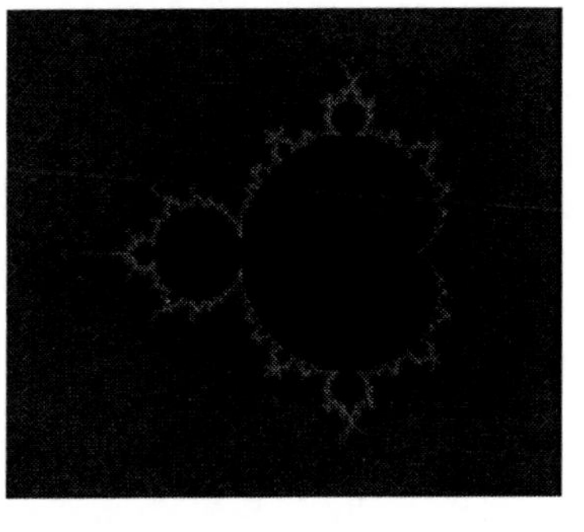

Figure 1. Mandelbrot Set

IOP Conf. Series: Journal of Physics: Conf. Series **1085** (2018) 032034 doi:10.1088/1742-6596/1085/3/032034

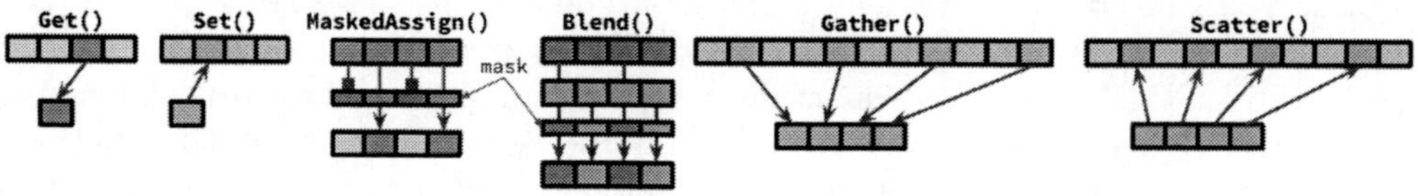

Figure 2. Illustration of VecCore API operations

```
template<typename T>
void mandelbrot(T xmin, T xmax, size_t nx,
                T ymin, T ymax, size_t ny,
                size_t max_iter, unsigned char *image)
{
    T dx = (xmax - xmin) / T(nx);
    T dy = (ymax - ymin) / T(ny);

    for (size_t i = 0; i < nx; ++i) {
        for (size_t j = 0; j < ny; ++j) {
            size_t k = 0;
            T x = xmin + T(i) * dx, cr = x, zr = x;
            T y = ymin + T(j) * dy, ci = y, zi = y;

            do {
                x  = zr*zr - zi*zi + cr;
                y  = 2.0 * zr*zi + ci;
                zr = x;
                zi = y;
            } while (++k < max_iter && (zr*zr+zi*zi < 4.0));

            image[ny*i+j] = k;
        }
    }
}
```

Listing 2: Scalar implementation

```
template<typename T>
void mandelbrot_v(Scalar<T> xmin, Scalar<T> xmax, size_t nx,
                  Scalar<T> ymin, Scalar<T> ymax, size_t ny,
                  Scalar<Index<T>> max_iter,
                  unsigned char *image)
{
    T iota;
    for (size_t i = 0; i < VectorSize<T>(); ++i)
        Set<T>(iota, i, i);

    T dx = T(xmax - xmin) / T(nx);
    T dy = T(ymax - ymin) / T(ny), dyv = iota * dy;

    for (size_t i = 0; i < nx; ++i) {
        for (size_t j = 0; j < ny; j += VectorSize<T>()) {
            Scalar<Index<T>> k{0};
            T x = xmin + T(i) * dx,       cr = x, zr = x;
            T y = ymin + T(j) * dy + dyv, ci = y, zi = y;

            Index<T> kv{0};
            Mask<T> m{true};

            do {
                x = zr*zr - zi*zi + cr;
                y = T(2.0) * zr*zi + ci;
                MaskedAssign<T>(zr, m, x);
                MaskedAssign<T>(zi, m, y);
                MaskedAssign<Index<T>>(kv, m, ++k);
                m = zr*zr + zi*zi < T(4.0);
            } while (k < max_iter && !MaskEmpty(m));

            for (size_t k = 0; k < VectorSize<T>(); ++k)
                image[ny*i+j+k] = (unsigned char) Get(kv, k);
        }
    }
}
```

Listing 3: VecCore implementation

The two implementations are quite similar to each other, the main differences can be seen inside the innermost `do/while` loop, that in the vectorized implementation must account for the differences when treating several points at the same time. The vectorized implementation requires using a mask to assign only to the right indices, and checking for termination becomes more complex due to that as well, since care must be taken to not return from the inner loop until all points in the SIMD vector have been properly computed.

		Intel Core i7 6700			Intel Xeon Phi 7210	
Runtime (ms)		GCC-7.2	Clang-5.0	ICC-18.0	GCC-7.2	ICC-18.0
Single Precision	Scalar	550	549	677	3415	3609
	Scalar Backend	570	569	677	3353	3510
	Vc 1.3.3	110	110	126	1064	1160
	UME::SIMD 0.8.1	117	117	131	–	543
Double Precision	Scalar Algorithm	548	548	672	3409	3602
	Scalar Backend	571	571	674	3348	3502
	Vc 1.3.3	267	267	257	2101	2087
	UME::SIMD 0.8.1	421	421	422	–	846
Speedup vs Scalar						
Single Precision	Scalar Backend	0.96	0.96	1.00	1.02	1.03
	Vc 1.3.3	5.00	4.99	5.37	3.21	3.11
	UME::SIMD 0.8.1	4.70	4.69	5.17	–	6.65
Double Precision	Scalar Backend	0.96	0.96	0.99	1.02	1.03
	Vc 1.3.3	2.05	2.05	2.61	1.72	1.73
	UME::SIMD 0.8.1	1.30	1.30	1.59	–	4.25

Table 1. Runtime in milliseconds of Mandelbrot set example for various configurations.

Runtime and speedups are shown in Table 1. It is important to note the difference in frequency between the Core i7 (3.4 GHz) and the Xeon Phi (1.3 GHz), hence the difference in absolute times. Moreover, performance comparisons using Julia sets (similar to Mandelbrot, but fixing c and varying z) show that the speedup strongly depends on branch divergence—that is, how many times the vectorized loop can return early. Complex regions of the fractal decrease the maximum speedup that can be obtained with vectorization. Therefore, these results should be used only as a relative measurement of performance between backends, and not as a measurement of the overall vectorization efficiency.

4. Electromagnetic Physics Models

Collision events at the LHC frequently produce energetic photons and electrons that lead to electromagnetic particle showers containing a large number of particles. Since particles generated in these showers constitute the vast majority of particles generated in a collision event, the simulation of electromagnetic processes is computationally expensive. Therefore, these processes are the natural first candidates for vectorization.

In the GeantV project [1, 2], VecCore has been used in the implementation of vectorized versions of electromagnetic physics models. Figure 3 shows the vectorization speedups for each model on an Intel Xeon Phi (Knights Corner) for the main electromagnetic processes. For 64 or more tracks, the speedup is between 5x and 6.5x in double precision, when the maximum speedup is 8x. Considering that these models rely on gather/scatter operations that are not so fast compared with other SIMD operations such as arithmerics, masking, etc, this result is very promising. Nevertheless, these speedups are expected to increase once a vectorized pseudo-random number generator is implemented in the framework.

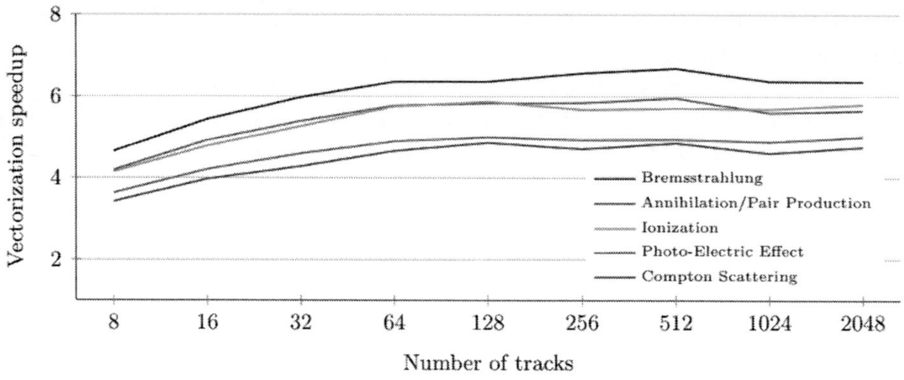

Figure 3. Vectorization speedup of electromagnetic models on Intel Xeon Phi

5. Geometry Algorithms

VecGeom[7, 8], the geometry library used in GeantV [1, 2, 9] and recent versions of Geant4 [10, 11, 12], has introduced a vectorized multi-particle API based on VecCore to perform ray casting, distance calculations, and navigation in sets of particles within detector geometries. Performance gains on Intel Knights Landing for VecGeom shapes compared to previous implementations are shown in Figure 4. Since Vc does not support KNL, the scalar backend has been used (i.e. speedup comes from compiler auto-vectorization only). As expected, in this case the UME::SIMD backend outperforms Vc by a significant margin.

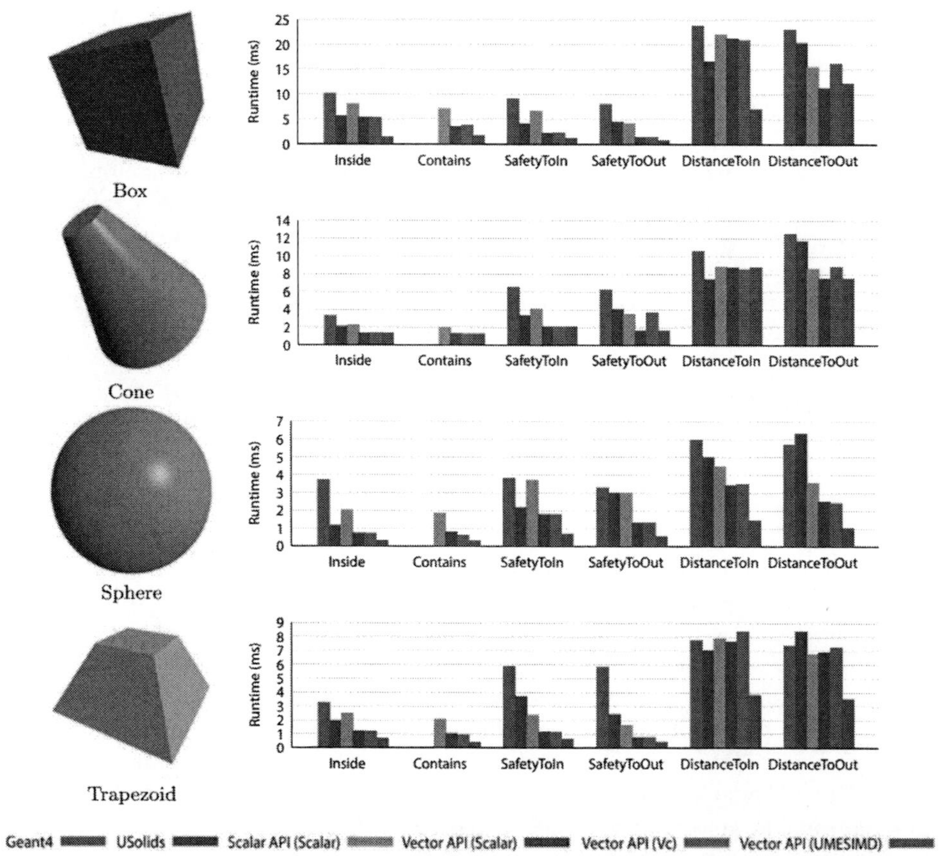

Figure 4. Vectorization speedups of selected shapes on Intel Knights Landing (KNL)

6. Conclusions

VecCore introduces a new unified abstraction layer on top of existing SIMD libraries that lets users write generic vectorized code that can be used with different instruction sets and hardware architectures. It also allows the user to change libraries to take advantage of the strengths of each library, or avoid drawbacks like lack of support for some architectures (e.g. KNL, ARM, or PowerPC), since VecCore has a simple scalar backend that works across architectures. If new backends are added, users of VecCore can take advantage of them without needing to change their code, by only changing the backend configuration. These features make VecCore a useful library for writing SIMD-enabled algorithms that are expected to perform well on a variety of architectures.

7. Acknowledgements

VecCore development has been funded by an Intel Parallel Computing Center project at São Paulo State University (UNESP), Fermilab, and CERN, as part of the GeantV project. VecCore is now maintained by the ROOT Team.

References

[1] Amadio G, Apostolakis J, Bandieramonte M, Bhattacharyya A, Bianchini C, Brun R, Canal P, Carminati F, Duhem L, Elvira D, de Fine Licht J, Gheata A, Iope R L, Lima G, Mohanty A, Nikitina T, Novak M, Pokorski W, Seghal R, Shadura O, Vallecorsa S and Wenzel S 2015 *Journal of Physics: Conference Series* **664** 072006 URL http://stacks.iop.org/1742-6596/664/i=7/a=072006

[2] Apostolakis J, Brun R, Carminati F and Gheata A 2012 *International Conference on Computing in High Energy and Nuclear Physics 2012 (CHEP2012)* **396** 022014 URL http://iopscience.iop.org/1742-6596/396/2/022014/pdf/1742-6596_396_2_022014.pdf

[3] ROOT Data Analysis Framework URL https://root.cern

[4] Kretz M and Lindenstruth V 2011 *Software: Practice and Experience* **42** 1409–1430 URL http://onlinelibrary.wiley.com/doi/10.1002/spe.1149/abstract

[5] Karpiński P and McDonald J 2017 A high-performance portable abstract interface for explicit SIMD vectorization *Proceedings of the 8th International Workshop on Programming Models and Applications for Multicores and Manycores* PMAM'17 (New York, NY, USA: ACM) pp 21–28 ISBN 978-1-4503-4883-6 URL http://doi.acm.org/10.1145/3026937.3026939

[6] Vallecorsa S and Amadio G Auto-vectorization: recent progress *21st Geant4 Collaboration Meeting* (2016) URL https://agenda.infn.it/contributionDisplay.py?contribId=125&sessionId=22&confId=11196

[7] Apostolakis J, Brun R, Carminati F, Gheata A and Wenzel S 2014 *Journal of Physics: Conference Series* **513** 052038 URL http://stacks.iop.org/1742-6596/513/i=5/a=052038

[8] Apostolakis J, Bandieramonte M, Bitzes G, Brun R, Canal P, Carminati F, Cosmo G, Licht J C D F, Duhem L, Elvira V D, Gheata A, Jun S Y, Lima G, Nikitina T, Novak M, Sehgal R, Shadura O and Wenzel S 2015 *Journal of Physics: Conference Series* **608** 012023 URL http://stacks.iop.org/1742-6596/608/i=1/a=012023

[9] Apostolakis J, Brun R, Carminati F, Gheata A and Wenzel S 2014 *Journal of Physics: Conference Series* **523** 012004 URL http://stacks.iop.org/1742-6596/523/i=1/a=012004

[10] et al S A 2003 *Nuclear Instruments and Methods in Physics Research Section A: Accelerators, Spectrometers, Detectors and Associated Equipment* **506** 250 – 303 ISSN 0168-9002 URL http://www.sciencedirect.com/science/article/pii/S0168900203013688

[11] Asai M 2012 *International Conference on Computing in High Energy and Nuclear Physics 2012 (CHEP2012)* **396** 052007 URL http://iopscience.iop.org/1742-6596/396/5/052007/pdf/1742-6596_396_5_052007.pdf

[12] et al J A 2006 *Nuclear Science, IEEE Transactions on* **53** 270–278 ISSN 0018-9499 URL http://dx.doi.org/10.1109/TNS.2006.869826

IOP Conf. Series: Journal of Physics: Conf. Series **1085** (2018) 032035 doi:10.1088/1742-6596/1085/3/032035

A scalable new mechanism to store, query and serve the ATLAS detector description through a REST web API

R M Bianchi[1,†] **and I Vukotic**[2,†]

[1] University of Pittsburgh, Department of Physics and Astronomy, 100 Allen Hall, 3941 O'Hara St, Pittsburgh PA 15260, US

[2] University of Chicago, High Energy Physics, 5640 S Ellis Ave, Chicago IL 60637, US

[†]on behalf of the ATLAS Collaboration

E-mail: riccardo.maria.bianchi@cern.ch

Abstract. Until now, geometry information for the detector description of the ATLAS experiment was only defined in C++ code, stored in online relational databases integrated into the experiment's frameworks or described in files with text-based markup languages. In all cases, to build and use the complete detector geometry, a full software stack was needed. In this paper, we present a new and scalable mechanism to store the geometry data and to query and serve the detector description data through a web interface and a REST API. This new approach decouples the geometry information from the experiment's framework. Moreover, it provides new functionalities to users, who can now search for specific volumes and get partial detector description, or filter geometry data based on custom criteria. We present two approaches to build a REST API to serve geometry data, based on two different technologies used in other fields and communities: The graph database Neo4j and the search engine ElasticSearch. We describe their characteristics, and we compare them in a HEP context.

1. Introduction

Every modern particle physics experiment is based on a sophisticated particle detector, usually composed of many different sub-detectors, each of them targeting a specific measurement. The so-called *Detector Description* is the piece of data containing the detailed description of all the sub-detectors, their position in the 3D space and their relationships with the other components of the whole detector.

The Detector Description of the ATLAS Experiment [1] is based on the GeoModel [2] library, which lets the developers define the geometry in the C++ code, using *nodes* (shapes, transformations, materials, and so forth) to describe the structure and the characteristics of the detector geometry.

Until now, the ATLAS geometry was stored in-memory only, built on the fly by the software framework when a client asked for it, taking the required information both from C++ code and from online databases. Hence, the geometry data of the experiment could only be accessed and interpreted through the experiment's software framework using scripts or compiled code, making it impossible to interactively query and explore the detector geometry and to use it in stand-alone or cross-platform applications.

Building upon the earlier work [3] on a persistent representation of the geometry data, this paper describes the efforts recently made to design and develop a new web-based REST API service to store, query and serve the Detector Description of the ATLAS Experiment, in order to make the experiment's geometry data accessible and usable outside the experiment's framework.

2. The ATLAS detector description and the GeoModel library

The ATLAS experiment uses the GeoModel library to describe the volumes and structure of the geometry tree in C++ code. Figure 1 shows two examples of basic GeoModel trees. A *logical volume* (Figure 1(a)) defines a fundamental entity within the tree: A logical volume has a name, and it references a *shape node* and a *material node*, to define a single abstract type of object. This abstract entity is then used to build a concrete *physical volume*, placed in the 3D space. The same logical volume can be used to define multiple physical volumes, and other kinds of nodes can be used to place volumes and to set attributes and properties (Figure 1(b)).

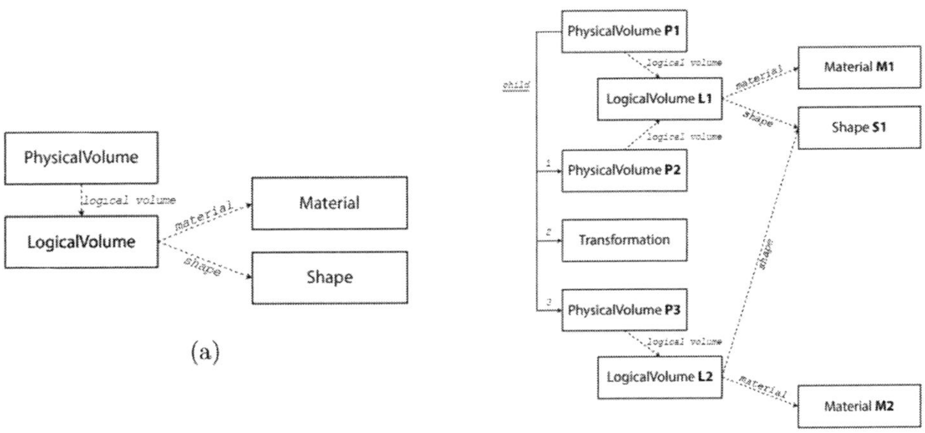

(a) (b)

Figure 1: (a) the definition of a logical volume, and its usage in the definition of a physical volume, as described in Ref. [3]; (b) an example of a GeoModel tree, where both shared and nested nodes are used: The Physical Volume P2, the Transformation node, and the Physical Volume P3 are all children of P1; both P1 and P2 share the same Logical Volume L1 (a *shared node*), which means they use the same material M1 and the same shape S1; P3, on the contrary, references the logical volume L2, which uses the shape S1 (another *shared node*, because it is used by L1, too) and the material M2. The position of the nodes within the tree is important: In this example, the Transformation node (child 2) affects only the P3 node (child 3).

The hierarchy of the nodes and their attributes are defined in different C++ files, typically related to specific sub-detectors. In addition to that, parameters which can change over time are stored in a dedicated online database, and retrieved through the experiment's software framework. The resulting GeoModel tree is then built on-the-fly upon request, through the experiment's framework, and stored in memory. At that time the experiment's geometry is ready to be used by applications running inside the framework.

3. The GeoModel structure

The GeoModel library offers many types of nodes, used to build objects, place volumes, and set attributes; as seen above, nodes can be shared within the tree. Moreover, physical volumes

IOP Conf. Series: Journal of Physics: Conf. Series **1085** (2018) 032035 doi:10.1088/1742-6596/1085/3/032035

can be nested, through a *parent-child* relationship: The *parent* can act as an outer container, containing several *children volumes*. Figure 1(b) shows an example of a more complex tree, where both shared and nested nodes are used. In GeoModel the order of children and nodes is critical: Nodes like space transforms affect only the nodes after them, like the Transformation node in the example tree; moreover, transformations are accumulated: When traversing the sub-tree, if a transform node is found, its transformation matrix is combined with the transformation matrix of the other transform nodes before that in the sub-tree.

The detector description of a modern HEP experiment uses a very complex hierarchy of sub-trees to define every single piece of the geometry. In ATLAS roughly 80,000 physical volumes are used, built with nearly 50,000 logical volumes. The shape and material nodes used to define the logical volumes are about 100,000 and 380, respectively; and approximately 200,000 transform nodes are used to place all the pieces of the ATLAS detector correctly. The number of the transform nodes is higher than the number of the physical volumes because a same physical volume can be used to define different pieces of the detector (*shared volumes*).

4. Serving the detector description

The resulting ATLAS description is a huge and complex hierarchical structure of roughly 500,000 GeoModel nodes. In addition, another piece of data is part of the detector description: The relationships among all the nodes and the position of each node in each sub-tree. There are around 665,000 of these entries and they are part of a complete description of the ATLAS GeoModel tree.

In the earlier paper, which the work described in this article is built upon [3], the focus has been placed on the design of a new data model to efficiently dump and store this complex data structure independently of the experiment's software framework, and on the development of a new mechanism to make a persistent local copy of it. Using the server-less database SQLite [4] we could store a persistent copy of the full GeoModel structure into a small file[1], which can be easily distributed to end users. The new mechanism allows a straightforward restore of the GeoModel, but it has some limitations: The whole detector description tree is restored, without the possibility of building sub-trees only; there is no way to easily inspect and examine the detector geometry; and using the new SQLite database, asking for nodes based on properties and relationships requires very complicated SQL queries.

To efficiently query and retrieve our complex tree-structured geometrical data, two open-source architectures and their relative platforms have been identified: The graph-based database Neo4j [6, 7], and the Elasticsearch search engine and its graphical interface Kibana [5]. The two platforms have a different purpose, and they have been chosen to address two different use cases: The first as development and debug tool, mainly to examine the actual detector description; the latter for an easy and straightforward use of the detector geometry by third-party applications which are not interested in the description structure itself, but only in the final outcome, that is the *rendered geometry*[2].

5. Graph-based data structure to store detector description data

The graph-based format used by Neo4j is ideal to store the GeoModel data: Neo4j has no concept of data tables, it only works with nodes, relationships, and properties. Therefore, the complex structure of the GeoModel tree is efficiently stored in full detail, without using auxiliary intermediate link tables like in SQL-based data formats. Also, all sorts of properties

[1] With the current implementation, the complete ATLAS geometry fits into a 46 MBytes SQLite file. For comparison, the corresponding in-memory footprint is about 130 MBytes [3].

[2] In this context, the expression *"rendered geometry"* refers to the geometry computed by traversing the GeoModel tree.

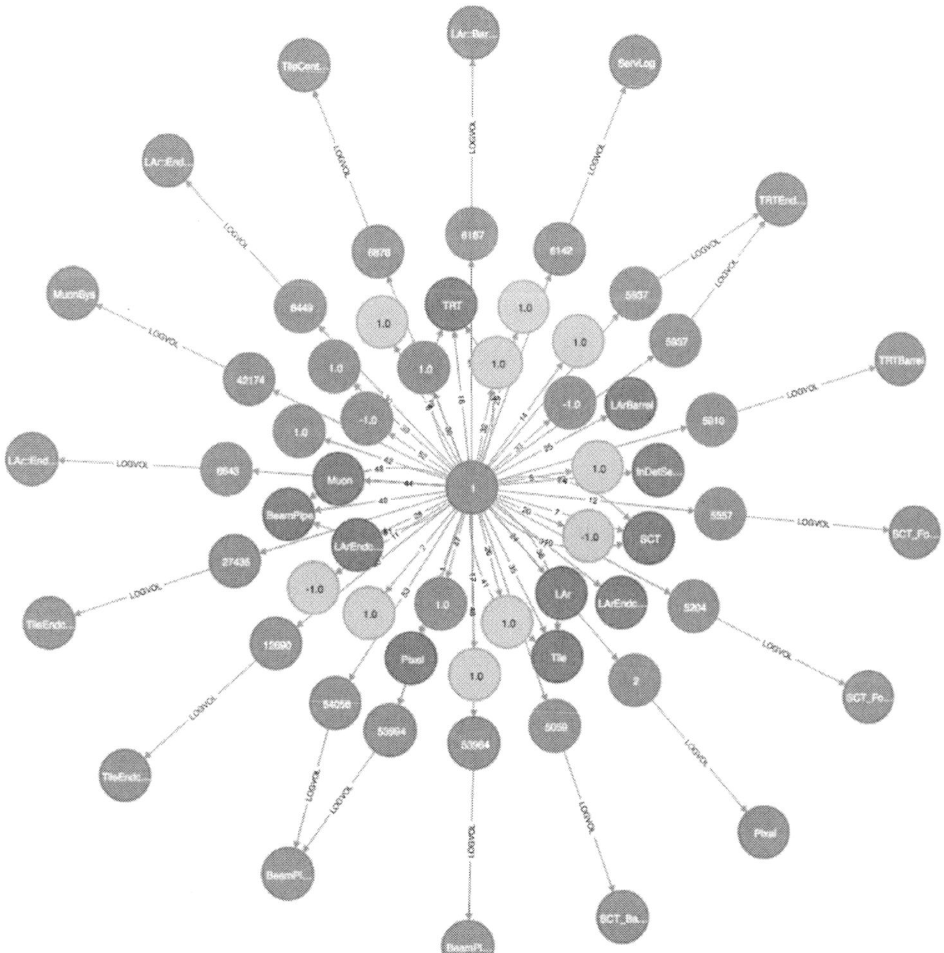

Figure 2: The graphical output of a Neo4j query retrieving the top logical volumes of the ATLAS geometry; Physical volumes are rendered in blue, Logical volumes in green, Transformations in grey and yellow. The node at the center is the World Volume. The numbers on the arrows state the children's position in the subtree. The red Name Tags define the global name of subtrees defining the geometry of the ATLAS sub-detectors: *Pixel* detector, *Muon* spectrometer, *LAr* and *Tile* calorimeters, and so forth. In this graph, only the nodes of the tree satisfying the example query are shown.

can be attached to each node or relationship in an independent way, without having to create additional tables for a single property or entity.

The first step required the export of the GeoModel data stored in SQLite tables (as described in Ref. [3]) into comma-separated files (CSV). Then, the CSV data was imported into Neo4j though a series of Cypher queries that create nodes and relationships. *Cypher* is the native query language of Neo4j, which is focused on graph-based data and optimized to perform queries and operations involving nodes, relationships, links, paths, and properties.

Cypher can be used to query the GeoModel graph as well, once the data are imported in the Neo4j database. Its syntax is very powerful, letting developers accurately inspect the detector description while developing or debugging it. Listing 1 shows an example of actual Cypher queries, used to inspect the structure of the ATLAS detector description. Figure 2 shows an example of the graphical result of a Neo4j query.

```
/*retrieve a specific logical volume filtering on its name*/
match (n:LogVol) where n.name = "TileEndcapNeg" return n;
/*as above, but retrieve the GeoPhysLogVol which uses it, too*/
match (n:PhysVol)-[r:LOGVOL]->(m:LogVol) where m.name = "TileEndcapNeg" return m,
    labels(m),n,r;
/*get all "Box" Shape nodes used together with "Air" as Material*/
match (s:Shape {type: "Box"})<-[rs:SHAPE]-(l:LogVol)-[rm:MATERIAL]->(m:Material {
    name: "Air"}) return s,rs,l,rm,m;
/*get all \emph{root volume} children and sort them by the 'position' property of
    the 'CHILD' relationship*/
match (n:RootVolume)-[r:CHILD]->(m) with r,m order by r.position return r.
    position, m;
```

Listing 1: Example of Cypher queries which can be used with the Neo4j interface

6. A search engine to serve the rendered geometry

The data returned by Neo4j reflects the raw structure of the GeoModel tree, and that is useful for developing and debugging the detector description. But, as briefly introduced in Section 4, other kinds of applications are interested in the final geometry volumes instead—for instance, all tools that are interested in the visual representation of the ATLAS detector, such as event displays and applications for outreach and education. To be able to provide that kind of data to clients, we chose to use the search engine Elasticsearch. The goal is to let applications quickly retrieve geometry objects without having to traverse the geometry tree.

Data from the SQLite file are converted to JSON, then inserted into Elasticsearch, traversing the tree data and pre-computing the geometry data at index time. Also, geometry data are *flattened*: All the attributes are collected along the way, and all the space transforms are computed and accumulated, in order to get all the attributes and the absolute position of each volume directly associated with the object itself. Ultimately, Elasticsearch does not store the actual GeoModel structure anymore; it stores *flattened objects* instead, which represent final usable objects of the geometry. The loss of information about the GeoModel structure is compensated by an increased speed in getting the final geometry volumes and in a greater convenience in querying them. The data can be interactively inspected using Kibana front-end, or queried and retrieved using Elasticsearch REST API. Listing 2 shows an example of Lucene queries (Lucene is the query language used by Elasticsearch); while Figure 3 shows flattened geometry data retrieved through the Kibana web-based interface.

```
/*retrieve a specific volume*/
name:TileEndcapNeg
/*retrieve all Pixel volumes in the EndCap side C*/
tags:Pixel AND tags:EndcapC
/*retrieve all box-shaped volumes whose material is ""Air*/
shape:box AND material:Air
```

Listing 2: Example of Lucene queries which can be used with the Ealsticsearch/Kibana interface

7. Comparison and conclusions

The main issues with the current implementation with the ATLAS Geometry system are the inability to access the detector geometry outside the experiment's framework, the lack of a way to query and inspect the detector description, and the lack of a tool to easily retrieve the final

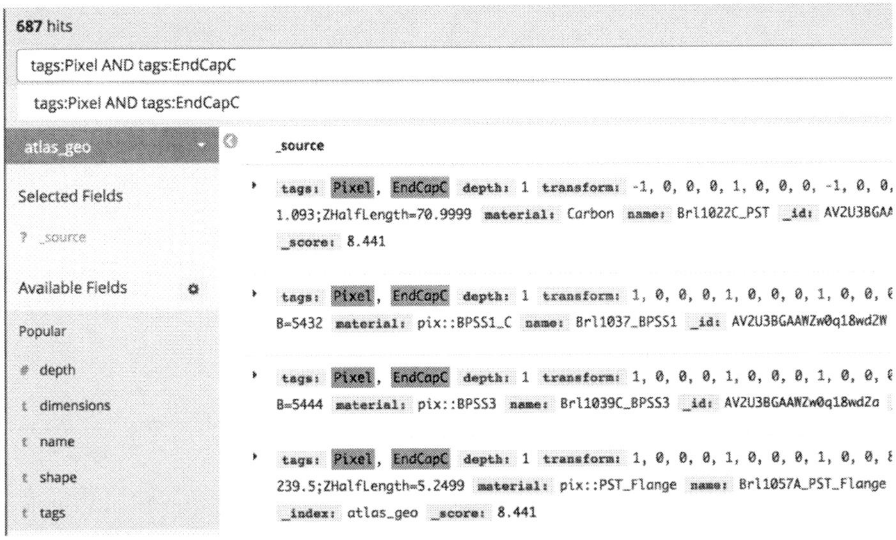

Figure 3: *Flattened* geometry data retrieved through the Kibana web-based interface; each line gives the information of a single volume, collected traversing the geometry tree: Name, space position, dimensions, material and so forth.

rendered geometry or usable partial data. The first issue has been solved in the earlier work [3], defining a new model to store the complete geometry data in a SQLite database in an storage-wise efficient way. But the SQL-based data model could not solve the remaining two issues: Easily querying and serving the geometry. Those were addressed in the work presented in this paper.

As said, GeoModel is composed by nodes interconnected by many types of relationships; however, relationships are not natively stored as objects in SQL-based databases; link tables and external refs are used to connect two items in two different tables. Therefore, SQL queries aimed at retrieve nodes and the relationships between them get very complicated very quickly. Here is where a graph-DB comes to rescue, by natively storing data as nodes and relationships. That lets users easily interact with the geometry data, examining and visualizing all nodes and relationships with convenient graph-aware queries. In addition to that, a search engine like Elasticsearch, let us easily store the rendered geometry into a searchable data structure, to easily retrieve specific final volumes ready to be used in external applications.

In conclusion, while GeoModel has been designed to minimise the memory footprint and to ease the definition of complex geometries by using efficient C++ code, the new mechanism using Neo4j and Elasticsearch has been designed and optimised to easily and interactively query the geometry data and to get the final volumes by just using simple text-based online queries.

Table 1 summarizes all the features offered by the new systems, compared to the current implementation. The original *in-framework* mechanism is designed to store all the latest and versioned data describing the ATLAS geometry. The SQLite-based persistification mechanism is designed to create a persistent copy of specific versions of the ATLAS geometry. On the other hand, the newest mechanisms presented here are designed to easily store and query the persistified geometry data. As one can infer from the table, the two new systems are complementary, since they address different targets and use-cases. Neo4j stores and exposes the full GeoModel structure, which can be queried, inspected and visualized. While Elasticsearch/Kibana serves only the final, computed geometrical volumes, ready to be queried and used by client applications interested in the visualization of the final *rendered* geometry.

IOP Conf. Series: Journal of Physics: Conf. Series **1085** (2018) 032035 doi:10.1088/1742-6596/1085/3/032035

	Current implementation (*in-framework*)	Persistent SQLite DB	Neo4j	ElasticSearch Kibana
Out-of-framework access to complete geometry data	No	**Yes**	**Yes**	**Yes**
Local copy of complete geometry data	No	**Yes**	**Yes**	No
Interactive query of GeoModel nodes and relationships	No	*Yes*‡	**Yes**	No
Visualization of GeoModel nodes and relationships	No	No	**Yes**	No
Partial retrieval of geometry data (subtrees)	No	*Yes*‡	**Yes**	**Yes**
Fast retrieval of final physical volumes with global positions	No	No	No	**Yes**

‡*with complex SQL queries*

Table 1: Comparison of different tools used in ATLAS to store, query and retrieve the geometry and data description data. The first column shows the current *in-framework* system; the second column refers to the mechanism described and presented in Ref. [3]; the third and the fourth columns refer to the newest mechanism which is the object of this paper.

Together, they offer a complete suite of tools to store, handle, inspect, retrieve detector description data for a large HEP experiment like ATLAS.

References

[1] ATLAS Collaboration, *The ATLAS Experiment at the CERN Large Hadron Collider*, JINST 3 (2008) S08003.
[2] Boudreau J and Tsulaia V 2004 *The GeoModel Toolkit for Detector Description*, CHEP '04, Book of Abstracts, `https://indico.cern.ch/event/0/contributions/1294152/`
[3] Bianchi R M, Boudreau J and Vukotic V, *A new experiment-agnostic mechanism to persistify and serve the detector geometry of ATLAS* (2017), CHEP 2016, `https://cds.cern.ch/record/2243141/`
[4] Hipp D R *et al.*, "*SQLite*" [software], `https://www.sqlite.org`
[5] Elasticsearch BV, "*Kibana*" [software], `https://github.com/elastic/kibana`
[6] Neo Database AB, *The Neo Database—A Technology Introduction* [White paper], (2006), `http://dist.neo4j.org/neo-technology-introduction.pdf`
[7] "*Neo4j*" [software], `https://github.com/neo4j/neo4j`

IOP Conf. Series: Journal of Physics: Conf. Series **1085** (2018) 032036 doi:10.1088/1742-6596/1085/3/032036

Virtualization of the ATLAS software environment on a shared HPC system

A J Gamel[1,2], U Schnoor[1], K Meier[1,2], F Bührer[1], M Schumacher[1]
Albert-Ludwigs-Universität Freiburg,
[1]Physikalisches Institut, Hermann-Herder-Strasse 3
[2]Rechenzentrum, Hermann-Herder-Strasse 10
79104 Freiburg im Breisgau, Germany

E-mail: anton.gamel@physik.uni-freiburg.de

Abstract. The shared HPC cluster NEMO at the University of Freiburg has been made available to local ATLAS users through the provisioning of virtual machines incorporating the ATLAS software environment analogously to a WLCG center. This concept allows to run both data analysis and production on the HPC host system which is connected to the existing Tier2/Tier3 infrastructure. Schedulers of the two clusters were integrated in a dynamic, on-demand way. An automatically generated, fully functional virtual machine image provides access to the local user environment. The performance in the virtualized environment is evaluated for typical High-Energy Physics applications.

1. Introduction

High-Performance Computing (HPC) and other research cluster computing resources provided by universities can be useful supplements to the ATLAS [1] collaboration's own WLCG [2] computing resources for data analysis and production of simulated event samples. Known difficulties to use these opportunistic resources are incompatibilities in network, batch schedulers, operating system, software installed and no access to external resources required. There exist a vast variety of different and very individual approaches to make opportunistic resources available, mainly for ATLAS production [3] [4] [5]. In order to use the University of Freiburg HPC-cluster NEMO [6] (388/TOP500 [7], 287280 HEP-SPEC) we further pursued the local approach already started and described here [8], taking advantage of the preexisting OpenStack [9] layer, allowing virtual machines (VMs) wrapped as jobs to be started on top of the NEMO bare metal installation.

2. Tasks

A transparent setup that makes NEMO resources available for users and that is as well smooth and flexible for administrators had to include the following:

- Full ATLAS / WLCG analysis and production Tier2/Tier3 environment
- Full local user environment including access to dCache [10] and cluster filesystem BeeGFS [11]
- Easy generation of up-to-date VM images and if possible in an automated procedure
- Integration of local Tier2/Tier3 batch system slurm [12] and NEMO scheduler Torque/Moab [13] [14]
- Start VMs on-demand using separate batch queues on Tier2/Tier3 slurm scheduler.

3. Generation of Virtual Machines using Packer

For the diskless installations of workernodes on the Tier2/Tier3 cluster a template machine is used that is a complete and fully functional workernode installation residing on VMware [15]. This template is configured by Puppet [16]. Exactly this Puppet configuration can be used to create virtual images with the identical setup the current workernodes on the Tier2/Tier3 cluster also have. Packer [17] installs a minimal SL6.8 from iso-file in a nested KVM on the OpenStack Service-Environment. After software update and reboot, Packer injects a valid Puppet certificate into the VM and connects the Puppet client with the Puppet master. Full configuration by Puppet takes some time since all packages are installed one by one. At this stage also the kernel modules are built to access the cluster filesystem later. After configuration has finished a rootfs-resize [18] optimizes the size of the image and Packer creates a qcow2 image under version control ready to be uploaded to NEMO OpenStack.

4. Integration of batch systems

The VMs that are now available via OpenStack have to be started on request i.e. if a job is queued in the corresponding slurm partition. Unfortunately the Elastic Computing feature of slurm that may start and stop physical machines on-demand only offers very limited control over VMs on remote systems. We therefore adapted ROCED [19] which is a modular python program interfacing different batch systems to different cloud sites and managing virtual machines on these sites. Moreover ROCED provides the administrator with handles to monitor the status of the queues as well as to analyze the usage. It was initially designed for HTCondor [20] and was now adapted to be used with slurm.

The ATLAS user (figure 1) submits the Tier3 job as usual but to a dedicated queue that reflects for fairshare reasons the group affiliation (1). ROCED demons, on a separate VM, monitor for incoming jobs (2) and trigger (3) the start of a job-wrapper in the NEMO UI & Moab environment (4). Once the job is launched by Moab, the job wrapper requests a VM from OpenStack (5) that starts into the job-wrappers space (6). After the start of the VM, the most important features of the environment are automatically checked, e.g. file system access and mounts as well as network access. If the tests are passed, the VM contacts the slurm control demon on the slurm server and is ready to receive jobs from the slurm queue (7). Several ROCED services can be run without interference in order to serve different slurm partitions. Watchdog processes shutdown the VM if idle for a certain period or set it to be drained in time before the end of walltime approaches. Since the VMs have no cvmfs cache available neither on virtual disk nor on the host, within the NEMO network a dedicated squid proxy VM was installed that handles all the requests.

IOP Conf. Series: Journal of Physics: Conf. Series **1085** (2018) 032036 doi:10.1088/1742-6596/1085/3/032036

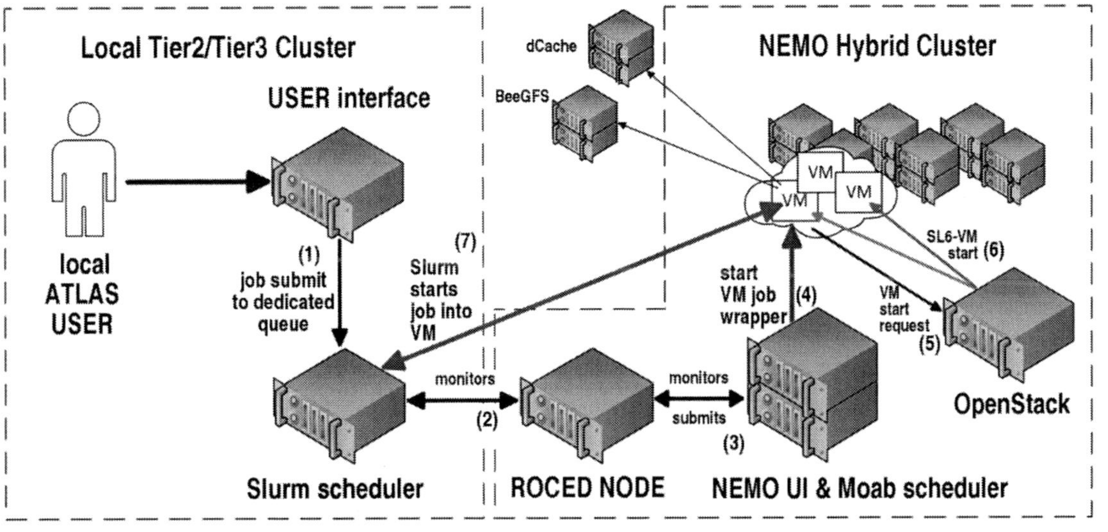

Figure 1. Overview: NEMO cluster receives jobs from local Tier2/Tier3 cluster via ROCED node.

5. Performance in the virtualized environment

NEMO and the Tier2/Tier3 cluster partly share the identical hardware configuration: 4in1 INTEL S2600KPR board with 2x INTEL CPU E5-2630v4 2.20GHz 40cores hyperthreaded, 128GB RAM, SSD. It was therefore possible to compare the performance of jobs in the VMs (4-core, CentOS7 host) with the performance of jobs running on bare metal (multicore, diskless install). For the evaluation two applications were chosen. On the one hand a Powheg/Pythia8 [21] [22] event generation that used ATLAS software via cvmfs (figure 2), on the other hand the HEP-SPEC06 benchmark application [23] [24]. As a reference HEP-SPEC06 benchmarks were also run on reserved machines on bare metal using different numbers of cores (figure 3). In both applications the jobs in the VMs perform better than the jobs on bare metal: no loss of performance in the VMs can be determined (figure 2 and 3). The difference in performance, VMs versus bare metal, was unexpected and is under further investigation.

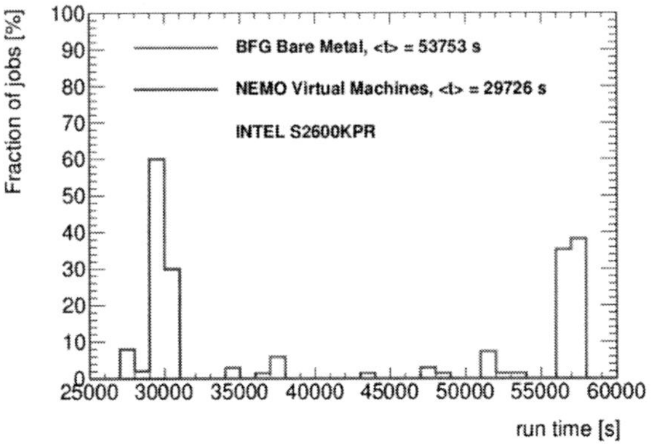

Figure 2. VM nodes better performing compared to bare metal.

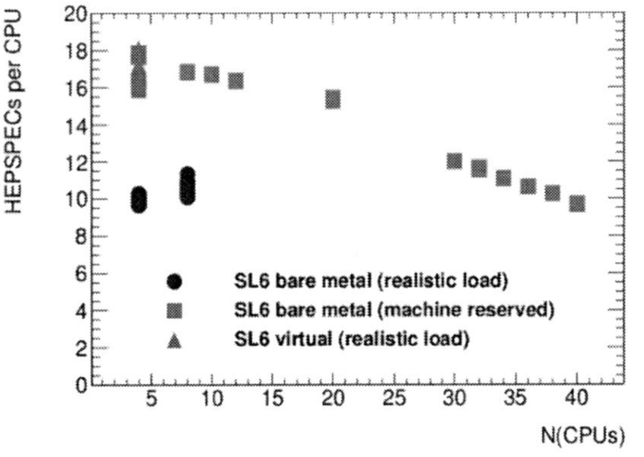

Figure 3. HEP-SPEC06 on 4-core VMs similar to optimal conditions on bare metal.

6. Outlook

Meanwhile NEMO Moab received an improvement and is able to start a job with a VM in it directly from OpenStack without the need of a job wrapper. The job monitoring information ROCED makes available to Slurm or to the user is still sparse and has to be enhanced. In the described setup a queue for Tier2 production can be established as well. For the transition from SL6 to CentOS7 a template to generate VMs with CentOS7 is in preparation.

7. References

[1] ATLAS Collaboration 2008 J. Inst. 3 S08003
[2] Shiers J 2007 Comput. Phys. Commun. 177 219-223
[3] Kennedy J A, Kluth S, Mazzaferro L and Walker R 2015 Bringing ATLAS production to HPC resources - A use case with the Hydra supercomputer of the Max Planck Society. In: 21st International Conference On Computing in High Energy and Nuclear Physics (Chep2015), Parts 1-9, Vol. 664, 092019
[4] Cameron D, Filipčič A, Guan W, Tsulaia V, Walker R and Wenaus T 2016 Exploiting opportunistic resources for ATLAS with ARC-CE and the Event Service. 22nd International Conference on Computing in High Energy and Nuclear Physics, (CHEP 2016), San Francisco, USA, 10 - 14 Oct 2016
[5] Barreiro M, Fernando H, De K, Klimentov A, Nilsson P, Oleynik D, Padolski S, Panitkin S and Wenaus T 2016 Integration of Titan supercomputer at OLCF with ATLAS production system. 22nd International Conference on Computing in High Energy and Nuclear Physics, (CHEP 2016), San Francisco, USA, 10 - 14 Oct 2016
[6] http://www.hpc.uni-freiburg.de/nemo [accessed 2017-09-27]
[7] https://www.top500.org/list/2017/06/?page=4 [accessed 2017-09-27]
[8] Meier K *et al* 2016 J. Phys.: Conf. Ser. 762 012012
[9] OpenStack [software] https://www.openstack.org/ [accessed 2017-09-27]
[10] dCache [software] https://www.dcache.org/index.shtml [accessed 2017-09-27]
[11] BeeGFS [software] https://www.beegfs.io/content/ [accessed 2017-09-27]
[12] Slurm [software] https://slurm.schedmd.com/ [accessed 2017-09-27]
[13] Torque [software] http://www.adaptivecomputing.com/products/open-source/torque/ [accessed 2017-09-27]
[14] Moab [software] http://www.adaptivecomputing.com/products/hpc-products/moab-hpc-basic-edition/ [accessed 2017-09-27]
[15] https://www.rz.uni-freiburg.de/services-en/serverdienste_en/virtualisierung?set_language=en [accessed 2017-09-27]
[16] Puppet [software] https://puppet.com/ [accessed 2017-09-27]
[17] Packer [software] https://www.packer.io/ [accessed 2017-09-27]
[18] rootfs-resize [software] https://github.com/ctyler/rootfs-resize [accessed 2017-09-27]
[19] Erli G, Fleig G, Hauth T and Riedel, S "ROCED" [software] Available from https://github.com/roced-scheduler/ROCED [accessed 2017-09-27]
[20] HTCondor [software] http://research.cs.wisc.edu/htcondor [accessed 2017-09-27]
[21] Nason P 2004 JHEP11 040, hep-ph/0409146
[22] Sjöstrand T, Mrenna S and Skands P 2008 Comput. Phys. Comm. 178 852-867
[23] SPEC® CPU2006 [software] https://www.spec.org/cpu2006/ [accessed 2017-11-13]
[24] http://w3.hepix.org/benchmarks/benchmarking.html [accessed 2017-11-13]

IOP Conf. Series: Journal of Physics: Conf. Series **1085** (2018) 032037 doi:10.1088/1742-6596/1085/3/032037

GeantV alpha release

G Amadio[1], Ananya[2], J Apostolakis[1], M Bandieramonte[1], S Behera[3], A Bhattacharyya[3], R Brun[1], P Canal[4], F Carminati[1], G Cosmo[1], V Drogan[5], L Duhem[6], D Elvira[4], K Genser[4], A Gheata[1], M Gheata[1,7], I Goulas[1], F Hariri[1], V Ivantchenko[1], S Jun[4], P Karpinski[1], G Khattak[1], D Konstantinov[1], H Kumawat[3], G Lima[4], J Martínez-Castro[11], P Mendez Lorenzo[1], A Miranda-Aguilar[11], K Nikolics[1], M Novak[1], E Orlova[9], W Pokorski[1], A Ribon[1], R Sehgal[3], R Schmitz[8], S Sharan[1], O Shadura[1], S Vallecorsa[1,10], S Wenzel[1]

[1] CERN - European Organization for Nuclear Research, Geneva, Switzerland
[2] IIT- Indian Institute of Technology, India
[3] Bhabha Atomic Research Center, India
[4] Fermi National Accelerator Laboratory, USA
[5] National University of Kyiv, Ukraine
[6] Intel Corporation
[7] Institute of Space Sciences, Romania
[8] University of Minnesota, USA
[9] Higher School of Economics, Moscow, Russia
[10] Gangneung-Wonju National University, Gangneung, South Korea
[11] Centro de Investigación en Computación, Mexico City, Mexico

andrei.gheata@cern.ch

Abstract. In the fall 2016, *GeantV* went through a thorough community evaluation of the project status and of its strategy for sharing the R&D results with the LHC experiments and with the HEP simulation community in general. Following this discussion, *GeantV* has engaged onto an ambitious 2-year road-path aiming to deliver a beta version that has most of the final design and several performance features of the final product, partially integrated with some of the experiment's frameworks. The initial *GeantV* prototype has been updated to a vector-aware concurrent framework, which is able to deliver high-density floating-point computations for most of the performance-critical components such as propagation in field and physics models. Electromagnetic physics models were adapted for the specific *GeantV* requirements, aiming for the full demonstration of shower physics performance in the alpha release at the end of 2017. We have revisited and formalized *GeantV* user interfaces and helper protocols, allowing to: connect to user code, provide recipes to access efficiently MC truth and generate user data in a concurrent environment.

1. Introduction

Improving the CPU performance is a major objective of simulation R&D and the related program of work is connected to the LHC schedule for the high luminosity phase. Adapting the simulation workflow and algorithms to better profit from the FLOPS potential of modern architectures is very important. The R&D undertaken by the *GeantV* project to increase the data and instruction locality has shown [1] that important performance improvements are within reach even in highly complex frameworks such as particle transport simulation.

The *GeantV* project aims at speeding-up particle transport simulation by a factor between 2 and 5 by recasting the traditional approach into a more cache and vector-friendly form, targeting instruction-level parallelism. Tracks are regrouped per-step into "baskets" collecting particles that go through a common processing phase, allowing to perform repetitive work and to use SIMD registers even for algorithms that are not internally vectorizable.

A first R&D phase revealed an important speed-up potential coming from geometry computation. This led to a vectorization-capable geometry package demonstrating excellent performance in both scalar and multi-particle modes. The scalar version of the *VecGeom* [2] package is now part of the official releases of *Geant4* [3], while the vectorization abstraction layer *VecCore* [4] developed in the context of *GeantV* was released as an independent package, adopted for vectorization by *ROOT* [5]. The major *GeantV* challenges after this first period of R&D were to extend the basket approach to other processing stages and to develop the first version of the electromagnetic physics package, preparing the field for a fully vectorized EM shower simulation.

A thorough review of the project goals revealed the need to expose *GeantV* to the community for early integration and testing with experiment frameworks, targeting high-luminosity LHC for the production phase. As a result, *GeantV* went through deep transformations touching the core scheduler, the interfaces and the physics package, aiming to deliver an alpha version at the end of 2017 and a beta version in 2018. The following sections briefly describe these transformations and the general features to be expected in the first release, and on-going R&D works expected to be delivered in 2018.

2. A generic vector scheduling framework

Making use of the SIMD pipelines available today even in commodity PC's has been already investigated in *GeantV* [6]. An important lesson resulting from this R&D is that vectorizing only some of the computing intensive algorithms is not sufficient for enabling significant overall gains from the SIMD pipelines. The entire simulation workflow needs to be optimized to sustain a low-overhead continuous vector data flow. *GeantV* prototyped and put in production an approach that splits the stepping procedure for tracks into stages, to accumulate several particles before actually performing the actions involved by each stage.

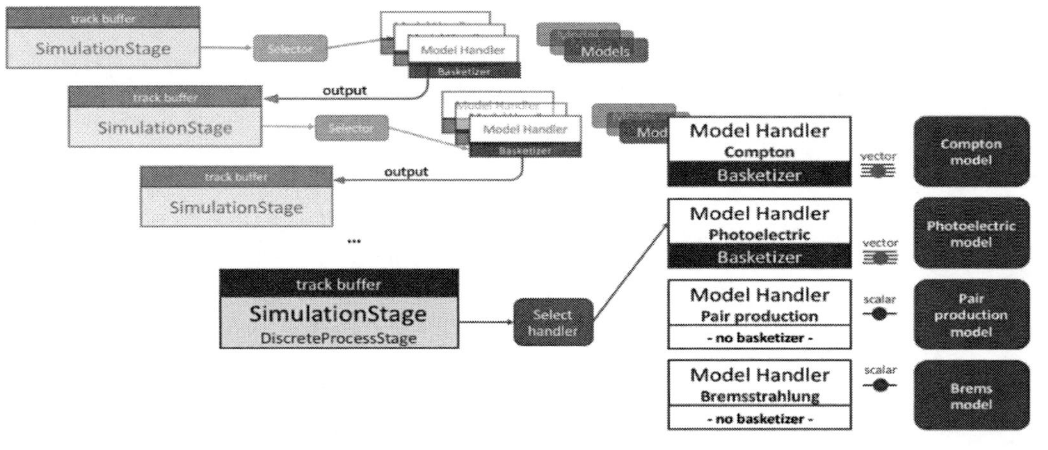

Figure 1 The GeantV stepping procedure decomposed in stages. Each stage loops over input tracks, selecting among a set of models corresponding to the stage. Tracks may be either "basketized" with others and sent to vectorized algorithms, or processed by the selected model in scalar mode. This approach improves the locality and cache coherency compared to the classical stack-based approach.

IOP Conf. Series: Journal of Physics: Conf. Series **1085** (2018) 032037 doi:10.1088/1742-6596/1085/3/032037

As illustrated in Figure 1, the generation of the final states from discrete physics processes is an example of a simulation stage, but the same mechanism is suitable for propagation in magnetic field or geometry queries. Particles entering a stage are directed to different physics model handlers that accumulate suitable tracks into baskets, executing the different model algorithms in multi-particle mode.

This kind of design changes the traditional stack-based execution into a multi-stream execution pattern. It creates the premises for increasing code and data locality while allowing to feed multiple data to vectorized code.

The new version of the *GeantV* scheduler (version 3) corrects many of the drawbacks of the previous implementations. The memory management has been fully re-written to prevent bloating due to event mixing in the most challenging conditions: high number of threads, geometry of large complexity or low production cuts producing many secondary particles. The implementation uses a special buffer that prioritizes the transport of particles of older generation, allowing to "consume" previous particle showers before generating and transporting new ones. Figure 2 shows the maximum resident memory as function of the number of threads in the case of a complex geometry setup and in the case of very large EM showers. Even if the scalability of the new scheduler is not improved much, the single-threaded *GeantV* applications are 40%-80% faster now due to the increased locality induced by the generalized vector flow, as shown in Figure 3. Vectorization benefits are not considered in this comparison and they are expected to increase the performance gain when available for physics and field propagation.

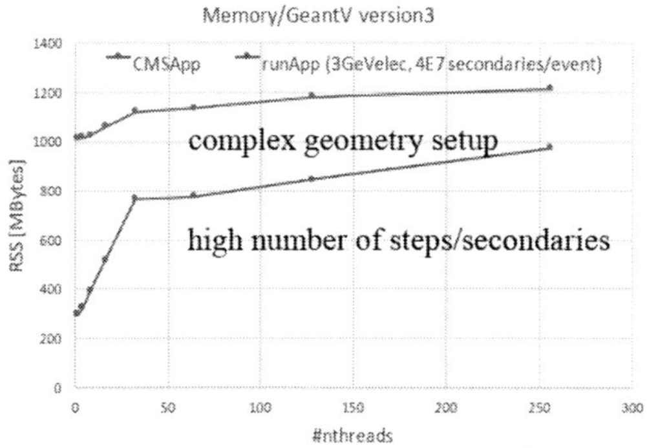

Figure 2 Resident Set Size (RSS) vs number of threads. Memory in GeantV scheduler (version 3 is controlled by applying a policy prioritizing the transport of older generation secondary particles, while re-using the containers released by completed tracks.

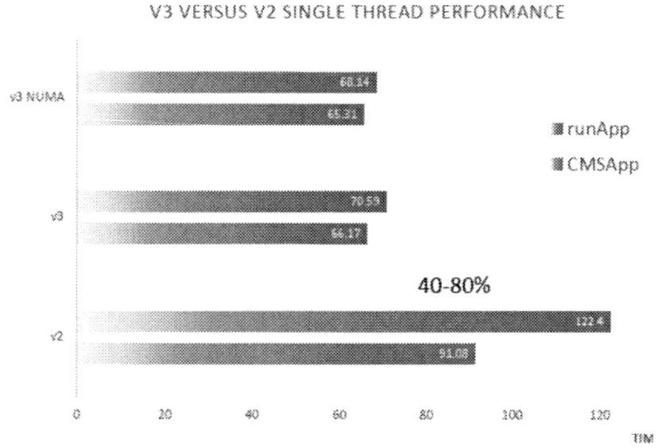

Figure 3 Single thread performance improvements compared to the previous version, induced by increased locality. Tests are shown for both a simple sampling calorimeter (upper) and a realistic CMS setup (middle), compared to previous version (v2)

The coming *GeantV* prototype release will be "hardware topology"-aware. The increase of the number of cores on modern hardware leads to more complex cache and memory hierarchies (Non-Uniform Memory Architecture or NUMA) with un-avoidable extra costs for non-local data access. Programs having frequent cache misses are more penalized by non-uniform memory accesses than the others. *GeantV* can run NUMA-aware on demand, activating a topology discovery service based on the *hwloc* library [7]. The application can configure its concurrency settings to deploy separate propagators on each locality node, which uses threads and data pinned to the node and minimizes the need of data transfers among different propagators.

IOP Conf. Series: Journal of Physics: Conf. Series **1085** (2018) 032037 doi:10.1088/1742-6596/1085/3/032037

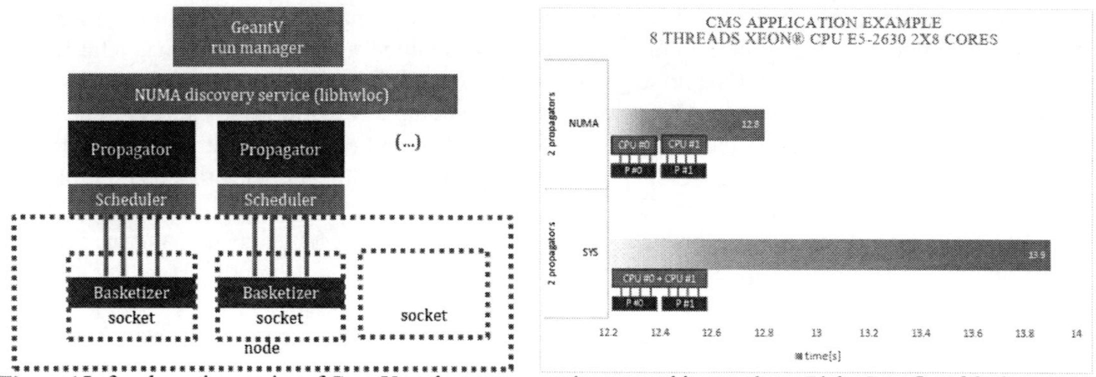

Figure 4 Left: schematic mapping of GeantV track transport services to machine topology. Right: Benefits of deployment of a number of propagators matching the number of locality nodes (sockets) of the machine. The alpha release of *GeantV* can be configured NUMA-aware, detecting the hardware topology and confining the data and processing flow to local resources as much as possible.

As shown in Figure 4, each propagator is pinned to a node, and it behaves almost like a separate process on its own locality, while different propagators are work-balanced by a single event server service. Compared to the scheduling done by the OS, this mode can bring benefits of up to 10% or more depending on the hardware.

3. GeantV physics in the alpha release

The development of electromagnetic showers in calorimeters plays a major role for the CPU performance of simulations. This is also true for jets because hadronic showers, induced by the hadron components of jets, have electromagnetic components coming from the decays of neutral pions. This implies that the development of vectorized EM physics models is one of the main areas where the effort should be focused in the short and medium term for developing a high-performance detailed simulation.

An important objective of the *GeantV* physics development is therefore to review the physics algorithms and their implementation in order to improve their accuracy and better adapt them to a multi-particle flow. Most of the EM physics models in *GeantV* were revisited with respect to the theory and rewritten with state-of-the-art implementations. The resulting improvements were adapted also to *Geant4*, which is of utmost importance for both validating and making available the new *GeantV* developments to the community.

The *GeantV* strategy is to first provide scalar versions having vectorizable data structures and general helpers (such as sampling methods), then to proceed with the vectorization of these models one-by-one, offering a testbed where the scalar and vector versions can be switched on demand. This will also facilitate the validation of the new developments and the benchmarking of their performance in a basket flow environment.

The alpha release will deliver vectorized geometry, but mainly scalar physics fully describing EM showers. The different *GeantV* models were tested and validated by test applications implemented in both *GeantV* and *Geant4*. Tests are showing results matching at per mil level for sensitive observables, such as visible energy per primary, number of steps or number of produced secondary particles of different types. Table 1 presents such a comparison for electron-induced showers in the ATLAS simplified calorimeter, while Figure 5 presents the validation of different options of the multiple scattering model in two different setups. The available physics models in the alpha release are: Bremsstrahlung, Ionisation and Coulomb scattering (with multiple scattering) for electrons/positrons and Compton, Conversion and Photoelectric for gammas, in the range [100eV-100TeV].

Table 1 Total energy deposited and cumulative track length for electron-induced showers in simplified ATLAS sampling calorimeter. The corresponding GeantV and Geant4 simulation results are matching within statistical errors.

10^5 1 [GeV] e- in ATLAS bar. simpl. cal. : 50 layers of [2.3 mm Pb + 5.7 mm lAr]; p.cut = 0.7 [mm]								
	e^-/e^+: **ionisation, bremsstrahlung, msc;** γ: **Compton, conversion**							
	GeantV				**Geant4**			
material	E_d[GeV]	rms [MeV]	tr.l. [m]	rms [cm]	E_d[GeV]	rms [MeV]	tr.l. [m]	rms [cm]
Pb	0.69450	15.198	51.015	1.189	0.69448	15.234	51.016	1.192
lAr	0.22792	14.675	106.11	7.592	0.22796	14.656	106.13	7.582

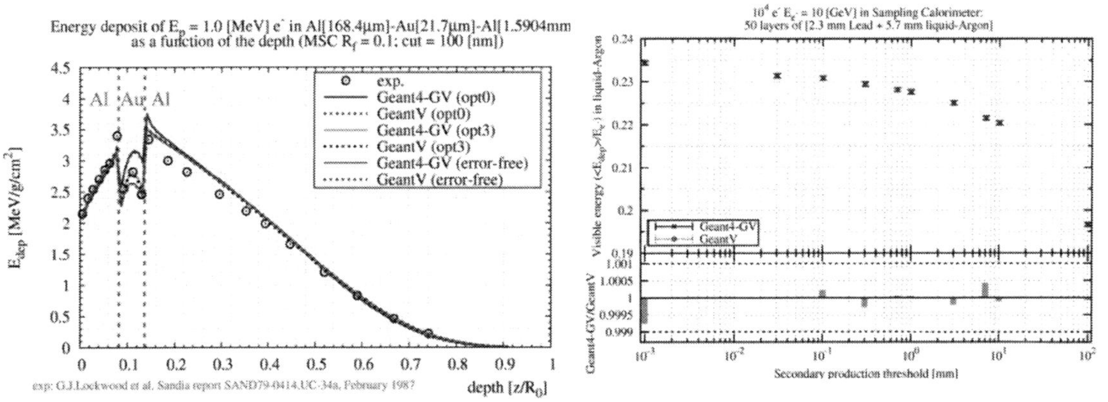

Figure 5 GeantV EM physics validation against Geant4 10.4 beta. Left: Energy deposit depending on depth for 1 MeV electrons in multi-layered Al-Au-Al target. Different options of the multiple scattering model have been compared for GeantV, Geant4 and data. Right: visible energy per primary for 10 GeV electrons in liquid-Argon for multi-layered sampling calorimeter. The Geant4-GV physics list matching the physics settings used in GeantV was consistently used in Geant4-based simulations.

4. User interfaces and examples

The *GeantV* recast of the simulation approach has several practical implications, several of which are affecting the user code. For instance, track mixing from different events is important for sustaining the basketized flow, and this will affect the user scoring methods. It is interesting to note, however, that several current or planned experiment frameworks will also support track-level parallelism. This will affect the data management and I/O on the user side, requiring concurrent bookkeeping and management of multiple events in flight. Extending the user API with vector signatures will not affect existing scalar user code, but it will allow new vectorized user code to exploit the particle-level parallelism offered by the framework.

The alpha release will introduce a set of interfaces defining the interaction with the user application, conceptually very similar to the *Geant4* ones. The user application will still be notified at the beginning/end of the run/event/primary tracks, and will be able to score in sensitive detectors, using similar methods as before. Some new paradigms derived from running concurrently a limited number of events will have impact on the data management. The signal produced in sensitive elements by different threads needs to be merged per event before being digitized. The user scoring data will need to have thread local state to prevent expensive locking required by thread safety.

GeantV provides on-demand services to deal with the new concurrency paradigm and ease the transition of user applications to the new environment. A series of simple to more complex examples demonstrating the complete workflow will be delivered with the first release. There will also be examples of the integration of *GeantV* with user applications providing their own event loop. We are targeting a first level of integration with task-based frameworks such as *CMSSW* [8] and *GAUDI* [9].

5. A ML approach for integrating fast simulation

The required increase in CPU performance for the LHC high-luminosity program cannot be achieved only with algorithmic improvements exploiting SIMD, structural changes and multi-particle flow. It is also necessary to develop fast sampling algorithms as well as fast simulation approaches that replace the simulation of full showers. The project aims to investigate further the ways in which full and fast simulation capabilities can be integrated in a single detector simulation framework.

The target is to integrate a simple example in the alpha release demonstrating fast simulation of a calorimeter using ML-based inference. This is the result of an ongoing R&D [10] testing a generic Generative Adversarial Networks (GAN) [11] based approach via computer vision techniques using as input 3D calorimeter data. The initial tests are able to reproduce the shower profiles and single cell response for mono-energetic particles coming from the same direction, in the 25x25x25 cells calorimeter (Figure 6).

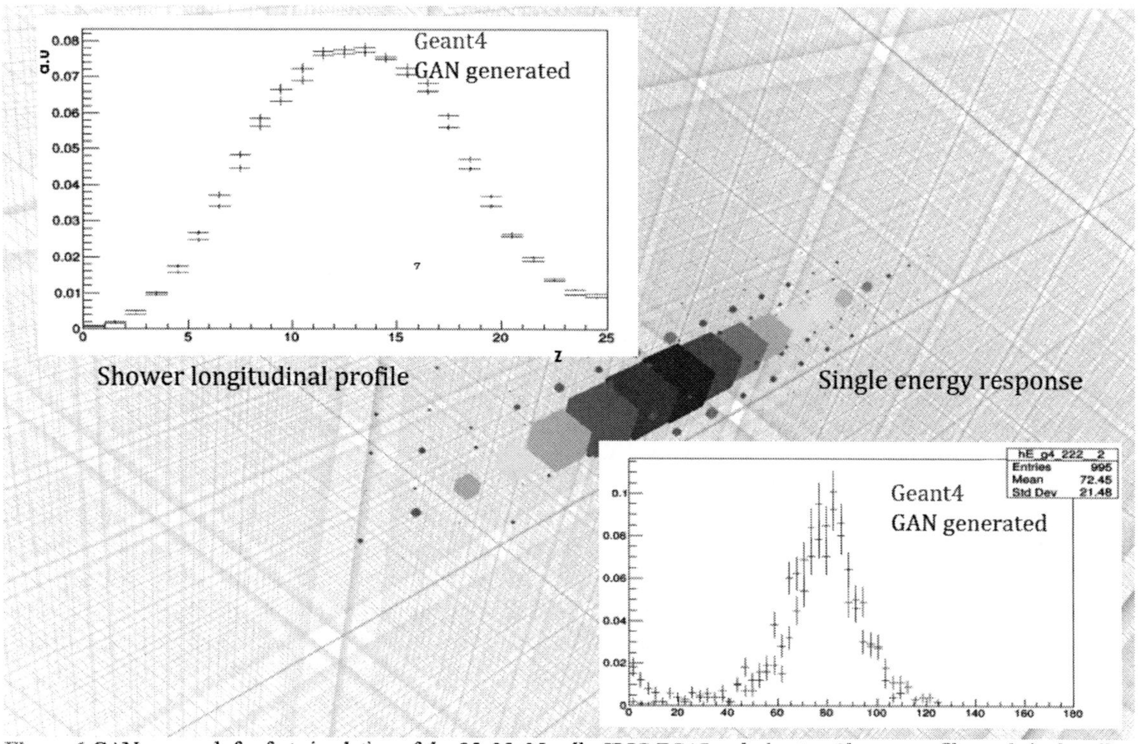

Figure 6 GAN approach for fast simulation of the 25x25x25 cells CLIC ECAL calorimeter. Shower profiles and single cell response for the trained network describe well the input data.

After a first phase trying to understand the limitations of such approach (such as training time versus re-usability of the trained network), we will try to generalize it, using multi-objective regression (momentum and angular distributions, particle type, ...). The next phase aims at generalizing this approach to other types of detectors in an adaptive manner, optimizing for example the network topology for the problem to solve.

6. Conclusions

After a first phase of intensive R&D, the *GeantV* prototype has tagged an alpha release [12]. This version delivers only part of the design performance, but allows the community to have a first look at a new technology featuring track-level parallelism. This will benefit both *GeantV* and the applications using it. The alpha tag features the specific *GeantV* interfaces, and several examples of different degrees of complexity demonstrating the complete workflow, each one having a correspondent *Geant4* application.

The alpha release features full EM physics for electrons, positrons and gamma particles in scalar mode, demonstrated in different concurrency scenarios, using either internal event loop and static threads or external event loop in a task-based approach.

We expect a very close collaboration and feedback from experiments during 2018, including early tests of framework integration with *GeantV* for the beta release at the end of the year. The beta release is expected to feature full *GeantV* performance for vectorized EM physics, and a production-quality geometry.

Acknowledgements
The authors wish to acknowledge the contribution of Intel to the *GeantV* project through the Intel Parallel Computing Centre (IPCC) program. We also want to acknowledge the very useful technical contributions of CERN openlab, which allowed performing several studies and performance optimizations.

References

[1] J Apostolakis et al 2015 Adaptive track scheduling to optimize concurrency and vectorization in GeantV J Phys: Conf Ser **608** 012003
[2] J Apostolakis et al 2015 J. Phys.: Conf. Ser. **608** 012023
[3] S Agostinelli et al 2003 *Nuclear Instruments and Methods* **A 506** (53pp)
[4] G Amadio et al 2017 Speeding up experiments software with VecCore, a portable SIMD library *ACAT 2017*
[5] ROOT project, "ROOT" [software] version 6.13/02, 2018. Available from https://root.cern.ch/content/release-61302 [accessed 2018-04-09]
[6] Apostolakis J, Brun R, Carminati F and Gheata A 2012 *J.Phys: Conf. Ser.* **396** 022014 http://iopscience.iop.org/1742-6596/396/2/022014
[7] Portable Hardware Locality project, "hwloc" [software], version 2.0.1, 2018. Available from https://www.open-mpi.org/projects/hwloc [accessed 2018-04-09]
[8] CMSSW project, "CMSSW Application Framework" [software], 2018. Available from https://twiki.cern.ch/twiki/bin/view/CMSPublic/WorkBookCMSSWFramework [accessed 2018-04-09]
[9] Gaudi project, "Gaudi" [software] version v30r2, 2018. Available from http://gaudi.web.cern.ch/gaudi [accessed 2018-04-09]
[10] S Vallecorsa et al 2017 A Machine Learning tool for fast simulation *ACAT 2017*
[11] Goodfellow I J et al 2014 Generative Adversarial Networks. *ArXiv e-prints*: 1406.2661
[12] GeantV project, "GeantV" [software] version 0.3.0, 2018. Available from https://gitlab.cern.ch/GeantV/geant/tree/alpha [accessed 2018-04-09]

Event Display in the JUNO Experiment

Jiang Zhu[a,1], Zhengyun You[a,2*] and Yumei Zhang[b,3],
on the behalf of JUNO Collaboration

[a] School of Physics, Sun Yat-sen University, Guangzhou 510275, China
[b] Sino-French Institute of Nuclear Engineering and Technology, Sun Yat-sen University, Guangzhou 510275, China

Email: [1]zhuj38@mail2.sysu.edu.cn, [2]youzhy5@mail.sysu.edu.cn, [3]zhangym26@mail.sysu.edu.cn

Abstract. The current event display system in the offline software of Jiangmen Underground Neutrino Observatory Experiment(JUNO) is based on the ROOT EVE package. We use Unity, a renowned game engine, to improve its performance and make it available on different platforms. Compared to ROOT, Unity provides a more vivid demonstration for high energy physics experiments and can be ported to different platforms easily. We build a tool for event display in JUNO with Unity. It provides us an intuitive way to observe the detector model, the particle trajectories and the hit distributions.

1. Introduction

An event display visualizes a high energy physics experiment, showing the detector structure and event information. It can also be used to improve the reconstruction algorithm and help physics analyses. With the event display system, physicists can understand the detector structure quickly and can observe some special events intuitively.

Jiangmen Underground Neutrino Observatory Experiment (JUNO) is a next-generation reactor neutrino experiment. It is located in Jiangmen City, Guangdong Province, China, which is in a distance of 53km from both of Yangjiang and Taishan nuclear power plants [1]. The main scientific purpose of JUNO is to measure the mass hierarchy of the three generations of neutrinos using inverse beta decay [2]. The central detector of JUNO is a 35.4m-diameter acrylic sphere with 20000 tons liquid scintillator inside and more than 17000 Photo Multiple Tubes(PMTs) around the sphere's surface. The whole detector is surrounded by water and there is a muon tracker on the top of it [1].

Based on the demands of the JUNO experiment, the collaboration develops the offline software system for JUNO based on the software framework SNiPER [3], (Software for Non-collider Physics ExpeRiment). The current event display system of JUNO is integrated in the JUNO offline software system and it is based on ROOT [4]. We use the Event Visualization Environment(EVE) package [5] to build this application for JUNO to show the detector structure, to animate the event hits, to visualize the reconstruction vertex and other event information. At the same time, we are also developing a new event display system based on the Unity engine to make the event display system less dependent on the JUNO offline software system so that the users can run the event display on their own PCs.

* Supported by National Natural Science Foundation of China (11405279, 11675275) and the Strategic Priority Research Program of Chinese Academy of Sciences (XDA10010900).

2. Event display based on ROOT

ROOT is a data analysis framework for high energy physics [4], based on C++. The EVE package is a visualization library in ROOT. It can manage the geometry objects and GUI components at a high level and render the image with OpenGL [5].

In the framework of ROOT, we use the EVE package to build the current event display system SERENA, which stands for Software for Event display with ROOT EVE in Neutrino Analysis. Figure 1 shows the current scheme of the event display in JUNO offline software.

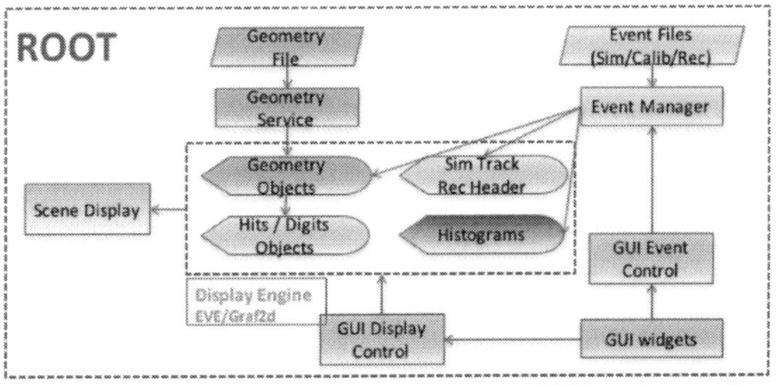

Figure 1. The current scheme of the event display in JUNO offline software

SERENA loads the geometry file of the JUNO detector to get the structure information and uses the EVE package to generate the visual objects. Then it reads the event information from the simulation/calibration/reconstruction files generated by the JUNO offline software. After getting the PMT hits information in each event, it changes the visual effects of the objects generated with the geometry file, builds the reconstruction vertex and shows the detailed information of the event. Users are able to select the event that they want to display with the GUI interface. In addition, it can read the data from the event file directly and project the hit distributions into azimuth angle to get the 2D plot so that users can observe the distribution in 2D projection view.

Figure 2 shows the GUI of SERENA. In the middle is the display windows, showing the PMT hit distributions in the central detector. The left side has the visual object selector, with which users can select the specific structure components they want to display or not, such as hiding the water pool, the acrylic ball or the veto PMTs. The right side has the widgets to control the event, where users can choose the specific event and demonstrate the PMT hit animation. With SERENA we can show the detector structure, animate the event hit, draw the construction vertex and make a comparison between simulation and reconstruction for reconstruction algorithm tuning.

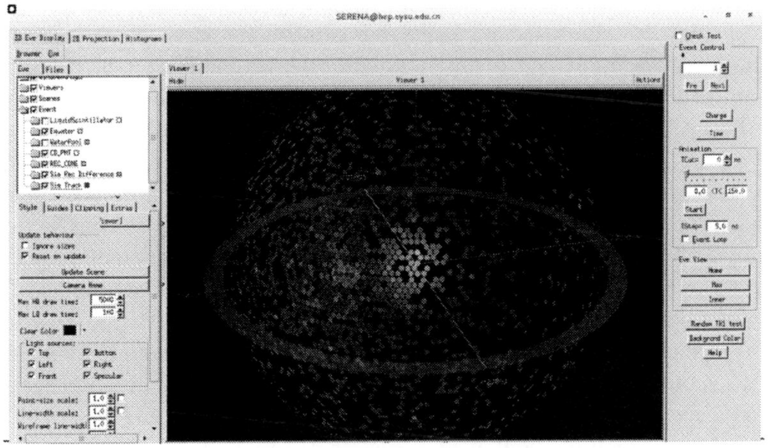

Figure 2. GUI of SERENA

For now, SERENA has satisfied most of the demands of JUNO experiment for event display. It is integrated in the JUNO offline software system as a package. Usually the JUNO offline software will be set up on servers with Scientific Linux system. If users want to run the display system on their own laptops, they will have to install some plugins like XQuartz or VNC to display the screen locally. The situation is even more complex as different users use different operation systems. An alternative event display system, which is less dependent on the JUNO offline software, might be helpful for users. We consider using the Unity engine as suitable for building such an independent event display system.

3. Event display based on Unity

3.1. Introduction to Unity

Unity [6] is a renowned game engine. Many interesting games like the Firewatch, Monument Valley, and Disney Crossy Road is developed with the Unity engine. Unity is known for its support for multiple platforms, which means an application developed with Unity can be built for other platforms very easily. The developers do not need to spend too much time on porting their application to make it available on Windows, Linux, macOS or in the browser. Not only famous game companies, but also many independent developers like using Unity to build their games. The personal edition of Unity is free for people who do not make sizable profits with it and the personal edition includes most of the functions of Unity. What's more, Unity is not just for games and there are many successful projects in education, simulation and visualization. In the field of high energy physics, there is a project developed by CERN called CAMELIA (Cross-platform Atlas Multimedia Educational Lab for Interactive Analysis) [7]. It can be used to visualize the detector of ATLAS experiment and analyze the collider events in LHC. When developing with Unity, developers script with C# or JavaScript. There are many useful modules for developers to build their GUI very quickly, like button, slider, toggle and so on.

3.2. Requirements analysis

In contrast to the event display system based on ROOT, the new event display system based on Unity is built as a client so that the system can be less dependent on the JUNO offline software and can run on users' own PCs only after importing the event data. As an event display system in high energy physics experiment, it should do the visualization of the detector structure, restore the process of event hits, load reconstruction data and make comparisons between reconstruction and simulation. At the same time, we try to build this system as a client to make it easy to upgrade into the online event display system for further development.

3.3. Scheme and data flow

The new event display system based on Unity has almost the same data flow as the one based on ROOT, as figure 3 shows. In JUNO offline software, the original detector description file is in text format. Then the offline software builds the detector structure with Geant4 and runs the simulation process. After that, it can output the geometry file in GDML format or store it in a ROOT file as ROOT geometry objects, together with the simulation data. In the new event display system, we also get the detector structure from the geometry file, read the event data from the event file generated by JUNO offline and obtain the reconstruction vertex and energy from the reconstruction file. All the data files generated by JUNO offline are in ROOT format, including the geometry, simulation, calibration and reconstruction files. Without the ROOT framework and the event data model of JUNO offline [8], the data files cannot be read by the new event display system. It is inefficient to add the whole ROOT library and JUNO data model into the new application as it would consume disk space and computing resources. It also should be noted that JUNO offline is still in the development stage and the data model may be adjusted in the coming versions. Instead, we write a ROOT macro to output the data we need into text format and then the new display system can read the text files to get the event data and detector structure. The macro will be available for users in JUNO offline. It is more efficient that we do the data conversion in the computing node, where we set up the JUNO offline software.

Geometry File -> Detector structure

Event & MC truth File -> PMT hits and simulation info

Reconstruction File -> Reconstructed vertex and energy

Figure 3. Data flow of the event display based on Unity

3.4. Interface and functions

With the data flow in figure 3, we realize three main functions in the new event display system, including the visualization of detector structure, animation of event hits and comparison between simulation and reconstruction. Figure 4 is the visualization of the JUNO detector. From outside to inside, the cylinder on the outermost layer is the water pool that contains the central detector and there are about 1700 veto PMTs in the water pool. The next layer is the central detector PMTs around the surface of the acrylic sphere (only half is shown in figure 4). And the inner layer is the liquid scintillator.

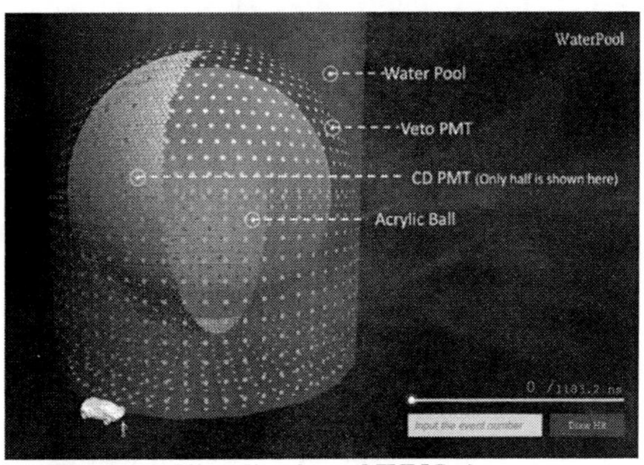

Figure 4. Visualization of JUNO detector

Figure 5 shows a PMT hit distribution in the detector. Users can input the event number they want to display in the lower right corner of the window, and then click the draw button to trigger the animation of the central detector PMTs. The circles that appear in figure 5 represent the PMTs that have already been hit and the color of the circles stands for the number of photons they receive. Users are also able to freeze time to observe the specific distribution at that moment by dragging the slider in the lower right corner. Furthermore, the system provides detailed information about detector components if users move their mouse cursor onto a specific component. Figure 5 shows the identifier, the coordinates and the first hit time of the PMT.

It also provides visualization of the reconstruction vertex and energy. Figure 6 shows the initial point, the energy deposition point and the reconstruction point. Users can observe the difference between simulation and reconstruction intuitively and use it for reconstruction algorithm tuning.

Different to the ROOT based event display, to maximize the view of users, we hide the detailed settings from the main interface. A shortcut is set up to call out the option menu of this system, where users can select the objects to display, change the lighting effect and view more information about the event file. In addition, there is also a camera system for users to observe the detector and events from different angles as they want.

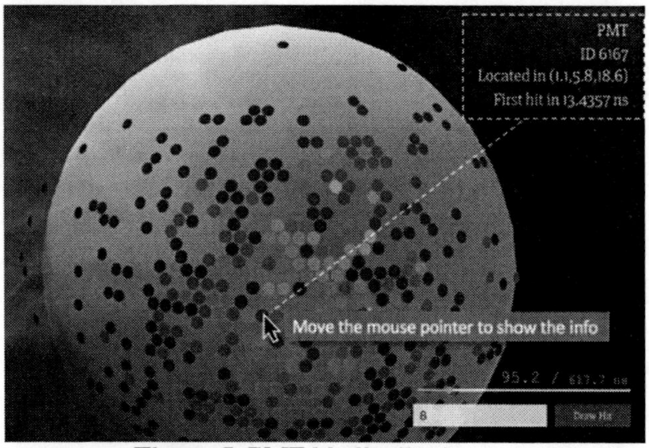

Figure 5. PMT hit distribution

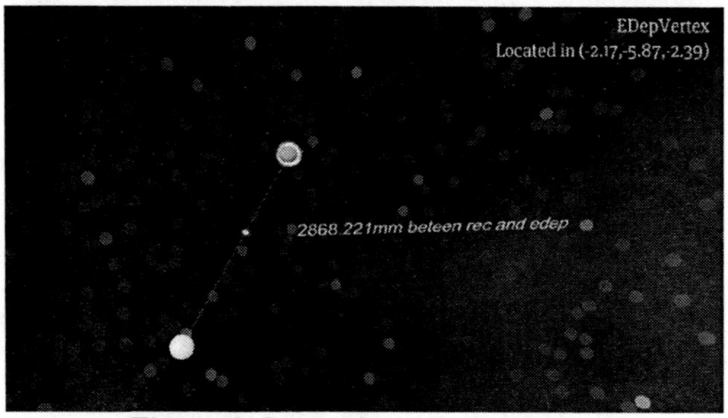

Figure 6. Comparison of vertexes

3.5. Test of portability

We have built the new event display system on different platforms to test the portability of the application developed by Unity and to see if the software can run successfully in the corresponding operating systems. Considering the operation system that the members of JUNO usually use, we have built the event display in Windows 8.1, Scientific Linux 7.2 and macOS 10.12. The application is available on these platforms. What's more, we also build it into html format and use Chrome 58.0 to run it with a lower display resolution, so it's possible for us to build an online event display system with Unity.

4. Comparison of the two event display systems

Currently there are two event display systems for JUNO. The first one, SERENA, is in the framework of ROOT and the second one is based on the Unity engine.

The event display system based on ROOT is approaching full functionality. It has realized the visualization of the JUNO detector and users can use it for reconstruction algorithm tuning and physic analysis. Compared to the Unity one, it has the advantage that it can easily read the data in ROOT format since it is in the framework of ROOT. On the other hand, SERENA is integrated in the JUNO offline software so users need to set up the whole offline software to run the event display. If the users

IOP Conf. Series: Journal of Physics: Conf. Series **1085** (2018) 032038 doi:10.1088/1742-6596/1085/3/032038

want to receive the screen of event display remotely, they need to install some other plugins. And the visual effects of SERENA are also limited by the ROOT EVE package itself.

The biggest advantage of the event display based on Unity is that it can be deployed onto different platforms quickly after development. And it is built as a client so users can run the event display on their own PCs without setting up the whole offline software. Besides, as a game engine, it is easier for the developer to realize fancier visual effects. On the contrary, the shortcoming of the system is obvious. The extra data conversion is necessary to get the event data from the ROOT format file generated by the offline software.

5. Conclusion

Considering the demands of the JUNO experiment, we have built two event display systems to meet the requirements of users. The original event display based on ROOT, SERENA, is approaching full functionality and the new event display system based on Unity is essentially available. With the Unity engine, we can easily realize better visualization effects. The new event display system is highly portable and it is possible to be upgraded into the online event display system.

The authors would like to thank the members of JUNO for their valuable discussions and suggestions, and thank the support from National Natural Science Foundation of China (11405279, 11675275) and the Strategic Priority Research Program of Chinese Academy of Sciences (XDA10010900).

References

[1] Adam T, An F, An G, et al. JUNO conceptual design report. *arXiv preprint arXiv:1508.07166*.

[2] An F, An G, An Q, Antonelli V, Baussan E, Beacom J et al. Neutrino physics with JUNO. *Journal of Physics G: Nuclear and Particle Physics*. 2016;**43**(3):030401.

[3] Zou J, Huang X, Li W, Lin T, Li T, Zhang K et al. SNiPER: an offline software framework for non-collider physics experiments. *Journal of Physics: Conference Series*. 2015;**664**(7):072053.

[4] https://root.cern.ch

[5] Tadel M. Overview of EVE – the event visualization environment of ROOT. *Journal of Physics: Conference Series*. 2010;**219**(4):042055.

[6] https://unity3d.com

[7] http://medialab.web.cern.ch/content/camelia

[8] Li T, Xia X, Huang X, Zou J, Li W, Lin T et al. Design and development of JUNO event data model. *Chinese Physics C*. 2017;**41**(6):066201.

IOP Conf. Series: Journal of Physics: Conf. Series **1085** (2018) 032039 doi:10.1088/1742-6596/1085/3/032039

Muon g-2 reconstruction and analysis framework for the muon anomalous precession frequency

K S Khaw on behalf of the Muon g-2 collaboration
University of Washington, Department of Physics, Box 351560, Seattle, WA 98195, USA
E-mail: khaw84@uw.edu

Abstract. The Muon g-2 experiment at Fermilab, with the aim to measure the muon anomalous magnetic moment to an unprecedented level of 140 ppb, has started beam and detector commissioning in Summer 2017. To deal with incoming data projected to be around tens of petabytes, a robust data reconstruction and analysis chain based on Fermilab's *art* event-processing framework is developed. Herein, I report the current status of the framework, together with its novel features such as multi-threaded algorithms for online data quality monitor (DQM) and fast-turnaround operation (nearline). Performance of the framework during the commissioning run is also discussed.

1. Introduction

The Muon g-2 experiment [1] at Fermilab aims to measure the muon anomalous magnetic moment, a_μ to an unprecedented precision of 140 ppb after several years of data taking. In the experiment, a longitudinally-polarized 3.1 GeV/c μ^+ beam will be delivered by an accelerator complex (after upgrading and modifying the existing antiproton ring) at Fermilab. This beam will be injected into the muon storage ring, which is a superconducting dipole magnet, through a superconducting inflector magnet. After traversing one-quarter of the ring, the beam is "kicked" by three pulsed magnetic kickers onto the intended storage orbit. There are four electrostatic quadrupoles installed symmetrically around the ring to focus the beam vertically. Due to the $B = 1.45$ T dipole magnetic field in the ring, the muon will undergo cyclotron motion with frequency ω_c and the spin of the muon will precess in the magnetic field with frequency ω_s. The difference between the two is the anomalous precession frequency, $\omega_a = \omega_s - \omega_c$, and it can be extracted by fitting the time modulation of the number of positrons (coming from muon decay) above an energy threshold detected by the calorimeters [2, 3] installed around the inside of the ring, with a fit function $N(t) = N_0 e^{-t/\gamma\tau_\mu}[1 + A\cos(\omega_a t + \phi)]$. Here, N_0 is the normalization factor, $\gamma\tau_\mu$ the dilated lifetime of the muon, A the average asymmetry of the decay and ϕ the initial phase. About 400 NMR probes are installed around the ring to measure the magnetic field around the ring. The field is expressed in terms of free proton's Larmor frequency in the field after convoluted with the muon beam distribution, $\tilde{\omega}_p$. a_μ can then be extracted using the formula $a_\mu = \frac{g_e}{2}\frac{\omega_a}{\tilde{\omega}_p}\frac{m_\mu}{m_e}\frac{\mu_p}{\mu_e}$, where $\frac{g_e}{2}$ is the electron g-factor, $\frac{m_\mu}{m_e}$ the muon-to-electron mass ratio and $\frac{\mu_p}{\mu_e}$ the proton-to-electron magnetic moment ratio. This article will focus on the data reconstruction and analysis framework for the ω_a frequency measurement.

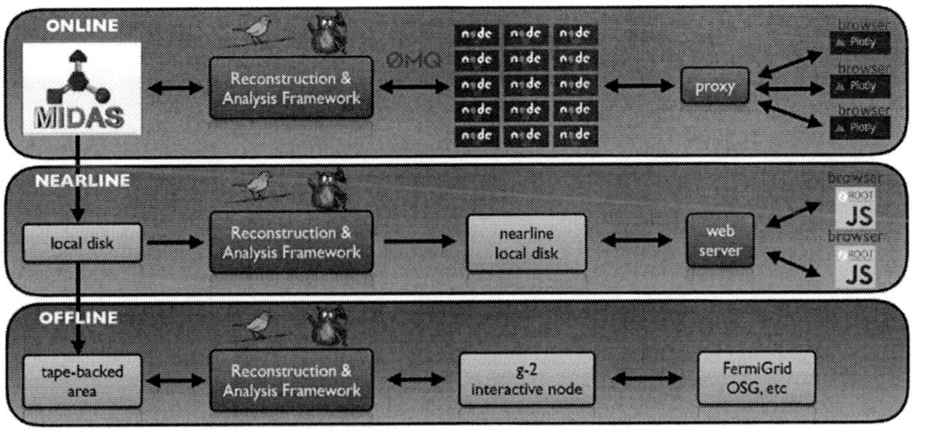

Figure 1. Data flow from online to nearline to offline for the Muon g-2 experiment.

2. Data processing flow

The Muon g-2 event building process is done using a modern data acquisition system, MIDAS [4] developed at PSI and TRIUMF. Data from twenty-four calorimeters, three tracking stations, four fiber harps, Inflector Beam Monitoring System (IBMS), T-zero (T0) detector, three kickers and so on are streamed into a backend machine running MIDAS DAQ. It is worth noting that due to the high incoming data rate (20 GB/s), calorimeter frontend machines are equipped with NVIDIA K40 GPUs to pre-select over-threshold short pulses from 54 PbF_2 crystals per calorimeter each recorded at 800 MSPS in 700 μs long raw waveforms. The DAQ system is triggered by an FC7-based clock and control system which takes input from the Fermilab accelerator trigger system. system. The average rate at which muon bunches are injected (each bunch is defined as a *fill* in subsequent paragraphs) and hence the DAQ rate is 12 Hz. The raw data is stored in the MIDAS binary format in a 40 TB RAID6 local disk before being transferred to a tape-backed area in Fermilab for permanent storage.

The data quality monitoring (DQM) system for the experiment is a custom-built framework built on top of software stacks of MIDAS, *art* [5], ZeroMQ [6], Node.js [7] and Plotly [8]. Communication between MIDAS and *art* is enabled using a custom-built module in *art* called *midastoart*. It is developed based on the MIDAS *mserver* utility. The reconstructed and analyzed data are then aggregated using the Node servers, as shown in Fig. 1(top). Finally, the data are plotted using Plotly, an online data analytics and visualization tool. A reverse proxy is also implemented to balance the query traffic.

In addition to the DQM, the experiment also has a nearline operation (see Fig. 1 (middle)) that provides live feedback regarding the quality of the data files and the quality of the reconstructed data. It also produces the latest calibration data for the DQM and the offline data processing. It operates on a multi-core machine utilizing the availability of the multi-threaded modules. Reconstructed and analyzed data are stored in the *art*/ROOT [9] format in the nearline local disk. Histograms of high-level physics variables are also stored in separate ROOT files. Users can browse the histograms through a web server running JSROOT [10].

Once the raw data files are transferred to the tape-backed area, they will be processed using a combination of FermiGrid and Open Science Grid (OSG) as shown in Fig. 1 (bottom). A web interface named POMS [11], which was initially developed by a Fermilab offline production team, is utilized to automate and to book-keep the production process.

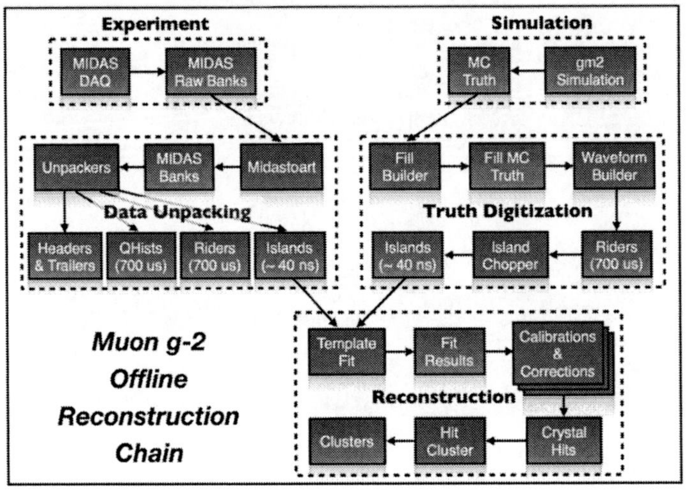

Figure 2. Data reconstruction and analysis chain implemented in the *art* framework for the ω_a measurement.

3. Data reconstruction and analysis framework

The Muon g-2 offline reconstruction and analysis chain is depicted in Fig. 2. Each stage of data processing (data unpacking, digitization, and reconstruction) consists of a series of algorithms and data products. For example, for the simulation chain, the digitization stage starts with a *fill builder* which aggregate 10^4 muon decay events and converts them into a single g-2 fill event. A waveform building module with the implementation of SiPM responses and digitizer behaviors is used to simulated the digitizer waveforms. Finally, an *island chopping* module is used to filter out over-threshold short pulses (also called *islands*) in the 700 μs long waveforms, similar to the island chopping done by the GPUs in the frontend machines. A typical reconstruction stage for the threshold-method (T-method) ω_a analysis starts with the pulse fitting (using a template pulse shape, χ^2 minimization is done using Eigen linear algebra library [12]) and ends with the hit clustering. Each algorithm will output a data product that serves as an input to the next algorithm. By having this modularity, it is very convenient to study alternative reconstruction algorithms developed by different collaborators.

The difference between simulated and experimental data in the reconstruction stage is that, in the simulation, all the PbF_2 crystals and the silicon photomultipliers (SiPMs) are assumed to be identical (e.g. in terms of crystal light yields and SiPM gains). Quality control of the crystals and the SiPMs have indicated that they could differ by about 20% each regarding light transmission and breakdown voltage. Hence in practice, additional steps have to be taken to equalize them as much as possible in the experiment. A gain calibration procedure using fast pulsed laser and neutral-density filters is implemented to extract gain calibration constants (from fitted pulse integral to the number of photo-electron, *npe*), and a laser monitoring system is used to monitor the SiPM gain fluctuations. Additional corrections involve monitoring the laser intensity fluctuation using pin diodes and photo-multiplier tubes (PMT). Finally the energy calibration constants (from *npe* to MeV) are extracted using either the minimum ionizing particle (MIP) signals from the muons or the end-point matching of the decay energy spectrum. All these calibration and correction constants are applied to the fitted energy in sequence, and the final number is a proxy for the particle energy. On the other hand, the hit time information is extracted from the template-fitting technique, and it has a resolution of about 20 ps. A pulse-pileup separation of 100% can be achieved for a time difference of larger than 5 ns between two pulses.

The ω_a offline framework also has extensive interfaces to the database of the experiment for reconstruction and analysis purposes. They are the *MIDAS online database (ODB) parser* module and the *database service* module. The ODB is written into the raw MIDAS file and includes most of the run information needed for data reconstruction. A dedicated ODB parser based on the boost library is implemented to extract information like the digitizer channel to crystal number mapping and the number of enabled digitizer channels. Calibration and correction constants extracted by *art* modules are stored in a FNAL central database for g-2. A dedicated database service that can talk to both databases is implemented using libraries like *pqxx* (Postgres C++ API) [13] and *boost::asio* (URL request library) [14].

In addition to the feature mentioned above, we have also utilized the Intel Threading Building Blocks (TBB) to parallelize data processing within an art module. This parallelization is possible because our data products are in a nested structure and data from each calorimeter is independent from each other. We have implemented such parallelization in the unpacking and the reconstruction stages. A fill event processing rate of 8 – 10 Hz is achievable using a 16-core CPU machine, and such a speed up enabled us to run DQM and nearline operation without missing many events.

4. Performance during the Summer 2017 commissioning run

During the 5-week commissioning run in Summer 2017, the event rate was about 0.15 Hz. The DAQ uptime was about 60% and the online DQM and nearline uptimes were close to 100%. Around 20 TB of data was recorded. The nearline data processing rate was limited by the ROOT I/O. To catch up with the DAQ rate, the ROOT compression level had to be reduced.

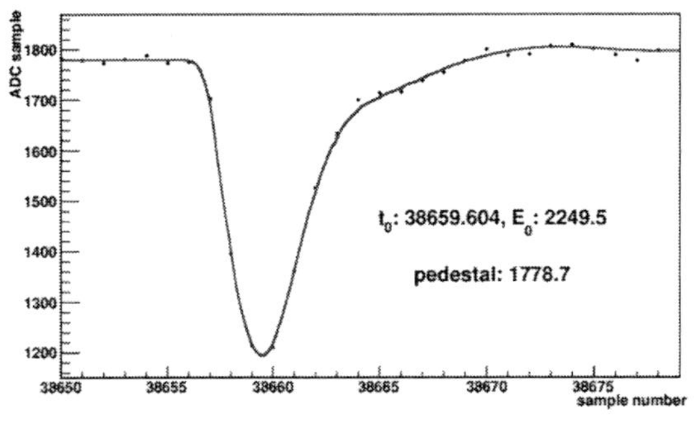

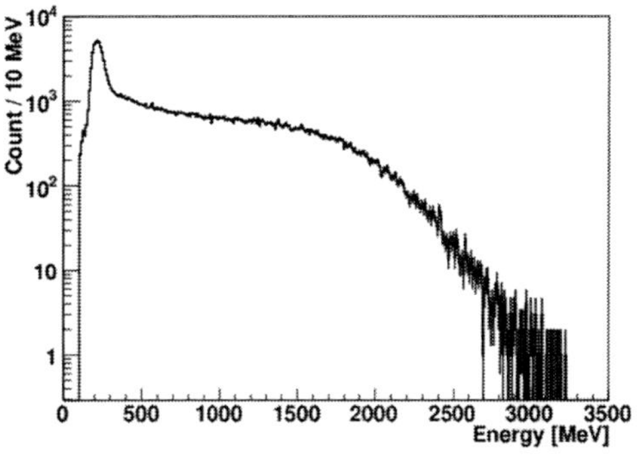

Figure 3. (left) Template fit of a digitized SiPM pulse. Red curve is the template pulse shape and black dots are the digitized samples for a typical positron impact. (right) Distribution of the reconstructed particle energy of a typical run.

As part of the upstream beamline that eliminates the proton and the pion beam was not ready during the commissioning run, the incoming beam was not purely muon but a mixture of the three species (p, π^+, μ^+). An example of the fitted pulse is shown in Fig. 3(left). A typical energy distribution of the detected particles is displayed in Fig. 3(right). The low energy peak ($\approx$ 200 MeV) is dominated by the $p/\pi^+/\mu^+$ MIP signals whereas the higher energy part is coming from mainly muon-decay positrons. A typical hit distribution is shown in Fig. 4(left). It is in good agreement with the simulation [15] where most of the hits are expected to be concentrating on the left-hand side of the calorimeter which is closer to the storage region of the beam. A preliminary result showing the modulation of the high energy positron rate is depicted in Fig. 4(right).

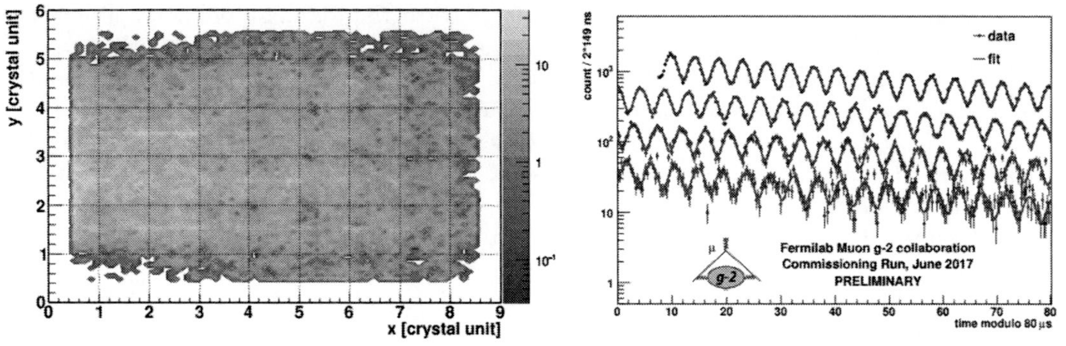

Figure 4. (left) Distribution of the reconstructed hit positions on calorimeters. (right) The number of positron with $E > 1.86$ GeV as a function of time detected by the calorimeters. The modulation is at the muon anomalous precession frequency.

5. Conclusions

In summary, the reconstruction and analysis framework for the ω_a analysis of Muon g-2 experiment is implemented into a highly-integrated system from online to offline using modern standardized tools. The raw detector data was first processed using GPU farms to reduce the incoming rate by a factor of 100, and the reconstruction and analysis of the data are performed using the algorithms implemented in the *art* framework. The reconstruction stage is a chain of modules performing pulse fitting, energy calibration and hit clustering. The system performed very well during the Summer 2017 commissioning run and is currently being optimized based on the data collected during the commissioning run.

6. Acknowledgments

The author thanks A. Fienberg for useful discussions in preparing this manuscript. This work was supported, in part, by the U.S. Department of Energy Office of Science, Office of Nuclear Physics under award number DE-FG02-97ER41020. This manuscript has been authored by Fermi Research Alliance, LLC under Contract No. DE-AC02-07CH11359 with the U.S. Department of Energy, Office of Science, Office of High Energy Physics.

References

[1] J Grange *et al.* [Muon g-2 Collaboration], Muon (g-2) Technical Design Report," arXiv:1501.06858 [physics.ins-det].

[2] A T Fienberg *et al.*, "Studies of an array of PbF_2 Cherenkov crystals with large-area SiPM readout," Nucl. Instrum. Meth. A **783**, 12 (2015) doi:10.1016/j.nima.2015.02.028 [arXiv:1412.5525 [physics.ins-det]].

[3] J Kaspar *et al.*, "Design and performance of SiPM-based readout of PbF_2 crystals for high-rate, precision timing applications," JINST **12**, no. 01, P01009 (2017) doi:10.1088/1748-0221/12/01/P01009 [arXiv:1611.03180 [physics.ins-det]].

[4] MIDAS project, "MIDAS" [software], version 2.1, available from https://bitbucket.org/tmidas/midas [accessed 2017-10-18]

[5] C Green, J Kowalkowski, M Paterno, M Fischler, L Garren and Q Lu, "The Art Framework," J. Phys. Conf. Ser. **396**, 022020 (2012). doi:10.1088/1742-6596/396/2/022020

[6] ZeroMQ project, "ZeroMQ" [software], version 4.2.2, Available from http://zeromq.org/intro:get-the-software [accessed 2017-10-18]

[7] Node.js project, "Node.js" [software], version 6.11.4, Available from https://nodejs.org/en/download/ [accessed 2017-10-18]

[8] Plotly Technologies Inc., "Collaborative data science," Plotly Technologies Inc., Montrëal, QC, 2015, https://plot.ly

[9] R Brun and F Rademakers, "ROOT: An object oriented data analysis framework," Nucl. Instrum. Meth. A **389**, 81 (1997). doi:10.1016/S0168-9002(97)00048-X
[10] JavaScript ROOT project, "JSROOT" [software], version 5.2.1, Available from https://root.cern.ch/root/htmldoc/guides/JSROOT/JSROOT.html#javascript-root [accessed 2018-04-09]
[11] D Box *et al.*, "The FIFE Project at Fermilab : Computing for Experiments," PoS ICHEP **2016**, 176 (2016).
[12] Eigen project, Eigen [software], version 3.3.4, Available from https://bitbucket.org/eigen/eigen/ [accessed 2017-10-18]
[13] pqxx project, "pqxx" [software], version 5.0.1, Available from https://github.com/jtv/libpqxx [accessed 2017-10-20]
[14] J Torjo, "Boost.Asio C++ Network Programming," http://www.packtpub.com/boost-asio-cplusplus-network-programming/book
[15] gm2ringsim project, "gm2ringsim" [software], version 7.0.4, Available from https://cdcvs.fnal.gov/redmine/projects/gm2ringsim [accessed 2017-10-20]

A Conditions Data Management System for HEP Experiments

P J Laycock[1], D Dykstra[2], A Formica[3], G Govi[2], A Pfeiffer[1], S Roe[1], R Sipos[4]
[1] CERN, [2] Fermilab, [3] CEA Saclay, [4] MTA-ELTE Lendület CMS Particle and Nuclear Physics Group, Eötvös Loránd University, Budapest, Hungary
E-mail: paul.james.laycock@cern.ch

Abstract. Conditions data infrastructure for both ATLAS and CMS have to deal with the management of several terabytes of data [1, 2]. Distributed computing access to this data requires particular care and attention to manage request-rates of up to several tens of kHz. Thanks to the large overlap in use cases and requirements, ATLAS and CMS have worked towards a common solution for conditions data management with the aim of using this design for data-taking in Run 3. In the meantime other experiments, including NA62, have expressed an interest in this cross-experiment initiative. For experiments with a smaller payload volume and complexity, there is particular interest in having simple payload storage. The conditions data management model is implemented in a small set of relational database tables. A prototype access toolkit consisting of an intermediate web server has been implemented, using standard technologies available in the Java community. Access is provided through a set of REST services for which the API has been described in a generic way using standard Open API specifications, implemented in Swagger. Such a solution allows the automatic generation of client code and server stubs and further allows changes in the backend technology transparently. An important advantage of using a REST API for conditions access is the possibility of caching identical URLs, addressing one of the biggest challenges that large distributed computing solutions impose on conditions data access, avoiding direct DB access by means of standard web proxy solutions.

1. Conditions in HEP experiments

In the context of the HEP Software Foundation community white paper [3], a cross-experiment working group of experts in conditions data met to discuss best practices in conditions data management. This short paper represents a concise summary of the recommendations of that working group, which will be documented in more detail elsewhere.

Broadly speaking, conditions data is defined as the non-event data required by data-processing software to correctly simulate, digitise or reconstruct the raw detector event data. The non-event data required to maintain, operate and optimise detectors can be broken down into the following categories:

(i) Configuration
(ii) Detector Control System (DCS), from Process Visualization and Control System (PVSS)
(iii) Monitoring
(iv) Calibrations, alignments

IOP Conf. Series: Journal of Physics: Conf. Series **1085** (2018) 032040 doi:10.1088/1742-6596/1085/3/032040

Conditions data thus largely consists of (iv), together with the subset of (i) and (ii) that are required for data processing. Other non-event data may also be required for data processing, for example machine parameters, and thus this list only attempts to give an indication of the scope. In practice, any non-event data from any source that are required for optimal data processing can be considered as conditions data. In general, conditions data vary with time but with a granularity much coarser than the event, ranging from one year to one run, down to a granularity of the order of one minute for a small subset. For those exceptional cases where granularity is even higher, great care must be taken to keep payload volumes under control. Finally, careful versioning of groups of conditions data for production use cases is a critical item to guarantee reproducibility.

1.1. Workflows
Given the definition of conditions data, the workflows that need to be supported are predominantly those of offline data processing. However, both CMS and ATLAS high level (software) triggers use the same software framework as offline data reconstruction and must also be supported, as must any workflow that requires conditions data.

A non-exhaustive list of important workflows follows:

(i) Subsystem calibration: conditions determination and testing (including the ability to access a copy of the conditions stored locally); uploading conditions to the production database.

(ii) Prompt data processing, with conditions updates adhering to strict protocols

(iii) Offline data processing, including Monte Carlo simulation, using pre-determined conditions.

1.2. Data volumes, read and write rates
Conditions data tend to scale in a controlled way during the lifetime of an experiment, typically producing data volumes of gigabyte to terabyte scale. It is worth noting that on ATLAS, where the same database instance was used for conditions data as well as the DCS and trigger data, the DCS and trigger data dominated the offline and online database instances, respectively. It is therefore strongly recommended by this working group to factorise conditions data from other use cases, even where these are related. Typically write-rates for conditions data must support of the order of 1 Hz (to ensure good support for several independent systems writing conditions data every minute), with the majority of conditions data being updated much less frequently than this. On the other hand, read-rates up to several kHz must be supported for distributed computing workflows, where thousands of jobs needing the same conditions data may start up at the same time. Conditions data are typically written once and read frequently.

2. Conditions Database Archetype
The archetypal solution to a conditions database management system (CDMS) is shown in figure 1. The conditions data payloads are stored in a master database and are accessed using a client-server design through a REST interface, achieving a high degree of separation between client and server. This allows for simple clients which all of the experiments found to be very desirable. Due to the read-rate requirements, caching is also extremely important and good experience was seen when using web-proxy caches, e.g. the Squid cache shown here. An advanced prototype called Crest exists [4] and some key design principles learned from its implementation are detailed in the following.

2.1. Payload serialisation technology
Experiments will inevitably choose their favourite payload serialisation technology. This working group recommends placing most emphasis on homogeneity and long-term maintenance when

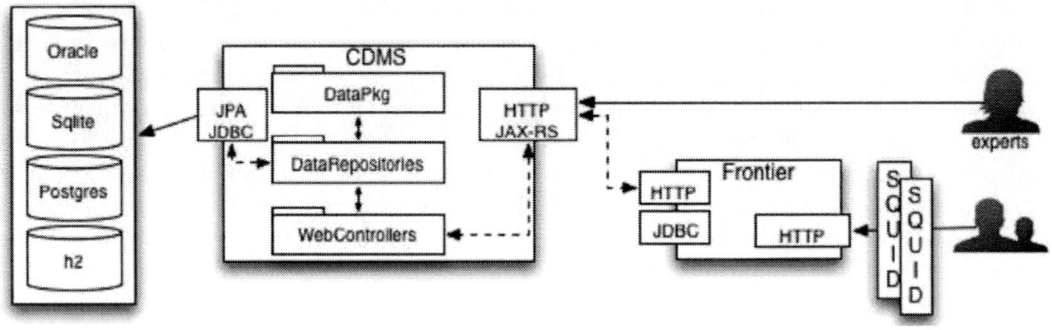

Figure 1. The conditions database archetype described in the text.

making this choice. Inhomogeneity and home-grown solutions all place additional burden on projects that typically lack the resources to support this after the initial build and commissioning phase of an experiment. CMS has very good experience of removing choice and only supporting boost-serialised C++ objects, with all classes belonging to one package in the CMSSW framework. Such a strategy lends itself more readily to long-term maintenance and minimises hurdles to data preservation. It is noted that there are payload formats used more in industry which would lend themselves more to higher-level functionality without the need of the software framework, but, while this is attractive, the choice of format tends to be driven by software framework developers.

2.2. Database backend

One of the key features of the design is that it is agnostic to particular choices of database backend. This was also one of the key features of COOL [5, 6], but due to the lack of caching in COOL, the queries themselves had to become more complicated. Thus on ATLAS, which uses an Oracle back-end, a significant amount of Oracle DBA effort was required to tune the queries and make them performant. It is therefore important to realise that real flexibility with respect to choices in database backends only comes when the system as a whole is simplified.

2.3. Client-side requirements

The client layer should be as simple as possible and should be as agnostic of the rest of the architecture as it is possible to be in order to improve maintainability. The database insertion tools in particular benefit from adopting a simple e.g. REST interface. The client layer needs to take care of payload deserialisation, as the remaining architectural components will deal with serialised objects. Experience also shows that clients should be able to manage multiple proxies and servers to provide robustness against server failures.

2.4. Caching layer

CMS and ATLAS require an intermediate layer between the client and server to provide caching capabilities. Considering the distributed use case, where thousands of jobs start at the same time and require the same conditions data, this is a clear requirement and one that can be well met using web-proxy technologies. An alternative solution would be to use a distributed filesystem with good caching capabilities, and the ALICE experiment is gaining experience using CVMFS [7]. The simplicity of this solution makes it very attractive and it has thus been adopted as a strategy by the NA62 experiment. This in turn suggests that, as a design requirement, it should be possible to represent a conditions database on a filesystem. The primary challenge is

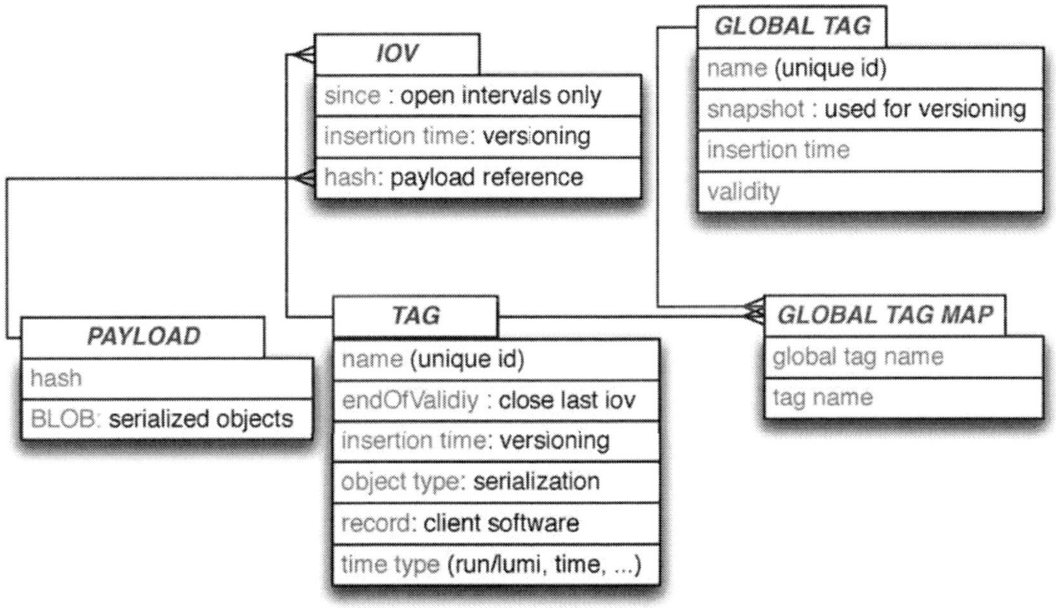

Figure 2. The data model for conditions data management described in the text.

to make the filesystem mapping use the CVMFS caching layers efficiently. Several experiments also have experience using SQLite replicas which are attractive for workflows where the exact subset of conditions is known in advance (some MC workflows), but a performant caching layer is preferable for general use cases.

2.5. Data Model

The data model for conditions data management is an area where the experiments have converged on something like a best practice shown in figure 2. A global tag is the top-level configuration (version) of all conditions data needed for a particular data processing use case. For a given system and a given interval of validity, a global tag will resolve to one, and only one, conditions data payload. The Global Tag resolves to a particular system Tag via the Global Tag Map table. A system Tag consists of many intervals of validity (IOV) or entries in the IOV table. Finally, each entry in the IOV table maps to a payload via its unique hash key in the Payload table. A relational database is a good choice for this design.

This design has several key features. Firstly, conditions data payloads are uniquely identified by a hash which is the sole reference to any given conditions data payload. The payload data has been separated from the data management metadata and could in principle be placed in a separate storage system. This could also be important for data preservation, as the entire metadata component will occupy a trivial data volume and could exist in e.g. an SQLite file, while the payload storage could be handled separately. Secondly, IOVs are resolved independently of payloads and are also cacheable. Efficient caching is a key design requirement for any conditions database that must support high rate data access.

2.6. A git-based approach

For workflows which are completely offline and asynchronous with respect to data-taking, LHCb has adopted a different approach. Using git as the versioning system, conditions are placed in a directory structure, one for each type of condition. A file is used to map timestamps to payload

IOP Conf. Series: Journal of Physics: Conf. Series **1085** (2018) 032040 doi:10.1088/1742-6596/1085/3/032040

files, and a simple format is used to allow a level of indirection to improve performance. With the algorithm used, payload look-up on timestamp is linear within the file, so large numbers of IOVs cause performance issues; the file format allows a timestamp to point either directly to a payload file or to a directory, thus allowing partitioning of the lookup. Versioning is then taken care of by creating a git tag, equivalent to a global tag.

3. Future and roadmap

The Conditions Data Management System described here is expected to meet the needs of CMS and ATLAS into the HL-LHC era [8]. Data volumes are not expected to exceed a terabyte per year, and the rate of requests (determined by the computing resources of the experiments) are expected to peak at tens of kHz. Assuming the solutions presented here work in Run 3, they are expected to continue to work beyond that. The most important issue for those experiments will be maintenance and operation in the face of evolving hardware and infrastructure, which makes consolidation to a simple and modular design, such as that presented here, crucial for the experiments' continued success. Full scale tests of this design are expected in the coming years, but based on experience with similar systems (the current CMS approach) the outlook is positive. Nevertheless, there are still open questions that need to be addressed. The functionality to produce a filesystem-based replica of a conditions database, that can be accessed transparently to the client, needs to be prototyped. The NA62 experiment will look at this problem in reverse, moving from a filesystem-based approach to a relational database, requiring a suitable filesystem layout. Meanwhile, on a time-scale of Run 3, the large data rates will force LHCb and ALICE to only reconstruct data online. While they will potentially be more exposed to data quality losses, their offline conditions data access requirements are greatly reduced.

3.1. Conditions for analysis

Despite the plans for LHCb and ALICE to be effectively conditions-free, there will nevertheless be higher-level calibrations applied to the physics data, and this will require some management. Belle II expects that this will be part of the analysis model from the outset, and there is wider interest in the analysis community in using systems like those described here to deliver analysis conditions data. Given the plans of Belle II, and the growing interest in the analysis community, feedback on the suitability of these solutions is expected within the next few years.

4. Conclusion

Following a decade of experience with the LHC experiments, conditions data management is an important component of HEP software and one that is often considered relatively late in the software design cycle. While the data volumes are easily accommodated by several database technologies, the workflows can be demanding. In particular, access rates at the level of tens of kHz is critical (and non-trivial) to support, and the importance of a caching system like Frontier [9, 10] or CVMFS, in particular for CMS and ATLAS, cannot be overstated.

Several experiments have converged on similar solutions for solving conditions data management problems, based on the good performance of the CMS approach. All of the experiments want a high degree of separation between the client and server side of the problem, with a client that is relatively simple (in contrast to the original COOL solution). An intermediate layer with caching capability is needed to support high rate requests and web-proxies have performed very well. REST interfaces and industry standard components, together with a relational data model to represent the global tag concept, can support the wide variety of workflows used in HEP while being sufficiently modular to evolve easily. For purely offline workflows, CVMFS is attractive and could be produced from the master database. This would resemble the git-based approach used by LHCb, where the IOV structure is encapsulated by a simple file. Analysis conditions data management could also benefit from using these solutions.

IOP Conf. Series: Journal of Physics: Conf. Series **1085** (2018) 032040 doi:10.1088/1742-6596/1085/3/032040

References

[1] Barberis D, Formica A, Gallas E J, Govi G, Miotto G L and Pfeiffer A 2015 *J. Phys. Conf. Ser.* **664** 042015
[2] Guida S D, Govi G, Ojeda M, Pfeiffer A and Sipos R 2015 *J. Phys. Conf. Ser.* **664** 042024
[3] Alves Jr A A *et al.* 2017 (*Preprint* 1712.06982)
[4] *Crest homepage* https://github.com/HSF/Crest
[5] Valassi A, Basset R, Clemencic M, Pucciani G, Schmidt S A and Wach M 2008 *Proceedings, 2008 IEEE Nuclear Science Symposium, Medical Imaging Conference and 16th International Workshop on Room-Temperature Semiconductor X-Ray and Gamma-Ray Detectors (NSS/MIC 2008 / RTSD 2008): Dresden, Germany, October 19-25, 2008* pp 3021–3028
[6] *Persistency homepage* https://twiki.cern.ch/twiki/bin/view/Persistency
[7] *CVMFS homepage* https://cernvm.cern.ch/portal/filesystem
[8] *DB Futures Workshop* https://indico.cern.ch/event/615499/timetable/?view=standard
[9] Dykstra D 2011 *J. Phys. Conf. Ser.* **331** 042008
[10] *Frontier homepage* http://frontier.cern.ch/

The LHCb Online system in 2020: trigger-free read-out with (almost exclusively) off-the-shelf hardware

T Colombo[1], A Amihalachioaei[1,2], K Arnaud[3], F Alessio[1], L Brarda[1], J-P Cachemiche[3], D Cámpora[1,4], S Cap[5], L Cardoso[1], F Cindolo[6], M Daoudi[1], P Durante[1], P-Y Duval[3], C Faerber[1], P Fernández[1], M Frank[1], D Galli[6], C Gaspar[1], F Hachon[3], M Jevaud[3], B Jost[1], R Le Gac[3], M Manzali[7], U Marconi[6], H Mohammed[1], N Neufeld[1], F Pisani[1,6], L Promberger[1,8], C Quast[1], F Réthoré[3], F Sborzacchi[9], R Schwemmer[1], S T'jampens[5], S Valat[1], J Viana Barbosa[1], B Vőneki[1] and G Vouters[5]

[1] CERN, Geneva, Switzerland
[2] Gheorghe Asachi Technical University, Iași, Romania
[3] Centre de Physique des Particules de Marseille (CPPM), Marseille, France
[4] Universidad de Sevilla, Sevilla, Spain
[5] Laboratoire d'Annecy-le-Vieux de Physique des Particules (LAPP), Annecy-le-Vieux, France
[6] Università di Bologna and Sezione INFN, Bologna, Italy
[7] Università di Ferrara and Sezione INFN, Ferrara, Italy
[8] Hochschule Karlsruhe – Technik und Wirtschaft, Karlsruhe, Germany
[9] Laboratori Nazionali di Frascati dell'INFN, Frascati, Italy

E-mail: `Tommaso.Colombo@cern.ch`

Abstract. The LHCb experiment at CERN has decided to optimise its physics reach by removing the first level hardware trigger for 2020 and beyond. In addition to requiring fully redesigned front-end electronics this design creates interesting challenges for the data-acquisition and the rest of the online computing system. Such a system can only be realized within realistic cost using as much off-the-shelf hardware as possible. Relevant technologies evolve very quickly and thus the system design is architecture-centred and tries to avoid to depend too much on specific technologies. In this paper we describe the design, the motivations for various choices and the current favoured options for the implementation, and the status of the R&D. We will cover the back-end readout, which contains the only custom-made component, the event-building, the event-filter infrastructure, and storage.

IOP Conf. Series: Journal of Physics: Conf. Series **1085** (2018) 032041 doi:10.1088/1742-6596/1085/3/032041

1. Introduction

LHCb [1] is one of the four major experiments at CERN's Large Hadron Collider (LHC). It is a single-arm forward spectrometer, designed for precision measurements of B-meson decays. In 2019 and 2020 the LHCb experiment will be substantially upgraded in order to reach unprecedented precision on the main observables of the b and c-quark sectors. Critical parts of this upgrade are the trigger-less readout system and the full software trigger.

The experiment observes proton-proton inelastic collision events delivered by the LHC at a rate of 30 MHz. Due to limitations in the current readout system, this rate must be reduced to 1.1 MHz using a fixed-latency hardware trigger system [2], which can only base its selection on basic physics signatures available at this stage (high-energy or high-momentum particles). The selection is further refined by a software event filter (also called high-level trigger).

The hardware trigger is currently the largest source of event selection inefficiencies for the experiment. Therefore, one of the main goals of the LHCb upgrade is its replacement with a full software event filter. This requires a new trigger-less readout system, a new event builder, and a substantially upgraded event filter system in order to acquire, assemble, and select events at the full rate of 30 MHz. The upgrade enables applying event selections that are as similar as possible to those used in offline analyses, so that selection efficiencies are maximised and systematic uncertainties are minimised [3].

The building blocks of the new system are shown in Figure 1. The front-end electronics of the various detectors are synchronised with the LHC collision frequency by the timing and fast control system (TFC). After zero-suppression, the acquired data fragments are pushed to around 500 event builder servers via almost 10000 point-to-point optical links. Event builder nodes are all interconnected by a dedicated high-performance network used to assemble event fragments into full events. They are divided in small groups, with each group sending the events it assembles to a subdivision of the event filter computer farm (called sub-farm). The event filter nodes run the selection algorithms. They have access to a large disk-based temporary storage for events, that is used to relax the throughput requirements on the event filter and to maximise the farm utilisation.

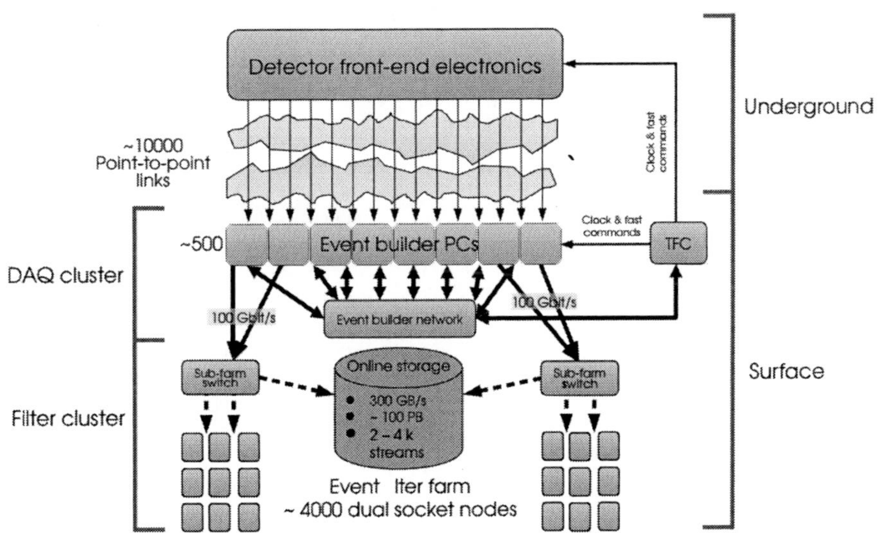

Figure 1. Layout of the new LHCb DAQ & event filter.

IOP Conf. Series: Journal of Physics: Conf. Series **1085** (2018) 032041 doi:10.1088/1742-6596/1085/3/032041

2. Readout, timing, and fast control

Each detector in the upgraded experiment will be equipped with front-end electronics capable of acquiring its signals at the LHC maximum collision rate of 40 MHz [1]. The interface between the detectors and the readout system consists of almost 10000 simplex optical links. The links use a custom protocol (GBT [4]) developed to be implementable with radiation-hardened electronics, a necessary feature to operate in close proximity with the detectors. Each optical link can carry up to 4.48 Gb/s of data from the underground experimental area to the surface where it is received by a custom PCI Express readout board, called TELL40, housed in an event-builder node. The received event fragments are buffered in the node's main memory waiting to be assembled into a full event. A dedicated supervisor board, called SODIN, centrally manages the readout of events. It distributes the LHC clock to the rest of the system, it ensures that all front-ends and readout boards are synchronous, and it regulates the data-flow from the front-ends to the readout boards. Dedicated interface boards, called SOL40, fan-out the timing and synchronisation commands from the supervisor board to the front-ends and readout boards, while also acting as the interface between the front-ends and the experiment's control system. The flow of event data and commands is shown in Figure 2. More details on the clock and timing distribution are available in [5].

The functionality of the readout boards (TELL40), supervisor board (SODIN), and interface boards (SOL40) is actually implemented on the same hardware platform: a PCI Express Gen 3.0 x16 add-in card, shown in Figure 3, based on an Intel Arria 10 FPGA and high-density optical

[1] In LHCb, one every four "collision" events delivered by the LHC is actually empty, so the real inelastic collision rate corresponds to the 30 MHz quoted in section §1.

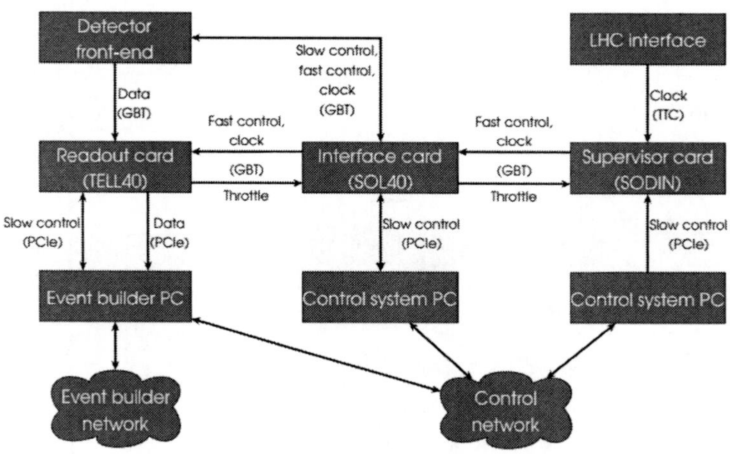

Figure 2. Readout, timing, and fast control data-flow.

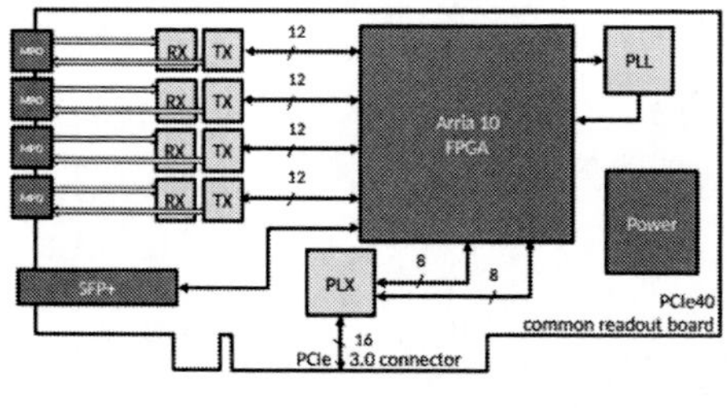

Figure 3. The common LHCb readout, timing, and control board.

I/O with up to 48 duplex GBT ports. The functional differentiation is achieved by changing the card's firmware. These cards are the only pieces of custom-built hardware in the system. In order to read out the entire LHCb experiment, 500 readout boards, less than a hundred interface boards, and a few supervisor boards are necessary. Having all three functions share a single hardware platform lowers their production and maintenance costs.

In its readout incarnation, the FPGA is programmed with a high-performance DMA engine capable of copying the received event fragments into the main system memory at over 100 Gb/s. More details on its architecture and performance are available in [6, 7].

3. Event builder

The main goal of the upgraded LHCb data-acquisition system is enabling the event filter to efficiently select events based on the full data readout from the detectors. This requires designing an event builder system capable of collecting and aggregating data fragments from the various detectors at the full collision rate of 30 MHz. The data size of a full event is projected to be as high as 150 kB. Therefore, taking into account protocol overheads and other potential inefficiencies, the design goal for the new event builder was set to a sustained throughput of up to 40 Tb/s.

An event builder is essentially made up of three components: the readout units, the builder units, and the network interconnecting them. Each readout unit receives the event fragments from the detector point-to-point links and makes them available on the network. For each event, one of the builder units gathers all the event fragments from all the readout units, assembles them into a complete event, and makes it available to the event filter. Since the data flows in a single direction (from the readout units to the builder units), it is possible to double the network utilisation and halve the total number of nodes in the event builder system by "folding" the builder units onto the readout units. Essentially, a single event builder node can host both a readout unit, which outputs event fragments onto the network, and a builder unit, which receives event fragments from the same network. An obvious problem with this architecture is that, if the event builder network is already close to saturation, a separate network is necessary to transfer the complete events from the builder units to the event filter nodes. However, this is not necessarily a disadvantage: being single-purpose, the event-builder network can be based on any high-performance network technology: the choice depends solely on the price/performance ratio for this specific application. Moreover, the traffic pattern is known in advance and it can be tuned to maximise network usage. For these reasons, the development effort for the new LHCb event builder is focused on the folded architecture.

In principle, the basic readout/builder unit can be implemented by a standalone FPGA-based readout board equipped with the appropriate number of inputs, to receive data from the experiment and from the event-builder network, and outputs, to send data to the event-builder network. Ostensibly, dispensing with the requirement for a server to host the readout boards would lead to significant cost savings. As a matter of fact, this solution comes with significant limitations. High-performance network technologies, such as InfiniBand and OmniPath, or a full TCP/IP over Ethernet stack are generally too complex or expensive to implement on a FPGA. While there are numerous commercially available integrated circuits implementing these technologies, these are commonly designed to interface with a complete computer and operating system via PCI Express links. Moreover, purely FPGA-based network stacks must be relatively limited, to save on FPGA resources and complexity, and most likely cannot include a network congestion control mechanism. This must be compensated by adequately large buffers in the network switches interconnecting the boards. Deep-buffer network switches are a niche product and as such tend to be very expensive compared to their shallow-buffered counterparts. These and other considerations, explained in [8], led to the adoption of an event builder architecture based on commercially available servers and network interfaces.

The basic building block for the implementation is an event-builder node housing three PCI Express x16 add-in cards: a TELL40 readout board, a 100 Gb/s network adapter for the event builder network, and another 100 Gb/s network adapter to send the assembled events to the event filter nodes for processing. The total number of nodes is essentially determined by the number of TELL40 readout boards needed by the various detectors, currently estimated to be around 500. In order to reach the required 40 Tb/s aggregate throughput, each node must be able to sustain a throughput of at least 80 Gb/s. Two measurements are crucial to achieving that goal: internal node throughput and aggregate network throughput.

Each node must be capable of handling a 80 Gb/s flow from the TELL40 to its main memory and from there to the event-builder network adapter, plus another 80 Gb/s flow from the event-builder network adapter to the main memory and from there to the event filter network adapter. This is certainly possible on current-generation server hardware, as shown in [9]. However, at those rates, network protocols implemented in software, as is usually the case for the TCP/IP suite for example, introduce a very significant CPU usage both for packet processing and kernel-space/user-space data copies [9]. Both these problems can be avoided using hardware-assisted remote DMA (RDMA) technologies. The InfiniBand and OmniPath network stacks support this by design; many Ethernet network adapters also support either iWARP or RoCE, two competing approaches to RDMA over Ethernet.

Having established that a single node is capable of sustaining the required throughput is not enough: the event-builder network must be able to sustain the aggregate traffic to and from the nodes. The traffic pattern is particularly challenging. To maximise throughput, every node builds a different event in parallel with all the other nodes. This results in a continuous network-wide scatter/gather operation: the readout unit on each node scatters the fragments corresponding to each event to their respective builder units; the builder unit on each node gathers all the fragment corresponding to a single event from the readout units that have them. Naturally, a bottleneck-free network would be best suited to handle this all-to-all traffic. Ideally, if the nodes are partitioned in two arbitrary halves, the bandwidth available between the partitions should always correspond to the sum of the bandwidths of one half of the nodes. The most commonly network topology with this property is the so-called fat tree: a multi-layer network with equal link capacity between the layers, implemented by building a tree with multiple roots. The simplest incarnation of a fat-tree network, consisting of two layers of 4-port switches, is shown in Figure 4.

Even in the absence of bottlenecks, maximising the network usage is not an easy task: collisions, i.e. two or more senders trying to use the same network link at the same time, cause inevitable performance degradation. An extensive measurement campaign is currently underway to demonstrate the feasibility of a 40 Tb/s event builder with a fat-tree network. Fortunately, many existing supercomputers employ a fat-tree network, so a proof-of-concept requires only the development of event builder benchmark software, but no hardware investment. Two such benchmarks are currently in development: DAQPIPE [9] and LSEB [10]. They can probe a large parameter space, including varying communication sizes, number of in-flight communications,

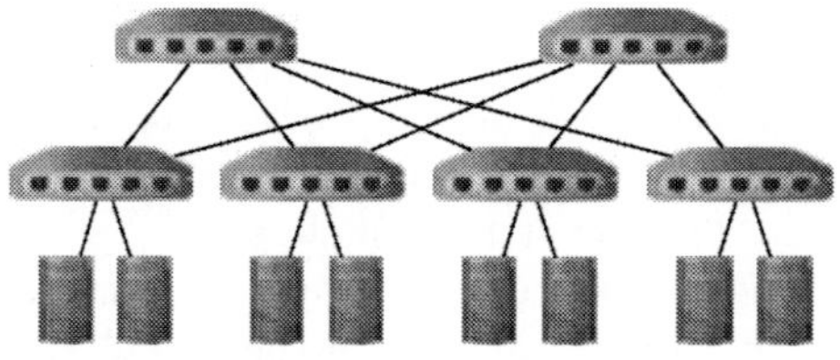

Figure 4. Fat tree network with 4-port switches.

and different communication scheduling patterns. Among those, the most relevant

The results of the largest-scale test performed so far, on a 72-node partition of an 100 Gb/s InfiniBand-based supercomputer with 18 nodes per switch, are shown in Figure 5. Two different communication scheduling patterns were used: a "shift" pattern (each node H_i sends a block of data to node H_{i+1}, then the next block to H_{i+2}, and so on until all other nodes have received data from H_i), and a "random" pattern (where each node sends a block of data to a randomly chosen destination until all other nodes have received data from it). With the random pattern, and an optimised selection of its other parameters, the DAQPIPE benchmark can achieve a per-node bandwidth of around 72 Gb/s, 8 Gb/s short of the stated objective of 80 Gb/s per node.

While the benchmark results are promisingly close to the goal, they also cast a doubt on the scalability of the system: Figure 5 shows that, as the number of nodes in the system increases, the per-node throughput decreases. This is most likely due to the increased probability of communication collisions as the network grows. Performing a continuous scatter/gather collective operation on a fat-tree network while completely avoiding collisions is possible [11], but requires all nodes to schedule their communications in a synchronised fashion. A proof-of-concept test of this idea was performed on 32 nodes interconnected by a 1 Gb/s Ethernet fat-tree network with 4 nodes per switch. With the nodes' clocks synchronised using the NTP protocol and a scheduling granularity of 100 ms, large enough to absorb the effects of operating system jitter, each node can reach 90% of the maximum throughput of its network interface. These results suggest that this distributed scheduling is a viable way of increasing the event builder's performance. However, the tests should be repeated at a larger scale with high-performance network interfaces to fully validate the concept.

4. Event filter

The event filter computing cluster will consist of as many as 4000 dual-processor servers. It will employ a two-level filtering strategy already in use and validated with the current experiment. A first filter level, called HLT1, will perform a fast reconstruction and selection of the events at the full data-acquisition rate of 30 MHz. It is expected to reduce the event rate to around 1 MHz. The events selected by HLT1 will be saved to a temporary disk buffer. A second filter level, called HLT2, will then asynchronously read the saved events and perform a full reconstruction and selection, reducing the event rate to a final output of around 100 kHz.

The intermediate disk buffer allows the maximal exploitation of the event filter cluster. The LHC isn't constantly providing collisions: after a collision period that can last up to a day, a few hours are needed to prepare for the next collision period. Therefore, when the accelerator is

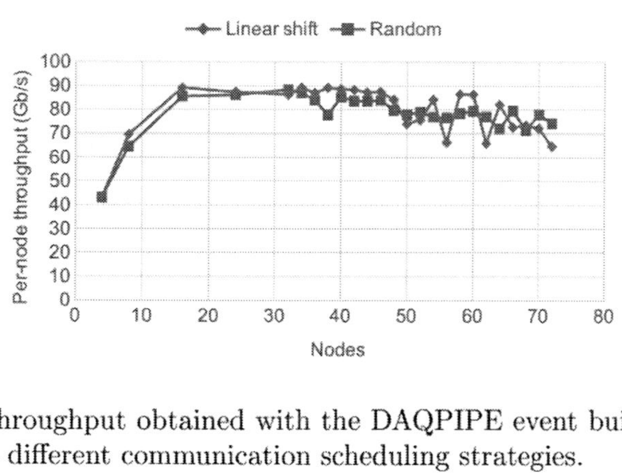

Figure 5. Per-node throughput obtained with the DAQPIPE event builder benchmark on up to 72 nodes, using two different communication scheduling strategies.

IOP Conf. Series: Journal of Physics: Conf. Series **1085** (2018) 032041 doi:10.1088/1742-6596/1085/3/032041

running, most of the servers will be dedicated to HLT1 to ensure a timely handling of the new experimental data. During the LHC downtime, all the nodes will instead run HLT2 to finish processing the fraction of events that couldn't be handled in real-time. This strategy keeps the event filter cluster occupied at all times, but requires a large, high-performance storage system: it must sustain a maximum throughput of 150 GB/s in input and 150 GB/s in output. Moreover, to absorb long LHC collision periods, its capacity should be in the order of tens of petabytes. Its definitive implementation is not yet defined and will be driven by cost considerations.

5. Conclusion

The LHCb experiment will undergo an ambitious upgrade in 2019 and 2020. Its readout system will be upgraded to support trigger-less data acquisition at 30 MHz, which will require an event builder system capable of a throughput of 40 Tb/s.

The execution of the upgrade plans is proceeding well. The readout boards, their firmware, and associated control software are already well advanced in development. The event builder benchmarks present no show-stoppers and tests suggest that there is room for further improvement. Implementation evaluations are underway for the event filter cluster and its associated network and storage.

Big implementation challenges remain. The upgrade project is entering its most interesting phases yet: the finalisation of designs and finally the construction of the system.

References

[1] LHCb Collaboration 2008 *JINST* **3** S08005
[2] LHCb Collaboration 2003 *LHCb trigger system Technical Design Report* CERN-LHCC-2003-031 (Geneva: CERN)
[3] LHCb Collaboration 2014 *LHCb Trigger and Online Upgrade Technical Design Re*port CERN-LHCC-2014-016 (Geneva: CERN)
[4] Moreira P, Marchioro A and Kloukinas K 2007 *Proc. Top. Workshop Electron. Part. Phys.* (Geneva: CERN) p 332
[5] Alessio F *et al* 2015 *JINST* **10** C02033
[6] Bellato M *et al* 2014 *J. Phys.: Conf. Ser.* **513** 012023
[7] Durante P *et al* 2015 *IEEE Trans. Nucl. Sci.* **62** 1752
[8] Schwemmer R and Neufeld N 2015 *IEEE Trans. Nucl. Sci.* **62** 1747
[9] Valat S *et al* 2017 *IEEE Trans. Nucl. Sci.* **64** 1480
[10] Manzali M *et al* 2017 *IEEE Trans. Nucl. Sci.* **64** 1486
[11] Zahavi E *et al* 2011 *Concurr. Computat.: Pract. Exp.* **22** 217

IOP Conf. Series: Journal of Physics: Conf. Series **1085** (2018) 032042 doi:10.1088/1742-6596/1085/3/032042

Using containers with ATLAS offline software

J Elmsheuser[1], L Heinrich[2], G Stewart[3], M Vogel[4] on behalf of the ATLAS Collaboration

[1] Brookhaven National Laboratory, PO Box 5000, Upton, NY 11973, USA
[2] New York University, 70 Washington Square South, New York, NY 10012, USA
[3] University of Glasgow, University Avenue, Glasgow Metropolitan Area G12 8QQ, UK
[4] University of Wuppertal, Gausstrasse 20, 42119 Wuppertal, Germany

E-mail: `marcelo.vogel@cern.ch`

Abstract. This paper describes the deployment of ATLAS offline software in containers for software development. For this we are using Docker, which is a lightweight virtualization technology that encapsulates a piece of software inside a complete file system snapshot. The deployment of offline releases via containers removes the strict requirement of compatibility between the runtime environment needed for job execution and the configuration of worker nodes at computing sites. If these two are decoupled from each other, sites can upgrade their nodes whenever and however they see fit. In this work, ATLAS software is distributed in containers either via the CernVM File System (CVMFS) or by means of a full ATLAS offline release installation. In software development, separating the build and runtime environment from the development environment allows users to take advantage of many modern code development tools that may not be available in production runtime setups like SLC6. It also frees developers from depending on resources like lxplus at CERN, allowing the use of an average laptop for ATLAS code development. We include in this document a basic comparison in performance of the two deployment options in two popular host operating systems: Ubuntu and OS X.

1. Introduction

Docker is a light weight, software based virtualization technology [1], which encapsulates a piece of software inside a complete file system that contains everything that applications need to run: code, runtime, system tools, system libraries, etc. (see Figure 1, and 2). Processes run natively in the host yet isolated in containers, differentiating Docker from virtual machines, which simulate both hardware and software. This isolation guarantees that the software will always run in the same way regardless of the host's configuration, thus providing a uniform runtime environment ideal for distributed computing. The ATLAS Computing Group of the ATLAS Collaboration [2] is actively engaged in the deployment of container technologies for software development.

2. ATLAS software in Docker images

The ATLAS Offline Software Group currently hosts a fairly comprehensive body of documentation with detailed instructions for building, running and managing containers with ATLAS software [3]. At the core of the 'containers' section of the documentation is the so-called Dockerfile, which is a configuration file used by the Docker engine to build images in a completely automatic way. In a nutshell, Dockerfiles consist of a set of sequential instructions with each one adding a new layer to the image (see Figure 3).

IOP Conf. Series: Journal of Physics: Conf. Series **1085** (2018) 032042 doi:10.1088/1742-6596/1085/3/032042

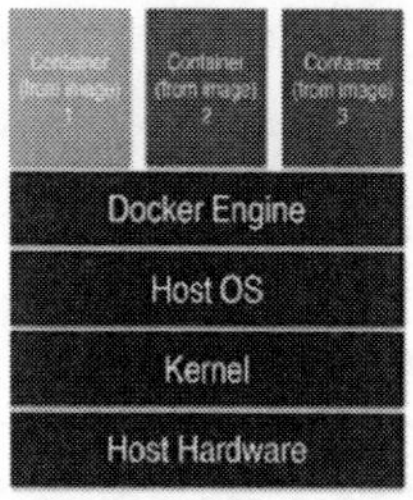

Figure 1. Docker will share the host OS kernel, but isolate different environments or 'containers' with respect to data, code, networking, etc.

Figure 2. Docker images can be thought of as archived, executable file system snapshots.

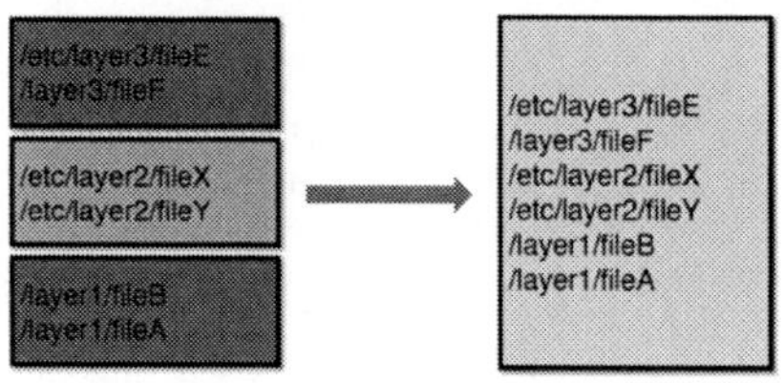

Figure 3. Docker images are built from layers.

In the containers sections of the ATLAS Software Documentation, two workflows are described for deploying ATLAS software in containers running images built via Dockerfiles. In the first option, a light-weight docker image with a basic Scientific Linux OS (see Figure 4) delivers ATLAS software through mounted CVMFS volumes [4]. In the second option, a complete release is installed in a docker image, which only depends on external CVMFS volumes for conditions and geometry data (see Figure 5). The CVMFS volume shown in Figures 4 and 5 is bind mounted to unprivileged containers via the CVMFS Docker Volume plugin supported by CERN IT [5]. CVMFS is installed in the host system where the cache can be shared across containers.

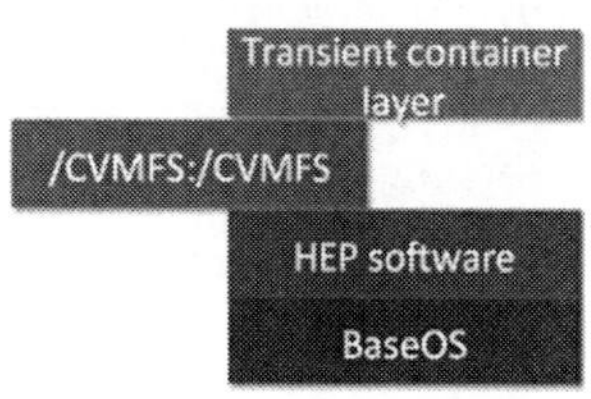

Figure 4. Light image (~2GB). ATLAS software available through CVMFS.

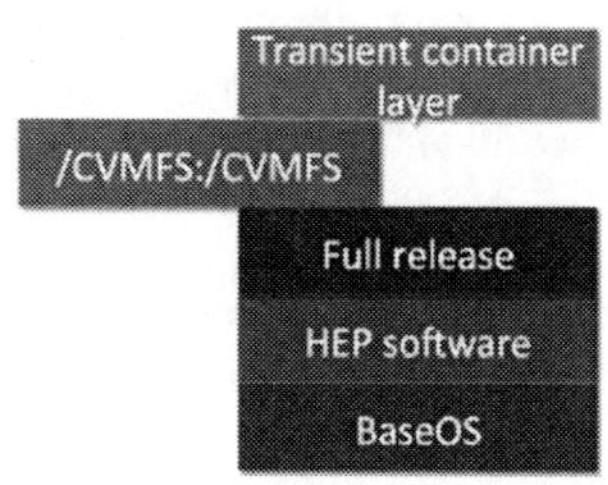

Figure 5. Full release (~23GB). CVMFS is used only for conditions data.

The ATLAS software documentation provides detailed information on how to install Docker on a personal Mac and Ubuntu machine. The deployment option (light vs. full-release) depends of course on the particular use case at hand. The light image is best suited for raw data reconstruction, which relies heavily on CVMFS for conditions and calibrations; while the stand-alone image is best fitted for environments with no connectivity to the Internet and for analysis and simulation jobs. External mounts guarantee small image sizes, but make it difficult to add a user code layer. On the other hand stand-alone images are large, but provide the best encapsulation of software. Given their individual advantages, they both represent the current strategy for delivering ATLAS offline and/or analysis software on more modern OS platforms than Scientific Linux 6. In fact, there is currently an effort underway to produce the light Docker image as part of the nightlies production system.

Dockerfiles corresponding to the two build options shown in Figures 4 and 5 can be obtained from the ATLAS Gitlab docker repository using the 'git clone' command:

```
git clone https://:@gitlab.cern.ch:8443/atlas-sit/docker.git
```

As mentioned before, the Docker engine is instructed through the Dockerfile to execute instructions in a command-line fashion. The same could be done manually, of course, as when installing packages directly on a Linux box; however, a Dockerfile facilitates image building by providing a clean and standardized way of installing and configuring software on top of more basic images and/or prebuilt layers. The light image option differs little from the current strategy for delivering experiment-specific software to sites connected to the WLCG, that is, through mounted CVMFS volumes. The most important difference is that the whole runtime comes packaged in a container, which can in principle run on any minimally configured host. On the other hand, in the case of the stand-alone option, the ultimate goal is to provide complete independence from the network and network file systems. In this case the Dockerfile executes the commands necessary to build and configure an entire release from RPMs, much like the releases served through CVMFS are built. We currently provide examples of both types of images in a public registry at Docker Hub [6]. To pull the images, execute:

```
docker pull atlas/atlas_external_cvmfs # for the light image
docker pull atlas/athena:<RELEASE NUMBER> # for the full-release image
```

whether the light or full-release option is used, the resulting image can be used for software development by first cloning the Git Athena repository inside the container, and then follow the standard development workflow described in the ATLAS Software Documentation [7]. This tutorial describes the current framework to install packages, which is based on the Git version control system and the CMake build system.

3. Performance of Containers Running ATLAS Software

We document here a basic comparison in performance of the two deployment options discussed in section 2 in two platforms frequently used by developers: Ubuntu and OS X. The metrics measured are the wall-time and the CPU-time during a raw data reconstruction and a full simulation job. For each type of transformation the job is executed using both the light image, where ATLAS software is available through CVMFS, and the full-release image, where CVMFS is only used for conditions and geometry. As a reference, the jobs are also run on bare metal on a SLC6 test-bed machine.The general technical specifications of the machines used in the tests are listed in Table 1.

The reconstruction test job ran on 25 raw data events. Figures 6 and 7 show the performance of containers relative to bare metal when using the light image and the full-release image, respectively. Table 2 summarizes the results of the reconstruction test. The full-simulation test job ran on 5 events. Figures 8 and 9 show the performance of containers relative to bare metal, and Table 3 summarizes the results of the full-simulation test. All measurements were taken with a 'warm' CVMFS cache.

IOP Conf. Series: Journal of Physics: Conf. Series **1085** (2018) 032042 doi:10.1088/1742-6596/1085/3/032042

Table 1. Technical specifications of testing systems.

Test Machine	Description
`Mac laptop`	MacBook Pro, i7 2.2 GHz, macOS Sierra 10.12.6 Docker for Mac 17.06.0 configured for 2 cores and 4GB CVMFS 2.3.3, FUSE 3.5.4
`Linux desktop`	Intel Core 2 DUO CPU E7500, 2.93GHz, Memory 3.6GB Ubuntu 14.04, Docker 17.05.0, CVMFS 2.3.5
`Linux test-bed`	Intel Xeon CPU E5645 2.40GHz, Memory 2.4GB SLC6 6.8, bare metal, CVMFS 2.4.1

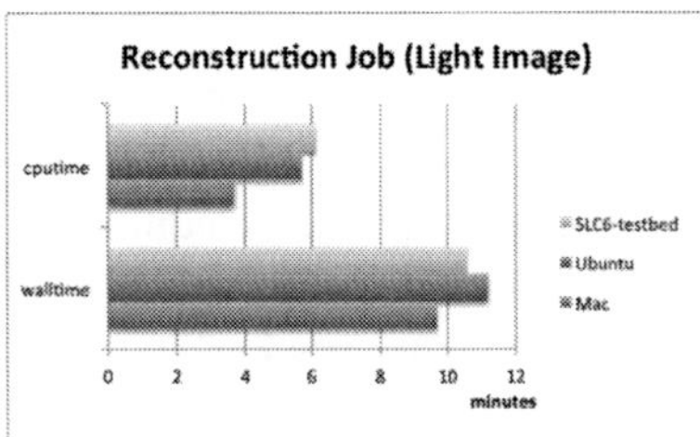

Figure 6. Light image.

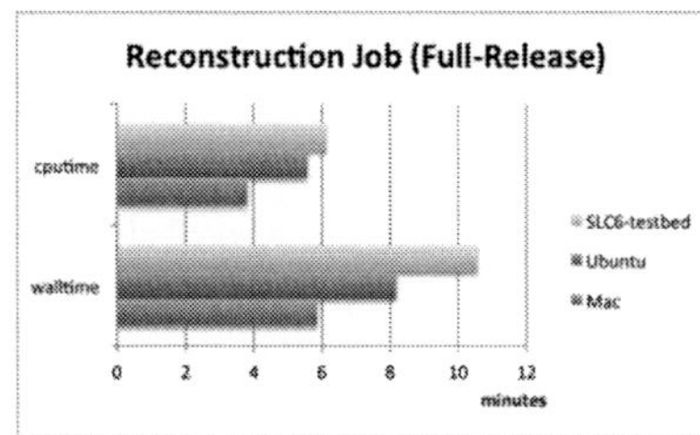

Figure 7. Full release image.

Table 2. Reconstruction test job (times in minutes).

	Light image		Full release	
System	wall-time (σ)	CPU-time (σ)	wall-time (σ)	CPU-time (σ)
Mac	9.66 (0.23)	3.66 (0.05)	5.85 (0.50)	3.79 (0.03)
Ubuntu	11.19 (0.20)	5.72 (0.01)	8.18 (0.35)	5.58 (0.01)
SLC6-testbed	10.56 (0.56)	6.13 (0.08)	10.56 (0.56)	6.13 (0.08)

σ is the sample standard deviation

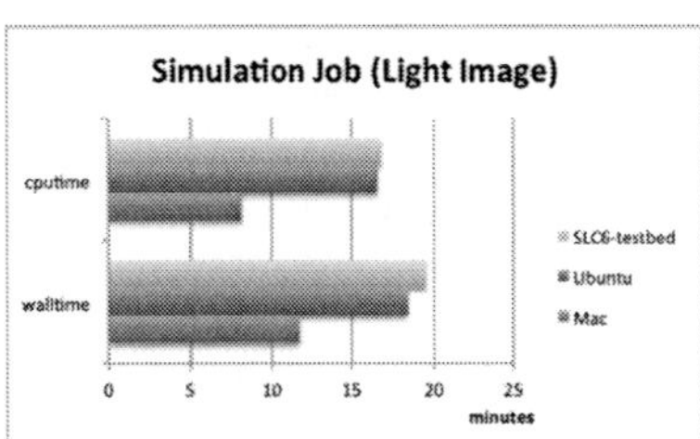

Figure 8. Light image.

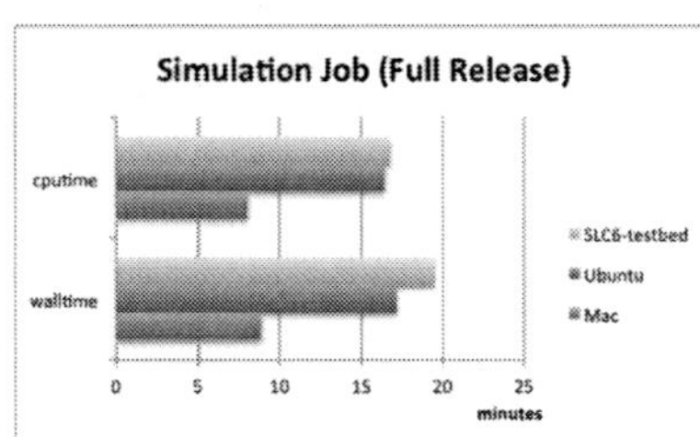

Figure 9. Full release image.

4. Consistency requirement of bind mounts in macOS

The current implementation of bind mounts in macOS provides a way of relaxing a default requirement of perfect consistency between the views of the mounted filesystem as seen from the host and from the container. The following flags can be added to the bind mount Docker arguments:

Table 3. Simulation test job (times in minutes).

	Light image		Full release	
System	wall-time (σ)	CPU-time (σ)	wall-time (σ)	CPU-time (σ)
Mac	11.75 (0.37)	8.12 (0.17)	8.84 (1.14)	8.00 (0.11)
Ubuntu	18.42 (0.28)	16.58 (0.03)	17.18 (0.24)	16.44 (0.03)
SLC6-testbed	19.51 (2.00)	16.77 (0.31)	19.51 (2.00)	16.77 (0.31)

σ is the sample standard deviation

- cached: delays are allowed before updates in the host appear in the container
- delegated: delays are allowed before updates in the container appear in the host

The same metrics used in section 3, wall-time and CPU-time, were compared for the three options specifying levels of consistency: consistent, cached and delegated in a reconstruction test. The degradation in performance seen when consistency is required (see Figure 10) is practically removed when the full-release image is used, regardless of the level of consistency (see Figure 11).

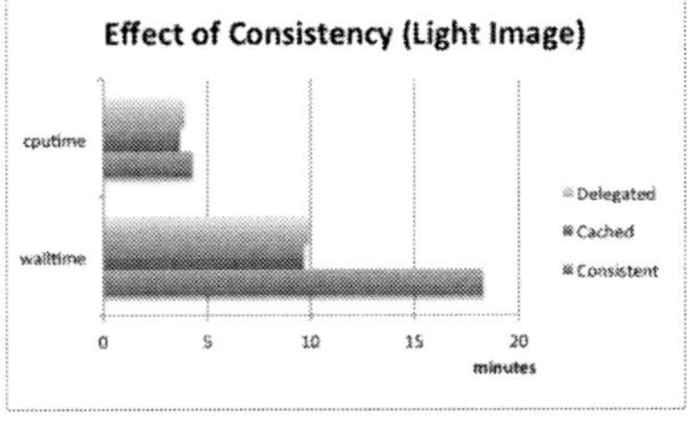

Figure 10. Light image. Relaxing consistency improves performance.

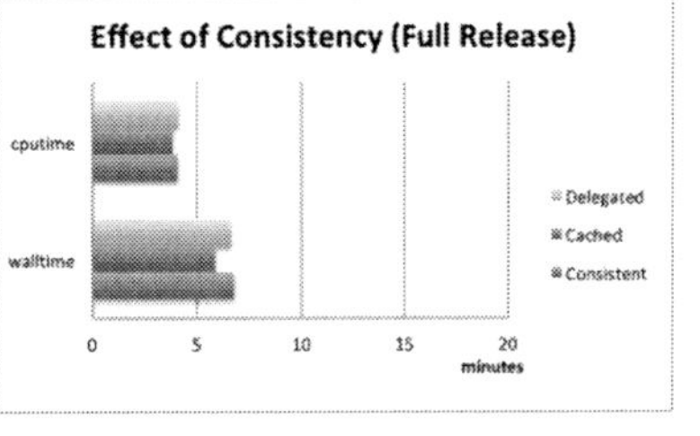

Figure 11. Full release image. Minimal use of cvmfs.

5. Summary and outlook

The preliminary tests carried out in this study showed that jobs running in containers with ATLAS software performed as well as the same jobs running on bare metal. Although the immediate objective here was to provide a platform for ATLAS software development, the long term goal is to exploit the advantages of containers in physics analyses and in production workloads such as event generation, simulation, reconstruction and physics derivations. The unique feature of containers of packaging and isolating software inside an entire file system represents an efficient implementation of a uniform runtime environment ideal for distributed computing.

References

[1] Docker: https://www.docker.com
[2] ATLAS Collaboration 2008 *J. Inst. 3* S08003.
[3] How to use Docker containers in ATLAS: https://atlassoftwaredocs.web.cern.ch/athena/intro-docker
[4] CernVM File System: http://cernvm.cern.ch/portal/filesystem
[5] CVMFS Docker Volume plugin: https://gitlab.cern.ch/cloud-infrastructure/docker-volume-cvmfs
[6] ATLAS Docker Hub image repository: https://hub.docker.com/u/atlas/dashboard
[7] ATLAS Software Git workflow: https://atlassoftwaredocs.web.cern.ch/gittutorial

ATLAS BigPanDA monitoring

A Alekseev[1], A Klimentov[2], T Korchuganova[1], S Padolski[2*], T Wenaus[2]
on behalf of the ATLAS Collaboration

[1]Tomsk Polytechnic University, 30 Lenin av., Tomsk, 634050, Russia
[2]Brookhaven National Laboratory, P.O. Box 5000, Upton, NY 11973-5000, USA

*Presenter, primary author

E-mail: siarhei.padolski@cern.ch

Abstract. BigPanDA monitoring is a web application that provides various processing and representation of the Production and Distributed Analysis (PanDA) system objects states. Analysing hundreds of millions of computation entities, such as an event or a job, BigPanDA monitoring builds different scales and levels of abstraction reports in real time mode. Provided information allows users to drill down into the reason of a concrete event failure or observe the broad picture such as tracking the computation nucleus and satellites performance or the progress of a whole production campaign. PanDA system was originally developed for the ATLAS experiment. Currently, it manages execution of more than 2 million jobs distributed over 170 computing centers worldwide on daily basis. BigPanDA is its core component commissioned in the middle of 2014 and now is the primary source of information for ATLAS users about the state of their computations and the source of decision support information for shifters, operators and managers. In this work, we describe the evolution of the architecture, current status and plans for the development of the BigPanDA monitoring.

1. Introduction

The ATLAS experiment [1] uses the PanDA [2] (Production and Data Analysis) Workload Management System for managing workflows for all data and Monte Carlo processing on over 170 WLCG [3] data centers and opportunistically used resources such as commercial clouds, supercomputers, volunteer machines. As of the middle of 2017, it manages executing of 2M jobs daily and run 300k simultaneous jobs.

To perform monitoring and effectively debug described computation payload a dedicated Web application was developed. The BigPanDA monitoring produces a set of aggregated reports, dashboards and representations of entities manipulated by the PanDA Workload Management system. There are four functional roles distinguished among the BigPanDA monitoring users: physicists, production managers, shifters and developers. Despite there being some common requirements all groups focus on different views provided by the system. For example, production managers need to have high-level aggregates which report the progress of particular payloads while developers should have an access to job logs to trace possible issues. All of this information is accessible in the BigPanDA monitoring in real-time mode along with two years' history of job execution.

While the development of the monitoring system is primarily driven by the ATLAS requirements, other experiments, for example, COMPASS [4], also utilize their instances of BigPanDA to monitor custom computations.

2. Architecture

Basically, two mutually exclusive requirements defined the BigPanDA architecture: the system should work as a Web application analyzing data in a real-time mode and on the other hand it is expected to perform some accounting with aggregates of historical data. To find an appropriate balance between them the following approach was applied:

- Hotspots in Python source code were identified and these code fragments were reimplemented in PL/SQL in order to improve the processing-over-data performance.
- Web pages containing few independent data sources were separated into asynchronous fragments loaded into a web page by AJAX [5] or AngularJS [6] libraries.
- The system contains many SQL queries which involve large amounts of data in the processing algorithm and therefore they could not be executed in real-time mode. To cover such cases, we developed a proactive caching approach which is described later.

The general architecture diagram is presented in figure 1.

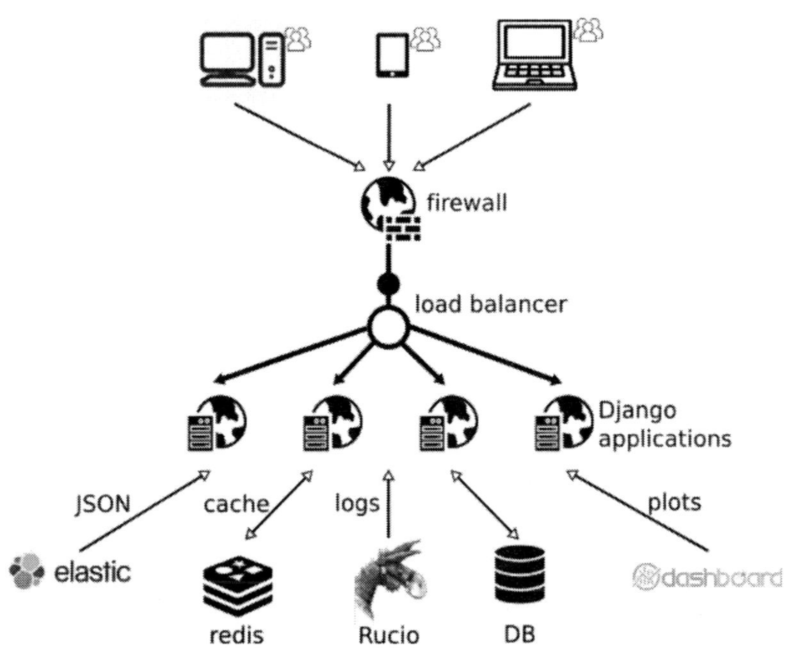

Figure 1. BigPanDA system architecture diagram

The Load Balancer (LB) distributes incoming traffic across the Apache Server instances which handle user requests. Currently, we are using the Nginx HTTP server [7] for this purpose.

Continuous development following the growing community needs requires the efficient utilization of frameworks. We are using Django [8] as a basic framework for BigPanDA monitoring because of its versatility, flexibility, and number of ready to use components. The dynamic data load provided by AJAX and AngularJS and Web page design is based on ZURB Foundation [9] components.

BigPanDA collects data from following sources as shown in Figure 1:

IOP Conf. Series: Journal of Physics: Conf. Series **1085** (2018) 032043 doi:10.1088/1742-6596/1085/3/032043

- ElasticSearch [10] instance providing access to parsed and indexed log records of PanDA server [11], JEDI [12] and Pilot [13].
- Rucio [14] provides access to log files for individual jobs from the distributed file storage to BigPanDA web servers which make the files accessible via ordinary URLs.
- CERN dashboard [15] service provides summary plots incorporated into the main page of the BigPanDA monitoring.
- Oracle DB [16] used by PanDA to keep all transactions is the main sources of data for BigPanDA monitor. Thanks to the Django DB Middleware [17] BigPanDA also supports MySQL [18] backend for installations beyond ATLAS.

As it was mentioned above a cache system is actively utilized in the monitoring. It has two principal purposes:

- To reduce the extra load on the Django part of the BigPanDA application and to the database. When a user requests particular information the response from the monitoring system is kept in the Redis cache [19] for 20 minutes and can be reused for the same requests in the future.
- Another advantage of data caching is the proactive preparation of aggregated data in order to reduce the waiting time and rise up the system's responsiveness. Some views require to process large volumes of data and these requests cannot be completed within a few seconds. For example, searching for jobs using a wildcard filter pattern may require processing almost 900 million database rows. To make such aggregates always available for display, a special crawler passes through a dynamically generated list of views with different filter conditions and then stores the outputs in the Redis cache.

As of the middle of 2017 BigPanDA monitoring servers receive 20k requests daily, with 35% of them conimg from automatic scripts or other third-party systems developed by users. Such custom applications can interact with the monitoring system using the JSON data format.

3. User-oriented content

BigPanDA monitoring generates around 70 different views providing information for the specific user roles described above. The distribution of requests for the most demanded views is shown in Figure 2.

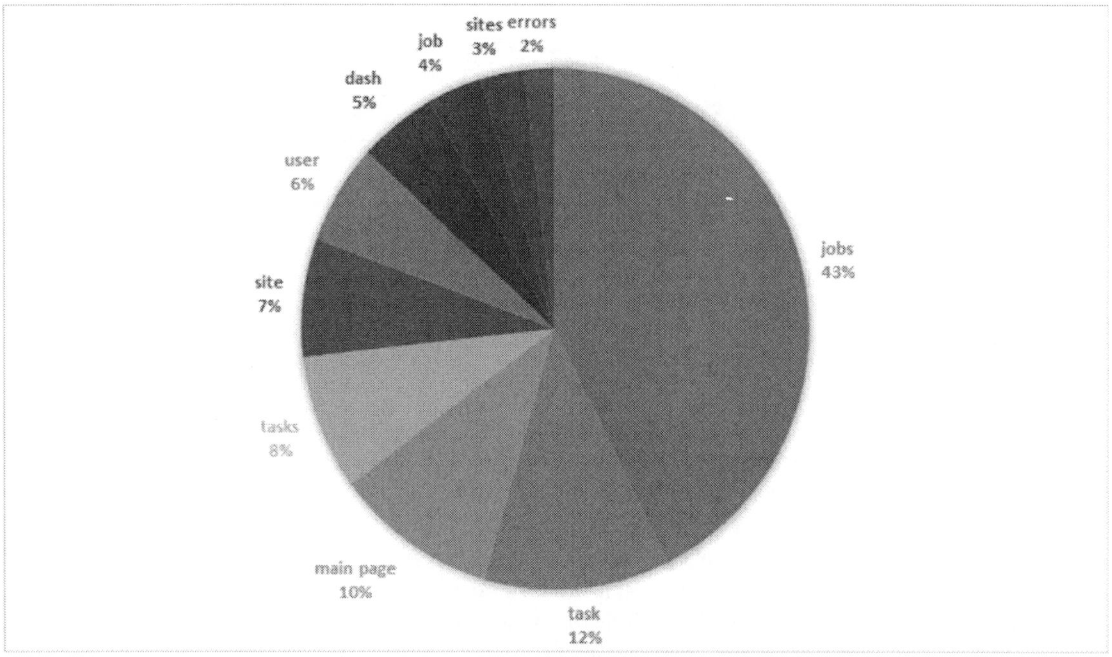

Figure 2. Distribution of the most frequently accessed views

The most important objects managed by the PanDA workload system are presented in the BigPanDA monitoring in two forms – as an individual (e.g. job, task, site) and as a part of a summary (e.g. jobs, tasks, sites). The second representation provides aggregates for key indicators in a compact form such as computation accounting, payload distribution over sites, or its internal structure etc. We will describe some of the most frequently accessed views below.

3.1. Jobs view
The jobs view provides three summaries on the top level: Job attribute summary, Overall error summary and Job list. In the first view the 26 most informative jobs attributes are aggregated by values and are available for the following drilling-down. Overall error summary accounts the errors codes of selected jobs and also provides some textual description of them. The Job list is the table containing a subset of selected jobs. There are several different filters available for this view: jobs modification time, assigned computing site, the name of the user who submitted the job, simulation software version, etc.

3.2. Tasks view
The tasks view has a similar structure as the one described above with the only difference being that the aggregated objects are the tasks.

3.3. User view
This view is the starting point used by most of ATLAS physicists to monitor execution of custom payloads. It contains tasks and jobs summaries filtered by a particular user, as well as the totals of the CPU time consumed. A recommender system is also integrated into this view. It analyses prior visits of BigPanDA monitoring and suggests links to views and objects that can probably be the focus of the user's attention.

3.4. Failures analysis
BigPanDA monitoring has advanced capabilities for error analysis. There are four basic inputs processed by the system: individual job logs, DB records, PanDA server logs and other infrastructure software logs. The last one produces nearly 400 GB of records daily. In order to manage and aggregate this data flow in an efficient way a special setup based on ELK stack [20] was developed [21]. BigPanDA Monitoring also has a special querying interface that allows one to compose custom requests to Kibana and deliver a response to the corresponding views.

4. Conclusion
BigPanDA monitoring is a Web application developed to cover the various information needs of PanDA users such as access to large scope analytics dashboards, jobs and tasks summaries, results of logs processing, and payload objects state. To provide a smooth user experience of working with BigPanDA, a different acceleration of the following approaches has been introduced: proactive caching, processing-over-data and dynamic data loading. The described system serves not only human requests, but is also used as a data processing engine with almost 7,000 JSON requests per day. The future plans for BigPanDA monitoring include further acceleration of processing speed, following the growing needs of the users and a deeper content personalization.

Acknowledgements
This material is based upon work supported by the U.S. Department of Energy, Office of Science, Office of Advanced Scientific Computing Research, under Award No. DE-SC-0012704. This work is partly funded by the Russian Federal budget (Government Task "SCIENCE") project No. 2.9223.2017/8.9

References

[1] ATLAS Collaboration 2008 The ATLAS Experiment at the CERN Large Hadron Collider, JINST 3 S08003
[2] Maeno T for the ATLAS Collaboration 2008 PanDA: Distributed production and distributed analysis system for ATLAS, *J. Phys.: Conf. Series* 119 062036
[3] Bird I 2011 Computing for the Large Hadron Collider *Annual Review of Nuclear and Particle Science* vol 61 pp 99-118
[4] Abbon P et al. 2007 The COMPASS experiment at CERN *Nucl. Instrum.Meth.* A577 455
[5] AJAX project, "AJAX" [software], version 3.2, 2017. Available from http://api.jquery.com/category/ajax/ [accessed 2017-09-30]
[6] AngularJS project, "AngularJS" [software], version 1.2, 2017. Available from https://angularjs.org/ [accessed 2017-09-30]
[7] Nginx project, "nginx" [software], version 1.12.1, 2017. Available from https://nginx.org/en/download.html [accessed 2017-09-30]
[8] Django project, "Django" [software], version 1.11.5, 2017. Available from https://docs.djangoproject.com/en/1.11/ [accessed 2017-09-30]
[9] ZURB Foundation project, "ZURB Foundation" [software], version 6, 2017. Available from http://foundation.zurb.com/sites/docs/ [accessed 2017-09-30]
[10] ElasticSearch project, "Elastic Search" [software], version 5.0, 2017. Available from https://www.elastic.co/products/elasticsearch [accessed 2017-09-30]
[11] Maeno T et al. 2014 Evolution of the ATLAS PanDA workload management system for exascale computational science *J. Phys.: Conf. Ser.* 513 032062
[12] De K, Golubkov D, Klimentov A et al. 2014 Task management in the new ATLAS production system. *J. Phys.: Conf. Ser.* 513 032078
[13] Nilsson P et al 2011 *J. Phys.: Conf. Ser.* 331 062040
[14] Garonne V et al 2014 *J. Phys.: Conf. Ser.* 513 042021
[15] Andreeva J, Campana S, Karavakis E et al. 2012 ATLAS job monitoring in the Dashboard Framework *J. Phys.Conf. Ser.* 396
[16] Oracle DB project, "Oracle" [software], version 11g, 2017. Available from http://www.oracle.com/technetwork/database/database-technologies/express-edition/downloads/index.html [accessed 2017-09-30]
[17] Django Databases Middleware project, "Django Databases" [software], version 1.11, 2017. Available from https://docs.djangoproject.com/en/1.11/ref/databases/ [accessed 2017-09-30]
[18] MySQL project, "MySQL" [software], version 5.7, 2017. Available from https://www.mysql.com/ [accessed 2017-09-30]
[19] Nelson J 2016 Mastering Redis. *Birmingham:Packt Publishing* p 340
[20] ELK stack project, "ELK" [software], version, 2017. Available from https://www.elastic.co/products [accessed 2017-09-30]
[21] Alekseev A, Barreiro Megino F, Klimentov A, Korchuganova T, Maeno T and Padolski S 2017 Applying Big Data solutions for log analytics in the PanDA infrastructure *Symp. on Nuclear Electronics and Computing* (Montenegro: Budva) https://indico.jinr.ru/getFile.py/access?contribId=199&sessionId=18&resId=0&materialId=slides&confId=151 [accessed 2017-10-02]

IOP Conf. Series: Journal of Physics: Conf. Series **1085** (2018) 032044 doi:10.1088/1742-6596/1085/3/032044

Toward real-time data query systems in HEP

Jim Pivarski[1], **David Lange**[1], **Thanat Jatuphattharachat**[2]
[1] Physics Department, Princeton University, Princeton, NJ 08544, USA
[2] Computer Engineering Chulalongkorn University, Krung Thep Maha Nakhon 10330, Thailand

E-mail: pivarski@princeton.edu

Abstract. Exploratory data analysis tools must respond quickly to a user's questions, so that the answer to one question (e.g. a visualized histogram or fit) can influence the next. In some SQL-based query systems used in industry, even very large (petabyte) datasets can be summarized on a human timescale (seconds), employing techniques such as columnar data representation, caching, indexing, and code generation/JIT-compilation. This article describes progress toward realizing such a system for High Energy Physics (HEP), focusing on the intermediate problems of optimizing data access and calculations for "query sized" payloads, such as a single histogram or group of histograms, rather than large reconstruction or data-skimming jobs. These techniques include direct extraction of ROOT TBranches into Numpy arrays and compilation of Python analysis functions (rather than SQL) to be executed very quickly. We will also discuss the problem of caching and actively delivering jobs to worker nodes that have the necessary input data preloaded in cache. All of these pieces of the larger solution are available as standalone GitHub repositories, and could be used in current analyses.

1. Benefits of a query service

Despite everything that's changed in computing since the first AIHENP/ACAT workshop, one aspect of today's High Energy Physics (HEP) analyses that a physicist 27 years ago would recognize is the use of private skims. Rather than analyzing the reconstructed data directly, physicists skim it (dropping events) and slim it (dropping particle attributes), copy this summary closer to where they will be working, and then analyze the summary. The reason is performance: smaller data with fewer steps between the analyst and the data result in quicker exploratory plots, which allow the human analyst to be more engaged in the investigation. Long processing times are acceptable for established, push-button procedures, but a new analysis requires creative work and interactive discovery.

However, producing such a skim can take a long time— weeks or months before seeing the first plot— and is fraught with tradeoffs. Excluding important events or (more often) important particle attributes can be costly, and including unnecessary data only makes the process take longer and use more disk. Moreover, the copy uses extra resources (disk and processing), which will only get tighter as data volumes outpace computing budgets, and it can effectively "price out" small analysis groups. Since it is a distinct copy, it does not benefit from improvements or corrections to the source data until refreshed, and can often be a year or more out of date. Finally, it forces physicists to take time and attention away from physics and statistical issues to solve an IT problem. If the analysis can be performed without this skim, so much the better.

IOP Conf. Series: Journal of Physics: Conf. Series **1085** (2018) 032044 doi:10.1088/1742-6596/1085/3/032044

Table 1. Rate of processing a "query sized" payload: filling one histogram of jet p_T for all jets in a $t\bar{t}$ sample, illustrating the orders of magnitude lost to providing a full framework for heavy event processing (all single-threaded).

rate	operation
0.018 MHz	full framework (CMSSW, single-threaded C++)
0.029 MHz	load all 95 jet branches in ROOT
2.8 MHz	load jet p_T branch (and no others) in ROOT
12 MHz	allocate C++ objects on heap, fill, delete
31 MHz	allocate C++ objects on stack, fill histogram
250 MHz	minimal "for" loop in memory (single-threaded C)

Real-time analysis on primary datasets is possible, and has been demonstrated in industry, usually in the form of large, distributed databases used internally by a company's data analytics team. These systems are built from Apache Drill, Impala, Kudu, Hawk, or Dremel and tuned to provide SQL access to petabytes of data in seconds. Unlike Hadoop and Spark, which are optimized for high throughput, these systems are optimized for low latency, and they use database-style indexes and query planning to avoid churning through all data with every request.

We can learn from these systems, adopting their techniques to provide similar access to HEP data, but there are some important differences between industry needs and HEP needs. A physicist's analysis is both simpler and more complex than a data scientist's analysis: it is simpler in that HEP does not require all-to-all processing, such as transposing a table of customer purchases to make product recommendations— all of the information that is needed to process one physics event is contained in that event or small auxiliary tables, not all the other events. Once collected, a HEP dataset is immutable— new events do not accumulate while the physicist is analyzing it. However, physics analyses are more complex in the depth of processing applied to that one event, examining many combinations of particles to search for relationships among the particles. To help with the bookkeeping, physicists are inclined to think in terms of objects, iterating over collections of particle objects, rather than exploding and aggregating tables as in SQL.

Because of the differences, we do not expect an off-the-shelf database to solve our problem directly, but we can adapt established techniques to work for HEP.

2. Optimizing for "query sized" payloads

An analysis query is the smallest unit of exploratory analysis, in which the physicist asks one question, such as "What is the distribution of muon η for the second-highest p_T muon in each event, subject to track-fit χ^2 constraints?" The conditions may be very complex, involving mass calculations, vertex fits, generated/reconstructed particle matching, computed with deeply nested loops and external function calls, but it is a request for one plot. The request and the plot are small enough to be transmitted through a slow network, and the data processing touches at most a dozen particle attributes out of thousands. Even a petabyte-sized dataset would only need to yield a terabyte of data, distributed among hundreds of workers, to service this request.

HEP frameworks are not designed for this style of data access. Table 1 shows how four orders of magnitude in processing rate are lost to provide all the services of a full event framework (CMSSW in this example). These include creating full C++ particle objects with all attributes loaded, allocated randomly on the heap for convenient memory management, compared to a sequential, contiguous, and possibly vectorized histogram fill over a simple array. These services are essential for "heavy" processing tasks like event reconstruction, which touch most attributes of each particle, but are inappropriate for "query sized" tasks like filling a histogram.

Table 2. Illustration of exploding nested, hierarchical objects. The data below are logically conceived as a list of lists of character/integer pairs, but stored in memory as four flat arrays describing the structure of the outer list, the inner list, the 1st and 2nd attributes.

logical data	`[[(a,1), (b,2), (c,3), (d,4)], [], [(e,5), (f,6)]], [], [[(g,7)]]`
outer offsets	`[0, 3, 3, 4]`
inner offsets	`[ 0, 4, 4, 6, 7]`
1st attribute	`[ a, b, c, d, e, f, g ]`
2nd attribute	`[ 1, 2, 3, 4, 5, 6, 7 ]`

Private skims strip most of this away for lightweight processing in a physicist's personal analysis environment; there is no reason a lightweight framework could not also be provided centrally.

Similar to low-latency databases used in industry, ROOT stores objects in an "exploded" form (called "splitting" in ROOT), with all values of each attribute in a separate array. This is even possible for hierarchically nested data, containing arbitrary length lists of objects, though multiple arrays with different lengths are required to encode it. This allows ROOT to only read attributes required for the analysis function— such as a terabyte of a petabyte dataset.

However, databases like Apache Drill go one step further by leaving the data in their exploded form during processing, without materializing rows as runtime objects[1]. Selective reading accounts for the first two orders of magnitude in Table 1, while eliminating object materialization accounts for most of the last two by skipping memory allocation, non-sequential memory scans, and permitting compiler optimizations like loop vectorization.

To permit this kind of access for ROOT data, Brian Bockelman and Zhe Zhang added a method to ROOT I/O that skips the `GetEntry` (event materialization) process[2]. Query sized calculations on the resulting arrays (computing momentum magnitudes from components) run 5 times faster than a streamlined `GetEntry` loop and 10 times faster than `TTree::Draw` or `TTreeReader` for uncompressed or LZ4-compressed input data.

3. Code transformation and compilation

For interactive analysis, it must be possible to define analysis functions at runtime and then compile them to run at full speed on arrays. Drill converts SQL to Java bytecode[1], while Brian and Zhe's test cases use Numba to JIT-compile Python[2]. What these tools lack is a mechanism to convert a physicist's object oriented view of the data into array operations. Below, I will describe how we can do this and how the transformed code performs.

Given the exploded data shown in Table 2 and user code such as

```
for outerlist in dataset:
    for innerlist in outerlist:
        for pair in innerlist:
            compute(pair.first, pair.second)
```

we want to convert it to a form that doesn't reference "`dataset`," "`outerlist`," "`innerlist`," or "`pair`" as objects, instead deferring to "`outeroffsets`," "`inneroffsets`," "`first`" and "`second`," where the data are stored as arrays.

Such a transformation can be performed algorithmically on the user code's Abstract Syntax Tree (AST), converting the above into

```
for (i = 0;  i < 3;  i++)
   for (j = outeroffsets[i];  j < outeroffsets[i+1];  j++)
      for (k = inneroffsets[j];  k < inneroffsets[j+1];  k++)
         compute(first[k], second[k]);
```

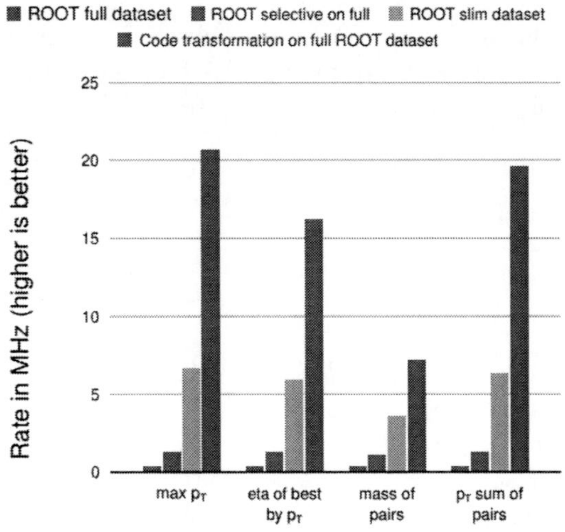

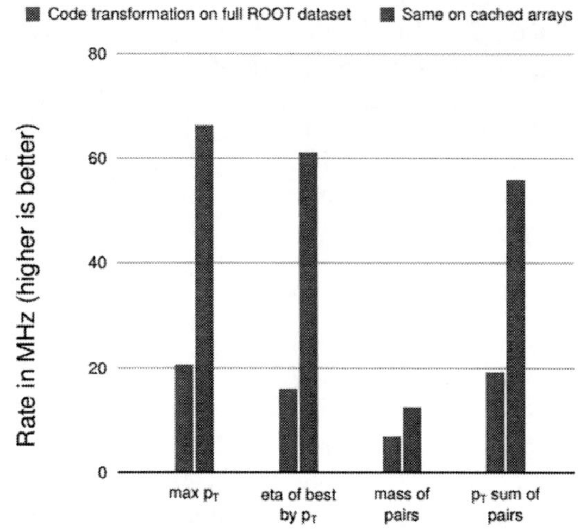

Figure 1. Processing rate of four analysis functions (Table 3) implemented in C++ with `GetEntry` and our Python code transformation/compilation. "ROOT full dataset" loads all particle attributes, including unnecessary ones, "selective on full" uses selective reading (`SetBranchStatus`), "slim dataset" uses a specially preselected input sample, "Code transformation on full ROOT dataset" uses Brian and Zhe's array extraction and our code transformation. Though uncompressed in warm cache, ROOT file reading is still the bottleneck; transformed code on raw arrays cached in memory is several times faster (all single-threaded).

by replacing each "`outerlist`" AST node with its corresponding "`outeroffsets[i]`" and each "`pair.first`" with its corresponding "`first[k]`." The first transformation rule eliminates references to list objects and the second eliminates references to record objects.

This transformation is like a type-inferring compilation pass, in which the types of dataset substructures must be propagated through the code, including assignments to new variables, and syntactic structures such as "for" loops have special transformation rules. Functions such as "len" (requesting the length of a list) must be overloaded as well (e.g. "`offsets[j+1] - offsets[j]`").

In special cases such as this one, we can even convert the above into

```
for (k = 0;  k < inner[outer[3]];  k++)
   compute(first[k], second[k]);
```

by recognizing that the outer and inner loops are total and sequential. Thus, when the transformed function is passed to a bytecode compiler such as Numba or Clang, the non-nested "for" loop may be more highly optimized, possibly vectorized (depending on "`compute`"). This code transformation technique is discussed in more detail in[3].

The performance of the transformed code was studied with a simulated Drell-Yan dataset containing 5.4 million collisions in the CMS detector. Calculations were performed in single-threaded processes on a `i2.xlarge` Amazon Web Services (AWS) instance, files were read from warmed cache, and the contents were uncompressed. Results are presented in Figure 1.

4. Distributed processing with cache

The optimizations discussed above focus on increasing single-threaded throughput, but of course the main benefit of a centralized query service is parallelization. If we attain our latency goal of no more than a second per plot and a hundred physicists are online, submitting a query every ten seconds, then each physicist would get a tenth of the whole cluster at a time.

Table 3. Analysis functions tested in Figure 1: "max p_T" is a per-event aggregation, "eta of best" adds the complication of maximizing one attribute while plotting another, "mass of pairs" iterates over distinct pairs of particles per event, computing an essential HEP function, and "p_T sum of pairs" has the same loop structure without the expensive cosh and cos calls.

max p_T

```
for event in dataset:
  maximum = 0.0
  for muon in event.Muon:
    if muon.pt > maximum:
      maximum = muon.pt
  fill_histogram(maximum)
```

eta of best by p_T

```
for event in dataset:
  maximum = 0.0
  best = None
  for muon in event.muons:
    if muon.pt > maximum:
      maximum = muon.pt
      best = muon
  if best is not None:
    fill_histogram(best.eta)
```

mass of pairs

```
for event in dataset:
  n = len(event.muons)
  for i in range(n):
    for j in range(i+1, n):
      m1 = event.muons[i]
      m2 = event.muons[j]
      mass = sqrt(
        2*m1.pt*m2.pt*(
        cosh(m1.eta - m2.eta) -
        cos(m1.phi - m2.phi)))
      fill_histogram(mass)
```

p_T sum of pairs

```
for event in dataset:
  n = len(event.muons)
  for i in range(n):
    for j in range(i+1, n):
      m1 = event.muons[i]
      m2 = event.muons[j]
      s = m1.pt + m2.pt
      fill_histogram(s)
```

HEP analysis functions are embarrasingly parallel, which would scale linearly if data can be delivered to them efficiently. However, this depends critically on caching, since data processing is considerably faster than disk access, and caching is local. An input dataset in memory on one machine is only useful if subsequent jobs requiring that input are sent to the same machine.

Rather than dispatch subtasks round-robin or to the least busy compute node, we want compute nodes to pull subtasks with a preference for input data they already have in cache. Figure 2 shows a scheme for accomplishing this, using Apache Zookeeper to advertise new subtasks and globally mark them as in progress and delete them when done. Compute nodes pull subtasks rather than having work pushed to them because only the compute nodes know which input datasets are locally in cache. Pulling work also reduces the effect of stragglers for the same reason as work-stealing. Additionally, the work-pulling algorithm should not strictly

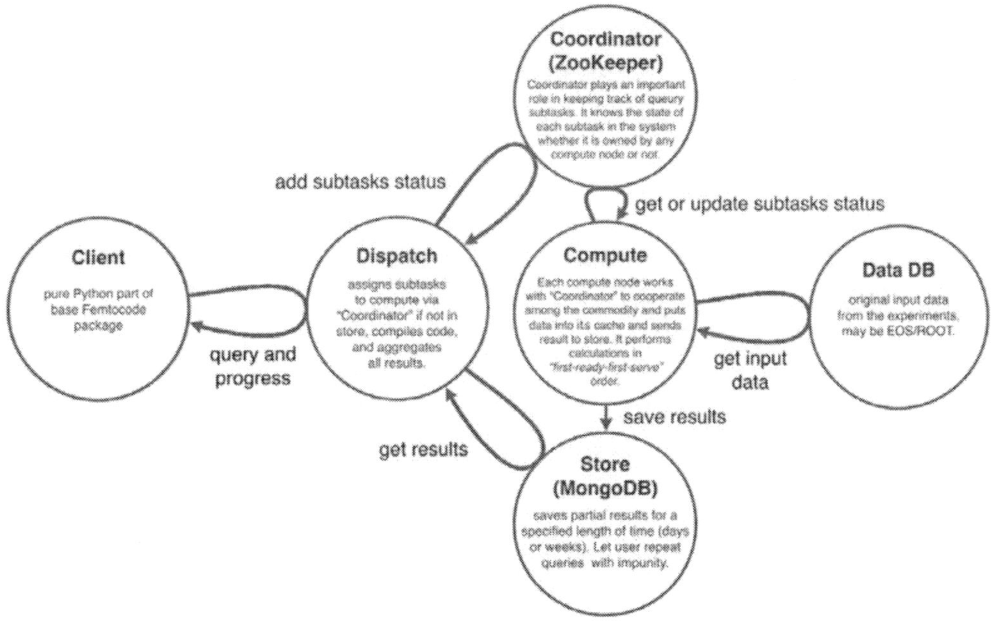

Figure 2. Schematic for distributed query processing to minimize cache misses (see text).

IOP Conf. Series: Journal of Physics: Conf. Series **1085** (2018) 032044 doi:10.1088/1742-6596/1085/3/032044

require input data to be in cache, but should check for work in two rounds: the first takes only cache-local work, but if there is no cache-local work to do, compute nodes will take any work after a sub-second delay. This allows the compute nodes that can best service the request to have first dibs, but if there's too much for them to do, new compute nodes will download the input data so that the handling of a popular dataset elastically scales with demand.

For the most common type of query, histogram aggregation, subtasks represent partial histograms that must be aggregated centrally. For responsiveness, we imagine storing partial histograms in a document database like MongoDB and aggregating whatever is available at regular intervals. That way, the user would see results accumulate interactively and can cancel malformed queries. The range of supported tasks can be extended by adopting generalized aggregation with Histogrammar[4].

This system was developed in draft form by one of us (Thanat)[5], to serve as a model for development. This draft system lacks only realistic work payloads, substituting them with random delays.

5. Summary

A centralized query service that is as fast and convenient as private skims would have a profound impact on the working habits of thousands of physicists. It doesn't represent an incremental change of existing infrastructures, and as such, needs quite a few new components to make it work.

Rather than trying to introduce this system in a "big bang," we are rolling out individual components, even if they tend to increase reliance on private skims in the short term. For instance, array access to ROOT data and Python code transformation/compilation can accelerate and simplify analysis using skims. We are considering ways of implementing a distributed system as a virtual ROOT file, to provide physicists with a familiar interface, even if it isn't used with the accelerated query engine. This way, we can get feedback about parts of the system before the whole is in place. This method also encourages spin-off technologies.

The following software is either available now or soon.

- Array access to ROOT data, via Numpy and `TTreeReaderFast`, is scheduled for **ROOT 6.14**[6]. Users willing to compile ROOT may try the feature early using a custom fork[7].
- The same interface is available now in a Python package called **uproot**[8].
- The code transformation techniques described here are implemented in **OAMap**[9], which accelerates functions on data in Apache Arrow[10] format, with on-the-fly conversion from ROOT to Arrow. The purpose of targeting Arrow rather than ROOT directly is to attract contributions from Big Data projects that rely on Arrow.

Readers are encouraged to try out these packages and report any usability issues, as this will guide the process of developing the query system.

Acknowledgements

This work was supported by the National Science Foundation under grants ACI-1450377 and PHY-1624356.

References

[1] M Hausenblas and J Nadeau, "Apache drill: Interactive ad-hoc analysis at scale," *Big Data*, vol. 1(2), pp. 100–104, 2013. [Online]. Available: https://doi.org/10.1089/big.2013.0011

[2] B Bockelman, Z Zhang, and J Pivarski, "ROOT-BulkIO and Numpy interface," *Journal of Physics: Conference Series*, 2017. [Online]. Available: https://indico.cern.ch/event/567550/contributions/2627167/

[3] J Pivarski, P Elmer, B Bockelman, Z Zhang, "Fast Access to Columnar, Hierarchical Data via Code Transformation," *http://arxiv.org/abs/1708.08319*, Aug. 2017.

[4] J Pivarski, A Svyatkovskiy, F Schenck, and B Engels, "histogrammar-python: 1.0.0," Sep. 2016. [Online]. Available: https://doi.org/10.5281/zenodo.61418
[5] T Jatuphattharachat. (2017) femto-mesos. [Online]. Available: https://github.com/JThanat/femto-mesos
[6] R Brun and F Rademakers, "ROOT: An object oriented data analysis framework," *Nucl. Instrum. Meth.*, vol. A389, pp. 81–86, 1997. [Online]. Available: https://doi.org/10.1016/S0168-9002(97)00048-X
[7] B Bockelman. (2017) root-bulkapi-fastread-v2. [Online]. Available: http://github.com/bbockelm/root/tree/root-bulkapi-fastread-v2
[8] J Pivarski. (2017) "Uproot: Minimalist ROOT I/O in pure Python and Numpy.". [Online]. Available: https://github.com/scikit-hep/uproot
[9] ——. (2017) "OAMap: Toolset for Computing Directly on Hierarchically Nested, Columnar Data". [Online]. Available: https://github.com/diana-hep/oamap
[10] The Apache Arrow team. (2016) "Apache Arrow: Powering Columnar In-Memory Analytics". [Online]. Available: https://arrow.apache.org/

IOP Conf. Series: Journal of Physics: Conf. Series **1085** (2018) 032045 doi:10.1088/1742-6596/1085/3/032045

Designing and prototyping the control system for the Cherenkov Telescope Array

I Oya[1], M Füßling[1], P O Antonino[2], M Araya[3], J Borkowski[4], A Bulgarelli[5], J Castroviejo[6], J Colomé[6], V Conforti[5], A Garcia-Piquer[6], J Guárdia[6], L Hagge[7], E Lyard[8], A Morgenstern[2], M Mayer[9], D Melkumyan[1], T Murach[1], L Pizarro[3], I Sadeh[1], T Schmidt[1], U Schwanke[9], G Spengler[9], J Schwarz[10], G Tosti[11], R Walter[8] and P Wegner[1] for the CTA Consortium[12]

[1] DESY, Zeuthen, Germany
[2] Fraunhofer IESE, Kaiserslautern, Germany
[3] UTFSM, Valparaíso, Chile
[4] CAMK, Torun, Poland
[5] I.A.S.F. di Bologna, Italy
[6] IEEC-CSIC, Spain
[7] DESY, Hamburg, Germany
[8] University of Geneva - Departement d'Astronomie, Switzerland
[9] Humboldt-Universitaet zu Berlin, Germany
[10] INAF - Osservatorio Astronomico di Brera, Italy
[11] University of Perugia, Italy
[12] Full consortium author list at http://cta-observatory.org

E-mail: igor.oya.vallejo@desy.de

Abstract. The Cherenkov Telescope Array (CTA) is the next-generation atmospheric Cherenkov gamma-ray observatory. The Observation Execution System (OES) team within the CTA project is designing and prototyping the software to execute the observations and to handle the acquisition of scientific data at GB/s rates. In this contribution we show the OES system as it is being designed using the Unified Modeling Language (UML) and Systems Modeling (SysML) formalisms. In addition, we present the status of the associated prototyping activities.

1. Introduction

The Cherenkov Telescope Array (CTA) [1] will consist of two facilities, one in the southern (Cerro Armazones Chile) and the other in the northern hemisphere (La Palma, Spain). The two sites will contain dozens of telescopes of different sizes, constituting one of the largest astronomical installations under development.

CTA will implement simultaneous automatic operation of multiple sub-arrays of telescopes. It will be capable of quick re-scheduling of observations (within a few seconds), in order to allow observations of elusive transient events. The operation, control, and monitoring of the distributed multi-telescope CTA arrays is inherently complex. As such, they pose new challenges in scientific instrumentation control systems and in particular in the context of ground-based gamma-ray astronomy. The system providing the functionality to execute the observations and to handle the acquisition of scientific data in CTA is the Observation Execution System (OES).

In the the following we will first introduce the model-driven methodology adopted to design the OES system (Section 2), then present a brief description of the main sub-systems of OES and describe the status of the associated prototypes (Section 3), and finally offer some concluding remarks (Section 4).

2. Model Driven System Design

The OES architecture is designed using the Software Platform Embedded System (SPES) methodology [2] implemented via the Unified Modeling Language (UML) and Systems Modeling (SysML) formalisms.

The architecture is composed of two type of elements: *architecture drivers* and *architecture design*. The architecture drivers specify functional and quality requirements for the system. The architecture design is expressed by different viewpoints. Figure 1 illustrates via an example the main viewpoints we use in the OES architecture, which are the following:

- In the *context* viewpoint the OES system is considered to be a black-box interacting with a set of external systems and *stakeholders* (human actors). The interaction is performed via the exchange of information through the flow of *functional data*; in the example in Fig. 1 we show a system and a stakeholder exchanging information with the OES system.
- In the *functional* viewpoint we derive a network of functions that exchange data with the external entities outside the system as well as between them. The functions are grouped hierarchically, down to the elementary elements named *behavior functions*.
- In the *logical* viewpoint, we create a hierarchical decomposition of the systems, the higher level elements corresponding to the sub-systems contained within the OES (see also Fig. 2). These components are decomposed in a hierarchy of lower level components, down to the elementary implementation units (*behavior components*), that realize the functions previously defined in the functional viewpoint. This viewpoint contains the provided and required interfaces of the components and the concrete (*logical*) data exchanged. In addition to the UML component diagrams which show the static description of the OES system, this viewpoint uses UML activity and sequence diagrams to specify the behavior of the system.
- We describe the deployment specification of the logical components in the *technical* viewpoint (not shown in the example in Fig. 1).

The application of the architecture model serves the following purposes in the teamwork to create the OES system:

- Specification of the architecture design, including interfaces, behavior and data model.
- Relationships (traces) from drivers (functional and quality requirements) to model elements, ensuring consistency.
- Definition of specifications of software to be created by the development teams, which can be either within the CTA organization or external software development companies.
- Vision sharing and scoping (including interface identification and definition), two important ingredients for collaborative software engineering.
- A foundation to implement the project management to build the OES system. The work package product breakdown structure (PBS) and the work breakdown structure (WBS) can be built based on the functional and logical viewpoints. The model can help to identify risks, internal and external dependencies, and to plan staging scenarios in the construction of the OES software.

IOP Conf. Series: Journal of Physics: Conf. Series **1085** (2018) 032045 doi:10.1088/1742-6596/1085/3/032045

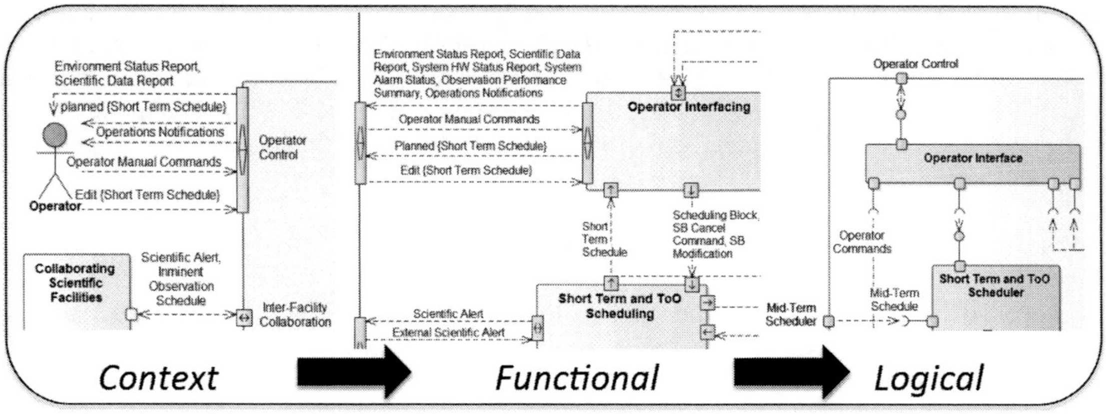

Figure 1. Example to illustrate the main viewpoints of the OES architecture. The arrows exemplify the process of going from an abstract description to a more concrete level of detail, from the Context (left) to the Functional (middle) and then to the Logical viewpoints (right). The diagram utilizes the SysML and UML notation. See text for details.

The OES will be implemented as a distributed software application using the ALMA Common Software (ACS) CORBA-based framework [3, 4]. One instance of the OES will be deployed in each CTA site. Figure 2 shows a diagram indicating the main components of the OES system, described in the next section.

3. OES Subsystems and their Prototypes

The OES design presented here takes as input many concepts from the Array Control and Data Acquisition (ACTL) idea discussed within the CTA Consortium [5, 6]. We have produced advanced prototypes for most OES sub-systems during the past few years, implemented in the ACS framework. The scope of these sub-systems, together with the status of the prototypes is briefly summarized below:

- The **Short Term and ToO Scheduler** is responsible for deciding, at real time, how to group and use the telescopes of a CTA installation to perform nightly operations, based on a mid-term schedule supplied to it before the beginning of the night operations. This sub-system is also responsible for reacting at real time to changing environmental conditions and external or internal scientific alerts (scientific alerts are also known as ToOs; target of opportunity alerts). The scheduler prototype is implemented in C++, and uses advanced artificial intelligence algorithms (meta-heuristic optimization) [10, 11].
- The **Manager and Central Control** is responsible for the execution of the scheduling blocks provided by the scheduler by sending corresponding commands to the telescopes, while supervising the ongoing operations, bookkeeping the allocation of telescopes to sub-arrays, and overseeing the Data Handling System. A prototype of this sub-systems exists, implemented in Java and using a supervision tree architecture. The prototype includes a dedicated Python environment to run operations scripts. The latter are the high level procedures implementing the actions to be performed by the telescopes in a sub-array during a particular operation.
- The **Data Handling System** is responsible for getting the stream of data from the Cherenkov cameras of the CTA telescopes and handling these data, which arrive at a rate of the order of GB/s, according to the sub-array they participate in. As such, it contains the central trigger, data acquisition, local storage, and a pipeline to provide scientific alerts

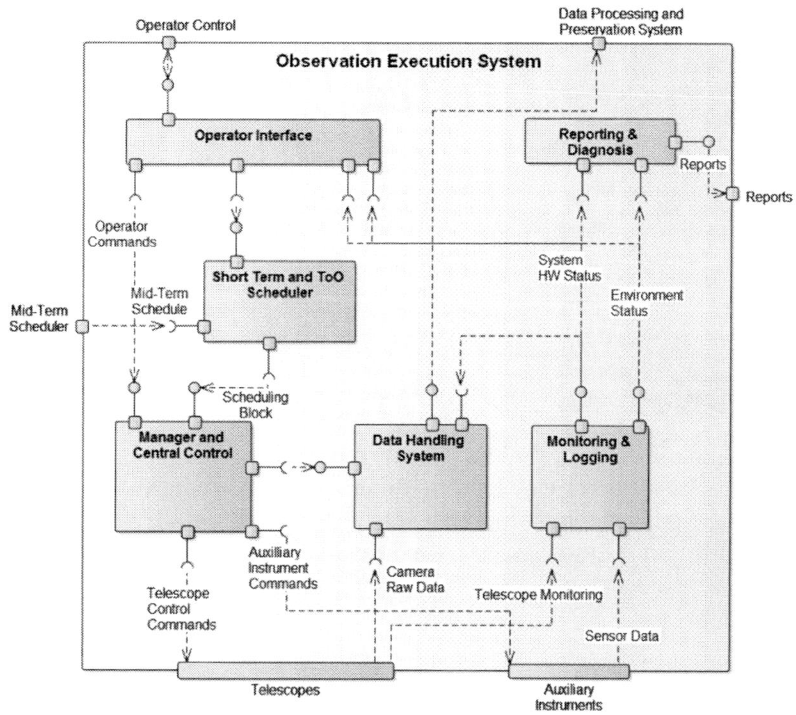

Figure 2. Logical view of OES, representing the main components of the system. Only the highest level components, and the most relevant data elements and interfaces are shown. The diagram uses the UML component notation, and the basic entities are the software components, depicted as pink boxes. The dashed lines show the flow of data elements and the arrows the direction and sign of the flow.

at real time. More details on the existing prototype of the data acquisition, implemented in C/C++, is provided in [7]. The concept and design ideas of the science alerts pipeline is presented in [8].

- The **Operator Interface** offers to the operator located in the control room of the CTA installation a comprehensive view of the status of the observations and the hardware and means to interact with the system. Our prototype [9] is based on web technologies, with the front-end implemented in JavaScript and the back-end in Python (Pyramid Web Framework). Figure 3 shows a few examples of existing prototype panels.
- The **Monitoring System** is responsible for monitoring data items from the telescope and other devices deployed at the CTA sites, and making those data available for the operator interface, for quick-look quality checking, as well as for detailed inspection. In our estimates, we assume a total of 150,000 data items to be monitored at a rate of about 1 Hz. Based on the assumed number of data items, we are prototyping solutions that are using redis, MongoDB and Cassandra nosql data base technologies.
- The **Reporting and Diagnosis** sub-system is responsible for gathering the relevant data from the other OES subsystems in order to produce status and quality reports during the night for the Operator Interface and for other systems outside the OES. This system has not yet been prototyped.

A preliminary integration of the OES sub-system prototypes mentioned above is underway, starting with integration of the Central Control with the Short Term Scheduler and the Data

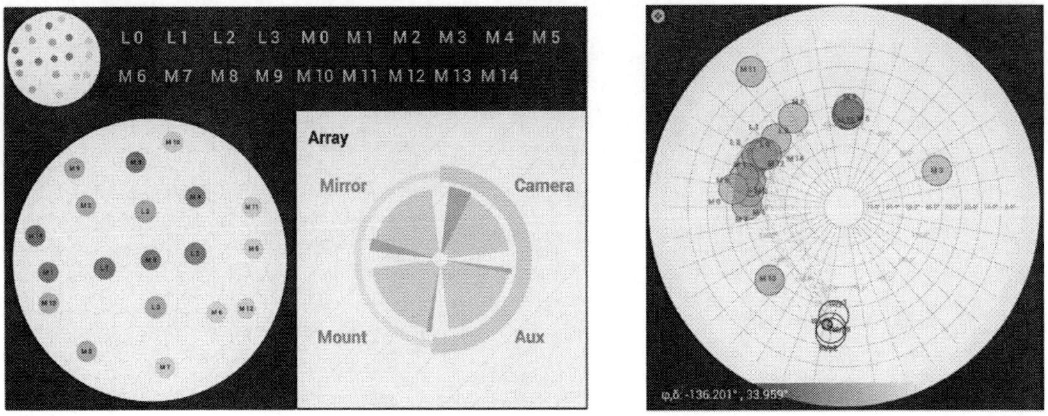

Figure 3. Two panels from the operator User Interface prototype. The *array zoomer* (left panel) shows the status of array elements at different levels of detail by implementing multi-scale navigation with semantic zooming concepts (see [9] for details). The figure in the right panel shows the position of telescopes (blue/red circles) on the sky in polar coordinates.

Handling System.

4. Conclusions

We have set up a model-driven approach to design the OES software. The system has been decomposed in several sub-systems, most of them already having associated prototypes. The pre-integration of the sub-system prototypes is the first step in the construction phase of the OES system. Our experience during the last couple of years is that the usage of the formal architecture approach presented here improves the communication of expectations and interface definitions between the different development teams, which are physically distributed around the world. Following this, we conclude that such model-driven approach will be instrumental for the success of the OES implementation in the future.

Acknowledgments

The authors acknowledge the work of the ACTL team members of the CTA consortium during the last years, which was instrumental for the definition of the OES system. We gratefully acknowledge financial support from the agencies and organizations listed here: http://www.cta-observatory.org/consortium_acknowledgments

References

[1] Acharya B, Actis M, Aghajani T, et al. 2013 *Astroparticle Physics* **43**, 3
[2] Pohl K, Hönninger H, Achatz R, Broy M (Eds.) 2012 *Model-Based Engineering of Embedded Systems: The SPES 2020 Methodology* (Berlin: Springer)
[3] Chiozzi G, Jeram B, Sommer H, et al. 2014 *Proc. SPIE* 9149, 91490H
[4] Schwarz J, Sommer H and Farris A 2004 *ASP Conference Series* 314, 634
[5] Oya I, Füßling M, Antonino PO, et al 2016 *Proc. SPIE* 9913, 991303
[6] Füßling M, Oya I, Balzer A, et al 2016 *Proc. SPIE* 9913 99133C
[7] Lyard E, Walter R 2017 *End-to-end data acquisition pipeline for CTA*, *Proc. ICRC* (*Preprint* arXiv:1709.04203)
[8] Bulgarelli A, Fioretti V, Zoli A, et al 2014 *Proc. SPIE* 9145 91452X
[9] Sadeh I, Oya I, Schwarz J and Pietriga E 2016 *Proc. SPIE* 9913, 99130X
[10] Colomé J, Colomer P, Campreciós J, et al 2012, *Proc. SPIE* 5496, 205-218
[11] Garcia-Piquer A, Morales JC, Ribas I, et al 2017 *Astronomy & Astropysics* **604**, A87.

IOP Conf. Series: Journal of Physics: Conf. Series **1085** (2018) 032046 doi:10.1088/1742-6596/1085/3/032046

Cross-architecture Kalman filter benchmarks on modern hardware platforms

Cámpora Pérez D H[1,2], Awile O[1], Bouizi O[3] and Neufeld N[1]

[1] CERN, CH-1211 Geneva 23, Geneva, Switzerland
[2] Universidad de Sevilla, C/San Fernando, 4, C.P. 41004, Sevilla, Spain
[3] Intel France, 2, rue de Paris, 92196 Meudon Cedex, France

E-mail: dcampora@cern.ch

Abstract.
The 2020 upgrade of the LHCb detector will vastly increase the rate of collisions the online system needs to process in software in order to filter events in real-time. 30 million collisions per second will pass through a selection chain where each step is executed conditional to its prior acceptance.

The Kalman filter is a process of the event reconstruction that, due to its time characteristics and early execution in the selection chain, consumes 40% of the whole reconstruction time in the current trigger software. This makes it a time-critical component as the LHCb trigger evolves into a full software trigger in the upgrade.

The algorithm Cross Kalman allows performance tests across a variety of architectures, including multi and many-core platforms, and has been successfully integrated and validated in the LHCb codebase. Since its inception, new hardware architectures have become available exposing features that require fine-grained tuning in order to fully utilize their resources.

In this paper we present performance benchmarks and explore the Intel® Skylake and Intel® Knights Landing architectures in depth. We determine the performance gain over previous architectures and show that the efficiency of our implementation is close to the maximum attainable given the mathematical formulation of our problem.

1. Introduction

The projected upgrade of the LHCb detector at CERN in 2020 [1] will increase the estimated collision rate to 30 MHz, requiring data processing at 40 Tbit/s. As part of the upgrade, the first stage of filtering known as hardware level trigger will be discontinued in favor of a full software trigger [2]. Consequently, the processing rate required in the software trigger will increase by a factor 40, due to both the increase in rate of events processed in software, and the influx of larger events.

The Kalman filter is a linear quadratic estimator invented by Rudolf E. Kálmán in 1960 [3], that allows accurate estimation of the state of a system. In its discrete formulation it consists of a *predict* phase where the state is propagated according to its physical model, and an *update* phase where the measurements observed are integrated in the state prediction. Since its inception, it has successfully been applied to a range of applications including space tracking [4] and meteorology [5].

In high-energy physics experiments, the Kalman filter is a fundamental part of the reconstruction of particle trajectories coming from particle collisions. In LHCb, the Kalman filter consumes 40% of the reconstruction time, rendering it a critical component in the reconstruction. In order to cope with the increase in data rate, the software in the LHCb reconstruction is being updated to current hardware standards.

IOP Conf. Series: Journal of Physics: Conf. Series **1085** (2018) 032046 doi:10.1088/1742-6596/1085/3/032046

Cross Kalman is a cross-architecture Kalman filter optimized for a variety of SIMD processors [6] [7]. It is optimized for low-rank Kalman filters on multi and many-core processors, and we have successfully integrated it in the LHCb reconstruction software. It also serves as a self-contained package to evaluate different architectures in the context of the LHCb upgrade.

In this paper, we focus on the evaluation of modern Intel® platforms, including Xeon Phi™ and Xeon® Scalable Processors. Our software takes into account features like the support of the AVX-512F ISA extension and high bandwidth MCDRAM. We evaluate the scalability and price performance of Cross Kalman in all platforms. In addition, we evaluate our Kalman filter integrated in the multithreaded LHCb framework *Gaudi* [8]. We show the ability of Cross Kalman to scale to all considered configurations, and draw conclusions over the platforms under analysis.

2. A vectorized Kalman filter

In the LHCb software level trigger, particle trajectory reconstruction (*track reconstruction*) involves a complex selection chain where the trajectory of each particle is determined in several steps, due to the various subdetectors that compose the LHCb tracker [9]. Several of these stages involve simplified fits, in the form of square root fits or simplified Kalman filters. In contrast, Cross Kalman performs a fit in three steps: For each track, a fit is performed in the *forward* direction, followed by the *backward* direction, and finally a *smoother* is applied to average both states [10].

Additionally, a track is composed of two kind of nodes. *Signal* nodes are produced by detector signals, whereas virtual *reference* nodes are placed at certain positions in the track in order to determine track parameters up-front for later stages of the reconstruction. While signal nodes trigger a Kalman filter step composed of *predict* and *update* phases, reference nodes only trigger the *predict* phase in either direction. Figure 1 depicts two particles traversing the LHCb detector. The particle at the top would trigger three *predicts* before the first *update*, in the forward direction.

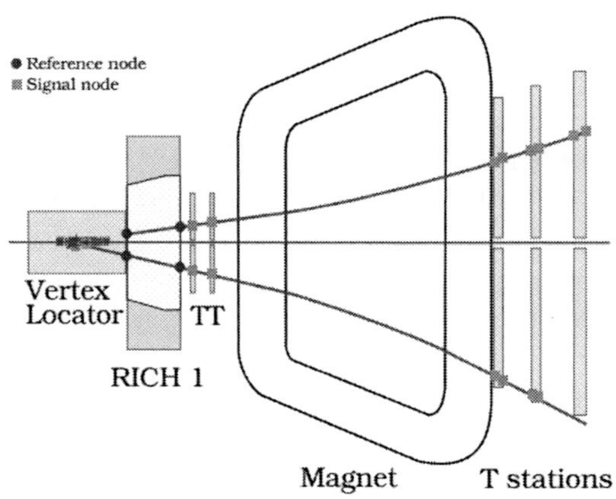

Figure 1: Schematic of two particle trajectories traversing LHCb detectors (in blue). The collision at their origin (in red) takes place in the Vertex Locator subdetector. The tracks are composed of signal nodes, produced in the Vertex Locator, TT and the three tracking stations T1, T2 and T3. A magnet bends the tracks between the TT and the tracking stations. Reference nodes are placed at predefined locations in order to determine their track parameters.

In order to adapt the inherent complexity of the problem to any SIMD architecture, we have developed Cross Kalman following three design aspects: *flow control*, *data structures* and *backend implementation*.

IOP Conf. Series: Journal of Physics: Conf. Series **1085** (2018) 032046 doi:10.1088/1742-6596/1085/3/032046

SIMD processors require non-divergent workflows to operate at peak performance. In Intel® processors, vector units perform the same operation over an array of data. This requirement has become more evident as the vector units have become wider. Therefore, we employ a static scheduler to assign computations to SIMD lanes. Since the execution of nodes in different tracks is independent, we execute them in a horizontally parallel scheme. We allow for a configurable lane width, which makes our scheduler scale to the considered SIMD platforms.

The LHCb Kalman filter needs to be processed in the backward direction as well. Our scheduler guarantees this requirement trivially by traversing the scheduled object in the reverse direction.

The *data structure* that we use in our application is Array of Structures of Array (AOSOA). This data structure has been shown to work well with SIMD architectures [11], as it maximizes locality. In Cross Kalman, the data is assumed to be already in this format in the order we require to process it. This assumption is safe, as the schedule can be generated prior to populating these data structures (cf. section 2.1).

The data is allocated in contiguous chunks of aligned memory. In order to avoid overheads from allocating big chunks of memory, *data stores* are created on demand and can store a configurable number of elements. In order to be able to traverse the data stores, iterators and reverse iterators that are aware of the store size are provided.

Finally, several *backend implementations* are provided to perform the fit computations. A scalar implementation, alongside Vectorclass [12] and UMESIMD [13] vectorized implementations are provided. We also provide OpenCL and CUDA implementations in our proto-application *Cross Kalman Mathtest*, showing that we can run on many types of SIMD architectures.

2.1. Integration

Our software has been integrated and validated in the Gaudi framework. Following the pre- and post-condition requirements of the current Kalman filter application *TrackMasterFitter* (TMF), we have developed *TrackVectorFitter* (TVF) as a SIMD aware drop-in replacement for TMF. TVF implements the design of Cross Kalman, with the necessary code modifications and data adaptations.

In order to minimize the impact of data transposition, we generate the static schedule prior to any data generation. This way, memory references are populated as the Cross Kalman filter would expect them. Then, the data is populated in-place. This method improves the data generation performance, and allows for future optimizations by vectorizing the methods for data generation. Additionally, the output data is required to adhere to Gaudi's data layout for the processes that follow in the selection chain after the Kalman filter. We transpose the data only once, at the end of TVF's execution. These optimizations reduce the overhead of changing the data structure from AOS to AOSOA. Nevertheless, we have observed a performance penalty which currently is 3.4% of the total execution time of TVF.

Finally, a version of Gaudi with multithreaded support is currently under development. We have made our application reentrant in order to support this version of the framework.

3. Results

The results in this section were obtained with Cross Kalman v1.7, compiled with `gcc 6.2.0` and flags `-O2 -march=native`. The events used, were generated with the official LHCb Monte Carlo event generator, and the results were validated against the existing Kalman filter implementation. NUMA pinning and TBB tasks were used to efficiently load the nodes. Table 1 describes the nodes used for these tests. Xeon® E5-2630 v3 (Haswell) is the current production platform at LHCb. Xeon® E5-2683 v4 and Xeon® Platinum 8170 represent medium-range and high-range processors from the Broadwell and Skylake family respectively. Xeon Phi™ 7210 (Knights Landing) is a many-core alternative to the more traditional Xeon® based solutions.

	Xeon® E5-2630 v3	Xeon® E5-2683 v4	Xeon Phi™ 7210	Xeon® Platinum 8170
# cores	8	16	64	26
base frequency	2.40 GHz	2.10 GHz	1.30 GHz	2.10 GHz
Cache	20 MB L3	40 MB L3	32 MB L2	35.75 MB L3
TDP	85 W	120 W	215 W	165 W
DRAM configuration	64 GB	64 GB	96 GB 16 GB MCDRAM	192 GB
STREAM bandwidth	41 GB/s	76 GB/s	77 GB/s (DRAM)	101 GB/s
Recommended price (ark.intel.com, August 2017)	\$667.00 - \$671.00	\$1846.00	\$1881.00	\$7405.00 - \$7411.00

Table 1: Machine configurations under study.

In Figure 2a we show a performance comparison across the architectures under consideration. The leftmost bar represents the scalar performance of Cross Kalman on the production platform. When running the vectorized implementation on the same processor, we observe a 1.36× speedup. We note also a 1.27× speedup of the Skylake processor over Knights Landing. A price performance comparison is shown in Figure 2b. The Xeon Phi™ Knights Landing processor is the most cost-effective solution for Cross Kalman, being 40% cheaper than the production platform. Medium and high-range CPUs are not cost-effective for our workload.

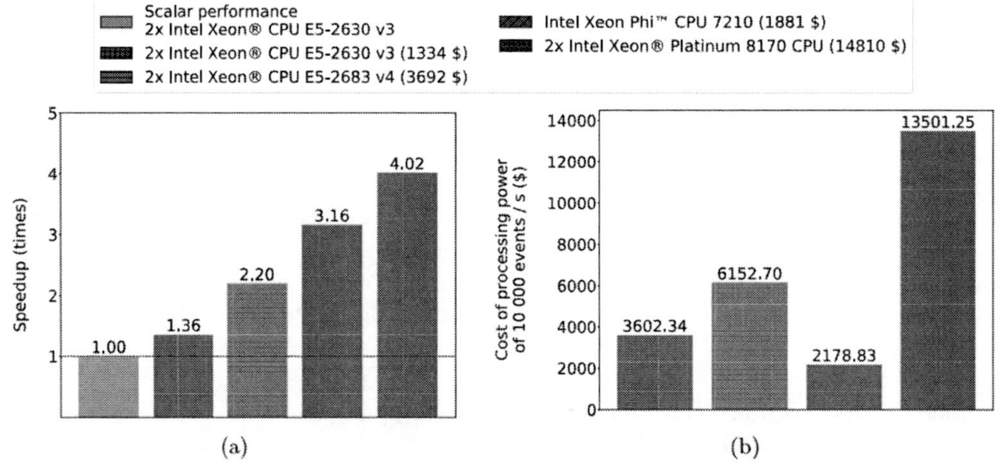

Figure 2: (a) Performance of Cross Kalman relative to scalar implementation across several architectures (higher is better). The best configuration for each architecture in terms of active threads is used, shown in Figure 3a. The leftmost bar represents Cross Kalman's scalar implementation, running on a dual-socket Xeon® E5-2630 v3 processor. (b) Price required to obtain a throughput of 10 000 events per second (lower is better). The Xeon Phi™ processor shows a better *price for throughput* ratio than its competitors.

Figure 3a shows the scalability of Cross Kalman in the architectures under study. Our workflow saturates Xeon® processors when configured with all physical processors, and we do not gain by adding extra Hyperthreads (HTs). The notable exception is the Knights Landing platform, where we gain an additional 31% with 2 HTs and 13% with 4 HTs. We attribute this behavior on Knights Landing to its higher bandwidth MCDRAM, which we believe is in the case of our application only saturated when using all available Hyperthreads.

Figure 3b shows the scalability of the Gaudi algorithm *TrackBestTrackCreator*, which contains TVF. The Gaudi framework is run in multithreaded mode with as many active

IOP Conf. Series: Journal of Physics: Conf. Series **1085** (2018) 032046 doi:10.1088/1742-6596/1085/3/032046

processors as indicated in the X axis. The behavior of this workload is very different than Cross Kalman. The application benefits from the HTs in both Xeon® architectures, however not in Xeon Phi™. Both in Xeon Phi™ and Skylake, the peak performance is observed when configured with 80 cores. We conclude Gaudi is not yet optimized for many-core execution.

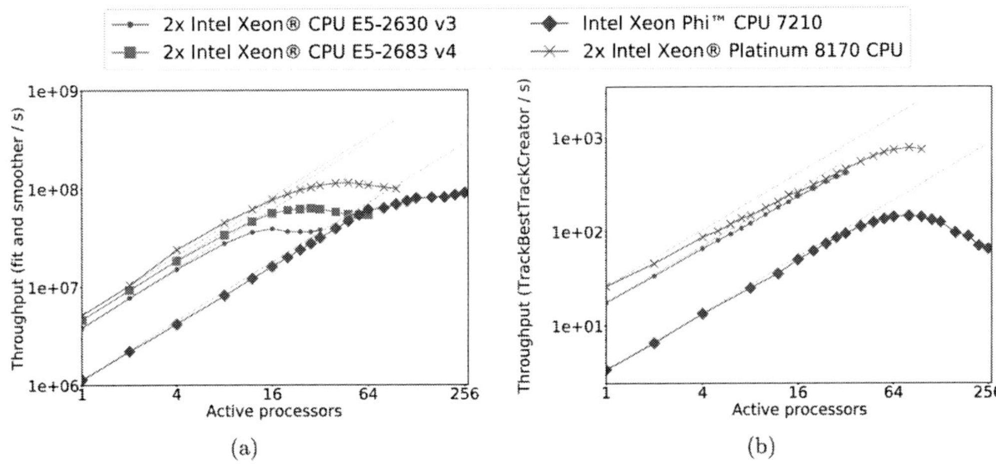

Figure 3: (a) Throughput of Cross Kalman across architectures. All processors show a degradation of performance when using HTs with the exception of Xeon Phi™. (b) Throughput of algorithm executing TVF in the Gaudi framework. Xeon® processors scale better than in Cross Kalman, in contrast to Xeon Phi™.

Finally we present a Roofline model [14] of our fit and smoother programs in Figure 4. The scalar implementation is provided by an emulated vector mode in UMESIMD, and we observe the overhead introduced by this emulation causes the scalar fit and smoother to underperform. We saturate the Skylake processor with our fit implementation when configured with SIMD lane width 8. Even though the smoother formulation shows a higher arithmetic intensity, we observe there may still be room for optimization.

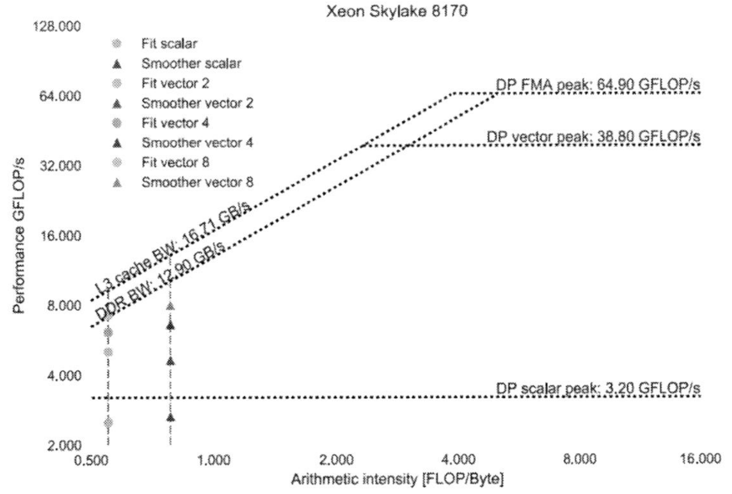

Figure 4: Roofline model of fit and smoother processes under Xeon® Platinum 8170 (Skylake). Different double precision vector width configurations are shown.

4. Conclusions

We have presented results of our evaluation of modern hardware architectures using Cross Kalman. We compared three dual-socket Xeon® architectures including a production Haswell LHCb server, and a many-core Xeon Phi™ architecture. The Skylake processor outperforms the other considered alternatives in raw computational performance. At the same time, the Xeon Phi™ processor under study is the most cost-effective solution for the Cross Kalman workload, offering the same performance as our production dual-socket Haswell system for 60% of its cost.

We have successfully ported our algorithm to the Gaudi multithreaded framework. We observe a different behavior in the scalability of Cross Kalman and Gaudi versions. We believe this is due to the inner workings of the framework, currently in development in its multithreaded form. In particular, many-core architectures do not seem to scale. We will further investigate the causes in the framework and try to mitigate this effect.

Cross Kalman saturates the Skylake platform for the fit process, while the arithmetic intensity of the smoother process seems to leave room for improvement, which we plan to study further. We will continue to track the performance of modern hardware architectures and adapt our software to it, and observe the evolution of the different platforms.

Acknowledgements

The authors would like to thank the High-Throughput Computing Collaboration at CERN openlab for fruitful discussions through the process of designing and writing the presented software. Thanks to C. Potterat for his contributions on the validation of the software. Thanks to F. Lemaitre for his contribution of the vectorized transposition code. In addition, thanks to W. Hulsbergen and R. Aaij for the mathematical discussions and data structure design. Finally, thanks to A. Riscos Núñez for his guidance and support.

References

[1] The LHCb Collaboration 2012 Framework TDR for the LHCb Upgrade: Technical Design Report Tech. Rep. CERN-LHCC-2012-007. LHCb-TDR-12 URL `https://cds.cern.ch/record/1443882`

[2] The LHCb Collaboration 2014 LHCb Trigger and Online Upgrade Technical Design Report Tech. Rep. CERN-LHCC-2014-016. LHCB-TDR-016 URL `https://cds.cern.ch/record/1701361`

[3] Kalman R E 1960 *Journal of Basic Engineering* **82** 35–45 ISSN 0098-2202 URL `http://dx.doi.org/10.1115/1.3662552`

[4] Mcgee L A and Schmidt S F 1985 Discovery of the Kalman filter as a practical tool for aerospace and industry Tech. rep. URL `https://ntrs.nasa.gov/search.jsp?R=19860003843`

[5] Houtekamer P L and Mitchell H L 1998 *Monthly Weather Review* **126** 796–811 ISSN 0027-0644 URL `http://journals.ametsoc.org/doi/abs/10.1175/1520-0493%281998%29126%3C0796%3ADAUAEK%3E2.0.CO%3B2`

[6] Cámpora Pérez D H *Journal of Physics: Conference Series* to appear

[7] Cámpora Pérez D H, Awile O and Potterat C *Euro-Par 2017: Parallel Processing Workshops* to appear

[8] Barrand G *et al.* 2001 *Comput. Phys. Commun.* **140** 45–55

[9] Schiller M 2011 *Track reconstruction and prompt K_S^0 production at the LHCb experiment* Dissertation University of Heidelberg

[10] Hulsbergen W 2009 *Nuclear Instruments and Methods in Physics Research Section A: Accelerators, Spectrometers, Detectors and Associated Equipment* **600** 471 – 477 ISSN 0168-9002 URL `http://www.sciencedirect.com/science/article/pii/S0168900208017567`

[11] Gou C, Kuzmanov G and Gaydadjiev G N 2010 *Proceedings of the 24th ACM International Conference on Supercomputing* ICS '10 (New York, NY, USA: ACM) pp 179–188 ISBN 978-1-4503-0018-6 URL `http://doi.acm.org/10.1145/1810085.1810111`

[12] Fog A 2012 VCL C++ vector class library URL `http://www.agner.org/optimize`

[13] Karpiński P and McDonald J 2017 *Proceedings of the 8th International Workshop on Programming Models and Applications for Multicores and Manycores* PMAM'17 (New York, NY, USA: ACM) pp 21–28 ISBN 978-1-4503-4883-6 URL `http://doi.acm.org/10.1145/3026937.3026939`

[14] Williams S, Waterman A and Patterson D 2009 *Communications of the ACM* **52** 65 ISSN 00010782 URL `http://portal.acm.org/citation.cfm?doid=1498765.1498785`

Continuous software quality analysis for the ATLAS experiment

A Washbrook, on behalf of the ATLAS collaboration
SUPA, School of Physics and Astronomy, The University of Edinburgh, James Clerk Maxwell Building, Peter Guthrie Tait Road, Edinburgh, EH9 3FD, United Kingdom

E-mail: `awashbro@ph.ed.ac.uk`

Abstract.
The software for the ATLAS experiment on the Large Hadron Collider at CERN has evolved over many years to meet the demands of Monte Carlo simulation, particle detector reconstruction and data analysis. At present over 3.8 million lines of C++ code (and close to 6 million total lines of code) are maintained by an active worldwide developer community. To run the experiment software efficiently it is essential to maintain a high level of software quality standards. The methods proposed to improve software quality practices by incorporating checks into the new ATLAS software build infrastructure will be discussed.

1. Introduction

The regular application of software quality tools in large collaborative projects is required to keep software defects to an acceptable level. Defects such as redundant code paths and errors of omission are often viewed as minor transgressions from the perspective of the individual developer and may not be flagged by compilers as part of the software build process. However if left unchecked the accumulation of defects invariably results in performance degradation at scale and leads to problems with the long-term sustainability of the software.

A wide range of software quality tools are used by the developer community in the ATLAS collaboration to identify, track and resolve any defects in their code. A limited selection of static code analysis tools (namely `cppcheck` [2] and `Coverity` [3]) are used to perform periodic scans across main development branches to provide scheduled notifications to code maintainers on any defects that require urgent resolution. More general code quality indicators, coverage testing tools and code formatting checkers are also used as part of the development and build process.

Although software quality tools have proven effective for defect identification, there remains a non-trivial sociological challenge to resolve defects in a timely manner. High priority defects (such as the existence of uninitialised variables or potential sources of memory leaks) are usually dealt with promptly, but lower priority defects can often remain unresolved. This is especially problematic in sections of code subjectively considered to be non-critical by maintainers and leads to a backlog of legacy defects that are likely to be unaddressed. The issue is compounded by areas of code where responsibility and provenance are unrecorded and prevalent where developer effort is reorganised or not retained.

To provide a high level of code quality a two-fold approach is therefore required. Firstly, defects over a certain age threshold should be periodically re-evaluated on their validity and importance and be disregarded if their impact is viewed as marginal. Secondly, to reduce defect accumulation it is necessary to identify and address defects *before* they are introduced into a widely used software release. It is this latter approach that is of interest in the studies described here.

2. Continuous Software Quality Evaluation

Recent wholesale changes to the ATLAS software build infrastructure have provided an ideal opportunity to apply software quality evaluation as an integral part of the development workflow. Source code control is now provided by `git`, thereby enabling the use of the `Gitlab` [4] repository manager to provide a framework for code review (or *merge requests*) before any modifications are included in the main development branches. Any submitted merge request (or subsequent code modifications within the same merge request) triggers an ordered sequence of build correctness checking and unit testing jobs to be run by a continuous integration (CI) service (`Jenkins` [5]). A test result summary is then pushed to the `Gitlab` merge request discussion via the `Gitlab` Application Programming Interface (API). All test results are monitored by a dedicated rota of code review shifters to validate changes and to enforce experiment software guidelines before a merge request is accepted.

By extension, this framework can be leveraged to include the preemptive checking of defects by the addition of software quality tests within the CI test chain. Any new defects left unresolved after the code review can be attributed to the developer initiating the request and thus an audit trail of known defects is automatically provided. Given the high volume of merge requests it is necessary for CI-based software quality tests to be quick (typically less than 5 minutes) to reduce load on the build systems and to provide a reasonable response time for input into the review process. It is therefore preferable to perform checks *only on code directly affected by any changes in a given merge request.* The results from software quality tools should remain open to interpretation from code reviewers to avoid erroneous or trivial quality concerns blocking urgent merge requests from being accepted.

3. Feasibility Studies

The feasibility of continuous software quality evaluation was tested by applying an example static code analysis tool to run within the ATLAS CI framework.

3.1. Development Testbed

The inclusion of software quality CI tests without full validation was considered to be too disruptive to the ongoing review process. This motivated the construction of a test development environment hosted at multiple sites and isolated from the operational software build infrastructure.

The use of `Docker` [6] containers enabled the rapid emulation of key services in the ATLAS software build infrastructure, thereby creating a sandbox to develop new CI tests and to explore how to interpret and report software quality test results. An illustration of the distributed test-bed can be seen in Figure 1.

The merge request workflow was simulated by importing a snapshot of the ATLAS software repository. Mock merge requests were then generated to inject a range of code modification patterns that could test the response of the software quality tools and reporting mechanisms. The mock requests generated trivial changes (such as the addition of a comment in the code) and modifications that were more representative of developer contributions. Parallel development and reproducible results across test-bed instances was made possible by placing the deployed container instances under version control in a registry hosted in a common `Gitlab` project area. A stable version of the service and CI test configuration states could then be tagged for inclusion into the main ATLAS software build infrastructure.

IOP Conf. Series: Journal of Physics: Conf. Series **1085** (2018) 032047 doi:10.1088/1742-6596/1085/3/032047

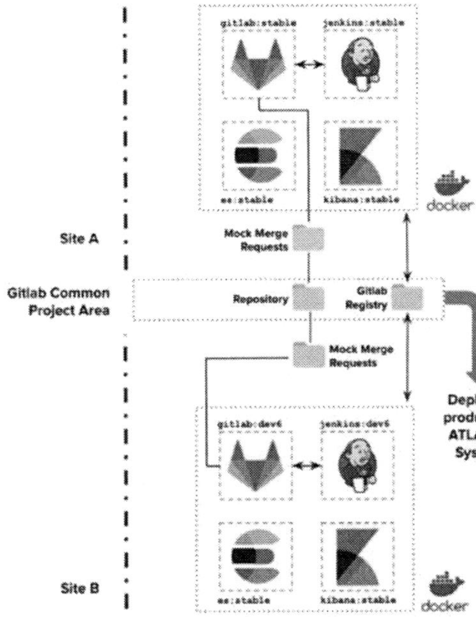

Figure 1. Illustration of the distributed container testbed used to develop CI software quality tests.

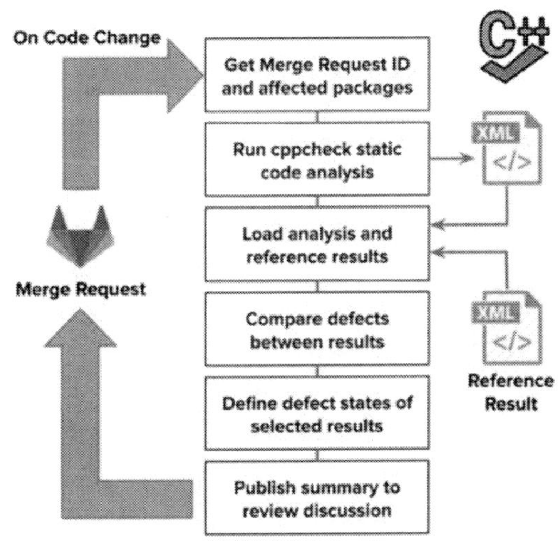

Figure 2. Illustrated workflow for the `cppcheck` CI test job.

3.2. Defect Identification and Reporting

The software quality application `cppcheck` was used to generate a representative set of defect results within a CI test job. This application was chosen because of the relatively quick time to generate results from a large code base (approximately 40 minutes for the entire ATLAS release) and for the minimal effort needed to apply checks to a selected subset of the code base.

The illustrated workflow is shown in Figure 2. It can be seen from the figure that payload checking is only small part of the overall workflow. The merge request data are first retrieved before the list of files affected by the proposed modification is extracted and passed to `cppcheck` for processing. This filtering step considerably reduced the execution time of `cppcheck` compared to running over the full development branch by default. It was reasonably expected that `cppcheck` would take less than a minute to run when only a small amount of code modifications were made.

In earlier iterations of this model it was anticipated that the results could be passed directly to the merge request for review. However, it was decided that this would pollute the review discussion. For example, a merge request may contain minor changes to numerous sections of code containing historical defects unrelated to the modification under review. It was therefore more beneficial to provide shorter feedback indicating a state change. Defects are either *introduced*, *removed* or remain *unresolved* against a reference result generated from the main development branch. The role of the software quality CI test after results gathering was therefore to identify and label differences between the merge request and reference result, and to publish a summary report of the state changes in the merge request discussion.

The summary report was constructed in a markdown syntax which could be parsed in a consistent format by `Gitlab`. An example report is shown in Figure 3. Some care was taken in applying defect prioritisation and in truncating the report to a maximum length where necessary. Any defects introduced by the proposed code modification were given more prominence to motivate action to be taken and to halt further increases to the overall number of defects.

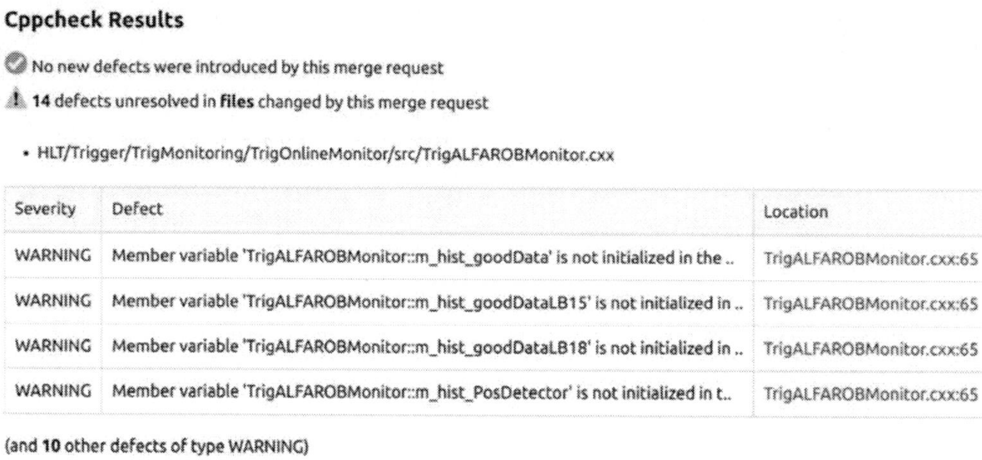

Cppcheck Results

No new defects were introduced by this merge request

14 defects unresolved in **files** changed by this merge request

- HLT/Trigger/TrigMonitoring/TrigOnlineMonitor/src/TrigALFAROBMonitor.cxx

Severity	Defect	Location
WARNING	Member variable 'TrigALFAROBMonitor::m_hist_goodData' is not initialized in the ..	TrigALFAROBMonitor.cxx:65
WARNING	Member variable 'TrigALFAROBMonitor::m_hist_goodDataLB15' is not initialized in ..	TrigALFAROBMonitor.cxx:65
WARNING	Member variable 'TrigALFAROBMonitor::m_hist_goodDataLB18' is not initialized in ..	TrigALFAROBMonitor.cxx:65
WARNING	Member variable 'TrigALFAROBMonitor::m_hist_PosDetector' is not initialized in t..	TrigALFAROBMonitor.cxx:65

(and **10** other defects of type WARNING)

Figure 3. Sample formatted report generated by the cppcheck CI test job.

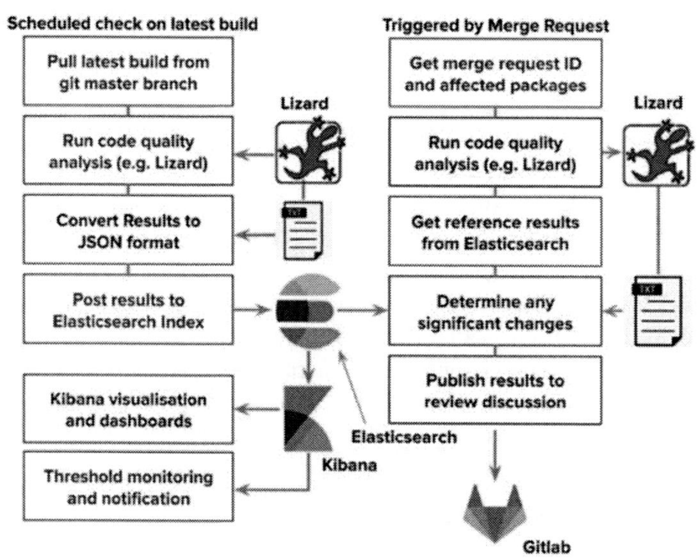

Figure 4. Illustrated workflow of the code quality trend analysis CI test job.

3.3. Trend Analysis

Defect identification is only one aspect of an overall software quality evaluation strategy. Code quality indicators can also be used to signify the quality and relative maintainability of a given piece of software. Indicators can be relatively simple metrics (e.g. the number of lines of code with comments) or more complex interpretations (e.g. function decision depth [8]).

Indicator values taken in isolation add limited information to the code review process. For example, a file with a reported *cyclomatic complexity* value (i.e. the number of possible logical paths through a section of code) needs to be placed into context by supplementing the value with trend information (e.g. whether this has significantly changed as a consequence of a code change) and a threshold value given at which affirmative action should be taken.

The application of code quality indicators within a CI framework was tested using the `Lizard` [7] code quality tool, which provided a relatively fast method of generating several key quality indicator metric values for code contained in the main development branch. The CI test workflow was similar to the `cppcheck` feasibility test described in Section 3.2 and is shown in Figure 4.

Reference results were converted into a `JSON` format and stored in a `Elasticsearch` [9] index. `Kibana` [9] was then applied to the curated dataset to provide a dashboard visualisation of indicator trends across successive versions of the code.

The response of this proposed workflow was tested by storing the code quality indicator results for each defined `C++` function from several historical snapshots of the main development branch (where each branch contains over 219,000 functions). In the most recent snapshot a highly branched code section was inserted into a sample function to determine if changes in cyclomatic complexity could be easily identified through `Elasticsearch` queries and `Kibana` dashboards.

Despite the low latency found in isolating potential quality issues across a large code base, some care will be needed managing the size of `Elasticsearch` indices if the number of stored snapshots and code quality indicators is increased.

4. Future Improvements

Once the evaluation and deployment of a CI-based software quality check is completed, experience will be gathered from code review shifters and developers to help optimise testing coverage and results presentation. Some areas of potential development are described below.

4.1. Alternative Code Analysis Tools

The CI software quality test workflow was designed to have minimal dependency on the underlying static code quality application. Further development of the workflow could adopt a plugin approach to accommodate results gathering and comparison from other tools. `Coverity` is now being evaluated to provide more comprehensive code analysis coverage. As before, only the modifications covered by an individual merge request should be considered for analysis to reduce the overall time of the CI-based test. An incremental build method is one such more lightweight method being considered.

4.2. Triage Methods

New processes to assign defect resolution action through *reviewer-led* triage are being tested to help reduce the overall number of defects in the main development branches. Any new approach would need to naturally fit within the code review workflow. One such implementation provides a connection service between the `Coverity` triage store and `Gitlab`. For example, `Gitlab` *webhooks* could monitor triage actions initiated in the merge request discussion. Reviewer requests (e.g. assigning a particular defect to a responsible group) could be converted into a API call to the `Coverity` triage store via a third party service. An immediate benefit would be the suppression of unimportant or incorrectly identified defects at source to avoid repetition of future code review effort.

5. Outlook

The feasibility tests described above have demonstrated that it is possible to provide continuous software quality evaluation by the inclusion of lightweight tests in the new code review process.

IOP Conf. Series: Journal of Physics: Conf. Series **1085** (2018) 032047 doi:10.1088/1742-6596/1085/3/032047

Before significant effort is invested in providing additional features there needs to be an assessment of whether related, but more established, tools and services that provide similar functionality are preferable. The balance between the benefits of enhanced test coverage to the additional overhead of deployment (and user education) of a service external to the chosen ATLAS build infrastructure tools would have to be addressed.

The reduction of software defects can be helped further by developer engagement. For example, software quality considerations can be applied by promoting analysis tools integrated into code development environments. This will help catch defects before the code is committed for review.

Finally, the recent migration from legacy and bespoke elements in the build process has meant that the development of software quality CI tests has been independent of the underlying code. With growing experience the practices described above could be applied outside of the ATLAS experiment to help improve the code quality in other software projects in the high energy physics community and elsewhere.

References

[1] ATLAS Collaboration, The ATLAS Experiment at the CERN Large Hadron Collider, 621 JINST 3 (2008) S08003.

[2] Cppcheck - A tool for static C/C++ code analysis, "Cppcheck" [software], version 1.80, 2017. Available from http://cppcheck.sourceforge.net/ [accessed 2017-10-01]

[3] Synopsys Static Analysis, "Coverity" [software], version 8.7.1, 2017. https://www.synopsys.com/software-integrity/security-testing/static-analysis-sast.html [accessed 2017-10-01]

[4] Gitlab Community Edition [software], version 9.2.4, 2017. Available from https://packages.gitlab.com/gitlab/gitlab-ce [accessed 2017-10-01]

[5] Jenkins open source automation server [software], version 2.46.2, 2017. Available from https://jenkins.io/download/ [accessed 2017-10-01]

[6] Docker [software], version 1.12.6-11, 2017. Available from https://www.docker.com/get-docker [accessed 2017-10-01]

[7] Lizard [software], Available from https://github.com/terryyin/lizard [accessed 2017-10-01]

[8] Software Quality Metrics Resources [Online]. Available from https://www.imagix.com/links/software-metrics.html [accessed 2017-10-01]

[9] Elastic Open Source Search and Analytics [software], version 5.5.1, 2017. Available from https://www.elastic.co/downloads [accessed 2017-10-01]

IOP Conf. Series: Journal of Physics: Conf. Series **1085** (2018) 032048 doi:10.1088/1742-6596/1085/3/032048

Parallelized JUNO simulation software based on SNiPER

T Lin[1], J H Zou[1], W D Li[1], Z Y Deng[1], G F Cao[1], X T Huang[2] and Z Y You[3]

(On Behalf of the JUNO Collaboration)

[1]Institute of High Energy Physics, Chinese Academy of Sciences, Beijing, China
[2]Shandong University, Jinan, China
[3]Sun Yat-sen University, Guangzhou, China

E-mail: `lintao@ihep.ac.cn, zoujh@ihep.ac.cn`

Abstract. The Jiangmen Underground Neutrino Observatory (JUNO) is a neutrino experiment to determine the neutrino mass hierarchy. It has a central detector used for neutrino detection, which consists of a spherical acrylic vessel containing a 20 kt liquid scintillator (LS) and about 18,000 20-inch photomultiplier tubes (PMTs) to collect light from LS. Around the central detector, there is a water pool to provide shielding from background radioactivity. The outer water pool is also equipped with about 2000 PMTs to measure cosmic ray muons by detecting Cherenkov light.

As one of the important parts in JUNO offline software, the serial simulation framework is developed based on SNiPER. It manages physics generator, detector simulation, event mixing and digitization. However, Geant4 based detector simulation of such a large detector is time-consuming and challenging. It is necessary to take full advantages of parallel computing to speed up simulation. Starting from version 10.0, Geant4 supports event-level parallelism. Even though this is based on pthread, it can be extended with other libraries such as Intel TBB and MPI. Therefore it is possible to parallelize JUNO simulation framework by integrating Geant4 and SNiPER.

In this paper, our progress in developing parallelized simulation software is presented. The SNiPER framework can run in sequential mode, Intel TBB mode or other modes. The SNiPER task component is in charge of the event loop, which is like a simplified application manager. Two types of tasks are introduced in the simulation framework, one is a global task and another is a worker task. The global task will run only once to initialize detector geometry and physics processes before any other tasks spawned. Later it is accessed by other tasks passively. The worker tasks will be spawned after the global task is done. In each worker task, a Geant4 run manager is invoked to do the real simulation. In this way the simulation framework and the underlying TBB have been decoupled. Finally, the software performance of parallelized JUNO simulation software is also presented.

1. Introduction

JUNO [1,2] is an underground neutrino experiment to measure the neutrino mass hierarchy and oscillation parameters. It will be located in southern China, about 53 km away from two nuclear power plants. Figure 1 shows the schematic view of the JUNO detector. To detect neutrinos,

IOP Conf. Series: Journal of Physics: Conf. Series **1085** (2018) 032048 doi:10.1088/1742-6596/1085/3/032048

the innermost part, which is called the central detector, is filled with a 20 kt liquid scintillator (LS). When neutrinos react with the LS, light is produced and then collected by the surrounding photomultiplier tubes (PMTs). To suppress background radioactivity from PMTs and rocks, water surrounds the spherical acrylic vessel. The outer water pool is also equipped with PMTs to veto cosmic ray muon events by detecting Cherenkov light. On the top of the water pool, a top tracker is used to measure muons.

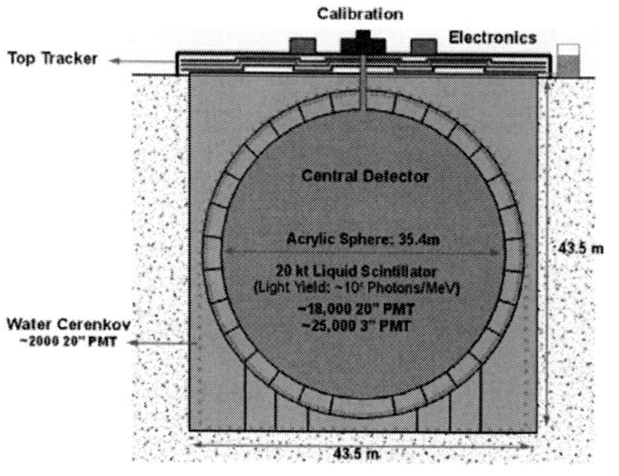

Figure 1. Schematic view of JUNO detector

Offline software is an important part of the JUNO experiment. Simulation software is used for detector design and optimization, algorithm tuning and physics studies. PMT waveform reconstruction is used to calibrate charge and time for each PMT. Reconstruction algorithms are used to reconstruct events and get physics objects for physics analysis. These algorithms are built upon a software framework called SNiPER [3], and compose a full chain of data processing [4], as shown in Figure 2.

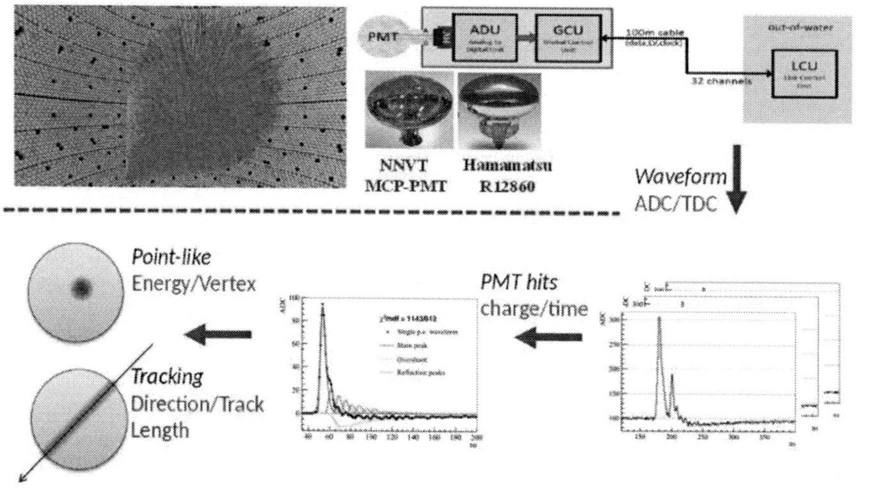

Figure 2. Offline data processing chain

IOP Conf. Series: Journal of Physics: Conf. Series **1085** (2018) 032048 doi:10.1088/1742-6596/1085/3/032048

2. Towards parallelized detector simulation framework

A serial detector simulation framework [5] is implemented to integrate SNiPER and Geant4 [6,7]. Figure 3 shows the core class diagram. It is lightweight with only a few classes in the design. Simulation software can be easily migrated from a standalone application into simulation framework. However, simulation of a large detector is not only time consuming, but requires considerable memory and I/O resources as well. After the construction of the full geometry, the physical memory allocation is about 700 MB. More memory is required during the execution of event loop. Especially to the cosmic ray muons, millions of optical photons are produced and propagated, which is a challenge to the memory. Meanwhile, the waveforms are generated with 1 GHz sampling rate and written to files during digitization simulation, which is I/O consuming.

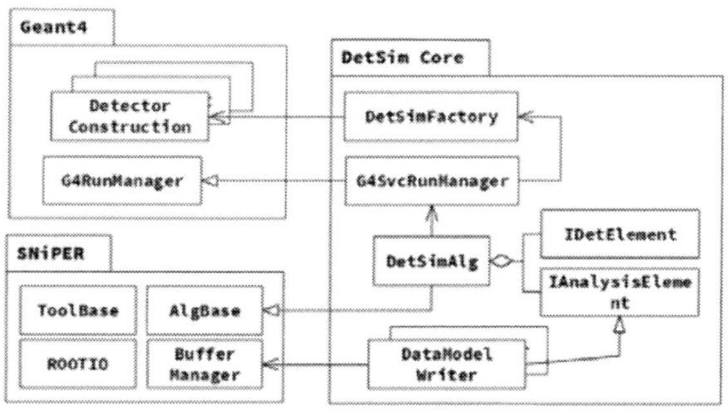

Figure 3. Design of detector simulation framework

A parallelized simulation software is necessary for Monte-Carlo production in the future. It is foreseeable that more and more cores will be integrated in a simple chip, such as the Intel Knights Landing (KNL) processor. But it is not economical to equivalently increase the amount of memory. The memory can be a bottleneck in this situation. According to the results of ATLAS simulation on the Intel KNL, less than 200 workers are used in multi-process jobs due to the limited memory, while 288 threads are used in multi-threaded jobs [8]. In order to use computing and memory resources efficiently, the underlying frameworks adopt the multi-threading model for both the ATLAS and CMS experiments [9–11].

In the SNiPER framework, a parallelized solution based on Intel Threading Building Blocks (TBB) [12] is proposed. Intel TBB supports task-based programming, which simplifies the thinking of parallel computing. In SNiPER, `Task` is an important component, which is like a lightweight application manager. In sequential mode, a `Task` is configured by the user and then invokes registered algorithms. To support parallel computing, SNiPER Muster (Multiple SNiPER Task Scheduler) is designed and implemented to integrate the SNiPER `Task` component and Intel TBB worker. Multiple instances of `Task` components are created before execution. The SNiPER Muster does not execute these `Tasks` directly. It creates corresponding TBB-based workers to execute the `Task` component. For better performance, each worker object is recycled rather than allocated for each event, which is called *contiuation* in the Intel TBB terminologies. At the beginning of each recycling, the worker will takes over an idle SNiPER `Task` component in priority order. As shown in Figure 4, the first worker takes over an I/O `Task`, and it will not execute a regular `Task` until the next recycling.

Starting from version 10.0, Geant4 supports multi-threaded simulation [13]. Events are simulated in different threads. With the evolution of Geant4, it is possible to run simulation with Intel TBB, MPI and so on [14]. A parallelized simulation framework is developed to integrate the latest Geant4 and SNiPER Muster, which takes full advantages of parallel computing. Based

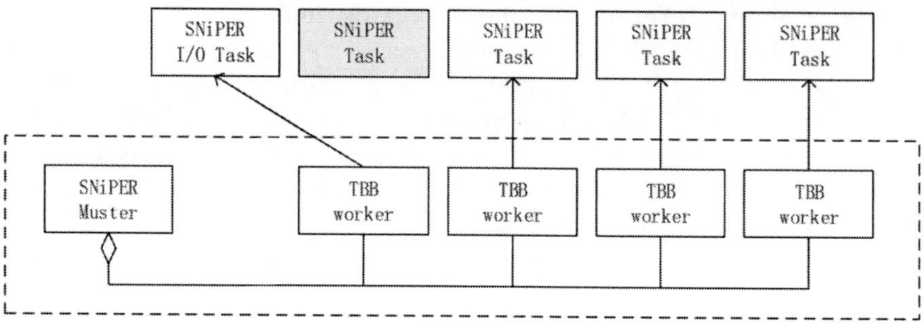

Figure 4. SNiPER Muster

on SNiPER Muster, the simulation framework is decoupled from the underlying Intel TBB, and becomes more flexible and simpler.

3. SNiPER Muster based simulation framework

The parallelized simulation framework developed is based on SNiPER Muster. Several new classes are introduced, compared with the serial simulation framework. The event loop is still controlled by SNiPER instead of Geant4, however, the control flow is rather different from the serial simulation framework. As shown in Figure 5, the simulation framework consists of a global task and several worker tasks. The global task is in charge of initialization of the detector geometry and physics lists. A customized master run manager is derived from Geant4's `G4MTRunManager` to take control of the simulation. A corresponding service is created, so that SNiPER can initialize run manager automatically.

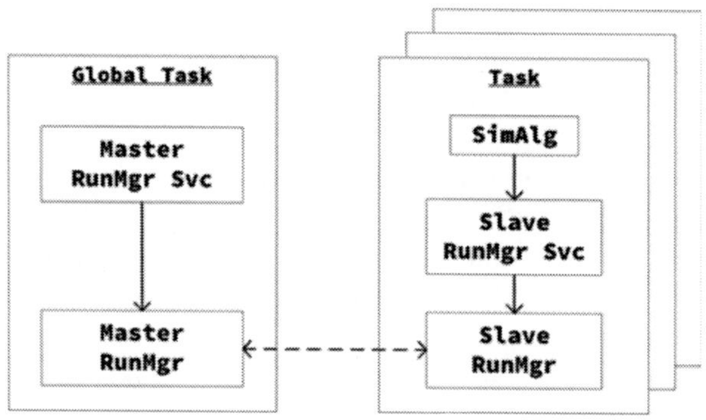

Figure 5. Design of parallelized simulation framework

After the initialization of the global task, SNiPER Muster starts the worker tasks. Each worker manages a simulation algorithm and a slave run manager service. The simulation algorithm can invoke the slave run manager via the corresponding service. The slave run manager is derived from Geant4's `G4WorkerRunManager`, which breaks the original event loop and simulates one event each call. Instead of executing a fixed number of events in each worker, the execution of worker is controlled by SNiPER Muster. If there are events to be simulated, the worker will invoke the simulation algorithm. Otherwise, the worker will be stopped. Hence, this event dispatching procedure can make full usage of all available CPUs.

IOP Conf. Series: Journal of Physics: Conf. Series **1085** (2018) 032048 doi:10.1088/1742-6596/1085/3/032048

4. Performance measurements

In these performance measurements, 1000 events of single γs with an energy of 2.2 MeV are generated at detector center. The simulation runs on a blade server with Intel Xeon CPU E5-2680 v3 @ 2.5 GHz and 64 GB of memory. The operating system is Scientific Linux 6.5 with GCC 4.9.4. The Geant4 version is 10.03.p01, built with `global-dynamic` TLS (Thread Local Storage) model. The physics lists `G4EmStandardPhysics` and `G4OpticalPhysics` are used in simulation. An average of three measurements is used for each case. In order to eliminate I/O interference, the events are not saved into ROOT files.

Figure 6 shows speedup ratio versus number of threads. The dotted line represents the ideal result. The gray line with full square represents the result without any optimization. Up to four threads, the speedup is good. However, when increasing the number of threads to 24, the speedup ratio is stuck around 8.

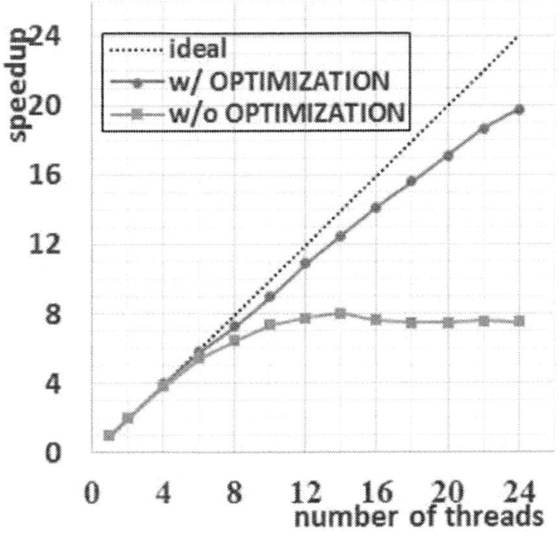

Figure 6. Speedup versus number of threads

The Intel's VTune Amplifier profiler is used to analyze this performance issue. We find that the hotspot comes from a mutex used in Geant4's `G4MaterialPropertiesTable`. This mutex is used when each optical photon calculates its group velocity. The problem is that there are a lot of optical photons in JUNO simulation, so this mutex is accessed frequently in order to get the velocity. When there are multiple threads accessing the mutex, the lock becomes a bottleneck. The optimization is moving this part into initialization stage. When refractive index is set, the group velocity is precalculated during initialization. As shown in Figure 6, the blue line with full circle represents the result with optimization. After this optimization, better performance scaling is achieved.

5. Conclusions

In this paper, we show how JUNO simulation software is parallelized based on SNiPER Muster. By introducing several new classes, Geant4 10 is integrated into SNiPER Muster. The global task initializes detector geometry and physics processes while the worker is invoked by SNiPER Muster to simulate one event. Events are dispatched dynamically to different workers, which

IOP Conf. Series: Journal of Physics: Conf. Series **1085** (2018) 032048 doi:10.1088/1742-6596/1085/3/032048

improves CPU utilization. The software performance studies show that there are performance issues due to a mutex in material properties table. After an optimization, the simulation software achieves a close to linear speedup, which fulfills requirements.

To speedup events such as cosmic ray muons, we are investigating how to use track-level parallelism in the simulation framework. The idea is to use MPI to dispatch tracks from the master node to different worker nodes. In the future, MPI and TBB can be applied simultaneously in different levels of the JUNO parallelized simulation software.

Acknowledgments

This work is supported by Joint Large-Scale Scientific Facility Funds of the NSFC and CAS (U1532258), the Strategic Priority Research Program of the Chinese Academy of Sciences, Grant No. XDA10010900, National Natural Science Foundation of China (11575224, 11405279, 11675275).

References

[1] An F *et al.* (JUNO) 2016 *J. Phys.* **G43** 030401 (*Preprint* 1507.05613)
[2] Djurcic Z *et al.* (JUNO) 2015 (*Preprint* 1508.07166)
[3] Zou J H, Huang X T, Li W D, Lin T, Li T, Zhang K, Deng Z Y and Cao G F 2015 *J. Phys. Conf. Ser.* **664** 072053
[4] Huang X T, Li T, Zou J H, Lin T, Li W D, Deng Z Y and Cao G F 2016 *PoS* **ICHEP2016** 1051 URL https://pos.sissa.it/282/1051
[5] Lin T, Zou J, Li W, Deng Z, Fang X, Cao G, Huang X and You Z (JUNO) 2017 *22nd International Conference on Computing in High Energy and Nuclear Physics (CHEP 2016) San Francisco, CA, October 14-16, 2016* (*Preprint* 1702.05275) URL http://inspirehep.net/record/1514060/files/arXiv:1702.05275.pdf
[6] Agostinelli S *et al.* (GEANT4) 2003 *Nucl. Instrum. Meth.* **A506** 250–303
[7] Allison J *et al.* 2006 *IEEE Trans. Nucl. Sci.* **53** 270
[8] Farrell S, Calafiura P, Leggett C, Tsulaia V and Dotti A (ATLAS) 2017 *J. Phys. Conf. Ser.* **898** 042012
[9] Calafiura P, Lampl W, Leggett C, Malon D, Stewart G and Wynne B 2015 *J. Phys. Conf. Ser.* **664** 072031
[10] Stewart G A *et al.* (ATLAS) 2016 *J. Phys. Conf. Ser.* **762** 012024
[11] Sexton-Kennedy E, Gartung P, Jones C D and Lange D 2015 *J. Phys. Conf. Ser.* **608** 012034
[12] Intel threading building blocks https://www.threadingbuildingblocks.org/ [Online; accessed 19-March-2018]
[13] Allison J *et al.* 2016 *Nucl. Instrum. Meth.* **A835** 186–225
[14] Dotti A, Asai M, Barrand G, Hrivnacova I and Murakami K 2016 *Proceedings, 2015 IEEE Nuclear Science Symposium and Medical Imaging Conference (NSS/MIC 2015): San Diego, California, United States* p 7581867 (*Preprint* 1605.01792) URL http://inspirehep.net/record/1456144/files/arXiv:1605.01792.pdf

IOP Conf. Series: Journal of Physics: Conf. Series **1085** (2018) 032049 doi:10.1088/1742-6596/1085/3/032049

The LHCb Software and Computing Upgrade towards LHC Run 3

S Roiser[1] and C Bozzi[1,2]
on behalf of the LHCb Computing Project
[1] European Organization for Nuclear Research (CERN), Geneva, Switzerland
[2] Sezione INFN di Ferrara, Ferrara, Italy

E-mail: stefan.roiser@cern.ch, concezio.bozzi@cern.ch

Abstract. LHCb is planning major changes for its data processing and analysis workflows for LHC Run 3. Removing the hardware trigger, a software only trigger at 30 MHz will reconstruct events using final alignment and calibration information provided during the triggering phase. These changes pose a major strain on the online software framework which needs to improve significantly. The foreseen changes in the area of the core framework include a re-design of the event scheduling, introduction of concurrent processing, optimizations in processor cache accesses and code vectorization. Furthermore changes in the areas of event model, conditions data and detector description are foreseen. The changes in the data processing workflow will allow an unprecedented amount of signal events to be selected and therefore increase the load on the experiments simulation needs. Several areas of improvement for fast simulation are currently being investigated together with improvements needed in the area of distributed computing. Finally the amount of data stored needs to be reflected in the analysis computing model where individual user analysis on distributed computing resources will become inefficient. This contribution will give an overview of the status of those activities and future plans in the different areas from the perspective of the LHCb computing project.

1. Introduction

The LHCb experiment will be upgraded for data taking in Run 3 and after [1]. Tu fully profit from the increase in instantaneous luminosity from 4×10^{32} to $2 \times 10^{33} cm^{-2} s^{-1}$ the current L0 hardware trigger will be removed, and a full software trigger will be deployed, with the goal of sustaining trigger capabilities up to the inelastic event rate of 30 MHz. The full read-out of the detector at this rate has a major impact on software and computing systems.

A study of the trigger output rates for different physics scenarios is reported in the LHCb Trigger and Online Upgrade TDR [2]. In summary, output bandwidths between 2 and 5 GB/s are expected, to be compared with 700 MB/s in Run 2. If the current computing model and software framework are kept, the data storage capacity and computing power required to process data at this rate, and to generate and reconstruct equivalent samples of simulated events, will exceed the current capacity by at least an order of magnitude. It is therefore mandatory to study how the current model can be changed to sustain this scenario.

The introduction of the split High Level Trigger (HLT) concept in the Run 2 data taking [3] has allowed the integration of real-time calibration and alignment into the data taking process [4]. The online reconstruction is therefore equivalent to the offline one, thus enabling analyses to be

IOP Conf. Series: Journal of Physics: Conf. Series **1085** (2018) 032049 doi:10.1088/1742-6596/1085/3/032049

performed on physics objects produced directly out of the trigger. Furthermore, the introduction of the `Turbo` data stream, where only a subset of the event information is saved, allows to decrease the event size, and therefore the output bandwidth and the need for additional offline computing resources. The physics program at LHCb will be clearly maximized by further exploiting these concepts, which ultimately means to move the event reconstruction and selection as close as possible to the online processing, and to implement data streaming as early as possible in the computing model.

A major challenge to be faced in the upgrade era is therefore the efficient usage of computing resources. Multi-core and many-core architectures, as well as coprocessors such as GPGPUs and FPGAs, offer significant speedups, making them particularly suited for the event filter farm. The efficient use of these new processors requires a paradigm shift in the LHCb core software framework, scheduling, and event model. The R&D activities related to this topic are presented in Section 2.

The events accepted by the trigger will be distributed for analysis in a similar way as today through the Worldwide LHC Computing Grid (WLCG). Section 4 details the challenges to be dealt with in this domain. Given that event reconstruction and selection will happen at the trigger level, the event format will be compact, and comparable to the ones currently used (*Data Summary Tape, DST* or *Micro Data Summary Tape, μDST*). Offline data processing will be very limited, and the storage costs for the recorded data will be driven by the HLT output rate. The vast majority of offline CPU work will be therefore dedicated to the production of simulated events. As the number of simulated events is proportional to the number of data events, it follows that the work needed for simulation will exceed the expected resource increase considerably. It is therefore necessary to speedup simulation and pursue alternative ways, where faster or parameterized simulations are employed where possible. Section 3 summarizes the work plan for these activities. A summary is given in Section 5.

2. Core software framework

The stringent requirements on the data processing throughput of the LHCb upgrade imply that the utilization of computing capacities of current and emerging hardware has to be improved both quantitatively and qualitatively.

2.1. Framework

The data processing model of the `Gaudi` framework [5, 6] used in LHCb currently implies sequentially processing of events. To mitigate these limitations, several techniques and models are considered, such as simultaneous multithreading, a task-based programming model [7], by which the basic computation units, *i.e.* the `Gaudi` algorithms, are represented and handled as tasks, and concurrent data processing, with inter- and intra-event, and (optionally) intra-task concurrency. The sustainability and conformity of the above-mentioned principles in the `Gaudi` framework were demonstrated by a prototype [8], which constitutes the building block of the framework for the upgrade.

Another wide-ranging change will concern the mechanisms of declaration of data dependencies between the `Gaudi` Algorithms. In particular, their input and output data requirements must be explicitly declared to the framework. This is needed by the `Gaudi` task scheduler for task concurrency control.

The concurrent paradigm places several stringent constraints on the event data model. First and foremost, once data is made available and written to the Transient Event Store (TES), it must remain immutable.

A functional framework was developed with the above requirements in mind, including a read-only TES, in the context of `Gaudi`. About one hundred algorithms have been adapted, some of them were partially ported to be used in the current framework, with significant gain

IOP Conf. Series: Journal of Physics: Conf. Series **1085** (2018) 032049 doi:10.1088/1742-6596/1085/3/032049

in performance. An application was built (*Mini-Brunel*), that includes a Kalman filter for charged track reconstruction. and full photon reconstruction in the RICH detectors. Preliminary measurements of the performance show a good scalability in terms of multi-threading, a very low and scalable memory footprint. Fig 1 shows a comparison of memory usage where the multi-threaded jobs uses a factor 40 less memory than the multi-processing one.

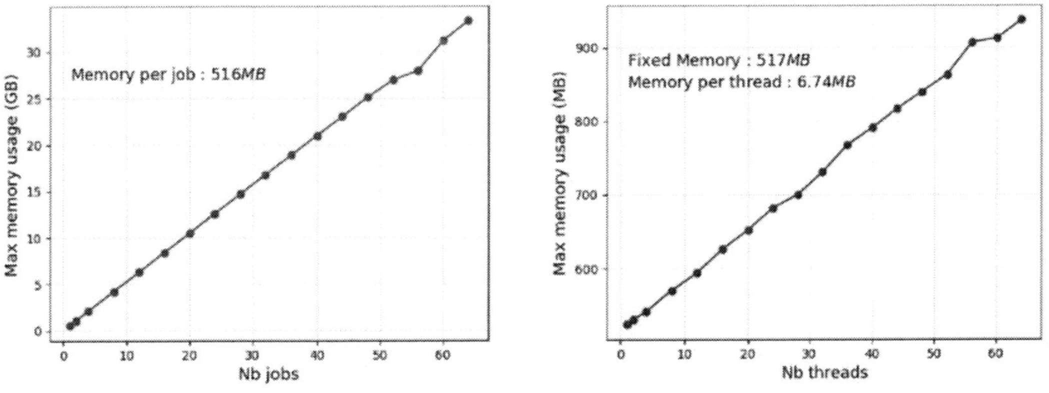

Figure 1. Memory consumption for multi-job and multi-thread approaches

2.2. Event Model

The LHCb event data model [9] has been very successful in enabling developers to efficiently write reconstruction or selection code, by providing *e.g.* an object representation of raw data, reconstructed objects and decay trees, and limiting the number of changes in the interface to the data model. However, as a result of the Array of Structures (AoS) design of the event model classes in its current form, the exploitation of SIMD instruction is more difficult. In addition, current event model objects are not composable, so if information needs to be added to objects without modifying them, they need to be copied. This costs significant amounts of memory, which in turn leads to sub-optimal usage of resources.

As a result of the changes to the core framework, the event model objects are therefore required to become read-only after their initial creation, to be composable, to allow the choice of memory layout (SoA, AoS, etc.) and to use single precision where possible.

2.3. Detector Description and Conditions Database

The LHCb Detector Description (`LHCbDD` [10]) is based on a home-made framework, developed along the lines of the Geant 4 geometry, with special extensions to implement generic active volumes. The persistent format chosen was XML. The current implementation has several limitations, in particular it is not thread-safe, and therefore has negative impact on the throughput performance of a multithreaded software framework.

The `DD4Hep` toolkit [11] has been investigated as a replacement. It has been shown[12] that using `DD4Hep` from `Gaudi` is relatively easy. The full LHCb geometry has been automatically converted and tested. In addition, a geometry where the detailed structure of the LHCb detector is averaged out over a small number of elements, has been defined which e.g. improves the throughput for algorithms such as the Kalman filter.

The LHCb conditions data are currently stored in a database managed with the `Cool/Coral` libraries [13], developed by CERN/IT in collaboration with LHCb and ATLAS. The level of thread-safety of the `Cool/Coral` library is unclear. Moreover, the XML format used for

IOP Conf. Series: Journal of Physics: Conf. Series **1085** (2018) 032049 doi:10.1088/1742-6596/1085/3/032049

		SSE4		AVX2	
		time (s)	Speedup	time (s)	Speedup
double	scalar	233.462		228.752	
double	vectorized	122.259	1.90	58.243	3.93
float	scalar	214.451		209.756	
float	vectorized	55.707	3.85	26.539	7.90

Table 1. Performance of vectorized Rich photon reconstruction

persistency is slow to read and not compact. The transient representation of conditions data are also likely to change, in order to speed up code performance.

In order to adapt the conditions interfaces to a multithreaded environment, a proposal and a prototype have been developed in the context of `Gaudi`. A prototype to manage the conditions backend with `git` is also available, that offers better performance and easier maintenance.

2.4. Optimization

The task-based framework described above has been successfully used on multi- and many-cores architectures. Further studies have been performed on alternative architectures such as GPGPUs and FPGAs, with promising results in terms of throughput [14, 15]. An example for performance improvement in the RICH photon reconstruction via code vectorsation is shown in Table 1. However, a large investment in code rewrite should be taken into account and a cost-benefit analysis should be performed before taking decisions, as well as the reproducibility of results obtained on these alternative architectures.

Performance improvements are expected by taking advantage of wide processing units and improved scheduling in `Gaudi`, and by minimizing cache misses. Examples of parallelism in the LHCb reconstruction software are given in [16, 17]. Monitoring and in-depth measurements in these domains are challenging but nevertheless needed in order to make substantial progress.

3. Simulation

In the current offline processing, the majority of CPU work (about 70%) is spent for Monte Carlo simulation. The current simulated samples correspond to about 15% of the total data statistics. Given that in the LHCb upgrade the trigger purity will increase, it is expected that a larger fraction of events will have to be simulated. This, together with the increased luminosity and the more complex nature of the events, results in an increase between one and two orders of magnitudes of needed CPU work.

Two avenues are being pursued in order to mitigate the required computing resources: the usage of fast simulations, where speed-up is obtained with either fully parameterized, or fast detector response, or reuse of events, and the usage of parallelized simulation frameworks (multi-threading, multi-processor) and of geometries of different complexity.

The flexibility of the LHCb Simulation framework, `Gauss` [18], allows to implement, in addition to the full `Geant4` simulation [19, 20], a variety of predefined safe and easy to use alternative configurations. Many options are possible including simulating only part of an event, replacing a full `Geant4` simulation with faster versions for a given detector or disabling specific processes (*e.g.* Cherenkov effect in RICH detectors), stopping particles at a specific stage, reducing the geometry, re-using the underlying event or merging a simulated signal with real

IOP Conf. Series: Journal of Physics: Conf. Series **1085** (2018) 032049 doi:10.1088/1742-6596/1085/3/032049

data, or even using a fully-parametric simulations providing reconstructed objects and using as much as possible available packages like Delphes [21].

Another important aspect is enabling concurrency in the simulation application. This would allow the usage of multi- and many-core architectures. The main player here is `Gaudi` with its multi-threaded version, see above, and its proper coexistence with the multi-threaded version of external tools like `Geant4`.

In any simulation with complicated geometry, the majority of time is spent navigating the geometry itself. New geometry packages with improved performances have become available for `Geant4`, like `USolid` [22]. This alternative version of `Geant4` geometry should be tested in the `Gauss` framework. In addition, the outcome of the evaluation on the change in the detector description (see Section 2) would have an impact on how `Gauss` transfers the geometry information to `Geant4`. A fully vectorized geometry package, `VecGeom`, is being developed in the context of `GeantV` [23], the vectorized version of `Geant4`. These newer options should be tested as soon as they will become available.

On a broader view, LHCb started a collaboration with the software developers of the Future Circular Collider (FCC) project to create `Gaussino`, an experiment-independent version of `Gauss`, with the intention that `Gauss` will be based on it and provide the LHCb specific functionality. The development of `Gaussino` should allow an easier way to test new design and package options with simpler settings but in an environment very similar to that of LHCb.

4. Distributed computing and data analysis

In the data flow model used in Run 1, the full raw event information is kept up until the end in the processing steps, the stripping, in which a selected subset of triggered events is provided to users for physics analysis. This model implies a heavy use of offline CPU, for reconstruction and stripping campaigns, and storage resources for raw data. Moreover it requires data to be moved from tape to disk and vice-versa at each stripping or reconstruction campaign to allow processing. One important advantage of this model is that the full event information is always saved (either on disk or on tape), thereby allowing for improvements in the reconstruction and selection criteria to be introduced at any moment.

Although very robust and well oiled, the Run 1 data flow model cannot be sustained in the upgrade era. The projected computing resources will not allow for the storage of the raw event information for the entire collected luminosity. Therefore the whole data processing, from the data acquisition to the final physics objects, needs to be changed.

The concepts of split HLT and `Turbo` stream, introduced in the Run 2 data taking and already described in Section 1, will be further exploited in Run 3, where the fully software trigger system will give a dramatic increase in efficiency for most physics channels and the current stripping step will effectively happen online, thereby allowing for quick analysis turnaround time and resource optimization.

The building blocks for distributed computing and analysis in the upgrade era are detailed below. All of them can be already investigated with the current software framework.

4.1. Turbo stream to become the default

The capabilities of the `Turbo` stream should be extended in order to produce different output types, as discussed in Section 4.2. The benefit of this approach is essentially the removal of unnecessary information earlier in the workflow, with a clear benefit for online and offline resources optimization.

4.2. New event formats

Currently, the output of the stripping can be either in the μDST or the DST formats. For both of them, parts of the raw event can be selectively saved as needed by analysts. In the past years, μDST has become the most widely used format (90% of the events available to physics analysis). In the μDST only the selected decay candidate and related information (such as PV and multiplicities) are saved in the output. In addition, other quantities can be calculated during the stripping job and saved in μDST but no additional information can be extracted after stripping. On the other hand, in the DST all tracks and calorimeter clusters are saved. On average, the μDST occupies ten times less disk space than a DST. It is known that, although several analyses cannot be performed on μDST, full event information is still not needed in most cases. Therefore it is necessary to investigate new data formats in between μDST and DST where, for example, only a cone of tracks around the selected candidate is stored such that storage usage is optimized without compromising the physics analysis.

4.3. Centralised ntuple production and alternative approaches
It is foreseen that ROOT [24] ntuples will still be heavily used to perform data analysis which are produced by users in an unscheduled activity by running over `Turbo` or stripping output. The possibility to have a scheduled, centralized ntuple production is being investigated, based on the work done within ALICE [25]. These productions would be centrally managed in a timescale of weeks. It should also be noted that (μ)DSTs can also be directly used for physics analysis, but this requires at the moment the entire LHCb software stack.

4.4. Distributed computing
Infrastructures such as `Dirac` [26] will still be the main tool through which data is processed and distributed. The modular architecture of `Dirac` allows for adiabatic improvements of components, in order to cope with the upgrade conditions. In parallel, more dynamic and flexible ways of data placement will be investigated, for example by monitoring the accesses to a given dataset.

5. Conclusions
The LHCb experiment will be upgraded for the Run3 data taking in 2021 onwards. A review of the ongoing R&D work in the domains of software and computing for the upgrade has been given in this paper.

The performance of the software trigger in terms of data throughput will be optimized by changing significantly basic building blocks such as the core framework, that will be task-based, the event model, that will be adapted to efficiently use wide processing units, and non-event data. Such changes will also be useful to improve the simulation framework.

In other areas, such as the distributed computing, the analysis model, the implementation of alternative fast simulations there will be a natural evolution towards the upgrade era, with current Run 2 data taking to be used as testbed.

Acknowledgments
The authors would like to acknowledge the contributions of R. Aaij, M. Cattaneo, R. Cenci, P. Charpentier, M. Clemencic, A. Contu, G. Corti, B. Couturier, C. Haen, G. Raven, I. Shapoval, F. Stagni in the development of the work presented in this paper.

References
[1] Aaij R *et al.* (LHCb collaboration) 2012 Framework TDR for the LHCb Upgrade: Technical Design Report LHCb-TDR-012
[2] Aaij R *et al.* (LHCb collaboration) 2014 LHCb Trigger and Online Upgrade Technical Design Report LHCb-TDR-016
[3] Michielin E (LHCb) 2016 *PoS* **ICHEP2016** 996. 4 p URL `https://cds.cern.ch/record/2287601`

[4] Martinelli M and Collaboration L 2017 *Journal of Physics: Conference Series* **898** 032039 URL http://stacks.iop.org/1742-6596/898/i=3/a=032039
[5] Mato P 1998 GAUDI-Architecture design document Tech. Rep. LHCb-98-064 CERN Geneva URL https://cds.cern.ch/record/691746
[6] Barrand G *et al.* 2001 *Comput. Phys. Commun.* **140** 45–55
[7] Task-based Programming https://software.intel.com/en-us/node/506100
[8] 2012 The Concurrent Framework Project (CF4Hep) http://concurrency.web.cern.ch/GaudiHive
[9] Roiser S 2003 *Event data modelling for the LHCb experiment at CERN* Ph.D. thesis Vienna, Tech. U. URL https://cds.cern.ch/record/692288
[10] Ponce S, Mato Vila P, Valassi A and Belyaev I 2003 *eConf* **C0303241** THJT007 (*Preprint* physics/0306089)
[11] https://github.com/AIDASoft/DD4hep
[12] Clemencic M and Karachaliou A 2015 *J. Phys. Conf. Ser.* **664** 072012
[13] Valassi A, Basset R, Clemencic M, Pucciani G, Schmidt S A and Wach M 2008 COOL, LCG conditions database for the LHC experiments: Development and deployment status *Proceedings, 2008 IEEE Nuclear Science Symposium, Medical Imaging Conference and 16th International Workshop on Room-Temperature Semiconductor X-Ray and Gamma-Ray Detectors* pp 3021–3028
[14] Faerber C 2017 *Journal of Physics: Conference Series* **898** 032044 URL http://stacks.iop.org/1742-6596/898/i=3/a=032044
[15] Gallorini S, Lucchesi D, Gianelle A, Amerio S and Corvo M 2017 *Journal of Physics: Conference Series* **898** 032029 URL http://stacks.iop.org/1742-6596/898/i=3/a=032029
[16] Prez D H C 2017 *Journal of Physics: Conference Series* **898** 032052 URL http://stacks.iop.org/1742-6596/898/i=3/a=032052
[17] Stahl M 2017 *Journal of Physics: Conference Series* **898** 042042 URL http://stacks.iop.org/1742-6596/898/i=4/a=042042
[18] Clemencic M *et al.* 2011 *J. Phys. Conf. Ser.* **331** 032023
[19] Agostinelli S *et al.* (Geant4 collaboration) 2003 *Nucl. Instrum. Meth.* **A506** 250
[20] Allison J, Amako K, Apostolakis J, Araujo H, Dubois P *et al.* (Geant4 collaboration) 2006 *IEEE Trans.Nucl.Sci.* **53** 270
[21] Selvaggi M 2014 *J. Phys. Conf. Ser.* **523** 012033
[22] Apostolakis J *et al.* 2015 *J. Phys. Conf. Ser.* **608** 012023
[23] Amadio G *et al.* 2015 *J. Phys. Conf. Ser.* **664** 072006
[24] Brun R and Rademakers F 1997 *Nucl. Instrum. Meth.* **A389** 81–86
[25] Zimmermann M (ALICE) 2015 *J. Phys. Conf. Ser.* **608** 012019 (*Preprint* 1502.06381)
[26] Tsaregorodtsev A *et al.* 2010 *J. Phys. Conf. Ser.* **219** 062029

IOP Conf. Series: Journal of Physics: Conf. Series **1085** (2018) 032050 doi:10.1088/1742-6596/1085/3/032050

Last developments of the INFN CNAF Long Term Data Preservation (LTDP) project: the CDF data recover and safekeeping

P P Ricci[1,3], A Cavalli[1], S Dal Pra[1], A Falabella[1], E Fattibene[1], M Pezzi[1] and S Amerio[2]

[1] INFN CNAF, viale Berti Pichat 6/2 40127 Bologna, Italy.
[2] INFN Sezione di Padova, Via Marzolo 8, 35131 Padova, Italy.

E-mail: pierpaolo.ricci@cnaf.infn.it

Abstract. The INFN CNAF Tier-1 has become the Italian national data center for the INFN computing activities since 2005. As one of the reference sites for data storage and computing provider in the High Energy Physics (HEP) community it offers resources to all the four LHC experiments and many other HEP and non-HEP collaborations. The CDF experiment has used the INFN Tier-1 resources for many years and, after the end of data taking in 2011, it faced the challenge to both preserve the large amount of scientific data produced and give the possibility to access and reuse the whole information in the future using the specific computing model. For this reason starting from the end of 2012 the CDF Italian collaboration, together with the INFN CNAF and Fermilab (FNAL), introduced a Long Term Data Preservation (LTDP) project with the purpose of preserving and sharing all the CDF data and the related analysis framework and knowledge. This is particularly challenging since part of the software releases is no longer supported and the amount of data to be preserved is rather large. The first objective of the collaboration was the copy of all the CDF RUN-2 raw data and user level ntuples (about 4 PB) from FNAL to the INFN CNAF tape library backend using a dedicated network link. This task was successfully accomplished during the last years and, in addition, a system to implement regular integrity check of data has been developed. This system ensures that all the data are completely accessible and it can automatically retrieve an identical copy of problematic or corrupted file from the original dataset at FNAL. The setup of a dedicated software framework, which allows users to access and analyse the data with the complete CDF analysis chain, was also carried out with the addition of users and system administrators detailed documentation for the long-term future. Furthermore a second and more ambitious objective emerged during 2016 with a feasibility study for reading the first CDF RUN-1 dataset now stored as an unique copy in a huge amount (about 4000) of old Exabyte tape cartridges. With the installation of compatible refurbished tape drive autoloaders an initial test bed was completed and the first phase of the Exabyte tapes reading activity started. In the present article, we will illustrate the state of the art of the LTDP project with a particular attention to the technical solutions adopted in order to store and maintain the CDF data and the analysis framework, and to overcome the issues that have arisen during the recent activities. The CDF model could also prove useful for designing new data preservation projects for other experiments or use cases.

1. The INFN CNAF Tier-1 CDF use case

The INFN CNAF Tier-1 was involved in the CDF collaboration since the very beginning of its activity as computing center. At present the CNAF Tier-1 offers resources, in terms of computing, storage and general IT services to all the four LHC experiments and ~30 others non-LHC collaborations (including Astroparticle Physics). The CDF experiment (FNAL Tevatron) was one of the first CNAF Tier-1 users and after the end of the official CDF data taking activity (2011) we started the CDF long term data preservation LTDP collaboration with a close connection with FNAL, to preserve the amount of data

[3] Corresponding Author

produced during the last years of production of the detector. The technical term for defining this kind of activity is *"bit preservation"* since it usually only includes the maintenance of the data merely. The ability to access and reuse them in the future is generally defined as *"framework preservation"* and it includes all the software services that grant the accessibility and usability of the preserved datasets to the scientific communities.

The first major task that we granted to the CDF collaboration was the maintenance (bit-preservation) of the whole 2 PByte (raw data and analysis-level ntuples) that made up the RUN-2 first dataset collected during 2001-2011. The idea was copying them from the FNAL storage facilities "master copy" to the CNAF Tier-1 as a backup copy. We have decided, as an extended goal, to preserve a complete copy of the CDF RUN-2 whole amount of data and Montecarlo samples at CNAF in addition to the software transfer services (access, data analysis), for a total of roughly 4 PByte of data and associated software. To accomplish these two tasks, a geographical connection on a dedicated 10 Gbps link from CNAF to FNAL was made available from the GARR network association [1]. Consequently, the copy process was split in two separated phases:

- PHASE 1: end 2013 - early 2014 → All data and MC user level ntuples (2.1 PB)
- PHASE 2: starting from mid 2014 → All remaining raw data (1.9 PB)

The PHASE 1 also includes the installation and configuration of the necessary software services for the data transfer as outlined in the following section.

2. The CDF RUN-2 bit preservation activity

2.1. The PHASE 1 and PHASE 2 data transfer activity

The Sequential Access via Metadata (SAM) data handling tool (developed at FNAL) [2] was installed on dedicated servers at CNAF for orchestrating the data transfer that is actually made using the GridFTP protocol. In Figure 1 an overall schema of the FNAL/CNAF data transfer system is reported. The CDF user data request triggers a pre-staging process in the FNAL disk cache from the tape backend. This operation is made in parallel with the data transfer and permits a direct disk-to-disk transfer of only the pre-staged data at FNAL. Data are copied via GridFTP protocol, using the "third party transfer" option.

The SAM station performs a real time validation of the checksum (stored in a dedicated Oracle database) on the transferred data, in order to grant the reliability, as soon as it is saved to the CNAF Tier-1 hierarchical storage management (HSM) system [3]. Once the data are in the CNAF disk cache, they are automatically migrated to the HSM tape backend using custom integration of SAM and GridFTP commands. The Oracle CDF database also stores information about the specific dataset locations and metadata and it is the authoritative source of all the query requests about the CDF stored data. A specific dedicated copy script was created in order to automatize the data transfer process. The whole copy process (PHASE 1 and PHASE 2) was completed during spring 2015. This means that all CDF RUN-2 data are replicated at CNAF HSM facility (tape library) and fully checked. Furthermore, the SAM station was upgraded with the SAMWeb tool which now uses HTTP protocol for accessing the CDF database [2].

2.2. The CDF RUN-2 dataset check script

During the last phase of the copy process the idea of an automated system for regular checking of the data integrity has arisen. For this reason the copy script has evolved into a check script for verifying data and maintaining alignment from the FNAL "master copy" to the CNAF secondary one. The idea was to obtain a system for implementing regular integrity check on data and provide automatic correction of corrupted or unavailable datasets. This check script can be executed periodically (using automated scheduling tools or user commands) over specific CDF datasets, ensuring that all data are completely accessible. It can also automatically retrieve an identical copy of problematic or corrupted file from the original dataset at FNAL.

IOP Conf. Series: Journal of Physics: Conf. Series **1085** (2018) 032050 doi:10.1088/1742-6596/1085/3/032050

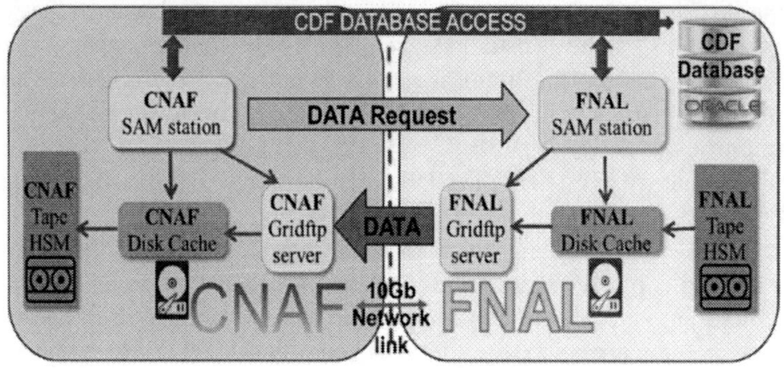

Figure 1. The FNAL/CNAF data transfer system.

The script performs three main actions as detailed below:

- Get information for the specific dataset.
- Check the file information checksum (CRC).
- Eventually heal the dataset and create a report of all the fulfilled operations.

Firstly, the script checks if the files which make up the dataset are correctly migrated to the CNAF HSM tape system. If some files are missing from a specific dataset, the copy is automatically triggered from FNAL to CNAF using the CNAF SAM station. After that, the script checks whether the file information checksum (CRC) is present as file extended attribute. If the CRC is absent or incorrect, the stage-in from the CNAF HSM tape system is triggered and the CRC is calculated on the retrieved file once available on the disk cache. Finally the script checks if the CRC is exactly the same stored in the CDF Oracle database info. If the CRC is different, the CNAF file (as secondary copy) is considered invalid and a new copy is triggered from the FNAL master copy by the use of the CNAF SAM station. It is clear that at the moment the FNAL master copy of the data is supposed to be proof of corruption and thereby represents the CDF data integrity model. In the future it will be probably possible to implement a "two way" system where in case of FNAL data corruption or unavailability, the CNAF "backup copy" could be used for recovering the original dataset.

3. The CDF RUN-1 bit preservation activity

3.1. The data recovery layout

The RUN-1 of CDF (1992-1995) contains events that demonstrate the "top quark" existence and it represents unique dataset in physics since the D0 collaboration did not keep the data at that time. The energy and conditions were different from RUN-2 but still proton-antiproton collisions [4]. This Tevatron conditions uniqueness brings both scientific and educational value for the RUN-1 datasets. After an initial review it was found that all valuable data are stored in ~4000 Exabyte 8 mm (Data8) tape cartridges with a maximum capacity of 5GB each. The old text database and Fortran software code (for analysis, simulation and visualization) are still available. Some preliminary tests using the RUN-2 CDF software were carried out at FNAL and indicated that probably the RUN-1 dataset can still be used with modern software. For this reason a bunch of 20 test tapes was delivered to CNAF in early 2016 in order to check the feasibility of a large scale data retrieve operation. Preparatory attempts showed that the tape data on the Data8 tape cartridges could be accessed using old Exabyte compatible tape drives connected via SCSI to modern O.S. servers (Linux S.L.6). Since a single tape contains variable amount of data (from 10Mbyte to the nominal capacity of 5Gbyte) the reading speed rates from 400Kbyte/s to 1.2MB/s and this translates in a reading time of 3-4 hours for a full tape. This means that the usage of single tape drives should clearly be excluded giving preference to tape drive autoloaders with 7-10 slots capacity each that could speed up the whole process. Furthermore, Exabyte latest generation Mammoth drives are backward compatible (read-only) with our tapes and this means that last generation autoloaders should be preferred since they are easier to find on the refurbished market. Technically speaking, the

RUN-1 data retrieval was considered possible and therefore the full amount of ~4000 tapes were packed and sent to CNAF at end of 2016. We acquired in the refurbished market one EXB-210 10-slots Exabyte Autoloader and two EZ17 7-slots Mammoth Autoloaders (all Linux compatible) with one year "swap warranty". Using old Dell 1950 with PCI-X servers, dismissed SCSI cards and cabling recycled from INFN Tier-1, we set up a hardware system whose layout is depicted in Figure 2.

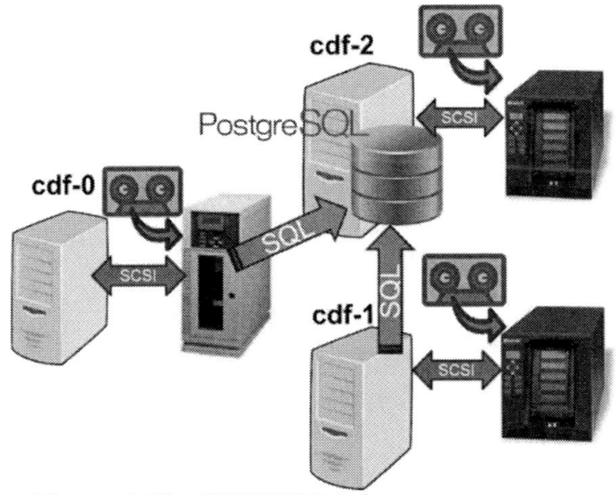

Figure 2. The CDF RUN-1 data recovery system.

The figure shows that three servers are directly connected via SCSI to the three autoloaders. The original text database (with number of files per tape and other relevant info about stored data) was imported into a PostgreSQL database running on one of the servers and this database also keeps track of the reading progress (success/failure statistics, size, CRC, etc…) into a table populated by the script operating the tape drives.

3.2. First results of the RUN-1 data recovery process

We decided to divide the RUN-1 data recovery operation in phases similarly to the RUN-2 data copy. PHASE 1 was defined with a target of successfully reading a subset of 10% (400) data tapes, followed by an evaluation of the success and failure ratio in order to estimate the sustainability of the operation. Specific scripts were also designed for operating the autoloaders, managing read operation and dealing with error conditions (i.e. partially read tapes). After the end of PHASE 1 we realized that two different tape media were used at the time: the Sony Data type QG-112M (~1000 tapes) and the Fuji "consumer" type P6-120 (the remaining ~3000 tapes). The Fuji brand tapes tend to slow down and block the tape drives as well as dirty the drive head. They also showed a considerably high *completely unreadable* status (15-20%) from all drives. Focusing on the Sony tapes we got better results, as we read 440 tapes with only 4 *completely* and 8 *partially unreadable* tapes. Anyway we noticed that, after a few months of continuous usage, all drives tended to malfunction and they stopped working permanently. Since the on-site assistance is not available due to lack of expertise, the only solution was to send the autoloaders to the refurbish expert sites worldwide, with a significant loss of time. In Figure 3, a graph describing the read operation trend during the first two months of activity is reported. As shown, a total of ~1350Gbyte over ~580 tapes were successfully read which means roughly 10 tapes/day with an average of 2.3Gbyte per tape. The occasional interruptions in the increasing trend confirm that the autoloaders drives need constant human intervention to retry read of failed tapes or to deal with struck drives. For this reason, the sustainability of the whole read operation should be seriously reconsidered. Therefore, we defined and started PHASE 2, which consists of reading the subset of all the Sony tapes (~1000 cartridges roughly 25% of the total) with the addition of the rebuilding of the RUN-1 data framework software. Since the remaining 3000 Fuji "consumer" tapes showed a higher failure rate (almost unsustainable) compared to the Sony ones, it will be hardly possible to complete the whole operation given the limits of the current hardware.

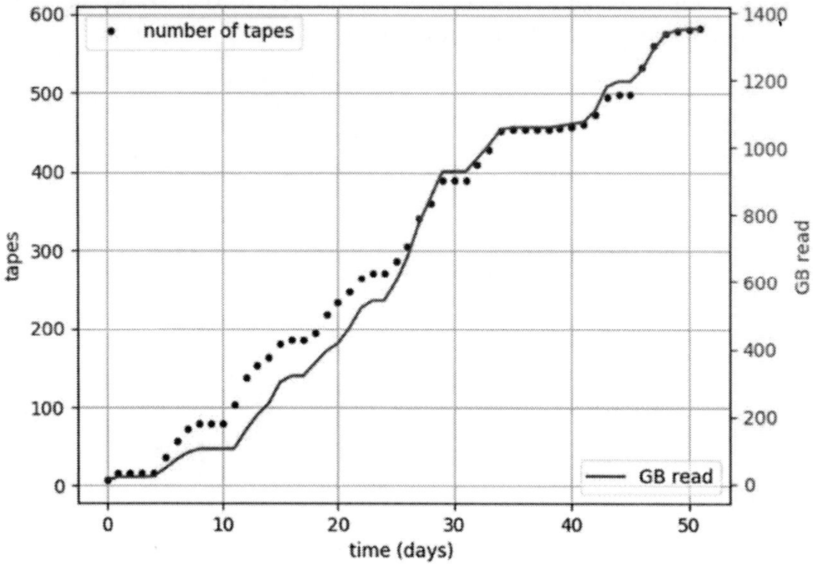

Figure 3. The results of two months of CDF RUN-1 read activity.

4. The CDF Framework preservation use case

4.1. The software and job submission system replica

The primary goal of the framework preservation is to safeguard the full infrastructure for using the experiment software and data in the long term future. We based the implementation on the instantiation of virtual machines (VMs) that runs the necessary services and can be put to "sleep" and resumed as needed. There are 3 principal service nodes in the CDF production framework:

- **cdfsam:** the SAM Station (for accessing the CDF datasets)
- **cdfdata:** the CDF users area (for user-level data e.g. *home* directory or custom software)
- **cdfheadsrv:** the CDF job submission frontends (head node)

Currently the SAM station software code is based on SL6 and therefore all cdfsam stations will remain SL6 virtual machines. The CDF analysis software has been using SL5 but a new software legacy release based on SL6 has been officially released (Feb. 2015) and therefore must be preserved [5]. In addition the CDF software is distributed using the CernVM File System [6] (CVMFS) which uses a server currently located only at FNAL for distributing the software to the end users.

The CDF job submission system is now based on JobSub [7] a system already in use at FNAL from other communities. In Figure 4 a scheme of the CDF software framework is represented. As reported in the figure for a "FNAL-independent" long term future job submission and software analysis system we need to replicate at CNAF the job submission head node (using JobSub), the CVMFS server and the CDF Oracle database. In particular a preliminary study for the CDF database replication was carried out and it is described in the following sub-section.

4.2. *The CDF Oracle database replica*

The CDF Oracle database is composed by two main instances: the *offline* instance (information for offline data processing and bookkeeping) and the *online* instance (data taking condition). We decided to use for the CNAF dedicated installation the Oracle version 11gR2 running on Oracle O.S. Linux 7.3. This is different from the O.S. used at FNAL (SUN Solaris) but aligned to the other database in production at our site. A single VM machine was installed with two Oracle database instances: the instances were named ***cdfofdp*** for the *offline* database and ***cdfondp*** for the *online* one. The VM is hosted on a four nodes KVM virtualization cluster, with the virtual machine images residing on a GPFS filesystem. The main reason was that the Oracle VM can be managed with the usage of GPFS *file clones* capability, a sort of snapshots where we can easily go back to a clean state in case of serious problems during the database import test. Using the Oracle Data Pump technology two export files were produced on February 2017 and copied from FNAL to CNAF containing a replica of both the *offline* and *online* complete databases with a size of 230GB each. The files contains a huge number of entries to be imported into our installation in particular:

- the number of table rows in the *offline* DB is ~3.3 x 10^9 rows on 958 tables
- the number of table rows in the *online* DB is ~4.2 x 10^9 rows on 766 tables

For this reason we decided to limit the import test to only two tables with ~9.1 x 10^6 rows with the addition of the relative indexes. The imported tables contain data that seems to be reasonable, however a cross-check with the original data is mandatory to demonstrate the real data alignment between the two databases. As a conclusion, the import from the Oracle database dump on a new installation with a different O.S. requires some special investigation and adjustment of the data definition statements included in the exports, in order to fit the data into the destination environment.

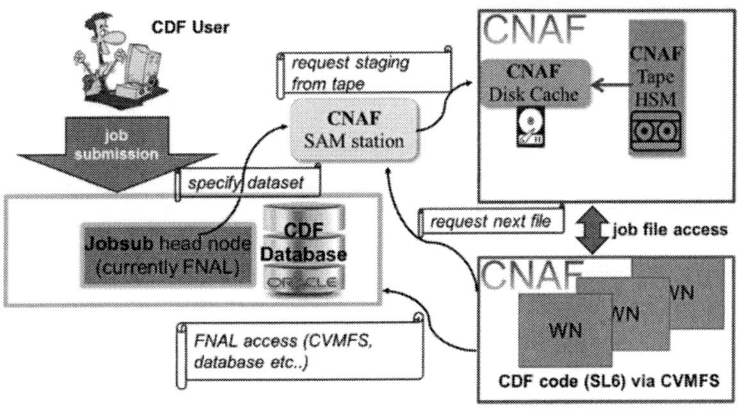

Figure 4. CDF software analysis framework.

5. Conclusion and future plans

This paper describes the latest developments in the INFN CNAF Tier-1 LTDP project. During the last years the complete CDF RUN-2 dataset was successfully copied to the INFN CNAF Tier-1. Nevertheless the integrity of the stored data should be regularly verified. For this reason, we are going to improve the check script described in section 2 in order to validate files on the CNAF HSM tape backend with periodically scheduled stage-in operation from tape to disk. About the CDF RUN-1 datasets, the PHASE 2 data retrieval should be completed soon, but some uncertainties still remains about the remaining consumer Fuji tapes that contains the residual 75% amount of data. Furthermore, the RUN-1 Fortran software framework could be revived using VMs and perhaps the RUN-2 software could also be adapted to work with the RUN-1 dataset. Regarding framework preservation, the job submission system could be fully replicated at CNAF ensuring full autonomy for these activities from FNAL. This includes replicating the FNAL CDF Oracle database, which can be carried out using FNAL database export files imported at CNAF. However, the problem of maintaining over time the synchronization between the two databases over two different sites (FNAL/CNAF) still remains. A dedicated website is under completion for collecting all relevant documents and could contain dedicated pages with specific information about CNAF archived datasets. Finally, and in conclusion, the CDF use case model will certainly be useful for designing future data preservation projects for other experiments.

Acknowledgments

We would like to thank the whole FNAL CDF Data Preservation team and all the CDF experiment people in particular Ray Culbertson, Eugene Schmidt Jr, Jonathan Lewis, Giorgio Bellettini and Donatella Lucchesi for the valuable support in the development of the present work.

References

[1] Amerio S et al., The Long Term Data Preservation (LTDP) project at INFN CNAF:CDF use case *J. Phys. Conf. Ser.* ***608*** *(2015) 012012, proceedings of the 2014 ACAT Conference.*

[2] Illingworth R A, A data handling system for modern and future Fermilab experiments *J.Phys. Conf. Ser.* ***513*** *(2014) 032045, proceedings of the 2013 CHEP Conference.*

[3] Bonacorsi D et al., The Grid Enabled Mass Storage System (GEMSS): the Storage and Data management system used at the INFN Tier1 at CNAF *J. Phys. Conf. Ser.* ***396*** *(2012) 042051 proceedings of the 2012 CHEP Conference.*

[4] Abe F et al., Evidence for top quark production in $\bar{p}p$ collisions at $\sqrt{s}$ =1.8 TeV *Phys. Rev.* ***Vol.50****, Iss. 5, September 1994 D 50, 2966.*

[5] Amerio S et al., Data preservation at the Fermilab Tevatron *Elsevier Nuclear Instruments and Methods in Physics Research Section A* ***Vol. 851****, 11 April 2017*, pp 1–4.

[6] Buncic P et al., CernVM - a virtual appliance for LHC applications *PoS(ACAT08)****012****, proceedings of the 2008 ACAT Conference.*
Info about CVMFS also avaliable online: https://cernvm.cern.ch/portal/filesystem

[7] Box D et al., Progress on the Fabric for Frontier Experiments Project at Fermilab *J. Phys. Conf. Ser.* ***664*** *(2015) 062040, proceedings of the 2015 CHEP Conference.*
Info about JobSub also avaliable online: https://cdcvs.fnal.gov/redmine/projects/jobsub/wiki

Predictive analytics tools to adjust and monitor performance metrics for the ATLAS Production System

F H Barreiro Megino[1], M Borodin[2], D Golubkov[3,4], M Grigorieva[4,5], M Gubin[5], A Klimentov[4,6], T Korchuganova[5], T Maeno[6], S Padolski[6], M Titov[4] on behalf of the ATLAS Collaboration

1 University of Texas at Arlington (US)
2 University of Iowa (US)
3 Institute for High Energy Physics (RU)
4 National Research Centre Kurchatov Institute (RU)
5 National Research Tomsk Polytechnic University (RU)
6 Brookhaven National Laboratory (US)

E-mail: maksim.gubin@cern.ch

Abstract. Having information such as an estimation of the processing time or possibility of system outage (abnormal behaviour) helps to assist to monitor system performance and to predict its next state. The current cyber-infrastructure of the ATLAS Production System presents computing conditions in which contention for resources among high-priority data analyses happens routinely, that might lead to significant workload and data handling interruptions. The lack of the possibility to monitor and to predict the behaviour of the analysis process (its duration) and system's state itself provides motivation for a focus on design of the built-in situational awareness analytic tools.

1. Introduction

The ATLAS [1] Production System (ProdSys) [2] is an automated scheduling system that is responsible for central production of Monte-Carlo data, highly specialized production for physics groups, as well as data pre-processing and analysis using such facilities as grid infrastructures, clouds and supercomputers. Its current processing rate is around 2M tasks per year (that corresponds to 365M jobs per year). ProdSys evolves to accommodate a growing number of users and new requirements from the ATLAS Collaboration (the ATLAS experiment at the Large Hadron Collider, CERN), physics groups and individual users. ATLAS Distributed Computing in its current state is the aggregation of large and heterogeneous facilities, running on the WLCG (Worldwide LHC Computing Grid) [3], academic and commercial clouds, and supercomputers.

There are two key components of ProdSys that represent the core of the system: i) Database Engine for Tasks (DEfT) is used to formulate the tasks, chains of tasks and task groups (i.e., production request), complete with all necessary parameters; ii) Job Execution and Definition Interface (JEDI) is a subsystem (also part of the workload management system PanDA [4]) that performs task-level workload management (i.e., brokerage and execution), dynamic job definition and execution (for resource usage optimization) [5].

IOP Conf. Series: Journal of Physics: Conf. Series **1085** (2018) 032051 doi:10.1088/1742-6596/1085/3/032051

Having information such as an estimation of the processing time or possibility of system outage (abnormal behavior) helps to monitor system performance and to predict its next state. The current cyber-infrastructure of ProdSys presents computing conditions in which contention for resources among high-priority data analyses happens routinely, that might lead to significant workload and data handling interruptions. The lack of the possibility to monitor and to predict the behavior of the analysis process (its duration) and system's state itself caused us to focus on design of the built-in situational awareness analytic tools.

The proposed suite of tools aims to estimate completion time (so-called "Time To Complete", TTC) for every (production) task, completion time for a chain of tasks, and to predict the failure state of the system based on "abnormal" task processing times. Its implementation is based on Machine Learning methods and techniques, and besides the historical information about finished tasks it uses ProdSys job execution information and the state of resource utilization (real-time parameters and metrics). The next planned step after task duration prediction is task profiling, detecting anomalous task behavior, and predicting task chain execution duration.

2. Approaches to estimating completion time for tasks

Several approaches were considered for predicting task execution times (i.e., task execution durations) on the basis of simulation modeling, and on the basis of machine learning.

Within the simulation modeling approach, tools for modeling distributed computing systems such as SimGrid, CloudSim, SPECI, CDOSIM and DCSim were considered for use. [6, 7] Based on the provided functionality, the SimGrid (a toolkit that provides core functionalities for the simulation of distributed applications in heterogeneous distributed environments) was chosen for further consideration.

Within the framework of the machine learning approach, modeling based on deep neural networks and ensemble methods, such as random forests and gradient-boosted trees, were considered. After performing the preliminary analysis of the approaches, we decided to implement the methods in the following order: apply gradient-boosted trees and random forest regression methods for task duration prediction (use minimal set of parameters), refine predictions by developing a framework allowing to concurrently use several different predictive models (that are built with different "extended" sets of parameters of used methods) and compare their precision, adding the deep learning-based model to the set of models, and choosing the ones that give better precision while keeping acceptable performance. [8]

The decisive factor in choosing the machine learning-based approach over the simulation-based approach was the rate of growth and change of the WLCG and rapid increase of its computation load, making simulations impractical, as the scale and complexity of ProdSys exceeds the capabilities of the best grid modeling systems currently available. [2, 9]

A machine learning-based approach, unlike a simulation-based one, uses historical data to describe functioning of the system instead of its exact structure description for model training, and therefore allows for gradual increase of prediction precision with increase of the size of data sample, in turn allowing the researcher to choose a desired trade-off between model scale and precision, while a simulation-based approach features sharp precision drop when the used model is incomplete or outdated, necessitating using a full scale model of the system.

3. Framework for prediction of the task execution duration

The imprecise nature of models used to predict task durations, as well as the need to provide users with task duration estimates for all steps of the task lifecycle, preferably starting before the task is completely defined, to simplify the planning of data processing, necessitated creating a task duration prediction framework that allows using several models simultaneously, monitoring their precision and performance, adding and removing new models without impact on the framework

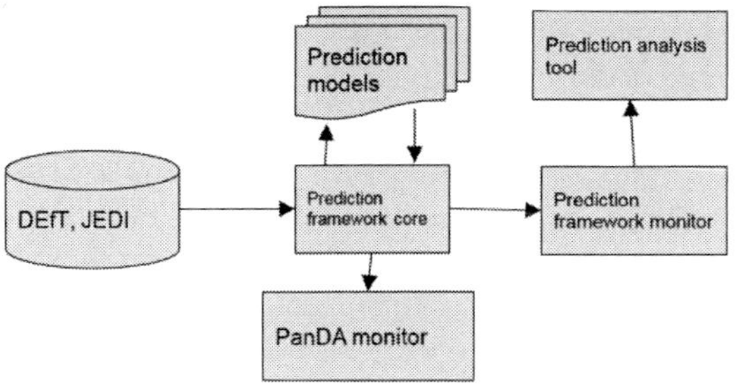

Figure 1. Task duration prediction framework structure

functionality, updating models according to a predefined schedule, and gathering statistics to be used in further improving of the framework.

3.1. Task duration prediction framework components

The structure of the task duration prediction framework is displayed on Figure 1.

The *prediction framework core component* provides the system's core functionality. It maintains a connection to a set of prediction models (by the following blocks: collector and predictor) and allows to operate with new models (e.g., extend the set with newcomers) and control their status in the system, serves as a source of prediction data for PanDA monitor, as well as generates and sends logs of its functioning to the prediction framework monitor. Input data for model creation is collected from ProdSys components (i.e., DEfT/JEDI).

Prediction framework monitor accumulates the logs provided by the prediction framework core, and allows operators to review logs, as well as to configure conditional alerts to be sent out in case of certain conditions being met.

Prediction analysis tool implements several metrics that can be used to control the stability, performance and precision of various models based on the logs gathered by the prediction framework monitor, predictions made by the models, and factual task durations.

Generating a prediction involves the following steps: the framework core component collects data with task parameters from the DEfT/JEDI database, converts them into a format accepted by the models, and then sends it to the prediction models using the task duration predictor API. After receiving predictions, the prediction framework core sends the results to the PanDA monitor to be displayed. Every step of the process (e.g., receiving a request for prediction generation, polling the models, the model response times and predicted values) is stored using the prediction framework monitor.

Results of the prediction process conducted by the framework are displayed by the PanDA Monitor [10] (that shows the profile of the corresponding task with produced predictions).

3.2. Prediction models

The requirement of providing users with an estimate of task execution duration for every step of task lifecycle forces use of a tiered approach to task duration prediction that means using a certain type of the model for the particular stage of the task lifecycle.

3.2.1. Preliminary prediction. It is often useful for the data processing planning to have a rough estimation of data processing duration even during the preliminary planning stage, when the user does not yet know how much resources can be allocated to the task, when the task will be scheduled for for execution, and the exact size of the input data. Initial "rough" estimation of the task execution duration is produced by the classifier component, that returns an average duration of tasks of the same class as the requested one (based on historical data). The class of the task is defined by the following properties: *project name*, *production step*, and *provenance*.

3.2.2. "Cold" model. Models of the "cold" category use numerous task parameters for their creation. Those parameters are known at the stage of task definition, such as type of the input data, size of the input data (expressed as the number of processing jobs), type of processing to be performed on the input data, and type of the processing queue that will be used by the task. Limiting the model to the parameters that are known at the time of task definition allows the model to give prediction at the early stage of task lifecycle, but it also means that its prediction will lack the precision, as the model does not account for the state of processing environment.

3.2.3. "Warm" model. After the execution of the task begins, it launches up to 10 "scout" jobs to measure several important performance metrics, such as amount of RAM an average job uses, and an average job duration, that allows to better gauge the load that the task will affect on the WLCG, as well as the state of WLCG itself. The knowledge of these metrics allows for significant improvement of task duration prediction precision. "Warm" model category is based on utilization of these metrics. Models of this category make use of all parameters available to the previously defined category, as well as metrics provided by running scout jobs.

3.2.4. "Hot" model. Models of the "hot" category are aimed aimed at making predictions of the remaining runtime for a task that is being processed. These models are created on (and can make use of) every metric available to models of "cold" and "warm" categories, as well as parameters of task jobs, such as average job duration, trends in job duration, and occurrence of failed jobs, to predict task duration with the highest precision.

3.3. Implemented and planned model types

The current implementation of the preliminary prediction model is based on a simple classifier. It is planned to switch over to a machine learning based model if the classifier's precision proves insufficient for practical use. The "cold", "warm" and "hot" models currently in use are based on the Gradient-Boosted Trees regression method from Spark MLlib machine learning library. [11]

During the next step of the framework development, random forest and deep learning-based models will be added to the models pool for performance and precision evaluation, followed by selecting the most precise models with acceptable performance for use in predicting task durations.

3.4. Planned extensions of task duration prediction framework

The preliminary research has been performed to gauge feasibility of detecting anomalous behaviour of running tasks. We have found that tasks of the same type tend to have similar patterns of job execution. The execution profile of a typical task is shown in Figure 2. The red line shows when the jobs comprising the task finish, the green dots show the start times of the corresponding jobs. By viewing the task profile it is often possible to notice anomalies visually long before a task becomes overdue.

Another promising direction of the framework development is task chain duration prediction. Due to the way tasks are implemented in ProdSys, the next task in the chain can often starts execution before the previous task finishes. This leads to the fact that duration of the task chain is shorter than the sum of tasks' durations, and preliminary tests with Gradient-Boosted Trees-based models showed that predicting task chain duration with high precision is possible, conditional upon knowing task durations with high precision. As currently task duration prediction has low precision for tasks that are not currently running, the implementation of task chain duration prediction was delayed for now.

IOP Conf. Series: Journal of Physics: Conf. Series **1085** (2018) 032051 doi:10.1088/1742-6596/1085/3/032051

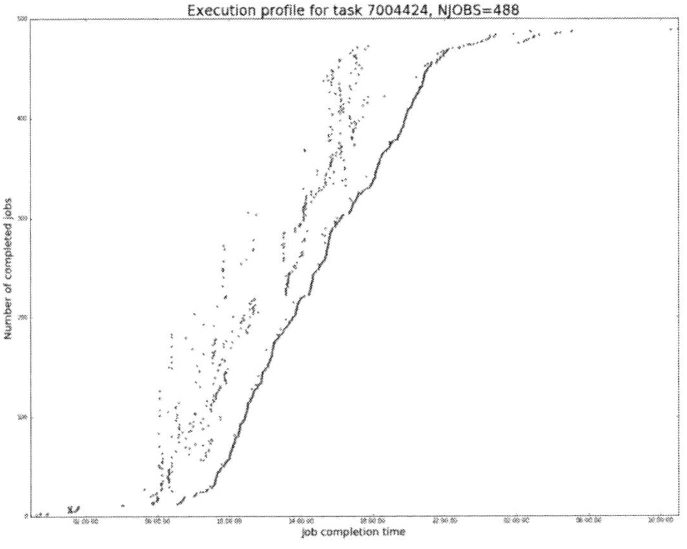

Figure 2. Execution profile of a typical task

4. Conclusion

Adding predictions of the task execution duration to the features of the ProdSys monitoring will allow users/operators to notice task execution delays as they happen, and take measures to resolve them. Combined with the anomaly detection framework, this will further improve efficiency of ProdSys data processing and reduce the data handling interruptions.

Acknowledgments

We wish to thank all our colleagues from ATLAS experiment, PanDA team and CERN IT.

TPU team research was funded by the Russian Science Foundation under contract 16-11-10280.

NRC KI work was funded by the the Russian Ministry of Science and Education under contract 14.Z50.31.0024.

References

[1] Aad G, Abat E, Abdallah J, Abdelalim A, Abdesselam A, Abdinov O, Abi B, Abolins M, Abramowicz H, Acerbi E *et al.* 2008 *Journal of Instrumentation* **3** S08003–S08003

[2] Borodin M, De K, Garcia J, Golubkov D, Klimentov A, Maeno T, Vaniachine A *et al.* 2015 *Journal of Physics: Conference Series* vol 664 (IOP Publishing) p 062005

[3] Bird I 2011 *Annual Review of Nuclear and Particle Science* **61** 99–118

[4] Maeno T, De K, Klimentov A, Nilsson P, Oleynik D, Panitkin S, Petrosyan A, Schovancova J, Vaniachine A, Wenaus T *et al.* 2014 *Journal of Physics: Conference Series* vol 513 (IOP Publishing) p 032062

[5] De K, Golubkov D, Klimentov A, Potekhin M, Vaniachine A *et al.* 2014 *Journal of Physics: Conference Series* vol 513 (IOP Publishing) p 032078

[6] Casanova H 2001 *Cluster computing and the grid, 2001. proceedings. first ieee/acm international symposium on* (IEEE) pp 430–437

[7] Malhotra R and Jain P 2013 *International Journal* **3**

[8] Caruana R and Niculescu-Mizil A 2006 *Proceedings of the 23rd international conference on Machine learning* (ACM) pp 161–168

[9] Fujiwara K and Casanova H 2007 *Proceedings of the 2nd international conference on Performance evaluation methodologies and tools* (ICST (Institute for Computer Sciences, Social-Informatics and Telecommunications Engineering)) p 12

[10] Klimentov A, Nevski P, Potekhin M and Wenaus T 2011 *Journal of Physics: Conference Series* vol 331 (IOP Publishing) p 072058

[11] Meng X, Bradley J, Yavuz B, Sparks E, Venkataraman S, Liu D, Freeman J, Tsai D, Amde M, Owen S *et al.* 2016 *The Journal of Machine Learning Research* **17** 1235–1241

IOP Conf. Series: Journal of Physics: Conf. Series **1085** (2018) 032052 doi:10.1088/1742-6596/1085/3/032052

Experience with SPLUNK for archiving and visualisation of operational data in ATLAS TDAQ system

A Kazarov[1], G Avolio[2], A Chitan[3] and M Mineev[4]

[1]NRC "Kurchatov Institute" - PNPI, St. Petersburg, Russian Federation
[2]CERN, Geneva, Switzerland
[3]National Institute of Physics and Nuclear Engineering, Bucharest, Romania
[4]Joint Institute for Nuclear Research, JINR Dubna, Russian Federation

E-mail: Andrei.Kazarov@cern.ch

Abstract. The ATLAS Trigger and Data Acquisition (TDAQ) is a large, distributed system composed of several thousands interconnected computers and tens of thousands software processes (applications). Applications produce a large amount of operational messages at the order of 10^4 messages per second, which need to be reliably stored and delivered to TDAQ operators in a quasi real-time manner, and also be available for post-mortem analysis by experts.

We have selected SPLUNK, a commercial solution by Splunk Inc, as an all-in-one solution for storing different types of operational data in an indexed database, and a web-based framework for searching and presenting the indexed data and for rapid development of user-oriented dashboards accessible in a web browser.

The paper describes capabilities of the Splunk framework, use cases, applications and web dashboards developed for facilitating the browsing and searching of TDAQ operational data by TDAQ operators and experts.

Trigger and Data Acquisition of the ATLAS experiment at LHC

ATLAS Experiment

A Toroidal LHC ApparatuS (ATLAS) [1] is a particle physics experiment at the Large Hadron Collider (LHC) at CERN. The LHC is producing proton-proton head-on collisions with center-of-mass energy equal to 13 TeV at 40 MHz collision rate. The ATLAS detector comprises more than 140 million electronic channels which deliver raw event data at the rate of order of TB/s.

Trigger and Data Acquisition system (TDAQ)

TDAQ is one of the core ATLAS systems [2] with the following key characteristics :

- Manages filtering and transfer of ATLAS experiment data from the detector to large-scale mass-storage, handles the flow of 1.5 MB events at rate up to 100 kHz
- A distributed computing system with more than 40000 applications running on a cluster of 2300 nodes
- Non-stop operation 24 hrs/day, 7 days/week during LHC Run 1 and Run 2
- Includes applications and frameworks for Configuration, Control and Monitoring the overall ATLAS data taking activity

One of the TDAQ applications is described in this paper.

Motivations for the use of SPLUNK for archiving and visualization

TDAQ applications produce a huge number of operational monitoring data on-line, which is necessary to gather, to archive, to process, to analyse and finally to present to the ATLAS operations crew and to experts to facilitate on-line supervision of the system and also for post-mortem analysis. This includes, in particular, a large amount of operational messages (at the order of $O(10^4)$ messages per second), and other types of data like statistics of the busy fraction induced by ATLAS detectors.

SPLUNK, a commercial solution by Splunk Inc [3] was selected as an all-in-one solution for storing different types of operational data in an indexed database, and a web-based framework for searching, analysing and presenting the data. An important feature of Splunk is that it allows rapid development and easy maintenance of web applications, including user-oriented and task-oriented dashboards accessible in a web browser.

Splunk is developed as a system for archiving and processing of applications logs produced by processes running on large-scale clusters, which nicely can be extended to any type of textual operational data present in the TDAQ system.

SPLUNK workflow

The Splunk server (also called *indexer*) can receive data from files, TCP sockets or from the standard output of user scripts. It is capable to index any type of textual data it receives, and data are immediately available in search queries and in dashboards. To achieve the desired search performance, one can provide Splunk with some information about the structure of inserted data, e.g. fields which need to be indexed. TDAQ data are well-structured 'field=value' events, and for each type of events we developed a 'properties' configuration file, that gives Splunk an idea of which fields need indexing.

A typical event produced by the ERS (Error Reporting Service [4]) TDAQ application looks like:

```
t=1386776531, rn=224190, part=ATLAS, uname=crrc, msgID=rc::OngoingTransition, host=pc-tdq-onl
-77, app=RootController, sev=INFO, text="Transition INITIALIZE from IGUI is ongoing.", context="
PACKAGE_NAME: RunController. FILE_NAME: ../src/lib/RootController.cc. FUNCTION_NAME: virtual
void daq::rc::RootController::receive(const char*, const char*, const char*). LINE_NUMBER: 316.
DATE_TIME: 1386776531.", params="trans: INITIALIZE from IGUI. ", quals="RunController ", chained
="0", gh=1755720156
```

Which is indexed by Splunk with the help of this indexing configuration:

```
[ers]
TIME_PREFIX = ^t=
MAX_TIMESTAMP_LOOKAHEAD = 20
SHOULD_LINEMERGE = true
BREAK_ONLY_BEFORE = ^t=\d{10},\srn=\d+,\s
SEGMENTATION = ers
EXTRACT-fields = ,\stext="(?<text>.*)",\scontext="(?<context>.*)",\sparams="(?<params>.*)",\
squals="(?<quals>.*)",\schained=(?<chained>.+),\sgh=\d+$
```

After the data are indexed, a process of getting results out of Splunk can be presented as a pipeline:

```
search(filter) | process(analyze, transform) | present(chart, dashboards)
```

where the (search | process | present) chain is literally a typical Splunk search string - a pipeline of commands, passing results to each other. A search is usually a selection of interesting fields by value. For processing the search results Splunk offers a great spectrum of functions: statistical, aggregation, transformation etc. Finally, selected and processed data are presented to the user in the form of tables, different forms of charts and plots which can be integrated into web pages as ready-to-use dashboards. The user is also allowed to work with the raw 'search' Splunk application, and to analyse data with a pipeline of Splunk commands. Below is an example search string from one of the TDAQ Splunk application demonstrating usage of Splunk search and processing capabilities:

```
search index=daqeff sourcetype=daq_eff Fill=* | dedup LumiBlock RunNumber | eval eDurRFP=
Duration*ReadyForPhysics*Eff | eval DurSB=Duration*StableBeams | eval eRTSB=Duration*StableBeams
*Eff | eval eRT=Duration*Eff | eval endT=_time+(Duration/1000) | eval SB1=case(StableBeams=1,
_time) | eval R4P1=case(ReadyForPhysics=1,_time) | eval inEff=(1-Eff)*Duration | chart eval(
round(sum(DurSB)/3600000,2)) as "Stable Beam Time [h]", eval(round((sum(eDurRFP)/sum(DurSB))
*100,1)) as "Physics Efficiency [%]" by RunNumber | sort RunNumber
```

Typically a Splunk application includes a number of dashboards which hide the complexity of data processing from the end users. Dashboards can be developed in a simple XML format, also they can be converted to the HTML format where the full power of JavaScript can be used to add more functionality if needed. A very useful feature for operational monitoring are real-time search forms and dashboards, where information is refreshed automatically as soon as fresh data get indexed.

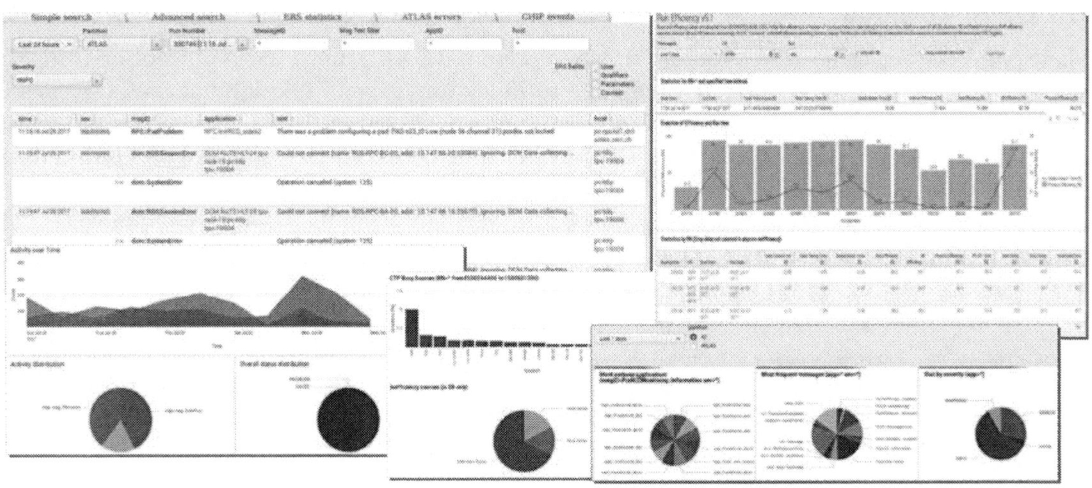

Figure 1: Dashboards from TDAQ Splunk applications.

DAQ applications implemented in SPLUNK

A single Splunk server can host a number of independent applications, thus reducing the deployment and maintenance efforts. The TDAQ Splunk server hosts four applications, each handling specific type of data coming from different sources:

ERS web browser Provides access to millions of ERS messages produced by TDAQ applications in a web browser. It allows searching the messages based on run number, application name, host, severity, message text etc. Results are available in a table and can be exported. More advanced dashboards with statistics of messages (per application, severity and message type) are available for experts. ERS messages are collected from all running TDAQ applications and stored in files in an intermediate buffer on a shared filesystem before being indexed by Splunk.

Run efficiency dashboards Detailed information about ATLAS data-taking efficiency (fraction of dead-time induced by ATLAS detectors per lumi-block) is stored by the Central Trigger in an Oracle database. It is being indexed by Splunk, and a number of dashboards are available for experts, allowing to examine the efficiency evolution by run and to analyze sources of inefficiency by subsystem and by type.

CHIP actions dashboard CHIP [5] is a central expert-system like application, aiming to automate different operational and recovery procedures. This Splunk application includes few dashboards which present the history and distribution of actions performed by CHIP.

Access Manager logs browser This is a classical Splunk application, where log files from distributed Access Manager [6] server instances are collected and indexed by Splunk, and a number of dashboards are provided with filtered information relevant for different use cases. This application is under development.

Screenshots of the dashboards from first three applications are presented on Figure 1.

Performance

A single Splunk server running on a virtual machine node (four cores, 7GB RAM) can handle the rates of events coming from TDAQ data sources as presented on Figure 2. In total more then 26M of messages were indexed, taking 16GB of the disk space.

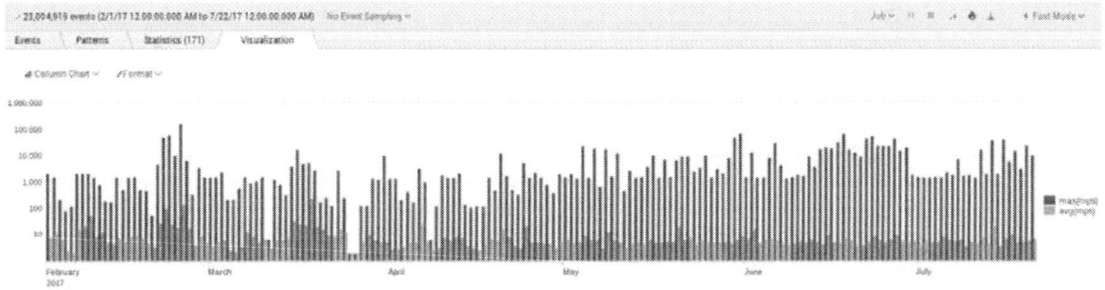

Figure 2: Maximum and average number of messages indexed per second, for all data collected in course of 2017. This is a visualization of the results of a Splunk query.

Conclusions and outlook

Splunk is successfully used as an all-in-one solution framework for archiving and visualizing of the TDAQ operational data, providing easy integration, fast web applications development and low maintenance costs. An open-source solution, also adopted by CERN, like combination of Elastic Search and Kibana are considered for evaluation as a possible alternative.

References

[1] ATLAS Collaboration 2008 *Journal of Instrumentation, vol.* **3** S08003
[2] M Abolins et al 2016 *JINST 11* **06** P06008
[3] Splunk web page `www.splunk.com`
[4] S Kolos, A Kazarov and L Papaevgeniou 2015 *J. Phys.: Conf. Ser.* **608** 012004
[5] G Anders, G Avolio, G Lehmann Miotto and L Magnoni 2015 *J.Phys.Conf.Ser.* **608** 1, 012007
[6] M Valsan et al 2011 *J.Phys.Conf.Ser.* **331** 022042

IOP Conf. Series: Journal of Physics: Conf. Series **1085** (2018) 032053 doi:10.1088/1742-6596/1085/3/032053

Multi-threading in the ATLAS High-Level Trigger

Adam Edward Barton

Physics Department, Lancaster University, Lancaster, United Kingdom, LA1 4YB

E-mail: abarton@cern.ch

Abstract. Over the next decade of LHC operation the instantaneous luminosity will reach up 7.5 times the design value with over 200 interactions per bunch-crossing. This will pose unprecedented challenges for the ATLAS trigger system.

We report on a prototype of the High Level Trigger software that maximizes commonality with offline reconstruction software while retaining the key aspects of trigger functionality including regional reconstruction and early event rejection. We report on the first experience of migrating trigger algorithms to this new framework and present the next steps towards a full implementation of the ATLAS trigger within AthenaMT.

1. Introduction

The ATLAS trigger consists of two trigger levels. The first level trigger, L1, is implemented in hardware and the High Level Trigger (HLT) is implemented in software executed on a farm of processors. The HLT selection software runs online (i.e. in real-time) in the HLT processor farm located at the ATLAS experimental area.

With the evolution of the CPU market to many-core systems, both the ATLAS offline reconstruction and High-Level Trigger software will have to transition from a multi-process to a multithreaded processing paradigm in order not to exhaust the available physical memory of a typical compute node. The new multithreaded ATLAS software framework, AthenaMT, has been designed from the ground up to support both the offline and online use-cases with the aim to further harmonize offline and trigger algorithms. The latter is crucial both in terms of maintenance effort and to guarantee the high trigger efficiency and rejection factors needed for the next two decades of data-taking. A second goal is to provide greater harmonisation of Trigger and Offline code that will ease the interchange of algorithms between the two environments and thereby reduce maintenance effort.

2. Athena and AthenaMP

The ATLAS software framework, Athena, was originally designed to process events serially and run algorithms sequentially on each event. In order to reduce memory usage a concurrent framework, AthenaMP [1, 2], was developed that creates worker processes from a common parent process. The workers independently process events in parallel, but processing within each worker is sequential. To minimize the total memory requirement, AthenaMP uses a "copy-on-write" mechanism to share read-only data (initially memory is shared between processes, however a write operation causes the corresponding page of memory to be copied to become thread-specific). This substantially reduces the total memory usage. A limitation of the copy-on-write mechanism is that a whole memory page (currently 4kB) is copied as soon as any bits in that page are changed.

A framework that allows the CPU cores to be utilized in a more flexible way is desirable to limit the scaling of memory with CPU core count. To address this, the AthenaMT framework is being developed that permits parallelisation of algorithms within an event in addition to event-wise parallelisation. This offers the potential to decrease memory usage by reducing the number of events processed in parallel (events in-flight) to less than the number of threads. A further memory reduction can be achieved if algorithms can be made reentrant, such that a single algorithm instance can be executed concurrently in different threads, rather than having separate clones of the algorithm. This provides a more efficient use of memory than the copy-on-write mechanism used in AthenaMP since the latter is limited by the finite size of the memory page.

3. AthenaMT

AthenaMT will use components of the multi-threaded framework GaudiHive [3]. AthenaMT will use the Intel Thread Building Blocks library to execute algorithms on available CPU threads [4]. Algorithms using data from one event/collision can be parallelized and multiple events can also run in parallel (Intra-event and Inter-event parallelism).

In the current Athena framework algorithms are scheduled in a manually defined sequence, this would be very sub-optimal in a MT environment yet data-dependencies must be respected. AthenaMT [5, 6, 7] will implement a data-dependent scheduler where each algorithm will explicitly declare its dependencies ("DataHandles"). The dependency graph will be constructed and verified during initialization. Figure 1 illustrates how the AthenaMT scheduler exploits TBB to schedule execution of algorithms whose data dependencies have been met in available threads. The framework permits different implementations of schedulers and the designs are still being adjusted according to emerging criteria.

Algorithms executed concurrently in different threads may be from the same event or different events (multiple events in flight). Algorithms will be either reentrant (purely procedural) or cloneable but singleton algorithms may be permitted as a migration step. Singleton services will be required to be thread-safe.

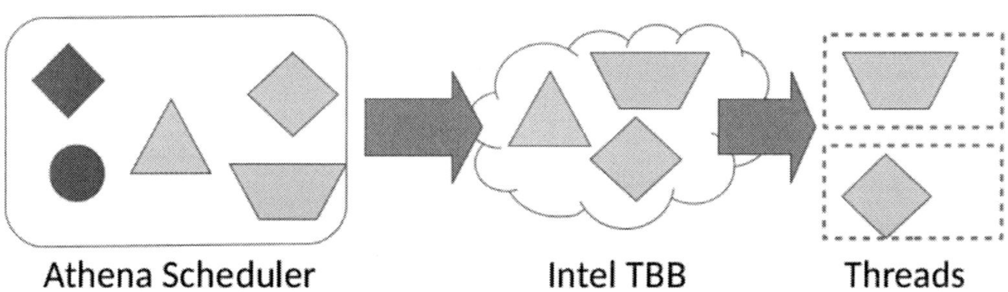

Figure 1. Illustration of the how the AthenaMT scheduler exploits TBB to schedule algorithms in available threads. Red shapes represent algorithm instances that have unmet data dependencies while green shapes represent algorithms for which all data dependencies have been met

4. AthenaMT and the trigger

In the current framework, there is only one HLT algorithm that is part of the athena workflow, the HLT Steering. The HLT reconstruction and selection algorithms are scheduled by the HLT

Steering that also provides them with a Region of Interest (RoI) context. An RoI is a geometrical sub-region of the event, most often created from L1 information, in which HLT reconstruction is performed.

In the existing HLT design, data is retrieved and reconstructed in limited geometrical RoIs. This reduces bandwidth and CPU requirements: a trigger decision can be reached without needing to examine the entire event. This differs from offline reconstruction, where all data retrieval and processing are performed for entire events. The differences between online and offline processing have led to different algorithm interfaces. While offline algorithms access the central storage location for event data (StoreGate), HLT algorithms receive Trigger Element (TE) objects as arguments to their main execute() method. Trigger Elements contain links to the relevant input data for the algorithm that is usually specific to the RoI that the algorithm is set to process. The Trigger Steering is the HLT-specific framework component responsible for executing algorithms for particular RoIs, and passing them the appropriate trigger elements.

We intend to remove the HLT-specific algorithm interface in the new framework, making it easier to share code between online and offline environments. Instead of using Trigger Elements to present RoI data, this will be handled in a transparent way using a new framework component, EventViews. In the current HLT, a sequence of one or more reconstruction algorithms (RecoAlg) and a hypothesis algorithm (HypoAlg) is executed in an RoI by the Trigger Steering, passing data using Trigger Elements. In the AthenaMT framework the RecoAlgs receive their data from EventViews that have the same interface as the Data Store but only provide access to objects added to the EventView at creation or by algorithms executed in the EventView. Figure 2 illustrates the creation by an EventViewCreatorAlg of three EventViews with three algorithms scheduled to be executed in each EventView. A MergerAlg running in the event context combines the output objects from the three views in the event Data Store. The implementation for EventViews is still in the process of being designed and may be integrated into the scheduler in the future.

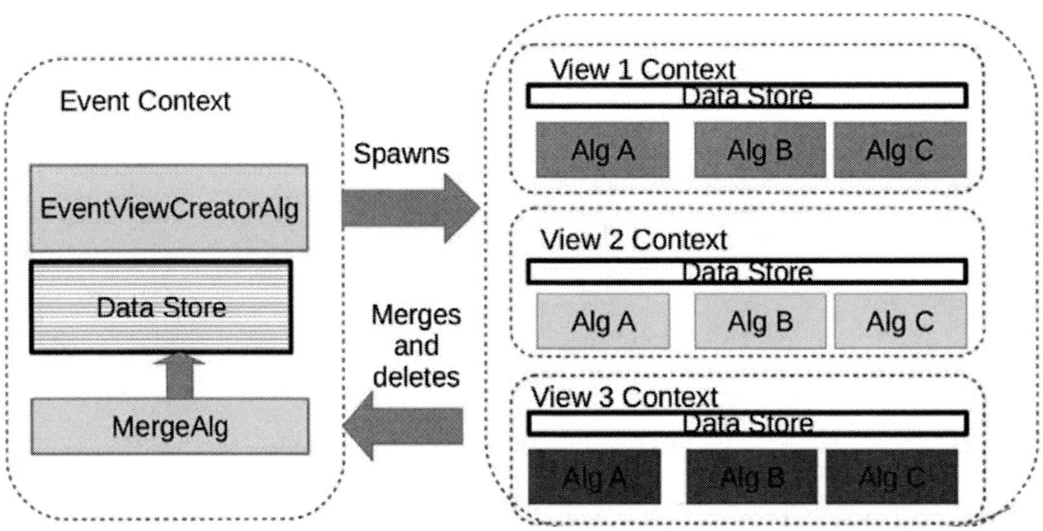

Figure 2. A schematic diagram showing the relationship of the Event View algorithms and dataflow in relation to the event-wide algorithms.

IOP Conf. Series: Journal of Physics: Conf. Series **1085** (2018) 032053 doi:10.1088/1742-6596/1085/3/032053

5. Control Flow

The trigger menu is composed of a list of trigger requirements each of which corresponds to a "chain" of reconstruction and selection algorithms required to determine if the trigger requirement is met. The decision taking infrastructure has to ensure that the list of trigger requirements, described in the trigger menu as chains, are evaluated with a minimum of processing. Therefore, the processing is divided into "steps", each represented by a sequence of reconstruction followed by hypothesis testing. Two steps are illustrated in Figure 3. The division into steps allows further processing to be avoided if a fast negative decision is made in an earlier step. It also allows eventwide early rejection if there are no chains passing a given step. This is implemented using Control Flow dependencies. The scheduler first determines if an algorithm is control ready before evaluating if its data dependencies are met. The Control Flow dependencies are implemented by two framework components shown as dashed circles in Figure 3, the sequence AND (blue) that sets algorithms control-ready in sequence conditional on the status of the previous algorithm and the Parallel OR (red) that makes algorithms control-ready at the same time. In the simple example shown in Figure 3, the upper Sequence AND enables step 2 after step 1 and the lower Sequence AND first enables execution of the criteria filter and only if the filter is passes, enables the step 2 Parallel OR that enables the step 2 algorithms.

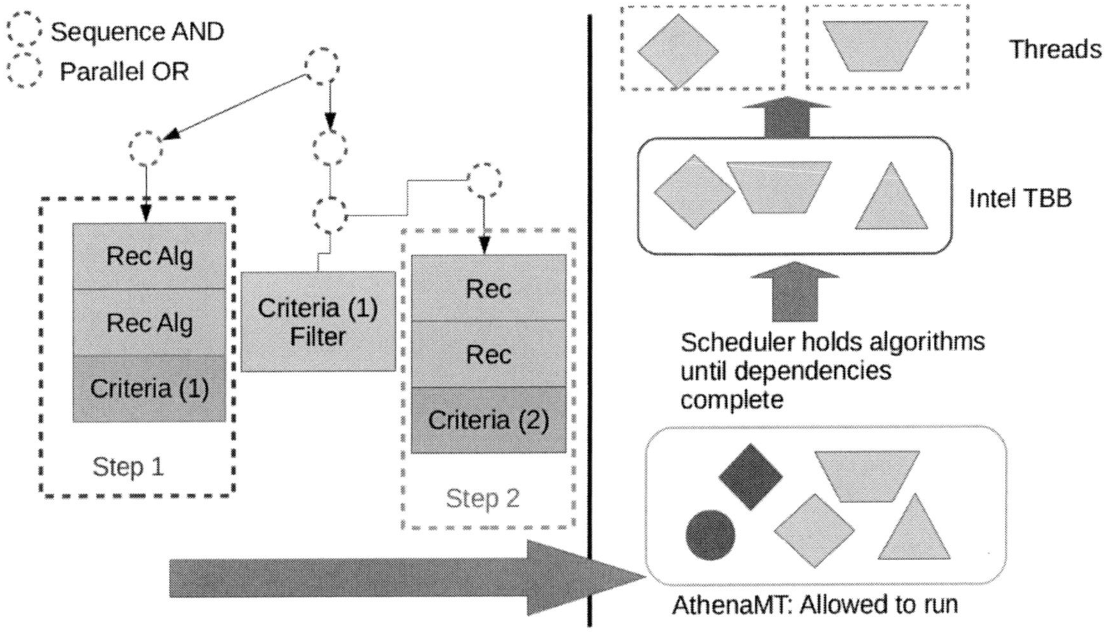

Figure 3. An illustration of the control flow mechanisms used in the trigger. Chains are ran in step 1 initially, if they meet the requirements of the criteria filter then step two is executed. If not and no other chains are accepted the event could be rejected early.

6. Conclusion

ATLAS is redesigning the trigger, reconstruction and analysis framework to have the flexibility required to cope with increasing demands and exploit future hardware. The trigger will also benefit from greater unification of the online and offline code-bases. A basic set of algorithms are now functional. The full set of algorithms will be migrated over the next 3 years. We expect to have a first implementation by the end of 2018 and a full implementation by 2020.

IOP Conf. Series: Journal of Physics: Conf. Series **1085** (2018) 032053 doi:10.1088/1742-6596/1085/3/032053

References

[1] P. Calaura, W. Lavrijsen, C. Leggett, M. Marino and D. Quarrie, The Athena control framework in production, new developments and lessons learned, CHEP 2004 Conf. Proc. C04-09-27 pp 456-458 (2005)
[2] Binet S et al., 2012 Multicore in production: Advantages and limits of the multiprocess approach in the ATLAS experiment J. Phys.: Conf. Series 368 012018 (ACAT2011 proceedings)
[3] M. Clemencic, B. Hegner, P. Mato and D. Piparo, Introducing concurrency in the Gaudi data processing framework, J. Phys. Conf. Ser. 513, 022013 (2014).
[4] Intel Threading Building Blocks Documentation https://software.intel.com/en-us/tbb-documentation
[5] C. Leggett et. al. Development of a Next Generation Concurrent Framework for the ATLAS Experiment ATL-SOFT-PROC-2015-017
[6] The Atlas Collaboration ATLAS Future Framework Requirements Group Report ATL-SOFT-PUB-2016-001
[7] C. Leggett et. al. AthenaMT: Upgrading the ATLAS Software 1 Framework for the Many-Core World with Multi-Threading, ATL-SOFT-PROC-2017-019

IOP Conf. Series: Journal of Physics: Conf. Series **1085** (2018) 032054 doi:10.1088/1742-6596/1085/3/032054

Test Management Framework for the Data Acquisition of the ATLAS Experiment

A Kazarov[1], A Corso Radu[2], G Avolio[3], G Lehmann Miotto[3], I Soloviev[2] and G Unel[2]

[1]NRC "Kurchatov Institute" - PNPI, St. Petersburg, Russian Federation
[2]University of California, Irvine, USA
[3]CERN, Geneva, Switzerland

E-mail: Andrei.Kazarov@cern.ch

Abstract. Trigger and Data Acquisition (TDAQ) of the ATLAS experiment is a large distributed and heterogeneous system: it consists of thousands of interconnected computers and electronics devices that operate coherently to read out and select relevant physics data. Advanced testing and diagnostics capabilities of the TDAQ control system are a crucial feature which contributes significantly to smooth operation and fast recovery in case of problem and, finally, to the high efficiency of the whole experiment.

The base layer of the verification and diagnostic functionality is a test management framework. We have developed a flexible test management system that allows experts to define and configure tests for different components, indicate follow-up actions to test failures and describe inter-dependencies between TDAQ or detector elements. This development is based on the experience gained with the previous test system that was used during the first three years of data taking. We discovered that more emphasis needed to be put on the flexibility and configurability of the verification and diagnostics functionality by the many people that are, each, knowledgeable and expert on individual components of the experiment.

In this paper we describe the design and implementation of the test management system and also some aspects of its exploitation during the ATLAS data taking in the LHC Run 2.

1. Trigger and Data Acquisition of the ATLAS experiment at LHC

ATLAS Experiment

A Toroidal LHC ApparatuS (ATLAS) [1] is a particle physics experiment at the Large Hadron Collider (LHC) at CERN. The LHC is producing proton-proton head-on collisions with center-of-mass energy equal to 13 TeV at 40 MHz collision rate. The ATLAS detector comprises more than 140 million electronic channels which deliver raw event data at the rate of order of TB/s.

Trigger and Data Acquisition system

TDAQ is one of core ATLAS systems [2] which manages filtering and transfer of experiment data from the ATLAS detectors to large-scale mass-storage. The TDAQ system is composed of a large number of distributed hardware and software components (about 3000 machines and more than 40000 concurrent processes) which, in a coordinated manner, provide the data-taking functionality of the overall system. The system is required to handle data coming in parallel from the detector readout over some 1800 point-to-point readout links forming a flow of 1.5 MB events at rate up to 100 kHz. During LHC runs, TDAQ is maintained in non-stop operation mode by a group of operators and experts on call.

Online Software
Given the complexity of the system and the probability of failures of hardware and software components, having advanced testing and diagnosing capabilities is an essential requirement for the TDAQ Online Software framework.

The Online Software [3] encompasses the software to configure, control and monitor the TDAQ system. It is based on a number of services which provide essentially the glue that holds the various sub-systems together. The Test Management is one of these services and is devoted to the verification of the functioning of the TDAQ system by executing tests on request.

2. Test Management Framework

Use Cases
The Test Management (TM) is a framework allowing from one side for experts: to develop individual tests for system components, to define dependencies between testable components, and to configure test dependencies and follow-up actions.

From another side, the testing expertise stored in TM is used (Figure 1): by the Run Control (RC) system that periodically verifies the functioning of the components it is in charge of; by the Central Hint and Information Processor (CHIP) [4] that analyses tests results to diagnose problems and applies follow-up actions; by the DAQ Operator or Expert who can have a view on testable hierarchy of TDAQ configuration components and perform testing of a subset of it via a dedicated graphical user interface. The Test Management utilizes other components of the Online Software framework. The most important ones are: Configuration Database (DB) [5] which provides the description of the TDAQ system (and tests) configuration, and Process Manager (PMG) [6] which offers a service to create, control and monitor the status of all the processes in the TDAQ system.

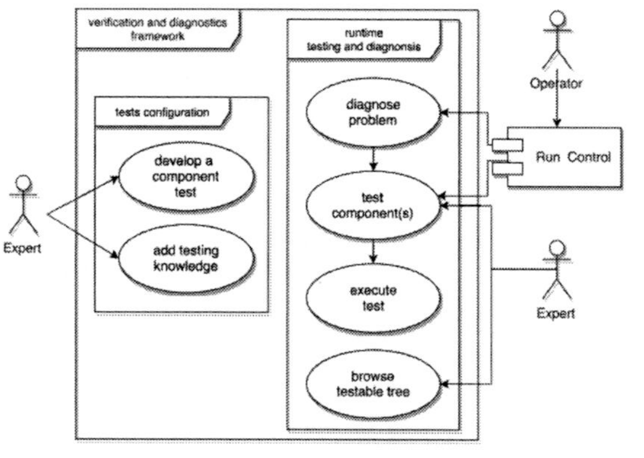

Figure 1: Test Management Framework use cases.

New requirements
The Test Management and Diagnostics service [7] was revised after the first data taking period of ATLAS. A set of new functional requirements were added:

- Experts shall be able to define the order in which tests should be executed for a component; the sequence may dynamically change based on the result of completed tests.
- Experts shall be able to define the order with which inter-related components shall be tested; the test sequence may change depending on the result obtained for the components.
- Experts shall be able to define what should be done upon failure of a test or a component to further diagnose the issue or recover.

All these requirements point towards an increased configurability of the system by ATLAS experts. The expert knowlege, like description of the testing behaviour listed above, must be completely contained in the TDAQ configuration database, which can be populated and modified by the system experts using standard database editing tools.

An additional constraint was added, based on the evolution of other parts of the TDAQ system:

- The test management functionality should be provided in Java (in addition to C++), in order to be available for Java-based TDAQ applications like CHIP

Based on those extension of requirements a complete re-implementation was carried out.

Design and implementation

Typically tests are small programs or scripts which can be executed remotely on a host in the distributed TDAQ system, depending on configuration of a particular test. A single test can also be implemented as two or more programs started on different hosts, e.g. to verify the functionality of a link between source and destination elements in the system. A test returns a single result from a predefined enumeration `{Passed, Failed, Unresolved}`. Every component in the system may have a number of tests defined, where each individual test verifies a particular functionality of the component. In this case the test result for a component is a combination of individual tests resuls.

The test management framework is modular and is implemented as client libraries (Java, C++) and a Qt based application for the graphical user interface. In particular the package includes:

- a database schema to describe tests, components, dependencies and follow-up actions
- a client library (Java, C++) to perform tests on individual components
- a client library (C++) to handle the testing of components taking into account the other components they depend on (e.g. before testing an application, test that the computer it should run on works correctly)
- a GUI to allow the operator to browse a tree of testable TDAQ components and to request the execution of tests for a subset of it

The C++ and Java implementations address the same requirements and are based on the same configuration schema but they where developed independendly, as the need in Java implementation arised in course of the evolution of the TDAQ software. Both implementations achieve a good, comparable performance and are capable of launching many tests in parallel without imposing any significant overhead: for example a test for configuration of 1100 components (computer nodes) requires launching of 4500 individual tests on the tested nodes which is done in 32 parallel threads in total takes about 8 seconds. The Java implementation is, at the time being, capable of optimally parallelizing execution using less threads than the C++ implementation. Its threading model may thus be ported to the C++ implementation as well in the future.

The handling of dependencies between components is implemented at present only in the C++ library. A forward chaining inference engine is used to launch the testing of every component at the appropriate time, based on the tests policies and the test results obtained for other components.

Test Configuration schema

The Figure 2 represents an object schema for configuring all aspects of the test configuration. It can be divided into 3 main functional groups:

(1) Classes `Test`, `Test4Object`, `Test4Class` and `Test4CORBAServer` allow to describe a Test and to associate it to an object or class of objects in configuration. An `ExecutableTest` can be

IOP Conf. Series: Journal of Physics: Conf. Series **1085** (2018) 032054 doi:10.1088/1742-6596/1085/3/032054

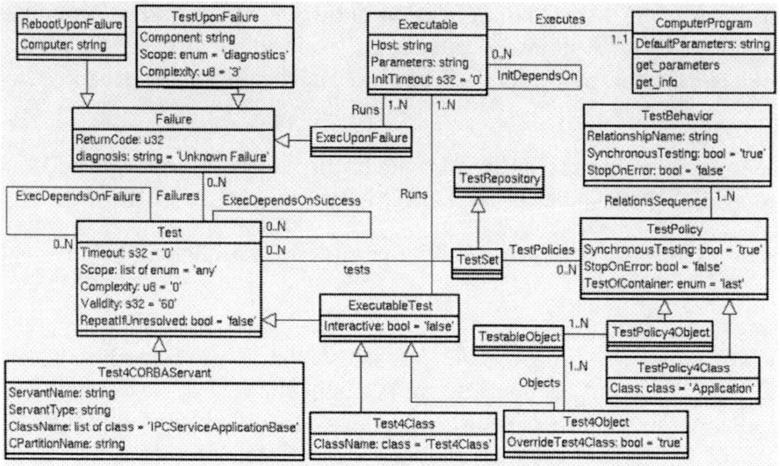

Figure 2: Test Configuration object schema.

launched on any host described in the system, whereas **Test4CORBAServer** is a special ping-like test which is executed locally as a library call in the caller thread, avoiding the overhead of launching of a process.

(2) Classes **TestPolicy**, **TestPolicy4Object**, **TestPolicy4Class** and **TestBehaviour** allows to build a hierarchical tree of testable components, following TDAQ configuration schema and to define policies (order, synchronicity) for testing the groups of components in the tree.

(3) Classes **Failure**, **RebootFailure**, **ExecUponFailure** and **TestUponFailure** define the possible actions which can be executed when test fails with a particular error code.

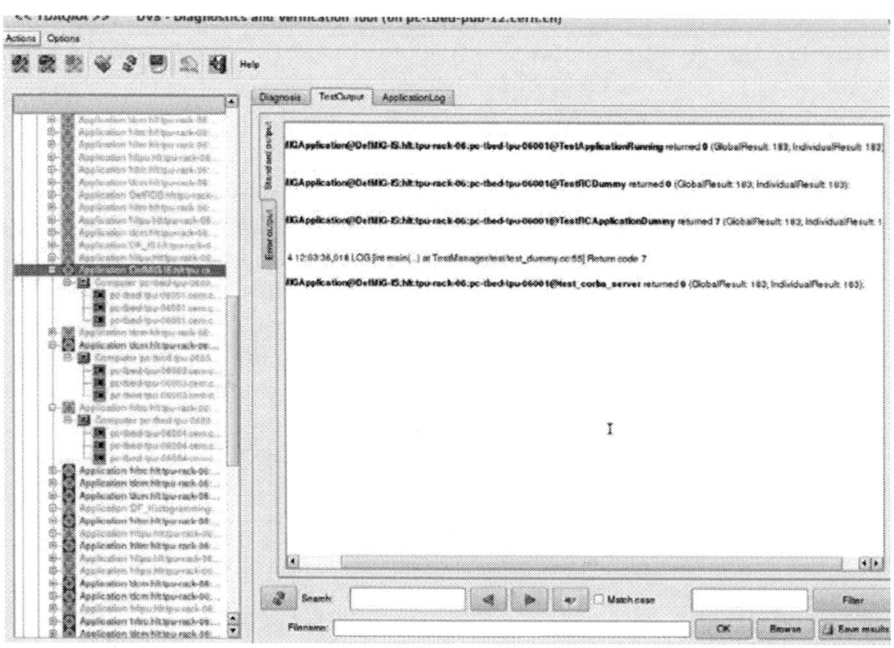

Figure 3: A schreenshot of the Testing GUI with a testable tree on the left and some tests results on the right panel.

IOP Conf. Series: Journal of Physics: Conf. Series **1085** (2018) 032054 doi:10.1088/1742-6596/1085/3/032054

GUI

The GUI (Figure 3) allows an operator to browse a testable TDAQ configuration, to select and to test one or more components, or to run individual tests for one component. The tree-like representation of a configuration is built according to the schema described before and indicates their interdependencies (e.g. an application relies on a working computer to run on). The operator can browse the detailed output of the tests, as well as the diagnosis that has been created based on the test results and the follow-up suggestions that have been configured by the experts in the configuration database.

3. Conclusions and outlook

A highly-configurable, modular and performant framework was developed for handling complex testing and diagnostics scenarios in Online Software for the ATLAS data acquisition system. It allows to store expert's knowledge about testing capabilities of the systems and to use it during TDAQ operations in order to maintain high data taking efficiency of the system in LHC Run 2. Some development is foreseen in C++ implementation in order to maximize utilization of available resources.

References

[1] ATLAS Collaboration 2008 *Journal of Instrumentation, vol.* **3** S08003
[2] M. Abolins et al. 2016 *JINST 11* **06** P06008
[3] Lehmann Miotto G. et al. 2010 *Nucl.Instrum.Meth.* **A623** 549-551
[4] Anders G. et al. 2015 *J.Phys.Conf.Ser.* **608** 1, 012007
[5] Lehmann Miotto G. et al. 2008 *J.Phys.Conf.Ser.* **119** 022004
[6] Avolio G. et al. 2008 *IEEE Trans.Nucl.Sci.* **55** 399-404
[7] Kazarov A. et al. 2007 *IEEE Trans.Nucl.Sci.* **54** 604-608

IOP Conf. Series: Journal of Physics: Conf. Series **1085** (2018) 032055 doi:10.1088/1742-6596/1085/3/032055

Exploiting Apache Spark platform for CMS computing analytics

M Meoni[1], V Kuznetsov[2], L Menichetti[3], J Rumševičius[4], T Boccali[5] and D Bonacorsi[6]

[1] University of Pisa & INFN Pisa, Italy
[2] Cornell University, NY, U.S.A.
[3] CERN, Switzerland
[4] Vilnius University, Lithuania
[5] INFN Pisa, Italy
[6] University of Bologna and INFN, Italy

E-mail: marco.meoni@cern.ch, vkuznet@gmail.com, luca.menichetti@cern.ch, justinas.rumsevicius@gmail.com, tommaso.boccali@pi.infn.it, daniele.bonacorsi@unibo.it

Abstract. The CERN IT provides a set of Hadoop clusters featuring more than 5 PBytes of raw storage with different open-source, user-level tools available for analytical purposes. The CMS experiment started collecting a large set of computing meta-data, e.g. dataset, file access logs, since 2015. These records represent a valuable, yet scarcely investigated, set of information that needs to be cleaned, categorized and analyzed. CMS can use this information to discover useful patterns and enhance the overall efficiency of the distributed data, improve CPU and site utilization as well as tasks completion time. Here we present evaluation of Apache Spark platform for CMS needs. We discuss two main use-cases CMS analytics and ML studies where efficient process billions of records stored on HDFS plays an important role. We demonstrate that both Scala and Python (PySpark) APIs can be successfully used to execute extremely I/O intensive queries and provide valuable data insight from collected meta-data.

keywords: CERN, CMS, HEP, Big Data, Hadoop, Spark, Analytics, Machine Learning.

1. Introduction

The need for agile data analytics platforms is increasing in High Energy Physics (HEP). The current generation of HEP experiments has globally surpassed the Exa-Byte of storage with deployed computing resources exceeding one million computing cores. On top of this, computing infrastructures are highly distributed, and data transfer between sites is a major activity in the computing operations. Systems of such complexity require a careful monitoring in order to be maintained in a healthy state. At this scale processing TBytes of data on a daily basis become a real challenge. Big Data derived approaches have started to be introduced in order to get a better understanding of the computing operations and utilization of computing resources. Among others, the Elasticsearch, Hadoop Map-Reduce, and Spark are attractive solutions to handle large data sets. CERN itself is moving most of its computing dashboards from a standard RDBMS oriented approach (current dashboards use Oracle as a back-end) towards Hadoop based solutions [1]. The expected outcome of this transition is to efficiently process large data sets and aggregate information across distributed data providers. Such studies are becoming more and more important due to the need of reducing computing resources and the associated cost with respect to general increase of physics data collected by the LHC experiments. Funding agencies and review committees are, in fact, asking for increasingly optimized computing operations which can only benefit from a deeper understanding of computing resource utilization.

Since 2015, the Compact Muon Solenoid (CMS) experiment [2] has stored large sets of computing logs ("meta-data") about user accounting, jobs life-cycle, resources utilization, sites performance, software versions, and file access, on Hadoop file systems. The Apache Spark open-source cluster-computing framework has been evaluated as a valuable candidate to handle large amount of this meta-data stored on Hadoop file system (HDFS). Spark provides a scalable and high-performance data processing solution based on APIs offered in a variety of languages.

The benefit of using Spark for CMS experiment is two-fold. On one hand, it reduces the processing time on large datasets (DS) and manages effectively heterogeneous meta-data. On the other, it allows to perform Machine Learning (ML) studies and/or computing analytics over large DS from distributed data sources. Starting from unstructured raw data, the data preparation and analysis steps are completely defined in Hadoop infrastructure. This opens up a possibility to offload production RDBMS workload onto a Hadoop+Spark environment which is more suitable for large data aggregations and distributed joins.

In this paper we address two main use cases:

- CMS data popularity which provides various metrics about CMS user activities based on data available at CERN HDFS system. In particular, we discuss the aggregation pipeline from multiple data-streams and present useful information about user access patterns, such as number of accesses, CPU metrics versus data-tiers, sites, in visualization dashboards;
- Intelligent data placement based on modeling (via ML) and predicting popular CMS datasets to be used in analysis jobs. This activity includes fast queries for reprocessing monitoring time intervals, file-to-DS aggregation and correlation, as well as the evaluation of predictive models.

The paper is organized as follows. Sect. 2 outlines the related works, with focus on efforts applied to Big Data analytics in HEP. Then we provide description of system setup in Sect. 3 where we discuss technical parts of the data flow and underlying technologies. Then, we address the CMS data popularity use case in Sect. 4. Finally, we demonstrate how Hadoop and Spark platforms can be used to analyze large data sets in order to predict user behavior in Sect. 5 followed by discussion of using these predictions in sites caching strategies.

2. Related Work

After the introduction of Spark platform [3] for the general public in 2010 and following adaptation of SQL [4] and ML [5] within Spark, many businesses realized its power within the domain of data analytics. Small and large companies successfully applied this framework to different use cases which required processing large datasets. Up until now, the HEP uses GRID infrastructure to process individual jobs at distributed sites. In order to leverage Map-Reduce paradigm and distributed nature of Hadoop+Spark platform many aspects should be modified. Among the others, changes to middle-ware in order to read underlying ROOT data-format, adaptation of data workflow scheduling and submission tools as well as overall design of data processing [6]. Such effort is undergoing within recently funded projects such as DIANA-HEP [7]. The analytics use cases, e.g. extracting data insight from available meta-data (logs), is very well studied and adapted in a business world. Recently, HEP experiments realized that they can explore their meta-data to gain additional insight about user activities and apply this knowledge to improve their computing models [1]. In particular, exploiting Machine Learning techniques on computing meta-data is a hot topic in the HEP community at large [8]. The data popularity problem was studied by all major HEP experiments, e.g. in CMS [9, 10], LHCb [11], and

ATLAS [12]. It was found that ML approach can be useful addition to custom data placement solutions adopted by experiments. In this work, we extend previous CMS attempts [9, 10] to use CMS meta-data in the context of dataset popularity and extend discussed techniques to Spark platform.

3. System Setup and Implementation

The European Organization for Nuclear Research (CERN) has deployed two large Hadoop clusters featuring nearly 5 PBytes of raw storage. Currently, they consist of around 50 nodes equipped with 64 GBytes of RAM and 32 cores/node on a mix of Intel/AMD CPUs and CentOS7/SLC6 operating systems.

The Spark cluster is tuned and optimized to reach maximum scalability and parallelism. Every job is calibrated in favor of local processing within the same executor and reducing the data exchange between cluster nodes. The Spark paradigm consists of reading data once and processing them as much as possible in memory, minimizing data shuffling, network communication between executors, and disk I/O operations. In addition, together with the cluster resource manager (YARN in our case), Spark jobs can profit from dynamic allocation of their executors, automatically scaling in and out their number when necessary. This allows users to run the same job against folders of the same dataset without adapting different parameters according to their individual sizes.

The CERN Hadoop service provides scientists with a variety of tools and Big Data frameworks, such as Spark, Pig, Hive, which are available in different clusters. These are distinct but not isolated environments where data stored in one HDFS namespace can be accessed by jobs running indiscriminately in whatever cluster. Obviously jobs reading the local namespace will be faster, and if necessary data can be moved from one cluster to another using Map-Reduce jobs.

CMS data is stored in daily folders in CSV, JSON, or Avro format. Spark 2.0 framework, via the SparkSQL context, is equipped to read these formats. These components take as input an HDFS path and may require additional parameters depending on the kind of the format, e.g. a header of CSV format, a data schema or quotation symbols, and so on. All of them return a DataFrame used in forthcoming computations.

Since Spark is written mainly in Scala and Java, we originally used the former language for work discussed in Sect. 5. Later we found that the Scala programming language is not (yet) well adopted in HEP community. Therefore, we made significant effort to adopt our code base to PySpark framework, as explained in Sect. 3.2.

In next Sections 3.1 and 3.2, we provide details of our experience with Scala and PySpark APIs. We mostly outline pros and cons of both languages based on our practice with them.

3.1. Scala usage

Our original attempt to process large DS on HDFS was done in the Scala programming language and it reflects results described in Sect 5. Here we would like to highlight our observation of using Scala for this task. Fig. 1 shows the processing time required for DS popularity monitoring queries. The first plot shows the amount of time, in seconds, that was needed by a number of monitoring queries. The names on the x-axis are borrowed from the corresponding materialized views (MV) implemented in the current Oracle-based monitoring infrastructure. The second plot shows the reprocessing time to reproduce the results of an Oracle DB view. We found that regardless of the Oracle view or selected time period the average processing time remains in a range of few minutes[1]. This makes Spark and Scala a very effective combination for reprocessing

[1] In order to exploit built-in distribution and persistence, we relied on `repartition()` and `persist()` API to split data uniformly across the computing nodes and cache them during the aggregation steps.

IOP Conf. Series: Journal of Physics: Conf. Series **1085** (2018) 032055 doi:10.1088/1742-6596/1085/3/032055

popularity queries on Hadoop. In fact, the same results take continuous running of incremental MV updates in Oracle back-end.

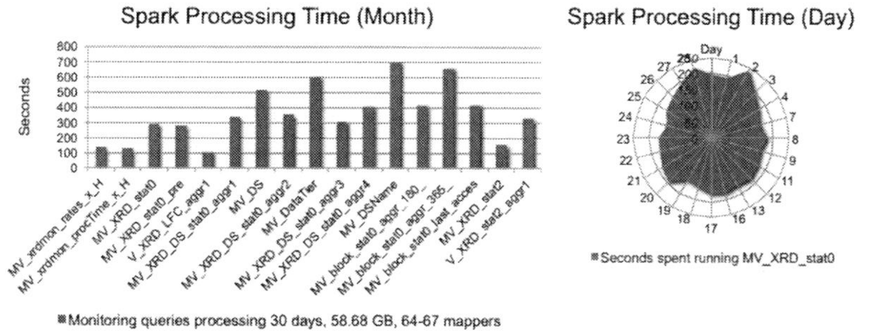

Figure 1. Fast reprocessing of DS popularity monitoring query on Apache Spark. The left plot shows monthly processing time for individual Oracle DB views implemented in Spark. The plot on a right shows processing time of one Oracle DB view for daily records.

3.2. PySpark usage

The transition from Scala to PySpark was trivial. But we found a few particularly interesting issues every programmer should know about when developing code using Python (PySpark) APIs.

Careful crafting of data is required to fit it into worker's node memory. For instance, most of the run-time errors we experienced were related to java.lang.OutOfMemoryError, Java heap, and GC errors issues. A typical example would be the temptation to use *df.collect*() PySpark function to collect intermediate results. Instead, the code should perform series of operations over the data and either write them out to HDFS or run aggregation steps. Further, when the data are skewed not all workers obtain the right amount of workload. For instance, grouping or shuffling operations of non equally distributed data can cause the slowness of entire job within executor. To take advantage of parallelism the code should not iterate over the containers, instead, optimized PySpark APIs should be used, e.g. $df.foreach$ in favor of $for\ item\ in\ df.collect()$ pattern or $df.treeReduce(lambda\ x, y : x + y)$ instead of local non cluster-distributed functions $np.sum(df.collect())$.

Users are advised to rely on PySpark functions to avoid slowness of built-in Python functions over DataFrame operations, e.g. $pyspark.sql.functions.split$ should be used instead of *split* Python built-in function.

To avoid these types of pitfalls we developed our own framework called CMSSpark [13] which provided all necessary tools for code submission to Spark platform, set up necessary Java libraries for data processing (e.g. support for CSV or Avro data formats) as well as set up proper run parameters optimized for CERN Spark cluster.

This framework significantly helped new developers to easily adopt their business logic into framework and quickly produce desired results. For instance, any user with access to the CERN Hadoop cluster can quickly launch distributed (Py)Spark job to process CMS data, focusing on what operation to run rather than struggling on how to properly submit YARN-based batch job.

4. CMS Data Popularity

The common approach of data analytics is based on RDBMS solutions. At CERN, the Oracle DB is a de-facto standard for many bookkeeping and analytics services. Even though it works extremely well we see that the amount of information HEP experiments want to process is growing exponentially each year. We already found certain limitations on data processing pipelines within Oracle DB. For instance, the CMS most populated DBs, Data Bookkeeping System (DBS) and Data Location System (PhEDEx) [14], already contain a few hundred GBytes of data. Merging data only from those two DBs becomes an issue. Even though they reside in one physical DB, the cross-joins among them caused large latencies on complex queries due to large data volume. Moreover, growing number of collected meta-data about user jobs which resides outside of Oracle DB may provide an additional insight into data utilization by our infrastructure. Spark platform provides a valuable solution to overcome these obstacles.

The CMS, among other HEP experiments, has joined a new trend of storing various meta-data to HDFS starting in 2015. This effort was driven by individual users and groups willing to investigate a new technology to analyze unstructured data, such as logs from user based jobs. By 2017 we already collected billions of records from various data providers. Here we present a successful effort showing utilization of Hadoop+Spark platform for analytics purposes and modeling of DS popularity based on available meta-data. We start our discussion with CMS analytics use case where significant portions of these data were used.

4.1. Data Processing

We used Spark jobs to acquire various information about user activities such as total number of accesses to dataset and processing time of analysis jobs at specific sites. We identified four data-streams (in the form of logs) representing this information: AAA [15] logs of remote XrootD accesses; EOS [16] logs of file accesses on CERN EOS system; CMSSW [17] logs capturing information about CMS software framework, and CRAB [18] logs representing user analysis jobs. All of these data are accessible on HDFS.

Each data-stream contains various metrics associated with a job in question, e.g. name of the file used during a job, user Distinguished Name (DN), processing time of the job, and similar. In order to get insight into data present in these data streams they should be properly cleaned, sorted, filtered, and joined with data from other sources such as DBS and PhEDEx DBs which provide information about dataset, block and site names. The latter two DBs are quite large, their current size is at the level of a few hundred GBytes each. In order to answer simple questions about users accessing a particular dataset on a certain site(s), it was required to join various attributes from multiple streams along with DB tables from DBS and PhEDEx Oracle DBs. We found that processing time to join and aggregate data across DBS/PhEDEx DBs and data-streams is impossible to achieve in a reasonable amount of time using a vertical approaches based on RDBMS solutions[2]. At the same time a simple dump of DBS and PhEDEx DBs into HDFS and performing a regular Spark job to join desired data-streams with them allows quickly (on a scale of minutes) to get insight into data we were looking for. For instance, a daily snapshot consists of about 130M log records which were joined at a file level with DBS and PhEDEx DBs, grouped and aggregated to roughly a few thousand records in under an hour of Spark processing time. We measured that monthly stats can be obtained in about 12 hours. This allowed us to build a series of dashboards to look at data utilization in the CMS experiment.

[2] A typical job to aggregate information from Oracle DBs and a single data-stream required many hours to complete using materialized view across all data.

4.2. Data Analytics and Monitoring

There are different approaches to data monitoring. On one hand, someone can delegate pushing information from individual GRID sites and applications into the central monitoring system. On the other, such information can be obtained from an individual logs stored in a common place (e.g. central HDFS) via external jobs. The former approach requires proper instrumentation of monitoring metrics and coordination of data injection from all participating data providers. The latter relies on significant processing power to collect and aggregate information in a reasonable time.

The ability to efficiently process large data sets on HDFS via Spark jobs opens up a possibility to build various dashboards based on user access metrics extracted from various data-streams. For that purpose we used CERN MONIT system [1] based on ElasticSearch and Kibana tools. The data processed by Spark jobs were injected as JSON documents into CERN MONIT system via Apache ActiveMQ Stomp [19] protocol. We achieved a reduction factor of 5000 coming from meta-data records (logs) on HDFS to aggregated records pushed into CERN MONIT system. Fig. 2 shows one part of the CMS popularity dashboard we built.

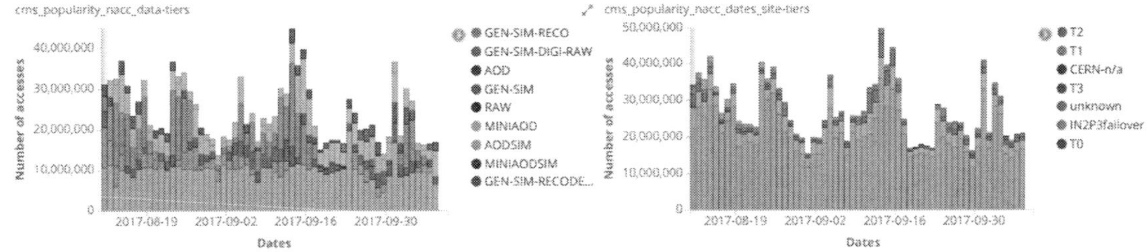

Figure 2. CMS popularity dashboard based on processing four CMS data streams: AAA, EOS, CMSSW and CRAB. The displayed part of the dashboard represents the number of accesses vs data-tiers and site-tiers as time series plots.

Even though various dashboard plots were useful to get a broader view on daily activities of CMS users, we were also able to address specific questions raised by various users. For example, how much data resides on all T2 sites with detailed breakdown of number of events and replica sizes of common used data-tiers.

While looking at the current status of resource utilization, which is a very important component of our daily operations, we oversee that it can be further enhanced by predicting user behavior and adjusting our resources accordingly.

5. Modeling CMS Popularity

Usage of Machine Learning in HEP was mostly used in physics analysis such as discrimination of signal versus background, clustering algorithms, object reconstructions, or similar tasks. Recently, ML ideas started penetrating into other domains of HEP such as helping physicists better understand computing resources. For instance, ML modeling has been discussed in the scope of dataset placement optimization, reduction of transfer latencies as well as in network-aware applications for identifying anomalies in network traffic, predicting congestions and WAN path optimization. Here we present results build on top of previous studies for predicting dataset popularity in context of better data placement strategies. We extend these results by demonstrating how to predict dataset popularity using larger datasets and use them to build strategies for better data placement at participating sites. In fact, DS placement at most WLCG experiments is based on DS popularity which should be considered to make an optimal choice of

IOP Conf. Series: Journal of Physics: Conf. Series **1085** (2018) 032055 doi:10.1088/1742-6596/1085/3/032055

replication to maximize data availability for processing and analysis needs. As shown in [9, 20], CMS dataset popularity can be successfully predicted using ML approach.

In this work, the generation of ML dataframes involved several processing steps on Spark and it was achieved through a Scala reduction pipeline shown in Fig. 3. It highlights the number of samples and the computing time at each step of the reduction chain. In the shown example, we start from two billion AAA file access logs, merge them with PhEDEx DB to extract dataset and site information and end-up with half a million records suited for ML studies.

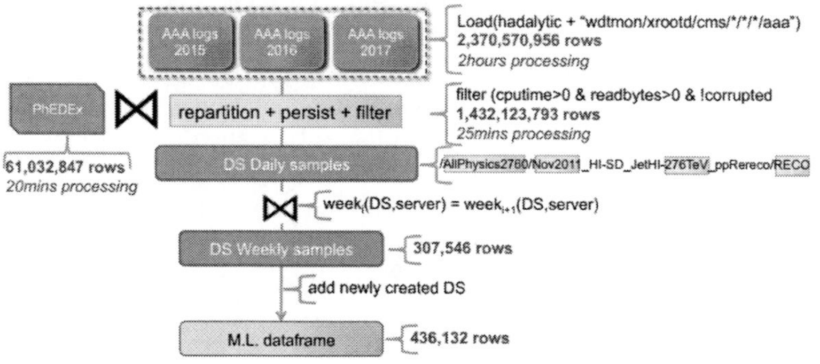

Figure 3. An example of reduction pipeline of Spark components from file AAA access logs level to a dataframe suitable for Machine Learning input.

The input dataframe for popularity predictions consisted of a set of attributes and a popularity label. This label is computed on the basis of the usage metric designed as popularity feature. When usage metric in some $time\_period_{i+n}$ is higher than a given threshold, the samples corresponding to $time\_period_i$ is marked as popular. Our choice for timespan n was chosen to be one week. This was justified by empirical observation. When the temporal window in the training set was increased to 3 weeks the number of positive samples decreases remarkably and prediction accuracy decreased. In fact, Fig. 4 shows that the number of DS accesses over 3 weeks time frame is about 50% lower than across 2 weeks, which explains that most DS have a time-limited access pattern.

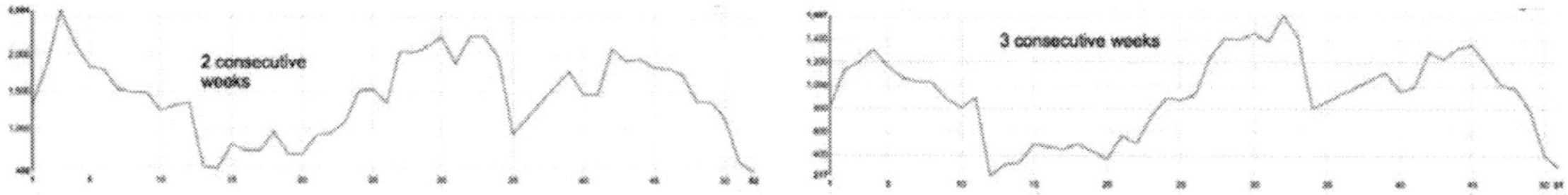

Figure 4. Dataset access patterns during 2 and 3 week period of time.

After applying a cut on number accesses to a dataset, we converted used dataset into classification problem which later was trained on Spark via various classifiers available in Spark MLLib. We achieve the same level of accuracy as reported in [9].

The outcomes from the classifiers are used to implement a data caching policy based on popularity predictions. The datasets are stored at various CMS Grid sites to minimize the network traffic and spread the load for analysis jobs. The current site placement policy is based on assumptions on the expected needs for datasets, mostly using the data-tier. The use of storage, at least for small-to-medium sites, is based on Least Recently Used (LRU) caching systems in order to select which elements should be removed when a site reaches its capacity. Instead, we design a different caching policy that relies on knowledge of DS popularity in the next week and keeps DS in the cache if it is still popular. If a given DS is selected for eviction but is marked as popular by the prediction model, then it is maintained inside the cache.

Preliminary results are detailed in [20], with the main difference w.r.t LRU observed at lower caching size threshold. We always assume a cold start at each site which generates a number of compulsory cache misses corresponding to the first reference to each distinct DS. Our caching strategy outperforms LRU by up to 20% when the cache size is limited. This makes it very effective in production sites with limited storage or bandwidth. It paves the way for further comparison with others caching strategies and for research aiming at enabling smart policies for dataset replica management based on demonstrated ML approach.

6. Conclusions and Outlook

Spark has proven to be a very solid framework for data analytics and ML. It is extremely effective for crunching large datasets available on HDFS and extracting aggregation metrics from underlying meta-data. It can be integrated into HEP workflow via PySpark APIs or stand-alone application written in Scala or Python. We discussed possible pitfalls programmers should be considered using either of the programming languages.

We demonstrated that Spark framework represents a valuable alternative to RDBMS solutions for analytics needs. It allows to (re-)process and aggregate various metrics from multiple data-streams efficiently in distributed nature. We were able to access logs from different data streams including AAA, EOS, CMSSW and CRAB and join them with the HDFS dumps of two largest CMS databases, DBS and PhEDEx. As a result the data can be easily aggregated into set of dashboards based on CERN MONIT system and be used for various monitoring tasks within collaboration.

We also implemented a scalable pipeline of Scala components to accomplish mission-critical data reduction and ML processing from billion of records available at CERN HDFS system and build ML models that can be used towards data-driven approach in CMS computing infrastructure. For example, dataset popularity predictions can play a significant role in intelligent data placement at GRID sites both for newly created datasets and holding dataset samples frequently accessible by CMS physicists. A clever caching strategies based on ML predictions can lead to cost-effective data placement across GRID sites and better utilization of available computing resources.

In conclusion, CMS considers the Spark platform as a potentially crucial component in the ecosystem that would enable LHC experiments to attack Big Data problems in the long run.

References

[1] Aimar A et al 2017 *Unified Monitoring Architecture for IT and Grid Services*, *J. Phys.: Conf. Series* **898** 092033

[2] Chatrchyan S et al 2008 *The CMS Experiment at the CERN LHC*, JINST **3** S08004T

[3] Matei Z et al 2010 *Spark: Cluster Computing with Working Sets*, HotCloud 10–10

[4] Armburst M et al 2015 *Spark SQL: Relational Data Processing in Spark*

[5] Meng X et al 2016 *MLlib: Machine Learning in Apache Spark*, J. Machine Learning Research **17** 1-7

[6] Gutsche O et al 2016 *Big Data in HEP: A comprehensive use case study*, *Preprint* arXiv:1703.04171

[7] DIANA HEP project, part of NSF's Software Infrastructure for Sustained Innovation (SI2) program, http://diana-hep.org/pages/activities.html

[8] HEP Software Foundation white papers: http://hepsoftwarefoundation.org/activities/cwp.html

[9] Kuznetsov V, Li T, Giommi L, Bonacorsi D and Wildish T 2016 *Predicting dataset popularity for the CMS experiment*, *J. Phys.: Conf. Series* **762** 012048

[10] Meoni M, Boccali T, Magini N, Menichetti L and Giordano D 2017 *XRootD popularity on Hadoop clusters*, *J. Phys.: Conf. Series* **898** **7** 072027

[11] Hushchyn M, Charpentier P and Ustyuzhanin A 2015 *Disk Storage Management for LHCb Based on Data Popularity Estimator*, *J. Phys.: Conf. Series* **664** 042026

[12] Beermann T et al 2014 *Popularity Prediction Tool for ATLAS Distributed Data Management*, *J. Phys.: Conf. Series* **513** 042004

IOP Conf. Series: Journal of Physics: Conf. Series **1085** (2018) 032055 doi:10.1088/1742-6596/1085/3/032055

[13] https://github.com/vkuznet/CMSSpark
[14] Giffels M et al *The CMS Data Management System*, *J. Phys.: Conf. Series* **513** 042052
[15] Xrootd project page: http://www.xrootd.org/
[16] Mascetti L et al 2015 *Disk storage at CERN*, *J. Phys.: Conf. Series* **664** 042035
[17] CMS software framework, https://github.com/cms-sw/cmssw
[18] Spiga D et al 2008 *CRAB: The CMS distributed analysis tool development and design* Nuclear Physics B - Proceedings Supplements 177-178 **1** 267-268
[19] Apache ActiveMQ Stomp, http://activemq.apache.org/stomp.html
[20] Meoni M, Perego R and Tonellotto N 2017 *Popularity-based caching of CMS datasets*, ParCo

IOP Conf. Series: Journal of Physics: Conf. Series **1085** (2018) 032056 doi:10.1088/1742-6596/1085/3/032056

Mastering Opportunistic Computing Resources for HEP

M J Schnepf, C Heidecker, M Fischer, M Giffels, E Kuehn, A Heiss, A Petzold and G Quast

KIT - Karlsruhe Institute of Technology, Germany

E-mail: {matthias.schnepf, christoph.heidecker, max.fischer, manuel.giffels, eileen.kuehn, andreas.heiss, andreas.petzold, guenter.quast}@kit.edu

Abstract. As results of the excellent LHC performance in 2016, more data than expected has been recorded leading to a higher demand for computing resources. It is already foreseeable that for the current and upcoming run periods a flat computing budget and the expected technology advance will not be sufficient to meet the future requirements. This results in a growing gap between supplied and demanded resources.

One option to reduce the emerging lack of computing resources is the utilization of opportunistic resources such as local university clusters, public and commercial cloud providers, HPC centers and volunteer computing. However, to use opportunistic resources additional challenges have to be tackled. At the Karlsruhe Institute of Technology (KIT) an infrastructure to dynamically use opportunistic resources is built up. In this paper tools, experiences, future plans and possible improvements are discussed.

1. Introduction

Computing resources for the LHC experiments are usually provided statically by the Worldwide LHC Computing Grid (WLCG) [1]. However, the demand for computing resources varies over the time due to periods of data taking and conferences. This results in either over- or under-commitment of available resources, both having its drawbacks. In addition, the current deployment strategy of WLCG sites is relying on a flat budget and the expected technology advance. However, this will not be sufficient to cover future resource needs for the LHC computing leading to a gap between supplied and demanded resources [2].

Dynamic integration of additional computing resources for High Energy Physics (HEP) could help to reduce this gap and would allow more flexibility of the computing models. Therefore, one of the research topics at the Karlsruhe Institute of Technology (KIT) is the development of an infrastructure for a transparent and dynamic provisioning of computing resources not dedicated to HEP, so-called opportunistic resources provided for example by High Performance Computing (HPC) Centers, scientific and commercial cloud providers.

In this paper the developed infrastructure, experiences, future plans and possible improvements are discussed.

2. Integration of Opportunistic Resources

The integration of opportunistic resources can be grouped into three different categories as shown in Fig. 1. One category is the extension of resources for a longer period of time as

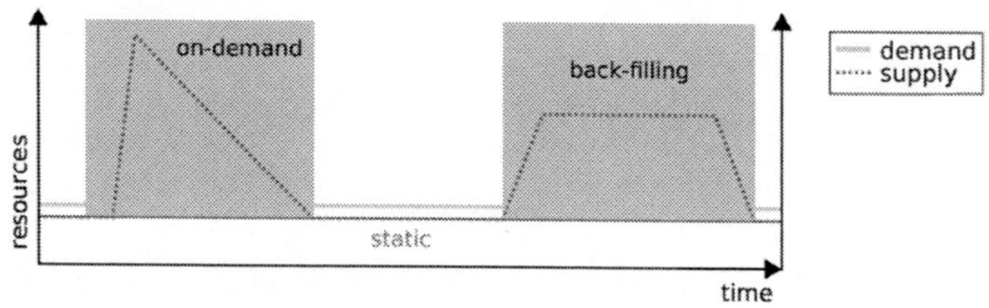

Figure 1: Three different ways to extend statically available HEP compute resources (green line). The constant capacity expansion (black line), the dynamic extension of resources (depicted on the left side) that depends on the current demand (orange line) and the back-filling approach (shown on the right side) which is independent of the current demand. The dotted blue line shows to total amount of compute resources available over the time.

static resources. The other two categories are the dynamic on-demand extension of resources to cover peak-loads and the back-filling approach. Since the extension with static resources is comparable to the deployment of new dedicated HEP computing resources, we will focus on the dynamic approaches in this paper. Typically HEP computing resources are provided to the experiments utilizing the concepts of traditional Grid computing. In other words, it means that resource providers have to maintain a complex framework with special dependencies in terms of the operating system and the HEP software requirements to make their resources available to the HEP community. However, this is usually not feasible on opportunistic resources, since it is often incompatible with the intended purpose of those resources.

To overcome this challenge it is necessary to utilize modern virtualization and container technology following the "Infrastructure as a Service" approach. This leads to a task sharing between the resource provider, who solely operates the infrastructure and HEP that manages its complex software environment using virtual machine images or containers. This approach allows HEP to utilize additional resources in a dynamic fashion from various providers like scientific and commercial clouds as well as High Performance Computing centers, given that a virtualized environment or a container solution is offered for resource provisioning.

Many tools utilized in HEP computing do already support modern cloud technologies and therefore provide a profound base for making use of opportunistic resources. The HTTP based CERN Virtual Machine File System (CVMFS) [3] is well suited for the world-wide distribution of experiment software. Using conventional caching HTTP proxies enables a scalable on-demand provisioning of software. This makes it to an adequate choice to deploy software on opportunistic resources. The XRootD protocol [4] and the data federations of the experiments completes this technology and provides transparent access to the experiment data located on traditional Grid storage systems. In addition, the batch system HTCondor [5] which is commonly used in HEP is also perfectly suited to integrate dynamic and opportunistic resources due to its pull workload mechanism as well as its native support of running jobs in `docker` [6] and `singularity` [7] containers.

3. Resource Management

In addition to the software stack describe above, the cloud scheduler `ROCED` (**R**esponsive **O**n-**D**emand **C**loud-**e**nabled **D**eployment) [8] has been developed to address one of the remaining challenges, namely the dynamic management of resources depending on the demand. `ROCED` monitors the demand for computing resources by checking the job queue of a batch system as well as the available resources. Based on this information it can either request new or release unused resources by starting or terminating virtual machines, respectively. The workflow of `ROCED` is depicted in Fig.2. `ROCED` provides interfaces to various cloud APIs and batch systems, to allocate

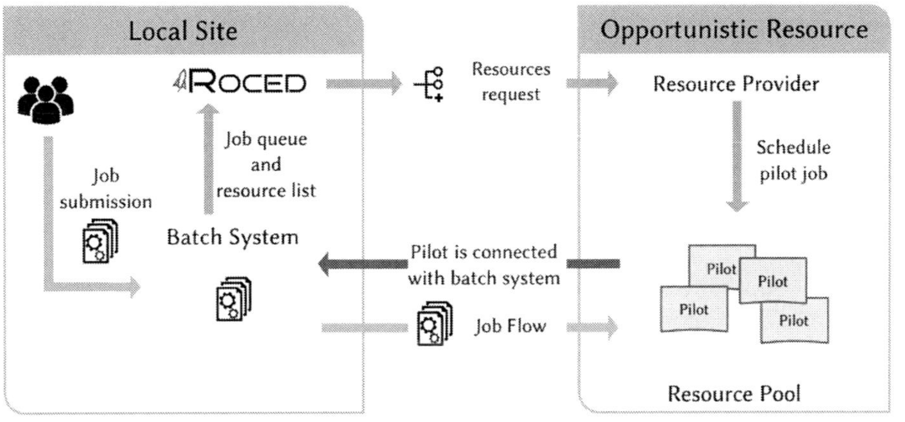

Figure 2: The resource manager ROCED monitors the job queue and the available resources of the batch system. Based on this information it can request additional resources from a supplier like a HPC center or a cloud provider if necessary. Once the resources are provisioned, the pilots connect to the batch system and new jobs can be assigned.

and integrate resources from different providers. Nevertheless, support for additional batch systems and cloud APIs can be easily added thanks to its modular design. ROCED can handle several different cloud resources at once and implements a basic cloud brokerage functionality based on cost.

4. Experiences with Opportunistic Resources

The development of ROCED started back in 2011, since then the group has gained longterm experience with the acquisition, integration, and utilization of different kind of opportunistic resources for HEP computing, which will be described in the following using illustrative examples. The following resources have been integrated transparently for the users into the same HTCondor batch system.

4.1. Desktop Cloud

The institutes desktop PCs are primarily used for office applications like email, web browsing, text editing, etc. For this purpose, the users rely on a modern operating system like Ubuntu Linux supporting state-of-the-art versions of office applications. On the other side, recently procured desktop PCs equipped with powerful eight core Intel i7 CPUs, 16 GB RAM and SSDs are well suited to run HEP applications. Bearing in mind the strict requirements of the HEP software stack based on a mature operating system, both use cases seem incompatible at the first glance. However, a recently added feature of the HTCondor batch system allows the execution of jobs inside docker containers. By utilizing this feature, modern container technology could be used to manage the balancing act between office use and HEP applications. In total around 300 additional job slots could be added to the batch system more than doubling the locally available computing power for HEP. In order to avoid interferences with the daily users of the desktop PCs, compute jobs are automatically suspended and resumed depending on the local utilization of the PCs. Altogether the utilization of the desktop PCs as well as the available HEP computing power could be increased by using modern container technologies without unreasonably great effort and expense.

IOP Conf. Series: Journal of Physics: Conf. Series **1085** (2018) 032056 doi:10.1088/1742-6596/1085/3/032056

4.2. Commercial Cloud Providers

The usage of commercial cloud providers to cover peak loads has been evaluated as well. In this case, the `ROCED` cloud scheduler was used for the on-demand resource provisioning and integration in the locally operated `HTCondor` batch system. The provisioning of resources on the infrastructure of commercial cloud providers still relies mostly on traditional virtualization techniques even more and more providers started to support container technologies as well. Therefore, virtual images based on SL6 and `CVMFS` were used in order to supply the HEP software environment. In order to allow efficient access to conditions data and `CVMFS` repositories at CERN, service infrastructure like HTTP proxies need to be deployed on the cloud sites as well. So far, the integration of resources from commercial cloud providers has been tested based on grants and the experiences gained are mixed. For global players, almost everything works out of the box, whereas for smaller vendors often special adaptions are needed. The cost of virtual resources procured from commercial providers is nearly dropping every year, however the cost of network traffic still remains very high. Nowadays the integration of commercial providers in scientific computing is driven by the Helix Nebula Science Cloud (HNSciCloud) project [10] Founded by the European Union. Further test as part of the HNSciCloud project are currently ongoing.

4.3. High Performance Computing Center NEMO

In the context of the bwHPC project [9] of the state of Baden-Württemberg, HEP is also entitled to use HPC resources at the bwFORCluster NEMO at the University of Freiburg [11]. NEMO is a shared HPC cluster between neuroscience, microsystem engineering and particle physics. Each community has different needs and requires a very different software stack. The only viable solution is to operate a classic HPC infrastructure on bare metal in combination with a virtualized infrastructure on the same cluster [12]. For this purpose a hypervisor is installed on each compute node and the virtual infrastructure is managed by OpenStack. However, the needs for an accounting and fairshare mechanism on the cluster leads to a more complex setup compared to the commercial cloud providers described above. The resource allocation needs to be done via the HPC batch system. Therefore, a batch job containing a small script is submitted, which takes care of blocking a job slot on a corresponding worker node, as well as requests the provisioning of a virtual machine on the same worker node via the OpenStack API. Using this setup up to 8000 cores can be requested in parallel and the scalability of the solution has been proven up to 11000 cores. The NEMO cluster is currently largest supplier for opportunistic resources integrated into HEP computing at KIT. The resource utilization of the KIT HEP group at NEMO is shown in Fig.3.

5. Outlook

The usage of large amounts of opportunistic resources comes along with new challenges. Unlike traditional Grid computing sites, suppliers of opportunistic resources usually do not provide permanent storage systems or do not offer access to dedicated WLCG networks. Thus, it is recommended to use opportunistic resources only for CPU intensive tasks so far. Otherwise the remote data access to Grid storage systems could overload the network and the storage systems as well. As a result I/O intensive tasks would suffer from a low CPU utilization. However, in the following concepts to use opportunistic resources also for I/O intensive tasks are presented.

One solution could be transparent caching of data locally at the site, for example by using shared storage available at HPC centers. This concept is described in more detail in [13].

Another supplementing approach could be the improvement of the job scheduling based on the network utilization and the available network bandwidth as shown in Fig. 4. HEP workflows usually consist of several thousands jobs running the same application with different input parameters or files. Thus, it can be assumed that the network access pattern of different

IOP Conf. Series: Journal of Physics: Conf. Series **1085** (2018) 032056 doi:10.1088/1742-6596/1085/3/032056

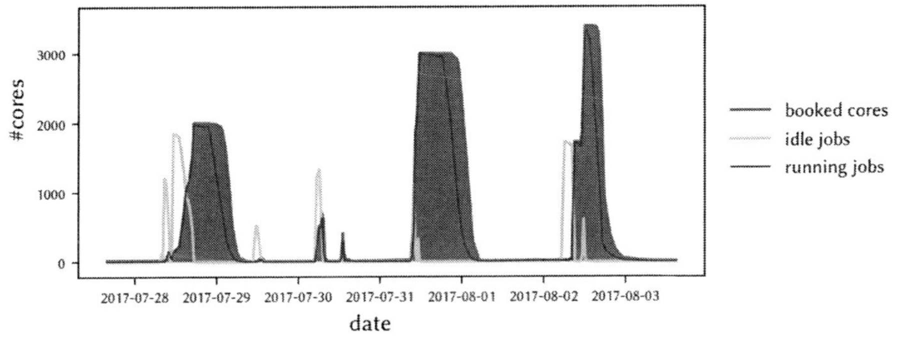

Figure 3: Resource utilization of the KIT HEP group at the NEMO HPC center. The orange line shows the number of idle jobs which corresponds to the demand for additional resources. The blue area represents the number of cores integrated in our batch system. The blue line shows the number of used CPU cores. Depending on the utilization of the NEMO cluster, the number of integrated cores could increases slower than the number of demanded cores. Due to draining effects idle job slots can only be released once all jobs running inside one VM are finished.

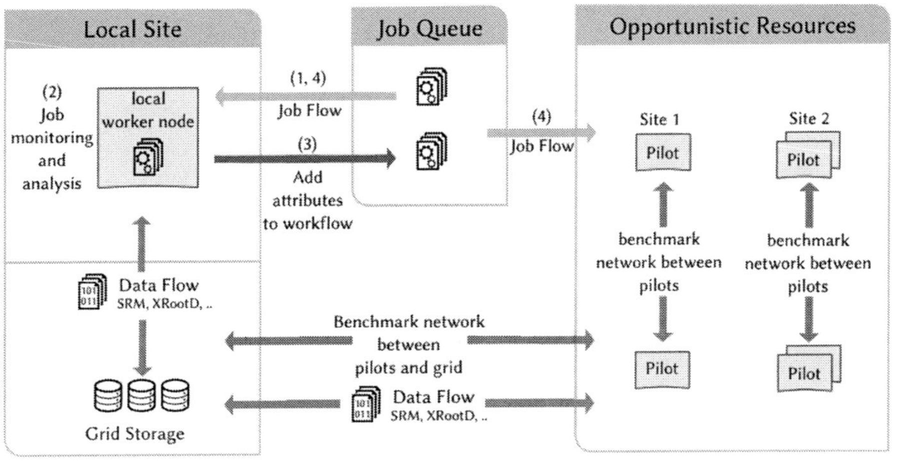

Figure 4: Improved scheduling using information about estimated network utilization of jobs and available bandwidth between sites. (1) Run sample jobs of a given workflow on local resources. (2) Monitor and analyze the network I/O behaviour of these jobs. (3) Add network I/O related attributes to the workflow description. (4) Evaluate additional attributes for the scheduling of jobs on opportunistic resources. In addition, coordinated network benchmarks between the local site and the oppportunistic resource are performed to determine the available network bandwidth.

jobs in the same workflow are very similar. The network I/O behaviour can be estimated by running sample jobs of a workflow on local resources. Afterwards network I/O related attributes are added to the remaining jobs of the workflow.

In addition, coordinated network benchmarks between the Grid sites and the opportunistic resources are performed to determine the available bandwidth using for example the perfSONAR [14] toolkit.

By taking into account the available bandwidth and the estimated network I/O requirements of the jobs in a given workflow, the scheduling process can be improved leading to a more efficient

IOP Conf. Series: Journal of Physics: Conf. Series **1085** (2018) 032056 doi:10.1088/1742-6596/1085/3/032056

utilization of the available network bandwidth and therefore to a better CPU utilization on the opportunistic resources. Both concept are currently in development and will be evaluated in the future.

6. Summary

This paper presents the current state of the integration of opportunistic resources into the KIT HEP computing resource pool. To deploy the required HEP software and operating system stack on those resources modern virtualization and container technologies are utilized. The dynamic on-demand resource management is done using the cloud scheduler `ROCED` developed at KIT. All acquired resources are integrated into the same `HTCondor` batch system to allow transparent access to the users. Further concepts to tackle the challenge of efficient remote data access arising from the usage of opportunistic resources by taking into account additional information about network utilization and available bandwidth during the job scheduling process are presented as well. Overall the ability of the dynamic integration of opportunistic resources into HEP computing has been demonstrated.

Acknowledgment

The authors acknowledge the support by the DFG-funded Doctoral School "Karlsruhe School of Elementary and Astroparticle Physics: Science and Technology", the German Helmholtz Association and the state of Baden-Württemberg through bwHPC.

References

[1] Eck C et al. 2005 LHC computing Grid : Technical Design Report *Technical Design Report LCG*, https://cds.cern.ch/record/840543

[2] Bird I et al. 2016 Status of the WLCG Project, including Financial Status, CERN-RRB-2016-124, https://cds.cern.ch/record/2210410 [accessed 2017-10-12]

[3] Buncic P et al. "CVMFS" [software], version 2.X.Y, http://iopscience.iop.org/1742-6596/219/4/042003

[4] XRootD project, "XRootD" [software], http://xrootd.org/

[5] HTCondor project, "HTCondor" [software], version 8.6.1, Available from https://research.cs.wisc.edu/htcondor/downloads/ [accessed 2017-10-12]

[6] Docker inc., "docker" [software], version 17.05.0-ce, Available from https://www.docker.com/ [accessed 2017-10-12]

[7] Kurtzer GM, Sochat V, Bauer MW, "singularity" [software], version 2.2.1, 11 May 2017. https://doi.org/10.1371/journal.pone.0177459

[8] ROCED project, "ROCED" [software], version 1.0.0, Available from https://github.com/roced-scheduler/ROCED [accessed 2017-10-12]

[9] bwHPC-C5: Coordinated Compute Cluster Competence Centers, http://www.bwhpc-c5.de/ [accessed 2017-10-12]

[10] Helix Nebula Science Cloud, http://www.hnscicloud.eu/ [accessed 2017-04-02]

[11] Meier K et al. 2016 Dynamic provisioning of a HEP computing infrastructure on a shared hybrid HPC system, *Journal of Physics: Conference Series* Volume 762 012012

[12] Fischer F et al. 2017 On-demand provisioning of HEP compute resources on cloud sites and shared HPC centers, *Journal of Physics: Conference Series* Proceedings of 22st International Conference on Computing in High Energy and Nuclear Physics (to be published)

[13] Heidecker C et al. 2017 Provisioning of data locality for HEP analysis workflows, *Journal of Physics: Conference Series* Proceedings of 18th International workshop of Advanced Computing and Analysis Techniques (to be published)

[14] perfSONAR collaboration, perfSONAR [software], Available from https://www.perfsonar.net/ [accessed 2017-10-12]

IOP Conf. Series: Journal of Physics: Conf. Series **1085** (2018) 042001 doi:10.1088/1742-6596/1085/4/042001

The ATLAS Electron and Photon Triggers

Samuel David Jones, on behalf of the ATLAS Collaboration

School of Mathematical and Physical Sciences, Pevensey II, University of Sussex, Falmer Campus, Brighton, BN1 9QH, United Kingdom

E-mail: `samuel.david.jones@cern.ch`

Abstract. Electron and photon triggers covering transverse energies from 5 GeV to several TeV are essential for signal selection in a wide variety of ATLAS physics analyses to study Standard Model processes and to search for new phenomena. The ATLAS trigger system is divided in a hardware-based Level-1 trigger and a software-based high-level trigger, both of which were upgraded during the LHC shutdown in preparation for Run 2 operation. To cope with the increasing luminosity and more challenging pile-up conditions at a centre-of-mass energy of 13 TeV, the trigger selections at each level are optimised to control the rates and keep efficiencies high. To achieve this goal multivariate analysis techniques are used. The ATLAS electron and photon triggers and their performance with Run 2 data are presented.

1. Introduction

The ATLAS [1] electron and photon (e/γ) triggers select events with electrons and photons from the huge amount of data produced by high energy proton-proton collisions at the Large Hadron Collider [2], only a fraction of which can be recorded. Electrons and photons are common to many interesting experimental signatures at the LHC. For example, $W \to e\nu$ was a key channel for the measurement of the W mass [3] and photon signatures played a major role in the Higgs boson discovery [4]. Electrons and photons are also present in many new physics scenarios.

In Run 2, the LHC has reached a peak instantaneous luminosity of 16.8×10^{33} cm^{2}s^{-1}, twice the level seen during Run 1. The increased rate of collision data offers enormous physics potential, presenting new challenges for the e/γ triggers in the form of increased rates and harsher pileup conditions. Many improvements have been made in Run 2 to the e/γ triggers to maintain efficiencies at Run 1 levels for as long as possible whilst keeping the rates at a manageable level.

These proceedings provide a brief overview of the design and recent upgrades to the ATLAS e/γ triggers as well their performance in 2016 and 2017 data taking.

2. The ATLAS Electron and Photon Triggers

e/γ triggers at ATLAS select events based on inputs from the calorimeter and the inner detector. The calorimeter system measures the energy deposits with an outer hadronic layer and an inner electromagnetic layer. These surround the inner detector, which provides the tracking of charged particles with a pixel detector and semiconductor tracker. A Transition Radiation Tracker (TRT) allows for additional separation of electrons and photons and adds particle identification capability based on transition radiation.

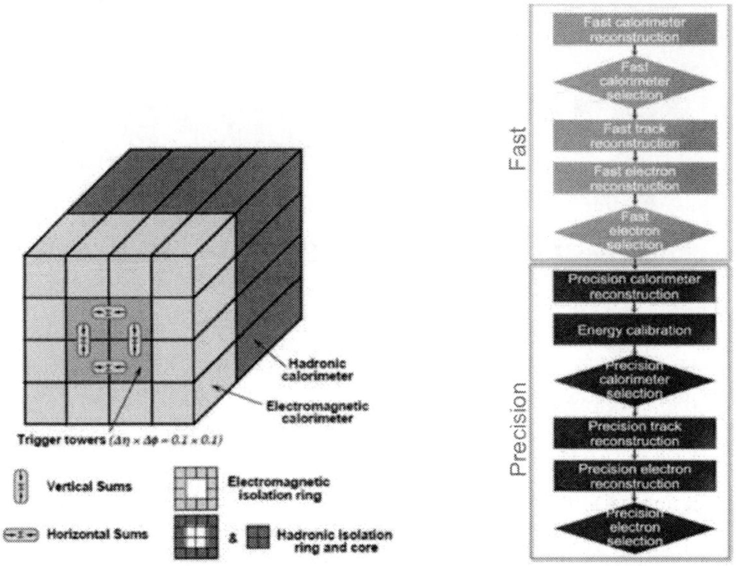

Figure 1. Schematic of the L1 calorimeter step (left) and the HLT trigger algorithm (right). At L1 a sliding-window algorithm uses the four overlapping trigger towers highlighted in green to locate a local energy maximum for energy reconstruction. The HLT process consists of a fast sequence for early rejection which precedes more precise offline algorithms for the final selection.

The ATLAS trigger system uses these inputs to provide a trigger "decision" on whether events are recorded, reducing the rate from 40 MHz to $\sim$ 1 kHz, around 20% of which is reserved for e/γ triggers. The e/γ trigger decision begins at Level 1 (L1) which identifies Regions-Of-Interest (ROIs) based on hardware information. The High Level Trigger (HLT) is then seeded by the ROI, using the full detector granularity in this region to provide the final trigger decision.

The L1 trigger decision uses calorimeter information to build an ROI consisting of 4×4 trigger towers (see figure 1), analogue summations of energy in the calorimeter cells, with granularity 0.1×0.1 in pseudorapidity (η) and azimuthal angle (ϕ). A transverse energy (E_{T}) threshold is applied, which is η dependent to account for energy losses and the geometry of the detector. A sliding-window algorithm identifies a local energy maximum using four overlapping towers within a 2×2 central region for EM energy reconstruction. The EM and hadronic isolation rings are formed from the 12 towers surrounding the central cluster in the EM and hadronic layers respectively. These rings are used for isolation tests to discriminate against hadronic activity. The hadronic core energy behind the EM cluster provides additional hadronic rejection.

At the HLT, the full detector granularity is used within the ROI for the final trigger decision. Photons and electrons are identified with EM clusters, with an additional matching track requirement for electrons. The calorimeter reconstruction is typically less expensive in terms of computing, and therefore precedes the track reconstruction to reject events early. The HLT trigger sequence is shown in figure 1 (right).

3. Upgrades to Electron and Photon Identification in Run 2

In 2017 the LHC is producing collisions at greater rates and with higher than ever average pileup per bunch crossing. These conditions have prompted a number of innovative upgrades to the trigger algorithms, with the aim of reducing the trigger rate without compromising the efficiency in order to reduce and delay the need to increase trigger E_{T} thresholds. Rate predictions are

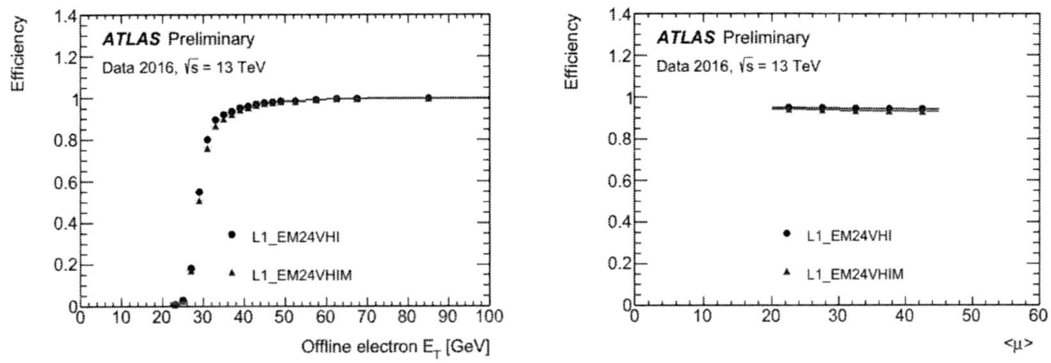

Figure 2. Performance of the new medium L1 isolation working point [9]. The efficiency is shown as a function of transverse energy (left) and the average number of interactions per bunch crossing (right). The naming convention is as follows: *L1* indicates a Level 1 trigger, *EM* indicates electromagnetic, the subsequent number indicates the transverse energy threshold, *V* indicates a pseudorapidity dependent transverse energy threshold, *H* indicates hadronic isolation and *I* indicates EM isolation. The *M* identifies the new medium working point implemented in the trigger menu for 2017 data taking.

measured using an enhanced bias dataset [5] and the electron efficiencies are measured using the *Tag and Probe* method [6] with $Z \to ee$ electrons. Photon efficiencies [7] are measured using the *bootstrap* method, where the efficiency is measured with respect to a looser L1 trigger.

3.1. L1 Reoptimisation
In 2017 the L1 isolation selection was optimised by considering variables based on the EM isolation energy compared with the cluster energy. The new 'medium' L1 isolation working point achieves a rate reduction of 10.8% with a minimal efficiency loss of 1.3% for a L1 E_{T} threshold of 24 GeV, currently the default L1 EM trigger E_{T} threshold. The medium working point is implemented for the trigger menu in 2017 data taking. The performance of the new L1 working point is shown in figure 2. The efficiency loss is mostly concentrated in the turn-on, with the plateau efficiency mostly unchanged.

3.2. Ringer Upgrade
Another Run 2 improvement to the electron trigger is the ringer upgrade to the fast calorimeter step [8]. The ringer algorithm is an alternative to the previous cut based selection which explores the conic geometry of the electron shower shape, building successive rings in the calorimeter. These are then used to build a vector of discriminating variables based on generalised shower shapes, which are fed into a neural network (NN) classifier. A Multilayer Perceptron is used with the hidden and output layer activation function set to tanh. The data is split into into a training set (60%) in which the NN weights are adjusted and a test set (40%) for verification and performance evaluation. The classifier performs particle identification targeting efficiency of the complete HLT trigger with a significant reduction to the number of calls to the inner tracking detector, which is typically more expensive in terms of computing.

The ringer algorithm achieves a ×2 improved background rejection compared with the older cut based selection used in the fast calorimeter step, with the efficiency almost unchanged. The performance of the ringer trigger in 2016 is shown figure 3. The ringer algorithm is now the default in electron triggers at ATLAS.

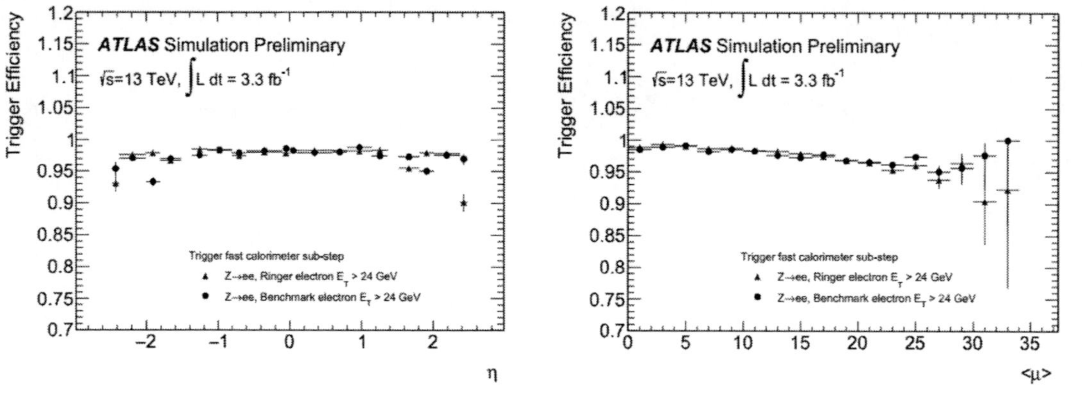

Figure 3. Performance of the ringer algorithm in the calorimeter fast-step of the HLT, compared with the previously default cut based method [9]. The efficiency is shown as a function of pseudorapidity (left) and the average pileup per bunch crossing (right).

3.3. Likelihood Based Electron Identification

A common set of variables is used for electron and photon identification [6][7]. These variables are based on properties of the EM energy cluster such as the shower shape and the energy ratios between cells within the cluster. In Run 1, cut based selections were used for both electron and photon identification, with the photon selection reoptimised in Run 2 for harsher pileup conditions, and higher instantaneous luminosity and centre-of-mass energy. In Run 2, a likelihood (LH) based approach was introduced for offline electron identification. The LH method is a multivariate analysis technique, simultaneously evaluating electron cluster variables to form a LH discriminant:

$$d_{\mathcal{L}} = \frac{\mathcal{L}_S}{\mathcal{L}_S + \mathcal{L}_B}, \qquad \mathcal{L}_{S(B)}(\vec{x}) = \prod_{i=1}^{n} P_{s(b),i}(x_i) \tag{1}$$

where $\mathcal{L}_{S(B)}$ is the signal (background) likelihood, $\vec{x}$ is a vector of the input variables and $P_{s(b),i}(x_i)$ are the signal (background) probability densities corresponding to each of the inputs. Three ID levels are formed by cutting on different values of the LH discriminant $d_{\mathcal{L}}$. These represent a tradeoff between purity and efficiency and are chosen such that the samples of electrons selected are subsets of looser IDs. The LH approach has now been introduced online, providing a better matching between online and offline identification and giving a 20% improvement in the online trigger rate for a similar efficiency. The LH identification is now the default for electron triggers in Run 2.

4. Electron and Photon Trigger Performance in 2016 and 2017

The trigger selection has been progressively tightened in Run 2 to control the trigger output rates. By tightening the identification level of the triggers the E_{T} thresholds can be maintained at a lower transverse energy than would otherwise be necessary. At L1, the trigger identification has moved from a non-isolated to an isolation based selection and the E_{T} threshold has been increased from 18 GeV at the beginning of Run 1 to 24 GeV in 2017. At HLT the likelihood based identification improved the trigger rate, and an additional 45% rate reduction was achieved in 2016 by moving from medium likelihood ID to tight likelihood ID for the single electron trigger. The single trigger E_{T} threshold has also been tightened from 24 GeV in 2015 to 28 GeV in

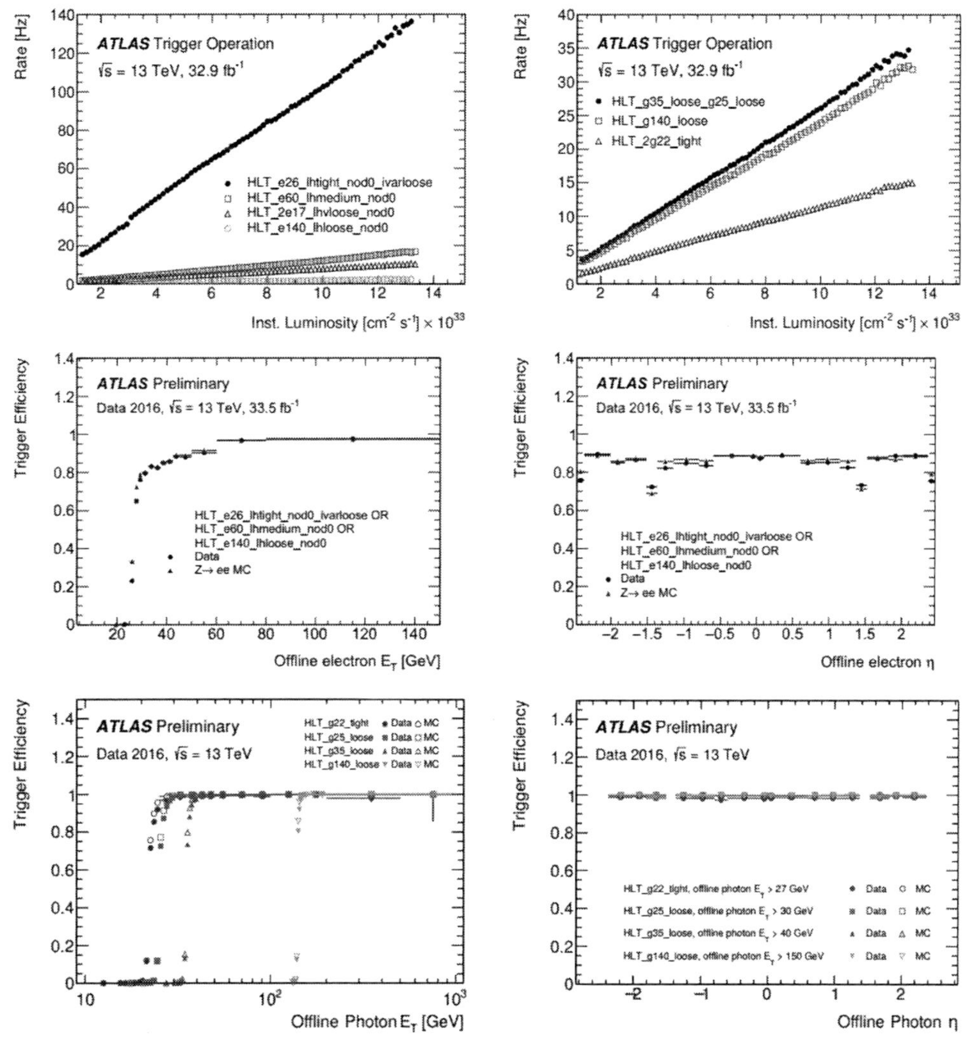

Figure 4. Trigger output rates (top) and electron (middle) and photon (bottom) performance in 2016 [9]. The efficiencies are shown as a function of transverse energy (left) and pseudorapidity (right). The electron efficiency is measured using the *Tag and Probe Method* using $Z \rightarrow ee$ decays for data and Monte Carlo. The photon efficiency is measured with the *bootstrap* method using events recorded with a looser Level 1 trigger.The naming convention is as follows: *HLT* indicates a High Level Trigger, *e (g)* indicates an electron (photon) trigger, the subsequent number indicates the transverse energy threshold, next the identification level is identified (*tight, loose* or *medium*), where *lh* is used to indicate a likelihood based trigger. *nod0* indicates that no transverse impact parameter cuts are required and *ivarloose* indicates a variable sized cone isolation requirement.

2017. For the dielectron trigger, a lower E_{T} threshold can be used. During Run 2 the dielectron trigger E_{T} threshold has been raised to 17 GeV from 12 GeV. The electron output rates for the triggers that make up the logical OR for the single electron trigger and the dielectron trigger are shown in figure 4 (top left). The electron trigger rate is dominated by $W \rightarrow e\nu$ events (around

IOP Conf. Series: Journal of Physics: Conf. Series **1085** (2018) 042001 doi:10.1088/1742-6596/1085/4/042001

55%), with subdominant contributions from multijet (35%) and $Z \to ee$ (10%) events.

For the photon triggers a similar strategy is employed, with the single photon E_T threshold progressing from 120 GeV to 140 GeV, and the 20 GeV diphoton E_T threshold moving to 22 GeV. For the asymmetric diphoton trigger, requiring 25 GeV and 35 GeV photons, the rate is controlled by tightening the trigger ID from a loose cut based ID to a medium cut based ID.

In 2016 and 2017 data taking variable track-isolation is used for the primary single electron trigger. At high transverse energies, track isolation losses become important. To mitigate the efficiency loss at high energies, a logical OR is used with high E_T threshold, looser, non-isolated triggers. The effect can be seen in the overall efficiency vs. transverse energy in figure 4 at 60 GeV where a discontinuity is seen in the trigger turn-on. A significant increase in efficiency in the plateau is achieved with minimal contribution to the electron trigger rate.

5. Conclusions

Improvements to the ATLAS e/γ triggers at L1 and HLT, particularly in rate reduction, have allowed the transition to a higher luminosity LHC without large increases from the Run 1 E_T thresholds and thus improving the physics potential of the experiment. The 2016 performance plots in figure 4 show excellent agreement between data and Monte Carlo simulation and a good overall performance, continuing in the most recent results for 2017 data taking. With the LHC running at design luminosity in 2018 and 2019 it is anticipated that around 80 fb^{-1} of collision data will be produced. The ATLAS e/γ triggers will continue to operate at close to the current energy thresholds, offering exciting prospects for electron and photon signatures.

References

[1] The ATLAS Collaboration 2008 *JINST* **3** S08003

[2] Evans L and Bryant P 2008 *JINST* **3** S08001

[3] The ATLAS Collaboration 2017 *Measurement of the W-boson mass in pp collisions at $\sqrt{s} = 7$ TeV with the ATLAS detector* `arXiv:1701.07240v1` [hep-ex]

[4] The ATLAS Collaboration 2012 *Observation of a new particle in the search for the Standard Model Higgs boson with the ATLAS detector at the LHC* `arXiv:1207.7214v2` [hep-ex]

[5] The ATLAS Collaboration 2016 *Trigger monitoring and rate predictions using Enhanced Bias data from the ATLAS Detector at the LHC* ATL-DAQ-PUB-2016-002

[6] The ATLAS Collaboration 2016 *Electron efficiency measurements with the ATLAS detector using the 2015 LHC proton-proton collision data* ATLAS-CONF-2016-024

[7] The ATLAS Collaboration *Measurements of the photon identification efficiency with the ATLAS detector using 4.9 fb1 of pp collision data collected in 2011* ATLAS-CONF-2012-123

[8] João Victor da Fonseca Pinto and ATLAS Collaboration 2016 J. Phys.: Conf. Ser. 762 012049

[9] The ATLAS Collaboration Public Egamma Trigger Plots for Collision Data `https://twiki.cern.ch/twiki/bin/view/AtlasPublic/EgammaTriggerPublicResults`

IOP Conf. Series: Journal of Physics: Conf. Series **1085** (2018) 042002 doi:10.1088/1742-6596/1085/4/042002

Computer simulation on homogeneity testing for weighted data sets used in HEP

Petr Bouř and Václav Kůs

Department of Mathematics, Faculty of Nuclear Sciences and Physical Engineering, Czech Technical University in Prague, Trojanova 13, 12000 Praha 2, Czech Republic

E-mail: `petr.bour@fjfi.cvut.cz`

Abstract. Modified statistical homogeneity tests on weighted data samples are commonly used in high energy physics applications. We do typically apply the tests in order to test homogeneity of weighted and unweighted samples, e.g. Monte Carlo simulations compared to the real data measurements. The asymptotic approximation of p-value of our weighted variants of homogeneity tests are investigated by means of simulation experiments. The simulation is performed for various probability sample distributions. We show that the asymptotic characteristics of the weighted homogeneity tests are valid for the specific distribution of weights.

1. Introduction

In high energy physics, homogeneity testing precedes many analysis and modeling techniques, particularly in machine learning (ML) applications. It is the case when we apply a data preprocessing procedure called data weighting. Via assigning weights $w_1, \ldots, w_n > 0$ to simulated observations $x_1, \ldots, x_n$, $n \in \mathbb{N}$, we are able to fine-tune our Monte Carlo (MC) simulation data set with respect to our requirements. However, statistical theory concerning homogeneity tests does not handle any weighting procedures, nor associates weights with observations, except for sporadic works on weighted histograms, as e.g. [1]. Therefore, the classical homogeneity tests must be adjusted for weighted data sets. Despite relatively straightforward incorporation of weights into the classical homogeneity tests and their modification, asymptotic properties of these tests can be no longer guaranteed. Thus, our goal is to investigate the validity of asymptotic properties of homogeneity testing for weighted observations through computer simulation.

The need for homogeneity testing may typically arise from a simple signal/backgrounds binary classification task. In this common ML application, we often use MC simulation for both ML classifier training and testing. We may then apply the trained classifier to real measured data set (DATA). Naturally, we expect both MC $\sim F$ and DATA $\sim G$ to be identically distributed: $F \equiv G$. Otherwise, the classification model will not perform well.

2. Weighted tests of homogeneity

Let us assume it is vital to guarantee homogeneity of DATA and MC distributions prior to subsequent utilization of some ML methods. For this purpose, we first define an analogy with empirical distribution function (EDF) for weighted data set. Let $\boldsymbol{X} = (X_1, \ldots, X_n)$ be iid random variables distributed by cumulative distribution function (CDF) $F(x)$ and let

$(w_1, \dots, w_n)$ be their corresponding weights, where $W = \sum_{i=1}^{n} w_i$. We define the weighted empirical distribution function (WEDF) to be

$$F_n^W(x) = \frac{1}{W} \sum_{i=1}^{n} w_i I_{(-\infty,x]}(X_i), \tag{1}$$

where $I_A(X)$ is the indicator of the set A. Note that in the case of $w_i = 1$ for all $i = 1, \dots, n$ (i.e. unweighted DATA), the definition of WEDF goes over to usual EDF.

In order to avoid an investigation of an unknown parametric family, we shall pursue our homogeneity testing only with nonparametric approaches. Thus, proceeding further in this section, we present the Kolmogorov-Smirnov test based upon EDFs of two data sets $\boldsymbol{X}_1 = \left(X_1^{(1)}, \dots, X_{n_1}^{(1)}\right)$, $\boldsymbol{X}_2 = \left(X_1^{(2)}, \dots, X_{n_2}^{(2)}\right)$, with respective distribution functions F, G. By the homogeneity hypothesis [3], as our null hypothesis is H_0, we understand

$$H_0 : F = G \quad \text{vs} \quad H_1 : F \neq G \quad \text{at significance level} \quad \alpha \in (0, 1). \tag{2}$$

We require our homogeneity tests to meet the condition $P(W_C|H_0) \leq \alpha$, where W_C is a critical region for the specific test statistic T. We reject hypothesis H_0 if $T \in W_C$. The nature of homogeneity testing prompted us to look for the p-value, i.e., the lowest significance level α for which we reject hypothesis H_0. Thus, for every $\alpha >$ p-value we may automatically reject hypothesis H_0.

2.1. Two sample Kolmogorov-Smirnov test

Let F_{n_1}, G_{n_2} denote the EDFs of two data samples $\boldsymbol{X}_1, \boldsymbol{X}_2$ with respective sample sizes n_1, n_2. We consider the test statistic

$$D_{n_1,n_2} = \sup_{x \in \mathbb{R}} |F_{n_1}(x) - G_{n_2}(x)|. \tag{3}$$

It is clear from the Glivenko-Cantelli lemma [2] that, under the true H_0, it holds that $D_{n_1,n_2} \xrightarrow{a.s.} 0$ for $n_1, n_2 \to \infty$. Furthermore, due to [4], it follows that for the true H_0 and $\lambda > 0$

$$\lim_{n_1,n_2 \to \infty} P\left(\sqrt{\frac{n_1 n_2}{n_1 + n_2}} D_{n_1,n_2} \leq \lambda\right) = 1 - 2\sum_{k=1}^{\infty} (-1)^{k-1} e^{-2k^2\lambda^2}. \tag{4}$$

Therefore, we obtain the approximate p-value as $2\sum_{k=1}^{\infty} (-1)^{k-1} e^{-2k^2\lambda_0^2}$, where $\lambda_0 = \sqrt{\frac{n_1 n_2}{n_1+n_2}} D_{n_1,n_2}$. However, for weighted data sample we are forced to replace EDFs F_{n_1}, G_{n_2}, and the numbers of entries n_1, n_2, with their respective WEDFs $F_{n_1}^{W_1}, G_{n_2}^{W_2}$, and the sums of weights W_1, W_2 in (3) and (4). Instead of (3), we thus obtain the test statistic

$$D_{n_1,n_2}^{W_1,W_2} = \sup_{x \in \mathbb{R}} \left|F_{n_1}^{W_1}(x) - G_{n_2}^{W_2}(x)\right|. \tag{5}$$

The definition (1) of WEDF makes it clear that the statistic $D_{n_1,n_2}^{W_1,W_2} \xrightarrow{a.s.} 0$ for $n_1, n_2 \to \infty$ and $W_1, W_2 \to \infty$. Nevertheless, it is important to notice some of the weaknesses inherent in the above approach. This modified test for the weighted data sample does not have to obey the asymptotic property (4). Let us emphasize that the p-value obtained using the statistic $D_{n_1,n_2}^{W_1,W_2}$ cannot be considered a regular approximate p-value without subsequent detailed research. At this point, we are not able to present a rigorous mathematical proof. However, we intent to supply the HEP community with some recommendations on the suitability of the weighted tests. This is why we propose partial validation of our approach in section 3. A key insight we provide is the stable numerical verification performed for a few fundamental data distributions.

IOP Conf. Series: Journal of Physics: Conf. Series **1085** (2018) 042002 doi:10.1088/1742-6596/1085/4/042002

3. Simulation experiments

We have already mentioned the problem of insecure asymptotic properties when applying weighted modifications of the standard tests. We now turn our attention at the numerical simulation. For our purposes here, the best way would be to validate the asymptotic properties using standard unweighted tests. This requires us to plug into the testing an unweighted data set (only instead of weighted MC; DATA is unweighted already). We shall do this by appropriate transformation of the weighted ensemble MC into the unweighted ensemble $\text{MC}_\dagger$ [5].

3.1. Rearranging technique

We make two requirements for the transformation. Firstly, we desire to preserve or exploit information contained in MC weighting since the weight assigned to an observation refer to what extent the distribution should be present in the neighbourhood of this observation. Secondly, we require that the sum of weights in MC corresponds to the number of observations in the unweighted $\text{MC}_\dagger$. Continuing in this manner, we now proceed as follows. Denote by $\boldsymbol{X} = \left(X_{(1)}, \ldots, X_{(n)}\right)$ the ordered sample in MC with weights $(w_1, \ldots, w_n)$ and let $W = \sum_{i=1}^{n} w_i$. Let $N = \lfloor W \rfloor$ denote the desired number of observations in the new transformed ensemble $\text{MC}_\dagger$. Given both our requirements regarding $\text{MC}_\dagger$, we are constructing special weighted averages from $\boldsymbol{X}$. For simplicity, we presume $0 \leq w_i \leq 1$ for all $i = 1, \ldots, n$. Into the set of the first weighted average we include the smallest possible number of observations $\left(X_{(1)}, \ldots, X_{(k_1)}\right)$ such that $1 \leq \sum_{i=1}^{k_1} w_i < 2$. Thereby, $\sum_{i=1}^{l} w_i < 1$ for all $l < k_1$. The portion of weight w_{k_1} of the observation $X_{(k_1)}$ which contributes above 1 to the sum $\sum_{i=1}^{k_1} w_i$ will not be included into the first weighted average. Hence, we denote this residual portion by $r_{k_1} = \sum_{i=1}^{k_1} w_i - 1$. Thereafter, the first observation $Y_{(1)}$ in $\text{MC}_\dagger$ can be defined as the following weighted average

$$Y_{(1)} = \frac{\sum_{i=1}^{k_1} X_{(i)} w_i - X_{(k_1)} r_{k_1}}{\sum_{i=1}^{k_1} w_i - r_{k_1}}. \tag{6}$$

From the definition of r_{k_1} we arrive at

$$Y_{(1)} = \sum_{i=1}^{k_1} X_{(i)} w_i - X_{(k_1)} r_{k_1} = \sum_{i=1}^{k_1 - 1} X_{(i)} w_i + X_{(k_1)} (w_{k_1} - r_{k_1}). \tag{7}$$

The residual portion r_{k_1} will be added to the next weighted average for $Y_{(2)}$. In general, for $Y_{(j)}$ we write

$$r_{k_j} = \sum_{i=k_{j-1}+1}^{k_j} w_i - r_{k_j - 1} - 1 \tag{8}$$

$$Y_{(j)} = X_{(k_{j-1})} r_{k_{j-1}} + \sum_{i=k_{j-1}+1}^{k_j - 1} X_{(i)} w_i + X_{(k_j)} (w_{k_j} - r_{k_j}). \tag{9}$$

Repeating the same steps we transform the original weighted ensemble MC with $\boldsymbol{X} = \left(X_{(1)}, \ldots, X_{(n)}\right)$ into the new unweighted ensemble $\text{MC}_\dagger$ with $\boldsymbol{Y} = \left(Y_{(1)}, \ldots, Y_{(\tilde{n})}\right)$. We have distributed the weights from the MC so that there is the unit weight for each observation $Y_{(j)}$. Therefore, we are authorized to apply standard homogeneity tests, which guarantees the asymptotic properties.

IOP Conf. Series: Journal of Physics: Conf. Series **1085** (2018) 042002 doi:10.1088/1742-6596/1085/4/042002

3.2. Generic p-value validation
We have verified an eligible usage of modified weighted tests in previous section with data sets originating from high energy physics [5]. Now, we provide a reader with more general verification. In our simulation, we consider several different distributions for $\boldsymbol{X} = (X_1, \ldots, X_n)$: Beta, Cauchy, Exponential, Laplace, Logistic, Lognormal, Normal, Uniform and Weibull, whilst the weights $\boldsymbol{W} = (W_1, \ldots, W_n)$ are taken from the Beta distribution, as we can easily tune the expected value,

$$W \sim Beta(\alpha, \beta) \implies E\left[W\right] = \frac{\alpha}{\alpha + \beta}. \tag{10}$$

The appropriate number of data points was determined by preliminary convergence studies. Otherwise, the simulation steps proceed as follows:

(i) Generate n random weighted data points $(\boldsymbol{X}, \boldsymbol{W})$, e.g. $n = 3500000$.

(ii) Estimate weighted distribution from all the observations $(\boldsymbol{X}, \boldsymbol{W})$ using kernel density estimator.

(iii) Repeat all the following items k times, e.g. $k = 1000$:

- (a) Choose $m_w = \frac{n}{k}$ weighted observations from $(\boldsymbol{X}, \boldsymbol{W})$ as the current MC sample, e.g. $m_w = 3500$.
- (b) Generate $m_u \approx \sum_{i=1}^{m_w} w_i$ unweighted observations from the estimated weighted distribution and consider them to be the current DATA sample, e.g. $m_u = 1000$.
- (c) Apply weighted homogeneity test MC vs DATA.
- (d) Rearrange MC into unweighted sample MC$_\dagger$ and apply standard unweighted test.

Thus, we obtain k individual p-values from the weighted tests and also another k corresponding p-values from the unweighted tests. We may now check asymptotic properties of both weighted and unweighted tests.

For all the distributions under consideration we arrived at two main results. First, the condition $P(W_C|H_0) \leq \alpha$ for significance level α is uniformly satisfied as it is shown in figure 1, i.e. both EDFs are located under diagonal line in the graph. Second, both weighted modifications and unweighted tests have the same resulting p-value distribution. This was tested via common classical homogeneity tests for unweighted data. Nevertheless, the extraordinary correspondence is already obvious from the graph in figure 1. Quite similar results were achieved for all other simulated distributions: Beta, Cauchy, Exponential, Laplace, Logistic, Normal, Uniform and Weibull.

4. Discussion
We carried out a numerical validation of the modified statistical homogeneity tests for generated data sets under Beta-weighting. Our simulation verifies that the approximate asymptotic properties remain the same for both weighted and unweighted tests. In practice, we may either utilize modified weighted tests or we may apply the rearranging technique from section 3.1 directly with the unweighted standard tests (where the asymptotics are theoretically proven). However, our verification was performed for the specific choice of weights distribution only. In the future research, we aim to investigate the effect of various homogeneity tests and different weights distribution on the overall significance and power. We also plan to explore the possibility of theoretic validation of some representative weighted tests for wide-ranging data distribution families. It may be reached by appropriate restrictions imposed on the distribution of weights, since, in practice, there exists only a limited number of high energy physics applications under rather specific weighting.

Please note we only aimed to provide empirical recommendations for weighted tests of homogeneity. A reader may use these only for an experimental setup with data and weights distributions similar to ours. On the other hand, this may cover large number of cases as HEP data often come from these fundamental distributions or their mixture. In any case, the reader may always utilize our validation procedures to verify whether the output of his test corresponds to a regular approximate p-value.

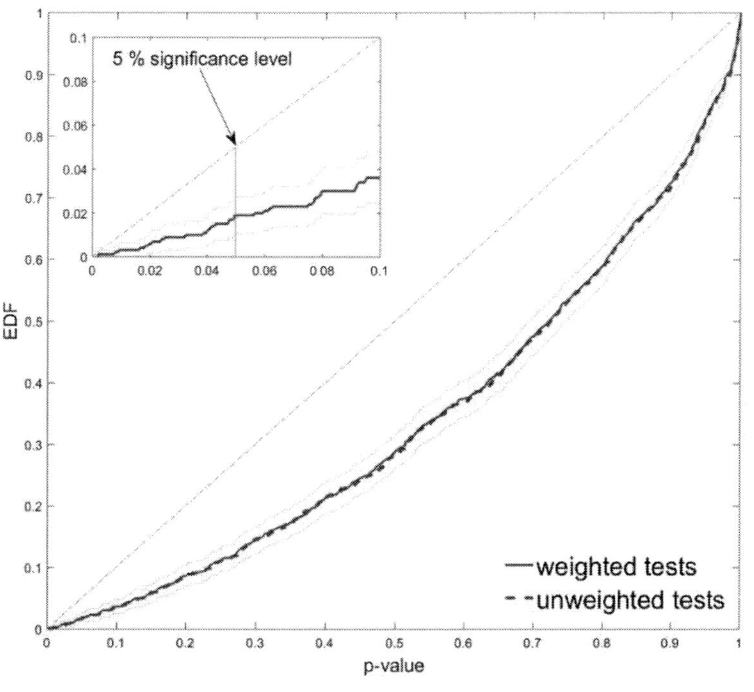

Figure 1. EDF of p-value for weighted and unweighted tests of homogeneity. Homogeneity of p-value distributions was tested via common classical homogeneity tests for unweighted data. Underlying data are taken from the lognormal distribution.

Acknowledgments
This work was supported by the grants LG15047 (MYES), LM2015068 (MYES), GA16-09848S (GACR), and SGS15/214/OHK4/3T/14 (CTU).

References
[1] Gagunashvili N D 2015 Chi-square goodness of fit tests for weighted histograms. Review and improvements *JINST* **10** P05004
[2] van der Vaart A W 1998 *Asymptotic Statistics* (Cambridge: Cambridge University Press) p 265
[3] DeGroot M H and Schervish M J 2010 *Probability and Statistics* (Boston: Pearson) p 647
[4] Smirnov N J 1944 *Usp. Mat. Nauk.* **10** 179–206
[5] Bouř P 2016 *Development of statistical nonparametric and divergence methods for data processing in D0 and NOvA experiments* (Prague: FNSPE CTU) p 66

IOP Conf. Series: Journal of Physics: Conf. Series **1085** (2018) 042003 doi:10.1088/1742-6596/1085/4/042003

Real-time alignment and reconstruction: performance and recent developments at the LHCb experiment

Agnieszka Dziurda[1], Lucia Grillo[2], Francesco Polci[3], Michael Sokoloff[4]

[1] European Organization for Nuclear Research (CERN), Geneva, Switzerland
[2] University of Manchester, Manchester, United Kingdom
[3] LPNHE, Paris, France
[4] University of Cincinnati, Cincinnati, OH, United States

E-mail: lucia.grillo@cern.ch

Abstract. The LHCb detector is a single-arm forward spectrometer designed for the efficient reconstruction decays of c- and b-hadrons. LHCb has introduced a novel real-time detector alignment and calibration strategy for LHC Run II. Data collected at the start of the fill are processed in a few minutes and used to update the alignment, while the calibration constants are evaluated for each run. This is one of the key elements which allow the reconstruction quality of the software trigger in Run-II, which fully includes the particle identification selection criteria, to be as good as the offline quality of Run-I. This approach greatly increases the efficiency, in particular for the selection of charm and strange hadron decays. We discuss strategy and performance of this novel approach, followed by a presentation of the recent developments implemented for the 2017 run of data taking, and with the performance and reconstruction quality achieved by the LHCb experiment in LHC Run-II.

1. Introduction

The LHCb detector [1][2] is a single-arm forward spectrometer covering the pseudorapidity range $2 < \eta < 5$, designed for the study of beauty and charm hadron decays with special attention to CP violating phenomena as well as searches for physics beyond the Standard Model through rare decays. The detector includes a high-precision tracking system consisting of a silicon-strip vertex detector (VELO) surrounding the *pp* interaction region [3], a large-area silicon-strip detector (TT) located upstream of a dipole magnet with a bending power of about 4 Tm, and three stations, indicated as T stations (T1-T3 in Fig. 1) of silicon-strip detectors (IT) in the region closer to the beam pipe and straw drift tubes (OT) [4] in the outer region, placed downstream of the magnet. The particle identification system is composed of two ring-imaging Cherenkov detectors (RICH1, RICH2)[5], electromagnetic (ECAL) and hadronic (HCAL) calorimeter, and a system of muon chambers (M1-M5 in Fig. 1)[6]. A schematic view of the LHCb detector and its sub-systems is shown in Fig. 1.

In Run-II (2015-2018) of data taking the LHCb experiment collects data with the bunch crossing rate of 30 MHz. The hardware trigger, based on information from the calorimeter system and muon chambers, reduces the rate of accepted candidates to about 1 MHz, at which the full detector can be read out. The candidates accepted by this stage are processed by the high-level software trigger, with two split stages (HLT1 and HLT2). The real-time reconstruction needs to be achieved within 35 (650) ms in the HLT1 (HLT2). All events passing the first stage are

IOP Conf. Series: Journal of Physics: Conf. Series **1085** (2018) 042003 doi:10.1088/1742-6596/1085/4/042003

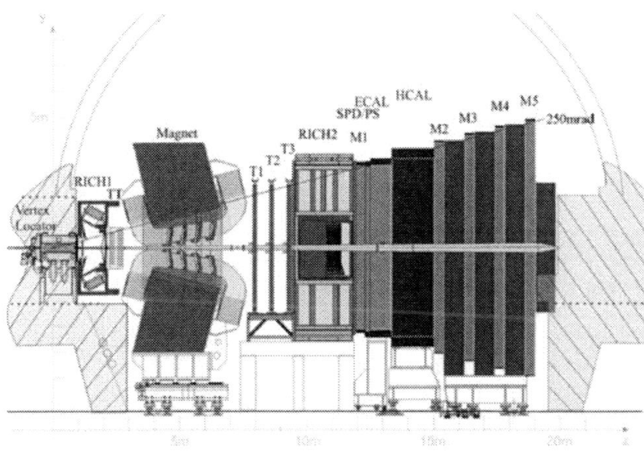

Figure 1. Schematic view of the LHCb detector. The Vertex Locator indicates the vertex detector (VELO). The labels RICH1 and RICH2 denote the Ring Imaging Cherenkov detectors, TT and T1-T3 are the tracking stations, the latter composed of IT and OT detectors in the inner and outer region respectively, M1-M5 indicate the muon chambers, SPD (PS) is the scintillator pad (preshower) detector. ECAL and HCAL mark the electromagnetic and hadronic calorimeters, respectively.

buffered on the Event Filter Farm (EFF). Once available, a fraction of these events is used for the run-by-run real-time alignment and calibrations of the detector. The detector conditions determined with the alignment and calibration tasks can be used immediately in HLT1 and/or in HLT2, resulting in a full, best performing event reconstruction, exploiting the full detector information, achieved as output of HLT2. The level of reconstruction quality that previously was achieved after the offline processing, in Run-II is achieved at the software trigger level, without any need of further offline processing. Fig. 2 illustrates the event reconstruction and selection strategy used by LHCb in Run-II, including the track and primary vertex (PV) reconstruction algorithms, further explained in Sec.3.

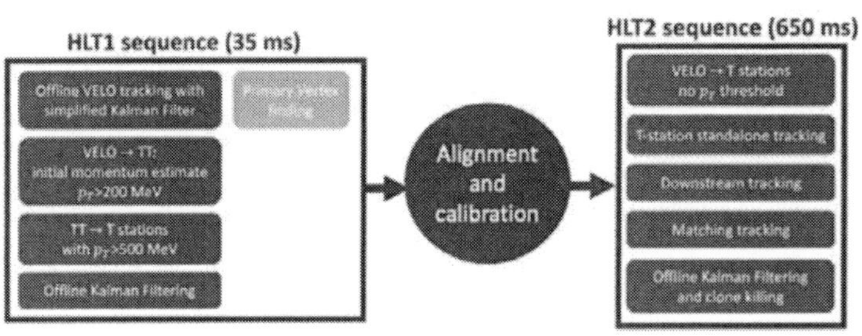

Figure 2. Schematic view of the event reconstruction strategy at LHCb in Run-II. See Sec. 3 for the description of the track and vertex reconstruction algorithms, and Fig. 3 for the track types.

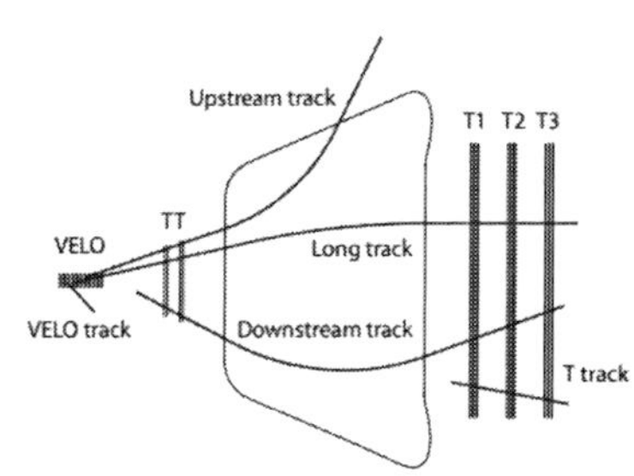

Figure 3. LHCb track types.

2. Real-time alignment and calibration

The spatial alignment of a detector and the accurate calibration of its subcomponents are essential elements to achieve the best physics performance. Pressure and/or temperature changes, together with physical movements either induced by operational conditions, i.e. magnet polarity switch, or mechanical interventions, can modify the position of the detectors and/or their response. The position and orientation of detector elements in the global reference frame must be known with an accuracy significantly better than the single hit resolution. A real-time alignment and calibration procedure is crucial for detector conditions changing with time.

This is particularly important for the VELO, which is retracted and re-inserted for stable beam conditions in each fill to be centred on the primary vertex position in the transverse plane and is therefore sensitive to position changes on fill-by-fill basis. Precise alignment of the VELO is necessary for an accurate identification of primary and secondary vertices, and thus better impact parameter (IP), defined as the minimum distance of a track to a primary vertex, PV, and decay-time resolutions. A more precise alignment of the entire tracking system directly implies a better invariant mass resolution. The time calibration of the OT improves the track reconstruction efficiency in the software trigger [9]. An accurate calibration and alignment of RICH detectors [10] improves the particle identification and allows to increase the sample purity. The calibration of the calorimeter systems [10] is crucial for the energy measurement of neutral particles, affecting the mass resolution of radiative b-hadron decays. The information from the first two stations of the muon system is used to estimate the momentum of the muons passing through the detector, used in the hardware trigger. To guarantee high trigger efficiency without a significant charge asymmetry, an alignment precision better than 1 mm is required.
The alignment of the tracking system uses an iterative procedure of minimizing the residuals of a Kalman filter and takes into account multiple scattering, energy loss in the material and magnetic field information [11]. The usage of primary vertex and mass constraints minimizes global distortions [12]. The alignment and calibrations are performed at regular intervals, which can be at the beginning of the run, fill or less frequently. The resulting constants are updated only if they differ significantly from the constants used to take data. Automatic monitoring procedures are used to make sure the variation of the constants is always within the expected range, and is stable during the fill. Each automatic alignment procedure runs at the start of each fill when the required sample of candidates has been selected, saved in the buffer and reconstructed within the EFF. The VELO alignment is performed first since different beam position with respect to the retractible halves of the VELO directly impact the physics performance of the experiment, the sample needed for the alignment can be collected rapidly and other sub-detectors are aligned relative to the VELO. Figure 4 shows the stability of the tracking system alignment for translations in x and y (VELO), z (Tracker) and x (Muon system, M1-M5 in Fig.1) directions during the first part of 2017 data taking. The horizontal dashed lines indicate the threshold within which the alignment is not updated. Each point denotes the difference between the new constant and the parameter from the previous alignment. The constants are updated every few fills and an overall good stability is found.

3. Real-time track reconstruction

The LHCb tracking system has three main sub-detectors: VELO, TT and T stations. Based on the information from them, several type of tracks can be reconstructed: VELO tracks with hits in the VELO; upstream tracks with hits in the VELO and TT; T tracks with hits in the T stations; downstream tracks with hits in the TT and T stations, which most likely correspond to daughters of long-lived particles. Finally long tracks with hits in at least the VELO and the T stations. Since long tracks are formed using hits before and after the magnetic field, they have the most accurately measured momentum. Due to their properties downstream and long tracks are the most useful for physics analyses. The track types are shown in Fig. 3.
The track reconstruction consists of three main parts. Firstly, the signatures (hits) produced in the detector by charged particles are found and the tracks are formed combining them. Then, all tracks found are fitted using a Kalman filter which obtains the best possible estimate of the true trajectory. This procedure includes corrections due to energy loss and multiple scattering. Finally, duplicated tracks are removed and a container with the best unique tracks is created.
The strict time budget allowed for the reconstruction in HLT1 and HLT2 stages, 35 and 650 ms respectively, requires the usage of different track reconstructions. In HLT1, all VELO tracks are reconstructed and fitted with a simplified Kalman filter. These tracks serve as an input to the

IOP Conf. Series: Journal of Physics: Conf. Series **1085** (2018) 042003 doi:10.1088/1742-6596/1085/4/042003

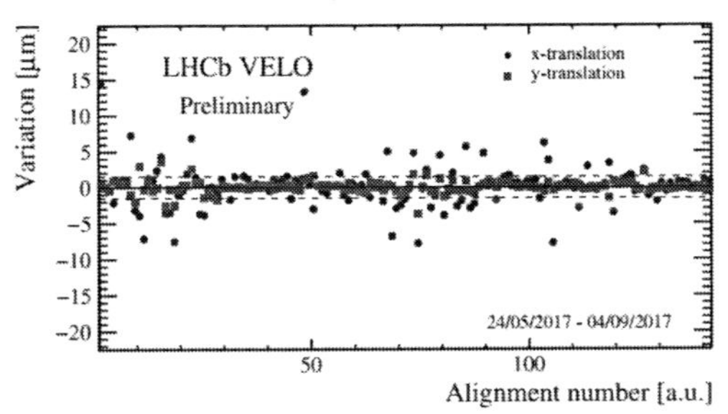

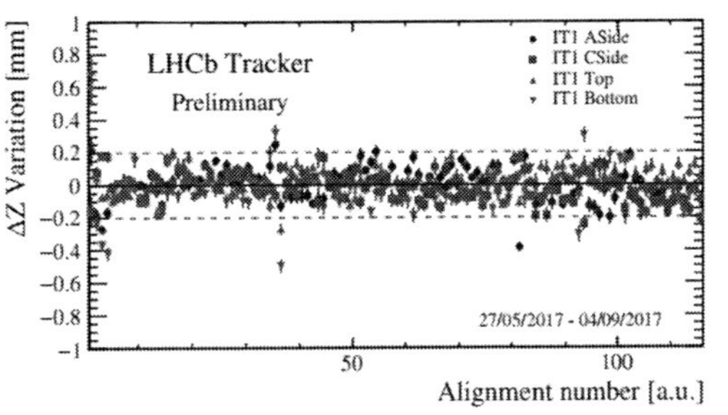

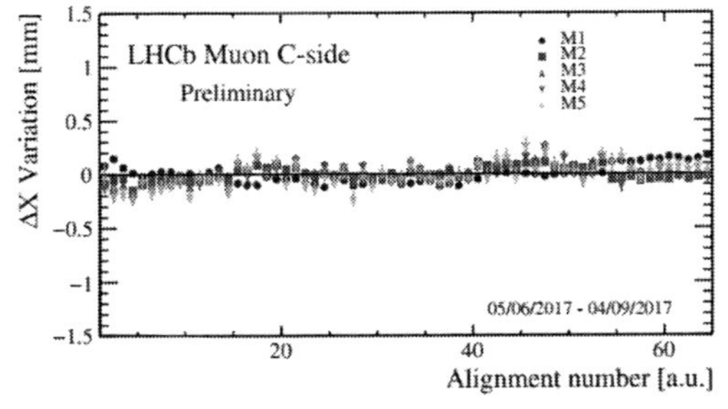

Figure 4. Stability of the alignment constants for VELO (x and y translations), Tracker, IT elements (translation in z direction) and Muon system, right detector side (x translations).

primary vertex finding, some of them are extended to the TT stations creating upstream tracks with an initial momentum estimation. The upstream tracks are extended further to long tracks by looking for hits in the T stations. The resulting long tracks are Kalman filtered and selected with a transverse momentum threshold above 500 MeV/c. They serve as the necessary input to most of the calibration and alignment algorithms. The second stage, HLT2, runs with looser requirements. The VELO or upstream tracks are extended further to long tracks by looking for hits in the T stations, this time without a transverse momentum threshold as well as without requiring clusters in the TT. In addition, T tracks are reconstructed and combined with VELO tracks (long tracks) or with clusters in TT (downstream tracks). All track candidates are fitted with a Kalman filter and clones from different algorithms are removed.
The LHCb experiment intensively uses machine learning in the tracking algorithms. Neural networks are used in the forward tracking (responsible for finding long tracks by extending upstream or VELO tracks) [7] as well as a dedicated fake track rejection algorithm [8]. In 2017 a new tuning for the downstream tracking has been performed exploiting two different multivariate techniques. Firstly, the algorithm filters T tracks using a bonsai Boosted Decision Tree with 11 dimensional discretized space improving the fake T tracks rejection. Then the remaining T track candidates are matched with TT hits. Finally, the good track candidates are selected based on a neural network decision. Overall signal efficiency improvement of about O(3 - 5%) has been found, together with O(3 - 5%) fake reduction.

4. Performance
Excellent reconstruction performance is achieved as shown in Fig 5 - 8. A decay-time resolution of about 45 fs is found for a 4-track vertex. The primary vertex resolution for 25 tracks in z direction is estimated to be 77 μm, while the impact parameter resolution in the x direction is 13 μm for tracks with high transverse momentum. An average tracking efficiency larger than 96% is achieved. In all the relevant physical quantities, the reconstruction performance achieved by LHCb in Run-II at trigger level is comparable or better than the performance of the offline reconstruction in Run-I.

IOP Conf. Series: Journal of Physics: Conf. Series **1085** (2018) 042003 doi:10.1088/1742-6596/1085/4/042003

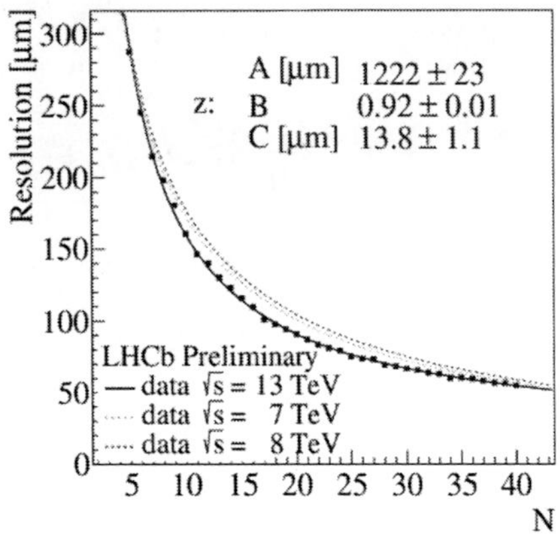

Figure 5. PV resolution along the beam axis (z) versus number of tracks used. Data at $\sqrt{s} = 7$ and 8 TeV collected in Run-I, in 2011 and 2012 respectively are compared to the Run-II, $\sqrt{s} = 13$ TeV dataset.

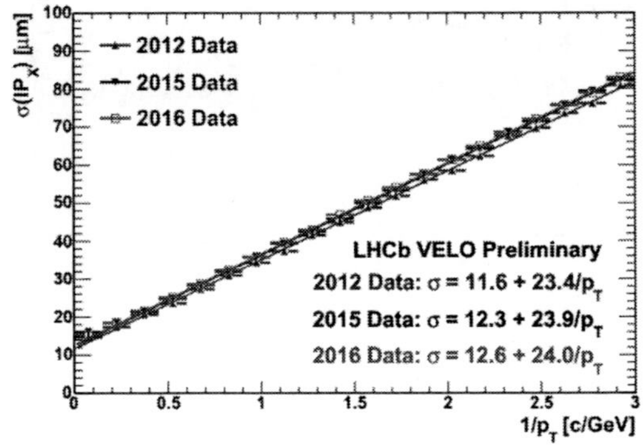

Figure 6. IP resolution as function of the inverse of the transverse momentum. The Run-I dataset has been collected in 2012 while separate Run-II datasets acquired in 2015 and 2016 are compared.

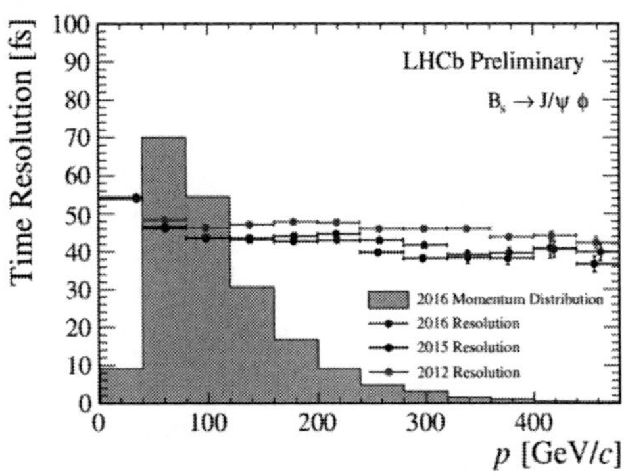

Figure 7. Decay-time resolution for decays with a 4-track secondary vertex as function of momentum. The Run-II datasets (2015 and 2016) are compared to Run-I (2012).

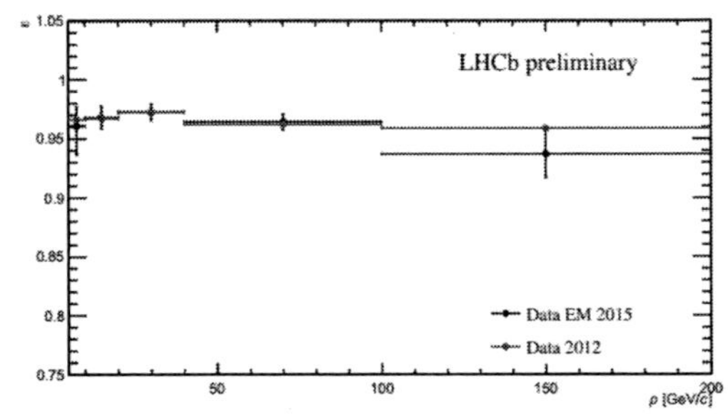

Figure 8. Long track (see Fig 3) reconstruction efficiency as function of the momentum of the track. Performance on data acquired in the first period of Run-II (Data EM) is compared to the Run-I performance. The method is described in Ref. [13]

5. References

[1] Alves Jr. A. A. *et al.* (LHCb collaboration), *JINST* **3** (2008) S08005
[2] Aaij R. *et al.* (LHCb collaboration), *Int. J. Mod. Phys* A **30** (2015) 1530022, arXiv:1412.6352
[3] Aaij R. *et al.* (LHCb VELO group), *JINST* **9** (2014) P09007, arXiv:1405.7808
[4] Arink R. *et al.* (LHCb Outer Tracker group), *JINST* **9** (2014) P01002, arXiv:1311.3893
[5] Adinolfi M. *et al.* (LHCb RICH group), *Eur. Phys. J.* C **73** (2013) 2431 arXiv:1211.6759
[6] Alves Jr. A. A. *et al.*, *JINST* **8** (2013) P02022, arXiv:1211.1346
[7] Stahl M., LHCb-PROC-2017-013, CERN-LHCb-PROC-2017-013.
[8] Seyfert P. *et al*, LHCb-PUB-2017-011.
[9] d'Argent Ph. *et al.* (LHCb Outer Tracker group), *JINST* **12** (2017) no.11, P11016, arXiv:1708.00819
[10] Borghi S., *Nucl. Instrum. Meth. A*, A **845** (2017) 560-564, https://doi.org/10.1016/j.nima.2016.06.050
[11] Hulsbergen W., *Nucl. Instrum. Meth.* A **600** (2009) 471 - 477, arXiv:0810.2241
[12] Amoraal J. , *et al.*, *Nucl. Instrum. Meth.* A **712** (2013), 48-55 arXiv:1207.4756
[13] Aaij R. *et al.* (LHCb collaboration), *JINST* **10** (2015) P02007, arXiv:1408.1251

IOP Conf. Series: Journal of Physics: Conf. Series **1085** (2018) 042004 doi:10.1088/1742-6596/1085/4/042004

Robust circle reconstruction with the Riemann fit

R Frühwirth[1] **and A Strandlie**[2]

[1] Institute of High Energy Physics of the Austrian Academy of Sciences, Nikolsdorfer Gasse 18, A-1050 Wien, Austria

[2] NTNU Norwegian University of Science and Technology, P. O. Box 191, N-2802 Gjøvik, Norway

E-mail: rudolf.fruehwirth@oeaw.ac.at, are.strandlie@ntnu.no

Abstract. Finding and fitting circles from a set of points is a frequent problem in the data analysis of high-energy physics experiments. In a tracker immersed in a homogeneous magnetic field, tracks are close to perfect circles if projected to the bending plane. In a ring-imaging Cherenkov (RICH) detector, circles of photons around the crossing point of charged particles have to be found and their radii estimated. In both cases, non-negligible background may be present that tends to complicate the pattern recognition and to bias the circle fit. In this contribution we present a robust circle fit based on a modified Riemann fit that removes or significantly reduces the effect of background points. As in the standard Riemann fit, the measured points are projected to the Riemann sphere or paraboloid, and a plane is fitted to the projected points. The fit is made robust by replacing the usual least-squares regression by a least median of squares (LMS) regression. Because of the high breakdown point of the LMS estimator, the fit is insensitive to background points. The LMS plane is used to initialize the weights of an M-estimator that refits the plane in order to suppress eventual remaining outliers and to obtain the final circle parameters. The method is demonstrated on three sets of artificial data: points on a circle plus a comparable number of background points; points on two overlapping circles with additional background; and points obtained by the simulation of tracks in a drift chamber with mirror points and additional background. The results show high circle finding efficiency and small contamination of the final fitted circles.

1. Introduction

The fitting of circular arcs or full circles to a discrete set of measurements is an important problem in high-energy physics experiments. In ring-imaging Cherenkov (RICH) detectors rings of photons have to be reconstructed; from their estimated radii the Cherenkov emission angle of the crossing particles can be computed. Significant background can be expected, and there may be overlapping rings from other particles in the same event. Tracking detectors often are embedded in homogeneous magnetic fields, resulting in a nearly circular track in the bending plane of the particle trajectories. There is usually background from other tracks in the event, and in a drift chamber there is additional irreducible background because of the left-right ambiguity of the position measurement.

Methods for finding rings in RICH detectors include, amongst others, the Hough transform [1, 2, 3, 4], track-based maximum likelihood [5], global maximum likelihood [6], trackless ring finding via Markov chain Monte Carlo [7], and deconvolution [8]. Methods in the pattern recognition literature include learning automata [9], genetic algorithms [10] and center-based clustering [11]. It should be noted, however, that some of these are not relevant to the problem studied here, as

IOP Conf. Series: Journal of Physics: Conf. Series **1085** (2018) 042004 doi:10.1088/1742-6596/1085/4/042004

they are designed to the recognition of entire circular objects, for instance coins or balls, in an image.

In this contribution we propose a novel method for finding and fitting rings: a robust Riemann circle fit for tagging the noise points, followed by an adaptive least-squares estimator for obtaining the best possible precision. The Riemann fit [12, 13, 14] has recently been extended to a scale and location invariant version [15], following a recommendation in [16], including the full error propagation from the measured points to the final circle parameters. In this contribution it is shown how to make the circle fit robust against background. The least-squares regression, which is the core of the Riemann fit, is replaced by a robust regression algorithm. The robust fit is used to identify the points on the circle, and the found circle serves as the starting point for the final adaptive least-squares regression. If track information is available, it can be used to define a window for preselecting the points included in the robust fit.

The paper is organized as follows. After a brief recapitulation of the basic concept of the Riemann track fit the robust least median of squares (LMS) regression is briefly described (Section 2). The full chain of the LMS fit followed by the adaptive fit is evaluated first on full circles with background (Section 3), then on overlapping circles with background (Section 4), and finally on a simulation of points in a drift chamber with mirror points and additional background (Section 5). The paper is concluded by a summary and an outlook to further work.

2. Robust Riemann circle fit

We assume that the measured points (u_i, v_i), $i = 1, \ldots, N$, in the (x, y)-plane are given in Cartesian coordinates and have already been centered and scaled (see [15]). They are mapped to the "Riemann paraboloid", a second-degree surface with the equation $z = x^2 + y^2$, by the following projection:

$$x_i = u_i, \quad y_i = v_i, \quad z_i = u_i^2 + v_i^2.$$

In this way the circle in the plane with the equation

$$(u - u_0)^2 + (v - v_0)^2 = \rho^2$$

is mapped to the plane in 3D space with the equation

$$z - 2xu_0 - 2yv_0 = \rho^2 - u_0^2 - v_0^2.$$

The observed points $\boldsymbol{r}_i = (x_i, y_i, z_i)^\mathsf{T}$, $i = 1, \ldots, N$, are fitted to the plane $\boldsymbol{n}^\mathsf{T}\boldsymbol{r} + c = 0$, where $\boldsymbol{r} = (x, y, z)^\mathsf{T}$, $\boldsymbol{n}$ is the unit normal vector and c is the signed distance of the plane from the origin. Let $d_i(\boldsymbol{n}, c) = \boldsymbol{n}^\mathsf{T}\boldsymbol{r}_i + c$ denote the distance of the point $\boldsymbol{r}_i$ from the plane. The standard Riemann fit is orthogonal LS regression, in which the (possibly weighted) sum of the squared distances d_i of the points from the plane is minimized:

$$(\boldsymbol{n}_{\mathrm{LS}}; c_{\mathrm{LS}}) = \arg\min_{(\boldsymbol{n};c)} \textstyle\sum_{i=1}^{N} w_i\, d_i^2.$$

The weights reflect the precision of the measured points. For the sake of simplicity, we assume equal weights $1/N$ from now on. The generalization to arbitrary weights is straightforward.

It is well known that LS regression is sensitive to background or outlying measurements. A highly robust alternative to LS regression is LMS regression [17], in which not the sum but the median of the squared distances is minimized:

$$(\boldsymbol{n}_{\mathrm{LMS}}; c_{\mathrm{LMS}}) = \arg\min_{(\boldsymbol{n};c)} \mathrm{med}_{i=1}^{N}\, d_i^2.$$

The breakdown point of the LMS regression is 0.5. This means that up to 50% of the measurements can be moved to arbitrary positions without introducing a bias into the fit.

The price to pay for the extraordinary robustness of the LMS regression is threefold: it is more difficult to compute than LS regression; it is less efficient in terms of its asymptotic variance; and there is no simple prescription for computing the covariance matrix of the estimated parameters. We have therefore used the LMS regression only for identifying the background measurements. Computing the LMS regression parameters is essentially a combinatorial problem: one has to find the subset of size $\lfloor N/2 \rfloor$ that minimizes the sum of the squared orthogonal distances to the plane. As it is hardly possible to check all such subsets, triples of points are selected randomly, defining a plane. The squared distances of all points to this plane are computed, and the plane with the smallest median of the squared distances is taken as an approximation of the LMS plane. This plane is used to compute the initial weights of the final fit, which is a redescending M-estimator with deterministic annealing [18, 19, 20], implemented as an iterated re-weighted LS regression. The final temperature of the annealing process is set to $T = 0.25$, resulting in a hard discrimination between inliers and outliers. Points having a final weight above 0.5 are tagged as inliers and are included in the final circle.

In the following sections the robust Riemann circle fit will be compared to randomized Hough transform (RHT). As in the robust Riemann fit, triples of points are selected randomly, from which the center and the radius of the circle defined by the triplet are computed and stored. After finding the most populated bin in the 2D histogram of the center coordinates, the found circle is defined as the medoid of all centers and radii in this bin. Points are tagged as inliers or outliers by a cut on their weighted radial distance to the found circle.

3. Performance with full circles plus background

We have simulated 10000 circles with superimposed background. The signal consists of points on a circle with radius $\rho = 1$, which is contained in a window of size 1.5×1.5 around the origin. The points are distributed uniformly on the circumference of the circle. In analogy to a RICH detector, where the number of photons follows a Poisson distribution, the number of points is drawn from a Poisson distribution with mean 20. Gaussian position errors with $\sigma_x = \sigma_y = 0.02$ are added to their (x, y)-positions. The background points are uniformly distributed in the window, and their number is drawn from a Poisson distribution with mean 15. This means that in about 22.3% of the cases there are at least as many background points as signal points in the window.

The performance can be summarized by the efficiency of the circle finding and by the purity and the completeness of the reconstructed circles. The efficiency of the circle finder can be measured by counting the outliers in the residual distributions of the circle parameters, which are the coordinates of the center and the radius. In this context, an outlier is defined as a residual which is outside the interval $[m - 4\,s, m + 4\,s]$, where m and s are robust estimates of the location and the spread of the residual distribution, respectively. We have chosen m as the median and s as the median of the absolute deviations from the median (MAD) times 1.485, which is very close to one for large standard normal samples. According to this criterion, the efficiency is about 98%. The average bias μ, the standard error σ and the root-mean-square error RMSE of the estimated circle parameters (x_c, y_c, ρ) are listed in Table 1, along with the corresponding results of the RHT. The robust Riemann fit gives smaller RMSE for all three parameters.

The purity of a found circle is defined as the ratio of the signal inliers over all inliers. Its average over all 4899 found circles is 94.6%. The completeness is the ratio of the signal inliers over all signal points. Its average is 98.8%. The corresponding figures for the RHT are again 98% for the efficiency, 92.9% for the average purity, and 99.7% for the average completeness. The robust fit is almost twice as fast as the RHT.

Figure 1 shows two examples of circle fits. The true circle is drawn in cyan, the LMS circle

IOP Conf. Series: Journal of Physics: Conf. Series **1085** (2018) 042004 doi:10.1088/1742-6596/1085/4/042004

in magenta, and the final circle in black. In Example 1 on the left, which is representative for the overwhelming majority of cases, the three circles are virtually identical, although two background points that are close to the true circle are included. In Example 2 on the right, the circle is biased by the inclusion of two background points by the LMS regression (magenta dotted circle), from which the M-estimator (black circle) cannot recover.

Table 1. Average bias μ, standard error σ and root-mean-square error RMSE of the estimated circle parameters. Left: robust Riemann fit, right: randomized Hough transform.

	x_c	y_c	ρ
$\mu \cdot 10^3$	0.027	0.037	0.815
$\sigma \cdot 10^2$	0.784	0.781	0.546
RMSE $\cdot 10^2$	0.784	0.781	0.552

	x_c	y_c	ρ
$\mu \cdot 10^3$	−0.028	0.151	0.173
$\sigma \cdot 10^2$	1.140	1.147	0.811
RMSE $\cdot 10^2$	1.140	1.147	0.811

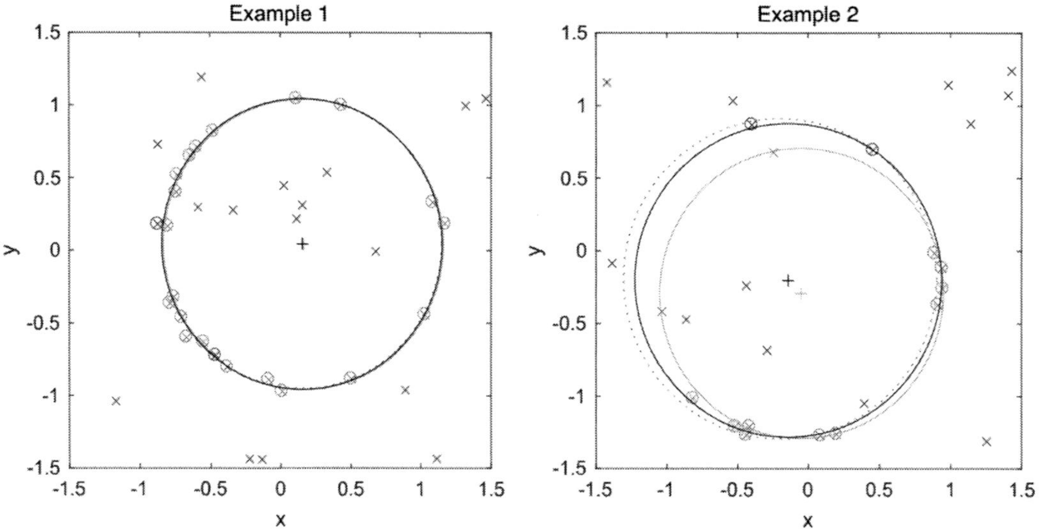

Figure 1. Two examples of fitted circles. Cyan: true circle with true center, green: signal points, red: background points, circled: points tagged as inliers by the M-estimator, magenta dotted: LMS circle, black: final circle with final center

4. Performance with overlapping circles plus background

We have simulated 10000 pairs of overlapping circles that are fully contained in the square $[-2.5, 2.5] \times [-2.5, 2.5]$. The primary circle is centered at the origin and has the radius $\rho_1 = 1$. The number of points on this circle is drawn from a Poisson distribution with mean $N_1 = 20$. The center of the secondary circle is uniformly distributed in the square $[-1.5, 1.5] \times [-1.5, 1.5]$, its radius ρ_2 is drawn from a uniform distribution in $[0.5\,\rho_1, \rho_1]$, and the number of points is drawn from a Poisson distribution with mean $N_2 = 15$. The points on the circles are smeared with Gaussian errors as before. The background points are uniformly distributed in the window, and their number is drawn from a Poisson distribution with mean $N_b = 10$.

After the fit, for each of the circles the sum of the weights of its point are computed, and the circle with the larger sum is tagged as found. The efficiency of the circle finding as well as the purity and completeness of the found circles are defined as in Section 3. The total circle finding efficiency, measured as above, is 94.7%. Of the 4735 found circles inside the efficiency cut, 2498 are primary circles and 2237 are secondary circles. The purity averaged over all circles within the efficiency cut is 94.2%, the completeness is 98.7%. There is no significant difference between primary and secondary circles. The corresponding figures for the RHT are 97.1% for the efficiency, 92.8% for the average purity, and 99.8% for the average completeness. The average bias μ, the standard error σ and the root-mean-square error RMSE of the estimated circle parameters (x_c, y_c, ρ) are listed in Table 1, along with the corresponding results of the RHT. Again, the robust Riemann fit gives smaller RMSE for all three parameters, but the RHT is slightly more efficient. As the robust fit needs more iterations with larger background, the RHT is slightly faster in this case.

Figure 2 shows an example of a successful fit. The left hand panel shows all points including the background, the right hand panel shows which circle has been found and which points are tagged as inliers. After removing the points on the found circle the procedure can be repeated in order to find the other circle.

Table 2. Average bias μ, standard error σ and root-mean-square error RMSE of the estimated circle parameters. Left: robust Riemann fit, right: randomized Hough transform.

	x_c	y_c	ρ
$\mu \cdot 10^3$	0.168	0.081	0.736
$\sigma \cdot 10^2$	0.810	0.818	0.579
$\text{RMSE} \cdot 10^2$	0.810	0.818	0.583

	x_c	y_c	ρ
$\mu \cdot 10^3$	0.253	0.031	0.291
$\sigma \cdot 10^2$	1.112	1.131	0.813
$\text{RMSE} \cdot 10^2$	1.113	1.131	0.814

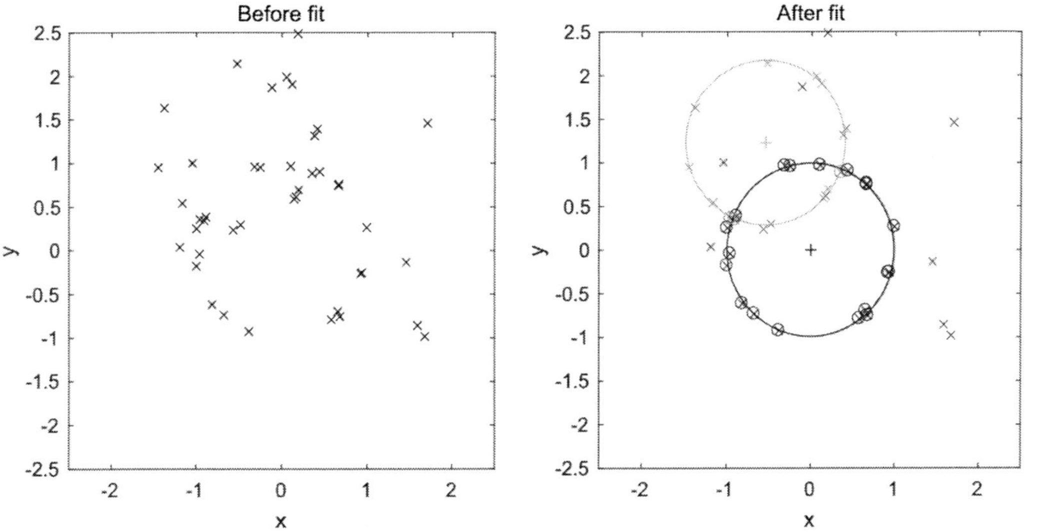

Figure 2. An example of a successful fit with overlapping circles. Left: all points before the fit. Right: after the fit; cyan: true circles with true centers, blue: signal points of the primary circle, green: signal points of the secondary circle, red: background points, circled: points tagged as inliers by the M-estimator, black: final circle with final center

IOP Conf. Series: Journal of Physics: Conf. Series **1085** (2018) 042004 doi:10.1088/1742-6596/1085/4/042004

5. Performance with mirror points

We have simulated tracks in a simplified and down-scaled model of the central drift chamber of the Belle II experiment [21], without the stereo layers. The model has 40 layers of sense wires, the cell size is 2 cm, and the point resolution is 150 µm. Each true point has a mirror point. In each layer, there is the probability p_{N} of an additional noise point plus its mirror point in either the signal wire itself or in the two adjacent wires on each side of the signal wire. The LMS regression is the same as in the previous sections. The M-estimator has been modified to include competition between a point and its mirror point, which is effectively an EM algorithm with deterministic annealing [22] and a special case of the Elastic Planes algorithm in [12].

Table 3 shows the average purity and the average completeness of the reconstructed track for three values of p_{N}. The completeness is close to one in all cases, the purity drops to about 94% at the highest noise level, as more noise points are included in the tracks. An example of a track at the highest noise level is shown in Figure 3. This track has 17 noise points and their mirror points. In layer 2, the signal point is missing and its mirror point is included. Three noise points are included in layers 6, 15 and 28.

Table 3. Average purity and average completeness of tracks with mirror points at three noise levels.

p_{N}	0.0	0.2	0.4
purity	0.994	0.966	0.942
completeness	0.994	0.998	0.998

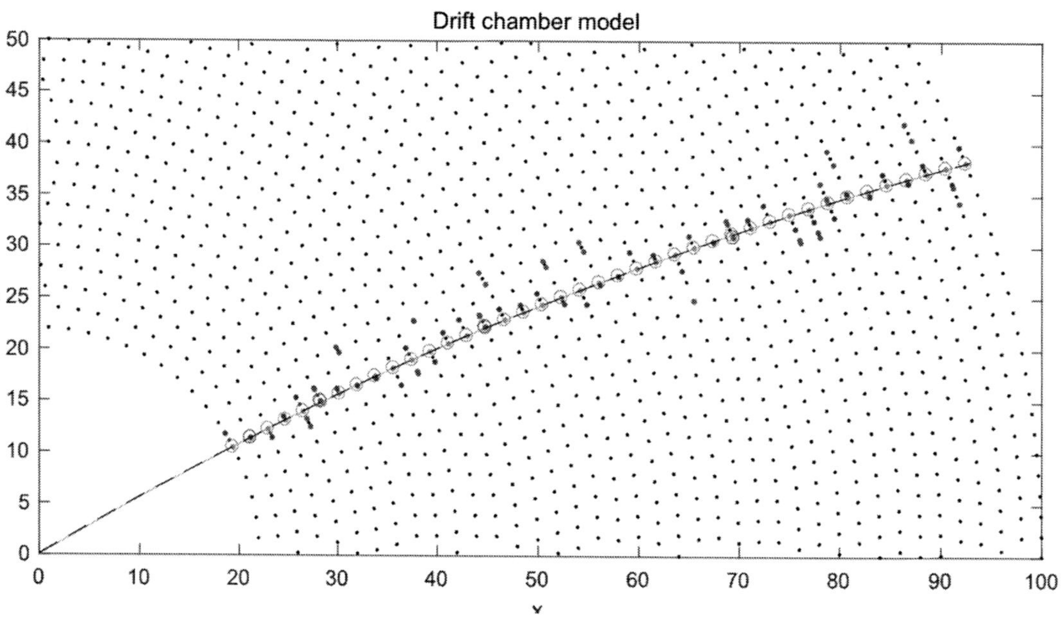

Figure 3. An example of a track at noise level $p_{\mathrm{N}} = 0.4$. Cyan: true track, green: true points, red: mirror points, magenta: noise points, circled: points included in the track, black: final track.

6. Summary and outlook

It has been shown that circle finding against a substantial background can be implemented by a robust Riemann circle fit based on LMS regression in combination with an M-estimator. Its performance is comparable to that of the randomized Hough transform. As LMS regression is computationally quite costly, further work will explore other robust regression methods such as least trimmed squares [17] or L_1 regression. Alternatively, the results of the randomized Hough transform can serve as the initial values for the M-estimator, in particular in the case of several overlapping circles. Finally, the fitting of tracks with mirror points will be extended to the case of crossing or overlapping tracks.

References

[1] Adamczewski J et al 2014 *Nucl. Instr. Meth. Phys. Res.* A **766** 250
[2] Volpe G on behalf of the ALICE Collaboration 2014 *Nucl. Instr. Meth. Phys. Res.* A **766** 259
[3] Abbon P et al 2008 *Nucl. Instr. Meth. Phys. Res.* A **595** 233
[4] Wilkinson G 2008 *Nucl. Instr. Meth. Phys. Res.* A **595** 228
[5] Starič M 2008 *Nucl. Instr. Meth. Phys. Res.* A **595** 237
[6] Buszello C P on behalf of the LHCb RICH Collaboration 2008 *Nucl. Instr. Meth. Phys. Res.* A **595** 245
[7] Lester C G 2006 *Nucl. Instr. Meth. Phys. Res.* A **560** 621
[8] Morháč M, Matoušek V 2011 *Nucl. Instr. Meth. Phys. Res.* 2011 A **639** 249
[9] Cuevas E et al 2012 *IET Image Processing* **6** 1124
[10] Ayala-Ramirez V et al 2006 *Pattern Recognition Letters* **27** 652
[11] Scitovski R and Marošević T 2015 *Pattern Recognition Letters* **52** 9
[12] Strandlie A, Wroldsen J, Frühwirth R and Lillekjendlie B 2000 *Comput. Phys. Commun.* **131** 95
[13] Strandlie A and Frühwirth R 2002 *Nucl. Instr. Meth. Phys. Res.* A **480** 734
[14] Strandlie A, Wroldsen J and Frühwirth R 2002 *Nucl. Instr. Meth. Phys. Res.* A **488** 332
[15] Strandlie A and Frühwirth R 2017 *Nucl. Instr. Meth. Phys. Res.* A **867** 72
[16] Chernov N 2010 *Circular and linear regression* (Boca Raton: CRC Press)
[17] Rousseeuw P J and Leroy A M 1987 *Robust Regression and Outlier Detection* (New York: Wiley & Sons)
[18] Huber P J 2004 *Robust Statistics: Theory and Methods* (New York: Wiley & Sons)
[19] Li S Z 1996 *Pattern Recognition* **29** 159
[20] Frühwirth R and Waltenberger W 2008 *Austrian J. Stat.* **37** 301
[21] Taniguchi N on behalf of Belle II collaboration 2017 *J. Instrum.* **12** C06014
[22] Ueda N and Nakano R 1998 *Neural Netw.* **11** 271

IOP Conf. Series: Journal of Physics: Conf. Series **1085** (2018) 042005 doi:10.1088/1742-6596/1085/4/042005

Performance studies of GooFit while estimating the global statistical significance of a new physical signal

Alexis Pompili and Adriano Di Florio
(on behalf of CMS Collaboration)

Dipartimento Interateneo di Fisica di Bari and I.N.F.N.-Sezione di Bari
via G. Amendola 173, 70126 Bari, Italy

E-mail: alexis.pompili@ba.infn.it, adriano.diflorio@ba.infn.it

Abstract. Graphical Processing Units (GPUs) represent one of the most sophisticated and versatile parallel computing architectures that are currently entering the High Energy Physics field. GooFit is an open source tool interfacing ROOT/RooFit to the CUDA platform on *nVidia* GPUs that acts as an interface between the MINUIT minimization algorithm and a parallel processor which allows a Probability Density Function to be evaluated in parallel.

In order to test the computing capabilities of GPUs with respect to traditional CPU cores, a high-statistics pseudo-experiment method has been implemented both in ROOT/RooFit and GooFit frameworks with the purpose of estimating the local statistical significance of an already known signal. The optimized GooFit application running on GPUs provides striking speed-up performances with respect to the RooFit application parallelized on multiple CPU workers by means of the PROOF-Lite tool.

This method is extended to situations when, dealing with an unexpected signal, a global significance must be estimated. The Look-Elsewhere-Effect is taken into account by means of a scanning technique in order to consider - within the same background-only fluctuation and everywhere in the relevant mass spectrum - any fluctuating peaking behavior with respect to the background model. The execution time of the fitting procedure for each MC toy can considerably increase, thus the RooFit-based approach gets so time-expensive that may become unreliable while GooFit is an excellent tool to carry reliably out this p-value estimation method.

1. Introduction to GooFit

The word *GPU-accelerated computing* refers to an enhancement of application performances that can be obtained by offloading compute-intensive portions to the GPU, while the remaining code still runs on the CPUs. The computing capabilities are enhanced once a sequence of elementary arithmetic operations are performed in parallel on a huge amount of data. In the context of High Energy Physics (HEP) analysis applications, GooFit is an under development open source data analysis tool that is being optimized to allow the HEP analyst to perform complex fit tasks on GPUs or multithreaded CPU backends. Its description and details can be found elsewhere [1]; it has been very recently updated to a version 2.0 [2] that is characterized by a set of new features. However the results shown here have been obtained by using the previous main version. GooFit is used in applications for parameter estimation, that interfaces ROOT [3]/RooFit [4] to the CUDA [5] parallel computing platform on nVidia's GPUs (it also supports OpenMP). Parameter estimation and interpolation are a crucial part of many physics analyses. The evaluation of Probability Density Function (PDF) on large datasets is

IOP Conf. Series: Journal of Physics: Conf. Series **1085** (2018) 042005 doi:10.1088/1742-6596/1085/4/042005

usually the bottleneck in the MINUIT [6] algorithm. `GooFit` acts as an interface between the MINUIT minimization algorithm and a parallel processor which allows a PDF to be evaluated in parallel. Specifically the `FitManager` object [1] forms the interface between MINUIT (running on CPU) and a GPU which enables the evaluation in parallel of the PDF representing the physical model (`GooPdf` object). Fit parameters are estimated at each negative-log-likelihood (NLL) minimization step on the *host side* (CPU) while the PDF/NLL is evaluated on the *device side* (GPU) [7] until fit convergence and consequently a cyclical intense memory transfer takes place between CPU and GPU. The `FitControl` object [1] allows to switch between Maximum Likelihood (ML) and χ^2 fits, either unbinned or binned.

Preliminary tests performed with an unbinned ML fit, either by using a single CPU and by using an additional GPU, indicate amazing speed-up values of several hundreds when the involved number of events is greater than few thousands and the considered task is enough compute-intensive (a convolution of a Breit-Wigner and a Gaussian resolution function was considered in our tests). As expected, for a binned ML fit, the speed-up ranges from few units to few dozens (increasing with the number of bins). Applications using, even recursively, a series of several fits with complicated PDFs, can evidently take advantage of the GPU acceleration by using `GooFit`. Once implemented within `GooFit`, Monte Carlo pseudo-experiments represent a good example of an application with these characteristics, as discussed in the next section.

2. `GooFit` performances for Monte Carlo toys

Monte Carlo pseudo-experiments (MC toys) are used to estimate the probability (*p-value*) that background fluctuations would - alone - give rise to a signal as much significant as that seen in the data. To test the computing capabilities of GPUs with respect to CPU cores, a high-statistics MC toys technique was implemented both in ROOT/RooFit and `GooFit` frameworks [8] with the aim to estimate a *p-value* and specifically the local statistical significance of the structure observed by CMS close to the kinematical threshold of the $J/\psi\phi$ invariant mass in the $B^+ \to J/\psi\phi K^+$ decay [9].

The used hardware setup consists in two servers, one equipped with two nVidia TeslaK20 and 32 cores (16 + 16 by Hyper-Threading) and the other with one nVidia TeslaK40 and 40 (20 + 20) cores [10]. To efficiently run RooFit MC toys on the 72 CPUs available on the two servers hosting the GPUs, the PROOF-Lite [11] tool is used. On the other hand the *nVidia* Multi Process Service tool [12] allows the execution of - up to 16 - simultaneous processes on the same GPU acting as a scheduler and allowing a balanced full usage of the GPU. The optimized

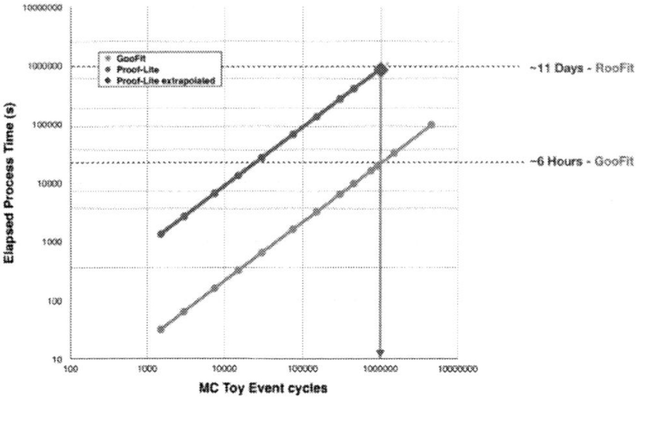

Figure 1. Comparison for the elapsed time employed with two TeslaK20 and one TeslaK40 together as a function of the number of MC toys; `GooFit`/MPS runs 48 concurrent processes while RooFit/PROOF-Lite runs on 72 CPUs. For 1M toys the red diamond point shows the extrapolated time (about 11days) for the RooFit application.

IOP Conf. Series: Journal of Physics: Conf. Series **1085** (2018) 042005 doi:10.1088/1742-6596/1085/4/042005

GooFit application running on GPUs has provided [8] striking speed-up performances with respect to the RooFit application parallelized on multiple CPUs by means of PROOF-Lite tool. In particular, from the point of view of the end-user analyst having at its own disposal all the 72 CPU cores and the three GPUs, it has been measured that 1M of MC toys can be produced in about 11days with RooFit/PROOF-Lite and in about 6 hours only with GooFit/MPS (Fig.1).

The extension of this method when a new unexpected signal is reconstructed will be presented in the next section.

3. Clustering approach to address the LEE

When a new unexpected signal is reconstructed in HEP, the *global* significance of the associated peak needs to be estimated and the Look Elsewhere Effect [13] must be taken into account. This implies to consider, within the same background-only fluctuation and everywhere in the relevant mass spectrum, any random peaking behaviour with respect to the expected background model. Thus a scanning technique based on a clustering approach has been developed.

Beforehand a pseudo-data invariant mass distribution of 15K candidates in a generic region of interest, namely [1, 18]GeV, has been generated according to a fictitious 7th order polynomial background model on the top of which any desired amount of a *significant* signal, mimicked by a Voigtian model, can be artificially added close to 8GeV (as for instance in Fig.2). At this mass value a 60MeV mass resolution is considered. The fits to the pseudo-data distribution of Fig.2 are performed accordingly: the background-only model (the *Null Hypothesis* H_0) is a 7th order polynomial function whereas the background+signal model (the *Alternative Hypothesis* H_1) is obtained by adding a Voigtian function. The resolution values in the latter are reasonably increased as a function of the increasing invariant mass while satisfying the 60MeV constraint at 8GeV. By performing the H_0 and H_1 fits, the (local) statistical significance of this peak is $Z\sigma = 5.5\sigma$ with Z approximately estimated by:

$$Z \simeq \sqrt{-2[ln(L_{H_1}) - ln(L_{H_0})]} \tag{1}$$

where L_{H_0} (L_{H_1}) is the likelihood evaluated for H_0 (H_1) hypothesis [14].

The MC toys method is configured as follows. As first step of each toy iteration, a distribution based on the background-only model is generated over the whole mass spectrum and the H_0 fit is performed. As a second step the clustering technique acts on each generated pseudo-experiment as follows:

(i) search for a *seed* bin, namely for a bin whose content fluctuates more than $x\sigma$ strictly above the value of the background function in the center of that bin (σ is the statistical error associated to the considered bin).

(ii) Add any *side* bin to the *seed* bin if it holds a content that fluctuates more than $z\sigma$ strictly above the value of the background function in the center of that bin, otherwise the *seed* bin forms a 1-bin cluster.

(iii) Check also for *light* seeds, namely bins that fluctuate more than $y\sigma$ with $z < y < x$ and with at least a *side* bin fluctuating more than $z\sigma$. In case of positive result a cluster is formed.

In the third step, a series of independent H_1 fits is performed by cycling on the clusters collected in the clustering step. At the end of this step the fit with the best ΔNLL (the test statistic) is chosen. On the whole a ΔNLL distribution is obtained over all the processed MC toys.

A set of *baseline* clustering parameters $(x, y, z) = (2.25, 1.50, 1.00)$ has been chosen in order to satisfy two concurrent requirements: not missing any possible interesting fluctuation and avoiding selecting too many irrelevant fluctuations. This *baseline* configuration has been run on about 76M pseudo-experiments and the ΔNLL distribution is shown in Fig.3, with the superimposed red line indicating the ΔNLL_{data} value for the pseudo-data.

IOP Conf. Series: Journal of Physics: Conf. Series **1085** (2018) 042005 doi:10.1088/1742-6596/1085/4/042005

Table 1. Mean number of alternative hypothesis fits per toy ($< fit_{H_1} >$) and fraction of toys with no fit (f_{nofit}) for the three different clustering configurations described in the text.

Clustering configs.	$< fit_{H_1} >$	f_{nofit}
Tight (3.00, 1.75, 1.00)	2.2	~10%
Baseline (2.25,1.50, 1.00)	4.5	~1%
Loose (2.00, 1.25, 1.00)	6.6	0.1%

The global *p-value* is then estimated by:

$$p = \int_{\Delta NLL_{data}}^{\infty} f(\Delta NLL) d(\Delta NLL) \simeq \frac{9.820 \cdot 10^2}{7.584 \cdot 10^7} \simeq 1.295 \cdot 10^{-5} \quad (2)$$

This corresponds to the *global* statistical significance $Z\sigma = \Phi^{-1}(1-p)\sigma \simeq 4.22\sigma$, through the inverse function of the cumulative distribution of the standard Gaussian. As expected by considering the LEE, the *global* significance is relevantly lower than the estimated *local* one.

4. Evaluation of the possible systematic uncertainty

In order to test the behavior of the method and to estimate the possible systematic uncertainty associated to the clustering technique, three sets of configuration parameters, i.e. three values for the (x, y, z) parameters, have been carefully considered. After some tests with different cuts two further configurations are chosen besides the baseline clustering cuts: a set of tighter values $(3.00, 1.75, 1.00)$ and a set of looser values $(2.00, 1.25, 1.00)$. The Tab.1 reports details about these three clustering configurations such as the average number of H_1 fits per toy and the fraction of toys with no fit.

These three configurations have been run on a same common set of 45M fluctuations and the three corresponding ΔNLL distributions are shown superimposed in Fig.4.

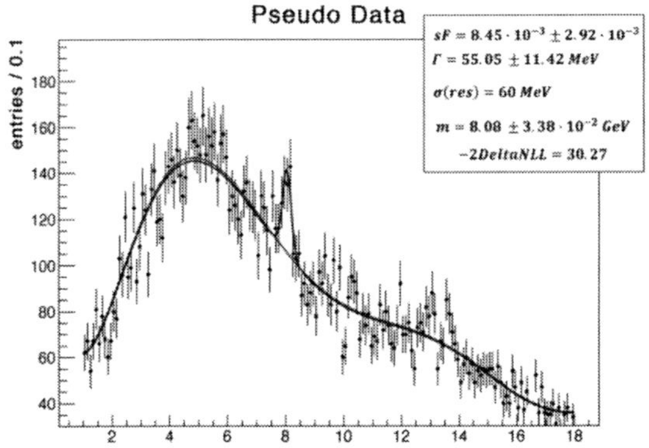

Figure 2. Simulated invariant mass distribution (pseudo-data). $H_0(H_1)$ fit is in red (blue); in the top right box the best values for the estimated parameters of the H_1 model are given.

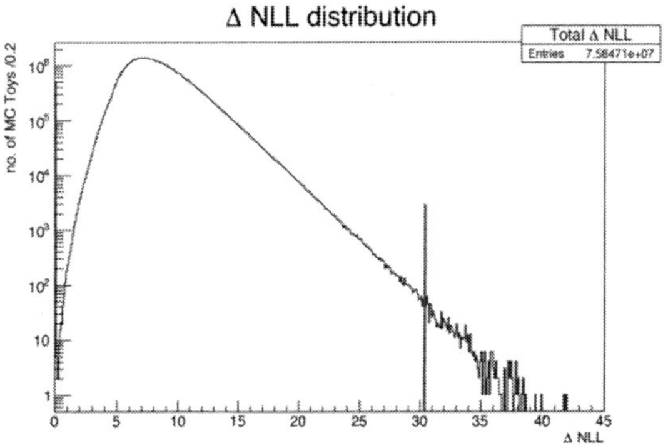

Figure 3. ΔNLL distribution for about 76M toys for the baseline configuration of clustering technique. The red line indicates the $\Delta NLL_{data} \simeq 30.27$ value for the pseudo-data distribution in Fig.2.

IOP Conf. Series: Journal of Physics: Conf. Series **1085** (2018) 042005 doi:10.1088/1742-6596/1085/4/042005

Table 2. Estimated *global* significances for the 3 clustering configurations with respect to different *local* significance values estimated by Eq.(1).

Local Significance	**4.0σ**	**4.5σ**	**5.0σ**	**5.5σ**	**6.0σ**
Tight (3.00, 1.75, 1.00)	2.21	2.91	3.58	4.22	4.87
Baseline (2.25,1.50, 1.00)	2.20	2.91	3.58	4.22	4.87
Loose (2.00, 1.25, 1.00)	2.20	2.91	3.58	4.22	4.87

By focusing on the region of interest for the estimation of the statistical significance, i.e. the tail of the ΔNLL distribution ($\Delta NLL > 20$), it is evident that there is no relevant difference (Fig.5) among the three configurations. This can be appreciated by inspecting, in Figg.4, 5, the normalized deviations $(x - y)/(x + y)$ of the other two distributions with respect to the baseline distribution. This is also confirmed by examining the estimated *global* significances for the *p-values* corresponding to different values of *local* significances, as reported in Tab.2. It can be concluded that the systematic uncertainty on the *p-values* associated to the method is negligible.

5. Conclusions

A high-statistics pseudo-experiment method, based on a scanning and clustering approach, has been implemented and tested within `GooFit` framework with the purpose to estimate the *global* statistical significance of an unexpected new signal. The presented results clearly indicate that the systematic uncertainty associated to the method is negligible and that the *p-value* estimation is not affected by the clustering configuration. This kind of validation studies has been performed by exploiting the high performance of an optimized `GooFit` application running on GPU-equipped servers.

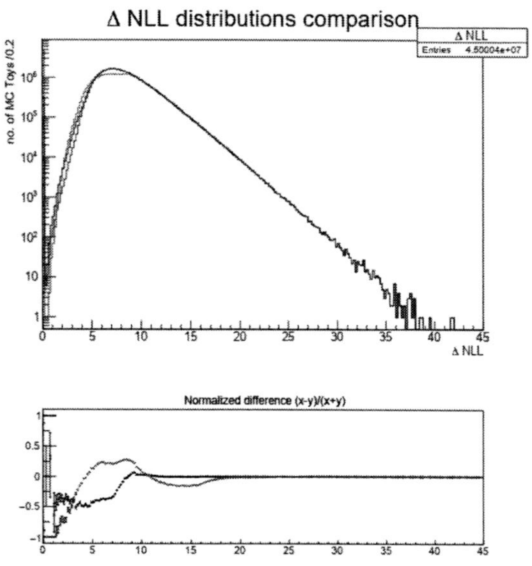

Figure 4. ΔNLL distributions for 45M of common fluctuations for the 3 configurations: baseline (black), tight (red) and loose (blue).

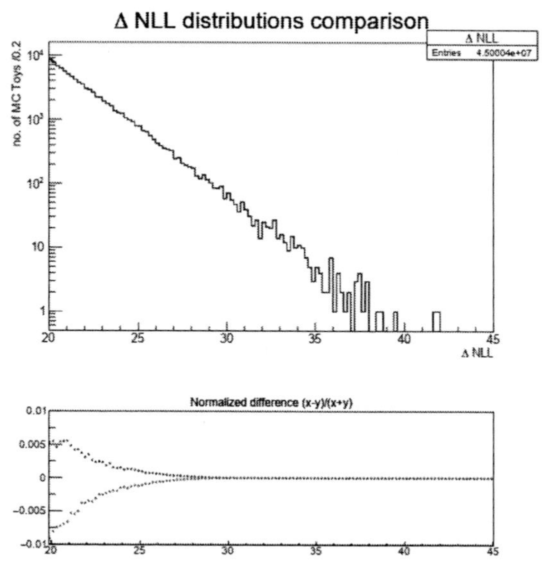

Figure 5. The same 3 ΔNLL distributions of Fig.4 once zoomed into the range 20.0-45.0 to inspect their tail behaviour.

IOP Conf. Series: Journal of Physics: Conf. Series **1085** (2018) 042005 doi:10.1088/1742-6596/1085/4/042005

Aknowledgements
We are grateful for valuable support from G.Donvito, manager of the High Performance Cluster hosted by the ReCas [15] Data Center.

References
[1] Andreassen R, Meadows B T, de Silva M and Sokoloff M D *J. Phys.: Conf. Series* **513**, 052003 (2014)
[2] Schreiner H F et al. *arXiv:1710.08826* (2017)
[3] Brun R and Rademakers F *Nucl. Instrum. Meth.* **A 389**, 81-86 (1997)
[4] Verkerke W and Kirkby D P *eConf* **C0303241** MOLT007 (*Preprint* physics/0306116) (2003)
[5] `https://docs.nvidia.com/`; for these studies CUDA version 6.5(7.0) were used for Tesla K20(40) boards.
[6] James F and Roos M CERN Program Library routine D506 (1989) (long write-up)
[7] This can be checked by nVidia Visual Profiler (`nvvp`)).
[8] Pompili A and Di Florio A, *J. Phys. Conf. Ser.* **762**, 012044 (2016) [ACAT 2016]
[9] CMS Collaboration, *Phys. Lett.* **B 734**, 261 (2014)
[10] Tesla K20(K40) has 5(12)GB GDDRs; the 16(20) cores are E5-2640 v2 @ $2.0GHz$ with 64(256)GB of RAM
[11] `https://root.cern.ch/proof`; see for specific PROOF-Lite applications on multi-core servers: Barbone L, Donvito G and Pompili A, *J. Phys.: Conf. Series* **396** 042017 (2012)
[12] `https://docs.nvidia.com/deploy/pdf/CUDA_Multi_Process_Service_Overview.pdf`
[13] Gross E and Vitells O *Eur. Phys. J.* **C70**, 552 (2010)
[14] I.Narsky and F. C. Porter, *Statistical Analysis Techniques in Particle Physics: Fits, Density Estimation and Supervised Learning*, Wiley, 2013.
[15] ReCas is a project financed by the italian MIUR (*PONa3_00052, Avviso 254/Ric.*); its web page is `http://www.recas-bari.it/index.php/en/`.

IOP Conf. Series: Journal of Physics: Conf. Series **1085** (2018) 042006 doi:10.1088/1742-6596/1085/4/042006

Weakly Supervised Classification For High Energy Physics

Lucio Mwinmaarong Dery, Benjamin Nachman, Francesco Rubbo, Ariel Schwartzman

E-mail: ldery@stanford.edu, bnachman@cern.ch, rubbo@slac.stanford.edu, sch@slac.stanford.edu

Abstract. We present a classification algorithm that applies the machine learning paradigm of Learning from Label Proportions (LLP) [1] to enable learning on unlabelled data. Our algorithm, Weakly Supervised Classification, receives as its only input the class proportions of batches of data but makes per-instance classification decisions matching the performance of fully supervised approaches. We apply our model to the problem of Quark-Gluon tagging and show that it is robust to underlying mismodelling of the simulated data unlike fully supervised learning.

1. Introduction

Classification problems abound in High Energy Physics; from Quark-Gluon tagging [2] and b-tagging [3] to boosted W tagging [4], there is a strong need to discriminate the types of particles produced by collision events. Traditionally, these classification problems are approached by training a model that takes in individual feature-label pairs and learns to generalize from the supervision provided to unseen instances. The feature-label pairs used for training are generated via high fidelity Monte-Carlo (MC) simulations that model physical processes at distances ranging from 10^{-25} meters all the way to the macroscopic dimensions of detectors.

Since these simulations are only approximate, the underlying physical processes are not always perfectly model and as a consequence, the performance of classifiers trained via this approach and applied to actual data on collision events are in some cases sub-optimal. Comparisons of tagging efficiencies between data and simulation show large $(10\%-30\%)$ differences for b-tagging [5] quark/gluon tagging [6], boosted W tagging [7], and high p_T top quark tagging [8]. Since using collider data for supervised training is not an option due to the absence of labels, there is the need to develop unsupervised or weakly supervised approaches that directly leverage collider data.

In some cases, though the per-instance class labels for data are unknown, we know the relative proportions in which the classes occur in. For example, at a fixed order in perturbation theory, the probability for an outgoing parton to be a quark or a gluon depends on well-known parton distribution functions and matrix elements. Relying only on the weak supervision of class proportions and leveraging insights from the areas of Multiple Instance Learning (MIL) [9]

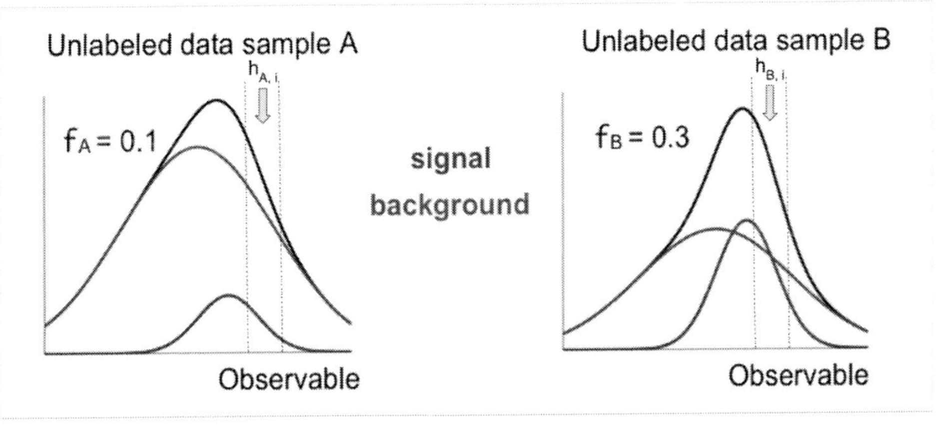

Figure 1: A rendering of the setup in section 2.1. The black distribution is obtained by binning the features of all examples, both signal and background, in a batch. f_A, f_B give the background to signal proportion in batches A and B respectively. The same bin i is chosen for in both histograms for resolving the true distributions of classes A and B

and LLP we propose an algorithm that matches the performance of fully supervised classifiers. More specifically, given a batch of data where only the relative proportions of classes are known, we learn a per-instance mapping from features to labels under the constraint that statistics of the predicted labels in a batch match the expected proportions.

2. Method

Though proportions are a much weaker learning signal than per-instance labels, we will prove in the section that follows that it is a sufficient signal in finding a decision boundary that optimally classifies the data.

2.1. Probability Density Functions from Fractions

We show the feasibility of our approach by providing an analytic solution to a simple 2-batch problem where the class proportions are known. Consider two batches of data, A and B, each of which contains a mixture of two classes {0,1} in proportions f_A and f_B respectively. Each input is a feature vector $x_i \in \mathcal{R}^d$. In order to identify the two components, we need to resolve the Probability Density Functions (PDFs) corresponding to the individual classes. To do this, we first build the PDF corresponding to each of the individual batches. By considering m bins for each dimension d, we adopt a binning strategy, creating a 1-dimensional histogram (with $d \times m$ bins) representing a density function over feature bins. Since we know that each batch is a mixture of inputs from both classes, the height of a particular bin i is a linear combination of the true heights $h_{0,i}$ and $h_{1,i}$ for each class in their relative proportions. Figure 1 illustrates the setup. Using $h_{A,i}$ and $h_{B,i}$ as the height of bin i in batches A and B respectively, formally

$$h_{A,i} = f_A h_{0,i} + (1 - f_A) h_{1,i}$$
$$h_{B,i} = f_B h_{0,i} + (1 - f_B) h_{1,i}$$

Given the system of linear equations above, we can solve for the independent heights of each class in each bin i, $h_{0,i}$ and $h_{1,i}$ respectively. This allows us to extract the PDFs for each of the individual classes. Having obtained the PDFs for each class, we can easily build an optimal classifier that distinguishes the two classes based on the likelihood ratios.

The above procedure provides an analytic approach to learning from batch proportions. It is however not a practical approach. Specifically, the binning strategy is intractable when the feature space is large. The system of equations set up also becomes over-constrained when more than two batches of proportions are provided, which means there might not exist an exact solution. We get around both these problems by using a neural network to approximate both the binning strategy and the optimal decision function.

2.2. *Weakly Supervised Classifiers*

The supervised binary classification problem is set up as follows. Given a training set $\mathcal{T}$, consisting of M_{train} tuples $(x_i, y_i)^{M_{train}}$ where x_i is a feature vector of an instance and y_i is the corresponding true label, we learn a classifier $\mathcal{F}_{sup}$ which generalizes to a unseen set of M_{test} examples $\mathcal{D} = \{x_j\}^{M_{test}}$ of unknown labels. To learn $\mathcal{F}_{sup}$ we set up a loss function l_{sup} that is independently evaluated for each member of a batch of data N.

$$\mathcal{F}_{sup} = argmin_{\mathcal{F}} \sum_i l_{sup}\Big(\mathcal{F}(x_i), y_i\Big)$$

In the weakly supervised setup however, the training set consists of tuples (X_B^j, y_j) where X_B^j is the jth batch of examples containing M_{batch} instances. The label y_j in this case is the proportions of members of X_B^j that belong to class A. We still need the classifier to produce a per-instance binary decision though it is fed batched data. To achieve this, we introduce a loss that ties together the predictions of a batch. That is, $\mathcal{F}_{weak}$ makes a prediction for each member of the batch, but the loss, l_{weak} is evaluated on how well the distribution of batch predictions match the expected proportions y_j. More formally,

$$\mathcal{F}_{weak} = argmin_{\mathcal{F}} \sum_j l_{weak}\left(\frac{\sum_{x_i \in X_b^j} \mathcal{F}(x_i)}{M_{batch}}, y_j\right)$$

3. Results

To test our approach, we apply weak supervision to the quark-gluon tagging problem. We generate a dataset of $2 \rightarrow 2$ quark-gluon scattering (dijet) events, simulated using the Pythia 8.18 event generator [10]. Jets are clustered using the $anti-k_t$ algorithm [11] with distance parameter R = 0.4 via the FastJet 3.1.3 [12] package. Jets are classified as quark- or gluon-initiated by considering the type of the highest energy quark or gluon in the full generator event record that is inside a 0.3 radius of the jet axis. For simplicity, one transverse momentum range is considered: 45 GeV $< p_T <$ 55 GeV. Additionally, there is a pseudo-rapidity requirement that mimics the usual detector acceptance for charged particle tracking: $|\eta| < 2.1$. Heuristically, gluons have twice as much strong-force charge as quark jets, resulting in more constituents and a broader radiation pattern. More precisely, at leading logarithm the ratio of the average gluon and quark jet constituent multiplicity is equal to the ratio of the QCD color-factor (Casimir) associated with quarks (CF) and gluons (CA): $CA/CF = 9/4 \approx 2$ [13], [14]. Due to Casimir scaling, the ROC curve for any p_T and angle-weighted moments of the jet radiation pattern is set by CA/CF [15]. The following variables are useful for quark/gluon discrimination: the

IOP Conf. Series: Journal of Physics: Conf. Series **1085** (2018) 042006 doi:10.1088/1742-6596/1085/4/042006

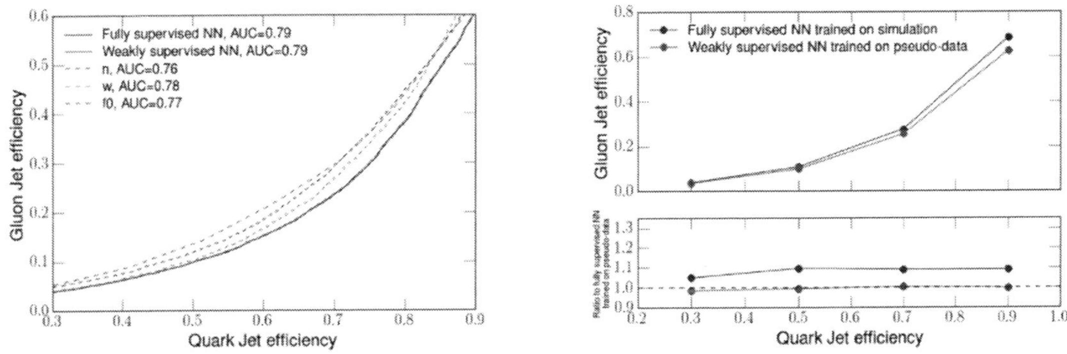

(a) (b)

Figure 2: In (a) Fully and Weakly supervised classifiers are trained on identical simulated data and evaluated on a test sample drawn from the same population. The weakly supervised classifier matches the performance of the fully supervised one. The curves corresponding to the three input observables used as discriminant are shown as reference. In (b) Fully supervised classifier (blue line) is trained on a labeled simulated training sample. The weakly supervised classifier (red line) is trained on an unlabeled pseudo-data training sample. In both cases, the performance is evaluated on the same pseudo-data test sample. The ratios to the performance of a fully supervised classifier trained on a labeled pseudo-data sample are shown in the bottom pad

number of jet constituents n, the first radial moment in p_T (jet width) w, and the fraction of the jet p_T carried by the leading $anti-k_T$ R = 0.1 subjet f_0. The constituents i considered for computing n and w are the hadrons in the jet with $p_T > 500MeV$. A weakly supervised classifier with one hidden layer of size 30 is trained by considering 12 bins of the distribution of the absolute difference in pseudorapidity between the two jets [16]. The proportion of quark initiated jets varies between 0.21 and 0.32

As can be seen from Figure 2(a) our classifier achieves the same performance as a fully supervised classifier. The key application we envision for Weak Supervision is the circumvention of the use of imperfect simulation data to build classifiers for high energy physics. To highlight this, we setup an experimental scenario to evaluate the performance of a fully supervised classifier trained on simulation but evaluated on pseudo-data against a weakly supervised classifier trained directly on pseudo-data. We build the pseudo-data samples by distorting the probability distributions of n and w in the training sample to emulate the difference in efficiency measured in [2] between simulation and data. As can be seen from Figure 2(b) weak supervision outperforms fully supervised, demonstrating ability to circumvent mis-modeling issues presented by training on simulated data.

The demonstrations provided so far have been on relatively low dimensional inputs. Weak supervision has also been applied to high dimensional jet images for the Q-G problem, achieving competitive results as in [17]

If you wish to replicate our findings, please find the code for Weak Supervision and the experiments described at this link : `https://doi.org/10.5281/zenodo.322813`

4. Conclusion

We have presented a new approach to classification in High Energy Physics in cases where class proportions are known but individual labels are not readily available. This weakly supervised classification has broad applicability and has been demonstrated in one important discrimination task in high energy physics: quark versus gluon jet tagging. Our method serves to lay the groundwork for circumventing the mis-modelling issues that plague training on data from simulations, allowing us to train directly on collider data.

References

[1] Giorgio Patrini, Richard Nock, Paul Rivera, and Tiberio Caetano. (almost) no label no cry. In *Advances in Neural Information Processing Systems*, pages 190–198, 2014.

[2] Melissa Franklin, John Huth, Valerio Ippolito, David Lopez Mateos, Kevin Mercurio, Masahiro Morii, William Spearman, Andy Yen, and G Zevi Della Porta. Light-quark and gluon jet discrimination in pp collisions at $\sqrt{s} = 7$TeV with the ATLAS detector. 2014.

[3] ATLAS Collaboration et al. Performance of b-jet identification in the ATLAS experiment. *Journal of instrumentation*, 11(04):P04008, 2016.

[4] Georges Aad, B Abbott, J Abdallah, O Abdinov, R Aben, M Abolins, OS AbouZeid, H Abramowicz, H Abreu, R Abreu, et al. Identification of boosted, hadronically decaying w bosons and comparisons with ATLAS data taken at $\sqrt{s} = 8$TeV. *The European Physical Journal C*, 76(3):154, 2016.

[5] CMS collaboration et al. Identification of b-quark jets with the CMS experiment. *Journal of Instrumentation*, 8(04):P04013, 2013.

[6] CMS collaboration et al. Performance of quark/gluon discrimination in 8 tev pp data. *CMS Physics Analysis Summary CMS-PAS-JME-13-002, CERN*, 2013.

[7] CMS collaboration et al. Identification techniques for highly boosted w bosons that decay into hadrons. *arXiv preprint arXiv:1410.4227*, 2014.

[8] CMS collaboration et al. Boosted top jet tagging at CMS. Technical report, CMS-PAS-JME-13-007, 2013.

[9] Thomas G Dietterich, Richard H Lathrop, and Tomás Lozano-Pérez. Solving the multiple instance problem with axis-parallel rectangles. *Artificial intelligence*, 89(1-2):31–71, 1997.

[10] Torbjörn Sjöstrand, Stephen Mrenna, and Peter Skands. A brief introduction to pythia 8.1. *Computer Physics Communications*, 178(11):852–867, 2008.

[11] Matteo Cacciari, Gavin P Salam, and Gregory Soyez. The anti-kt jet clustering algorithm. *Journal of High Energy Physics*, 2008(04):063, 2008.

[12] Matteo Cacciari, Gavin P Salam, and Gregory Soyez. Fastjet user manual. *The European Physical Journal C*, 72(3):1896, 2012.

[13] Jason Gallicchio and Matthew D Schwartz. Quark and gluon tagging at the lhc. *Physical review letters*, 107(17):172001, 2011.

[14] Keith A Olive, Particle Data Group, et al. Review of particle physics. *Chinese Physics C*, 38(9):090001, 2014.

[15] Andrew J Larkoski, Jesse Thaler, and Wouter J Waalewijn. Gaining (mutual) information about quark/gluon discrimination. *Journal of High Energy Physics*, 2014(11):129, 2014.

[16] Georges Aad, B Abbott, J Abdallah, O Abdinov, B Abeloos, R Aben, M Abolins, OS AbouZeid, NL Abraham, H Abramowicz, et al. Measurement of the charged-particle multiplicity inside jets from $\sqrt{s} = 8$TeV pp collisions with the ATLAS detector. *The European Physical Journal C*, 76(6):322, 2016.

[17] Patrick T Komiske, Eric M Metodiev, Benjamin Nachman, and Matthew D Schwartz. Learning to classify from impure samples. *arXiv preprint arXiv:1801.10158*, 2018.

IOP Conf. Series: Journal of Physics: Conf. Series **1085** (2018) 042007 doi:10.1088/1742-6596/1085/4/042007

Machine learning for antihydrogen detection at ALPHA

A Capra for the ALPHA Collaboration
TRIUMF, 4004 Wesbrook Mall, Vancouver BC V6T 2A3, Canada
E-mail: acapra@triumf.ca

Abstract. The ALPHA experiment at CERN is designed to produce and trap antihydrogen to the purpose of making a precise comparison with hydrogen. The basic technique consists of driving an antihydrogen resonance which will cause the antiatom to leave the trap and annihilate. The main background to antihydrogen detection is due to cosmic rays. When an experimental cycle extends for several minutes, while the number of trapped antihydrogen remains fixed, background rejection can become challenging. Machine learning methods have been employed in ALPHA for several years, leading to a dramatic reduction of the background contamination. This allowed ALPHA to perform the first laser spectroscopy experiment on antihydrogen.

1. Introduction

Antihydrogen is a tool to explore the fundamental symmetries [1] that lie at foundation of the Standard Model, i.e., the CPT invariance, and of the General Relativity, i.e., the Equivalence Principle. The CPT symmetry can be tested by comparing the spectra of hydrogen and antihydrogen. The ALPHA collaboration observed for the first time a transition between Zeeman levels of the antihydrogen ground state [2], placed the most stringent upper limit to the antihydrogen electric charge [3], performed the first laser spectroscopy experiment [4] and, more recently, measured the hyperfine splitting of the ground state [5]. An experiment to measure the gravitational acceleration of antihydrogen is currently being built.

In the next section, the ALPHA apparatus is described in some detail with the outline of typical experimental cycle. Successively, the techniques to identify antihydrogen are presented. Finally, the measurement of the 1S-2S transition, depicted in Fig. 1, is described as the prototypical case of using *supervised Machine Learning* for antihydrogen identification. This measurement is a test of CPT invariance with a relative precision of 2×10^{-10}.

2. The ALPHA antihydrogen trapping apparatus

The ALPHA apparatus [6] is located at the CERN Antiproton Decelerator (AD) and its main component are shown in Fig. 2.

Antihydrogen is formed by mixing together a positron and an antiproton plasma. These non-neutral plasma are confined and manipulated in a so-called Penning-Malmberg trap, which consists of an *electrodes* stack and a 1 T solenoid (not shown in the picture).

Cold antihydrogen is trapped near the minimum of the magnetic field gradient produced by the superposition of the field due to the *mirror coils* and to the *octupole*. The depth of

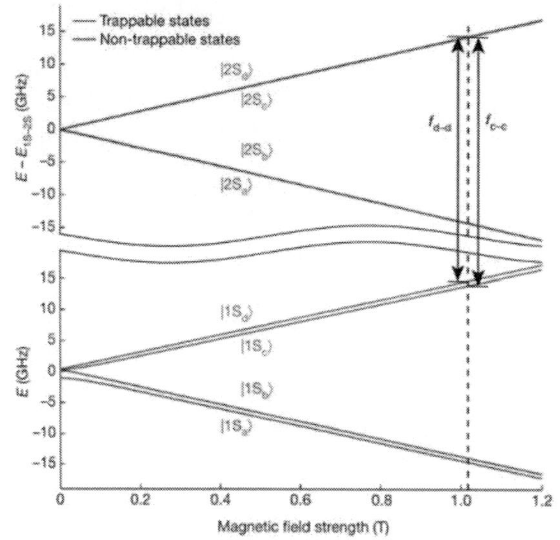

Figure 1. Ground and first excited state of the antihydrogen atom. Since antihydrogen is confined in a magnetic field, Zeeman splitting of the hyperfine levels occur. Only substates where the total spin of the antiatom is anti-parallel to the magnetic field are trappable (blue), the others (red) are not confined in the ALPHA trap. The frequency of radiation is such that it drives a transition between trappable states, as indicated by the black arrows, at a given magnetic field strength.

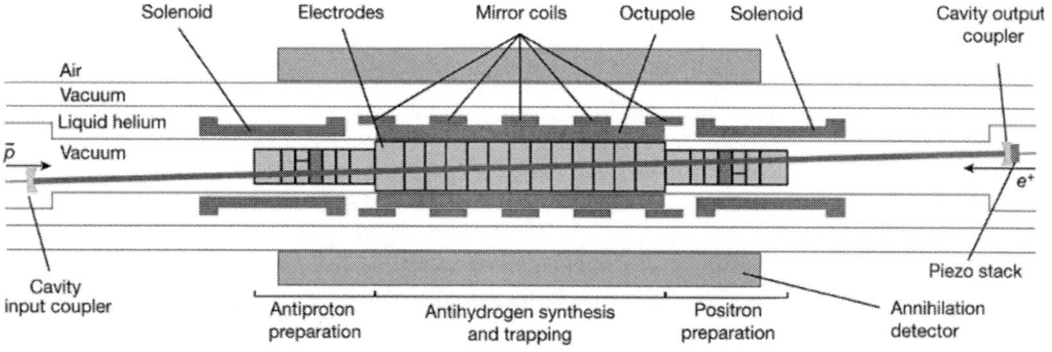

Figure 2. View of the main components of the ALPHA apparatus for the measurement of the 1S-2S transition. The text in italics in the present section refers to the labels in this figure.

the "minimum-B trap" is approximately $50\,\mu$eV, or 0.56 K in temperature units, and is largely limited by the maximum current that can circulate in the magnets.

One of the key features of ALPHA is its capability of detecting antihydrogen annihilation. The *annihilation detector* is a silicon strip tracking detector. It is composed of 72 double-sided modules, arranged in three layers. The "silicon detector" permits to monitor the antihydrogen production and to perform the analysis of the response of the antiatoms to the electromagnetic radiation, through the determination - or reconstruction - of the their annihilation location.

The laser system produces 150 mW of 243 nm laser light. In order to increase the light intensity in each direction, allowing the antiatoms to be excited in a reasonable amount of time, the apparatus includes a *power build-up cavity*.

IOP Conf. Series: Journal of Physics: Conf. Series **1085** (2018) 042007 doi:10.1088/1742-6596/1085/4/042007

2.1. A typical experimental cycle
In the following the main steps to produce and study antihydrogen are listed with a brief explanation.

Positrons and antiprotons are loaded at centre of the ALPHA apparatus (see Fig. 2), after being cooled by several order of magnitude.

Antihydrogen synthesis in the minimum-B trap. The antiproton plasma is slowly, i.e, on the time scale of 1 s, put into contact with the positron plasma, producing antihydrogen. Most of the antihydrogen created is not magnetically confined, either because has too much kinetic energy or because the bound state is too weak. This phase is referred as "mixing". There are ten expected *cold* antihydrogen atoms confined per mixing. Overall the antihydrogen production takes approximately four minutes and can be repeated several times to trap the desired number of antiatoms.

Trapped antihydrogen is illuminated with electromagnetic radiation, e.g., 243 nm or 121 nm laser light, or 28 GHz microwave, for several minutes. The laser interrogation, for instance, lasts for 600 s.

Trap shutdown where the current in the magnets is ramped down in approximately 1.5 s and all the trapped antiatoms are released from the confinement.

During the mixing phase all the events recorded with the silicon detector are due to the annihilation of "hot" antihydrogen (i.e., antiatoms with too much kinetic energy to be confined) with a negligible background contamination. During the laser (or microwave) irradiation, one expects to see antihydrogen annihilation occurring due to various loss mechanisms. This is the so-called "appearance channel" for antihydrogen spectroscopy measurements. During the first few hundreds of milliseconds of the trap shutdown all the antiatoms, that have not interacted with the electromagnetic radiation, annihilate. This is the "disappearance channel".

The dataset for the experiment under consideration follows a simple protocol, in which three types of trials are conducted: "on resonance", where the build-up cavity is locked to the hydrogen transition frequency, "off resonance", where the laser is detuned by 200 kHz (at 243 nm), and "no laser", where no radiation is present during the 600 s the antiatoms are held in the trap.

3. Detection and identification of antihydrogen
An antihydrogen atom is detected when it annihilates on the trap's wall. On the one hand, the annihilation of positron (a pair of 511 keV gamma rays) is not observed in the current setup. On the other hand, the antiproton annihilation with a nucleon produces mostly charged and neutral π mesons. The π^0 decays rapidly into two high energy gamma rays, which may subsequently produce electron-positron pairs in the surrounding material. The $\pi^{\pm}$ have an average total momentum in the range of 100 - 300 MeV and are tracked in the silicon detector.

Each annihilation event is *reconstructed* through the following - rather standard - steps.

(i) The three-dimensional position of the hits due to the charged particles impinging on the detector is obtained from the charge profile on the strips.

(ii) Tracks are identified by searching through all possible combinations of three hits, one for each of the three layers.

(iii) The identified tracks are fit to helices, since the detector is immersed in an uniform magnetic field.

(iv) The helices are selected so that only the ones most likely due to $\pi^{\pm}$ mesons are used the in the final stage.

(v) The annihilation location, called the *vertex*, is the point where the helices pass closest to one another.

IOP Conf. Series: Journal of Physics: Conf. Series **1085** (2018) 042007 doi:10.1088/1742-6596/1085/4/042007

Annihilation events are not the only type of events that trigger the detector readout. The detector occasionally records events due to cosmic rays at rate of $\approx$ 10 Hz. These events constitute the main background to antihydrogen annihilation identification and, as such, must be removed from the data samples. A "cuts analysis" using two variables, as described in [7], achieves a background suppression of 99.9%, which corresponds to a cosmic rays event rate of 47 mHz. This rate is equivalent to 28 background events in the "appearance channel". If one considers that the typical trapping rate is of 10 antiatoms per mixing cycle, the annihilation signal is washed out without a stronger background suppression.

3.1. Background rejection: a classification problem

The difficulty of further suppressing the background arises from events like the ones shown in Fig. 3. The event labelled as "annihilation" (a) was detected during the mixing phase. The cosmic rays events (b,c) were detected when no antiprotons were present in the ALPHA trap. The event (b) is due to a single particle and can be easily classified as a cosmic ray (likely a muon). Even though the event (c) is originates outside the ALPHA apparatus, it is not trivial to classify it as a cosmic ray event.

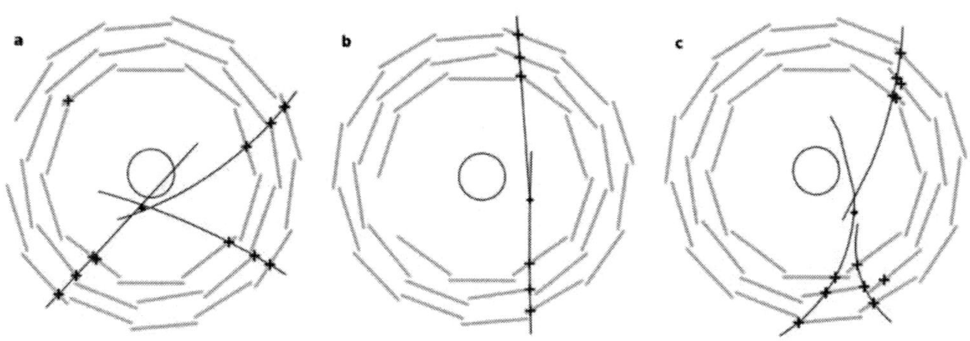

Figure 3. Cross-section view of real events in the ALPHA apparatus. The grey rectangles identify the position of the silicon detector modules. The blue crosses are the hits on the detector and the blue lines represent the reconstructed tracks (helices). a) Antihydrogen annihilation: three tracks event with reconstructed vertex near the trap wall (blue circle). b) Cosmic ray: reconstructed as a "two-tracks" event, even if it is due to single incoming particles. c) Event due to a cosmic ray that produces multiple tracks in the detector and that can be misidentified as an antihydrogen annihilation.

The reduction of the background events to an acceptable rate is only possible by exploiting more sophisticated methods of event classification.

3.2. Training, training samples and discriminating variables

A Machine Learning tool from the SPR package [8] has been employed to classify detected events, achieving high statistical significance in the first spectroscopy experiment on antihydrogen. The classifier of choice is a *Bagged Decision Tree* because of its immediate interpretation.

In ALPHA it is possible to train the classifier on actual data, i.e., without resorting to the generation of a sample of Monte Carlo data. Indeed, the background data sample is collected by operating the silicon detector without antiprotons in the trap, while the signal data sample is derived from the mixing phase, i.e., the annihilation events recorded during the $\overline{\text{H}}$ synthesis. In the present analysis, the background sample consists of $\approx$1.6M events, while $\approx$208k antihydrogen annihilation events compose the signal sample.

It must be stressed at this point that quantities that relate to the antihydrogen annihilation position along the trap axis, i.e., the z coordinate of a Cartesian reference system, cannot be used in the training. This is due to the fact that the antihydrogen annihilation events during the mixing are centred around where the antiproton and positron plasma are confined (see Fig. 1), while the ones due to the interaction with the laser light (or the microwave radiation) depend on the spatial profile of the electromagnetic wave and on the magnitude of the trapping magnetic field.

The training samples are split equally into three subsamples: two for optimization and validation of the training and one for testing, which will be used to evaluate the performance of the classifier.

The following discriminating variables were chosen based on how their distributions for the signal and background event samples can be differentiated.

(i) number of hits N_{hit} (large for annihilation events),

(ii) number of reconstructed tracks N_{helices} (large for annihilation events),

(iii) number of reconstructed tracks used in vertexing $\tilde{N}_{\text{helices}}$ (large for annihilation events),

(iv) ϕ component of the reconstructed vertex (uniform for annihilation events),

(v) r component of the reconstructed vertex (ideally all the annihilation events occur at the trap radius however, due to the finite resolution of the detector, there is a broad peak centred on the expected value, while vertexes due to the cosmic rays do not have a preferred radial position),

(vi) squared residual δ (this is a measure of the tracks straightness, which is greater for the cosmic rays),

(vii) ϕ component of sphericity tensor eigenvector Σ_ϕ (see Eq. (1) and Fig. 4),

(viii) combination of two largest sphericity tensor eigenvalues $\sqrt{\lambda_1^2+\lambda_2^2}$,

(ix) z component of sphericity tensor eigenvector Σ_z.

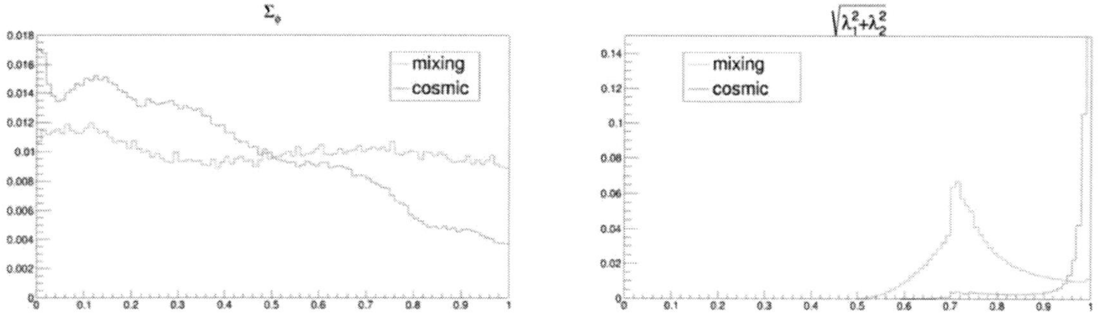

Figure 4. Distributions of the ϕ component (left) and of the combination of two largest eigenvalues (right) of the sphericity tensor, given by Eq. (1) for the mixing and the cosmic rays samples. The distributions are normalized to the total number of events in each sample.

The topology of the detected event can be characterized by the *sphericity tensor*, given by

$$S_{ab} = \frac{1}{N_{\text{helices}}} \sum_{i=1}^{N_{\text{helices}}} \frac{p_a^i\, p_b^i}{|\boldsymbol{p}_i|^2}, \tag{1}$$

where $\boldsymbol{p}_i$ is the momentum of the ith reconstructed track and $a, b = x, y, z$. By standard diagonalization of S_{ab}, one finds the eigenvalues λ_j, with $j = 1, 2, 3$, such that $\lambda_1 \geq \lambda_2 \geq \lambda_3$. The eigenvector corresponding to λ_1 is labelled $\boldsymbol{\Sigma}$.

3.3. Result of the training for the appearance channel

The classifier training described above is general enough that is adopted for different analysis channels. In this section the results of the training for the appearance channel are presented, since this analysis is the most challenging one, given the long observation time.

The output of the trained classifier[1] on the test sample is shown in Fig. 5. The cut on the classifier output is placed by maximizing the *Punzi significance* [9]

$$\frac{S}{\frac{3}{2}n_\sigma + \sqrt{B}}, \tag{2}$$

where S and B are the number of events classified as annihilation (signal) and background, respectively, and $n_\sigma = 3$ is the number of standard deviations that separate the signal from the background. This yields a background, or "false positive", event rate of 4 mHz ($\pm 7\%$) and the total number of signal events accepted by the classifier is $(37.6 \pm 0.2)\%$ of the mixing sample. The signal acceptance is not too low since the background is suppressed by a factor of twelve and the statistical significance calculated with Eq. (1) is equal to 2.5.

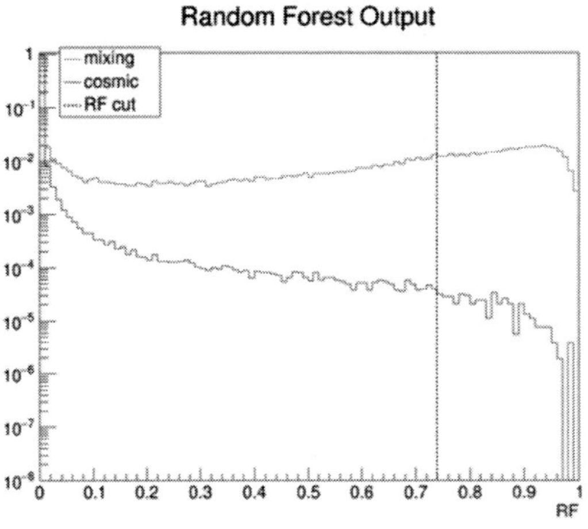

Figure 5. Random Forest classifier (Bagged Decision Tree) with 100 decision tree replicas, 4 variables per tree, 16 events per leaf. The dashed line corresponds to the optimal value of Eq. (2). The distributions are normalized to the total number of events in each sample.

As previously mentioned, it is important to study the z distribution of the reconstructed annihilation vertex. In Fig. 6 it is shown that the z distribution is not distorted by the classifier selection.

4. Laser spectroscopy of trapped antihydrogen

With the trained classifiers at hand, one for the disappearance channel and one for the appearance channel, it is possible to count the antihydrogen annihilation events. A summary of the number of antihydrogen annihilation detected in the two analysis channels is presented in Tab. 1. The on-resonance and off-resonance counts in the disappearance channel differ by 92 ± 15 counts: a clear signature that the resonant laser frequency removed $\approx 58\%$ of the trapped antiatoms. In the appearance channel, the counts in the off-resonance and no laser trials are consistent with the expected number of background events.

[1] The decision trees are grown by optimizing the Gini index.

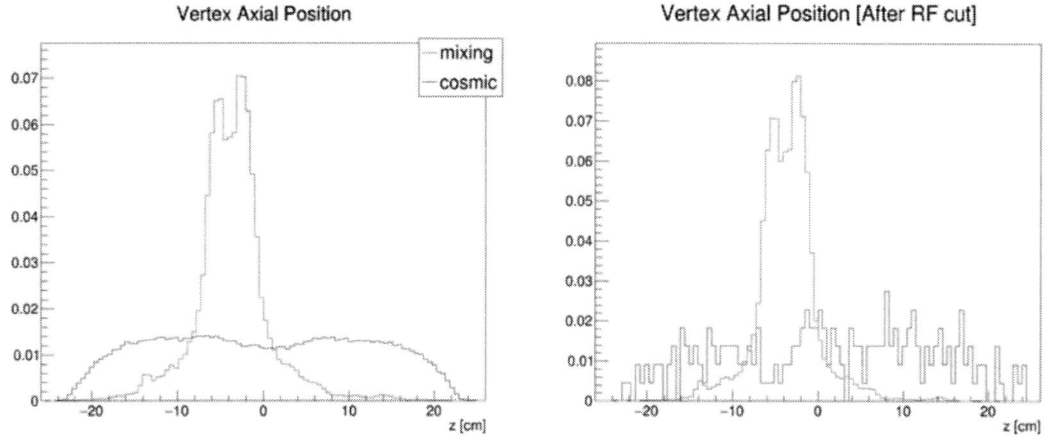

Figure 6. The distributions of the z component of the reconstructed vertex for the mixing and the cosmic rays samples. The distributions are normalized to the total number of events in each sample. The distribution of the mixing sample (left) is not altered after the classifier selection (right).

	Disappearance channel			Appearance channel		
	Background	$\overline{\mathrm{H}}$ events	Uncertainty	Background	$\overline{\mathrm{H}}$ events	Uncertainty
On-resonance	0.7	67	8.2	28.4	79	8.9
Off-resonance	0.7	159	13	28.4	27	5.2
No laser	0.7	142	12	28.4	30	5.5

Table 1. Summary table. The "background" columns indicate the expected number of cosmic rays event. The "$\overline{\mathrm{H}}$ events" columns report the actual number of events classified as antihydrogen annihilation. The "uncertainty" columns are the counting statistics errors.

Taking into account the laser linewidth, the line broadening due to the magnetic field and the detuning of the off-resonance trials, the frequency $\nu_{\mathrm{1S\text{-}2S}}$ of the transition is determined to a relative precision of $\delta\nu/\nu_{\mathrm{1S\text{-}2S}} \approx 2 \times 10^{-10}$.

5. Conclusion

The adoption of machine learning techniques allowed ALPHA to perform breakthrough experiments on antihydrogen. ALPHA relies upon annihilation position reconstruction for antihydrogen detection. The identification of the antihydrogen annihilation can be challenging due to the cosmic rays background. However, the classifier described here, which is trained on real data, strongly suppressed the false positive event rate, resulting in the first laser spectroscopy experiment on antihydrogen. The analysis presented constitutes a test of CPT invariance at 0.2 ppb level. The current experimental effort consists in determining the lineshape of the 1S-2S transition, where, once again, the use of machine learning classifiers is essential to perform a high precision measurement.

Acknowledgments
The author wishes to acknowledge the whole ALPHA collaboration for the great team effort that lead to this historical measurement: M Ahmadi, B X R Alves, C J Baker, W A Bertsche, E Butler, C Carruth, C L Cesar, M Charlton, S Cohen, R A Collister, S Eriksson, A Evans, N A Evetts, J Fajans, T Friesen, M C Fujiwara, D R Gill, A Gutierrez, J S Hangst, W N Hardy, M E Hayden, C A Isaac, A Ishida, M A Johnson, S A Jones, S Jonsell, L Kurchaninov, N Madsen, M R Mathers, D Maxwell, J T K McKenna, S Menary, J M Michan, T Momose, J J Munich, P Nolan, K Olchanski, A Olin, P Pusa, C Ø Rasmussen, F Robicheaux, R L Sacramento, M Sameed, E Sarid, D M Silveira, S Stracka, C So, G Stutter, T D Tharp, J E Thompson, R I Thompson, D P van der Werf, J S Wurtele. In particular, SS for introducing the machine learning techniques to ALPHA, and JTKM and AO for their tireless analysis efforts.

References
[1] Charlton M, Eades J, Horváth D, Hughes R J and Zimmermann C 1994 *Phys. Rep.* **241** 65–117
[2] Amole C *et al.* (ALPHA Collaboration) 2012 *Nature* **483** 439–443
[3] Ahmadi M *et al.* (ALPHA Collaboration) 2016 *Nature* **529** 373–376
[4] Ahmadi M *et al.* (ALPHA Collaboration) 2016 *Nature* **541** 506–510
[5] Ahmadi M *et al.* (ALPHA Collaboration) 2017 *Nature* **548** 66–69
[6] Amole C *et al.* (ALPHA Collaboration) 2014 *Nucl. Instrum. Meth.* **A735** 319–340
[7] Andresen G B *et al.* (ALPHA Collaboration) 2012 *Nucl. Instrum. Meth.* **A684** 73–81
[8] Narsky I 2005 *arXiv preprint* **arXiv:physics/0507143**
[9] Punzi G 2003 *arXiv preprint* **arXiv:physics/0308063v2**

IOP Conf. Series: Journal of Physics: Conf. Series **1085** (2018) 042008 doi:10.1088/1742-6596/1085/4/042008

Novel functional and distributed approaches to data analysis available in ROOT

G. Amadio[1], J. Blomer[1], P. Canal[2], G. Ganis[1], E. Guiraud[1,3], P. Mato Vila[1] L. Moneta[1], D. Piparo[1], E. Tejedor[1], X. Valls Pla[1,4],
[1]CERN, [2]FNAL, [3]University of Oldenburg, [4]University Jaume I

E-mail: `root-dev@cern.ch`

Abstract. The bright future of particle physics at the Energy and Intensity frontiers poses exciting challenges to the scientific software community. The traditional strategies for processing and analysing data are evolving in order to (i) offer higher-level programming model approaches and (ii) exploit parallelism to cope with the ever increasing complexity and size of the datasets. This contribution describes how the ROOT framework, a cornerstone of software stacks dedicated to particle physics, is preparing to provide adequate solutions for the analysis of large amount of scientific data on parallel architectures.

The functional approach to parallel data analysis provided with the ROOT TDataFrame interface is then characterised. The design choices behind this new interface are described also comparing with other widely adopted tools such as Pandas and Apache Spark. The programming model is illustrated highlighting the reduction of boilerplate code, composability of the actions and data transformations as well as the capabilities of dealing with different data sources such as ROOT, JSON, CSV or databases. Details are given about how the functional approach allows transparent implicit parallelisation of the chain of operations specified by the user.

The progress done in the field of distributed analysis is examined. In particular, the power of the integration of ROOT with Apache Spark via the PyROOT interface is shown.

In addition, the building blocks for the expression of parallelism in ROOT are briefly characterised together with the structural changes applied in the building and testing infrastructure which were necessary to put them in production.

1. Future challenges for analysis at the Energy and Intensity frontiers

The full exploitation of the LHC is the highest priority in the European Strategy for Particle Physics adopted as part of the ESFRI Roadmap [1]. A major upgrade of the LHC will take place in about three years from now. The luminosity delivered by the machine will be about ten times higher than the present, nominal one.

Such an evolution poses a series of challenges to us for what concerns all steps of the HEP data processing chain: triggering, reconstruction, simulation, digitisation and analysis.

Assuming the performance of the software presently in use for HEP data processing and a reasonable evolution of hardware technologies, the amount of resources the LHC community will need to cope with the aforementioned computation requirements will have to be nearly about ten times bigger than what it is today [2]. This is clearly not realistic.

In addition, a sizable contribution to the need of computing resources is to be accounted for the High Intensity Frontier program: about 10% of the budget presently required by a LHC experiment today [3].

IOP Conf. Series: Journal of Physics: Conf. Series **1085** (2018) 042008 doi:10.1088/1742-6596/1085/4/042008

From the point of view of the software, a better exploitation of resources goes hand in hand with parallelisation. However, for this approach to succeed, the intrinsic complexity of software parallelisation needs to be hidden from the user.

This paper proposes a new technique to optimise the usage of resources of the last step of the HEP data processing: analysis. A declarative approach offered by the ROOT [4] framework is discussed: TDataFrame [5]. Its cornerstones are the exploitation of parallel architectures, including accelerators, as well as a straightforward interface for scientists: the effort expended on achieving high parallel processing rates is wasted unless accompanied by outstanding achievements in the simplification of the programming model.

2. A declarative approach to HEP data analysis: TDataFrame

Starting from release 6.10, ROOT provides a tool to interact with its columnar datasets in a declarative way: TDataFrame.

The approach draws inspiration from other widely adopted data analysis frameworks such as Pandas [6] or Spark [7]. In the context of TDataFrame, an analysis is made of two main kinds of operations: *transformations* and *actions*. Examples of *transformations* are the application of a filter to select entries, the creation of a new column, also based on the content of other existing columns, the caching of the dataset in memory or its writing on disk. Examples of *actions* are the creation of a histogram, counting entries and extracting the content of columns. Transformations and actions are chained to achieve the desired output results. The approach is similar to the one of the more traditional data processing frameworks such as Gaudi [8]: the event loop is not controlled by the user but the data processing units are orchestrated by the framework to manipulate the data (see listings 1).

```
// Traditional approach, explicit loop
TTreeReader data(tree);
TTreeReaderValue<A> x(data, "x");
TTreeReaderValue<A> y(data, "y");
TTreeReaderValue<A> z(data, "z");
while(data.Next()) if(IsGoodEvent(*x, *y, *z)) h.Fill(*x);
h.Draw();

// Hidden event loop, declarative style
TDataFrame d(tree);
auto h = d.Filter(IsGoodEvent, {"x","y","z"}).Histo1D("x");
h.Draw();
```

Listing 1. A TDataFrame usage example comparing to the previously existing handle to access columnar datasets: TTreeReader. The code illustrates how straightforward it is to access the data, to apply a filter and to create a histogram.

Historically, TDataFrame was born as a tool to implement *functional chains*. It was soon realised that the tool could have been much more powerful if it had been able to represent "functional graphs". For example, the code snippet 2 can be graphically represented as figure 1:

(i) an interface is created to the dataset: the TDataFrame,

(ii) a filter, expressed in the form of a lambda, is applied to it,

(iii) a new column, z is added for the entries passing the filter,

(iv) two histograms are created depending on this last node.

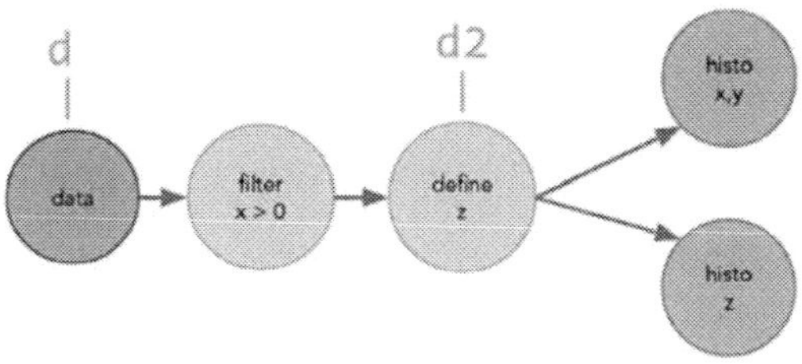

Figure 1. Complex control flows, which for example incarnates a series of cuts on collision events, can be concisely and expressively represented with TDataFrame.

```
// d2 is a new data-frame, a transformed version of the d
auto d2 = d.Filter([](const A& x){return x > 0;}, {"x"})
           .Define("z", "x*x + y*y");

// make multiple histograms out of it
auto hz = d2.Histo1D("z");
auto hxy = d2.Histo2D("x","y");
```

Listing 2. Intermediate nodes can be saved into temporary variables for the comfort of the user. Predicates associated to a transformation such as a filter can be expressed as a *callable* or a C++ string which is jitted by the ROOT interpreter and associated to the necessary columns on behalf of the user provided that the variables have the same name in the dataset.

It should be noted that all transformations and actions are executed in the same loop over all entries: it is therefore possible to create an arbitrary number of histograms or other quantities in a single event loop, in contrast with more traditional approaches such as the one offered by ROOT `TTree::Draw`.

One of the ideas behind TDataFrame is therefore to express an analysis as a "pipeline" of steps to be performed on the input data: the user chooses *what* has to be run while ROOT takes care of *how* the calculation is structured, for example for what concerns optimisations. For instance, the parallel runtime of ROOT is exploited so to run in parallel the analysis on behalf of the user transparently: the same code written for the sequential case is executed on all the cores available by the ROOT parallel runtime.

A serious investment was made in the programming model offered to the scientists by the TDataFrame interface. The APIs were designed respecting modern C++ best practices. In addition, the Python ROOT interface, PyROOT [9], was also considered during every stage of the design in order to provide seamless access to the functionality offered in the two languages by construction. In order to guarantee top performance, template metaprogramming techniques were aggressively adopted in the implementation. As a result, desirable behaviours such as absence of virtual function calls as well as type-safe access to the datasets are guaranteed.

3. Benchmarks

Besides usability, the principal added value of an approach like the one discussed in this contribution is performance and scaling on many cores architectures.

Figure 2 shows the behaviour of TDataFrame when dealing with a real-life study of the effect of parton distribution functions adopted by the Monte Carlo generator on the measured mass of the W boson. In this particular case, eight kinematic and quality cuts were applied and more than a thousand tridimensional histograms (70x10x10 bins) were filled. The input dataset was a flat n-tuple contained in a compressed 2.5 GB ROOT file. The times measured to calculate the scaling are *real times* needed by the program to execute, therefore including startup time, histogram merging as well as tear-down of the application: in other words, the time a user would have to wait to see the result of the study.

IOP Conf. Series: Journal of Physics: Conf. Series **1085** (2018) 042008 doi:10.1088/1742-6596/1085/4/042008

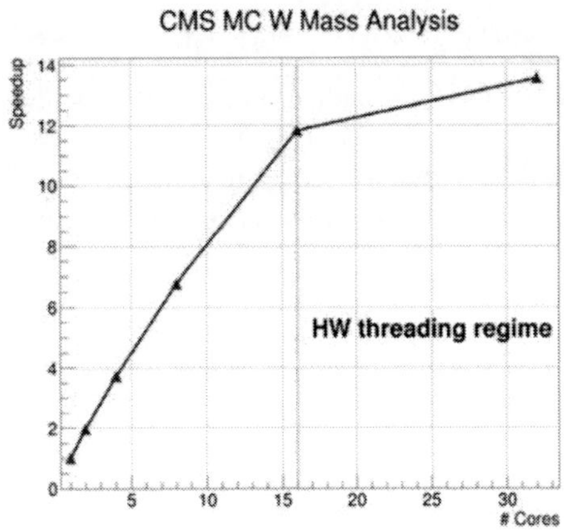

Figure 2. Scaling of a real-life application based on TDataFrame, running on an a machine equipped with Intel Xeon CPU E5-2650 v2 @ 2.60GHz equipped with 32 logical cores. Despite the presence of NUMA domains an encouraging scaling is achieved. Thanks to M. Dunser for providing the original code of this benchmark.

A comparison of different approaches to analysis of HEP data stored in ROOT columnar format has been performed. A sample study on a LHCb open dataset was examined, considering three different ways of accessing the data, compressed and uncompressed, namely: the usage of the bare TTree interface, the adoption of the TTreeReader tool, the exploitation of TDataFrame, in sequential and parallel mode. The results are shown in figure 3.

The TDataFrame approach is slightly slower than the TTreeReader one. The former, in its implementation, relies on the latter and adds some overhead (which is planned to be removed in the 6.12 ROOT release). The parallelised approach, possible only with TDataFrame, shows that with no additional effort required by the user, the full exploitation of the resources of the machine can be achieved.

Parallelisation can be achieved also with multiple processes, for instance running on a multicore computer or a batch farm, at the price of imposing a final, sequential, merging of the results. The kind of multithreading made accessible by TDataFrame avoids this last onerous step.

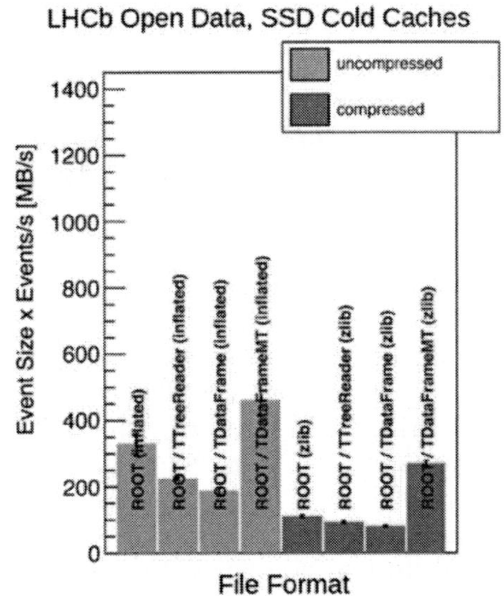

Figure 3. Amount of collision data processed for different file formats on a commodity laptop featuring an Intel i7 CPU with eight logical cores. The machine is equipped with an SSD. Considering the simplifications introduced by the programming model of TDataFrame and the possibility to parallelise an existing analysis on all cores available with no effort, the advantage over more traditional methods is clear. The overhead imposed by TDataFrame and TTreeReader over the bare TTree usage is understood and planned to be removed in a forthcoming ROOT releases.

4. The PyROOTSpark Project

The declarative approach described above is not specific to TDataFrame. The PyROOTSpark R&D project [10] was initiated in order to be able to exploit Spark [7] clusters via a slim Python interface. The underlying principle is shown in listing 3, i.e. to allow the design of a ROOT based data analysis designed around the map-reduce [11] pattern.

```
import ROOT
from DistROOT import DistTree
dTree = DistTree(filelist = ["myFile1", "myFile2"],
                 treename = "myTree", npartitions = 8)
myHistos = dTree.ProcessAndMerge(fillHistos, mergeHistos)
```

Listing 3. The map and reduce functions are executed by Spark on ROOT datasets. The Spark engine is used as a scheduler for the work. PyROOT guarantees the communication between C++ and Python, whereas PySpark the connection between Python and the Spark interface.

With HEP workflows running on Spark clusters in mind, an advanced and elegant monitoring system has also been developed to inspect the status of the jobs [12].

5. Conclusions and Outlook

A novel, declarative approach to ROOT based data analysis has been described: TDataFrame. The conjunction of TDataFrame's programming model and the possibility to run the same analysis in sequential and parallel mode without any effort required by the user makes the approach particularly powerful. With a few lines of code a scientist can apply filters to a columnar dataset, create histograms, create new columns, cache in memory or save the dataset on disk. A well behaved scaling of real-life usecases has been demonstrated on many cores servers as well as commodity laptops. The declarative approach is also being pursued in the context of PyROOTSpark, which offers a minimal interface to run ROOT based analyses orchestrating the simultaneous usage of ROOT, Python and Spark thanks to PyROOT and PySpark.

In the future, TDataFrame will evolve in order to offer an even better integration with PyROOT, giving for example the possibility to specify cuts as Python functions, and more handles to orchestrate parallelism such as an efficient *task local* storage. PyROOTSpark will be made available to the community as part of the CERN notebook service [13] in order to allow CERN scientists to transparently take advantage of Spark resources with limited knowledge of the technology and a minimal interface.

References

[1] Commission E 2017 European strategy forum on research infrastructures URL https://ec.europa.eu/research/infrastructures/index_en.cfm
[2] Bird I 2016 Wlcg and experiment computing for run ii URL https://indico.cern.ch/event/563488
[3] Fuess S, Gutsche O, Kirby M, Kutschke R, Lyon A, Norman A, Perdue G and Sexton-Kennedy E 2015 *Journal of Physics: Conference Series* **664** 032012 URL http://stacks.iop.org/1742-6596/664/i=3/a=032012
[4] Brun R and Rademakers F 1997 *Proceedings AIHENP'96 Workshop* vol A389 ed in Phys N I M pp 81–86 URL http://root.cern
[5] Guiraud E, Naumann A and Piparo D 2017 TDataFrame: functional chains for ROOT data analyses URL https://doi.org/10.5281/zenodo.260230
[6] McKinney W 2010 *Proceedings of the 9th Python in Science Conference* ed van der Walt S and Millman J pp 51 – 56
[7] Zaharia M, Xin R S, Wendell P, Das T, Armbrust M, Dave A, Meng X, Rosen J, Venkataraman S, Franklin M J, Ghodsi A, Gonzalez J, Shenker S and Stoica I 2016 *Commun. ACM* **59** 56–65 ISSN 0001-0782 URL http://doi.acm.org/10.1145/2934664
[8] Barrand G, Belyaev I, Binko P, Cattaneo M, Chytracek R, Corti G, Frank M, Gracia G, Harvey J, Herwijnen E, Maley P, Mato P, Probst S and Ranjard F 2001 **140** 45–55
[9] Generowicz J, T L P Lavrijsen W, Marino M and Mato P 2005

[10] Tejedor E and Piparo D 2017 Pyrootspark Accessible at: https://github.com/etejedor/root- spark URL `https://doi.org/10.5281/zenodo.1010129`
[11] Dean J and Ghemawat S 2008 *Commun. ACM* **51** 107–113 ISSN 0001-0782 URL `http://doi.acm.org/10.1145/1327452.1327492`
[12] R K, Tejedor E and Piparo D 2017 Spark monitor 0.0.8 URL `https://doi.org/10.5281/zenodo.1012585`
[13] Piparo D, Tejedor E, Mato P, Mascetti L, Moscicki J and Lamanna M 2018 *Future Generation Computer Systems* **78** 1071 – 1078 ISSN 0167-739X

IOP Conf. Series: Journal of Physics: Conf. Series **1085** (2018) 042009 doi:10.1088/1742-6596/1085/4/042009

Speeding up prediction performance of BDT-based models

E. Khairullin[1,3] and A. Ustyuzhanin[2,3]

[1] Moscow Institute of Physics and Technology, Institutskiy per. 9, Dolgoprudny, Moscow Region, 141700, Russia

[2] National Research University Higher School of Economics, 20 Myasnitskaya st., Moscow 101000, Russia

[3] Yandex School of Data Analysis, 11/2, Timura Frunze st., Moscow 119021, Russia

E-mail: mikari@yandex-team.ru

Abstract. The outcome of a machine learning algorithm is a prediction model. Typically, these models are computationally expensive, where improving of the quality the prediction leads to a decrease in the inference speed. However it is not always tradeoff between quality and speed. In this paper we show it is possible to speed up the model by using additional memory without losing significat prediction quality for a novel boosted trees algorithm called CatBoost. The idea is to combine two approaches: training fewer trees and merging trees into a kind of hashmaps called DecisionTensors. The proposed method allows for pareto-optimal reduction of the computational complexity of the decision tree model with regard to the quality of the model. In the considered example the number of lookups was decreased from 5000 to only 6 (speedup factor of 1000) while AUC score of the model was reduced by less than 10^{-3}.

1. Introduction

Today, Machine Learning (ML) approaches are used in various fields ranging from the information retrieval in the Internet, to speech recognition, to optimizatimg steel compositions, to the search for the New Physics. The resources (memory, CPU, network etc.) can be very limited when solving tasks in production settings.

For most ML algorithms, improving the quality of the model usually leads to higher levels of resource consumption. For example, when using tree-based boosting algorithms, the increase in the number of trees from tens to thousands usually leads to significant increases in the quality, the model size and inference time. Higher demands for the computational resources of the more accurate models are usually satisfied by various technical optimizations of the model storage mechanism. For example vectorisation of the tree weights storage can be used. Another approach is to modify the training process so that it creates models with certain properties that will accelerate their application without maintaining the required level of quality.

For example, the Topological Trigger in the CERN-based LHCb experiment has very strict application time restrictions. In [2], the original MatrixNet [1] model allowed for significant improvements in quality. However, to get acceptable applying time the model was coarsed and converted to a Bonsai Boosted Decision Tree format [4].

In this paper, we consider a modification to the training process in the decision tree-based algorithm CatBoost. CatBoost is a successor of MatrixNet [3]. Our approach is also applicable

IOP Conf. Series: Journal of Physics: Conf. Series **1085** (2018) 042009 doi:10.1088/1742-6596/1085/4/042009

to other boosted decision tree-based algorithms.

2. CatBoost

CatBoost is the implementation of a tree-based gradient boosting in which oblivious decision trees are used [3]. The values of each feature i are discretized into baskets using the boundaries $B_i = b_{i1}, ..., b_{ip_i}$, where p_i is the number of boundaries. The boundaries themselves are determined in advance using various approaches. The original feature vector with real values is replaced by a vector of binary values 0 and 1 ($false$ and $true$). The value of the i-th attribute f_i is replaced with a binary vector $G_i = g_{i1}, ..., g_{ip_i}$ where $g_{ij} = f_i > b_{ij}$. Finally, all G_i are concatenated into one vector. Consecutively, the resulting decision will a make bunch of boolean checks instead of comparing some real feature value with the thresholds.

The number of baskets used for the discretization can be controlled by specifying the appropriate parameter during the training procedure.

The learning rate is another important training characteristic in boosted decision-tree methods. The smaller it is, the smaller the contribution of each next tree in the ensemble. However, this has to be compensated by a large number of trees and vice versa. The learning rate can be chosen by the cross-validation process given the fixed number of trees. Usually, the more trees the higher the model quality.

CatBoost shows already acceptable quality with default parameters, so we variate only number of trees (and learning rate to compensate) because number of trees affects time. The other parameters were left at their default values.

3. Decision Tensor

One of the simplest ways to accelerate computational performance is the preliminary calculation. The feature vector obtained after the discretization takes a limited number of values. Therefore, we can precompute the predictions for all possible value combinations.

Let n be the number of features. Let us consider some model Z, trained by some algorithm, containing a set of trees T. For convenience, we assume that the values of every feature belongs to the interval $[0; 1]$. In other words, every feature vectors lies in n-dimensional cube $[0; 1]$.

Let B_i be the a of basket boundaries for the i-th feature, which was used in T. Then we can split the hyper-cube $[0; 1]$ into small hyper-parallelepipeds. The prediction of the model is constant within each hyperparallelepiped. In fact, the model Z allows you to calculate the value at any point of our hypercube. If we pre-calculate the value for each piece, then we will accelerate the application of our model, effectively reducing it to the time required to search a piece that contains the point. We called this approach a Decision Tensor. The asymptotic complexity of the application of the tree-based mode is $(h \cdot t)$ where h is the height of the trees, t is the total number of the trees. Asymptotic complexity of Tensor is $(n \cdot \log(b))$ where n is the number of the features and b is the number of cuts per feature (this complexity is estimated using a binary search). It should be noted for a small b a simple linear search could be slightly faster that a binary search. With a small number of features the tensor-based prediction may be computed much faster than the tree-based model.

However, Decision Tensors have a larger memory footprint. It needs $O(b^n)$ bytes of memory and it may be too large for common computer. Therefore, the precomputed tensor can only be used for small b and n.

In the general, the size of one tensor D can be estimated as follows:

$$D = (B_1, ..., B_n),$$

$$B_i = \{b_{i1}, ..., b_{ip_i}\},$$

IOP Conf. Series: Journal of Physics: Conf. Series **1085** (2018) 042009 doi:10.1088/1742-6596/1085/4/042009

$$S(D) \propto \prod |B_i| = \prod p_i$$

Consider the case for small n. We can get a small b in two main ways: by limiting the number of cuts for each feature and training the model with a large enough tree count or by severely limiting the tree count. In the latter case, the total number of effective cuts will be exactly $h \cdot t$ without taking into account repeated use of the same cuts.

The first method was can be found in [2]: which used a trained MatrixNet (ancestor of CatBoost) model with a small number of cuts that has been precomputed into a Decision Tensor. The consumption of memory for a single tensor can be extremely huge provided that acceptable quality is achieved, so the natural extension of the idea is the construction of Decision Tensors ensemble but with some deterioration of applying time.

4. Tensor Similarity

Let's introduce the metric for the similarity of two tensors. Suppose, we have two tensors: $D^1 = (B_1^1, ..., B_n^1)$ and $D^2 = (B_1^2, ..., B_n^2)$. The sum $D^u = D^1 + D^2$ could be defined as $D^u = (f_1^u, ..., f_f^u)$, where $f_i^u = f_i^1 \cup f_i^2$. Now, let's define similarity as the difference between the sum of their sizes and the size of their union:

$$Sim(D^1, D^2) = S(D^1) + S(D^2) - S(D^1 + D^2)$$

5. Algorithm for combining set of Decision Trees to a set of Decision Tensors

Suppose we have T trees and we want to combine them into M Decision Tensors. The efficiency metric of the partition is the resulting memory footprint of the tensor set. In general, this is not an easy task and it requires a separate study. We used the following heuristic approach:

(1) Initialize M Tensors by placing in each one an arbitrarily chosen tree
(2) Choose any of the remaining trees
(3) Combine it with the most similar tensor
(4) Repeat steps 2-3 until there are no unused trees left.

Such reduction of the set of trees to a set of Tensors we call partitioning.

6. A complete algorithm for constructing an Decision Tensors Ensemble

There are two key approaches:

- We use a rough partitioning by a small number of cuts and train an arbitrarily large number of trees.
- The partitioning is done quite detailed, but at the same time we significantly reduce the number of trained trees, thereby reducing the number of actually used buckets.

The algorithm actively uses both of these ideas.

(1) Determine time and size constraints
(2) Determine the desired number of tensors (M) based on time constraints
(3) Train the model with a limited number of buckets
(4) Merge the trees from the obtained model into the first Decision Tensor from the ensemble
(5) Choose another set of hyperparameters using a random search
    - Train the model, using the previously trained model as a baseline

IOP Conf. Series: Journal of Physics: Conf. Series **1085** (2018) 042009 doi:10.1088/1742-6596/1085/4/042009

- Transform it into M-1 Decision Tensors

(6) Remove ensembles that do not fit the required size limits

(7) Choose the ensemble with the best quality on an additional data sample.

This algorithm identifies the Pareto-optimal ensemble of Decision Tensors.

7. Real problem illustration

We have used CatBoost as a main algorithm and also took the dataset and the model from [2]. Number of features in the dataset is 10, the model was trained with 5000 trees and 64 buckets. We compared our algorithm for constructing Decision Tensors Ensemble with the trivial approach: training a small number of trees but with high value of the training speed. For this we divided the data set into three parts: training (50%), test (25%) and validation (25%). We trained the models on the training sample, then used the test for the selection of the best models under given constraints.

We have limited the number of tensors to 6 and the total size of the models to one gigabyte, so every prediction requires 4 bytes (float). We have scanned only through the key CatBoost hyperparameters, i.e. number of trees and the learning rate.

The figure of merit we've taken also from [2], which is the area under the ROC curve where False Positive Rate less than 0.05. For the convenience, we have normalized this metric so that its maximum value (with an ideal model) is equal to one.

8. Discussion

We compared the CatBoost model with 5000 trees, the fast CatBoost models for 1-125 trees and the Decision Tensors Ensembles (DTE) with different memory limits and different numbers of tensors.

Figure 1 shows the dependence of the quality on the number of lookups. Figure 2 shows the quality versus the total inference time at the full validation dataset. Table 1 shows properties of the models trained in the process.

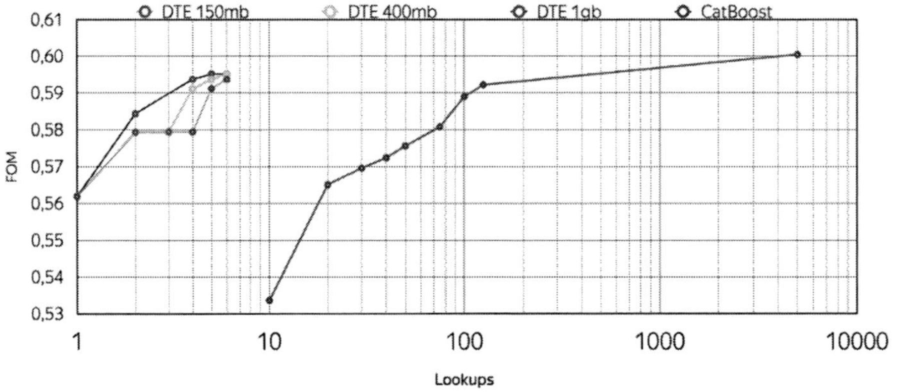

Figure 1. Lookups count versus FOM for different models sizes

It can be seen that at a comparable quality with fast CatBoost models (1-125 trees), DTE allows a 25 times reduction in the number of lookups. And in comparison with the best model (5000 trees), the number of lookups is 1000 times smaller with a small loss in quality (< 0.006 with the best quality of 0.6).

IOP Conf. Series: Journal of Physics: Conf. Series **1085** (2018) 042009 doi:10.1088/1742-6596/1085/4/042009

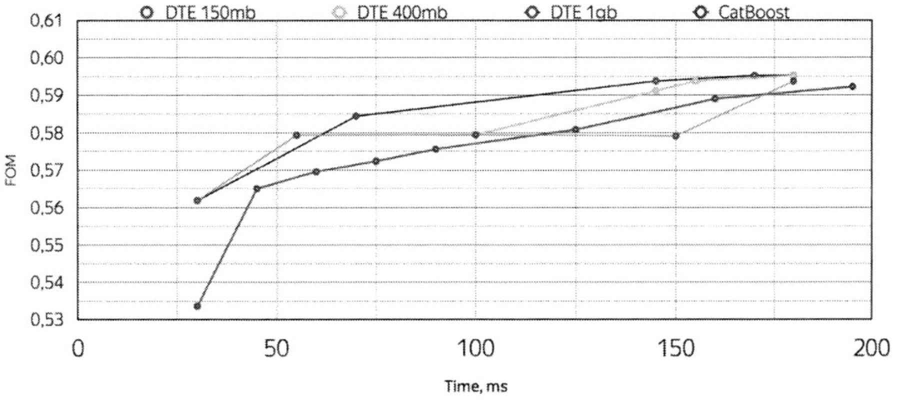

Figure 2. Application time versus FOM for different models sizes

Table 1. Properties of the some models trained in the process

Model	Roc	Time, ms	Size, Mb
catboost-5000	0.6004	10000	< 5
catboost-100	0.589	160	< 5
catboost-75	0.5808	125	< 5
catboost-10	0.5336	30	< 5
big dte-6	0.5952	180	1000
big dte-3	0.5938	145	1000
big dte-2	0.5844	70	1000
small dte-2	0.5794	55	150
dte-1	0.5618	30	150

However, the difference in working time is not so significant - 2.5 times and 50 times, respectively. This is due to the principles of the CPU caching and the size of our dataset. Since the Decision Tensor occupies significant amount of memory and does not fit into the CPU cache almost every lookup results in reading from the RAM, which bottlenecks the working time. CatBoost models are small enough to fit into the cache. There was a single operation type of model application in our tests, so the CatBoost models stayed persistently in the cache. In real applications, CatBoost model will most likely not stay in the cache alone and will be even slower because of this. On the contrary the performance of the Decision Tensor Ensemble will not degrade due to that fact.

9. Conclusion

Decision Tensor Ensembles allow you to speed up the inference by complex Boosted Decision Trees algorithms (such as CatBoost) in real-time systems. The algorithm considered for building tensor ensembles allows for additional tradeoff between memory and the inference speed without significant loss of the model quality. In addition, the use of the Ensemble can significantly improve the model quality in comparison with a single Decision Tensor and can be used in the LHCb experiment triggers.

IOP Conf. Series: Journal of Physics: Conf. Series **1085** (2018) 042009 doi:10.1088/1742-6596/1085/4/042009

References

[1] Gulin A, Kuralenok I, Pavlov D 2011 Winning The Transfer Learning Track of Yahoo!'s Learning To Rank Challenge with YetiRank *Yahoo! Learning to Rank Challenge* p 63–76

[2] Likhomanenko T, Ilten P, Khairullin E, Rogozhnikov A, Ustyuzhanin A, Williams M 2015 LHCb Topological Trigger Reoptimization (IOP Publishing) *Journal of Physics: Conference Series* Vol. 664, No. 8, p. 082025

[3] Dorogush A, Gulin A, Gusev G, Kazeev N, Ostroumova L, Vorobev A Fighting biases with dynamic boosting arXiv:1706.09516

[4] Gligorov V, Williams M 2013 Efficient, reliable and fast high-level triggering using a bonsai boosted decision tree *Journal of Instrumentation* 8 P02013

ACAT2017 IOP Publishing
IOP Conf. Series: Journal of Physics: Conf. Series **1085** (2018) 042010 doi:10.1088/1742-6596/1085/4/042010

Learning to Remove Pileup at the LHC with Jet Images

Patrick T. Komiske,[a] **Eric M. Metodiev,**[a] **Benjamin Nachman,**[b] **and Matthew D. Schwartz**[c]

[a] Center for Theoretical Physics, Massachusetts Institute of Technology, Cambridge, MA 02139, USA
[b] Physics Division, Lawrence Berkeley National Laboratory, Berkeley, CA 94720, USA
[c] Department of Physics, Harvard University, Cambridge, MA 02138, USA

E-mail: pkomiske@mit.edu, metodiev@mit.edu, bpnachman@lbl.gov, schwartz@physics.harvard.edu

Abstract. We present the Pileup Mitgation with Machine Learning (PUMML) algorithm for pileup removal at the Large Hadron Collider (LHC) based on the jet images framework using state-of-the-art machine learning techniques. We demonstrate that our algorithm outperforms existing methods on a wide range of jet observables up to pileup levels of 140 collisions per bunch crossing. We also investigate what aspects of the event our algorithms are utilizing by understanding the learned parameters of a simplified version of the model.

1. Introduction

The Large Hadron Collider (LHC) is operated at very high instantaneous luminosities to achieve the large statistics required to search for exotic Standard Model (SM) or beyond the SM processes as well as for precision SM measurements. Most collisions are soft, with the protons dissolving into mostly low-energy pions that disperse throughout the detector. A typical collision of this sort at the LHC will contribute about 0.6 GeV/rad^2 of energy [1, 2]. Occasionally, one pair of protons within a bunch crossing collides head-on, producing hard, high-energy radiation of interest. At high luminosity, this hard collision, or leading vertex (LV), is always accompanied by soft proton-proton collisions called pileup. The data collected thus far by ATLAS and CMS have approximately 20 pileup collisions per bunch crossing on average ($\langle \text{NPU} \rangle \sim 20$); the data in Run 3 are expected to contain $\langle \text{NPU} \rangle \sim 80$; and the HL-LHC in Runs 4-5 will have $\langle \text{NPU} \rangle \sim 200$. Mitigating the impact of this extra energy on physical observables is one of the biggest challenges for data analysis at the LHC.

The Pileup Mitigation with Machine Learning (PUMML) algorithm was introduced in Ref. [3], building on the jet images paradigm [4, 5] which has found successful application in jet tagging with convolutional neural networks [6, 7] and generation [8, 9]. To apply the convolutional neural network paradigm to cleaning an image itself, we exploit the finer resolution of the tracking detectors relative to the calorimeters of ATLAS and CMS and the fact that tracking information allows charged particles to be identified as coming from either pileup or the leading vertex. We give as input to our network three-channel jet images: one channel for the charged LV particles, one channel for the charge pileup particles, and one channel, at slightly

IOP Conf. Series: Journal of Physics: Conf. Series **1085** (2018) 042010 doi:10.1088/1742-6596/1085/4/042010

lower resolution, for the total neutral particles. We then train the network to reconstruct the unknown image for LV neutral particles.

We apply the algorithm to $R = 0.4$ anti-k_t jets. The jet image inputs are square grids in pseudorapidity-azimuth (η, ϕ) space of size 0.9×0.9 centered on the charged leading vertex transverse momentum (p_T)-weighted centroid of the jet. To simulate the detector resolutions of charged and neutral calorimeters, charged images are discretized into $\Delta\eta \times \Delta\phi = 0.025 \times 0.025$ pixels and neutral images are discretized into $\Delta\eta \times \Delta\phi = 0.1 \times 0.1$ pixels. We use the following three input channels:

`red` = the transverse momenta of all neutral particles

`green` = the transverse momenta of charged pileup particles

`blue` = transverse momenta of charged leading vertex particles

The output of our network is also an image:

`output` = the transverse momenta of neutral leading vertex particles.

Only charged particles with $p_T > 500$ MeV were included in the green or blue channels. Charged particles not passing this charged reconstruction cut were treated as if they were neutral particles. The different resolutions for charged and neutral particles initially present a challenge, since standard architectures assume identical resolution for each color channel. To avoid this issue, we perform a direct upsampling of each neutral pixel to 4×4 pixels of size $\Delta\eta \times \Delta\phi = 0.025 \times 0.025$ and divide each pixel value by 16 such that the total momentum in the image is unchanged.

The convolutional neural net architecture used in this study took as input 36×36 pixel, three-channel pileup images. Two convolutional layers, each with 10 filters of size 6×6 with 2×2 strides, were used after zero-padding the input images and first convolutional layer with a 2-pixel buffer on all sides. The output of the second layer has size $9 \times 9 \times 10$, with the 9×9 part corresponding to the size of the target output and the 10 corresponding to the number of filters in the second layer. In order to project down to a $9 \times 9 \times 1$ output, a third convolution layer with filter size 1×1 is used. This last 1×1 convolutional layer is a standard scheme for dimensionality reduction. A rectified linear unit (ReLU) activation function was applied at each stage. A schematic of the framework and architecture is shown in Fig. 1. All neural network implementation and training was performed with the python deep learning libraries Keras [10] and Theano [11]. The dataset consisted of 56k pileup images, with a 90%/10% train/test split. He-uniform initialization [12] was used to initialize the model weights. The neural network was trained using the Adam [13] algorithm with a batch size of 50 over 25 epochs with a learning rate of 0.001. The following loss function was used to train PUMML was a modified per-pixel logarithmic squared loss with a value of $\bar{p} = 10$ GeV:

$$\ell = \left\langle \log\left(\frac{p_T^{(\mathrm{pred})} + \bar{p}}{p_T^{(\mathrm{true})} + \bar{p}}\right)^2 \right\rangle. \tag{1}$$

The PUMML architecture is *local* in that the rescaling of a neutral pixel is a function solely of the information in a patch in (η, ϕ)-space around that pixel. The size of this patch can be controlled by tuning the filter sizes and number of layers in the architecture. Further, due to weight-sharing in convolutional layers, the same function is applied for all pixels. Building this locality and translation invariance into the architecture ensures that the algorithm learns a universal pileup mitigation technique, while carrying the benefit of drastically reducing the number of model parameters. Indeed, the PUMML architecture used in this study has only 4,711 parameters, which is small on the scale of deep learning architectures, but serves to highlight the effectiveness of using modern machine learning techniques (such as convolutional layers) in high energy physics without necessarily using large or deep networks.

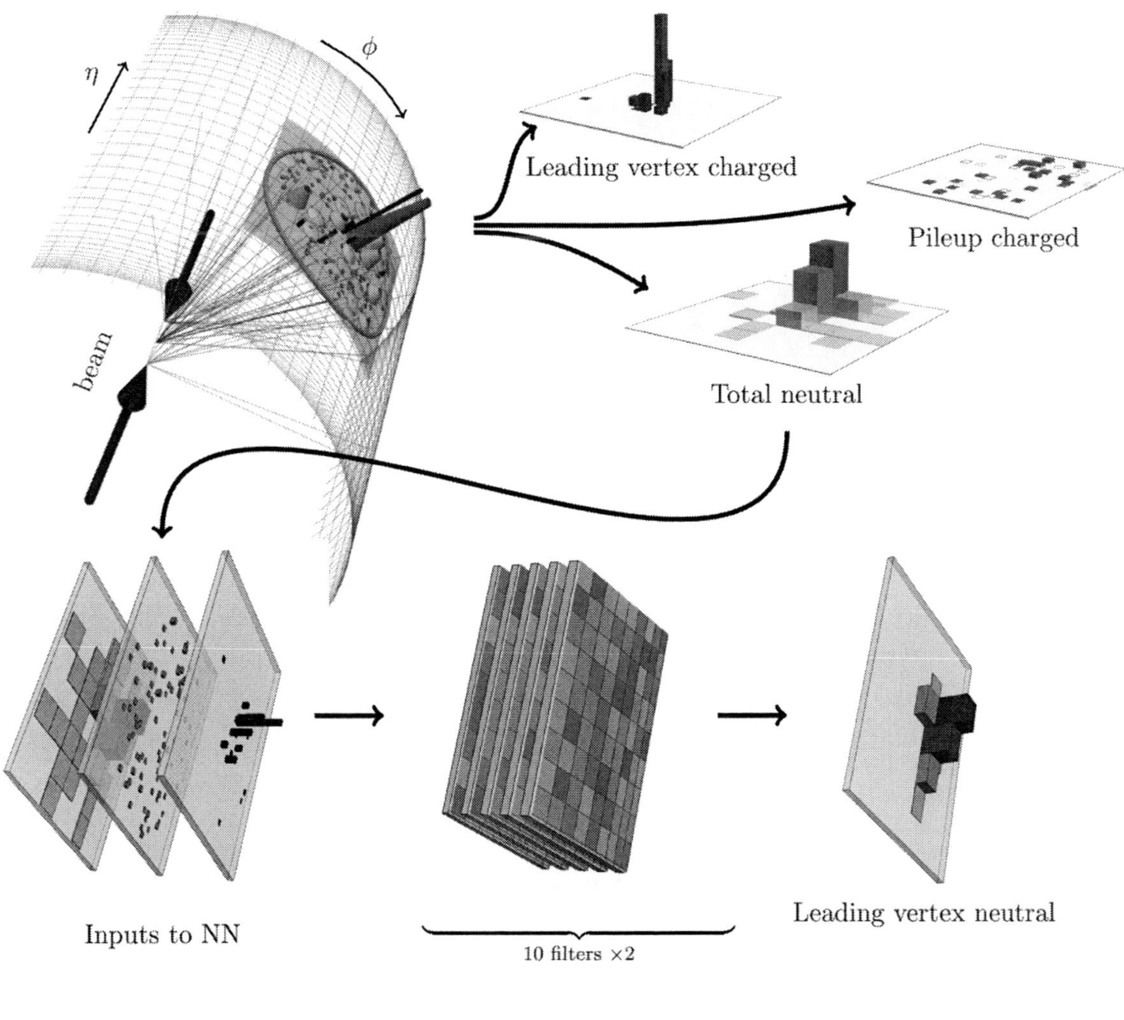

Figure 1: An illustration of the PUMML framework. The input is a three-channel image: blue/purple represents charged radiation from the leading vertex, green is charged pileup radiation, and yellow/orange/red is the total neutral radiation. The resolution of the charged images is higher than for the neutral one. These images are fed into a convolutional layer with several filters whose value at each pixel is a function of a patch around that pixel location in the input images. The output is an image combining the pixels of each filter to one output pixel.

2. Performance

To test the PUMML algorithm, we consider $q\bar{q}$ light-quark-initiated jets coming from the decay of a scalar with mass $m_\phi = 500$ GeV. Events were generated using Pythia 8.183 [14] with the default tune for pp collisions at $\sqrt{s} = 13$ TeV. Pileup was generated by overlaying soft QCD processes onto each event. Final state particles except muons and neutrinos were kept. The events were clustered with FastJet 3.1.3 [15] using the anti-k_t algorithm [16] with a jet radius of $R = 0.4$. A parton-level p_T cut of 95 GeV was applied and up to two leading jets with $p_T > 100$ GeV and $\eta \in [-2.5, 2.5]$ were selected from each event. All particles were taken to be massless.

Samples were generated with the number of pileup vertices ranging from 0 to 180. Since the model must be trained to fix its parameters, the learned model depends on the pileup distribution used for training. For our pileup simulations, we trained on a Poisson distribution of NPUs with mean $\langle\text{NPU}\rangle = 140$. For robustness studies, we also tried training with NPU$=140$ for each event or NPU$=20$ for each event. For comparison, we show the performance of two powerful and widely used constituent-based pileup mitigation methods: PUPPI [17] and SoftKiller [18]. In both cases, default parameter values were used: $R_0 = 0.3$, $R_{\text{min}} = 0.02$, $w_{\text{cut}} = 0.1$, $p_{\text{T}}^{\text{cut}}(\text{NPU}) = 0.1 + 0.007 \times \text{NPU}$ (PUPPI), grid size $= 0.4$ (SoftKiller). Variations in the PUPPI parameters did not yield a large difference in performance. Both PUPPI and SoftKiller were implemented at the particle level and then discretized for comparison with PUMML.

To evaluate the performance of different pileup mitigation techniques, we compute several observables and compare the true values to the corrected values of the observables. To facilitate a comparison with PUMML, which outputs corrected neutral calorimeter cells rather than lists of particles, a detector discretization is applied to the true and reconstructed events. Our comparisons focus on the following four jet observables:

- **Jet Mass**: Invariant mass of the leading jet.
- **Dijet Mass**: Invariant mass of the two leading jets.
- **Energy Correlation Functions**, $\text{ECF}_N^{(\beta)}$ [19]: Specifically, we consider the logarithm of the two- and three-point ECFs with $\beta = 4$.

As a measure of pileup mitigation effectiveness, we show the distributions of the per-event percent error in reconstructing the true values of the observables in Fig. 2. A distribution more peaked around zero indicates better reconstruction performance. PUMML outperforms the other pileup mitigation techniques by this metric, successfully reconstructing jet substructure observables such as the jet mass and the energy correlation functions.

3. Opening the black box

While it is generally very difficult to determine what a network is learning, one possible probe is to examine the weights of the filter layers in the convolutional network. For our full network, these weights are complicated and the subtractor that the network learns is difficult to probe analytically. Instead, we trained a simplified PUMML network with a single 12×12 pixel filter, which spans 3×3 neutral pixels, with no bias term. The different channels of this filter are shown in Fig. 3. The neutral filter clearly selects the relevant neutral pixel for subtraction, while the charged pileup filter is approximately uniform (with the value dependent on the specific choice of loss function and activation function), and the charged leading vertex filter does not significantly contribute.

The filter values motivate the following parameterization of what PUMML is learning:

$$p_T^{\text{N,LV}} = 1.0\, p_T^{\text{N,total}} - \beta\, p_T^{\text{C,PU}} + 0.0\, p_T^{\text{C,LV}}, \tag{2}$$

for some $\mathcal{O}(1)$ constant β, where $p_T^{\text{N,LV}}$, $p_T^{\text{N,total}}$, $p_T^{\text{C,PU}}$, and $p_T^{\text{C,LV}}$ are the neutral-pixel-level transverse momenta of the neutral leading-vertex particles, all neutral particles, charged pileup

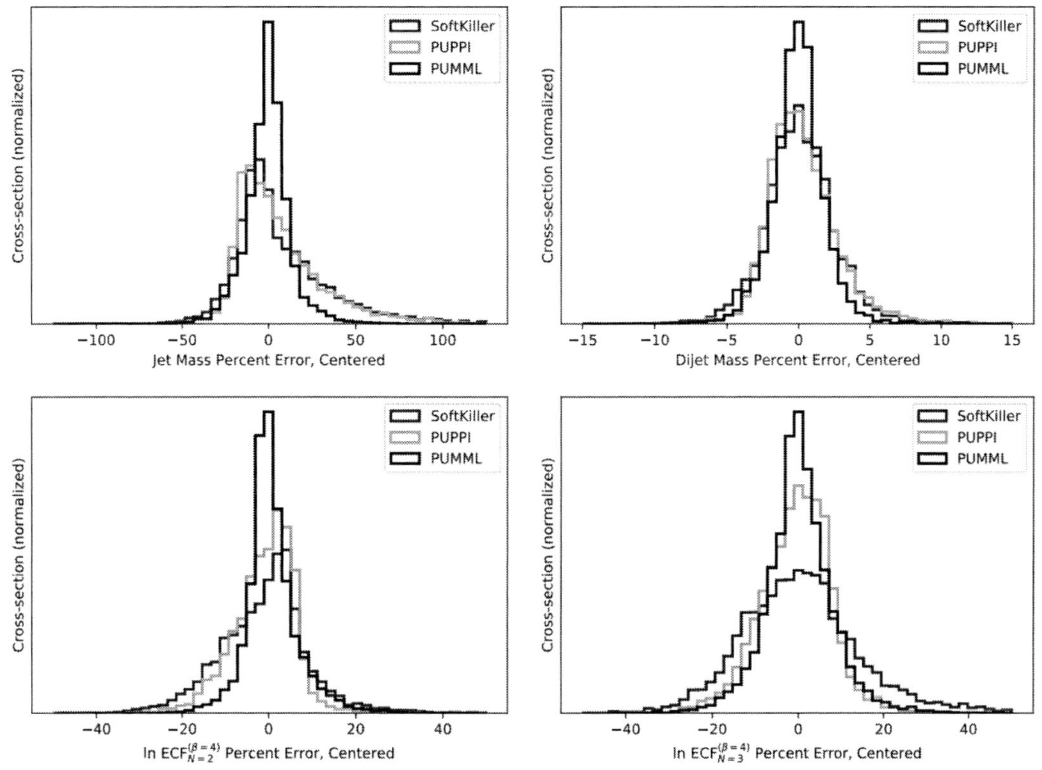

Figure 2: Distributions of the percent error between reconstructed and true values for leading jet mass (top left), dijet mass (top right), ln $\mathrm{ECF}^{(\beta=4)}_{N=2}$ (bottom left), and ln $\mathrm{ECF}^{(\beta=4)}_{N=3}$ (bottom right) for the considered pileup subtraction methods with Poissonian $\langle\mathrm{NPU}\rangle = 140$ pileup. All distributions are centered to have median at 0. The improved reconstruction performance of PUMML is highlighted by its taller and narrower peaks.

particles, and charged leading-vertex particles, respectively. The values 1.0 and 0.0 in Eq. 2 are stable (to the 0.05 level) under variations in the loss and activation functions. This is reassuring as the learned subtractor is thereby robust in the NPU $\to$ 0 limit despite begin trained on $\langle\mathrm{NPU}\rangle = 140$.

Eq. 2 is remarkably similar to the physically-motivated formula used in Jet Cleansing [20]. Cleansing is built on the observation that since pileup is the incoherent sum of many separate scattering events, its variance is smaller than the variance of the radiation from the leading-vertex. Thus, it is better to estimate $p_T^{\mathrm{N,PU}}$ from $p_T^{\mathrm{C,PU}}$ than to estimate $p_T^{\mathrm{N,LV}}$ from $p_T^{\mathrm{C,LV}}$. The simplest form of Cleansing (Linear Cleansing) gives the formula:

$$p_T^{\mathrm{N,LV}} = p_T^{\mathrm{N,tot}} - \left(\frac{1}{\overline{\gamma_0}} - 1\right) p_T^{\mathrm{C,PU}}, \tag{3}$$

where $\overline{\gamma_0}$ is the average ratio of charged p_T to total p_T in a subjet. Thus this simple one 12×12 filter PUMML network is learning a subtractor of the same parametric form as Linear Cleansing.

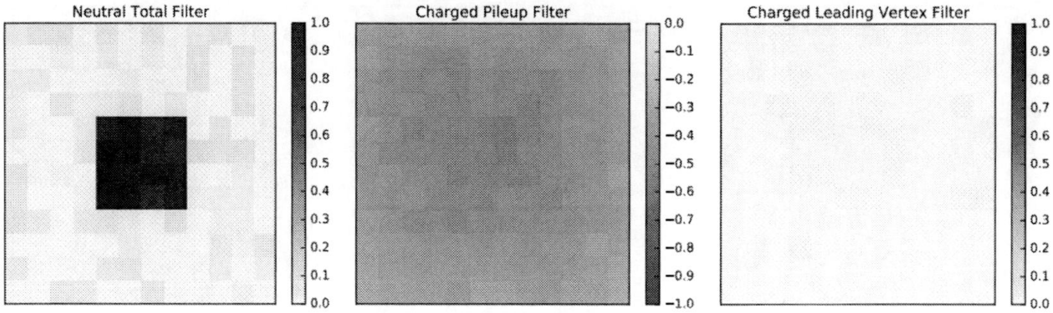

Figure 3: Filter weights for a simple PUMML network with a single 12×12 filter and a ReLU activation function trained with $\langle \text{NPU} \rangle = 140$. The network has selected the relevant neutral pixel, turned off the charged leading vertex contribution, and is using the charged pileup information uniformly.

4. Conclusions

In this paper, we have introduced the first application of machine learning to the critically important problem of pileup mitigation at hadron colliders. We have phrased the problem of pileup mitigation in the language of a machine learning regression problem. The method we introduced, PUMML, takes as input the transverse momentum distribution of charged leading-vertex, charged pileup, and all neutral particles, and outputs the corrected leading vertex neutral energy distribution. We demonstrated that PUMML works at least as well as, and often better than, the competing algorithms PUPPI and SoftKiller in their default implementations.

To prevent the model from learning simulation artifacts, it is preferable to train on actual data rather than simulation. In many machine learning applications in collider physics, obtaining truth-level training samples in data is a substantial challenge. To overcome this challenge in classification tasks, Refs. [21] and [22] introduce an approach to train from mixed samples with different class proportions. For PUMML and pileup mitigation more broadly, a more direct method to train on data is possible. To simulate pileup, we overlay soft QCD events on top of a hard scattering process, both generated with Pythia. Experimentally, there are large samples of minimum bias and zero-bias (i.e. randomly triggered) data. There are also samples of relatively pileup-free events from low luminosity runs. Thus we can construct high-pileup samples using purely data. This kind of *data overlay* approach, which has already been used by experimental groups in other contexts [23, 24], could be perfect for training PUMML with data. Therefore, an implementation of ML-based pileup mitigation in an actual experimental setting could avoid mis-modeling artifacts during training, thus adding more robustness and power to this new tool.

Acknowledgments

The authors would like to thank Philip Harris, Francesco Rubbo, Ariel Schwartzman, Nhan Tran, and Jesse Thaler for helpful conversations. PTK and EMM would like to thank the MIT Physics Department for its support. Computations in this paper were run on the Odyssey cluster supported by the FAS Division of Science, Research Computing Group at Harvard University. This work was supported by the U.S. Department of Energy, Office of Science under contracts DE-AC02-05CH11231 and DE-SC0013607. Additional support was provided by the Harvard Data Science Initiative. Cloud computing resources were provided through a Microsoft Azure for Research award.

IOP Conf. Series: Journal of Physics: Conf. Series **1085** (2018) 042010 doi:10.1088/1742-6596/1085/4/042010

References

[1] Khachatryan V *et al.* (CMS) 2017 *JINST* **12** P02014 (*Preprint* 1607.03663)
[2] Aaboud M *et al.* (ATLAS) 2017 (*Preprint* 1703.09665)
[3] Komiske P T, Metodiev E M, Nachman B and Schwartz M D 2017 (*Preprint* 1707.08600)
[4] Cogan J, Kagan M, Strauss E and Schwarztman A 2015 *JHEP* **02** 118 (*Preprint* 1407.5675)
[5] de Oliveira L, Kagan M, Mackey L, Nachman B and Schwartzman A 2016 *JHEP* **07** 069 (*Preprint* 1511.05190)
[6] Komiske P T, Metodiev E M and Schwartz M D 2017 *JHEP* **01** 110 (*Preprint* 1612.01551)
[7] Kasieczka G, Plehn T, Russell M and Schell T 2017 *JHEP* **05** 006 (*Preprint* 1701.08784)
[8] de Oliveira L, Paganini M and Nachman B 2017 (*Preprint* 1701.05927)
[9] Paganini M, de Oliveira L and Nachman B 2017 (*Preprint* 1705.02355)
[10] Chollet F 2015 Keras
[11] Bergstra J, Breuleux O, Bastien F, Lamblin P, Pascanu R, Desjardins G, Turian J, Warde-Farley D and Bengio Y 2010 *Proc. 9th Python in Science Conf* pp 1–7
[12] He K, Zhang X, Ren S and Sun J 2015 *Proceedings of the IEEE international conference on computer vision* pp 1026–1034
[13] Kingma D and Ba J 2014 *arXiv preprint arXiv:1412.6980*
[14] Sjöstrand T, Ask S, Christiansen J R, Corke R, Desai N, Ilten P, Mrenna S, Prestel S, Rasmussen C O and Skands P Z 2015 *Comput. Phys. Commun.* **191** 159–177 (*Preprint* 1410.3012)
[15] Cacciari M, Salam G P and Soyez G 2012 *Eur. Phys. J.* **C72** 1896 (*Preprint* 1111.6097)
[16] Cacciari M, Salam G P and Soyez G 2008 *JHEP* **04** 063 (*Preprint* 0802.1189)
[17] Bertolini D, Harris P, Low M and Tran N 2014 *JHEP* **10** 059 (*Preprint* 1407.6013)
[18] Cacciari M, Salam G P and Soyez G 2015 *Eur. Phys. J.* **C75** 59 (*Preprint* 1407.0408)
[19] Larkoski A J, Salam G P and Thaler J 2013 *JHEP* **06** 108 (*Preprint* 1305.0007)
[20] Krohn D, Schwartz M D, Low M and Wang L T 2014 *Phys. Rev.* **D90** 065020 (*Preprint* 1309.4777)
[21] Dery L M, Nachman B, Rubbo F and Schwartzman A 2017 *JHEP* **05** 145 (*Preprint* 1702.00414)
[22] Metodiev E M, Nachman B and Thaler J 2017 (*Preprint* 1708.02949)
[23] Marshall Z, Collaboration A *et al.* 2014 *Journal of Physics: Conference Series* vol 513 (IOP Publishing) p 022024
[24] Haas A (ATLAS) 2017 Atlas simulation using real data: Embedding and overlay Tech. rep.

IOP Conf. Series: Journal of Physics: Conf. Series **1085** (2018) 042011 doi:10.1088/1742-6596/1085/4/042011

Analysis Preservation and Systematic Reinterpretation within the ATLAS experiment

Kyle Cranmer, Lukas Heinrich on behalf of the ATLAS collaboration
New York University 726 Broadway, New York, NY 10003, USA
E-mail: lukas.heinrich@cern.ch

Abstract. The LHC data analysis software used in order to derive and publish experimental results is an important asset that is necessary to preserve in order to fully exploit the scientific potential of a given measurement. An important use-case is the re-usability of the analysis procedure in the context of new scientific studies such as the reinterpretation of searches for new physics in terms of signal models that not studied in the original publication (RECAST). We present the usage of the graph-based workflow description language *yadage* to drive the reinterpretation of preserved HEP analyses. The analysis software is preserved using Docker containers, while the workflow structure is preserved using plain JSON documents. This allows the re-execution of complex analysis workflows on modern distributed container orchestration systems and enables a systematic reinterpretation service based on such preserved analysis.

ATL-SOFT-PROC-2017-075
15 December 2017

1. Introduction

The discovery of the Higgs boson by the ATLAS and CMS collaborations[1, 2, 3] during first data-taking period ('Run-1') of the Large Hadron Collider (LHC) has completed the Standard Model of Particle Physics (SM). Besides the continued precision measurements of the Standard Model parameters such as the Higgs boson mass, the search for phenomena predicted by Beyond Standard Model (BSM) theories is a major focus of the LHC collaborations. While there is a large number of plausible BSM scenarios, constraints on both human and computational resources in the collaborations prohibit dedicated data analyses for each proposed extension of the Standard Model. Hence, the collaborations focus on a selection of data analyses targeting complementary final states whose design is guided either by concrete BSM scenarios or so-called "simplified models"[4] acting as proxies for a class of phenomenologically similar theories. No significant excess has been observed thus far in these analyses and typically results are presented as statistical inference results on the parameters of the guiding (simplified) model.

These published limits are not easily translatable into constraints on BSM theories that are not well-aligned with the assumptions of the original BSM model. However, often the analyses are still sensitive to such alternative models, even though they were not optimized for them. This provides an opportunity to still derive limits by *re-interpreting* (or *re-casting*) the analysis in a two-stage procedure:

(i) Generation of simulated events of the new candidate theory, including reconstruction and detector simulation ("Signal Generation")

(ii) A re-execution of the analysis (including event selection and statistical evaluation) of new signal paired with archived background estimates and observed collision data. ("Analysis")

The appeal of this approach is that it allows a broad survey of possible models by efficiently re-using the limited set of analyses. Notably such a reinterpretation is much less resource-intensive than a dedicated analysis for the new candidate theory: It only requires the relatively light-weight simulation of a new signal, while keeping the resource-intensive analysis procedure design as well as the data and SM background estimates fixed.

As most BSM searches are not unfolded, the above recipe can only be accurately implemented by the LHC collaboration itself as both the detector simulation (using Geant4[5]), the analysis pipeline (using the collaboration-specific Event Data Model and analysis software) as well as necessary data inputs are not readily accessible for non-collaboration members. While third-party implementations exist for both signal generation (e.g. via Delphes[6]) and analysis execution via a number of analysis catalogs provided by Rivet[7] and CheckMate [8], those can only provide approximations that rely on published performance and analysis-specific data, such as reconstruction efficiency tables and resolutions. As analyses adopt complex multivariate event selections or non-standard reconstruction objects, such re-implementations may not be feasible anymore. With the well-structured processes inside of the ATLAS collaboration to generate new signals, the barrier to reinterpretation has been in preserving the analysis pipeline in a reusable way.

2. Analysis Preservation via Declarative Workflows and Linux Containers

2.1. Continuous Software Preservation with Container Images

High energy physics analyses are characterized by their highly distributed development model, where individual processing stages such as event selection and the statistical evaluation are provided by sub-teams within the collaboration mostly without centralized knowledge of how to execute the entire analysis pipeline. The preservation problem thus factorized across into two sub-problems:

(i) preserving software and computing environment for individual processing stages

(ii) preserving the workflow logic how to assemble those stages into a coherent analysis pipeline

For the former, the recent advances in Linux Container technology from large industry vendors has enabled a realistic framework of capturing the highly diverse code bases used within an analysis — from systematically developed analysis and event processing frameworks to individual one-off shell scripts — in a standardized, archivable, and portable software environment.

A collaboration-wide effort is now underway to provide base container images adhering to the Open Container Initiative (OCI) image specification [9] along two broad paths: on the one hand as "lean containers" providing the base operating system that are then paired with domain specific high-energy physics (HEP) software taken from the global read-only filesystem CVMFS[10] and on the other hand as "standalone" containers that are independent from an external filesystem. These images are then distributed via a container image registry (such as Docker Hub).

With an increased usage of modern hosted version control services (such as GitHub and GitLab) and their integration with continuous integration services, those images then facilitate the continuous software preservation of individual analysis code that are built into analysis-specific (or sub-stage specific) container images — ensuring that preservation happens in a automated manner *during* the development of the analysis, avoiding the need for post-facto preservation that may be difficult to achieve.

2.2. Declarative Container-based Workflows

Once the individual software environment is preserved, the workflow can be conveniently described in the declarative workflow specification *yadage*. The specification is defined as pure JSON schemas for long-term archivability and native interoperability with the digital library

IOP Conf. Series: Journal of Physics: Conf. Series **1085** (2018) 042011 doi:10.1088/1742-6596/1085/4/042011

software Invenio[11] that is widely used within HEP. The language, described in more detail in reference [12], provides a extensible framework for describing arbitrary, run-time dependent directed acyclic graphs (DAGs) of individual container workloads. Within that framework, individual processing stages are defined as "packaged activities" (packtivities) parametrized such that they can be executed on variable inputs that are provided as a JSON document (see Figure 2), while the workflow DAG is modeled as a number of *stages* that extend a directed acyclic graph by scheduling one or more instances of packtivities (graph nodes) and defining their execution dependencies (graph edges) as soon as all required inputs are known.

2.3. Workflow Archival in the CERN Analysis Preservation Portal

Once the software and workflows are preserved within the collaboration they are ready to be archived within the CERN Analysis Preservation Portal[13] (CAP), a in-progress project for the long-term analysis archival led by the CERN Scientific Information Services division. The portal is designed to hold a wide range of data regarding an analysis, including metadata such as a list of contributors, static data assets in the form of e.g. background estimates (that can be re-used for reinterpretation) as well as source code and software environments. The portal, based on the Invenio platform, will provide a web user interface to search and view preserved analyses.

3. Distributed Workflow Execution using modern Container Orchestration Engines

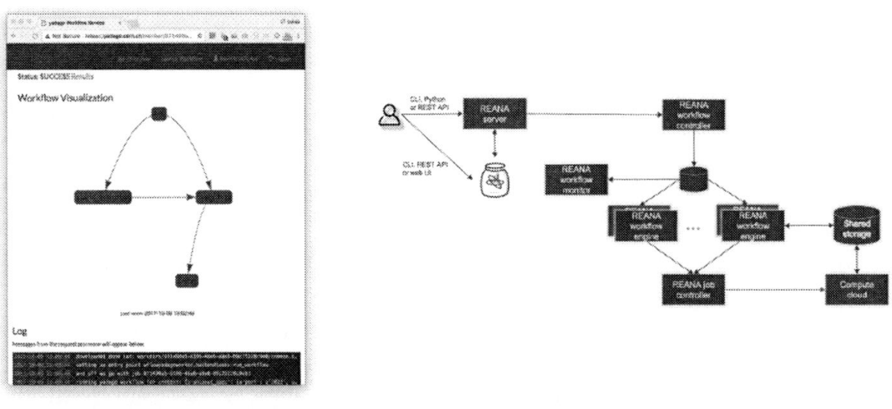

Figure 1. REANA workflow monitor and REANA architecture

With analysis software captured as standard OCI containers the workflows archived in CAP can be executed on distributed computing clusters with a variety of container orchestration tools, such as Kubernetes. These can be conveniently set up both on commercial cloud providers or using private clouds managed by OpenStack such as the CERN OpenStack Installation. In order to facilitate the execution of the workflows stored in the CERN Analysis Preservation Portal, a dedicated workflows-as-a-service platform REANA is being developed as a collaboration between CERN, DASPOS[14] and DIANA-HEP[15], offering collaboration members an easy entry-point to re-execute preserved analysis without managing a container cluster on their own. The platform will handle e.g. access control, storage (e.g. CephFS, EOS), CVMFS access (if needed), monitoring and logging based on industry and community standards. An overview of the architecture is shown in Figure 1.

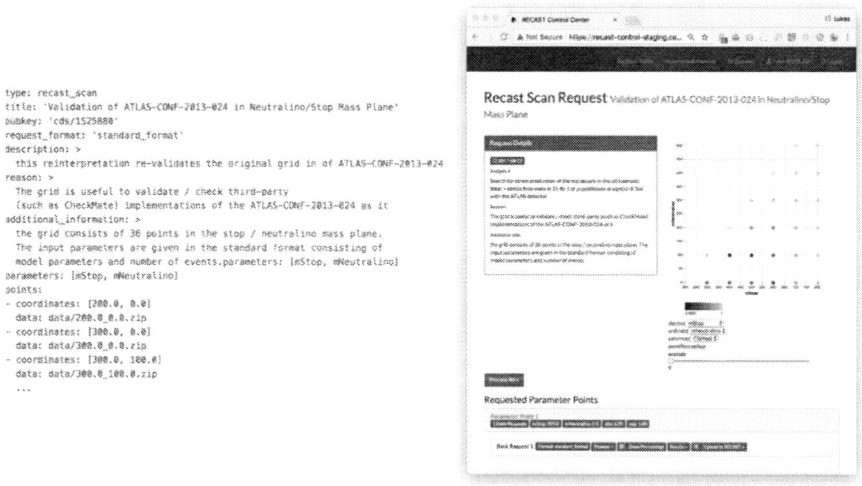

Figure 2. RECAST request and resulting web-based presentation

4. RECAST — leveraging preserved workflows for reinterpretation

Based on CAP as an analysis archive and REANA as an analysis execution engine, the collaboration-internal reinterpretation service can feasibly be implemented. This additional layer connects physics research questions to the generic but physics-unaware backends provided by CAP and REANA through a reinterpretation-specific interface and tools. Reinterpretations are well-posed scientific questions about BSM models that have statistical inference results as their answers. That basic structure is largely invariant to the specific implementing workflows for signal generation and analysis.

Therefore, a web-based interface is being developed to organize both models and inference results of reinterpretations. Reinterpretation requests can be uploaded via both the web interface as well via a command line tool that specifies the details of the reinterpretation such as the analysis to be reinterpreted and input data in the form of a file archive that holds parameter-point specific information. The specific format of the model input can be chosen between one of several options such as dataset identifiers for pre-generated signals available from collaboration-internal storage or Monte Carlo generation input data such as parameter cards and UFO model descriptions[16].

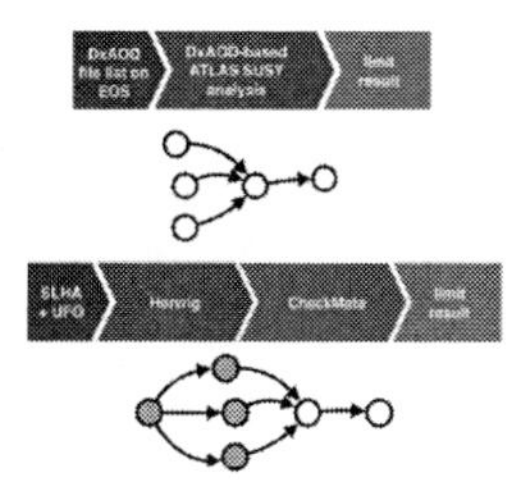

Figure 3. Workflows with and without signal generation.

Depending on the input provided as well as the targeted analysis, the service then selects suitable implementations through a single workflow stored in CAP or a combination of workflows. The former is suitable if the input format is a pre-generated signal such that the analysis can be run directly on the new input, while the latter is chosen for the case where the analysis inputs need to be generated on the fly. In this case the composability of the workflow engine enables combining independently developed workflows for signal generation steps (different Monte Carlo generators, variable simulation settings) with analysis implementations that may include fully preserved ATLAS analyses but also re-implementations such as Rivet or collaboration internal truth-based analyses. The two main modes are visualized in Figure 3.

The desired combination can then be submitted to REANA for processing, where the workflow definitions are retrieved from a workflow repository like CAP and upon completion

IOP Conf. Series: Journal of Physics: Conf. Series **1085** (2018) 042011 doi:10.1088/1742-6596/1085/4/042011

the reinterpretation results such as limits or signal region event yields are transferred back to RECAST. Similarly to the model inputs, results may be produced by the workflows in a variety of ways (text files, JSON documents, ROOT files); RECAST converts those outputs into a standardized format, and presents them on the web interface and the associated command line tools.

5. Future Developments

Containerized workflows have been used for a number of ATLAS reinterpretations[17, 18] and based on these, the described system is being designed for streamlining the internal reinterpretation efforts of the ATLAS experiment by interfacing with the CERN Analysis Preservation portal. As the collaboration gains more experience with both analysis preservation and reinterpretation, it may be possible to open such a service to the wider high-energy community as envisioned by the RECAST proposal [19, 20]. In such a scheme phenomenologists could then upload reinterpretation requests. Depending on resource requirements and availability as well as scientific merit, the collaboration may then choose to fulfill such requests by reusing its preserved analysis and publishing the results on archives such as HepData[21].

References

[1] Aad G *et al.* (ATLAS Collaboration) 2012 *Phys. Lett.* **B716** 1–29 (*Preprint* 1207.7214)

[2] Aad G *et al.* (ATLAS Collaboration) 2008 *Journal of Instrumentation* **3** S08003 URL http://stacks.iop.org/1748-0221/3/i=08/a=S08003

[3] Chatrchyan S *et al.* (CMS) 2012 *Phys. Lett.* **B716** 30–61 (*Preprint* 1207.7235)

[4] Alves D (LHC New Physics Working Group) 2012 *J. Phys.* **G39** 105005 (*Preprint* 1105.2838)

[5] Agostinelli S *et al.* 2003 *Nuclear Instruments and Methods in Physics Research Section A: Accelerators, Spectrometers, Detectors and Associated Equipment* **506** 250 – 303 ISSN 0168-9002 URL http://www.sciencedirect.com/science/article/pii/S0168900203013688

[6] de Favereau J, Delaere C, Demin P, Giammanco A, Lematre V, Mertens A and Selvaggi M (DELPHES 3) 2014 *JHEP* **02** 057 (*Preprint* 1307.6346)

[7] Buckley A, Butterworth J, Lonnblad L, Grellscheid D, Hoeth H, Monk J, Schulz H and Siegert F 2013 *Comput. Phys. Commun.* **184** 2803–2819 (*Preprint* 1003.0694)

[8] Drees M, Dreiner H, Schmeier D, Tattersall J and Kim J S 2015 *Comput. Phys. Commun.* **187** 227–265 (*Preprint* 1312.2591)

[9] OCI 2017 Open Container Initiative URL https://www.opencontainers.org

[10] Blomer J, Buncic P and Fuhrmann T 2011 *Proceedings of the First International Workshop on Network-aware Data Management* NDM '11 (New York, NY, USA: ACM) pp 49–56 ISBN 978-1-4503-1132-8 URL http://doi.acm.org/10.1145/2110217.2110225

[11] Kuncar J, Nielsen L H and Simko T 2014 Invenio v2.0: A Pythonic Framework for Large-Scale Digital Libraries Tech. Rep. ATLAS-CONF-2016-033 CERN Geneva URL http://urn.fi/URN:NBN:fi-fe2014070432294

[12] Cranmer K and Heinrich L 2017 (*Preprint* 1706.01878)

[13] Chen X, Dallmeier-Tiessen S, Dani A, Dasler R, Fernández J D, Fokianos P, Herterich P and Šimko T 2016 *CERN Analysis Preservation: A Novel Digital Library Service to Enable Reusable and Reproducible Research* (Cham: Springer International Publishing) pp 347–356 ISBN 978-3-319-43997-6

[14] DASPOS 2017 DASPOS Project Website URL https://daspos.crc.nd.edu/

[15] DIANA-HEP 2017 DIANA-HEP Project Website URL http://diana-hep.org/

[16] Degrande C, Duhr C, Fuks B, Grellscheid D, Mattelaer O and Reiter T 2012 *Comput. Phys. Commun.* **183** 1201–1214 (*Preprint* 1108.2040)

[17] Aad G *et al.* (ATLAS Collaboration) 2015 *JHEP* **10** 134 (*Preprint* 1508.06608)

[18] Aaboud M *et al.* (ATLAS Collaboration) 2016 *JHEP* **09** 175 (*Preprint* 1608.00872)

[19] Cranmer K and Yavin I 2011 *Journal of High Energy Physics* **2011** 38 ISSN 1029-8479 URL http://dx.doi.org/10.1007/JHEP04(2011)038

[20] 2015 ATLAS Data Access Policy Tech. Rep. ATL-CB-PUB-2015-001 CERN Geneva URL https://cds.cern.ch/record/2002139

[21] Maguire E, Heinrich L and Watt G 2017 *22nd International Conference on Computing in High Energy and Nuclear Physics (CHEP 2016) San Francisco, CA, October 14-16, 2016* (*Preprint* 1704.05473) URL https://inspirehep.net/record/1592380/files/arXiv:1704.05473.pdf

IOP Conf. Series: Journal of Physics: Conf. Series **1085** (2018) 042012 doi:10.1088/1742-6596/1085/4/042012

Identification of hadronically decaying W bosons and top quarks using multivariate techniques at ATLAS

Tatsumi Nitta, On behalf of the ATLAS Collaboration
Waseda Research Institute for Science and Engineering, Waseda University, Tokyo, Japan
E-mail: tatsumi.nitta@cern.ch

Abstract. By colliding protons and examining particles emitted from the collisions, the Large Hadron Collider aims to study the interactions of quarks and gluons at the highest energy accessible in a controlled experimental way. In such collisions, W bosons or top quarks which have TeV scale momentum can be accessible. Reconstructing such boosted jets are becoming important. In particular, the ability to identify original particle that decays to quarks against normal QCD jets plays a central role in various searches at high energy scale. This is typically done by the use of a single physically motivated observable constructed from the constituents of the jet. In this contribution, multiple complementary observables are combined using boosted decision trees and neural networks to increase the ability to distinguish W bosons and top quarks from light quark jets in the ATLAS experiment.

Since hadronic decay ratios of W boson and top quark are higher than leptonic decay, these are essential probes for rare phase space, especially for heavy resonance search at the Large Hadron Collider (LHC). For instance, W and top quark jets are used in diboson resonance searches [1], $t\bar{t}$ resonance search [2], and measurement of differential cross section of high momentum top quarks [3] in ATLAS [4]. In this regime, frequently W and top quark jets have much higher momenta than their rest mass, thus these can be reconstructed by a single large radius jet. Such a large radius jet tends to have similar features to QCD jets produced by the single quark or gluon at much higher rate at LHC. In order to take into account radiation pattern of the jets, physically motivated observables such as mass and the number of core energy deposition area inside the jets have been used to effectively tag W or top quark jets in ATLAS [5, 6]. Applying machine learning techniques with several high level features of the large radius jet is a natural way to expand their performance by analogy to a multivariate b-tagging technique [7] . In this study, two kinds of machine learning techniques, Deep Neural Network (DNN) implemented by Keras [8] and Boosted Decision Tree (BDT) implemented by TMVA [9] are studied.

Table 1. Summary of training and testing sample statistics for each signal and background processes.

Purpose	Training (Top Quark)		Training (W Boson)		Testing		
Sample	Signal	Background	Signal	Background	Top Signal	W Signal	Background
Number of jets	1×10^6	1×10^6	7×10^5	7×10^5	4×10^5	3×10^5	1×10^6

IOP Conf. Series: Journal of Physics: Conf. Series **1085** (2018) 042012 doi:10.1088/1742-6596/1085/4/042012

1. Training Sample

All simulation samples are made by PYTHIA8 [10], W bosons are collected from $W' \rightarrow WZ$ process, top quarks are collected from $Z' \rightarrow t\bar{t}$ and quark and gluon jets are collected from QCD dijets events. The number of events used in this study is summarized in table 1. To avoid any bias due to difference of p_T distribution, training samples are reweighted to obtain flat p_T spectrum. While testing samples are reweighted to background p_T spectrum. Input variable candidates for BDT and DNN are gathered from high level feature proposed by a theoretical community shown in table 2 [11].

Table 2. Summary of candidate variables. The definition of variables are discussed in Ref. [11].

Observable	Variable	Used For
Jet mass	m^{comb}	top,W
Jet p_T	p_T	top,W
Energy Correlation Ratios	e_3, C_2, D_2	top,W
N-subjettiness	τ_1,τ_2,τ_3 ,τ_{21},τ_{32}	top,W
Center of Mass Observables	Fox Wolfram (R_2^{FW})	W
Splitting Measures	Z_{CUT} ,$\sqrt{d_{12}}$,$\sqrt{d_{23}}$	top,W
Planar Flow	P	W
Angularity	a_3	W
Aplanarity	A	W
KtDR	$KtDR$	W
Qw	Q_w	top

Table 3. Top-quark tagging inputs groups for DNN as used in figure 2.

Group	variables
1	C_2,D_2,τ_{21},τ_{32}
2	1, m^{comb}
3	2, p_T
4	τ_1,τ_2,τ_3,e_3,m^{comb},p_T
5	1, $\sqrt{d_{12}}$,$\sqrt{d_{23}}$,Q_W
6	2, $\sqrt{d_{12}}$,$\sqrt{d_{23}}$,Q_W
7	4, $\sqrt{d_{12}}$,$\sqrt{d_{23}}$,Q_W
8	3, $\sqrt{d_{12}}$,$\sqrt{d_{23}}$,Q_W
9	1, 4, $\sqrt{d_{12}}$,$\sqrt{d_{23}}$,Q_W

2. Optimization

Input variables are selected from the candidate variables for each BDT and DNN independently to achieve the best performance for each technique. To find the optimal set of BDT input variables, the single input variable that gives the largest discrimination power is sequentially added to the BDT input list. Similar to the BDT training, the DNN is trained on different sets of input variables in order to find the optimal set of input variables. Unlike the BDT, sets of input variables are not defined by successively adding variables but are defined by grouping the input variables related to the corresponding signal shown in table 3. Input variables selection is shown in figure 1 and 2.

IOP Conf. Series: Journal of Physics: Conf. Series **1085** (2018) 042012 doi:10.1088/1742-6596/1085/4/042012

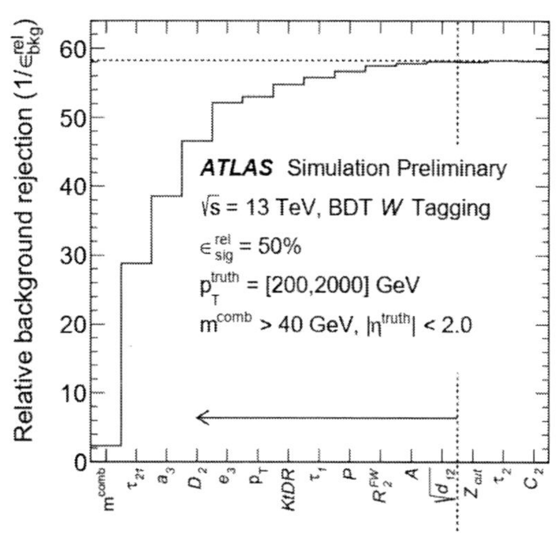

Figure 1. BDT background rejections for each variable set at the 50% W tagging efficiency. 12 variables which are at left side of dashed line are chosen for BDT W-tagging.

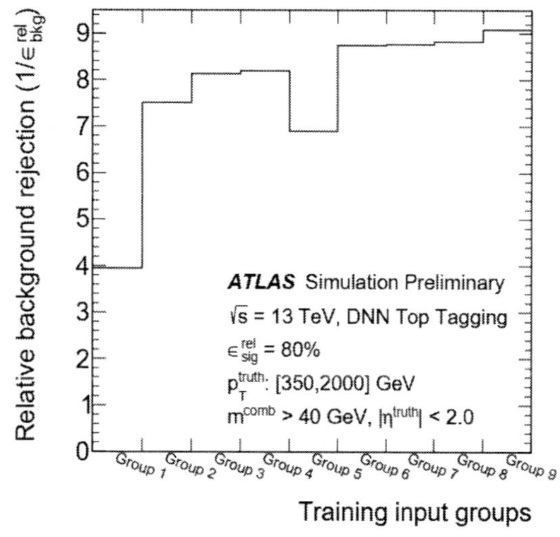

Figure 2. DNN background rejections for each variable group at 80% top-quark tagging efficiency. Input variable groups are described in table 3. Group 9 is chosen as a input set for DNN tagger.

Hyper-parameters for each BDT and DNN have been optimized by a grid-scan approach and the detail is described in elsewhere [11].

3. Result

The discrimination power of the taggers are quantitatively evaluated by receiver operating characteristic (ROC) curves as shown in figure 3 and 4. Both BDT and DNN taggers improve background rejection by 20% (200%) at 50% W (80% top) jet efficiency compared to 2-variables optimized tagger currently used in ATLAS [6].

These taggers are validated by real data where signal and background are separately enriched by analysis-level event selections. The data used were taken in 2015 and 2016 at a center of mass energy of $\sqrt{s}$ = 13 TeV corresponding to integrated luminosities between 36.1 and 36.7 fb^{-1}. These data are ensured that all sub-detector systems and trigger system were properly functional. From this dataset, three different samples are analyzed to study the performance of W-boson and top-quark tagging algorithms as follows:

- $t\bar{t}$ signal sample, enriched in hadronically decaying top quarks,
- quark or gluon jet from γ + jet background sample, covering the p_{T} range from 200 GeV to about 2 TeV,
- multijet background sample, covering the p_{T} range from 500 GeV to about 3.5 TeV.

Signal modelings for each BDT-W tagger and DNN-Top tagger are shown in figure 5 and 6. Background rejections are compared between data and MC simulation by Pythia8 [10] and Sherpa [12] as shown in figure 7 and 8. Differences between simulation and data are less than 10% within statistical uncertainties.

IOP Conf. Series: Journal of Physics: Conf. Series **1085** (2018) 042012 doi:10.1088/1742-6596/1085/4/042012

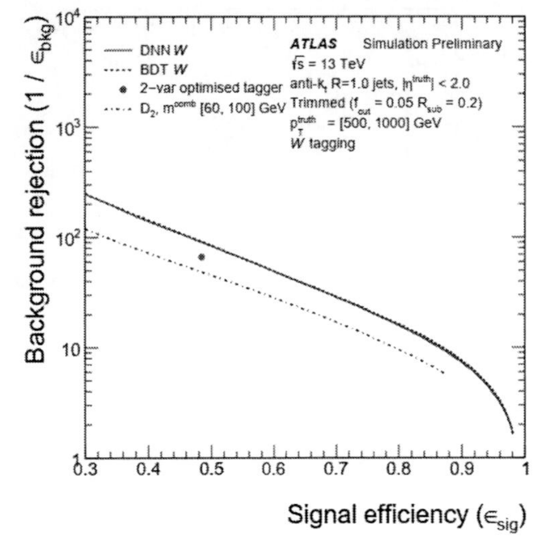

Figure 3. The performance comparison of the W tagger in a $p_{\mathrm{T}} = [500, 1000]$ GeV as a function of signal efficiency.

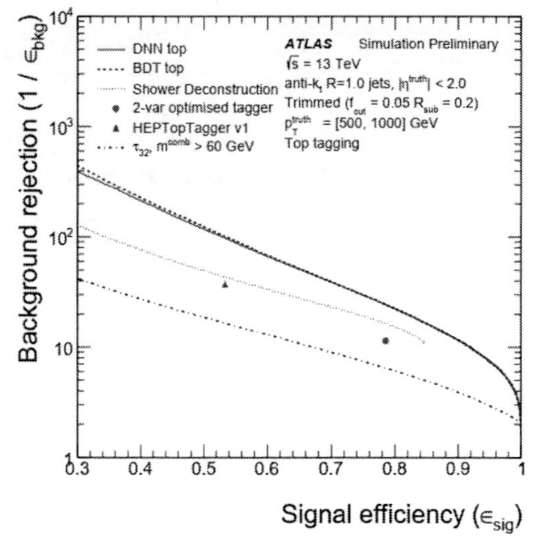

Figure 4. The performance comparison of the top tagger in a $p_{\mathrm{T}} = [500, 1000]$ GeV as a function of signal efficiency.

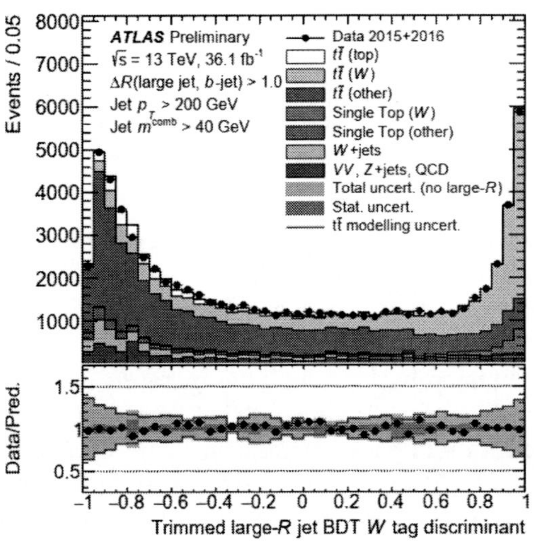

Figure 5. Comparison of the observed data and predicted MC distributions of the BDT discriminants for W-boson tagging. The $t\bar{t}$ MC sample is decomposed using generator level information. The systematics uncertainties for these distributions do not include large radius jet uncertainties.

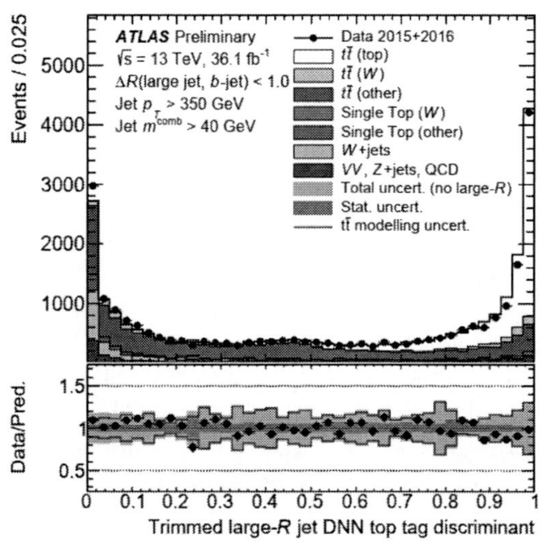

Figure 6. Comparison of the observed data and predicted MC distributions of the DNN discriminant for top-quark tagging. The $t\bar{t}$ MC sample is decomposed using generator level information. The systematics uncertainties for these distributions do not include large radius jet uncertainties.

4. Conclusion

W and top quark jet taggers with BDT and DNN are constructed in the ATLAS experiment. Both BDT and DNN taggers improve background rejections by 20% (200%) at 50% W (80%

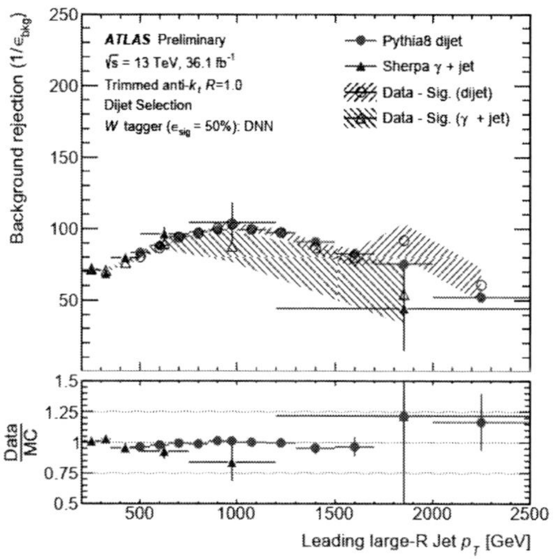

Figure 7. Background rejection as a function of leading jet p_{T} for the DNN W-boson taggers. γ + jet simulation (data) are shown as solid blue (hollow black) triangles and dijet simulation (data) are shown as red (hollow black) circles. For each bin in data, the predicted signal contribution is subtracted prior to computing the background rejection. Only the statistical uncertainties are shown.

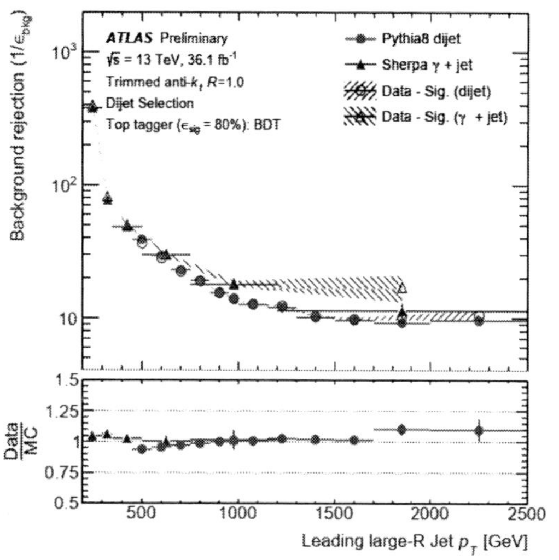

Figure 8. Background rejection as a function of leading jet p_{T} for the BDT top-quark taggers. γ + jet simulation (data) are shown as solid blue (hollow black) triangles and dijet simulation (data) are shown as red (hollow black) circles. For each bin in data, the predicted signal contribution is subtracted prior to computing the background rejection. Only the statistical uncertainties are shown. Last point of γ+jets in ratio panel is not shown since it is outside of scale.

top) jet efficiency compared to 2-variables optimized tagger currently used in ATLAS. There is no difference between BDT and DNN taggers in terms of signal efficiency and background rejection power. Signal and background modelings are studied with $t\bar{t}$, γ+jet and dijet events. The differences between data and simulation are within statistical uncertainties.

References

[1] ATLAS Collaboration 2016 *JHEP* **09** 173 (*Preprint* 1606.04833)
[2] ATLAS Collaboration 2016 Search for heavy particles decaying to pairs of highly-boosted top quarks using lepton-plus-jets events in proton–proton collisions at $\sqrt{s}$ = 13 TeV with the ATLAS detector ATLAS-CONF-2016-014 URL https://cds.cern.ch/record/2141001
[3] ATLAS Collaboration 2016 *Phys. Lett. B* **756** 52 (*Preprint* 1512.06092)
[4] ATLAS Collaboration 2008 *JINST* **3** S08003
[5] ATLAS Collaboration 2016 *Eur. Phys. J. C* **76** 154 (*Preprint* 1510.05821)
[6] ATLAS Collaboration 2017 Performance of Top Quark and W Boson Tagging in Run 2 with ATLAS ATLAS-CONF-2017-064 URL https://cds.cern.ch/record/2281054
[7] ATLAS Collaboration 2015 Expected performance of the ATLAS b-tagging algorithms in Run-2 ATL-PHYS-PUB-2015-022 URL https://cds.cern.ch/record/2037697
[8] F Chollet 2015 Keras URL https://github.com/fchollet/keras
[9] A Hoecker et al 2007 TMVA: Toolkit for Multivariate Data Analysis PoS ACAT (2007) 040
[10] Sjostrand T, Mrenna S and Skands P Z 2008 *Comput. Phys. Commun.* **178** 852–867 (*Preprint* 0710.3820)
[11] ATLAS Collaboration 2017 Identification of Hadronically-Decaying W Bosons and Top Quarks Using High-Level Features as Input to Boosted Decision Trees and Deep Neural Networks in ATLAS at $\sqrt{s}$ = 13 TeV ATL-PHYS-PUB-2017-004 URL https://cds.cern.ch/record/2259646
[12] Gleisberg T, Hoeche S, Krauss F, Schonherr M, Schumann S *et al.* 2009 *JHEP* **0902** 007 (*Preprint* 0811.4622)

IOP Conf. Series: Journal of Physics: Conf. Series **1085** (2018) 042013 doi:10.1088/1742-6596/1085/4/042013

Hydra: a C++11 framework for data analysis in massively parallel platforms

A A Alves Júnior[1] and M D Sokoloff[2]
Physics Department - University of Cincinnati
400 Geology/Physics Building PO Box 210011 Cincinnati, OH 45221-0011 United States
E-mail: aalvesju@cern.ch[1] and sokoloff@ucmail.uc.edu[2]

Abstract. Hydra is a header-only, templated and C++11-compliant framework designed to perform the typical bottleneck calculations found in common HEP data analyses on massively parallel platforms. The framework is implemented on top of the C++11 Standard Library and a variadic version of the Thrust library and is designed to run on Linux systems, using OpenMP, CUDA and TBB enabled devices. This contribution summarizes the main features of Hydra. A basic description of the overall design, functionality and user interface is provided, along with some code examples and measurements of performance.

1. Introduction

Despite the ongoing efforts of modernization, a large fraction of the software used in High Energy Physics (HEP) is legacy. It mostly consists of libraries assembling single threaded, Fortran and C++03 mono-platform routines. Concomitantly, HEP experiments keep collecting samples with unprecedented large statistics, while data analyses become increasingly complex. Even deploying powerful computers, it is common to spend days performing calculations to reach a result, and they often need re-tuning anyway.

On the other hand, computer processors will not increase clock frequency anymore in order to reach higher performance. Indeed, the current road-map to improve overall performance is to deploy different levels of concurrency. One effect of which is the proliferation of multi-thread friendly and multi-platform environments among HPC data centers. Unfortunately, HEP software is not completely prepared yet to fully exploit concurrency and to deploy more opportunistic computing strategies.

Hydra proposes a computing model to approach these issues. The Hydra framework provides collection of parallelized high-level algorithms, addressing some of of typical computing bottlenecks commonly found in HEP, and a set of optimized containers and types, through a modern and functional interface, allowing to enhance HEP software productivity and performance, keeping the portability between NVidia GPUs, multi-core CPUs and other devices compatible with CUDA, TBB and OpenMP computing models.

2. Design highlights

Hydra is a C++11 template framework organized using a variety of static polymorphism idioms and patterns. This ensures the predictability of the stack size at compile time, which is critical for stability and performance when running on GPUs and minimizes the overhead introduced by the user interface when engaging the actual calculations. Furthermore, the combination of static polymorphism and templates allows exposure of the maximum amount of code to the compiler, in the context in which the code will be used, contributing to activate many compile time optimizations that could not be accessible otherwise. Hydra's interface and implementation details extensively deploy patterns and idioms that enforce thread-safety and efficient memory access and management. The following list summarizes some of the main design choices adopted in Hydra:

- Hydra provides a set of optimized STL-like containers that can store multidimensional data-sets using SoA[1]
- Data handled using iterators and all classes manage allocated resources using RAII idiom.
- The framework is type and thread-safe.
- No limitation on the maximum number of dimensions that containers and algorithms can handle.

The types of devices in which Hydra can be deployed are classified by back-end type, according to the device compatibility with certain computing models. Currently, Hydra supports four back-ends, which are CPP, OpenMP, CUDA and TBB. Code can be dispached and executed in all supported back-ends concurrently and asynchronously in the same program, using the suitable policies represented by the symbols `hydra::omp::sys`, `hydra::cuda::sys`, `hydra::tbb::sys` , `hydra::cpp::sys`, `hydra::host::sys` and `hydra::device::sys`. These policies define the memory space where resources should be allocated to run algorithms and store data.

For mono-back-end applications, source files written using Hydra and standard C++ compile for GPU and CPU just exchanging the extension from .cu to .cpp and one or two compiler flags. There is no need for refactoring code.

3. Basic features

Currently, Hydra provides a collection of parallelized high-level algorithms, addressing some computing-intensive tasks commonly found in data analyses in HEP. The available high-level algorithms are listed below:

- Multidimensional p.d.f. sampling.
- Parallel function evaluation on multidimensional data-sets.
- Five fully parallelized numerical integration algorithms: Genz-Malik, self-adaptive and static Gauss-Kronrod quadratures, plain, self-adaptive importance sampling and phase-space Monte Carlo integration.
- Phase-space Monte Carlo generation and sampling.
- Interface to `ROOT::Minuit2`[1] minimization package, allowing to accelerate maximum likelihood fits over multidimensional large data-sets.
- Parallel implementation of the S-Plots[2] technique prescription, for statistical unfolding data distributions.

Hydra also provides two multidimensional containers optimized to store large POD data-sets with dimensions represented by numbers with the same or different types. A significant number of facilities are also available to filter, copy and perform other operations. In the following subsections, some of the basic features are described.

[1] Structure of arrays or SoA is a layout separating elements of a structure into one parallel array per field. This ease the data manipulation with SIMD instructions and, if only a specific field of the structure is needed, only this field can to be iterated over, allowing more data to fit onto a single cache line.

IOP Conf. Series: Journal of Physics: Conf. Series **1085** (2018) 042013 doi:10.1088/1742-6596/1085/4/042013

```
struct Gaussian: public hydra::BaseFunctor<Gaussian,double,2>
{
  //implementing the Evaluate() method of the Gaussian functor.
  //unsigned int n : number of arguments
  //double *x: pointer to the array of arguments
   __host__ __device__
  inline double Evaluate(unsigned int n, double* x){
    double _m2 = x[0]*_par[0]*_par[0];
    double _s2 = _par[1]*_par[2];

    return exp(-m2/(2.0 * _s2 ))/( sqrt(2.0*_s2*pi));
  }
};
```

Listing 1. Example of implementation of a functor representing a Gaussian function in Hydra. The access to the Hydra interface and type information is implemented deriving the functor from the `hydra::BaseFunctor<Functor, ReturnType, NParameters>` base class. The functor's parameters, in this case the mean and sigma of the Gaussian, are accessible in the scope of the functor via the symbols `_par[i]` or in the functor instantiation scope via accessors `SetParameter(...)` and `GetParameter(...)`.

3.1. Functors and arithmetic operators

In all situations where custom calculations need to be performed, Hydra framework instantiates algorithms to call the user's code that is passed using functors or C++11 lambdas. Hydra adds features and type information to generic functors using the CRTP idiom. To be compliant with Hydra's interface, the functors need to derive from the `hydra::BaseFunctor<Functor, ReturnType,NParameters>` class and to implement the `__host__ __device__ Evaluate(...)` method. Functor's values can be cached. The Listing 1 shows how to implement a simple functor to calculate a Gaussian function.

3.2. C++11 Lambdas

The user can define a C++11 lambda function and convert it into a Hydra functor using the function `hydra::wrap_lambda()`. The wrapped lambda will have access to all functionality of Hydra. The Listing 2 shows a basic example of usage for this functionality. Hydra also supports the usage of C++11 lambdas as fit models. In this case, the lambda function needs to have a different signature and the list of fit parameters needs to be passed to `hydra::wrap_lambda()`.

3.3. Arithmetic operators

Hydra overloads the basic arithmetic operators for all objects deriving from `hydra::BaseFunctor`. Composition of functors is supported as well. For example, the lines shown in Listing 3 are completely legal C++11 code.

3.4. Containers

Hydra framework provides one dimensional STL-like vector container for each supported back-end, aliasing the underlying Thrust types. Beyond this, Hydra implements two native multidimensional containers: `hydra::multivector` and `hydra::multiarray`. In these containers, the data corresponding to each dimension is stored in contiguous memory regions

```
...
double two = 2.0;
//define a simple lambda and capture "two"
auto my_lambda = [=] __host__ __device__ (unsigned int n, double* x)
    {
          return two*sin(x[0]);
    };
//convert it into a Hydra functor
auto my_lambda_wrapped  = hydra::wrap_lambda(my_lambda);
...
```

Listing 2. Example of usage of `hydra::wrap_lambda()` to wrap a simple lambda function multiplying sin(x) by two.

```
...
//basic arithmetic operations
auto A_plus_B  = A + B;
auto A_minus_B = A - B;
auto A_times_B = A * B;
auto A_per_B   = A/B;
//any composition of basic operations
auto any_functor = (A - B)*(A + B)*(A/C);
// C(A,B) is represented by:
auto compose_functor = hydra::compose(C, A, B)
...
```

Listing 3. Using overloaded arithmetic operators to build up expressions using functors. If `A`, `B` and `C` are Hydra functors, the code above is completely legal.

and when the container is traversed, each entry is accessed as a `hydra::tuple`, where each field holds a value corresponding to a dimension. Both classes implement an interface completely compliant with a STL vector and also provides constant and non-constant accessors for the single dimensional data. The container `hydra::multivector` is suitable to store data-sets where the dimensions are represented by entries with different POD types. `hydra::multiarray` is designed to store data-sets where all dimensions are represented by the same type. Data is copiable across different back-ends.

4. Code examples and performance measurements

4.1. Multidimensional numerical integration

The VEGAS algorithm[3] samples the integrand and adapts itself, so that the sampling points are concentrated in the regions that make the largest contribution to the integral. Hydra's implementation follows the structure of the corresponding GSL algorithm, but parallelizes the integrand evaluations and accumulations in each iteration. There is no limit in the number of dimensions the integrator can handle. Figure 1 shows the performance of Hydra's VEGAS implementation.

4.2. Phase-Space Monte Carlo

Phase-Space Monte Carlo simulates the kinematics of a particle with a given four-momentum decaying to a n-particle final state, without intermediate resonances. Samples of phase-space

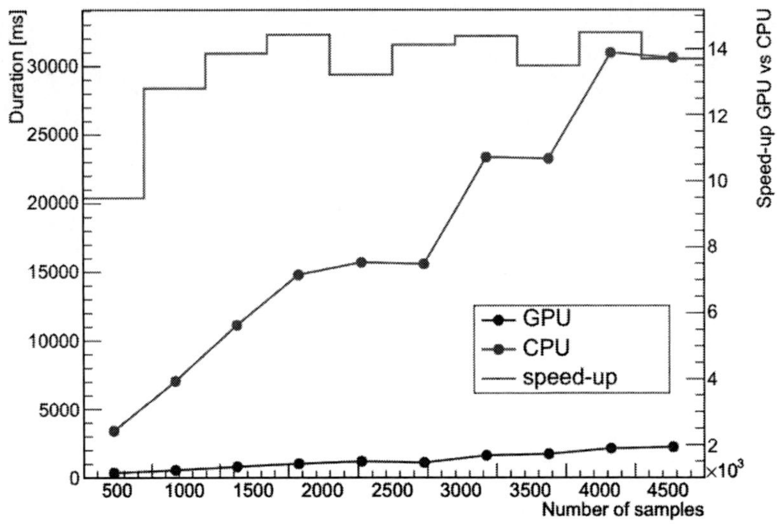

Figure 1. Performance of Hydra's VEGAS implementation as a function of the number of integrand calls per iteration. The integrand is a 10-dimensional Gaussian distribution. The GPU model is Tesla K40c and the CPU is Intel Xeon(R) CPU E5-2680 v3 @ 2.50GHz (one thread).

```
//Masses of the particles
hydra::Vector4R Mother(mother_mass, 0.0, 0.0, 0.0);
double Daughter_Masses[3]{ daughter1_mass, daughter2_mass,
    daughter3_mass };
//Create PhaseSpace object
hydra::PhaseSpace<3> phsp(Mother_mass, Daughter_Masses);
//Allocate the container for the events
hydra::Events<3, device> events(ndecays);
//Generate
phsp.Generate(Mother, events.begin(), events.end());
```

Listing 4. This snippet shows how to use the Hydra's phase-space Monte Carlo generator to produce a sample with `ndecays` three-body decays.

Monte Carlo events are widely used in HEP studies where the calculation of phase-space volume is required as well as a starting point to implement and to describe the properties of models with one or more resonances or to simulate the response of the detector to decay's products[4]. Hydra provides an implementation of the Raubold-Lynch method[4] and can generate the full kinematics of decays with any number of particles in the final state. Sequential decays, evaluation of model, production of weighted and unweighted samples and many other features are also supported. Listing 4 shows how to use the phase-space Monte Carlo generator to produce a sample with `ndecays` three-body decays. Figure 3 shows the phase-space generator performance as a function of the sample size.

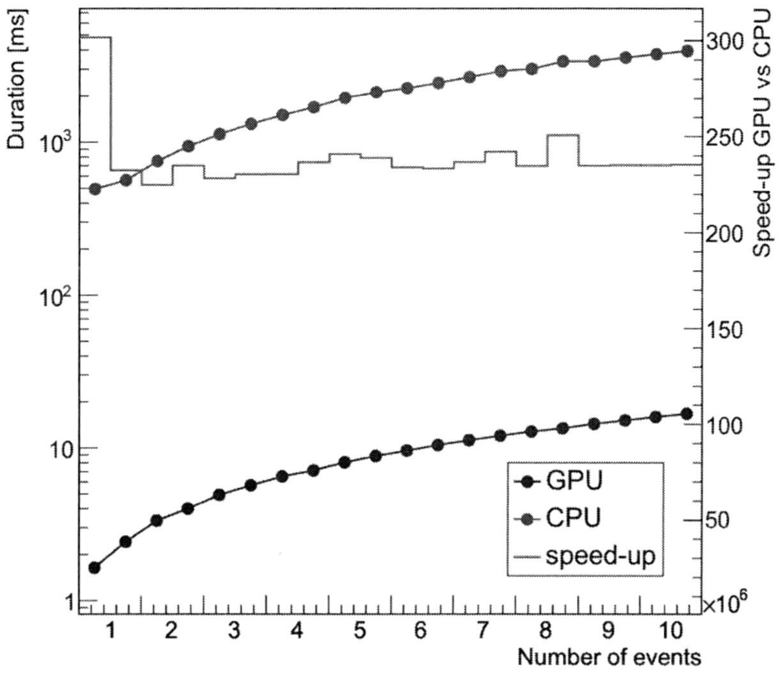

Figure 2. The performance of the phase-space generator as a function of the sample size. The GPU model is Tesla K40c and the CPU is Intel Xeon(R) CPU E5-2680 v3 @ 2.50 GHz (one thread).

4.3. Interface to Minuit2

Hydra implements an interface to `ROOT::Minuit2` that parallelizes the FCN calculation [1]. This dramatically accelerates the calculation over large data-sets. Hydra normalizes the pdfs on-the-fly using analytical or numerical integration algorithms provided by the framework and handles data using iterators.

Listing 5 shows how to build a simple model composed of two pdfs, respectively representing a Gaussian and an exponential distributions. The model is used to perform an extended likelihood fit. The FCN object provided by Hydra parallelizes the function calls in the back-end the data is stored in. The processing of a sample with 20 million events takes 4.8 seconds in a Tesla K40c and 299.8 seconds in a Intel Xeon(R) CPU E5-2680 v3 @ 2.50 GHz (one thread), resulting in a speed-up of a factor ~62x.

5. Summary

The basic design, performance and functionality of the header-only, C++11-compliant framework Hydra have been introduced. Some of the basic interfaces and algorithms are discussed in Section 3. The performance measurements for running some of Hydra's algorithms using CUDA and one CPU thread are discussed in Section 4, and show that Hydra can be

IOP Conf. Series: Journal of Physics: Conf. Series **1085** (2018) 042013 doi:10.1088/1742-6596/1085/4/042013

```
//analytical integral functors
GaussAnalyticIntegral GaussIntegral(min, max);
ExpAnalyticIntegral   ExpIntegral(min, max);
//build the pdfs
auto Gaussian_PDF    = hydra::make_pdf(Gaussian, GaussIntegral);
auto Exponential_PDF = hydra::make_pdf(Exponential, ExpIntegral);

//make a extended pdf model
std::array<hydra::Parameter*, 2> yields{NGaussian, NExponential};
auto Model = hydra::add_pdfs(yields, Gaussian_PDF, Exponentia_PDF );
//get the FCN
auto Model_FCN = hydra::make_loglikehood_fcn(Model, data_d.begin(),
   data_d.end());
//pass the FCN to Minuit2
...
```

Listing 5. This code snippet shows how to build simple model composed by two pdfs, respectively representing a Gaussian and an exponential distributions. The model is used to perform a extended likelihood fit, able to predict the yields of each component. The FCN object provided by Hydra, will parallelize the function calls in the back-end the data is stored.

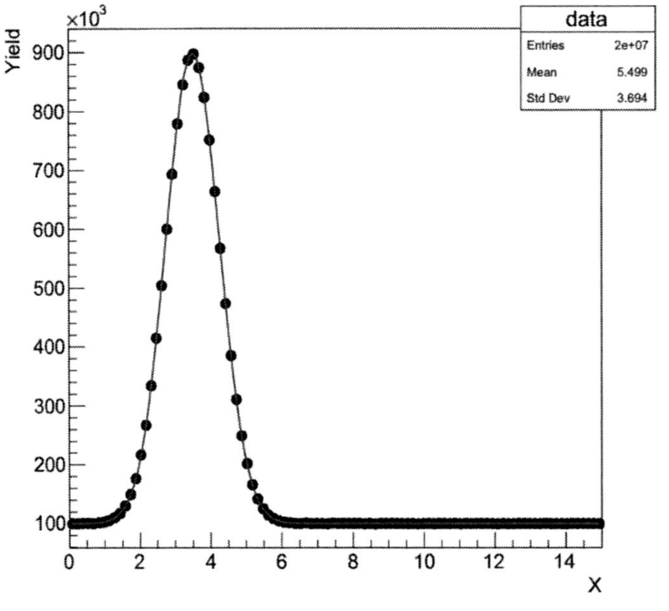

Figure 3. Result of the extended unbined likelihood fit performed over a 20 million events sample.

up to 250 times faster than conventional software, depending on the graphics card. For CPU multi-threads back-ends, TBB and OpenMP, if the problem size is large and calculations pay the cost of thread creation and destruction, the algorithms scale linearly with the number of threads deployed. Since Hydra is header only, no additional building process needs to be done beyond the inclusion of the required headers. Hydra has been presented at several conferences, including NVidia's GTC 2017[5]. It is currently being used in a data analysis aiming to measure the charged Kaon mass at the LHCb Experiment at CERN[6]. The initial development of Python bindings for Hydra was one of the projects of Google Summer of Code 2017[7].

Hydra is open source and released under GPL version 3 license. The project is hosted on GitHub at `https://github.com/MultithreadCorner/Hydra`.

6. Acknowledgments

This work was performed with support from NSF Award PHY-1414736. NVidia provided K40 GPUs for our use through its University Partnership program.

References

[1] Minuit2 Package `https://root.cern.ch/root/html/MATH_MINUIT2_Index.html`

[2] Pivk M and Le Diberder F R 2005 *Nucl. Instrum. Meth.* **A555** 356–369 (*Preprint* `physics/0402083`)

[3] Lepage G P 1978 *Journal of Computational Physics* **27** 192 – 203 ISSN 0021-9991 URL `http://www.sciencedirect.com/science/article/pii/0021999178900049`

[4] James F E 1968 (CERN) URL `https://cds.cern.ch/record/275743`

[5] Alves Junior A A 2017 *GTC2017* (San José, US) Presentation ID S7340

[6] Alves Junior A A and Contu A 2017 *Rare n Strange 2017: strange physics at LHCb* (Santiago de Compostela, Spain)

[7] Hydra.Python: python bindings for Hydra `https://summerofcode.withgoogle.com/dashboard/project/6669304945704960/overview/`

GooFit 2.0

Henry Schreiner[1], Christoph Hasse[2], Bradley Hittle[3], Himadri Pandey[1], Michael Sokoloff[1] and Karen Tomko[3]

[1] University of Cincinnati, 2600 Clifton Ave, Cincinnati, OH 45220, USA
[2] Technical University of Dortmund, Emil-Figge-Straße 50, 44227 Dortmund, Germany
[3] Ohio Supercomputer Center, 1224 Kinnear Rd, Columbus, OH 43212, USA

E-mail: henry.schreiner@uc.edu

Abstract. The GooFit package provides physicists a simple, familiar syntax for manipulating probability density functions and performing fits, and is highly optimized for data analysis on NVIDIA GPUs and multithreaded CPU backends. GooFit was updated to version 2.0, bringing a host of new features. A completely revamped and redesigned build system makes GooFit easier to install, develop with, and run on virtually any system. Unit testing, continuous integration, and advanced logging options are improving the stability and reliability of the system. Developing new PDFs now uses standard CUDA terminology and provides a lower barrier for new users. The system now has built-in support for multiple graphics cards or nodes using MPI, and is being tested on a wide range of different systems. GooFit also has significant improvements in performance on some GPU architectures due to optimized memory access. Support for time-dependent four-body amplitude analyses has also been added.

1. Introduction

Multidimensional fits to large datasets, with tens of millions of events, are becoming increasingly common in High Energy Physics (HEP). The models can be computationally intense, with hundreds of parameters. In preparing a model and fitting the data, the most natural description is usually to prepare components,[1] and then combine these components for a specific fit. This is the approach taken by RooFit [1], one of the most commonly used physics fitting packages. With the advent of highly parallel computing architectures, the need arose for a system that would make implementing fitting efficiently in parallel as intuitive to physicists as RooFit.

Graphical Processing Units (GPUs) provide an alternative to traditional computing, with the ability to do computations in a massively parallel compute engine, with thousands of floating point operations processed at the same time (see Table 1). Since most of the time spent in fitting a probability distribution to a dataset is composed of independently computing the value of a function at each point in the dataset, this maps well to the processing capabilities of GPUs.

There are several complications however; programming on a graphics card uses a different language (such as CUDA for NVIDIA) and requires data to be managed between memory in the host and the device. GPUs have a simpler instruction set, no branch prediction, and an emphasis on single precision compute, especially for the cheaper "gamer"-class cards. They also must compute the exact same instruction across groups of data at the same time. Even a

[1] Although commonly called PDFs, these components include complex quantum mechanical amplitudes whose magnitude-squared values represent PDFs.

Table 1. The advertised floating point operations per second (FLOP/s) for several common graphics cards, for both single precision (SP) and double precision (DP), and the number of Streaming Multi-processors (SMPs). Some approximate FLOPs of Intel CPUs for comparison include 0.0864 GFLOP/s for a 2015 i5 dual core processor or 0.461 GFLOP/s for a 12 core Xeon from a similar year.

	Name	Cores	Clock	GFLOP/s SP	GFLOP/s DP	Cost
Gamer	GTX 1050 Ti	768	1290 Mhz	1980	62	$150
	GTX 1080 Ti	3584	1596 Mhz	11300	330	$850
Server	Tesla K40	2880	745 Mhz	4290	1430	$3000
	Tesla P100	3584	1329 Mhz	9300	4700	$10000

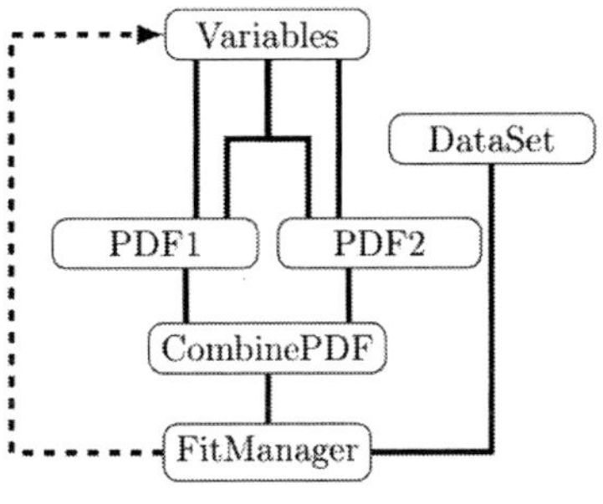

Figure 1. Illustration of the structure of a GooFit program. The `Variables` represent both observables and parameters, and are input to PDFs. The PDFs can be combined using combination PDFs. The final PDF is fed into a `FitManager`, which also takes a binned or unbinned `DataSet` with data to attach to an observable. The `FitManager` performs a fit, and sets the parameter variables accordingly.

simple if statement will force a mask to be applied and both sides of the if statement must be computed separately. These factors combine to make writing GPU code substantially different from traditional CPU code.

One of the first fitting systems designed for parallel fits on a GPU was the GooFit package, first released in 2013 by Rolf Andreassen [2]. The project was designed using the CUDA language from NVIDIA, and used their Thrust library [3] to manage kernel launches. By the end of the year, GooFit was expanded with an Thrust supported OpenMP backend to enable parallel fitting on CPU devices "with the flip of a (compiler) switch".

GooFit 1.0 had an impressive list of built-in PDFs, including a specialized system for amplitude analyses of three body particle decays (often called Dalitz plot analyses). Combination PDFs provided ways to build more complex PDFs out of simpler building blocks. Several examples were provided. Simple composition of PDFs using existing building blocks was GooFit's design target (see Figure 1), and it succeeded in that. The comparative performance for different systems can be seen in Table 2, and scalability when changing the number of OpenMP threads can be seen in Figure 2 and Figure 3.

However, providing new PDFs was non-trivial, and many of the more advanced Physics PDFs had large sets of custom code that was not generalized or shared with the rest of GooFit, such as complex return values or signal generation. GooFit was hard to build and had specific system requirements, and the development was fragmented across several Git repositories on GitHub. The GooFit 2.0 project was undertaken to make GooFit easier to build and develop with, and to combine the development effort. Future work is addressing other aspects of the design to make PDFs easier to author and simpler to maintain.

IOP Conf. Series: Journal of Physics: Conf. Series **1085** (2018) 042014 doi:10.1088/1742-6596/1085/4/042014

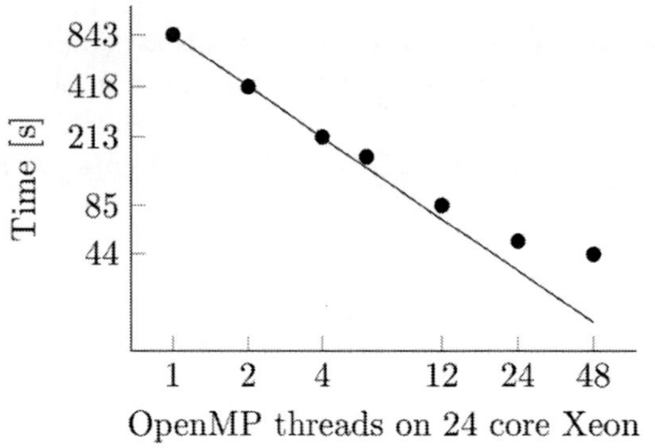

Figure 2. Performance of $\pi^+\pi^-\pi^0$ with 16 time dependent amplitudes on an Intel Xeon E5-2680 dual-chip system (24 cores). The line illustrates ideal scaling.

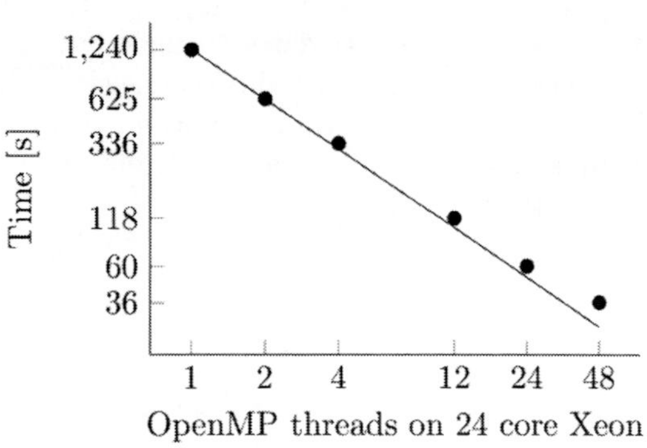

Figure 3. Performance of the Zach fit (D^{*+} and D^0 mass difference measurement with BaBar data) with 16 time dependent amplitudes.

Table 2. Comparison of performance of example code from the GooFit 2.0 examples on different systems. $\pi^+\pi^-\pi^0$ on the left, Zach fit on the right.

Type	Device	Time
2 Cores	Core 2 Duo	1159. s
GPU	GeForce GTX 1050 Ti	86.4 s
GPU	Tesla K40	64.0 s
MPI	Tesla K40 ×2	39.3 s
GPU	Tesla P100	20.3 s

Type	Device	Time
2 Cores	Core 2 Duo	738. s
GPU	GeForce GTX 1050 Ti	60.3 s
GPU	Tesla K40	96.6 s
MPI	Tesla K40 ×2	54.3 s
GPU	Tesla P100	23.5 s

2. New build system

The build system has been completely redesigned using CMake, the most popular cross-platform build system. This was done in a three step procedure, allowing an evolutionary change to the traditional build system to support new folder structures and code reorganization while the new system was being built (described in more detail in [4]).

The new CMake-powered build added a plethora of new features to GooFit. Support for IDEs, such as XCode and QtCreator is now available; IDE-centric features such as header listings and folder organization are included. More backends are now supported, such as the single threaded "CPP" backend, and a limited form of support for Intel's Threaded Building Blocks (TBB). Support for Intel and LLVM compilers was also added, including the Apple LLVM compiler on macOS for the first time. Data files are automatically downloaded from GitHub by the build system for the examples. All of the required libraries are now downloaded through Git submodules if not discovered, including several new header-only libraries. The helper files that make these features possible are available in a separate Git repository, for reuse by other non-GooFit packages. A new packaging system was added as well, allowing a user to easily develop code without forking the GooFit project.

An initial set of unit tests were added, primarily focusing on parts of GooFit that were refactored during the 2.0 development process, such as the `FitManager` and `Variables`. The examples now have a script that runs all of them sequentially, and checks the validity of the results. Continuous integration for the OpenMP backend builds, tests, and runs the examples on the Travis CI service. This system also builds the Doxygen comments in the source into documentation for every version. Code coverage was added with GCov and the CodeCov service

IOP Conf. Series: Journal of Physics: Conf. Series **1085** (2018) 042014 doi:10.1088/1742-6596/1085/4/042014

to estimate the percentage of the code base covered by the test cases, and to report changes on new pull requests.

The Travis CI system presented several difficulties. The GooFit library required the ROOT library to install [5], but building ROOT takes more than the alloted time on a Travis worker node utilizing both available threads. This was overcome by using a prebuilt binary and caching it on Travis. CMake is trivial to download and run for any system, so a recent copy is obtained on Travis and used as well. A standalone copy of Minuit 2 was made available for GooFit 2.0, so there is longer a requirement that ROOT be available, although since it is still used in most of the examples and a few optional tests, ROOT remains part of the standard Travis build.

GooFit Docker images for both OpenMP and CUDA (using NVIDIA-Docker) were added to provide a simple way for new users to start using GooFit. An exact set of commands to prepare a basic GooFit environment, developed and tested using pristine operating system Docker containers, was developed for CentOS7, OpenSUSE, Alpine, and Ubuntu systems.

3. Modernization

GooFit was originally designed for CUDA 4.0 and NVIDIA CUDA compute architecture 2.0. Several of the GooFit forks were already using C++11 features available in CUDA 7+, so the decision was made to target CUDA 7 and higher for the GooFit 2.0 upgrade. Code cleanup and modernization was initially performed manually, but was slow and laborious for such a large code base. Some changes, such as renames, were done using regular expressions in Python using the `ModernizeGooFit.py` script. This was instrumental in removing custom terminology that would be unfamiliar to new developers. Where possible, standard CUDA or Thrust terminology was used for all backends, and several spellings were normalized to match ROOT. To facilitate rewriting GooFit using newer language constructs, the Clang-Tidy program was used. It processed all of the source code using the built-in CMake 3.6+ support, and changed a specified set of features. One feature was processed at a time, using Git to view the changes. Some of the most useful changes were the use of override for all overriding virtual functions, the use of range-style `for` loops, and the use of `nullptr` vs. `0` or `NULL`.

The usage of a custom GooFit class for complex numbers was replaced with the new Thrust complex number class. This also required supplying an external addition that enhanced the `ldg` CUDA intrinsic[2] to support non-intrinsic types to retain recent performance gains on midrange NVIDIA hardware [6]. Logging is now under a unified interface using the fmt library to provide simple Python-like formatted messages [7]. Compile time settings in CMake allow debug and trace logging to be added to GooFit with no runtime cost if disabled. Unified errors are provided with a custom exception subclass that also utilizes the fmt library. Timing statistics have been added to the standard output. A modified version of the small FeatureDetector library checks for missed compiler optimization and prints warnings as needed [8].

One of the features commonly needed for the examples and for analyses was the addition of a Command Line Interface parser, CLI11 [9]. This library was designed to provide a completely general utility for creating command line interfaces for complex and performance dependent applications, with the ability to provide customized behavior for specific toolkits. GooFit provides a customized subclass for the main application, adding standard GooFit specific options, backend information, color printing through the Rang library [10], checks for missing compiler optimization, and a few other features. The GooFit version sets defaults that are designed to provide a natural one-to-one mapping of a command line interface and the standard models described by fits, while CLI11 defaults remain true to standard Unix programs.

[2] In CUDA, `__ldg()` reads from global memory using the texture-path, which is a read-only memory path providing faster global memory access.

4. New software features

One of the key features of GooFit is the caching of partial results inside PDFs to be reused in the computations, taking advantage of the fact that many parameters remain the same between calls. The `Variable` change detection system was improved to support multiple datasets. The caching system for specific PDFs was modified to support a "Structure of Arrays" format which gives better performance.

Another recent addition is preliminary MPI support. This is included as an option in the CMake build, and the standardized CLI11 application class allows setup and teardown to conditionally be included. The provided MPI support divides the dataset for the application by the number of processes involved in the calculation. The compilation and execution of MPI is supported for both OpenMP and CUDA. If multiple GPU processes run on the same node, they will each select a different GPU, if available.

The default fitter in GooFit has been changed to the newly supported Minuit 2 fitter, with the Minuit 1 fitter still provided if ROOT is present. The Minuit 2 fitter was factored out of ROOT, and a new CMake build was added. If ROOT is not found, GooFit will default to the standalone Minuit 2. As part of this change, the Minuit 1 fitter was completely redesigned to provide a more consistent interface. The new version avoids global variables and has automated variable synchronization. The Minuit 2 fitter provides direct access to the FCN and Minuit 2 variables, improved logging output, and direct access to all the Minuit 2 controls and output.

The other non-PDF core classes in GooFit, `DataSets` and `Variables`, were also rewritten. Input to a dataset is much simpler and faster internally, improving code transparency and maintainability. The inheritance design was improved; it is now possible to write generic code with `DataSet` and select the binned or unbinned variations at runtime, simplifying several examples. `Variables` now have a much more tightly controlled API, allowing GooFit to catch errors more reliably at compile time. Further runtime checks were added to warn unsuspecting users who load data out of range.[3] Several operator overloads were added to make manipulation of the `Variable` value as convenient as direct access.

5. New physics features

Several physics analyses have used new physics features from the various GooFit forks that were merged into GooFit 2.0.

Support for three-body time-dependent amplitude analyses was added. This was used to measure mixing in the decay $D^0 \to \pi^+\pi^-\pi^0$ [11]. The code for this analysis has been provided in both GooFit as an example, and as the original RooFit based package. The data for this example were made public for the first time by the *BaBar* collaboration just before the release of GooFit 2.0. Another use of this feature can be seen in the LHCb mixing and CP violation search in $D^0 \to K^0_S\pi^-\pi^+$ [12].

Four-body time-integrated and time-dependent amplitude analyses support was added, as well. This was developed for and used in a mixing parameter search in $D^0 \to K^+\pi^-\pi^+\pi^-$ [13].

Toy Monte Carlo generation for multi-body systems was added using the MCBooster library [14]. This is used in the three- and four-body amplitude analysis PDFs for integration, signal generation, and coordinate transformations. Physics analyses that use this include the mixing parameter search previously mentioned as well as a model independent partial wave analysis (MIPWA) of $D^+ \to h^-h'^+h'^+$ [15].

6. GooFit future developments

One of the most requested features, Python bindings, has been developed as a proof-of-principle for an example, a simple exponential. Only the minimal set of tools needed for that one example

[3] The author was one such unsuspecting user.

is provided in GooFit 2.0, but it is relatively straight forward to extend to the rest of GooFit. The bindings were constructed with PYBIND11 [16], and work with OpenMP or GPU backends. The Python interface is disabled by default, but is included in the test builds on Travis.

In development for GooFit 2.1 are drastically expanded Python bindings, with support for most of GooFit's features and all of the PDFs. The current development version is available from the PyPI system through PIP (Package Installer for Python).

GooFit development continues at a rapid pace. A new indexing system is being designed and implemented across all PDFs, which will simplify PDF authoring and enable new optimizations to the backend in the future. The Hydra package is being considered for inclusion, a new framework for data analysis with massively multi-threaded platforms, developed by the same author as MCBooster as a replacement [17].

7. Conclusions

GooFit has undergone a major code structure transformation. The changes behind the 2.0 design is enabling new features to be added, is reducing the burden on analysts building and using GooFit, and is encouraging contributions to a unified code base. The code runs faster and is better at catching a user's mistakes. It runs on more systems than ever and supports IDEs and debuggers. It provides a set of tools to assist analysts in writing GooFit code.

The future of GooFit looks bright. The Python bindings will make PDF composition even easier to access. The PDF indexing and redesign efforts will make developing in GooFit easier, and could potentially improve GooFit's already exciting performance. Hydra inclusion may further improve the performance and abilities of analysts.

References

[1] Verkerke W and Kirkby D P 2003 *eConf* **C0303241** MOLT007 (*Preprint* physics/0306116)
[2] Andreassen R E, de Silva W M, Meadows B T, Sokoloff M D and Tomko K A 2014 *IEEE Access* **2** 160–176
[3] Hoberock J and Bell N 2010 *Thrust* Version 1.8.2, accessed 2017-10-20 URL http://thrust.github.io/
[4] Schreiner H 2017 Modernizing GooFit: A Case Study *Proceedings of the Practice and Experience in Advanced Research Computing 2017 on Sustainability, Success and Impact* PEARC17 (ACM) pp 30:1–30:5 ISBN 978-1-4503-5272-7 URL http://doi.acm.org/10.1145/3093338.3093364
[5] Brun R and Rademakers F 1997 *Nucl. Instrum. Meth.* **A389** 81–86
[6] NVIDIA 2013 *Generics* Accessed 2017-10-20 URL https://github.com/bryancatanzaro/generics
[7] Zverovich V 2012 *fmt* Version 4.0.0, accessed 2017-10-20 URL https://github.com/fmtlib/fmt
[8] Yee A 2015 *FeatureDetector* Accessed 2017-10-20 URL https://github.com/Mysticial/FeatureDetector
[9] Schreiner H 2017 *CLI11* Version 1.2, accessed 2017-10-20 URL https://github.com/CLIUtils/CLI11
[10] Gauniyal A 2016 *Rang* Version 2.1, accessed 2017-10-20 URL https://agauniyal.github.io/rang
[11] Lees J P *et al.* (BaBar) 2016 *Phys. Rev.* **D93** 112014 (*Preprint* 1604.00857)
[12] Reichert S 2015-12-18 *Measurement of the mixing parameters of neutral charm mesons and search for indirect CP violation with $D^0 \to K_S^0\pi^+\pi^-$ decays at LHCb* Ph.D. thesis Dortmund U. URL http://inspirehep.net/record/1503629/files/CERN-THESIS-2015-348.pdf
[13] Hasse C, Albrecht J, Alves Jr A A, d'Argent P, Evans T D, Rademacker J and Sokoloff M D 2017 Amplitude analysis of four-body decays using a massively-parallel fitting framework *22nd International Conference on Computing in High Energy and Nuclear Physics (CHEP 2016) San Francisco, CA, October 14-16, 2016* (*Preprint* 1702.06735)
[14] Alves Jr A A and Sokoloff M D 2017 (*Preprint* 1702.05712)
[15] Sun L, Aoude R, dos Reis A C and Sokoloff M D 2017 Model-independent partial wave analysis using a massively-parallel fitting framework *22nd International Conference on Computing in High Energy and Nuclear Physics (CHEP 2016) San Francisco, CA, October 14-16, 2016* (*Preprint* 1703.03284)
[16] Jakob W 2015 *pybind11* Version 2.2.1, accessed 2017-10-20 URL https://github.com/pybind/pybind11
[17] Alves Jr A A 2017 Hydra: A Framework for Data Analysis in Massively Parallel Platforms (San José: GTC2017) Presentation ID S7340

IOP Conf. Series: Journal of Physics: Conf. Series **1085** (2018) 042015 doi:10.1088/1742-6596/1085/4/042015

Deep learning for inferring cause of data anomalies

V. Azzolini[1], M. Borisyak[2,3], G. Cerminara[4], D. Derkach[2,3], G. Franzoni[4], F. De Guio[5], O. Koval[6,3], M. Pierini[4], A. Pol[7], F. Ratnikov[2,3], F. Siroky[4], A. Ustyuzhanin[2,3] and J-R. Vlimant[8]

[1] Massachusetts Institute of Technology, Cambridge, USA
[2] National Research University Higher School of Economics, Moscow, Russia
[3] Yandex School of Data Analysis, Moscow, Russia
[4] CERN, European Organization for Nuclear Research, Geneva, Switzerland
[5] Texas Tech University, Lubbock, USA
[6] Skolkovo Institute of Science and Technology, Moscow, Russia
[7] University of Paris-Saclay, Paris, France
[8] California Institute of Technology, Pasadena, USA

E-mail: `fratnikov@hse.ru`

Abstract. Daily operation of a large-scale experiment is a resource consuming task, particularly from perspectives of routine data quality monitoring. Typically, data comes from different sub-detectors and the global quality of data depends on the combinatorial performance of each of them. In this paper, the problem of identifying channels in which anomalies occurred is considered. We introduce a generic deep learning model and prove that, under reasonable assumptions, the model learns to identify 'channels' which are affected by an anomaly. Such model could be used for data quality manager cross-check and assistance and identifying good channels in anomalous data samples. The main novelty of the method is that the model does not require ground truth labels for each channel, only global flag is used. This effectively distinguishes the model from classical classification methods. Being applied to CMS data collected in the year 2010, this approach proves its ability to decompose anomaly by separate channels.

1. Introduction

Data quality monitoring is a crucial task for every large scale high energy physics experiment. The challenge is driven by a huge amount of data. A considerable amount of person power is required for monitoring and classification. Previously [1], we designed a system, which automatically classifies marginal cases in general: both of 'good' and 'bad' data, and use human expert decision to classify remaining 'grey area' cases.

Typically, data comes from different sub-detectors or other subsystems, and the global data quality depends on the combinatorial performance of each of them. In this work, we aim instead to determine which sub-detector is responsible for anomaly in the detector behaviour, knowing only global flag. For this study, we use data [2] acquired by CMS detector [3] at CERN LHC. A proposed system can indicate affected channels and draw the attention of human experts to other channels. Data from the channels, which are reconstructed relying primarily on normally operating sub-detectors, can be used for further specific physics analysis.

IOP Conf. Series: Journal of Physics: Conf. Series **1085** (2018) 042015 doi:10.1088/1742-6596/1085/4/042015

2. Data and feature extraction

Data preprocessing procedure is the same as in the previous work [1], and described in detailes there.

All data is divided into time quanta - LumiSections[1], which are labelled as 'good' or 'bad'. Physics object is a proxy for a particle recorded in the detector which is constructed from recorded raw data collected by several sub-detectors. The information coming from four channels per each used LumiSection. These channels are equivalent to physics objects: muons, photons, particle flow jets (PF) or calorimeter jets (calo). Objects (equivalent to physics particles) are quantiled by their transverse momentum to have fixed number of features in each event. Then every selected object is characterized by its reconstructed physics properties: mass, spatial location, kinematics. And statistics for each feature for the entire lumisection is computed (5 percentiles, mean and variance). For instance, 'muons_mass_q4_p25' is one of the features in muon channel. Additionally, features like instant luminosity, number of particles in event and others were also introduced.

In total we have almost 16k data points (LumiSections) 25% of which were labeled as anomalous. There are 267 features in muon, 232 in photon, 126 in PF, 266 in calo channel. We hold 10% of data for tests. The final validation for our system is a computing correlations between the method predictions and experts labels for CMS sub-detectors.

3. Method

In order to predict a probability of anomaly in different channels separately we build a special multi-head neural network (NN) configuration presented in Fig. 1. The model was implement with Lasagne and Theano [4-5].

NN consist of four branches and each sub-networks has in input features from corresponding channel. Each branch returns a score for its channel. At the end, sub-networks are connected and the whole network is trained to recover global labels.

In this paper we present results when sub-networks are connected with kind of 'Fuzzy AND' operator:

$$\exp\left[\sum_{i=1}^{4}(f_i - 1)\right] \tag{1}$$

where f_i- is an output of the last layer of sub-networks.

It is proved for this operator, that under reasonable assumptions, the model learns to identify 'channels' which are affected by an anomaly.[2] As network connection operator, logistic regression and min operator with dropout could also be used.

The main benefit is that proposed approach uses only aggregated global quality tag for training, but allows to predict anomalies for separate channels.

In this way, each subnetwork returns score:

- close to 1 for good lumisections;
- close to 1 for anomalies invisible from subnetworks channel data, but visible to other subnetworks channels in the NN method;
- close to 0 for anomalies visible from subnetworks channel data.

Thus NN decomposes anomalies by channels.

'Fuzzy AND' approach assumes that there are anomalies not seen from all channels. They look like 'good' data points in the channel, surrounded by them and are classified by this

[1] Segment of data corresponding to 23 seconds of data taking

[2] Proof for 'Fuzzy AND' operator decomposition properties and code of the systems with different operators for network connection are available in `https://github.com/yandexdataschool/cms-dqm/`

IOP Conf. Series: Journal of Physics: Conf. Series **1085** (2018) 042015 doi:10.1088/1742-6596/1085/4/042015

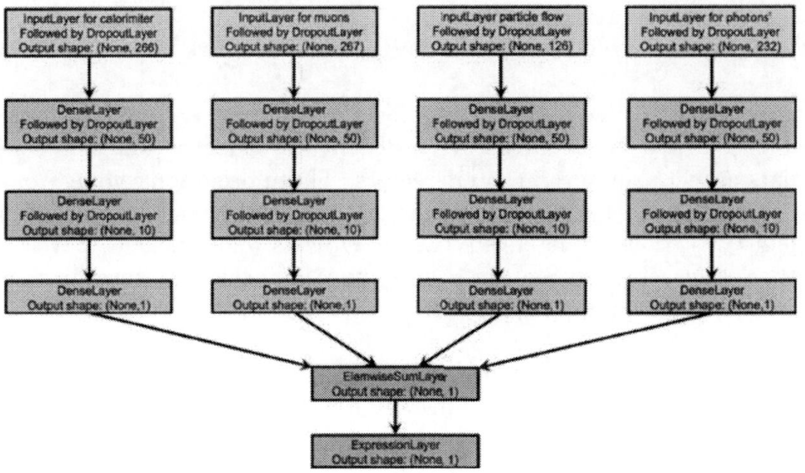

Figure 1. NN architecture with four sub-networks for each channel

channel as 'good' too. It is desired setting, but it causes a problem of small gradients of 'Fuzzy AND' function for close to the hyperplane data points during training. Long time is needed to change the hyperplane slightly when anomalous data points, which are potentially visible from particular channel, have already 'bad' labels in other channels. A simple cross-entropy loss for 'Fuzzy AND' output of the whole network is not sensitive enough in such cases. To resolve the problem and to accelerate the convergence we use a dynamic loss function:

$$L' = (1 - C) \cdot L + C \cdot (L_1 + L_2 + L_3 + L_4)/4, \tag{2}$$

where $L-$cross-entropy loss for 'Fuzzy AND' output of the network; L_i- so called 'auxiliary' losses, cross-entropy of corresponding sub-network scores against global labels; $C-$ decreasing along iterations constant to regulate amount of 'pretraining'.

With such 'soft pretraining' dynamic loss function we can force sub-networks to be more accurate and to take care about ambiguous samples during the first training iterations, but then to pay more attention to the predictive power of the whole NN against global labels. Thus, simple enough separation hyperplane is constructed during training, and problem of the small gradient, which is mentioned above, is avoided.

4. Results and discussions

Being applied to CMS data collected in 2010, method proves its ability to decompose anomaly by separate channels. Distributions of predictions in each NN branch are shown in Fig. 2. As expected, we can see scores close to one for 'good' samples. And 'bad' data has two options, it could be visible from channel (score close to zero) or not. We can think about the second cases, as data is not affected by an anomaly and it is still useful for for further physical analysis. This part of data is needed to be reviewed by CMS experts channel wise.

Thus, method suggests that most of anomalies occurred (or at least better identified) by calo channel. And there are some data from others channels, which does not look like anomalous (predicted labels close to one when the global label is "bad") and can be saved.

In these experiments global predictive power of the whole network on validation set is rather high, ROC AUC score equals to 0.96. To verify obtained results we calculate correlations between sub-network predictions and experts' labels for CMS subsystems, presented in Fig. 3. Performance on the test part of the data is presented. Inclusive global labels were used for

IOP Conf. Series: Journal of Physics: Conf. Series **1085** (2018) 042015 doi:10.1088/1742-6596/1085/4/042015

training, and all others were not. All ROC AUC scores a higher than 0.5. It means that there is a clear correlation between sub-networks outputs and corresponding subsystem labels or some of them are almost independent (scores close to 0.5), but there is no anti-correlations, as it is expected. For example, Muon branch predictions are similar to Muon expert's labels more than other branches. And it is less correlated with Jets, Photons/Electron and ECAL than others. A surprising observation is that NN don't sensitive CSC muon chambers anomalies.

Anomalies in 2010B CMS OpenData usually happens in several sub-detectors simultaneously. We selected lumisections with anomalies in the only one sub-detector to get more precise comparison, and computed percentage of such exclusive anomalies identified by NN branches, results are shown in Fig. 4. Unfortunately, there are only four such sub-detectors, and they don't contain enough number of anomalies for reliable interpretation. We observe perfect behavior of PF branch, which is not depended on muon chambers, but information from ECAL and HCAL is used for the reconstruction of PF physical properties. And one of the artifacts is that anomalies in RPC muon chambers are visible for calo branch.

So, there are still various subtle effects that need to be studied and taken into account in order to get production quality. On this stage the system suits for expert cross-check assistant. And, as an automated approach and the expert statistics are not mutually exclusive, injection of expert statistics into the feature set is a good starting point for the improvement of the system.

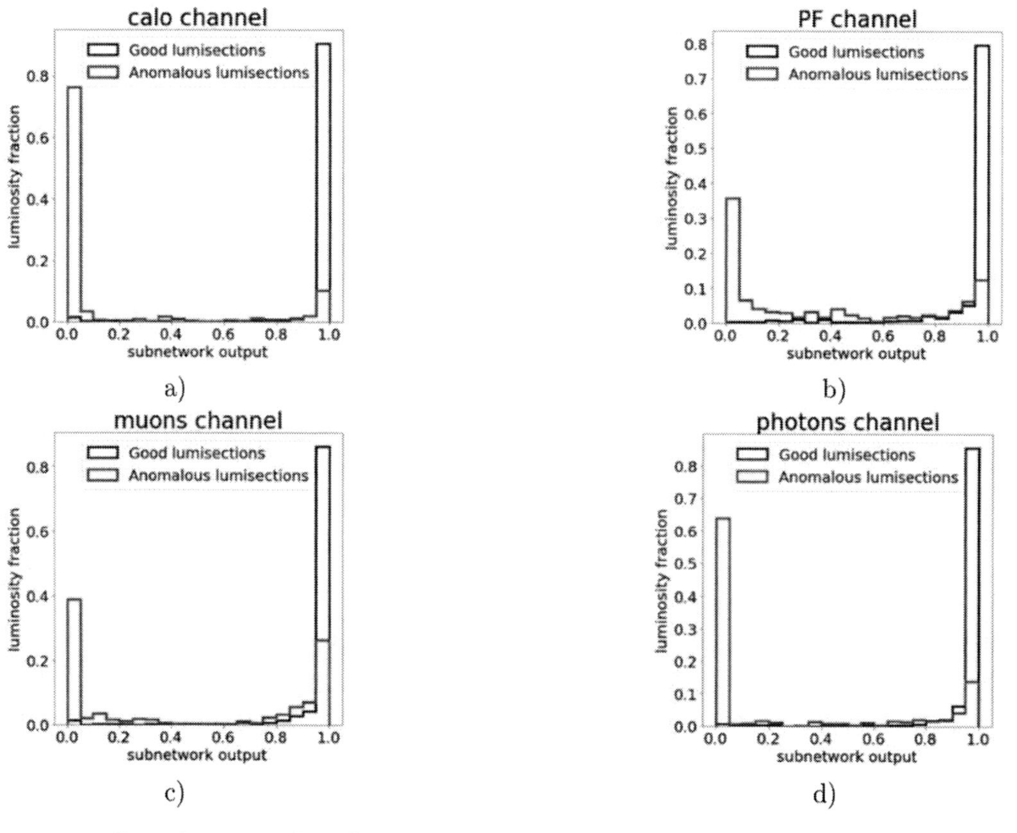

Figure 2. Distributions of predictions returned by NN branches build on features from a) calorimiter, b) particle flow jets, c) muons, d) photons channels.

IOP Conf. Series: Journal of Physics: Conf. Series **1085** (2018) 042015 doi:10.1088/1742-6596/1085/4/042015

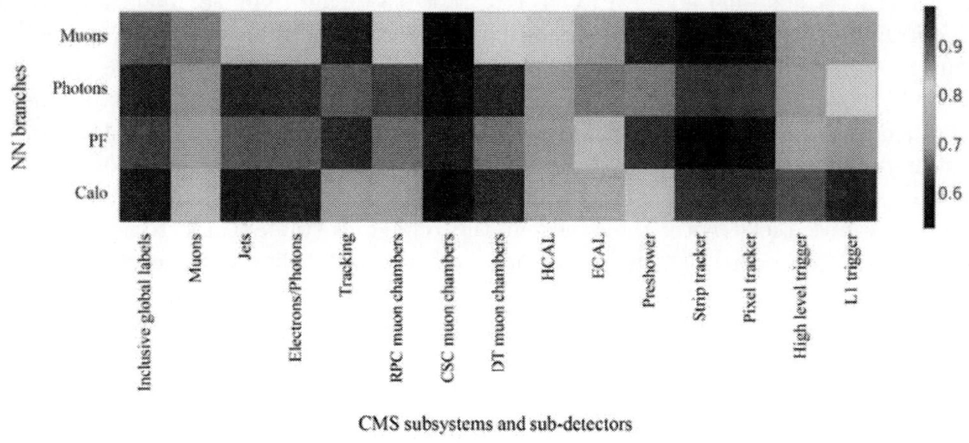

Figure 3. ROC AUC scores of NN branches scores against experts' labels for CMS subsystems. Number of anomalies in each subsystem is given in the brackets.

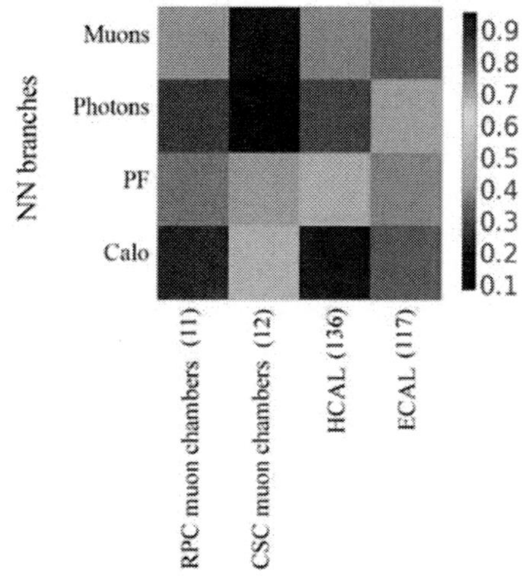

Figure 4. Fraction of identified exclusive anomalies in CMS subsystems by NN branches. Number of exclusive (when anomaly is happened in only one detector) anomalies is given in the brackets.

5. Conclusions

In this work, we described a deep learning approach for inferring cause of data anomalies. While developed with the CMS experiment at CERN in mind, we use an agnostic approach which allows the straightforward adaptation of the proposed algorithm to different experimental setups. Method shows its ability to decompose anomalies by separate channels, being applied to data collected by the CMS experiment in 2010. Although only global quality labels were used for training, we got correlation between sub-networks outputs and corresponding true subsystem labels, what proves the principles of the approach.

IOP Conf. Series: Journal of Physics: Conf. Series **1085** (2018) 042015 doi:10.1088/1742-6596/1085/4/042015

Acknowledgements
The research leading to these results was partly supported by Russian Science Foundation under grant agreement № 17-72-20127.

References
[1] Borisyak M, Ratnikov F, Derkach D and Ustyuzhanin A. *Towards automation of data quality system for CERN CMS experiment.* arXiv:1709.08607 [physics.data-an], September 2017
[2] Calderon A *at al. Open access to high-level data and analysis tools in the CMS experiment at the LHC* 2015 *J. Phys.: Conf. Ser.* **664** 032027
[3] CMS Collaboration, The CMS experiment at the CERN LHC, 2008 *JINST* **3** S08004
[4] Theano Development Team, *Theano: A Python framework for fast computation of mathematical expressions.* arXiv:1605.02688 [cs.SC], May 2016
[5] Dieleman S *at al.* Lasagne: First release. 10.5281/zenodo.27878, August 2015.

IOP Conf. Series: Journal of Physics: Conf. Series **1085** (2018) 042016 doi:10.1088/1742-6596/1085/4/042016

Parallelized Kalman-Filter-Based Reconstruction of Particle Tracks on Many-Core Architectures

G Cerati[4], P Elmer[2], S Krutelyov[1], S Lantz[3], M Lefebvre[2], M Masciovecchio[1], K McDermott[3], D Riley[3], M Tadel[1], P Wittich[3], F Wurthwein[1], and A Yagil[1]

[1] UC San Diego, 9500 Gilman Dr., La Jolla, California, USA 92093
[2] Princeton University, Princeton, New Jersey, USA 08544
[3] Cornell University, Ithaca, New York, USA
[4] Fermi National Accelerator Laboratory, Batavia, USA

E-mail: Daniel.Riley@cornell.edu

Abstract. Faced with physical and energy density limitations on clock speed, contemporary microprocessor designers have increasingly turned to on-chip parallelism for performance gains. Algorithms should accordingly be designed with ample amounts of fine-grained parallelism if they are to realize the full performance of the hardware. This requirement can be challenging for algorithms that are naturally expressed as a sequence of small-matrix operations, such as the Kalman filter methods widely in use in high-energy physics experiments. In the High-Luminosity Large Hadron Collider (HL-LHC), for example, one of the dominant computational problems is expected to be finding and fitting charged-particle tracks during event reconstruction; today, the most common track-finding methods are those based on the Kalman filter. Experience at the LHC, both in the trigger and offline, has shown that these methods are robust and provide high physics performance. Previously we reported the significant parallel speedups that resulted from our efforts to adapt Kalman-filter-based tracking to many-core architectures such as Intel Xeon Phi. Here we report on how effectively those techniques can be applied to more realistic detector configurations and event complexity.

1. Introduction

Finding tracks is the most computationally complex of the steps reconstructing events in the CMS detector. The speed of online reconstruction has a direct impact on how effectively interesting data can be selected from the 40 MHz collisions rate, while the speed of the offline reconstruction limits how much data can be processed for physics analyses. The time spent in tracking will become even more important in the HL-LHC era of the Large Hadron Collider, as the increase in event rate will increase the detector occupancy (pile-up, PU), leading to the exponential increases in track reconstruction time, illustrated in Figure 1 [1].

At the same time, due to power density and physical scaling limits, the performance of single CPUs has slowed, with advances in performance increasingly relying on highly parallel architectures. In order to sustain the higher HL-LHC processing requirements without compromising physics performance or unduly delaying results, the LHC experiments must make full use of highly parallel architectures.

As a potential solution, we are investigating adapting the existing CMS track finding algorithms and data structures [2] to make efficient use of highly parallel architectures, such

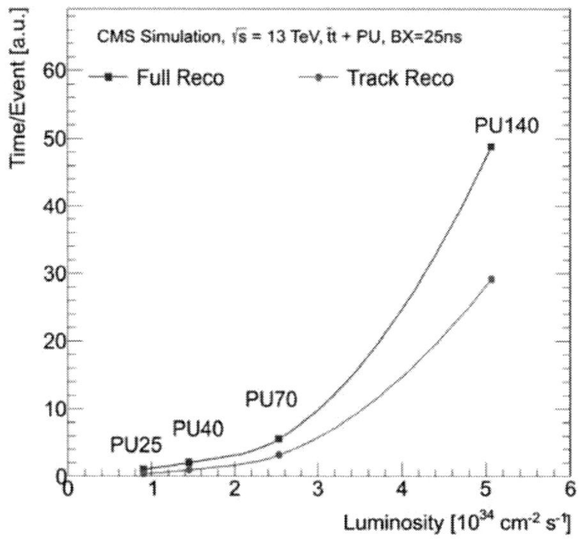

Figure 1. CPU time per event versus instantaneous luminosity, for both full reconstruction and the dominant tracking portion. Simulated data with pile-up of 25 primary interactions per event (PU25) corresponds to the data taken during 2012, while pile-up of 140 (PU140) corresponds to the low end of estimates for the HL-LHC era.

as Intels Xeon Phi and NVIDIA general-purpose graphics processing units (GPGPUs). In this investigation we have followed a staged approach, starting with Intel Xeon and Xeon Phi Knights Corner (KNC) architectures, an idealized detector geometry, and a series of simpler "warm-up" exercises such as track fitting. This simplified problem domain was used to develop our tools, techniques, and understanding of the issues scaling track finding to many cores. The warm-up exercises let us develop useful components while also allowing the physicists to become familiar with the computational tools and techniques, while the computational experts learned about the problem domain. Armed with the results of those initial investigations, we are now addressing more realistic detector geometries and event content, as well as adding new platforms. This paper gives an overview of our progress to date and assesses the effectiveness of our staged approach.

2. Kalman Filter Tracking

Our targets for parallel processing are track reconstruction and fitting algorithms based on the Kalman Filter [3] (KF). KF-based tracking algorithms are widely used to incorporate estimates of multiple scattering directly into the trajectory of the particle. Other algorithms, such as Hough Transforms and Cellular Automata [4][5], are more naturally parallelized. However, these are not the main algorithms in use at the LHC today. The LHC experiments have an extensive understanding of the physics performance of KF algorithms; they have proven to be robust and perform well in the difficult experimental environment of the LHC.

KF tracking proceeds in three main stages: seeding, building, and fitting. Seeding provides the initial estimate of the track parameters based on a few hits in the innermost regions of the detector; seeding is currently out of scope for our project. Track building projects the track candidate outwards to collect additional hits, using the KF to estimate which hits represent the most likely continuation of the track candidate. Track building is the most time consuming step, as it requires branching to explore multiple candidate tracks per seed after finding compatible hits on a given layer. When a complete track has been reconstructed, a final fit using the KF is performed to provide the best estimate of the track's parameters.

To take full advantage of parallel architectures, we need to exploit two types of parallelism: vectorization and parallelization. Vector operations perform a single instruction on multiple data (SIMD) at the same time, in lockstep. In tracking, branching to explore multiple candidates per

IOP Conf. Series: Journal of Physics: Conf. Series **1085** (2018) 042016 doi:10.1088/1742-6596/1085/4/042016

seed can interfere with lock-step single-instruction synchronization. Multi-thread parallelization aims to perform different instructions at the same time on different data. The challenge to track building is workload balancing across different threads, as track occupancy in a detector is not uniformly distributed on a per event basis. Past work by our group has shown progress in porting sub-stages of KF tracking to support parallelism in simplified detectors [6][7][8][9]. As the hit collection is completely determined after track building, track fitting can repeatedly apply the KF algorithm without branching, making this a simpler case for both vectorization and parallelization; results in implementing vectorized parallel KF tracking to Xeon Phi were shown previously [10].

2.1. Matriplex

The implementation of a KF-based tracking algorithm consists of a sequence of operations on matrices of sizes from 3×3 up to 6×6. In order to optimize efficient vector operations on small matrices, and to decouple the computational details from the high level algorithm, we have developed a new matrix library. The *Matriplex* memory layout uses a matrix-major representation optimized for loading vector registers for SIMD operations on small matrices, using the native vector-unit width on processors with small vector units or very large vector widths on GPGPUs. Matriplex includes a code generator for defining optimized matrix operations, with support for symmetric matrices and on-the-fly matrix transposition. Patterns of elements that are known by construction to be zero or one can be specified, and the resulting code will be optimized to eliminate unnecessary register loads and arithmetic operations. The generated code can be either standard C++ or macros that map to architecture-specific intrinsic functions.

3. Initial Test Scenario

In order to study parallelization with minimal distractions, we developed a standalone KF-based tracking code using a simplified ideal barrel geometry with a uniform longitudinal magnetic field, gaussian-smeared hit positions, a particle gun simulation with flat transverse momentum distribution between 0.5 and 10 GeV, no material interaction, and no correlation between particles nor decays. This simplified configuration was used to study vector and parallel performance and to study the performance of different choices of coordinate system and handling of KF-updates. These studies led to the use of a hybrid coordinate system, using global Cartesian coordinates for spatial positions and polar coordinates for the momentum vector.

3.1. Initial Results

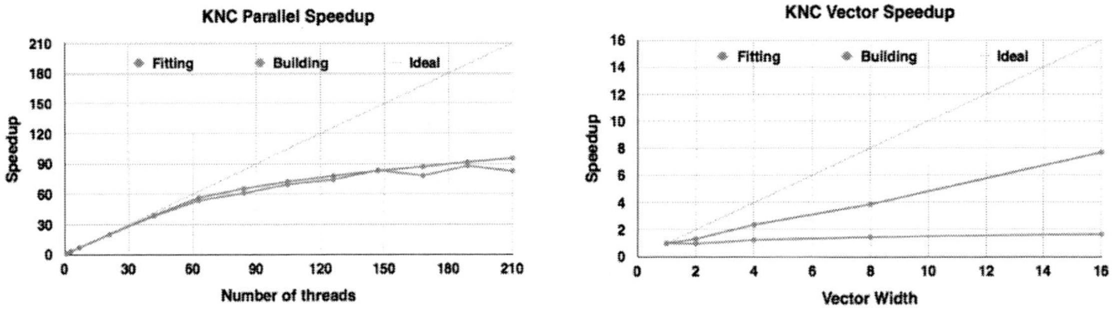

Figure 2. KNC parallel and vector speedups for track fitting and building.

Figure 2 illustrates the parallel and vector speedups on the KNC platform for the simple test scenario. We achieve good parallelization up to the number of physical cores, 61, with some continued improvements from hyper-threading (KNC CPUs change hyper-thread every clock cycle, so full utilization would be a speedup of 122). While both track fitting and building are "embarrassingly parallel", we believe the parallelization speedup is limited by memory bandwidth and cache sizes, so our scaling performance is somewhat dependent on the details of the memory subsystem. For the relatively simple track fitting we also see good vector performance. The combinatorial nature of the track building algorithm, which examines and adds a variable number of hit candidates in each layer, results in many branching operations that impede vectorization, as well as adding frequent repacking operations to keep the full vector width utilized. The larger and more complicated data structures used for selecting hit candidates also results in poorer data locality and higher bandwidth requirements. We are continuing to investigate strategies for improving track building vector utilization.

3.2. Initial Lessons Learned

Achieving acceptable vector and multi-thread parallel performance required careful attention to detail. Regular profiling with Intel VTune Amplifier and examination of the compiler optimization reports helped identify many obstacles, some relatively subtle. For multi-thread parallelism, the two most critical issues were memory management and workload balancing. To avoid memory stalls and cache conflicts, we reduced our data structures to the minimum necessary for the algorithm, optimized our data structures for efficient vector operations, and minimized object instantiations and dynamic memory allocations. Workload balancing for track building is complicated by the uncertain distribution of track candidates and hits, which can result in large tail effects for a naive static partitioning scheme. We found that using Intel Threaded Building Blocks [11] (TBB) tasks, with smaller units of work and dynamic "work-stealing", let us naturally express thread-parallelism in a way that allowed more dynamic allocation of resources and reduced the tail effects from uneven workloads.

4. Adding Complications

Recently we have nearly completed work on using more realistic detector geometries and realistic events. In addition, we have added new platforms, principally the NVIDIA Tesla K20/K40 and Pascal P100 GPGPUs, which present a very different programming model.

4.1. Realistic Geometries

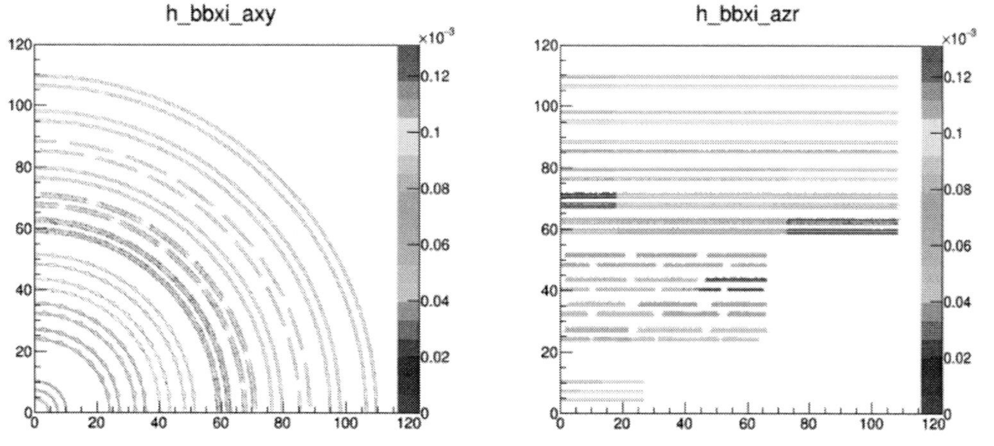

Figure 3. x-y and r-z views of simplified CMS geometry

IOP Conf. Series: Journal of Physics: Conf. Series **1085** (2018) 042016 doi:10.1088/1742-6596/1085/4/042016

We implemented two more realistic geometries. The first adds endcaps to our idealized geometry, yielding a "cylinder with lids" that was used for initial algorithm development. The second geometry implements a simplified form of the CMS tracker geometry, illustrated in Figure 3. For track building in the less regular geometry, track propagation is performed in two steps, an initial propagation to the average radius of the barrel or average Z of the endcap disk, followed by further propagation steps to the position of each candidate hit. KF updates are performed on the plane tangent to the hit radius. These choices avoid the complexities of the full detector description. We are currently validating the physics performance of our algorithms with the simplified CMS geometry, prior to completing computational performance studies.

4.2. Realistic Events

Event data from the full CMS simulation suite is translated into a reduced form that can be processed by our standalone tracking code. Seeds used are from the first iteration of the CMS iterative tracker [1]. Realistic events typically have lower occupancy and less uniformity in track distribution, leading to more challenges for workload balancing and thus to worse multi-thread scaling than our "toy" setup. To counter this effect we process multiple events at a time, filling in gaps caused by the varying levels of parallelism within a single event. Figure 4 shows that this strategy is effective at recovering good parallel scaling for more realistic events, overcoming the workload balancing issues from lower occupancy events.

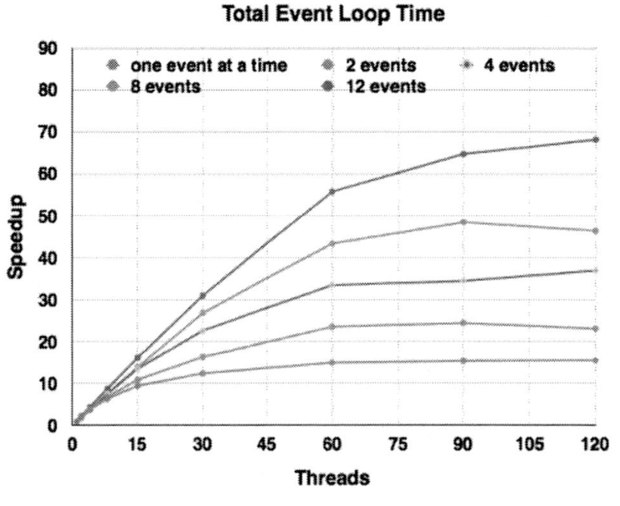

Figure 4. KNC track building speedup as a function of the number of threads, for events with occupancy similar to seeds from the first iteration of the CMS iterative tracker. For these sparser events, processing tracks from multiple events in parallel significantly improves scaling.

4.3. GPGPU Developments

We have also implemented the track fitting and building algorithms on NVIDIA GPGPUs using the CUDA toolkit. The large core count of GPGPUs and their thread scheduling policies force a large number of threads to be concurrently scheduled. This implies that coping with branching is critical as threads of a *warp* (a group of 32 consecutive threads) execute the same common instructions at the same time. Investigating the memory layout efficiency for GPGPUs, we found Matriplex to be the most efficient of the alternatives we tested as it naturally allows coalesced memory accesses, where memory requests from consecutive threads to consecutive locations can be coalesced into a single transaction. Figure 5 shows the results of these studies. Using the Matriplex data structure also facilitated code sharing, by templating low level routines such as KF propagation to accept Matriplex structures sized for CPU or GPGPU, while only high-level "steering" routines needed to be customized. To minimize branching, the GPGPU version allocates a fixed number of slots for each track seed to store the corresponding track candidates

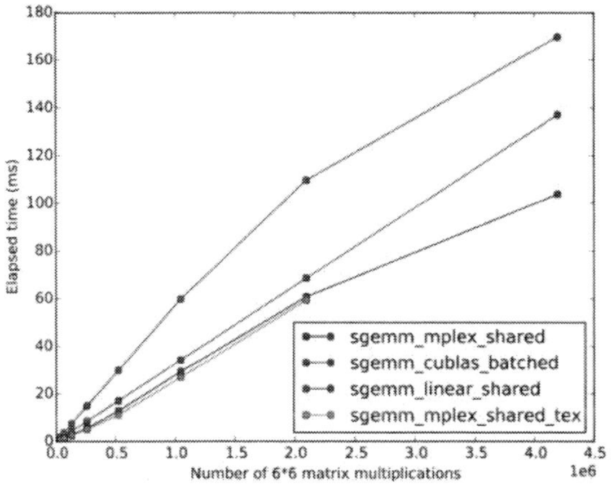

Figure 5. Performance comparison of different memory organizations for multiplying 6×6 matrices, showing the time as a function of the number of 6×6 matrix multiplications. Compared to arrays of matrices using standard libraries, Matriplex (labeled "mplex") shows superior performance.

from that seed. The candidates are organized as a heap, allowing rapid identification of the best candidates. As in the CPU version, we found that multiple events in flight was also necessary for scaling so that expensive copy operations could be overlapped with processing. We are currently investigating whether the design choices made for the GPGPU implementation can be applied to improve our vector performance on the Xeon platforms.

5. Conclusion and Outlook

We have made significant progress in parallelized and vectorized Kalman Filter-based end-to-end tracking R&D on Xeon and Xeon Phi architectures, with some work on GPGPUs. Through the use of a variety of tools we have developed a good understanding of bottlenecks and limitations of our implementation which has led to further improvements. With our own Matriplex package and TBB, we can achieve good utilization of unconventional highly parallel vector architectures; these results should also be applicable to newer architectures with large vector registers, such as Intel Skylake-SP. We are currently focusing on processing fully realistic data.

We found our staged approach to the problem to be effective in developing our understanding of the problem and promoting effective collaboration across domains. Despite some initial concerns about diluting our efforts, we have also found that addressing multiple architectures has been useful for developing approaches that can be applied across architectures.

Acknowledgments

This work is supported by the U.S. National Science Foundation, under the grants PHY-1520969, PHY-1521042, PHY-1520942 and PHY-1120138, and by the U.S. Department of Energy.

References

[1] Cerati G 2015 *PoS VERTEX* **2014** 037
[2] The CMS Collaboration 2014 *JINST* **9** P10009
[3] Fruehwirth R 1987 *NIM* A **262** 444-450
[4] Pantaleo F, Di Florio A, Carta A, Pierini M 2017 *Preprint* ACAT2017-78
[5] Emeliyanov D, Howard J 2012 *JoP: Conf. Series* **396** 012018
[6] Cerati G *et al.* 2015 *JoP: Conf. Series* **664** 072008
[7] Cerati G *et al.* 2016 *EPJ Web Conf.* **127** 00010
[8] Cerati G *et al.* 2016 *Preprint* physics.ins-det/1702.06359
[9] Cerati G *et al.* 2017 *Preprint* physics.comp-ph/1705.02876
[10] Cerati G *et al.* 2014 *JoP: Conf. Series* **608** 012057
[11] Reinders M 2007 *Intel Threading Building Blocks: Outfitting C++ for Multi-core Processor Parallelism* (Cambridge, MA: O'Reilly Media)

IOP Conf. Series: Journal of Physics: Conf. Series **1085** (2018) 042017 doi:10.1088/1742-6596/1085/4/042017

Controlling Physical Attributes in GAN-Accelerated Simulation of Electromagnetic Calorimeters

Luke de Oliveira
Vai Technologies & Lawrence Berkeley National Laboratory
E-mail: lukedeo@ldo.io

Michela Paganini
Yale University Department of Physics & Lawrence Berkeley National Laboratory
E-mail: michela.paganini@yale.edu

Benjamin Nachman
Lawrence Berkeley National Laboratory
E-mail: bnachman@cern.ch

Abstract.
High-precision modeling of subatomic particle interactions is critical for many fields within the physical sciences, such as nuclear physics and high energy particle physics. Most simulation pipelines in the sciences are computationally intensive – in a variety of scientific fields, Generative Adversarial Networks have been suggested as a solution to speed up the forward component of simulation, with promising results. An important component of any simulation system for the sciences is the ability to condition on any number of physically meaningful latent characteristics that can effect the forward generation procedure. We introduce an auxiliary task to the training of a Generative Adversarial Network on particle showers in a multi-layer electromagnetic calorimeter, which allows our model to learn an attribute-aware conditioning mechanism.

1. Introduction

Modeling the interactions of particles with media is critical across many physical sciences. Detailed simulation of particle collisions and subsequent interactions at the LHC experiments, along with simulating exact detector response, is very computationally expensive, requiring billions of CPU hours and roughly half of LHC computing resources [1, 2, 3].

Recently, deep learning-based generative models including Generative Adversarial Networks [4] and Variational Auto-Encoders [5] have been proposed and tested as a solution to significantly speed up scientific simulation in Oncology [6], Cosmology [7, 8, 9], High Energy Physics [10, 11], and many other basic science fields. We propose a simple extension to the CaloGAN model [10] allowing such a generation model to be conditioned on a vector of physically meaningful characteristics. We introduce a series of auxiliary tasks to encourage our model to learn $p(x|\xi)$, where ξ is a vector of meaningful characteristics which should guide the generation procedure.

2. The Dataset

The publicly available dataset from Ref. [12], composed of 500,000 e^+, 500,000 π^+, and 400,000 γ GEANT4-generated [13] showers, is used to construct and validate a proposed architecture for conditional modeling. The detector geometry present in the data consists of a cubic region along the z direction of $V = (480\text{ mm})^3$ of an ATLAS-inspired EM calorimeter, at a distance of $z_0 = 288$ mm from (0,0,0). The volume is radially segmented into three layers of depth 90 mm, 347 mm, and 43 mm composed of alternating layers of lead and liquid Argon.

In this work, all GEANT4 coordinates native to the simulation are transformed into ATLAS coordinates for consistency with practical use-cases in experiments.

3. Generative Adversarial Networks

Generative Adversarial Networks (GAN) [4] are a method to learn a generative model by recasting the generation procedure as a minimax game between two actors that are parameterized by deep neural networks. A *generator* network is tasked with building samples that are very similar to some target data distribution, and a *discriminator* network (the adversary) is tasked with learning to distinguish real-looking samples from fake-looking samples.

Formally, assume we have a data distribution we wish to model, $x \sim p_{\text{data}}(x) \in \mathcal{X}$. We construct a generator network G that maps a d-dimensional latent prior $z \sim p_z(z) \in \mathbb{R}^d$ to synthetic samples $G : \mathbb{R}^d \longrightarrow \mathcal{X}$. The map G implicitly defines a learned probability distribution from which we can sample, $p_G(x)$. In order to direct $p_G(x)$ towards the data distribution $p_{\text{data}}(x)$, a discriminator network D is tasked with taking a sample x and classifying it with a label y as originating from the data distribution ($y = 1$) or from the implicit synthetic distribution ($y = 0$), i.e., $D : \mathcal{X} \longrightarrow [0, 1]$.

In the original formulation of GANs [4] a loss $\mathcal{L}_{\mathsf{adv}}$ is constructed (Eqn. 1) to guide the learning towards equilibrium. When G and D are allowed to be drawn from the space of all continuous functions (i.e., they have infinite capacity), this system converges to a unique Nash Equilibrium [14] in which the implicit distribution $p_G(x)$ exactly recovers the target data distribution $p_{\text{data}}(x)$, and $D(\cdot)$ is 1/2 everywhere [4]. It can be shown [4] that this procedure minimizes the Jensen-Shannon divergence between $p_G(x)$ and $p_{\text{data}}(x)$, $\text{JSD}(p_G(x) \parallel p_{\text{data}}(x))$.

$$\mathcal{L}_{\mathsf{adv}} = \underbrace{\mathbb{E}_{z\sim p_z(z)}[\log(1 - D(G(z)))]}_{\substack{\text{term associated with } D \text{ classifying} \\ \text{a sample from } G \text{ as fake}}} + \underbrace{\mathbb{E}_{x\sim p_{\text{data}}(x)}[\log D(x)]}_{\substack{\text{term associated with } D \\ \text{perceiving a real sample as real}}} \tag{1}$$

In the original formulation, G minimizes $\mathcal{L}_{\mathsf{adv}}$ while D minimizes $-\mathcal{L}_{\mathsf{adv}}$, i.e., the game is zero sum. However, this formulation suffers from gradient saturation when a synthetic sample $x = G(z)$ is seen as very fake, i.e., when $D(G(z)) \approx 0$, stagnating the learning procedure due to near-zero gradients. To overcome this, Ref. [4] proposes the non-saturating heuristic, replacing the generator objective with Eq. 2, which it minimizes just as before. This formulation is utilized in our experiments, and we let $\mathcal{L}_D = \mathcal{L}_{\mathsf{adv}}$ represent the same quantity as in Eq. 1, except now it is only used for the discriminator.

$$\mathcal{L}_G = -\mathbb{E}_{z\sim p_z(z)}[\log D(G(z))] \tag{2}$$

Through architecture design [15, 16, 17, 18, 19, 20] and more theoretically-sound training objectives used [21, 22, 23, 24, 25, 26, 27], GANs have emerged as one of the most promising methods for neural networks to learn generative models of complicated and structured data spaces. We follow Ref. [28] and impose task-specific metrics which allow us to move away from likelihood-level notions of quality which do not directly translate to tasks of interest. We utilize the original heuristic loss formulation (Eq. 2) with label flipping [20] in lieu of more recent advances. The authors expect future work will experiment with effective conditioning mechanisms for other more popular GAN formulations such as the Wasserstein GAN [22] and the Cramér GAN [27].

To effectively condition on physical attributes, we add additional terms to the loss outlined in Eqn. 1, which we explore in-depth in Sec. 4. We modify the CaloGAN architecture [10] to perform the attribute-conditional task, leaning on inspiration from Ref. [17] for conditioning mechanisms.

4. The Conditional CaloGAN

To create a GAN based simulator that presents a useful solution for fast simulation, we need not only learn $p_{\mathsf{data}}(x)$, but we must also approximate $p_{\mathsf{data}}(x|\xi)$, where ξ is a vector of conditioning attributes. Given the dataset outlined in Sec. 2, ξ is chosen to be $\xi = (E, x_0, y_0, \theta, \phi)$, where x_0 and y_0 are the incident coordinates and θ and ϕ are incident angles. As per Ref. [10], an absolute-deviance is defined over requested and reconstructed energy from a simulated collision, encouraging the GAN to learn the conditional distribution. In this work, a generalization is presented. For quantities that do not have a direct mathematical formula given a shower (i.e., x_0, y_0, θ, and ϕ), we build a separate head on the discriminator to learn a specific submodel Ξ which focuses on learning to regress on said quantities given a shower as input. This estimate $\hat{\xi} = \Xi(x)$ is learned along with the normal alternating gradient step. To both $\mathcal{L}_G$ and $\mathcal{L}_D$ from Sec. 3, $\mathcal{L}_{\hat{\xi}}$ (Eq. 3) is added such that both players seek to minimize the attribute reconstruction error

$$\mathcal{L}_{\hat{\xi}} = \lambda^T \mathbb{E}_{x\sim p_{\mathsf{data}},\, \xi \sim p_\xi(\xi|x)} \left[||\xi - \Xi(x)|| \right]. \tag{3}$$

Note that in Eq. 3 a hyperparameter λ is included to allow us to individually control the contribution for each attribute reconstruction task. Combining this with the energy conditioning loss $\mathcal{L}_E$ and associated weight λ_E from the original CaloGAN formulation [10], our final optimization problem consists of the generator minimizing $\mathcal{L}_G + \mathcal{L}_{\hat{\xi}} + \lambda_E \mathcal{L}_E$ and the discriminator minimizing $-\mathcal{L}_D + \mathcal{L}_{\hat{\xi}} + \lambda_E \mathcal{L}_E$.

5. Sparsity Considerations

As in most scientific applications, sparsity plays an important role in determining how viable a generates sample is. In the original CaloGAN work [10], a raw sparsity (or occupancy) calculation is performed and fed as additional information to the discriminator. For an image $X \in \mathbb{R}^{m\times n}$, the sparsity is calculated as:

$$\mathsf{sparsity}(X) = \frac{1}{mn} \sum_{i<m} \sum_{j<n} \mathbb{I}[X_{i,j} \neq 0]. \tag{4}$$

Note that although this allows the discriminator to learn to reject generated samples that do not match the sparsity levels of real samples, this formulation does not allow for any gradient signal to propagate to the generator due to all-zero sub-derivatives of the indicator function $\mathbb{I}[\cdot]$. To ameliorate this deficiency, *soft sparsity* is introduced – a quantity that in the limit of tunable hyperparameters, converges to Eqn. 4. In particular, we define this quantity as

$$\mathsf{softsparsity}(X) = \left(\frac{1}{nm}\right) \left\| \frac{|X|^\alpha + |X|^{-\beta}}{|X|^\alpha + |X|^{-\beta} + 1} \right\|_1, \tag{5}$$

where $\alpha, \beta > 0$, all matrix powers and $|\cdot|$ operators are assumed to act *pointwise* on X, and $\|\cdot\|_1$ is the entry-wise 1-norm rather than the induced norm.

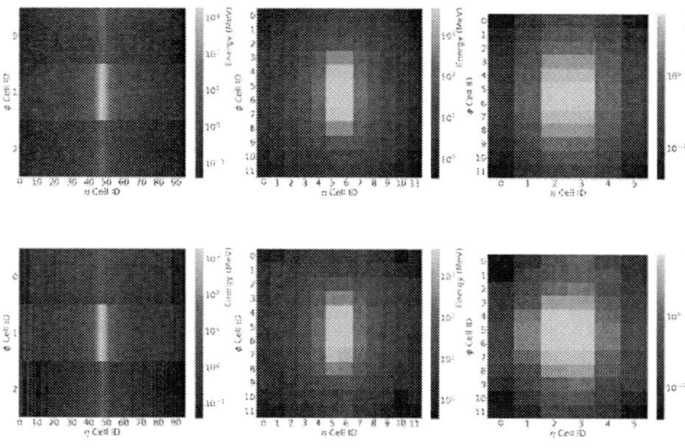

Figure 1. Average generated calorimeter image from GEANT (top) and our model (bottom) for e^+. From left to right, we proceed through layers of the segmented calorimeter.

Examining Eqn. 5, we note that $\mathsf{softsparsity} : \mathbb{R}^{m\times n} \longrightarrow [0, 1]$, and that the limit behavior for Eqn. 5 is consistent with Eqn. 4, i.e.,

$$\lim_{\alpha,\beta\to\infty} \mathsf{softsparsity}(X) = \frac{1}{mn}\sum_{i<m}\sum_{j<n}\mathbb{I}[X_{i,j} \neq 0]. \quad (6)$$

This information is added to the discriminator in raw form as well as a minibatch discrimination [20] step in order to allow our model to learn the correct distribution of sparsity.

6. Experiment & Results

6.1. Setup

A Conditional CALOGAN (Sec. 4) with soft sparsity information (Sec. 5) is trained on the dataset described in Section 2. We perform our experiments on the simulated e^+ showers.

To allow the model to learn to generate translation invariant samples, we replace all locally connected layers in the original CALOGAN formulation with normal convolution layers, as well as a fully-convolutional attention mechanism to mediate layer-to-layer dependence. In addition, convolutional kernels are added as blurring operations to specific detector layers for smearing effects, where weights are initialized to $\frac{1}{nm}$ for an $n\times m$ filter and are not updated during training, which was empirically found to improve generation quality. This explicit blur operation allows the model to learn a factored problem which marginalizes out smearing effects, and allows the model to focus on important factors in shower patterns.

Network trainings are performed using the KERAS [29] library with customized TensorFlow [30] operations across four Pascal Architecture NVIDIA® Titan X GPUs for a total of 150 epochs.

6.2. Results

As a first layer of validation, and commensurate with the validation performed in Ref. [10], average images per calorimeter layer between images from the dataset of GEANT4 images are compared to images generated from the proposed model, as shown in Figure 1. This visualization illustrates the increased energy dispersion compared to the original dataset presented in Ref. [10]. The comparison in Figure 1 shows a high level of visual agreement between GAN-generated and GEANT4-generated showers.

In addition to aggregate image pattern validation, nearest GAN neighbors[1] are retrieved

[1] By standard vectoral 2-norm.

IOP Conf. Series: Journal of Physics: Conf. Series **1085** (2018) 042017 doi:10.1088/1742-6596/1085/4/042017

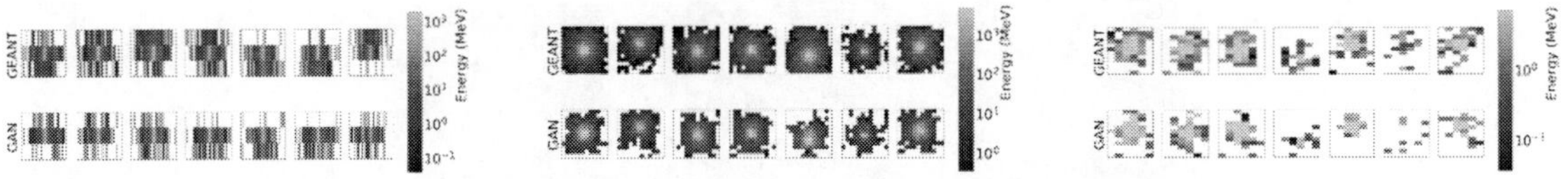

Figure 2. Nearest GAN-generated neighbors (top) for seven random GEANT4-generated e^+ showers (bottom) for the first layer (left), second layer (middle), and last layer (right) of the calorimeter.

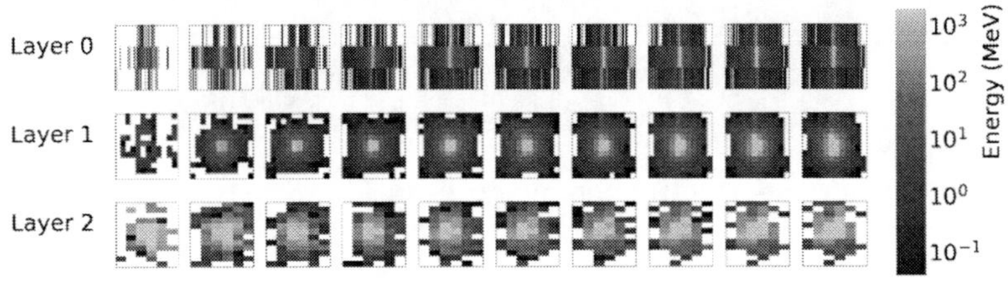

Figure 3. Interpolation across physical range of incident energy as a conditioning latent factor for e^+ showers, with energy increasing from 1 GeV to 100 GeV from left to right. Each point in the interpolation is an average of 10 showers, with each point along the traversal build from an identical latent prior z.

for seven GEANT4 images and used to validate that (a) our model does not memorize shower patterns, and (b) that the full space of displacements (both angular and positional) are explored.

At the nearest-neighbor level, the model produces convincing energy deposition patterns, as shown in Figure 2. The model does not appear to memorize the training dataset. In addition, positional variance (observed by noticing energy centroid deviations from the center of the calorimeter image) is well explored by the GAN, as shown by GAN-generated images matching all positions given by GEANT4.

To further verify our models ability to condition on physical attributes, the latent space for each conditioning variable is traversed, showing how the model learns about each conditioning factor. In any practical setting, such conditioning mechanisms will need to be tuned to a high level of fidelity.

To illustrate the model's internal representation, incident energy, x_0, and θ manifolds are traversed at regular intervals along the trained range. In Figure 3, incident energy is traversed, clearly showing more energetic behavior as the incident energy is increased from left to right.

Similarly, the latent space for x_0 is traversed, and the resulting impact on generated image is shown in Figure 4. We note that as x_0 increases, shower position shifts downward, which is consistent with the ATLAS coordinates used in the dataset described in Sec. 2.

Finally, as we traverse θ (Fig. 5) we illustrate the shower behavior dynamic using a difference between the middle point in interpolation space and each point along the θ traversal. As θ increases, we note that the width and dispersion decreases and the showers become significantly more centralized[2], which is consistent with the ATLAS definition of θ.

7. Conclusion

In this work, we explore the ability of GANs to be conditioned on physically meaningful attributes towards the ultimate goal of creating a viable, comprehensive solution for fast, high fidelity simulation of electromagnetic calorimeters. Clearly, GANs show great potential for

[2] In Figure 5, areas turning blue indicate that less energy is deposited in that particular section of the image at a given point in latent space.

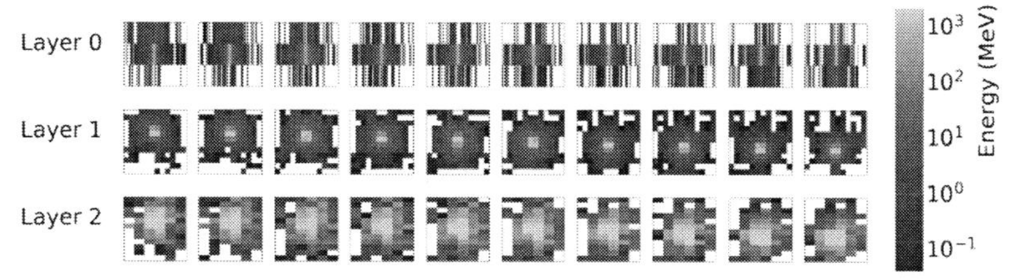

Figure 4. Interpolation across physical range of x_0 as a conditioning latent factor for e^+ showers. Note in the ATLAS coordinate system, x represents the vertical direction in this dataset. Each point in the interpolation is an average of 10 showers, with each point along the traversal build from an identical latent prior z.

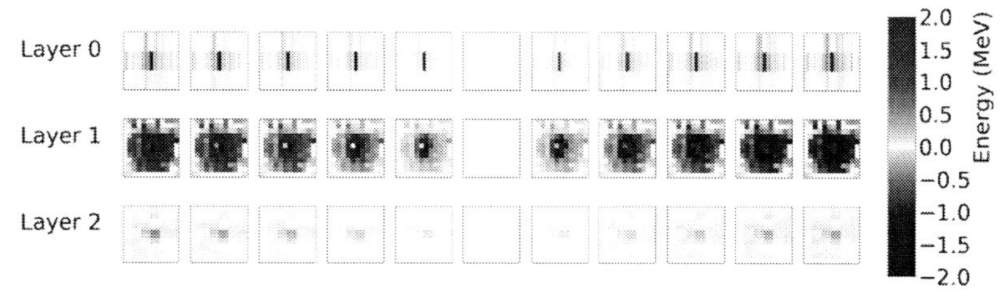

Figure 5. Interpolation across physical range of θ as a conditioning latent factor for $e+$ showers, with θ increasing from left to right. Each point in the interpolation is an average of 10 showers subtracted from the middle point along the interpolation path, with each point along the traversal build from an identical latent prior z.

controllability of generation procedures, but much future work remains. In particular, a thorough investigation around dynamics between the attribute estimation portion of the network, Ξ, and the overall training objective should be pursued, particularly as it relates to the final fidelity of the attribute estimates. In addition, future work should examine newer GAN formulations (as outlined in Sec. 3) and their ability to improve image quality.

[1] Flynn J 2015 Computing Resources Scrutiny Group Report Tech. Rep. CERN-RRB-2015-014 CERN Geneva URL `https://cds.cern.ch/record/2002240`
[2] Karavakis E, Andreeva J, Campana S, Gayazov S, Jezequel S, Saiz P, Sargsyan L, Schovancova J, Ueda I and the Atlas Collaboration 2014 *Journal of Physics: Conference Series* **513** 062024 URL `http://stacks.iop.org/1742-6596/513/i=6/a=062024`
[3] Bozzi C 2015 LHCb Computing Resource usage in 2014 (II) Tech. Rep. LHCb-PUB-2015-004. CERN-LHCb-PUB-2015-004 CERN Geneva URL `https://cds.cern.ch/record/1984010`
[4] Goodfellow I J, Pouget-Abadie J, Mirza M, Xu B, Warde-Farley D, Ozair S, Courville A and Bengio Y 2014 *ArXiv e-prints* (*Preprint* `1406.2661`)
[5] Kingma D P and Welling M 2013 *ArXiv e-prints* (*Preprint* `1312.6114`)
[6] Kadurin A, Aliper A, Kazennov A, Mamoshina P, Vanhaelen Q, Khrabrov K and Zhavoronkov A 2016 *Oncotarget* **8** ISSN 1949-2553
[7] Ravanbakhsh S, Lanusse F, Mandelbaum R, Schneider J and Poczos B 2016 (*Preprint* `1609.05796`)
[8] Schawinski K, Zhang C, Zhang H, Fowler L and Santhanam G K 2017 *Monthly Notices of the Royal Astronomical Society* **467** L110–L114 (*Preprint* `1702.00403`)
[9] Mustafa M, Bard D, Bhimji W, Al-Rfou R and Lukić Z 2017 *ArXiv e-prints* (*Preprint* `1706.02390`)
[10] Paganini M, de Oliveira L and Nachman B 2017 (*Preprint* `1705.02355`)

[11] de Oliveira L, Paganini M and Nachman B 2017 *Computing and Software for Big Science* **1** 4 ISSN 2510-2044 URL `https://doi.org/10.1007/s41781-017-0004-6`

[12] Paganini M, de Oliveira L and Nachman B 2017 Electromagnetic calorimeter shower images with variable incidence angle and position URL `http://dx.doi.org/10.17632/5fnxs6b557.2`

[13] Agostinelli S *et al.* (GEANT4) 2003 *Nucl. Instrum. Meth.* **A506** 250–303

[14] Goodfellow I J 2017 *CoRR* **abs/1701.00160** URL `http://arxiv.org/abs/1701.00160`

[15] Radford A, Metz L and Chintala S 2015 *ArXiv e-prints* (*Preprint* 1511.06434) URL `http://arxiv.org/abs/1511.06434`

[16] Reed S E, Akata Z, Mohan S, Tenka S, Schiele B and Lee H 2016 *ArXiv e-prints* (*Preprint* 1610.02454) URL `http://arxiv.org/abs/1610.02454`

[17] Odena A, Olah C and Shlens J 2016 *ArXiv e-prints* (*Preprint* 1610.09585)

[18] Zhang H, Xu T, Li H, Zhang S, Huang X, Wang X and Metaxas D 2016 *ArXiv e-prints* (*Preprint* 1612.03242)

[19] Mirza M and Osindero S 2014 *ArXiv e-prints* (*Preprint* 1411.1784) URL `http://arxiv.org/abs/1411.1784`

[20] Salimans T, Goodfellow I J, Zaremba W, Cheung V, Radford A and Chen X 2016 *ArXiv e-prints* (*Preprint* 1606.03498) URL `http://arxiv.org/abs/1606.03498`

[21] Chen X, Duan Y, Houthooft R, Schulman J, Sutskever I and Abbeel P 2016 *ArXiv e-prints* (*Preprint* 1606.03657) URL `http://arxiv.org/abs/1606.03657`

[22] Arjovsky M, Chintala S and Bottou L 2017 *ArXiv e-prints* (*Preprint* 1701.07875)

[23] Gulrajani I, Ahmed F, Arjovsky M, Dumoulin V and Courville A 2017 *ArXiv e-prints* (*Preprint* 1704.00028)

[24] Qi G 2017 *CoRR* **abs/1701.06264** URL `http://arxiv.org/abs/1701.06264`

[25] Nowozin S, Cseke B and Tomioka R 2016 *ArXiv e-prints* (*Preprint* 1606.00709)

[26] Lin M 2017 *ArXiv e-prints* (*Preprint* 1704.06191)

[27] Bellemare M G, Danihelka I, Dabney W, Mohamed S, Lakshminarayanan B, Hoyer S and Munos R 2017 *CoRR* **abs/1705.10743** URL `http://arxiv.org/abs/1705.10743`

[28] Theis L, van den Oord A and Bethge M 2015 *ArXiv e-prints* (*Preprint* 1511.01844)

[29] Chollet F 2017 Keras `https://github.com/fchollet/keras`

[30] Abadi M, Agarwal A, Barham P, Brevdo E, Chen Z, Citro C, Corrado G S, Davis A, Dean J, Devin M, Ghemawat S, Goodfellow I, Harp A, Irving G, Isard M, Jia Y, Jozefowicz R, Kaiser L, Kudlur M, Levenberg J, Mané D, Monga R, Moore S, Murray D, Olah C, Schuster M, Shlens J, Steiner B, Sutskever I, Talwar K, Tucker P, Vanhoucke V, Vasudevan V, Viégas F, Vinyals O, Warden P, Wattenberg M, Wicke M, Yu Y and Zheng X 2015 TensorFlow: Large-scale machine learning on heterogeneous systems software available from `https://www.tensorflow.org/` URL `http://tensorflow.org/`

IOP Conf. Series: Journal of Physics: Conf. Series **1085** (2018) 042018 doi:10.1088/1742-6596/1085/4/042018

Domain adaptation with gradient reversal for MC/real data calibration

A. Ryzhikov[1,2] **and A. Ustyuzhanin**[1,2]
[1] National Research University Higher School of Economics, 20 Myasnitskaya st., Moscow 101000, Russia
[2] Yandex School of Data Analysis, 11/2, Timura Frunze st., Moscow 119021, Russia
E-mail: artemryzhikov@gmail.com

Abstract.
It is quite common part of the data analysis in High Energy Physics to train a classifier for signal and background separation. In case the signal under investigation is a rare process, the signal sample is simulated and background sample is taken from the real data. Such setting create an unnecessary bias: the classifier might learn not the characteristic of the signal but the characteristic of the imperfect simulation. So the challenge is to train the classifier in such way that it picks up signal/background difference and doesnt overfit to the simulation-specific features. The suggested approach is based on cross-domain adaptation technique using neural networks with gradient reversal. The network architecture is a dense multi-branch structure. One branch is responsible for the signal/background discrimination, the second branch helps to avoid the overfitting on the Monte-Carlo training dataset. The tests showed that this architecture is a robust mechanism for choosing trade-offs between discrimination power and overfitting. So the resulting networks successfully distinguishes the signal from the background, but does not distinguish simulated events from the real ones. Moreover, such architecture could to be easily extended with more branches, and each one could be responsible for specific discrete and continuous domains. For example, the additional third network's branch could help to reduce the correlation between the classifier predictions and reconstructed mass of the decay, thereby making such approach highly viable for wide variety of physics searches. But such network's extensions weren't investigated during this work.

1. Introduction

Vast majority of data analyses in High Energy Physics relies on discrimination between signal and background events. Physics has pioneered [3] and adopted application of the machine learning (ML) approaches for that task. Methods like Decision Trees, Boosting and Neural Networks play significant role in the modern physics discoveries. Today, with the advance of the deep learning, new software frameworks and increased computational power neural networks are becoming more and more popular.

However the application of a large number of machine learning methods for the problems of High Energy Physics need to be considered with care. The reason for that is low generalization quality and certain number of physical restrictions for discriminative models. Besides demand for high values for corresponding figures of merit models should provide physically-sound result. In this research we propose a new approach for building signal/background discriminator based on neural networks for high energy physics problem. The method is based on cross-domain adaptation technique with gradient reversal [4]. It suggests the following architecture for the neural network: a common part (responsible for generation of important features) and three branches connected to it: the first branch is responsible for signal detection, the second branch helps to avoid overfitting on simulated events and and the third one avoids sculpting a false mass peak.

2. Problem

Usage of the simulated sample is fairly common approach in the High Energy Physics. However it is often hard to include all physics factors into Monte-Carlo (MC) simulation. Moreover, not all variables can be simulated accurately enough, so the discrepancies may lead either to:

(i) High simulation costs of signal and background
(ii) ML models trained on simulated sample tend to overfit to the simulation artifacts and work poorly on real data.

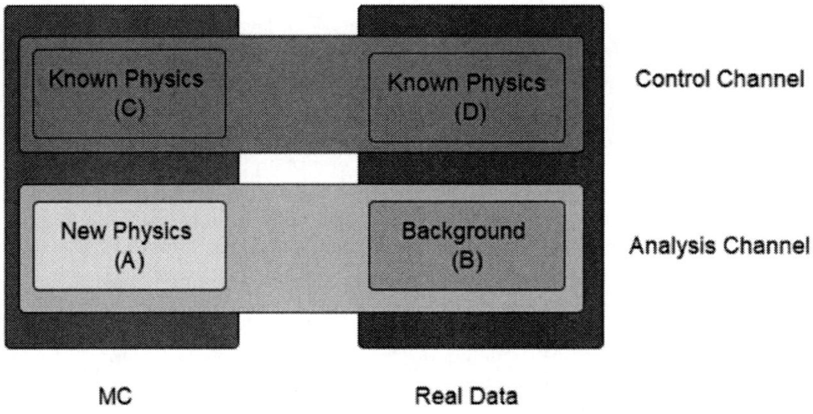

Figure 1. Training on the mixture of the simulated (MC) and the real samples [5]

In this research, we consider $\tau \rightarrow 3\mu$ decay that has been published at the Data Science challenge "Flavours of Physics" on *kaggle.com* [6]. The challenge is three-fold:

(i) The quality of the signal discrimination from the background should be as high as possible. The evaluation metric for the signal discrimination is Weighted Area Under the ROC Curve (truncated AUC) [6]
(ii) Due to the absence of signal events ($\tau \rightarrow 3\mu$) from data, the classifier should be trained using a mixture of simulated signal ("A", Fig. 1) and real data background ("B" , Fig. 1), it is possible to demonstrate high performance by exploiting the features that are specific to the simulation. It is required that the classifier should not have a large discrepancy when applied to real and simulated data. To verify this, we use much more frequent control channel, $D_s \rightarrow \phi\pi$ ("C" and "D", Fig. 1), that has a similar topology as say the, $\tau \rightarrow 3\mu$ (analysis channel), but contains both simulated ("C", Fig. 1) and real ("D", Fig. 1) signals. Thus the goal is to train a classifier able to separate "A" from "B" but not "C" from "D" (Fig. 1). A Kolmogorov-Smirnov (KS) test [10] [11] is used to evaluate the similarity between the classifier output distributions on C and D. In our problem KS is calculated between predictions distributions for real and simulated data for $D_s \rightarrow \phi\pi$ channel. The KS-value of the test should be less than 0.09 (according to the original competition's conditions [6]).

(iii) The classifier output should not be correlated with the reconstructed mass feature, i.e. its output distribution should not sculpt artificial peaks that could be interpreted as a (false) signal. To test the flatness we have used Cramer-von Mises (CvM) test that gives the uniformity of the distribution [7].

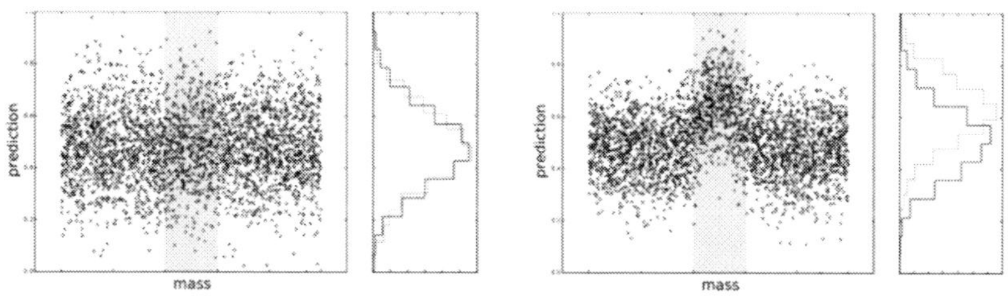

Figure 2. Illustration of the CvM correlation test [7]. On the left side there is no correlation with mass (low CvM values). On the right side, the model's predictions are highly correlated with mass (high CvM values)

3. Baseline

At the moment of this publication the *Data Doping* [2] was the best technique to deal with the classifier overfitting to MC-speficic features. We use this technique as the baseline.

The idea of Data Doping is to "dope" the training sample with a small fraction of MC events from the control channel (denoted by C at Fig. 1) labeled as background. Thus, it forbids the classifier to rely on features discriminating real and background events. The optimal fraction of the doping events was taken from [5].

4. Domain adaptation

As an alternative to *Data Doping* we applied method based on *Cross-Domain adaptation with gradient reversal* [4]. The concept is similar to GAN (Generative Adversarial Nets) [9]: we use an additional network branch that is trained to discriminate real from simulated (domain classifier) events. During the training process we reverse the gradient from the domain classifier in order to prevent the whole model from discrimination of the real signal from the simulation (Fig. 3).

Instead of reversing the gradient it is possible to use the domain classifier loss function with negative sign, since

$$-\frac{\partial L_d}{\partial \Theta_d} = \frac{\partial(-L_d)}{\partial \Theta_d} \tag{1}$$

effectively using the negative cross-entropy (between predicted and target domain) as the objective for *Domain classifier* (where L_d and Θ_d means it's cross-entropy loss and parameters respectively).

The network architecture has a three separately fitable dense branches with common dense part (Fig. 3) and consists of the following components (we will follow the definitions of the original article of [4] below):

- Feature extractor - responsible for feature preprocessing;
- Label predictor - responsible for the target prediction (signal/background discrimination with cross-entropy loss);

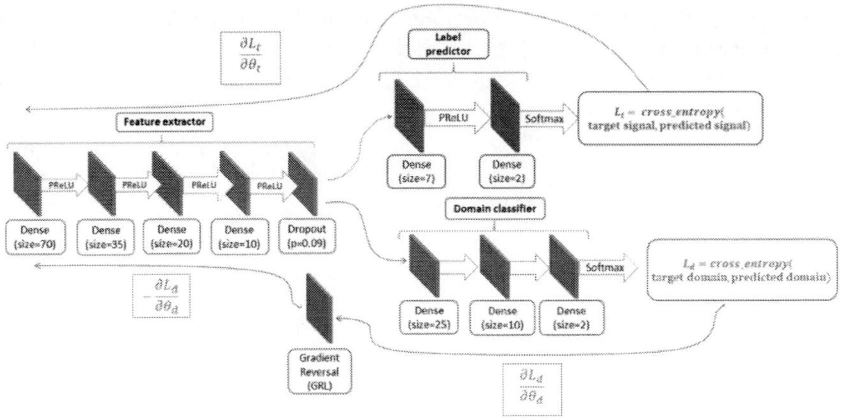

Figure 3. Cross-domain adaptation with gradient reversal [4]

- Domain classifier - responsible for cross-domain adaptation and prevents the network from overfitting to MC domain (with negative cross-entropy loss as described above);
- Mass predictor - helps to eliminate the correlation between classifier predictions and the reconstructed mass of the decay (It wasnt tested in this research and our architecture was tested without this part. It is omitted from the Figure 3. Theoretically it was designed as additional branch, similar to domain classifier, working along the same principle).

5. Data

Training and test datasets (Analysis channel) consists of 67000+ and 855000+ events of signal ($\tau \to 3\mu$) and background events respectively. Control channel consists of 71000+ events of signal ($D_s \to \phi\pi$) and background. All events are described by 46 features. It is taken from the "Flavours of Physics" challenge [6].

6. Training

The architecture was implemented in *Python 2.7* [1] using *Lasagne (ver. 2.1)* [2] framework.

The model was trained for 40 epoches with RMSProp optimizer. To achieve highest stability and reproducibility we trained only *Feature extractor* and *Label predictor* parts several (20) epoches with frozen *Domain classifier* and after that the *Domain classifier* was trained concurrently with the first two parts.

During the first step only *Feature extractor* and *Label predictor* were trained. *Domain classifier* is frozen, as mentioned above. Training procedure was 20-epoches RMSProp-optimization of categorical (2-classes) cross-entropy (between analysis channel's signal/background) loss. Batch size was 1000. Learning rate at first step was 0.01 and was decayed 10 times each 5 epoches.

In the second step we restored *Feature extractor* and *Label predictor* from previous step and trained all three parts. *Feature extractor* and *Label predictor* were trained in the same way (20 epoches, same loss, same batch size, same initial learning rate and the same learning rate decay policy), but during this step the *Domain classifier* was concurrently trained each 6 batches of *Feature extractor* and *Label predictor's* training. The *Domain classifier's* training parameters are the same except the batch size (batch size=300).

To eliminate metric in the *KS*-test we increased such Domain classifiers parameters as learning rate, corresponding batch size and batches frequency:

- learning rate ratios between branches (*learning rate multiplier*)

- batch size ratios. The best observed values were 1000 and 300 for *Label predictor* and *Domain classifier* respectively
- batches frequency. The best observed ratio for number of batches per epoch was 6:1 for *Label predictor* and *Domain classifier* respectively

Figure 4 represents such dependency from one of such parameters. It was observed that too small values of *KS* makes *CvM* values higher and *AUC* metric smaller. So the goal was to find balance between *KS*, *CvM* and *AUC* using parameters described above.

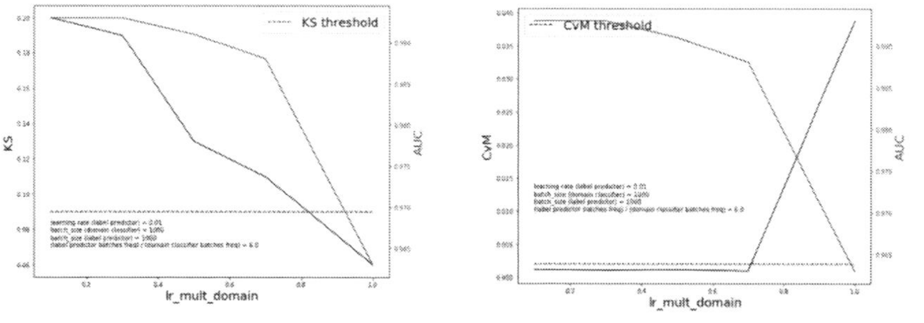

Figure 4. Metrics dependency from domain classifiers *learning rate multiplier*

7. Results

In the research we've compared the following models: *Baseline* (label predictor from Fig. 4 without Domain Adaptation), *Domain Adaptation* (our approach), *Data Doping*. Models were tested on 85000+ events of signal ($\tau \to 3\mu$) and background. KS and CvM thresholds were chosen 0.09 and 0.02 respectively

The tests outcome is shown in (Table 1) demonstrates that this architecture is a robust mechanism for choosing trade-off between discrimination power and overfitting.

As shown in the table below our method provides the best solution for signal detection problem ($\tau \to 3\mu$).

	$AUC(truncated)$	KS	CvM
Baseline	0.999	0.18	0.0008
Data Doping	0.974	0.09	0.0011
Domain adaptation	0.979	0.06	0.0008

Table 1. AUC, KS, CvM scores for different approaches

From the table above we can conclude that the Domain Adaptation approach can be used for training models on MC and real data mixture: it avoids using poorly simulated features while keeping the discrimination quality quite high.

8. Conclusion

Domain Adaptation approach addresses important problem of modern data analysis in the High-Energy Physics. It can be used without any special data pre-processing, giving flexibility to control signal/background discrimination quality versus real/simulation mismatch overfitting. The code is published on github [8] that demonstrate training/testing performed on public dataset [6]. This approach is much more flexible than *Data Doping* since it can be easily extended by additional restrictions (like mass-invariant predictions) by further network branching.

IOP Conf. Series: Journal of Physics: Conf. Series **1085** (2018) 042018 doi:10.1088/1742-6596/1085/4/042018

References

[1] Python project, “Python“ [software], version 2.7.13, 2016. Available from https://www.python.org/downloads/release/python-2713/ [accessed 2017-07-12]
[2] Lasagne project, "Lasagne" [software], version 0.2.dev1, 2017. Available from https://github.com/Lasagne/Lasagne [accessed 2017-07-12]
[3] Amidi E 1996 A Search for the Top Quark Using Artificial Neural Networks (Massachusetts: Northeastem University Boston)
[4] Ganin Y and Lempitsky V 2015 Unsupervised domain adaptation by backpropagation (Moscow: Skolkovo Institute of Science and Technology)
[5] Gaitan V 2016 Data Doping solution for “Flavours in Physics“ challenge *Heavy Flavour Data Mining workshop* (Zurich: University of Zurich, Irchel Campus)
[6] *Flavours of Physics Competition, https://www.kaggle.com/c/flavours-of-physics*
[7] Rogozhnikov A, Bukva A, Gligorov V, Ustyuzhanin A, Williams M 2015 New approaches for boosting to uniformity (Moscow: Yandex)
[8] Ryzhikov A, Ustyuzhanin A 2017 Source code for Domain Adaptation research *https://github.com/Leensman/Cross-domain-adaptation-on-HEP-HSE-course-work-*
[9] J. Goodfellow I, Pouget-Abadie J, Mirza M, Xu B, Warde-Farley D, Ozair S, Courville A, Bengio Y 2014 Generative Adversarial Networks (Montreal: Universite de Montreal)
[10] Kolmogorov A 1933 Sulla determinazione empirica di una legge di distribuzione
[11] Smirnov N 1948 Table for estimating the goodness of fit of empirical distributions

IOP Conf. Series: Journal of Physics: Conf. Series **1085** (2018) 042019 doi:10.1088/1742-6596/1085/4/042019

Automated proton track identification in MicroBooNE using gradient boosted decision trees

Katherine Woodruff
on behalf of the MicroBooNE collaboration
Physics Department, New Mexico State University
Las Cruces, NM 88003

E-mail: kwoodruf@nmsu.edu

Abstract. MicroBooNE is a liquid argon time projection chamber (LArTPC) neutrino experiment that is currently running in the Booster Neutrino Beam at Fermilab. LArTPC technology allows for high-resolution, three-dimensional representations of neutrino interactions. A wide variety of software tools for automated reconstruction and selection of particle tracks in LArTPCs are actively being developed. Short, isolated proton tracks, the signal for low-momentum-transfer neutral current (NC) elastic events, are easily hidden in a large cosmic background. Detecting these low-energy tracks will allow us to probe interesting regions of the proton's spin structure. An effective method for selecting NC elastic events is to combine a highly efficient track reconstruction algorithm to find all candidate tracks with highly accurate particle identification using a machine learning algorithm. We present our work on particle track classification using gradient tree boosting software (XGBoost) and the performance on simulated neutrino data.

1. Introduction

The three up and down valence quarks in the nucleon only account for a small percent of its mass. Gluons that bind the valence quarks split into quark-antiquark pairs of up, down, and strange flavor. This sea of quarks and gluons carries the remainder of the nucleon mass. The structure of the quark-gluon sea and how its elements combine with the valence quarks to give the nucleon its measured structure is not precisely known.

The net spin of the proton comes from a combination of the spin and orbital momentum of the quarks and gluons. The net contribution from the spin of strange quarks and antiquarks, Δs, is zero in the static quark model.

In the 1980s the European Muon Collaboration [1] and several subsequent experiments found that the Ellis-Jaffe Sum Rule was violated in polarized, charged-lepton, inclusive, deep inelastic scattering (DIS). The Ellis-Jaffe sum rule assumes that SU(3) flavor symmetry is valid and that $\Delta s = 0$. For the results to be consistent with exact SU(3) flavor symmetry, Δs must be *negative*. Follow-up measurements using semi-inclusive deep inelastic scattering have been consistent with $\Delta s = 0$, but these determinations of Δs are highly dependent on the fragmentation functions used [2].

An independent determination of Δs can be made using neutral-current (NC) elastic neutrino-proton scattering. The NC elastic cross section depends directly on Δs and no assumptions about SU(3) flavor symmetry or fragmentation functions are needed. Previous measurements using elastic neutrino-proton scattering [3, 4, 5, 6] have been able to resolve final state protons down to a kinetic energy of $T \sim$ 240 MeV which corresponds to a momentum transfer of $Q^2 = 0.45$ GeV^2. These measurements also found $\Delta s < 0$, but the results are highly dependent on the choice of the axial form factor Q^2 dependence. To extract Δs, G_A^s must be extrapolated to $Q^2 = 0$. Detecting events with lower momentum transfer would lessen the dependence on the choice of the model. Global fits to electron-proton and neutrino-proton elastic scattering data have found $\Delta s = -0.30 \pm 0.42$. Based on data from a simulation of the MicroBooNE detector and the Booster Neutrino Beam (BNB), the uncertainty on the global fit to Δs is estimated to decrease by a factor of ten when including MicroBooNE data [7].

2. Elastic neutrino-proton scattering

The elastic lepton-nucleon scattering cross section depends on the axial, electric, and magnetic form factors which represent the finite structure of the nucleon. The axial form factor, G_A, represents the spin structure, and the electric and magnetic form factors, G_E and G_M, represent the electric and magnetic structure, respectively.

2.1. Neutral-current elastic scattering

The NC elastic neutrino-proton cross section [8] can be written in terms of the neutral-current form factors G_A^{NC}, G_E^{NC}, and G_M^{NC} that represent the finite structure of the proton. The electric and magnetic form factors of the proton have been determined in a world-wide measurement program of elastic electron-proton scattering using hydrogen targets and quasi-elastic electron-nucleon scattering using light nuclear targets (specifically deuterium and helium) [9, 10]. In the limit where the momentum transfer (Q^2) goes to zero, the quark contributions to the axial form factor, G_A^{NC}, become the net contribution of individual the quark spin to the proton spin so that

$$G_A^{NC}(Q^2 = 0) = \frac{1}{2}(-\Delta u + \Delta d + \Delta s)\,.$$

The difference of the up and down spin contributions, $\Delta u - \Delta d$, is proportional to the axial vector coupling constant g_A measured in hyperon β decay [11], therefore a measurement of G_A^{NC} can determine Δs.

2.2. Experimental measurement

For free nucleon scattering, the final state of an NC elastic neutrino-proton interaction consists of a neutrino and a proton. Since it isn't possible to detect the outgoing neutrino, the signal is a single proton track. In order to extrapolate the axial form factor to zero, we need to detect very low energy protons. The kinematics of the interaction are determined entirely by the proton kinetic energy, T_P, by $Q^2 = 2T_P M_P$.

We estimate that MicroBooNE can detect NC elastic events down to a minimum of $Q^2 \sim 0.08$ GeV^2. The momentum transfer is determined by the kinetic energy of the proton in NC elastic interactions. MicroBooNE can detect protons with a track length of at least 1.5 cm which corresponds to a kinetic energy of ~40 MeV in liquid argon giving $Q^2 \sim 0.08$ GeV^2. The set of events with a neutrino-proton final state is a subset of NC elastic scattering events, since the neutrinos are interacting with argon atoms and not free nucleons.

3. MicroBooNE

The MicroBooNE detector [12] is a liquid-argon time projection chamber (TPC) located in the Booster Neutrino Beam at Fermilab. MicroBooNE is a high-resolution detector designed to investigate the apparent excess of low-energy electron neutrino interactions seen in the MiniBooNE experiment [13]. MicroBooNE's ability to detect low-energy neutrino interactions

IOP Conf. Series: Journal of Physics: Conf. Series **1085** (2018) 042019 doi:10.1088/1742-6596/1085/4/042019

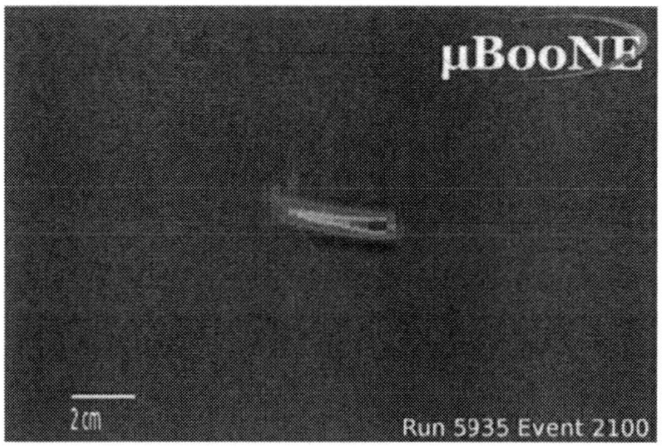

Figure 1. Proton track candidate in MicroBooNE data.

allows it to detect neutral current neutrino interactions with a low momentum transfer. It began taking data in October of 2015. Figure 1 shows an example of a proton track candidate in MicroBooNE.

The TPC [12] is 10 meters long in the beam direction, 2.3 meters tall, and 2.5 meters in the electron drift direction. It takes 2.3 ms for electrons to drift across the full width of the TPC in the electric field. Events are read out on three anode wire planes. In addition to the TPC, MicroBooNE has a light collection system with nanosecond timing resolution to collect scintillation light emitted by the liquid argon. The PMTs determine the initial time of the interaction to help with cosmic rejection. In order for an event to be read out, there must be an optical signal within a 23 μs window around the BNB spill.

3.1. Events in MicroBooNE
The data from each *event* in MicroBooNE can be visualized as a set of three high resolution images (one from each anode plane). Each image has approximately 20 million pixels (3,000 wires by 9,600 time ticks). It takes 30 MB of disk space to store one MicroBooNE event. In Fig. 1, the X-axis corresponds to wire number and the Y-axis corresponds to time tick. The dark blue background corresponds to no signal on the wire, and the colored pixels correspond to charge deposited on the wire. The color or intensity of the pixel corresponds to the amount of energy deposited in the TPC.

4. Automated event selection
MicroBooNE is close to the surface of the Earth, which results in a large cosmic ray background. This can be seen in the bottom image in Fig. 2. There are an average of twelve cosmic muon tracks per readout frame [14]. In addition, there are approximately five times as many event triggers caused by cosmic rays coincident with the BNB spill than actual neutrino interactions. During MicroBooNE's three year run, we expect to have ~200,000 neutrino interactions and ~1,000,000 cosmic interactions. This means that automated neutrino event reconstruction and identification algorithms are required. These algorithms are currently being developed for liquid argon TPCs.

4.1. Track reconstruction in LArSoft
Track reconstruction is handled in the Liquid Argon Software framework (LArSoft) [15]. One-dimensional hits are found by fitting Gaussian functions to noise-filtered [16] waveforms that are read out from the anode wires in the TPC. This is done for all of the wires on all three of the planes. The result is a two-dimensional image for each of the three wire planes, where the two dimensions are wire number and time. These 2D hits are used as inputs to the Pandora Software Development Kit [17]. Pandora contains pattern recognition algorithms that have been

IOP Conf. Series: Journal of Physics: Conf. Series **1085** (2018) 042019 doi:10.1088/1742-6596/1085/4/042019

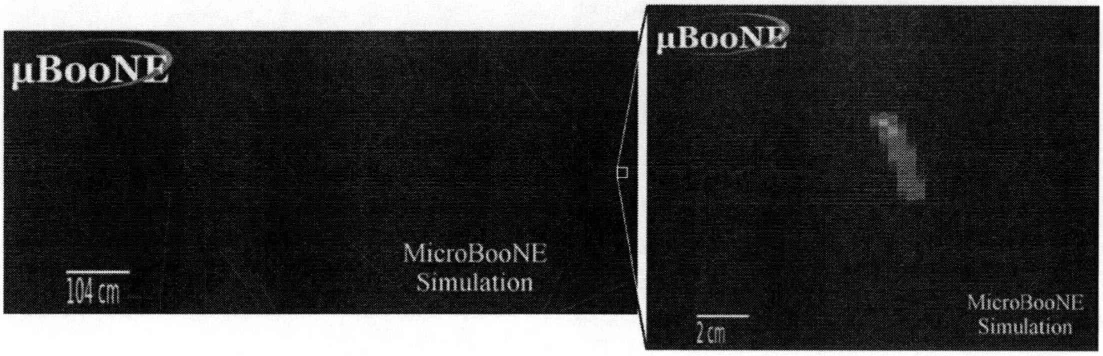

Figure 2. 2D event display of a simulated neutral-current elastic event in MicroBooNE that was successfully classified as a proton. The left image shows the side view of the entire MicroBooNE TPC. The right image is a close-up event display of the simulated proton track. All of the additional tracks are from cosmic rays.

optimized to reconstruct tracks from neutrino interactions in liquid argon TPCs at the BNB energy range. The Pandora algorithms take a set of hits and reconstruct neutrino interaction vertices.

At this point the size of each event has been reduced from millions of pixels to about 20 reconstructed track objects with very little information loss. We can attempt to identify the type of particle and interaction that produced the set of tracks. In the NC elastic case, we want to specifically select proton tracks.

4.2. Proton track identification

Neutral-current elastic interactions are the most difficult to detect automatically because there is typically only one visible particle coming from the interaction vertex. There is no unique topology separating these events from the cosmic background.

Each reconstructed track object in a MicroBooNE event has several reconstructed physics properties associated with it. These properties fall into the categories of geometric, calorimetric, and optical. The geometric properties are related to the position, shape and size of the track. This includes variables like whether the track is entering the TPC, how long the track is, and how curvy a track is. Calorimetric properties all have to do with the charge deposited along the track. We can use information about how much total charge was deposited by the track, the average charge deposited per cm, and the difference between the amount of charge deposited at the beginning or the end of the track. We can also create variables that represent the shape and scale of the dE/dx (or dQ/dx) curve at any point along the reconstructed track. Additionally, we can use optical information from the PMT system to help characterize tracks. In this work we use the distance between the reconstructed track and the closest optical flash that was in the beam time window.

To identify proton tracks, we use a gradient-boosted decision tree classifier. We chose to use decision trees because they are easily interpretable and the inputs can be a mix of numeric and categorical variables. A detailed description of gradient decision tree boosting can be found in the documentation for the XGBoost[18, 19] software library that was used.

4.2.1. The decision tree model We created a multi-class gradient-boosted decision tree classifier, using the XGBoost software library, to separate five different track types: any proton track, muons or pions from BNB neutrino interactions, tracks from electromagnetic showers from

IOP Conf. Series: Journal of Physics: Conf. Series **1085** (2018) 042019 doi:10.1088/1742-6596/1085/4/042019

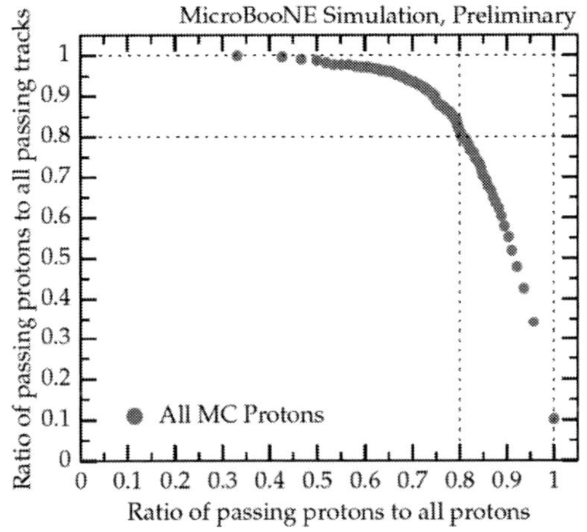

Figure 3. The efficiency versus the purity of simulated protons selected by the boosted decision tree classifier for a series of proton probability cuts between zero and one. The dashed horizontal and vertical lines at 0.8 are there to guide the eye and correspond to a proton probability cut of approximately 30%.

BNB interactions, and any non-proton track produced by a cosmic ray interaction. The classifier takes reconstructed track features as input and outputs a probability of the track having been produced by each of the given particle types. The reconstructed features are based on the track's geometric, calorimetric, and optical properties.

The training data that we use to make the decision trees comes from Monte Carlo simulation. The BNB interactions are simulated using the GENIE neutrino generator [20, 21], and cosmic interactions are simulated using the CORSIKA cosmic ray generator [22]. The particles generated by GENIE and CORSIKA are passed to Geant4 [23, 24] where they are propagated through a simulated MicroBooNE detector. For training and testing of the trees we only use tracks that were reconstructed in LArSoft.

Of the reconstructed test tracks that were input to the classifier, 84% of the protons from simulated neutrino interactions, and 63% of the protons from simulated cosmic interactions were classified correctly as protons. Figure 3 shows the efficiency versus purity for different proton probability cuts in the range from zero to one. Figure 4 shows the protons from simulated neutrino interactions as a function of proton kinetic energy. Of the reconstructed test tracks that were classified as protons, 89% were true simulated protons (22% neutrino induced protons and 67% cosmic induced protons). Figure 5 shows the breakdown of track types that are classified as protons.

5. Conclusions

Whether the strange quarks in the nucleon sea contribute negatively or not at all to the spin of the nucleon is an open question. Neutrino NC elastic scattering offers a unique way to determine Δs that is independent of the assumptions required by previous measurements. The MicroBooNE liquid argon TPC can detect low-Q^2 NC elastic events and is currently taking neutrino data at Fermilab. Automated event reconstruction and selection methods are being developed to analyze the large amount of high-resolution neutrino events in MicroBooNE. In these proceeding, we summarize the current status of MicroBooNEs automatic reconstruction and identification of proton tracks in neutral-current and charged-current interactions. The automated reconstruction chain in MicroBooNE successfully reduces the size of the events from millions of pixels to individual reconstructed track objects. This allows us to use boosted decision trees for proton identification. The boosted decision tree model presented here correctly classifies 84% of simulated protons from neutrino interactions, and can be improved with further tuning to the decision tree model and the simulation.

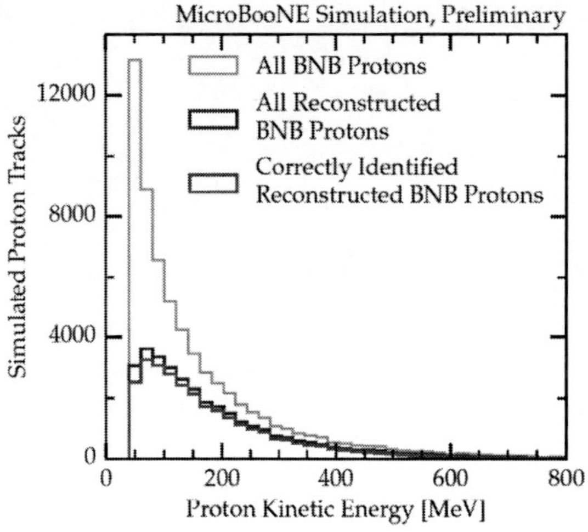

Figure 4. Simulated proton tracks as a function of kinetic energy. The light blue shows the full set of protons from simulated BNB neutrino interactions. The dark blue shows the number of those tracks that were reconstructed with the Pandora algorithms. The red line shows the subset of the reconstructed tracks that are classified as protons by the boosted decision trees.

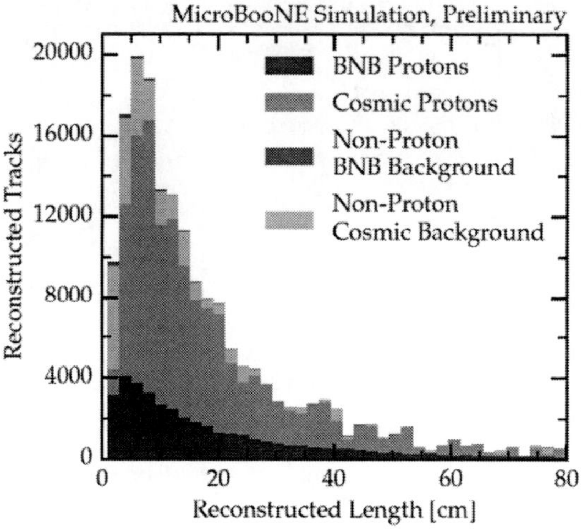

Figure 5. Simulated particle types that are classified as protons by the boosted decision trees as a function of reconstructed length. Light blue shows simulated cosmic-induced protons, and dark blue shows the protons from simulated BNB neutrino interactions. The tan area shows all other simulated cosmic tracks, and the red area shows all other tracks from simulated BNB neutrino interactions.

Acknowledgments

This work was supported by the US Department of Energy, Office of Science, Medium Energy Nuclear Physics Program.

References

[1] Ashman J *et al.* (European Muon) 1989 *Nucl. Phys.* **B328** 1
[2] Aidala C A, Bass S D, Hasch D and Mallot G K 2013 *Rev. Mod. Phys.* **85** 655–691 (*Preprint* 1209.2803)
[3] Ahrens L A *et al.* 1987 *Phys. Rev.* **D35** 785
[4] Garvey G T, Louis W C and White D H 1993 *Phys. Rev. C* **48**(2) 761–765
[5] Aguilar-Arevalo A A *et al.* (MiniBooNE) 2010 *Phys. Rev.* **D82** 092005
[6] Aguilar-Arevalo A A *et al.* (MiniBooNE) 2015 *Phys. Rev.* **D91** 012004
[7] Pate S and Trujillo D 2014 *EPJ Web Conf.* **66** 06018 (*Preprint* 1308.5694)
[8] Alberico W M, Bilenky S M and Maieron C 2002 *Phys. Rept.* **358** 227–308
[9] Armstrong D S and McKeown R D 2012 *Ann. Rev. Nucl. Part. Sci.* **62** 337–359 (*Preprint* 1207.5238)
[10] Cates G D, de Jager C W, Riordan S and Wojtsekhowski B 2011 *Phys. Rev. Lett.* **106** 252003
[11] Patrignani C *et al.* (Particle Data Group) 2016 *Chin. Phys.* **C40** 100001
[12] Acciarri R *et al.* (MicroBooNE) 2016 *Submitted to: JINST* (*Preprint* 1612.05824)
[13] Aguilar-Arevalo A A *et al.* (MiniBooNE) 2009 *Phys. Rev. Lett.* **102** 101802
[14] The MicroBooNE Collaboration 2016 *MicroBooNE Public Note* **1002** URL `http://microboone.fnal.gov/public-notes/`
[15] LArSoft Project, "LArSoft" [software], version 05_08_00 [accessed 2017–10–18] URL `http://scisoft.fnal.gov/scisoft/bundles/larsoft/v05_08_00/larsoft-v05_08_00.html`
[16] The MicroBooNE Collaboration 2016 *MicroBooNE Public Note* **1009**
[17] Acciarri R *et al.* 2017 *J. Phys. Conf. Ser.* **888** 012142
[18] XGBoost Project, "XGBoost" [software], version 0_6 [accessed 2017–10–18] URL `https://github.com/dmlc/xgboost/tree/v0.60`
[19] Chen T and Guestrin C 2016 *CoRR* (*Preprint* 1603.02754) URL `http://arxiv.org/abs/1603.02754`

[20] GENIE Project, "GENIE" [software], version 2_12 [accessed 2017–10–18] URL `http://genie.hepforge.org/svn/generator/branches/R-2_12_0/`
[21] Andreopoulos C *et al.* 2010 *Nucl. Instrum. Meth.* **A614** 87–104 (*Preprint* 0905.2517)
[22] Heck D, Schatz G, Thouw T, Knapp J and Capdevielle J N 1998
[23] Geant4 Project, "Geant" [software], version 4_10_1 [accessed 2017–10–18] URL `http://geant4.cern.ch/support/source_archive.shtml`
[24] Agostinelli S *et al.* (GEANT4) 2003 *Nucl. Instrum. Meth.* **A506** 250–303

IOP Conf. Series: Journal of Physics: Conf. Series **1085** (2018) 042020 doi:10.1088/1742-6596/1085/4/042020

How easily can neural networks learn relativity?

Kartik Chitturi and Peter Onyisi

Department of Physics, University of Texas at Austin, Austin, TX 78712, United States

E-mail: kartik.chitturi@utexas.edu, ponyisi@utexas.edu

Abstract. Relativistic invariants are key variables in high energy physics and are believed to be learned implicitly by deep learning approaches. We investigate the minimum network complexity needed to accurately extract such invariants. Doing so will help us understand how complex a neural network needs to be to obtain certain functions. We find that neural networks do well with predicting transverse momentum of a collision which illustrates the fact that non-linear functions can be learned. On the other hand, invariant mass was much more difficult to predict. Further work will be done to learn the reason why. However the non-linearity of the function can be ruled out as the sole reason.

1. Introduction

Roughly, the Universal Approximation Theorem states that a feed-forward neural network can approximate (most) continuous functions with a single hidden layer. This means that in theory relativistic invariants such as invariant mass, transverse momentum, and rapidity may be learned using only one hidden layer. However, this theorem does not constrain the size of the hidden layer. In fact, in practice often times fewer nodes may be needed with multiple hidden layers. On the other hand, networks with more than a single hidden layer are more difficult to train. This trade-off that occurs when determining the number of hidden layers needs to be more carefully examined, especially in the context of high energy physics.

We examine how adjusting the number of nodes, number of layers, and the activation function used affects the following problems:

- Finding $p_T(\mu\mu)$ for $Z \to \mu\mu$ given the 4-vectors,
- Finding $p_T(\mu\mu)$ of $t\bar{t} \to \mu\mu + X$ given the 4-vectors,
- Finding the invariant mass of $\mu\mu$ for $t\bar{t} \to \mu\mu + X$ given the 4-vectors.

This work was performed using Keras [1] (with a Tensorflow [2] backend) on a NVIDIA Tesla K80. The $t\bar{t} \to \mu\mu + X$ events are generated with the Powheg generator [3] using the Pythia 8 parton shower [4]; $Z \to \mu\mu$ events are generated with Sherpa 2.2 [5].

2. Finding Transverse Momentum

We examine the accuracy of finding transverse momentum of dimuon pairs in $Z \to \mu\mu$ and $t\bar{t} \to \mu\mu + X$ using single layer and two layer neural networks. Both the single layer and two layer models used rectified linear units (ReLU) as the activation function. For the two layer model, the number of nodes in the second layer is less than or equal to the number of nodes in the first layer. The 4-vectors, (E, p_x, p_y, p_z), of the two muons are given as the 8 inputs for the input layer.

We find that with a single hidden layer, the neural network trains quite well. As seen in Figure 1, with three nodes we obtain decent results. Figure 2 shows that with 10 nodes we achieve almost perfect accuracy.

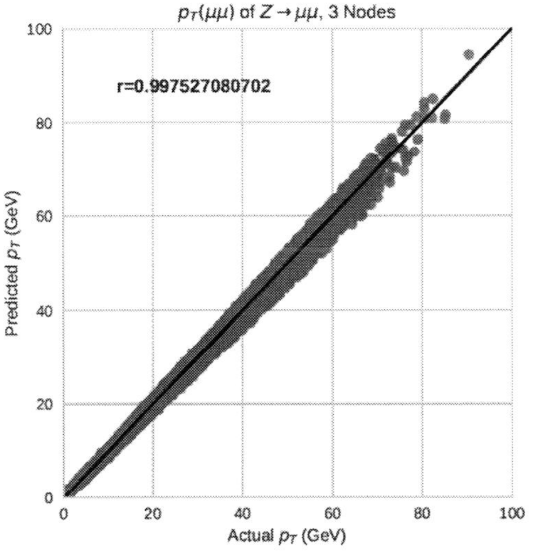

Figure 1. The predicted vs. actual $p_T(\mu\mu)$ of $Z \rightarrow \mu\mu$ for a neural network with a single layer with 3 nodes.

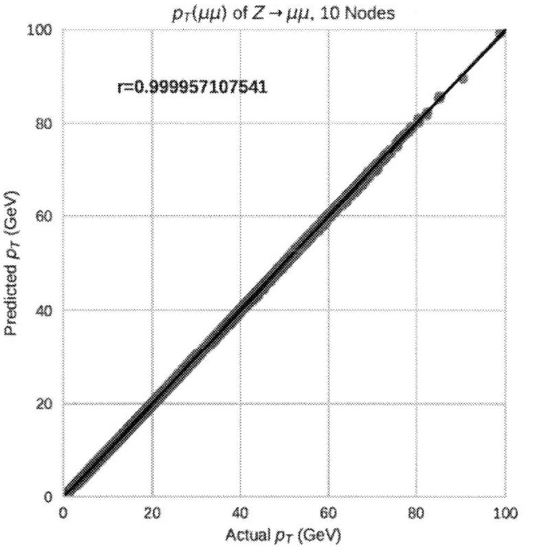

Figure 2. The predicted vs. actual $p_T(\mu\mu)$ of $Z \rightarrow \mu\mu$ for a neural network with a single layer with 10 nodes.

Sample independence was seen from the fact that the same neural network can also find p_T of dimuon production from $t\bar{t}$ with high accuracy. This is shown in Figure 3.

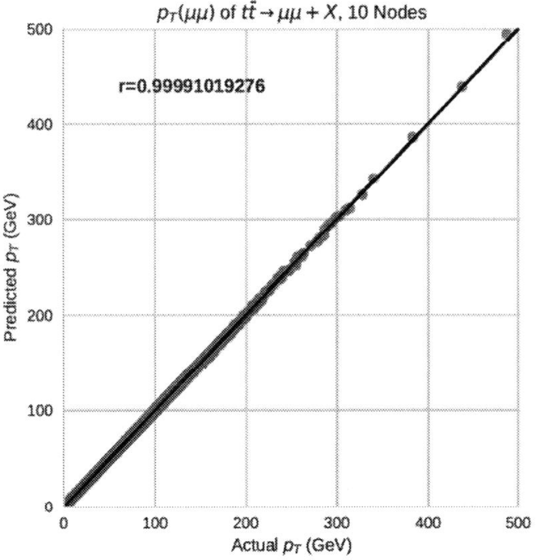

Figure 3. The predicted vs. actual $p_T(\mu\mu)$ of $t\bar{t} \rightarrow \mu\mu + X$ for the same neural network with a single hidden layer with 10 nodes.

IOP Conf. Series: Journal of Physics: Conf. Series **1085** (2018) 042020 doi:10.1088/1742-6596/1085/4/042020

Investigating the weights for this hidden layer we find that E and p_z in the two 4-vectors are given weights of essentially zero. Additionally, the weights for p_x and p_y are symmetric for both muons. This means that the neural network is learning a non-linear function of $(p_{x1} + p_{x2})$ and $(p_{y1} + p_{y2})$ where the numerical subscripts indicate the muon.

3. ReLU vs. LeakyReLU

We find that upon adding a second layer, accuracy did not increase. In fact, as Figure 4 shows, the accuracy actually dipped when we had only a few nodes in the second hidden layer.

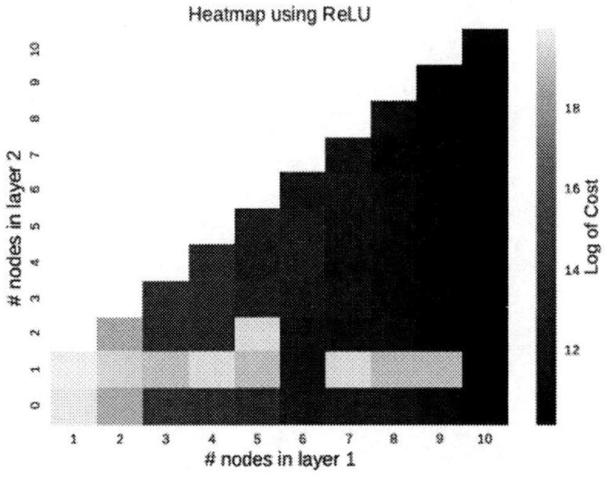

Figure 4. Logarithm of accuracy for a two hidden layer model predicting $pT(\mu\mu)$ of $Z \to \mu\mu$ using ReLU.

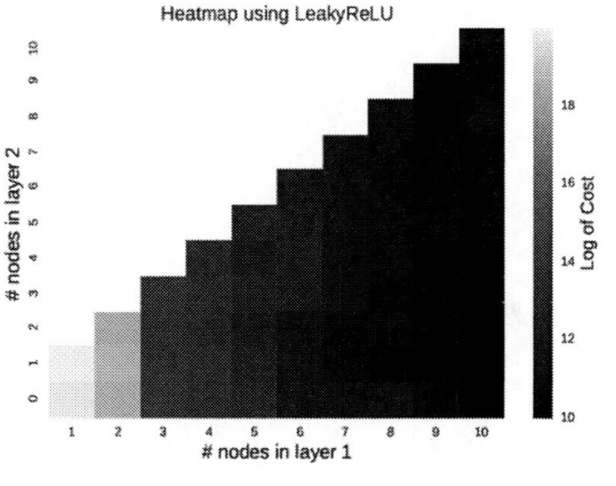

Figure 5. Logarithm of accuracy for a two hidden layer model predicting $pT(\mu\mu)$ of $Z \to \mu\mu$ using LeakyReLU.

This occurred when using both the softplus function and ReLU as the activation function. The problem is that ReLU and softplus suffer from a zero gradient for negative inputs. This means that during training, the neurons can "die" and constantly output zero. LeakyReLU (which is ReLU with a small slope for negative values) was created to fix this "dying ReLU" problem. Figure 5 demonstrates that using LeakyReLU gives a non-decreasing accuracy for the neural network with respect to the number of layers (and the number of nodes).

4. Invariant Mass

Dimuon invariant mass of $t\bar{t} \to \mu\mu + X$ was found using the same energy-momentum 4-vectors. However, as demonstrated in Figure 6, the accuracy for finding invariant mass was much lower than the accuracy for p_T. When we increase to 20 nodes in the single hidden layer, there is improvement but the neural network still does not achieve the same accuracy as p_T. This is illustrated in Figure 7.

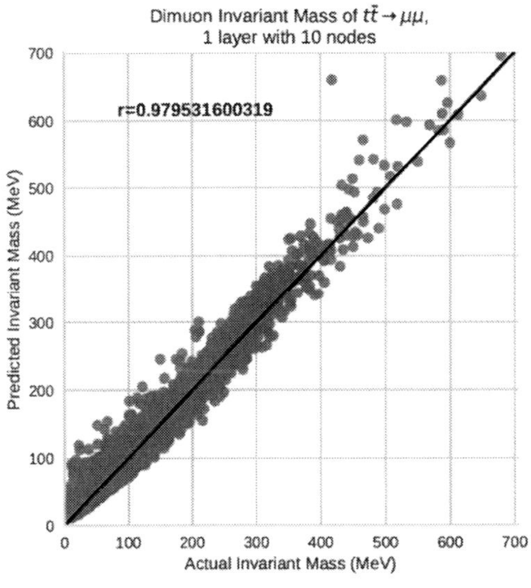

Figure 6. Predicted vs. actual dimuon invariant mass of $Z \rightarrow \mu\mu$ with 10 nodes in a single hidden layer.

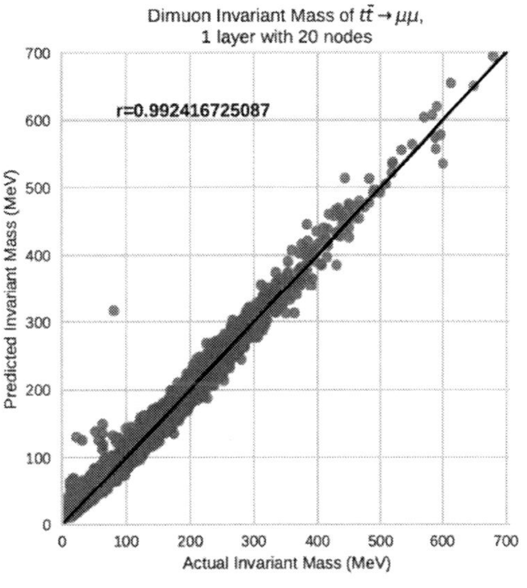

Figure 7. Predicted vs. actual dimuon invariant mass of $Z \rightarrow \mu\mu$ with 20 nodes in a single hidden layer.

We only begin to see similar accuracy when using two layers with 80 nodes each as shown in Figure 8.

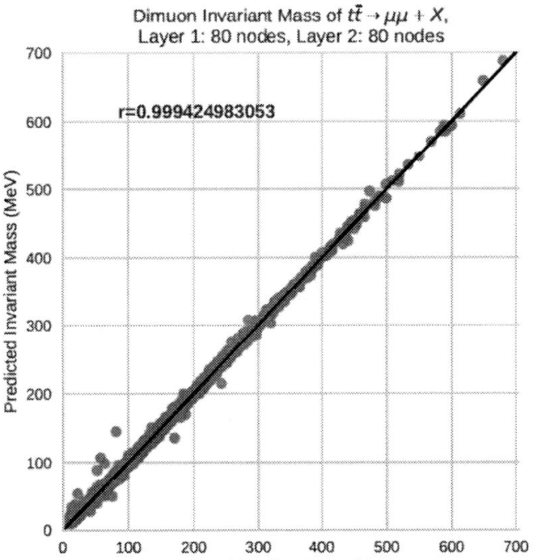

Figure 8. Predicted vs. actual dimuon invariant mass of $Z \rightarrow \mu\mu$ with 80 nodes in two layers.

Additionally, we see in Figure 9 that if there a small number of nodes there is almost no increase in accuracy when we add a second layer. Only when we use a large number of nodes, as exhibited in Figure 10, does adding a second layer improve accuracy.

IOP Conf. Series: Journal of Physics: Conf. Series **1085** (2018) 042020 doi:10.1088/1742-6596/1085/4/042020

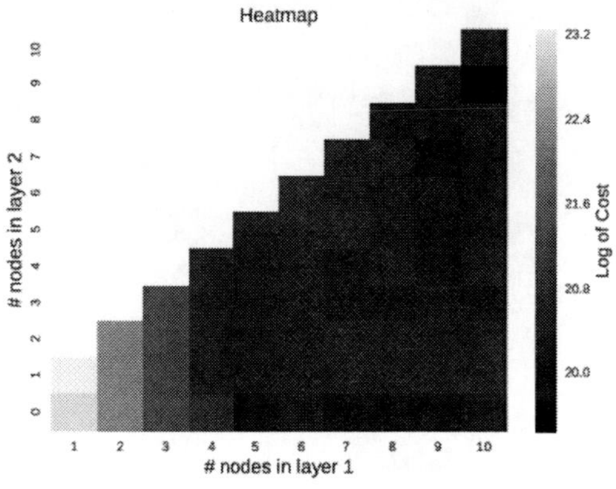

Figure 9. Logarithm of accuracy for a two hidden layer model predicting dimuon invariant mass of $Z \to \mu\mu$ using a low number of nodes.

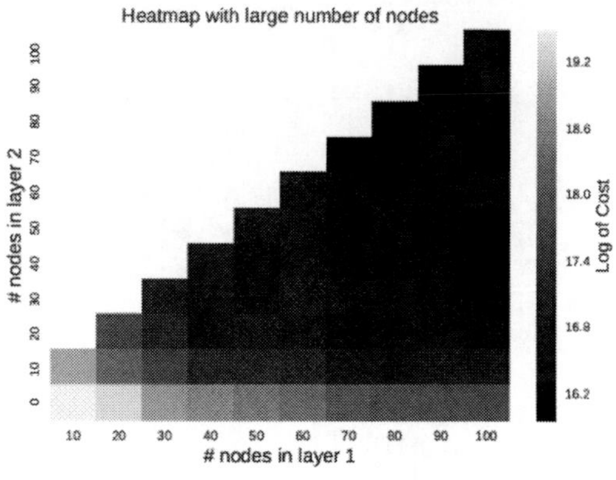

Figure 10. Logarithm of accuracy for a two hidden layer model predicting dimuon invariant mass of $Z \to \mu\mu$ using a high number of nodes.

The precise reason for the degradation of accuracy when investigating invariant mass (as opposed to transverse momentum) is unknown. To better understand the difference we compared the accuracies of $E^2 - p^2$ (invariant mass) and $E^2 + p^2$ (a physically meaningless variable) in a single layer model. This is illustrated in Figure 11.

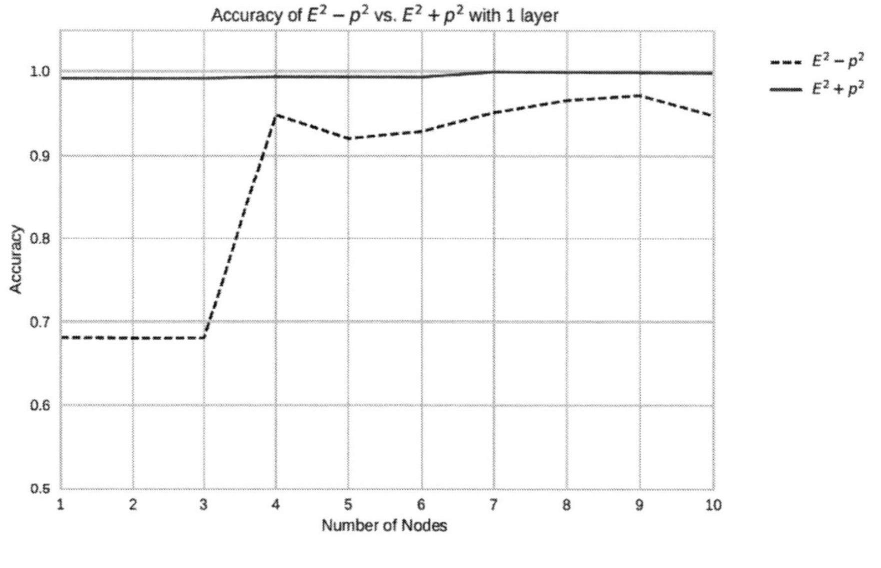

Figure 11. Comparison of invariant mass and $E^2 + p^2$ accuracies for a neural network with a single hidden layer.

We see that near perfect accuracy is achieved for $E^2 + p^2$. Since $E^2 + p^2$ and $E^2 - p^2$ have similar functional forms, it stands to reason that the invariant mass is difficult to train on because of its negative sign. This suggests that in general hyperbolic functions train worse than elliptic functions. However more research is required to understand the precise for this.

5. Conclusion

Neural networks can learn some non-linear relativistic invariants easily. For example, transverse momentum can be learned using only a single layer with 10 nodes. However other non-linear functions such as invariant mass are much more difficult to learn. It seems this is due the sign within the formula for invariant mass. Why this sign causes difficulty is unknown and further study of this could prove to be a fruitful avenue of research.

Acknowledgments

We wish to thank Jim Pivarski for useful discussions.

References

[1] Keras Project, "Keras" [software], version 2.0.5, Available from https://github.com/fchollet/keras/releases/tag/2.0.5 [accessed 2017-10-18]
[2] TensorFlow Project, "TensorFlow" [software], version 1.2.1, Available from https://github.com/tensorflow/tensorflow/releases/tag/v1.2.1 [accessed 2017-10-18]
[3] Frixione S, Nason P and Ridolfi G 2007 A Positive-weight next-to-leading-order Monte Carlo for heavy flavour hadroproduction *JHEP* **0709** 126 (*Preprint* 0707.3088 [hep-ph])
[4] Bierlich C et al., "Pythia" [software], version 8, Available from http://home.thep.lu.se/ torbjorn/Pythia.html [accessed 2018-2-16]
[5] Sherpa Project, "Sherpa" [software], version 2.2, Available from https://sherpa.hepforge.org/trac/wiki [accessed 2018-2-16]

IOP Conf. Series: Journal of Physics: Conf. Series **1085** (2018) 042021 doi:10.1088/1742-6596/1085/4/042021

A Toolkit to Study Sensitivity of the Geant4 Predictions to the Variations of the Physics Model Parameters

Laura Fields[1], Krzysztof Genser[1], Robert Hatcher[1], Michael Kelsey[2], Gabriel Perdue[1], Hans Wenzel[1], Dennis H. Wright[2], Julia Yarba[1]

[1] Fermi National Accelerator Laboratory†, Batavia, IL, USA

[2] SLAC National Accelerator Laboratory, Stanford, CA, USA

E-mail: `yarba_j@fnal.gov`

Abstract. Geant4 is the leading detector simulation toolkit used in high energy physics to design detectors and to optimize calibration and reconstruction software. It employs a set of carefully validated physics models to simulate interactions of particles with matter across a wide range of interaction energies. These models, especially the hadronic ones, rely largely on directly measured cross-sections and phenomenological predictions with physically motivated parameters estimated by theoretical calculation or measurement. Because these models are tuned to cover a very wide range of possible simulation tasks, they may not always be optimized for a given process or a given material. This raises several critical questions, e.g. how sensitive Geant4 predictions are to the variations of the model parameters, or what uncertainties are associated with a particular tune of a Geant4 physics model, or a group of models, or how to consistently derive guidance for Geant4 model development and improvement from a wide range of available experimental data. We have designed and implemented a comprehensive, modular, user-friendly software toolkit to study and address such questions. It allows one to easily modify parameters of one or several Geant4 physics models involved in the simulation, and to perform collective analysis of multiple variants of the resulting physics observables of interest and comparison against a variety of corresponding experimental data. Based on modern event-processing infrastructure software, the toolkit offers a variety of attractive features, e.g. flexible run-time configurable workflow, comprehensive bookkeeping, easy to expand collection of analytical components. Design, implementation technology, and key functionalities of the toolkit are presented and illustrated with results obtained with Geant4 key hadronic models.

1. Introduction

Geant4 [1] is the leading detector simulation toolkit used to design detectors and to optimize calibration and reconstruction software in such domains as high energy and nuclear physics, space science, reactor safety, and medical applications. It offers a set of carefully validated physics models to simulate interactions of particles with matter accross a wide range of interaction energies. Because Geant4 cannot offer a single modeling algorithm to cover the entire energy domain from zero to the TeV scale, for all known processes and particles, a combination of models is typically used to perform a simulation task. Each hadronic model relies on a substantial number of parameters involved in calculations associated with modeling the interactions. This naturally leads to the questions of how sensitive the Geant4 predictions are to the variations of

IOP Conf. Series: Journal of Physics: Conf. Series **1085** (2018) 042021 doi:10.1088/1742-6596/1085/4/042021

these parameters, and what uncertainties are associated with a Geant4 physics model, or a group of models, involved in simulation and optimization of a detector design. As of today, at least 3 Geant4 hadronic models, PreCompound [2], Bertini cascade [3] (Bertini) and Fritiof [4] (FTF), offer a configuration interface that allows to vary up to 39 parameters and 17 switches, in order to study the sensitivity of the simulated results. The Geant4 collaboration is currently working to expand the number of configurable hadronic models and to identify additional parameters and switches to include in the configuration interface. However, while studying the sensitivity of the Geant4 simulated results, it is important to prevent the use of physically incorrect settings of the model parameters or the neglect of correlations among parameters. Thus, an important part of advertising to the users community configurable parameters of hadronic models is documentation of the validity range of each parameters, and the correlations of parameters, where applicable. This challenge has motivated the Geant4 collaboration to start a new initiative and develop a toolkit to study the effects of varying model parameters on the simulated results, and to explore the associated uncertainties. The primary goal is to estimate the physically meaningful range of validity of the different model parameters and to determine their optimal values and uncertainties from a global fit to relevant data. Model parameters exposed to the users will be thoroughly documented. As an additional benefit, understanding variations in the parameters and how they can possibly impact sytematics may provide valuable input towards tuning Geant4 models by the authors.

2. Toolkit Features

We have designed and implemented a comprehensive, modular, user-friendly software package that allows the user to randomly sample in the multiparameter space and to run simulation with multiple configurations of the Geant4 models. Another feature is the possibility to run a collective analysis of multiple variants of the resulting physics observables of interest and simultaneous benchmark versus relevant experimental datasets (one or many, as applicable). For benchmarking simulated results versus relevant experimental data or to perform statistical analysis, programmatic (C++) access to the DoSSiER repository [7] is also implemented. Geant4 version 10.3.patch01 or later is required to utilize this software. The toolkit is based on a modern event-processing framework Art [5] and offers a variety of features, such as a user-friendly model configuration API, flexible run-time configurable workflow, geometry and sensitive detector setup in GDML [6] if needed, comprehensive bookkeeping, analysis of multiple variants of the resulting physics observables of interest, and an extensible design. The toolkit components and the workflow are schematically presented in figure 1.

3. Preliminary Results

In our initial study we explored sensitivity of the Bertini and FTF simulated results to the variations of their parameters. An example of the effects of varying selected Bertini parameters on the simulated energy deposit in the LArIAT [9] Liquid Argon detector is shown in figure 2. The simulation uses the QGSP_FTFP_BERT physics list [1, 8] that includes the Bertini covering the 0-9.9 GeV energy range. Predictions using the default values of Bertini parameters were compared with results obtained with several alternative settings of these parameters within a physically meaningful range. In this study the varied parameters were the nuclear radius and the "internal" hadron-nucleon cross-section for an incident particle traveling through the target nucleus ("Radius Scale" and "XSec Scale" in figures 2 and subsequent figures). The toolkit has also been used to model hadron-nucleus interactions with various settings of the Geant4 Bertini or FTF parameters and to perform initial statistical analysis of the simulated results by benchmarking them versus relevant data from HARP [10, 11] and ITEP-771 [12] experiments. Results are given as χ^2 per number of degrees of freedom (χ^2/ndf). The χ^2 is calculated for each simulated spectrum and the corresponding data spectrum. Figure 3 shows preliminary results

IOP Conf. Series: Journal of Physics: Conf. Series **1085** (2018) 042021 doi:10.1088/1742-6596/1085/4/042021

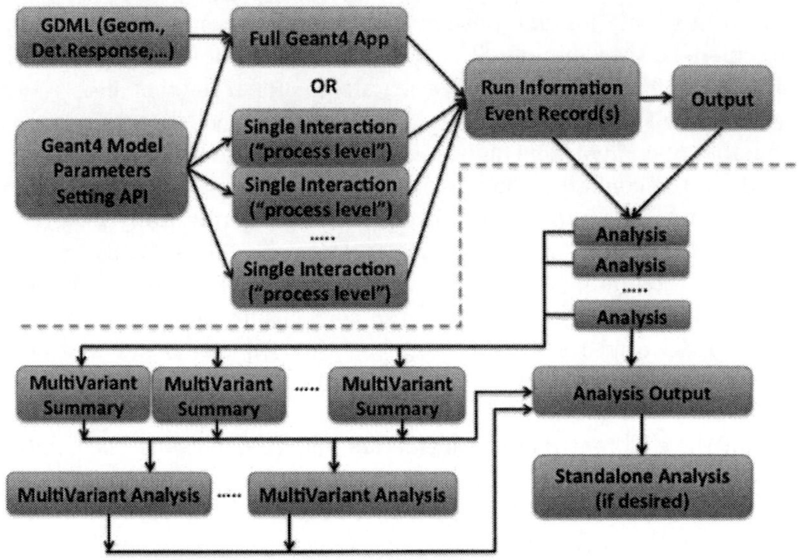

Figure 1. Software components and workflow.

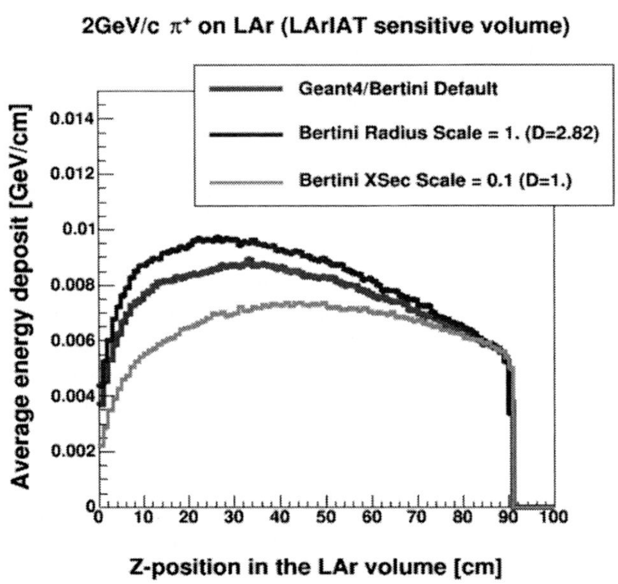

Figure 2. Longitudinal profile of a simulated hadronic shower induced by 2 GeV/c π^+ incident on a Liquid Argon detector volume (LArIAT) [9] with different settings of Bertini parameters.

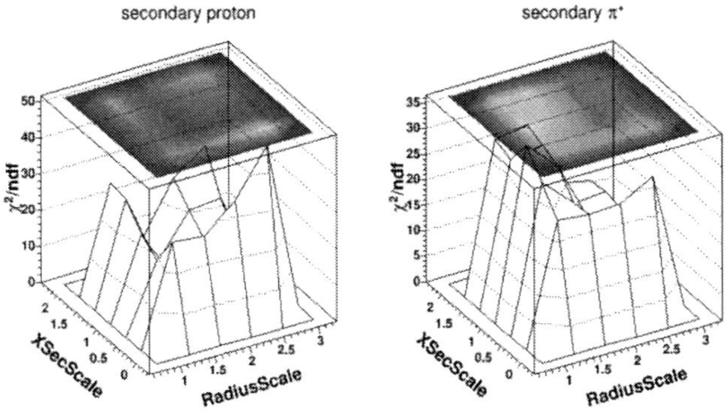

Figure 3. Statistical comparison of the simulated results for proton (left) and π^+ (right) production by a 5 GeV/c proton beam interacting with a Lead nucleus. Results are shown in a form of χ^2/ndf as a function of Bertini nuclear radius scale (Radius Scale) and the "internal" hadron-nucleon cross-section scale (XSec Scale) settings. The χ^2 is calculated for each simulated spectrum, as obtained for a given pair of radius scale and cross-section scale settings, and the relevant experimental data spectrum [10, 11, 12]. The ndf is the number of degrees of freedom.

on the χ^2/ndf as a function of the Bertini parameters values in the 2-parameters space.

We also varied the parameter that controls modeling of the trailing effect ("Trailing Radius"). The trailing effect takes into account the local depletion of the nuclear density following an intra-nuclear collision; it is roughly the radius (in fm) of the depleted region. In this study we randomly selected 100 settings of the parameter uniformly distributed within its validity range. Preliminary results are shown in figure 4 as χ^2/ndf versus the value of Bertini trailing radius. Results of modeling proton production in hadron-nucleus interactions with different variants of Bertini appear to favor the trailing radius setting of approximately 0.7 while by default this feature is disabled (i.e. trailing radius is set to zero). However, it is important to remember that there exist correlations among the Bertini parameters. Figure 5 illustrates the effect of varying nuclear radius on simulated hadron production in hadron-nucleus interactions as modeled with the traling radius set to zero (default) or to 0.7. While both sets of χ^2/ndf as a function of nuclear radius are sensitive to the nuclear radius value, they exhibit substantially different patterns. In figures 4 and 5 the χ^2 is calculated for each simulated spectrum and the relevant experimental data spectrum [10, 11, 12].

We have performed a similar study using FTF to model hadron production in hadron-nucleus interaction and to explore the sensitivity of the simulated observables. Preliminary results are shown in figures 6 and 7. Figure 6 illustrates how the χ^2/ndf depends on the value of one of the FTF parameters that control modeling of the nuclear destruction of the target nucleus [13] ("FTFP NUCDESTR_P1_TGT" in figure 6). The parameter in question may be set to a fixed value (e.g. 1.0 in the Geant4 release 10.3.patch01) or it can be expressed as a function of the number of nucleons in the target nucleus (as projected in future Geant4 releases). In this study we varied this parameter assuming that its value is the same for all types of the target nuclei, and we simulated hadron production in proton-Carbon or proton-Lead interactions at 5 GeV/c. The χ^2 is calculated for each simulated spectrum and the relevant experimental data spectrum [10, 11, 12]. As presented in figure 6, the distributions of χ^2/ndf as a function of NUCDESTR_P1_TGT obtained for light (Carbon) or heavy (Lead) nuclei exhibit substantially different patterns, which is especially clear for the secondary protons. This is likely to confirm that the parameter should depend on the number of nucleons in the target nucleus. However,

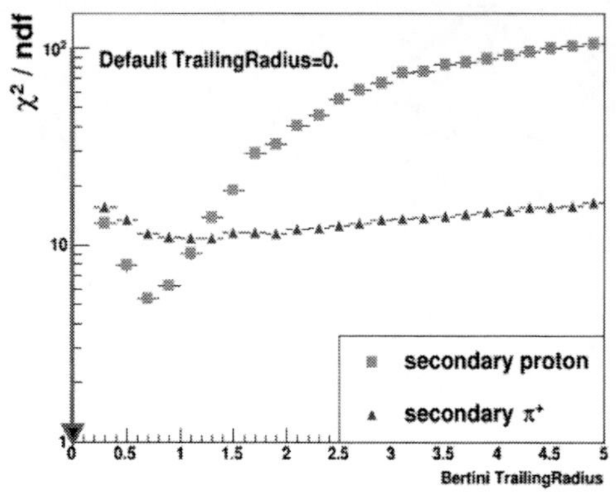

Figure 4. Statistical comparison of the simulated results for proton and π^+ production in proton-Lead interactions at 5 GeV/c. Results are shown in a form of χ^2/ndf as a function of Bertini trailing radius. Experimental data from HARP [10, 11] and ITEP-771 [12] are used to calculate the χ^2/ndf.

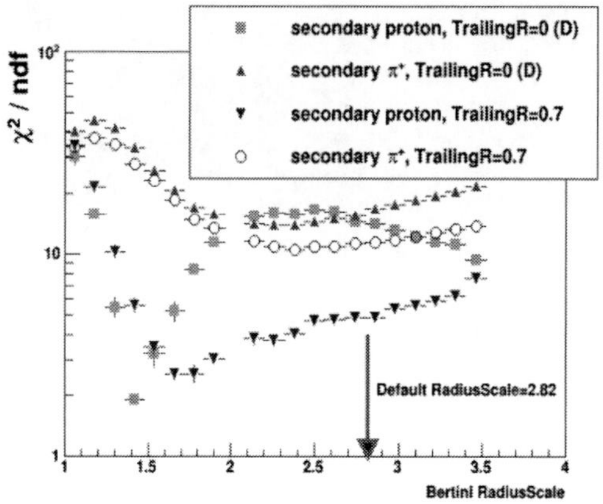

Figure 5. Statistical comparison of the simulated results on proton and π^+ production in proton-Lead interactions at 5 GeV/c. The χ^2/ndf versus Bertini nuclear radius is shown for different settings of the trailing radius. Experimental data are from HARP [10, 11] and ITEP-771 [12].

the correlation among FTF parameters should be thoroughly explored before a more definite statement is made. Figure 7 illustrates how the simulated results on hadron production in hadron-nucleus interactions depend on one of the parameters, average transverse momentum ("FTFP BARYON_AVRG_PT2" in figure 7) that is involved in modeling excitation of the participating hadron as a QCD string (the string is further decayed through the fragmentation mechanism). The χ^2 is calculated for each simulated spectrum and the relevant experimental data spectrum [14].

4. Future Plans

Our nearest-term task is to understand errors and correlations of the Geant4 hadronic models parameters via simulation and analysis in the multiparameter space, accross a wide range of energies, beam particle types, and target nuclei. In addition to the possibility to run various types of Geant4-based simulation in the multiparameters space, we have also developed an interface to the Professor package [15] that has been used for tuning a number of Monte Carlo event generators. Professor relies on the bin-wise parameterisation of the results simulated in the multiparameters to a low order polynomial function; then it uses such function to construct the $\chi^2 = \Sigma_{bin}((interpolation - data)^2/error^2)$ and to further numerically minimize it. The obvious benefit of such approach is that it substantially reduces the number of simulated samples required for the study. We plan to obtain more results from the Professor-based statistical analysis for such key Geant4 hadronic models as PreCompound, Bertini, and FTF.

5. Summary

In response to requests from the user community, we are developing a software toolkit to explore the impact of varying Geant4 model parameters on the simulated physics results, to refine the validity ranges of the parameters, and to numerically determine correlations among them. Comparison of simulated results versus experimental data, including statistical analysis, can be done via programmatic access to the DoSSiER repository. The toolkit was used to study the

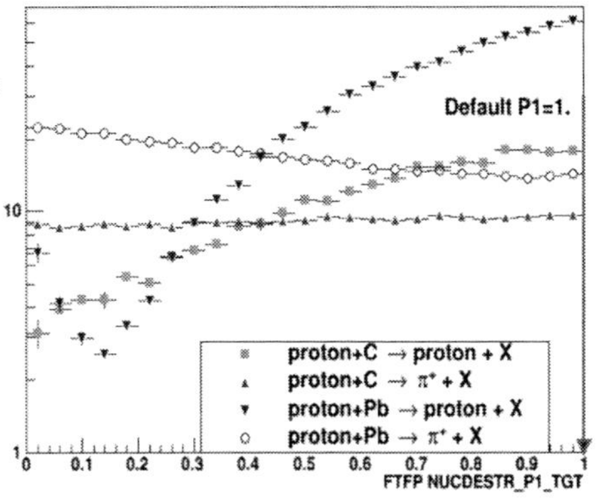

Figure 6. Statistical comparison of the simulated results for proton and π^+ production proton-Carbon or proton-Lead interactions at 5 GeV/c. Results are shown as χ^2/ndf versus one of the FTF parameters involved in modeling nuclear destruction of the target nucleus. The χ^2/ndf is calculated versus data from HARP [10, 11] and ITEP-771 [12].

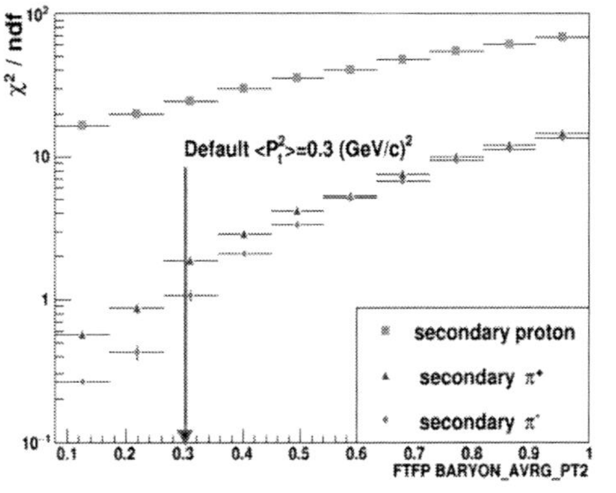

Figure 7. Statistical comparison of the simulated results for proton and pion production in proton-Carbon interactions at 158 GeV/c. Results are shown as χ^2/ndf versus one of the FTF parameters involved in modeling excitation of the participating hadron. The χ^2/ndf is calculated versus spectra measured by NA49 [14].

effects of varying Geant4 hadronic model parameters on simulated observables, with Geant4 Bertini and FTF models as the use-cases. Selected results are included in this presentation for illustration. Our near-term plans include further development of the toolkit and use of the Professor tuning toolkit for analysis in the multiparameter space.

Acknowledgments

† Operated by Fermi Research Alliance, LLC under Contract No. DE-AC02-07CH11359 with the United States Department of Energy.

References

[1] Allison J et al. 2016 *Nucl. Instrum. Meth.* A **835** 186
[2] `http://cern.ch/geant4/UserDocumentation/UsersGuides/PhysicsReferenceManual/fo/PhysicsReferenceManual.pdf`
[3] Wright D H and Kelsey M 2015 *Nucl.Instrum.Meth.* A **804** 175
[4] Uzhinsky V 2013 *International Conference on Calorimetry for the High Energy Frontier* 260
[5] `http://art.fnal.gov/`
[6] Chytracek R, McCormick J, Pokorski W and Santin G 2006 *IEEETrans.Nucl.Sci.* **53** 2892
[7] Wenzel H et al., 2017 DoSSiER: Database of Scientific Simulation and Experimental Results *Preprint* FERMILAB-CONF-16-508-CD
[8] `http://cern.ch/geant4/UserDocumentation/UsersGuides/ForApplicationDeveloper/html`
[9] `http://lariat.fnal.gov/`
[10] Catanesi M G et al. 2008 *Phys.Rev.* C **77** 05520
[11] Apollonio M et al. 2009 *Phys.Rev.* C **80** 035208
[12] Bayukov Yu et al. 1983 Preprint ITEP-83-148; Bayukov Yu et al. 1985 *Sov.J.Nucl.Phys.* **42** 116
[13] `http://geant4.web.cern.ch/geant4/UserDocumentation/UsersGuides/ForToolkitDeveloper/html`
[14] `https://spshadrons.web.cern.ch/spshadrons`
[15] `http://professor.hepforge.org`

IOP Conf. Series: Journal of Physics: Conf. Series **1085** (2018) 042022 doi:10.1088/1742-6596/1085/4/042022

End-to-End Event Classification of High-Energy Physics Data

M Andrews[1], M Paulini[1], S Gleyzer[2], B Poczos[3]

[1] Department of Physics, Carnegie Mellon University, Pittsburgh, USA
[2] Department of Physics, University of Florida, Gainesville, USA
[3] Machine Learning Department, Carnegie Mellon University, Pittsburgh, USA

E-mail: mandrews@cmu.edu, paulini@heps.phys.cmu.edu, sergei@cern.ch, bapoczos@cs.cmu.edu

Abstract. Feature extraction algorithms, such as convolutional neural networks, have introduced the possibility of using deep learning to train directly on raw data without the need for rule-based feature engineering. In the context of particle physics, such end-to-end approaches can be used for event classification to learn directly from detector-level data in a way that is completely independent of the high-level physics reconstruction. We demonstrate a technique for building such end-to-end event classifiers to distinguish simulated electromagnetic decays in a high-fidelity model of the CMS Electromagnetic Calorimeter.

1. Introduction & Motivation

An essential part of any new physics search at the Large Hadron Collider (LHC) at CERN involves event classification, or distinguishing signal events from the background. Traditional machine learning techniques have relied on high-level features in the form of particle 4-momenta - consistent with our understanding of particle physics phenomenology. However, such high-level features are the result of a rule-based casting of the raw detector data into progressively more physically-motivated quantities. While such approaches have proven to be useful, they are highly dependent on our ability to completely and effectively model all aspects of particle decay phenomenology. On the other hand, powerful image-based Convolutional Neural Network (CNN) algorithms have emerged, capable of training directly on raw data, learning the pertinent features unassisted - so-called *end-to-end deep learning classifiers*. In this paper, we explore the use of such classifiers on simulated data from the Electromagnetic Calorimeter (ECAL) of the Compact Muon Solenoid (CMS) detector at the LHC. Through the use of simulated toy decays, we demonstrate that these classifiers are able to effectively discern event phenomenology completely unassisted.

In this paper, we restrict ourselves to electromagnetic showers. A number of recent efforts have concentrated on image-based physics classification - for instance, jet shower [1,2] and neutrino classification [3], as well as traditional event classification

IOP Conf. Series: Journal of Physics: Conf. Series **1085** (2018) 042022 doi:10.1088/1742-6596/1085/4/042022

[4–6]. However, these approaches mostly rely on the output of rule-based high-level physics reconstruction algorithms and are thus subject to any mis-modeling contained therein. While a few have begun to use low-level data for event classification [7], in their current form, they still employ image construction techniques that depict the underlying detector geometry in rough approximation. As suggested in [6], such techniques potentially suffer from pixelization effects. In contrast, we construct our images in the detector coordinate system using only detector-level data. This minimizes our exposure to potential mis-modeling effects outside of the event classifier itself and ensures as high fidelity a representation of the detector as possible.

In section 2, we describe how CMS ECAL performs hit reconstruction. In section 3, we describe the use of end-to-end classifiers for electromagnetic shower classification. Then, in section 4, we discuss the construction of a full event classifier using various types of state-of-the-art deep learning models. We discuss our future plans in section 6 and summarize our conclusions in section 7.

2. Detector Reconstruction

2.1. The CMS ECAL Detector

The CMS experiment is one four large collider experiments at the LHC at CERN [8]. The focus of this paper is on the Electromagnetic Calorimeter, the sub-detector of CMS, responsible for resolving and localizing the energies of photons and electrons. It plays a major role in the detection of the Higgs boson and other Electroweak phenomena [4]. The ECAL is a hermetically-sealed cylinder composed of lead tungstate ($PbWO_4$) crystals packed together into a barrel section (EB) and two circular endcap (EE) sections. The barrel section is composed of 61200 crystals segmented 170-fold in pseudorapidity (η) and 360-fold in the azimuth angle (ϕ). Each circular endcap section contains 7324 crystals segmented in a rectilinear grid. The crystals are angled toward the interaction point but with a slight offset. This minimizes particles slipping through cracks but means the amount of an ECAL crystal a particles traverses is position-dependent.

2.2. ECAL Hit Reconstruction

2.2.1. Particle Shower Formation and Detection. On its way from the interaction point, a high-energy electron (e) or photon (γ) will interact with material from other subdetectors in CMS, causing it to shower and deposit its energy over a range of crystals in the ECAL. While shower formation is a stochastic process, on average, roughly 94% of the e/γ shower's energy is deposited within a 3 by 3 grid of crystals. Upon interacting with the ECAL crystals, the e/γ shower induces scintillating light which is measured by Avalanche Photodiodes (in EB) or Vacuum Phototriodes (in EE) at the other end of the crystal as a signal pulse. To first approximation, e/γ shower profiles are expected to be identical in the ECAL. However, due to its interaction with the magnetic field of the

CMS solenoid (B = 3.8 T), the charged electron emits bremsstrahlung, preferentially in ϕ. This introduces a higher-order perturbation to the e shower profile causing it to be more spread out and slightly asymmetric in ϕ compared to the photon shower.

2.2.2. The Digitized Hit. The signal pulse is amplified and shaped by a multi-gain preamplifier before being digitized at 40 MHz. The net effect is a signal pulse with a stepped profile that rises sharply to a peak before falling off gradually, with a full width of under 150 ns at half maximum. For every recorded event, ten such steps or *samples* are stored, giving ten *amplitude* readings per crystal, separated by 25 ns. These samples define the raw *digitized hit* or *digi*, and represent the lowest level, physically-sensible calorimeter data. The timing calibration is such that the pulse appears on the 4th time sample, such that the baseline electronics noise, known as the *pedestal*, can be estimated from the first three time samples.

2.2.3. The Reconstructed Hit. A fitting algorithm is applied to the digitized hit to determine the energy and time from the peak and shape of the pulse. A number of effects must be corrected for before arriving at a calibrated hit. Most notable is the crystal transparency loss and recovery of the crystals under the presence and absence of beam radiation, respectively. After correcting for this and other effects, a calibrated, *reconstructed hit* or *rec hit* is produced containing energy and timing information.

2.2.4. High-level Physics Features. Through a rule-based process known as high-level physics reconstruction [9], the reconstructed hits are transformed into higher-level features like particle IDs, 4-momenta, and shower shapes. Further in the analysis chain, even more complex features are engineered to achieve greater separation of signal from background. These features have their roots in particle physics theory. In machine learning parlance, these are the equivalent of *hand-engineered* features. In particle physics, as in computer vision, high-level features have traditionally served as inputs to machine learning algorithms like fully-connected neural networks and boosted decisions trees for event classification.

Having described ECAL hit reconstruction, we use the term *end-to-end classifier*, more precisely, to denote classifiers that use either digitized or reconstructed hits as inputs.

3. Shower Classification

As a first step towards event classification, we tackle the easier problem of e/γ shower classification. To begin, we use events generated with exactly one negatively charged e or one γ of fixed transverse momentum $p_T = 50\,\text{GeV}$ fired from the interaction point to a direction sampled uniformly from pseudorapidity $|\eta| < 1.4$ and azimuthal angle $-\pi < \phi < \pi$, effectively constraining the e/γ shower to the barrel section of ECAL.

3.1. Input Image

To construct the image of the shower, we take a grid of 32 by 32 crystals from ECAL centered around the maximum-energy shower deposit (the *shower seed*). Each pixel in the image grid will correspond exactly to one crystal, though not necessarily the *same* crystal from event to event, and will be filled with the relevant data for that crystal: amplitude, energy, etc. As the image is always centered on the shower seed, shower classification does not take full advantage of the feature translation abilities of CNNs. The classifier must still be able to generalize due to the stochastic nature of shower formation, and position-dependent crystal interaction. Images of averaged electron and photon showers are shown in Figure 1.

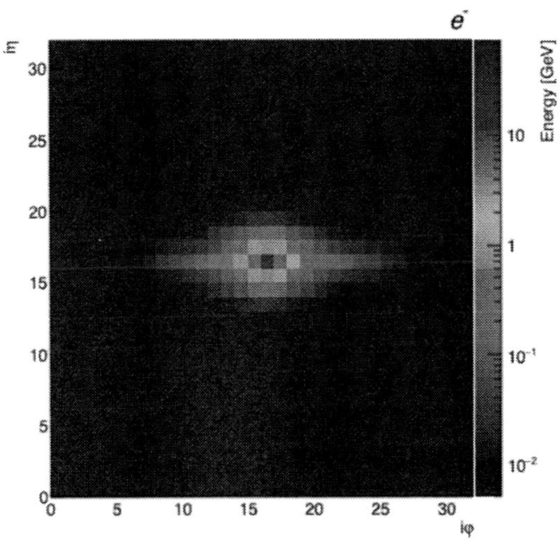

(a) Average electron shower

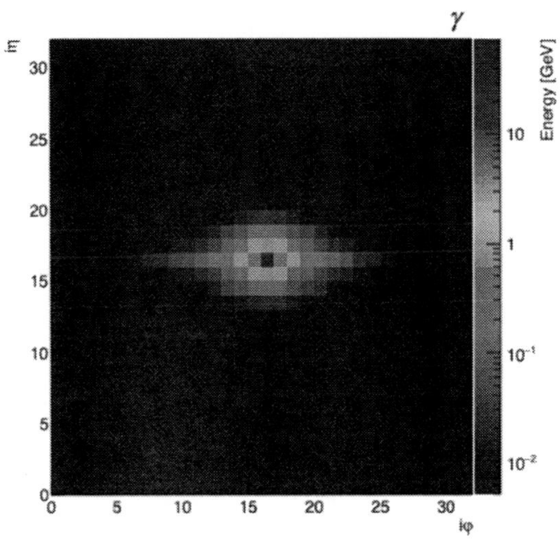

(b) Average photon shower

Figure 1: e/γ showers averaged over 50k showers. The e shower is slightly more spread out in ϕ-in addition to being slightly asymmetric-due to bremsstrahlung effects.

3.2. Preprocessing

Preprocessing is essential for ensuring numerical stability in the optimizer, especially with large images and deep networks. We experimented with linear and logarithmic transformations. For the latter, because the shower image is sparse, we attempted different schemes of shifting the log-transformed inputs relative to the sparse values.

For the digi amplitudes, there is the additional complication of neighboring crystals reading out the pedestal values due to the activation of the ECAL selective readout algorithm. We, therefore, tried different pedestal subtraction schemes in addition to applying linear or log transformations.

IOP Conf. Series: Journal of Physics: Conf. Series **1085** (2018) 042022 doi:10.1088/1742-6596/1085/4/042022

3.3. Network Architecture.

We experimented with deep neural network architectures that can be primarily grouped on the following categories: Convolutional Neural Networks (CNN) [10], Long Short-Term Memory (LSTM) [11], and Fully-Connected Networks (FCN) [12]. All of these were implemented using the Keras Deep Learning library [13].

The CNN category included simplified implementations of VGG [14], Inception [15], and Residual Net (ResNet) [16] architectures. We tried the following data combination schemes:

(i) energy only

(ii) energy and time stacked

(iii) digitized amplitudes stacked

(iv) energy, time, and digitized amplitudes stacked

(v) (ii) and (iii) concatenated at output of convolutional layers

(vi) (ii) and (iii) concatenated at output of fully-connected layers

The LSTM networks only used digi inputs and included the following architectures: (i) Time-distributed: each digi amplitude image is connected to its own CNN, flattened, and given as a sequence to LSTM layers. (ii) Convolutional: the LSTM layers act directly on a time sequence of convolutional layers [17].

The FCNs learned from the flattened energy and time images or the flattened digi amplitude images and consisted of 2-, 3-, 6-hidden layers with 256 units per layer.

3.4. Training Strategy.

The simulated sample consisted of 498k events evenly balanced between e and γ. Of these, 320k (64%) were used for training, 89.6k (18%) for validation, and 88.4k (18%) for a final test set. The validation set was used to evaluate competing architectures and hyper-parameters within the above three network categories. For the best network in each category, we evaluate their performance on the test set, with the results summarized in section 5.

For e/γ classification, we used binary cross-entropy loss with the ADAM adaptive gradient optimizer, both with and without explicit learning rate decay. The initial learning rate was set to 1×10^{-3} and we trained for 50 epochs by which point the validation loss had more than plateaued. A dropout of 20% for fully-connected layers and ReLU activation were used, with weights initialized from a truncated normal distribution where applicable. Finally, the area under the curve of the Receiver Operating Characteristic (ROC AUC) was chosen as a performance figure of merit, due it interpretability in terms of signal efficiency and background rejection.

IOP Conf. Series: Journal of Physics: Conf. Series **1085** (2018) 042022 doi:10.1088/1742-6596/1085/4/042022

4. Event Classification

Two studies are presented. The first is a generalization of the shower classification to full detector images. Recall that the e/γ particles are fired randomly, and so the showers will appear to move in the image from event to event, providing a real test of the feature-translation abilities of CNNs in a sparse image.

Then, we proceed to generalize to double particle decays-either an electron-positron pair e^+e^-, or a photon pair $\gamma\gamma$. The particles are allowed to have a range of transverse momenta $p_T = (20, 80)$ GeV but it is required to be the same for each particle. The pair decays back-to-back, such that the position of one is the inverse of the other in either e^+e^- or $\gamma\gamma$ decay. One can think of two ways a classifier might learn to distinguish multi-particle events: either from a difference in particle shower profiles, or from a difference in the spatial arrangement of the particle showers, i.e. event kinematics. Back-to-back decays provide a useful scenario for investigating the former due to the identical kinematics. This forces the network to learn only from the shower profiles, providing an unbiased test of the ability of CNNs to perform feature translation. Finally, we include pile-up of $\langle \text{PU} \rangle = 25$, but no underlying event simulation.

4.1. Input Image & Preprocessing.

In event classification, the image grid is 170 by 360 pixels corresponding to the full ECAL barrel geometry. There is again an exact correspondence between calorimeter crystals and image pixels. However, now the crystals underlying the detector image are always fixed.

4.2. Network Architecture & Training Strategy.

A VGG-type network proved to be cumbersome for event classification. The convolutional to fully-connected interface caused an explosion of model weights by several orders of magnitude. Coupled with the increased image sparsity ($> 90\%$), this made training unstable at learning rates above $\sim 10^{-3}$. Instead, we experimented primarily with ResNet-type architectures without any fully-connected layers. While the ResNets were stable even above learning rates of $\sim 10^{-3}$, we found that a lower learning rate of 5×10^{-4} worked best.

Interestingly, the original residual block worked better with sparse data compared to the bottleneck version of the residual block [16]. We found the best performance for the residual block *without* batch normalization, and with max pooling instead of average pooling, as might be expected. A total of three down-samples performed better compared to the typical five used in most ResNet implementations. Finally, with 320k training events, any ResNet deeper than about 23 convolutional layers showed no further gain in performance.

5. Results.

5.1. Shower Classification Results

The classification scores of each deep learning model evaluated on the test set are summarized in Table 1.

Table 1: Shower Classification Results.

Category	**Network**	**ROC AUC**
CNN	VGG, energy	0.807
LSTM	Conv-LSTM, digis	0.799
FCN	3-layers, digis	0.770

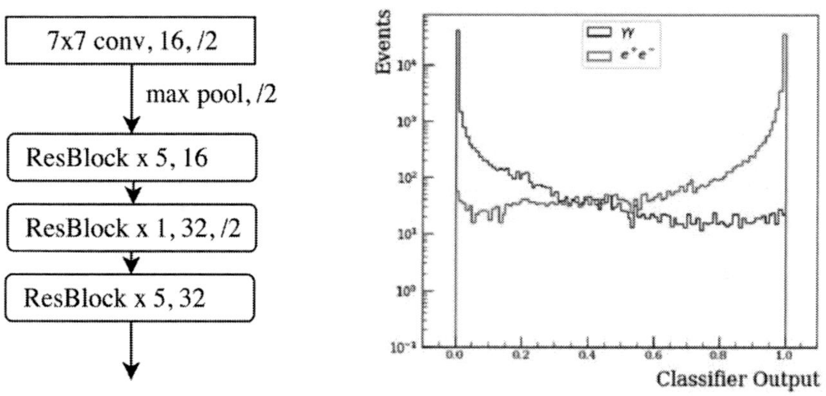

(a) ResNet-23 (b) e^+e^- vs. $\gamma\gamma$ Classifier Output

Figure 2: Event Classification Results for e^+e^- vs. $\gamma\gamma$ using a 23-layer Residual Net.

Within the CNN category, the different architectures all scored within 0.1% of each other. LSTM-based architectures, as expected, took considerably longer to train but showed no clear advantage over purely-CNN architectures. Within the FCN category, the different networks all scored within $< 1\%$ of each other. The FCN-based architectures consistently underperformed, even though feature translation did not play a major role. Linear preprocessing worked best in each category, though, in general, the results were not sensitive to the choice of preprocessing.

As noted in 2.2.1, in the ECAL, e showers are at first approximation expected to be identical to their γ counterparts. And, by eye, it is difficult to tell the two apart. Yet our results indicate that deep learning classifiers are able make very fine distinctions between the two, picking up on higher-order perturbations to deliver a good classification score.

5.2. Event Classification Results

The classification scores of the best deep learning model (ResNet-23) applied to single and di-particle events are summarized in Table 2.

Table 2: Event Classification Results.

Decay	**Network**	**ROC AUC**
e vs. γ	ResNet-23	0.788
e^+e^- vs. $\gamma\gamma$	ResNet-23	0.997

The single e/γ score compares favorably to the one from shower classification, although is slightly lower. Therefore, while CNNs handle feature translation well even for highly sparse data, the transition is not perfect. One important questions is - what advantage does event classification offer over simpler shower classification? A partial answer is provided by the di-particle result.

As shown in Table 2, the classifier score for double photon versus double electron events is considerably higher. While the boost in classification score seems a little surprising at first, it is to be expected. In particular, one *could* model the di-particle prediction using the product of the individual shower classifiers' predictions. If the mean single shower mis-classification accuracy is y'_{single}, then on average, one would accurately classify the di-particle pair at a rate of $y_{\text{pair}} = 1-(y'_{\text{single}})^2$. From the shower classification CNN, $y'_{\text{single}} \sim 0.26$, giving $y_{\text{pair}} \sim 0.93$. The observed event classifier accuracy of ~ 0.97 is even higher, suggesting that there are other effects present. Importantly, this accuracy is achieved over a range of momentum and with pile-up. Even though we expect even better performance compared to hand-engineered features in more complicated decay topologies, the fact that the classifiers perform so well even in the relatively simple two-body decays, is encouraging. As this example suggests, the strength of end-to-end event classifiers may lie in tackling complex, multi-particle decays, without the need to introduce explicit kinematics.

6. Future Work

For future work, we would like to extend our study in two directions. First, we plan to study more complex pp decays with non-trivial kinematics, with an eye towards a realistic physics analysis. There are additional challenges, such ensuring that the classifier is not correlated with an observable of interest. Second, we plan to extend the end-to-end approach to the full CMS detector. The next natural steps are to include the ECAL endcaps, the Hadronic Calorimeter and then the Tracker and Muon Systems. Building a full end-to-end classifier from all the available data from the CMS detector is the ultimate goal of this study.

IOP Conf. Series: Journal of Physics: Conf. Series **1085** (2018) 042022 doi:10.1088/1742-6596/1085/4/042022

7. Conclusions

In this paper, we have presented a technique for building end-to-end physics classifiers for single electromagnetic shower classification and event classification. First, we constructed high-fidelity images of electromagnetic showers from the CMS ECAL barrel using low-level detector data. Afterwards, we used these images to train various deep learning models to distinguish electron- and photon-induced showers and perform both shower and event classification. In both cases, the classifiers were able to exploit higher-order effects to obtain good discrimination. Finally, we distinguished pairs of identical particles in back-to-back decays, where a marked increase in performance was seen, suggesting the classifiers had learned to effectively model this type of a particle decay. While there are still challenges and more work needed to bring this approach to data, our results indicate a significant potential of end-to-end classifiers for high-energy physics data analysis.

References

[1] de Oliveira L. et al., *Jet-Images—Deep Learning Edition*, JHEP **07** 069 (2016).
[2] Kasieczka G. *Deep-learning Top Taggers or The End of QCD?*, JHEP 2017:6 (2017).
[3] Aurisano A. et al.,*A Convolutional Neural Network Neutrino Event Classifier*, JINST **11** p. P09001 (2016).
[4] The CMS Collaboration, *Observation of the diphoton decay of the Higgs boson and measurement of its properties*, Eur. Phys. J. C **74**:3076 (2014).
[5] Baldi P. et al., *Searching for Exotic Particles in High-Energy Physics with Deep Learning*, Nature Communications **5** 4308 (2014).
[6] Louppe G. et al., *QCD-Aware Recursive Neural Networks for Jet Physics*, arXiv:1702.00748.
[7] Bhimji W. et al., *Deep Neural Networks for Physics Analysis on low-level whole-detector data at the LHC*. https://indico.cern.ch/event/567550/contributions/2629673/
[8] The CMS Collaboration, *The CMS experiment at the CERN LHC*, JINST **3** p.S08004 (2008).
[9] The CMS Collaboration, *Particle-flow reconstruction and global event description with the CMS detector*, JINST **12** P10003 (2017).
[10] LeCun Y. et al. *Gradient-based learning applied to document recognition*. Proceedings of the IEEE. **86** (11): 2278-2324 (1998).
[11] Hochreiter S., Schmidhuber J., *Long Short-term Memory*. Neural Computation. **9** 1735-80, (1997)
[12] Rumelhart D. *Learning representations by back-propagating errors*. Nature. **323** (6088): 533-536 (1986).
[13] Keras Team, "Keras" [software], version 2.0.3, Available from https://github.com/fchollet/keras [accessed 2017-04-17].
[14] Simonyan K., Zisserman A., *Very Deep Convolutional Networks for Large-Scale Image Recognition*, arXiv:1409.1556.
[15] Szegedy C., *Going Deeper with Convolutions*, arXiv:1409.4842.
[16] Kaiming H. et al., *Deep Residual Learning for Image Recognition*, arXiv:1512.03385.
[17] Shi X. et al., *Convolutional LSTM Network: A Machine Learning Approach for Precipitation Nowcasting*, arXiv:1506.04214.

IOP Conf. Series: Journal of Physics: Conf. Series **1085** (2018) 042023 doi:10.1088/1742-6596/1085/4/042023

The HEP.TrkX Project: Deep Learning for Particle Tracking

Aristeidis **Tsaris**[1,a], Dustin **Anderson**[3], Josh **Bendavid**[3], Paolo **Calafiura**[2], Giuseppe **Cerati**[1], Julien **Esseiva**[2,4], Steven **Farrell**[2], Lindsey **Gray**[1], Keshav **Kapoor**[1], Jim **Kowalkowski**[1], Mayur **Mudigonda**[2], **Prabhat**[2], Panagiotis **Spentzouris**[1], Maria **Spiropoulou**[3], Jean-Roch **Vlimant**[3], Stephan **Zheng**[3], Daniel **Zurawski**[1]

[1] Fermi National Accelerator Laboratory, Batavia, Illinois, United States of America
[2] Lawrence Berkeley National Laboratory, Berkeley, California, United States of America
[3] California Institute of Technology, Pasadena, California, United States of America
[4] Haute école spécialisée de Suisse occidentale, Fribourg, Switzerland

E-mail: [a]atsaris@fnal.gov

Abstract. Charged particle reconstruction in dense environments, such as the detectors of the High Luminosity Large Hadron Collider (HL-LHC) is a challenging pattern recognition problem. Traditional tracking algorithms, such as the combinatorial Kalman Filter, have been used with great success in HEP experiments for years. However, these state-of-the-art techniques are inherently sequential and scale quadratically or worse with increased detector occupancy. The HEP.TrkX project is a pilot project with the aim to identify and develop cross-experiment solutions based on machine learning algorithms for track reconstruction. Machine learning algorithms bring a lot of potential to this problem thanks to their capability to model complex non-linear data dependencies, to learn effective representations of high-dimensional data through training, and to parallelize easily on high-throughput architectures such as FPGAs or GPUs. In this paper we present the evolution and performance of our recurrent (LSTM) and convolutional neural networks moving from basic 2D models to more complex models and the challenges of scaling up to realistic dimensionality/sparsity.

1. Introduction

The Large Hadron Collider (LHC) [1] is the highest energy collider ever constructed. The goal of the LHC is to probe fundamental physics questions, such as the basic building blocks of matter and their interactions. Two counter-circulating proton beams collide in the center of building-sized detectors, such as ATLAS [2] and CMS [3] and by measuring the energy and momentum of the escaping particles, fundamental laws of physics can be tested. The reconstruction of the trajectories of charged particles plays a key role in identifying particles and measuring their charge and momentum. A simplified tracking work-flow consists of track seeding, track building and track fitting. Typically, a Combinatorial Kalman Filter (CKF) algorithm is being used in both track fitting and track building with the latter to be the most CPU intense process in the reconstruction work-flow.

The High-Luminosity LHC era (HL-LHC) will bring significant challenges to particle tracking. With increasing the number of interactions per beam crossing ("pile up") the detector occupancy will also increase. Traditional tracking algorithms scale quadratically or worse [4] with detector occupancy and assuming a stable improvement in the CPU performance the next years, tracking algorithms will need to run faster and in parallel.

The goal of the HEP.TrkX project is to apply advanced machine learning algorithms to the HL-LHC track reconstruction problem. Deep learning architectures have been explored for vertex and track finding, as well as track fitting reconstruction challenges. We prototyped and developed the algorithms using datasets produced with "toy" simulations, while the robustness of the algorithms was tested with more realistic datasets. Machine learning algorithms have shown a great ability to learn effective representations of high dimensional data through training in various datasets. They also scale linearly with input data size and computationally are very regular, which makes them ideal to run efficiently on data parallel architectures on an HPC environment.

2. Deep Learning Models for Tracking with Toy Datasets

Long Short Term Memory (LSTM) [5] networks are state of the art Recurrent Neural Networks (RNN) that have seen great success in sequence prediction. We can imagine the problem of track finding as a hit prediction task, where a track has known hits in the first three layers (seed) and the neural network predicts the hit location on the other layers. The input to each LSTM layer is a detector layer in the format of pixel array followed by a single fully connected layer (FC). The FC is applied separately to each LSTM output and produces a pixel prediction for the same detector layer.

Figure 1 shows a sketch of this model architecture along with the input and the model prediction. The 2D toy dataset used here was generated using straight lines constrained within the detector volume. The target track has pixel hits in the first three layers compared to the incomplete background tracks. This model architecture was adjusted and tested to realistic dataset as described in Section 4.

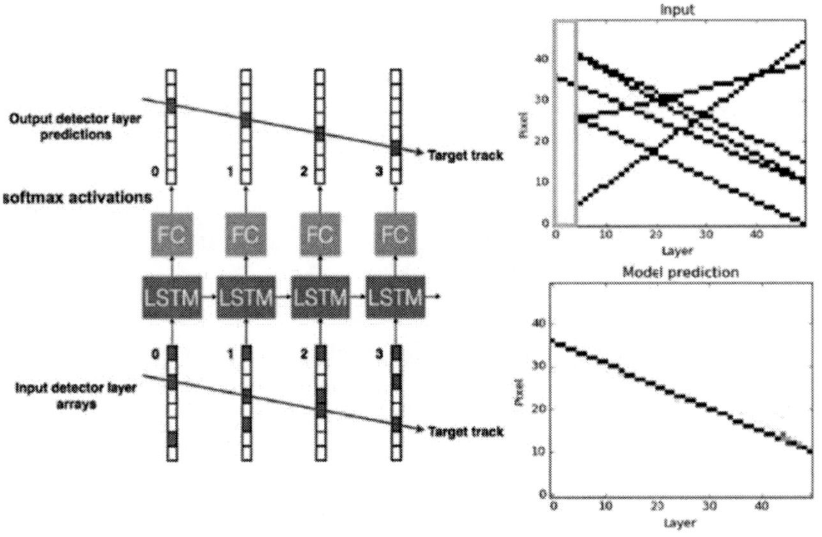

Figure 1. The plot on the left shows the LSTM model architecture [6] used to classify hits in one track. The plot on the right shows a 2D representation of the input and the output of the model. The cyan box in the upper right plot shows the seed of the target track in the first three layers.

S. Farrell et al. performed a detailed study of various model architectures and a comparison of the model predictions for 2D and 3D toy datasets [6]. The track finding problem was tested with bidirectional LSTM, deep-LSTM and next-layer LSTM (making prediction for the next detector layer). Also models that include Convolutional Neural Networks (CNNs) have been tested in toy datasets, either by modeling the track dynamics in a pixel level classification, or in an end-to-end approach where the model predicts the parameters of the track. The models that showed promising results will be tested against realistic datasets, as described in the next sections.

3. Hit Assignment to Tracks with TrackMLRamp Dataset

An alternative way to use RNN networks for track finding is to develop the model to assign hit to tracks. The input in this case is a set of hit positions (r, Φ and Z in this case) in the detector and the output is a matrix of probabilities of the hits assigned to tracks. In this case all input hits are fed in across all layers. The model architecture can be seen in the right plot of Figure 2 where three Bi-directional GRU (Gated Recurrent Units) layers were used with two dropout layers in between. GRU layers where used instead of LSTMs, since they appear to be trained faster for the shallow network architecture used.

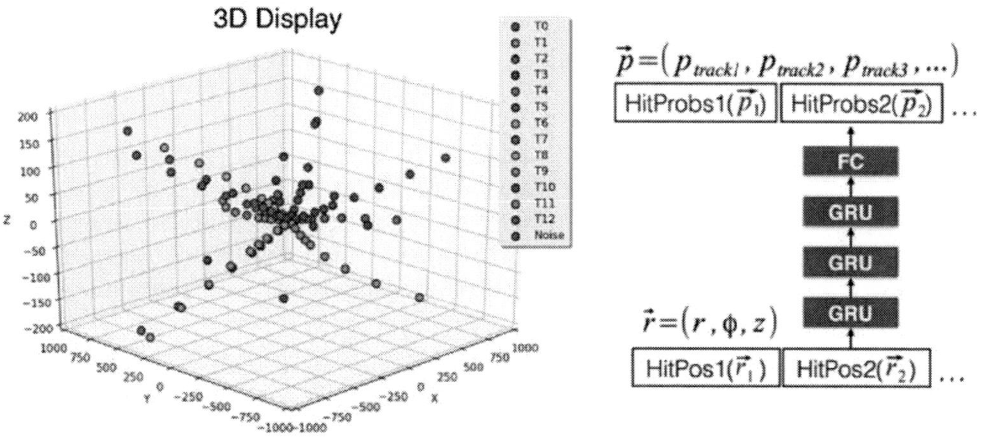

Figure 2. The plot on the left shows a 3D representation of the modified TrackMLRamp Dataset [7]. The plot on the right shows the GRU model used to assign hit to tracks.

A 3D dataset was generated to train this neural network based on the 2D dataset used in "TrackMLRamp hackathon" [7]. Events were generated with a constant magnetic field in perfectly circular detectors. A longitudinal dimension was added with respect to the beam, along with noise hits, distributed randomly across the layers. A 3D representation of an example event with 13 tracks can be seen in the left plot of Figure 2.

Figure 3 shows the accuracy as a function of the number of tracks, where the accuracy is defined as the fraction of hits in the event that is assigned to correct track. Events with one track have the lowest average accuracy, since with a flat noise rate across all events the ambiguity becomes larger compared to multi-track events. For events with higher number of tracks the accuracy seems relatively stable, and we clearly have not reached the point that the accuracy is saturated. To study this we will generate statistically larger dataset and possibly use a deeper network.

The overal hit assignment accuracy in the previous case was 87%, were detector hits assigned to only one track, the one with the highest probability. We also explored accepting multiple

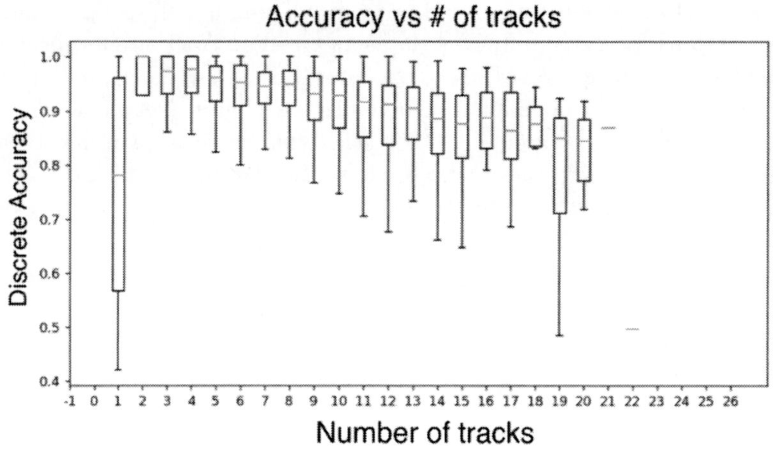

Figure 3. The plot shows the accuracy as a function of the number of tracks for the hit assignment GRU model.

hit assignments to tracks, for different threshold values of the prediction probabilities, in an effort to improve the accuracy. This study is shown in Figure 4 where the top left shows the probability of hit assignment to multiple tracks with at least a threshold value. With low enough threshold selection the probability is increased and the number of ambiguities appears not to be that high (right plot of Figure 4). For example for the particular dataset that is presented here, a threshold value of 0.7 would result a 10% more accurate hit assignment. We can still keep the events that are not covered with this threshold and resolve the ambiguities due to multiple hit assignment with another off-line algorithm downstream the reconstruction chain.

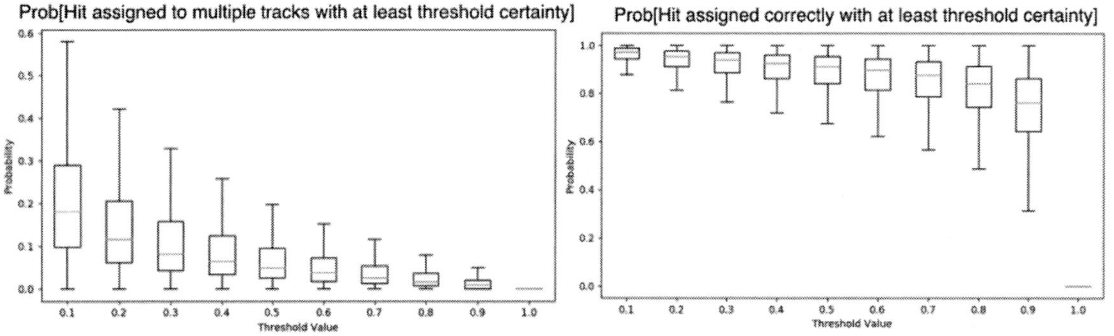

Figure 4. The plot on the left shows the probability of hit assignment to multiple tracks as a function of threshold value. The plot on the right shows the probability of correctly hit assigned to multi tracks as a function of threshold value.

4. Hit Classification and Vertex Finding with ACTS Dataset

After exploring various deep learning architectures in toy datasets, a set of those architectures is being tested against more realistic datasets simulated with A Common Tracking Software (ACTS) [8]. Deep learning algorithms described in the previous sections are tested on MC samples from a generic LHC-like detector simulation. By increasing the complexity and realism of the dataset, we test the robustness of the tracking algorithms against different configurations. Figure 5 shows the layout of the detector geometry generated with the ACTS framework. Only hits in the barrel volume detectors are used while the hits in the end-cap detector are also available for future studies.

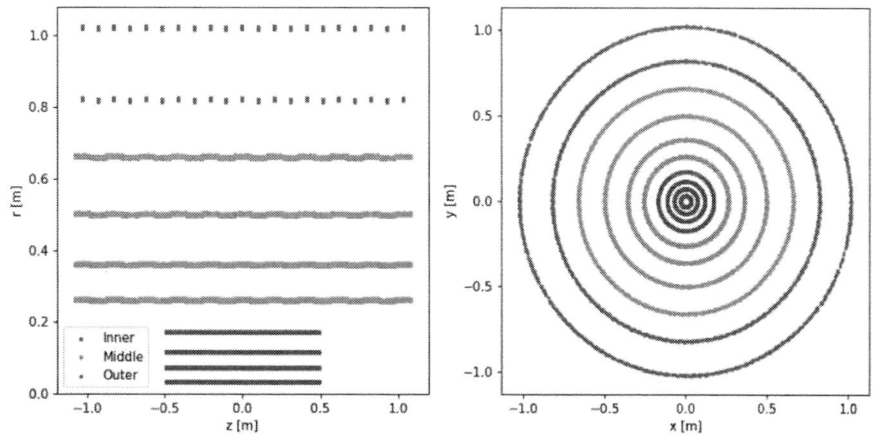

Figure 5. These scatter plots show the geometry of the ACTS dataset. Ten detector layers from three cylindrical barrel volumes are used. The left plot shows the longitudinal z coordinate vs. cylindrical radius r, while the right plot shows the x-y plane. Detector end-cap layers are not utilized and not shown.

Figure 6 shows a generated example event used by the LSTM hit classification model. The left and middle plots are projected 3D histograms which show the input data and the ground truth used to train the model. The gray color scale is setup so that dark represents the maximum number of hits. The hit classification model was used in this dataset, as described in section 2, adjusted for 3D classification using variable-sized layers and binning for the barrel-volume detector layers. As in the 2D case, the first three layers were used as a seed to constraint the target track. The right plots of Figure 6 shows the model prediction for both projections in the detector, Z and Φ and the gray scale represents the probability of the prediction. The binning was coarse and many events overwhelm with occupancy so overall accuracy was 70% but it was qualitatively seen to perform much better when occupancy is lower. This suggests that greater granularity (probably in smaller detector sub-regions) could scale up better.

The construction of track seeds from triplet-hit combinatorics can be constrained by an estimate of the interaction vertex position along the beamline (the Z axis). A deep learning solution was thus developed which takes as input the three innermost barrel pixel layer hits binned in Z and Φ as an RGB image and produces a prediction for the binned primary vertex Z position. The model is based on LeNet-5 [9] and is trained on ACTS data generated with five average interactions per sample. Figure 7 shows an example sample prediction and target as well as the thresholded accuracy as a function of bin size. An 82% accuracy can be achieved by covering 57% of the events with a resolution of 7.65 mm. Studies are ongoing to improve the performance of this model.

5. Conclusion

CNNs and LSTMs have shown great results in classification and sequential problems, but are still relatively unexplored for particle tracking problems. In this work, a variety of deep learning

IOP Conf. Series: Journal of Physics: Conf. Series **1085** (2018) 042023 doi:10.1088/1742-6596/1085/4/042023

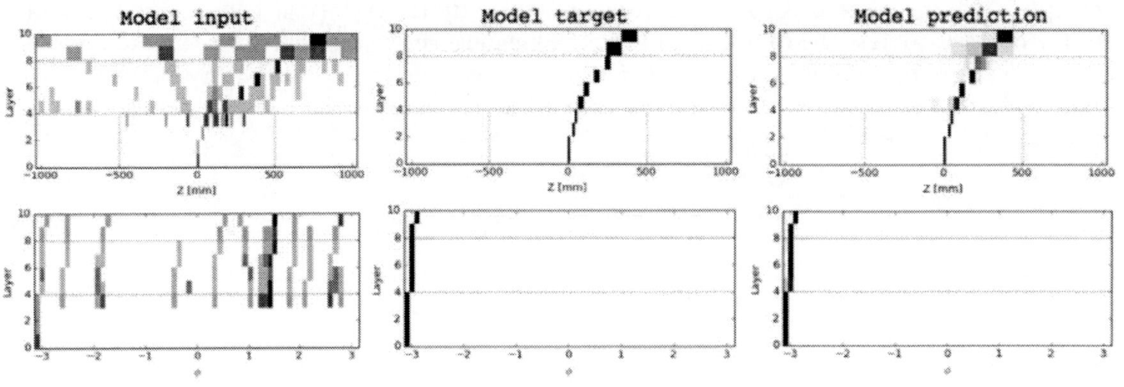

Figure 6. The plots show the input hit distribution to the LSTM network, the target tracks and the model prediction. The top plots are a projection in the Z coordinate of the detector (illustrated in Figure 5) and the bottom plots are a projection in the azimuthal angle of the detector.

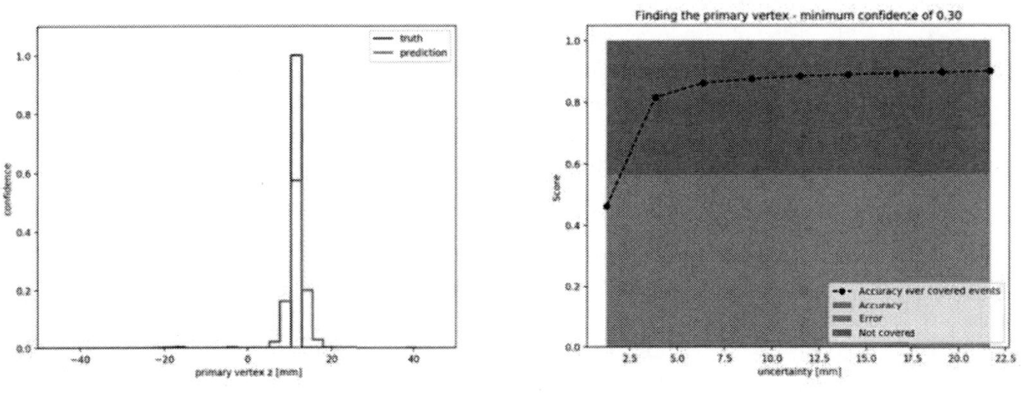

Figure 7. The plot on the left illustrates the binned Z-position of the target vertex (blue) and the output of the model (red) with a bin size of $2.55mm$. The plot on the right shows the accuracy and the error rate as a function of resolution (bin-size) with a 0.3 threshold in the confidence level of the model. The dashed line is the accuracy for events that pass this threshold (57% of the total events).

architectures was tested in semi-realistic datasets and where it was possible an optimization of the output was performed for higher efficiency of the model prediction. We have already started exploring our models with datasets generated with the ACTS framework. There are numerous architectures that have been tested with toy datasets but not yet fully adopted by the latest examples. We would like to fix this dataset (or a variation of it) and make comparisons with common tracking metrics between the different approaches described above. Also a comparison will follow with a traditional tracking algorithm such as the CKF.

The authors would like to thank the funding agencies DOE ASCR and COMP HEP for supporting this work, as well as the numerous tracking experts from ATLAS and CMS who have shared insights and experience.

IOP Conf. Series: Journal of Physics: Conf. Series **1085** (2018) 042023 doi:10.1088/1742-6596/1085/4/042023

References

[1] L. Evans, P. Bryant, JINST 3, S08001 (2008).
[2] G. Aad et al. (ATLAS), JINST 3, S08003 (2008)
[3] S. Chatrchyan et al. (CMS), JINST 3, S08004 (2008)
[4] G. Cerati et al., J. Phys. Conf. Ser. 664, 072008 (2015)
[5] S. Hochreiter, J. Schmidhuber, Neural Comput. 9, 1735 (1997)
[6] S. Farrell et al., EPJ Web of Conferences 150, 00003 (2017)
[7] TrackMLRamp hackathon, a 2D tracking challenge, https://indico.cern.ch/event/577003/contributions/2476446/attachments/1423512/2183608/tr170307_davidRousseau_CTDWIT2017_trackML.pptx.pdf
[8] A Common Tracking Software Project, http://acts.web.cern.ch/ACTS/index.php
[9] Y. LeCun et al., Proc. IEEE 86 (11) (1998) 22782324

IOP Conf. Series: Journal of Physics: Conf. Series **1085** (2018) 042024 doi:10.1088/1742-6596/1085/4/042024

The ATLAS Event Index: The Architecture of the Core Engine

Dario Barberis[1], Zbigniew Baranowski[2], Andrea Favareto[1], Julius Hřivnáč[3], Fedor Prokoshin[4], Grigori Rybkin[3], Rainer Toebbicke[2] and Ruijun Yuan[3] on behalf of the ATLAS Collaboration

[1] Universita di Genova and INFN Sezione di Genova, Via Dodecaneso 33, 16146 Genova, Italy
[2] CERN, Route de Meyrin 385, 1217 Meyrin, Switzerland
[3] Linear Accelerator Laboratory, Batiment 200, 91440 Orsay, France
[4] Federico Santa Maria Technical University Valparaiso, Avenida de, Av. Espana 1680, Valparaiso, Region de Valparaiso, Chile

E-mail: `Julius.Hrivnac@cern.ch`

Abstract.
The global view of the ATLAS Event Index system has been presented in the 17th ACAT Conference. This article concentrates on the architecture of the system core component. This component handles the final stage of the event metadata import. It organizes its storage and provides a fast and feature-rich access to all information. A user is able to interrogate metadata in various ways, including by executing user-provided code on the data to make selections and to interpret the results. A wide spectrum of clients is available, from a set of Linux-like commands to an interactive graphical Web Service. The stored event metadata contain the basic description of the related events, the references to the experiment event storage and the full trigger record and can be extended with other event characteristics. Derived collections of events can be created. Such collections can be annotated and tagged with further information.

1. ATLAS Event Index

The ATLAS [1] Event Index [2] (presented in the 17th ACAT Conference [3]) is a system which provides experts and users with event-level metadata services. A user can find information about specific events from the Event Index data rather than from event dataset files. Event Index stores data GUIDs (Global Unique IDentifiers) as well as a variety of other event-wise information. All data are organized in TagFiles, which are currently implemented as Hadoop [4] MapFiles. A set of TagFiles is a TagSet, which can be accessed in the same way as a TagFile. All TagFiles are registered in a Catalog. Each operation on the Event Index database is implemented as a transformation of TagFiles, its result is again a standard TagFile subject to common interface. TagFiles can be processed in the most general way by applying any valid (Java) code to them. All operations are also registered in a Journal.

The system is integrated in the ATLAS offline environment and is available from all its distributions and via an Apache/Tomcat Web Service, which runs on two load-balanced servers. It runs on a Hadoop cluster provided by CERN IT. Currently, it handles over 80 billion events, both data and Monte Carlo.

IOP Conf. Series: Journal of Physics: Conf. Series **1085** (2018) 042024 doi:10.1088/1742-6596/1085/4/042024

2. Interfaces

External access to data is available via several interfaces:

- *Linux-like command line* which requires a small local client.
- *Interactive Graphical Interface* is just a simple graphical version of the command line interface for various commands (like Event Lookup command `el` (Figure 2)). Results are represented in a graphical view (like dataset overlaps table (Figure 4) or trigger statistics table and histogram (Figure 3)). The data oriented interface (Figure 1) presents all existing datasets with actions available to each of them.
- *Stateless Web Service* serves as an interface to other ATLAS services.

All interfaces implement the same functionality. External access is protected by the CERN Shibboleth [5] system.

3. Commands

The most generic access is possible by using two general commands available as Linux commands and as Web Service):

- `catalog` command allows searching and modification of the TagFile Catalog.
- `ei` command gives full searching access to all data. Result of each request is a new TagFile. User can specify search and assembly criteria using any valid (Java) code. This command is very rich, but more complex to use. There are several special auxiliary processors available to users, allowing more complex operations, like generating statistics histograms for selected TagFile variables.

Some frequently-used operations are available as special wrapper commands. The most important are:

- `el` is Event Lookup command to locate of events based on their event numbers.
- `ti` is Trigger Interface command to create special statistics tables about event triggers.
- `inspect` gives more direct access to TagFile content. It is useful for detailed study and for TagFiles with non-standard schema.

4. Auxiliary HBase Tables

Some data are assembled in auxiliary HBase [6] tables.

Catalog table keeps all information about TagFiles, their content, format, genealogy and basic characteristics. Catalog search is in most cases the first stage of general search.

Journal keeps track of all non-trivial operations. It allows to search for already delivered results, helping to investigate possible problems and to find usage patterns and usage statistics.

Event Lookup tables (one for real data and one for Monte Carlo data) give very fast mapping of event numbers to containing files. Tables are generated from imported HDFS files and keep references to them to allow more detailed searches and reporting. They are used by the `el` command.

5. Future improvements

5.1. Query Spaces

All TagFiles can be considered as entities in the TagFile space. All operations (derivations, merging, etc.) then create relations between those entities. We can develop a measure allowing to assess distances between entities. A user then can get an immediate approximate (or sometimes exact) result to the request based on distance to nearest already available TagFiles because results themselves are represented by TagFiles.

IOP Conf. Series: Journal of Physics: Conf. Series **1085** (2018) 042024 doi:10.1088/1742-6596/1085/4/042024

5.2. Universal Command Line Client
A downloadable application running on all usual platforms (Linux, MS Windows, Mac, Android) is under development. It will give command-like interface, so far only available on Linux, to all users.

5.3. Conclusion
The ATLAS Event Index system is in production and provides a succesfull service to the community. Its modular architecture allows rapid adaptation to new user requirements. New features and improvements are under active development.

References

[1] The ATLAS Collaboration 2008 The ATLAS Experiment at the CERN Large Hadron Collider JINST 3 S08003
[2] Barberis D et al 2015 The ATLAS EventIndex: architecture, design choices, deployment and first operation experience J. Phys.: Conf. Ser. 664 042003
[3] D Barberis et al 2016 J. Phys.: Conf. Ser. 762 012028
[4] Hadoop http://hadoop.apache.org
[5] Ormancey E 2008 CERN single sign on solution J. Phys.: Conference Ser. 119 (8) 082008
[6] HBase http://hbase.apache.org

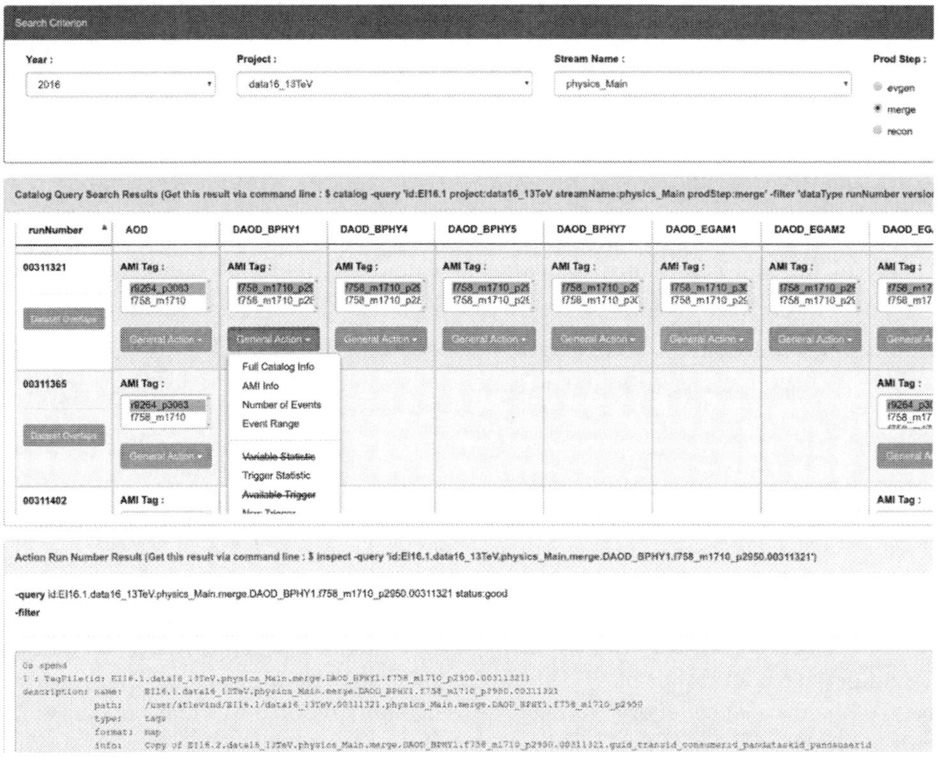

Figure 1. Data oriented Web Service proposes all actions available for a particular dataset.

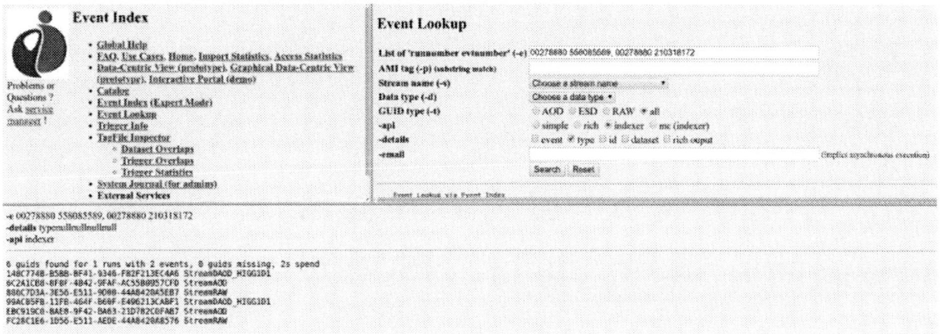

Figure 2. Generic Event Lookup Web Service.

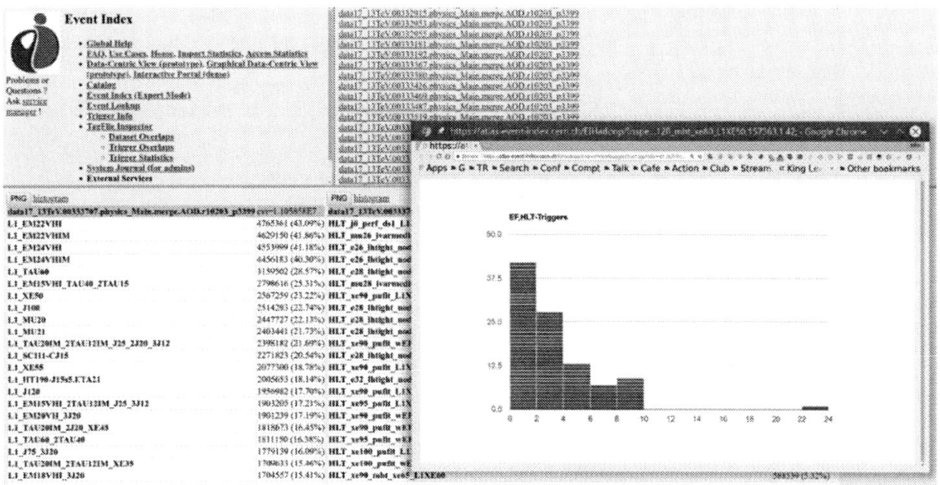

Figure 3. Trigger Statistics Web Service.

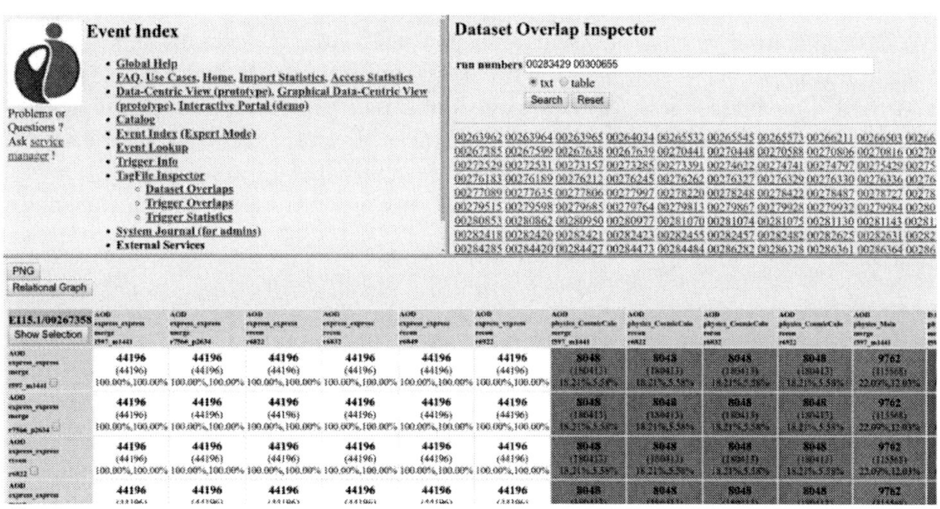

Figure 4. Web Service showing overlap between datases.

IOP Conf. Series: Journal of Physics: Conf. Series **1085** (2018) 042025 doi:10.1088/1742-6596/1085/4/042025

Machine Learning for electromagnetic showers reconstruction in emulsion cloud chambers

S.Shirobokov[1,2], A.Filatov[1,2], V.Belavin[1,2,3], A.Ustyuzhanin[1,2]

[1]National Research University Higher School of Economics, Myasnitskaya 20, 101000 Moscow, Russia
[2]Yandex School of Data Analysis, Timura Frunze 11 bld 2, 119034 Moscow, Russia
[3]Moscow Institute of Physics and Technology, Institutskiy per. 9, Dolgoprudny, Moscow Region, 141700, Russia

E-mail: shirobokov@yandex-team.ru, afilatov@hse.ru, belavin@phystech.edu

Abstract. Traces of electromagnetic showers in the neutrino experiments may be considered as signals of dark matter particles. For example, SHiP experiment is going to use emulsion film detectors similar to the ones designed for OPERA experiment from dark matter search. The goal of this research is to develop an algorithm that can identify traces of electromagnetic showers in particle detectors, so it would be possible to analyse and compare various dark matter hypothesis. Both real data and signal simulation samples for this research come from OPERA experiment. Also we have used exploited algorithm for electromagnetic showers identification as a baseline. Although in this research we have used no hints about shower origin.

1. Introduction

Neutrinos are invisible in the detector, but a very small fraction of them will interact with material in the detector and produce an electromagnetic shower in case of a scattered electron and also a hadron shower in case of a scattered nucleus. Atomic electrons can also be scattered by light dark matter which is searched for by other experiments, including those at CERN, for example at SHiP [1] which is going to use very similar neutrino detection techniques as OPERA experiment [2]. The goal of the research is to find traces of electromagnetic showers within a very dense background sample. Baseline algorithm used by OPERA [3] relies on the known origin of the shower, but in our case we are not going to use any a priori knowledge of the shower origin. Several solutions to the problem are suggested.

2. Data sample

The OPERA experiment data samples are structured corresponding to physical bricks of photo emulsion (emulsion cloud chamber). Each brick consists of 57 emulsion films. Each of the film plays the role of calorimeter and is used for track detection from electromagnetic shower, originated from the point of neutrino interaction. Each track[1] is described by 3 coordinates and 2 direction variables (slopes of the track in X/Y projections, assuming that xy-plane is parallel to emulsion films) as well as goodness of fit parameter χ^2. The total background studied in

[1] Mostly we are using word "track" to describe hits in emulsion films, but sometimes we refer to "track" as real particle trace, especially when discussing shower origin and energy estimation.

IOP Conf. Series: Journal of Physics: Conf. Series **1085** (2018) 042025 doi:10.1088/1742-6596/1085/4/042025

this work is $\sim 2 \cdot 10^6$ tracks, which constitute one tenth of the background tracks per event comparing to the real background sample. Signal sample representing electromagnetic samples is simulated.

3. Artem's solution

The algorithm proceeds in two steps: at the first step we would like to clean the brick saving as much signal as possible. At the second step, the algorithm classifies remaining track sequences and detects the background which was not detected at the first step.

3.1. SVM step

At the first step of the algorithm we train Support Vector Machine (SVM) [4] to classify pairs of tracks: SVM should indicate whether the pair belongs to signal or background class. We compute the following features: impact parameters to both direction, euclidean distance between tracks, tangent of angle, difference in projection angles to X and Y, χ^2 for both tracks.

The algorithm can be described by the following pseudo-code.

Algorithm 1: SVM step

```
for every layer in the brick do
    for every track in a layer do
        Find neighbors on the next layer;
        Find the probability of each neighbor to be a continuation of the track using SVM;
        if the highest probability is bigger than threshold or the track has more than h
          ancestors then
            Leave the track;
        else
            Delete the track;
```

The neighbors on the next layer are tracks that lie inside the square of the given area with the center in the continuation of the predecessor track.

3.2. CRF step

After the SVM step we get a cleared brick where only sequence-like structures are left. In order to clean the brick from the background which forms linear structures we run the Conditional Random Field (CRF) [5] algorithm to classify the sequences.

It is important to note that CRF learns the sequences of the edges, not the sequence of tracks. We use this approach, because coordinates of tracks do not contain any information which could help us to classify the sequences, whereas transition features between tracks can help us a lot. From the cleared brick we have created sequences using SVM: if classifier indicates that two tracks are a part of one sequence with the high probability, then we concatenate them to a single sequence.

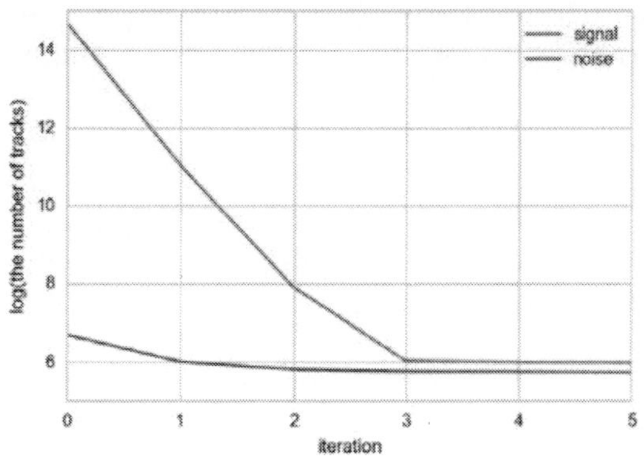

Figure 1. Dynamic of signal and background in the SVM step.

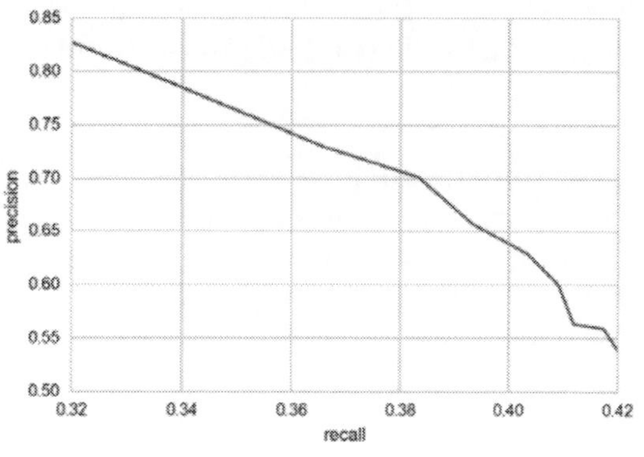

Figure 2. Precision/recall curve for SVM step (Full brick on last iteration).

3.3. Experimental results

In the experiment we applied SVM step multiple times to facilitate the work at the CRF step. On the figure 1, one can see that after multiple SVM steps number of noise decreases dramatically, whereas the signal decreases very slowly. The figure 2 demonstrates the trade-off between precision and recall for the SVM step. Applying CRF step we got the following results. The mean precision/recall for the second step is about 0.48/0.9. Energy resolution is 0.43.

4. Sergey's solution

4.1. Location of shower origin. Initial classification

Since the original dataset is highly imbalanced - highly enriched with background, the main purpose of the first stage is to discard it as much as possible. For that, the new features are created.

Each layer of emulsion is considered with its two neighbor layers. For each track a projection of its direction is obtained for both layers. All the tracks belonging to a small rectangle of size R over the track projections to both layers are selected. Then, the Impact Parameter (IP) projection between given track and selected tracks are calculated, and K tracks with smallest IP are selected. In addition to given features, χ^2 differences and χ^2 itself are added as features. In total there are 81 features for each track. After that, XGBoost classifier (XGB) [6] with weighted penalty function is trained. The weights are shifted towards signal tracks, to penalize classifier more, if it makes mistakes on signal events.

4.2. Topological filter

After the first stage, there is still a large amount of background left. However, the key point here is that in signal events the tracks are clustered near the real shower tracks, whereas background tracks are distributed evenly over the brick volume. Using this fact, this stage of the algorithm provides topological filter, by deleting all unclustered tracks. To do this, the median of X, Y coordinates in each layer of brick are calculated. A typical plot of median value is shown on fig. 3. As one can see there is a plateau, corresponding to the tracks, that form the real shower. Then the algorithms proceeds to a couple of stages, firstly detecting plates, corresponding to the plateau, and then deleting tracks in each layer if it lies away from the median. After that, only tracks in the center of shower are selected. By calculating series of center of selected tracks for each layer, and fitting Principal Component Analysis (PCA) on them, selecting first component, one gets the direction of the shower.

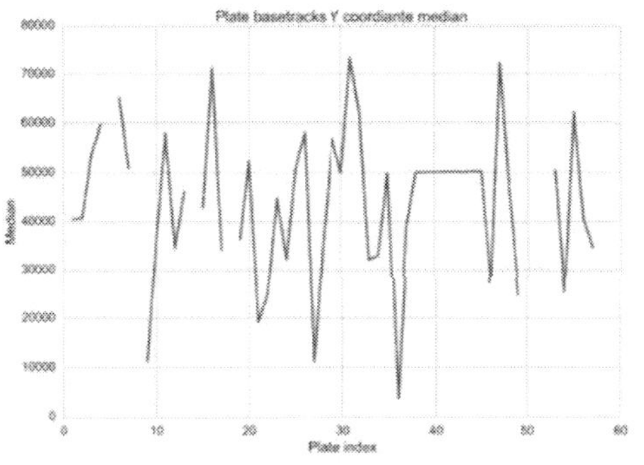

Figure 3. Plate tracks Y median values as a function of plate number.

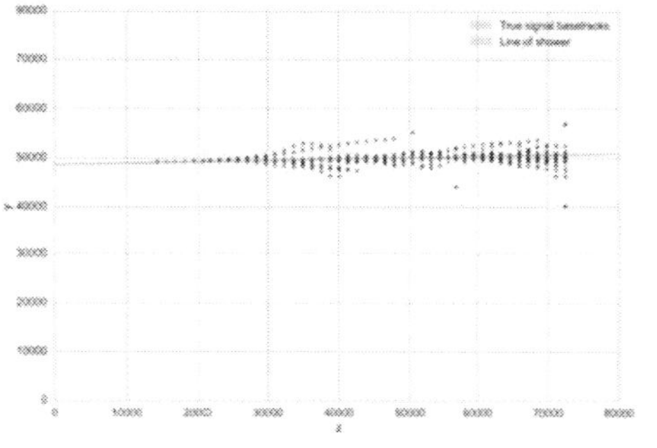

Figure 4. Typical (Y,Z) view of brick after clearance. Red points are selected as signal. Blue is the true signal. Green is shower line.

4.3. Shower origin search

Now, algorithm proceeds iteratively. On each iteration it is assumed that we know tracks at plate number N. Then, building line on known tracks, one finds intersection point of line and plates N-2 and N-1. The algorithm supposes that shower origin is at plate N-2, finds tracks in smaller region near the point at plate N-1 and calculate IP to "supposed" origin. Tracks with IP smaller than $threshold_1 = 150\mu m$ are considered. If there are no such tracks, tracks with bigger IP, but with small angle w.r.t. "origin" are considered. If there are such tracks, plate N-1 assumed to contain signal tracks and plate N-2 marked as non-origin. The process is repeated, starting from plate N-1. The iterations stop once there were no suitable tracks found at plate N-2.

The idea of the search above is that IP feature of the signal tracks near shower origin is distributed highly distinguishable from the IP of the background tracks.

The algorithm has run over 400 events and more than 200 tracks were selected. The mean distance error in XY plane is 0.4 mm, and mean distance error in Z coordinate is 2.4 mm. About 85% of the showers origins are detected within 2 plates distance from the true origin.

5. Vlad's solution

Shower identification is conducted in two separated steps. Firstly, LightGBM [7] classifier (gradient boosting framework that uses tree based learning algorithms) is applied in order to reduce background/signal ratio. Secondly, conditional Random Field (CRF) model is trying to find patterns in the data.

5.1. First step. Background Filtering

First of all for each track T_0 in the layer N we are searching for two possible successive tracks T_1, T_2 in the next layer $N+1$. To measure likelihood that some track T_i in the layer $N+1$ is successive to the track T_0 from the layer N the integral distance between pair of track is calculated (fig. 5), thus two tracks with the lowest integral distance are chosen as successive.

Secondly, for the track T_0 a set of features is constructed: integral distances $d(T_0, T_1)$ and $d(T_0, T_2)$; $|\theta_{X,T_0} - \theta_{X,T_1}| + |\theta_{y,T_0} - \theta_{y,T_1}|, |\theta_{X,T_0} - \theta_{X,T_2}| + |\theta_{y,T_0} - \theta_{y,T_2}|$, where $\theta_{x,T}$ is a slope between of track T in X projection; $\chi^2_{T_0}, \chi^2_{T_1}, \chi^2_{T_2}$.

LightGBM classifier is trained on the generated data. For this model Precision Area Under Curve (PR-AUC) in a brick equals 0.30. By varying threshold, we are changing both efficiency and background/signal ratio, which is crucial at the second step.

IOP Conf. Series: Journal of Physics: Conf. Series **1085** (2018) 042025 doi:10.1088/1742-6596/1085/4/042025

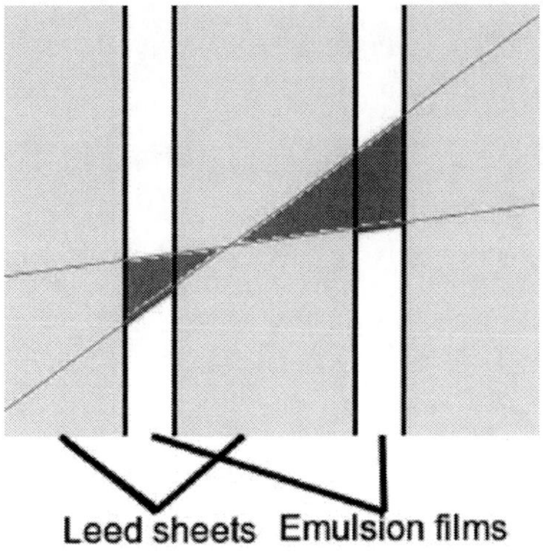

Figure 5. Red segments represent tracks. Area of the blue segment is a measure of similarity between two tracks.

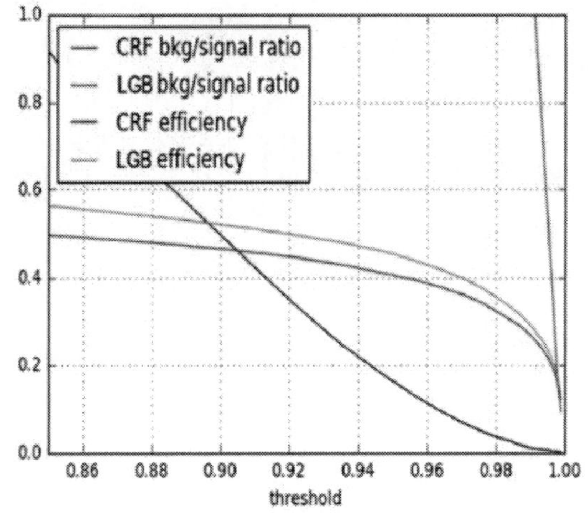

Figure 6. Background to signal ratios and efficiencies as a function of threshold. LGB stands for LightGBM.

5.2. Second step. Find shower patterns using CRF

At the second step, CRF model is applied to the filtered data. The main advantage of the CRF is an ability to take into account relations between tracks.

The underlying idea of the CRF model is an energy minimization of the following potential:

$$E = \sum_i (t_i\theta_i(1) + (1-t_i)\theta_i(0)) + \sum_{(i,j)\in Edges} (t_i(1-t_j)\theta_{ij}(1,0) + t_j(1-t_i)\theta_{ij}(0,1)),$$

where t_i is a class label of track T_i; $Edges$ is a set of such pairs of indices (i,j) such that if T_i lies in N plate, then T_j should lie in $N+1$ or $N+2$ plate and $d(T_i,T_j) < 5\cdot 10^5$.

To set up CRF model one needs to define unary potentials $\theta_i(1)$ and $\theta_i(0)$ that represent prior knowledge whether the track is a signal or background. For this problem we define them in the following way: $\theta_i(0) = 0$, $\theta_i(1) = -(\log p(t_i=1) - C) + C\cdot\log(1-N_i)$, where C is a 25th percentile over $\{p(t_i=1)\}_{i=1}^N$, N_i is a number of tracks T_j that lie in one of the two following layers and satisfy constraint $d(T_i,T_j) < 5\cdot 10^5$, and $p(t_i=1)$ is a probability obtained from GBT classifier trained at the first step. Pairwise potentials $\theta_{ij}(1,0), \theta_{ij}(0,1)$ represent similarity between two tracks, thus $\theta_{ij}(0,1) = \theta_{i,j}(1,0) = 5\cdot 10^5/d(T_i,T_j)$.

Graph construction and energy minimization are done by PyMaxflow library [8]. On the (fig. 6) one can see the background reduction effect of CRF model as a function of threshold applied at the background filtering stage. CRF model significantly reduces background/signal ratio with the comparably small decrease in the efficiency.

6. Energy Resolution

6.1. Shower tracks selection

Once the direction and initial point of shower are approximately found, the Hosseni algorithm [3] is applied. Mean average precision (area under curve) is 0.81 ± 0.07.

6.2. Energy resolution
For energy resolution one builds regression of the form

$$E_{rec} = a \cdot N_{st} + b, \quad ER = \sigma\left(\frac{E_{rec}}{E_{true}}\right),$$

where $E_{rec}-$ estimated energy of initial particle, $N_{st}-$ number of selected tracks, $ER-$ is energy resolution, $E_{true}-$ is true energy of initial particle.

Looking at tracks with less than 20 GeV energy of initial particle, and fitting the above regression, one obtains energy resolution about 0.27.

7. Conclusion
Constructed algorithms are tested with one tenth of real background in an event. The algorithms are optimized to work well with showers containing more than 200 tracks. The best energy resolution obtained is at the level of 0.27 that is comparable to the resolution of the baseline algorithm [3], but our algorithms do not rely on the information of the shower origin. With some modifications these algorithms could be scaled for multi-shower setting, which is crucial for SHiP experiment.

8. Acknowledgements
We would like to thank OPERA collaboration and Prof. Giovanni De Lellis, spokesperson of the collaboration, for sharing the data and simulated signal samples.

References
[1] Alekhin S et al 2016 A facility to Search for Hidden Particles at the CERN SPS: the SHiP physics case *Reports on Progress in Physics* 79(12)
[2] Acquafredda R et al 2009 The OPERA experiment in the CERN to Gran Sasso neutrino beam *Journal of Instrumentation* 4.04 P04018
[3] Hosseini B 2015 Search for Tau Neutrinos in tau to e Decay Channel in the OPERA Experiment *Napoli University*
[4] Cortes C Vapnik V 1995 Support vector machine *Machine learning* 20(3) pp 273-297
[5] Lafferty J McCallum A Pereira F 2001 Conditional random fields: Probabilistic models for segmenting and labeling sequence data
[6] Chen T and Guestrin C 2016 Xgboost: A scalable tree boosting system. *Proceedings of the 22nd acm sigkdd international conference on knowledge discovery and data mining* (ACM) pp 785-794
[7] Ke, Guolin and Meng, Qi and Finley, Thomas and Wang, Taifeng and Chen, Wei and Ma, Weidong and Ye, Qiwei and Liu, Tie-Yan 2017 LightGBM: A Highly Efficient Gradient Boosting Decision Tree *Advances in Neural Information Processing Systems* 30 pp 3149-3157
[8] Boykov Y and Kolmogorov V 2004 An Experimental Comparison of Min-Cut/Max-Flow Algorithms for Energy Minimization in Vision *In IEEE Transactions on PAMI* 26(9) pp 1124-1137

IOP Conf. Series: Journal of Physics: Conf. Series **1085** (2018) 042026 doi:10.1088/1742-6596/1085/4/042026

Background Suppression with the Belle II Neural Network Trigger

Sebastian Skambraks, Sara Neuhaus, Christian Kiesling
Max Planck Institute for Physics, Munich, Germany
E-mail: sskambra@mpp.mpg.de

Abstract. Neural networks are going to be used in the pipelined first level trigger of the upgraded flavor physics experiment Belle II at the high luminosity B factory SuperKEKB in Tsukuba, Japan. An instantaneous luminosity of $\mathcal{L} = 8 \times 10^{35}\mathrm{cm}^{-2}\mathrm{s}^{-1}$ is anticipated, 40 times larger than the world record reached with the predecessor KEKB. Background tracks, with vertices displaced along the beamline (z-axis), are expected to be severely increased due to the high luminosity. Using input from the central drift chamber, the main tracking device of Belle II, the online neural network trigger provides 3D track reconstruction within the fixed latency of the first level trigger. In particular, the robust estimation of the z-vertices allows a significantly improved suppression of the machine background. Based on a Monte Carlo background simulation, the high event rate faced by the first level trigger is analyzed and the benefits of the neural network trigger are evaluated.

1. Introduction

Belle II [1] at SuperKEKB [2] is a high luminosity upgrade of Belle [3], which was successful in confirming the CKM mechanism as the main source of CP violation in the standard model. In asymmetric e^+e^- collisions at the $\Upsilon(4S)$ resonance (e^+ with 4 GeV and e^- with 7 GeV), Belle II will be used to search for new physics at the precision frontier. A key feature that will enable the high instantaneous luminosity of $\mathcal{L} = 8 \times 10^{35}\mathrm{cm}^{-2}\mathrm{s}^{-1}$ is the nano-beam scheme: at the Interaction Region (IR), the vertical beam size will be squeezed to only $\mathcal{O}(60\,\mathrm{nm})$. Besides $\Upsilon(4S)$ events, with a large average track multiplicity (≈ 10), Belle II has access to important low multiplicity physics decay channels like τ pair production, initial state radiation, dark photons searches, etc. [4].

The downside of the luminosity upgrade is a severely increased background rate. This background manifests itself in noise and tracks. All events suffer from background noise (single hits), which increases the occupancy in all detectors and thus requires stable algorithms. Background tracks, found at the first trigger level, are a dangerous threat to low multiplicity physics events. Fortunately, many background tracks have their vertices displaced in the longitudinal coordinate z. The expected z vertex distribution of background trigger tracks is shown in figure 1 and will be discussed in section 3.

A 3D track reconstruction at the first trigger level will enable a significant reduction of the background track rate. The neural network trigger [5] is a noise robust solution to this 3D tracking problem that will be implemented on FPGA hardware [6]. After reviewing the neural network trigger and its integration into the first level trigger in section 2, the simulation of the expected backgrounds and its possible suppression are discussed in section 3.

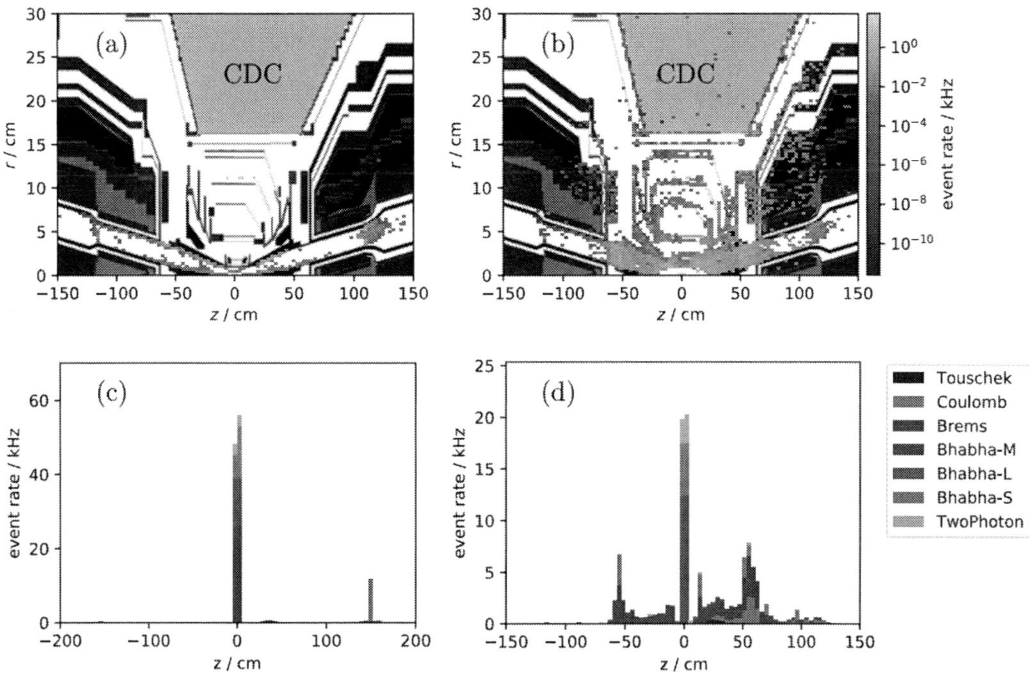

Figure 1. Rate of background events with at least one 2D trigger track. (a) & (b) Longitudinal (z) and radial (r) components of the track vertices on top of the material (detectors, beam-pipe, etc.). (c) & (d) z vertex distribution for different background types. (a) & (c) Generated particles before the simulation of the material interaction with Geant4. (b) & (d) Vertices of the simulated particles with a match to the tracks found by the 2D trigger.

2. Neural Network Trigger

At the first trigger level of Belle II, events are processed online using only information from detectors with a fast readout time. The main sub-triggers use the Central Drift Chamber (CDC) and the electromagnetic calorimeter (ECL). The neural network trigger is integrated in the CDC subtrigger [7]. Detectors with a longer readout time, especially the precise silicon detectors (silicon vertex detector - SVD & pixel detector - PXD), are only read out if a trigger signal is received. All the sub-trigger systems and the combination of their results in the global decision logic (GDL) are operated as deadtime-free pipelines using FPGAs. Required is a fixed latency of under 5 μs while providing a trigger rate of 30 kHz [1].

2.1. CDC Trigger

The CDC is a drift chamber consisting of 56 layers of sense wires, with high voltage field wires in between. The measurement of the drift time provides a distance information of a charged track to a sense wire. The 56 layers are grouped to 9 Super Layers (SL) of wires with the same orientation: 5 axial SL with wires parallel to the z axis (direction of the 1.5 T magnetic field) and 4 stereo SL inclined with a stereo angle w.r.t. the z axis that enables 3D track reconstruction [1].

In the CDC subtrigger, the first component is the Track Segment Finder (TSF): it is a noise suppression and data reduction element that combines all layers within a SL to Track Segments (TS) [8]. A TS is active if at least four out of five layers within the TS pattern have a hit. An active TS contains one priority wire with its drift time and indicates whether the track

IOP Conf. Series: Journal of Physics: Conf. Series **1085** (2018) 042026 doi:10.1088/1742-6596/1085/4/042026

passed left or right of the wire. The TSF is followed by a 2D track finder based on a Hough transformation [9] using only the axial TS positions (no drift times) as input [5]. For each found track two parameters are provided: p_T and ϕ, the transverse momentum of the track and the azimuthal angle. Parallel to the 2D finder, the event time is determined to enable the use of the drift times. Next, each found track is extended with 3D information: z and θ, the longitudinal vertex position and the polar angle of the track. This 3D reconstruction is carried out by two alternative approaches, the neural network and a least square fit. Finally, in the GDL, the CDC results are combined with the other subtriggers and a final trigger decision is made.

2.2. Neural Network

The essential part in the neural network trigger is a multilayer perceptron with one hidden layer using the hyperbolic tangent as activation function. The neural network is applied on single tracks and trained to estimate z and θ, scaled to the codomain of the activation function. This scaling is realized by expecting tracks to have their z vertex within 1 m around the IR (i.e. $z \in [-50, 50]$ cm). The neural network trigger receives p_T and ϕ of the tracks found, the event time t_0 and all the TS (axial and stereo). At first, a preprocessing step associates each 2D track with the corresponding TS and thus enables the neural network to process single tracks. The input to a network is one TS per SL per track, e.g. 9 TS for a track with a TS in every SL. In order to increase the efficiency, a dedicated pretrained network is selected, based on the actual number of TS per track. To form the input, each TS is represented with three values: two for the position of the TS (ϕ_{rel}, α) and the drift time t. The position is calculated relative to the 2D track, ϕ_{rel} is the distance of the 2D track to the TS and α is the arc length along the 2D track to the TS divided by the radius of the 2D track. Altogether, a network for a track with 9 TS has 27 inputs and 2 outputs. The size of the hidden layer is set to 127 nodes. A detailed description of the current neural network structure and its inputs is provided in [5].

The networks are trained with simulated single muon tracks with overlaid background noise. The track parameters are uniformly distributed in z, $\cos(\theta)$, ϕ and p_T^{-1} and the vertices are located on the z axis. The training is carried out using the iRPROP$^-$ [10] backpropagation algorithm within the library FANN [11]. The size of the training data set was chosen to be 10× the number of weights in the network.

Figure 2(a) shows the neural network reconstructed z vertices of the background tracks. The true vertex positions in figure 1(d) are smeared, with a smaller total rate (at least 3 stereo TS are required). The limited z range is clearly visible and tracks with vertices outside of the z

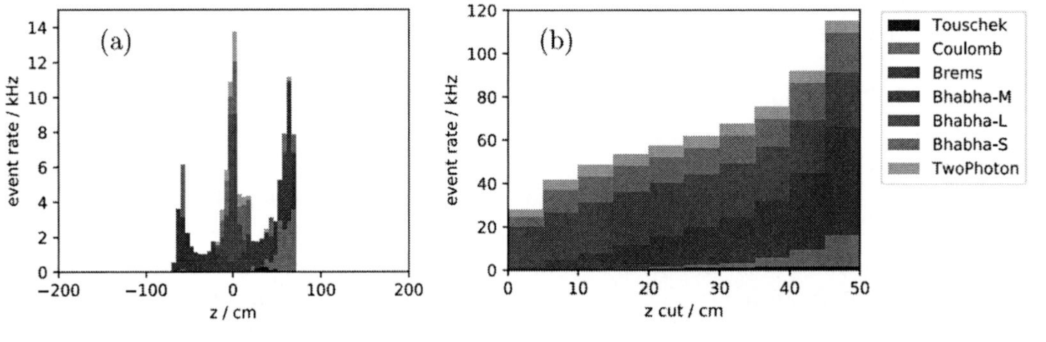

Figure 2. (a) Neural network estimates of the z vertices. (b) Integrated background rate for cuts based on the neural network z estimates.

range are nicely predicted at the boundaries. Figure 2(b) is an integrated version of 2(a) for cuts in z (z_{cut}) varied in 5 cm steps. It shows the remaining background event rate, if only tracks are triggered where the estimated z value z_{NN} satisfies: $-z_{\text{cut}} < z_{\text{NN}} < z_{\text{cut}}$.

3. Background

Two main classes of background can be distinguished: luminosity and machine background. Luminosity backgrounds are e^+e^- events ($z = 0$ cm) from uninteresting QED processes. The main luminosity backgrounds are 2γ events ($e^+e^- \to e^+e^-e^+e^-$) and (radiative) Bhabha ($e^+e^- \to e^+e^-\gamma$). Machine background is caused by particle loss from a single beam, i.e. when particles hit the beam wall at any position along the beamline ($z \neq 0$ cm). The main sources of machine background [12] are the Touschek effect [13] (intra beam scattering) and scattering of beam particles on residual gas nuclei (elastic/Coulomb and inelastic/Brems). Touschek background will increase due to the nano-beam scheme, beam-gas scattering due to the bad vacuum conditions coming along with the small beam-pipe radius. The machine background rate in the IR is reduced by collimators [14]. Via material scattering, the vertex positions of the background trigger tracks may deviate from the primary background vertices, i.e. luminosity background can also produce tracks with $z \neq 0$ cm. The background types considered are listed in table 1.

Table 1. Background types considered in the simulation with their physics generators and their expected rate of events containing at least one 2D trigger track.

background	process	generator	2D trigger rate
TwoPhoton	$e^+e^- \to e^+e^-e^+e^-$	AAFH [15]	6 kHz
Bhabha-S		BBBrem [16]	20 kHz
Bhabha-M	$e^+e^- \to e^+e^-\gamma$	BHWide [17]	52 kHz
Bhabha-L		BHWide	26 kHz
Touschek	intra bunch	SAD [18]	2 kHz
Coulomb	elastic beam-gas	SAD	15 kHz
Brems	inelastic beam-gas	SAD	1 kHz

3.1. Background Simulation

In machine background processes, beam particles experience a small deviation in momentum or energy and have to be tracked along the collider ring for several turns until they eventually get lost. In the simulation, a sample of deviated beam particles, weighted by the physics probability of the deviation, is tracked with SAD. Here all generated particles (weighted) are simulated 10× with Geant4, i.e. low rated events are included. The rates are used directly in the evaluation.

In luminosity backgrounds, the number of generated (unweighted) events corresponds to a given time equivalent. The radiative Bhabha events are divided into three parts (S/M/L), because the cross section strongly depends on the polar scattering angle of one outgoing lepton with the z axis ($\theta_{e^\pm}$) and the whole phase space should be represented. In Bhabha-S (small), at least one scattering angle is required to be small $\theta_{e^\pm} < 0.5°$. Bhabha-L (large) contains the events, where for both scattering angles holds $\theta_{e^\pm} > 1°$ and at least one angle has $\theta_{e^\pm} > 10°$. Bhabha-M (medium) covers the remaining phase space.

IOP Conf. Series: Journal of Physics: Conf. Series **1085** (2018) 042026 doi:10.1088/1742-6596/1085/4/042026

3.2. Material Scattering

The products of a scattering process with the material might be in the acceptance region of the CDC trigger. Therefore, generated background particles not in the acceptance produce detectable secondaries and tracks from the IR have secondaries with vertices at $z \neq 0\,\text{cm}$. Figure 1 shows the vertices of the initial background particles and the vertices of the particles that are actually found by the trigger, i.e. after material scattering. The particles in figure 1(a) which are generated on the beam-pipe can be identified with machine background in figure 1(c). Figure 1(b) shows the vertices after material scattering at various positions and figure 1(d) shows that both luminosity and machine background undergo scattering. Figure 1(d) is the expected z vertex distribution of events with at least one 2D trigger track.

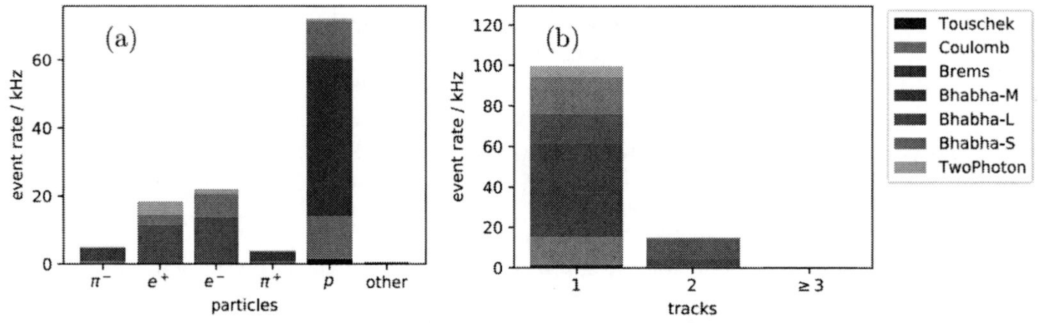

Figure 3. Properties of the background events. (a) Particle spectrum of the simulated Monte Carlo particles matched to the 2D trigger tracks. (b) Trigger track multiplicity in the background events.

Figure 3 shows the spectrum of the particle types causing trigger tracks and the track multiplicity. The spectrum is dominated by protons (p), produced by nuclear spallation of the material. The final states of inelastic scattering processes (p, π^+, π^-) are produced by all background types, while e^+ and e^- tracks without scattering are observed only from the luminosity backgrounds.

3.3. Background Suppression

The goal of the z vertex reconstruction is to reject background tracks with z vertices outside of the IR. From the perspective of the neural network trigger alone, the background sample splits into two parts: the reducible ($z \neq 0\,\text{cm}$) and the irreducible ($z = 0\,\text{cm}$) background. Substantially, the irreducible backgrounds are e^+e^- tracks from the IR caused by the luminosity backgrounds. Countermeasures are installed in the GDL of the trigger, where the results of all subtriggers are available (Bhabha vetos). Basically, ECL and CDC information is combined in order to identify electron and positron tracks.

Figure 4(a) shows the Monte Carlo z vertices of the reducible background events, here defined as $|z| \geq 1\,\text{cm}$, with an event rate of 81.4 kHz. The largest contribution to the reducible background are scattered Bhabha-M events. In Figure 4(b), the integrated rates for different z_{cut} values show that a huge suppression of the background rate is enabled by the neural network z vertex trigger. This allows to open up the other trigger conditions and thus to enhance the sensitivity for new physics. In order to improve the efficiency for low multiplicity physics, even a single track trigger can be considered. Given the maximum trigger rate of 30 kHz, a multiplicity dependent z_{cut} and pre-scaling may be used.

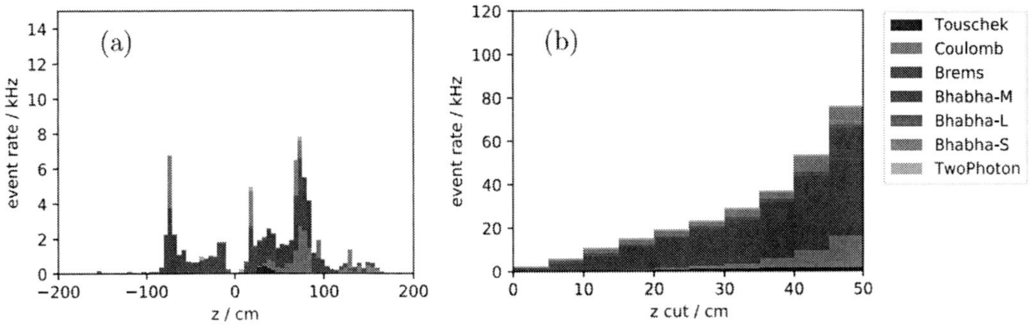

Figure 4. Suppression of the reducible background with vertices $|z| \geq 1\,\mathrm{cm}$. (a) Monte Carlo z distribution of the simulated particles matched to 2D trigger tracks. (b) Integrated background event rate of the reducible background at different cut values (z_{cut}) on the neural network z estimates.

4. Conclusion

A neural network tracking algorithm, robust against noise, has been developed for the Belle II first level trigger. It allows a significant suppression of the background rate and enables even a single track trigger. Luminosity and machine background tracks, with a total event rate of $\approx 120\,\mathrm{kHz}$, undergo material scattering. About $80\,\mathrm{kHz}$ is reducible by a z vertex trigger, the remaining $\approx 40\,\mathrm{kHz}$ from QED events (at $z = 0\,\mathrm{cm}$) have to be suppressed by a special veto in combination with the calorimeter. The exact determination of z_{cut} will depend on the efficiency of the other trigger logics in the GDL and has to be the subject of a combined study.

References

[1] Abe T *et al.* (Belle-II) 2010 Belle II Technical Design Report (*Preprint* 1011.0352)
[2] Hashimoto S *et al.* 2004 LoI for KEK Super B Factory, Part I: Physics
[3] Abashian A *et al.* 2002 *Nucl. Instrum. Methods A* **479** 117–232 ISSN 0168-9002
[4] Bevan A J *et al.* 2014 *The European Physical Journal C* **74** 3026 ISSN 1434-6052
[5] Neuhaus S, Skambraks S and Kiesling C 2017 *EPJ Web Conf.* **150** 00009
[6] Baehr S, Skambraks S, Neuhaus S, Kiesling C and Becker J 2017 *Journal of Instrumentation* **12** C03065
[7] Skambraks S *et al.* 2015 *IEEE Transactions on Nuclear Science* **62** 1732–1740 ISSN 0018-9499
[8] Iwasaki Y, Cheon B, Won E, Gao X, Macchiarulo L, Nishimura K and Varner G 2011 *IEEE Transactions on Nuclear Science* **58** 1807–1815 ISSN 0018-9499
[9] Hough P V C 1959 *Conf. Proc.* **C590914** 554–558
[10] Igel C and Hüsken M 2000 *Proc. 2nd Int. Symp. Neural Computation* pp 115–121
[11] Nissen S "FANN" [software], version 2.2.0, Available from https://github.com/libfann/fann
[12] Nakayama H, Iwasaki M, Kanazawa K, Nakano H, Ohnishi Y, Tanaka S and Tsuboyama T 2012 *Proc. of IPAC* pp 1825–1827
[13] Bernardini C, Corazza G F, Di Giugno G, Ghigo G, Haissinski J, Marin P, Querzoli R and Touschek B 1963 *Phys. Rev. Lett.* **10**(9) 407–409
[14] Nakayama H, Suetsugu Y, Kanazawa K, Ohnishi Y, Funakoshi Y, Ohmi K and Zou D a 2012 *Proc. of IPAC* pp 1104–1106
[15] Berends F, Daverveldt P and Kleiss R 1985 *Nuclear Physics B* **253** 441 – 463 ISSN 0550-3213
[16] Kleiss R and Burkhardt H 1994 *Comput. Phys. Commun.* **81** 372–380 (*Preprint* hep-ph/9401333)
[17] Jadach S, Placzek W and Ward B 1997 *Physics Letters B* **390** 298–308 ISSN 0370-2693
[18] Hirata K 1988 *Proceedings of the Second Advanced ICFA Beam Dynamics Workshop* CERN 88-04

IOP Conf. Series: Journal of Physics: Conf. Series **1085** (2018) 042027 doi:10.1088/1742-6596/1085/4/042027

Update of the BESIII Event Display System

Shuhui Huang, Zhengyun You

Sun Yat-sen University, Guangzhou, 510275, China

E-mail: huangshh28@mail2.sysu.edu.cn, youzhy5@mail.sysu.edu.cn

Abstract. The BESIII spectrometer is located at the Beijing Electron-Positron Collider (BEPCII). Recently, the endcap parts of the Time-Of-Flight system (TOF) have been upgraded and consequently, an upgrade of the BESIII visualization software (BesVis) is necessary. The Event Display visualizes particle interactions in the detector and plays an important role for the data acquisition system (DAQ), reconstruction algorithms tuning and physics analyses. The graphical interface of Event Display is based on ROOT GUI. The detector description is stored in GDML files and is converted into the ROOT geometry system.

1. Introduction

The BEPCII[1] is a double-ring symmetric e^+e^- collider with a center-of-mass energy between 2.0 and 4.6 GeV and a maximum luminosity of 10^{33} $cm^{-2}s^{-1}$. Large data samples have been collected in the τ-charm region.

BESIII[2] is a multipurpose spectrometer working at BEPCII. The detector covers 93 percent of the full solid angle and provides an excellent momentum resolution of charged tracks and good energy resolution of neutral particles. As shown in Figure 1, the BESIII detector consists of four subdetectors:

- Main Drift Chamber (MDC): a 43 layers drift chamber with helium-based gas surrounds the beam pipe. It is used to measure ionization energy loss and to precisely determine charged particle tracks. The single wire resolution is 115 μm. The momentum resolution for charged tracks at 1 GeV/c is 0.5%.

- Time-Of-Flight system (TOF): a Time-Of-Flight system is built up for particle identification which consists of a barrel part with 2 layers including 88 pieces of 5 cm thick, 2.4 m long plastic scintillators and two end-cap parts with 36 modules based on new Multigap-Resistive-Plate-Chamber (MRPC). The time resolution of TOF system is 68 ps and 60 ps in the barrel and endcaps respectively, corresponding to a $2\sigma K/\pi$ separation for momentum under about 1.0 GeV/c. The solid-angle coverage is $|cos\theta| < 0.83$ in barrel and $0.85 < |cos\theta| < 0.95$ in the endcap.

- Electromagnetic Calorimeter (EMC): an electromagnetic calorimeter made of CsI(Tl) with one cylindrical shape barrel and two endcaps. The energy resolution for showers in the EMC is 2.3% for 1 GeV photons and the position resolution is 6 mm in barrel and 9 mm in endcaps. The barrel EMC covers $|cos\theta| < 0.83$ and the endcaps cover $0.83 < |cos\theta| < 0.93$.

IOP Conf. Series: Journal of Physics: Conf. Series **1085** (2018) 042027 doi:10.1088/1742-6596/1085/4/042027

- Muon Counter system (MUC): a Muon Counter system is made of 1600 m^2 of resistive plate chamber arranged in 9 layers in barrel and 8 layers in endcaps, which is used to identify muons with momentum greater than 500 MeV/c. The position resolution is about 2 *cm*.

- Superconducting Solenoid Magnet (SSM): A superconducting solenoid magnet with a central magnetic field of 1.0 Tesla is located in the area between EMC and MUC, providing an optimum magnetic field for precise momentum measurements of charged particles around 1 GeV/c.

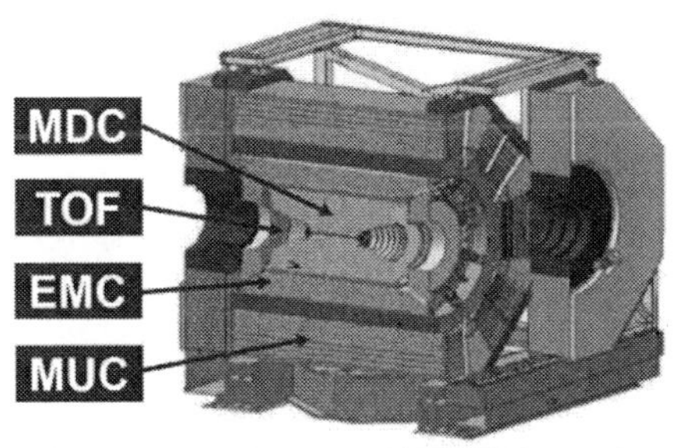

Figure 1. The BESIII detector and its 4 subdetectors.

2. Upgrade of the Time-Of-Flight System

For the old TOF detector, each part of it includes 48 fan-shaped, 5 *cm* thick, plastic scintillators and single read-out system for each scintillator. Compared to the old TOF detector, the new endcap of the TOF system[3] is replaced with Multigap-Resistive-Plate-Chamber (MRPC)[4]. As shown in Figure 2, each part of the new TOF detector endcap consists of 36 modules and 2 layers, each containing 18 modules with 12 gas layers and 24 read-out strips. The read-out system is a new version with single end but double sided. With the upgrade of MRPC, the time resolution of endcap TOF has been improved from 100 *ps* to 60 *ps*.

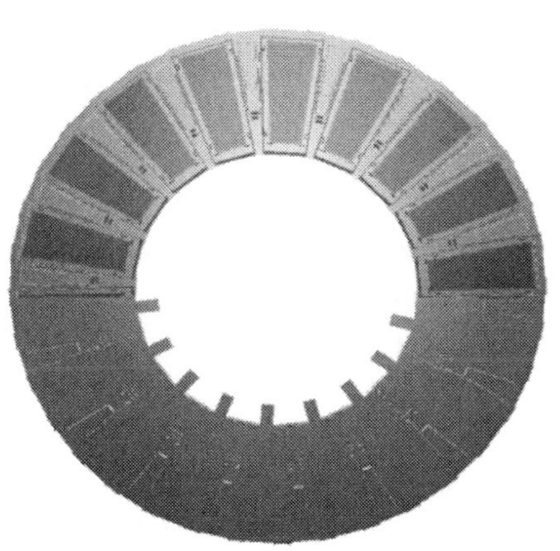

Figure 2. Detector model of the endcap TOF system with display in ROOT.

IOP Conf. Series: Journal of Physics: Conf. Series **1085** (2018) 042027 doi:10.1088/1742-6596/1085/4/042027

3. The Event Dispaly System

The BESIII event display system is based on ROOT[5]. In the event display system, the geometry of the detector is described with GDML[6] files originally, which has also been converted into ROOT geometry file for faster data read-in[7]. ROOT GUI is used to build the graphical user interface for convenient interactive display.

3.1. Display of the new endcap TOF detector

In order to display the new endcap TOF detector, we converted the original GDML file into ROOT geometry. The 2D view and 3D view of the new TOF detector are shown in Figure 3 and Figure 4, respectively.

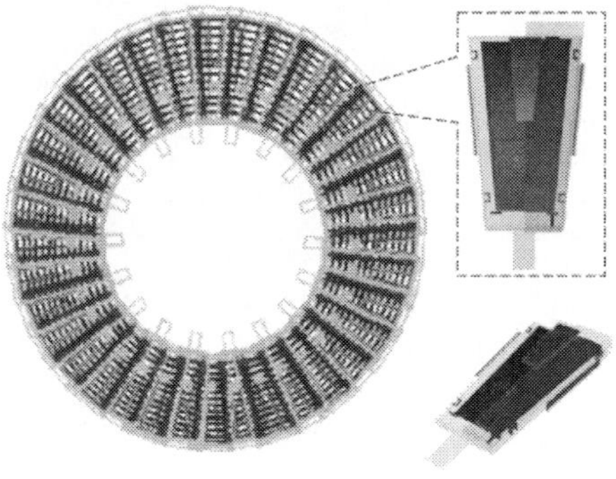

Figure 3. East endcap TOF with 36 modules and each module with 12 read-out strips.

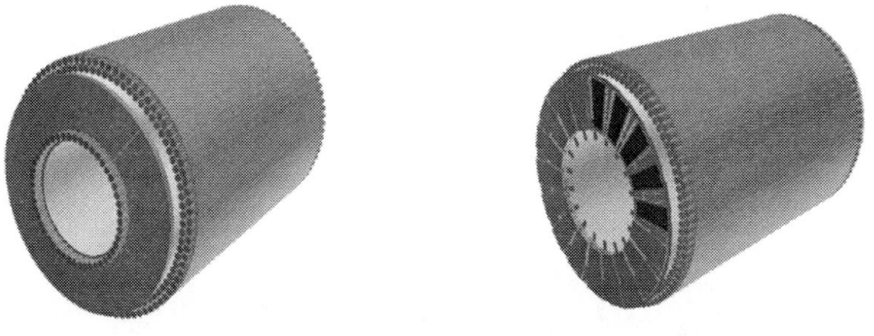

(a) (b)

Figure 4. (a) 3D view of the old TOF with single layer scintillator endcap, (b) 3D view of the new TOF geometry with MRPC module endcap.

3.2. The current Event Display System

The GUI of BESIII Event Display System consists of several parts, as shown in Figure 5. The main window is the XY View and ZR View of BESIII detector geometry, and in each view window, users can zoom in or out as they like. On the top of the display window is a status window in blue, which shows the information of each event displayed below. On the right is a toolbar, where users can rotate the detectors in XY View and ZR View, choose which detector to be displayed and which event to be shown, etc. 3D display of BESIII Event Display System is shown in Figure 6.

IOP Conf. Series: Journal of Physics: Conf. Series **1085** (2018) 042027 doi:10.1088/1742-6596/1085/4/042027

In order to display the detector more precisely, each endcap TOF is shown as 36 ladder-shaped MRPC modules and each module has 12 ladder-shaped strips. As shown in Figure 7, one module has been divided into 12 ladder-shaped strips. If one strip is fired, it will be tagged and shown in red for highlight.

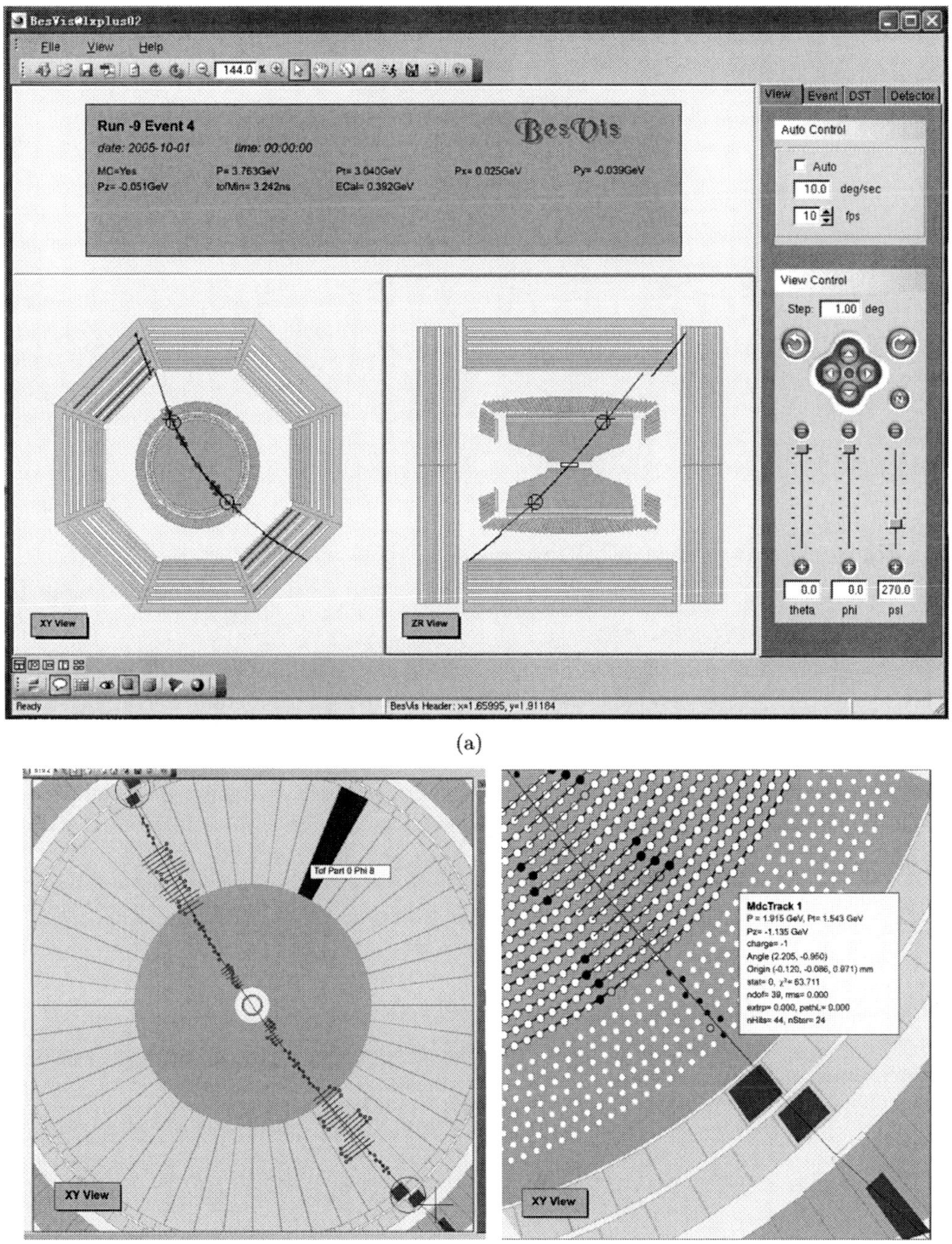

(a)

(b) (c)

Figure 5. The Graphical User Interface of BESIII event display and 2D projection of a physics event.

IOP Conf. Series: Journal of Physics: Conf. Series **1085** (2018) 042027 doi:10.1088/1742-6596/1085/4/042027

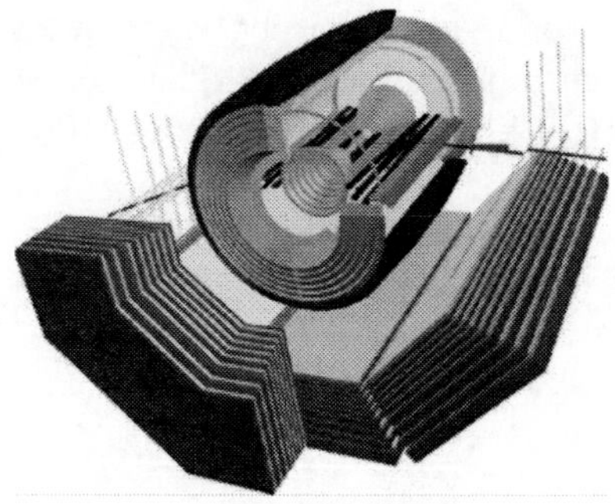

Figure 6. 3D view of a physics event in BESIII detector.

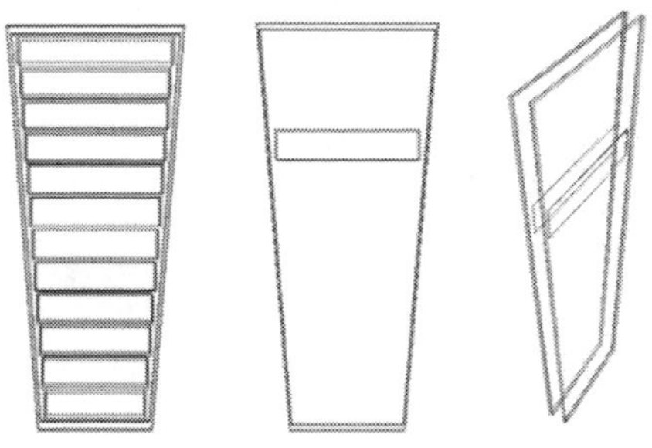

Figure 7. One module of MRPC on endcap TOF with fired strips dispalyed.

4. Summary
BESIII experiment provides a good chance to study τ-charm physics at the energy region of 2.0-4.6 GeV. In the BESIII detector, the TOF system plays an important role in particle identification, vertex reconstruction and so on. The upgrade of the endcap TOF system with MRPC provides a more precise Time-Of-Flight measurement on BESIII. Therefore, a better particle identification is possible and improvements for physics analyses are expected. The update of the Event Display system with the new detector geometry is necessary and the work is currently in progress.

The authors would like to thank the BESIII collaborators for the valuable discussions and suggestions in the work, the support from National Natural Science Foundation of China (11405279, 11675275) and The Test Plan of Top-Notch Students Training in Basic Disciplines at YAT-SEN School, Sun Yat-sen University.

5. References
[1] Zhang, C. and Wang, J. Proceedings of EPAC (Luzern, Switzerland, 2004), 230232
[2] M. Ablikim et al. Design and Construction of the BESIII Detector. *Nucl. Instrum. Meth.*, A614:345–399, 2010.
[3] Zhi Wu et al. First results of the new endcap TOF commissioning at BESIII. *JINST*, 11(07):C07005, 2016.
[4] Yong-Jie Sun et al. A prototype MRPC beam test for the BESIII ETOF upgrade. *Chin. Phys.*, C36:429–433, 2012.
[5] Rene Brun and Fons Rademakers. ROOT – An object oriented data analysis framework. *Nucl. Instrum. Meth.* , A389:81–86, 1997
[6] You Zheng-Yun, Liang Yu-Tie, and Mao Ya-Jun. A method for detector description exchange among root geant4 and geant3. *Chinese Physics C*, 32(7):572, 2008.
[7] Yu-Tie Liang et al. A uniform geometry description for simulation, reconstruction and visualization in the BESIII experiment. *Nucl. Instrum. Meth.*, A603:325–327, 2009.

IOP Conf. Series: Journal of Physics: Conf. Series **1085** (2018) 042028 doi:10.1088/1742-6596/1085/4/042028

A study of the applicability of recommender systems for the Production and Distributed Analysis system PanDA of the ATLAS Experiment

M Titov[1], G Záruba[2], K De[2], A Klimentov[3], S Jha[4] on behalf of the ATLAS Collaboration

[1] National Research Centre "Kurchatov Institute", Moscow, Russia
[2] University of Texas at Arlington, Arlington, TX, USA
[3] Brookhaven National Laboratory, Upton, NY, USA
[4] Rutgers, The State University of New Jersey, New Brunswick, NJ, USA

E-mail: mikhail.titov@cern.ch

Abstract. Scientific computing has advanced in the ways it deals with massive amounts of data, since the production capacities have increased significantly for the last decades. Most large science experiments require vast computing and data storage resources in order to provide results or predictions based on the data obtained. For scientific distributed computing systems with hundreds of petabytes of data and thousands of users it is important to keep track not just of how data is distributed in the system, but also of individual users' interests in the distributed data (reveal implicit interconnection between user and data objects). This however requires the collection and use of specific statistics such as correlations between data distribution, the mechanics of data distribution, and mainly user preferences. This work focuses on user activities (specifically, data usages) and interests in such a distributed computing system, namely PanDA (Production ANd Distributed Analysis system). PanDA is a high-performance workload management system originally designed to meet production and analysis requirements for a data-driven workload at the Large Hadron Collider Computing Grid for the ATLAS Experiment hosted at CERN (the European Organization for Nuclear Research). In this work we are going to investigate whether data collection that was gathered in the past in PanDA shows any trends indicating that users could have mutual interests that would be kept for the next data usages (i.e., data usage patterns), using data mining techniques such as association analysis, sequential pattern mining, and basics of the recommender system approach. We will show that such common interests between users indeed exist and thus could be used to provide recommendations (in terms of the collaborative filtering) to help users with their data selection process.

1. Introduction

For a distributed data analysis system to work, the user-selected data has to be available together with the user's executables at a compute node; more precisely, this distribution of executables and data is one of the most important tasks of such systems. In this paper we are going to focus on PanDA (Production ANd Distributed Analysis system). PanDA is a high-performance workload management system originally designed to meet production and analysis requirements for a data-driven workload at the Large Hadron Collider Computing Grid for the ATLAS (A Toroidal LHC ApparatuS) experiment [1] hosted at CERN (the European Organization for Nuclear Research) in Switzerland.

For a system like PanDA to work efficiently, data (usually large) with high user demand should be preloaded at several computing clusters so that job executables (usually small) can be scheduled with fewer restrictions and shorter turnaround times. It is thus to the advantage of PanDA to explicitly collect information on how "hot" (with high demand) datasets are in order for it to be able to correlate the number of replicas over the network. One way to predict the need for an individual dataset is to determine whether it is being used by a user who has shared interests (in other datasets) with a number of other users. Thus, the goal of such analysis would be to perform an evaluation of correlations between users' interests and similarities in data usage between correlated users. The outcome of such analysis could reveal what impact a recommender system would have if it was integrated into PanDA. Currently in ATLAS, sharing interest in datasets is usually done by word of mouth or over emails and blogs.

An earlier study [2] of data popularity with stochastic methods based on information from PanDA (job parameters of processed data from the ATLAS experiment) required a highly complex model design; yet it only showed insignificant correlation between datasets and actual user interests. The methodology did not consider relationships between different users' past datasets but was based on pure popularity analysis.

1.1. PanDA system

Production and Distributed Analysis system PanDA is a pilot-based workload management system. This means, that workload is assigned based on the feedback from successfully activated and validated *pilot jobs*, which are lightweight processes that probe the environment (on compute nodes and clusters) and act as *"smart wrappers"* for the payload. (Generally, pilots manage data and executables on compute nodes they are running on.) In PanDA, an independent subsystem manages the delivery of pilot jobs to all worker nodes via a number of cluster and grid scheduling systems (e.g., Condor-G). Pilot based systems like PanDA also enable the integration of non-grid based resources and thus can scale and work simultaneously on large clusters as well as individual computers (by using local *pilot submission factories*).

Jobs are defined by a set of specifications, their associated datasets and the input/output files within them. The PanDA server places such jobs into a global job queue, upon which a brokerage module operates to prioritize and assign work on the basis of job type, priority, input data and its locality, available CPU resources, and other brokerage criteria. Allocation of job sets to computer centers is followed by the dispatch of corresponding input datasets to those centers or nearby ones, handled by a data service system located within the ATLAS distributed data management system.

2. Data Analysis Techniques

2.1. Data Mining

Data mining, in general, is considered as the analysis step of *knowledge discovery in databases* (KDD), and is defined as the application of specific algorithms for extracting patterns from data [3]. Its functionality is used to specify the kind of patterns to be found in data mining tasks [4]. Such data mining tasks can be classified into two categories: descriptive and predictive. *Descriptive* mining tasks characterize the general properties of the data in the database. *Predictive* mining tasks perform inference on the current data in order to make predictions [4].

In this paper, we focus on descriptive mining tasks, and thus the goal of our mining process is to capture, model, and analyze the behavioral patterns and profiles of users interacting with data (i.e., the discovery of associations and correlations within PanDA data).

2.2. Data Modeling and Representation

Transactional form is used to represent data for analysis. Set of transactions are as following $\mathbb{T} = \{t_1, t_2, \ldots, t_M\}$, where t_m is a transaction that aggregates objects from the following sets: set of users $\mathbb{U} = \{u_1, u_2, \ldots, u_K\}$ and set of items (used data) $\mathbb{I} = \{i_1, i_2, \ldots, i_N\}$. Each transaction t_m is characterized by its owner, set of used data, starting time, and duration. Thus $t_m = \{eid, u_k, I(t_m) \mid u_k \in \mathbb{U},\ I(t_m) \subseteq \mathbb{I}\}$, where *eid* (*eventid*) is a transaction start time, u_k is transaction's owner, $I(t_m)$ is a nonempty set of items that were used by user u_k within the transaction t_m. Since a transaction is supposed to describe which items are requested during a certain time window by a user, we restrict the duration of transactions to a common transaction time window Δt. (At the creation of the transaction database, if an activity is longer than Δt then it gets split into several Δt long transactions.)

Transactions can then be grouped by their users; inside each such group, transactions are ordered by their start time (*eid*) thus representing a data sequence s_k for user u_k. The set of all data sequences then is: $\mathbb{S} = \{s_1, s_2, \ldots, s_K\}$, where $s_k = \{u_k, T(u_k) \mid u_k \in \mathbb{U},\ T(u_k) \subseteq \mathbb{T}\}$, and $T(u_k)$ is an ordered subset of transactions for user u_k.

The cumulative set of user features and preferences can be used to define a user-specific function that calculates the degree of interest of a user in a specified item and can be seen as the *user profile*. The set of user profiles is thus $\mathbb{UP} = \{up_1, up_2, \ldots, up_K\}$ and $interest\_degree(k, n) = up_k(i_n)$.

2.3. Association Analysis

The task of association analysis is to discover association rules by mining frequent itemsets. In general, the goal of association analysis is to discover hidden relationships in large data sets. Thus our objective is to identify item pairs in that co-occur more frequently than an analyst-predetermined value in the transactional database.

A rule is defined as an implication of the form $X \Rightarrow Y$ (a.k.a., the *if-then* rule), where $X, Y \subseteq \mathbb{I}$ and $X \cap Y = \varnothing$. Association rules are rules that surpass analyst-specified *minimum support* and *minimum confidence* thresholds. The support $supp(X)$ of the set of items X determines how often the rule is applicable to a given data set, and is defined as the fraction of transactions that contain the defined set of items: $supp(X) = |\{t_m \mid X \subseteq I(t_m),\ t_m \in \mathbb{T}\}| \times |\mathbb{T}|^{-1}$; the confidence of rule $X \Rightarrow Y$ determines how frequently items in Y appear in transactions that contain X: $conf(X \Rightarrow Y) = supp(X \cup Y) \times supp(X)^{-1}$. Therefore $X \Rightarrow Y$ will be used as an association rule if it satisfies analyst-predefined minimum support (σ) and minimum confidence limits (δ):

$$supp(X \cup Y) \geq \sigma \quad \text{and} \quad conf(X \Rightarrow Y) \geq \delta \tag{1}$$

2.4. Association Rule Mining Algorithms

A common strategy for association rule mining algorithms is to decompose the problem into two major subtasks [5]: i) *Frequent Itemset Generation*, where the objective is to find all the sets of items that satisfy the minimum support threshold (these sets are then called frequent itemsets); ii) *Rule Generation*, where the objective is to extract all the high-confidence rules from the frequent itemset generation step; these rules are then referred to as *strong rules*.

Frequent Itemset Generation is based on the property of itemsets that every sub-itemset of a frequent itemset is also frequent, i.e. if $X, Y \subseteq \mathbb{I}$ and $X \subseteq Y$ and Y is a frequent itemset then X is frequent itemset as well; this implies that $supp(X) \geq supp(Y)$ (*anti-monotone* property of support). This property helps to eliminate the need to evaluate supersets that contain non-frequent itemsets. A frequent itemset is called a *maximal itemset* if it does not have a frequent superset.

Most associative rule mining algorithms are based on the *Apriori* algorithm [6]. Apriori is an iterative algorithm based on the above described properties and uses so-called *prior knowledge* discovered at each iteration. It operates by constructing *candidate* k-itemsets, filtering these candidates based on frequency, and using the results to explore and construct *(k+1)*-itemsets (supersets over the previous candidate k-itemsets). These steps repeat until there are no more new candidates.

IOP Conf. Series: Journal of Physics: Conf. Series **1085** (2018) 042028 doi:10.1088/1742-6596/1085/4/042028

2.5. Sequential Pattern Mining

The sequential pattern mining technique encompasses the mining of frequently occurring patterns ordered by time (i.e., ordered events). In our case these are the so-called paths that users follow (*data-sequences*). The problem is to find all and/or the longest (with the most number of items and/or with the most number of transactions) sequential patterns with an analyst-specified *minimum support*, where the support of a sequential pattern relates to the number of data-sequences that contain the pattern (patterns are subsequences).

Association rules indicate intra-transaction relationships, while sequential patterns represent the correlation between transactions. They help to reduce the potentially large number of sequences into the most interesting sequential patterns. To meet different user requirements, it is important to use a minimum support which prunes sequential patterns of no interest.

In our case, performance (i.e., how long it takes to identify patterns) was not of utmost importance as our goal was to determine if there are significant patterns in our data. Algorithm SPADE [7] was identified to best suit our needs as it lends itself to easy modifications (to satisfy our custom criteria) and shows linear scalability with respect to the number of sequences. SPADE is based on lattice search techniques and provides the possibility to impose constraints on the mined sequences. The key features of SPADE include the layout of the database in a vertical id-list database format (the rows of the database consist of object-timestamped pairs associated with an event) with the search space decomposed into sub-lattices which are processed independently in main memory thus enabling the database to be scanned a maximum of only three times (sometimes just once on some pre-processed data).

3. Problem Statement

Given the per user data usage of items in PanDA, our task is to find correlations among users and items. We represent data in a transactional form, i.e., each sequence of transactions is associated with a user, and represents the sequence of itemsets that the particular user used during the transaction time window.

The first step to reach our goal is then to find associations within items. Such associations will state that there is a relation between two items if the usage of the first item tends to indicate the usage of the second item (regardless of relations between users). Having associated items will provide some indication that there are frequent sequences of itemsets. Establishing then the presence of sequential patterns (relying on these associations found) will provide confirmation that relationships between user data usages may exist. More precisely, sequential patterns for which the support is greater than the analyst-specified minimum support will be considered *frequent* and thus show that there are associations between users' data usages.

Strong correlations between data usages will provide indication that the future employment of data mining techniques for classifying user preferences and to predict users' future activities (data usage) will likely lead to a usable recommender system for users.

4. Data Study

The previous section does not only provide a statement of the problem but also outlines a path leading to a solution. Distributed computing systems provide information about users, their jobs, and data that is used as the input during the processing of these jobs. As mentioned earlier, PanDA collects such information as records of jobs with references to user (i.e., job's owner, *prodUserId*) and data patterns (i.e., job's input data, *prodDBlock*). In PanDA, the historical data for the year 2011 (when there was no decision making subsystem of PanDA in production) contains about 220M records. A preprocessing application we have written extracted 1,814 users (1,597 relevant users) and 380,111 items (220,867 relevant items). (Relevant users are defined as users who had analysis jobs that terminated successfully; relevant items have a similar definition related to success and parsability.)

IOP Conf. Series: Journal of Physics: Conf. Series **1085** (2018) 042028 doi:10.1088/1742-6596/1085/4/042028

The corresponding statistical data showed that 95.6% (1,526 users) of all relevant users had requested at least two distinct items; that 68.9% of relevant users (1,100 users) had requested more than 10% of the average number of requested items; and that 47.6% of relevant items (105,097 items) were requested by at least two distinct users. In this paper we ignore items that have only been used by one user; however, in a future recommender system such items may be considered if they show similarities with other, more popular, items.

The average number of (distinct) items used per user is 500 with a standard deviation of 1,758. This large standard deviation indicates strong skew; 39 users heavily exceed the average (by more than a standard deviation). As users with such large numbers of items used would overwhelm with their recommendations, we choose to remove them from this analysis leaving us with 1,558 users. (Indeed keeping these users in our analysis would show a much larger number of frequent data sets but may be misleading as to how useful such recommendations would be. Furthermore, our analysis of these users indicates that many of them either represented artificial job requesting agents or used each item only once. In general, data of such users would introduce too much noise into the analysis.)

4.1. Transaction Time Window Estimation

To get a reasonable evaluation by the item association analysis, data usage should be investigated first to find a good estimate for the transaction time window. Thus potential recommendations of items, which were not used by users before, would be simulated (these recommendations will be stored and cross referenced against real item use). The recommendations would be based on similarity between users, i.e., similarity in data usage.

4.1.1. Problem Definition Every user $u_k \in \mathbb{U}$ has three associated subsets:

- Used items $UI(u_k) = \{i_j : [(tu_{kj}, n_{kj}), \ldots] \mid i_j \in \mathbb{I}\}$, where tu_{kj} is the date when item i_j was used by user u_k, and n_{kj} is the number of times (i.e., number of jobs) the corresponding item was used during that day. (This implies that a specific item may have a sequence of (tu_{kj}, n_{kj}) pairs associated with it.)
- Similar users $SU(u_k) = \{u_m : [(ts_{km}, sc_{km}), \ldots] \mid u_m \in \mathbb{U}\}$, where ts_{km} is timestamp, time when similarity between two users was evaluated by the analysis system; sc_{km} is a similarity coefficient between users u_k and u_m. (When sc_{km} is updated then ts_{km} is also updated.)
- Recommended items $RI(u_k) = \{i_j : \{u_m : [(tr_{jm}, sc_{km}), \ldots] \mid u_m \in SU(u_k)\} \mid i_j \in \mathbb{I}\}$, where tr_{jm} is a timestamp, i.e., time when user u_m had used item i_j; and sc_{km} is a similarity coefficient between users u_k and u_m at the moment of tr_{jm}. Thus this set contains items that user u_k may receive as recommended based on similar interests.

The goal of our simulation-based analysis is to find an average time difference between potential recommendation and actual data usage. Again, this will help us determine a proper transaction time window needed for the sequential pattern mining.

4.1.2. Evaluation As described previously the processing time period for used items is set to calendar days. For every user during each processing period the following actions have been applied: i) Maintain $UI(u_k)$: list of used items is created and inserted into $UI(u_k)$, if one of these items was recommended earlier (i.e., the same item was already in $RI(u_k)$) then move corresponding item from recommended items to the *certain used item object* (a list that is maintained in the simulation); ii) Maintain $SU(u_k)$: maintain the sets of similar users, i.e., evaluate every user-pair and create corresponding objects at $SU(u_k)$ with the Jaccard index (Formula 2) as the similarity coefficient, if this coefficient exceeds a threshold value; iii) Maintain $RI(u_k)$: create new recommendations by inserting all used items of just established similar users into $RI(u_k)$.

$$jsc(u_k, u_m) = \frac{|UI(u_k) \cap UI(u_m)|}{|UI(u_k) \cup UI(u_m)|} = (\frac{num\ common\ used\ items}{num\ total\ used\ items}) \tag{2}$$

Figure 1 shows the per day average (per user) number of items used (and its standard deviation), the average number of recommended items in $RI(u_k)$, and the true positive recommendations (related to the certain used object list size). PanDA does not currently have a recommender system; thus the data on which our simulation analysis is based was not influenced by recommendation-induced item usages. As we are interested in establishing possible relationships between interests of different users to see if a recommender system would add value to their work, in this analysis we are not going to consider false positives and false negatives among our data sets.

Further analysis revealed that the average time difference between a potential recommendation and actual data use among the true positive simulated recommendations is ~28.7 days with a standard deviation of ~33.5 days. For the sake of simplicity, we elected to use a transaction time window of 30 days for the following analysis.

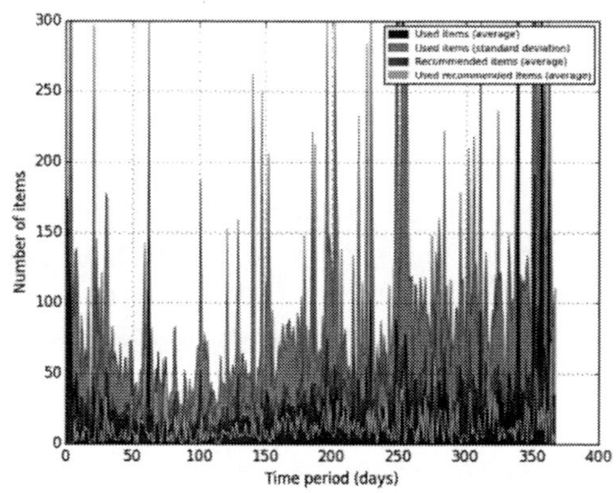

Figure 1. Number of items per day during the analysis period (y-axis is truncated to 300 items)

4.2. Association Rule Mining

For most of our association analysis we used **R**, a language and environment for statistical computing and graphics. Association rules were generated with *arules* [8] (a computational R library for mining association rules and frequent itemsets) by using the Apriori algorithm. Table 1 shows the results for applying the *arules* tool; data in columns headed with italics were used as inputs to arules. We varied the minimum support to show its impact on the number of associated items and rules, and elected to insert the more meaningful values into the table.

With a transaction time window of 30 days most associated items do not have a support greater than 0.036 (3.6% of transactions contain associated items); with smaller transaction time windows this upper threshold for support becomes even smaller. The results show the existence of associations for 30 day windows at a valid minimum support.

4.3. Frequent Sequences in Data

The association analysis in Section 4.2 confirmed that certain subsets of used items are correlated as far as user interests are considered; however this is not enough to conclude that sequences of used items are correlated as well. As described earlier, sequential patterns between related users would show to what extent their usage activities overlap. Users' data usage activities are represented as data sequences with the previously determined transaction time window of 30 days (the maximum number of transactions per user is 12 for the provided PanDA data).

There exist several implementations of sequential pattern mining algorithms; for our purposes the most applicable representatives were identified as: i) the R package "arulesSequences" - mining frequent sequential patterns with the cSPADE algorithm [9]; and ii) SPMF (Sequential Pattern Mining Framework) - an open-source data mining library written in Java, specialized in pattern mining [10]. Both tools showed good results with certain portions of our data, but not with the entire data set. The number of sequences in conjunction with their sheer sizes were

IOP Conf. Series: Journal of Physics: Conf. Series **1085** (2018) 042028 doi:10.1088/1742-6596/1085/4/042028

Table 1. Numbers for association rules (generated with R library "arules").

Transaction time window (TTW), days	Number of transactions established	*Minimum support*	*Minimum confidence*	Number of associated items found	Number of associated rules created
		0.035	0.5	71	43,446
30	7,299	**0.034**	0.5	150	15,652,534
		0.0335	0.5	166	39,144,905
		0.023	0.5	55	1,629
5	20,276	**0.022**	0.5	141	9,434,742
		0.0215	0.5	161	39,563,180

causing frequent crashes of the above tools. Thus we took it upon us to create our own sequential data mining tool, specifically tailored to handle the intricacies due to the dimensions of our data. As described before, our algorithm of choice for sequential pattern mining is SPADE [7] (Apriori-based). In addition to an implementation of SPADE, we extended it with CMAP (Co-occurrence MAP) [11] and bit sequence representation; this tailoring of SPADE was done to increase the computational efficiency. Our implementation in Python is available at Ref. [12].

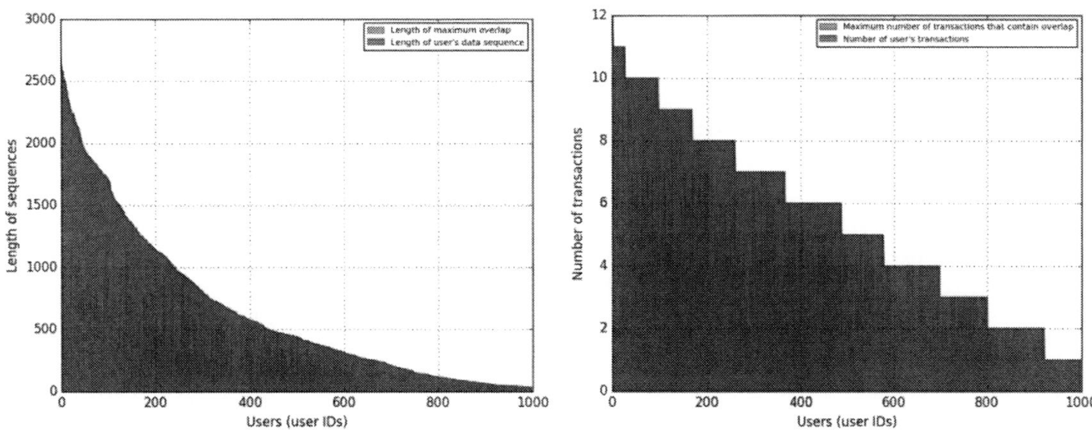

(a) Length of sequences per user (b) Number of transactions per user

Figure 2. Maximum data usage overlap per user

Our custom SPADE tool was applied to every pair of users to detect the degree of their correlation. The average sequence length (the average sum of lengths of transactions) in our transactional database is about 700. Sequential pattern mining reveals that for 998 users (64% of relevant users) the sequences share a significant portion of the timeline of used items with at least one other user, i.e., there is a sequence pattern (for every pair of sequences) that is a subsequence for each of the paired sequences. More precisely, we found that the average length of a sequential pattern is 43.58% of the lengths of the original sequences (with a standard deviation of 27.25%). To avoid biases coming from short, completely overlapping sequences, we removed the outliers and only considered sequences with lengths greater than 5% of the average length. (The number of users with sequence lengths less than 5% of the average length is 269 or 16.8%

IOP Conf. Series: Journal of Physics: Conf. Series **1085** (2018) 042028 doi:10.1088/1742-6596/1085/4/042028

out of the total number of relevant users.) Figure 2 shows histograms of the lengths of users' data usage sequences with the corresponding maximum overlap. This significant average length of sequential patterns (43.58%) indicates a strong correlation between users' data needs and thus validates our belief that a recommender system would enable users to find interesting data more readily and rapidly for their experiments.

5. Conclusion

In this work we investigated whether data that was gathered in the past in PanDA shows any trends that indicate that users could have mutual interests.

Initial analysis of data usages confirmed that a certain percentage of recommendations from similar users are actually followed despite the low quality of those recommendations. Thus user activity can follow some usage pattern of the group of similar users and be correlated with specific user interests. Association analysis showed that correlation between items existed; however, even to spot small signs of these correlations took significant computational cost. To be able to analyze PanDA data further, we have implemented a custom version of the SPADE sequential pattern mining algorithm with extensions that accommodate the dimensionality of said data. Deeper analysis, that included considerations for relationships between items in relation to users, presented correlations between users based on items relations. Indeed we found that data usage activity shows about 44% overlap for 64% of all relevant users. We also found that another about 17% of all relevant users had overlapping, but not significant data usage correlations.

These findings indicate a strong correlation between users' data needs, validating our belief that a recommender system would enable users to find interesting data more readily and rapidly for their experiments. Indeed our results have helped us raise awareness of the potential benefits of a recommender system to PanDA. Our current efforts are focused on an actual implementation of a recommender system that will be closely integrated with the PanDA system thus enabling ATLAS users to benefit early and frequently from these identifications of similarities.

Acknowledgments

We would like to thank the PanDA team for providing the data for our analysis and for their continued support, and to express our appreciations to CERN, UTA and NRC KI for the provided facilities. In Addition, this work has been carried out using computing resources of the federal collective usage center Complex for Simulation and Data Processing for Mega-science Facilities at NRC "Kurchatov Institute", http://ckp.nrcki.ru/. This work was funded in part by the Russian Ministry of Science and Education under contract No. 14.Z50.31.0024.

References

[1] ATLAS Collaboration 2008 *JINST* **3**
[2] Titov M, Záruba G, Klimentov A and De K 2012 *Journal of Physics: Conference Series* **396**
[3] Rajaraman A and Ullman J D 2011 *Mining of Massive Datasets* (Cambridge University Press)
[4] Han J and Kamber M 2006 *Data Mining: Concepts and Techniques* 2nd ed (Morgan Kaufmann Publishers)
[5] Tan P N, Steinbach M and Kumar V 2005 *Introduction to Data Mining* (Pearson)
[6] Agrawal R and Srikant R 1994 *Proceedings of the 20th International Conference on Very Large Data Bases* VLDB '94 (Morgan Kaufmann Publishers Inc.)
[7] Zaki M J 2001 *Machine Learning* **42**
[8] Hahsler M, Grün B and Hornik K 2007 *SIGKDD Explorations* **2**
[9] Hahsler M, Chelluboina S, Hornik K and Buchta C 2011 *Journal of Machine Learning Research* **12**
[10] Fournier-Viger P, Gomariz A, Gueniche T, Soltani A, Wu C W and Tseng V S 2014 *Journal of Machine Learning Research* **15**
[11] Fournier-Viger P, Gomariz A, Campos M and Thomas R 2014 *Advances in Knowledge Discovery and Data Mining* (Springer International Publishing) chap Fast Vertical Mining of Sequential Patterns Using Co-occurrence Information
[12] Titov M Data relations discovery package https://github.com/mtitov/pyrexplorer

IOP Conf. Series: Journal of Physics: Conf. Series **1085** (2018) 042029 doi:10.1088/1742-6596/1085/4/042029

Deep learning in jet reconstruction at CMS

Markus Stoye[1] on behalf of the CMS collaboration
CERN[1]

E-mail: markus.stoye@cern.ch

Abstract. Deep learning has led to several breakthroughs outside the field of high energy physics, yet in jet reconstruction for the CMS experiment at the CERN LHC it has not been used so far. This report shows results of applying deep learning strategies to jet reconstruction at the stage of identifying the original parton association of the jet (jet tagging), which is crucial for physics analyses at the LHC experiments. We introduce a custom deep neural network architecture for jet tagging. We compare the performance of this novel method with the other established approaches at CMS and show that the proposed strategy provides a significant improvement. The strategy provides the first multi-class classifier, instead of the few binary classifiers that previously were used, and thus yields more information and in a more convenient way. The performance results obtained with simulation imply a significant improvement for a large number of important physics analysis at the CMS experiment.

1. Introduction

The reconstruction of jets, sprays of particles that originate from the hadronization process of a quark or gluon, is a central element in high energy physics collider experiments. Jets can originate from the following particles: b quarks, c quarks, light quarks (u, d, s quark), and gluons. The b and c quarks are heavier that the other aforementioned quarks. If the focus is on identifying b and c quarks we speak of (heavy) flavor tagging. If we want to classify gluons, we cannot speak of flavor tagging as they do not have a physical flavor. We speak either of generic tagging if the gluon class is added or of quark vs. gluon tagging, if it is only a binary classifier. Recently several studies, using simplified simulations, were made on applying deep neural networks in jet tagging. It was proposed to use the analogy of the calorimeter cells to pixels in images to apply convolutional or dense networks that are often used for computer vision [1, 2, 3, 4]. The results ranged from no significant improvement to slight improvements with respect to established methods and did not include jets originating from heavy flavor quarks. Also, the use of recurrent neural networks was proposed [5, 6] in context of heavy flavor and other jet tagging.

In section 2, we present the jet tagging inputs in more detail in terms of the intrinsic data hierarchy and complexity. We present in section 3 the currently recommended flavor tagger of the CMS experiment [7] as well as the new proposal that is built on deep learning in the sense that layers for feature engineering are added that use the intrinsic hierarchy of the data. Finally, in section 5, we present results of the former standard CMS flavor tagger (CSVv2), the currently recommended flavor tagger (DeepCSV) and the generic new proposal (DeepJet). We also compare the generic tagging capabilities to other deep neural network architectures for quark vs. gluon classification.

IOP Conf. Series: Journal of Physics: Conf. Series **1085** (2018) 042029 doi:10.1088/1742-6596/1085/4/042029

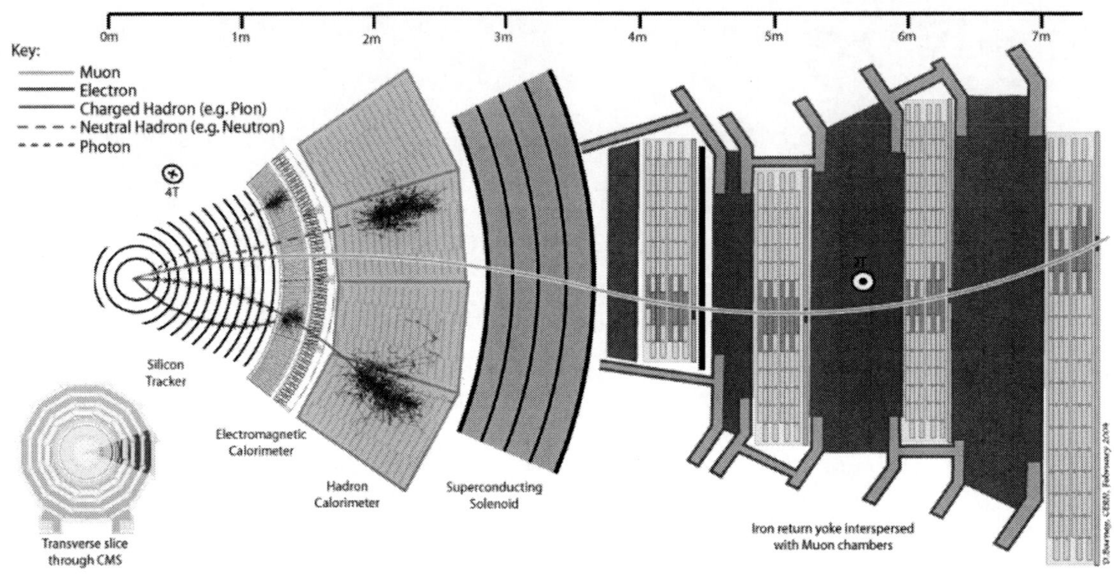

Figure 1. Slice of the CMS detector. It illustrates the different detector types and their different response to different particle types, such as muons or hadrons.

2. Particle flow jets

Particle flow jets consist of a list of particles reconstructed by the particle flow algorithm [12]. A single particle leaves various individual signals in the CMS subsystems. The subsystems are composed of different detector technologies as can be seen in Figure 1, which all have different signal responses to different particles. So, the particle data is very heterogeneous. As different particles carry different information, we split the particles, that are reconstructed by the particle flow algorithm, into two categories; charged particles which we measure in the tracker, and neutral that are only measured in the calorimeter. The object lists are sorted in descending order of a measure of their displacement significance, which for heavy flavor tagging is known to be a key observable.

There is a natural data hierarchy as jets are composed of particles, all of which have features. To build these lists we use the clustering algorithm anti-k_T [13] that identifies particles belonging to one particle spray initiated by a quark or gluon. Typically, fewer than 50 particles are found in such lists. The clustering algorithm has a parameter that is related to the radius of the cone of the particles spray which is set to 0.4. Tracks, i.e. the trajectories of charged particles are used to find vertices from where several tracks originated. Light quarks and gluons hadronize dominantly to short lived hadrons and their tracks originate directly from the proton-proton collision point, called the primary vertex. Secondary vertices are displaced from the primary vertex and are produced by heavy flavor hadrons that do not decay immediately and stem from b or c qaurks. If secondary vertices are found inside the cone of the jet, we add them as additional information to the jet.

3. Neural network designs

For the DeepCSV tagger we first apply quality criteria to tracks. Then we use the first six most displaced tracks and build seven features by hand that are traditionally used for b tagging [14]. Also, information from the most displaced vertex is used and again the features are similar to

those in [14] as well as the seven global features that we use, such as the transverse momentum of the jet or number of charged particles in the jet. All these features (around 60) are the input to a dense neural network with six layers and 100 nodes per layer.
For the deep learned DeepJet tagger the input is significantly extended. No quality criteria are applied for the charged particles of the jet. We use the features that are used for the heavy flavor tagging of DeepCSV and, in addition we also use features that are likely useful for jet classification, e.g. features that are related to the quality of tracks. All together this makes 18 features per charged particle. For neutral particles, we only use six features; some of which are calculated with respect to the secondary vertex, e.g. the opening angle between the particle and the first secondary vertex, if present. Finally, we use up to four secondary vertices that were reconstructed in the jet cone and use again the same features as in DeepCSV and a few additional ones, adding up to 12 features per vertex. These lists are then input to the neural network architecture. The first step is to engineer features per particle or vertex. For this purpose, several 1D 1x1 convolutional layers are applied to the lists of objects. For charged particles and secondary vertices, four layers with 64, 32, 32, and eight nodes are used. For the neutral particles with considerably less information content only 3 layers are used with 32, 16 and 8 nodes. This strategy allows us to use many input features per particle, which is important as it allows to use rather complete information for each particle or vertex. Altogether the number of input features is around 650. The output of the convolutional layer is then given to a LSTM [15] recurrent neural network. The output of the three recurrent networks that correspond to the charged particles are 150, 50, and 50 intermediate features that correspond to the charged particles, the neutral particles and the secondary vertices, respectively. These intermediate features are concatenated with eight global features, that are used traditionally in flavor tagging, and given to a dense neural network with six layers with 200 nodes for the first and 100 nodes for the remaining layers. For both DeepCSV and DeepJet the ReLU [16] activation function is used, and the softmax function for the last layer. We add DropOut [17] (rate 0.1) layers and batch normalization [18]. For optimization, the Adam optimizer [19] is used with a learning rate of $\epsilon = 10^{-8}$ without automatic learning rate decay. Instead, the learning rate is decreased manually during training time if the loss remains constant. For the technical implementation, we use Keras [20] and Tensorflow [21].

4. Training strategy

Both DeepCSV and DeepJet differ from previous taggers also in the training strategy. In both cases for the first time in CMS a multi-class classification approach is used for the classes. For DeepCSV the light quark class and the gluon class are merged, as usually done in flavor tagging. For DeepJet though these are individual classes to provide a generic tagger.
Another change with respect to previous taggers is that a larger and more diverse sample is used for the training. We use simulated events from two processes, namely QCD and top quark pair production. The more diverse samples reduce the danger of obtain a classifier too specific to a process. The larger sample sizes allow to increase the complexity of the tagger without danger of overfitting. Depending on the simulation version we use 40 to 100 million jet samples. Typically, we reserve 10% of the samples for both, validation and development. The training needs about 30 minutes on a NVIDIA GTX 1080 GPU for DeepCSV and 24 hours for DeepJet.

5. Performance comparison in real data and simulations

Figure 2 compares the DeepCSV to the old tagger (CSVv2) [14] of the CMS experiment and a significantly better performance is observed. At the same true positive rate (b jet efficiency) of about 65%, the false positive rate (misidentification probability) for uds quarks and gluon is reduced from 1% to 0.6%. The 1% false positive rate is a typical working point used for classification. The improvement originate from the changes in feature selection, new training samples, and a new machine learning algorithm. The efficiencies (true positive rate) of the tagger above a certain threshold on the discriminator are evaluated in real collision data as well as in simulation, details of the procedure are described in Ref. [14]. Figure 3 illustrates the so-called scaling factors that measure the difference in tagging efficiencies in simulation and real collision data. DeepCSV and CSVv2 show the similar agreement in simulation and data, which means that DeepCSV is also more performant in real data collisions. Also for other classes DeepCSV outperformed the old CMS taggers, specifically the binary c quark vs. light quark and c quark vs. b quark taggers.

Figure 4 shows the comparison of DeepCSV to DeepJet (labeled DeepFlavour as it is used for

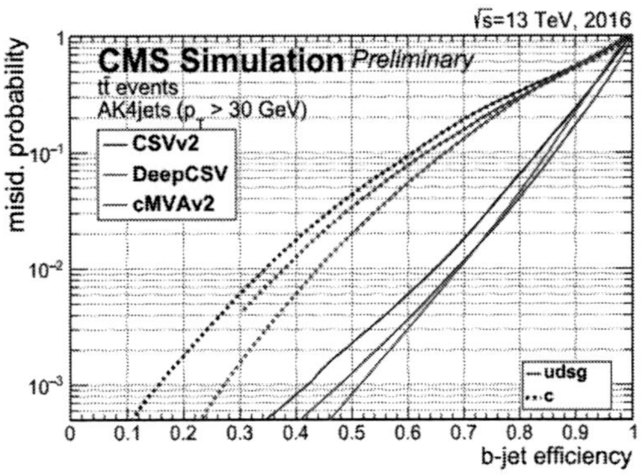

Figure 2. [8] True positive (b jet efficiency) rate vs. false positive rate of light jets and c quarks in simulation with 2016 conditions for jets with more than 30 GeV transverse momentum from top quark pair production. DeepCSV and CSVv2 are explained in the text. cMVA uses muon information, which for validation purposes is not used by other taggers.

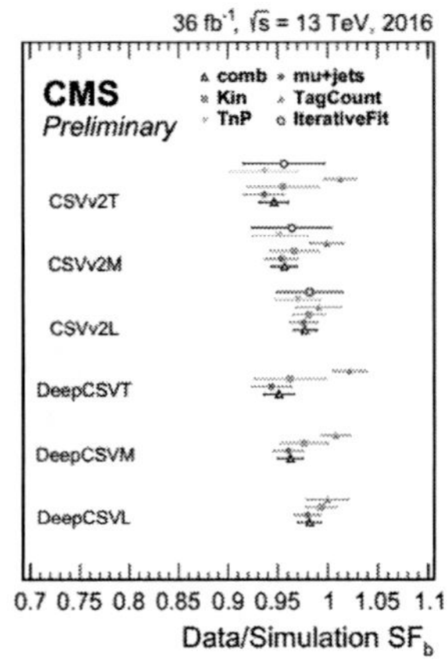

Figure 3. Difference in b jet efficiency between real data and simulation for different working points (loose (L), medium (M), and tight (T) with 10%, 1%, 0.1% false positive rate, respectively) for DeepCSV and CSVv2. Several complementary methods are used as described in [14, 8].

flavor tagging in this case). A further significant gain is observed; the false positive rate is reduced from 1% to 0.7% for 78% true positive rate. We also tested using the extended input of DeepJet as the input to an extended dense neural network. The resulting performance is labeled noConv in Figure 4 and shows that the more sophisticated custom architecture is needed. We also tested using the input of DeepCSV as input for the DeepJet architecture and effectively reproduced the DeepCSV result. We tested in several steps increasing the input from DeepCSV to DeepJet and found that a significant improvement was achieved by not using the quality criteria for particle selection of DeepCSV and CSVv2. The second biggest impact were the additional features per charged particles. The additional secondary vertices and neutral particles gave only modest gain, but are also important for other classes of the generic DeepJet tagger. In Figures 6 and 5 DeepCSV and DeepJet are compared at very high transverse momentum of the b jets. The gain is very significant; the false positive rate is reduced from 1% to only 0.12% at the usual working point. Also here a main limitation was the charged particle quality selection and as well the upper limit of only six charged particles (tracks) that were used in DeepCSV. We thus

learned that in order to reduce the input feature dimension, useful information was lost in the past flavor taggers of CMS.
For quark vs. gluon separation we compare DeepJet to custom deep neural networks. In [3] it was shown that these slightly outperform the classical approaches, namely using only a few handcrafted features that are input machine learning tools. As reference we used a neural network as described in [3], i.e. a 2D convolution layers working on a jet image. To produce images the continuous particle positions are made discrete, i.e. they are pixelized. We also built a slimmed down version of DeepJet, as for light quark vs. gluon separation it is evident from first principle arguments that just a fraction of the input to DeepJet is relevant. We used only 4 features per particles and removed the 1x1 convolutional layers. The secondary vertices were dropped as well. The slimmed down version of DeepJet is labeled recurrent in the Figure. Figure 7 compares the three networks, that all were trained with the same samples. DeepJet and the slimmed down version of DeepJet performed similarly well and only marginally better than the 2D convolution. The 2D convolution works well because quark vs. gluon separation mostly dependent on particle densities and energies densities, which can well be represented in images. Heavy flavor tagging however is much more complex.

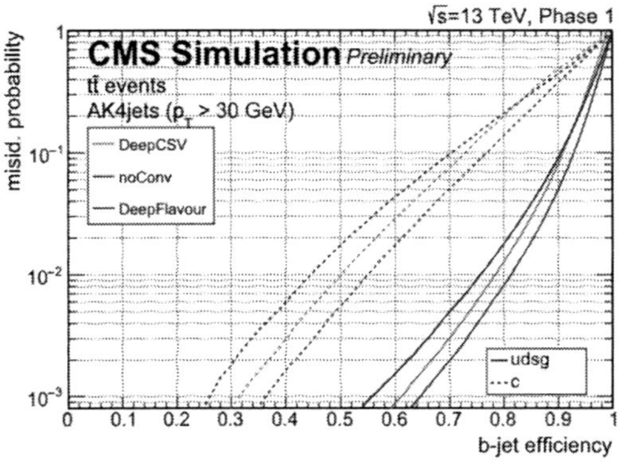

Figure 4. [9] True positive rate (b jet efficiency) vs. false positive rate of light jets and c quarks in simulation with 2017 conditions for jets with more than 30 GeV transverse momentum from top pair production. The jet-taggers DeepJet, DeepCSV, and noConv are described in the text.

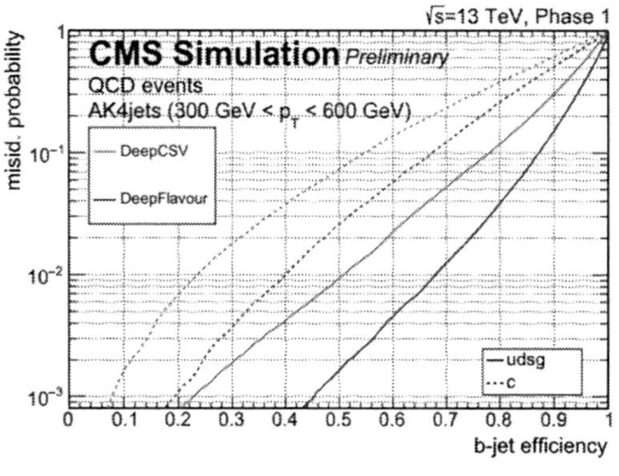

Figure 5. [9] True positive rate (b jet efficiency) vs. false positive rate of light jets and c quarks in simulation with 2017 conditions for jets in the range of 300-600 GeV of transverse momentum from QCD processes. The DeepJet and DeepCSV taggers are described in the text.

6. Conclusion

We present a new network architecture for jet tagging for the LHC experiments. DeepCSV, the currently recommended tagger of CMS based on a deep neural network, led to significant gain in tagging performance and was validated in real collision data. An even more recent development is DeepJet, a generic tagger for all hadron classes, which is based on deep learning. It outperforms the other taggers and especially at large transverse momenta of the jets the gain is very significant, nearly on order of magnitude less false positive rate for the standard threshold definitions. If the expectations from simulations are confirmed in real data that would yield to a significant improvement of the scientific output of the CMS experiment.

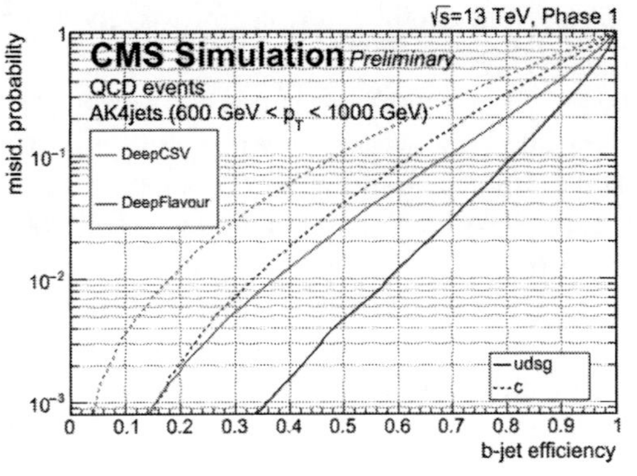

Figure 6. [9] True positive rate (b jet efficiency) vs. false positive rate of light jets and c quarks in simulation with 2017 conditions for jets in the range of 600-1000 GeV of transverse momentum from QCD processes. The DeepJet and DeepCSV taggers are described in the text.

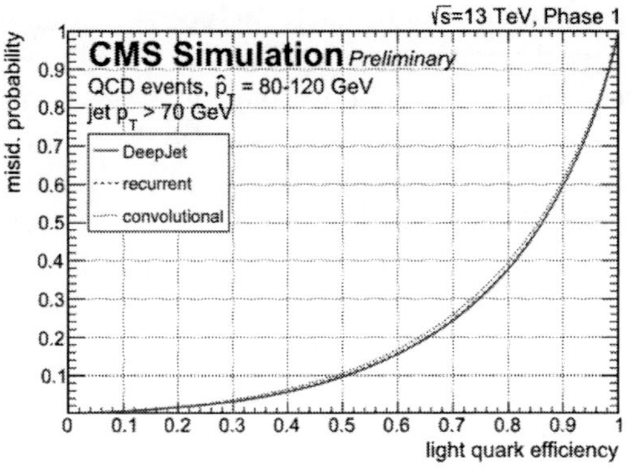

Figure 7. [10] True positive rate (light jet efficiency) vs. false positive rate of gluon jets in simulation with 2017 conditions for jets in the range of 80-120 GeV of transverse momentum from QCD processes. Three different (DeepJet, recurrent, convolutional) neural network based taggers are tested. Only DeepJet is a multi-class classifier that can also be used for flavor tagging.

References

[1] J. Cogan, M. Kagan, E. Strauss and A. Schwarztman, "Jet-Images: Computer Vision Inspired Techniques for Jet Tagging," JHEP **1502**, 118 (2015)

[2] G. Kasieczka, T. Plehn, M. Russell and T. Schell, "Deep-learning Top Taggers or The End of QCD?," JHEP **1705**, 006 (2017)

[3] P. T. Komiske, E. M. Metodiev and M. D. Schwartz, "Deep learning in color: towards automated quark/gluon jet discrimination," JHEP **1701**, 110 (2017)

[4] P. Baldi, K. Bauer, C. Eng, P. Sadowski and D. Whiteson, "Jet Substructure Classification in High-Energy Physics with Deep Neural Networks," Phys. Rev. D **93**, no. 9, 094034 (2016)

[5] D. Guest, J. Collado, P. Baldi, S. C. Hsu, G. Urban and D. Whiteson, "Jet Flavor Classification in High-Energy Physics with Deep Neural Networks," Phys. Rev. D **94**, no. 11, 112002 (2016)

[6] G. Louppe, K. Cho, C. Becot and K. Cranmer, "QCD-Aware Recursive Neural Networks for Jet Physics," arXiv:1702.00748 [hep-ph] (2017)

[7] CMS Collaboration, "The CMS Experiment at the CERN LHC," JINST **3**, S08004 (2008)

[8] CMS Collaboration, "Heavy flavor identification at CMS with deep neural networks," Detector Performance Summary: CMS-DPS-17-005, http://cds.cern.ch/record/2255736, (2017)

[9] CMS Collaboration, "CMS Phase 1 heavy flavour identification performance and developments," Detector Performance Summary: CMS-DPS-17-013, http://cds.cern.ch/record/2263802, (2017)

[10] CMS Collaboration, "New developments for jet substructure reconstruction in CMS," Detector Performance Summary: CMS-DPS-17-027, https://cds.cern.ch/record/2275226, (2017)

[11] ATLAS Collaboration, "Identification of Jets Containing b-Hadrons with Recurrent Neural Networks at the ATLAS Experiment," ATLAS note: ATL-PHYS-PUB-2017-003, http://cds.cern.ch/record/2255226, (2017)

[12] CMS Collaboration, "Particle-flow reconstruction and global event description with the CMS detector," JINST **12**, no. 10, P10003 (2017)

[13] M. Cacciari, G. P. Salam and G. Soyez, "The Anti-k(t) jet clustering algorithm," JHEP **0804**, 063 (2008)

[14] CMS Collaboration, Identification of b-quark Jets at the CMS Experiment in the LHC Run2 Startup," CMS public analysis summary BTV-15-001, (2015)

[15] S. Hochreiter and J. Schmidhuber, "Flat Minima," Neural Computation 9(1):1-42, http://people.idsia.ch/ juergen/fm/ (1997)

[16] V. Nair and G. E. Hinton, "Rectified linear units improve restricted boltzmann machines," Proceedings of the 27th international conference on machine learning (ICML-10), pp. 807–814. (2010)

[17] N. Srivastava *et al.*, "Dropout: a simple way to prevent neural networks from overfitting," Journal of machine

IOP Conf. Series: Journal of Physics: Conf. Series **1085** (2018) 042029 doi:10.1088/1742-6596/1085/4/042029

learning research **15**, no. 1, 1929–1958. (2014)
[18] S. Ioffe and C. Szegedy, "Batch normalization: Accelerating deep network training by reducing internal covariate shift," International Conference on Machine Learning, pp. 448–456. (2015)
[19] D. P. Kingma and J. Ba, "Adam: A Method for Stochastic Optimization," arXiv: 1412.6980 [cs.LG] (2014)
[20] Chollet, François et al., "Keras," https://github.com/fchollet/keras, (2015)
[21] Martín Abadi et al., "TensorFlow: Large-Scale Machine Learning on Heterogeneous Systems," https://www.tensorflow.org (2015)

IOP Conf. Series: Journal of Physics: Conf. Series **1085** (2018) 042030 doi:10.1088/1742-6596/1085/4/042030

CMS Analysis and Data Reduction with Apache Spark

Oliver Gutsche[2], Luca Canali[1], Illia Cremer[4], Matteo Cremonesi[2], Peter Elmer[5], Ian Fisk[3], Maria Girone[1], Bo Jayatilaka[2], Jim Kowalkowski[2], Viktor Khristenko[1], Evangelos Motesnitsalis[1], Jim Pivarski[5], Saba Sehrish[2], Kacper Surdy[1], Alexey Svyatkovskiy[5]

[1]European Organization for Nuclear Research CERN, Geneva, Switzerland
[2]Fermi National Accelerator Laboratory, Batavia, IL, USA
[3]Flatiron Institute of the Sions Foundation, New York, NY, USA
[4]Intel Corp.
[5]Princeton University, Princeton, NJ, USA

E-mail: `gutsche@fnal.gov`

Abstract. Experimental Particle Physics has been at the forefront of analyzing the world's largest datasets for decades. The HEP community was among the first to develop suitable software and computing tools for this task. In recent times, new toolkits and systems for distributed data processing, collectively called "Big Data" technologies have emerged from industry and open source projects to support the analysis of Petabyte and Exabyte datasets in industry. While the principles of data analysis in HEP have not changed (filtering and transforming experiment-specific data formats), these new technologies use different approaches and tools, promising a fresh look at analysis of very large datasets that could potentially reduce the time-to-physics with increased interactivity. Moreover these new tools are typically actively developed by large communities, often profiting of industry resources, and under open source licensing. These factors result in a boost for adoption and maturity of the tools and for the communities supporting them, at the same time helping in reducing the cost of ownership for the end users. In this talk, we are presenting studies of using Apache Spark for end user data analysis. We are studying the HEP analysis workflow separated into two thrusts: the reduction of centrally produced experiment datasets and the end analysis up to the publication plot. Studying the first thrust, CMS is working together with CERN openlab and Intel on the CMS Big Data Reduction Facility. The goal is to reduce 1 PB of official CMS data to 1 TB of ntuple output for analysis. We are presenting the progress of this 2-year project with first results of scaling up Spark-based HEP analysis. Studying the second thrust, we are presenting studies on using Apache Spark for a CMS Dark Matter physics search, investigating Spark's feasibility, usability and performance compared to the traditional ROOT-based analysis.

1. Introduction

Experimental Particle physics has been at the forefront of analyzing the world's largest datasets for decades. These datasets are produced by sophisticated detector systems that observe particle interactions. In this paper, we discuss the analysis of data from the field of High Energy Physics (HEP), where known particles are made to collide at very high energies to probe the fundamental constituents of nature. The most basic concept of how data in HEP is organized is an event: all detector signals associated with a single beam crossing and subsequent high-energy collision.

IOP Conf. Series: Journal of Physics: Conf. Series **1085** (2018) 042030 doi:10.1088/1742-6596/1085/4/042030

Events are the natural unit of HEP data and may be processed separately, which is why the computational problems of particle physics can be easily parallelized. The underlying data organization holds for all subfields of experimental particle physics.

Events must be reconstructed to convert detector signals into measurements of particles produced in collisions. This is usually done centrally for each detector. The reconstructed events are then input to the final analysis done by individual researchers or groups of researchers exploring a multitude of physics questions. This process is very idiosyncratic, as individual researchers or groups are searching for different physics phenomena in the same data. This results in a challenging computational problem, where as many as thousands of physicists analyze the same datasets to extract different physics results.

The analysis process uses properties of the event such as energy or momentum of particles produced in the collisions. It is based on comparing the distribution of properties measured in many events with theoretical predictions of the same property, either calculated from first principles or simulated using Monte Carlo techniques. Particle physics is a statistical science. Statistically significant numbers of recorded and simulated events are required to make claims.

Analysis is also an iterative process by nature, meaning that an analyst wants to explore the data and see his refinements of selection and information extraction preferrably instantaneously. With ever increasing data volumes in particle physics, interactivity is harder to achieve, simply due to the time required to read and transform the data. In the future, exponentially growing datasets will make this problem acute. New techniques will be required if we are to continue exploring the nature of matter and the universe effectively.

Recently, new toolkits and systems have emerged outside of the HEP community to analyze Petabyte and Exabyte datasets in industry, collectively called "Big Data". These new technologies use different approaches and promise a fresh look at analysis of extremely large datasets. Our goal in investigating these new technologies is to improve the efficiency and turn-around times of HEP analysis. At the same time, using industry tools would educate our graduate students and postdoctoral researchers in these techniques and improve their chances on the job market outside academia. It would also make the HEP community part of an even larger community of data analysts.

In this paper, we present our experience applying one of these technologies, Apache Spark [1], to the HEP analysis problem. We incorporate lessons learned from our previous investigations [2] and present new approaches and tools developed to enable HEP analysis in Apache Spark.

2. The Traditional Analysis Workflow

The particle physics community was the first to develop suitable software and computing tools for the HEP analysis task. The requirement from the point of the analyst is to minimize the "time-to-insight", to get from data to physics results as quickly as possible. Analysis is an iterative process, repeated in response to discoveries and mistakes. Interactivity is very important to facilitate an efficient conversation with the data. Asking scientific questions and answering them using data needs to be as user friendly and direct as possible to be successful.

Many different physics topics can be under investigation concurrently, each looking at different characteristics and even subsets of the data. For all individual analyses, the same programmatic analysis steps are used to minimize "time-to-insight": (a) Skimming (dropping events in a disk-to-disk copy), (b) Slimming (dropping variables in a disk-to-disk copy), (c) Filtering (selectively reading events into memory), (d) Pruning (selectively reading variables into memory).

The traditional user analysis workflow for CMS data applies two C++ frameworks to the centrally produced dataset: CMSSW [3], specially designed for processing CMS data, and ROOT [4], which is a general, experiment-independent C++ toolkit. The ROOT framework provides statistical tools and a serialization format to persist reconstructed and transformed C++ objects in files and is the underlying basis of CMSSW.

IOP Conf. Series: Journal of Physics: Conf. Series **1085** (2018) 042030 doi:10.1088/1742-6596/1085/4/042030

Although the CMSSW C++ framework can be very efficient, its data organization based on serialization of C++ objects in files can be difficult for end users to understand. Moreover, it operates at a low level of abstraction, and even if a user knows how to process data efficiently, the difficulty in setting up a manual procedure may be outweighed by the time required to do so.

Most data analysts or analysis groups start by translating the class structure of the data into a "flat ntuple", in which events are rows of a table with primitive numbers or arrays of numbers as columns. These ntuples are written to files using the ROOT framework as well. They may then be analyzed without the CMSSW framework, with minimal dependencies.

This "ntupling" step requires significant computing resources as it accesses the centrally produced samples. Grid resources are used for this step to exploit parallelization. Workflow management systems take care of orchestrating the whole analysis workflow. Significant burden of bookkeeping and failure resubmission is put on the individual analyst or analysis groups though.

Often, these resulting ntuples are still too big for interactive analysis. Most analysis groups therefore introduce additional steps in which the ntuples themselves are skimmed and slimmed.

In the last step of the analysis, quantities from the final ntuple are aggregated and plotted as histograms.

The time scale of a complete traditional analysis workflow can range from days to weeks, depending on how many events are needed for the analysis. The first step is repeated about four times a year and the produced ntuples can be used by more than one analysis. The skimming and slimming step is executed between once and four times a month. The actual analysis step, producing plots, is repeated many times a day, since it represents the iterative and interactive part of the analysis.

To give an example of data volumes of the different steps in a traditional analysis, we use a typical analysis of CMS [5], one of the 4 experiments at the Large Hadron Collider (LHC) at CERN in Geneva, Switzerland [6]. For an analysis of the 2015 dataset, the analysis ntuples have a size of about 2 Terabytes from a communal dataset of 200 Terabytes. These are then skimmed and slimmed down to several Gigabytes that are used for exploration and producing plots.

3. The Spark Analysis Workflow

In our previous usability study [2] of Apache Spark, we implemented the traditional analysis workflow by converting data into the AVRO [7] format and uploading it to the HDFS [8] file system of our development Spark cluster. The analysis code itself was reimplemented in the scala [9] language from it's original version written in C++. This resulted in a much simpler structure of the analysis itself due to the functional programming approach used in scala. It also resulted in much better portability to other facilities and technologies than in the traditional case.

While using Apache Spark to analyze data, analysts noticed that especially the tools to aggregate data in histograms were not suited for the new analysis paradigm. HEP physicists are used to control the event loop themselves and iterate through events programatically. Apache Spark manages the concurrency of the analysis and does not expose the event loop to the analysts.

A new toolkit was developed to aggregate data in histograms especially made for the map-reduce environment of Apache Spark and similar technologies. This package, called **Histogrammar** [10], provides a functional interface to fill histograms in Apache Spark by passing lambda functions and uses them in the same way as transformations are used in Apache Spark. The actual plot of a histogram is produced afterwards using a wide variety of plotting tools.

The biggest impediment in the first study to use the new technology as identified by the analysts was the need to convert the data in a new format that Apache Spark can use efficiently. This step was considered time consuming and not providing benefits over the traditional analysis workflow. To enable Apache Spark to understand the data structures of HEP experiments stored in ROOT files more directly, we developed a package called **spark-root** [11]. It is based on a Java-only implementation of the ROOT I/O libraries and connects ROOT to Apache Spark to be able to read ROOT object collections and automatically infer their class schema. The data is then available from within Apache Spark as DataFrames/Datasets/RDDs. This enables Apache Spark to read ROOT files directly from HDFS and also from the EOS [12] storage system through a new **Hadoop-xrootd connector** [13].[1] The connector was developed and tested with EOS. It is planned to expand its functionality gradually and support other storage systems that allow access through the xrootd protocol [14].

In the following, we present performance tests of the full analysis workflow producing plots of histograms by reading ROOT files from the EOS storage system, which were conducted on CERN facilities.

A stand-alone test was performed using a dedicated test cluster of 3 nodes (1 name node, 2 data nodes) each with an Intel(R) Xeon(R) CPU E5-2650 @ 2.00GHz with 32 cores, 128 GB RAM and 10 Gb/s network connection. Files in different formats (text, parquet [15], ROOT) were read either from the public or CERNBOX [16] EOS instances at CERN, where the public EOS instance shows decreased performance characteristics because of its physical size and access pattern optimization compared to the CERNBOX instance. The results are shown in Table 1.

Table 1. Stand-alone test of reading files from the EOS storage system using a dedicated test cluster. Shown are copy and Spark access of different file formats. The access test of files from Spark used 2 executors and 4 cores.

	EOS to HDFS copy	**EOS from Spark**
Text (200 GB from CERNBOX)	1 Gbit/s	300 Mbit/s
PARQUET (200 GB from CERNBOX)	800 Mbit/s	6-9 Gbit/s
ROOT (200 GB from public EOS)	400 Mbit/s	2.3 Gbit/s

The hadoop-xrootd connector and spark-root performed reasonably well in this stand-alone test. The copy test was limited by the performance of the public EOS instance performance and the characteristics of the HDFS setup of the test cluster, which copied a file first to the name node and then distributed it across the cluster. The Apache Spark test was limited by the uneven file size distribution of the input ROOT files and the fact that reading text files from Apache Spark is much less optimized then for example reading parquet files. The stand-alone tests didn't show any significant problems, although effects from different parts of the infrastructure (Apache Spark, local disk, memory, number of files, ...) are not easy to disentangle. Further tuning and optimization will be needed in the future.

We performed larger scale tests reading 0.5 TB of input ROOT files using the CERN analytix cluster of currently 36 nodes. A single node/single executor test was performed to compare reading ROOT files from HDFS and EOS. Reading 0.5 TB ROOT files from HDFS finished in about 4 hours where reading the same files from EOS only finished in about 9.5 hours. We concluded that the hadoop-xrootd connector is currently slower by a factor 2-3 than HDFS. It is expected that further tuning and optimization will be able to close the performance gap.

To test the parallel scaling by using more executors and cores of the analytix cluster, we distributed equal number of ROOT files over 160 tasks (40 executors with 4 cores per executor).

[1] EOS [12] is a storage system developed and deployed at CERN and other LHC grid sites.

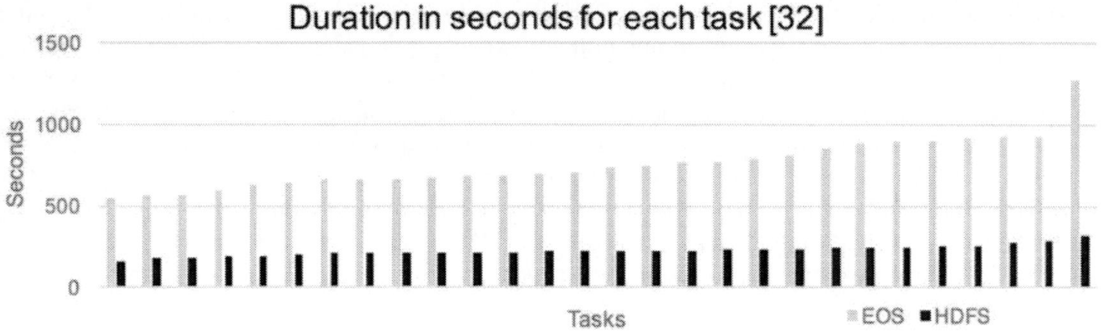

Figure 1. The parallel scalability test showed a larger difference between the HDFS and EOS read performance than the single executor/task test because of the tail of slow executors being longer for the xrootd-hadoop connector reading from EOS.

This is sub optimal because of the uneven distribution of file sizes, but sufficient for an initial large scale test. Table 2 shows the results reading ROOT files from HDFS and EOS.

Table 2. Cluster test of reading ROOT files from HDFS and EOS. 160 tasks were used (40 executors with 4 cores per executor) to analyze 0.5 TB of ROOT files in Apache Spark. The total read time is calculated by subtracting the total CPU time from the total executor time.

	HDFS from Spark	**EOS from Spark**
Total executor time	5.8 h	11.7 h
Total CPU time	2.9 h	2.9 h
Total Read time	2.9 h	8.8 h
Run time	5 min	19 min

As expected we see a significant speed up compared to the single executor/core case. The difference between HDFS and EOS read performance is larger than in the single case though. This is due to the uneven distribution of file sizes per task which results in the tail of slow executors being longer for the xrootd-hadoop connector (see for example Fig. 1 in a different executor/core configuration with 32 tasks).

The average reading performance from HDFS is about 70 MB/s per task and from EOS about 20 MB/s per task. This confirms the conclusion that reading from EOS is currently about 2-3 times slower than from HDFS. The plan for the performance tuning is to first look at the impact of the network and optimize the task fragmentation.

4. Conclusions and Next Steps

We presented studies of executing the traditional HEP analysis workflow on Apache Spark. The two largest caveats from the initial studies were solved by developing the histogrammar package to fill histograms in map-reduce style technologies and to read ROOT files directly from Apache Spark, even from the CERN EOS storage system. This removed the need to convert the input data into a format that Apache Spark understands natively.

The performance of the complete analysis workflow reading ROOT files is about 2-3 times slower on EOS than on HDFS, but further tuning and optimizations are expected to close the gap.

The scaling behavior results are promising to investigate larger and larger input volumes and reach the goal to perform interactive analysis on very large datasets. We hope the developed technologies and packages will enable the CERN openlab/Intel project [17] to develop a integrated analysis facility based on Apache Spark to reduce 1 PB of data in 5 hours to 1 TB for further analysis.

5. Acknowledgments

We would like to thank the CMS collaboration and the LHC to provide the data for the use case and the ROOT based workflow. We would also like to thank CERN openlab for enabling wide ranging collaboration on this research topic with international laboratories, universities and industry partners and being the spark for innovation. And we would like to thank Intel, our industry partner, in being interested in our challenges and helping us solving them. This manuscript has been co authored by Fermi Research Alliance, LLC under Contract No. DE-AC02-07CH11359 with the U.S. Department of Energy, Office of Science, Office of High Energy Physics, and by the National Science Foundation under grant ACI-1450377 and Cooperative Agreement PHY-1120138. The United States Government retains and the publisher, by accepting the article for publication, acknowledges that the United States Government retains a non-exclusive, paid-up, irrevocable, world-wide license to publish or reproduce the published form of this manuscript, or allow others to do so, for United States Government purposes.

References

[1] Zaharia M, Chowdhury M, Franklin M J, Shenker S and Stoica I 2010 *Proceedings of the 2Nd USENIX Conference on Hot Topics in Cloud Computing* HotCloud'10 (Berkeley, CA, USA: USENIX Association) pp 10–10 URL `http://dl.acm.org/citation.cfm?id=1863103.1863113`
[2] Gutsche O, Cremonesi M, Elmer P, Jayatilaka B, Kowalkowski J, Pivarski J, Sehrish S, Surez C M, Svyatkovskiy A and Tran N 2017 *CoRR* **abs/1703.04171** URL `http://arxiv.org/abs/1703.04171`
[3] Elmer P, Hegner B and Sexton-Kennedy L 2010 *J. Phys. Conf. Ser.* **219** 032022
[4] Brun R and Rademakers F 1997 *Nuclear Instruments and Methods in Physics Research Section A* **389** 81 – 86 ISSN 0168-9002 new Computing Techniques in Physics Research V URL `http://www.sciencedirect.com/science/article/pii/S016890029700048X`
[5] Chatrchyan S e a (CMS Collaboration) 2008 *JINST* **3** S08004
[6] Evans L and Bryant P 2008 *Journal of Instrumentation* **3** S08001 URL `http://stacks.iop.org/1748-0221/3/i=08/a=S08001`
[7] Apache avro URL `http://avro.apache.org`
[8] Shvachko K, Kuang H, Radia S and Chansler R 2010 *Proceedings of the 2010 IEEE 26th Symposium on Mass Storage Systems and Technologies (MSST)* MSST '10 (Washington, DC, USA: IEEE Computer Society) pp 1–10 ISBN 978-1-4244-7152-2 URL `http://dx.doi.org/10.1109/MSST.2010.5496972`
[9] Scala URL `http://www.scala-lang.org`
[10] Pivarski J, Svyatkovskiy A, Schenck F and Engels B 2016 histogrammar-python: 1.0.0 URL `https://doi.org/10.5281/zenodo.61418`
[11] Khristenko V and Pivarski J 2017 diana-hep/spark-root: v0.1.14_pre1 release URL `https://doi.org/10.5281/zenodo.1019880`
[12] Eos: Large disk storage at cern URL `https://eos.web.cern.ch`
[13] Motesnitsalis V hadoop-xrootd-connector URL `https://gitlab.cern.ch/awg/hadoop-xrootd-connector`
[14] Dorigo A, Elmer P, Furano F and Hanushevsky A 2005 Xrootd- a highly scalable architecture for data access WSEAS Transactions on Computers
[15] Apache parquet URL `https://parquet.apache.org/`
[16] Mascetti L, Labrador H G, Lamanna M, Mocicki J and Peters A 2015 *Journal of Physics: Conference Series* **664** 062037 URL `http://stacks.iop.org/1742-6596/664/i=6/a=062037`
[17] Cern openlab/intel cms big data project URL `https://cms-big-data.github.io`

IOP Conf. Series: Journal of Physics: Conf. Series **1085** (2018) 042031 doi:10.1088/1742-6596/1085/4/042031

Machine Learning Algorithms for b-Jet Tagging at the ATLAS Experiment

Michela Paganini
on behalf of the ATLAS Collaboration

Yale University, Department of Physics, 217 Prospect Street, New Haven, CT 06511-8499

E-mail: michela.paganini@cern.ch

Abstract. The separation of b-quark initiated jets from those coming from lighter quark flavors (b-tagging) is a fundamental tool for the ATLAS physics program at the CERN Large Hadron Collider. The most powerful b-tagging algorithms combine information from low-level taggers, exploiting reconstructed track and vertex information, into machine learning classifiers. The potential of modern deep learning techniques is explored using simulated events, and compared to that achievable from more traditional classifiers such as boosted decision trees.

1. Introduction

With the increase in the expected integrated luminosity delivered in the remaining years of Run 2 by the Large Hadron Collider [1] (LHC), a growing emphasis is put on algorithmic improvements at the reconstruction level [2] to leverage the full sensitivity potential for measurements and searches of interest to the ATLAS [3] physics program.

The task of identifying jets that contain b-hadrons and separating them from jets initiated by lighter quark flavors is of particular significance due to the prominence of b-hadrons in final states of interesting physics processes, and to the abundant background light-flavored-jet production at hadron colliders such as the LHC. This task, commonly referred to as b-tagging, is broadly applied to both precise standard model measurements and searches for signatures of new physics phenomena. b-tagging can be cast as a classification problem with the objective of correctly assigning jet flavors using the trajectories of charged particles (tracks) reconstructed

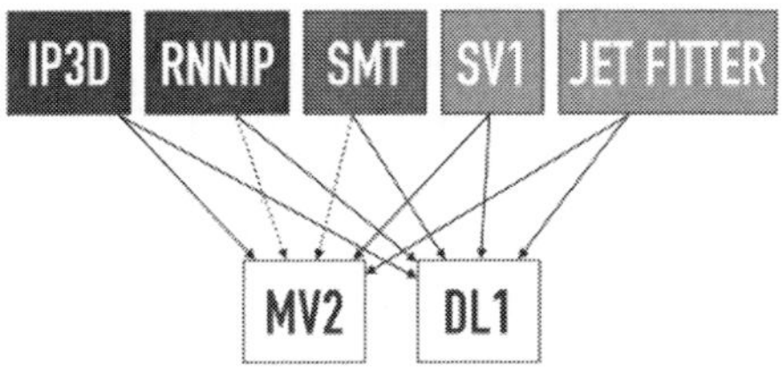

Figure 1: The set of b-tagging algorithms currently in use in the ATLAS experiment at the LHC includes five low-level taggers (IP3D, RNNIP, SMT, SV1, and JetFitter), as well as two high level taggers (MV2, and DL1) which take as inputs combinations of the outputs of the low-level taggers.

in the inner detector [4], and then associated to jets reconstructed from clusters of energy in the electromagnetic [5] and hadronic [6] calorimeters.

We present the current hierarchy of algorithms used for b-tagging in ATLAS (Figure 1), and outline the latest Monte Carlo-based optimization results [7], with specific focus on deep learning-powered improvements, which include the use of fully-connected and recurrent neural networks. These are the algorithms to be used to analyze the data recorded in 2017-2018.

2. Heavy hadron topology

The typical heavy hadron topology presents one or more vertices that are displaced from the hard-scatter interaction point. The longer lifetimes of heavy-flavor hadrons correspond to macroscopic decay lengths that can be resolved by the ATLAS vertexing detectors. The hadron's decay is governed by its heavy quark decay chain. A b-hadron will usually decay through a cascade to a c-hadron, strange hadron, etc. Therefore, in a b-initiated event, the goal is to reconstruct not only the primary vertex whence the event initiates, but also a secondary, and possibly tertiary vertex along the decay cascade. Finally, the heavy-flavor hadrons' non-negligible branching ratio to semi-leptonic decays ($\approx 21\%$) suggests using the presence of soft muons within jets as yet another discriminating feature for flavor tagging.

Jets used for b-tagging are reconstructed using the anti-k_t algorithm [8] with radius $R = 0.4$, and pass the following selection criteria [7]:

- $p_{\mathrm{T}}^{\mathrm{jet}} > 20$ GeV
- $|\eta^{\mathrm{jet}}| > 2.5$
- if $|\eta^{\mathrm{jet}}| < 2.4$ and $p_{\mathrm{T}}^{\mathrm{jet}} < 60$ GeV: $\mathrm{JVT}^{\mathrm{jet}} > 0.59$

where JVT is the output of the Jet Vertex Tagger algorithm [9] used to suppress jets from pile-up interactions. An additional overlap removal requirement is imposed to discard electrons reconstructed as jets.

Tracks are associated to the nearest jet based on an angular separation criterion that varies as a function of the jet p_{T} [10].

At truth level, using the kinematic properties of the jet under investigation and the hadrons in the event, the logic of the devised labeling scheme goes as follows:

if $(\Delta R(\mathrm{jet},\ B_{\mathrm{hadron}}) < 0.3)\ \&\ (p_{T_{B\ \mathrm{hadron}}} > 5\ \mathrm{GeV})$ ***then***
  jet flavor $\leftarrow b$
else if $(\Delta R(\mathrm{jet},\ C_{\mathrm{hadron}}) < 0.3)\ \&\ (p_{T_{C\ \mathrm{hadron}}} > 5\ \mathrm{GeV})$ ***then***
  jet flavor $\leftarrow c$
else if $(\Delta R(\mathrm{jet},\ \tau_{\mathrm{lepton}}) < 0.3)\ \&\ (p_{T_{\tau\ \mathrm{lepton}}} > 5\ \mathrm{GeV})$ ***then***
  jet flavor $\leftarrow \tau$
else
  jet flavor $\leftarrow$ *light*
end if

where $\Delta R(a,b) \equiv \sqrt{(\phi_a - \phi_b)^2 + (\eta_a - \eta_b)^2}$ [1] and p_{T} identifies the component of the momentum that is perpendicular to the beam direction.

3. Low-level taggers

Given the unique topology of heavy hadron decays (Section 2), ATLAS deploys a suite of low-level b-tagging algorithms that aim at exploiting different discriminative aspects of b-quark jets.

[1] ATLAS uses a right-handed coordinate system with its origin at the nominal interaction point in the centre of the detector and the z-axis along the beam pipe. The x-axis points from the interaction point to the centre of the LHC ring, and the y-axis points upwards. Cylindrical coordinates (r, ϕ) are used in the transverse plane, ϕ being the azimuthal angle around the z-axis. The pseudorapidity is defined in terms of the polar angle θ as $\eta = -\ln\tan(\theta/2)$.

3.1. Secondary vertex-based b-tagging algorithms

Physically grounded algorithms have been designed to look for displaced vertices. ATLAS uses an inclusive secondary vertex algorithm, SV1 [11], which looks for displaced tracks and groups them in a secondary vertex, and a multi-vertex finding algorithm, JetFitter [12], which reconstructs the topology and fits the entire decay chain along the hadron line of flight.

3.2. Impact parameter-based b-tagging algorithms

Another low-level b-tagging approach exploits the impact parameter properties of tracks and how they differ in the heavy flavor and light flavor scenarios. The goal is to compute the likelihood of tracks associated to jets originating at the primary vertex.

We define the transverse impact parameter, d_0, as the distance of closest approach of the track to the primary vertex in the $r - \phi$ plane (the plane perpendicular to the beam direction $\hat{z}$), and the longitudinal impact parameter, z_0, as the distance from the primary vertex to the intersection of the track projection onto the longitudinal plane with the beam axis [13]. The sign of the impact parameters indicates whether the track crosses the jet axis in front of or behind the primary vertex with respect to the jet direction of flight. While tracks in light jets have impact parameter significances that are close to zero, those generated from b-hadron decays and consistent with the secondary vertex hypothesis are more likely to deviate from zero [13].

The tracks used for impact parameter-based tagging satisfy:

- $p_{\mathrm{T}}^{\mathrm{track}} > 1$ GeV
- $|d_0| < 1$ mm and $|z_0 \sin\theta| < 1.5$ mm
- $N_{\mathrm{Si}}^{\mathrm{hits}} \geq 7$ and $N_{\mathrm{Si}}^{\mathrm{holes}} \leq 2$ with $N_{\mathrm{pixel}}^{\mathrm{holes}} \leq 1$

where a hole is defined as a missing expected hit and the subscripts on N identify different inner detector components.

3.2.1. IP3D Traditionally, impact parameter tagging uses a binned 2D likelihood method in different track grade categories using the transverse and longitudinal impact parameter significances. The IP3D tagger is based on a log likelihood ratio discriminant, computed – assuming no correlation among tracks – as the following sum of per-track contributions:

$$\text{IP3D LLR} = \sum_{i=1}^{N} \log \frac{p_{b_i}}{p_{u_i}} \tag{1}$$

where N is the number of tracks associated to the jet, and p_b, p_u are the template PDFs for the b and light flavor hypotheses.

However, in a b-hadron decay, several charged particles with large impact parameters can emerge from the secondary (or tertiary) vertex. In this scenario, the tracks impact parameters are intrinsically correlated: once a track with large impact parameter is found, finding a second track with large impact parameter becomes more likely. On the other hand, such a correlation should not be observed if there is no displaced decay, which is often the case for light-flavor jets. IP3D's treatment of tracks as independent limits its power as its formulation (Eq. 1) disregards known track correlations.

3.2.2. RNNIP The traditional IP3D tagger has been augmented by the introduction of RNNIP [14], a Recurrent Neural Network-based b-tagging classifier.

Recurrent Neural Networks (RNNs) are generally used to learn sequence-based dependencies for arbitrary-length input sequences. The fundamental unit is a cell that holds an internal state vector. To allow information to persist, RNNs use loops that feed the output of the

IOP Conf. Series: Journal of Physics: Conf. Series **1085** (2018) 042031 doi:10.1088/1742-6596/1085/4/042031

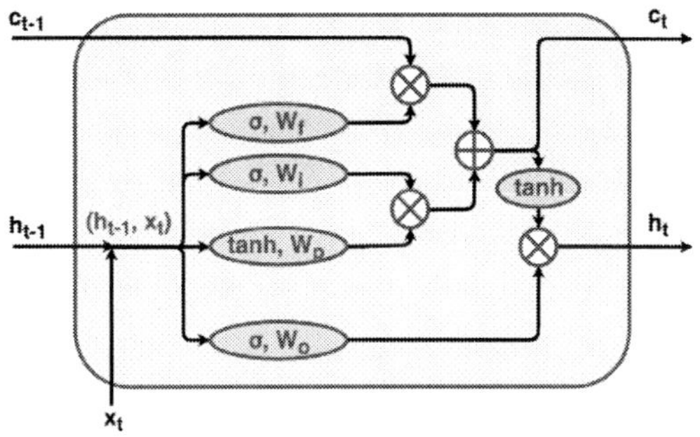

Figure 2: Anatomy of an LSTM cell with output, input, and forget gates.

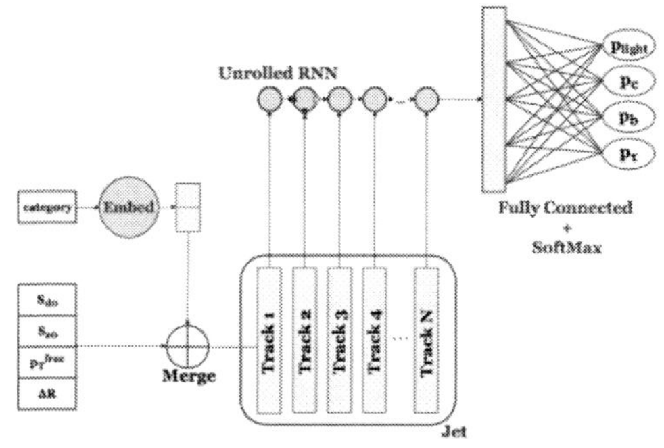

Figure 3: Diagram of the RNNIP tagger from input (left) to output (right).

cell back into the cell itself at the following step. Much of the recent success of RNNs comes from the formulation of Long Short-Term Memory [15] (LSTM) units and later variants such as Gated Recurrent Units [16] (GRUs). These modifications at the cell level mitigate issues related to vanishing and exploding gradients, to improve the knowledge persistence of long-term dependencies. They do so with different internal gating mechanisms that modify the cell state in order to regulate the relative importance of long-term versus short-term information, allowing the unit to learn when to read, write and reset the memory. The inner structure of an LSTM unit is depicted in Figure 2, while Figure 3 shows its usage in the context of the tagger.

This architecture is the natural candidate for inputs in sequence format. RNNIP represents jets as a sequence of tracks ordered by absolute transverse impact parameter significance, as track ordering studies suggest. Each track is described by a vector of features: transverse and longitudinal impact parameter significances, jet p_T fraction carried by the track, distance between the track and the jet axis, and a learned 2D embedding of the track quality.

This four-class tagger predicts a jet flavor probability associated with each flavor (b, c, τ, and light), which could be used to select a class using a maximum a posteriori estimation. For the purpose of calibrating multiple efficiency points, this information is instead combined into the discriminant

$$\mathrm{RNNIP}(f_c, f_\tau) = \log\left(\frac{p_b}{f_c \cdot p_c + f_\tau \cdot p_\tau + (1 - f_c - f_\tau) \cdot p_u}\right) \tag{2}$$

where each background importance can be varied by varying their fraction (f_c, f_τ) after training.

3.3. Muon-based b-tagging algorithm

The Soft Muon Tagger [7] (SMT) relies on information from the reconstructed muons from semi-leptonic decays of heavy-flavor hadrons. SMT provides orthogonal information to impact parameter- and vertex-based taggers. This tagger uses a group of 6 hand-engineered variables and combines them in a boosted decision tree to obtain a discriminant.

4. High-level taggers

4.1. MV2

The recently improved MV2 tagger is a gradient boosted decision tree trained using the `ROOT` `TMVA` library [17] to perform a binary classification of jets into b and non-b. Three variants are investigated which differ in the quantity of input variables.

MV2: this baseline model uses 24 input variables from lower level taggers (IP2D, IP3D, SV1, JetFitter) and kinematic properties. The full list is available in Ref. [18];

MV2Mu: adds the output of the SMT BDT to the list of inputs;

IOP Conf. Series: Journal of Physics: Conf. Series **1085** (2018) 042031 doi:10.1088/1742-6596/1085/4/042031

MV2MuRnn: adds the output of RNNIP to the list of inputs.

Importance sampling is performed to achieve a flat 2D distribution in p_T and η. The hyper-parameters of the model were optimized for the 2016 Run [13] to 1000 trees with maximum depth of 30, and minimum node size of 0.05% of the training sample.

4.2. DL1

DL1 [19] is deep neural network whose architecture is a mixture of fully-connected, maxout [20], and batch normalization [21] layers with rectified linear units [22] as activation functions. Dropout [23] is used as a stochastic regularization technique.

As with RNNIP, DL1 is trained using `Keras` [24], and incorporated into the ATLAS Athena code base [25] via the custom C++ network evaluator provided through the `LWTNN` library [26]. `rootpy` [27] and `root_numpy` [28] were used to convert the input file format.

The network is trained to minimize the cross-entropy loss using the Adam optimizer [29]. A thorough grid search is performed to select the network structure and hyper-parameters such as the number of hidden layers, number of nodes per layer, and the learning rate.

The three output nodes calculate the probabilities associated with each jet flavor (here: p_b, p_c, p_u) as highly non-linear functions of the input features. A final discriminant for b-tagging is obtained by combining the outputs into one tunable function of the fraction of c-jets in the background, f_c:

$$\mathrm{DL1}(f_c) = \log\left(\frac{p_b}{f_c \cdot p_c + (1 - f_c) \cdot p_u}\right) \tag{3}$$

A similar equation can be derived for c-tagging, by swapping the b and c labels. Although the c-fraction can be varied depending on the background composition of each analysis, in order to facilitate the comparison with MV2, f_c is herein set to the natural occurring fraction in a typical $t\bar{t}$ sample, *i.e.* 7%. This parameter regulates the trade-off between c and light rejections at a given b tagging efficiency.

While DL1 and MV2 have been found to have similar performance, DL1 has emerged out of need for improved flexibility and power in future R&D cycles. The neural network can be easily trained with minimal standalone code; it is GPU-enabled; it is modular and can be effortlessly extended to include more input variables. It is also amenable to be trained using an adversarial loss to minimize data-Monte Carlo discrepancies [30], and it grants the possibility for future end-to-end trainings in conjunction with RNNIP.

5. Performance in Simulation

The high level taggers are trained on a newly introduced hybrid sample which combines $t\bar{t}$ with a broad Z' sample at $m_{Z'} = 4$ TeV to increase the representation of high p_T jets at training time.

Figure 4 shows that RNNIP improves light rejection by a factor of 2 and c rejection by a factor of 1.2 compared to IP3D, and outperforms it across the entire p_T spectrum. A sizable fraction of b-jets is tagged correctly by only one of the two algorithms, where IP3D is more efficient at low track multiplicity and short decay length, while RNNIP is superior at high multiplicity.

The three MV2 versions are evaluated on exclusive $t\bar{t}$ and Z' samples; results from the latter are presented in Figure 5. Including muon information improves b-to-light jet separation, but the improvement vanishes at high p_T. On the other hand, b-to-c separation does not strongly benefit from the muon-in-jet information for efficiencies $> 50\%$ because of lepton presence in both b and c decays. MV2MuRnn further improves performance, especially in the high-p_T region where the RNNIP tagger is most performant. For example, when evaluating on Z' events (Figure 5), at $p_T^{\mathrm{jet}} = 1$ TeV the RNN improves the light-jet rejection by a factor ~ 1.5 and the c-jet rejection by a factor of ~ 1.3, compared to the baseline MV2 2017 configuration.

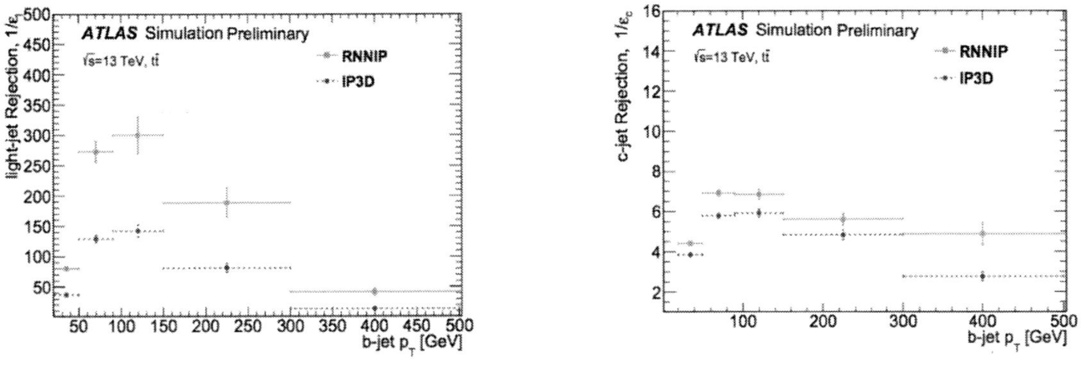

(a) (b)

Figure 4: Performance comparison in bins of p_{T}, as it appeared in Ref. [7], between IP3D and RNNIP for light jet (a) and c-jet rejection (b) in simulated $t\bar{t}$ events at a constant b-efficiency of 70%.

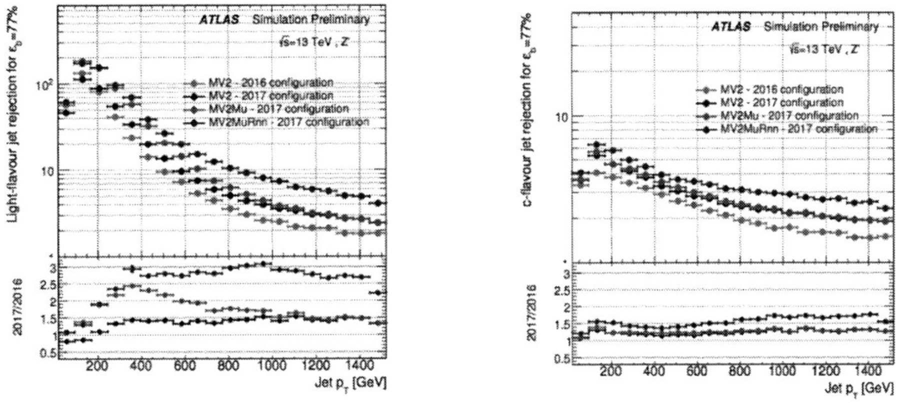

(a) (b)

Figure 5: Performance comparison in bins of p_{T}, as it appeared in Ref. [7], between MV2 in the 2016 configuration (brown), MV2 in the 2017 configuration (black), MV2Mu (red) and MV2MuRnn (blue) for light jet (a) and c-jet rejection (b) in simulated Z' events at a constant b-efficiency of 77%. The ratio plots use the 2016 MV2 configuration as baseline.

6. Conclusion

ATLAS developed new algorithms to identify b-jets using machine learning techniques. Their performance was investigated using Monte Carlo simulation of $p\bar{p}$ events with $\sqrt{s} = 13$ TeV in the ATLAS detector. These methods augment the already rich suite of b-tagging algorithms adopted by the ATLAS collaboration, by extracting complementary features and enhancing the overall flavor tagging performance.

ATLAS analysts can also expect an increased level of agreement between the train (Monte Carlo) and test (data) domains due to improvements in the tracking simulation with respect to the 2016 configuration, which reduces the performance losses associated with out-of-domain model application.

Acknowledgements
This work was supported in part by the Office of High Energy Physics of the U.S. Department of Energy under contract DE-FG02-92ER40704.

References
[1] Evans L and Bryant P 2008 *JINST* **3** S08001
[2] Mitrevski J 2015 *Journal of Physics: Conference Series* **664** 072034
[3] ATLAS Collaboration 2008 *JINST* **3** S08003
[4] ATLAS Collaboration 2010 *The European Physical Journal C* **70** 787–821 ISSN 1434-6052
[5] 1996 ATLAS liquid argon calorimeter: Technical design report Tech. Rep. CERN-LHCC-96-41 CERN Geneva
[6] 1996 ATLAS tile calorimeter: Technical design report Tech. Rep. CERN-LHCC-96-42 CERN Geneva
[7] 2017 Optimisation and performance studies of the ATLAS *b*-tagging algorithms for the 2017-18 LHC run Tech. Rep. ATL-PHYS-PUB-2017-013 CERN Geneva URL `http://cds.cern.ch/record/2273281`
[8] Cacciari M, Salam G P and Soyez G 2008 *Journal of High Energy Physics* **2008** 063
[9] 2014 Tagging and suppression of pileup jets with the ATLAS detector Tech. Rep. ATLAS-CONF-2014-018 CERN Geneva URL `https://cds.cern.ch/record/1700870`
[10] ATLAS Collaboration 2016 *JINST* **11** P04008
[11] 2017 Secondary vertex finding for jet flavour identification with the ATLAS detector Tech. Rep. ATL-PHYS-PUB-2017-011 CERN Geneva URL `http://cds.cern.ch/record/2270366`
[12] Piacquadio G and Weiser C 2008 *Journal of Physics: Conference Series* **119** 032032
[13] 2016 Optimisation of the ATLAS *b*-tagging performance for the 2016 LHC Run Tech. Rep. ATL-PHYS-PUB-2016-012 CERN Geneva URL `http://cds.cern.ch/record/2160731`
[14] 2017 Identification of Jets Containing *b*-Hadrons with Recurrent Neural Networks at the ATLAS Experiment Tech. Rep. ATL-PHYS-PUB-2017-003 CERN Geneva URL `http://cds.cern.ch/record/2255226`
[15] Hochreiter S and Schmidhuber J 1997 *Neural Comput.* **9** 1735–1780 ISSN 0899-7667
[16] Chung J, Gulcehre C, Cho K and Bengio Y 2014 (*Preprint* `cs.NE/1412.3555`)
[17] Hoecker A, Speckmayer P, Stelzer J, Therhaag J, von Toerne E and Voss H 2007 *PoS* **ACAT** 040 (*Preprint* `physics/0703039`)
[18] 2015 Expected performance of the ATLAS *b*-tagging algorithms in Run-2 Tech. Rep. ATL-PHYS-PUB-2015-022 CERN Geneva URL `https://cds.cern.ch/record/2037697`
[19] ATLAS Collaboration 2017 URL `http://cds.cern.ch/record/2274065`
[20] Goodfellow I J, Warde-Farley D, Mirza M, Courville A and Bengio Y 2013 Maxout networks *Proceedings of the 30th International Conference on International Conference on Machine Learning - Volume 28* ICML'13 (JMLR.org) pp III–1319–III–1327
[21] Ioffe S and Szegedy C 2015 Batch normalization: Accelerating deep network training by reducing internal covariate shift *Proceedings of the 32nd International Conference on International Conference on Machine Learning - Volume 37* ICML'15 (JMLR.org) pp 448–456
[22] Nair V and Hinton G E 2010 Rectified linear units improve restricted boltzmann machines *Proceedings of the 27th International Conference on International Conference on Machine Learning* ICML'10 (USA: Omnipress) pp 807–814 ISBN 978-1-60558-907-7
[23] Srivastava N *et al.* 2014 *J. Mach. Learn. Res.* **15** 1929–1958 ISSN 1532-4435
[24] Chollet F *et al.* 2015 Keras URL `https://github.com/fchollet/keras`
[25] 2005 ATLAS Computing: technical design report URL `http://cds.cern.ch/record/837738`
[26] Guest D H, Smith J W, Paganini M, Kagan M, Lanfermann M and Krasznahorkay A 2017 lwtnn/lwtnn: Version 2.4 DOI: 10.5281/zenodo.848256
[27] Dawe N, Waller P, Friis E K *et al.* 2015 rootpy: 0.8.0 DOI: 10.5281/zenodo.18897
[28] Dawe N, Ongmongkolkul P and Stark G 2017 *The Journal of Open Source Software* **2**
[29] Kingma D P and Ba J 2014 (*Preprint* `cs.LG/1412.6980`)
[30] Ganin Y *et al.* 2015 (*Preprint* `stat.ML/1505.07818`)

IOP Conf. Series: Journal of Physics: Conf. Series **1085** (2018) 042032 doi:10.1088/1742-6596/1085/4/042032

Tools for Trigger Rate Monitoring at CMS

Geoffrey N Smith[1], Charles N Mueller[1], Andrew S Wightman[1] on behalf of CMS

[1] University of Notre Dame, Notre Dame, IN (US)

E-mail: gsmith15@nd.edu, charles.mueller@cern.ch, andrew.steven.wightman@cern.ch

Abstract. In 2017, the LHC delivered an instantaneous luminosity of roughly $2.0 \times 10^{34}\text{cm}^{-2}\text{s}^{-1}$ to the Compact Muon Solenoid (CMS) experiment, with about 60 simultaneous proton-proton collisions ($<\mu>$) per event. In these challenging conditions, it is important to be able to intelligently monitor the rate at which data are being collected (the trigger rate). It is not enough to simply look at the trigger rate; it is equally important to compare the trigger rate with expectations. We present a set of software tools that have been developed to accomplish this. The tools include a real-time component - a script that monitors the rates of individual triggers during data-taking, and activates an alarm if rates deviate significantly from expectation. Fits are made to previously collected data and extrapolated to higher $<\mu>$. The behavior of triggers as a function of $<\mu>$ is then monitored as data are collected - plots are automatically produced on an hourly basis and uploaded to a web area for inspection. This same set of tools can also be used offline in data certification, as well as in more complex offline analysis of trigger behavior.

1. Introduction

The trigger system is extremely important to data acquisition at the Compact Muon Solenoid (CMS) [1] experiment at CERN. Roughly 500 separate algorithms combine to filter the approximately 40 MHz rate of collisions down to 1 kHz of data (corresponding to 1 GB/s) collected for offline analysis. Without an effective trigger system, physics at CMS would not be possible.

As the LHC pushes to higher beam intensities, CMS has to be ready to respond to emergencies if the trigger rates go out of expected range. The trigger rate is very sensitive to all aspects of the detector and how they operate, so it often provides the first indication of a detector problem. Therefore, it is very important to be able to intelligently monitor, characterize and visualize trends in trigger rates.

The trigger Field Operation Group at CMS has developed a set of software tools to accomplish this task, consisting of a suite of python modules and scripts that perform a variety of functions. Fits are made to the trigger rates in previous runs using linear and non-linear regression. These fits are then compared to the instantaneous trigger rate as data are being collected, in order to spot small (unexpected) deviations in rate. As well as this real-time component, the software provides a variety of additional features that are used in offline analysis.

2. Fits

First, fits are made to the trigger rate in previous runs as a function of the average number of collisions in an LHC bunch crossing, $<\mu>$ (also known as average in-time pile-up). Runs to be

IOP Conf. Series: Journal of Physics: Conf. Series **1085** (2018) 042032 doi:10.1088/1742-6596/1085/4/042032

fit are selected from a list of known good runs, and the fits are performed using the ROOT [2] software.

The rate information is obtained from a central SQL database at CMS. Before fitting, the raw trigger rate is first corrected for deadtime, Level-1 (L1) and High Level Trigger (HLT) prescales, as well as the number of colliding bunches in the LHC. Performing this correction facilitates comparisons and extrapolations between runs with different conditions, and allows a smooth function to be fit between runs. For each fit, several fit functions are attempted, and the final function is selected based on a χ^2 minimization.

Ideally, trigger rates should depend linearly on instantaneous luminosity. However, some have nonlinear behavior, mainly due to pile-up effects. Candidates for multiple object triggers may be found in independent collisions, and triggers that rely on measuring calorimeter deposits (such as jets and jet sums) increase in rate as the number of contributions from independent collisions increases. After exploring several options, we found that the behavior of most triggers is well-described by either linear, quadratic, or exponential (sinh) functions. The triggers with linear and quadratic behavior are primarily those that trigger on single and double objects, respectively. The triggers with the highest sensitivity to pile-up are the jet sum triggers, and they are described best by exponential functions. Examples of fits are shown in Figure 2.

3. Online Monitoring

3.1. Real-time Component

Using fits from previous runs, the trigger rates can be monitored during data-taking by comparing the instantaneous rate to the rate predicted by the fit. A list of approximately 20 L1 and HLT triggers is used for real-time online monitoring in the CMS control room. This list is selected such that all CMS subdetectors are monitored and all physics objects are represented. An automated script running continuously during data-taking checks current trigger rates against the prediction from the fit for each trigger in the list. Rate information is displayed in the trigger shifter terminal. The rate information is updated every two minutes— updating more frequently can cause errors as there is a lag associated with the time it takes to propagate new rate information to the database (this is an aspect of the software we are working to improve). When the rate of a given trigger is at least five times higher than the uncertainty on the fit for a significant period of time, the script prints a warning, and an email summarizing the problematic triggers is automatically sent to on-call experts. In order to decrease false alarms due to noise, the warning and email are only activated when three consecutive queries of the database indicate a higher than expected rate. The script also activates audible alarms in the control room in the event that rates exceed more extreme, fixed thresholds; in this case, warnings are tailored to help diagnose problems with individual subsystems. For example, if the rate of a certain L1 e/gamma trigger is higher than 25kHz, the warning "critical L1 e/gamma trigger rate: please check that ECAL and calorimetric triggers are behaving correctly" is given.

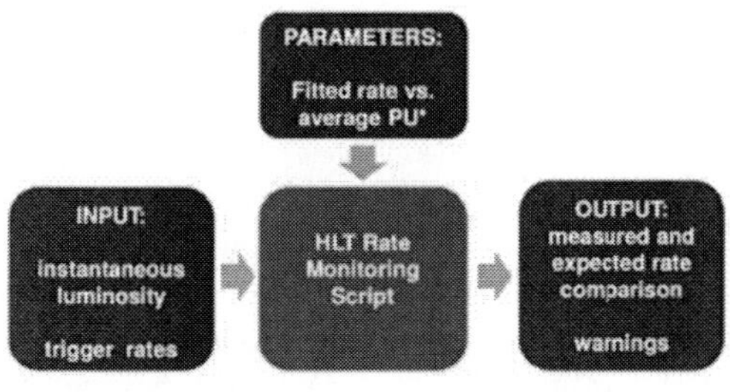

Figure 1. Flowchart for the real-time component of the rate-monitoring software.

3.2. Summary Plots

In addition to the real-time component, the software produces plots that compare the trigger rate versus <μ> to the expected rate for each fill. These plots have been integrated into the

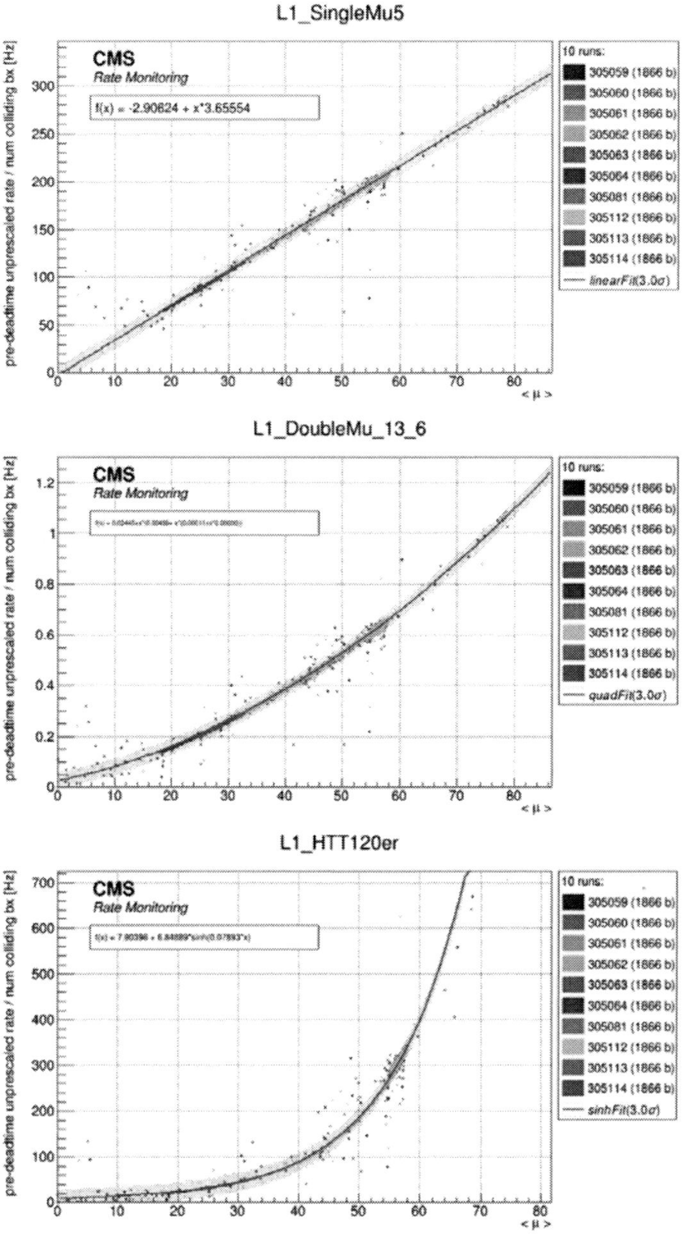

Figure 2. Examples of adjusted trigger rate versus $<\mu>$. The points correspond to the per-lumisection rate, color-coded for different runs, and the pink band is the 3σ error band from the fit. From top to bottom: examples of triggers whose behavior as a function of $<\mu>$ is best described by linear, quadratic and hyperbolic sine functions, respectively.

central CMS Web Based Monitoring service (WBM); a dedicated page is linked from each Fill Report page on WBM, containing plots with rates for that fill (Figure 3). The page contains rate versus $<\mu>$ plots for all the triggers in the monitored trigger list. Plots for all HLT and L1 triggers, as well as stream and dataset plots, are available via links from this main page.

IOP Conf. Series: Journal of Physics: Conf. Series **1085** (2018) 042032 doi:10.1088/1742-6596/1085/4/042032

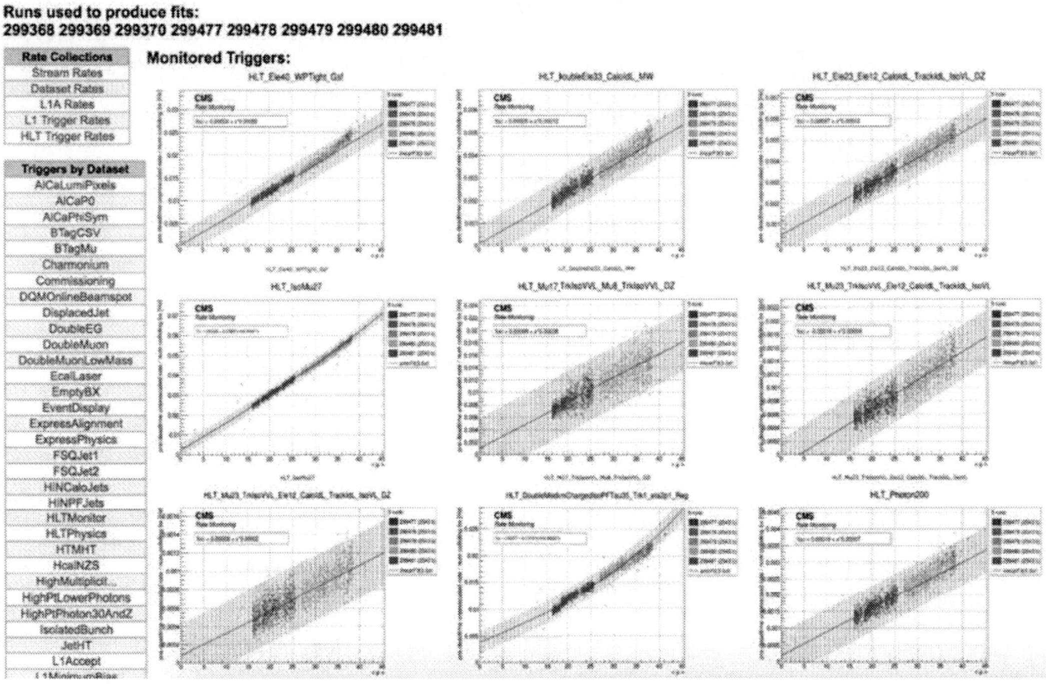

Figure 3. Screenshot of WBM page produced by the rate-monitoring software, showing plots of rate versus $<\mu>$ for LHC fill 5976. Selected triggers are shown on the main page, with links on the left-hand column to further pages containing plots for every L1 and HLT trigger in the trigger menu, as well as dataset and stream rates.

Additional links directly to the plots of the individual triggers are available from the HLT and L1 summary pages. The plots are produced by a cron job run on a dedicated machine that updates the plots on an hourly basis (for the current fill only). Work is ongoing to develop a better implementation which would allow the plots to be updated in real time.

4. Offline Use

The software is also used for offline data certification. For a given list of runs and list of triggers, rate-versus-lumisection plots can be produced for each run and trigger (one lumisection equals 2^{18} orbits of the LHC beam, or about 23 seconds). An overlay of the rate prediction from the fit enables easy comparison by offline validators (Figure 4), and a text summary is also produced highlighting the runs and lumisections where triggers deviated significantly from expectation. The software has a modular organization, and can be extended to perform additional functions by other collaborators.

5. Summary

We have described a set of software tools currently in use at CMS to monitor trigger rates. This software is used for real-time monitoring of trigger rates by shifters and on-call experts, as well as offline for data certification and other studies. In particular, the trigger shifter script together with the plots on WBM have allowed fast identification and diagnosis of detector issues. The sofware has proven to be a critical tool to the successful operation of the trigger and to successful data-taking at CMS.

For further information about the rate-monitoring software, see the Github repository here [3].

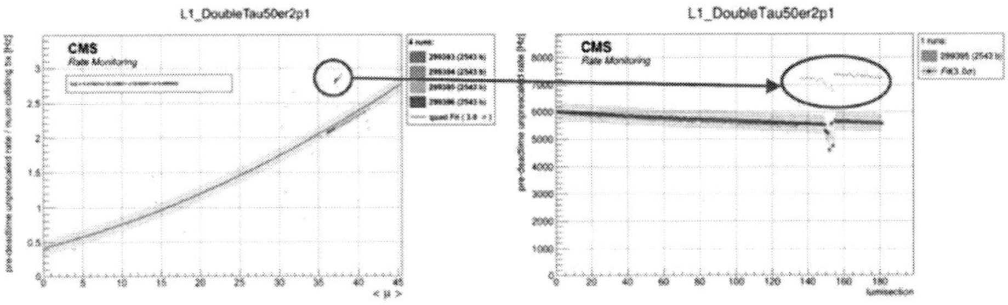

Figure 4. Examples of plots used for run certification. Runs where the trigger rate deviated significantly from the fit are first identified in rate-versus-<μ> plots (left). Rate-versus-lumisection plots are then used to identify the affected lumisections (right).

6. References

[1] CMS Collaboration 2008 *The CMS experiment at the CERN LHC* JINST 3 S08004
[2] Rene Brun and Fons Rademakers, *ROOT - An Object Oriented Data Analysis Framework*, Proceedings AIHENP'96 Workshop, Lausanne, Sep. 1996, Nucl. Inst. & Meth. in Phys. Res. A 389 (1997) 81-86. See also http://root.cern.ch/.
[3] A. Wightman et. al, *Repository with various tools to monitor HLT and L1 rates.* https://github.com/cms-tsg-fog/RateMon.

High-statistics and GPU Accelerated Data Analysis

Philipp Eller (for the IceCube-Gen2 Collaboration)
Penn State University, Dept. of Physics, 104 Davey Lab, 16802 PA University Park
E-mail: pde3@psu.edu

Abstract. We present methods to perform high statistics data analyses to investigate fundamental neutrino properties in large volume neutrino detectors, fast and with modest computational resources. The introduced measures are threefold: speeding up computations using graphics processors, evaluating the underlying physics processes on a grid instead of treating every event individually and lastly applying smoothing methods to quantities obtained from Monte Carlo simulations. We show that with our method we can get reliable analysis results using significantly less simulation than what is usually needed, and that the timing to run an analysis with our method is independent of sample size.

1. Introduction

Proposed very large volume neutrino telescopes (VLVnTs), such as PINGU [1, 2] or ORCA [4], will collect unprecedented amounts of atmospheric neutrinos (in a typical event selection, several hundred thousand events a year). Said neutrinos can be used to perform various measurements, such as the atmospheric oscillation angle θ_{23}, the mass squared splitting Δm^2_{31} and its sign (neutrino mass hierarchy, NMO) and the rate of ν_τ appearance just to name a few. Also limits on non-standard interactions, sterile neutrinos and other exotic phenomena can be extracted.

What all of these analyses have in common, is that they compare distributions of simulated neutrino events with given model parameters to the observed data. The typical procedure is to perform a maximum likelihood analysis of the reconstructed energy and arrival direction of the observed neutrinos to find the maximum likelihood estimators of the model parameters. To do so, a large parameter space needs to be explored, and every time the distributions of the reconstructed event properties (templates) generated to compare them to the data. The distributions that are compared to the data should be of significantly better statistical precision (typically an order of magnitude) than the data itself.

The canonical way of achieving this is by simulating large enough samples, we refer to this in the following as *direct histogramming*. A way to alleviate statistical fluctuations of otherwise too small samples is to use smoothing techniques, such as kernel density estimation (KDE) [5] instead of simple histograms, we refer to this as *direct KDE*. Our proposed method will be explained in the following and referred to as *PISA*. The used techniques and methods are explained in much greater detailed in [3].

2. The PISA staged approach

Instead of directly using simulated data by putting the events into a histogram or KDE to obtain the final distributions that can directly be compared to the data, we take a different route. The simulated data is rather used to characterize the experiments properties, here its acceptance and resolutions. These quantities together with the atmospheric neutrino flux and its flavor oscillation probabilities are used to generate high-statistics samples of reconstructed event properties in the following way: The initial flux (1) multiplied together with the oscillation probabilities (2) and the experiment's acceptance (3) yield truth-level distributions (see figure 1). In a convolution with point spread and energy resolution function (4) these spectra are translated into the space of reconstructed variables.

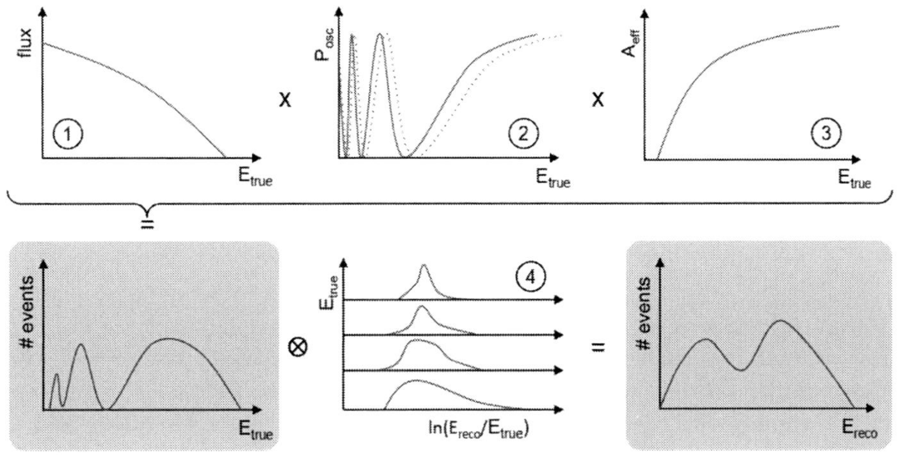

Figure 1: Illustration of the PISA staged approach for obtaining event templates, here for simplicity using a characterization in one dimension (energy) only. Steps 1, 2, and 3 are in true energy (E_{true}); the product of these yields the expected event distribution (lower left). Smearing this spectrum with energy-dependent energy resolution functions (step 4) gives the reconstructed event rate spectrum (lower right). The different horizontal axes on panel 4 are used to show resolution kernels of different true energies. Note that the dotted green line in step 2 shows a hypothetical change of oscillation parameters, affecting only stage 2.

2.1. Validation

Table 1 shows, that in the limit of dense enough grids on which the individual properties are evaluated this method is equivalent to direct histogramming. When using 320 x 320 grid points or more, the maximum χ^2_{max} distance of all bins is more than an order of magnitude smaller than the average $\langle\chi^2\rangle = 1.0$ that is expected when comparing to data. In the next section it will be demonstrated that this way is faster when using large sample sizes. But more importantly, this approach allows to apply smoothing to quantities that are well understood and behaved.

Grid ($M \times N$)	40×40	80×80	160×160	320×320	640×640	1280×1280
$\langle\chi^2\rangle$	0.01067	0.00253	0.00060	0.00014	0.00003	0.00001
χ^2_{max}	1.45906	0.46930	0.19718	0.04974	0.00634	0.00172

Table 1: Average and maximal χ^2 deviations between non-smoothed staged approach and direct histogramming, for different grid point densities in (E_{true}, $\cos\vartheta$), using an MC dataset of 10^6 events.

IOP Conf. Series: Journal of Physics: Conf. Series **1085** (2018) 042033 doi:10.1088/1742-6596/1085/4/042033

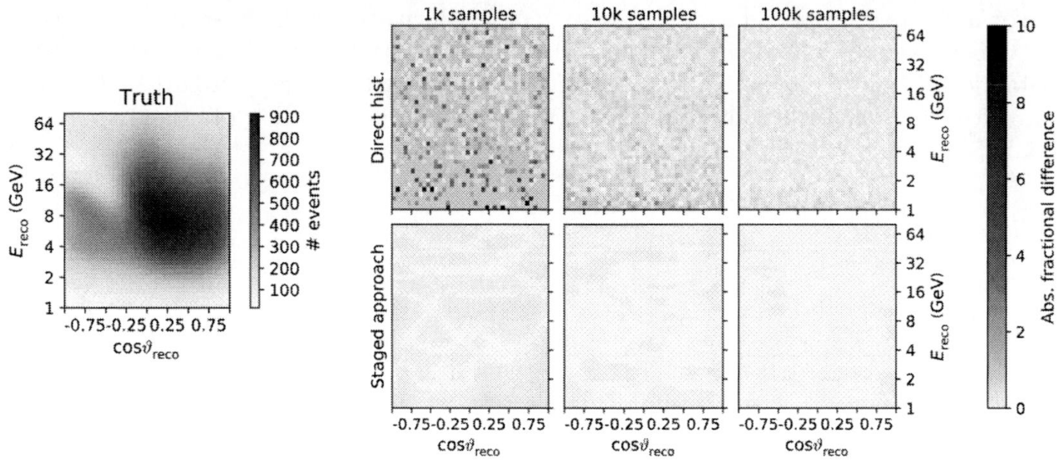

Figure 2: Parametric reference distribution for the ν_μ CC channel in ($\cos\vartheta_{reco}$, E_{reco}) (left panel) and relative residuals for the direct histogramming (right panel, top row) and our proposed method (right panel, bottom row) using different amounts of simulated events. The three columns in the right panel represent different MC sample sizes of 10^3, 10^4, and 10^5 events, respectively. The samples are drawn from an toy model distribution. Note that the residuals in the 1k-samples plot for direct histogramming go up to 31 but the scale is clipped at 10.

2.2. Smoothing

Instead of applying smoothing techniques to the final event distributions (as shown in *direct KDE*), we take the advantage that the underlying properties (acceptance and resolutions) of the experiment are inherently easier to deal with – the shape of the acceptance for example is completely decoupled from rapidly changing oscillation probabilities. We apply a simple spline smoothing to the acceptance and adaptive KDEs to the resolution functions. So this method is general and does not make assumptions of a parametric form of the resolution functions. A comparison of final level distributions obtained with histogramming and with the PISA method are compared for different sample sizes in figure 2 A more qualitative comparison of final level event distributions are shown in figure 3. These are obtained from a generic toy detector model, and therefore the underlying true distributions are known and also shown.

3. Results

Running a typical analysis (here the determination of the neutrino mass hierarchy) gives an idea of the severity of lacking statistical precision in the templates. Figure 4 shows the result one would obtain using a given sample size of simulated events and using a given technique[1]. For direct histogramming, relatively large sample sizes need to be used to obtain unbiased, realistic answers. The KDE applied to the final events can certainly help here, but it will be shown later that this techniques (apart from being inferior to our method) is extremely slow. The proposed PISA method outperforms the other two methods across the board and even for the smallest tested sample sizes, the obtained answers are still compatible with truth.

[1] The result is shown in terms of χ^2 difference between the best fit of both neutrino mass orderings to a given pseudo data trail.

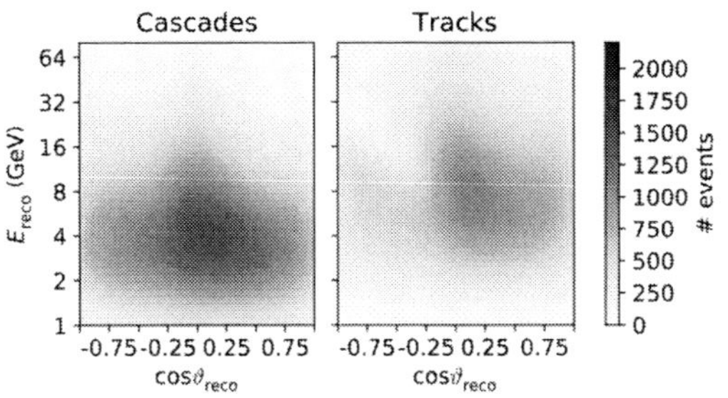

(a) Truth

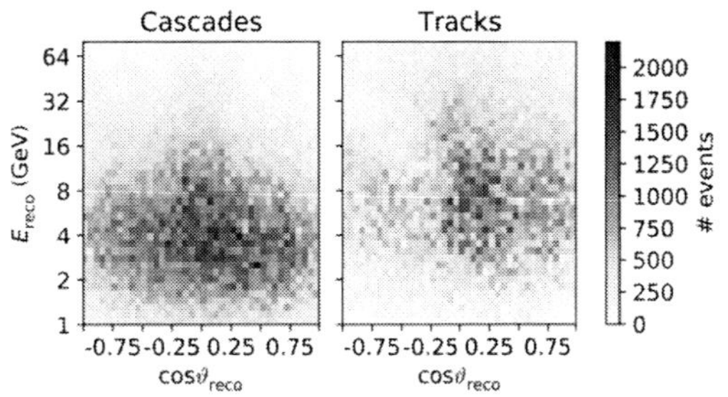

(b) Direct Histogramming

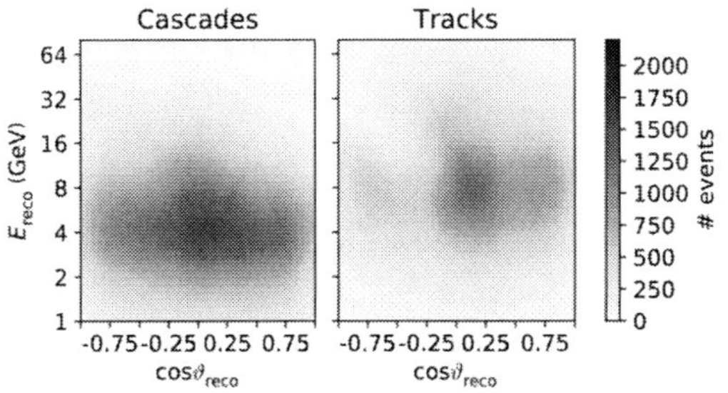

(c) Direct KDE

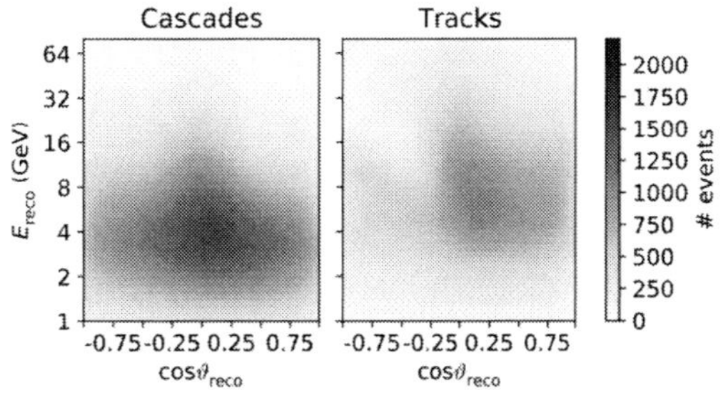

(d) PISA method

Figure 3: Final level templates used for the example data analysis. The reference distributions (truth) obtained directly for a toy detector model parameterizations are shown in panel (a). Given the same sample of 10^4 events the estimated distributions using histograms are shown in panel (b), using KDEs in panel (c), and using the PISA staged approach with a ($400x400$) grid in panel (d).

3.1. Benchmark

In figure 5 we show the average time necessary to generate a template, for a typical analysis this is done many times. While the direct histogramming method scales roughly linear with sample size, the direct KDE gets slower faster than linear as the sample size growth, making it impractical for usage with sample sizes much larger than $\approx O(10^5)$. The proposed PISA method is by construction independent of input sample size (excluding initial start-up time which is growing with sample size, but negligible when many (typically thousands) of templates need to be generated).

Furthermore, we implemented all of the methods, or part thereof, to run on graphics co-processors (GPUs) to achieve speed ups. The gain for the two direct methods is in the order of a magnitude faster, for the PISA approach not all steps have yet been ported to harness the GPUs power and therefore the speed improvement are less.

However, the PISA method is as fast or faster then the other method when sample sizes are considered that deliver useful physics answers as outlined in the previous section.

4. Summary

For-high statistics data analyses, as needed to analyze the data from new large neutrino telescopes, high-precision expected event distributions from given model assumptions are needed to compare to the data. The challenge is to keep the computational burden low to facilitate data analysis. While simple histogramming of events suffers from large inherent statistical fluctuations, state for the art smoothing techniques like KDE can help, but at the same time

IOP Conf. Series: Journal of Physics: Conf. Series **1085** (2018) 042033 doi:10.1088/1742-6596/1085/4/042033

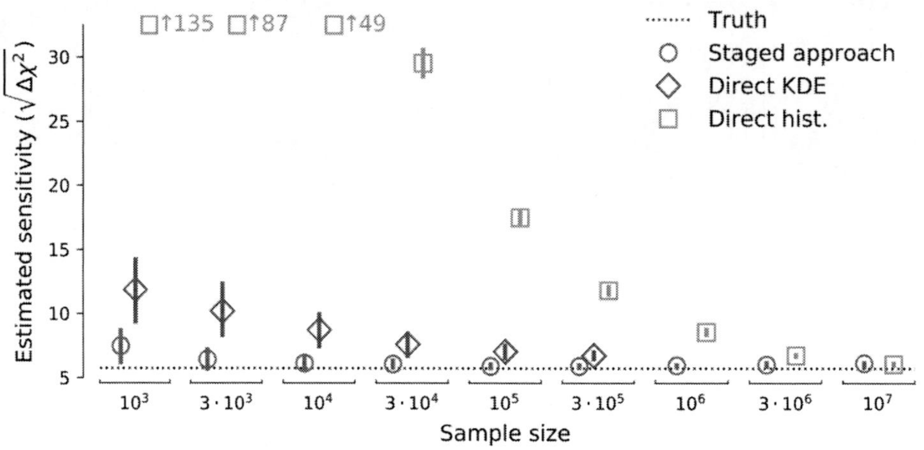

Figure 4: Estimated sensitivity ($\sqrt{\Delta\chi^2}$) to the NMO vs. sample size for direct histogramming, direct KDE, and the proposed PISA staged smoothing methods applied to multiple (between 50 and 200) statistically independent toy MC datasets. Vertical lines indicate central 68% quantiles. The dashed horizontal line shows the significance obtained from truth templates based on the parametric toy detector model. The staged approach outperforms the other methods—both in terms of bias and variance—for sample sizes through $3 \cdot 10^6$, with direct histogramming only matching the staged approach using 10^7 samples. Note that no data points exist for direct KDE and sample sizes above $3 \cdot 10^5$, as computational processing times become impractically large. Also note that direct histogramming is off-scale high for fewer than $3{\cdot}10^4$ events (mean values are indicated to the right of the corresponding markers).

can be slow. Our approach uses simulated events and smoothing techniques to characterize generic detector properties. The expected distributions of neutrino event properties are then constructed from these generic detector properties. The main advantages are higher precision due to reduced fluctuations and better computational performance.

References

[1] M. G. Aartsen et al. Letter of intent: the Precision IceCube Next Generation Upgrade (PINGU). 2017.
[2] M. G. Aartsen et al. PINGU: a vision for neutrino and particle physics at the South Pole. *J. Phys.*, G44(5):054006, 2017.
[3] M. G. Aartsen et al. Computational Techniques for the Analysis of Small Signals in High-Statistics Neutrino Oscillation Experiments. 2018.
[4] S. Adrian-Martinez et al. Letter of intent for KM3NeT 2.0. *J. Phys.*, G43(8):084001, 2016.
[5] Kyle S. Cranmer. Kernel estimation in high-energy physics. *Comput. Phys. Commun.*, 136:198–207, 2001.

IOP Conf. Series: Journal of Physics: Conf. Series **1085** (2018) 042033 doi:10.1088/1742-6596/1085/4/042033

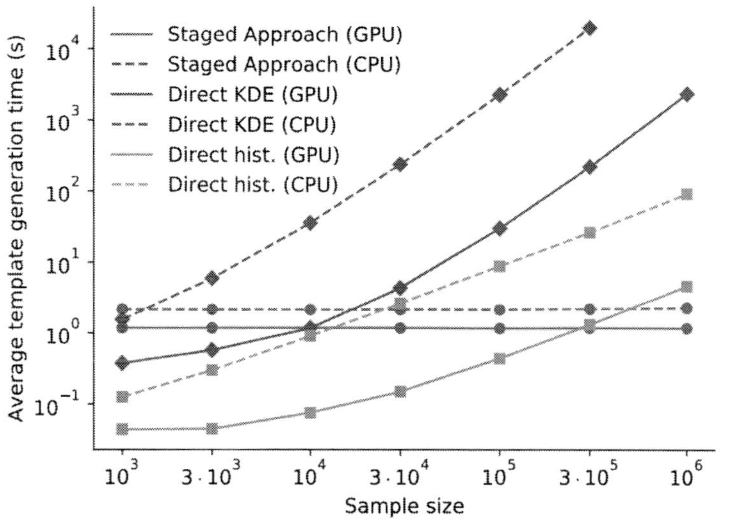

Figure 5: Average template generation time during a typical analysis for input data sets of varying size, shown for the direct histogramming, the direct KDE, and the staged approach. Solid lines represent timings based on (partial) GPU acceleration, whereas the dashed ones are for CPU-only calculations.

IOP Conf. Series: Journal of Physics: Conf. Series **1085** (2018) 042034 doi:10.1088/1742-6596/1085/4/042034

Deep Neural Networks for Physics Analysis on low-level whole-detector data at the LHC

Wahid Bhimji[1], Steven Andrew Farrell[1], Thorsten Kurth[1], Michela Paganini[1,2], Prabhat[1], Evan Racah[1]

[1] Lawrence Berkeley National Laboratory, Berkeley, CA 94720 USA
[2] Department of Physics, Yale University, New Haven, CT 06520, USA

E-mail: wbhimji@lbl.gov

Abstract. There has been considerable recent activity applying deep convolutional neural nets (CNNs) to data from particle physics experiments. Current approaches on ATLAS/CMS have largely focussed on a subset of the calorimeter, and for identifying objects or particular particle types. We explore approaches that use the entire calorimeter, combined with track information, for directly conducting physics analyses: i.e. classifying events as known-physics background or new-physics signals.

We use an existing RPV-Supersymmetry analysis as a case study and explore CNNs on multi-channel, high-resolution sparse images: applied on GPU and multi-node CPU architectures (including Knights Landing (KNL) Xeon Phi nodes) on the Cori supercomputer at NERSC.

We compare statistical performance of our approaches with selections on high-level physics variables from the current physics analyses, and shallow classifiers trained on those variables. We also compare time-to-solution performance of CPU (scaling to multiple KNL nodes) and GPU implementations.

1. Introduction

A major aim of experimental high-energy physics (HEP) is to find rare signals of new particles produced in large numbers of collisions at accelerators such as the Large Hadron Collider (LHC) and the ATLAS and CMS experiments. Improvements in classifying these collisions would improve sensitivity to discoveries that could overturn our understanding of the universe at the most fundamental level. Neural Networks have been used in high energy physics for some time. Recently attention has focused on deep learning to improve sensitivity, as well as tackle the increase in detector resolutions and data rates in HEP experiments.

Here we evaluate the use of deep Neural Network's (NN) on low-level detector data directly for physics analysis without reconstruction of physics objects like jets; without tuning of analysis variables; and using data from the entire calorimeter together with other sub-detectors. We use modern deep learning methods for performance; relate performance to that of physics variables; and ensure results are robust to factors such as pileup and alternative physics models.

The second goal of this work is to ensure these NN architectures run efficiently with popular deep-learning software frameworks on the Cori supercomputer at the National Energy Research Scientific Computing Center (NERSC). This is a major computing resource that is predominately composed of Intel Knights Landing (KNL) XeonPhi CPUs. We seek to improve single node performance on these CPUs, providing timings, optimisations and recipes, as well as distributed training up to the full scale of the machine (10k KNL nodes).

We provide background, in Section 2.1, on the physics use-case studied and benchmark selections; in Section 2.2, the CNN architecture we employ; and, in Section 2.3 the computing resources at NERSC. We then provide physics performance results in Section 3.1 together with interpreting how the learned network compares to the physics variables used in the benchmark analysis (in Section 3.2). Then in Section 3.3 we show timing results for implementations in various frameworks and on GPU and CPU resources at NERSC, before concluding in Section 4.

2. Background

2.1. Physics and datasets

For this study, we take as a use-case a particular analysis searching for new massive supersymmetric ('RPV-Susy') particles in multi-jet final states at the LHC [1]. We focus on 'gluino-cascade decays' with gluino and neutralino masses of 1400 GeV and 850 GeV by default (and varied for performance studies).

We use the Pythia event generator [2] interfaced to the Delphes fast detector simulation [3] and using the default Delphes ATLAS detector configuration ('card'). We generate events for two classes: the RPV-Susy signal and the most prevalent background ('QCD'). In order to have sufficient quantities of background events at high transverse momentum (p_T) that are harder to separate, the QCD sample is generated in ranges of p_T and then weighted according to the amount expected (cross-section) in both training loss and the evaluation.

Before training our network we apply some of the physics selections used in [1] to filter images to those more challenging to discriminate, resulting in a training sample of around 400k events. We compare the performance of our deep network to our own implementation of the full selections in [1] as a baseline benchmark. To produce the jet variables we use the same jet algorithm as in the physics analysis with parameters given in Appendix A and applied using 'FastJet' [4] within Delphes. The preselection and benchmark selections are also summarized in Appendix A. We have verified that our samples and baseline selections give performance comparable to that obtained in [1] providing a meaningful benchmark even though those selections were not tuned for exactly these datasets. We also compare our analysis to shallow classifiers based on these physics variables. As inputs to those classifiers, we use the full four-momentum of five jets with highest transverse momentum as well the jet variables used in the physics analysis selections: namely the sum of jet mass, number of jets, and $\Delta\eta$ between the leading 2 jets.

Data from the surface of the cylindrical detector is represented as a 2D image with coordinates corresponding to azimuthal angle ϕ and pseudorapidity η. For the pixel intensity in this image, we use either the overall energy deposited in the combined calorimeter or split the energy deposited in the electromagnetic and hadronic calorimeters, and add the number of tracks formed from the inner detector in that region to provide a three channel image. This is similar to the approach of [5] [6] except that we use large images covering the entire detector, and use these directly for classifying entire events rather than individual objects. We choose to bin the energy (and number of tracks) into uniform 64x64 bins,which correspond to the approximately 0.1x0.1 (η x ϕ) resolution of the ATLAS hadronic calorimeter, but also use larger 224x224 images for large-scale studies that could capture the granularity of the ATLAS electromagnetic calorimeter and future detectors.

2.2. Neural Net Architecture

We perform binary classification on the resulting images using a Convolutional Neural Net (CNN) [7] comprised of 3 convolution+pooling units (or 4 for larger images) with rectified linear unit (ReLU) activation functions [8]. These layers output into two fully connected layers which project into a two-dimensional vector on which a softmax function is applied to determine the

IOP Conf. Series: Journal of Physics: Conf. Series **1085** (2018) 042034 doi:10.1088/1742-6596/1085/4/042034

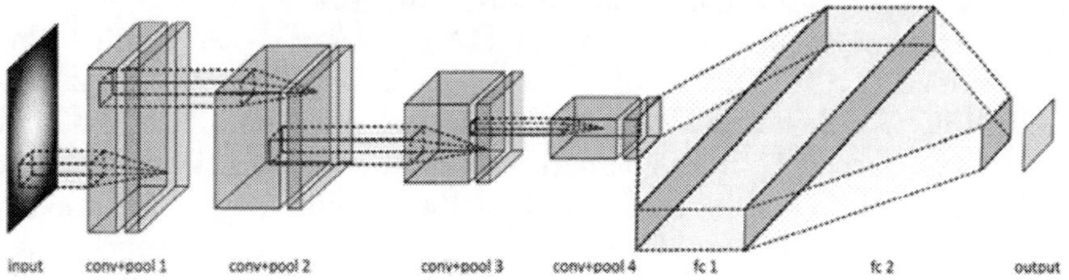

Figure 1: CNN architecture

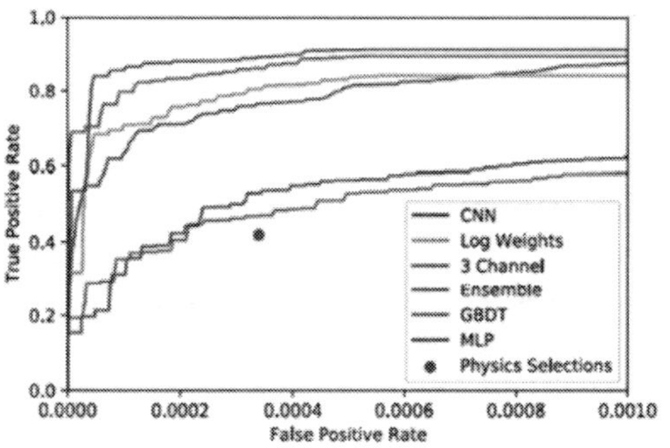

Figure 2: ROC curve comparison of different CNN implementations with physics selections and shallow classifiers

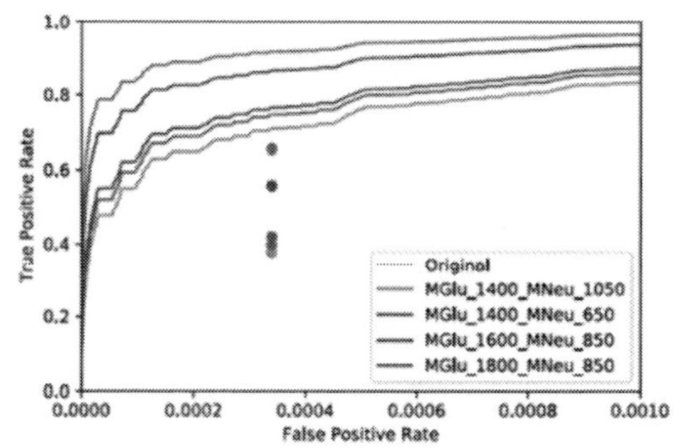

Figure 3: ROC curves for application of CNN to different signal mass samples

signal and background class probabilities. We use softmax with cross-entropy as the loss function and the ADAM optimizer [9]. The network employed is shown in figure 1.

2.3. Computational Resources
The CPU timing and scaling results reported here used the NERSC Cori system: a Cray XC40 with 2388 Intel (E5-2698 v3) Haswell and 9688 Intel XeonPhi 7250 Knight's Landing (KNL) compute nodes. Each KNL processor includes 68 1.4GHz cores with 4 HyperThreads and peak performance of 6 TeraFLOP/s (single-precision). Each Haswell node has dual 16-core 2.3 GHz processors with peak performance of 2.4 TeraFLOP/s. We compare performance to a standalone machine containing Titan X (Pascal) GPUs each capable of 10.2 TeraFLOP/s peak.

3. Results
3.1. CNN Performance
In order to obtain good significance to new physics it is necessary to combine high signal efficiency (True Positive Rate (TPR)) with very high background rejection (low False Positive Rate (FPR)). In figure 2 we show the ROC curve of TPR vs FPR for the CNN architecture described in 2.2 with a single channel for calorimeter energy (labeled 'CNN'). We compare to the physics selections described in 2.1. We achieve increased signal efficiency at same background rejection without using physics-based jet variables. We also compare the AMS (approximate median significance) [10] which gives a single measure accounting for initial pre-selection efficiency and relative signal/background cross-sections, and achieve a 1.8x improvement in AMS at the same background rejection as our implementation of the selections described in 2.1 and Appendix A.

IOP Conf. Series: Journal of Physics: Conf. Series **1085** (2018) 042034 doi:10.1088/1742-6596/1085/4/042034

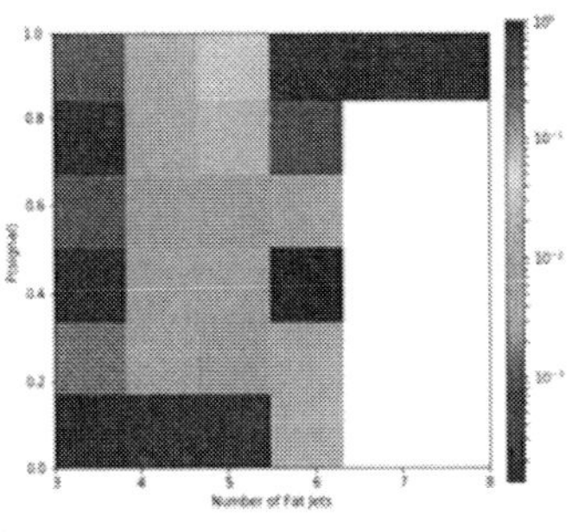

Figure 4: Comparison of CNN output with Number of Fat Jets

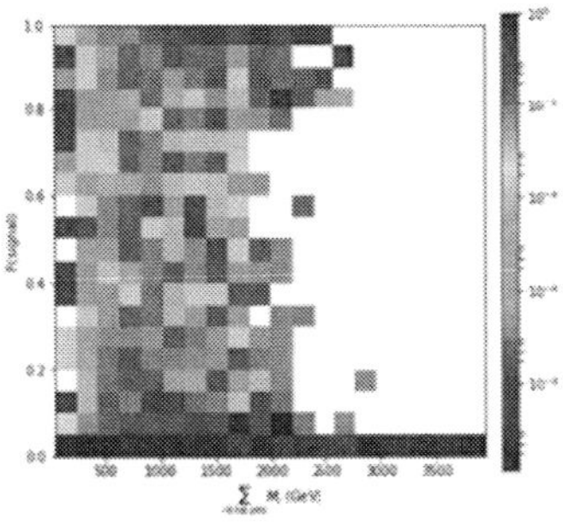

Figure 5: Comparison of CNN output with sum of jet mass

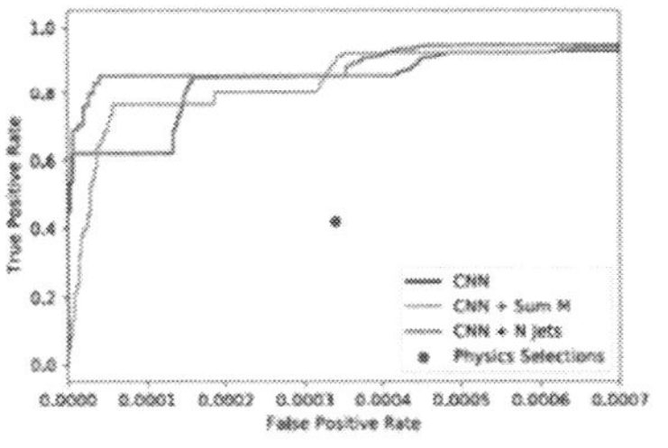

Figure 6: Change in performance on combining CNN output with jet variables

Comparison to shallow classifiers: We also compare to two shallow machine-learning approaches using high-level physics variables as inputs as detailed in Section 2.1 implemented with the scikit-learn package [11]. The first is a gradient boosted decision tree (GBDT) classifier with 100 estimators and a max depth of 3. The second is a single-hidden-layer fully-connected NN (MLP) with 100 hidden units. As shown in figure 2, these models outperform the traditional selections but under-perform relative to the CNN.

Further improving performance: In figure 2 we show further improvements to performance from modifications to our network or training. Firstly we alter the cross-section based weights applied to the training loss (described in section 2.2). The background cross-section weights are in some cases are seven orders of magnitude larger than that of the signal sample. This range of values causes some instability for training in some of the deep learning frameworks we use, so we also train the network with applying the natural logarithm of these normalized positive weights. This slightly improves the performance in the false-positive region of interest, though it performs worse than the simple cross-section for higher false-positive rates.

Next, we explore using three input channels in our CNN: namely to separate the energy deposited in the electromagnetic and hadronic calorimeters, and to add the number of tracks reconstructed in the same η/ϕ bin. Figure 2, shows the further performance improvements.

Finally a slight further improvement can be seen in figure 2 from creating an ensemble of the networks trained with original weights and the log weights (with 3 channels). This is done by taking the mean of the two network outputs as a prediction.

Robustness to different signal samples and pileup: The model described in preceding sections was trained on a specific cascade decay with gluino mass (MGlu) of 1400 GeV, and neutralino mass (MNeu) of 850 GeV. However, in figure 3 we show that applying this network to other signal samples with different input masses, without retraining, still offers good performance that exceeds that of the benchmark selections for each sample.

The above studies used simulated samples without a contribution from additional interactions per beam crossing (pile-up). We also generate samples using the default Delphes ATLAS detector configuration 'card' with added pile-up ($\mu = 20$). The physics selections benchmark has a higher false-positive rate for this sample (0.001) with a TPR of 0.6 at that point. The CNN still performs well with a TPR of around 0.9 at that FPR.

3.2. Relation between CNN and physics variables

In figures 4 and 5 we compare the predicted class probability of the CNN with two of the jet variables used in the physics analysis and a clear correlation can be seen. We also train a one-layer network combining the CNN output with each of these physics variables in turn.

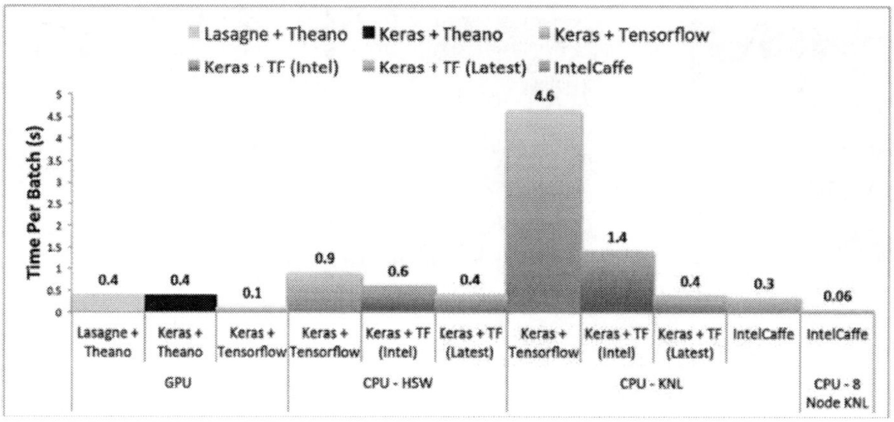

Figure 7: Comparison of time per batch (batch size=512) for different deep-learning frameworks on GPU, Haswell (HSW) and Knights Landing (KNL) CPU

The performance is shown in figure 6 and there is little or no further improvement within the statistics of the sample used, suggesting that the CNN is capturing much of the discriminating power of these jet variables, despite not being explicitly trained on them.

3.3. Training Time Performance on CPU and GPU architectures
In figure 7 we compare the training time for implementations of our network in different popular deep-learning frameworks and on different architectures. We report the time per batch of 512 examples excluding I/O. Not all implementations have been optimized so the aim is not to provide detailed comparisons of frameworks but rather to drive improved CPU performance. It can clearly be seen that the default Tensorflow (v1.2) [12] implementation performed very poorly on the KNL architecture. Several optmizations were recently introduced by Intel for Tensorflow on CPU [13] building on the Intel Math Kernel Libary (MKL) [14], such as multi-threaded convolutional layers, vectorizing over channels or filters and cache blocking. These offer substantial improvements as shown in the times labeled 'TF (Intel)' in figure 7. Further optimizations have been carried out in a collaboration between NERSC and Intel using the architecture and data from this paper, as well as other use-cases, resulting in the 'TF (Latest)' measurement in figure 7. These include MKL implementation of element-wise operations to avoid MKL to Eigen conversions. This version is currently available for use on NERSC systems and will be included into future releases of Tensorflow. Code, datasets and recipes are provided in Appendix B.

Performance improvements for CPU were also made in IntelCaffe [15]. Figure 7 shows this implementation also performs well. In addition a multi-node version of IntelCaffe using the Intel MLSL library [16] is available at NERSC. Figure 7 shows that using 8 nodes (with a constant batch size per node) one can train this network approximately 6x faster than a single node. With other collaborators, we have scaled this Caffe implementation on the same problem, but with large 224x224 images, to 9600 KNL nodes on Cori. That study is reported in detail in [17].

4. Conclusions
We have implemented deep convolutional networks on large whole-detector images directly for physics analysis. We find this offers improved sensitivity than physics-variable based selections, and shallow classifiers using those variables, without the need for jet reconstruction. Further improvements come, in particular, from adding multiple sub-detectors as channels. This network is robust to pileup and can be applied to other masses of the signal sample without retraining.

Furthermore, we have used this implementation and data to benchmark and improve popular deep learning libraries on CPU architectures including XeonPhi/KNL on the NERSC Cori supercomputer as well as to demonstrate distributed training up to 9600 KNL nodes.

References

[1] ATLAS Collaboration 2016 Search for massive supersymmetric particles in multi-jet final states produced in pp collisions at $\sqrt{s}$ = 13 TeV using the ATLAS detector at the LHC Tech. Rep. ATLAS-CONF-2016-057 CERN Geneva URL `http://cds.cern.ch/record/2206149`

[2] Sjöstrand T, Mrenna S and Skands P 2008 *Computer Physics Communications* **178** 852 – 867

[3] de Favereau J, Delaere C *et al.* 2014 *JHEP* **2014** 57

[4] Cacciari M, Salam G P and Soyez G 2012 *The European Physical Journal C* **72** 1896

[5] de Oliveira L, Kagan M, Mackey L, Nachman B and Schwartzman A 2016 *JHEP* **07** 069

[6] Komiske P T, Metodiev E M and Schwartz M D 2017 *JHEP* **01** 110

[7] LeCun Y, Bottou L, Bengio Y and Haffner P 1998 *Proceedings of the IEEE* **86** 2278–2324

[8] Nair V and Hinton G E 2010 *Proceedings of ICML-10* pp 807–814

[9] Kingma D P and Ba J 2014 *CoRR* **abs/1412.6980** URL `http://arxiv.org/abs/1412.6980`

[10] Adam-Bourdarios C, Cowan G, Germain C, Guyon I, Kegl B and Rousseau D 2014 URL `http://higgsml.lal.in2p3.fr/documentation`

[11] Pedregosa F *et al.* 2011 *Journal of Machine Learning Research* **12** 2825–2830

[12] 2017 TensorFlow version 1.2.0 `https://github.com/tensorflow/tensorflow/releases/tag/v1.2.0`

[13] 2017 TensorFlow Optimizations on Modern Intel Architecture `https://software.intel.com/en-us/articles/tensorflow-optimizations-on-modern-intel-architecture`

[14] 2017 Introducing DNN primitives in Intel Math Kernel Library `https://software.intel.com/en-us/articles/introducing-dnn-primitives-in-intelr-mkl`

[15] 2017 Intel distribution of Caffe* `https://github.com/intel/caffe`

[16] 2017 Intel Machine Learning Scaling Library for Linux OS `https://github.com/01org/MLSL`

[17] Kurth T *et al.* 2017 *Accepted for SC17, arXiv:1708.05256* URL `http://arxiv.org/abs/1708.05256`

Acknowledgements

We thank Ben Nachman and Brian Amadio for many valuable discussions and physics input, and Mustafa Mustafa and the Intel Tensorflow team for CPU optimizations. This research used resources of the National Energy Research Scientific Computing Center (NERSC), a DOE Office of Science User Facility supported by the Office of Science of the U.S. Department of Energy under Contract No. DE-AC02-05CH11231.

Appendix A. Benchmark analysis and pre-selection

The selections applied for the benchmark analysis were (following [1]):

Fat-jet object selection:
AntiKt R=1.0 trimmed ($R_{trim} = 0.2$, $P_{Tfrac} = 0.05$)
$P_T > 200$ GeV , $|\eta| < 2.0$
Preselection:
Leading Fat-jet $P_T > 440GeV$
$N_{Fat-Jet} > 2$
Analysis Selection:
$|\Delta\eta_{12}|$ between leading two Fat-jets < 1.4
($N_{Fat-Jet} >= 4$ AND $\Sigma M_{Fat-jet} > 800$ GeV)
OR ($N_{Fat-Jet} >= 5$ AND $\Sigma M_{Fat-jet} > 600$ GeV)

Appendix B. Datasets, Code and Recipes

Example code for implementing the networks used in this study, together with datasets and recipes for running at NERSC are provided at `http://www.nersc.gov/users/data-analytics/data-analytics-2/deep-learning/deep-networks-for-hep/`

IOP Conf. Series: Journal of Physics: Conf. Series **1085** (2018) 042035 doi:10.1088/1742-6596/1085/4/042035

Striped Data Server for Scalable Parallel Data Analysis

Jin Chang, Oliver Gutsche, Igor Mandrichenko, James Pivarski

ivm@fnal.gov

Abstract. A columnar data representation is known to be an efficient way for data storage, specifically in cases when the analysis is often done based only on a small fragment of the available data structures. A data representation like Apache Parquet is a step forward from a columnar representation, which splits data horizontally to allow for easy parallelization of data analysis. Based on the general idea of columnar data storage, working on the FNAL LDRD Project FNAL-LDRD-2016-032, we have developed a striped data representation, which, we believe, is better suited to the needs of High Energy Physics data analysis. A traditional columnar approach allows for efficient data analysis of complex structures. While keeping all the benefits of columnar data representations, the striped mechanism goes further by enabling easy parallelization of computations without requiring special hardware. We will present an implementation and some performance characteristics of such a data representation mechanism using a distributed no-SQL database or a local file system, unified under the same API and data representation model. The representation is efficient and at the same time simple so that it allows for a common data model and APIs for wide range of underlying storage mechanisms such as distributed no-SQL databases and local file systems. Striped storage adopts Numpy arrays as its basic data representation format, which makes it easy and efficient to use in Python applications. The Striped Data Server is a web service, which allows to hide the server implementation details from the end user, easily exposes data to WAN users, and allows to utilize well known and developed data caching solutions to further increase data access efficiency. We are considering the Striped Data Server as the core of an enterprise scale data analysis platform for High Energy Physics and similar areas of data processing. We have been testing this architecture with a 2TB dataset from a CMS dark matter search and plan to expand it to multiple 100 TB or even PB scale. We will present the striped format, Striped Data Server architecture and performance test results.

1. Introduction

High Energy Physics data analysis is an iterative process of distilling large amounts of data to extract physical information and presenting it in a most optimal way. Typically, the physicist repeats the process of refining the data filtering, data compilation and information representation many times, and reducing the time of individual iteration reduces the time of the overall analysis process and allows the physicist to deliver the results sooner.

Currently, HEP data is stored in files and the analysis is essentially a repeating process of running the analysis software over a large set of files. In order to reduce the iteration time, the physicists reduce the initial dataset (set of files) to a smaller set by using 2 methods:

- Skimming is a process of pre-filtering potentially interesting events so that there are fewer events to process during each run.

IOP Conf. Series: Journal of Physics: Conf. Series **1085** (2018) 042035 doi:10.1088/1742-6596/1085/4/042035

- Slimming is removing the elements of data structure which are not relevant for the particular analysis the physicist is performing. Typical dataset used for analysis before slimming contains ~1000 columns, whereas typical analysis uses only one or two dozens of these columns.

Once the dataset is reduced, the physicist can run their analysis code over smaller amounts of data in a more efficient way, but data reduction contributes to the overall analysis process and slows it down. In addition, because each physicist or group of physicists have their own way to reduce data, they all tend to replicate data multiple times between all the reduced datasets they create over time.

What we are proposing is a fundamentally different way of storing data (see Fig. 1), which would support a much more efficient and fast data analysis process. We propose to store analysis data in a database with efficient and scalable access and largely eliminate the need to reduce and duplicate data by providing direct and immediate access to the needed data and only to the needed data.

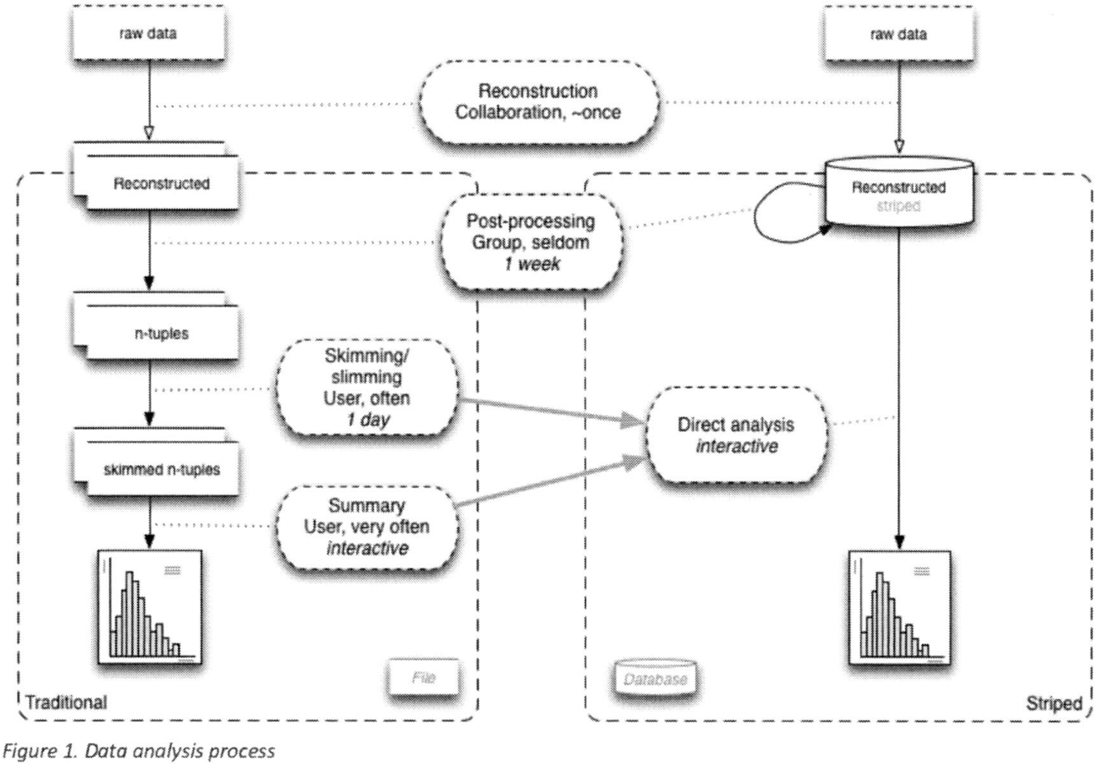

Figure 1. Data analysis process

In addition, we propose to move from the event loop style of analysis to vector-based calculations, which, when combined with functional and/or declarative algorithm representation, can be moved easily from CPU to GPU or other SIMD processing platforms as the underlying computing fabric.

2. Striped Data Representation

High Energy event data structures can be viewed as a list of uniform tuples, some elements of which can in turn be lists of uniform tuples, recursively. We developed a format of representing these kind of data structures, based on a columnar approach, suitable to store data in a non-relational database or a BLOB storage. As in columnar representations, the data is represented as a set of columns, each column is essentially a vector of values for a single data attribute for the whole dataset (see Fig. 2). Then we cut columns into segments called "stripes". Each stripe contains data from a set of HEP events, typically for 1K to 10K or more events. All columns are split along the same event boundaries. For example, if we have columns A and B, we split them in such a way that if we have a stripe for column A with data for events ranging from 112000 to 112999, then there will be a stripe for column

IOP Conf. Series: Journal of Physics: Conf. Series **1085** (2018) 042035 doi:10.1088/1742-6596/1085/4/042035

B with data for exactly the same events 112000 to 112999 in the same order. These event ranges are called Event Groups or Row Groups.

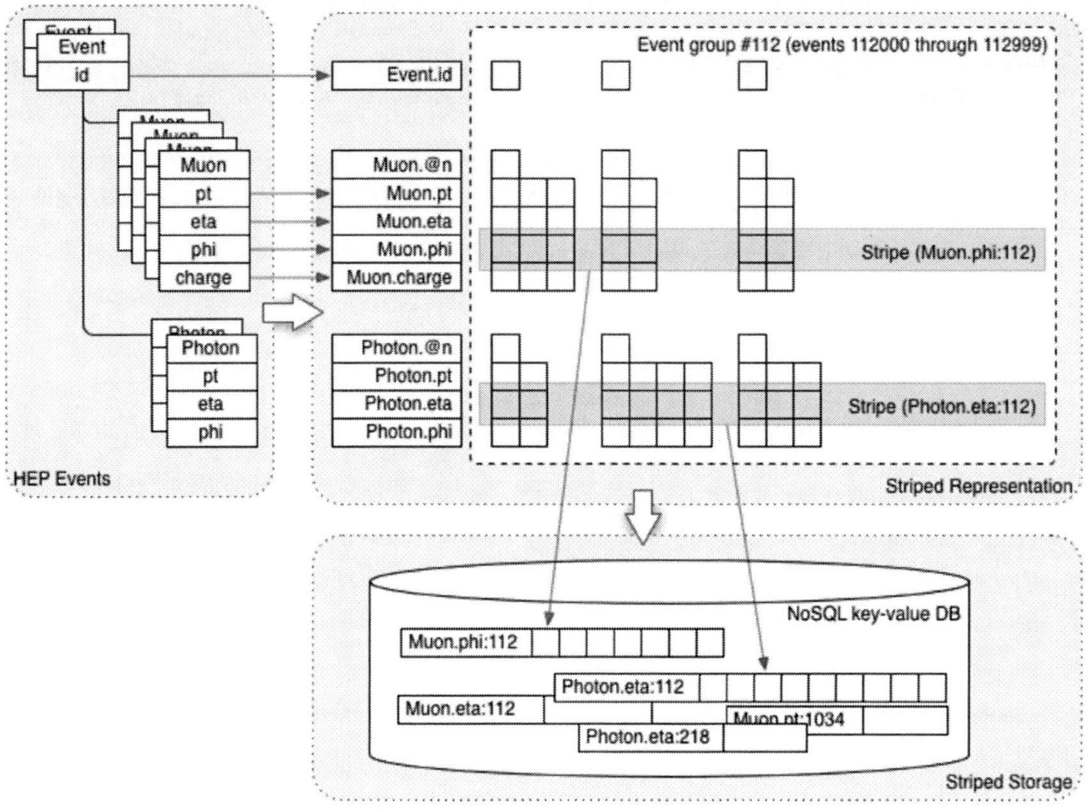

Figure 2. Striped Data Representation

This format of data representation works very well with no-SQL, key-BLOB data storage systems. Each stripe is stored as a BLOB, indexed by the combination of the dataset name, the column name and the row group number. The BLOB contents is in fact a numpy[1] array immediately importable into numpy without any data conversion, which makes the use of such storage very efficient.

Here are the advantages of using strd data representation for HEP data analysis:

1. We can use almost any storage capable of storing key-BLOB pairs. Those include a wide range of no-SQL databases. We can also use relational databases, or even a UNIX file system to store this kind of data and the details of the actual data representation can be hidden behind a uniform data access interface. Of course, each particular type of underlying storage has its own performance and other characteristics, and not all of them will be practically useful, but functionally any key-BLOB storage can support the striped format.
2. Columnar representation allows the user to retrieve only columns they need for their analysis and never touch unused columns.
3. Because data is represented in a numpy-friendly way, significant parts of calculations can be performed over stripes as vector operations instead of operations over individual event properties, thus moving large portion of calculations, if not all of them, from the event loop to vector calculations (see Fig. 3). On the other hand, we are not proposing to eliminate the event loop. The idea is to move as much as possible from event loop to vector calculations.

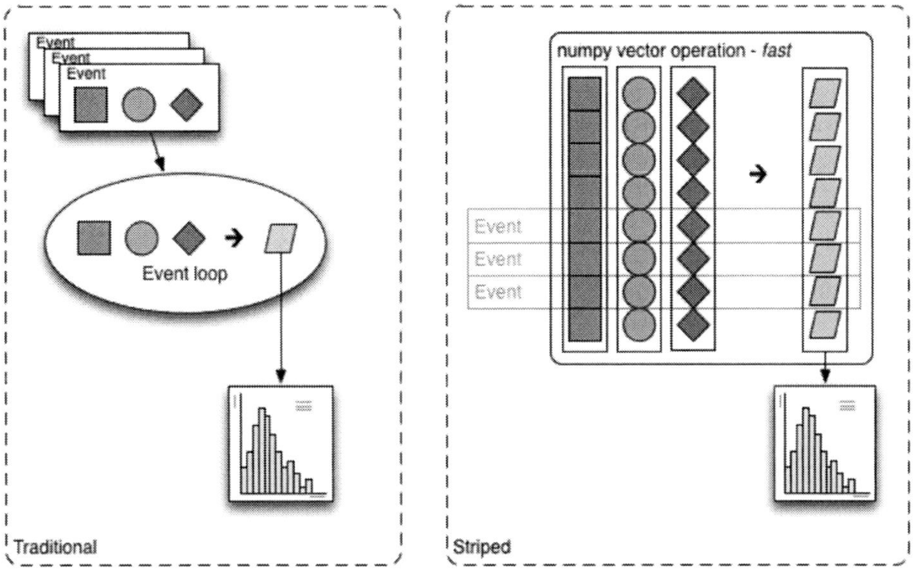

Figure 3. Moving calculations from event loop to vectors

4. Because stripes are immutable (except for rare cases when the entire dataset needs to be updated in the database, e.g. to correct an error), web services and data caching techniques can be leveraged to significantly increase efficiency of the data management and delivery and reduce development and maintenance cost of the system. This way we can take advantage of the fact that typical data analysis reuses the same data repeatedly multiple times.
5. Using web services as the data delivery mechanism makes it easy to leverage Internet and cloud technologies and provides great deal of flexibility in building distributed data centers, helping members of large international collaborations contribute their resources to the collaboration.

3. Implementation

We implemented a test instance of the striped storage system and a small analysis computing cluster at FNAL (see Fig. 4)

IOP Conf. Series: Journal of Physics: Conf. Series **1085** (2018) 042035 doi:10.1088/1742-6596/1085/4/042035

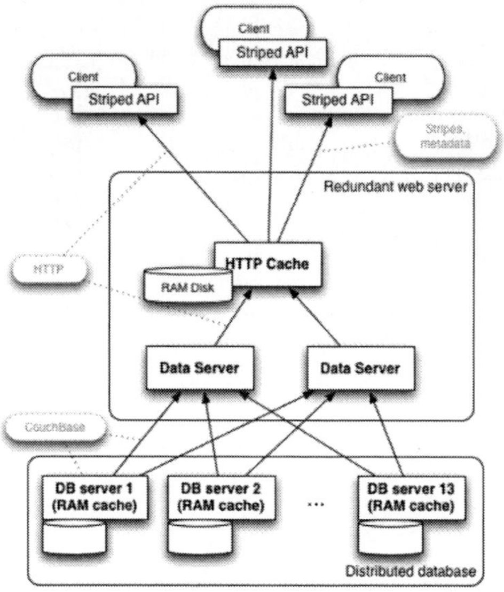

Figure 4. Demo instance of the striped storage

The data storage part was implemented using 13 servers. Each data server was a 8-core computer with 0.9TB data disk. We chose CouchBase[2] as the database software to store data. CouchBase is a rather simple, distributed no-SQL database with memory cache. On top of the CouchBase cluster, we put 2 redundant web servers and then a web cache computer with 20GB RAM-based cache. We used nginx[3] as the web server engine.

We loaded about 2TB, 0.5 billion events, about 700 columns of CMS[4] dark matter search n-tuples data into the system.

The computing portion was implemented as a set of independent worker processes in Python, running on 2 16-core computers, so we had 30 workers in total. The data was uniformly distributed among all available workers. Each worker has its own data cache. The algorithm of data distribution among workers always sends the same data to the same worker to make sure they can reuse their local cache. In total, we had 3 layers of RAM cache in the system – the CouchBase own memory cache, web server cache and then the worker local cache.

The analysis platform is also Python based. Currently, it is written in Python 2.7. The user front end is implemented as a iPython/Jupyter notebook with plotting using Matplotlib. The user notebook communicates with the workers via TCP sockets, so the user notebook can run anywhere, as long as it can reach the workers.

To measure performance of the demo system we built, we ran simple di-muon invariant mass analysis, building the distribution of Z-boson mass. On this our cluster, we managed to achieve the performance of about 1 million events per second using only 30 worker processes.

To demonstrate the flexibility of the system architecture, we created a Docker container with the worker code, and ran the analysis successfully running the workers on the iMac desktop, the analysis 2-node cluster described above and the AWS cloud. All the workers were accessing the data via cached web server located at FNAL.

IOP Conf. Series: Journal of Physics: Conf. Series **1085** (2018) 042035 doi:10.1088/1742-6596/1085/4/042035

4. Conclusion

We have developed a format to represent HEP data in a way suitable to be stored in a non-relational database. The data representation format and the database storage mechanism we developed allow to store and analyse data in much more efficient way it is done currently, by eliminating the need to slim and duplicate data. Also storing data in a database with scalable and efficient interfaces allow for scalable and massively parallel data analysis, which largely eliminates the need to skim data. The architecture we propose is highly flexible and does not require co-location of data and computing components. Wide use of Internet and web services technologies allows to deploy the storage as well as the computational components on the cloud. In addition, it allows to have the computing fabric to be distributed over WAN, heterogeneous and dynamic in terms of individual compute node availability. This allows a science collaboration to build distributed computing and storage facilities and have individual members of the collaboration flexibly contribute their resources to the collaboration.

The proposed computational model moves bulk of the computations from the event loop to the vector-based paradigm, which allows to move to functional and/or declarative programming models, which, in turn, allow to move computations easily from CPU to GPU and other SIMD platforms.

Utilization of data caching techniques significantly increases the efficiency of the analysis process, taking advantage of the repeating nature of the HEP analysis process.

We built a demo HEP data analysis platform based on iPython/Jupyter as the user interface.

We were able to build a small 30 cores demo analysis cluster and test its performance on a real 2TB dataset from CMS and achieved a performance of 1 million events per second with all the components written in Python.

The proposed approach has a great potential of replacing existing file-based data analysis models and can significantly increase the efficiency of the analysis process, reduce time to insight, add flexibility in building storage and analysis facilities, and open new areas of HEP research, currently not possible due to a relatively slow analysis process.

5. References

[1] S. van der Walt, S. C. Colbert, G. Varoquaux, The NumPy Array: A Structure for Efficient Numerical Computation, Computing in Science Engineering, 2011, Vol. 3

[2] http://couchbase.com

[3] http://nginx.org

[4] Chatrchyan, S. et. al., The CMS experiment at the CERN LHC, JINST, 2008, Vol. 3

IOP Conf. Series: Journal of Physics: Conf. Series **1085** (2018) 042036 doi:10.1088/1742-6596/1085/4/042036

Machine Learning approach to γ /π^0 separation in the LHCb calorimeter

Viktoria Chekalina and Fedor Ratnikov [1]

National Research University Higher School of Economics,
20 Myasnitskaya Str., Moscow, Russian Federation
Yandex School of Data Analysis,
16 Lva Tolstogo Str., Moscow, Russian Federation

E-mail: vicusya22@gmail.com

Abstract. Reconstruction and identification of particles in calorimeters of modern High Energy Physics experiments is a complicated task. Solutions are usually driven by a priori knowledge about expected properties of reconstructed objects. Such an approach is also used to distinguish single photons in the electromagnetic calorimeter of the LHCb detector at the LHC from overlapping photons produced from decays of high momentum π^0 . We studied an alternative solution based on first principles. This approach applies neural networks and classifier based on gradient boosting method to primary calorimeter information, that is energies collected in individual cells of the energy cluster. Mutial application of this methods allows to improve separation performance based on Monte Carlo data analysis. Receiver operating characteristic score of classifier increases from 0.81 to 0.95, that means reducing primary photons fake rate by factor of two or more.

1. Introduction

Reconstruction and identification in calorimeters of modern High Energy Physics experiments is a complicated task. The particular problem is to separate prompt photons produced in collisions and photons produced by π^0 which decay into two photons. The most confusing case is when two photons from the π^0 decay have so close directions, that they appear similar to a single cluster produced by prompt photons. Direct solutions are usually driven by a priori knowledge about expected properties of reconstructed objects. Currently used baseline solution of this problem for data collected by the LHCb experiment [1, 2] relies on variables characterising shape of the cluster produced in the electromagnetic calorimeter. In this paper we describe an exercise to separate those two signals without prior knowledge of particular discriminating features of the analysed cluster but rather train algorithm on plain energy deposits in the cluster cells around the cluster seed. This is an attempt to test if modern Machine Learning algorithms may be applied to low level experimental data. We call it *ab initio* approach.

2. Baseline Approach

LHCb Electromagnetic Calorimeter system [1] consists of Scintillating Pad Detector, Preshower Detector (PS), and shashlik-type electromagnetic calorimeter *per se* (ECAL). Energies of PS

[1] On behalf of the LHCb Collaboration

and ECAL cells in vicinity of the high energy "seed" ECAL cell are used to discriminate merged π^0 from prompt photons.

It is assumed that the main discriminating feature separating photon and merged π^0 clusters is a cluster shape. Merged π^0 clusters are expected to be elongated and asymmetric due to residual offset between two photons, while genuine photon clusters are expected to be more symmetrical.

Considering a cluster of N cells, with (x_c, y_c) to be coordinates of the cluster center of gravity, e_i to be energy of the ith cell, and (x_i, y_i) to be the cell coordinates. Then the matrix describing second moments of the cluster shape is defined as :

$$S_{xx} = \frac{\sum_{i=1}^{N} e_i(x_i - x_c)^2}{\sum_{i=1}^{N} e_i}, \; S_{yy} = \frac{\sum_{i=1}^{N} e_i(x_i - x_c)^2}{\sum_{i=1}^{N} e_i}, \; S_{xy} = S_{yx} = \frac{\sum_{i=1}^{N} e_i(x_i - x_c)(y_i - y_c)}{\sum_{i=1}^{N} e_i}$$

Features which are used for the baseline separation approach are defined as follows :

- $r_2 = \langle r \rangle = S_{xx} + S_{yy}$
- $r_2, r_4 = 1 - \frac{\langle r^2 \rangle^2}{\langle r^4 \rangle}$
- $k = \sqrt{(1 - 4\frac{S_{xx}S_{yy} - S_{xy}^2}{(S_{xx} + S_{yy})^2}}$
- $asym = \frac{S_{xy}}{\sqrt{S_{xx}S_{yy}}}$
- $\frac{E_{seed}}{E_{cl}}$
- $\frac{E_{seed} + E_{snd}}{E_{cl}}$

A 2-layer perceptron was applied to separate photon and π^0 clusters by these features built using energies in 3×3 ECAL cells area around the cluster seed . Full description of the baseline approach may be found in [3].

The training photon sample was done by K*gamma photons and set of π^0 from B decays. The performance of the baseline approach on the reference data described below is presented as dashed line in Fig. 1.

3. Ab initio approach

To separate photons and merged π^0 we consider energies in 5×5 ECAL and PS cells around the cell seed. However we do not build any sophisticated features based on physics considerations, but rather consider plain values of energy allocated in every cell of the 5×5 matrix in both ECAL and PS as a feature. These are 50 plain features to be used by the classifier.

We start with using 2-layer Neural Network which has 2 parallel branches of 2 full-connected layers for preprocessesing information from electromagnetic calorimeter and preshower. For decision tree approach we use XGBoost classifier with 6000 trees. To prevent overfitting, the tree's depth is restricted by 3.

Photons and π^0 used for training are produced in $B^0 \to K\pi\gamma$ and $B^0 \to K\pi\pi^0$ processes respectively, and cluster information is extracted from corresponding MC samples. To prove stability of obtained numbers we also use another π^0 sample: $B^0 \to J\psi K^\star$ with $K^\star \to K\pi^0$. For this study we only consider high energy clusters with transversal energy $E_T > 2\ GeV$. We also selected potentially merged π^0 by requiring for true MC photons from π^0 decay to be separated by not more than two calorimeter cells apart on the face of the ECAL.

The procedure of selecting and tuning architecture of the classifier is described in Sec. 4. To find enough number of trained epoch for neural network we plot train-test score curve for every step and noticed when it came out on the plateau. BDT-based classifiers demonstrate better performance for our problem, so we use the XGBoost [4] tool to train and test classifier. Fig. 1 demonstrates performance of the new approach in comparison with the baseline one. The score under Receiver operating characteristic curve improves from 0.89 for the baseline up to 0.97 in new approach. Considering 98% photon efficiency, new approach reduces fake rate from about 60% to about 30%.

IOP Conf. Series: Journal of Physics: Conf. Series **1085** (2018) 042036 doi:10.1088/1742-6596/1085/4/042036

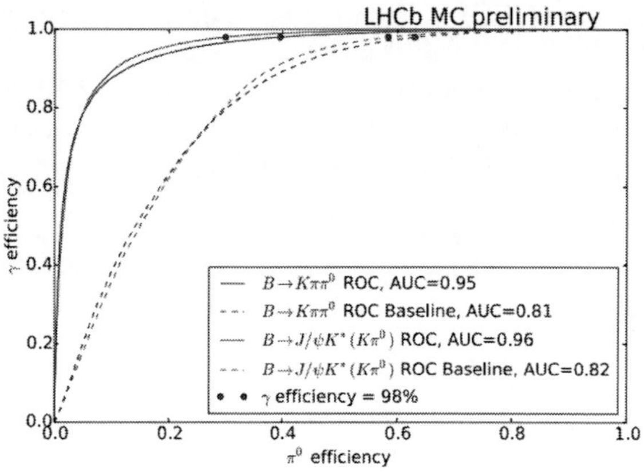

Figure 1. Classificator ROC curves for the baseline (dashed line) and new method (solid line). Different colours refer to different test samples.

Flatness is an important characteristics of the classifier, as it directly affects systematic uncertainties for physics analyses. Fig.2 demonstrates that new approach also have very good flatness in E_T.

3.1. Edge effects

LHCb electromagnetic calorimeter consists of three regions with cell sizes of 4×4, 6×6 and 12×12 cm^2 for inner, middle and outer calorimeter regions respectively. Regions are rectangular, thus the most of the calorimeter area contains the regular grid, and 5×5 clusters can be easily selected. However, every region still has edges which breaks regular structure. For two-sided borders we build complete cluster by upscaling or downscaling energy responses in the adjacent region. Clusters adjacent to innermost and outermost calorimeter remain incomplete. Generally, we have 3 different type of objects: full, reconstructed and incomplete. As properties of diversed cluster's type are expected to be different, a dedicated classifiers for border objects, reconstructed and incomplete, are trained. The full classifier discussed in the previous section, thus consists of 9 individually trained classifiers corresponding to three calorimeter regions, serving regular area, inner and outer edges for each region.

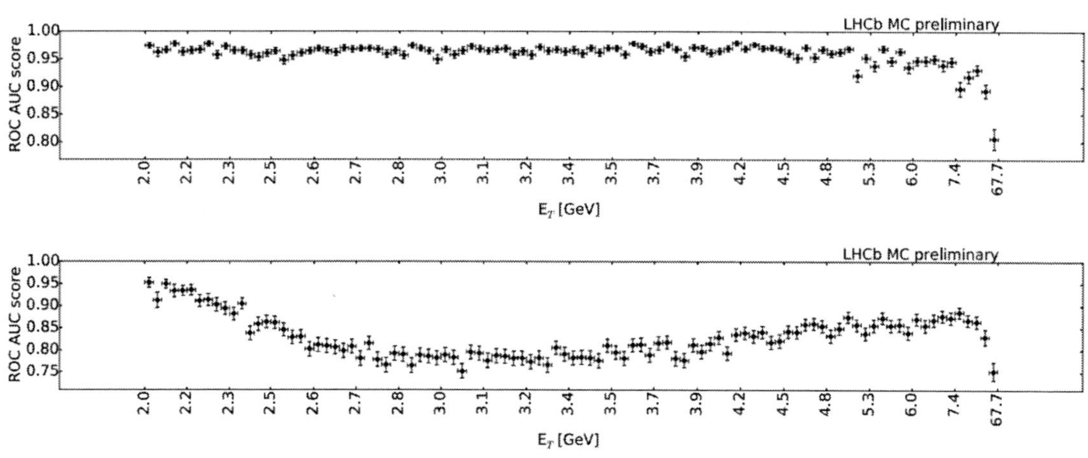

Figure 2. ROC AUC scores in different E_T intervals for baseline (bottom) and new (top) approaches.

IOP Conf. Series: Journal of Physics: Conf. Series **1085** (2018) 042036 doi:10.1088/1742-6596/1085/4/042036

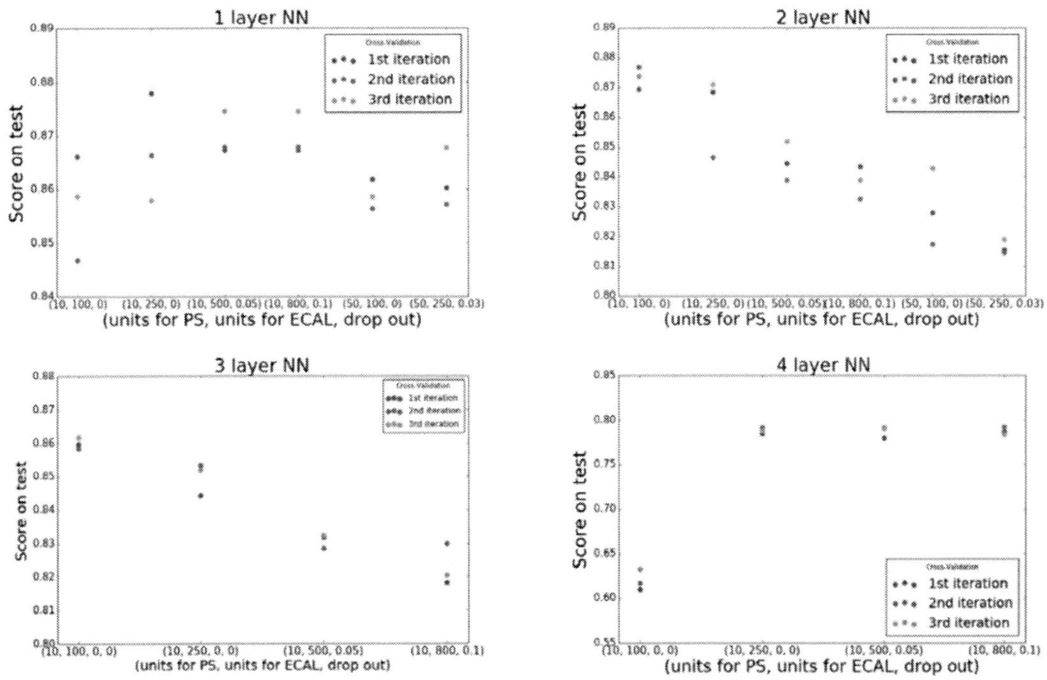

Figure 3. Cross Validation for network configurations with 1, 2, 3, 4 hidden layers. Quality for different iterations is presented as a function of (units for PS branch, units for ECAL branch, dropout).

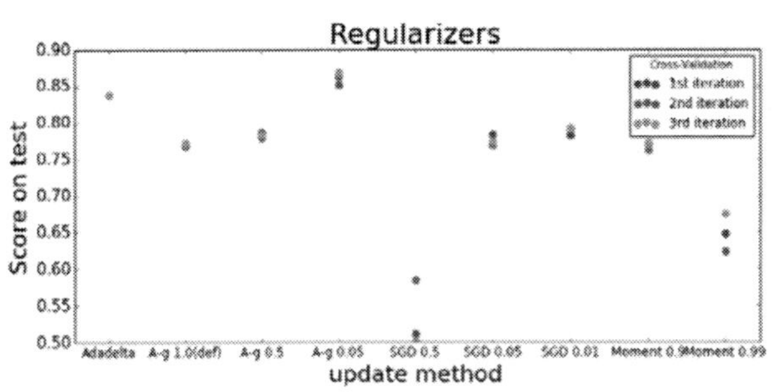

Figure 4. NN classifier quality for various kinds of update functions for 2-layer NN with (units for PS branch, units for ECAL branch, dropout)=(10, 100, 0)

4. Optimisation of the classifier architecture

To select the best classifier architecture we first have to compare Neural Network (NN) and Decision Tree (BDT) approaches. Regular area of the inner calorimeter is used as a reference for quantitive comparison of different classifiers

We start with optimising the possible NN architecture. As behaviours of PS and ECAL detectors are different, and ECAL is stacked after PS concerning the direction of the incoming particles, we build network consisting of two parallel branches of two full-connected layers separately preprocessesing information from ECAL and PS. We found that PS information is not as informative as ECAL. This could easily lead to overtraining. However we also found that using PS branch with smaller number of units could improve quality of the classifier.

When we consider classifiers structure, there are many hyper-parameters which should be optimised to get better quality. For neural networks, besides network topology hyper-parameters include: number of layers and number of units in each layer, method of regularisation, activation

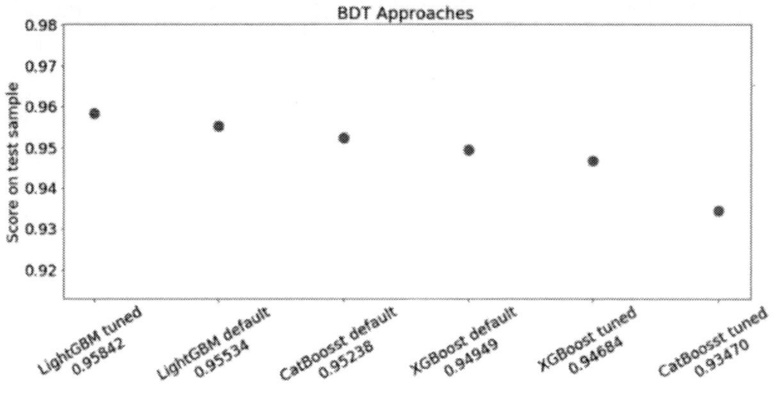

Figure 5. BDT classifiers quality for different BDT implementations with default and tuned set of hyper-parameters.

function, *etc.* Assuming that NN topology described above is fixed, we tune other hyper-parameters to get the best quality. We use 3-fold cross-validation and train network during 500 epoch. The later is about where the plateau for quality improvement is obtained.

Fig. 3 presents cross validation results for different NN configurations with 1-4 hidden layers. Despite of dropout, quality for 3 and 4 layers configuration degrades, that we explain by not too complicated structure of input features.

Results for final variating classifier optimisation method to get the best quality for the given number of steps is presented in Fig. 4.

For the BDT approach, we assayed XGBoost, CatBoost and LightGBM classifiers. For each classifier we found best configuration through tuning boosting parameters by ModelGym [5]. Fig. 5 presents results for both default or tuned configurations. The obtained quality for different classifiers is found to be very similar and significantly exceeds quality obtained with the best NN configuration, so we use default XGBoost (estimators=2000, learning rate=0.05, max depth = 5, min child weight = 2) to build new, ab initio classifier which was used in Sec. 3.

5. Conclusions

We introduced a new, ab initio method for γ/π^0 separation in calorimeter LHCb. It uses calorimeter cell energies in 5×5 area around cluster seed cell. Several neural network configurations and gradient boosting methods are tested, and XGBoost is used to evaluate results. After test on Monte Carlo samples, the method demonstrates an ability to significantly improve separation quality, and has no significant energy dependence.

Acknowledgements

We thank Alexander P. Ezhov, Andrey Ustyuzhanin, Miriam Calvo Gomez, and Max Chefdeville for fruitful discussions, suggestions and comments on the subject of this study.

References

[1] LHCb collaboration, A. A. Alves Jr.*et al.*, *The LHCb detector at the LHC*, JINST **3** (2008) S08005.
[2] LHCb collaboration, R. Aaij *et al.*, *LHCb detector performance*, Int. J. Mod. Phys. **A30** (2015) 1530022, arXiv:1412.6352.
[3] M. Calvo, E. Cogneras, O. Deschamps, M. Hoballah, *A tool for* γ π^0 *separation at high energies*, LHCb-PUB-2015-016.
[4] XGBoost: A Scalable Tree Boosting System. `arXiv:1603.02754 [cs.LG]`, March 2016
[5] Model Gym, (2017), GitHub repository, https://github.com/yandexdataschool/modelgym

IOP Conf. Series: Journal of Physics: Conf. Series **1085** (2018) 042037 doi:10.1088/1742-6596/1085/4/042037

Deep Learning for Signal and Background Discrimination in Liquid based Neutrino Experiment

Rui Li[a,1], Zhengyun You[a,2], Yumei Zhang[b,3]

[a] School of Physics, Sun Yat-sen University, Guangzhou 510275, China

[b] Sino-French Institute of Nuclear Engineering and Technology, Sun Yat-sen University, Guangzhou 510275, China

Email: [1]lirui47@mail2.sysu.edu.cn, [2]youzhy5@mail.sysu.edu.cn, [3]zhangym26@mail.sysu.edu.cn

Abstract. In high energy physics experiments, efficient data analysis tools are required to extract interesting information from the massive data. For large-scale liquid-based neutrino experiments, neutrino signals are usually overwhelmed in huge backgrounds. By constructing a liquid neutrino detector toy model, we generate simulation data in Geant4 [1] and run reconstruction for signal background discrimination. The low-level Photo Multipliers (PMT) hits are also projected to a 2D plane to create visualization outputs for classification. With the 2D images as input, we use the Convolutional Neural Network (CNN) as a specific application, which has shown remarkable performance in signal and background discrimination and outperforms those with high-level reconstruction outputs. The method is expected to be used in the neutrino experiments such as JUNO with further study.

1. Introduction

In high-energy physics (HEP) experiments, a huge number of events are recorded in the detectors in every second. The signals are usually overwhelmed by huge backgrounds. To search for the interesting physics in the huge data, it is essential to discriminate the signal events from backgrounds with some characteristic features being extracted. Neutrino experiments are good examples in HEP experiments where signal and background discrimination play an important role. Due to the extremely low cross section of neutrinos' interactions with matters, neutrino experiments are known for their low signal rates and high backgrounds, which include the cosmic ray muons, radioactive backgrounds from detector materials and environments, as well as accidental backgrounds, such as two independent signal events coinciding within the same event. To increase the statistics of signal events, neutrino experiments are usually built with large-scale liquid detectors, such as the Super-Kamiokande [2], DayaBay [3] and JUNO [4] experiments.

The traditional algorithms in HEP experiments often require extracting features from raw event data so that we can get better performance in event classification and physics analysis. For example, some critical physical quantities, such as the transverse momentum, secondary vertex and energy of the tracks in an event will be reconstructed. However, even if physicists spend a great deal of time on manually constructing the features, the result is still not promising because these algorithms are not good at solving

* Supported by National Natural Science Foundation of China (11405279, 11675275), the Strategic Priority Research Program of Chinese Academy of Sciences (XDA10010900) and the Recruitment Program of Global Experts.

non-linear problems. More or less, some information in the raw event data will be lost after data reconstruction.

Deep learning is a newly flourishing machine learning algorithm which shows superior performance on solving non-linear problems [5]. Much efforts have been taken in various kinds of HEP experiments and software developments [6, 7]. In all these applications, CNN is a powerful tool in image processing and recognition, which makes it very suitable for event classification. Some implementations in HEP experiments have been realized in the recent years [8, 9].

In this paper, we focus on signal and background discrimination in the large-scale liquid-based neutrino experiments and application of CNN on the image-based event output for classification. In section 2, we introduce Neural Network and CNN in deep learning. In section 3, the implementation of a toy liquid neutrino detector model is introduced, as well as the production of simulation data and classification with reconstruction outputs. In section 4, we discuss the 2D imaging of the PMT hits distribution in event outputs and application of CNN with the image input for signals and radioactive backgrounds discrimination. The classification results are compared with the results from the traditional Multilayer Perceptron (MLP) method using reconstruction data. Finally, in section 5, we make a summary and discuss the potential application of the method in future neutrino experiment.

2. Deep Learning

2.1. Neural Network

Neural network is a machine learning algorithm intellectively inspired by human brain. MLP [10, 11] is one of the traditional neural networks. In MLP, the basic element is a perceptron, which does exactly what a biological neuron do - turn to a specific state due to the input. For each input variable, the perceptron has a corresponding weight. Each perceptron has its own activation function and bias. Say the bias of the perceptron is b and the activation is the sigmoid function, a typical output of the perceptron with three inputs is $\left[e^{(w_1x_1+w_2x_2+w_3x_3+b)}\right]^{-1}$.

Neural network is the combination of connected perceptrons. As shown in the left plot of Figure 1, perceptrons are fully connected layer by layer. The left layer in purple represents the input layer, the number of the perceptron is decided by the number of inputs. Layers in blue represent the hidden layers. The total amount of hidden layers and perceptron can be adjusted. While training the neural network, different combinations need to be tried to find out which one outperforms the others. The green layer represents the output layer, which gives prediction of the neural network.

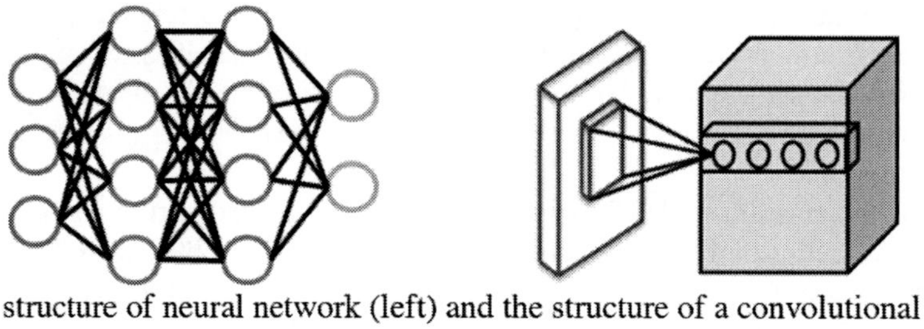

Figure 1. Basic structure of neural network (left) and the structure of a convolutional layer (right).

2.2. Convolution Neural Network

Convolution neural network [12] is inspired by the biological process in visual cortex. Instead of processing the information pixel by pixel, visual cortex does it in a part by part manner. It is widely used in the field of computer vision and has shown extraordinary performance.

In a convolutional layer, the neurons are no longer fully connected to its previous layer. There are a set of learnable filters in the convolution layer. During the forward process, the filter slides across the input and computes the dot product between the weights of filter and the input. When we slide the filter over the whole picture, we get something like a feature map of the original input. The basic structure of a convolutional layer is shown in the right plot of Figure 1.

3. Implementation in a Neutrino Detector Model

3.1. Detector Modelling

Neutrino experiments usually use a large amount of materials as the detector. In measurement of the neutrinos/antineutrinos from accelerators or reactors in nuclear power plants, inverse beta decay (IBD) $\overline{\nu_e} + p \to e^+ + n$ is one the most commonly used reaction for neutrino detection. The detecting materials are usually liquids, such as pure water (Super-Kamiokande, DayaBay), or liquid scintillator (KamLAND, JUNO). To detect the single photons from Cherenkov light or scintillation light, photon-multiplier (PMT) are widely used in neutrino experiments.

A toy neutrino detector model has been designed to simulate the large-scale liquid-based neutrino experiments. In this toy model, the detector is a spherical container with the diameter of 30 meters and is filled with liquid scintillator (linear alkylbenzene) for neutrino detection. The surface of the sphere is covered by 20-inch PMTs to measure the scintillation light from the liquid in the container. The number of the PMTs is about 10,000. A sketch of the detector model is shown in Figure 2. (Please note that the size of the PMTs and the size of container are not of the same scale but just for illustration.)

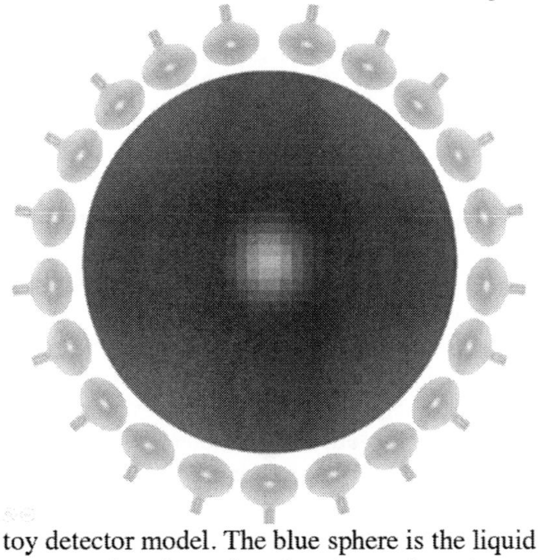

Figure 2. Sketch of the toy detector model. The blue sphere is the liquid container with grey PMTs arranged on its surface for photon detection.

3.2. Signals and Backgrounds

In this study, we focus on the discrimination of IBD signals and radioactive backgrounds. The signals are from neutrinos' IBD interaction with two final particles, a positron and a neutron. They both deposit energy in the liquid scintillator and produce optical photons, which will be received by the PMTs to reconstruct the vertex and energy of the event.

Due to the low signal counting rates in most neutrino experiments, background control is one of the most important and difficult works. The backgrounds events may come from different kinds of sources, including cosmic ray muons, radioactive background and accidental backgrounds. Neutrino experiments are usually build in deep underground with heavy overburdens above. Usually only high energy muons can penetrate the rocks on top of the detector, so the characteristics of cosmic ray muons are quite different from signals. In this study, we focus on discrimination of signals and radioactive backgrounds. In neutrino experiments, the level of radioactive backgrounds is greatly dependent on the purity of the environments and detector materials. The most common radioactive backgrounds come from the radioactive elements such as ^{238}U, ^{232}Th, ^{40}K and ^{222}Rn, which widely exist in the environment rocks,

PMT glasses, detector liquid due to impurity and the containers. According to different environment of a specific neutrino experiment and materials being used in a specific detector, radioactive backgrounds can be greatly different.

3.3. Simulation and Reconstruction
The detector model has been implemented in a Geant4 based detector simulation program. The signal IBD events are generated with event vertex randomly distributed in the container of liquid scintillator. Its final particles, the positron and neutron deposit energy after flying a short distance and produce scintillation lights. The photons transport in the liquid container and then produce hits in the PMT on the surface of the container. The backgrounds events are produced not only in the liquid scintillator, but also come from the container itself and the PMTs surrounding it.

The hits from all fired PMTs in an event will be written out for reconstruction. In neutrino experiments, event vertex and energy of the neutrino are two important variables for further physics analysis. In vertex reconstruction, we use the Charge Centered algorithm, in which the weighted position center of all fired PMTs after correction is set to be the vertex of the neutrino interaction point. For energy reconstruction, a Maximum Likelihood algorithm is used to fit the PMT hits distribution and get the most probable value of the reconstructed energy.

3.4. Classification with Reconstruction outputs
The reconstructed event vertex and event energy, together with other characteristic variables such as the total number of the PMT hits, can be used to discriminate IBD signals from radioactive backgrounds. The IBD signals can only be produced with their vertices in the liquid scintillator, while the radioactive background can also be produced from the container and PMTs near surface. The radioactive contributions from PMTs are greatly dependent on the materials of PMT glasses. In this way, the event vertex information can be used to separate signals and backgrounds, although there may be some overlaps in their distributions. The energy spectrum of reactor neutrinos is also different from those from radioactive elements decays, which makes it possible to use the event energy as a variable in discrimination. So long as the event vertex and energy are accurately reconstructed, the signal and background can be classified to some extent.

To test the performance of signal and background discrimination with event reconstruction outputs, we generate 10,000 signal IBD events and 10,000 radioactive background events, run through the detector simulation and event reconstruction to produce the variables for classification. The 5 reconstruction output variables to be used as input for classification are: reconstructed energy, number of total hits and the three-dimensional Euclidean coordinate of reconstructed vertex.

A traditional neural network MLP is used to study the discrimination power with the reconstruction outputs. The architecture of the network is shown in Figure 3. The input layer has 5 inputs, which correspond to the 5 reconstruction output variables. The number of hidden layer is 3, with the number of neurons in each layer being 50 or 100. The output has only one neuron to determine whether this event is signal or background.

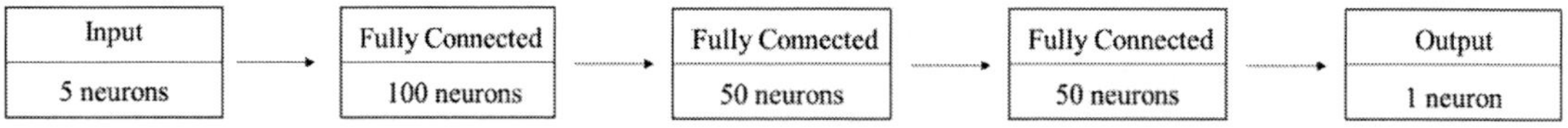

Figure 3. Architecture of the traditional neural network MLP.

The total 20,000 events of signals and backgrounds are separated into two samples, 18,000 events for training and 2,000 events for verifying. In training, we choose the Area Under the Curve (AUC) as the index of performance. The training result for AUC is 0.856, which will be used as a baseline for comparison with the CNN classification result.

4. Application of CNN for classification

4.1. 2D event display

The distribution of the fired PMT hits on surface of the spherical container can be projected to a θ-φ 2D plane to construct the image of the hits distribution of an event. For the PMTs with multiple hits, the total charges of all hits on the PMT are accumulated and the hit time is set to the earliest arrival hit's time. According to the difference on the production vertex of the event and the characteristic energy due to different particles in signal and background, the 2D projection event display can be used as graphical inputs for classification.

Figure 4 shows the typical 2D event display of a signal IBD event (left) and a radioactive background event (right). The scale of energy deposition is at MeV level. The signals are usually produced in the liquids in the container with a flatter hits distribution on the surface, while the backgrounds are produced closer to the surface with a sharper hits distribution on the surface. With the characteristics in 2D images, they can be used as inputs of CNN for classification. The images are low-level data with all raw information being kept. In comparison to the method using reconstruction outputs, no event reconstruction processes are required. It not only saves the time and computing resources, but also independent of the precision of any reconstruction algorithm.

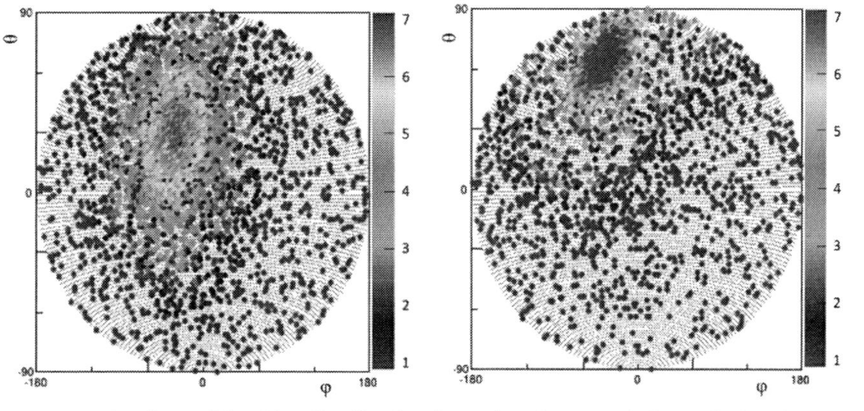

Figure 4. 2D projection of the hits distribution in a signal event (left) and a background event (right). The deposit energy in the events is on the scale of MeV.

4.2. CNN architecture

The size of input image is 314 by 329 pixels, which fully map the PMT arrangements on the surface of container. A CNN is constructed for classification with the image inputs. The architecture of the network is shown in Figure 5. The original 2D projection pictures are converted into RGB format and feed into it. There is a total of 5 hidden layers in the CNN. The first three hidden layers' structures are the same, they are all composed of two convolutional layers with 10 5 x 5 filters and one max-pooling layer with a 2 x 2 filter and its stride is set to be 2. The last two hidden layers are fully connected and each of them has 50 neurons. The output layer's activation function is the sigmoid function, the threshold is set to be 0.5 to determine whether the input is signal or background.

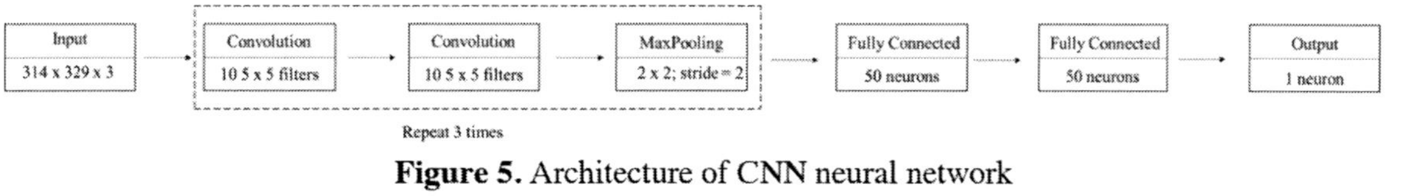

Figure 5. Architecture of CNN neural network

4.3. Training

The CNN model is trained on a GPU server with 4 NVIDIA Tesla K80 video cards. The same 20,000 event data samples in previous MLP tests are used in the CNN training. Instead of using the reconstruction output as input variables, the 2D hits projection images from each event are used as inputs of CNN network. In training, we use the Batch-Normalization [13] and Dropout [14] strategy to speed up the training and prevent overfitting. We also adopt ADAM as the optimizer [15]. The training lasts for 30 epochs before it converges. The training curve for CNN is shown in Figure 6.

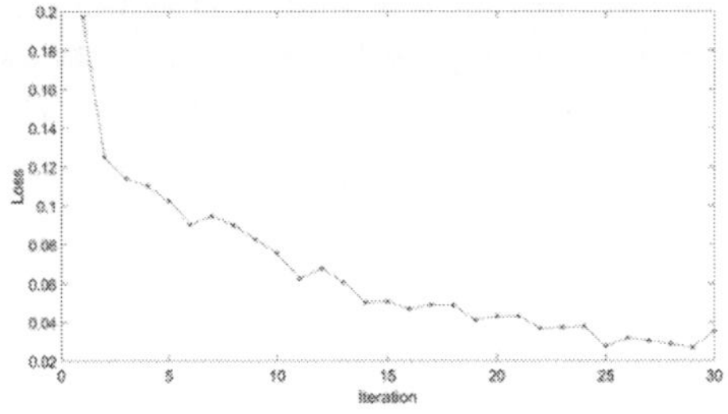

Figure 6. Convergence curve of the CNN in training.

4.4. Performance
The event classification results from MLP and CNN are listed in Table 1. We adopt AUC as the evaluation metrics to compare the classification power in different methods. As shown in Table 1, the CNN achieves much higher accuracy than MLP, which means that using CNN can significantly improve the performance of signal background discrimination without requiring any high-level feature. The classification accuracy is 97.4% for signals and 96.0% for backgrounds, respectively.

Table 1. Performance of event

	AUC	Variance
MLP	0.856	<0.001
CNN	0.975	<0.001

5. Summary
Deep learning has shown its powerful learning ability in different fields. It is skilful in handling massive data and finding the non-linear relationship between inputs and outputs. By applying the Convolutional Neural Networks in signal and background classification for a liquid neutrino detector model, we demonstrate that the low-level hits output can be transformed into images for event classification with CNN. Without knowing the high-level features of events from reconstruction, the signal and background discrimination power of CNN outperforms the traditional MLP method with reconstruction results. The toy detector model is designed to be a simplified version of the neutrino experiment JUNO. With further study, this method is expected to be applied in the realistic neutrino experiments in the future.

The authors would like to thank the supports from National Natural Science Foundation of China (11405279, 11675275), the Strategic Priority Research Program of Chinese Academy of Sciences (XDA10010900) and the Recruitment Program of Global Experts.

References
[1] S. Agostinelli et al. "GEANT4-a simulation toolkit", *Nucl. Instrum. Meth.*, A506.3 (2003):250-303.
[2] Y. Fukuda et al. (Super-Kamiokande Collaboration), "The Super-Kamiokande detector", *Nucl. Instrum. Meth.*, A501 (2003) 418-462.
[3] F.P. An et al. (Daya Bay Collaboration), "A side-by-side comparison of Daya Bay antineutrino detectors", *Nucl. Instrum. Meth.*, A685 (2012) 78-97.
[4] Z. Djurcic et al., (JUNO Collaboration), "JUNO Conceptual Design Report", *arXiv*:1508.07166.
[5] P. Baldi, P. Sadowski, and D. Whiteson. "Searching for exotic particles in high-energy physics with deep learning." *Nature Communications* (2014):4308.

[6] E. Racah et al., "Revealing Fundamental Physics from the Daya Bay Neutrino Experiment using Deep Neural Networks", *arXiv*:1601.07621.
[7] A. Bagoly et al., "Machine learning developments in ROOT", *J. Phys. Conf. Ser.* 898 (2017) no.7, 072046.
[8] K. Simonyan, A. Zisserman, "Very Deep Convolutional Networks for Large-Scale Image Recognition", *arXiv*:1409.1556.
[9] A. Aurisano. et al., "A Convolutional Neural Network Neutrino Event Classier", *JINST* 11 (2016), P09001.
[10] F. Rosenblatt, "Principles of Neurodynamics: Perceptrons and the Theory of Brain Mechanisms", Spartan Books, 1961.
[11] R. Reed and R. Marks, "Neural Smithing: Supervised Learning in Feedforward Artificial Neural Networks", A Bradford book, MIT Press, 1999.
[12] Y. LeCun, et al., "Backpropagation applied to handwritten zip code recognition", *Neural Comput.* 1 (1989) 541–551.
[13] Ioffe, Sergey, and C. Szegedy. "Batch Normalization: Accelerating Deep Network Training by Reducing Internal Covariate Shift", *arXiv*:1502.03167.
[14] Krizhevsky, Alex, I. Sutskever, and G. E. Hinton. "ImageNet classification with deep convolutional neural networks" *International Conference on Neural Information Processing Systems* Curran Associates Inc. 2012:1097-1105.
[15] Kingma, Diederik P, and J. Ba. "Adam: A Method for Stochastic Optimization" *Computer Science* (2014).

IOP Conf. Series: Journal of Physics: Conf. Series **1085** (2018) 042038 doi:10.1088/1742-6596/1085/4/042038

Machine-Learning-based global particle-identification algorithms at the LHCb experiment

Denis Derkach[1,2], Mikhail Hushchyn[1,2,3], Tatiana Likhomanenko[2,4], Alex Rogozhnikov[2], Nikita Kazeev[1,2], Victoria Chekalina[1,2], Radoslav Neychev[1,2], Stanislav Kirillov[2], Fedor Ratnikov[1,2]
on behalf of the LHCb collaboration

[1] National Research University — Higher School of Economics, Moscow, Russia
[2] Yandex School of Data Analysis, Moscow, Russia
[3] Moscow Institute of Physics and Technology, Moscow, Russia
[4] National Research Center Kurchatov Institute, Moscow, Russia

E-mail: `Denis.Derkach@cern.ch`

Abstract. One of the most important aspects of data analysis at the LHC experiments is the particle identification (PID). In LHCb, several different sub-detectors provide PID information: two Ring Imaging Cherenkov (RICH) detectors, the hadronic and electromagnetic calorimeters, and the muon chambers. To improve charged particle identification, we have developed models based on deep learning and gradient boosting. The new approaches, tested on simulated samples, provide higher identification performances than the current solution for all charged particle types. It is also desirable to achieve a flat dependency of efficiencies from spectator variables such as particle momentum, in order to reduce systematic uncertainties in the physics results. For this purpose, models that improve the flatness property for efficiencies have also been developed. This paper presents this new approach and its performance.

1. Introduction

Particle identification (PID) algorithms play a crucial part in any high-energy physics analysis. A higher performance algorithm leads to a better background rejection and thus more precise results. In addition, an algorithm is required to work with approximately the same efficiency in the full available phase space to provide good discrimination for various analyses.

PID at the LHCb experiment relies on several subsystems [1]. Two Ring Imaging Cherenkov (RICHs) sub-detectors provide charged hadron identification over a wide momentum range, from 2 to 100 GeV/c. Muons are identified mainly thanks to the muon chambers, while electron and photon identification is assured by the calorimeters. The information can be combined by simply computing Log Likelihoods (LL) separately for each sub-detector and combining them according to the mathematical definition [2], or by using a more elaborated method like an artificial neural network (see details in Section 3). The latter outperforms the LL approach and is used as a baseline in our studies. The above particle identification approaches are used by most of the analyses completed by LHCb collaboration.

In this paper, new models for particle identification that efficiently combine the diverse information available from all sub-detectors using advanced machine learning approaches are presented. The proposed models are based on Deep artificial Neural Network (DNN)

IOP Conf. Series: Journal of Physics: Conf. Series **1085** (2018) 042038 doi:10.1088/1742-6596/1085/4/042038

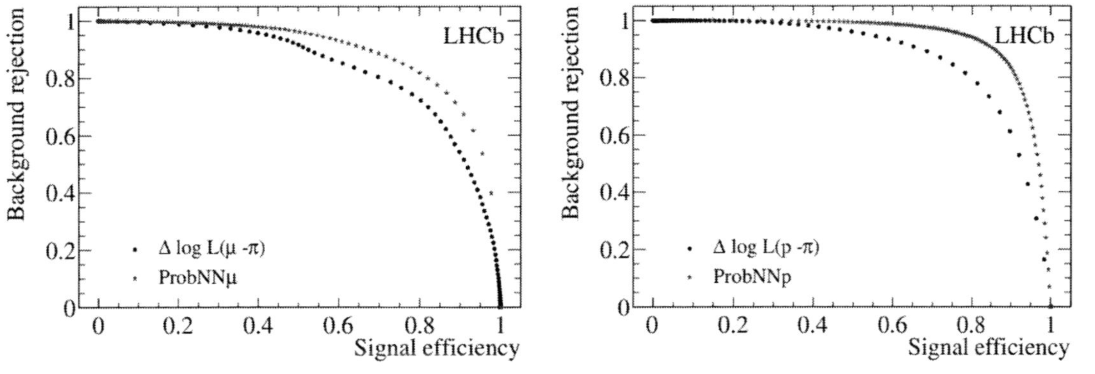

Figure 1. Background misidentification rates versus muon (left) and proton (right) identification efficiency, as measured in the $\Sigma^+ \to p\mu^+\mu^-$ decay study. The variables $\Delta\mathcal{L}(X-\pi)$ (black) and ProbNN (red), are compared for $5-10$ GeV/c muons and $5-50$ GeV/c protons, using data sidebands for backgrounds and simulated samples for the signal. The data sample used corresponds to 2012 sample collected at center-of-mass energy 8 GeV.

and gradient boosting over (ordinary and oblivious) decision trees. An additional class of algorithms providing flat efficiencies along several spectator observables (momentum, transverse momentum, pseudorapidity, number of particles in the event) is explored.

2. Problem Statement

The problem consists in identifying the charged particle type associated with a given track. There are five relevant particle species, namely, electron, muon, pion, kaon, proton, and ghost track (charged tracks that do not correspond to a real particle which passed through the detector) making a total of six hypotheses. Therefore, this is a multiclass classification problem. The information from RICHs, the electromagnetic and hadronic calorimeters and muon chambers are combined together with information provided by the tracking system. Apart from pre-aggregated into likelihood like observables subdetector responses [3], we also use track geometry variables and different detector flags. In addition to this, we used the muon identification [4] and calorimeter information about neutral clusters [5], which proved to be very useful to suppress fake tracks. In this paper, for the new methods we present only the result obtained with fully simulated events [6] used both for training and testing.

3. The current solution

The first machine learning algorithm used for the PID in LHCb is a fully-connected neural network (multilayer perceptron) with one hidden-layer implemented using the TMVA package [7]. The model, called ProbNN, was trained separately for each particle in the binary one-vs-rest classification mode, thus creating several separate models. The misidentification rates versus efficiency curves for the Log Likelihood, $\Delta\mathcal{L}(X-\pi)$, and ProbNN are shown in Figure 1. The improvement due to machine learning is clearly visible for both muons and protons.

4. New models

Two classes of algorithms are considered: first, 'non-uniform' algorithms, which have similar training target compared to the existing ProbNN; second, 'uniform' algorithms, which in addition is trained to have flat efficiencies along chosen kinematic variables.

IOP Conf. Series: Journal of Physics: Conf. Series **1085** (2018) 042038 doi:10.1088/1742-6596/1085/4/042038

Table 1. Relative increase of the 1-AUC scores for different particles species and ghosts (lower is better). The statistical uncertainty is lower than 1%.

Model	ghost	electron	muon	pion	kaon	proton
DNN	−29%	−41%	−52%	−37%	−20%	−17%
XGboost	−24%	−37%	−50%	−34%	−18%	−15%
CatBoost	−30%	−43%	−54%	−37%	−20%	−18%
flat 4d	−21%	−4%	−13%	−20%	+10%	+25%

We developed new PID models based on deep neural networks, from the `keras` library [8], and boosted decision trees, from the `CatBoost` [9] and `XGBoost` [10] libraries. Along with a general increase of information provided to the classifier, we used `XGBoost` library in the multiclass mode, which brought some additional improvement. To compare the algorithms' performances, we used the Area Under Receiver Operating Characteristic Curve (ROC AUC) for one-vs-rest classification (6 numbers in total). As can be seen from Table 1, the new models give encouraging results in comparison to the baseline. All numbers in Table 1 were found statistically siginficant using several methods proposed in [11, 12], paired t-test on cross-validation sample (see, for example [13]).

An important improvement that may lead to reduction of systematic uncertainties of physics annalyses is the construction of the PID model that does not depend on kinematic observables of the track. In order to flatten the efficiency dependence, we used special training procedure. This procedure effectively takes into account non-flatness of the output distribution by using the modified loss function as described in [14]. The corresponding loss function in this case looks like:

$$\mathcal{L} = \mathcal{L}_{ExpLoss} + \alpha \mathcal{L}_{FL}, \tag{1}$$

where $\mathcal{L}_{ExpLoss}$ corresponds to the classification loss function, the $\mathcal{L}_{FL}$ corresponds to the uniformity loss, and α is a parameter to control the trade-off between classification quality and uniformity. The $\mathcal{L}_{FL}$ is taken to be similar to the Cramer-von Mises measure:

$$\mathcal{L}_{FL} = < \int (F_{global}(s) - F_{local}(s))^2 \, ds >, \tag{2}$$

where $F_{global}(s)$ and $F_{local}(s)$ is a classifier predictions cumulative distribution function for the full sample or in a given interval, respectively. In order to reduce the efficiency dependency on the kinematic properties of the track we use the linear combination of flatness losses for momentum, transverse momentum, number of tracks in event and pseudorapidity:

$$\mathcal{L}_{FL_{4d}} = \mathcal{L}_{FL_p} + \mathcal{L}_{FL_{p_T}} + \mathcal{L}_{FL_{nTracks}} + \mathcal{L}_{\eta}. \tag{3}$$

This loss function is implemented using the `Decision Train` package [15], which uses fast oblivious decision trees for model construction. The results of flatness boosting approach are shown in Table 1 and in Fig. 2 with the label "flat 4d". The comparison of efficiency dependence on $1/p_T$ shows a significant improvement in flatness for each particle type. The lower performance of this model is the result of the trade-off between quality and flatness controlled by parameter α in Eq. 2. In these proceedings, we choose values of α to be 10 for track momentum, track transverse momentum components of the loss function and 6 for pseudorapidity and number of tracks components. The choice of α is motivated by the optimal trade-off between flatness and performance and is optimised using cross-validation sample. The final choice of the model depends on the particular physics analysis and should be made after

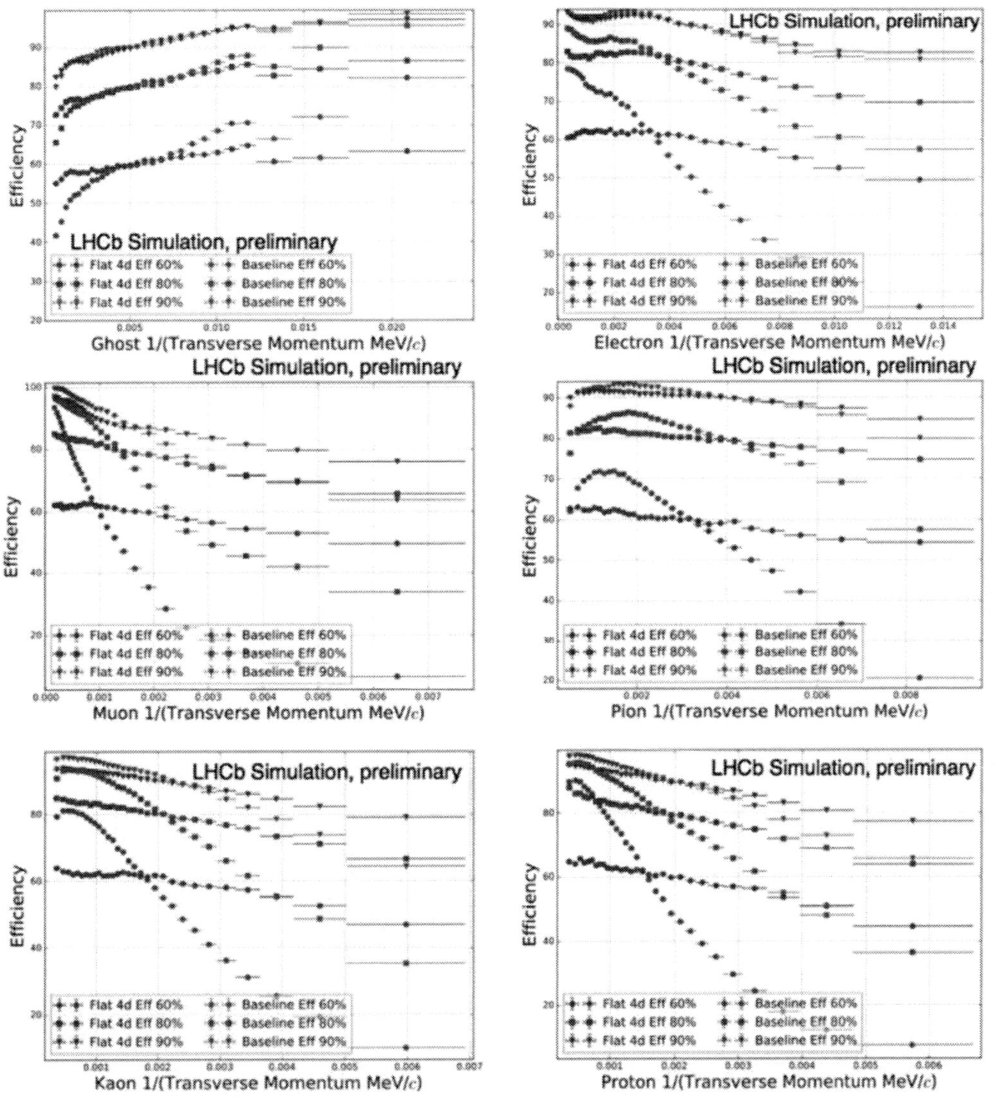

Figure 2. Efficiency of the "Flat 4d" model as a function of the inverse of the transverse momentum for each particle type, for different global efficiency cuts.

evaluation of the main sources of systematic uncertainties. Other approaches to flatten the efficiencies are also available for neural networks [16, 17].

5. Conclusions

Several new models based on state-of-the-art machine learning methods are developed and shown encouraging results in comparison to the baseline. Models drastically improving the flatness of the output with the respect to the spectator variables are built. The corresponding loss of overall efficiency is show to be small.

IOP Conf. Series: Journal of Physics: Conf. Series **1085** (2018) 042038 doi:10.1088/1742-6596/1085/4/042038

Acknowledgments
The research leading to these results has received funding from Russian Science Foundation under grant agreement n° 17-72-20127.

References
[1] Ignacio B *et al.* (LHCb collaboration) 2013 LHCb-TDR-014
[2] Patrignani C *et al.* (Particle Data Group) 2016 *Chin. Phys.* **C40** 100001
[3] Powell A 2011 5 p URL `https://cds.cern.ch/record/1322666`
[4] Archilli F *et al.* 2013 *JINST* **8** P10020 (*Preprint* `1306.0249`)
[5] Deschamps O, Machefert F P, Schune M H, Pakhlova G and Belyaev I 2003 Photon and neutral pion reconstruction Tech. Rep. LHCb-2003-091 CERN Geneva URL `https://cds.cern.ch/record/691634`
[6] Clemencic M *et al.* 2011 *J. Phys. Conf. Ser.* **331** 032023
[7] Hocker A *et al.* 2007 *PoS* **ACAT** 040 (*Preprint* `physics/0703039`)
[8] Chollet F *et al.* 2015 Keras `https://github.com/fchollet/keras`
[9] Dorogush A V, Gulin A, Gusev G, Kazeev N, Prokhorenkova L O and Vorobev A 2017 **abs/1706.09516** URL `http://arxiv.org/abs/1706.09516`
[10] Chen T and Guestrin C 2016 **abs/1603.02754** URL `http://arxiv.org/abs/1603.02754`
[11] DeLong E R, DeLong D M and Clarke-Pearson D L 1988 *Biometrics* **44** 837–845 ISSN 0006341X, 15410420 URL `http://www.jstor.org/stable/2531595`
[12] Hanley J A and McNeil B J 1983 *Radiology* **148** 839–843 pMID: 6878708 (*Preprint* `https://doi.org/10.1148/radiology.148.3.6878708`) URL `https://doi.org/10.1148/radiology.148.3.6878708`
[13] Demsar J 2006 *Journal of Machine Learning Research* **7** 1–30
[14] Rogozhnikov A, Bukva A, Gligorov V V, Ustyuzhanin A and Williams M 2015 *JINST* **10** T03002 (*Preprint* `1410.4140`)
[15] Rogozhnikov A 2015 Decision train classifier `http://arogozhnikov.github.io/` [Online; accessed 03-November-2017]
[16] Shimmin C, Sadowski P, Baldi P, Weik E, Whiteson D, Goul E and Sgaard A 2017 *Phys. Rev.* **D96** 074034 (*Preprint* `1703.03507`)
[17] Louppe G, Kagan M and Cranmer K 2016 (*Preprint* `1611.01046`)

IOP Conf. Series: Journal of Physics: Conf. Series **1085** (2018) 042039 doi:10.1088/1742-6596/1085/4/042039

Dynamic sharing of tape drives accessing scientific data

A. Cavalli, D. Cesini, E. Fattibene, L. Morganti, A. Prosperini, P. P. Ricci and V. Sapunenko

INFN CNAF, viale Berti Pichat 6/2 40127 Bologna, Italy

E-mail: enrico.fattibene@cnaf.infn.it

Abstract. The data management infrastructure operated at CNAF, the central computing and storage facility of INFN (Italian Institute for Nuclear Physics), is based on both disk and tape storage resources. About 40 Petabytes of scientific data produced by LHC (Large Hadron Collider) at CERN and other experiments in which INFN is involved are stored on tape. This is the highest latency storage tier within HSM (Hierarchical Storage Management) environment. Writing and reading requests on tape media are satisfied through a set of Oracle-StorageTek T10000D tape drives, shared among different scientific communities. In the next years, the usage of tape drives will become more intense not only due to the growing amount of scientific data to manage but also due to general trend to use tapes as "slow disk", announced by the main user communities. In order to reduce hardware purchases, a key point is to minimize the inactivity periods of tape drives. In this paper we present a software solution designed to optimize the efficiency of the shared usage of tape drives in our environment.

1. Tape-based facility at CNAF

CNAF is the major Data Center of INFN, offering resources and services to communities involved in scientific collaborations. As INFN participates to the LHC, the largest and most powerful particle accelerator in the world, CNAF is one of the 11 Tier-1 centers of the WLCG (Worldwide LHC Computing Grid), that receive data produced by the LHC experiments (ALICE, ATLAS, CMS, LHCb). Data coming from LHC are of the order of 1GB/sec on a monthly average, with peaks of 3 GB/s or more. Moreover, CNAF Data Center provides computing and storage facilities for 30 other experiments in which INFN is involved, belonging to Astrophysics, Astro-particle Physics and High Energy Physics domains. Data are stored on both disk and tape storage resources. At the time of writing, ~20 PB of data reside on disk and ~44 PB on tape.

1.1. Infrastructure and services

CNAF mass-storage infrastructure is based on a tape library Oracle-StorageTek SL8500 equipped with 17 T10000D tape drives used for scientific data and 9 T10000C drives used only for backup and recovery service. The overall capacity of the SL8500 library is 10000 slots, so ~85 PB could be stored with the existing technology. Tape-based storage is the highest latency storage tier within a HSM (Hierarchical Space Management) environment. In order to allow data access to scientific communities, the Storage Management group operates services based on a set of software packages:

- IBM Spectrum Scale [1]: formerly GPFS (General Parallel File System), a high-performance clustered file system developed by IBM. File systems can be partitioned into a number of *storage pools* implementing file placement policies and data migration rules from one pool to another according to some user-defined criteria.

- ISP (IBM Spectrum Protect) [2]: formerly TSM (Tivoli Storage Manager), a proprietary software designed by IBM, one of the leaders in data protection solutions. It offers a HSM extension to manage migrations from disk to tape and recalls from tape to disk of data hosted on Spectrum Scale file systems.
- StoRM (Storage Resource Manager) [3]: a software released by INFN based on SRM (Storage Resource Management) interface to access storage resources.
- GEMSS (Grid Enabled Mass Storage System) [4]: a software developed by INFN that provides a full HSM integration of Spectrum Scale, ISP and StoRM. It has been designed to optimize migration and recall operations.

Migrations and recalls are managed through HSM servers equipped with Fiber Channel connections to both storage disk and tape drives. Each server can be configured to handle one or more Spectrum Scale file systems. For each file system, an active HSM server is running and another standby server is configured and can be turned on in case of unavailability of the active one. At the moment, CNAF operates 6 active and 6 standby HSM servers.

1.2. GEMSS recalls

ISP HSM software can recall data from tape to disk using two possible methods: selective and transparent recalls. In case of selective recalls, the user (or a specific service on his/her behalf) asks for a file to be recalled from tape before submitting a job to a worker node, i.e. before making the first file access. This typically happens in the WLCG world, where the recall request is made via SRM commands, i. e. through StoRM service. Only when all the needed files have been recalled, the access is performed. Transparent recalls are triggered by a read operation (usually from user jobs) of a migrated file, i. e. in case only the stub file is present on disk. When the recall is finished and the file is accessible on disk, the control is given back to the user's process.

The standard ISP HSM behaviour consists in recalling files as soon as they are requested by users, following the order of the requests. As users have no knowledge of where the files are stored, and in particular of the way the files are ordered within a tape, such a procedure ends up in an inefficient usage of the tape resources. To overcome this limitation, GEMSS implements its own aggregation and reordering of tape recalls before submitting them to ISP.

GEMSS can handle both selective (triggered by a periodic scan of StoRM bring-online table or requested through GEMSS command *yamssEnqueueRecall*) and transparent recalls (triggered accessing the files). GEMSS server is able to transform transparent recalls in selective ones.

Figure 1 shows the selective tape-ordered recall system of GEMSS. First, requests are enqueued by a FIFO (First In First Out) method. The *yamssReorderRecall* process builds, for each tape, a list of files to recall sorted according to tape ordering. A recall process (*yamssProcessRecall*) can start for each tape file, according to the GEMSS configuration: for each file system, the maximum number of recall threads to send to ISP server is defined by the parameter *RECALL_RUNNING_THREADS*. Each running recall thread corresponds to a tape drive devoted to read the requested files. In the same way, the *MIGRATE_RUNNING_THREADS* parameter stabilizes the maximum number of running migration threads for each file system. Once the number of running recall threads hits the value of *RECALL_RUNNING_THREADS* parameter, all the other tape files are put in a queue. In case of new requests, *yamssReorderRecall* can add new files to the existing lists in the correct order. The *yamssMonitor* service is the supervisor of the reorder and recall phases. It discovers managed Spectrum Scale file systems on HSM nodes, reads the configuration file for each file system and triggers the needed actions, e.g. starting other processes. It loops continuously in background.

Within a file system, the criterion to assign priority to tapes to recall is given by the *RECALL_MAX_RETENTION* parameter (default value is 1800 seconds). In case pending recall threads waiting time is equal or lower than *RECALL_MAX_RETENTION*, priority is given to those tapes containing the largest number of files. Instead, if pending recalls threads waiting time is greater than *RECALL_MAX_RETENTION*, then priority is given following a FIFO method.

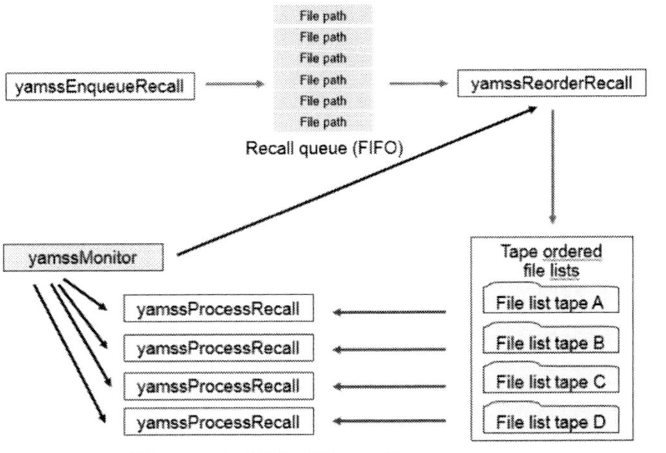

Figure 1. GEMSS recall system

2. Current status in tape drive usage

With the current configuration, the 17 T10000D tape drives dedicated to scientific data are shared among experiments, i. e. any of them can be used by migration or recall threads running for each file system. A maximum number of drives for recalls or migrations, statically defined in GEMSS, can be exploited by each file system. In case of scheduled massive recall or migration activity, these parameters are manually changed by administrators. There is no way to automatically change them.

Sometimes we notice a certain number of free drives that could actually serve pending recall threads, as shown in Figure 2 (plot on the left). In several cases, a subset of free drives could be profitably used to reduce the queue of pending recalls. However, the maximum number of possible running threads is limited by the HSM server throughput capacity. Currently each HSM server is equipped with a single FC8 (Fiber Channel 8 Gbit/sec) connection to the Tape Area Network, so it is capable of handling 800 MB/s simultaneously for inbound and outbound traffic. Given each T10000D tape drive can reach ~200-250 MB/s of throughput, at the moment each HSM server is able to support up to 4 migration and 5 recall processes, considering some observed inefficiencies in recalls due to the not-subsequent placement of files on tape. The HSM connection is planned to be upgraded to FC16 for each server next year.

In case of concurrency in the usage of drives, i. e. when recall or migration threads for one or more file systems cannot become running because of the lack of free drives, there is no way to dynamically change GEMSS parameters to give more priority to file systems that less used the system in the recent past.

Figure 2 (plot on the right) shows duration of recall and migration processes for the overall infrastructure, aggregated by day. The total usage is never greater than 8 days. This means that for the period of the plot, if we do not consider peaks, i. e. intervals of time with a drive usage above average, we could perform migration and recall activity with only 8 drives.

Moreover, in the next years, the usage of tape drives is expected to become more intense due to the growing amount of scientific data and to the trend, already disclosed by the main user communities, to use tapes as near-line (or "slow") disk, thereby increasing the reading traffic rate.

All of these considerations, together with the need to reduce hardware purchases, moved us to reflect on a drive usage optimization that would reduce their inactivity periods.

IOP Conf. Series: Journal of Physics: Conf. Series **1085** (2018) 042039 doi:10.1088/1742-6596/1085/4/042039

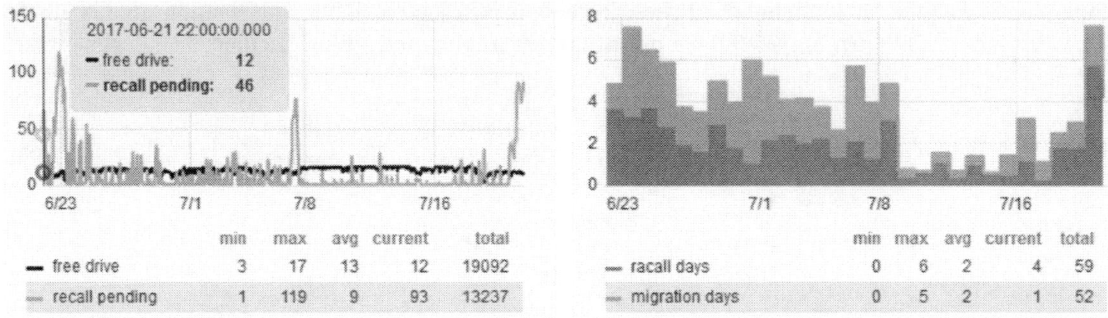

Figure 2: On the left: total number of recall threads pending and free drives. On the right: total duration (in days) of recall and migration processes, stacked plot aggregated by day. June-July 2017.

3. Dynamic sharing of tape drives

Given the weaknesses observed in the static-priority assignment of tape drives, we designed a software solution, hereafter named *Orchestrator*, which allows to dynamically allocate additional drives to file systems, in case free drives are there, and to improve on the management of concurrent recall accesses from different file systems.

Two new GEMSS parameters have been defined: *RECALL_MAX_RUNNING*, representing the maximum number of possible running recalls for each file system, taking into account the FC connection limits of the relevant HSM server; *RECALL_DEFAULT_RUNNING*, representing the value that should have the *RECALL_RUNNING_THREADS* parameter in normal conditions.

The *Orchestrator* uses the library InfluxDB-Python as a client for accessing InfluxDB and read its data. Indeed, monitoring information, essential for the *Orchestrator* to operate, is stored in InfluxDB and updated every five minutes. More specifically, such information includes the number of drives that are currently free or in use (taken from the ISP server), the number of running recall and migration threads, the number of pending recalls and the value of *RECALL_MAX_RUNNING* parameter for each file system (taken from the HSM servers). As a first step of the algorithm, the *Orchestrator* extracts all these relevant quantities from InfluxDB (Figure 3). Like every priority-driven algorithms, the *Orchestrator* performs on-line scheduling, so it makes decision without any knowledge about the kind and amount of workload that will come in the future.

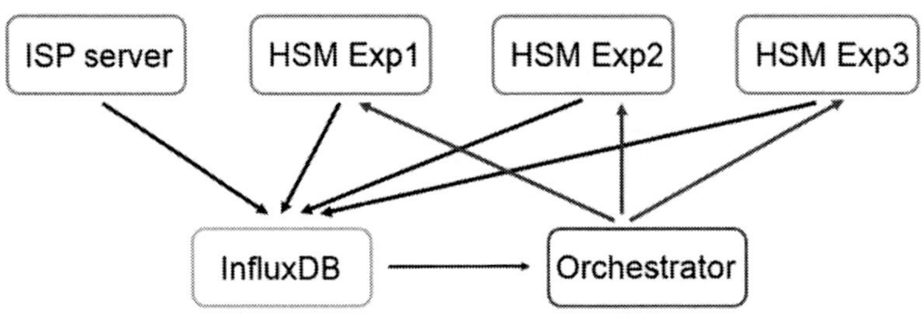

Figure 3. Monitoring information flows from ISP server and HSM servers to InfluxDB (black arrows). The *Orchestrator* reads it and updates GEMSS parameters in the HSM servers (red arrows).

Every five minutes, the *Orchestrator* inspects monitoring information. In case there are free drives and pending requests waiting for drives for any handled file system, it establishes the number of drives that can be assigned to each interested file system, comparing the number of actual running recalls with the value of *RECALL_MAX_RUNNING* parameter. Moreover, in order to exclude situations of ongoing rearrangements of tape drives, it checks whether the number of running threads equals the number of drives in use.

IOP Conf. Series: Journal of Physics: Conf. Series **1085** (2018) 042039 doi:10.1088/1742-6596/1085/4/042039

Whenever there is no concurrency among different file systems for the available resources, i.e. the number of available free drives is large enough to satisfy all the requests, the *Orchestrator* mitigates the pending requests. This is performed by means of modifying on the relevant HSM server the GEMSS parameter *RECALL_RUNNING_THREADS* for each interested file system, and by raising it to the value of *RECALL_MAX_RUNNING*. Of course, ideally one would like to lower the number of pending recall and migration threads by filling all the available free drives except for a reasonable reserve (which we set to 2, given the total of 17 drives).

Instead, a more complicated situation can happen in case the pool of free drives is not sufficient to satisfy all the requests. Moreover, it is also possible that pending recalls for a certain file system can not become running due to all the drives being busy with other operations. All these cases of concurrent access to tape drives are managed by the *Orchestrator* by computing and setting a dynamic priority for each file system on the basis of the following formula:

$$FSpriority = FSshare/(\alpha(usage\_time) + \beta(1 + run\_recall))$$

where *FSshare* is a static priority given to each file system, *usage_time* is the total recall time used by the file system in a fixed period of recent past (e.g. last 24 hours), *run_recall* is the number of recall running threads, and finally α and β are adjustable coefficients which allow to differently weight resources usage time in the past (*usage_time*) and current usage time (*run_recall*).

Once the file systems are placed in one common priority queue according to the values of *FSpriority*, the available free drives are assigned going through the sorted list. Then, when all the drives are occupied and new requests become pending, the value of *RECALL_RUNNING_THREADS* can be lowered for a given file system and increased for the file system with pending recalls. In practice, this procedure increases the priority of getting a tape drive for those file systems that did not extensively use resources in the recent past, and who are not currently performing many recalls and migrations.

In any case, when the actual number of running recall threads for a file system is equal or lower than *RECALL DEFAULT_RUNNING*, the parameter *RECALL RUNNING THREADS* is brought again to the default value (*RECALL_DEFAULT_RUNNING*) by the *Orchestrator*.

As noticed in paragraph 2.1, by setting *RECALL_MAX_RETENTION* parameter it would possible to give priority to certain pending recalls. Of course, it would be interesting to consider such procedure for future *Orchestrator* implementations in order to provide a different priority method to dedicated recall threads.

4. Conclusions

CNAF mass-storage infrastructure is handling tens of PB of scientific data. Data movements from disk to tape and vice versa are optimized by means of GEMSS software. In order to overcome the static assignment of a maximum number tape drives both for migration and recall processes, we designed a software solution to dynamically allocate additional drives to file systems and to manage concurrent requests. This solution is expected to optimize the tape drives usage, reducing migration and recall waiting time, that would be an important enhancement for CNAF mass-storage facility in view of the future growth of writing and reading rate. Moreover, the ability to maximize the drive exploitation would help CNAF in lowering hardware purchase, by reducing the need to purchase more tape drives in the future.

References

[1] IBM Spectrum Scale web page: www.ibm.com/systems/storage/spectrum/scale
[2] IBM Spectrum Protect web page: www.ibm.com/systems/storage/spectrum/protect
[3] R. Zappi R et al, *An efficient Grid data access with StoRM*, S.C. Lin and E. Yen (eds.), Data

Driven e-Science: Use Cases and Successful Applications of Distributed Computing Infrastructures (ISGC 2010), Springer Science + Business Media, LLC 2011
[4] Bonacorsi D et al., *The Grid Enabled Mass Storage System (GEMSS): the Storage and Data management system used at the INFN Tier1 at CNAF*, 2012 J. Phys. Conf. Ser. 396 042051 Proceedings of 2012 CHEP conference.

IOP Conf. Series: Journal of Physics: Conf. Series **1085** (2018) 042040 doi:10.1088/1742-6596/1085/4/042040

Convolutional Neural Network for Track Seed Filtering at the CMS High-Level Trigger

Adriano Di Florio
Dipartimento Interateneo di Fisica di Bari and I.N.F.N.-Sezione di Bari
via Amendola 173, 70126 Bari, Italy
E-mail: adriano.diflorio@ba.infn.it

Felice Pantaleo
CERN, European Organization for Nuclear Research
Geneva 23, CH-1211 Geneva, Switzerland
E-mail: felice.pantaleo@cern.ch

Antonio Carta
Dipartimento di Informatica, Università di Pisa
Largo Bruno Pontecorvo 3, I-56127 Pisa, Italy
E-mail: antonio.carta@di.unipi.it

on behalf of the CMS collaboration

Abstract.
Starting with Run II, future development projects for the Large Hadron Collider will constantly bring nominal luminosity increase, with the ultimate goal of reaching a peak luminosity of $5 \cdot 10^{34} cm^{-2} s^{-1}$ for ATLAS and CMS experiments planned for the High Luminosity LHC (HL-LHC) upgrade. This rise in luminosity will directly result in an increased number of simultaneous proton collisions (pileup), up to 200, that will pose new challenges for the CMS detector and, specifically, for track reconstruction in the Silicon Pixel Tracker. One of the first steps of the track finding work-flow is the creation of track seeds, i.e. compatible pairs of hits from different detector layers, that are subsequently fed to higher level pattern recognition steps. However, the set of compatible hit pairs is highly affected by combinatorial background resulting in the next steps of the tracking algorithm to process a significant fraction of fake doublets. A possible way of reducing this effect is taking into account the shape of the hit pixel cluster to check the compatibility between two hits. To each doublet is attached a collection of two images built with the ADC levels of the pixels forming the hit cluster. Thus the task of fake rejection can be seen as an image classification problem for which *Convolutional Neural Networks* (CNNs) have been widely proven to provide reliable results. In this work we present our studies on CNNs applications to the filtering of track pixel seeds. We will show the results obtained for simulated event reconstructed in CMS detector, focusing on the estimation of efficiency and fake rejection performances of our CNN classifier.

IOP Conf. Series: Journal of Physics: Conf. Series **1085** (2018) 042040 doi:10.1088/1742-6596/1085/4/042040

1. The CMS Trigger System

The Compact Muon Solenoid (CMS) [1] is a general purpose detector designed for the precision measurement of leptons, photons, and jets, among other physics objects, in proton-proton as well as heavy ion collisions at the CERN LHC [2]. The LHC is designed to collide protons at a center-of-mass energy of 14 TeV and a luminosity of $10^{34} cm^{-2} s^{-1}$. At design luminosity, the pp interaction rate is about 1 GHz but only a small fraction of these collisions contains events that can be interesting for CMS physics analyses and can be stored to be accessible offline. The trigger system is devoted to the selection of these events from the totality of the inelastic collision events. In order to accomplish this task, the CMS trigger system utilizes two levels [3]. The first one (L1) is implemented in custom hardware (FPGAs and ASICs). Within $4\mu s$ of a collision, it makes selection of candidates objects based on raw data from calorimeters and muon detectors and it restricts the output rate to 100 kHz. Events are then passed to the High-Level Trigger (HLT) that further refines the purity of the physics objects, and selects an average rate of 400 Hz for final offline storage.

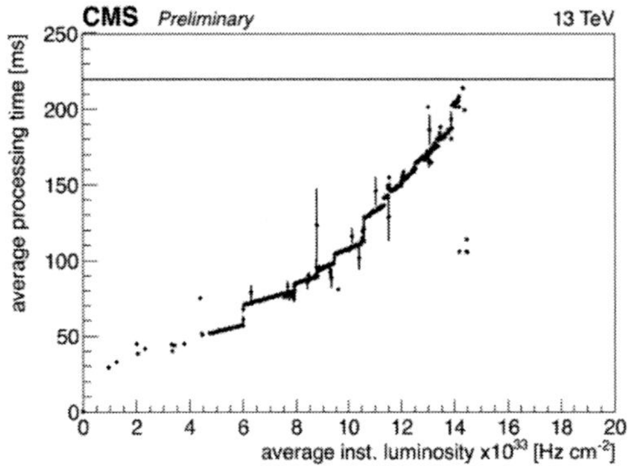

Figure 1. Average processing time per event with respect to the average instantaneous luminosity. [4]

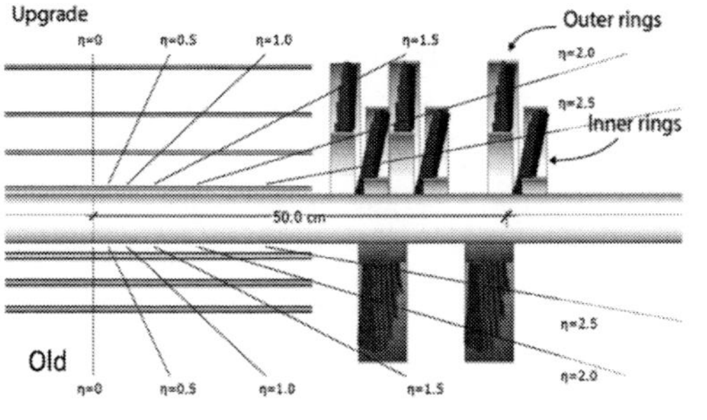

Figure 2. The Run 2 updated Silicon Pixel Tracker geometrical layout, optimized to offer full 4-hit tracking coverage up to a value of the pseudo-rapidity of 2.5 [5]

2. Track reconstruction at HLT

The HLT hardware consists of a processor farm, the Event Filter Farm (EVF) with about 26000 Intel Xeon cores [3].The HLT gathers information from all the CMS detectors and among them the Silicon Pixel Tracker has a crucial role in two of the first steps for the selection: the track reconstruction and the primary vertex finding. The track reconstruction software used at HLT is practically identical to that used for offline reconstruction with the difference that it has to fulfill stringent CPU timing constraints: the event selection has to be fulfilled with an average latency smaller than 260 ms. With the instantaneous luminosity increasing up to the ultimate goal of $5 \cdot 10^{34} cm^{-2} s^{-1}$ for the High Luminosity LHC (HL-LHC) upgrade, this time constraint is rapidly approaching (see Figure 1). Beside this, the number of simultaneous proton collisions per bunch crossing ("pileup") will continue to increase: the highest pileup reached averaged over a data run in 2011 and 2012 was ≈ 16 and ≈ 34, respectively, while in 2016 it exceeded 50 [6].

The track reconstruction procedure can be roughly split in four steps [3]. The first step is the construction of seeds (seeding), i.e. few (2-4) Pixel hits that act as the building blocks for track candidates and define the initial estimate for trajectory parameters. These candidates are then propagated by means of Kalman Filter techniques to find further compatible hits and build a track (building). The tracks are then fitted and marked with quality flags (fitting and selection).

In the current configuration, the very first step of seeding consists in collecting and mapping all the valid hits on the Silicon Pixel Detector layer. The Run 2 updated Silicon Pixel Tracker is arranged in four barrel layers (BPIX) and three forward disks (FPIX) in each end cap region (see Figure 2). Then, iterating on all the couples of consecutive *seeding layers*, the algorithm selects and collects all the compatible hit pairs (*doublets*), i.e. the hit pairs that could belong to the same particle and track. The compatibility between two hits is evaluated only on the basis of geometrical considerations, such as cuts in η, ϕ and r.

The doublet production is thus highly dominated by combinatorial background and constitutes a bottleneck for the subsequent steps. For example, for $t\bar{t}$ simulated events at energy of the center of mass of $\sqrt{s} = 13$ TeV, with average pileup $< PU >= 35$ and bunch time spacing of 25 ns, about 10^6 doublets are produced per each event. Therefore the reduction of the track reconstruction time hinges on the rejection of seeding combinatorial background. In the next sections a method for filtering doublet seeds based on CNNs is proposed.

3. Convolutional Neural Networks for Doublet Seeds Filtering

A traditional deep neural network is a machine learning model whose goal is to approximate a function $\hat{f}$ by composing a sequence of simpler functions. It consists of an input layer, one or more hidden layers and an output layer [7]. To a first approximation, each layer acts on the input as a composition of a matrix multiplication, tuned with a set of biases (b) and weights (w), and an activation function. Convolutional Neural Networks [8] are a specialized kind of neural network for processing data that has a grid-like structure, such as 2D images. The building block of a CNNs is a layer that uses *discrete convolution* in place of general matrix multiplication [7]. A convolutional layer takes in input a $n \times m \times r$ two dimensional image where r is the number of channels or depth, e.g. an RGB image has $r = 3$. The layer is characterized by k filters of size $k \times l \times q$ where k and l are smaller than the dimensions (n, m) of the image and typically $q = r$ (the filter has full depth receptive field). Each filter is devoted to detect some specific type of feature or shape at some spatial position in the input. The output of one or more stacked convolutional layers is usually passed to pooling layers that perform dimensionality reduction. A $p \times p'$ *pooling layer* replaces a $p \times p'$ region of the image with a single value, e.g. the maximum

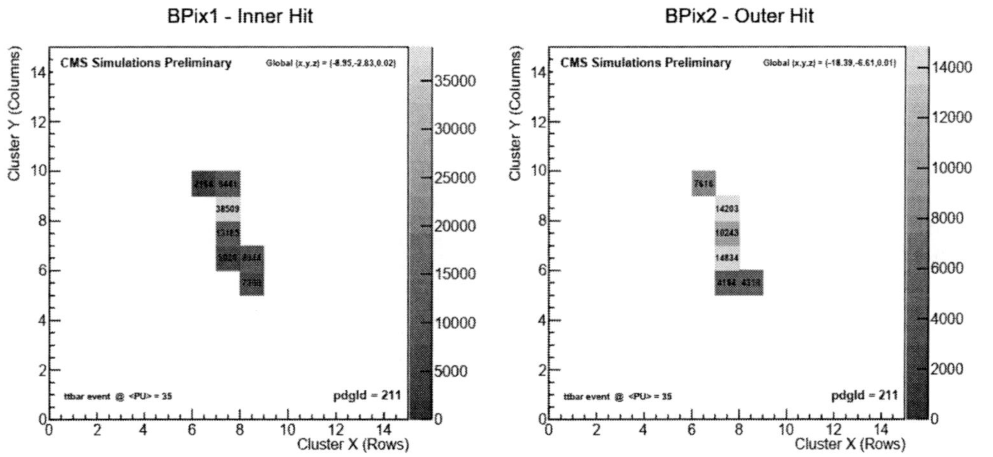

Figure 3. Example of a BPix1-BPix2 true doublet generated from $t\bar{t}$ simulated events corresponding to a π^+ track.

IOP Conf. Series: Journal of Physics: Conf. Series **1085** (2018) 042040 doi:10.1088/1742-6596/1085/4/042040

value in that region for a max pooling layer.

3.1. Doublet Hits Cluster Shape

As described above, while building seeds, the compatibility between two hits is evaluated only on the basis of geometrical considerations. A possible way of reducing the doublet fake ratio is taking into account that each hit is actually a cluster of pixels with its own shape. Each pixel is characterized by its 16-bit A.D.C. level ($\max_{ADC} = 2^{16} - 1$) and its local position (x, y) on the layer, where x is the azimuthal direction in the barrel detector and radial direction in the forward detectors. Then a single hit can be considered as a collection of three vectors: the x, the y and the ADC levels of each of its pixels. For each hit a 15×15 squared matrix M is built with the pixel local x on the rows and the local y on the columns. The matrix center is matched with the hit center of charge and each element m_{ij} is set to the A.D.C. level of the corresponding (x_i, y_j) pixel. Those pixels that stride over the hit cluster boundaries are set to zero (zero-padding technique).With this procedure each doublet can be considered as a collection of two 15×15 matrices, an example is shown in Figure 3, or as a two channel 2D image ($n \times m \times r = 15 \times 15 \times 2$). Thus the rejection of fake doublets is reduced to an image/pattern recognition task, perfectly suitable for being dealt with CNNs.

3.2. Doublet dataset: generation and features

To test the feasibility of this kind of approach, the generation (via PYTHIA 8 [9]) and the reconstruction of $t\bar{t}$ events at energy of the center of mass of $\sqrt{s} = 13$ TeV, with average pileup $< PU >= 35$ and bunch time spacing of 25 ns has been simulated within the CMS software framework (CMSSW [10]). For each event, both all the doublets produced and all the MC matched reconstructed tracks, i.e. associated with a tracking particle, are collected. A doublet is then labeled as *true* only if it is formed by pixel hits belonging to the same MC matched track. About 10^6 doublets are produced per each event and the ratio between *true* and *fake* doublets is between 300-400. For each doublet 537 parameters are stored:

- 225 + 225 *pixels* for the inner and the outer hit;
- 63 *doublet features* defined for each doublet and that include detector information and further hit and cluster characteristics;

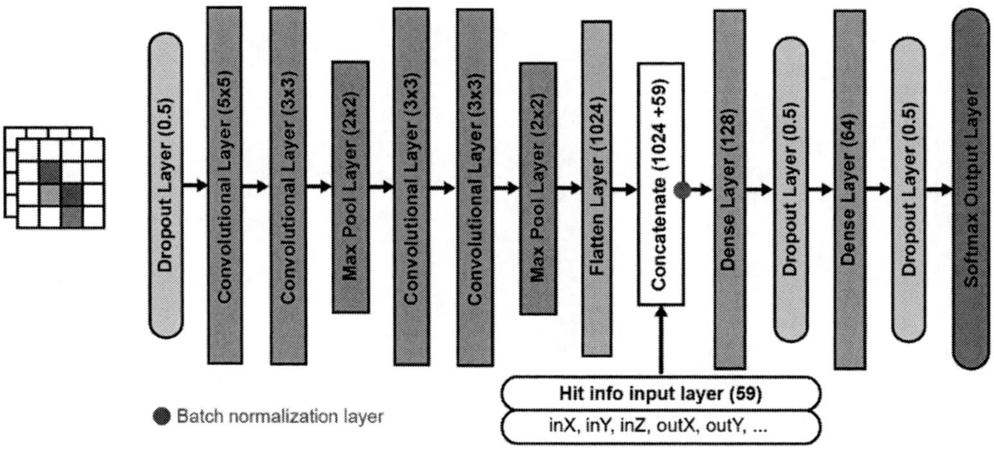

Figure 4. Layer map model architecture

IOP Conf. Series: Journal of Physics: Conf. Series **1085** (2018) 042040 doi:10.1088/1742-6596/1085/4/042040

- 24 *track labels* defined only for MC matched doublets, e.g. the corresponding track vertex coordinates, pt and eta;

On the whole, 1000 events are simulated, 800 for the training dataset, 150 for testing and 50 for validation. The training and the validation set are balanced so that the ratio between fake and true doublet is one. The whole balanced training dataset is composed of approximately 2.5 millions doublets.

4. CNN Classifier: the layer map model

The architecture for the doublet filtering classifier (*layer map model*) is shown in Figure 4 and it concatenates:

(i) *CNN block*: conventional stack of convolutional layers (4) with 5×5 or 3×3 filters and max pooling layers (2) whose output is reduced to a one dimensional structure through a flatten layer that returns a 1024 elements vector.

(ii) *Dense block*: stack of two fully connected layers that are fed with the one dimension reduced images from the previous block and 59 further doublets info (such as hits' detectors and coordinates).

The architecture is completed by a batch normalization layer inserted between the two blocks, *linear rectifier* activations for the convolutional and dense layers, three dropout layers and a *softmax* output layer for binary classification [7]. The two output neurons return the complementary probabilities that a doublet is fake or true ($p_{true} + p_{fake} = 1.0$). The 15×15 doublets pads, before being fed to the first block, are pre-processed as follows.

- Each pixel content is standardized with the mean and the standard deviation pre-computed on the whole 2.5 millions doublet dataset: $\mu = 13382.00$, $\sigma = 10525.13$.
- The input is split in 4 *angular channels*, 2 for the inner hit and 2 for the outer hit. From each hit pixel matrix $M_{i,o}$ two matrices are extracted as extra channels: one ($M_{i,o}^{cos}$) resulting from the multiplication of the original matrix by the cosine of the hit incident angle and one by the sine of the angle ($M_{i,o}^{sin}$) (θ_i for the inner and θ_o for the outer hit, see Figure 5).
- The input is split in 20 additional *detector channels*, 10 for the inner hit and 10 for the outer hit. Every channel corresponds to a single layer of the detector (4 barrel and 6 endcap) and is initialized to a 15×15 zero matrix. Then only the two channels corresponding to the inner and the outer layer are set, respectively, to the inner hit and the outer hit cluster matrix. This approach allows us to separate the hit shape clusters based on the specific layer and to apply a different transformation.

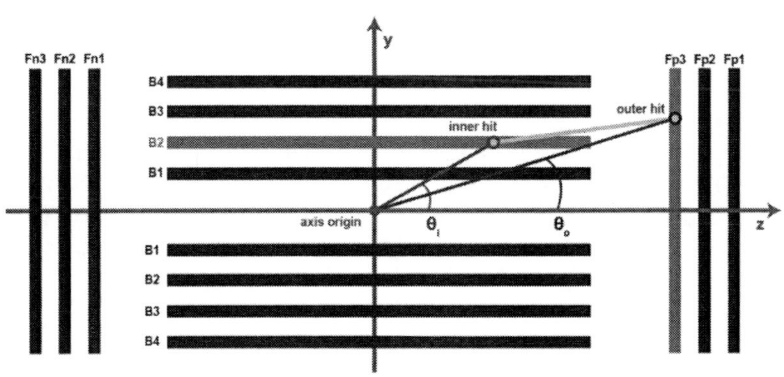

Figure 5. Example of an hit doublet with the inner hit on the third barrel layer (B3) and the outer hit on the first forwared layer (Fp1).

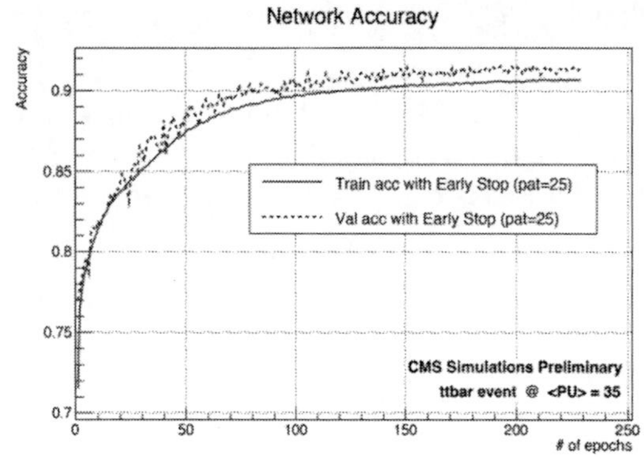

Figure 6. Accuracy for train (red ——) and validation datasets (blue - - - -) with early stopping patiance=25.

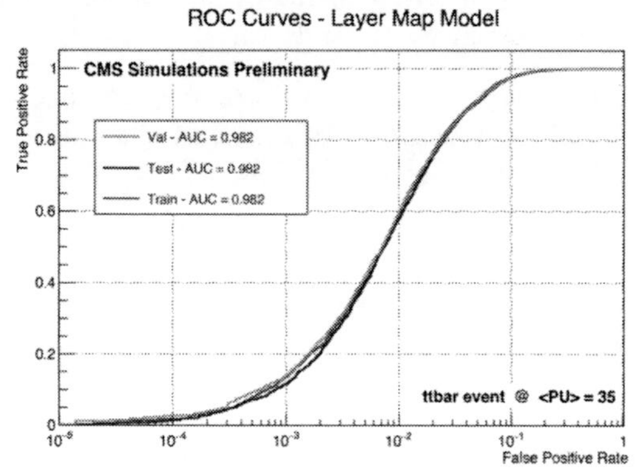

Figure 7. ROC curves for the train,test and validation dataset.

To sum up, the input dataset for the *layer map* model is composed of 24 channels 15×15 images standardized with their mean and their standard deviation. The model has been developed and trained with the `Keras` package [11], a high-level neural networks API, running on the top of `TensorFlow` library [12]. The network has been trained on NVIDIA Tesla K20 nodes at Bari Physics Department Tier 2 (ReCaS) [13] and on NVIDIA GTX1080 Ti GPUs at CERN.

5. Model testing and results

Once tested and tuned on a smaller sample, the model has been trained on the whole dataset (2.5M doublets) split in ten batches of 250k doublets in order to fit the available memory. The model has been trained with a *categorical cross entropy* loss function [7], using *Adam* [14] optimizer and accuracy as evaluation metric. At each batch training iteration the weights and the parameters of the trained network are passed to the next step. This procedure has been carried out for 20 global epochs on the whole dataset and each iteration is run with an *early stopping* callback, that is a form of regularization used to avoid over-fitting. It stops the network training when the selected metric (validation accuracy in our case) does not improve for a given number of consecutive epochs, denoted as *patience* (p=25 for the *layer map* setup). The training took about 5 days on a NVIDIA Tesla K20 node.

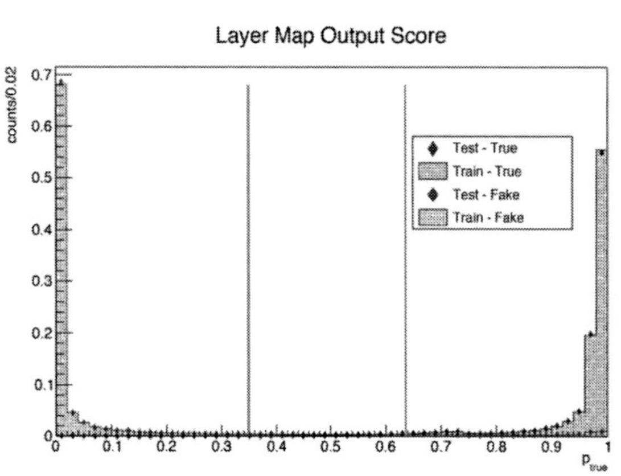

Figure 8. Network score for true (blue) and fake (doublets) for train (filled histogram) and test dataset (diamond markers). In purple the 0.94 accuracy threshold. In green the 0.99 efficiency threshold.

Table 1. Layer map network scores for train,validation and test dataset.

			TN/N at fixed TP/P		TP/P at fixed TN/N	
	AUC	**Acc**	**TP/P = 0.99**	**TP/P= 0.999**	**TN/N=0.99**	**TN/N=0.5**
Train	0.982	0.940	0.85426	0.6706	0.5896	0.9997
Test	0.982	0.941	0.8542	0.6709	0.5899	0.9996
Val	0.982	0.939	0.8525	0.6707	0.5948	0.9996

Figure 6 shows the model accuracy for train and validation dataset versus the number of training epoch for the first 250k batch. Validation curves follow the same trend as the training ones and this indicates that the network is not over-fitting on the training data. Note that the validation accuracy is always greater than the training one. This behavior depends on the fact that dropout layers are turned off when the network process the validation data. Therefore the network has more connections and neurons active thus is more complete and accurate.

The ROC curves for validation, test and training dataset, shown in Figure 7, completely overlay each other, and the area under the curves (AUCs) is more than 0.98. While assuring a 0.99 efficiency (true positive rate), network's sensitivity (true negative rate) reaches 0.85. The highest accuracy reached is about 0.94 for all the three datasets. See Table 1 for further network performance results. The normalized output score, namely the network estimated probability that a doublet is true (p_{true}) shows optimal separation between fake and true doublets sample. Both train and test p_{true} distributions are plotted in Figure 8 and the cut for an efficiency of 0.99. In order to compare them a two sided Kolmogorov-Smirnov test has been performed. This tests whether two samples are drawn from the same distribution [15]. For both true and fake histograms, the resulting score is $KS \approx 0.070$ corresponding to a p-value of $p_{val} \approx 0.961$, that assures us that the two histogram come from the same distribution with a very high level of confidence.

6. Conclusions and acknowledgments

In conclusion, the results described show that CNN techniques for mitigating combinatorial explosion look very promising and need to be further explored. Ongoing work includes the verification of the effect on the downstream track reconstruction, the exploration of different hardware architectures for fast inference and the final integration in the CMS reconstruction framework.

References

[1] CMS collaboration, *The CMS experiment at the CERN LHC*, 2008 JINST **3** S08004.
[2] L. Evans and P. Bryant, *LHC machine*, 2008 JINST **3** S08001.
[3] CMS collaboration, The CMS trigger system, JINST **12** (2017) P01020.
[4] https://twiki.cern.ch/twiki/pub/CMSPublic/HLTplotsCHEP2016/ .
[5] CMS Collaboration, *CMS Technical Design Report for the Pixel Detector Upgrade*, CERN-LHCC-2012-016, CMS-TDR-011 (2012).
[6] https://twiki.cern.ch/twiki/bin/view/CMSPublic/LumiPublicResults .
[7] I. Goodfellow, Y. Bengio, A. Courville, *Deep Learning*, The MIT Press (2016).
[8] Y. LeCun, et al. (1989), *IEEE Communications Magazine*, **27**(11), 4146.
[9] T. Sjstrand, S. Mrenna and P. Skands, *JHEP05* (2006) **026**, *Comput. Phys. Comm.* 178 (2008) **852**.
[10] https://twiki.cern.ch/twiki/bin/view/CMSPublic/WorkBookCMSSWFramework .
[11] Chollet, François and others, Keras, 2015, GitHub, https://github.com/fchollet/keras .
[12] TensorFlow: Large-Scale Machine Learning on Heterogeneous Systems, http://tensorflow.org .
[13] ReCas is a project financed by the italian MIUR (*PONa3_00052, Avviso 254/Ric.*); its web page is `http://www.recas-bari.it/index.php/en/`.

IOP Conf. Series: Journal of Physics: Conf. Series **1085** (2018) 042040 doi:10.1088/1742-6596/1085/4/042040

[14] D. P. Kingma, J. Ba, *Adam: A Method for Stochastic Optimization*, arXiv:1412.6980.
[15] W.T. Eadie et al. *Statistical Methods in Experimental Physics*, North-Holland, Amsterdam (1971).

IOP Conf. Series: Journal of Physics: Conf. Series **1085** (2018) 042041 doi:10.1088/1742-6596/1085/4/042041

Building a scalable python distribution for HEP data analysis

David J Lange

Physics Department, Princeton University, Princeton, NH 08544, USA

E-mail: dlange@princeton.edu

Abstract. There are numerous approaches to building analysis applications across the high-energy physics community. Among them are Python-based, or at least Python-driven, analysis workflows. We aim to ease the adoption of a Python-based analysis toolkit by making it easier for non-expert users to gain access to Python tools for scientific analysis. Experimental software distributions and individual user analysis have quite different requirements. Distributions tend to worry most about stability, usability and reproducibility, while the users usually strive to be fast and nimble. We discuss how we built and now maintain a python distribution for analysis while satisfying requirements both a large software distribution (in our case, that of CMSSW) and user, or laptop, level analysis. We pursued the integration of tools used by the broader data science community as well as HEP developed (e.g., histogrammar, root_numpy) Python packages. We discuss concepts we investigated for package integration and testing, as well as issues we encountered through this process. Distribution and platform support are important topics. We discuss our approach and progress towards a sustainable infrastructure for supporting this Python stack for the CMS user community and for the broader HEP user community.

1. Introduction

Use of the Python programming language has become ubiquitous in high-energy physics (HEP) and more broadly across data science. Python, as well as its user community, has evolved significantly since it was introduced into experiment software frameworks more than a decade ago. For example, the initial motivations to introduce Python as a dependency of the CMS software (CMSSW) distribution [1] were for runtime job configuration management, and for analysis using a light-weight version of the CMS framework (FWLite) [2] and/or the PyROOT toolkit [3].

These proceedings describe an effort to build a modern and maintainable Python toolkit matching the needs of both HEP frameworks and HEP analysts. The primary project goals include

1. **Reduce the startup burden for HEP analysts** interested in applying a new Python analysis tool to their project. If up to date Python-based tools are available in the same experimental software distributions that are used to access common HEP tools such as ROOT as well as experimental specific software, then the time needed for analysts to gain access to these tools can be considerably reduced.

IOP Conf. Series: Journal of Physics: Conf. Series **1085** (2018) 042041 doi:10.1088/1742-6596/1085/4/042041

2. **Facilitate use of a standard software environment** within analysis groups or experimental collaborations. HEP working groups typically consist of a geographically distributed community, typically using local computing resources for day-to-day work. The work of these groups can be simplified if they use a standard software platform, such as that typically maintained by central experimental teams. This approach reduces the chance to obtain different, or even incorrect, results due to inconsistent software versions or installations. Another aspect is the interoperability of software versions. Instead of small analysis groups each needing to understand the requirements and interdependencies of Python toolkits, a common software stack can be maintained and verified by an expert before put into use.
3. **Usability of distributed computing systems**. There is an increasing number of reasons why running Python-based toolkits on "the Grid" or more generically on a non-local set of computing resources is interesting. This will continue as the suite of computationally efficient data science tools available continues to increase.

These are some of the reasons why we felt a maintained Python toolkit is needed and of interest to the HEP community. In the next sections we will describe our approach and current state of implementation. We used CMSSW and its software distribution system as the basis for our work. The concepts and much of the implementation are not specific to CMS and can be easily ported. We will discuss this approach as well as how we have made this distribution easily available to developers not using CMSSW or otherwise working outside CMS.

2. Our approach

The CMSSW distribution has always included distributions of packages that form a relatively complete set of dependencies in addition to the build of the experiment specific codebase itself. This ensures that the entire software stack is built in a consistent manner. It was natural to expand this distribution to include our desired set of Python tools. Our aim was to do this in a simple yet maintainable way. Our approach was to adopt standard tools and make them easier for collaborators to use, rather than to build new capabilities.

2.1. PIP-based approach

As the number of Python packages expanded, we quickly realized that we needed to evaluate standard solutions for integrating Python tools without having to create and maintain code for each tool. After some evaluation, we selected PIP [4] to handle the Python tool management for CMSSW. PIP is a standard tool for managing Python packages that nearly all experienced Python users are familiar with.

Using PIP and the PYPI database [5] provides CMSSW a community standard mechanism for managing Python tools. We use source distributions on PYPI whenever possible to ensure that distributed toolkit is consistent with the full software stack that CMSSW distributes. We have only a few exceptions where a source based distribution is required.

This change also allowed us to standardize and simplify the code needed each time a Python tool is to be added to CMSSW. We can specify just the package, version and its dependencies to create the new tool in CMSSW. As CMSSW is RPM based, our implementation is driven by SPEC files, which look like this:

IOP Conf. Series: Journal of Physics: Conf. Series **1085** (2018) 042041 doi:10.1088/1742-6596/1085/4/042041

```
### RPM external py2-numba 0.33.0
## INITENV +PATH PYTHONPATH %{i}/${PYTHON_LIB_SITE_PACKAGES}

Requires: py2-funcsigs py2-enum34 py2-six
Requires: py2-singledispatch py2-llvmlite py2-numpy

## IMPORT build-with-pip
```

The last line is a small helper function (build-with-pip) that handles all of the common functionality. This is primarily to retrieve the sources from PYPI, to check that all dependencies are satisfied, to do the actual installation (including RPM creation in the case of CMSSW), and to provide hooks to handle some special cases (i.e., when options or environment variables are needed by the installation). We chose not to use PIP itself to discover, download and install dependencies. This choice was made to retain control over the full set of packages in CMSSW to help ensure that unwanted dependencies are not added by mistake.

Other advantages include the ability to discover the available Python packages distributed by CMSSW and their versions using a standard mechanism (PIP):

```
dlange> pip list
appdirs (1.4.3)
bleach (2.0.0)
Bottleneck (1.2.1)
certifi (2017.4.17)
chardet (3.0.4)
click (6.7)
climate (0.4.6)
```

2.2. *Allowing users to be nimble*

The most notable difference between the requirements of experimental distributions and user analysts is the expected turnaround time when software changes are needed. We added the VirtualEnv toolkit [6] to make this easy to do using the same tool (PIP) as used to put together the distribution itself. This approach also makes it possible to use the already existing mechanisms in CMS for making software changes automatically distributed to analysis jobs run on the CMS distributed computing system.

We use as an example the case where a user that wants to use Theano version 0.9.0 in their analysis whereas CMSSW currently uses 0.8.2 in the version of the software the user has. First the user will check what is current available in CMSSW (using two different mechanisms):

```
dlange> pip list | grep Theano
Theano (0.8.2)

dlange> python
Python 2.7.11 (default, Apr 28 2017, 13:50:27)
>>> import theano
>>> print theano.__version__
0.8.2
```

Next, a new working area for the VirtualEnv toolkit is created, activated and its corresponding path added to the Python path.

```
dlange> virtualenv updateTheano
New python executable in /build/dlange/CMSSW_9_3_X_2017-08-07-2300/updateTheano/bin/python
Installing setuptools, pip, wheel...done.
dlange> source updateTheano/bin/activate.csh

[updateTheano] dlange> setenv PYTHONPATH $PWD/updateTheano/lib/python2.7/site-
packages/:$PYTHONPATH
```

Then PIP is used to download and install the desired new version of Theano as well as any new dependencies. The CMSSW distribution is checked first for the dependencies before they are downloaded. In this case, the existing distribution already satisfies all of the needed dependencies.

```
[updateTheano] dlange> pip install Theano==0.9.0
Collecting Theano==0.9.0
Requirement already satisfied: scipy>=0.14
Requirement already satisfied: numpy>=1.9.1
Installing collected packages: Theano
Found existing installation: Theano 0.8.2
Not uninstalling theano at /cvmfs/cms-ib.cern.ch/.......... /lib/python2.7/site-packages, outside environment
/build/dlange/CMSSW_9_3_X_2017-08-07-2300/updateTheano
Successfully installed Theano-0.9.0
```

Finally, we can confirm that the new version of Theano is properly configured for use:

```
[updateTheano] dlange> pip list | grep Theano
Theano (0.9.0)

[updateTheano] dlange> python
Python 2.7.11 (default, Apr 28 2017, 13:50:27)
>>> import theano
>>> print theano.__version__
0.9.0
```

2.3. Installing outside of CMSSW

Using PIP also provides a natural mechanism to distribute our set of Python tools outside of CMSSW. We created a simple PYPI package [7] that can be installed in any environment using PIP. One important use case in CMS is since few collaborators run Scientific Linux on their personal laptops, this PIP-based mechanism is one way to obtain the same Python environment locally as is provided by the CMSSW distribution. The motivation of this work was to ease the standardization of tools across users in a working group. That is, working groups could adopt this PYPI package to easily provide the baseline versions of tools shared by the working group.

3. Python Tools

There is a broad range of Python toolkits of potential interest for data science and HEP, including an increasing number of toolkits developed specifically for HEP analysis use cases. We have attempted to add a broad range of tools that are in use across the HEP community. We have prioritized some packages simply based on CMS user interests and requests. A full list is beyond the scope of document, but can be found in our GitHub repository [8]. Example packages that we have included can be roughly grouped into five broad areas [9]:

1. **HEP specific toolkits**: Examples include rootpy, histogrammar, root_pandas, xrootdpyfs and root_numpy
2. **Notebook support**: The Jupyter stack

3. **Data science toolkits**: Examples include Bottleneck, pandas, downhill, sympy, numba, pytables and scipy.
4. **Interfaces to data storage formats**: Examples include HDF5, AVRO and ThriftPy.
5. **Machine learning toolkits**: Examples include Keras, tensorflow, scikit-learn, theano, and theanets.

4. Conclusion

We have described a new Python distribution mechanism meant to satisfy needs of both HEP analysis users and HEP experimental frameworks. It is based on PIP to provide ease of maintenance as well as a standard user interface to the distribution. Our development has been performed within the CMSSW distribution but does not strongly depend on it. We continue to extend this work with the aim to keep the distribution up to date as the data science Python tool kit evolves.

References

[1] G. Eulisse, et. al., "Rise of the Build Infrastructure", *J. Phys. Conf. Ser.* **513** 052009 (2014) (https://doi.org/10.1088/1742-6596/513/5/052009).

[2] C.D. Jones et al. 2008 Analysis environments for CMS, *J. Phys. Conf. Ser.* **119** 032027 (2008) (https://doi.org/10.1088/1742-6596/119/3/032027).

[3] https://root.cern.ch/pyroot

[4] https://pip.pypa.io/en/stable/

[5] https://pypi.python.org/pypi

[6] https://virtualenv.pypa.io/en/stable/.

[7] https://pypi.python.org/pypi/pyCMSSW currently version 0.1.

[8] https://github.com/cms-sw/cmsdist.

[9] Packages and their documentation can be found following this URL pattern: https://pypi.python.org/pypi/<package name>.

IOP Conf. Series: Journal of Physics: Conf. Series **1085** (2018) 042042 doi:10.1088/1742-6596/1085/4/042042

Boosted Decision Trees in the Level-1 Muon Endcap Trigger at CMS

Darin Acosta[1], Andrew Brinkerhoff[1], Elena Busch[2], Andrew Carnes[1], Ivan Furic[1], Sergei Gleyzer[1], Khristian Kotov[1], Jia Fu Low[1], Alexander Madorsky[1], Jamal Rorie[2], Bobby Scurlock[1], and Wei Shi[2] on behalf of the CMS Collaboration

[1] University of Florida
[2] Rice University

Abstract. The first implementation of a Machine Learning Algorithm inside a Level-1 trigger system at the LHC is presented. The Endcap Muon Track Finder (EMTF) at CMS uses Boosted Decision Trees (BDTs) to infer the momentum of muons in the forward region of the detector, based on 25 different variables. Combinations of these variables representing 2^{30} distinct patterns are evaluated offline using regression BDTs. The predictions for the 2^{30} input variable combinations are stored in a 1.2 GB look-up table in the EMTF hardware. The BDTs take advantage of complex correlations between variables, the inhomogeneous magnetic field, and non-linear effects – like inelastic scattering – to distinguish high momentum signal muons from the overwhelming low-momentum background. The new momentum algorithm reduced the background rate by a factor of three with respect to the previous analytic algorithm, with further improvements foreseen in the coming year.

1. Introduction

The Compact Muon Solenoid (CMS) is a detector at the Large Hadron Collider (LHC) located near Geneva, Switzerland. The LHC collides bunches of protons every 25 ns at a center of mass energy of 13 TeV. The CMS experiment detects the resulting particles and measures their kinematics using various subdetectors working in concert. With 40 million proton bunch crossings per second amounting to roughly 40 TB of data each second, saving the information from every event is not feasible. As such, the CMS trigger system chooses the interesting events to save to disk, operating in two stages [1]. The Level-1 (L1) trigger runs in hardware online reducing the throughput of data from 40 MHz to 100 KHz. From there, the High Level Trigger (HLT) operates in software online reducing the rate from 100 KHz to 1 KHz. In the end, about 1 GB/s is saved to disk.

With 40 MHz of input, the L1 Trigger has only 4 μs to decide whether to keep the information for an event. The Endcap Muon Track Finder (EMTF) – part of the L1 Trigger dedicated to muons – has only about 500 ns to determine the location, tracks, and momentum of the muons passing through the Cathode Strip Chambers (CSC) and Resistive Plate Chambers (RPC) in the endcaps of CMS [2]. High momentum muons are an important object for many physics analyses at CMS. As such, an accurate momentum assignment distinguishing low momentum muons (background) from high momentum muons (signal) is key to the EMTF trigger. In order to meet the timing requirements, the EMTF's logic is implemented in Field Programmable Gate

Arrays (FPGAs), a type of reprogrammable hardware that allows vast parallelization and speeds much greater than even the best CPUs.

To improve the transverse momentum (p_t) assignment for muons in the endcaps at Level-1, the EMTF team trained Boosted Decision Trees (BDTs) offline using TMVA [3], and stored the prediction scheme into a 1.2 GB Look-Up Table (LUT). The FPGAs then use the LUT online to assign the p_t in a single operation. Using the LUT to turn the BDT p_t assignment into a simple look-up enables the EMTF to utilize the power of a robust machine learning algorithm for its momentum predictions while still operating at the required time scale. Putting a parallelized version of the BDTs directly into the FPGAs, while hypothetically possible, would require more than the available number of logic gates. Such an implementation would still be slower than the LUT method, and changing the p_t assignment would require reprogramming the FPGA logic each time. The LUT method provides a simple way to run any machine learning evaluation at high speed by turning the evaluation into a single operation.

2. Metrics of Success

Two metrics are used to measure the success of the EMTF: the rate and the efficiency. The rate at X GeV is defined as the number of muons with a predicted p_t greater than X GeV. In other words, the rate consists of both true and false positives above the p_t threshold. The efficiency at X GeV is defined as the number of muons with both predicted p_t and true p_t greater than X GeV divided by the number of muons with true p_t above X GeV. Put another way, the efficiency measures the percentage of muons with true p_t above X GeV that were correctly predicted above X GeV. When evaluating the efficiency in data, the true p_t is typically provided by the offline reconstruction. When evaluating the efficiency in simulation, the true p_t is typically provided by the Monte Carlo truth. A good trigger will minimize the data saved without losing the interesting high p_t events where unexplored physics lies, i.e. it will minimize rate while maximizing the efficiency.

3. The EMTF Regression Project

A muon traveling through the endcap detectors has a chance to leave hits in four sequential stations labeled 1, 2, 3, and 4. The specific combination of hits like 1,3,4 is called the mode. Each station records the ϕ and θ location of a hit, among other information. The CSCs have better spatial resolution, so the ϕ and θ information is taken from the CSCs by default, but the RPC measurements for the station are used if the CSCs missed the hit in the same station. The charged muons travel through a magnetic field following curved paths due to the Lorentz force. The force causes the high p_t muons in a magnetic field to bend less and the low p_t muons to bend more. The difference in ϕ and θ between stations i and j, $\Delta\phi_{ij}$ and $\Delta\theta_{ij}$, quantify the curvature of the track. With most of the curvature accounted for by the $\Delta\phi$ variables, the $\Delta\phi$s provide the majority of the p_t discrimination.

A major difficulty in minimizing the rate is the steeply falling p_t distribution. A typical interesting event has p_t greater than 25 GeV, and there are about one thousand 5 GeV muons for every 25 GeV muon. With so many more low p_t events, predicting the low momentum muons poorly will drastically increase the rate. Moreover, in addition to the large number of low p_t muons, there are other notable difficulties: the muons travel through a non-uniform magnetic field, some scatter between detector stations, and those with high p_t often shower charged particles upon interacting with the detector material. Moreover, low p_t muons may spiral completely before getting to the next station. The scattering, showering, and spiraling add noise to the underlying true behavior, while the number of low p_t muons requires that the regression focus on the low momentum regime to prevent an explosion in the rate.

In order to assign p_t in a robust way and deal with the aforementioned difficulties, a BDT is trained for each possible mode using the discretized values for the features of Table 1. The

IOP Conf. Series: Journal of Physics: Conf. Series **1085** (2018) 042042 doi:10.1088/1742-6596/1085/4/042042

loss function and weights are chosen to focus on the low p_t events and minimize the rate while maintaining acceptable efficiency. Features are chosen for each mode to give the BDT the information needed to predict the p_t while dealing with the non-uniform magnetic field and the problematic scattering and showering effects.

The $\Delta\phi$ variables available for each mode are used as features to determine the curvature and get most of the p_t discrimination. However, the power of these variables depends largely on the track position in θ. The magnetic field varies as a function of θ affecting the magnitude of the curvature for a given p_t, thus correlating $\Delta\phi$, p_t, and θ. The link between these three makes θ the next most important training feature.

Variables modeling the mean and RMS of the available $\Delta\phi$s for the mode are also used as features in order to identify scattering and showering effects. If a muon were to scatter or shower between stations the recorded hit in a station may not be the true hit of the muon. Any $\Delta\phi$ involving this station will be an outlier. To determine the severity of the deviation and the likelihood of scattering/showering, the idea is to identify the outlier station and to compare the mean and RMS $\Delta\phi$ with and without the outlier station. The greater the difference the greater the severity. The nominal mean and RMS $\Delta\phi$ features are calculated using all available $\Delta\phi$s for the mode. The exclusive mean and RMS are calculated using all available $\Delta\phi_{ij}$ for the mode with i or j $\neq S_{out}$, where S_{out} is the outlier station. S_{out} is the excluded station such that leaving it out of the sum minimizes the mean and RMS. The outlier station, S_{out}, is also used as a feature. Including the nominal mean and RMS of $\Delta\phi$, the exclusive RMS and mean of $\Delta\phi$, and S_{out} as features helps the BDT differentiate scattering, showering, and normal events.

The features described above are the most important features, but not the whole collection. The front-rear (FR) bit designates whether the muon hit a front or rear CSC chamber in the station, and it is also included. The $\Delta\theta$s provide additional curvature information, and these are included as well. The B feature for each station is included as well, and it flags whether the ϕ, θ information for the station came from the CSCs or the RPCs. If there are bits available for the B_i feature it also includes information about the single station $\Delta\phi$ bend angle within a CSC chamber. Lastly, the $+/-$ feature stores the signs of the later $\Delta\phi$s relative to the first $\Delta\phi_{ij}$ for the mode.

4. Putting the BDTs into a Look-up Table

After training BDTs for each mode, the mode and the fundamental features from which the others can be derived are discretized and fit into a 30 bit word. The discretization scheme is different for each mode, detailed in Table 1. With the feature space compressed into 30 bits, there are 2^{30} possibilities that need to be assigned a p_t. A LUT is created by looping over all 2^{30} possible bit words, decoding each word into the fundamental features, deriving the secondary features, and sending the values to the BDT to assign the p_t prediction. Using 9 bits for the p_t, this amounts to a 1.2 GB LUT where each bit word value is an address and the p_t is the value in memory. Discretizing the feature space and creating a LUT turns the p_t assignment into a single operation. The LUT is then used by the FPGA logic online to assign p_t to muon tracks in the EMTF. The LUT method is a simple way to run any machine learning method quickly, but compressing the features into 30 or so bits may not always be feasible for the application.

Table 1. The feature discretization scheme for each mode.

Four Station Modes

Mode	**Feature**	$\Delta\phi_{12}$	$\Delta\phi_{23}$	$\Delta\phi_{34}$	+/-	$\Delta\theta_{14}$	B_1	B_2	B_3	B_4	FR_1	θ	Mode
1-2-3-4	**Bits**	7	5	4	2	2	2	1	1	1	1	3	1

Three Station Modes

Mode	**Feature**	$\Delta\phi_{12}$	$\Delta\phi_{23}$	+/-	$\Delta\theta_{13}$	B_1	B_2	B_3	FR_1	FR_2	θ	Mode
1-2-3	**Bits**	7	5	1	3	2	1	1	1	1	5	3
Mode	**Feature**	$\Delta\phi_{12}$	$\Delta\phi_{24}$	+/-	$\Delta\theta_{14}$	B_1	B_2	B_4	FR_1	FR_2	θ	Mode
1-2-4	**Bits**	7	5	1	3	2	1	1	1	1	5	3
Mode	**Feature**	$\Delta\phi_{13}$	$\Delta\phi_{34}$	+/-	$\Delta\theta_{14}$	B_1	B_3	B_4	FR_1	FR_3	θ	Mode
1-3-4	**Bits**	7	5	1	3	2	1	1	1	1	5	3
Mode	**Feature**	$\Delta\phi_{23}$	$\Delta\phi_{34}$	+/-	$\Delta\theta_{24}$	B_2	B_3	B_4	FR_2	–	θ	Mode
2-3-4	**Bits**	7	5	1	3	2	1	1	1	–	5	4

Two Station Modes

Mode	**Feature**	$\Delta\phi_{XY}$	$\Delta\theta_{XY}$	B_X	B_Y	FR_X	FR_Y	θ	Mode
X-Y	**Bits**	7	3	3	3	1	1	5	7

X-Y runs through the possible two station combinations: 1-2, 1-3, 1-4, 2-3, 2-4, 3-4.

5. Results and Conclusions

The LUT scheme utilizing the BDT predictions has been implemented in the EMTF for 2016 and 2017 data taking. As seen in Figure 1, the upgraded system – compared to the legacy system – reduces the rate at 25 GeV by a factor of three with no loss in efficiency. The legacy system was used in the endcaps until 2015 and assigned p_t using a maximum likelihood fit with up to three $\Delta\phi$s. More detailed information about the legacy system can be found in [1].

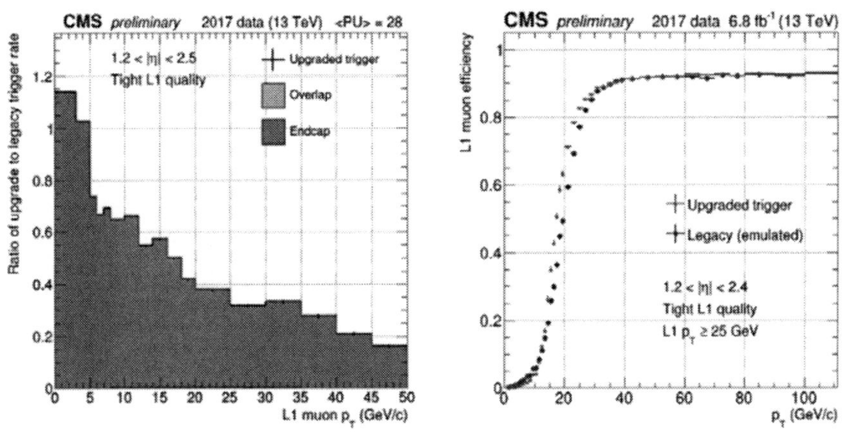

Figure 1. On the left, the upgraded EMTF rate divided by the legacy rate is shown for a variety of p_t thresholds. On the right, the upgraded and legacy efficiencies are presented for a 25 GeV threshold. The upgraded EMTF has a 3x lower rate than the legacy system at 25 GeV with virtually no difference in plateau efficiency for the same threshold. Plots are taken from [4].

References

[1] Khachatryan V *et al.* (CMS) 2017 *JINST* **12** P01020 (*Preprint* 1609.02366)
[2] Tapper A and Acosta D (CMS collaboration) 2013 CMS Technical Design Report for the Level-1 Trigger Upgrade Tech. Rep. CERN-LHCC-2013-011. CMS-TDR-12 additional contacts: Jeffrey Spalding, Fermilab, Jeffrey.Spalding@cern.ch Didier Contardo, Universite Claude Bernard-Lyon I, didier.claude.contardo@cern.ch URL https://cds.cern.ch/record/1556311

IOP Conf. Series: Journal of Physics: Conf. Series **1085** (2018) 042042 doi:10.1088/1742-6596/1085/4/042042

[3] Hoecker A, Speckmayer P, Stelzer J, Therhaag J, von Toerne E and Voss H 2007 *PoS* **ACAT** 040 (*Preprint* physics/0703039)
[4] CMS Collaboration 2017 URL https://cds.cern.ch/record/2286327

ACAT2017 IOP Publishing
IOP Conf. Series: Journal of Physics: Conf. Series **1085** (2018) 042043 doi:10.1088/1742-6596/1085/4/042043

The ATLAS Trigger Simulation with Legacy Software

Catrin Bernius
SLAC National Accelerator Laboratory, Menlo Park, California
E-mail: Catrin.Bernius@cern.ch

Gorm Galster
The Niels Bohr Institute, University of Copenhagen, Copenhagen, Denmark
E-mail: Gorm.Galster@cern.ch

Andy Salnikov
SLAC National Accelerator Laboratory, Menlo Park, California
E-mail: Salnikov@slac.stanford.edu

Joerg Stelzer
CERN, Geneva, Switzerland
E-mail: Joerg.Stelzer@cern.ch

Werner Wiedenmann
University of Wisconsin, Madison, Wisconsin
E-mail: Werner.Wiedenmann@cern.ch

Abstract. Physics analyses at the LHC require accurate simulations of the detector response and the event selection processes, generally done with the most recent software releases. The trigger response simulation is crucial for determination of overall selection efficiencies and signal sensitivities and should be done with the same software release with which data were recorded. This requires potentially running with software dating many years back, the so-called legacy software, in which algorithms and configuration may differ from their current implementation. Therefore having a strategy for running legacy software in a modern environment becomes essential when data simulated for past years start to present a sizeable fraction of the total. The requirements and possibilities for such a simulation scheme within the ATLAS software framework were examined and a proof-of-concept simulation chain has been successfully implemented. One of the greatest challenges was the choice of a data format which promises long term compatibility with old and new software releases. Over the time periods envisaged, data format incompatibilities are also likely to emerge in databases and other external support services. Software availability may become an issue, when e.g. the support for the underlying operating system might stop. The encountered problems and developed solutions will be presented, and proposals for future development will be discussed. Some ideas reach beyond the retrospective trigger simulation scheme in ATLAS as they also touch more generally aspects of data preservation.

1. Introduction

To analyse the data taken with the ATLAS detector [1] at the Large Hadron Collider (LHC), many analyses require a corresponding sample of simulated Monte Carlo (MC) data. This means that the production of new MC data needs to be maintained for all data taking periods. While MC data are generally produced with the newest software release, the simulated trigger response needs to be in agreement with when the data was taken by ideally using the same trigger algorithms and selections. This is necessary to reproduce selection efficiencies and trigger response as close as possible to those in the accumulated real data sample. Therefore the accurate re-simulation (e.g. because of improved detector response description, simulation of new physics processes etc.) of in particular the trigger response for data taking periods from several years ago poses a challenge. A solution to this is to use legacy software, the trigger software and conditions data that match the simulated data-taking period dating potentially many years back.

2. The ATLAS Monte Carlo Simulation Chain

A simplified view of the ATLAS MC simulation chain [2] is shown in Figure 1. Generated physics processes are put through a detailed simulation of the detector response (*Hits*). They are then available after the digitization processes as simulated raw data which are the input to the trigger simulation, adding the trigger response record to the event data. The MC simulation is completed by further event reconstruction. Data exchange between digitization, simulation and further reconstruction in form of *Raw Data Object (RDO)* [3], a *ROOT/POOL* [4, 5] based data format.

3. Considerations and Options for Simulation of the Trigger Response

There are several reasons that will require to re-simulate the trigger response, e.g. an improved description and/or understanding of the detector response, improved software for offline reconstructions in terms of algorithms and methods, increase of the MC sample size for future studies and the introduction of new event generators.

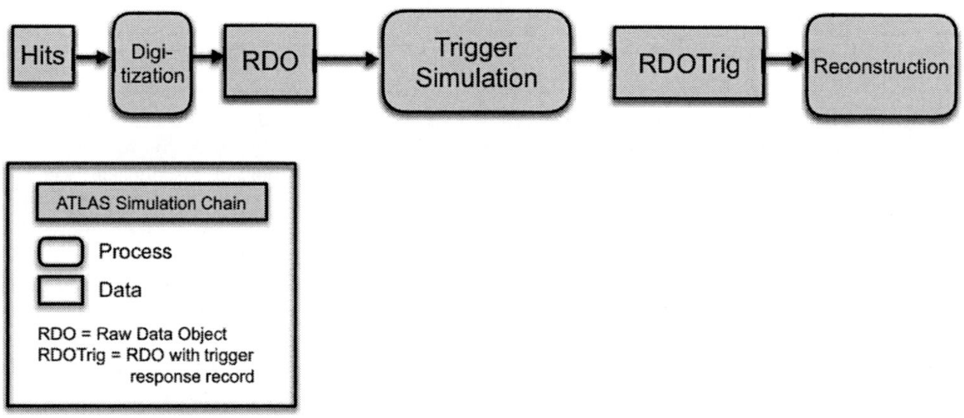

Figure 1. Simplified view of the ATLAS MC simulation chain. The physics processes that are generated are first put through a detailed simulation of the detector response (*Hits*), then digitized. The data is then passed on to the trigger simulation and further reconstruction. For the entire chain the same common simulation release is used. The data are exchanged in form of *Raw Data Object (RDO)*.

IOP Conf. Series: Journal of Physics: Conf. Series **1085** (2018) 042043 doi:10.1088/1742-6596/1085/4/042043

While detector response simulation and event reconstruction should be done with the newest software, the version used for the simulation of the trigger response needs to match that used for data taking. Mismatch between the software versions imposes a number of challenges with the most direct challenge being the data format compatibility with respect to the trigger response simulation. Therefore forward compatibility of format and content produced by new detector simulation or the possibility of the detector response conversion to an older format readable by the old trigger simulation needs to be guaranteed. Similarly backward compatibility or a conversion step need to be considered for the reconstruction that has to be able to read the old trigger response record.

Besides the data format challenges, there are the changes to hardware architectures, operating systems, core components and compiler changes which make it impossible to run the old trigger software "as is" today. The option of porting old trigger selection code to new simulation releases poses various problems as it would be necessary to keep old selection lines, algorithms, their configuration and conditions data operational along with the most recent trigger selections. This is not only problematic as it requires significant maintenance and manpower efforts but because of conflicting requirements in the trigger selections, the preservation of knowledge and the maintenance of the the infrastructure services.

The reuse of unmodified legacy trigger selection code from old releases can provide the most accurate trigger simulation of past data taking periods. The option to rerun legacy code allows for a strategy where the representative releases for a data taking period are chosen when all the required information and expertise is still available. The same selection algorithms and trigger configurations as during data taking can be applied and no or very little maintenance effort for conservation of the legacy trigger selection lines is required. Due to the advantages of this option, it has been further investigated and a simulation chain has been developed.

4. Trigger Simulation with Legacy Trigger Selection Software

To realise running the trigger legacy trigger selection code, there are various considerations that have to be addressed.

To be able to run with legacy releases, the long term conservation of trigger software releases, configuration data and conditions data becomes a necessity. In ATLAS, the software distribution which also includes all required external software components, the compilers with their run time environments and software configuration and development tools is done on the *cvmfs* file system [6].

Furthermore the ATLAS simulation chain requires a split into sub-steps which can use different software release versions. In the present simulation chain data are exchanged between the simulation modules as RDO data containing MC truth information and meta data with data processing parameters using a single software release. This guarantees the data compatibility between all steps, which is not given when running the trigger simulation with a different (legacy) software. In this case the data compatibility becomes an issue with respect to forward compatibility (output data from newer detector simulation need to be readable with older trigger releases) and backward compatibility (output data from old trigger release needs to be readable with newer reconstruction release).

Data which are directly read out from the detector hardware are available as *byte stream (BS)* data. This format is based on containers of 32 bit integers with a simple payload structure and described by the ATLAS raw data format [7]. It is tightly coupled to the detector readout hardware with only very few changes over time. Since it is a requirement that all ATLAS software releases can read data in BS format from all data taking periods, backward compatibility is guaranteed. Due to the very simple structure of the data format providing forward compatibility is much easier. The scripts to convert RDO data to BS data are already available in simulation releases and some ATLAS detectors already provide forward compatibility. The BS data format is therefore well suited to reach compatibility between different software release versions. However it does not provide structures to hold MC truth information or simulation meta data like the RDO data. The trigger simulation step only needs the detector raw data as input without any additional MC information and provides as output the trigger decision record in BS format. For the subsequent event reconstruction step the output can be merged with the other event reconstruction input data.

The additionally introduced sub-steps of RDO to byte stream data conversion before the trigger simulation and data merging after the trigger simulation which integrate the trigger simulation with a legacy release in the simulation chain are shown in Figure 2. In the data merging step the trigger decision record is converted from BS format to RDO format and added to the simulated data.

5. Use of Virtualisation - An Outlook

In the medium term the use of older releases on new operating systems can be achieved with a compatibility layer. However in the long term, using legacy code "as is" will be impossible on modern computing platforms. New computing hardware technologies, operating system changes and updates to the compilers and core run time libraries require changes to the legacy trigger releases. Virtualization enables the abstraction of the hardware and software run time layer from the underlying computing platform but comes at the expense of computational and resource overhead. Furthermore, external infrastructure services, e.g. for data input and output will most likely undergo important changes on a time scale of 10 years, requiring adaptations for the legacy trigger release. Patch releases to the legacy code should collect all these changes and allow for the integration with the environment external to the virtual machine. Encapsulation of the legacy trigger selection code in a virtual machine image extends the time span for of using legacy trigger selection code.

6. Conclusion

The challenges of simulating the trigger response precisely in MC simulation campaigns for data-taking periods dating back many years have been discussed and a strategy how to overcome

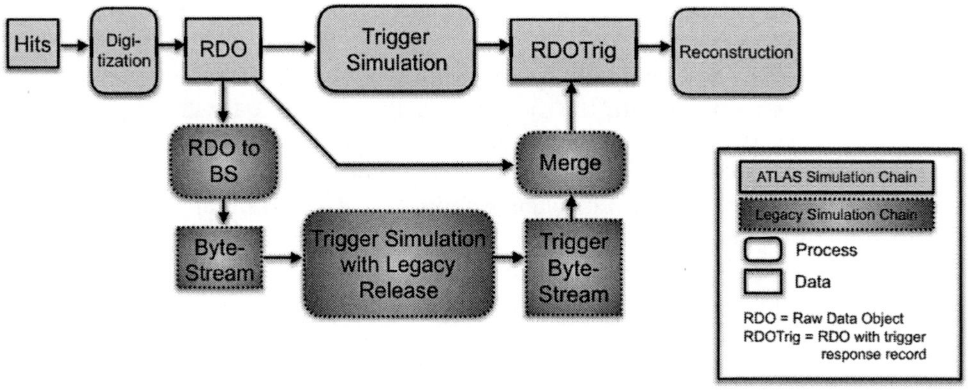

Figure 2. Simplified view of the ATLAS MC simulation and legacy simulation chain using a different, legacy software release for the trigger simulation. For the legacy simulation chain the data are converted from RDO to byte stream format and passed through the trigger simulation using a legacy software release. The resulting byte stream with trigger response record is then merged with information from the initial RDO and converted back to RDO format.

these difficulties has been laid out. While detector simulation and reconstruction use the newest available software releases, the simulation of the trigger response is done with older software which was used for this particular data-taking period that is simulated. Issues with data format compatibility between the different simulation steps have been studied and a modified simulation chain has been identified and implemented. Further work is still needed to completely integrate this modified simulation chain in the production workflow for large scale simulation campaigns in ATLAS.

References

[1] ATLAS Collaboration, JINST **3**, S08003 (2008).
[2] ATLAS Collaboration, Eur. Phys. J. C **70**, 823 (2010).
[3] P. van Gemmeren, D. Malon, for the ATLAS Collaboration, *Supporting High-Performance I/O at the Petascale: The Event Data Store for ATLAS at the LHC*, Proccedings of IEEE Cluster 2010, Grece, September 2010.
[4] *ROOT - A C++ framework for petabyte data storage, statistical analysis and visualization*, Computer Physics Communications, Anniversary Issue, Volume 180, Issue 12, Dec. 2009, P. 2499-2512.
[5] *POOL - Persistency Framework*, [http://pool.cern.ch].
[6] *CernVM Software Appliance*, [http://cernvm.cern.ch/portal].
[7] C.P.Bee, D.Francis, L.Mapelli, R.McLaren, G.Mornacchi, J.Petersen, F.J.Wickens, ATLAS-DAQ-98-129 (2004), [https://cds.cern.ch/record/683741].

IOP Conf. Series: Journal of Physics: Conf. Series **1085** (2018) 042044 doi:10.1088/1742-6596/1085/4/042044

The VISPA internet-platform in deep learning applications

Martin Erdmann, Benjamin Fischer, Robert Fischer, Erik Geiser, Christian Glaser, Gero Müller, Thorben Quast, Marcel Rieger, Martin Urban, Florian von Cube, David Walz and Christoph Welling

III. Physics Institute A, RWTH Aachen University, Germany

E-mail: erdmann@physik.rwth-aachen.de

Abstract. Latest developments in many research fields indicate that deep learning methods have the potential to significantly improve physics analyses. They not only enhance the performance of existing algorithms but also pave the way for new measurement techniques that are not possible with conventional methods. As the computation is highly resource-intensive both dedicated hardware and software are required to obtain results in a reasonable time which poses a substantial entry barrier. We provide direct access to this technology after a revision of the internet platform VISPA to serve the needs of researches as well as students. VISPA equips its users with working conditions on remote computing resources comparable to a local computer through a standard web browser. For providing the required hardware resources for deep learning applications we extend the CPU infrastructure with a GPU cluster consisting of 10 nodes with each 2 GeForce GTX 1080 cards. Direct access through VISPA, preinstalled analysis software and a workload management system allowed us on one hand to support more than 100 participants in a workshop on deep learning and in corresponding university classes, and on the other hand to achieve significant progress in particle and astroparticle research. We present the setup of the system and report on the performance and achievements in the above mentioned usecases.

1. Introduction

The project Visual Physics Analysis (VISPA)[1, 2] implements a development environment for data analyses in a web browser. It supports the typical development cycle of designing, executing, verifying and redesigning an analysis.

For analyses using deep learning techniques the working environment consisting of hardware and software components is essential for quick turnarounds. Throughout this proceeding this working environment is called workspace. The VISPA web platform gives its users access to computing resources which is considered as a key feature. The VISPA cluster is provided as default workspace but more computing resources can be added by the user. We present the setup of the system and focus on how the interplay between VISPA platform and cluster makes analysis resources available through a web browser.

2. The VISPA Server

The VISPA platform allows its users to work through a web-browser on a remote computing system, see figure 1. The only requirement to use the system is a modern web browser. If the user

IOP Conf. Series: Journal of Physics: Conf. Series **1085** (2018) 042044 doi:10.1088/1742-6596/1085/4/042044

wants to access additional workspaces than they must be accessible via SSH. Clients connect to the VISPA server via HTTPS and the connection between the server and the computing resource is established via remote procedure call (RPC) over SSH. So the VISPA server serves as a mediator between the user and the workspace by executing commands issued from client side and by returning the feedback from workspace side to the client.

All parts of the graphical user interface are written in HTML5 and JavaScript. This allows access from many different devices like desktop computers, tablets or mobile phones. The server uses the Python web framework CherryPy [3]. Code needed on workspace side e.g. to access files is written in Python [4] and copied automatically to the computing resource upon connection.

User credentials are stored in a SQL database like MariaDB [5]. For privacy reasons passwords to workspaces are not saved.

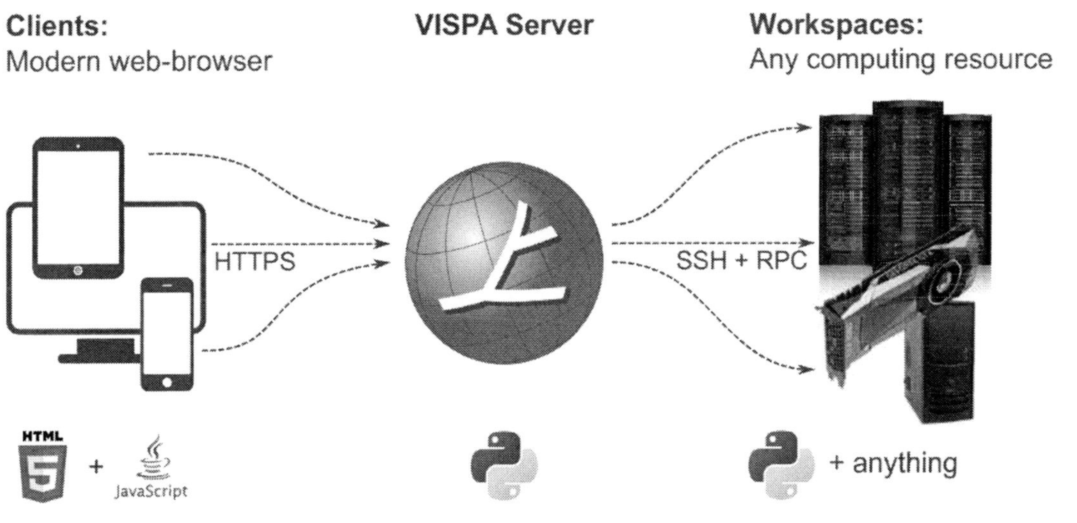

Figure 1. The VISPA Server as mediator between user and computing resources.

3. Extensions

The VISPA server provides different functionality by extensions. Default extensions are a file manager, a terminal and a code editor with execution and image preview capabilities. Further extensions are a browser for data of the Pierre Auger Observatory [6] or a parameter scan extension [7]. The latter helps to explore high-dimensional file system structures which appear naturally when naming output files of parameter scans by their parameters. An extension helping to work with various job submission tools like LSF or CRAB is currently under development.

4. The VISPA Cluster

The VISPA cluster operates since 2012 and consists of multiple computing nodes with in total 212 CPU cores and 20 NVIDIA GeForce GTX 1080 graphic cards (GPU). CPU only nodes are equipped with 16 GB of RAM, nodes assembled with a GPU provide 64 GB of RAM. The latter nodes were bought in 2016 and are dedicated to teaching purposes and analyses using deep learning techniques.

User space is provided by a network file system (NFS) server where home directories are stored on SSD and thus kept small. Data is stored on HDDs on the NFS server allowing for larger file sizes. The hard disks of the local computing nodes are used to cache the network file system.

IOP Conf. Series: Journal of Physics: Conf. Series **1085** (2018) 042044 doi:10.1088/1742-6596/1085/4/042044

Analysis software, e.g. TensorFlow [8], is installed and managed centrally which allows users to concentrate on analysis properties.

5. User Authentication

To provide a default computing resource, users registered on the VISPA server are able to access the VISPA cluster without additional password. For this purpose the MariaDB database is synchronized with a LDAP [9] database. The password to login to the VISPA server will also be used as cluster password allowing for an automatic login which gives users an interactive look and feel. Note that in previous publications, i.e. [10] the user authentication was achieved via a direct lookup into the SQL database. During the automatic login a round robin procedure is used to distribute the users to different login nodes which decreases the load on each individual node.

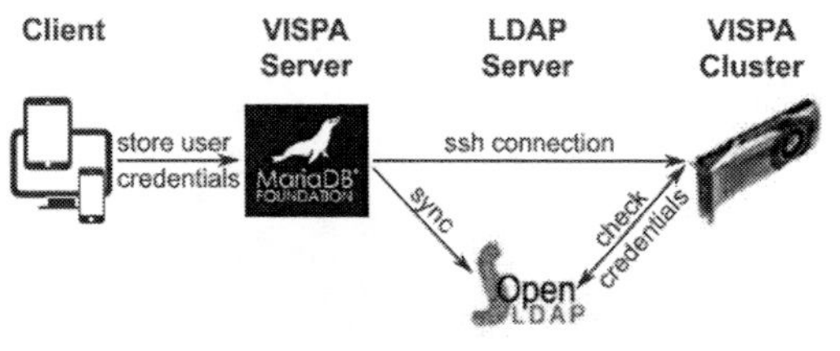

Figure 2. User authentication using an OpenLDAP database synchronized to the VISPA Server database

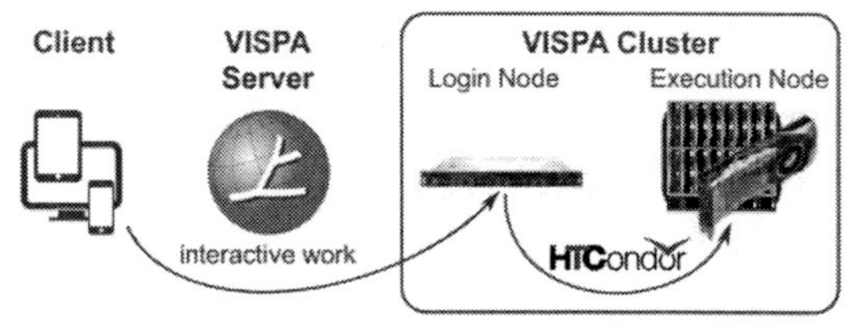

Figure 3. Job scheduling: Small analyses can be executed interactively on login node, longer jobs are managed by HTCondor.

6. Job scheduling

Computing resources should always be distributed on a fair share basis. Due to this reason only small interactive processes are allowed on the respective login node. More expensive jobs need to be send to the batch system software HTCondor [11], see also figure 3. The submission can be done directly from the code editor which lowers the entry barrier for inexperienced users. Graphic cards are not available interactively. Using HTCondor to schedule processes allows to distribute GPU resources on a fair share and not a first come first serve basis. Note that HTCondor is able to schedule jobs on a per GPU basis and even allows to request two GPUs.

7. Experience

The VISPA system consisting of server and cluster is used for education, outreach and research, see [12]. In 2017 it first served as a basis for a deep learning workshop in astroparticle physics with ~ 70 participants who needed a simultaneous access to the resources. Secondly, the execution of exercises of a university class on deep learning with ~ 90 students lead to high load peaks. The setup of the system facilitated the realization as everyone had the required hardware and software at their hands. Furthermore, the transparent approach to hide system details helped the participants to concentrate on deep learning techniques and physics problems.

VISPA is used to develop new deep learning methods in high energy and astroparticle physics [13, 14]. It allows several users to work on the same project using the same resources which streamlines the development process.

8. Conclusion

The VISPA web server and the VISPA computing cluster were presented. We focused on technical steps to connect webserver and cluster to obtain a seamless access to deep learning infrastructure. User authentication and distribution of resources are revealed as key points to provide an interactive look and feel. We have shown that the VISPA project helps to concentrate on physics problems. This increases the learning curve for students and the productivity for scientists.

Acknowledgments

This work is supported by the Ministerium für Wissenschaft und Forschung, Nordrhein-Westfalen and the Bundesministerium für Bildung und Forschung (BMBF)

References

[1] Bretz H.-P. et al., A Development Environment for Visual Physics Analysis, JINST 7 (2012) T08005 and http://vispa.physik.rwth-aachen.de
[2] VISPA software [software], 2017, Available from https://forge.physik.rwth-aachen.de/projects/vispa-web/repository
[3] CherryPy - A Minimalist Python Web Framework [software], version 10.1.1, 2017. Available from https://github.com/cherrypy/cherrypy/releases/tag/v10.1.1 [accessed 2017-10-12]
[4] Python project, Python [software], version 2.7.3, 2012. Available from https://www.python.org/downloads/release/python-273/ [accessed 2017-10-12]
[5] MariaDB [software], 2017. Available from https://github.com/MariaDB/server/releases/tag/mariadb-10.1.27 [accessed 2017-10-12]
[6] vispa-augeroffline event browser [software], 2017, Available from https://forge.physik.rwth-aachen.de/projects/vispa-augerofflineeventbrowser
[7] vispa-pscan [software], 2017, Available from https://forge.physik.rwth-aachen.de/projects/vispa-pscan
[8] TensorFlow [software], 2017, Available from https://github.com/tensorflow/tensorflow, [accessed 2017-10-13]
[9] OpenLDAP [software], version 2.4.40, Available from https://www.openldap.org/software/repo/openldap.git [accessed 2017-10-12]
[10] van Asseldonk D. et al., The VISPA internet platform for outreach, education and scientific research in various experiments, 2015, J. Phys. Conf. Ser., 664, 032031
[11] HTCondor [software], Available from https://research.cs.wisc.edu/htcondor/index.html
[12] Erdmann M. et al., A field study of data analysis exercises in a bachelor physics course using the internet platform VISPA, 2014 *Eur. J. Phys.* 3 **35** 035018
[13] Erdmann M. et al., A Deep Learning-based Reconstruction of Cosmic Ray- induced Air Showers, 2017 *preprint submitted to APJ*, arXiv:1708.00647
[14] Erdmann M. et al., Jet-parton assignment in ttH events using deep learning, 2017 *JINST* **12** P08020

IOP Conf. Series: Journal of Physics: Conf. Series **1085** (2018) 042045 doi:10.1088/1742-6596/1085/4/042045

Event Reconstruction and Simulation in PandaRoot for the $\overline{\text{P}}$ANDA Experiment

D. Steinschaden[1], on behalf of the $\overline{\text{P}}$ANDA Collaboration[2]
[1]Stefan Meyer Institute, Austrian Academy of Science, Vienna, Austria
[2]https://panda.gsi.de/author-list
E-mail: dominik.steinschaden@oeaw.ac.at

Abstract. The $\overline{\text{P}}$ANDA experiment, which is currently under construction at the Facility for Antiproton and Ion Research (FAIR) in Darmstadt, Germany, will address fundamental questions in hadron and nuclear physics via interactions of antiprotons with nuclei. It will be installed at the High Energy Storage Ring (HESR), which will provide an antiproton beam with a momentum range of 1.5 - 15 GeV/c and a high average interaction rate on the fixed target of 2×10^7 events/s. The $\overline{\text{P}}$ANDA experiment will adopt a continuous data acquisition and the expected data rate transmitted to a high-bandwidth computing network will be in the order of 200 GB/s. However, in order to be able to select many different and rare physics processes, an indiscriminate hardware trigger would not suffice. Instead, an online software-based data selection system will be used to select only relevant data and thereby reduce the data rate by a factor of up to 1000. Due to the high interaction rate a highly advanced online analysis must be developed to deal also with overlapping events. Scalability and parallelization of the reconstruction algorithms are therefore a particular focus in the development process. A simulation framework called PandaRoot is used to further optimize the detector performance and develop and evaluate different reconstruction algorithms for event building, tracking and particle identification. An overview about PandaRoot and the requirements on the event reconstruction algorithms is presented and algorithms for the event time reconstruction currently under development are discussed.

1. The $\overline{\text{P}}$ANDA experiment

The $\overline{\text{P}}$ANDA experiment (Antiproton Annihilation at Darmstadt) [1] will be one of the major experiments of the future international Facility for Antiproton and Ion Research [2][3] (FAIR), currently under construction at Darmstadt, Germany. The high-quality antiproton beam with a momentum range from 1.5 GeV/c to 15 GeV/c is provided by HESR (High Energy Storage Ring) [4]. The accelerator can be operated in two complementary modes. The high luminosity mode with an average luminosity of up to $L = 2 \times 10^{32}\,\text{cm}^{-2}\,\text{s}^{-1}$ uses stochastic cooling with a momentum resolution of $\Delta p/p < 10^{-4}$. The high resolution mode with $\Delta p/p < 4 \times 10^{-5}$ is achieved by using additional electron cooling and can be operated up to a momentum of $p = 8.9\,\text{GeV/c}$. The scientific program [5] includes charmonium and open-charm spectroscopy, searches and investigations of exotic states, e.g. multi quark states, hybrids and glueballs, the study of modifications of hadrons in nuclear matter, hyperon physics and γ-ray spectroscopy of hypernuclei. To measure and study the $\bar{\text{p}}\text{p}$ reactions comprehensively, a simultaneous measurement of all produced leptons, hadrons as well as photons is required. Therefor the $\overline{\text{P}}$ANDA experiment covers the full solid angle and is capable of measuring the energy and

IOP Conf. Series: Journal of Physics: Conf. Series **1085** (2018) 042045 doi:10.1088/1742-6596/1085/4/042045

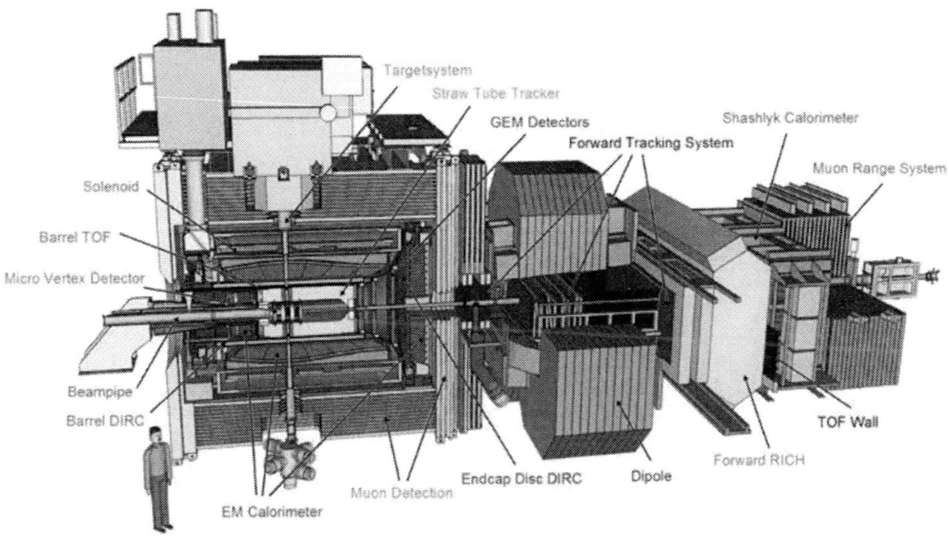

Figure 1. The $\overline{\text{P}}$ANDA spectrometer

momentum of all reaction products as well as being able to identify all particle types over the full momentum range. Figure 1 shows the full $\overline{\text{P}}$ANDA detector which consists of the Target Spectrometer covering the interaction point and the Forward Spectrometer with a maximum angular acceptance of 10 degrees horizontally and 5 degrees vertically in beam direction. A cluster-jet or pellet target system can be operated to scatter the antiprotons with an interaction rate of up to 2×10^7 Hz. The experiment will perform a trigger-less continuous read out, with raw data rates of about 200 GB/s, to provide the necessary flexibility for the complex physics program of $\overline{\text{P}}$ANDA, which will study a diverse range of channels with cross sections varying by many order of magnitudes.

2. PandaRoot

PandaRoot [6] is the software framework for the $\overline{\text{P}}$ANDA experiment which covers full simulations, reconstruction and analysis. It is based on the FairRoot [7] framework which is a project to provide a common computing structure for the future FAIR experiments. FairRoot handles the basic features, such as the interfaces with simulation, tasks, parameter database and the I/O. It is built on ROOT [8] and Virtual Monte Carlo [9]. PandaRoot is supported by various C++ compilers and several Linux distributions as well as macOS. Detector specific geometry, reconstruction and particle identification code is developed within PandaRoot and can be run in the respective stages as shown in the left part of figure 2. In the simulation stage specific initial state particle distributions, physics channels and antiproton-proton background reactions are produced. Therefore, several complementary event generators can be called, e.g. EvtGen [10], DPM [11], UrQMD [12] and Pythia [13]. Particle transport through the detector material is simulated with Virtual Monte Carlo, Geant4 [14] and Geant3 [15]. In the following digitization stage the generated MC data is processed to simulate the realistic response of the subdetectors. In the reconstruction stage information provided by the tracking detectors are combined to reconstruct charged tracks and propagate these to the outer subdetectors. Finally probability density functions (p.d.f.) are computed for every track based on different detectors and various particle identification (PID) concepts. These are combined to receive a global identification probability using Bayes theorem [6]. For the following analysis stage various fitting algorithms

IOP Conf. Series: Journal of Physics: Conf. Series **1085** (2018) 042045 doi:10.1088/1742-6596/1085/4/042045

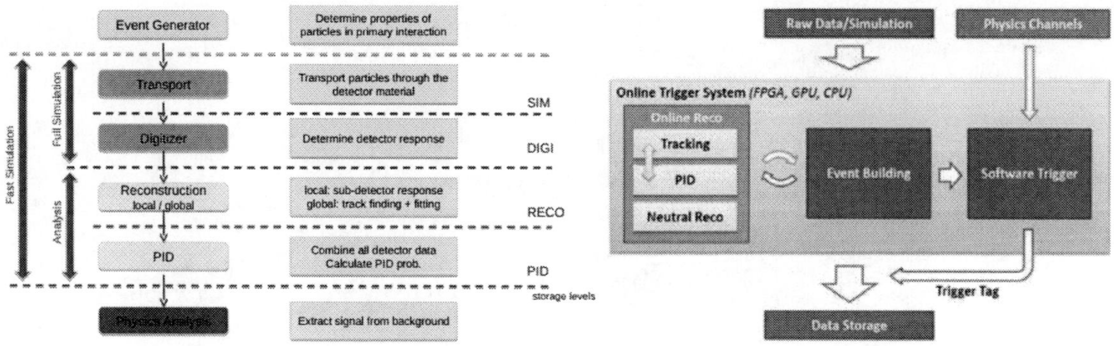

Figure 2. Left: Simulation stages within the framework PandaRoot. Right: Outline of the online reconstruction and trigger system for the continuous, time based read out of $\overline{\text{P}}$ANDA

for the four momentum and position of the particles as well as particle selection and combination mechanisms are provided.

3. Time based reconstruction

The beam from the HESR is Poission distributed, with a high average interaction rate of 20 MHz. Due to this, the $\bar{p}p$ annihilation events will overlap in time for slow subdetectors such as the Straw Tube Tracker [16]. Furthermore the $\overline{\text{P}}$ANDA detector will not use a hardware trigger since it would not be sophisticated enough to achieve the broad scientific program of the $\overline{\text{P}}$ANDA experiment. Instead every sub detector will operate in a self triggering mode. Therefore, the data of every sub system is marked with time stamps and streamed continuously to processing compute nodes. The data are processed directly before storage by the so called online reconstruction. This advanced online event sorting, reconstruction, and software filter is necessary to reduce the raw data rate of about 200 GB/s by a factor of up to 1000 before storage and fully replaces a common hardware trigger. The available operation time for this process will depend on the buffer memory and processing power, and is still under evaluation. The right-hand schematic in figure 2 shows an outline of the planned execution. In an iterative process the event wise sorting, called event building, followed by a fast track reconstruction and PID will be done. The obtained information will be used to enhance the event building and tracking further to a level where the software filter can decide whether to discard or store the respective data. A full and, in terms of computing power, expensive so called offline reconstruction and analysis is only performed at a later stage on the stored data and can be redone with different algorithms if necessary.

4. T_0 reconstruction

A rough sorting of the detector signals, in so called event packages, can be done as an initial step of the online reconstruction, without tracking and PID information. These event packages must contain all data pertinent to a $\bar{p}p$ annihilation event. However, the number of mismatched signals must be reduced to a minimum. Various algorithms exploiting different sub detectors for this first stage are tested currently within the collaboration. One suitable detector to support such algorithms is the Barrel-TOF due to its good time resolution which is below 75 ps [17][18][19]. The left part of figure 3 illustrates the basic algorithm. The received time stamps from the detector are corrected according to the detection position assuming straight line propagation of the particles from the interaction point with the speed of light. Time stamps of signals coming from the same $\bar{p}p$ annihilation will cluster in time to within 4 ns. The first signal of a cluster is

IOP Conf. Series: Journal of Physics: Conf. Series **1085** (2018) 042045 doi:10.1088/1742-6596/1085/4/042045

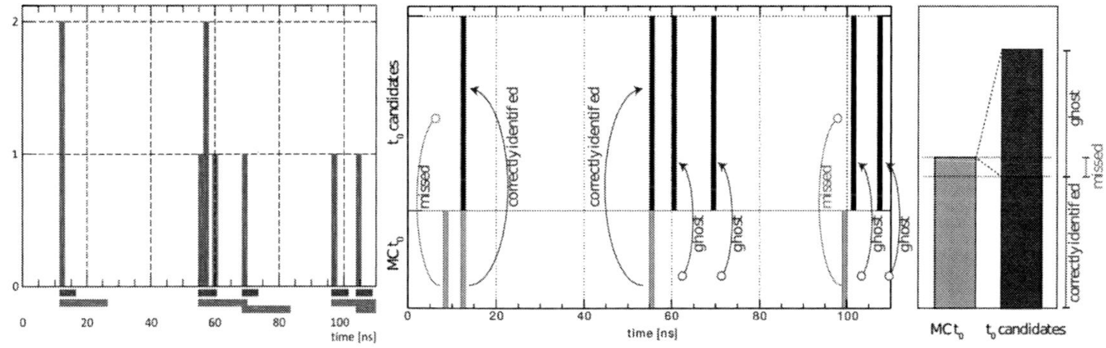

Figure 3. Left: basic principle of a Barrel TOF based event sorting algorithm. Every time stamp (blue) can potentially be the trigger for an event candidate. After a trigger has been accepted, there is a dead-time of 4 ns (red) where no other trigger is accepted. All timestamps after an accepted trigger and within a window of 15 ns are assumed to belong to a single event (green). These event time windows potentially overlap to ensure the completeness of the data. Center: Results for an average event rate of 20 MHz. The black lines indicate the triggered event times, the green lines the MC true values of the $\bar{p}p$ annihilation. Right: The diagram visualize the ratio between MC events, correctly identified and missed events as well as triggered ghost events.

used as an estimate of the $\bar{p}p$ annihilation time (t_0). The right part of figure 3 summarizes the simulation results for an average interaction rate of 20 MHz. 93% of the MC events are identified within 4 ns and the resolution of the estimated t_0 for these is about 0.55 ns. Additionally ghost events are triggered by signals from late arriving secondary particles with a rate of about 0.66 per MC event. First tests showed that crosschecking a potential trigger signal with signals of other sub detectors suppresses the ghost rate below 0.3.

After the first track reconstruction a relative time-of-flight algorithm improves the t_0 resolution and provides time-of-flight based PID information[20]. The basic principle is to iterate through all possible mass configurations (i.e. e, μ, π, K, p) and calculate the expected time of creation for every single track using the reconstructed momentum and flight path information provided by the tracking system. The conformities of the track creation times for all reconstructed tracks and mass assumptions are rated using a χ^2 probability weight based on the comparison of the measured time-of-flight and the calculated expected time-of-flight of the tracks. This task can be reduced to the minimization of the function

$$\chi^2_{(m_1,\dots,m_N)} = \sum_{i=1}^{N} \frac{(t_{i,0} - t_0)^2}{\sigma^2_{TOF}} \quad (1)$$

which is summing over all tracks. The annihilation time t_0 is the free parameter to minimize the function. m_i, $t_{i,0}$ and $\sigma_{i,TOF}$ are the mass assumption, calculated track creation time and the time of flight resolution for track i. Figure 4 summarizes the basic principle and the results for simulated events with more than 2 reconstructed charged tracks detected in the Barrel TOF. The presented algorithm enhances the event time resolution to a value of $\sigma = 0.12$ ns and provides a first time-of-flight based particle hypothesis.

5. Conclusion

PandaRoot is the $\overline{\text{P}}$ANDA software framework for full simulation, reconstruction and analysis. Various particle generators for different physics channels are integrated as well as different

IOP Conf. Series: Journal of Physics: Conf. Series **1085** (2018) 042045 doi:10.1088/1742-6596/1085/4/042045

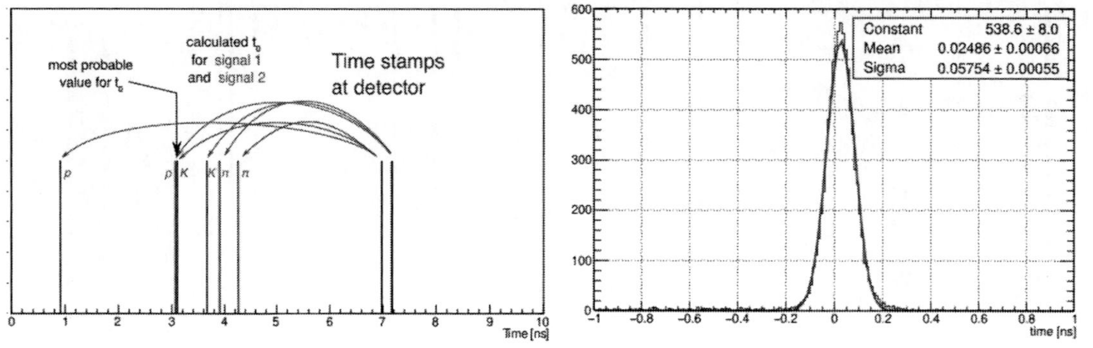

Figure 4. Left: For the detected signals in the Barrel TOF(blue) the corresponding possible track creation times according to a certain mass assumption are calculated (green and red). The combination providing the best conformity is equivalent to the most probable mass configuration. Right: The resulting t_0 distribution for events with more than 2 detected charged tracks shows a time resolution below $\sigma = 60\ ps$.

transport engines to simulate a realistic detector response. The event based simulation is in an advanced state and further improvements of tracking and PID algorithms are in development.

A strong effort is put into the time based simulation of the trigger-less data acquisition system. The continuous data stream must be sorted and reconstructed online on an event-by-event basis. This is done with an iterative and flexible algorithm before a sophisticated decision on discarding or storing the data is done. Currently algorithms are under development to reconstruct the $\bar{p}p$ annihilation time and the event structure. With a fast algorithm based on the time-of-flight system of $\overline{\text{P}}$ANDA a first event building is achieved with a t_0 resolution of $\sigma = 0.55$ ns. This value is improved using more advanced relative time-of-flight algorithms to the order of $\sigma = 0.12$ ns to support the online reconstruction. After passing the software trigger the full offline reconstruction chain will further enhance the reconstruction.

Acknowledgments

This work is supported by the Austrian Science Fund FWF under the Doctoral Program W1252-N27 Particles and Interactions.

References

[1] PANDA Collaboration 2005 Technical Progress Report, FAIR-ESAC/Pbar
[2] Spiller P and Franchetti G 2006 *Nuclear Instruments and Methods in Physics Research Section A: Accelerators, Spectrometers, Detectors and Associated Equipment* **561** 305 – 309 ISSN 0168-9002
[3] Henning W F 2007 *Journal of Physics G: Nuclear and Particle Physics* **34** S551
[4] Prasuhn D 2014 (PANDA meeting)
[5] Lutz M F M *et al.* (PANDA) 2009 (*Preprint* `arXiv:0903.3905`)
[6] Spataro S 2012 *Journal of Physics: Conference Series* **396** 022048
[7] Al-Turany M, Bertini D, Karabowicz R, Kresan D, Malzacher P, Stockmanns T and Uhlig F 2012 *Journal of Physics: Conference Series* **396** 022001
[8] B Rene F R 1997 *Nucl. Intr. Meth. A* **389** 81–86
[9] Hrivnacova I, Adamova D, Berejnoi V, Brun R, Carminati F, Fasso A, Futo E, Gheata A, Caballero I G and Morsch A 2003 *CoRR* **cs.SE/0306005**
[10] Lange D J 2001 *Nuclear Instruments and Methods in Physics Research Section A: Accelerators, Spectrometers, Detectors and Associated Equipment* **462** 152 – 155 ISSN 0168-9002
[11] Capella A, Sukhatme U, Tan C I and Van J T T 1994 *Physics Reports* **236** 225 – 329 ISSN 0370-1573
[12] Bleicher M, Zabrodin E, Spieles C, Bass S A, Ernst C, Soff S, Bravina L, Belkacem M, Weber H, Stcker H and Greiner W 1999 *Journal of Physics G: Nuclear and Particle Physics* **25** 1859

[13] Sjstrand T, Mrenna S and Skands P 2008 *Computer Physics Communications* **178** 852 – 867 ISSN 0010-4655
[14] Agostinelli S 2003 *Nucl. Inst. and Meth. A* **506** 250–303
[15] Brun R, Urban L, Carminati F, Giani S, Maire M, McPherson A, Bruyant F and Patrick G 1993 Geant: Detector description and simulation tool Tech. rep. CERN
[16] Erni W e a 2013 *The European Physical Journal A* **49** 25 ISSN 1434-601X
[17] Zimmermann S *et al.* 2017 *Journal of Instrumentation* **12** C08017
[18] Steinschaden D *et al.* 2015 *PoS* **EPS-HEP2015** 259
[19] Brunner S E, Gruber L, Marton J, Orth H and Suzuki K 2014 *Journal of Instrumentation* **9** C03010
[20] Basile M, Romeo G C, Cifarelli L, D'Ali G, Cesare P D, Giusti P, Massam T, Palmonari F, Sartorelli G, Valenti G, Contin A, Favale L, Zichichi A and Esposito B 1981 *Nuclear Instruments and Methods* **179** 477 – 485 ISSN 0029-554X

Modeling of modifications induced by jets in the relativistic bulk nuclear matter

Marcin Słodkowski[(1)], **Patryk Marcinkowski**[(1)], **Patryk Gawryszewski**[(1)], **Daniel Kikoła**[(1)], **Joanna Porter-Sobieraj**[(2)]
[(1)] Faculty of Physics, Warsaw University of Technology (PL)
[(2)] Faculty of Mathematics and Information Science, Warsaw University of Technology (PL)
E-mail: `m.slodkowski@if.pw.edu.pl`

Abstract. This work is focused on the influence of energy deposited by jets in the medium on the behavior of bulk nuclear matter. In the heavy ion reactions jets are widely used as probes in the study of Quark-Gluon-Plasma (QGP). Modeling using relativistic hydrodynamics with jets perturbation is employed to extract the properties of the QGP. In order to observe a modification of the collective characteristic of the matter, we use our (3+1) relativistic hydrodynamic code and jet energy loss algorithm implemented on the Graphics Cards (GPU). The program uses 7th order WENO algorithm and the Cartesian coordinate system to provide high spatial resolution and high accuracy in hydrodynamic simulations required to analyze the propagation of jets in the nuclear matter. We present how the propagation of jets in the medium could affect the measurements of the properties of a strongly interacting nuclear matter.

1. Introduction

The subject of this study is connected with investigation properties of Strongly Interacting Matter (SIM) and its phases and transition between them. In order to produce the hot nuclear matter on such extreme conditions the heavy nuclei are collided. The evolution of a high energy nucleus-nucleus collision can be distinguished a several stages: pre-equilibrium phase, equilibrium stages - formation of Quark-Gluon Plasma (QGP) phase, hadron gas and finally after freeze-out free hadron. During this stages the jets are produced and affected on expanding bulk nuclear matter. Detected particles have anisotropy in measured momenta in result of the heavy ion experiments. The general question is understand the origin of this momentum anisotropy of the produced particles. We known about two group of effects: flow and non-flow which can be influenced on changes in the final distribution of measured hadrons. Study jet-medium interaction and jet-induced flow can provide information about properties of the QGP phase. Fluctuation in the pre-equilibrium phase might have an impact of the dynamics of the colliding nucleus-nucleus system. Angular triggered two-particle correlations are shown in Fig. 1. ALICE and ATLAS experiments at CERN explan in Ref. [1] the shape of double peak on away side can be naturally explained by sum of measured anisotropic flow Fourier coefficients. This question is still under consideration if this correlation can be assign only to flow effects. Thus there is another open question if this structure is due to jet-medium interaction which show up in two-particle flow measurement. The (3+1) hydrodynamic code + jet energy loss algorithm may help answer of this question. Moreover odd flow harmonics (v_3, v_5, ...) can be generate only due to initial state fluctuations. Fourier flow coefficients as a function of transverse momentum

p_T are shown in Fig. 2. Fast and efficient hydrodynamic code is needed to study event-by-event initial state fluctuations and flow. For this purpose hydrodynamic implementation on the Graphics Cards (GPU) may be useful. Nevertheless approximately studies using averaged fluctuations, are also possible but full event-by-event simulations give more flexibility.

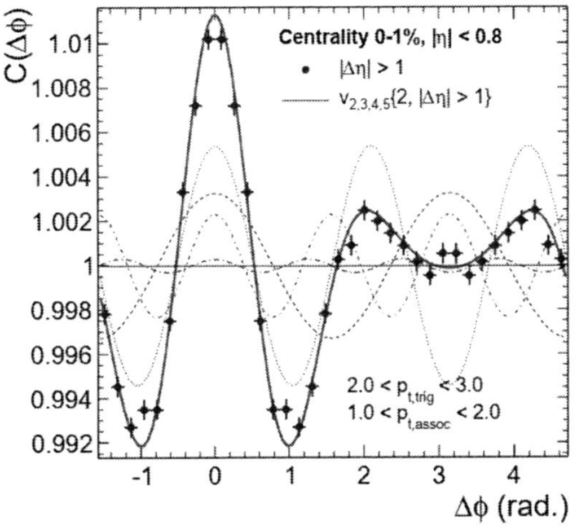

Figure 1. The two-particle azimuthal correlation, measured in $0 < \Delta\phi < \pi$ and shown symmetrized over 2π, between a trigger particle with $2 < p_T < 3$ GeV/c and an associated particle with $1 < p_T < 2$ GeV/c for the 0–1% centrality class. The solid red line shows the sum of the measured anisotropic flow Fourier coefficients v_2, v_3, v_4 and v_5 (dashed lines). Figure taken from [1].

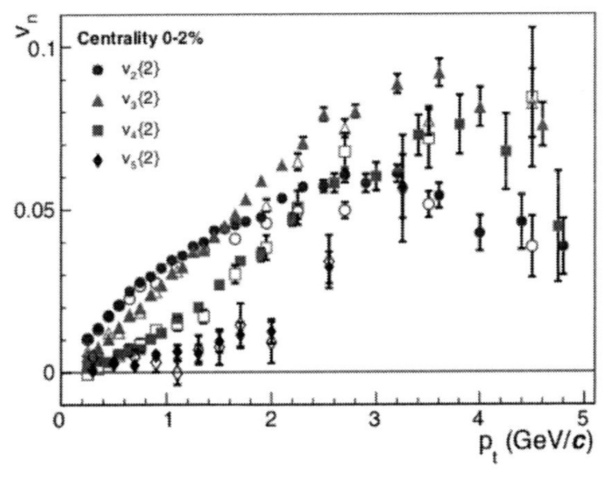

Figure 2. (color online) v_2, v_3, v_4, v_5 as a function of transverse momentum. The full, open symbols. are for $\Delta\eta > 0.2$ and $\Delta\eta > 1.0$, respectively. 0–2% centrality percentile. Figure taken from [1].

2. Hydrodynamics on the Graphics Cards (GPU)

Our hydrodynamic code can describe collective behaviors bulk nuclear matter. The Cartesian coordinate system is needed to study modification in the medium which affected by jets with high spatial resolution and high accuracy. Event-by-event flow fluctuations fast enough for good statistic are needed thus the fast and efficient hydrodynamic code can help. With this assumption we need a lot of computer power. Our GPU multi thread approach gives significant speed-up more than 2 orders of magnitude compared to single thread simulation [2, 3, 4, 5].

Our hydrodynamic program includes 4 major components:

(i) Initial condition for further hydrodynamic simulation,
(ii) Solving partial differential equations - hydrodynamic evolution of the system,
(iii) Freezee-out condition when the matter is not in the local thermal equilibrium.
(iv) Computing freeze-out hypersurface and particle emission function.

We begin our hydrodynamic simulation from generation initial condition based on UrQMD model [6] in order to prepare event-by-event simulation with fluctuating initial conditions what is also the point of interest in this investigation. Relativistic hydrodynamics simulations are based on universal hyperbolic equations of the conservation laws.

$$\frac{\partial U}{\partial t} + \frac{\partial F(U)}{\partial x} + \frac{\partial G(U)}{\partial y} + \frac{\partial H(U)}{\partial z} = 0 \tag{1}$$

with a determined Equation of State (EoS) $p = e^2/3$ where U is the vector of conserved quantities such as energy density (E), momenta density (M), net charge density (R) in the laboratory rest frame. Vectors of fluxes F, G, H in the x; y; z directions are defined as:

$$F(U) = \begin{bmatrix} (E+p)v_x \\ M_x v_x + p \\ M_y v_x \\ M_z v_x \\ R v_x \end{bmatrix}, \quad G(U) = \begin{bmatrix} (E+p)v_y \\ M_x v_y \\ M_y v_y + p \\ M_z v_y \\ R v_y \end{bmatrix}, \quad H(U) = \begin{bmatrix} (E+p)v_z \\ M_x v_z \\ M_y v_z \\ M_z v_z + p \\ R v_z \end{bmatrix}. \tag{2}$$

where v is the velocity, and p is pressure, defined by the energy e and charge density n in the local fluid rest frame, where velocity v vanishes ($v = (0,0,0)$). Additionally, the following relations occur:

$$\begin{aligned} E &= (\varepsilon + p)\gamma^2 - p \\ M_i &= (\varepsilon + p)\gamma^2 v_i, \qquad i = x, y, z \\ R &= n\gamma \end{aligned} \tag{3}$$

where ε is the energy density and $\gamma = \frac{1}{\sqrt{1-v^2}}$ is the Lorentz factor. Eq. 3 defines the transformation from local fluid rest frame variables to laboratory rest frame. There is the set of hyperbolic partial equations in the laboratory frame quantities defined in Eq. 2 In order to integrate them we use a finite-difference scheme on a Cartesian grid as follow:

$$\begin{aligned} U^{n+1}_{i,j,k} = U^{n}_{i,j,k} + \tfrac{\Delta t}{\Delta x}\left(F_{i-\frac{1}{2},j,k} - F_{i+\frac{1}{2},j,k}\right) + \\ + \tfrac{\Delta t}{\Delta y}\left(G_{i,j-\frac{1}{2},k} - G_{i,j+\frac{1}{2},k}\right) + \tfrac{\Delta t}{\Delta z}\left(H_{i,j,k-\frac{1}{2}} - H_{i,j,k+\frac{1}{2}}\right). \end{aligned} \tag{4}$$

For time propagation the standard 3th order Runge-Kutta methods are employed [7]. The main part of our software is the numerical implementation of 7th order WENO algorithm [8, 9] on the Graphics Cards (GPU) using nVidia CUDA framework [10]. The WENO is very accurrate algorithm. Numerical oscillation and numerical diffusion are balanced in this type of algorithm. The 7th order of WENO scheme tests gives very good performance in our implementation. We use energy loss $\frac{dE}{dx}$ procedure as a source in the hyperbolic equations.

$$\partial_\mu T^{\mu\nu} = J^\nu(x) \tag{5}$$

More details about Eq. 5 are described in Ref. [11]. The algorithm uses two mechanisms of jet energy loss which include gluon radiation and collision procedure of parton in dense medium. We used formula which is presented below:

$$\left(-\frac{\mathrm{d}E}{\mathrm{d}x}\right) = \kappa_{rad}\frac{C_R}{C_F}T^3 x + \kappa_{coll}\frac{C_R}{C_F}T^2 \tag{6}$$

where T is local temperature and $\kappa_{rad}, \kappa_{coll}, C_R, C_F$ coefficients depend on jet flavor (quark or gluon) and its energy [12]. In the simulations, we assumed $\kappa_{rad} = 4, \kappa_{coll} = 2.5, C_R/C_F = 1$. We assume that $\frac{\mathrm{d}E}{\mathrm{d}x}$ is small compared to the jet energy and the jet is not modified by the medium.

In order to accelerate of simulation we decided to use graphics processor GPU. The WENO algorithms have been implemented on a GPU with using the CUDA parallel programming model.

IOP Conf. Series: Journal of Physics: Conf. Series **1085** (2018) 052001 doi:10.1088/1742-6596/1085/5/052001

Our Hydro program starts on CPU in serial host mode. During this stage the initial data are copied to GPU coprocessor. Functions with numerical algorithm executed on a GPU are called kernels in parallel mode. Each kernel executed by launching blocks of threads. Threads form in block run together on the same stringing multiprocessor. Finally after parallel computation data are copied back to host and the program finished all tasks on CPU in serial mode.

Last stage of our simulation focuses on transformation of numerical grid quantities to kinetic description of particles produced during the nucleus-nucleus collision. Thus we use Cooper-Frye formula to obtain hypersurface in order to generate particles multiplicity spectra and momenta distribution.

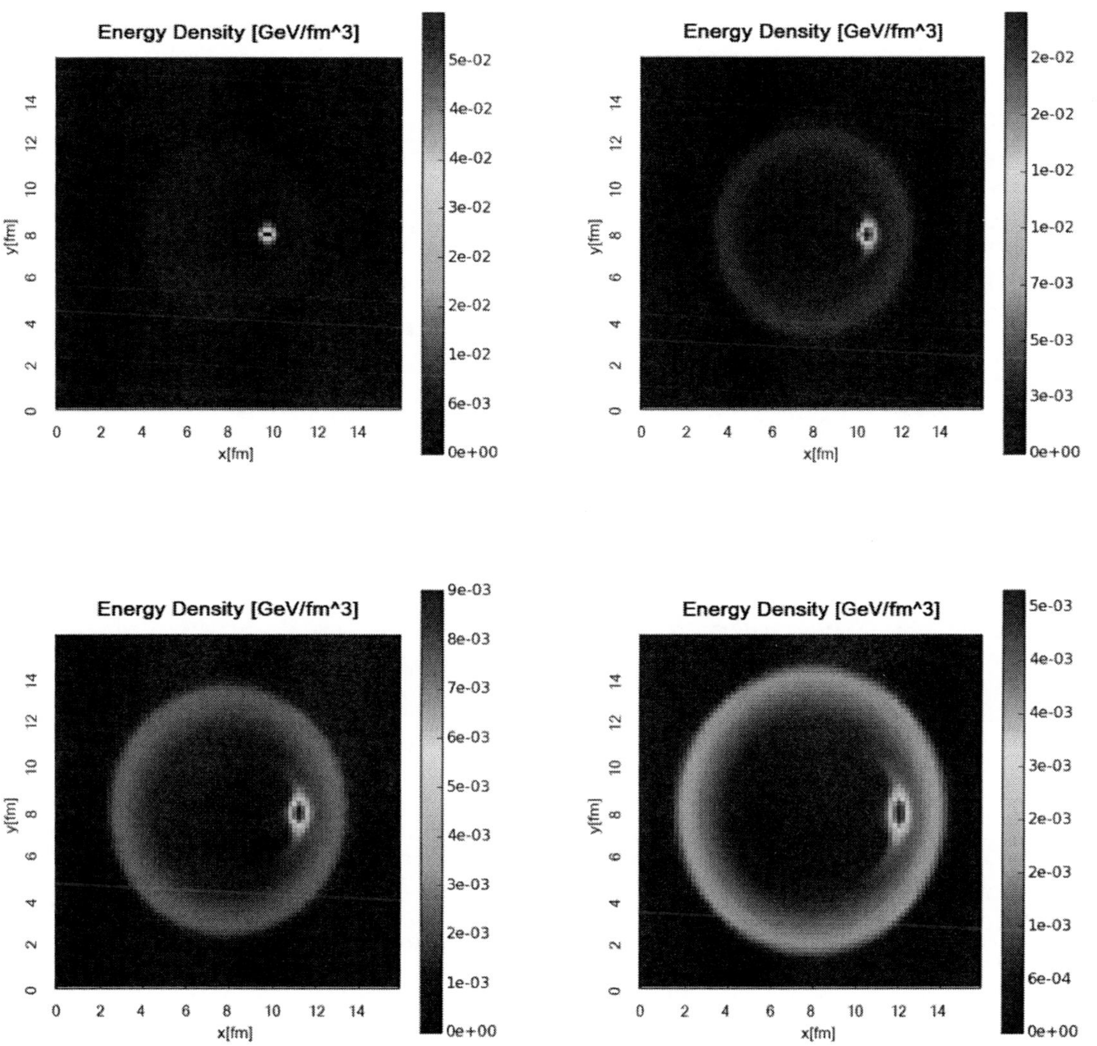

Figure 3. Energy density cross section with jet traveling trough the medium in xy plane at $z = 0$ after $t = 2.0$ fm/c see panel (a), $t = 2.8$ fm/c see panel (b), $t = 3.6$ fm/c see panel (c), $t = 4.4$ fm/c see panel (d).

3. Results

Our simulations are concentrated in modeling of modification induced by jets in the relativistic bulk nuclear matter [13]. In this section we present propagation of jets during the evolution of the medium. We simulate evolution of bulk matter together with energy loss $\frac{dE}{dx}$ of jet traveling through the medium. We performed ellipsoidal test in order to illustrate and validate our hydrodynamic code with jet energy loss algorithm. The initial condition of ellipsoidal test base on Ref. [14]. Response signal of the medium which is caused by jet propagation is observed in Fig. 3. During the matter evolution the Mach cone in the medium is visible. Energy profiles with time ($t = 2.0$ fm/c, $t = 2.8$ fm/c, $t = 3.6$ fm/c, $t = 4.4$ fm/c) are shown Fig. 4. It is clearly visible the propagation of jets in the medium could affect the properties of bulk nuclear matter. Surrounding matter where jet interacted with medium significantly increase due to jet energy loss. Employed realistic jet energy loss $\frac{dE}{dx}$ algorithm takes into account the decrease ionization of medium in time caused by jets. Initial condition from UrQMD model was prepared and preliminary hydrodynamic tests were performed in order to simulate comparable nucleus-nucleus interaction measured in the heavy ion experiments. Stability tests using implemented 7th order WENO were performed. We plan to finish Freeze-out implementation which is in progress. It is necessary to employ commonly used Cooper-Frey formula in order to obtain hypersurface. In the future we plan to use existing hadron Monte-Carlo generator such as TERMINATOR2 [15]. The interface between our hydro program and THERMINATOR2 generator is being developed. After this efforts there is possible to study elliptic flow, higher harmonic flow and other observables in order to compare to simulations experimental results such as $Pb + Pb$ and $Au + Au$ interactions at RHIC and LHC energies.

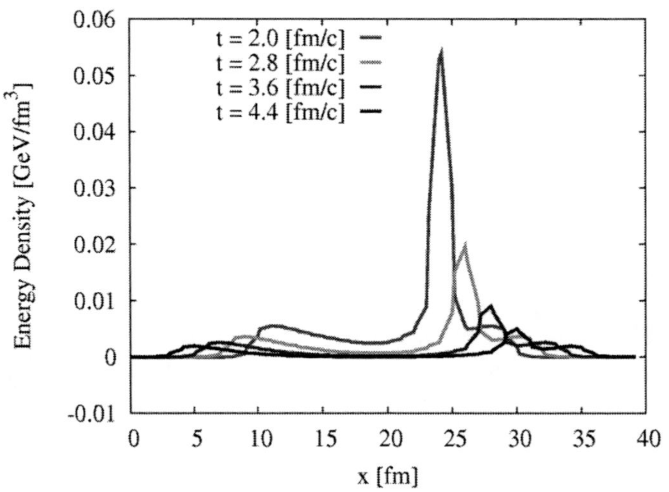

Figure 4. $1D$ projection $z = y = 0$ of energy density evolution in time with jet propagation

References

[1] K. Aamodt *et al.* [ALICE Collaboration], Phys. Rev. Lett. **107** (2011) 032301 doi:10.1103/PhysRevLett.107.032301 [arXiv:1105.3865 [nucl-ex]].

[2] J. Porter-Sobieraj, S. Cygert, D. Kikoła, J. Sikorski, M. Słodkowski *Concurrency and Computation: Practice and Experience* **27**(6) 1591-602, 2015

[3] S. Cygert, D. Kikoła,J. Porter-Sobieraj, J. Sikorski, M. Slodkowski M *Lect. Notes Comput. Sc.* **8384** 500-9, 2014

[4] J. Sikorski, S. Cygert, J. Porter-Sobieraj, M. Słodkowski, et al. . *J. Phys. Conf. Ser.* **509** 012059, 2014

[5] S. Cygert, J. Porter-Sobieraj, D. Kikoła D, J. Sikorski M. Słodkowski, *Fed. Conf. on Computer Science and Information Systems (Krakw)* pp 441-6, 2013
[6] S. A. Bass *et al.*, Prog. Part. Nucl. Phys. **41** (1998) 255 [Prog. Part. Nucl. Phys. **41** (1998) 225] doi:10.1016/S0146-6410(98)00058-1 [nucl-th/9803035].
[7] E.Constantinescu, A. Sandu, Computational ScienceICCS 2009, 293–301
[8] D. Balsara and C. W. Shu. Journal of Computational Physics, **160**(2):405-452, (2000).
[9] C. W. Shu. Fluid Dynamics, **17**:107-118, (2001).
[10] NVIDIA Corporation: NVIDIA CUDA Programming Guide Version 9.0, 2017,Available from: http://docs.nvidia.com/cuda/cuda-c-programming-guide/index.html [accessed 2017-08-17]
[11] Y. Tachibana, T. Hirano *Phys. Rev.* C **90** 021902, 2014
[12] Casalderrey-Solana J, Gulhan D C, Milhano J G, Pablos D and Rajagopal K 2014 *J. High Energy Phys.* JHEP10(2014)019
[13] P. Marcinkowski, M. Słodkowski, D. Kikoła, J. Sikorski, J. Porter-Sobieraj, P. Gawryszewski and B. Zygmunt, J. Phys. Conf. Ser. **668** (2016) no.1, 012115 doi:10.1088/1742-6596/668/1/012115 [arXiv:1511.02137 [nucl-th]].
[14] Y. M. Sinyukov and I. A. Karpenko, nucl-th/0505041.
[15] M. Chojnacki, A. Kisiel, W. Florkowski and W. Broniowski, Comput. Phys. Commun. **183** (2012) 746 doi:10.1016/j.cpc.2011.11.018 [arXiv:1102.0273 [nucl-th]].

IOP Conf. Series: Journal of Physics: Conf. Series **1085** (2018) 052002 doi:10.1088/1742-6596/1085/5/052002

Direct numerical computation and its application to the higher-order radiative corrections

K Kato[1], E de Doncker[2], T Ishikawa[3] and F Yuasa[3]

[1] Kogakuin University, Shinjuku, Tokyo 163-8677, Japan
[2] Western Michigan University,1903 West Michigan Avenue, Kalamazoo, MI 49008, United States
[3] KEK, Tsukuba, Ibaraki 305-0801, Japan

E-mail: kato@cc.kogakuin.ac.jp

Abstract. The direct computation method(DCM) is developed to calculate the multi-loop amplitude for general masses and external momenta. The ultraviolet divergence is under control in dimensional regularization. In this paper we report on the progress of DCM to several scalar multi-loop integrals after the presentation in ACAT2016. Also the discussion is given on the application of DCM to physical 2-loop processes including numerator functions.

1. Introduction

For the investigation of physics in the current and future collider experiments, a precise evaluation of higher order corrections in perturbative quantum field theory(QFT) is required. The calculation of the higher order radiative corrections turns out to be a large scale computation since the required number of Feynman diagrams is huge and the integral of each diagram is sometimes very complicated. The role of an automated system for the calculation of perturbative series in QFT is not only to manage such a large scale computation but to avoid possible errors caused by humans.

The starting point of the study in high-energy physics is QFT, i.e., the symbolic representation of the theory and the final output is the predicted numerical values to be compared with experimental results. So the specific feature of the system is depicted at which point one switches from the symbolic treatment to the numerical one. In this sense, our system would be described as the maximally numerical method.

The library of the multi-loop integrals is an important component of the automated system and the libraries are working well in the 1-loop calculations[1, 2, 3, 4, 5, 6, 7]. Beyond 1-loop, we have worked on the development of a computational method for Feynman loop integrals with a fully numerical approach. It is based on numerical integration and extrapolation techniques. In this paper, we describe the status and new developments in our techniques for the numerical computation of Feynman loop integrals.

2. Direct computation method

The multi-loop integral is essential for the higher-order radiative corrections. The integral can have singularities originating from the physics.

Table 1. The progress of DCM at ACAT2017. The number shown as 'x-dim.' stands for the maximum dimension of integrals.

	2-point(self-energy)	3-point(vertex)	4-point(box)
2-loop	massless, massive ref.[13, 22] 4-dim.	massive ref.[14, 15, 20] 5-dim.	massive ref.[16, 17] 6-dim.
3-loop	massless, massive ref.[20, 21] 7-dim.	massless ref.[18, 19] 6-dim.	
4-loop	massless ref.[20] 8-dim.		

– The function in the denominator can vanish, which is normally avoided by the analytic continuation with $m \to m - i0$ in the analytical method.
– The ultraviolet(UV) divergence can appear when the integral is divergent for the large momentum region when it is calculated in 4 space-time dimension.
– The infrared(IR) divergence can appear when a massless particle is included in the integral and the integral is divergent in the soft momentum region.

The singularity in the integrand disappears if we introduce the regularization, e.g., $m^2 \to m^2 - i\rho$, taking spacetime dimension to be $4 - 2\varepsilon$, or the introduction of the fictitious mass λ for a massless particle. With non-zero ρ, ε and λ the integral can be computed numerically and the physical value can be obtained by extrapolation to the limit $\rho \to 0$ and so forth.

It is already shown that DCM can handle these singularities numerically. For the scalar integrals the status is shown in Table.1. After the last ACAT[20], we have filled the box for 3-loop vertex functions. The complexity of numerical integration increases with the dimension of the integral and we have computed integrals up to dimension 8, or a diagram with 9 propagators. For the numerical integration, we use robust integration software in ParInt[8] or the double exponential transformation method[9] and use MPI[10] or other parallel environments for accelerating the computation.

An important feature of DCM is that one does not need to separate terms by hand as is done in the analytic treatment. In a large-scale computation, manual operation is the point where some error can happen. Suppose an integral I has a UV singularity as

$$I = \frac{C_{-K}}{\varepsilon^K} + \cdots + \frac{C_{-1}}{\varepsilon} + C_0 + C_1\varepsilon + C_2\varepsilon^2 + \cdots \tag{1}$$

where we take the spacetime dimension to be $n = 4 - 2\varepsilon$. Then, we compute I as the integral whose integrand includes the numerical value of ε. From a set of numerical values, we can obtain all values of the leading coefficients using a linear solver or Wynn's algorithm[11]. When the most singular term is $1/\varepsilon^K$, in the first case, we solve the linear equation

$$I(\varepsilon_j) = \sum_{k=-K}^{-K+N-1} C_k \varepsilon^k \qquad (j = 1, \ldots, N) \tag{2}$$

using an appropriate linear solver such as `dgefs.f` from the SLATEC Common Mathematical Library[12], and in the second case the following iteration is performed

$$a(j, k+1) = a(j+1, k-1) + \frac{1}{a(j+1,k) - a(j,k)} \qquad a(j,0) = \varepsilon^{-n} I(\varepsilon_j), a(j,-1) = 0 \tag{3}$$

to obtain approximations to the coefficient C_n(in the columns with k odd).

3. Application

The 2-loop amplitude including the numerator is processed in the following way. As an explicit example, we consider the calculation of the 2-loop electroweak self-energy function, $\Pi(s)$. The function is the sum of a number of diagrams, i.e., $\Pi(s) = \sum \Pi_j(s)$ and later we do not write the index j explicitly unless it is necessary. First, GRACE[1, 23, 24, 25] automatically generates the diagrams. Then for each diagram its numerator and denominator are given by a symbolic code in REDUCE to compute the integral:

$$\Pi(s) = \int [d\ell_1][d\ell_2] \frac{\mathbf{N}(\ell_1, \ell_2)}{P_1 P_2 \cdots P_N} = \int dx_1 dx_2 \cdots dx_N \delta(1 - \sum x_k) G \tag{4}$$

where $\mathbf{N}$ is the numerator and $P_k = p_k^2 - m_k^2 + i\rho$. Here,

$$G = \Gamma(N) \int [d\ell_1][d\ell_2] \frac{\mathbf{N}(\ell_1, \ell_2)}{\Delta^N} \tag{5}$$

where $\Delta = {}^{\mathrm{t}}\vec{\ell} A \vec{\ell} + 2 {}^{\mathrm{t}}\vec{\ell}\vec{b} + C$. Here $\vec{\ell} = \begin{pmatrix} \ell_1 \\ \ell_2 \end{pmatrix}$. After a sequence of the variable transformation, we have

$$\Delta = \ell_1^2 + \ell_2^2 - V, \quad \mathbf{N} = f^{00} + f^{10}\ell_1^2 + f^{01}\ell_2^2 + f^{20}(\ell_1^2)^2 + f^{11}\ell_1^2\ell_2^2 + f^{02}(\ell_2^2)^2 , \tag{6}$$

and G is given by

$$G = \sum \frac{(-1)^{N+k+m}}{(4\pi)^n} \frac{\Gamma(N-n-k-m)\Gamma(n/2+k)\Gamma(n/2+m)}{(\Gamma(n/2))^2} \frac{f^{km}}{U^{n/2} V^{N-n-k-m}}. \tag{7}$$

where U and the product of U and V are polynomials of x's,

$$U = \det A, \quad UV = -\det \begin{pmatrix} A & \vec{b} \\ {}^{\mathrm{t}}\vec{b} & C \end{pmatrix} . \tag{8}$$

The derivative function $d\Pi(s)/ds$ is also computed in a similar manner. The system generates the FORTRAN code for G automatically. However, at the present status, one must prepare a part of REDUCE code dependent on the topology of the diagram manually. This part needs more work for the complete automation. This point can be automated since the number of the topologies in 2-loop diagrams is rather limited, so that we can prepare all possible code for these in the system.

There is a variant of the treatment of the numerator used in the calculation of the 2-loop electroweak correction to muon $g-2$ in ref.[26]. In the future, we would also like to implement this method since it is preferable to have another method for the automated computation to confirm the results.

When the integral has UV divergence, we keep ε finite. Also when the denominator in the integrand vanishes in the integration region, we keep ρ finite. For an integral with both singularities, we execute a double extrapolation: First, we fix the value of ε and compute the integral for several values of ρ to estimate the limit as $\rho \to 0$ using either of the methods explained in the previous section. Then, we calculate the C coefficients of series in ε. A detailed discussion on the double extrapolation is found in ref.[27, 28].

As an instance, we consider the Higgs 2-point function of the 2-loop order in the electro-weak theory with the non-linear gauge fixing[25]. Then GRACE generates 3082 diagrams for the function including counter-term diagrams. So the system should perform well to handle such size of computation. Some test computations have shown that this example can be calculated within realistic computer time.

IOP Conf. Series: Journal of Physics: Conf. Series **1085** (2018) 052002 doi:10.1088/1742-6596/1085/5/052002

4. Summary

The DCM is well developed as an important tool to calculate the radiative corrections in 2-loop order when it is combined with GRACE system. Since it is based on a fully numerical method, once it is proved to work for the scalar integral, it can compute the physical amplitude without any special extension of the computational method.

As was already discussed, in the large scale calculation for the higher-order correction it is desirable to perform the calculation automatically, so that one can avoid possible human error. From this view point, the DCM still needs two points to be upgraded. One is to control the selection of the series of numerical values of $\{\varepsilon_j\}$ using some iterative test to find the proper values, i.e., a variation of a machine-learning method. Second is to introduce a system to analyze the structure of diagrams to provide the topology-dependent part of the numerator handler. In a future publication we plan to present explicit results for 2-loop processes.

Acknowledgments

We acknowledge the support from the National Science Foundation under Award Number 1126438, and the Center for High Performance Computing and Big Data at Western Michigan University. This work is further supported by Grant-in-Aid for Scientific Research (15H03668 and 17K05428) of JSPS, and the Large Scale Simulation Program No.16/17-21 of KEK.

References

[1] P. H. Khiem, J. Fujimoto, T. Ishikawa, T. Kaneko, K. Kato, Y. Kurihara, Y. Shimizu, T. Ueda, J.A.M. Vermaseren and Y. Yasui, Eur. Phys. J. **C 73** (2013) 2400.
[2] T. Hahn, Proceedings of Science **093** (2011) PoS(ACAT2010)078 (11 pages).
[3] J.-X. Wang, Nucl. Instr. Meth. **A534** (2004) pp.241-245.
[4] A. van Hameren, Comput. Phys. Commun. **182** (2011) pp.2427-2438.
[5] G. Cullen, J. P. Guillet, G. Heinrich, T. Kleinschmidt, E. Pilon, T. Reiter and M. Rodgers, Comput. Phys. Commun. **182** (2011) 2276.
[6] A. Denner, S. Dittmaier and L. Hofer, Comput. Phys. Commun. **212** (2017) 220.
[7] S. Carrazza, R. K. Ellis and G. Zanderighi, Comput. Phys. Commun. **209** (2016) 134.
[8] ParInt `http://www.cs.wmich.edu/parint`
[9] H. Takahasi and M. Mori, Publications of the Research Institute for Mathematical Sciences **9** (1974) pp.721-741.
[10] Open-MPI `http://www.open-mpi.org`
[11] P. Wynn, Mathematical Tables and Aids to Computing **10** (1956) pp.91-96.
[12] SLATEC `http://www.netlib.org/slatec/`
[13] F. Yuasa, T. Ishikawa, J. Fujimoto, N. Hamaguchi, E. de Doncker, Y. Shimizu, Proceedings of Science **070** (2009) PoS(ACAT08)122 (5 pages).
[14] E. de Doncker, Y. Shimizu, J. Fujimoto and F. Yuasa, Comput. Phys. Commun. **159** (2004) pp.145-156.
[15] E. de Doncker, F. Fujimoto, Y. Kurihara, T. Ishikawa, N. Hamaguchi, Y. Shimizu, F. Yuasa, Springer Lecture Notes in Computer Science, LNCS 6017 pp.139-154.
[16] E. de Doncker and F. Yuasa, Journal of Physics Conference Series **523** (2014) 012052.
[17] F. Yuasa, E. de Doncker, N. Hamaguchi, T. Ishikawa, K. Kato, Y. Kurihara, J. Fujimoto, Y. Shimizu, Comput. Phys. Commun. **183** (2012) pp.2136-2144.
[18] E. de Doncker and F. Yuasa, Procedia Computer Science **108C** (2017) pp.1773-1782.
[19] E. de Doncker, F. Yuasa, K. Kato and T. Ishikawa, Journal of Physics Conference Series **920** (2017) 012005.
[20] K. Kato, E. de Doncker, T. Ishikawa, J. Kapenga, O. Olagbemi and F. Yuasa, Journal of Physics Conference Series **762** (2016) 012070.
[21] E. de Doncker, F. Yuasa, F. Olagbemi, Procedia Computer Science **51** (2015) pp.1333-1342.
[22] E. de Doncker, F. Yuasa, K. Kato, T. Ishikawa and O. Olagbemi, Journal of Physics Conference Series **608** (2015) 012071.
[23] T. Ishikawa, et al., The GRACE manual v1.0, KEK Report **92-19** (1993).
[24] J. Fujimoto, T. Ishikawa, M. Jimbo, T. Kaneko, K. Kato, S. Kawabata, T. Kon, M. Kuroda, Y. Kurihara, Y. Shimizu, and H. Tanaka, Comput. Phys. Commun., **153** (2003) 106.
[25] G. Bélanger, F. Boudjema, J. Fujimoto, T. Ishikawa, T. Kaneko, K. Kato and Y. Shimizu, Phys. Rep. **430** (2006) 117.

[26] T. Ishikawa, N. Nakazawa and Y. Yasui, Journal of Physics Conference Series **920** (2017) 012009.
[27] E. de Doncker, F. Yuasa, and Y. Kurihara, Journal of Physics Conference Series **368** (2012) 012060.
[28] E. de Doncker and F. Yuasa, "Measurements in Quantum Mechanics", edited by Mohammad Reza Pahlavani, ISBN 978-953-51-0058-4, Chapter 14 (2012) pp.333-348.

IOP Conf. Series: Journal of Physics: Conf. Series **1085** (2018) 052003 doi:10.1088/1742-6596/1085/5/052003

Loopedia, a Database for Loop Integrals

C. Bogner[1], S. Borowka[2], T. Hahn[3], G. Heinrich[3], S.P. Jones[3], M. Kerner[3], A. von Manteuffel[4], M. Michel[5], E. Panzer[6], V. Papara[3]

[1] Institut für Physik, Humboldt-Universität zu Berlin, D–10099 Berlin, Germany
[2] Theoretical Physics Department, CERN, Geneva, Switzerland
[3] Max-Planck-Institut für Physik, Föhringer Ring 6, D–80805 Munich
[4] Dept. of Physics & Astronomy, Michigan State University, East Lansing, MI 48824, USA
[5] CP3, Université Catholique de Louvain, B–1348 Louvain-la-Neuve, Belgium
[6] All Souls College, University of Oxford, OX1 4AL, Oxford, UK

Abstract. Loopedia is a new database at `loopedia.org` for Feynman integrals, providing both bibliographic information as well as results made available by the community. Its bibliometry is complementary to that of INSPIRE or arXiv, as it allows to search for integrals by graph-theoretical specifications, e.g. their topology.

1. Introduction

Searching and accessing bibliographic information in high-energy physics is generally solved satisfactorily through databases such as INSPIRE and arXiv. The search within these databases is indexed by 'traditional' metrics: author, title, year of publication, etc. However, for the practitioner in loop calculations it is not easy to find e.g. publications which refer to a particular loop integral, where the integral is specified in some graph-theoretical way, say by its topology.

Loopedia [1] attempts to fill this gap, while not being limited to bibliographic information. The description field of each record can hold any kind of textual information (e.g. URLs to software), and in addition arbitrary files can be uploaded, for example Fortran programs or Maple worksheets.

2. Integrals and Graphs

Scalar Feynman loop integrals correspond to graphs and as such can be identified by a representation of the corresponding graph. A common choice is the *edge (adjacency) list:* each propagator (edge) is specified by the pair of vertices (nodes) it connects – external nodes need not be distinguished to identify the topology. For example, the following two-loop box graph has the edge list (e,0) (0,1) (0,2) (e,1) (1,3) (2,3) (2,4) (3,5) (e,4) (4,5) (e,5):

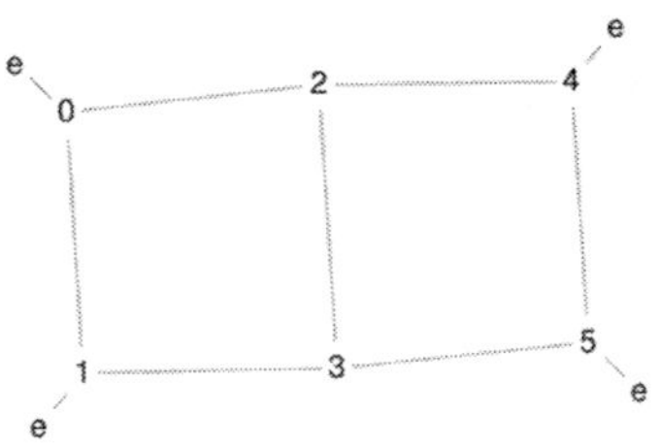

The Nickel index [2, 3, 4] provides an alternative representation which is constructed as follows: start at a vertex 0 and write down all vertices connected to 0 (`e` for external legs), then insert a '`|`'. Repeat the procedure for the other vertices but omit edges already included. For the two-loop box graph above this procedure leads to

vertex	0	1	2	3	4	5
connected to	$e12$	$e[0]3$	$[0]34$	$[12]5$	$e[2]5$	$e[34]$

and hence its Nickel index is `e12|e3|34|5|e5|e|`.

The Nickel index provides a compact identifier and has therefore been chosen as the basic object by which integrals are indexed in Loopedia. Neither the edge list nor the Nickel index is unique, however, i.e. depending on the chosen labelling the same graph may have different edge lists/Nickel indices. Loopedia therefore canonicalizes all input.

Besides this 'bare' Nickel index, which represents just the topology of the graph, Loopedia introduces an augmented form, the colored Nickel index, *CNickel* for short, which additionally captures the configuration of masses and external q^2. Loopedia color-codes them on the screen:

Nickel `e12|e3|34|5|e5|e|`
CNickel `e12|e3|34|5|e5|e|:110|10|00|0|11|1|`

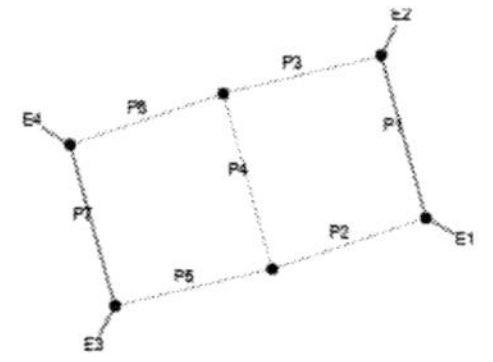

For each 'edge identifier' of the Nickel (`e` or a digit), the coloring has a mass identifier:

`0` for zero,
{`z`, `n`, `s`} for {any, non-zero, special} mass scale,
`1...9`, `a...y` (not `n`, `s`) for a definite non-zero mass scale.

The '`s`' option is for cases like thresholds or pseudo-thresholds, and the exact meaning of 'special' should be detailed in the description of the integral.

The 'definite' identifiers also represent arbitrary scales (though not zero) but unlike the '`zns`' choices they express equality of scales within the graph, similar to named patterns in a computer-algebra system. The graph in the example above has several massive propagators/legs, but all with the same scale '`1`'.

3. User Guide

Loopedia is located at `loopedia.org` and when accessed, the Start Page is displayed. One enters a graph in the input field and/or chooses constraints for the search with the lower control bars and then hits the Search button.

The graph is entered via its edge list. The user has some freedom there: generally the input is understood as long as a pairing of vertices into propagators can reasonably be inferred from it. This means that e.g. FeynArts [5] or QGRAF [6] notation can directly be pasted into the input form. An edge list is internally converted to a Nickel index right away. Alternatively, the Nickel index can also be entered directly.

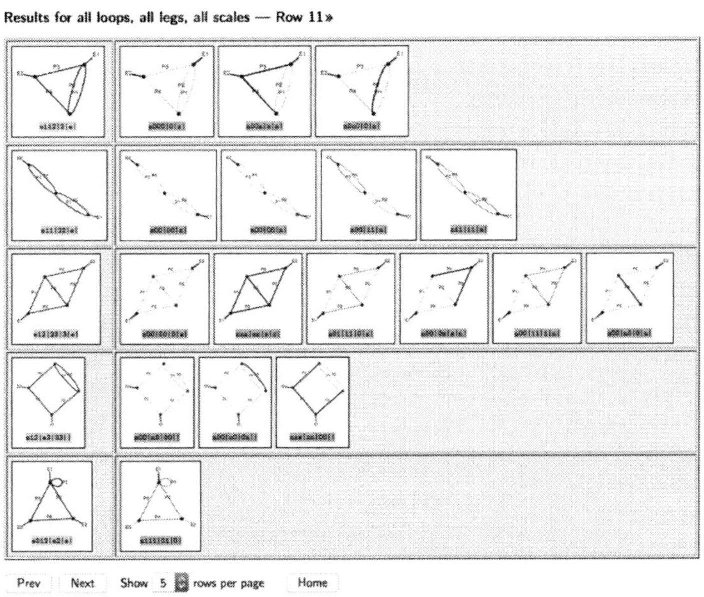

Figure 1. The Loopedia Graph Browser

3.1. Browsing Graphs

The search results are displayed in Loopedia's Graph Browser (Fig. 1). The matches are organized such that the topology appears on the left-hand side and the configurations for which information is available are on the right. Every graph icon is clickable.

Clicking on one of the graph icons opens the single-graph display for that graph. For icons in the left column it starts the Configuration Editor (Fig. 2), for icons in the right column it opens the Record Display (Fig. 3). The Graph Browser is skipped and the Configuration Editor is started immediately if the search delivers no matches and an explicit graph was entered, to commence adding a new record.

3.2. Viewing Single Graphs

The navigation panel at the top has buttons to edit the configuration [Edit • : •], the topology [Edit •], to browse the configurations of the topology [Browse • : •], and to return to the start page [Home]. Depending on the situation not all buttons may be active.

In the main part of the page, a picture of the graph is shown underneath its various textual representations (edge list, [C]Nickel index, database path). To the right the Configuration Chooser is displayed, with the defaults preset to the current values (if any).

IOP Conf. Series: Journal of Physics: Conf. Series **1085** (2018) 052003 doi:10.1088/1742-6596/1085/5/052003

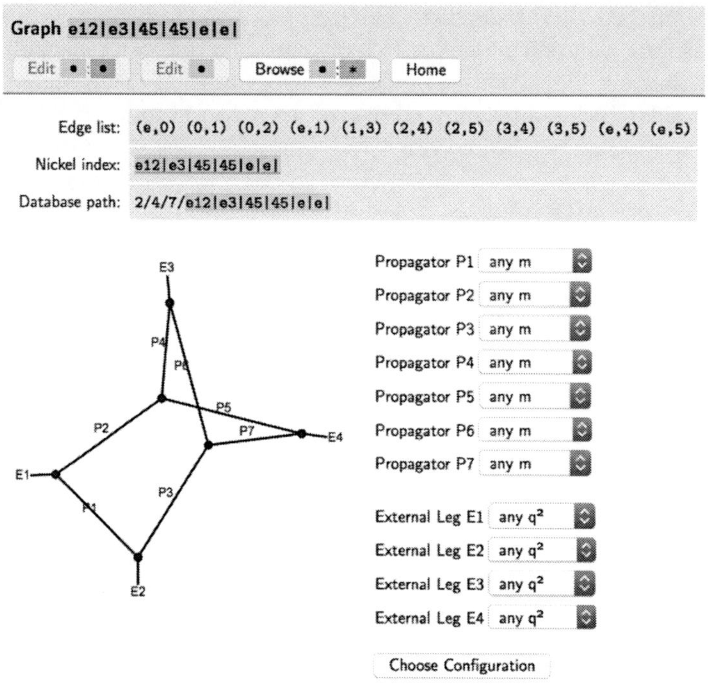

Figure 2. The Loopedia Configuration Editor

When editing a topology (bare Nickel index only) the page ends here, i.e. in order to proceed one needs to choose a configuration first. Otherwise the page continues with the available records and the New Record Form (Fig. 4).

3.3. Entering New Records

To enter a new record one must first navigate to a graph and a configuration as outlined above. The New Record Form is found beneath the available records.

Upon successful submission the submitted record is displayed again for review, with controls for editing and deletion (Fig. 5). An e-mail is sent to the submitter's address with an individualized URL that allows to access the review page again for future editing.

The submitted records are not visible publicly before they have been confirmed by a moderator. This is a safety measure we implemented currently to prevent situations like submitters deleting each other's records; this may change in the future. The moderators are notified together with the submitter and will usually make new records public in a timely manner.

4. Internal structure

Loopedia is realized as a bash script in a CGI environment which uses the Unix file system as database and is indexed by the `mlocate` utility.

The database contains (in the real filesystem sense) one directory for each bare Nickel index, with subdirectories for the configuration index (the second part of the CNickel) of the same

IOP Conf. Series: Journal of Physics: Conf. Series **1085** (2018) 052003 doi:10.1088/1742-6596/1085/5/052003

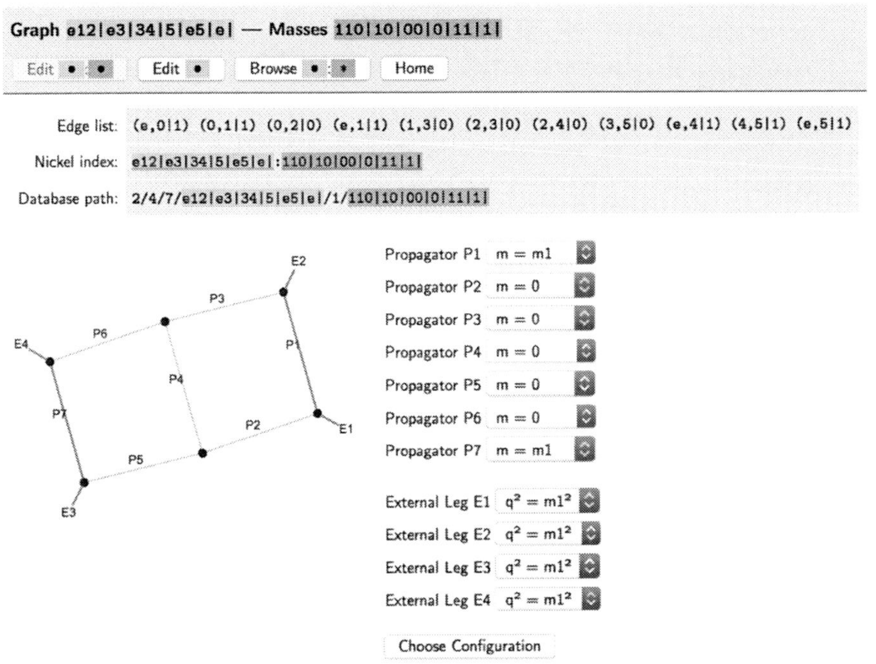

Figure 3. The Loopedia Record Display

graph. Underneath comes another layer of subdirectories for the visibility (`public`, `private`, `deleted`), and below that, one directory for each record. The actual implementation inserts a few more directory layers for indexing and performance reasons so that the full database path becomes

$$\texttt{db}/L/\ell/p/\text{Nickel}/s/\text{Config}/\text{Visibility}/\text{Record}$$

where L, ℓ, p, s are the number of loops, legs, propagators, and scales, respectively. The database root '`db`' is included in the path to be able to make anchored searches. Records have identifiers similar to the maildir format in Unix, e.g. `1482239373.Z1Fv`. (Maildir stores each incoming e-mail in a separate file, whose name includes a timestamp, to avoid race conditions. For the same reason, we store each record in a separate folder.)

Almost all graph-theoretical operations in Loopedia are taken care of by the Python library GraphState [4]. GraphState allows to supplement a graph with configurations, i.e. add extra integers to either the nodes or the edges. The colored Nickel index used by Loopedia implements a single edge coloring which encodes masses and external q^2.

To draw the graph, a Python function translates the graph into Graphviz's DOT language [7], from which neato produces an SVG image. SVG is an XML-based vector format rendered by all modern browsers and scales well in the icon-size plaquettes of the Graph Browser.

5. Summary

We presented a new database which allows to search for bibliographic and other information on loop integrals based on topology-related information. The intention of Loopedia is to collect the information available in the literature and to make it searchable by graph. It relies

IOP Conf. Series: Journal of Physics: Conf. Series **1085** (2018) 052003 doi:10.1088/1742-6596/1085/5/052003

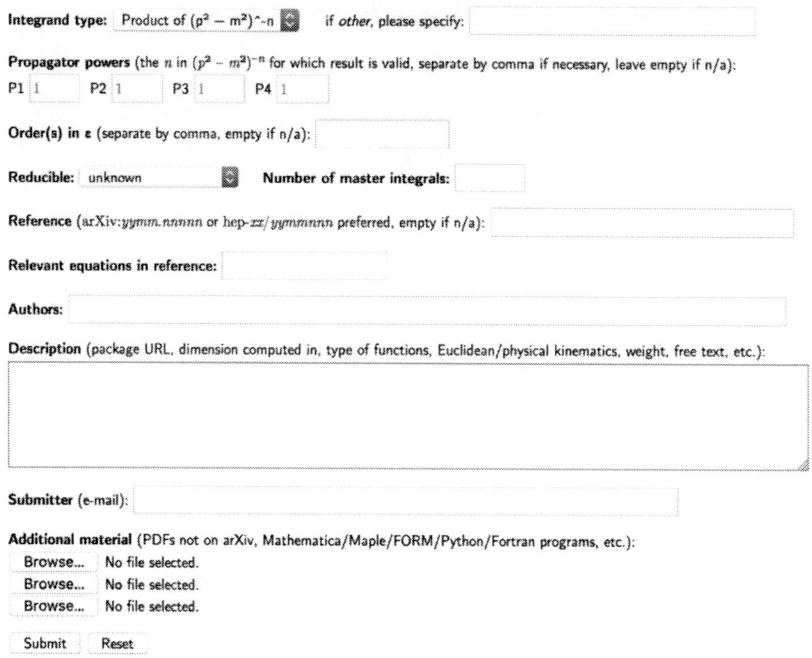

Figure 4. The New Record Form

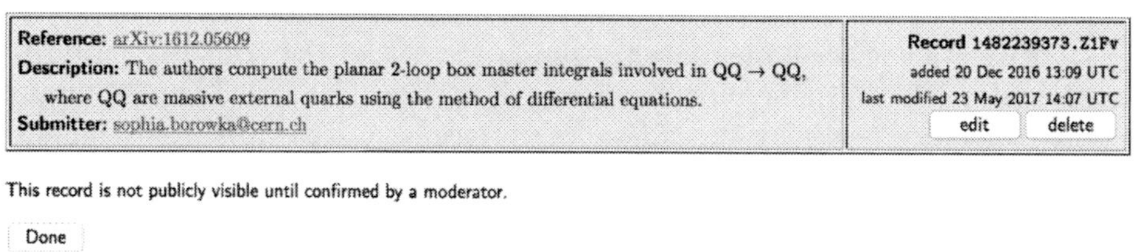

Figure 5. The Loopedia Record Review

on the concept that the community will actively contribute to the database by submitting new integral information. The database is hosted at the Max Planck Institute for Physics in Munich with broadband internet access and a daily backup at `loopedia.org`.

[1] C. Bogner *et al.*, "Loopedia, a Database for Loop Integrals," arXiv:1709.01266 [hep-ph].
[2] J.F. Nagle, *J. Math. Phys.* **7** (1966) 1588.
[3] B.G. Nickel, D.I. Meiron, G.A. Baker, Compilation of 2-pt and 4-pt graphs for continuous spin model, University of Guelph report, 1977.
[4] D. Batkovich, Y. Kirienko, M. Kompaniets, S. Novikov, arXiv:1409.8227.
The GraphState library is at `https://pypi.python.org/pypi/GraphState`.
[5] J. Küblbeck, M. Böhm, A. Denner, *Comput. Phys. Commun.* **60** (1990) 165;
T. Hahn, *Comput. Phys. Commun.* **140** (2001) 418 [hep-ph/0012260].
[6] P. Nogueira, *J. Comput. Phys.* **105** (1993) 279
[7] E.R. Gansner, S.C. North, *Software - Practice and Experience* **30** (2000) 1203, `graphviz.org`.

A development of an accelerator board dedicated for multi-precision arithmetic operations and its application to Feynman loop integrals II

H Daisaka[1], N Nakasato[2], T Ishikawa[3], F Yuasa[3], K Nitadori[4]

[1] Hitotsubashi University, 2–1, Naka, Kunitachi, Tokyo, 186–0801, Japan
[2] University of Aizu, Aizu-wakamatsu, Fukushima, 965–8580, Japan
[3] High Energy Accelerator Research Organization (KEK), 1–1, Oho, Tsukuba, Ibaraki, 305–0801, Japan
[4] RIKEN Advanced Institute for Computational Science, 7–1–26, Minatojima-minami-machi, Chuo-ku, Kobe, Hyogo, 650–0047, Japan

E-mail: daisaka@phys.science.hit-u.ac.jp

Abstract. Evaluation of a wide variety of Feynman diagrams with multi-loop integrals and physical parameters and its comparison with high energy experiments are expected to investigate new physics beyond the Standard Model. We have been developing a direct computation method of multi-loop integrals of Feynman diagrams. One of features of our method is that we adopt the double exponential rule for numerical integrations which enables us to evaluate loop integrals with boundary singularities. Another feature is that in order to accelerate the numerical integrations with multi-precision calculations, we develop an accelerator system with Field Programmable Gate Array boards on which processing elements with dedicated logic for quadruple/hexuple/octuple precision arithmetic operations are implemented. In addition, we also develop a programming interface designed for easy use of the system. The development is continued for practical use of the system. We present the current development of our system, and the numerical results of higher-loop diagrams performed using our system.

1. Introduction

The discovery of the Higgs boson in 2012 opened the door to a new era of elementary particle physics. While the Standard model has been successful, we have good enough reasons to expect new phenomena beyond it. In these investigations, precise theoretical predictions are crucial.

Recently, methods of evaluating Feynman integrals have expanded greatly in both analytic and numerical approaches. When we go into the higher order calculation, we have to include over thousands of diagrams and thus an automatic computation system is strongly required. Enormous efforts for developing such systems have been dedicated by many authors. For 1-loop integrals, several automated systems have shown successful results [1–7]. For multi-loop integrals, a lot of methods have been developed and the current status is reviewed in [8]. However, for evaluation of a wide variety of integrals including propagators with arbitrary mass scale and external momentum, there is still room for improvement.

We have been developing a fully numerical method for the evaluation of Feynman integrals which called *Direct Computation Method* (DCM) [9]. It is based on the numerical multi-dimensional integration and the extrapolation method. We can choose any kind of numerical

IOP Conf. Series: Journal of Physics: Conf. Series **1085** (2018) 052004 doi:10.1088/1742-6596/1085/5/052004

integration method in DCM if it gives integration results in a good enough precision. Here we use the *Double Exponential Formula* (DE) [10] which is one of the optimal methods in the case that there is an endpoint singularity. For the extrapolation, we use both linear and nonlinear extrapolation methods. So far, DCM have succeeded evaluating Feynman integrals with 2-, 3-, and 4-loops with 2, 3, and 4 legs [11].

However, DCM still has some difficulties for evaluating multi-loop diagrams. One is that the amount of calculation is very large, since a multi-loop diagram becomes an integration of multi-dimension. For example, a 3-loop self-energy diagram needs 7-th dimensional integration. Second is that an evaluation of multi-loop diagrams often requires calculation with high accuracy, higher than double precision, especially in the case that there is a strong singularity. Such a high accuracy calculation of Feynman loop integrals can be carried out by using high precision softwares (e.g., [12–14]). However, computation time will be much longer than the case using only double-precision. This is because in such softwares a high accuracy is realized by a combination of amount of arithmetic operations in ordinary CPUs, in other words, a large increase in operation count in the high precision case.

As one of the solutions to overcome these problems, we have been developing a dedicated accelerator system for multi-precision arithmetic operation. Based on the idea of GRAPE project(e.g., [15][16]), we have designed GRAPE-MPX architecture and the processor in which there are a number of processing elements (PE) with dedicated logic units for quadruple (hereafter MP4), hexuple (MP6), octuple-precision (MP8) arithmetic. So far, this processor has been implemented on a structured ASIC [17] and FPGAs [18]. The latest system is called GRAPE9-MPX [19][20].

In order to advance the evaluation of Feynman integrals, we continue to develop GRAPE9-MPX to have much computational power. We also continue to develop our programming interface, Goose and LSUMP, to fit to the current system, which supports the use of MPI and, furthermore, the use of GPU without changing our application program.

This paper organized as follows. The current status of GRAPE9-MPX system is explained in Section 2, and Section 3 presents the results of a 3-loop Feynman integral with two legs for the case propagators have two different mass scales as an application. The last section is dedicated for summary and discussion.

2. Overview of GRAPE9-MPX and programming interface

Figure 1 shows a schematic diagram and a picture of GRAPE9-MPX cluster system in KEK. This system is developed from the previous system with a single host and multiple FPGA boards [19][20] in order to enhance the computational power. The GRAPE9-MPX cluster system has 64 FPGA boards which are installed on 8 host computers. Each board is connected to a host via PCIe interface. For FPGA boards, we used Altera Arria V board from Intel Co. [21]. With the current implementation of our processor, the peak performance of the system achieves 403 GFlops for MP4, 185 GFlops for MP6, and 96 GFlops for MP8.

Our dedicated processor is designed by using VHDL (VHSIC Hardware Description Language) so that it is possible to implement our processor on another FPGA board, and also it is easy to make improvements. Our processor consists of two parts, MP processor and Control processor (CP). MP processor consists of PEs and broadcast memory units which form a SIMD processor. Each PE has a multiply unit, an adder unit, and register units for quadruple/hexuple/octuple-precision arithmetic. These arithmetic units perform in every clock cycle, but have 4 clock latency. Also, PE has an inverse square root (rsq) unit with a limited accuracy (19 bits for exponent and 32 bits for mantissa). This unit is used to create an initial guess for division arithmetic which is performed by Newton-Raphson method.

The role of CP is to control MP processor, by sending data and instruction necessary for calculation from a host computer to MP processor, and receiving calculation results from MP

IOP Conf. Series: Journal of Physics: Conf. Series **1085** (2018) 052004 doi:10.1088/1742-6596/1085/5/052004

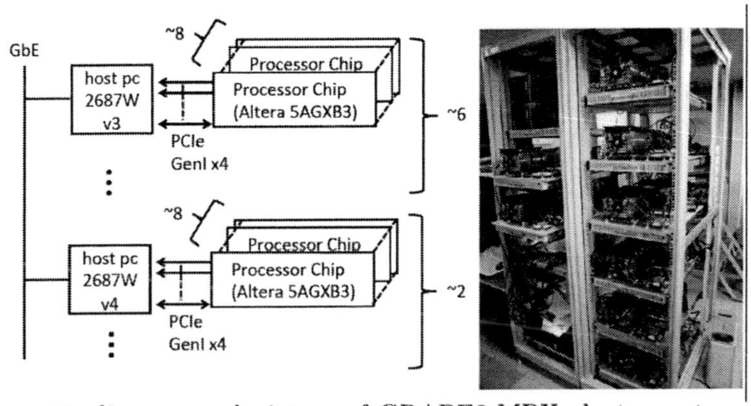

Figure 1. Schematic diagram and picture of GRAPE9-MPX cluster system. Each host has 8 FPGA boards and is connected with GbE. For host PCs, we use Intel Xeon E5-2687W v3(6 hosts) and E5-2687W v4(2 hosts) for CPU, 128GB memory, and X10 DRX MB from SuperMicro which has 11 PCIe slots.

processor and sending them to the host. It consists mainly of IO units (including PCIe and DRAM controllers), memory components for data and instruction, and their control unit. In the current implementation, the data memory has 32k words stored on onchip memory, whereas the instruction memory has 4k words on onchip memory, and 16M words on DRAM equipped on the FPGA board.

Table 1 shows a numerical representation for MP4, MP6, and MP8 used in PE, and a specification of our dedicated processor of the current implementation. Note that we used 19 bits for exponent in order to follow IEEE binary256 format for MP8, and the same exponent is also used for MP4 and MP6 for simplicity. Our numerical representation has the total bit width 4 bits wider than a standard format used in a host computer. Therefore, we cut 4 bits from mantissa or exponent when the data is sent back to CP and a host computer. The environmental variable decides from which we cut the bits.

	MP4 (quadruple)	MP6 (hexuple)	MP8 (octuple)
sign	1	1	1
exponent	19	19	19
mantissa	112	176	240
total	132	196	260
standard	128	192	256

	MP4	MP6	MP8
Number of PE	36	19	11
Clock(MHz)	88	78	68
Peak(Gflops)	6.3	2.9	1.5
logic utilization (%)	96	89	83
mem utilization (%)	52	55	62
DSP utilization (%)	64	69	62

Table 1. Numerical representation (length of bit) used in PE (left), and a specification and resource utilization of our processor currently implemented in the FPGA board (right). For logic synthesis, we use Quartus 13.1 by Altera.

As already stated in [19], one of applications suited for our system is an interaction type calculation as $f_i = \sum_{j=1}^{n_j} f(X_i, Y_j)$, where X_i is the i-th element of X, Y_j is the j-th element of Y, and n_j is the number of elements of Y. At calculation, data X_i is set on registers in i-th PE, and then data Y_j is sent from CP to all PEs. According to instructions sent from CP, i-th PE performs an evaluation of the function $f(X_i, Y_j)$ and a summation of the result. This process continues until data X_i and Y_i run out. In a program, this can be written as double loops. Note that a multi-dimensional integration can be expressed in double loops by using loop fusion technique that merges multiple loops into a single loop. Therefore, it is possible to accelerate a

IOP Conf. Series: Journal of Physics: Conf. Series **1085** (2018) 052004 doi:10.1088/1742-6596/1085/5/052004

calculation of Feynman loop integrals by using GRAPE9-MPX.

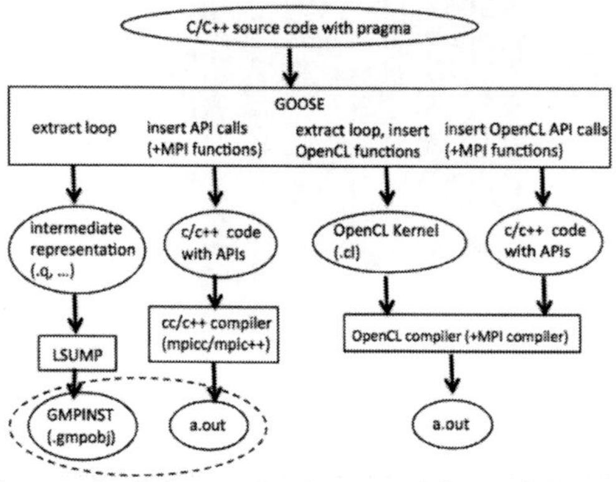


```
#pragma goose parallel for loopcounter(ixy, iz)
for(ixy = 0; ixy < ni; ixy++) {
  sumzG[ixy] = 0.0;
  for(iz = 0; iz < nj; iz++) {
    xx = dev_xx[ixy];
    yy = dev_yy[ixy];
    zz = x30_1[iz] * dev_cnt4[ixy];
    d = - xx * yy * s -
    tt * zz * (one - xx - yy - zz) +
    (xx + yy) * lambda2 +
    (one - xx - yy - zz) * (one - xx - yy) * fme2+
    zz * (one - xx - yy) * fmf2 ;
    sumzG[ixy] += gw30[iz] / (d * d);
  }
}
```

Figure 2. Schematic picture of flow of Goose and LSUMP and a sample code with the directive for a case of 1-loop box scalar integral.

Along with the development of the dedicated accelerator, we also have been developing our own compiler system, Goose and LSUMP, in order to use our system easily. Figure 2 shows the flow of compiling a program by Goose and LSUMP. Goose is a directive base compiler like OpenMP. Only procedure we have to take is to put a directive pragma (**`#pragma goose parallel`**) in an original C/C++ source code just before a double loop to be accelerated. LSUMP is our Domain Specific Language (DSL) compiler specially developed for GRAPE9-MPX. It generates a kernel executed in GRAPE9-MPX from the double loop extracted by Goose. For more detail, see Nakasato [22].

In order to use our cluster with multiple hosts and FPGA boards, we need to use a parallel computing interface such as MPI. We extended Goose to generate MPI API calls as well. In the current implementation, in the case that n hosts are used in calculation in parallel, the outer loop (i-loop) is divided into n sub loops, and each sub loop is assigned to each host. Also, we extended Goose to generate OpenCL API calls and kernels which enable us to accelerate our program by using GPGPU. For GPGPU, we use the Double-Double(DD) format [23]. Thus, by using Goose, we can accelerate our application program written in a simple C/C++ source code not only on GRAPE9-MPX but also on GPGPU with MPI.

3. Application to 3-loop Feynman integral

Here we present the numerical results of a 3-loop Feynman integral with 2 legs by DCM using GRAPE9-MPX system. Figure 3 shows a diagram of 3-loop self energy and the corresponding integral we evaluated. The number of the propagators is 8 as shown in Fig. 3 and the number of dimensions of the integration becomes 7 due to the δ-function. The masses of the propagator are given as $m_1 = m_2 = m_5 = m_6 = m_7 = m_8 = 1/2$ and $m_3 = m_4 = 1$, in order to compare results of Ghinculov [24]. We proceeded the evaluation of the diagram by changing the kinematic variable p not only in unphysical ($0 \leq p^2 < 1$) but also in physical ($1 < p^2 < 4$ and $4 < p^2 \leq 8$) region. For the physical region, we extrapolated a resultant value from results of the integral with different values of ρ. For the extrapolation, we used Wynn's ϵ-algorithm [25]. Table 2 lists a condition of calculation related to the DE formula and the extrapolation. In order to perform an integration that may be affected by singularity, we used a larger number of grids and small mesh sizes, and the extrapolation for calculations in physical region.

We computed a real part of I, I_{re}, in unphysical region, and both I_{re} and an imaginary part, I_{im}, in physical region. Figure 4 shows the behavior of our numerical results of I as a function of p^2. There are features which should be pointed out. First feature is that I diverges at $p^2 = 1$ and 4. Around them, the numerical computation becomes very hard. For example,

IOP Conf. Series: Journal of Physics: Conf. Series **1085** (2018) 052004 doi:10.1088/1742-6596/1085/5/052004

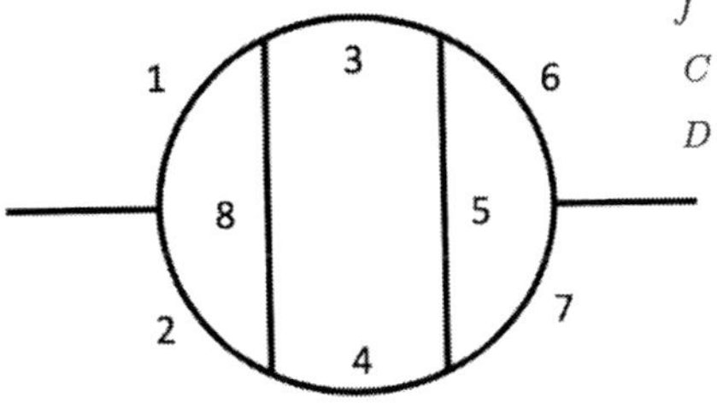

$$
\begin{aligned}
I &= \int dx_1 dx_2 dx_3 dx_4 dx_5 dx_6 dx_7 dx_8 \delta(1 - \sum_{l=1}^{8} x_l) \frac{1}{(D - i\rho C)^2} \\
C &= x_5(x_8 x_{123467} + x_{12} x_{3467}) + x_{67}(x_8 x_{1234} + x_{12} x_{34}) \\
D &= -p^2 \times (x_5(x_8 x_{136} x_{247} + x_1 x_2 x_{3467} + x_{12} x_{36} x_{47}) \\
&\qquad + x_8(x_6 x_7 x_{1234} + x_{13} x_{24} x_{67}) \\
&\qquad + x_{34}(x_1 x_2 x_{67} + x_{12} x_6 x_7) \\
&\qquad + x_{12} x_3 x_4 x_{67}) \\
&\quad + ({m_1}^2 x_1 + {m_2}^2 x_2 + {m_3}^2 x_3 + {m_4}^2 x_4 + {m_5}^2 x_5 \\
&\qquad + {m_6}^2 x_6 + {m_7}^7 x_2 + {m_8}^2 x_8) \times C \\
x_{ijk...} &= x_i + x_j + x_k + ...
\end{aligned}
$$

Figure 3. 3-loop self energy diagram and the corresponding integral in a massive case we evaluated. The values p and x_i denote a momentum of an external line and Feynman parameters, respectively. The term $i\rho$ prevents the denominator from vanishing in the integral domain.

	unphysical region(I_{re})	physical region (I_{re}, I_{im})
N	25	64
h	0.3	0.0875, 0.11718
ρ	0 (fixed)	1.15^{β}, $\beta = -51, -52, .., -74$
number of FPGA boards	1	64
precision	MP4	MP4
elapsed time per point	300 sec	11 hours

Table 2. Condition of calculation related to the DE formula and the extrapolation, as well as the number of FPGA boards used in calculation and precision, and elapsed time. N is the number of grids per dimension, h is a mesh size of the transformed variable in DE, and ρ is a parameter variable used for the extrapolation.

the extrapolated results, $I_{re}(p^2 = 3.99)$ is not well converged due to the severe singularity in the integration domain which causes the cancellation between positive and negative values. Second feature is that there are points which a sign for I changes at $p^2 \sim 2$ and 2.5 for I_{re}, and $p^2 \sim 3.5$ for I_{im}. Last feature is that I converges to zero for larger p^2. These features are similar to those obtained in Ghinculov [24]. [1]

Table 2 also shows the number of FPGA boards, precision used in calculation, and elapsed time per point p. By using our system with 64 FPGA boards, for physical region, a set of I (I_{re}, I_{im}) with 24 different values of β which is required for the extrapolation can be calculated in 11 hours which is acceptable. In order to increase the efficiency of calculation, we performed integrations of I_{re} and I_{im} with 8 different values of β at the same time (that is, 16 integrations at the same time). The reason why such calculation is possible is that PE has registers enough to store data, and C and D are common so that they can be reused to save computation. The calculation time of this method is about 4 times longer than that of a single integration case in which I_{re} or I_{im} is individually integrated with a single value of β. Thus, in using this method, four times higher computational efficiency can be obtained.

[1] The plot of I_{re} and I_{im} shown in Figure 5 of Ghinculov [24] seems to replace each other, because I_{re} should not be zero if $p^2 < 1$.

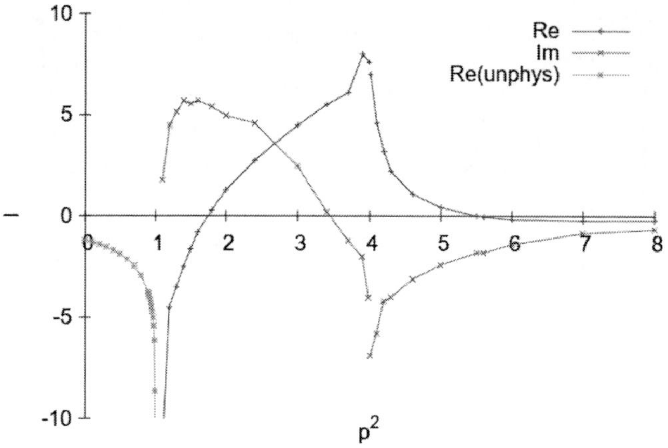

Figure 4. Numerical results of $I(I_{re}$ and $I_{im})$ as a function of p^2. Purple and green denote I_{re} and I_{im} for physical region, and blue denotes I_{re} for unphysical region.

4. Summary and discussion

We have been developing an accelerator system dedicated for multi-precision arithmetic operators. The current system is GRAPE9-MPX cluster system with 64 FPGA boards on 8 host computers. Our dedicated processor, including PE with arithmetic logic units for MP4/MP6/MP8, is implemented on the FPGA boards. We also developed a compiler system which enables us to accelerate our application program on GRAPE9-MPX without any change in the program except for putting pragma just before double loops to be accelerated. It should be noted that the current version of Goose can generate MPI API, and OpenCL API and kernels which has the DD format, so that we can use a cluster system and GPGPU. We performed an evaluation of a 3-loop self energy diagram with mass parameters used in Ghinculov [24] and showed that our numerical method (DE-DCM + GRAPE9-MPX) can reproduce the behavior of I seen in Ghinculov [24].

Note that we need to discuss about error of numerical results in Fig. 4. For unphysical region, the error is reasonable, but for physical region, the numerical error estimated from the extrapolation is 3-digit even for the best case and 1-digit for the worst case. The reason for the large error in physical region is probably due to the severe singularity appearing in the integral domain which causes large cancellation. One way to improve the accuracy is to tune parameters of the integration method, for example, increasing the number of grids and/or reducing mesh size. Another way is to divide the integral domain into sub domains and perform integration for each sub domain. In both cases, a large scale computation in multi-precision arithmetic such as quadruple/hexuple/octuple precision will play an important role. The results will appear in future paper.

Acknowledgment

We thank Prof. Kiyoshi Kato for fruitful discussion and comments. This work is further supported by Grant-in-Aid for Scientific Research (15H03602, 15H03668, and 17K05428) of JSPS, and the Large Scale Simulation Program No. 16/17-21 of KEK.

References

[1] Hahn T and Pérez-Victoria M 1999 *Computer Physics Communications* **118** 153 – 165
[2] Wang J X 2004 *Nucl.Instrum.Meth.* **A534** 241–245
[3] Bélanger G, Boudjema F, Fujimoto J, Ishikawa T, Kaneko T, Kato K and Shimizu Y 2006 *Phys. Rept.* **430** 117–209

IOP Conf. Series: Journal of Physics: Conf. Series **1085** (2018) 052004 doi:10.1088/1742-6596/1085/5/052004

[4] Cullen G, Guillet J P, Heinrich G, Reiter T and Rodgers M 2011 *Computer Physics Communications* **182** 2276–2284
[5] van Hameren A 2011 *Computer Physics Communications* **182** 2427–2438
[6] Carrazza S, Ellis R K and Zanderighi G 2016 *Computer Physics Communications* **209** 134–143
[7] Denner A, Dittmaier S and Hofer L 2017 *Computer Physics Communications* **212** 220–238
[8] Borowka S, Heinrich G, Jahn S, Jones S P, Kerner M and Schlenk J 2017 *Journal of Physics: Conf. Series* **920**
[9] Yuasa F, de Doncker E, Hamaguchi N, Ishikawa T, Kato K, Kurihara Y and Fujimoto J 2012 *Computer Physics Communications* **183** 2136–2144
[10] Takahasi H and Mori M 1974 *Publications of the Research Institute for Mathematical Sciences* **9** 721741
[11] de Doncker E, Shimizu Y, Fujimoto J and Yuasa F 2018 *Computer Physics Communications* **224** 164–185
[12] GMP URL `https://gmplib.org/`
[13] High-Precision Software Directory URL `http://crd-legacy.lbl.gov/~dhbailey/mpdist/`
[14] MPFR URL `http://www.mpfr.org/`
[15] Sugimoto D, Chikada Y, Makino J, Ito T, Ebisuzaki T and Umemura M 1990 *Nature* **345** 33–35
[16] Makino J 2006 *Computing in Science & Engineering* (8) 30–40
[17] Daisaka H, Nakasato N, Makino J, Yuasa F and Ishikawa T 2011 *Procedia Computer Science* **4** 878–887
[18] Nakasato N, Daisaka H, Fukushige T, Kawai A, Makino J, Ishikawa T and Yuasa F 2012 *Embedded Multicore Socs (MCSoC), 2012 IEEE 6th International Symposium on* 75–83
[19] Motoki S, Daisaka H, Nakasato N, Ishikawa T, T, Kawai A and Makino J 2014 *Journal of Physics: Conference Series* **608**
[20] Daisaka H, Nakasato N, Ishikawa T and Yuasa F 2015 *Procedia Computer Science* **51** 1323–1332
[21] Altera Co *Arria V Device Handbook* URL `www.altera.co.jp/literature/hb/arria-v/arriav_handbook.pdf`
[22] Nakasato N and Makino J 2009 *IEEE International Conference on Cluster Computing and Workshops* pp 1–9
[23] Hida Y, Li X and Bailey D 2001 *Proceedings of the 15th Symposium on Computer Arithmetic* pp 155–162
[24] Ghinculov A 1996 *Physics Letters B* **385** 279–283
[25] Wynn P 1956 *Mathematical Tables and Aids to Computing* **10** 91–96

IOP Conf. Series: Journal of Physics: Conf. Series **1085** (2018) 052005 doi:10.1088/1742-6596/1085/5/052005

High-speed evaluation of loop integrals using lattice rules

E de Doncker[1], **A Almulihi**[1] **and F Yuasa**[2]

[1] Department of Computer Science, Western Michigan University, Kalamazoo MI 49008, U. S. A.
[2] High Energy Accelerator Research Organization (KEK), Oho 1-1, Tsukuba, Ibaraki, 305-0801, Japan

E-mail:
{elise.dedoncker, ahmedhassan.almulihi}@wmich.edu, fukuko.yuasa@kek.jp

Abstract. Lattice rules are implemented in CUDA for many-core computations on GPUs. A high-speed evaluation of Feynman loop integrals is presented, based on lattice rules and suitable transformations. The accuracy and efficiency of the method are compared for higher order $\sin^m$-transformations. Extensive results are reported for classes of diagrams including 2-loop box and 3-loop self-energy diagrams with massive internal lines. The method is further combined with an extrapolation with respect to the dimensional regularization parameter.

1. Introduction

The goal of this work is to perform accurate loop integral computations, and to develop computer programs which evaluate multi-loop Feynman integrals in a fully numerical manner. Previously (e.g., in [1, 2, 3, 4]), we reported precise numerical results for 2-, 3- and 4-loop Feynman integrals using adaptive integration methods and extrapolation. In this paper we present a non-adaptive technique, applied to the calculation of 2- and 3-loop integrals, where the bulk of the computation, based on lattice rules with transformations by Sidi [5, 6], is executed on a GPU accelerator board. Lattice rules were used with Sidi's transformations in programs for 2- and 3-dimensional integration in (*r2d2lri* [7], *elrint3d* [8]). Lattice rules on GPU were recently applied to integrals resulting from sector decomposition [9].

A Feynman integral with L loops and N internal lines is given by

$$\mathcal{F} = \frac{\Gamma\left(N - \frac{\nu L}{2}\right)}{(4\pi)^{\nu L/2}}(-1)^N \int_0^1 \prod_{r=1}^{N} dx_r\, \delta(1 - \sum x_r) \frac{C^{N-\nu(L+1)/2}}{(D - i\varrho C)^{N-\nu L/2}}, \tag{1}$$

where C and D are polynomials (arising from determinants) depending on the topology of the Feynman diagram and physical parameters, and $\nu = 4$ is the space-time dimension. However, in case the integral diverges with $\nu = 4$, the space-time dimension can be considered as a function of the dimensional regularization parameter, $\nu = \nu(\varepsilon)$, for a regularization where $\mathcal{F}$ is expanded as a series in ε. Ultra-violet (UV) and infra-red (IR) singularities give rise to integrand singularities on the boundaries of the domain. We let $\nu(\varepsilon) = 4 - 2\varepsilon$ or $\nu(\varepsilon) = 4 + 2\varepsilon$ for a UV or IR singularity, respectively. For a finite integral computation, the expansion can also serve at the basis of a convergence acceleration or extrapolation to the limit as $\varepsilon \to 0$. Furthermore, we set $\varrho = 0$ in Eq (1) unless D vanishes in the domain.

By eliminating the δ-function in Eq (1), the integral is re-written as $\mathcal{F} = (4\pi)^{-\nu L/2}\mathcal{F}_{L,N}$ with

$$\mathcal{F}_{L,N} = \Gamma(N - \frac{\nu L}{2})(-1)^N \int_{\mathcal{S}_{N-1}} \frac{C^{N-\nu(L+1)/2}}{(D - i\varrho C)^{N-\nu L/2}}\, d\mathbf{x}, \tag{2}$$

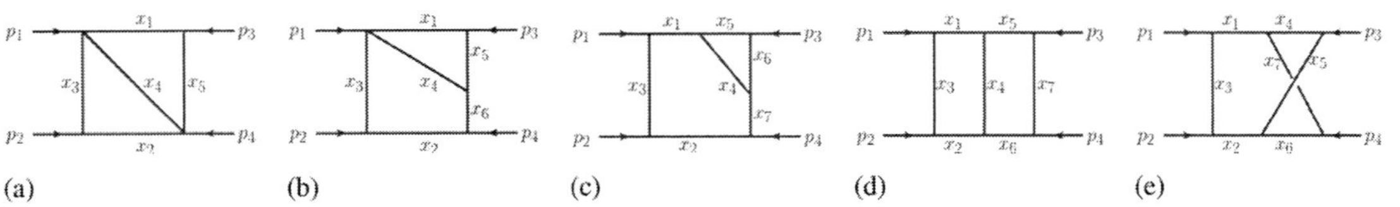

Figure 1. [2LB] Sample 2-loop box diagrams [10, 11]: (a) $N = 5$ (double-triangle), (b) $N = 6$ (tetragon-triangle), (c) $N = 7$ (pentagon-triangle), (d) $N = 7$ ladder, (e) $N = 7$ crossed box

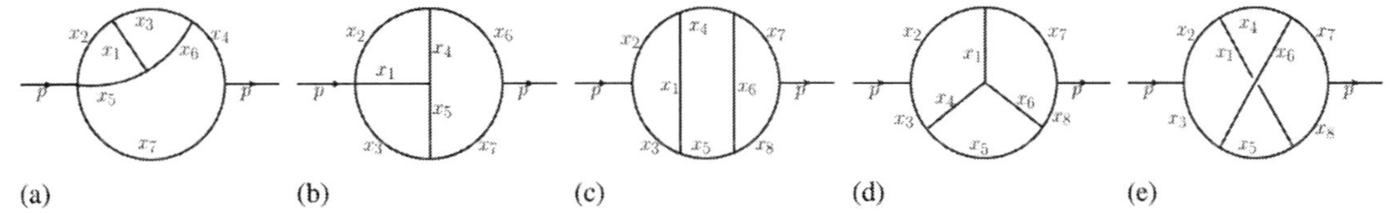

Figure 2. [3LS] Sample 3-loop self-energy diagrams with massive internal lines, cf., Laporta [10]: (a) $N = 7$, (b) $N = 7$, (c) $N = 8$, (d) $N = 8$, (e) $N = 8$.

Table 1. Integrals $\mathcal{F}_{L,N}$

DIAG	L	N	$\mathcal{F}_{L,N}$	VALUE	DIAG	L	N	$\mathcal{F}_{L,N}$	VALUE
Fig 1(a)	2	5	$-\int_{S_4} \frac{1}{CD}\,d\mathbf{x}$	$0.9509235623171^{(*)}$	Fig 2(a)	3	7	$-\int_{S_6} \frac{1}{CD}\,d\mathbf{x}$	$1.32644820827^{(*)}$
Fig 1(b)	2	6	$\int_{S_5} \frac{1}{D^2}\,d\mathbf{x}$	0.276209225359	Fig 2(b)	3	7	$-\int_{S_6} \frac{1}{CD}\,d\mathbf{x}$	$1.34139924145^{(*)}$
Fig 1(c)	2	7	$-2\int_{S_6} \frac{C}{D^3}\,d\mathbf{x}$	$0.1723367907503^{(*)}$	Fig 2(c)	3	8	$\int_{S_7} \frac{1}{D^2}\,d\mathbf{x}$	0.2796089232826
Fig 1(d)	2	7	$-2\int_{S_6} \frac{C}{D^3}\,d\mathbf{x}$	$0.1036407209893^{(*)}$	Fig 2(d)	3	8	$\int_{S_7} \frac{1}{D^2}\,d\mathbf{x}$	0.1826272375392
Fig 1(e)	2	7	$-2\int_{S_6} \frac{C}{D^3}\,d\mathbf{x}$	$0.0853513981538^{(*)}$	Fig 2(e)	3	8	$\int_{S_7} \frac{1}{D^2}\,d\mathbf{x}$	0.14801330396

where the domain is the $d = (N-1)$-dimensional unit simplex, $\mathcal{S}_d = \{\mathbf{x} \in \mathcal{C}_d \mid 0 \leq \sum_{k=1}^{d} x_k \leq 1\}$, and $\mathcal{C}_d = [0,1]^d$ is the unit cube in $\mathbb{R}^d$. We will further transform $\mathcal{F}_{L,N}$ to an integral over $\mathcal{C}_d$.

For $\nu = 4$ and $\rho = 0$, the integrals $\mathcal{F}_{L,N}$ corresponding to the diagrams in Fig 1 and 2 are listed in Table 1. The numerical values are from [10] – except for the crossed box integral value, that resulted from an extensive computation with large numbers of integrand evaluations using the adaptive ParInt code in long double precision [3]. The functions C and D, given in Table A1 of Appendix A, are derived according to [12]. The cases indicated by $^{(*)}$ in Table 1 are taken with positive sign in subsequent test results. The computations assume unit masses and $p^2 = 1$.

2. Rank-1 lattice rules

A rank-1 lattice rule approximation of an integral $\mathcal{I}f = \int_{\mathcal{C}_d} f(\mathbf{x})\,d\mathbf{x}$ is represented by

$$\mathcal{Q}f = \frac{1}{n}\sum_{j=0}^{n-1} f(\{\frac{j}{n}\mathbf{z}\}) \tag{3}$$

where $\mathbf{z}$ is a generator vector with integer components $z \in \mathcal{Z}_n = \{1 \leq z < n,\ gcd(z,n) = 1\}$. Here $\{\mathbf{x}\}$ denotes the vector in the half open cube $\mathcal{U}_d = [0,1)^d$ obtained by taking the fractional part of each component of $\mathbf{x}$. Classically n is prime (i.e., $\mathcal{Z}_n = \{1, 2, \ldots, n-1\}$); this is relaxed in [13, 14]. The generators used for the results in this paper are computed with the Fast CBC (component-by-component) algorithm by Nuyens and Cools [14, 15], which runs in $\mathcal{O}(d\,n\,\log n)$ time and $\mathcal{O}(n)$ space, and improves on the time of the CBC algorithm as introduced by Sloan and Reztsov [16].

IOP Conf. Series: Journal of Physics: Conf. Series **1085** (2018) 052005 doi:10.1088/1742-6596/1085/5/052005

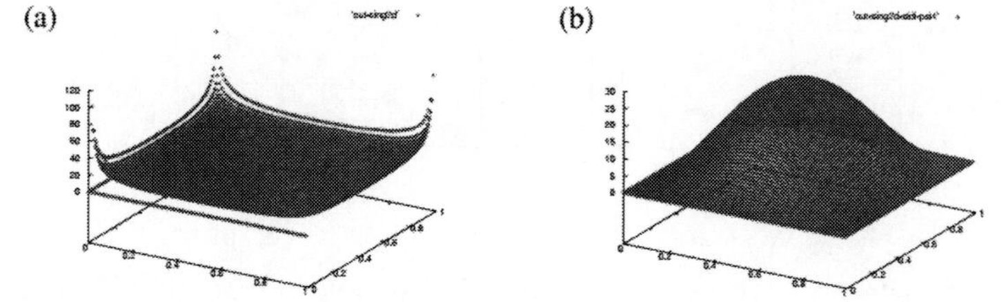

Figure 3. Ψ_4 transformation of $1/\sqrt{x_1x_2(1-x_1)(1-x_2)}$ over unit square: (a) without; (b) with transformation

Based on the theory of periodic functions, good lattice points (or **z** with optimal coefficients) as defined by Korobov [17, 18], and error bounds for lattice rule integration have been studied extensively in the literature (see, e.g., [13, 19, 20]), which further motivates the use of periodizing transformations.

3. Transformations

Consider the one-dimensional integral $If = \int_0^1 f(x)\,dx$ and a transformation $x = \varphi(t)$ such that $\varphi(0) = 0$, $\varphi(1) = 1$, and $\varphi'(t)$ has a number of derivatives that vanish at 0 and 1. Thus the transformed integrand in $If = \int_0^1 f(\varphi(t))\,\varphi'(t)\,dx$ can be extended periodically with period 1. Periodizing transformations include the transformations by Korobov [21], versions of the $\tanh$-transformation [22], the IMT [23, 24] and double exponential transformations [25]. For the $\tanh$ and double exponential type transformations one has to exercise considerable caution in their implementation, to avoid overflows and other run-time exceptions resulting from integrand evaluations near the boundaries.

In this paper we will rely on a class of transformations by Sidi [5, 6], $\varphi(t) = \Psi_m(t)$ given by

$$\Psi_m(t) = \frac{\theta_m(t)}{\theta_m(1)}, \qquad m = 1, 2, \ldots \tag{4}$$

with $\theta_m(t) = \int_0^t \sin^m(\pi u)\,du$. The transformations used for the work in this paper are listed below.

$\Psi_2(t) = t - \sin(2\pi t)/(2\pi), \quad \Psi_2'(t) = 2\sin^2(\pi t)$
$\Psi_4(t) = t + (-8\sin(2\pi t) + \sin(4\pi t))/(12\pi), \quad \Psi_4'(t) = \frac{8}{3}\sin^4(\pi t)$
$\Psi_6(t) = t - (45\sin(2\pi t) - 9\sin(4\pi t) + \sin(6\pi t))/(60\pi), \quad \Psi_6'(t) = \frac{16}{5}\sin^6(\pi t)$
$\Psi_8(t) = t + (-672\sin(2\pi t) + 168\sin(4\pi t) - 32\sin(6\pi t) + 3\sin(8\pi t))/(840\pi), \quad \Psi_8'(t) = \frac{128}{35}\sin^8(\pi t)$

Properties of $\Psi_m(t)$ are shown in [5, 6], also with respect to functions with end-point singularities. We will apply the transformations for periodization and also with the goal of dealing with integrand boundary singularities. Fig 3 illustrates the smoothening abilities of $\Psi_4(t)$, where the function $1/\sqrt{x_1x_2(1-x_1)(1-x_2)}$ is displayed before and after the transformation in both coordinate directions over the unit square. The jacobian "knocks out" the singular behavior in the transformed integrand.

4. Results

4.1. Lattice rule integration on GPU

The lattice rule integrations are implemented in CUDA C and executed on the *thor* cluster (of the Center for High Performance Computing and Big Data at WMU), where the host process runs on a node with Intel Xeon E5-2670, 2.6 GHz dual CPU and the lattice rule is evaluated on a (Kepler-20) GPU. For the CUDA program we used 64 blocks and 1024 threads per block on the GPU. The lattice generator vector is precomputed and is the only data communicated from the main program to the CUDA kernel, as a one-dimensional array of length d. The kernel sums the function values in Eq (3) over each block using the synchronization pattern for a reduction (see, e.g., [26]). A sample implementation of the integrand as a CUDA device function is given in Appendix C for the integral of Fig 1(b) in Table 1.

IOP Conf. Series: Journal of Physics: Conf. Series **1085** (2018) 052005 doi:10.1088/1742-6596/1085/5/052005

Table 2. Results for 2-loop box diagrams using no transformation or Ψ_2, Ψ_4, Ψ_6, and $(N-1)$-dim. lattice rule with 100M points. Listed are abs. accuracy (E_a) and execution times T in milliseconds (ms).

DIAG.		NO TRANS.		Ψ_2		Ψ_4		Ψ_6	
	N	E_a	T[ms]	E_a	T[ms]	E_a	T[ms]	E_a	T[ms]
2LB (a)	5	4.2 e-04	59	1.6 e-08	124	4.9 e-12	178	2.9 e-14	208
2LB (b)	6	1.6 e-04	58	3.0 e-07	154	3.6 e-10	224	6.2 e-11	258
2LB (c)	7	4.4 e-04	76	4.5 e-07	181	1.2 e-09	266	2.3 e-09	307
2LB (d)	7	4.0 e-05	76	3.7 e-11	181	6.6 e-11	266	1.1 e-11	310
2LB (e)	7	1.6 e-05	76	6.0 e-11	183	1.3 e-13	267	2.6 e-13	312

Table 3. Results for 3-loop self-energy diagrams using no transformation or Ψ_2, Ψ_4, Ψ_6, and $(N-1)$-dim. lattice rule with 100M points. Listed are abs. accuracy (E_a) and execution times T in milliseconds (ms).

DIAG.		NO TRANS.		Ψ_2		Ψ_4		Ψ_6	
	N	E_a	T[ms]	E_a	T[ms]	E_a	T[ms]	E_a	T[ms]
3LS (a)	7	5.7 e-02	77	3.4 e-04	184	1.0 e-05	267	2.5 e-07	309
3LS (b)	7	1.7 e-02	77	6.0 e-05	182	9.4 e-06	266	2.7 e-06	312
3LS (c)	8	2.8 e-03	87	1.2 e-05	214	2.4 e-06	316	5.3 e-07	366
3LS (d)	8	3.9 e-03	87	6.8 e-06	216	3.6 e-06	315	2.3 e-06	366
3LS (e)	8	1.7 e-03	86	2.6 e-06	218	1.2 e-07	315	6.9 e-08	367

Table 2 compares the accuracy and computation time for the 2-loop box diagrams of Fig 1(a-e), using a rank-1 lattice rule with 100,000,007 points generated for dimension $d = N - 1$, and Sidi's Ψ_2, Ψ_4, Ψ_6 transformations applied to the integrals; these are also compared to results obtained without a transformation (apart from mapping the unit simplex to the cube). The absolute accuracy E_a and the time in milliseconds are listed for each integration.

Similarly, Table 3 treats the 3-loop self-energy diagrams of Fig 2(a-e). These result in a more complicated structure and higher degree of the C and D polynomials, as well as difficult singularities on boundaries of the integration domain (cf., Table A1 in Appendix A). For example, D of 2(c) vanishes at $x_1 = x_2 = x_3 = 0$, and at $x_6 = x_7 = x_8 = 0$ (where $x_8 = 0$ corresponds to $\sum_{k=1}^{7} x_k = 1$ on $\mathcal{S}_7$). Results computed with Ψ_8 did not improve (or improved only slightly) on the results listed through Ψ_6. However, the accuracy improved significantly by the higher order transformations through Ψ_6.

Table 4 gives comparisons for lattice rules (LR) with 10M and 100M points, and adaptive integration (AD) with 100M points, using transformation Ψ_4 for the 2-loop box diagrams (2LB, Left), and Ψ_6 for the 3-loop self-energy diagrams (3LS, Right). The lattice rule with 100M points is clearly more accurate than the 10M rule, except for 3LS(d) where the accuracy is only marginally better (2.3 e-06 vs. 3.6 e-06).

For the sake of comparison, Table 4 also lists adaptive integration results with ParInt and 100M integrand evaluations, using 16 MPI processes on one 16-core node. The lattice integration method (LR) reaches considerably better accuracy with 100M evaluations and needs far less computation time than the adaptive integration (AD).

We further note that the (LR) GPU implementation achieves an excellent speedup (= the ratio of sequential to parallel time), which is around 456 for the crossed box integral 2LB (e) with 100M evaluations using the Ψ_4 transformation. The sequential time (around 122 seconds) was measured with *gettimeofday*, the parallel time (around 267.3 ms) with *cudaEventElapsedTime*.

4.2. Linear extrapolation

Extrapolation is tailored to an underlying asymptotic expansion, such as the expansions in [10] for loop integrals of the form $\mathcal{F}_{L,N}$ in Eq (2), with respect to the dimensional regularization parameter ε.

Table 4. Result comparisons for lattice rules (LR) with 10M and 100M points, and adaptive integration (AD) with 100M points, using transf. Ψ_4 for 2-loop box diagrams (2LB, Left), and tranf. Ψ_6 for 3-loop self-energy diagrams (3LS, Right). Listed are abs. accuracy (E_a) and execution times T in milliseconds (ms).

DIAG.	Ψ_4						DIAG.	Ψ_6					
2LB	LR 10M		LR 100M		Ad 100M		3LS	LR 10M		LR 100M		Ad 100M	
	E_a	T[ms]	E_a	T[ms]	E_a	T[ms]		E_a	T[ms]	E_a	T[ms]	E_a	T[ms]
2LB (a):	5.7 e-09	18	4.9 e-12	178	2.2 e-11	6495	3LS (a):	3.2 e-06	31	2.5 e-07	309	7.2 e-06	11782
2LB (b):	3.5 e-08	23	3.6 e-10	224	2.4 e-08	7971	3LS (b):	1.0 e-05	32	2.7 e-06	312	5.1 e-05	11757
2LB (c):	9.9 e-07	27	1.2 e-09	266	3.2 e-06	9765	3LS (c):	8.7 e-05	37	5.3 e-07	366	6.3 e-06	13497
2LB (d):	3.5 e-08	27	6.6 e-11	266	2.3 e-07	9721	3LS (d):	3.6 e-06	37	2.3 e-06	366	5.0 e-05	11707
2LB (e):	3.2 e-09	27	1.3 e-13	267	1.9 e-07	9730	3LS (e):	2.1 e-05	37	6.9 e-08	366	7.7 e-06	14000

Table 5. Results of linear extrapolation for loop integral of Fig 1(a), $\varepsilon_\ell = 2^{-\ell}$

ℓ	K	$I(\varepsilon)$	C_0	C_1	C_2	C_3
10	0	0.94969742599225604				
11	1	0.95030985357508291	0.9509222811579	-1.254251689629		
12	2	0.95061654757003522	0.9509235617007	-1.258185517020	2.685492832	
13	3	0.95077001482145529	0.9509235623195	-1.258189952916	2.694577546	-5.31585584
13	4	0.95084677853608013	0.9509235623197	-1.258189955234	2.694588624	-5.33530208
		Exact [10]:	0.9509235623171	-1.258189955235	2.694588643	-5.33534712

In general a linear extrapolation is performed for an underlying expansion

$$I(\varepsilon) \sim C_0\varphi_0(\varepsilon) + C_1\varphi_1(\varepsilon) + C_2\varphi_2(\varepsilon) + \ldots, \quad \text{as } \varepsilon \to 0$$

by creating an approximating sequence of $I(\varepsilon_\ell) \approx I(\varepsilon)$ such that

$$I(\varepsilon_\ell) \approx C_0\varphi_0(\varepsilon_\ell) + C_1\varphi_1(\varepsilon_\ell) + \ldots C_K\varphi_K(\varepsilon_\ell), \quad \ell = 1, \ldots, K+1,$$

and solving the resulting linear systems of orders $(K+1) \times (K+1)$, for increasing values of K and decreasing $\varepsilon = \varepsilon_\ell$. As an example, Table 5 shows extrapolated results for the leading term coefficients in the expansion of the 2-loop box integral of Fig 1(a). The functions $\varphi_k(\varepsilon) = \varepsilon^k,\ k \geq 0$, and $\varepsilon_\ell = 2^{-\ell}$.

5. Conclusions

We applied lattice rule integration to classes of 2-loop box and 3-loop self-energy Feynman diagrams, after a $\sin^m$-transformation, which was performed not only as a periodizing transformation but also to deal with integrand boundary singularities. We examined the effects of higher order transformations. The implementation in CUDA C was executed utilizing one Kepler 20 GPU for the integrand evaluations and summations over the blocks. The results were found far more accurate and efficient than parallel adaptive integration layered over MPI and using 16 CPU cores on one cluster node. The GPU lattice integration is promising for interaction cross section computations where thousands of integrals are needed. In future work it can be combined with other techniques, such as a suitable pre-partitioning of the domain.

Acknowledgments

We acknowledge the support from the National Science Foundation under Award Number 1126438, which funded the equipment used for the computations in this paper. This work is also funded by Grant-in-Aid for Scientific Research (17K05428) of JSPS, and the Large Scale Simulation Program No.16/17-21 of KEK. We further thank Dr. A. Sidi and Dr. S. Jones for pointing us to related papers.

IOP Conf. Series: Journal of Physics: Conf. Series **1085** (2018) 052005 doi:10.1088/1742-6596/1085/5/052005

Table A1. C and D functions in the integrands of Eq (2) and Table 1, for the diagrams in Fig 1 and 2

FIG	
1(a)	$C = x_{15}x_{234} + x_4x_{23}$
	$D = x_3^2x_4 + x_3x_4^2 + x_1^2x_{234} + x_3^2x_5 + 2x_3x_4x_5 + x_4^2x_5 + x_3x_5^2 + x_4x_5^2 - x_2^2x_{45}$
	$+x_2(x_4^2 + 2x_4x_5 + x_5^2 + x_3x_{45}) + x_1(x_2^2 + x_3^2 + x_4x_{45} + x_3(2x_4 + x_5) + x_2(x_{345} + x_4))$
1(b)	$D = (x_1^2 + x_5^2 + x_1x_5)x_{2346} + (x_2^2 + x_3^2 + x_6^2 + x_2x_3 + x_2x_6 + x_3x_6)x_{145} + x_4^2x_{12356}$
	$+3x_4x_5x_6 + 2x_1x_4x_{236} + 2x_4x_5x_{23}$
1(c)	$C = x_{1237}x_{456} + x_4x_{56}$
	$D = (x_1^2 + x_2^2 + x_3^2 + x_7^2 + x_1x_2 + x_{12}x_{37} + x_3x_7)x_{456} + x_4^2\, x_{123567}$
	$+(x_5^2 + x_6^2 + x_5x_6)x_{12347} + 3x_4(x_1x_5 + x_6x_7) + 2(x_{123}x_4x_6 + x_{237}x_4x_5)$
1(d)	$C = x_{1234}x_{4567} - x_4^2$
	$D = CM - x_2x_4x_5 - x_1x_4x_6 - x_{1234}x_5x_6 - x_3x_4x_7 - (x_{234}x_5 + x_1x_{45})x_7$
	$-(x_{134}x_6 + x_2x_{46})x_7 - x_1x_2x_{4567} - x_3(x_4x_5 + x_1x_{4567}) - x_3(x_4x_6 + x_2x_{4567})$
1(e)	$C = x_{12345}x_{12367} - x_{123}^2$
	$D = CM - x_2x_3x_4 - x_2x_3x_5 - x_2x_4x_5 - x_3x_4x_5 - x_2x_3x_6 - x_2x_4x_6 - 2x_3x_4x_6 - x_3x_5x_6$
	$-x_4x_5x_6 - x_2x_3x_7 - x_2x_4x_7 - x_3x_4x_7 - x_2x_5x_7 - x_3x_5x_7 - x_4x_5x_7 - x_2x_6x_7 - x_3x_6x_7$
	$-x_4x_6x_7 - x_5x_6x_7 - x_1(x_4x_5 + x_4x_6 + x_5x_6 + x_5x_7 + x_6x_7 + x_2x_{4567} + x_3x_{4567})$
2(a)	$C = -x_1^2x_{4567} + x_{125}x_{136}x_{4567} - x_{136}x_5^2 - 2x_1x_5x_6 - x_{125}x_6^2$
	$D = -x_1^2x_4^2 + x_{125}x_{136}x_4^2 - x_1^2x_{1567}x_{4567} + x_{125}x_{136}x_{1567}x_{4567} + x_{136}x_2^2x_{4567}$
	$+2x_1x_2x_3x_{4567} + x_{125}x_{4567}x_3^2 + 2x_{136}x_2x_4x_5 + 2x_1x_3x_4x_5 - x_{136}x_{1567}x_5^2 - x_3^2x_5^2$
	$+2x_1x_2x_4x_6 + 2x_{125}x_3x_4x_6 - 2x_1x_{1567}x_5x_6 + 2x_2x_3x_5x_6 - x_{125}x_{1567}x_6^2 - x_2^2x_6^2$
2(b)	$C = -x_{135}x_4^2 - x_1^2x_{4567} + x_{124}x_{135}x_{4567} - 2x_1x_4x_5 - x_{124}x_5^2$
	$D = -x_{13457}x_{135}x_4^2 - x_1^2x_{13457}x_{4567} + x_{124}x_{13457}x_{135}x_{4567} + x_{135}x_{4567}x_2^2 - 2x_1x_{13457}x_4x_5$
	$-x_{124}x_{13457}x_5^2 - x_2^2x_5^2 + 2x_{135}x_2x_4x_6 + 2x_1x_2x_5x_6 - x_1^2x_6^2 + x_{124}x_{135}x_6^2$
2(c)	$D = -x_{123}x_{13568}x_6^2 - x_2^2x_6^2 - x_1^2x_{13568}x_{678} + x_{123}x_{13568}x_{1456}x_{678} + x_{1456}x_2^2x_{678}$
	$+2x_1x_2x_4x_{678} + x_{123}x_4^2x_{678} + 2x_1x_2x_6x_7 + 2x_{123}x_4x_6x_7 - x_1^2x_7^2 + x_{123}x_{1456}x_7^2$
2(d)	$D = -x_{134568}x_{1678}x_4^2 - x_1^2x_{134568}x_{456} + x_{1234}x_{456}x_{134568}x_{1678} + x_{1678}x_2^2x_{456} - 2x_1x_{134568}x_4x_6$
	$-x_{1234}x_{134568}x_6^2 - x_2^2x_6^2 + 2x_1x_2x_{456}x_7 + 2x_2x_4x_6x_7 - x_4^2x_7^2 + x_{1234}x_{456}x_7^2$
2(e)	$D = -x_{13568}x_{2346}x_5^2 - x_{24}^2x_5^2 - x_{13568}x_{23}^2x_{5678} + x_{1235}x_{13568}x_{2346}x_{5678} + x_2^2x_{2346}x_{5678}$
	$-2x_2x_{23}x_{24}x_{5678} + x_{1235}x_{24}^2x_{5678} - 2x_{13568}x_{23}x_5x_6 - 2x_2x_{24}x_5x_6 - x_{1235}x_{13568}x_6^2 - x_2^2x_6^2$
	$-2x_2x_{2346}x_5x_7 + 2x_{23}x_{24}x_5x_7 - 2x_2x_{23}x_6x_7 + 2x_{1235}x_{24}x_6x_7 - x_{23}^2x_7^2 + x_{1235}x_{2346}x_7^2$

Table B1. Lattice rule generators

d	**z** components for $n = 10M$	**z** components for $n = 100M$
4	$(1, 2928962, 1859617, 3250721)$	$(1, 38278307, 43388112, 5988368)$
5	$(1, 3675449, 4456704, 3864450, 4934306)$	$(1, 38278307, 43388112, 48206750, 45950883)$
6	$(1, 2928962, 1859617, 4304839, 2689131, 2592686)$	$(1, 38278307, 34519177, 26711202, 33749025, 1281880)$
7	$(1, 2928962, 1859617, 4304839, 4474033, 273363, 1426318)$	$(1, 41883906, 22682973, 44229424, 29466837, 15518263, 42112409)$

Appendix A. C and D functions

The C and D functions of the loop integrals corresponding to the diagrams in Fig 1 and 2 are given in Table A1. We use the notation $x_{i_1 i_2 \ldots i_k} = x_{i_1} + x_{i_2} + \ldots + x_{i_k}$ for $1 \le i_k \le N$, and $M = \sum_r m_r^2 x_r$.

Appendix B. Lattice rule generators

Table B1 lists the components of the generator vector **z** in Eq (3) computed for dimensions $d = 4, 5, 6, 7$, with $n = 10,000,019$ and $n = 100,000,007$ points.

Appendix C. Sample integrand function

The program section below is a sample implementation of the integrand of Fig 1(b) as a CUDA device function. The integration is over the 5-dimensional unit cube obtained by a transformation from the unit simplex $\mathcal{S}_5$. Note that `pi12` $= 12\pi$ and `con` $= (\frac{8}{3})^5$ (from the Jacobian of the Ψ_4 transformation according to Eq (4)).

```
#include <math.h>

#define pi 3.14159265358979323846
#define pi12 37.69911184307751886152
#define con 134.84773662551440329049

const int dim = 5;

__device__ double f1(double x[]) { //Integrand function for
                                  //Fig 1(b)
  double x1,x2,x3,x4,x5,x6;
  double arg[5],dd,f0,phip,pixi;
  int i;

  f0 = 0;
  phip = 1.0;
  for(i = 0; i < dim; i++) { //Sidi Psi_4 transformation
     pixi = pi*x[i];
     arg[i] = x[i]+(-8.0*sin(2*pixi)+sin(4*pixi))/pi12;
     phip = phip*pow(sin(pixi),4);
  }
  if(phip == 0) return(0);

  x1 = arg[0];               //Simplex to cube transformation
  x2 = (1.0-x1)*arg[1];
  x3 = (1.0-x1-x2)*arg[2];
  x4 = (1.0-x1-x2-x3)*arg[3];
  x5 = (1.0-x1-x2-x3-x4)*arg[4];
  x6 = 1.0-x1-x2-x3-x4-x5;

  phip = phip*(1.0-x1)*(1.0-x1-x2)*(1.0-x1-x2-x3)*(1.0-x1-x2-x3-x4);
  if(phip == 0) return(0);

  dd = (x1*x1+x5*x5+x1*x5)*(x2+x3+x4+x6)
       +(x2*x2+x3*x3+x6*x6+x2*x3+x2*x6+x3*x6)*(x1+x4+x5)
       +x4*x4*(x1+x2+x3+x5+x6)+3.0*x4*x5*x6+
       2.0*x1*x4*(x2+x3+x6)+2.0*x4*x5*(x2+x3);

  if(fabs(dd) > 0.0) f0 = con*phip/dd/dd;
  return f0;
}
```

References

[1] de Doncker E, Fujimoto J, Hamaguchi N, Ishikawa T, Kurihara Y, Shimizu Y and Yuasa F 2011 *Journal of Computational Science (JoCS)* **3** 102–112 doi:10.1016/j.jocs.2011.06.003

[2] Kato K, de Doncker E, Ishikawa T, Kapenga J, Olagbemi O and Yuasa F 2016 *J. Physics: Conf. Ser.* **762** 012070, iopscience.iop.org/article/10.1088/1742-6596/762/1/012070

[3] de Doncker E, Yuasa F, Kato K, Ishikawa T, Kapenga J and Olagbemi O 2017 *J. Computer Phys. Communications* doi:10.1016/j.cpc.2017.11.001; Preprint at arXiv:[hep-ph] https://arxiv.org/abs/1702.04904

[4] de Doncker E and Yuasa F 2017 *Procedia Computer Science* **108** 1773–1782 https://doi.org/10.1016/j.procs.2017.05.253

[5] Sidi A 1993 *International Series of Numerical Mathematics* **112** 359–373

[6] Sidi A 2005 *Math. Comp.* **75** 327–343

[7] Robinson I and Hill M 2002 *Transactions on Mathematical Software* **28** 75–100 doi:10.1145/513001.513006
[8] Li T and Robinson 2010 *Transactions on Mathematical Software* **37** doi:10.1145/1824801.1824813
[9] Li Z, Wang J, Yan Q S and Zhao X 2016 *Chinese Physics C* **40** 033103 doi:10.1088/1674-1137/40/3/033103; Preprint at arXiv:1508.02512v1 [hep-ph]
[10] Laporta S 2000 *Int. J. Mod. Phys. A* **15** 5087–5159 arXiv:hep-ph/0102033v1
[11] de Doncker E and Yuasa F 2014 *Journal of Physics: Conf. Series (ACAT 2013)* **523** doi:10.1088/1742-6596/523/1/012052
[12] Yuasa F, de Doncker E, Hamaguchi N, Ishikawa T, Kato K, Kurihara Y and Shimizu Y 2012 *Journal Computer Physics Communications* **183** 2136–2144
[13] Niederreiter H 1978 *Monatshefte für Mathematik* **86** 203–219
[14] Nuyens D and Cools R 2006 *Journal of Complexity* **22** 4–28
[15] Nuyens D and Cools R 2006 *Math. Comp.* **75** 903–920
[16] Sloan I H and Reztsov A 2002 *Math. Comp.* **71** 263–273
[17] Korobov N M 1959 *Doklady Akademii Nauk SSSR* **124** 1207–1210 (Russian)
[18] Korobov N M 1960 *Doklady Akademii Nauk SSSR* **132** 1009–1012 (Russ.). Eng. trans. Soviet Math. Doklady, 1, 696-700
[19] Disney S A R and Sloan I H 1991 *Math. Comp.* **56** 257–266
[20] Sloan I and Joe S 1994 *Lattice Methods for Multiple Integration* (Oxford University Press, Oxford)
[21] Korobov N M 1963 *Number-Theoretic Methods in Approximate Analysis* (Fizmatgiz, Moscow (Russian))
[22] Sag T W and Szekeres G 1964 *Math. Comp.* **18** 245–253
[23] Iri M, Moriguti S and Takasawa Y 1970 *Kokyuroku of Res. Inst. for Math. Sci. Kyoto Univ.* **91** 82–118 (in Japanese); Engl. transl. in J. Comp. Appl. Math. 17 (1987), pp. 3-20
[24] Mori M 1978 *Publ. RIMS Kyoto Univ.* **14** 713–729
[25] Takahasi H and Mori M 1974 *Publications of the Research Institute for Mathematical Sciences* **9** 721–741
[26] Sanders J and Kandrot E 2010 *CUDA by Example: An Introduction to General-Purpose GPU Programming* (Pearson)

IOP Conf. Series: Journal of Physics: Conf. Series **1085** (2018) 052006 doi:10.1088/1742-6596/1085/5/052006

The R^*-operation and combinatorial challenges at five loops

Ben Ruijl[1], **Franz Herzog**[2], **Takahiro Ueda**[2], **J.A.M. Vermaseren**[2], **Andreas Vogt**[3]

[1] ETH Zurich, Wolfgang-Pauli-Str. 27, 8093 Zürich, Switzerland
[2] Nikhef Theory Group, Science Park 105, 1098 XG Amsterdam, NL
[3] Department of Mathematical Sciences, University of Liverpool, Liverpool L69 3BX, UK

E-mail: benrl@nikhef.nl, fherzog@nikhef.nl, tueda@nikhef.nl, t68@nikhef.nl, Andreas.Vogt@liv.ac.uk

Abstract. We sketch how the R^*-operation can be used to compute the poles of Feynman diagrams. Next, we identify computational difficulties when performing five-loop calculations. We provide four solutions that drastically reduce the number of generated terms.

1. Introduction

Many physically interesting quantities can be computed from just the pole parts of Feynman diagrams. For example, anomalous dimensions (such as the beta function), splitting functions, and decay rates. Often, it is much easier to compute the pole parts of diagrams than the finite part.

In this work we sketch the R^*-operation [1, 2, 3], which is able to compute the pole parts of L-loop diagrams by a computation of at most $L-1$-loop diagrams. Using a high-performance implementation of the R^*-operation combined with the Forcer program [4, 5, 6], we are able to compute the poles of five-loop massless propagator diagrams. Our code is all written in Form [7, 8, 9].

Performing efficient computations of five-loop diagrams is hard, since the Feynman rules create many terms. To alleviate these issues, we describe four methods to reduce the number of terms: (1) removal of propgator insertions, (2) delaying of Feynman rule substitution, (3) canonicalization of Feynman diagrams, and (4) efficient tensor reduction. These optimizations are not specific to the R^*-method.

Using these optimizations, we have computed the five-loop beta function for Yang-Mills theory with fermions in six days on one 32-core machine [10]. We have also computed the R-ratio, the Higgs decay to quarks, and to gluons [11].

The outline of this paper is as follows. In section 2 we briefly describe the R^*-operation. In section 3 we describe the four optimizations. Finally, we present the conclusion in section 4.

2. The R^*-operation

The R^*-operation can be used to compute the poles of Feynman diagrams [12, 13]. Recently, it has been extended to Feynman diagrams with arbitrary numerator structure [3]. In this section we briefly sketch how the R^*-operation works, focusing on UV-counterterms.

IOP Conf. Series: Journal of Physics: Conf. Series **1085** (2018) 052006 doi:10.1088/1742-6596/1085/5/052006

The basic object of the R^*-operation is the UV counterterm operation Δ acting on a graph G, which is defined as the poles of G in the limit of all loop momenta going to infinity with all contributions from subdivergences subtracted.

Additionally, we define the pole operator K on a Laurent series in ε as

$$K \sum_{i=-\infty}^{\infty} c_i \varepsilon^i = \sum_{i=-\infty}^{-1} c_i \varepsilon^i \,. \tag{1}$$

Then the R^* operation for some simple examples yields:

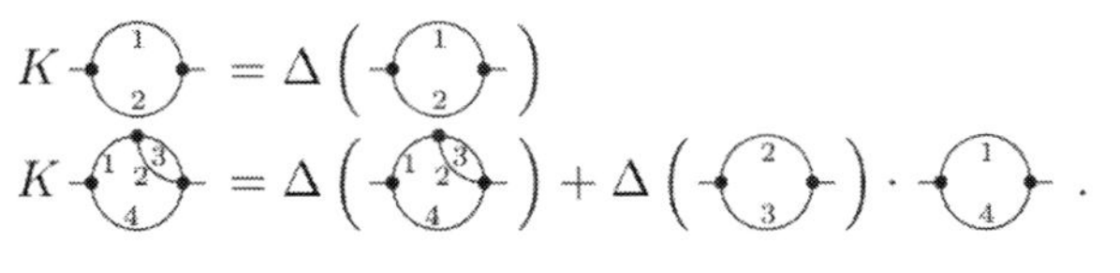

(2)
(3)

In general, all sets of non-overlapping divergent subdiagrams have to be considered:

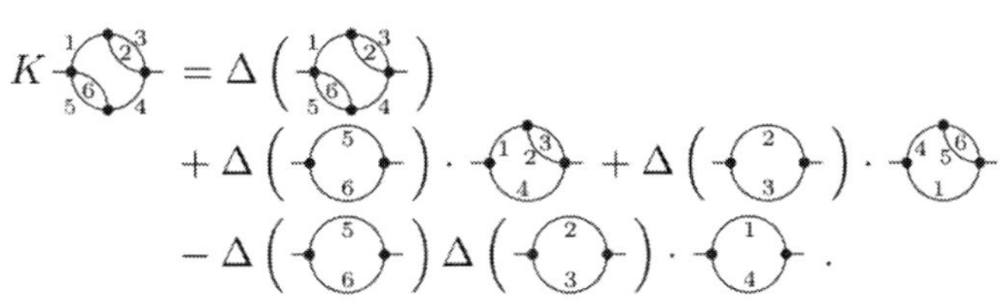

(4)

In the latter case, the contribution from two counterterms gets a minus sign, to prevent double counting.

For logarithmically divergent diagrams, Δ does not depend on external momenta or masses. Consequently, we can infrared rearrange (IRR) [14]:

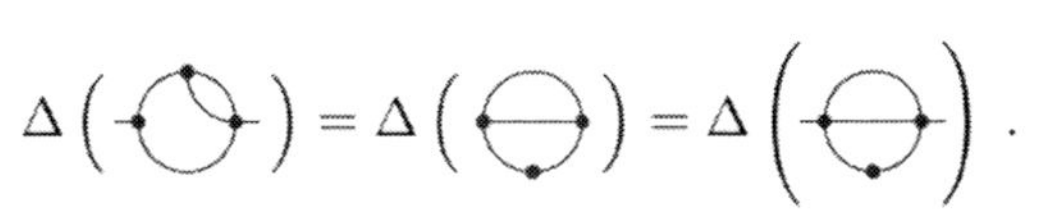

(5)

Using IRR and the definition of Δ, we can express the UV-counterterm of G in terms of simpler diagrams:

$$\Delta(G) \overset{\text{IRR}}{=} \underbrace{\Delta(G')}_{\text{Simpler than G}} = K(G') - \underbrace{\text{subdivergences}(G')}_{\text{Lower-loop diagrams}}$$

Using this setup, we can rewrite all L-loop input diagrams to $L-1$ scalar massless propagator integrals. For the applications at five loops, all those integrals can be computed using the Forcer program [4, 5].

3. Optimisations

Performing computations at five loops introduces at least four new bottlenecks compared to four loops: (1) the number of diagrams and their complexity grow exponentially, (2) the substitution of the Feynman rules is slow and creates millions of terms, (3) the number of counterterms grows exponentially, and (4) tensors of rank 10 have to be reduced, which involves solving large systems.

In this section we address these issues by presenting four optimisations, namely improved treatment of propagator insertions in section 3.1, delayed Feynman rule substitution in section 3.2, a canonical form algorithm for Feynman diagrams in section 3.3, and an efficient tensor reduction algorithm in section 3.4.

IOP Conf. Series: Journal of Physics: Conf. Series **1085** (2018) 052006 doi:10.1088/1742-6596/1085/5/052006

3.1. Treatment of propagator insertions

Many of the higher-loop corrections are self-energies of propagators in the diagram. Due to the local nature of the Feynman rules, these self-energies only depend on their external momentum (there are no contractions with other parts of the larger diagram), so they can be 'factorised' out:

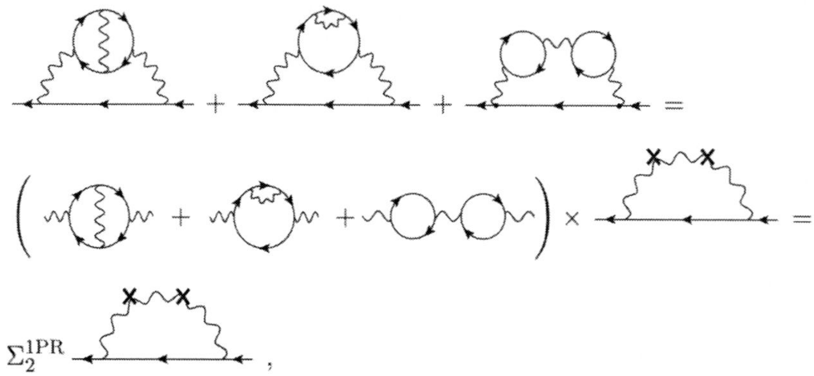

(6)

where the L-loop self-energy is replaced by $(p^2)^{-\varepsilon L}$ in the larger diagram (marked by L crosses). In a sense, the subdiagram is integrated out. The resulting simpler topology is multiplied by the one-particle-reducible L-loop self-energy. Since the L-loop subdiagram is of lower order, these quantities have already been computed and can easily be tabulated to prevent recomputations. For example, a five-loop diagram may contain the expensive 4-loop gluon propagator as a subdiagram.

For the R^*-operation, this representation has an issue: the non-integer power hides UV-divergent subdiagrams, which should be subtracted. However, since the exact contents of the $(p^2)^{-\varepsilon L}$ is factorised out, we may replace it with *any* L-loop subdiagram. Therefore we choose the simplest configuration: L scalar one-loop bubbles side by side.

Thus, for the R^*-operation we can remove propagator insertions by using the following relation:

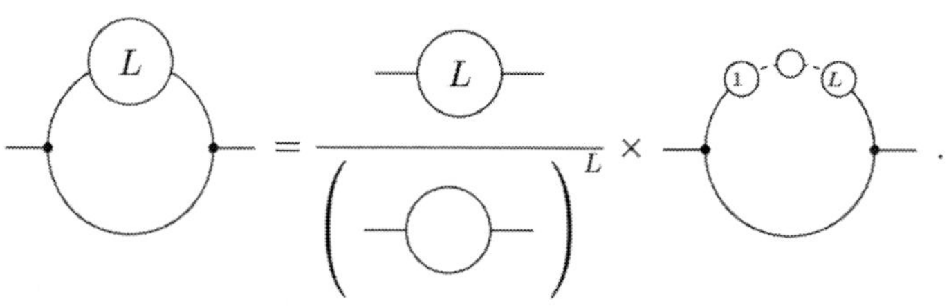

(7)

3.2. Delayed Feynman rule substitution

Substituting the Feynman rules creates many terms. For example, the following fully gluonic five-loop graph creates 12 029 521 scalar integrals in the Feynman gauge:

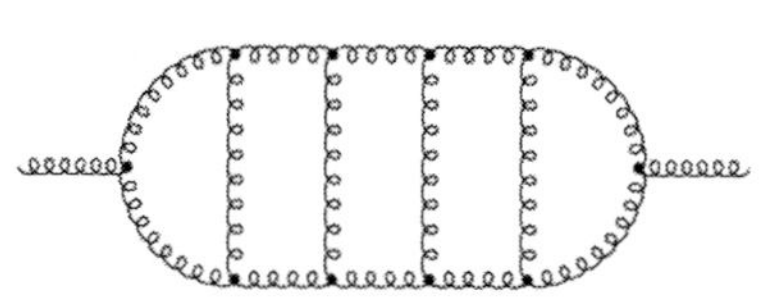

(8)

The source of the blow-up is the Feynman rule for the triple gluon vertex, which can be written in the following way:

$$v_{3g}(p_1^{\mu,a}, p_2^{\nu,b}, p_3^{\rho,c}) = -if^{abc}\left[(p_1 - p_2)^\rho g_{\mu\nu} + (2p_2 + p_1)^\mu g_{\nu\rho} + (-2p_1 - p_2)^\nu g_{\mu\rho}\right] . \quad (9)$$

Thus, for every vertex, six terms are created, of which some will evaluate to the same expression due to symmetries. For all these terms, expensive operations such as Taylor expansions and divergent subgraph recognitions have to be performed. However, these operations only depend on the momentum powers and are invariant under the way the momenta contract. So, we rewrite the triple gluon vertex in a way that exposes the momenta, but keeps all the contractions unsubstituted:

$$v_{3g}(p_1^{\mu,a}, p_2^{\nu,b}, p_3^{\rho,c}) = -if^{abc}p_1^{\sigma}t_3(\sigma,\nu,\rho,\mu) + if^{abc}p_2^{\sigma}t_3(\sigma,\mu,\rho,\nu)\,, \tag{10}$$

where

$$t_3(\mu,\nu,\rho,\sigma) = g_{\mu\rho}g_{\nu\sigma} + g_{\mu\sigma}g_{\nu\rho} - 2g_{\mu\nu}g_{\rho\sigma}\,. \tag{11}$$

After rewriting v_{3g} in terms of t_3, there are only $2^{10} = 1024$ terms for the Feynman diagram in eq. (8). We can keep our input in this compactified notation for as long as the actual contractions are not important, which is right until the tensor reduction.

We define the operation $\circ$ that applies the remaining Feynman rules to all components of the R^*-operation. For example:

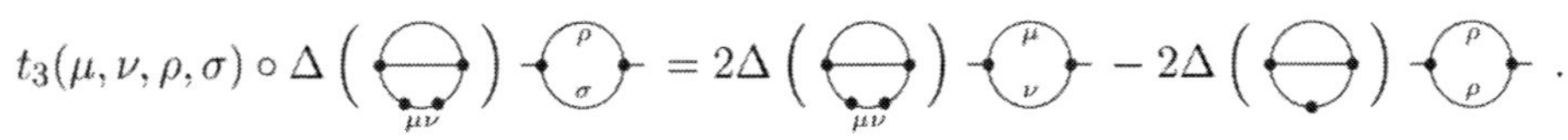
(12)

We stress that for this particular case contraction is *necessary*.

Similar rules can be devised for the other vertices and for the trace of gamma matrices. At five loops, the substitution of t_3 and similar structures is an expensive part of the calculation, since the number of generated terms is high.

3.3. Canonical forms for Feynman diagrams

The R^*-operation applied to five-loop diagrams will create many counterterms. In order to reduce computation time, it is important to compute the counterterms of a specific graph only once. In turn, this requires an efficient way to detect if two graphs are equal. One straightforward option is to keep a list of all the graphs that have already been processed and test for isomorphisms on every element of the list until one is found. If no match is found, the current graph can be added to the list. The two downsides of this method are that (1) an isomorphism test can be rather slow at five loops and (2) that the list of topologies grows rapidly.

A better alternative is to construct a *canonical form* of a graph. A canonical form is an isomorphism of the graph that is designated as the smallest by some yet to be defined measure. To test for equality, one can simply compare the canonical forms. Since isomorphy is first and foremost a property of the vertices, we give each vertex a label from 1 to n. For simplicity, let us consider a graph that has no dot products and only has edges with power 1.

We convert our graph to an edge representation:

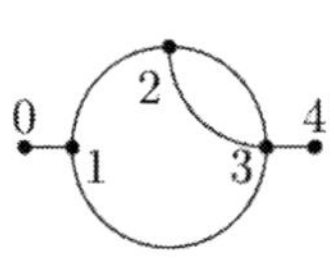

$$= e(0,1)e(1,2)e(2,3)e(2,3)e(1,3)e(3,4)\,. \tag{13}$$

Here, $e(n_1, n_2)$ is the edge function, in which we place the smallest vertex index as the first argument. The *edge list* is a lexicographically sorted list of edge functions, as is shown in eq. (13). Now we define the smallest isomorphism of a graph as the vertex labelling for which the edge list is lexicographically smallest.[1]

[1] In our program, we use the internal (deterministic) sorting order of FORM to determine the smallest isomorphism instead.

IOP Conf. Series: Journal of Physics: Conf. Series **1085** (2018) 052006 doi:10.1088/1742-6596/1085/5/052006

We can easily extend the graph notation to a graph where propagators can have different powers, by introducing a third argument to the edge function e:

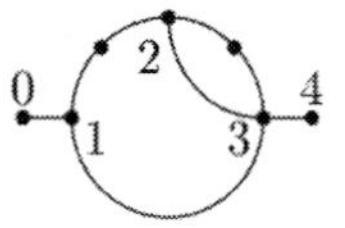

$$= e(0,1,1)e(1,2,2)e(2,3,1)e(2,3,2)e(1,3,1)e(3,4,1)\ , \tag{14}$$

where we again make sure that the first two arguments of $e(n_1, n_2, \ldots)$ are sorted. To add support for dot products and tensors, we extend the edge function even further:

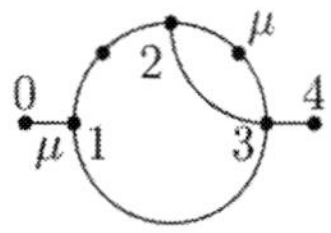

$$= e(0,1,1,\mu)e(1,2,2)e(2,3,1)e(2,3,2,\mu)e(1,3,1)e(3,4,1)\ . \tag{15}$$

We define the canonical signs of the momenta such that they always flow from the smallest vertex label to the highest. If a transformation changes the order, we flip the sign if the number of vectors in the momentum is odd:

$$e(2,1,n,\mu_1,\ldots,\mu_k) = (-1)^k e(1,2,n,\mu_1,\ldots,\mu_k)\ . \tag{16}$$

Finally, the momentum label p_i of each edge is uniquely defined by the position i of the edge in the edge list.

Now that most properties of the Feynman integral are captured in the extended edge list and we have defined which edge list is smallest, we use McKay's canonicalisation algorithm [15] to efficiently rewrite the complete Feynman integral to canonical form. A simplified version of this algorithm is implemented in FORM code.

3.4. Efficient tensor reduction

It can be shown that the tensor reduction of ultraviolet and infrared subtraction terms, required for the R^*-operation, is equivalent to the tensor reduction of tensor vacuum bubble integrals. In general tensor vacuum integrals can be reduced to linear combinations of products of metric tensors $g^{\mu\nu}$ whose coefficients are scalar vacuum integrals. Specifically a rank r tensor, $T^{\mu_1\ldots\mu_r}$, is written as a linear combination of $n = r!/2^{(r/2)}/(r/2)!$ combinations of $(r/2)$ metric tensors with coefficients c_σ, i.e.,

$$T^{\mu_1\ldots\mu_r} = \sum_{\sigma\in {}_2S_r} c_\sigma T_\sigma^{\mu_1\ldots\mu_r}\ , \qquad T_\sigma^{\mu_1\ldots\mu_r} = g^{\mu_{\sigma(1)}\mu_{\sigma(2)}}\ldots g^{\mu_{\sigma(r-1)}\mu_{\sigma(r)}}\ . \tag{17}$$

Here we define ${}_2S_r$ as the group of permutations which do *not* leave the tensor $T_\sigma^{\mu_1\ldots\mu_r}$ invariant. The coefficients c_σ can be obtained by acting onto the tensor $T^{\mu_1\ldots\mu_r}$ with certain projectors $P_\sigma^{\mu_1\ldots\mu_r}$, such that

$$c_\sigma = P_\sigma^{\mu_1\ldots\mu_r} T_{\mu_1\ldots\mu_r}\ . \tag{18}$$

From this it follows that the orthogonality relation,

$$P_\sigma^{\mu_1\ldots\mu_r} T_{\tau,\mu_1\ldots\mu_r} = \delta_{\sigma\tau}\ , \tag{19}$$

must hold, where δ is the Kronecker-delta. Since the projector $P_\sigma^{\mu_1\ldots\mu_r}$ of each tensor can also be written in terms of a linear combination of products of metric tensors, inverting an $n\times n$ matrix determines all the projectors. However, there are two issues. The first is that the size of the matrix grows rather rapidly as r increases. Instead of solving an $n\times n$ linear system, the

IOP Conf. Series: Journal of Physics: Conf. Series **1085** (2018) 052006 doi:10.1088/1742-6596/1085/5/052006

symmetry group of the metric tensors can be utilised to reduce the size of the system. From eq. (19) it follows that the projector P_σ is in the same symmetry group (the group of permutations which leave it invariant) as T_σ. For example, given a permutation $\sigma_1 = (123...(r-1)r)$,

$$T_{\sigma_1}^{\mu_1\ldots\mu_r} = g^{\mu_1\mu_2} g^{\mu_3\mu_4} \ldots g^{\mu_{r-1}\mu_r}\,. \tag{20}$$

The corresponding projector $P_{\sigma_1}^{\mu_1\ldots\mu_r}$ must be symmetric under interchanges of indices such as $\mu_1 \leftrightarrow \mu_2$, $(\mu_1,\mu_2) \leftrightarrow (\mu_3,\mu_4)$ and so on. Grouping the metric tensors by the symmetry leads to the fact that P_σ is actually written in a linear combination of a small number of m tensors instead of n $(m \le n)$,

$$P_\sigma^{\mu_1\ldots\mu_r} = \sum_{k=1}^{m} b_k \sum_{\tau\in A_m^\sigma} T_\tau^{\mu_1\ldots\mu_r}. \tag{21}$$

The set of groups $\{A_k^\sigma | k = 1..m\}$ must therefore each be closed under the permutations which leaves T_σ invariant and at the same time their union must cover once the group ${}_2S_n$. Contracting P_σ with a representative in each group gives an $m \times m$ matrix which can be inverted to yield the coefficients b_k. The number of unknowns m is $m = 5$ for $r = 8$ and $m = 22$ for $r = 16$, whereas we have $n = 105$ for $r = 8$ and $n = 2027025$ for $r = 16$. The comparison of these numbers illustrates that the exploitation of the symmetry of the projectors makes it possible to find the tensor reduction even for very large values of r, which could never have been obtained by solving the $n \times n$ matrix.

The second issue with tensors of high rank is the large number of intermediate terms that are created. Even though the system for the projector can be solved efficiently, $\mathcal{O}(n^2)$ terms will be created, of which some will merge due to symmetry. Let us consider rank 6, with 15 terms:

$$c_1 g^{\mu_1\mu_2} g^{\mu_3\mu_4} g^{\mu_5\mu_6} + c_2 g^{\mu_1\mu_3} g^{\mu_2\mu_4} g^{\mu_5\mu_6} + \ldots\,. \tag{22}$$

In most practical situations there is symmetry, both on the inside of the object that will be projected as on the outside. For example

$$A(p_1^{\mu_1} p_1^{\mu_2} p_1^{\mu_3} p_1^{\mu_4} p_2^{\mu_5} p_2^{\mu_6}) p_3^{\mu_1} p_3^{\mu_2} p_4^{\mu_3} p_4^{\mu_4} p_4^{\mu_5} p_4^{\mu_6} \tag{23}$$

is symmetric in exchanges of $\mu_1,\ldots,\mu_4$ and μ_5,μ_6 inside A, and is symmetric in μ_1,μ_2 and $\mu_3,\ldots,\mu_6$ outside A. The symmetry inside the object A will enforce that coefficient c_1 and c_2 (and others) will actually be the same. The symmetry on the outside will cause terms to merge. In the end, we could have used the symmetrised variant of eq. (22) instead:

$$c_1 \cdot (g^{\mu_1\mu_2} g^{\mu_3\mu_4} g^{\mu_5\mu_6} + 2 g^{\mu_1\mu_3} g^{\mu_2\mu_4} g^{\mu_5\mu_6}) + c_3 (2 g^{\mu_1\mu_2} g^{\mu_3\mu_5} g^{\mu_4\mu_6} + 10 g^{\mu_1\mu_5} g^{\mu_2\mu_6} g^{\mu_3\mu_4}). \tag{24}$$

We see that only two coefficients have to be computed instead of 15 and that there are only 4 terms in the output instead of 15. The challenge is to prevent these terms from being created in the first place by exploiting symmetry, instead of starting from eq. (22). We make use of the optimised FORM command `dd_`, which creates the tensor structure $T^{\mu_1,\ldots,\mu_r}$ without generating duplicates. If we evaluate `dd_(p1,p1,p1,p1,p2,p2)` and strip the coefficient we get `p1.p1^2*p2.p2+p1.p1*p1.p2^2`. These two terms represent the structure outside of c_1 and c_3 in (24). For each of these two terms, we solve for the coefficient. Next, we recreate the metric structures that would give this specific contraction.

A term generated by `dd_` consists of two different factors: $(p\cdot p)^a$ and $(p_1 \cdot p_2)^a$. For $(p\cdot p)^a$, we collect all possible indices involved with p. For eq. (23), this would be $\mu_1,\ldots,\mu_4$. Then we select all possible ways to get $2a$ elements from that list with `distrib_`. Next, we use *dd_* on those indices. Thus, for $p_1 \cdot p_1$ in the example we would get $g^{\mu_1\mu_2} + g^{\mu_1\mu_3} + g^{\mu_2\mu_3}$. For cases such

IOP Conf. Series: Journal of Physics: Conf. Series **1085** (2018) 052006 doi:10.1088/1742-6596/1085/5/052006

as $(p_1 \cdot p_2)^a$, we select a from the list of indices associated to p_1 and a from the list of p_2. Then we permute over the list of p_2. Using this algorithm, one can generate all possible contractions from the result without generating duplicates. To apply the outside symmetry, one can easily fill in the outside momenta associated to the indices instead of the indices themselves. `distrib_` and `dd_` will take the symmetry into account automatically.

4. Conclusion

We have sketched how the R^*-operation can be used to compute the poles of Feynman diagrams. Additionally, we have identified computational difficulties when performing five-loop calculations. We provide four solutions that reduce the number of generated terms.

Using these methods, we have computed the five-loop beta function for Yang-Mills theory with fermions in six days on one 32-core machine [10]. We have also computed the R-ratio, the Higgs decay to quarks, and to gluons [11].

Acknowledgements

This work is supported by the ERC Advanced Grant no. 320651, "HEPGAME".

References

[1] Chetyrkin K G and Tkachov F V 1982 *Phys. Lett.* **B114** 340–344
[2] Chetyrkin K G and Smirnov V A 1984 *Phys. Lett.* **B144** 419–424
[3] Herzog F and Ruijl B 2017 *JHEP* **2017** 37 ISSN 1029-8479 URL `http://dx.doi.org/10.1007/JHEP05(2017)037`
[4] Ruijl B, Ueda T and Vermaseren J 2017 Forcer URL `https://doi.org/10.5281/zenodo.1043944`
[5] Ruijl B, Ueda T and Vermaseren J A M 2017 (*Preprint* 1704.06650)
[6] Ueda T, Ruijl B and Vermaseren J A M 2016 *PoS* **LL2016** 070 (*Preprint* 1607.07318)
[7] Kuipers J, Ueda T, Vermaseren J A M and Vollinga J 2013 *Comput. Phys. Commun.* **184** 1453–1467 (*Preprint* 1203.6543)
[8] Tentyukov M and Vermaseren J A M 2010 *Comput. Phys. Commun.* **181** 1419–1427 (*Preprint* hep-ph/0702279)
[9] Ruijl B, Ueda T and Vermaseren J 2017 (*Preprint* 1707.06453)
[10] Herzog F, Ruijl B, Ueda T, Vermaseren J A M and Vogt A 2017 *JHEP* **02** 090 (*Preprint* 1701.01404)
[11] Herzog F, Ruijl B, Ueda T, Vermaseren J A M and Vogt A 2017 *JHEP* **08** 113 (*Preprint* 1707.01044)
[12] Chetyrkin K G and Tkachov F V 1982 *Phys. Lett.* **B114** 340–344
[13] Chetyrkin K G and Smirnov V A 1984 *Phys. Lett.* **B144** 419–424
[14] Vladimirov A A 1980 *Theor. Math. Phys.* **43** 417 [Teor. Mat. Fiz.43,210(1980)]
[15] McKay B D 1981 *Congressus Numerantium* **30** 45–87 URL `http://users.cecs.anu.edu.au/~bdm/nauty/pgi.pdf`

IOP Conf. Series: Journal of Physics: Conf. Series **1085** (2018) 052007 doi:10.1088/1742-6596/1085/5/052007

Minimisation strategies for the determination of parton density functions

Stefano Carrazza[1] **and Nathan P. Hartland**[2]

[1] Theoretical Physics Department, CERN, CH-1211 Geneva, Switzerland

[2] Department of Physics and Astronomy, Vrije Universiteit Amsterdam, NL-1081 HV and Nikhef Theory Group, Science Park 105, 1098 XG Amsterdam, The Netherlands

E-mail: stefano.carrazza@cern.ch[1], n.p.hartland@vu.nl

Abstract. We discuss the current minimisation strategies adopted by research projects involving the determination of parton distribution functions (PDFs) and fragmentation functions (FFs) through the training of neural networks. We present a short overview of a proton PDF determination obtained using the covariance matrix adaptation evolution strategy (CMA-ES) optimisation algorithm. We perform comparisons between the CMA-ES and the standard nodal genetic algorithm (NGA) adopted by the NNPDF collaboration.

1. Introduction

In perturbative QCD, parton distribution functions are used to describe the non-perturbative structure of hadrons. These functions are typically determined by means of a fit to a wide set of experimental data [1]. Such a fit is complicated by the non-trivial relationship between parton distributions and physical observables. Data on hadron-hadron collider processes is related to parton distributions by means of a double convolution with a kernel computed in perturbation theory. Given some representation of PDFs: f in terms of a set of model parameters: $\{a\}$ a typical hadron-hadron collider observable T may be be calculated as

$$T[f\{a\}] = \int \widetilde{T}_{ij}(x_1, x_2) f_{i/1}(x_1, \{a\}) f_{j/2}(x_2, \{a\}), \tag{1}$$

where $f_{i/1}$ and $f_{j/2}$ represent the parton distributions for PDF flavours i/j of the first and second beams respectively, and $\widetilde{T}$ is the relevant integration kernel for the observable T. The model parameters may then be determined by a fit to data by minimising some figure of merit, for example the correlated χ^2 between theory and data

$$\chi^2(\{a\}) = \sum_{i,j}^{N_{\rm dat}} (T_i[f] - D_i)(C^{-1})_{ij}(T_j[f] - D_j)\,, \tag{2}$$

where D_i represents the $i^{\rm th}$ data point in a set with covariance C. For the most part, in PDF determinations this minimisation procedure is performed with the use of standard gradient

[1] Preprint: CERN-TH-2017-241

descent methods. However the computation of the fitness gradient with respect to model parameters, i.e.

$$\frac{\partial \chi^2}{\partial a}, \tag{3}$$

is challenging due to the relationship between theory predictions and PDFs expressed in Equation 1. For fits of PDF models with relatively few parameters, such gradients may be either directly computed or reasonably approximated by the equivalent finite-differences. However when using models with a relatively large parameter space, such as those used in determinations in the NNPDF approach, the computation or approximation of these gradients can be a considerable computational burden. Consequently in NNPDF determinations of parton distributions, *gradient-free* minimisation methods have been employed. These typically take the form of a *Nodal Genetic Algorithm* (NGA) variant [2]. At each iteration of the minimisation, candidate parameter values are sampled around a search centre according to a Gaussian distribution, with the candidate minimising the χ^2 being selected for the search centre in the next iteration. Such an algorithm is *greedy* in that it always makes the locally optimal selection for the next iteration. For details of the algorithm see Ref. [3].

Despite the simplicity of the algorithm, it has proven to be an efficient way of exploring the complicated PDF parameter space in a manner that is relatively robust in terms of avoiding features such as local minima. However there are clear directions for improvement in terms of a minimisation algorithm. Firstly, the algorithm does not make use of any information on the structure of the parameter space in order to guide the exploration in subsequent steps. Secondly the current algorithm is rather sensitive to noise in the objective function, as it is constrained to only select the locally best candidate at each iteration.

In the NNPDF approach, the inherent susceptibility of the minimisation algorithm to noise is ameliorated by two mechanisms. Each PDF determination is performed as an ensemble of fits to different sub-samples of the complete dataset. This ensemble determination reduces the sensitivity of the overall result to fluctuations affecting individual fits, however there is the potential for PDF uncertainties to be inflated due to overly noisy ensemble members. Furthermore a form of cross-validated early stopping is applied to avoid over-fitting. This does not however completely prevent contamination of the results by noise.

In a recent application of the NNPDF methodology to the determination of fragmentation functions [4] an alternative minimisation strategy was employed, the active (μ, λ)-CMA-ES algorithm [5, 6]. The CMA-ES procedure includes mechanisms which overcome the two weaknesses in the standard NNPDF NGA minimisation described above. In the application to fragmentation functions, CMA-ES proved a reliable and efficient minimisation algorithm. Here we shall report results from a preliminary investigation into the application of the algorithm in the case of proton PDFs.

2. Comparison of minimisers

Two fits have been performed at NNLO in the NNPDF3.1 framework [7], one utilising the NGA minimiser, and the second with the CMA-ES algorithm with population $\mu = 80$ and initial step-size $\sigma = 0.1$. All other CMA-ES parameters are set as per the recommended defaults. The dataset and other fit parameters are identical to those used in the NNPDF3.1 NNLO determination.

In the first panel of Figure 1 we show the distribution of χ^2 fitness over the two PDF ensembles. The total χ^2 distribution demonstrates a significant difference between the results from the two minimisers. Ensemble members of the CMA-ES fit are typically better fit to their datasets than the NGA equivalent. Furthermore the results from the CMA-ES fit are considerably more consistent between ensemble members. To verify that this improvement in total fitness is not due to over-learning the training dataset, in the second panel we histogram

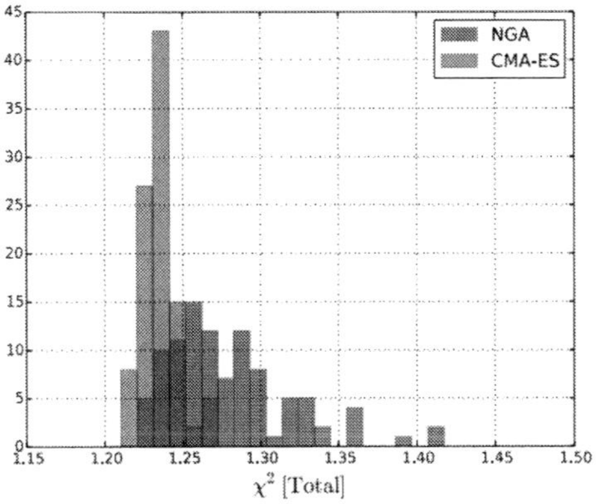

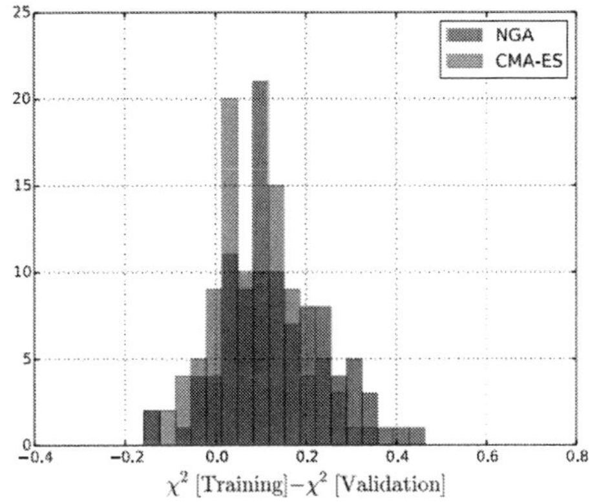

Figure 1. Histogram of χ^2 fitness to the total dataset (left) and the difference between training and validation χ^2 (right). Shown are the fitnesses of the CMA-ES ensemble (blue) and the NNPDF ensemble (green).

the differences between the χ^2 to the datasets used for training and validation. While results here are more consistent between the two minimisers, it is clear that the CMA-ES results have a slightly better balance between the training and validation fitnesses than in the NGA equivalent.

To investigate the greater ensemble consistency demonstrated in the CMA-ES χ^2 distribution, the arc-length defined as

$$L = \int \sqrt{1 + \left(\frac{df}{dx}\right)^2} dx, \tag{4}$$

is a useful quantity in that it offers a measure of the structural 'complexity' of a function. For a fit determining an underlying function that is expected to be largely smooth, results from fits suffering from contamination by noise in the objective function would be expected to have a typically larger arc-length than those less susceptible to noise.

In Figure 2 we compare the arc-lengths of the PDFs determined by the NGA minimiser and the CMA-ES . The figure demonstrates a clear and significant difference in both the PDF arc-lengths and their spread over the fit ensemble. The CMA-ES results demonstrate consistently lower and more regular arc-lengths, confirming the expectation that the results from the CMA-ES should be more resistant to noise in the objective function. Particularly striking is the difference in arc-length for the gluon PDF . A direct comparison of the gluon determination in the CMA-ES and NGA minimisers is shown in Figure 3, where even though the results are consistent between the two minimisers, the reduced structural complexity of the CMA-ES result can be clearly seen.

3. Conclusion

The CMA-ES minimiser has been successfully applied to the determination of parton distribution functions in the NNPDF approach. Results determined through the CMA-ES procedure exhibit greater consistency, improved agreement with data, and reduced complexity in comparison to results determined through the NGA minimiser. While these improvements are clear from performing fits to real data, the statistical consistency of the results are yet to be established. In terms of arc-lengths, while lower structural complexity may indicate a reduced sensitivity to noise in the objective function, it may also indicate an overly restrictive minimisation procedure whereby uncertainties may be underestimated. In order to ensure the statistical consistency of

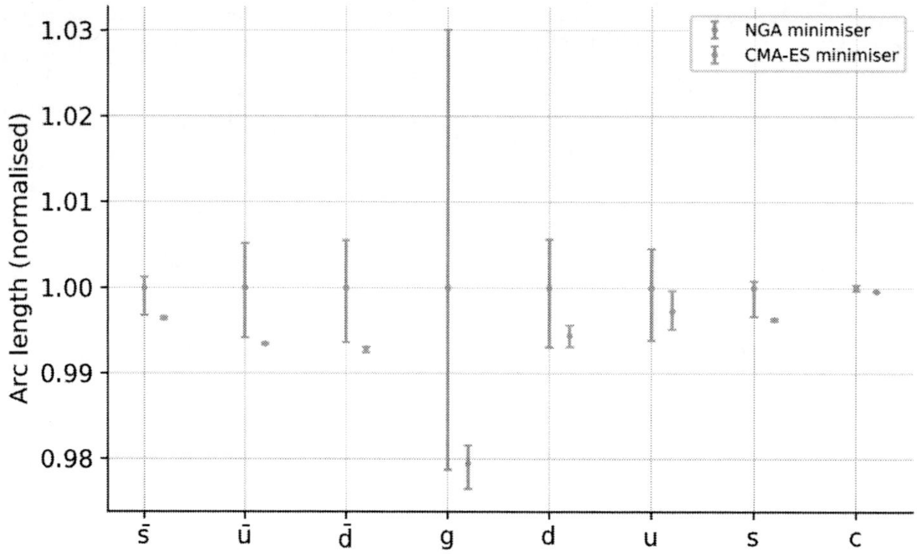

Figure 2. Arc-lengths L for PDFs determined via the NGA and CMA-ES minimisers . Results are normalised to the NGA value, and uncertainties are given as asymmetric 68% confidence intervals.

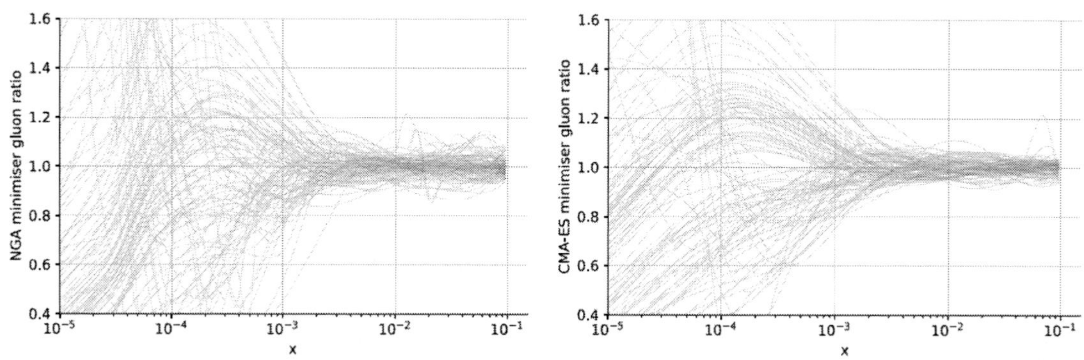

Figure 3. Gluon PDF ensemble distributions for the NGA (left) and the CMA-ES (right) minimisers. Each distribution is normalised to its central-value.

the results, further tests are needed. In Ref. [3] the consistency of the PDF uncertainties were determined by means of statistical closure tests. For the minimiser investigated here to provide reliable fits of PDFs, it must be able to demonstrate closure on a pseudo-dataset. These tests we leave for a future work.

Acknowledgements

S. C. is supported by the HICCUP ERC Consolidator grant (614577) and by the European Research Council under the European Union's Horizon 2020 research and innovation programme (grant agreement n° 740006). N. H. is supported by an European Research Council Starting Grant "PDF4BSM".

References

[1] J. Gao, L. Harland-Lang and J. Rojo, "The Structure of the Proton in the LHC Precision Era," arXiv:1709.04922 [hep-ph].

[2] D. J. Montana and L. Davis, *Training Feedforward Neural Networks Using Genetic Algorithms*, in *Proceedings of the 11th International Joint Conference on Artificial Intelligence - Volume 1*, IJCAI'89, (San Francisco, CA, USA), pp. 762–767, Morgan Kaufmann Publishers Inc., 1989.

[3] R. D. Ball *et al.* [NNPDF Collaboration], JHEP **1504** (2015) 040 doi:10.1007/JHEP04(2015)040 [arXiv:1410.8849 [hep-ph]].

[4] V. Bertone *et al.* [NNPDF Collaboration], Eur. Phys. J. C **77** (2017) no.8, 516 doi:10.1140/epjc/s10052-017-5088-y [arXiv:1706.07049 [hep-ph]].

[5] N. Hansen, A. Ostermeier, "Completely derandomized self-adaptation in evolution strategies," Evolutionary Computation (2001), no.2, 159-195 doi:10.1162/106365601750190398

[6] N. Hansen, "The CMA Evolution Strategy: A Tutorial," CoRR, 2016, [arXiv:1604.00772 [cs]].

[7] R. D. Ball *et al.* [NNPDF Collaboration], Eur. Phys. J. C **77** (2017) no.10, 663 doi:10.1140/epjc/s10052-017-5199-5 [arXiv:1706.00428 [hep-ph]].

High-precision calculation of the 4-loop contribution to the electron g-2 in QED[1]

Stefano Laporta

Dipartimento di Fisica e Astronomia, Università di Padova,
Istituto Nazionale Fisica Nucleare, Sezione di Padova,
Via Marzolo 8,
I-35131 Padova, Italy[2]

E-mail: stefano.laporta@pd.infn.it

Abstract. In this paper I show the result of the numerical evaluation of the mass-independent 4-loop contribution to the electron g-2 in QED with 1100 digits of precision. I show the semi-analytical fit to the numerical value, which contains harmonic polylogarithms of $e^{\frac{i\pi}{3}}$, $e^{\frac{2i\pi}{3}}$ and $e^{\frac{i\pi}{2}}$, one-dimensional integrals of products of complete elliptic integrals and six finite parts of master integrals, evaluated up to 4800 digits. I show also some information about the method and the program used.

1. Introduction

The QED mass-independent contribution to the electron g-2 can be expanded perturbatively in power series of the fine structure constant α:

$$A_1 = A_1^{(2)}\left(\frac{\alpha}{\pi}\right) + A_1^{(4)}\left(\frac{\alpha}{\pi}\right)^2 + A_1^{(6)}\left(\frac{\alpha}{\pi}\right)^3 + A_1^{(8)}\left(\frac{\alpha}{\pi}\right)^4 + A_1^{(10)}\left(\frac{\alpha}{\pi}\right)^5 + \ldots. \quad (1)$$

For further information about the theoretical prediction of electron g-2, see Ref. [1] in these proceedings. In Ref. [2] I have evaluated the 4-loop contribution $A_1^{(8)}$ up to 1100 digits of precision, finalizing a twenty-year effort [3–9] begun after the completion of the calculation of $A_1^{(6)}$ [10]. The first digits of the result are

$$A_1^{(8)} = -1.912245764926445574152647167439830054060873390658725345171329848\ldots. \quad (2)$$

The full-precision result is shown in table 1. The contribution to $A_1^{(8)}$ is given by 891 Feynman vertex diagrams, which can be obtained by inserting an external photon in 104 self-mass diagrams (depicted in Fig.1).

[1] Invited talk to ACAT 2017, University of Washington, Seattle, 21-25 August 2017
[2] since 1 September 2017

IOP Conf. Series: Journal of Physics: Conf. Series **1085** (2018) 052008 doi:10.1088/1742-6596/1085/5/052008

-1.9122457649264455741526471674398300540608733906587253451713298480060384439806517061427608927000036315837558415331473270056378514912854539 1
902804327050273822304345578957045562729309941296699760277782211578472033906415190816652709797086743811501215514797227432216427343192797595 86
074050057837384960701874328314024838025192249460742298558930463506140492252663431094424000235635688128062064549401322497759430042928883676 17
488992369151808780869897052635785337537769641170245361960134975744943612684861751626068323871867473038315059627418780153055148794005369777 98
369464278684326918431175889581159743566950433048349073613426586499531163878117434753854234883640855844418822372174567068710418233074305174 43
055739459611715508589611489952612660612469940731184039274723400234649695317354825848179982240973737107736574046451352112309124252811113721 53
021544537210148111211598489708842232798797204842014451228284515165852365617865945926009917330317213028654672123453405003491047007289244872 00
61604426132544906900043191519823004748818149431103849537829940629675867875385249781946989793132162197975750676701142904897962085050785592...

Table 1. First 1100 digits of $A_1^{(8)}$.

2. Gauge-invariant sets

The contributions of single Feynman diagrams may be I.R. or U.V. divergent, and depend on the gauge chosen. Therefore, it is convenient to regroup the diagrams in gauge invariant sets, which are finite. The number of diagrams at 2, 3 and 4 loop are 7, 72 and 891, respectively; they can be arranged in 3, 9(see Ref. [11]) and 25 gauge invariant sets, respectively. Some elements of the 25 sets at 4 loops are shown in Fig.2. The numerical contributions of each 4-loop set truncated to 40 digits are listed in Ref. [2].

3. The numerical fit

The contributions of all 4-loop vertex diagrams can be expressed by means of 334 master integrals belonging to 220 topologies. Due to the expected analytical complexity of the result, a completely analytical calculation of all the master integrals (similar to that at three-loop) has seemed out of reach. So I have used a different approach:

- Compute very high precision values of the ϵ−expansions of the master integrals in $D = 4-2\epsilon$ dimensions.
- Make the right analytical ansatz.
- Fit the rational coefficients of the ansatz by using the PSLQ algorithm [12–14]

As a result, one finds that the analytical expressions of the coefficients of the ϵ−expansions of the master integrals contain values of harmonic polylogarithms [15] with argument 1, $\frac{1}{2}$, $e^{\frac{i\pi}{3}}$, $e^{\frac{2i\pi}{3}}$ and $e^{\frac{i\pi}{2}}$, a family of one-dimensional integrals of products of elliptic integrals, and the finite terms of the ϵ−expansions of six master integrals. Work is still in progress to fit analytically these six unknown elliptical constants. The result of the analytical fit for $A_1^{(8)}$ can be written as follows:

$$A_1^{(8)} = T + \sqrt{3}V_a + V_b + W_b + \sqrt{3}E_a + E_b + U \; , \tag{3}$$

$$\begin{aligned}
T =& \frac{1243127611}{130636800} + \frac{30180451}{25920}\zeta(2) - \frac{255842141}{2721600}\zeta(3) - \frac{8873}{3}\zeta(2)\ln 2 + \frac{6768227}{2160}\zeta(4) + \frac{19063}{360}\zeta(2)\ln^2 2 \\
&+ \frac{12097}{90}\left(a_4 + \frac{1}{24}\ln^4 2\right) - \frac{2862857}{6480}\zeta(5) - \frac{12720907}{64800}\zeta(3)\zeta(2) - \frac{221581}{2160}\zeta(4)\ln 2 \\
&+ \frac{9656}{27}\left(a_5 + \frac{1}{12}\zeta(2)\ln^3 2 - \frac{1}{120}\ln^5 2\right) + \frac{191490607}{46656}\zeta(6) + \frac{10358551}{43200}\zeta^2(3) - \frac{40136}{27}a_6 + \frac{26404}{27}b_6 \\
&- \frac{700706}{675}a_4\zeta(2) - \frac{26404}{27}a_5\ln 2 + \frac{26404}{27}\zeta(5)\ln 2 - \frac{63749}{50}\zeta(3)\zeta(2)\ln 2 - \frac{40723}{135}\zeta(4)\ln^2 2 + \frac{13202}{81}\zeta(3)\ln^3 2 \\
&- \frac{253201}{2700}\zeta(2)\ln^4 2 + \frac{7657}{1620}\ln^6 2 + \frac{2895304273}{435456}\zeta(7) + \frac{670276309}{193536}\zeta(4)\zeta(3) + \frac{85933}{63}a_4\zeta(3) - \frac{142793}{18}a_5\zeta(2) \\
&+ \frac{7121162687}{967680}\zeta(5)\zeta(2) - \frac{195848}{21}a_7 + \frac{195848}{63}b_7 - \frac{116506}{189}d_7 - \frac{4136495}{384}\zeta(6)\ln 2 - \frac{1053568}{189}a_6\ln 2 \\
&+ \frac{233012}{189}b_6\ln 2 + \frac{407771}{432}\zeta^2(3)\ln 2 - \frac{8937}{2}a_4\zeta(2)\ln 2 + \frac{833683}{3024}\zeta(5)\ln^2 2 - \frac{3995099}{6048}\zeta(3)\zeta(2)\ln^2 2 \\
&- \frac{233012}{189}a_5\ln^2 2 + \frac{1705273}{1512}\zeta(4)\ln^3 2 + \frac{602303}{4536}\zeta(3)\ln^4 2 - \frac{1650461}{11340}\zeta(2)\ln^5 2 + \frac{52177}{15876}\ln^7 2 \; ,
\end{aligned} \tag{4}$$

IOP Conf. Series: Journal of Physics: Conf. Series **1085** (2018) 052008 doi:10.1088/1742-6596/1085/5/052008

Figure 1. The 4-loop self-mass diagrams.

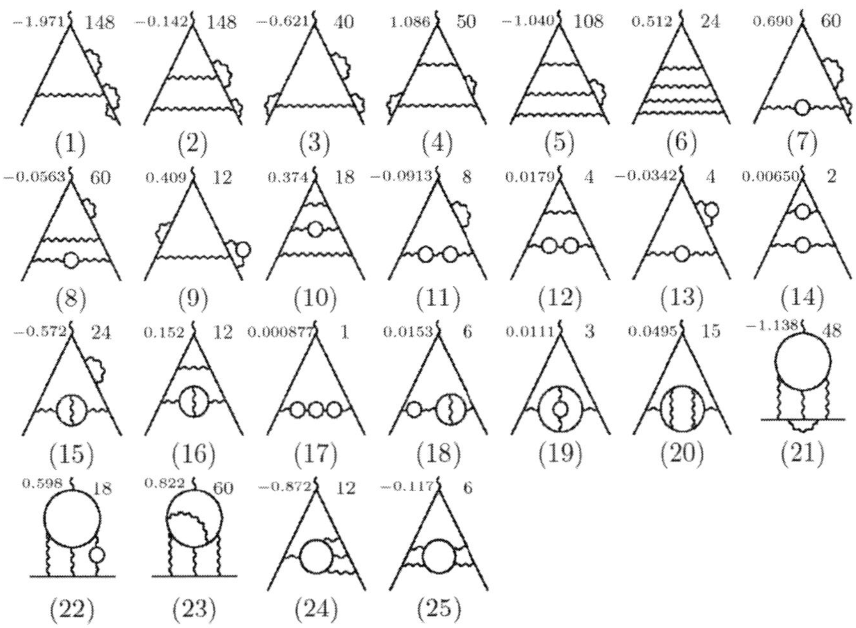

Figure 2. Typical representative diagrams of gauge-invariant sets. For each set only one diagrams is shown. Top left: contribution of the set to g-2; top right: number of diagrams of the set

$$\begin{aligned}
V_a =& -\frac{14101}{480}\mathrm{Cl}_4\left(\frac{\pi}{3}\right) - \frac{169703}{1440}\zeta(2)\mathrm{Cl}_2\left(\frac{\pi}{3}\right) + \frac{494}{27}\mathrm{Im}H_{0,0,0,1,-1,-1}\left(e^{i\frac{\pi}{3}}\right) + \frac{494}{27}\mathrm{Im}H_{0,0,0,1,-1,1}\left(e^{i\frac{2\pi}{3}}\right) \\
&+ \frac{494}{27}\mathrm{Im}H_{0,0,0,1,1,-1}\left(e^{i\frac{2\pi}{3}}\right) + 19\mathrm{Im}H_{0,0,1,0,1,1}\left(e^{i\frac{2\pi}{3}}\right) + \frac{437}{12}\mathrm{Im}H_{0,0,0,1,1,1}\left(e^{i\frac{2\pi}{3}}\right) + \frac{29812}{297}\mathrm{Cl}_6\left(\frac{\pi}{3}\right) \\
&+ \frac{4940}{81}a_4\mathrm{Cl}_2\left(\frac{\pi}{3}\right) - \frac{520847}{69984}\zeta(5)\pi - \frac{129251}{81}\zeta(4)\mathrm{Cl}_2\left(\frac{\pi}{3}\right) - \frac{892}{15}\mathrm{Im}H_{0,1,1,-1}\left(e^{i\frac{2\pi}{3}}\right)\zeta(2) \\
&- \frac{1784}{45}\mathrm{Im}H_{0,1,1,-1}\left(e^{i\frac{\pi}{3}}\right)\zeta(2) + \frac{1729}{54}\zeta(3)\mathrm{Im}H_{0,1,-1}\left(e^{i\frac{\pi}{3}}\right) + \frac{1729}{36}\zeta(3)\mathrm{Im}H_{0,1,1}\left(e^{i\frac{2\pi}{3}}\right) \\
&+ \frac{837190}{729}\mathrm{Cl}_4\left(\frac{\pi}{3}\right)\zeta(2) + \frac{25937}{4860}\zeta(3)\zeta(2)\pi - \frac{223}{243}\zeta(4)\pi\ln 2 + \frac{892}{9}\mathrm{Im}H_{0,1,-1}\left(e^{i\frac{\pi}{3}}\right)\zeta(2)\ln 2 \\
&+ \frac{446}{3}\mathrm{Im}H_{0,1,1}\left(e^{i\frac{2\pi}{3}}\right)\zeta(2)\ln 2 - \frac{7925}{81}\mathrm{Cl}_2\left(\frac{\pi}{3}\right)\zeta(2)\ln^2 2 + \frac{1235}{486}\mathrm{Cl}_2\left(\frac{\pi}{3}\right)\ln^4 2 \ ,
\end{aligned} \tag{5}$$

$$\begin{aligned}
V_b =& \frac{13487}{60}\mathrm{Re}H_{0,0,0,1,0,1}\left(e^{i\frac{\pi}{3}}\right) + \frac{13487}{60}\mathrm{Cl}_4\left(\frac{\pi}{3}\right)\mathrm{Cl}_2\left(\frac{\pi}{3}\right) + \frac{136781}{360}\mathrm{Cl}_2^2\left(\frac{\pi}{3}\right)\zeta(2) + \frac{651}{4}\mathrm{Re}H_{0,0,0,1,0,1,-1}\left(e^{i\frac{\pi}{3}}\right) \\
&+ 651\mathrm{Re}H_{0,0,0,0,1,1,-1}\left(e^{i\frac{\pi}{3}}\right) - \frac{17577}{32}\mathrm{Re}H_{0,0,1,0,0,1,1}\left(e^{i\frac{2\pi}{3}}\right) - \frac{87885}{64}\mathrm{Re}H_{0,0,0,1,0,1,1}\left(e^{i\frac{2\pi}{3}}\right) \\
&- \frac{17577}{8}\mathrm{Re}H_{0,0,0,0,1,1,1}\left(e^{i\frac{2\pi}{3}}\right) + \frac{651}{4}\mathrm{Cl}_4\left(\frac{\pi}{3}\right)\mathrm{Im}H_{0,1,-1}\left(e^{i\frac{\pi}{3}}\right) + \frac{1953}{8}\mathrm{Cl}_4\left(\frac{\pi}{3}\right)\mathrm{Im}H_{0,1,1}\left(e^{i\frac{2\pi}{3}}\right) \\
&+ \frac{31465}{176}\mathrm{Cl}_6\left(\frac{\pi}{3}\right)\pi + \frac{211}{4}\mathrm{Re}H_{0,1,0,1,-1}\left(e^{i\frac{\pi}{3}}\right)\zeta(2) + \frac{211}{2}\mathrm{Re}H_{0,0,1,1,-1}\left(e^{i\frac{\pi}{3}}\right)\zeta(2) \\
&+ \frac{1899}{16}\mathrm{Re}H_{0,1,0,1,1}\left(e^{i\frac{2\pi}{3}}\right)\zeta(2) + \frac{1899}{8}\mathrm{Re}H_{0,0,1,1,1}\left(e^{i\frac{2\pi}{3}}\right)\zeta(2) \\
&+ \frac{211}{4}\mathrm{Im}H_{0,1,-1}\left(e^{i\frac{\pi}{3}}\right)\mathrm{Cl}_2\left(\frac{\pi}{3}\right)\zeta(2) + \frac{633}{8}\mathrm{Im}H_{0,1,1}\left(e^{i\frac{2\pi}{3}}\right)\mathrm{Cl}_2\left(\frac{\pi}{3}\right)\zeta(2) \ ,
\end{aligned} \tag{6}$$

$$W_b = \zeta(2)\Bigg(-\frac{28276}{25}\mathrm{Cl}_2\left(\frac{\pi}{2}\right)^2 + 104\Bigg(4\mathrm{Re}H_{0,1,0,1,1}(i) + 4\mathrm{Im}H_{0,1,1}(i)\,\mathrm{Cl}_2\left(\frac{\pi}{2}\right) - 2\mathrm{Cl}_4\left(\frac{\pi}{2}\right)\pi + \mathrm{Cl}_2^2\left(\frac{\pi}{2}\right)\ln 2\Bigg)\Bigg) \ , \tag{7}$$

IOP Conf. Series: Journal of Physics: Conf. Series **1085** (2018) 052008 doi:10.1088/1742-6596/1085/5/052008

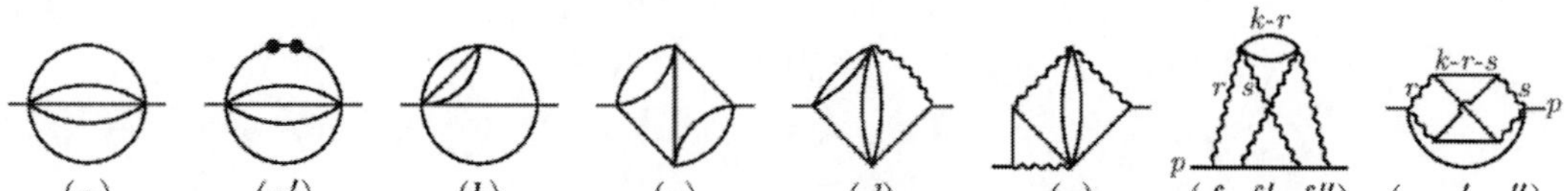

(a) (a') (b) (c) (d) (e) (f,f',f'') (g,g',g'')

Figure 3. Minimal set of master integrals which contain all the elliptic constants. The double dot in (a') means that denominator is raised to the power three. (f,f',f'') and (g,g',g'') have numerators equal to $(1,p.k,(p.k)^2)$, respectively.

$$
\begin{aligned}
E_a =&\pi\left(-\frac{28458503}{691200}B_3+\frac{250077961}{18662400}C_3\right)+\frac{483913}{77760}\pi f_2(0,0,1)+\pi\left(\frac{4715}{1944}\ln 2\, f_2(0,0,1)+\frac{270433}{10935}f_2(0,2,0)\right.\\
&\left.-\frac{188147}{4860}f_2(0,1,1)+\frac{188147}{12960}f_2(0,0,2)\right)+\pi\left(\frac{826595}{248832}\zeta(2)f_2(0,0,1)-\frac{5525}{432}\ln 2\, f_2(0,0,2)\right.\\
&+\frac{5525}{162}\ln 2\, f_2(0,1,1)-\frac{5525}{243}\ln 2\, f_2(0,2,0)+\frac{526015}{248832}f_2(0,0,3)-\frac{4675}{768}f_2(0,1,2)+\frac{1805965}{248832}f_2(0,2,1)\\
&\left.-\frac{3710675}{1119744}f_2(0,3,0)-\frac{75145}{124416}f_2(1,0,2)-\frac{213635}{124416}f_2(1,1,1)+\frac{168455}{62208}f_2(1,2,0)+\frac{69245}{124416}f_2(2,1,0)\right),
\end{aligned}
\tag{8}
$$

$$
\begin{aligned}
E_b =&\zeta(2)\left(\frac{2541575}{82944}f_1(0,0,2)-\frac{556445}{6912}f_1(0,1,1)+\frac{54515}{972}f_1(0,2,0)-\frac{75145}{20736}f_1(1,0,1)\right)\\
&-\frac{4715}{1458}\zeta(2)f_1(0,0,1),
\end{aligned}
\tag{9}
$$

$$
U =-\frac{541}{300}C_{81a}-\frac{629}{60}C_{81b}+\frac{49}{3}C_{81c}-\frac{327}{160}C_{83a}+\frac{49}{36}C_{83b}+\frac{37}{6}C_{83c}\,. \tag{10}
$$

C_{8xy} are the ϵ^0 coefficients of the ϵ−expansion of six master integrals (see f,f',f'',g,g',g'' of Fig.3); they appear only in the gauge-invariant sets 24 and 25. In the above expressions $\zeta(n)=\sum_{i=1}^{\infty}i^{-n}$, $a_n=\sum_{i=1}^{\infty}2^{-i}\,i^{-n}$, $b_6=H_{0,0,0,0,1,1}\left(\frac{1}{2}\right)$, $b_7=H_{0,0,0,0,0,1,1}\left(\frac{1}{2}\right)$, $d_7=H_{0,0,0,0,1,-1,-1}(1)$, $\mathrm{Cl}_n(\theta)=\mathrm{Im}\mathrm{Li}_n(e^{i\theta})$. $H_{i_1,i_2,\ldots}(x)$ are the harmonic polylogarithms. The integrals f_j are defined as follows:

$$
f_m(i,j,k) \quad = \quad \int_1^9 ds\, D_1(s)\mathrm{Re}\left(\sqrt{3^{m-1}}D_m(s)\right)\left(s-\frac{9}{5}\right)\ln^i(9-s)\ln^j(s-1)\ln^k(s)\ , \tag{11}
$$

$$
D_m(s) \quad = \quad \frac{2}{\sqrt{(\sqrt{s}+3)(\sqrt{s}-1)^3}}K\left(m-1-(2m-3)\frac{(\sqrt{s}-3)(\sqrt{s}+1)^3}{(\sqrt{s}+3)(\sqrt{s}-1)^3}\right)\ ; \tag{12}
$$

$K(x)$ is the complete elliptic integral of the first kind. The constants B_3 and C_3, defined in Ref. [8], admit the hypergeometric representations:

$$
B_3 =\frac{4\pi^{\frac{3}{2}}}{3}\left(\frac{\Gamma^2(\frac{7}{6})\Gamma(\frac{1}{3})}{\Gamma^2(\frac{2}{3})\Gamma(\frac{5}{6})}{}_4F_3\left({\frac{1}{6}\,\frac{1}{3}\,\frac{1}{3}\,\frac{1}{2} \atop \frac{5}{6}\,\frac{5}{6}\,\frac{2}{3}};1\right)+\frac{\Gamma^2(\frac{5}{6})\Gamma(-\frac{1}{3})}{\Gamma^2(\frac{1}{3})\Gamma(\frac{1}{6})}{}_4F_3\left({\frac{1}{2}\,\frac{2}{3}\,\frac{2}{3}\,\frac{5}{6} \atop \frac{7}{6}\,\frac{7}{6}\,\frac{4}{3}};1\right)\right), \tag{13}
$$

$$
C_3 =\frac{486\pi^2}{1925}\ {}_7F_6\left({\frac{7}{4}\,-\frac{1}{3}\,\frac{1}{3}\,\frac{2}{3}\,\frac{4}{3}\,\frac{3}{2}\,\frac{3}{2} \atop \frac{3}{4}\,1\,\frac{7}{6}\,\frac{11}{6}\,\frac{13}{6}\,\frac{17}{6}};1\right), \tag{14}
$$

Fig.3 shows the fundamental elliptic master integrals which contains irreducible combinations of B_3, C_3 and $f_m(i,j,k)$.

The PSLQ basis needed to fit T contains $F_{10}-1=54$ terms (F_i are the Fibonacci numbers, $F_1=F_2=1$). Due to the presence of harmonic polylogarithms of complex argument, the general basis needed to perform the fit of V_a and V_b contains $F_{17} - 1 = 1596$ terms. The practical maximum size of basis used in PSLQ fits is ≈ 500, so some guessing is needed in order to identify which elements are really needed in the fits (for example, only 93 for V_a and V_b). The precision of the calculation of elements of the basis is up to 9600 digits. The multi-pair parallel version of the PSLQ algorithm [13] publicly available from Ref. [14] has been essential to work out these difficult analytical fits in reasonable times.

4. Method of calculation

4.1. Diagrams generation

All the 104 4-loop self-mass diagrams are generated with a *C* program written *from scratch*; for each self-mass diagram, the corresponding vertex diagrams are generated by inserting a photon in all possible ways. For the sake of subsequent checks, one keeps also the diagrams which give no contribution because of Furry's theorem.

4.2. Extraction of the contribution to g-2

The contribution to g-2 of the diagram is extracted from the unrenormalized amplitude of each vertex diagram using suitable projectors, with a FORM [20, 21] program. The results are expressions which typically contain 100-30000 different Feynman integrals.

4.3. Identification of master integrals

The total number of topologically distinct master integrals is 334. 82 topologies have more than one master integral. For these topologies, products of scalar products are chosen as numerators, the choice depending on the momentum flow of the diagram. For the sake of subsequent checks, one processes *separately* the contributions from each one of the 104 self-mass diagrams. The master integrals are calculated separately for each self-mass diagram; therefore the total number of the master integrals actually calculated is 4607. Because of differences in momentum routing and therefore in the numerators, topologies with more than one master integral will have a slightly different set of master integrals in different self-mass diagrams. This provides a *lot of* useful checks.

4.4. Reduction to master integrals

The reduction is exact in D, and is performed by generating large systems of integration-by-parts identities and solving them with the algorithm implemented in `SYS` [3]. These large systems contain $4 \times 10^6 - 50 \times 10^6$ identities, with physical disk space ranging from $4GB$ to $100GB$. The reduction to master integrals has been repeated with 32 and 64 bit versions of `SYS`. In addition, the principal self-mass diagrams were reprocessed using a different momentum flow, checking that reduction to master integrals remained the same (after converting different sets of master integrals).

4.5. Numerical calculation of master integrals

For the numerical calculation of master integrals I have used combinations of the the difference equation method [3, 4] and the differential equation method [17–19]. An approach consists in calculating with difference equations (inserting the exponent n of the first electron propagator) the integral obtained putting the photon mass λ equal to the electron mass m, and integrating a differential equation in λ from $\lambda = m$ to $\lambda = 0$. A second approach consists in calculating with difference equations the diagram obtained by putting the the external momentum $p = 0$, and integrating a differential equation in p^2 from $p^2 = 0$ to $p^2 = -m^2$. The systems of difference and differential equations for master integrals are obtained by building systems of integration-by-parts identities and solving them by using the algorithm [3, 4], using rational arithmetic in D.

IOP Conf. Series: Journal of Physics: Conf. Series **1085** (2018) 052008 doi:10.1088/1742-6596/1085/5/052008

The sizes of the systems of difference or differential equations to be numerically solved are in the range $1MB-3GB$. Difference equations are solved by using the Laplace transformation method (integral representation of solutions and differential equation of the integrand). Differential equations are solved numerically, by using series expansion with truncated expansions in $\epsilon = (4-D)/2$ as coefficients. The minimum number of terms of the expansion in ϵ is 9 ($\epsilon^{-4} \ldots \epsilon^{4}$). There are cancellations of ϵ terms in intermediate steps; no care is used to avoid cancellations, as the corresponding numerical zeroes provide extremely useful checks. In the worst case 37 terms of expansions are needed. The standard precision of calculations is 4 kbit (1232 digits). About 130 digits are lost due to cancellations, so that the result has 1100 digits exact. Note that the analytical fit of some selected master integrals required a precision much higher, up to 16 kbit (9864 digits).

4.6. Renormalization

The renormalization counterterms are generated with two procedures written *from scratch* in *C* and FORM. Their expressions are reduced to master integrals already known in analytical form. I have chosen the Feynman gauge. I checked explicitly the (internal) gauge invariance of the contributions for arbitrary gauge of the photon line going into 1- ,2- or 3-loop vacuum polarization diagrams.

4.7. The program

In order to perform this calculation, in 1995 I begun writing 24000-line a comprehensive *C* program *from scratch*, `SYS` [3], containing all the necessary parts, among which:

- a simplified fast algebraic symbolic manipulator,
- a numerical solver of systems of difference and differential equations,
- and a library of multiprecision routines, operating on integer and rational numbers, polynomials in one and two variables, floating point numbers, series expansions with floating point coefficients.

Another important feature is the thorough protection of large buffers and I/O with checksums. Throughout the calculation, I have found several subtle corruptions, due to different causes, like marginal coupling of non-ECC RAM modules (error-rate: 1 bit per week), failing RAID systems (error-rate: one corrupted block of 64KB every 100GB), etc....) This has resulted in an highly reliable result.

References

[1] Laporta S "New results on g-2 calculation", *these proceedings*
[2] Laporta S 2017 *Phys. Lett.* B **772** 232
[3] Laporta S 2000 *Int. J. Mod. Phys.* A **15** 5087
[4] Laporta S 2001 *Phys. Lett.* B **504** 188
[5] Laporta S 2001 *Phys. Lett.* B **523** 95
[6] Laporta S 2003 *Acta Phys. Polon.* B **34** 5323
[7] Laporta S, Mastrolia P and Remiddi E 2004 *Nucl. Phys.* B **688** 165
[8] Laporta S 2008 *Int. J. Mod. Phys.* A **23** 5007
[9] Laporta S 2009 *Subnucl. Ser.* **45** 409.
[10] Laporta S and Remiddi E 1996 *Phys. Lett.* B **379** 283
[11] Cvitanovic P 1977 *Nucl. Phys.* B **127** 176
[12] Ferguson H R P and Bailey D H 1991 RNR Technical Report RNR-91-032.
[13] Bailey D H and Broadhurst D J 2001 *Math. Comput.* **70** 1719
[14] **ARPREC** [software] Bailey D H available from `http://crd-legacy.lbl.gov/~dhbailey/mpdist/` [accessed 29-12-2017]
[15] Remiddi E and Vermaseren J A M 2000 *Int. J. Mod. Phys.* **A15** 725

[16] Laporta S and Remiddi E 2005 *Nucl. Phys.* B **704** 349
[17] Kotikov A V, 1991 *Phys. Lett.* B **254** 158
[18] Remiddi E 1997 *Nuovo Cim.* A **110** 1435
[19] Gehrmann T and Remiddi E 2000 *Nucl. Phys.* B **580** 485
[20] Vermaseren J A M 2000 *preprint* arXiv:math-ph/0010025
[21] `FORM` [software] Vermaseren J A M available from `https://www.nikhef.nl/~form/` [accessed 29-12-2017]

IOP Conf. Series: Journal of Physics: Conf. Series **1085** (2018) 052009 doi:10.1088/1742-6596/1085/5/052009

Numerical Methods and the 4-point 2-loop Higgs amplitudes

S. P. Jones[a,1], **B. Ruijl**[b]

[a]Max Planck Institute for Physics, Föhringer Ring 6, 80805 München, Germany
[b]Institute for Theoretical Physics, ETH Zürich, 8093 Zürich, Switzerland

E-mail: `sjones@mpp.mpg.de`

Abstract. Some of the difficulties faced when calculating multi-loop amplitudes with several mass scales are reviewed. We then focus on one particular difficulty, the evaluation of the Feynman integrals, and introduce the program `pySecDec` which can be used to numerically compute such integrals. Some of the new features and in particular the sector symmetry finder, which can help to reduce the number of sectors to be numerically integrated after sector decomposition, are described.

1. Introduction

In recent years there has been significant progress in the calculation of multi-loop amplitudes and related higher order quantities. A prominent example is the computation of the N3LO QCD corrections to Higgs boson production in gluon-fusion [1]. However, processes with more legs but fewer loops, such as vector boson pair production at NNLO in QCD [2, 3, 4], have recently also been analytically computed. Furthermore, progress has been made in the calculation of processes which depend on many scales (i.e. with internal massive particles or massive external legs), for example the planar integrals relevant for Higgs boson plus jet production [5] as well as several integrals relevant for mixed EW-QCD corrections and QED corrections including fermion masses [6, 7, 8, 9]. Advances have also been made in the computation of multi-loop processes with 5 external legs [10, 11, 12] and very recently even 6 and 7 external legs [13].

Much of the recent progress has been driven by important insights into the analytical structure of Feynman integrals [14] as well as on-shell techniques, for example [15]. However, progress has also been made in the context of predominantly numerical calculations as demonstrated by the numerical computation of the NLO QCD corrections to Higgs boson pair production [16, 17]. An important open question is whether these numerical techniques can be used to compute further currently unknown processes such as the full Higgs boson plus jet amplitude or the Higgs plus Z-boson amplitude in gluon-fusion.

In these proceedings we briefly discuss some of the difficulties faced when computing multi-loop amplitudes with many mass scales. In Section 2 we discuss the various stages of a multi-loop calculation carried out using mostly traditional techniques and highlight where the current technological bottlenecks are. In Section 3 we focus on the particularly challenging issue of computing the relevant Feynman integrals and introduce a recent tool, `pySecDec` [18], which is

[1] Speaker.

IOP Conf. Series: Journal of Physics: Conf. Series **1085** (2018) 052009 doi:10.1088/1742-6596/1085/5/052009

capable of computing such integrals numerically. Finally, in Section 4 we focus in some detail on one new feature of `pySecDec`, the sector symmetry finder, and describe its implementation.

2. 2-loop Amplitudes

One standard method of computing multi-loop amplitudes consists of generating Feynman diagrams, inserting the Feynman rules (which produces integrals that must be evaluated), reducing the number of integrals using integration-by-parts identities [19, 20], then evaluating the remaining so-called master integrals. This method was used for the calculation of Higgs boson pair production [16, 17]. In more detail the steps taken (and compute time required) were:

(i) decompose the amplitude into form factors and construct projectors (minutes),
(ii) generate Feynman diagrams (seconds),
(iii) apply the projectors and compute the amplitude (hours/days),
(iv) perform integral reduction (6+ months),
(v) compute the master integrals (analytically: challenging, numerically: seconds/hours),
(vi) generate events and compute the differential cross-section (analytically: seconds?, numerically: hours/days).

Roughly the same timings are valid for the calculation of Higgs boson plus jet production at NLO, though the final step is currently in progress and could in principle take significantly longer than estimated above. The calculation of the 2-loop gluon-fusion contribution to Higgs plus Z-boson production is yet more complicated due to the additional mass scale (Z-boson mass), this will impact the time taken to perform the integral reduction symbolically, though, fixing the top-quark and Higgs boson masses to numerical values may make the integral reduction more tractable.

As can be seen from the timings above, by far the most time consuming step is the integral reduction. In the literature, several avenues are currently being explored to accelerate this step, see for example [21, 22, 23, 24]. Another major bottleneck for the analytic approach is the computation of the master integrals. The analytic evaluation of the master integrals is currently very challenging and so far only a subset of the integrals appearing in these processes is known [25, 5]. However, the numerical evaluation of the master integrals currently appears to be a practical alternative to the analytic approach and may be a promising research direction.

3. Overview of `pySecDec`

The program `pySecDec` [18] is designed to numerically compute dimensionally regulated parameter integrals. It is a complete rewrite of its predecessor `SecDec` using only open source software and is intended to be as modular and extensible as possible. The initial symbolic/algebraic steps performed by `pySecDec` are carried out using `python` [26], with extensive use of the packages `numpy` [27] and `sympy` [28]. Further symbolic manipulation and code generation are performed using `FORM` [29, 30, 31]. The numerical integration of the resulting integrand functions is carried out in `c++11` [33]. Extensive documentation for the code as well as tutorials and guides are provided using the Sphinx package [32].

The program `pySecDec` contains several new features in addition to many minor improvements, the key changes are:

- Support for an arbitrary number of regulators (not just ϵ).
- More flexible numerators (sums of contracted Lorentz vectors and inverse propagators).
- Improvements to the handling of integrals without a Euclidean region.
- Addition of a symmetry finder for detecting isomorphisms between sectors.

IOP Conf. Series: Journal of Physics: Conf. Series **1085** (2018) 052009 doi:10.1088/1742-6596/1085/5/052009

- Use of FORM [29, 30, 31] for code optimisation, this reduces the number of operations used to compute the integrand thus accelerating the numerical computation.
- Automatic generation of a c++ library which can be linked by external programs.

The input to pySecDec can be quite general parameter integrals. Much of what is discussed here and in Section 4 applies both to general parameter integrals and to the specific case of Feynman integrals. To rewrite a Feynman integral in terms of a dimensionally regulated parameter integral we first Feynman parametrise the integral. Concretely, an L-loop integral with N propagators, P_j, raised to arbitrary powers, ν_j, can be written as

$$G = \int_{-\infty}^{\infty} \left(\prod_{l=1}^{L} \frac{\mathrm{d}^D k_l}{i\pi^{\frac{D}{2}}} \right) \frac{1}{\prod_{j=1}^{N} P_j^{\nu_j}}$$
$$= (-1)^{N_\nu} \frac{\Gamma(N_\nu - LD/2)}{\prod_{j=1}^{N} \Gamma(\nu_j)} \int_0^\infty \left(\prod_{j=1}^{N} \mathrm{d}x_j \, x_j^{\nu_j - 1} \right) \delta(1 - \sum_{i=1}^{N} x_i) \frac{\mathcal{U}^{N_\nu-(L+1)D/2}(x_1, \ldots, x_N)}{\mathcal{F}^{N_\nu-LD/2}(x_1, \ldots, x_N; s_1, \ldots, s_m)}. \tag{1}$$

Here we have introduced the Feynman parameters $x_1, \ldots, x_N$ and carried out the momentum integration. The symbols $s_1, \ldots, s_m$ represent invariants or masses and $N_\nu = \sum_{j=1}^{N} v_j$. The functions $\mathcal{U}$ and $\mathcal{F}$ are the 1st and 2nd Symanzik polynomials respectively. If inverse propagators are present, or if the numerator of the integral contains scalar products involving the loop-momenta, there may be an additional polynomial $\mathcal{N}$.

When calculating loop integrals the first step taken within pySecDec is to integrate out the Dirac delta. For the decomposition strategies iterative and geometric_ku this is done by introducing

$$\int_0^\infty d^N x = \sum_{l=1}^{N} \int_0^\infty d^N x \prod_{j=1, j\neq l}^{N} \theta(x_l \geq x_j), \tag{2}$$

and performing one integration using the δ-distribution. After this step we have N integrals, which we refer to as primary sectors, each with one Feynman parameter set to one. For the decomposition strategy geometric the Cheng-Wu [34, 35] theorem is instead used to integrate out the Dirac delta, this amounts to replacing the δ-distribution in Eq. 1 by $\delta(1 - x_N)$. After this so-called primary decomposition, the full sector decomposition algorithm is applied to each primary sector, for a description of the various strategies see Refs. [36, 37, 18] and references therein. Finally, each sector is numerically integrated using one of the algorithms provided by CUBA [38] or, if the sector depends only on one variable, CQUAD [39, 40].

4. Sector Symmetry Finder

In pySecDec a sector consists of one or more (potentially exponentiated) polynomials in the integration parameters. In the case of loop integrals, a sector consists of decomposed $\mathcal{U}$ and $\mathcal{F}$ polynomials, which depend on the Feynman parameters, raised to the appropriate powers according to Eq. 1, and, if a numerator is present, an additional $\mathcal{N}$ polynomial.

If sectors are equivalent up to permutations of the integration parameters then it is sufficient to compute one such sector and multiply the result by the number of equivalent sectors, this saves compute time during the numerical integration step. Since all integration parameters x_i are integrated over the same domain $[0, 1]$ we may freely switch the labels of the integration parameters within each sector without changing the result after integration. We have therefore implemented algorithms, described below, which can identify equivalent sectors up to such a relabelling. These algorithms work not with sectors but polynomials. To convert a sector into

IOP Conf. Series: Journal of Physics: Conf. Series **1085** (2018) 052009 doi:10.1088/1742-6596/1085/5/052009

a single polynomial it suffices to multiply each polynomial within the sector by a unique label and sum the resulting polynomials.

To motivate the problem of identifying equivalent polynomials up to relabelling of variables consider the polynomials:

$$\begin{aligned} P_1 &= x_1^2 x_2^1 x_3^1 + x_1^1 x_2^1 x_3^2, \\ P_2 &= x_1^1 x_2^1 x_3^2 + x_1^1 x_2^2 x_3^1. \end{aligned} \tag{3}$$

In this case it is straightforward to see that the polynomials are equivalent up to relabelling of the variables x_i. By applying either permutation $S_1 = (2\ 1)$ or $S_2 = (2\ 3\ 1)$ to P_2 we obtain P_1. A brute force approach consists of applying all $n!$ permutations of the parameters $x_1, \ldots, x_n$ to one of the polynomials and then checking if the polynomials are equivalent. Such an algorithm has combinatorial complexity and becomes impractically slow for large problems.

In `pySecDec` there are four algorithms for identifying isomorphic sectors. Two of the algorithms, `iterative_sort` and `light_Pak_sort`, are relatively quick to run but do not find all isomorphisms. The remaining algorithms `dreadnaut` and `Pak_sort` find all identities but may take considerably longer to do so.

All algorithms presented in this section operate on the exponent list representation of multivariate polynomials. An exponent vector $\mathbf{e} = (e_1, e_2, \ldots, e_n) \in \mathrm{N}^n$ defines a monomial $\mathbf{x}^{\mathbf{e}} = x_1^{e_1} x_2^{e_2} \cdots x_n^{e_n}$. A term is the product of a non-zero coefficient c and a monomial, i.e. $c\mathbf{x}^{\mathbf{e}}$. The exponent list representation of a polynomial is a list of terms (coefficients and exponent vectors) appearing in the polynomial. For example, the first polynomial in Eq. 3 has exponent list representation:

$$P_1 = x_1^2 x_2^1 x_3^1 + x_1^1 x_2^1 x_3^2 \rightarrow \begin{array}{r} \\ \text{Term 1} \\ \text{Term 2} \end{array} \begin{array}{c} \text{Coeff} \\ \begin{pmatrix} 1 \\ 1 \end{pmatrix} \end{array} \begin{array}{c} \begin{matrix} x_1 & x_2 & x_3 \end{matrix} \\ \begin{pmatrix} 2 & 1 & 1 \\ 1 & 1 & 2 \end{pmatrix} \end{array} \tag{4}$$

In the exponent list representation simultaneously permuting the rows of the coefficient list and the exponent list corresponds to switching the order of terms in the polynomial, whilst permuting columns of the exponent list corresponds to switching the variable labels.

4.1. Graph-based (`dreadnaut`) method

We begin by discussing the `dreadnaut` symmetry finder algorithm. The algorithm works by constructing a graph representation of the multivariate polynomial, bringing the graph into a canonical form by relabelling the vertices, hashing the canonical graph and then comparing the canonical graphs of polynomials with identical hashes. If two polynomials have identical canonical graphs they are equivalent up to permutations of the variable labels. The use of the `dreadnaut` symmetry finder is disabled by default in `pySecDec 1.2.2`, it can be enabled by setting `use_dreadnaut = True` in the call to `make_package` or `loop_package`.

In Figure 1 (left panel) we show the conversion of the polynomial P_1 in Eq. 4 to a graph. The algorithm we utilise for converting a polynomial into a graph is based on that used in Ref. [41] for identifying isotopy of matrices. The following steps are taken:

(i) introduce a node of colour c for each variable (column of the exponent matrix),

(ii) introduce a node of colour r for each term (row of the exponent matrix),

(iii) introduce a node of colour e for each entry in the exponent matrix, connect the entry node to the node of the corresponding column and to the node of the corresponding row,

(iv) for each unique exponent $0, \ldots, p$ appearing in the exponent matrix introduce a node of a new colour $s_0, \ldots, s_p$ and connect the exponent node to any entry node whose corresponding entry in the exponent matrix has this exponent,

IOP Conf. Series: Journal of Physics: Conf. Series **1085** (2018) 052009 doi:10.1088/1742-6596/1085/5/052009

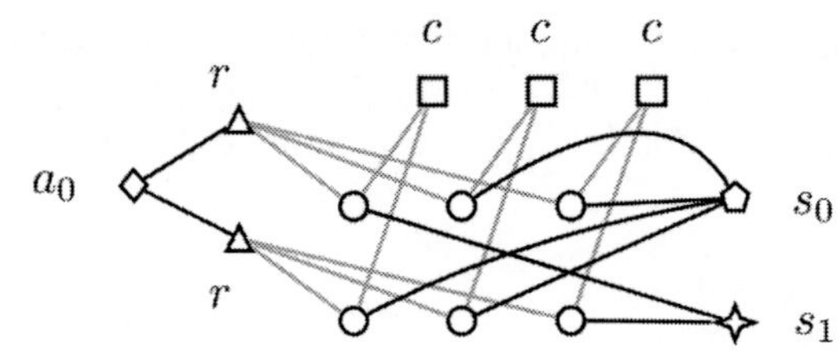


```
n=14 g
0: 5, 8; 1: 6, 9; 2: 7, 10; ! columns
3: 5, 6, 7; 4: 8, 9, 10;    ! rows
5: 12; 6: 11; 7: 11;
8: 11; 9: 11; 10: 12;
11: ; 12: ;                 ! exponents
13: 3, 4;                   ! coefficients
f=[0:2|3:4|5:10|11|12|13]
c x
z b
q
```

Figure 1. (Left panel) The graph corresponding to polynomial P_1 of Eq. 4 constructed as described in Section 4.1. (Right panel) The **dreadnaut** input corresponding to the graph.

(v) for each unique coefficient $0, \ldots, m$ in the coefficient list introduce a node of a new colour $a_0, \ldots, a_m$ and connect it to the row node of any term which has this coefficient.

To canonically label the graph we rely on the external program **dreadnaut** [41] which internally uses the graph algorithms of nauty and Traces. Usually the **dreadnaut** program can efficiently compute a canonical labelling of a graph, after which graphs that are isomorphic (the same except for vertex labels) become identical (exactly the same). Furthermore, after introducing a canonical labelling **dreadnaut** can output a graph hash (three 8-digit hex numbers). Isomorphic graphs always have the same hash, non-isomorphic graphs rarely have the same hash. These hashes can be used to quickly identify polynomials that may be identical. In Fig. 1 (right panel) we show the input to **dreadnaut** which performs these steps for the polynomial in Eq. 4; lines 1–8 define the graph, line 9 requests the canonical labelling, lines 10–11 instruct the program to output the graph hash and quit.

By default **dreadnaut** (and thus pySecDec) uses the McKay canonical graph labelling algorithm as implemented in nauty. Although this algorithm is known to have exponential complexity on some inputs [42, 43] it generally performs well and in practice we observe that it is typically much faster than a brute force algorithm.

4.2. Pak algorithm

An alternative method for identifying equivalent polynomials was described by Pak in Ref. [44]. Instead of mapping the polynomial to a graph, the algorithm works by permuting the columns of the exponent matrix (which corresponds to relabelling variables) in order to maximize some metric. The permutations which maximize the metric can be considered to produce a "canonical polynomial". The equivalence of polynomials up to relabelling variables can be established by comparing their "canonical" representation. The Pak algorithm is enabled by default in pySecDec 1.2.2, it can be disabled by setting `use_Pak = False` in the call to `make_package` or `loop_package`.

In principle this algorithm also has combinatorial complexity but beats a brute force algorithm by discarding more quickly obviously worse permutations.

4.3. Performance

In pySecDec the symmetry finding algorithms are applied first after primary sector decomposition and again after the complete sector decomposition. For benchmarking the symmetry finder we have selected 5 examples with between 2-4 loops and 3-4 legs. The examples triangle2L and triangle3L are taken from the pySecDec paper [18] and are described therein. In Figure 2 we depict the examples box2L, box3L and triangle4L which have been introduced for the present benchmark.

The number of symmetries identified by the symmetry finder depends strongly on the problem. In particular, with no numerator present, the number of unique primary sectors can

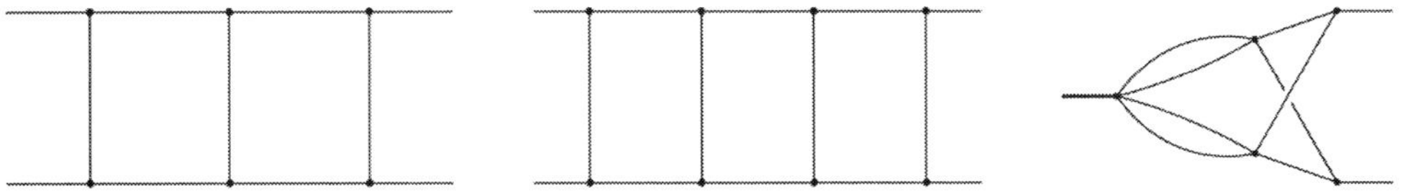

Figure 2. Examples box2L, box3L and triangle4L which are used to benchmark the performance of the symmetry finder algorithms. All internal propagators are taken to be massless and massive external legs are displayed in bold.

Table 1. The number of sectors before (total) and after (unique) applying the symmetry finder algorithms (`dreadnaut` or Pak) to each benchmark example. We show also the total wall time taken by each of the algorithms to identify all isomorphisms. All timings are taken using a single core of an Intel Core i7-7700 3.60 GHz CPU.

Example	Primary Sectors		Sectors		Time [s]	
	Total	Unique	Total	Unique	`dreadnaut`	Pak
triangle2L	6	4	34	21	1.9	0.1
triangle3L	7	3	448	124	7.0	0.54
triangle4L	8	3	2848	942	50	4.7
box2L	7	3	282	110	5.0	0.42
box3L	10	4	9414	3932	770	63

not be larger than the number of distinct graphs generated by pinching (removing) 1 propagator in each possible way. Many of the most complicated integrals appearing in Higgs boson pair production, Higgs boson plus jet production, and Higgs plus Z-boson production have few or no symmetries, though they typically have fewer sectors than the box2L example displayed here. For demonstration purposes, rather than selecting integrals from these processes, we have instead chosen integrals from the literature which have symmetries.

In Table 1, for each example, we display the total number of primary sectors after primary decomposition, the number of unique primary sectors (remaining after symmetries have been identified), the total number of sectors obtained without identifying any symmetries, and the the number of unique sectors (after symmetry finding). For all examples we use the default `iterative` sector decomposition strategy. When counting the total number of sectors without identifying symmetries we also do not identify symmetries for primary sectors. Both the graph-based `dreadnaut` algorithm and the Pak algorithm identify exactly the same sector symmetries and so the total number of sectors after symmetry finding does not depend on the method used.

In practice, within `pySecDec`, we usually first apply the quick `iterative_sort` algorithm followed by the `light_Pak_sort` algorithm then we apply a much slower full symmetry finder algorithm. In Table 1 we present timings produced running only the Pak or `dreadnaut` algorithm without first applying the quicker but imperfect algorithms. We observe that for these examples the Pak algorithm is between 10-20 times faster than the `dreadnaut` algorithm.

5. Conclusion

We have outlined some of the key difficulties obstructing the straightforward computation of 4-point 2-loop Higgs amplitudes. The most time consuming step in the methods that we have currently applied to the computation of these amplitudes is the use of integration-by-parts identities to reduce the number of Feynman integrals appearing. Another key difficulty

IOP Conf. Series: Journal of Physics: Conf. Series **1085** (2018) 052009 doi:10.1088/1742-6596/1085/5/052009

is the computation of the remaining integrals after this step. An overview of pySecDec, a program capable of computing such integrals numerically, was provided. The use of the recently introduced sector symmetry finder was described in detail. We have implemented two algorithms capable of identifying all sector symmetries, one is based on the canonical labelling of graphs and one, described previously by Pak, is based on the direct manipulation of the exponent list representation of multivariate polynomials. Although both algorithms identify precisely the same symmetries we find that the latter algorithm significantly outperforms the graph-based method for the examples considered here. The open source program pySecDec is publicly available and is provided with extensive documentation and usage guides.

Acknowledgements
We would like to thank the SecDec Collaboration: Sophia Borowka, Gudrun Heinrich, Stephan Jahn, Matthias Kerner, Johannes Schlenk and Tom Zirke for many interesting discussions and the fruitful collaboration. SPJ is supported by the Research Executive Agency (REA) of the European Union under the Grant Agreement PITN-GA2012316704 (HiggsTools).

References
[1] C. Anastasiou, C. Duhr, F. Dulat, et al. Phys. Rev. Lett. **114** (2015) 212001 [arXiv:1503.06056 [hep-ph]].
[2] T. Gehrmann, A. von Manteuffel and L. Tancredi, JHEP **1509** (2015) 128 [arXiv:1503.04812 [hep-ph]].
[3] A. von Manteuffel and L. Tancredi, JHEP **1506** (2015) 197 [arXiv:1503.08835 [hep-ph]].
[4] F. Caola, J. M. Henn, K. Melnikov, et al. JHEP **1506** (2015) 129 [arXiv:1503.08759 [hep-ph]].
[5] R. Bonciani, V. Del Duca, H. Frellesvig, et al. JHEP **1612** (2016) 096 [arXiv:1609.06685 [hep-ph]].
[6] R. Bonciani, S. Di Vita, P. Mastrolia and U. Schubert, JHEP **1609** (2016) 091 [arXiv:1604.08581 [hep-ph]].
[7] A. von Manteuffel and R. M. Schabinger, JHEP **1704** (2017) 129 [arXiv:1701.06583 [hep-ph]].
[8] S. Di Vita, P. Mastrolia, A. Primo and U. Schubert, JHEP **1704** (2017) 008 [arXiv:1702.07331 [hep-ph]].
[9] P. Mastrolia, M. Passera, A. Primo and U. Schubert, arXiv:1709.07435 [hep-ph].
[10] S. Badger, H. Frellesvig and Y. Zhang, JHEP **1312** (2013) 045 [arXiv:1310.1051 [hep-ph]].
[11] T. Gehrmann, J. M. Henn and N. A. Lo Presti, Phys. Rev. Lett. **116** (2016) 062001 Erratum: [Phys. Rev. Lett. **116** (2016) 189903] [arXiv:1511.05409 [hep-ph]].
[12] C. G. Papadopoulos, D. Tommasini and C. Wever, JHEP **1604** (2016) 078 [arXiv:1511.09404 [hep-ph]].
[13] D. C. Dunbar, J. H. Godwin, G. R. Jehu and W. B. Perkins, arXiv:1710.10071 [hep-th].
[14] J. M. Henn, Phys. Rev. Lett. **110** (2013) 251601 [arXiv:1304.1806 [hep-th]].
[15] S. Abreu, F. Febres Cordero, H. Ita, et al. Phys. Rev. Lett. **119** (2017) 142001 [arXiv:1703.05273 [hep-ph]].
[16] S. Borowka, N. Greiner, G. Heinrich, et al. Phys. Rev. Lett. **117** (2016) 012001 Erratum: [Phys. Rev. Lett. **117** (2016) 079901] [arXiv:1604.06447 [hep-ph]].
[17] S. Borowka, N. Greiner, G. Heinrich, et al. JHEP **1610** (2016) 107 [arXiv:1608.04798 [hep-ph]].
[18] S. Borowka, G. Heinrich, S. Jahn, et al. Comput. Phys. Commun. (2017) [arXiv:1703.09692 [hep-ph]].
[19] K. G. Chetyrkin and F. V. Tkachov, Nucl. Phys. B **192** (1981) 159.
[20] S. Laporta, Int. J. Mod. Phys. A **15** (2000) 5087 [hep-ph/0102033].
[21] A. von Manteuffel and R. M. Schabinger, Phys. Lett. B **744** (2015) 101 [arXiv:1406.4513 [hep-ph]].
[22] H. Ita, Phys. Rev. D **94** (2016) 116015 [arXiv:1510.05626 [hep-th]].
[23] K. J. Larsen and Y. Zhang, Phys. Rev. D **93** (2016) 041701 [arXiv:1511.01071 [hep-th]].
[24] P. Maierhoefer, J. Usovitsch and P. Uwer, arXiv:1705.05610 [hep-ph].
[25] R. Bonciani, V. Del Duca, H. Frellesvig, et al. JHEP **1508** (2015) 108 [arXiv:1505.00567 [hep-ph]].
[26] Python project, "Python" [software], version 3.6.3, 2017. Available from https://www.python.org/downloads/release/python-363/ [accessed 2017-11-10].
[27] Numpy package, "Numpy" [software], version 1.13.3, 2017. Available from https://pypi.python.org/pypi/numpy/1.13.3/ [accessed 2017-11-10].
[28] Sympy package, "Sympy" [software], version 1.1.1, 2017. Available from https://github.com/sympy/sympy/releases/tag/sympy-1.1.1/ [accessed 2017-11-10].
[29] J. A. M. Vermaseren, math-ph/0010025.
[30] J. Kuipers, T. Ueda and J. A. M. Vermaseren, Comput. Phys. Commun. **189** (2015) 1 [arXiv:1310.7007 [cs.SC]].
[31] B. Ruijl, T. Ueda and J. Vermaseren, arXiv:1707.06453 [hep-ph].
[32] Sphinx package, "Sphinx" [software], version 1.6.5, 2017. Available from https://pypi.python.org/pypi/Sphinx/1.6.5/ [accessed 2017-11-10].

[33] Programming Language C++, ISO/IEC 14882:2011.
[34] H. Cheng and T. T. Wu, Cambridge, USA: MIT-PR (1987) 285p
[35] V. A. Smirnov, Berlin, Germany: Springer (2006) 283 p
[36] G. Heinrich, Int. J. Mod. Phys. A **23** (2008) 1457 [arXiv:0803.4177 [hep-ph]].
[37] S. Borowka, G. Heinrich, S. P. Jones, et al. Comput. Phys. Commun. **196** (2015) 470 [arXiv:1502.06595 [hep-ph]].
[38] T. Hahn, Comput. Phys. Commun. **168** (2005) 78 [hep-ph/0404043].
[39] P. Gonnet, CoRR abs/1006.3962 (2010).
[40] M. Galassi et al., GNU Scientific Library Reference Manual - Third Edition. Network Theory Ltd., 3rd ed., 2009.
[41] B. D. McKay and A. Piperno, Journal of Symbolic Computation, 60, 94-112, 2014.
[42] M. Fürer, Tech. Rep. CS-87-36, Department of Computer Science, The Pennsylvania State University, University Park, 1987.
[43] T. Miyazaki, DIMACS Ser. Discrete Math. Theoret. Comput. Sci., 28, Amer. Math. Soc., Providence, RI, 239-256, 1997.
[44] A. Pak, J. Phys. Conf. Ser. **368** (2012) 012049 [arXiv:1111.0868 [hep-ph]].

IOP Conf. Series: Journal of Physics: Conf. Series **1085** (2018) 052010 doi:10.1088/1742-6596/1085/5/052010

On a Java library to perform S-expansions of Lie algebras

Carlos Inostroza[1], **Igor Kondrashuk**[2], **Nelson Merino**[3] **and Felip Nadal**[4]

[1] Departamento de Física, Universidad de Concepción, Casilla 160-C, Concepción, Chile
[2] Grupo de Matemática Aplicada, Departamento de Ciencias Básicas, Univerdidad del Bío-Bío, Campus Fernando May, Casilla 447, Chillán, Chile
[3] APC, Universite Paris Diderot, 10, rue Alice Domon et Leonie Duquet, 75205 Paris Cedex 13, France
[4] Instituto de Física Corpuscular (IFIC), Edificio Institutos de Investigación, c/ Catedrático José Beltrán, 2, E-46980 Paterna, España

E-mail: nemerino@gmail.com

Abstract. The S-expansion method is a generalization of the Inönü-Wigner (IW) contraction that allows to study new non-trivial relations between different Lie algebras. Basically, this method combines a Lie algebra $\mathcal{G}$ with a finite abelian semigroup S in such a way that a new S-expanded algebra $\mathcal{G}_S$ can be defined. When the semigroup has a zero-element and/or a specific decomposition, which is said to be resonant with the subspace structure of the original algebra, then it is possible to extract smaller algebras from $\mathcal{G}_S$ which have interesting properties. Here we give a brief description of the S-expansion, its applications and the main motivations that lead us to elaborate a Java library, which automatizes this method and allows us to represent and to classify all possible S-expansions of a given Lie algebra.

1. Introduction

As is well-known, the theory of Lie groups and algebras plays an essential role in Physics: it represents the mathematical tool allowing to describe the continues symmetries of a physical system and, via the Noether theorem, this is directly connected with the corresponding conservation laws of the system. Since the fifties, different mechanisms allowing to establish non-trivial relations between Lie algebras appeared and were very useful to understand interrelations between different physical theories. The original idea was introduced in Ref. [1], where it was pointed out that if two physical theories are related by means of a limit process (like Newtonian mechanics and special relativity which are related by the limit where the speed of light c goes to infinite), then the corresponding symmetry groups (the Galilean and Poincaré groups, in the example) under which those theories are invariant should also be related by means of a similar limit process. This process was formally introduced in Ref. [2, 3] and it is known nowdays as Inönü-Wigner (IW) contraction.

Many generalized contractions were introduced in the literature during the last decades. In particular, the Weimar-Woods (WW) contraction [4, 5, 6] is one of the most general realization of this method. Starting with an algebra which have a certain subspace structure, the WW contraction performs a suitable rescaling with a real parameter on the generators of each subspace. Then, a special limit for that parameter leads to the contracted algebra, which has the same dimension than the original one, but very different properties[1].

Another peculiar generalization of the contraction, known as *expansion* method, was parallelly introduced in the context of string theory [7] and supergravity [8]. This procedure not only is able to reproduce the WW contractions when the dimension is preserved in the process, but also may lead to expanded algebras whose dimension is higher than the original one. The main difference with the contraction method is that, using the dual description of a Lie algebra in terms of Maurer-Cartan (MC) forms, the rescaling is performed on some coordinates of the Lie group manifold and not on the generators of the Lie algebra.

Here we focus on an even more general procedure called *S-expansion*[2] [9], which combines the structure constants of the original algebra with the inner multiplication law of an abelian finite semigroup of order n to define a new *S-expanded algebra* $\mathcal{G}_S$. When the semigroup have a zero-element and/or a specific decomposition, which is said to be resonant with the subspace structure of the original algebra, then it is possible to extract smaller algebras, called *resonant subalgebras* and *reduced algebras*. In particular, the previous expansion method [7, 8] can be reproduced as a 0_S-reduction of the resonant subalgebra for an expansion with a special family of semigroups denoted by $S_E^{(N)}$.

During the last decade, many applications using the S-expansion has been performed [14, 15, 16, 17, 18, 19, 20, 21], mainly in the context of modified theories of gravity. At the begining, they considered only the family of semigroups $S_E^{(N)}$, which lead to the definition of the so called $\mathfrak{B}_N$ algebras. The use of other abelian semigroups to perform S-expansions of Lie algebras was first considered in Refs. [22], where it was pointed out that possible new applications could be performed if the following question is analyzed: *given two Lie algebras, is it possible to find a suitable semigroup that relates them by means of an S-expansion?* Of course, the answer depends on the specific algebras that we want to connect and involves to consider all possible finite abelian semigroups in each order[3] (for a brief review about that classification see the presentation [24], by C. Inostroza). In the same line, a study of the general properties S-expansion method (in the context of the classification of Lie algebras) was made in Ref. [25] and revealed that some properties of the semigroup allow to determine if a given property of the original algebra (like semisimplicity, compactness, etc) will be preserved or not under the expansion process. These results were shown to be useful as criteria to answer if two given algebras can be S-related. In particular, semigroups preserving semisimplicity were identified.

The results given in Refs. [22] and [25] (and in particular the use of semigroups preserving semisimplicity) were used in Ref. [26] to show that the semisimple version of the so called Maxwell algebra (introduced in [27]) can be obtained as an expansion of the AdS algebra. Later, this result was generalized in Refs. [28, 29] to new families of semigroups generating algebras denoted by $\mathfrak{C}_N$ and $\mathfrak{D}_N$ which have been useful to construct new (super)gravity models [30, 31, 32, 33, 34, 35, 36, 37, 38, 39, 40, 41, 42, 43, 44, 45, 46]. These several new applications show the importance of considering also semigroups outside of the $S_E^{(N)}$ family. Motivated by

[1] For example, if the original algebra is semisimple, then the contracted algebra is usually non-semisimple.

[2] The dual formulation of the S-expansion in terms of MC forms was performed in [10], while in Refs. [11, 12, 13] the method was also extended to other mathematical structures, like the case of higher order Lie algebras and infinite dimensional loop algebras.

[3] An example, it was considered the 2 and 3-dimensional that acts transitively on 2 and 3-dimensional spaces, i.e., the two 2-dimensional algebras $[X_1, X_2] = 0$, $[X_1, X_2] = X_1$ and the ten 3-dimensional algebras classified by Bianchi [23]. It was shown, that only four Bianchi algebras can be obtained as S-expansions of the the 2-dimensional algebras if one uses some specific semigroups of order 4, which do not belong in general to the $S_E^{(N)}$ family. Thus, this result cannot be obtained neither by a contraction nor by the expansion method [8]. The procedure used to construct by hand the multiplication table of those semigroups also made clear that, if a given problem involves the use of semigroups of higher order, then the use of computer programs is needed.

IOP Conf. Series: Journal of Physics: Conf. Series **1085** (2018) 052010 doi:10.1088/1742-6596/1085/5/052010

this, we have constructed a Java library [47, 48] that automatizes the S-expansion procedure and that is aimed to create a general picture about all possible S-relations between Lie algebras.

2. Brief review of the S-expansion method

Consider a Lie algebra $\mathcal{G}$ with generators $\{X_i\}$ and Lie product $[X_i, X_j] = C_{ij}^k X_k$ where C_{ij}^k are the structure constants. Consider also a finite abelian semigroup $S = \{\lambda_\alpha, \alpha = 1, \ldots, n\}$. An informal way (but useful for our purposes) of expressing the semigroup multiplication law is by means of quantities called *selectors*[4], denoted by $K_{\alpha\beta}^{\kappa}$ and defined by the relation $\lambda_\alpha \cdot \lambda_\beta = \lambda_{\gamma(\alpha,\beta)} = K_{\alpha\beta}^{\rho}\lambda_\rho$, where $K_{\alpha\beta}^{\rho} = 1$ if $\rho = \gamma(\alpha, \beta)$ and $K_{\alpha\beta}^{\rho} = 0$ if $\rho \neq \gamma(\alpha, \beta)$. Then, the S-expansion consists of the following steps.

Step I: Constructing S-expanded algebra. As shown in [9], the set $\mathcal{G}_S = S \otimes \mathcal{G}$ (with $\otimes$ being the Kronecker product between the matrix representation of S and $\mathcal{G}$) is also Lie algebra, which is called *expanded algebra*, if the basis elements are defined as $X_{(i,\alpha)} \equiv \lambda_\alpha \otimes X_i$ and Lie product by $\left[X_{(i,\alpha)}, X_{(j,\beta)}\right] \equiv \lambda_\alpha \cdot \lambda_\beta \otimes [X_i, X_j] = C_{(i,\alpha)(j,\beta)}^{(k,\gamma)} X_{(k,\gamma)}$. The structure constants of the expanded algebra $\mathcal{G}_S$ are fully determined by the selectors and the structure constants of the original Lie algebra $\mathcal{G}$, i.e., $C_{(i,\alpha)(j,\beta)}^{(k,\gamma)} = K_{\alpha\beta}^{\gamma} C_{ij}^k$.

Step II: Extraction of the resonant subalgebra. Consider the case where the original algebra has the subspace decomposition $\mathcal{G} = V_0 \oplus V_1$ with the following structure

$$[V_0, V_0] \subset V_0\,, \quad [V_0, V_1] \subset V_1\,, \quad [V_1, V_1] \subset V_0 \tag{1}$$

Suppose also that a given semigroup has a decomposition in subsets $S = S_0 \cup S_1$ satisfying

$$S_0 \cdot S_0 \in S_0\,, \quad S_0 \cdot S_1 \in S_1\,, \quad S_1 \cdot S_1 \in S_0 \tag{2}$$

which is called *resonant condition*, because of the similarity with the subspace structure (1) of the algebra[5]. Then, as shown in [9], the set $\mathcal{G}_{S,R} = (S_0 \otimes V_0) \oplus (S_1 \otimes V_1)$ is a subalgebra of the expanded algebra $\mathcal{G}_S = S \otimes \mathcal{G}$.

Step III: Extraction of the 0_S-reduced algebra. If semigroup contains an element 0_S satisfying $\lambda_\alpha \cdot 0_S = 0_S$ for any element $\lambda_\alpha \in S$, then this element is called a zero element. In that case, the sector $0_S \otimes \mathcal{G}$ can be removed from the expanded algebra in such a way that what is left is also a Lie algebra, called the *0_S-reduced algebra*. Remarkably, the reduced algebra is not necessarily a subalgebra of the expanded algebra.

3. Need of automizing the procedure

First we notice that the steps II and III are independent, but can also be applied simultaneously. This means that, depending on the semigroup that is going to be used in the expansion, the following algebras can be obtained:

(i) the expanded algebra $\mathcal{G}_S = S \otimes \mathcal{G}$,

(ii) the resonant subalgebra $\mathcal{G}_{S,R} = (S_0 \otimes V_0) \oplus (S_1 \otimes V_1)$, if the semigroup have at least one resonant decomposition,

(iii) the 0_S-reduced algebra if the semigroup have a zero element 0_S,

[4] In particular, the selectors provide a matrix representation for the semigroup $\lambda_\alpha \to (\lambda_\alpha)^{\rho}{}_{\beta} = K_{\alpha\beta}^{\rho}$, which is used in our library to represent a given semigroup.

[5] As it can be seen in [9], one can deal with algebras and semigroups having decompositions which are more general, but in the first version of our library we only consider the case given by (1) and (2).

(iv) the 0_S-reduction of the resonant subalgebra, if the semigroup has simultaneously a zero element and at least one resonant decomposition.

In order to study all possible S-expansions (i-iv) of a given algebra, one should consider the full set of abelian semigroups. As it will be reviewed in the presentation by C. Inostroza [24], the problem of enumerating the all non-isomorphic finite semigroups of a certain order is a non-trivial problem because the number of semigroups increases very quickly with the order of the semigroup. Thus, this task can be performed only up to a certain order and, in particular, we have used the program *gen.f* of Ref. [49] to generate the files *sem.2*, *sem.3*, *sem.4*, *sem.5* and *sem.6* which contain all the non isomorphic semigroups up to order 6. Those files can be used as input data for many of the programs that compose our library [47, 48].

For example, in the order 3 there are 18 non-isomorphic semigroups denoted by $S^a_{(3)}$ with $a = 1, ...18$, from which only 12 of them are abelian[6]. Thus, the different type of expansions that can be done with them are represented in the figure 1 where, for reasons of space, we use only the label 'a' to name the different semigroups in the horizontal axis, while in the vertical axis we represent the different kinds of expansions that can be performed with them. With a *gray number* it has also been identified the expansions that preserve the semisimplicity.

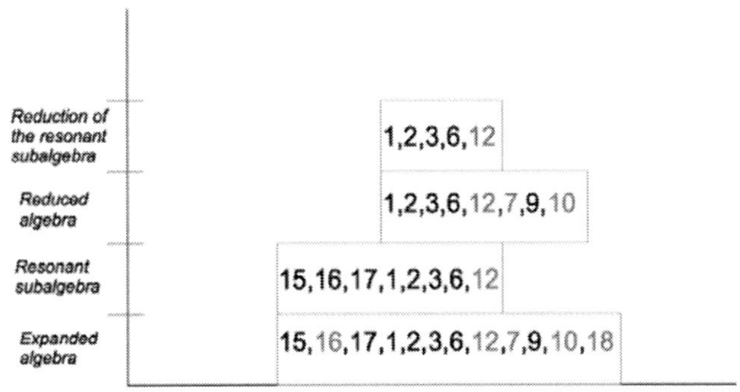

Figure 1. Representing different kind of expansions with semigroups of order 3.

To identify all possible resonant decompositions of all non-isomorphic semigroups in higher orders is not a trivial task. For this reason, we have constructed a Java library [47, 48] which allows us to study all possible S-expansions of the type (i-iv) using the mentioned lists *sem.n*.

[6] The non-isomorphic abelian semigroups of order 3 are:

$S^1_{(3)}$	λ_1	λ_2	λ_3
λ_1	λ_1	λ_1	λ_1
λ_2	λ_1	λ_1	λ_1
λ_3	λ_1	λ_1	λ_1

,

$S^2_{(3)}$	λ_1	λ_2	λ_3
λ_1	λ_1	λ_1	λ_1
λ_2	λ_1	λ_1	λ_1
λ_3	λ_1	λ_1	λ_2

,

$S^3_{(3)}$	λ_1	λ_2	λ_3
λ_1	λ_1	λ_1	λ_1
λ_2	λ_1	λ_1	λ_1
λ_3	λ_1	λ_1	λ_3

,

$S^6_{(3)}$	λ_1	λ_2	λ_3
λ_1	λ_1	λ_1	λ_1
λ_2	λ_1	λ_1	λ_2
λ_3	λ_1	λ_2	λ_3

,

$S^7_{(3)}$	λ_1	λ_2	λ_3
λ_1	λ_1	λ_1	λ_1
λ_2	λ_1	λ_2	λ_1
λ_3	λ_1	λ_1	λ_3

,

$S^9_{(3)}$	λ_1	λ_2	λ_3
λ_1	λ_1	λ_1	λ_1
λ_2	λ_1	λ_2	λ_2
λ_3	λ_1	λ_2	λ_2

,

$S^{10}_{(3)}$	λ_1	λ_2	λ_3
λ_1	λ_1	λ_1	λ_1
λ_2	λ_1	λ_2	λ_2
λ_3	λ_1	λ_2	λ_3

,

$S^{12}_{(3)}$	λ_1	λ_2	λ_3
λ_1	λ_1	λ_1	λ_1
λ_2	λ_1	λ_2	λ_3
λ_3	λ_1	λ_3	λ_2

,

$S^{15}_{(3)}$	λ_1	λ_2	λ_3
λ_1	λ_1	λ_1	λ_3
λ_2	λ_1	λ_1	λ_3
λ_3	λ_3	λ_3	λ_1

,

$S^{16}_{(3)}$	λ_1	λ_2	λ_3
λ_1	λ_1	λ_1	λ_3
λ_2	λ_1	λ_2	λ_3
λ_3	λ_3	λ_3	λ_1

,

$S^{17}_{(3)}$	λ_1	λ_2	λ_3
λ_1	λ_1	λ_2	λ_2
λ_2	λ_2	λ_1	λ_1
λ_3	λ_2	λ_1	λ_1

,

$S^{18}_{(3)}$	λ_1	λ_2	λ_3
λ_1	λ_1	λ_2	λ_3
λ_2	λ_2	λ_3	λ_1
λ_3	λ_3	λ_1	λ_2

,

4. On the maximal order of semigroups which the Java Library can perform

In some cases it is useful to have the full lists of non-isomorphic semigroups up to certain order. However, there is an intrinsic computational limitation to construct these lists for order 8 and higher. Indeed, the number of non-isomorphic semigroups in those orders has been evaluated only by using indirect techniques. For lower orders, instead, the construction of these tables is possible. For example, in Ref. [49] a fortran program gen.f was proposed, which, according to what is claimed in this reference, is able to generate these lists up to order 7. However, after running that fortran program we were able to obtain those lists only up to order 6. These lists are used as inputs for our library [47] although its methods are not restricted to the order 6. Indeed, the methods of our library also allow us to work with semigroup of higher orders. The only issue is that we do not have the full list of non-isomorphic tables for those higher orders. For example, our methods are able to read a multiplication table of order 20 and show in few seconds if that table is associative, commutative, to identify if there is a zero element and also allow to check if a given decomposition is resonant. The only task that involves more computations and time is the method *findAllResonances* that finds out all possible resonant decompositions of a given semigroup. In that case, one can obtain the answer in few minutes for semigroups of order 8, during about one hour for semigroups of order 9, and in our case we were unable to obtain an answer over 3 days for semigroups of order 10. All the computations above were made in usual laptop computers.

5. Conclusions

We have presented a brief description of the S-expansion method, its applications and the motivations to automatize this procedure. We have made this with a set of methods organized in the form of Java Library [47, 48] which is able not only to represent and perform basic operations with semigroups and Lie algebras but also to perform operations that are intrinsic of the S-expansion method, such as the characterization of resonances and the corresponding representation of the resonant subalgebra and the reduced algebras. The main features of this library will be described in the related talk by C. Inostroza [24].

Acknowledgments

We thank Andrei Kataev for his invitation to present this work at ACAT 2017. C.I. was supported by a Mecesup PhD grant and the Término de tesis grant from CONICYT (Chile). C.I. is very grateful to Local Organizing Committee and in particular to Gordon Watts for the financing his participation in ACAT. I.K. was supported by was supported by Fondecyt (Chile) grant 1050512 and by DIUBB (Chile) Grant Nos. 102609 and GI 153209/C. NM is supported by a Becas-Chile postdoctoral grant.

References

[1] Segal I E 1951 A class of operator algebras which are determined by groups *Duke Math. J.* **18** 221-65
[2] Inönü E and Wigner E P 1953 On the contraction of groups and their representations *Proc. Nat. Acad. Sci. USA* **39** 510-24
[3] Inönü E 1964 Contractions of Lie groups and their representations *Group theoretical concepts in elementary particle physics*, ed Gürsey F, (New York: Gordon and Breach) pp 391-402
[4] Weimar-Woods E 1991 Contractions of Lie algebras: generalized Inonu-Wigner contractions versus graded contractions *J. Math. Phys.* **32** 2028
[5] Weimar-Woods E 1995 Contractions of Lie algebras: generalized Inonu-Wigner contractions versus graded contractions *J. Math. Phys.* **36** 4519-48
[6] Weimar-Woods E 2000 Contractions, generalized Inönü and Wigner contractions and deformations of finite-dimensional Lie algebras *Rev. Math. Phys.* **12** 1505-29
[7] Hatsuda M and Sakaguchi M 2003 Wess-Zumino term for the AdS superstring and generalized Inonu-Wigner contraction *Prog. Theor. Phys.* **109** 853 (*Preprint* hep-th/0106114)

[8] de Azcarraga J A, Izquierdo J M, Picon M and Varela O 2003 Generating Lie and gauge free differential (super)algebras by expanding Maurer-Cartan forms and Chern-Simons supergravity *Nucl. Phys.* B **662** 185 (*Preprint* hep-th/0212347)
[9] Izaurieta F, Rodriguez E and Salgado P 2006 Expanding Lie (super)algebras through Abelian semigroups *J. Math. Phys.* **47** 123512 (*Preprint* hep-th/0606215)
[10] Izaurieta F, Rodriguez E, Perez A and Salgado P 2009 Dual Formulation of the Lie Algebra S-expansion Procedure *J. Math. Phys.* **50** 073511 (*Preprint* arXiv:0903.4712 [hep-th])
[11] Caroca R, Merino N and Salgado P 2009 S-Expansion of Higher-Order Lie Algebras *J. Math. Phys.* **50** 013503 (*Preprint* arXiv:1004.5213 [math-ph])
[12] Caroca R, Merino N, Perez A and Salgado P 2009 Generating Higher-Order Lie Algebras by Expanding Maurer Cartan Forms *J. Math. Phys.* **50** 123527 (*Preprint* arXiv:1004.5503 [hep-th])
[13] Caroca R, Merino N, Salgado P and Valdivia O 2011 Generating infinite-dimensional algebras from loop algebras by expanding Maurer-Cartan forms *J. Math. Phys.* **52** 043519
[14] Izaurieta F, Rodriguez E and Salgado P 2008 Eleven-dimensional gauge theory for the M algebra as an Abelian semigroup expansion of osp(32—1) *Eur. Phys. J.* C **54** 675 (*Preprint* hep-th/0606225)
[15] Izaurieta F, Rodriguez E, Minning P, Salgado P and Perez A 2009 Standard General Relativity from Chern-Simons Gravity *Phys. Lett.* B **678** 213 (*Preprint* arXiv:0905.2187 [hep-th])
[16] Quinzacara C A C and Salgado P 2012 Black hole for the Einstein-Chern-Simons gravity *Phys. Rev.* D **85** 124026 (*Preprint* arXiv:1401.1797 [gr-qc])
[17] Concha P K, Peñafiel D M, Rodríguez E K and Salgado P 2013 Even-dimensional General Relativity from Born-Infeld gravity *Phys. Lett.* B **725** 419 (*Preprint* arXiv:1309.0062 [hep-th])
[18] Concha P K, Peñafiel D M, Rodríguez E K and Salgado P 2015 Generalized Poincaré algebras and Lovelock-Cartan gravity theory *Phys. Lett.* B **742** 310 (*Preprint* arXiv:1405.7078 [hep-th])
[19] Quinzacara C A C and Salgado P 2013 Stellar equilibrium in Einstein-Chern-Simons gravity *Eur. Phys. J.* C **73** 2479
[20] Crisóstomo J, Gómez F, Salgado P, Quinzacara C, Cataldo M and del Campo S 2014 Accelerated FRW Solutions in Chern-Simons Gravity *Eur. Phys. J.* C **74** 3087 (*Preprint* arXiv:1401.2128 [gr-qc])
[21] Crisóstomo J, Gómez F, Quinzacara C and Salgado P 2016 Static solutions in Einstein-Chern-Simons gravity *J. Cosmol. Astropart. Phys.* JCAP 1606 (2016) no.06, 049 (*Preprint* arXiv:1601.06592 [gr-qc])
[22] Caroca R, Kondrashuk I, Merino N and Nadal F 2013 Bianchi spaces and their three-dimensional isometries as S-expansions of two-dimensional isometries *J. Phys.* A **46** 225201 (*Preprint* arXiv:1104.3541 [math-ph])
[23] Bianchi L 1898 Sugli spazi a tre dimensioni che ammettono un gruppo continuo di movimenti *Memorie di Matematica e di Fisica della Societa Italiana delle Scienze* Serie Terza **Tomo XI** 267-352
[24] Inostroza C, Kondrashuk I, Merino N and Nadal F 2017 On the algorithm to find S-related Lie algebras *Merged talk by C Inostroza at ACAT 2017*
[25] Andrianopoli L, Merino N, Nadal F and Trigiante M 2013 General properties of the expansion methods of Lie algebras *J. Phys.* A **46** 365204 (*Preprint* arXiv:1308.4832 [gr-qc])
[26] Díaz J D, Fierro O, Izaurieta F, Merino N, Rodríguez E, Salgado P and Valdivia O 2012 A generalized action for (2 + 1)-dimensional Chern-Simons gravity *J. Phys.* A **45** 255207 (*Preprint* arXiv:1311.2215 [gr-qc])
[27] Soroka D V and Soroka V A 2009 Semi-simple extension of the (super)Poincare algebra *Adv. High Energy Phys.* 234147 (*Preprint* hep-th/0605251)
[28] Salgado P and Salgado S 2014 $\mathfrak{so}(D-1,1)\otimes\mathfrak{so}(D-1,2)$ algebras and gravity *Phys. Lett.* B **728** 5
[29] Concha P K, Durka R, Merino N and Rodríguez E K 2016 New family of Maxwell like algebras *Phys. Lett.* B **759** 507 (*Preprint* arXiv:1601.06443 [hep-th])
[30] Salgado P, Szabo R J and Valdivia O 2014 Topological gravity and transgression holography *Phys. Rev.* D **89** 084077 (*Preprint* arXiv:1401.3653 [hep-th])
[31] Fierro O, Izaurieta F, Salgado P and Valdivia O 2014 (2+1)-dimensional supergravity invariant under the AdS-Lorentz superalgebra (*Preprint* arXiv:1401.3697 [hep-th])
[32] Concha P K, Peñafiel D M, Rodríguez E K and P. Salgado P 2014 Chern-Simons and Born-Infeld gravity theories and Maxwell algebras type *Eur. Phys. J.* C **74** 2741 (*Preprint* arXiv:1402.0023 [hep-th])
[33] Concha P K and Rodríguez E K 2014 Maxwell Superalgebras and Abelian Semigroup Expansion *Nucl. Phys.* B **886** 1128 (*Preprint* arXiv:1405.1334 [hep-th])
[34] Concha P K and Rodríguez E K 2014 N = 1 Supergravity and Maxwell superalgebras *J. High Energy Phys.* JHEP09(2014)090 (*Preprint* arXiv:1407.4635 [hep-th])
[35] González N, Salgado P, Rubio G and Salgado S 2014 Einstein-Hilbert action with cosmological term from Chern-Simons gravity *J. Geom. Phys.* **86** 339
[36] Concha P K, Rodríguez E K and Salgado P 2015 Generalized supersymmetric cosmological term in $N=1$ Supergravity *J. High Energy Phys.* JHEP08(2015)009 (*Preprint* arXiv:1504.01898 [hep-th])
[37] Concha P K, Fierro O, Rodríguez E K and Salgado P 2015 Chern-Simons supergravity in $D=3$ and Maxwell

IOP Conf. Series: Journal of Physics: Conf. Series **1085** (2018) 052010 doi:10.1088/1742-6596/1085/5/052010

superalgebra *Phys. Lett.* B **750** 117 (*Preprint* arXiv:1507.02335 [hep-th])
[38] Concha P K, Durka R, Inostroza C, Merino N and Rodríguez E K 2016 Pure Lovelock gravity and Chern-Simons theory *Phys. Rev.* D **94** 024055 (*Preprint* arXiv:1603.09424 [hep-th])
[39] González N, Rubio G, Salgado P and Salgado S 2016 Generalized Galilean algebras and Newtonian gravity *Phys. Lett.* B **755** 433 (*Preprint* arXiv:1604.06313 [hep-th])
[40] Durka R 2017 Resonant algebras and gravity *J. Phys.* A **50** 145202 (*Preprint* arXiv:1605.00059 [hep-th])
[41] Concha P K, Merino N and Rodríguez E K 2017 Lovelock gravities from Born-Infeld gravity theory *Phys. Lett.* B **765** 395-401 (*Preprint* arXiv:1606.07083 [hep-th])
[42] Ipinza M C, Concha P K, Ravera L and Rodríguez E K 2016 On the Supersymmetric Extension of Gauss-Bonnet like Gravity *J. High Energy Phys.* JHEP09(2016)007 (*Preprint* arXiv:1607.00373 [hep-th])
[43] Concha P K, Fierro O and Rodríguez E K 2016 Inönü-Wigner Contraction and $D = 2 + 1$ Supergravity *Eur. Phys. J.* C **77** 48 (*Preprint* arXiv:1611.05018 [hep-th])
[44] Peñafiel D M and Ravera L 2017 Generalized Inönü-Wigner Contraction as S-Expansion with Infinite Semigroup and Ideal Subtraction *J. Math. Phys.* **58** 081701 (*Preprint* arXiv:1611.05812 [hep-th])
[45] Artebani M, Caroca R, Ipinza M C, Peñafiel D M and Salgado P 2016 Geometrical aspects of the Lie algebra S-expansion procedure *J. Math. Phys.* **57** 023516 (*Preprint* arXiv:1602.04525 [math-ph])
[46] Ipinza M C, Lingua F, Peñafiel D M and Ravera L 2016 An Analytic Method for S-Expansion involving Resonance and Reduction *Fortsch. Phys.* **64** 854-80 (*Preprint* arXiv:1609.05042 [hep-th])
[47] https://github.com/SemigroupExp/Sexpansion/releases/tag/v1.0.0
[48] Inostroza C, Kondrashuk I, Merino N and Nadal F 2017 A Java library to perform S-expansions of Lie algebras *Preprint* arXiv:1703.04036 [cs.MS]
[49] Hildebrant J 2001 Handbook of Finite Semigroup Programs *LSU Mathematics Electronic Preprint Series Preprint* 2001-24

IOP Conf. Series: Journal of Physics: Conf. Series **1085** (2018) 052011 doi:10.1088/1742-6596/1085/5/052011

On the algorithm to find S-related Lie algebras

Carlos Inostroza[1], **Igor Kondrashuk**[2], **Nelson Merino**[3] **and Felip Nadal**[4]

[1] Departamento de Física, Universidad de Concepción, Casilla 160-C, Concepción, Chile
[2] Grupo de Matemática Aplicada, Departamento de Ciencias Básicas, Univerdidad del Bío-Bío, Campus Fernando May, Casilla 447, Chillán, Chile
[3] APC, Universite Paris Diderot, 10, rue Alice Domon et Leonie Duquet, 75205 Paris Cedex 13, France
[4] Instituto de Física Corpuscular (IFIC), Edificio Institutos de Investigación, c/ Catedrático José Beltrán, 2, E-46980 Paterna, España

E-mail: nemerino@gmail.com

Abstract. In this article we describe the Java library that we have recently constructed to automatize the S-expansion method, a powerful mathematical technique allowing to relate different Lie algebras. An important input in this procedure is the use of abelian semigroups and thus, we start with a brief review about the classification of non-isomorphic semigroups made in the literature during the last decades, and explain how the lists of non-isomorphic semigroups up to order 6 can be used as inputs in many of the methods of our library. After describing the main features of the classes that compose our library we present a new method called *fillTemplate* which tuns out to be very useful to answer whether two given algebras can be S-related.

1. Introduction

The S-expansion [1] is a powerful mathematical technique allowing to relate different Lie algebras, which contains as a particular case the Inönü-Wigner contraction [2, 3] and its generalizations [4, 5, 6]. This method also turns out to be a valuable tool to study the relations between different physical theories. Since we already gave a technical description of the S-expansion in the related proceeding contribution [7], as well as a review about its different physical applications, here we begin by summarizing the main ingredients needed to understand how the Java library [8] works. We have constructed this library to automatize the S-expansion procedure.

Basically, the S-expansion method combines the structure constants of a Lie algebra with the inner multiplication law of an abelian semigroup in such a way that leads to the definition of other Lie algebras, called *S-expanded* algebras. The inputs are:

- a Lie algebra $\mathcal{G}$ with generators $\{X_i\}$ and Lie product $[X_i, X_j] = C_{ij}^k X_k$ where C_{ij}^k are the structure constants (convention sum on repeated indices is assumed).
- an abelian finite semigroup $S = \{\lambda_\alpha, \alpha = 1, \ldots, n\}$, whose inner operation can be represented either by its multiplication table $A = (a_{\alpha\beta}) \equiv (\lambda_\alpha \cdot \lambda_\beta)$, with entries in λ_α, or by the *selectors* $K_{\alpha\beta}^\rho$ which are defined through the relation $\lambda_\alpha \cdot \lambda_\beta = \lambda_{\gamma(\alpha,\beta)} = K_{\alpha\beta}^\rho \lambda_\rho$, where $K_{\alpha\beta}^\rho = 1$ if $\rho = \gamma(\alpha, \beta)$ and $K_{\alpha\beta}^\rho = 0$ if $\rho \neq \gamma(\alpha, \beta)$. Thus, the components of a selector can only take the values 0 or 1.

As is well-known, the structure constants and the selectors provide respectively a matrix representation for $\mathcal{G}$ and S. This allows to show that the Kronecker product of those representations $\mathcal{G}_S = S \otimes \mathcal{G}$ is also a Lie algebra with generators $X_{(i,\alpha)} \equiv \lambda_\alpha \otimes X_i$, Lie product defined by $\left[X_{(i,\alpha)}, X_{(j,\beta)}\right] \equiv \lambda_\alpha \cdot \lambda_\beta \otimes \left[X_i, X_j\right] = C_{(i,\alpha)(j,\beta)}^{(k,\gamma)} X_{(k,\gamma)}$ and structure constants given by $C_{(i,\alpha)(j,\beta)}^{(k,\gamma)} = K_{\alpha\beta}^{\gamma} C_{ij}^{k}$. The explicit proof can be found in Ref. [1].

There are two cases where one can extract from $\mathcal{G}_S$ smaller algebras, with interesting properties. The first one happens when the original algebra has a subspace decomposition, e.g.[1], of the type $\mathcal{G} = V_0 \oplus V_1$ satisfying $[V_0, V_0] \subset V_0$, $[V_0, V_1] \subset V_1$, $[V_1, V_1] \subset V_0$. If the semigroup has a decomposition in subsets $S = S_0 \cup S_1$ satisfying the so called *resonant decomposition*:

$$S_0 \cdot S_0 \in S_0\,, \quad S_0 \cdot S_1 \in S_1\,, \quad S_1 \cdot S_1 \in S_0 \tag{1}$$

then, one can show that $\mathcal{G}_{S,R} \equiv (S_0 \otimes V_0) \oplus (S_1 \otimes V_1)$ is a subalgebra of $\mathcal{G}_S$, which is called the *resonant subalgebra*. The other case occurs when the semigroup have a zero element 0_S satisfying $\lambda_\alpha \cdot 0_S = 0_S$ for any element $\lambda_\alpha \in S$. In that case the whole sector $0_S \otimes \mathcal{G}$ can be erased from $\mathcal{G}_S$ in such a way that what remains is still a Lie algebra, which is called the *0_S-reduced algebra* $\mathcal{G}_{S,\mathrm{red}}$ (which is not necessarily a subalgebra of $\mathcal{G}_S$).

In summary, depending on the features of the abelian semigroup S that one uses in the S-expansion, the following types of algebra can be generated

(a) the expanded algebra $\mathcal{G}_S = S \otimes \mathcal{G}$,

(b) the resonant subalgebra $\mathcal{G}_{S,R}$, if S has at least one resonant decomposition of the type (1),

(c) the 0_S-reduced algebra $\mathcal{G}_{S,\mathrm{red}}$, if S has a zero element 0_S,

(d) the 0_S-reduction of the resonant subalgebra $\mathcal{G}_{S,R,\mathrm{red}}$, if S has both a zero element and a resonant decomposition.

Having introduced the main ingredients used in the S-expansion method, in the next sections we give a description of the Java library [8] that automatizes this procedure. In Section 2 we review some aspects about the classification of non-isomorphic semigroups, while in Section 3 we briefly describe the library. Then, in Section 4 we explain how the lists of all non-isomorphic abelian semigroups can be used to classify all possible S-expanded algebras (a-d). Finally, in Section 5, we will briefly describe a new method that extends our library and that is useful to answer if two given Lie algebras can be S-related.

2. Some aspects about the semigroup classification

As shown in the following table,

order	$Q = \#$ semigroups	
2	4	
3	18	
4	126	[Forsythe '54]
5	1,160	[Motzkin, Selfridge '55]
6	15,973	[Plemmons '66]
7	836,021	[Jurgensen, Wick '76]
8	1,843,120,128	[Satoh, Yama, Tokizawa '94]
9	52,989,400,714,478	[Distler, Kelsey, Mitchell '09]
10	**12,418,001,077,381,302,684**	[Distler, Jefferson, Kelsey, Kotthoff '16]

(2)

the problem of enumerating the all non-isomorphic finite semigroups of a certain order is a non-trivial problem, as the number Q of semigroups increases very quickly with the order

[1] As formulated in the original work [1], one can also deal with more general decompositions. However, on the first version of our library [8] we only focus in this case $\mathcal{G} = V_0 \oplus V_1$.

of the semigroup. This classification has been made by many different authors (see e.g. [9, 10, 11, 12, 13, 14, 15, 16, 17] and references therein). In lower orders, a list of the multiplication tables representing those semigroups can be explicitly constructed. For example, in Ref. [18] it has been claimed that its program *gen.f* allows to generate, in lexicographical ordering, the lists *sem.n* of these tables for the orders $n = 2, ..., 7$. After running that program, we got them only up to order 6 (for some reason the program stops when reaches the 835,927th of the 836,021 semigroups of order 7). In any case, we remark that similar lists are not explicitly known for orders higher than 8 and that the number of non-isomorphic semigroups in those orders has been reached only by using indirect techniques (see e.g., [19, 20]).

As we will see, the lists *sem.n* up to order $n = 6$ can be used as inputs in our library (although its methods are not restricted to that order[2]). In each list, a semigroup $S^a_{(n)}$ of order n is uniquely identified by the number $a = 1, ..., Q$ and the semigroup elements λ_α are represented by α with $\alpha = 1, ..., n$. The program *com.f* of [18] selects from those lists only the abelian semigroups (which are the ones of interest in the context of the S-expansion). For example, for $n = 2$ the semigroup elements are $\{1, 2\}$ and the program *com.f* gives:

$$\begin{array}{c|cc} S^1_{(2)} & 1 & 2 \\ \hline 1 & 1 & 1 \\ 2 & 1 & 1 \end{array}\ , \quad \begin{array}{c|cc} S^2_{(2)} & 1 & 2 \\ \hline 1 & 1 & 1 \\ 2 & 1 & 2 \end{array}\ , \quad \begin{array}{c|cc} S^4_{(2)} & 1 & 2 \\ \hline 1 & 1 & 2 \\ 2 & 2 & 1 \end{array}\ . \tag{3}$$

However, to avoid renaming the abelian semigroups in a new list, in our library we use directly the lists *sem.n* and select the abelian ones with a simple method when is needed[3].

3. Description of the Java library to perform S-expansions

To use the Java library it is necessary to download the linear algebra package *jama.jar*[4] from [21] and the following files from [8]:

File	**Brief description**
data.zip	Contains the the files *sem.n*
sexpansion.jar	Is the library itself
ReadMe.pdf	File with detailed installation instructions
examples.zip	Example programs explained in Section 5 of the Ref. [22]
Output_examples.zip	Output samples of the example programs

(4)

Our library is composed of 11 classes[5], each of them containing different methods allowing us to automatize the S-expansion procedure. As the full documentation of the library is available in [23], here we only describe the main features of these classes.

First, the class *SetS* allows to represent a set of integers, where none of them is repeated. It contains methods allowing to represent permutations, which are useful to determine isomorphisms between different semigroup tables. It also contains methods allowing to generate all possible subsets S_0 and S_1 of a semigroup S, with respectively $n_0 \leq n$ and $n_1 \leq n$ elements, such that $S = S_0 \cup S_1$ (notice that some elements can be simultaneously in S_0 and S_1). Thus, the methods of this class also provide the basics elements to the study of resonances.

[2] The methods of our library also allow us to perform semigroup calculations (like checking associativity, commutativity, finding zero element, isomorphisms and resonances) when the order n is higher than 6. The only issue is that we do not have the full list of non-isomorphic tables for those higher orders.

[3] It is also worth to mention that the lists *sem.n* are exhaustive, i.e., if one finds a semigroup S of order $n \leq 6$ whose multiplication table is not contained in the list *sem.n*, then S must be isomorphic to one and only one semigroup in that list. This can be easily checked with the methods implemented in our library and thus, we are able to work with arbitrary semigroups, i.e., not necessarily having a lexicographical ordering.

[4] In particular, we use the methods belonging to the class *Matrix* of the library *jama.jar*.

[5] To see the source code one only have to unzip the file *sexpansion.jar*.

IOP Conf. Series: Journal of Physics: Conf. Series **1085** (2018) 052011 doi:10.1088/1742-6596/1085/5/052011

The class *Semigroup* is used to represent discrete semigroups by means of the multiplication table $A = (a_{\alpha\beta}) \equiv (\alpha \cdot \beta)$ and contains methods allowing us to perform basic semigroup operations like checking associativity, commutativity, finding the zero element and loading the lists *sem.n* when is needed. It also contains methods that, when combined with those of the class *SetS*, allows to calculate isomorphisms and study all possible resonant decompositions of any given semigroup.

The class *Selector* allows to represent semigroups by means of their selectors of $K_{\alpha\beta}^{\rho}$ which, as explained in the introduction, are very convenient to describe S-expansions. It also contains methods allowing to determine if the semisimplicity is preserved under the S-expansion. The classes *SelectorReduced*, *SelectorResonant*, *SelectorResonantReduced* are constructed as a child of the class *Selector*, i.e., they extend that class for the cases where the semigroup have a zero element, a resonance and both simultaneously.

The class *StructureConstantSet* allows to represent the original algebra $\mathcal{G}$ by means of its structure constants C_{ij}^{k}. It also contains methods to perform basics calculations, like to determine the Killing-Cartan metric of $\mathcal{G}$.

The class *StructureConstantSetExpanded* allows to represent the Lie algebra resulting of performing the S-expansion of $\mathcal{G}$ with S. Finally, the child classes *StructureConstantSet-ExpandedReduced*, *StructureConstantSetExpandedResonant* and *StructureConstantSetExpande-dResonantReduced* allow to represent the expanded algebras $\mathcal{G}_{S,\text{red}}$, $\mathcal{G}_{S,R}$ and $\mathcal{G}_{S,R,\text{red}}$ when it corresponds.

Further details can be found in Ref. [22], which has been written as a handbook to use the library. Indeed, it explains in detail how the methods of our library has been constructed, how do they work and also explains the example programs that allow to check the results obtained in Refs. [24] and [25].

4. Classification of all possible S-expansions

The first motivation that lead us to construct our library was to provide the computational tool to study all possible S-expansions of the type (a-d) that one can perform with all the abelian non-isomorphic semigroups provided by the lists *sem.n*. On one side, this implies to identify all the abelian semigroups having a zero element, something that can be easily done not only by our library but also by the programs given in [18]. On the other side, a study of all possible resonant decompositions of these semigroups is also needed. To our knowledge, this was not done in the literature before our works [24] and [25] where we elaborated the first basic methods to study resonances, which we have improved recently and presented in the form of a Java library [8] in Ref. [22]. Thus, this is the first computational tool allowing to perform a full study of those resonances, as well as to classify and represent all possible S-expansions of the type (a-d) of a given Lie algebra. This result can be summarized in the following table,

	$n=2$	$n=3$	$n=4$	$n=5$	$n=6$
$\#\mathcal{G}_S$	3	12	58	325	2,143
#pss	2	5	16	51	201
$\#\mathcal{G}_{S,\text{red}}$	2	8	39	226	1,538
#pss	1	3	9	34	135
$\#\mathcal{G}_{S,R}$	1	8	48	299	2,059
#r	1	9	124	1,653	25,512
#pss	1	1	4	7	23
$\#\mathcal{G}_{S,R,\text{red}}$	0	5	32	204	1,465
#r	0	6	92	1,295	20,680
#pss	0	1	1	6	12

(5)

It gives for each order the number of S-expansions of the type (a-d) that can be performed, as well as the number *#pss* of S-expansions preserving semisimpliciy. Notice, a given semigroup can have more than one resonance, and with them the same semigroup may lead to different resonant subalgebras. Thus, in the table we count not only the number of abelian semigroups having at least one resonance, but also the total number of different resonances $\#r$.

5. A method to find S-related algebras

Imagine that we have to find a semigroup with some multiplications fixed, for example

$$\begin{array}{c|cccc} & \lambda_1 & \lambda_2 & \lambda_3 & \lambda_4 \\ \hline \lambda_1 & & \lambda_3 & & \lambda_4 \\ \lambda_2 & \lambda_3 & & \lambda_4 & \lambda_4 \\ \lambda_3 & & \lambda_4 & & \lambda_4 \\ \lambda_4 & \lambda_4 & \lambda_4 & \lambda_4 & \lambda_4 \end{array} \, . \tag{6}$$

As in Java an empty space cannot be left in an array, we choose to represent them by -1 in its place. For example, the template above will be represented by

$$\begin{pmatrix} -1 & 3 & -1 & 4 \\ 3 & -1 & 4 & 4 \\ -1 & 4 & -1 & 4 \\ 4 & 4 & 4 & 4 \end{pmatrix} . \tag{7}$$

Clearly, there are $4^4 = 256$ different ways to fill this template in a commutative way, where only some of them are associative so they are really semigroups. In general, for a template of order n where x independent component has been fixed, there are $n^{n(n+1)/2-x}$ different forms to fill it in a commutative way. This task is performed by the method *fillTemplate* of the class *Semigroup*, which reads the *template* and returns a *list* with all the tables that fill it. It is used as follow.

```
list = Semigroup.fillTemplate(template);
```

From all those tables one should select those which are associative and satisfy, if apply, a certain resonance. Applying the isomorphisms methods, we can finally identify the set of non-isomorphic semigroups that are solutions for the given template. As it will be explained in [26], this method turns out very useful to elaborate a general algorithm to answer if two given Lie algebras can be S-related.

6. Final Remarks

As our library has an open licence GNU, we expect that it can be improved and extended, for example to study other kind of decompositions like, e.g., $S = S_0 \cup S_1 \cup S_2$ where

$$\begin{array}{l} S_0 \times S_0 \subset S_0 \,, \quad S_0 \times S_1 \subset S_1 \,, \quad S_0 \times S_2 \subset S_2 \\ S_1 \times S_1 \subset S_0 \cap S_2 \,, \quad S_1 \times S_2 \subset S_1 \,, \quad S_2 \times S_2 \subset S_0 \cap S_2 \,. \end{array}$$

This could be useful for new physical applications because, as shown in [1], this type of decomposition can be used to study expansions of Lie super algebras (used in the context of supergravity and string theory), which have the subspace decomposition $\mathcal{G} = V_0 \oplus V_1 \oplus V_2$ with the following structure

$$\begin{array}{l} [V_0, V_0] \subset V_0 \,, \quad [V_0, V_1] \subset V_1 \,, \quad [V_0, V_2] \subset V_2 \\ [V_1, V_1] \subset V_0 \oplus V_2 \,, \quad [V_1, V_2] \subset V_1 \,, \quad [V_2, V_2] \subset V_0 \oplus V_2 \,. \end{array}$$

On the other hand, as mentioned in [27] it it would be interesting to analyze if the S-expansion could help to fit to the classification of solvable Lie algebras of a fixed dimension using S-expansions of semisimple Lie algebras of the same dimension. Thus, our library might also be useful to analyze that problem.

Acknowledgments

We thank Andrei Kataev for his invitation to present this work at ACAT 2017. C.I. was supported by a Mecesup PhD grant and the Término de tesis grant from CONICYT (Chile). C.I. is very grateful to Local Organizing Committee and in particular to Gordon Watts for the financing his participation in ACAT. I.K. was supported by was supported by Fondecyt (Chile) grant 1050512 and by DIUBB (Chile) Grant Nos. 102609 and GI 153209/C. NM is supported by a Becas-Chile postdoctoral grant.

References

[1] Izaurieta F, Rodriguez E and Salgado P 2006 Expanding Lie (super)algebras through Abelian semigroups *J. Math. Phys.* **47** 123512 (*Preprint* hep-th/0606215)
[2] Segal I E 1951 A class of operator algebras which are determined by groups *Duke Math. J.* **18** 221-65
[3] Inönü E and Wigner E P 1953 On the contraction of groups and their representations *Proc. Nat. Acad. Sci. USA* **39** 510-24
[4] Weimar-Woods E 2000 Contractions, generalized Inönü and Wigner contractions and deformations of finite-dimensional Lie algebras *Rev. Math. Phys.* **12** 1505-29
[5] Hatsuda M and Sakaguchi M 2003 Wess-Zumino term for the AdS superstring and generalized Inonu-Wigner contraction *Prog. Theor. Phys.* **109** 853 (*Preprint* hep-th/0106114)
[6] de Azcarraga J A, Izquierdo J M, Picon M and Varela O 2003 Generating Lie and gauge free differential (super)algebras by expanding Maurer-Cartan forms and Chern-Simons supergravity *Nucl. Phys.* B **662** 185 (*Preprint* hep-th/0212347)
[7] Inostroza C, Kondrashuk I, Merino N and Nadal F 2017 On the algorithm to find S-related Lie algebras *Merged talk by C Inostroza at ACAT 2017*
[8] https://github.com/SemigroupExp/Sexpansion/releases/tag/v1.0.0
[9] Forsythe G E 1955 SWAC computes 126 distinct semigroups of order 4 *Proc. Amer. Math. Soc.* **6** 443-447
[10] Motzkin T S and Selfridge J L 1956 Semigroups of order five *Bull. Amer. Math. Soc.* **62** 13-23
[11] Plemmons R 1967 There are 15973 semigroups of order 6 *Math. Algorithms* **2** 2-17
[12] Plemmons R 1970 Construction and analysis of non-equivalent finite semigroups *Computational Problems in Abstract Algebra: Proceedings of Conference (Oxford, 1967)* (Oxford: Pergamon) pp. 223-8
[13] Jürgensen H and Wick P 1977 Die Halbgruppen der Ordnungen ≤ 7 *Semigroup Forum* **14** 69-79
[14] Satoh S, Yama K, Tokizawa M 1994 Semigroups of order 8 *Semigroup Forum* **49** 7-29
[15] Distler A and Kelsey T 2009 The monoids of orders eight, nine and ten *Ann. Math. Artif. Intell.* **56** 3-21
[16] Distler A and T. Kelsey T 2008 The monoids of order eight and nine *9th Int. Conf. on Artificial Intelligence and Symbolic Computation AISC 2008: Proceedings of Conference (Birmingham, July 2008)* ed S Autexier, J Campbell, J Rubio, V Sorge, M Suzuki and F Wiedijk *(Lecture Notes in Computer Science vol 5144)* (Berlin: Springer) pp 61-76
[17] Distler A and Mitchell J D 2011 Smallsemi - a GAP package, version 0.6.4 http://tinyurl.com/jdmitchell/smallsemi/
[18] Hildebrant J 2001 Handbook of Finite Semigroup Programs *LSU Mathematics Electronic Preprint Series Preprint* 2001-24
[19] Distler A, Kelsey T and Mitchell J D http://www-circa.mcs.st-and.ac.uk/
[20] Distler A, Jefferson C, Kelsey T and Kotthof L 2012 The Semigroups of Order 10 *18th Int. Conf. on Principles and Practice of Constraint Programming CP 2012: Proceedings of Conference (Québec City, October 2012)* ed M Milano *(Lecture Notes in Computer Science vol 7514)* (Berlin: Springer Heidelberg) pp 883-899
[21] http://math.nist.gov/javanumerics/jama/
[22] Inostroza C, Kondrashuk I, Merino N and Nadal F 2017 A Java library to perform S-expansions of Lie algebras *Preprint* arXiv:1703.04036 [cs.MS]
[23] https://github.com/SemigroupExp/Sexpansion/wiki
[24] Caroca R, Kondrashuk I, Merino N and Nadal F 2013 Bianchi spaces and their three-dimensional isometries as S-expansions of two-dimensional isometries *J. Phys.* A **46** 225201 (*Preprint* arXiv:1104.3541 [math-ph])
[25] Andrianopoli L, Merino N, Nadal F and Trigiante M 2013 General properties of the expansion methods of Lie algebras *J. Phys.* A **46** 365204 (*Preprint* arXiv:1308.4832 [gr-qc])
[26] Inostroza C, Kondrashuk I, Merino N, Nadal F 2017 Algorithm to find S-related Lie algebras *in preparation*
[27] Nesterenko M 2013 S-expansions of three-dimensional Lie algebras *6th International Workshop Group analysis of differential equations and integrable systems GADEIS-VI: Proceedings of Conference (Protaras, June 2012)* ed O O Vaneeva, C Sophocleous, R O Popovych, P G L. Leach, V M Boyko and P A Damianou (Nicosia: Department of Mathematics and Statistics, University of Cyprus) pp 147-154 (*Preprint* arXiv:1212.1820[math-ph])

IOP Conf. Series: Journal of Physics: Conf. Series **1085** (2018) 052012 doi:10.1088/1742-6596/1085/5/052012

Go-HEP: writing concurrent software with ease and Go

S Binet[1]

[1] Laboratoire de Physique de Clermont-Ferrand, CNRS/IN2P3, FR

E-mail: `sebastien.binet@clermont.in2p3.fr`

Abstract. High Energy and Nuclear Physics (HENP) libraries are now required to be more and more multi-thread-safe, if not multi-thread-friendly and multi-threaded. This is usually done using the new constructs and library components offered by the `C++11` and `C++14` standards. These components are however quite low-level and hard to use and compose.

However, `Go` provides a set of better building blocks for tackling concurrency: goroutines and channels. This language is now used by the whole cloud industry: docker/moby, rkt, Kubernetes are obvious flagships for `Go`. But to be able to perform any meaningful physics analysis, one needs a set of basic libraries (matrix operations, linear algebra, plotting, I/O, ...) We present `Go-HEP`, a set of packages to easily write concurrent software to interface with legacy HENP `C++` physics libraries.

1. Introduction

With the end of Dennard Scaling, High Energy and Nuclear Physics (HENP) libraries and applications are now required to be multi-thread safe. This is to ensure the scaling of performances with new CPU architectures. The new `C++` standards introduced new constructs and library components: `std::thread`, `std::mutex`, `std::lock`, ...These components are however quite low-level and hard to use and compose, or easy to misuse. Moreover, `C++` is still plagued with scalability issues. Development speed is hindered by the interplay between the compilation model (based on `#include`) and `C++` templates. `C++` is a very large language, hard and subtle to teach and understand: this has an impact on code maintenance and the ability to bring newcomers up-to-speed on any given project. On the user side, `C++` is showing deficiencies as well: installing a project's many dependencies can be quite difficult. `Python` users will be surprised to learn there is still no `pip`-like mechanism to install, automatically and recursively, dependencies for a given project. Finally, most HEP software stacks rely on applications dynamically linked to hundreds shared libraries: packaging and deployment can be time consuming tasks.

Fixing the whole `C++` ecosystem sounds like a rather daunting task. It might be easier instead to start from a blank page, with a new language that addresses all the current deficiencies of `C++`, and improves the day-to-day life of a typical software programmer in a multi-core environment.

`Go` [1] was created in 2009 to address these issues. We will first describe briefly the `Go` programming language, its concurrency primitives and the features of its tooling that make `Go` a great fit for HENP applications. We will then introduce `Go-HEP`, a set of libraries and applications that aims to provide physicists with the basic tools to perform HENP analyses and integrate them in already existing `C++/Python` analysis pipelines.

IOP Conf. Series: Journal of Physics: Conf. Series **1085** (2018) 052012 doi:10.1088/1742-6596/1085/5/052012

2. Introduction to Go

Go is an open source language, released under the BSD-3 license in 2009. It is a compiled language with a garbage collector and builtin support for reflection. The language is reminiscent of C/C++: it shows a syntax similar to its older peers but adds first-class functions, closures and object-oriented programming via the concept of interfaces. Go is already available, via the gc toolchain, on all major platforms (Linux, Windows, macOS, Android, iOS, ...) and for many architectures (amd64, arm64, i386, s390x, mips64, ...) There is also a Go frontend to GCC, aptly named gccgo, that can can target all the platforms and architectures that the GCC toolchain supports. Griesemer, Pike and Thompson created Go to replace the multi-threaded C++ web servers that were hard to develop and slow to compile, even when using Google's massive infrastructure. As such, Go exposes two builtin concurrency primitives: the *goroutines* – very lightweight green threads – and the *channels* – typed conduits that connect goroutines together.

Launching goroutines is done by prepending a function (or method) call with the keyword go. Goroutines can be described as green threads: very lightweight execution contexts that are multiplexed on native threads. Each goroutine starts with a small stack (around $4KB$) that can grow and shrink as needed. This allows to write applications with thousands of goroutines without the need to buy high-end servers. Writing real world concurrent programs needs synchronization and communication: the ability to exchange data between goroutines. This is achieved *via* channels. Not only do channels allow to exchange data in a type safe manner but they are also a synchronization point: a goroutine trying to send a token of data through a channel will block until there is another goroutine on the other end of the channel trying to extract data from the channel, and *vice versa*. The last piece that makes building concurrent programs in Go efficient is the keyword select. This keyword is like a switch statement for controling the data flow between multiple channels and thus between multiple goroutines. The concurrency builtin tools of Go are easier to reason about and more composable than mutexes and locks.

Moreover, Go comes with a package system and a builtin tool to build Go code. There are no header files like in C++ as they require to recursively process all dependent headers and thus slow the build. Once compiled, Go packages are completely self-contained and do not require a client of package p1 – which itself depends on p2 and p3 – to know anything about these packages. The client only needs p1. This greatly improves the scalability of the build process as well as its speed. Thanks to the way third-party packages are imported in Go, *e.g.* `import "github.com/pkg/errors"`, a build tool only needs to parse and understand Go code to discover (recursively) dependencies of a given package. Coupled to the fact that package import strings contain the URL to the repository (GitHub, BitBucket, ...) holding the actual code, this allows a simple command – go get – to fetch code from the internet, (recursively) discover and fetch its dependencies and, finally build the whole artefact. The instructions to install code are valid on all platforms that are supported by a given Go toolchain: one just needs to point go get at a repository. Finally, Go code being very quick to compile makes static compilation manageable and enables simple deployment scenarii that boil down to scp-ing the resulting binary. Go exposes a productive development environment that is concurrency friendly. It is being used by many companies [1] beside Google, and in a variety of scenarii: from rocket telemetry to cloud systems to container orchestration. But for Go to be useful in HENP, physicists need libraries and tools to carry analyses. Moreover, these tools need to be interoperable with existing analyses pipelines.

3. Go-HEP

Go-HEP is a set of libraries and applications released under the BSD-3 license. They allow High Energy and Nuclear Physicists to write efficient analysis code in the Go programming language.

[1] A non-exhaustive list is maintained here: https://github.com/golang/go/wiki/GoUsers.

The `go-hep.org/x/hep` organization provides a set of pure-`Go` packages and building blocks to:

- write analyses in `Go`,
- write data acquisition and monitoring systems in `Go`,
- write control frameworks in `Go`.

As `Go-HEP` is written in pure-`Go`, the whole software suite and its dependencies are installable on all `Go` supported platforms (Linux, macOS, Windows, RPi3, ...) with:

```
$> go get go-hep.org/x/hep/...
```

The ellipsis (...) at the end instructs the `go` tool to compile and install all the libraries and applications that are part of the `go-hep.org/x/hep` repository.

The libraries provided by `Go-HEP` can be broadly organized around two categories:

- libraries that provide physics and statistical tools (Lorentz vectors, histograms and n-tuples, fits, jet clustering, fast detector simulation, plots, ...)
- libraries that provide low level interoperability with `C++` libraries, to allow `Go-HEP` users to integrate with existing analyses pipelines.

Indeed, analyses in HENP are mainly written in `C++` and `Python`. Even though `Go` has a native foreign function interface called `cgo` that allows `Go` code to use `C` code and vice versa, the libraries under `go-hep.org/x/hep` do not call any `C++` library (via a `C` shim library.) This decision allows to retain the quick edit-compile-run development cycle of `Go` and the easy deployment and cross-compilation of `Go` applications. Instead, the strategy for interoperating with `C++` is to integrate with *e.g.* ROOT [3] at the data file level. `Go-HEP` provides read/write access for LCIO, LHEF, HepMC, SLHA and YODA files. `Go-HEP` provides – at the moment – only read access to ROOT files via its `go-hep.org/x/hep/rootio` package.

In the following, we will describe two components of `Go-HEP`, `fads` and `hep/rootio`, that enable physics analyses.

4. fads

4.1. *fads-app*

`fads` is a "FAst Detector Simulation toolkit". This library is built on top of `Go-HEP`'s control framework, `fwk`. The control framework exposes a traditional API, with a `Task` type that can be started, process some event data and then stopped. Each event is processed by a goroutine. The framework runs these tasks in their own goroutine and lets the `Go` runtime deal with work stealing. Data between tasks are exchanged via an event store service which itself utilizes `channels` to ensure there are no data races. Besides an event store service, `fwk` also provides services for message logging, histogramming and n-tupling. N-tuple and histograms (1D and 2D) are provided by the `hep/hbook` and `hep/hbook/ntup` packages and are persistified using `hep/rio`, a binary file format that heavily draws inspiration from `SIO`, the binary file format the LCIO community uses.

Data dependencies between tasks are described at the job configuration level: each task declares what are its inputs (the type of input and a name that identifies it) and its outputs. `fwk` ensures there are no cycles between the tasks and connects the goroutines together via channels. `fwk` enables task-level parallelism and event-level parallelism via the concurrency building blocks of `Go`. For sub-task parallelism, users are required – by construction – to use the same building block, so everything is consistent and the `Go` runtime has the complete picture.

`fads` is itself a transliteration of Delphes [4] (v3.0.12) but with `fwk` underneath instead of `ROOT`. To assess the performances of `Go` and `fads` in a realistic settings, the whole ATLAS data card provided with Delphes has been implemented and packaged with `fads` under `hep/fads/cmd/fads-app`. This application is composed of:

IOP Conf. Series: Journal of Physics: Conf. Series **1085** (2018) 052012 doi:10.1088/1742-6596/1085/5/052012

- a HepMC file reader task,
- a particle propagator, a calorimeter simulator,
- energy rescalers, momentum smearers,
- electron, photon and muon isolation tasks,
- b-tagging and tau-tagging tasks, and a jet-finder task.

The jet finder task is based on a naive re-implementation of the `C++` FastJet [6] library: only the N^3 "dumb" clustering has been implemented so far. A small part of the directed acyclic graph of the resulting `fads-app` application can be found in figure 1.

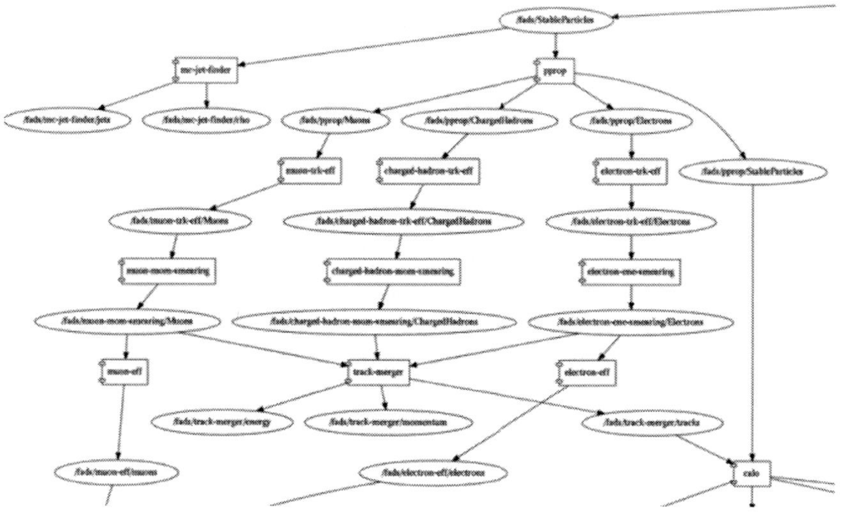

Figure 1. Part of the directed acyclic graph of data dependencies between tasks composing the `fads-app`, a `Go` transliteration of the Delphes' ATLAS data card example. Rectangles are tasks, ellipses are data collections.

Delphes was compiled with `gcc-4.8` and the N^3 clustering strategy hard-coded, `fads` was compiled with `Go-1.9`. Timings were obtained on a Linux Xeon CPU E5-4620@2.20GHz server with 64 cores and 128Gb RAM with a 10000 HepMC events input file.

The results for various numbers of threads are shown in figure 2. As Delphes is not thread-safe, we only show the data for one Delphes process and then draw a flat line for visual aid. The graph on the left shows a smaller memory footprint for `fads` that only matches that of Delphes' when 30 threads are used. The graph on the right shows that `fads` achieves a better event processing rate even when using only one thread, or when using only one thread but with the sequential event loop instead of the concurrent event loop (data point for $n_{thread} = 0$.)

`fads` achieves better performances than Delphes-3.0.12 on this data set. The output simulated data for `fads` was matched bit-by-bit with that of Delphes up to the calorimetry stage [2]. `fads` also achieves these performances without the need to merge output files.

4.2. `fads-rivet-mc-generic`

`fads` exports another application, `fads-rivet-mc-generic`, a reimplementation of the `MC_GENERIC` analysis from the Rivet [5] toolkit.

[2] We could not manage to match data further down the sequence because of Delphes' usage of PRNG to seed other PRNGs in the calorimeter. Control histograms agreed at the statistical level.

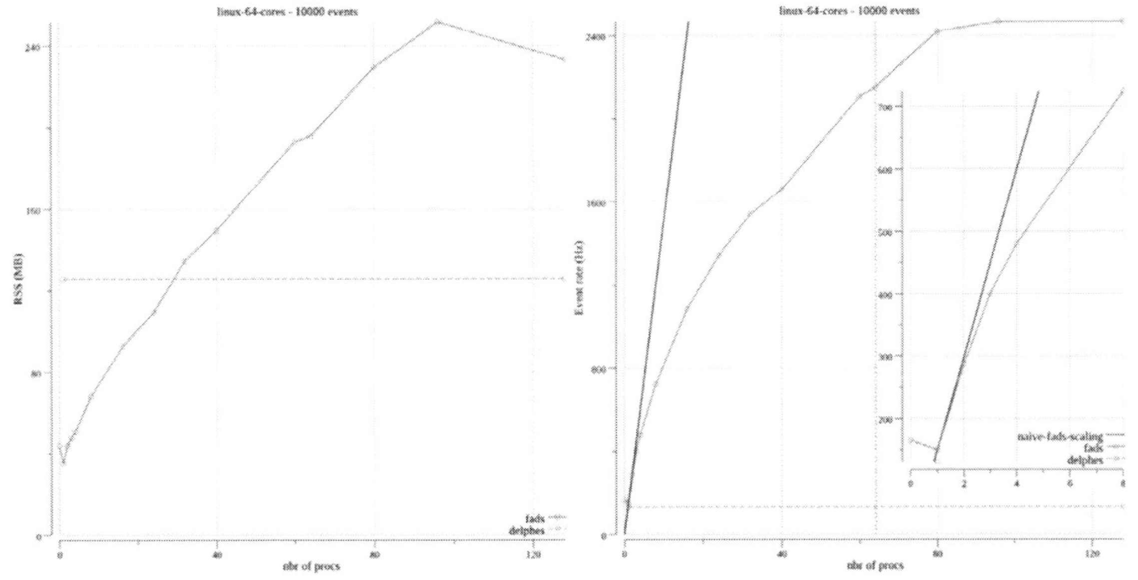

Figure 2. Memory usage (left) and event processing rate (right) for `fads` (solid red curve) and `Delphes` (green dash line).

	MaxRSS (Mb)	Real (s)	CPU (s)
`Rivet`	27	13.3	13.3
`fads`	23	5.7	5.7

Table 1. Runtime performances of `Rivet` and `fads` to read a file with Z events.

`fads` shows better performances than the `C++` application. As Rivet does not use ROOT for this application, the memory usage of Rivet is closer to that of `fads`. However, the event processing rate of `fads` with only one thread is twice as good as the one of Rivet. This is because the job steering and the event loop of Rivet are written in Python.

5. `hep/rootio`

The `hep/rootio` package is a pure-`Go` package that:

- decodes and understands the structure of `TFiles`, `TKeys`, `TDirectory` and `TStreamerInfos`,
- decodes and deserializes `TH1x`, `TH2x`, `TLeaf`, `TBranch` and `TTrees`.

At the moment, `hep/rootio` only concerns itself with reading ROOT files, although writing ROOT files is on the roadmap. Nonetheless, `hep/rootio` can already be useful and provides the following commands:

- `cmd/root-ls` lists the content of ROOT files,
- `cmd/root-print` extracts histograms from ROOT files and saves them as PDF or PNG,
- `cmd/root-dump` prints the contents of `TTrees`, event by event,
- `cmd/root-diff` prints the differences between the contents of two `TTrees`,
- `root-cnv-npy` converts simple `TTrees` to `NumPy` array data files,
- `cmd/root-srv` allows to interactively plot histograms from a file and branches from a tree. `root-srv` being pure-`Go`, it can be hosted on Google AppEngine.

IOP Conf. Series: Journal of Physics: Conf. Series **1085** (2018) 052012 doi:10.1088/1742-6596/1085/5/052012

`rootio` can read flat `TTrees` with `C++` builtins, static and dynamic arrays of `C++` builtins. `rootio` can also read `TTrees` with user defined classes containing `std::vector<T>` [3], another user defined class, `std::string` or `TString`, arrays of `C++` builtins and `C++ builtins`.

To assess the performances of `hep/rootio` with regard to the original `C++` implementation, we created two input files of 10^6 entries and 100 branches of `float64`. One file was compressed with the default settings of ROOT-6.10 ($686Mb$) and the other had no compression ($764Mb$).

`fCompression=0`	`VMem` (Mb)	`MaxRSS` (Mb)	Real (s)	CPU (s)
`ROOT`	517	258	6.7	6.6
`hep/rootio`	43	42	12.9	12.9

`fCompression=1`	`VMem` (Mb)	`MaxRSS` (Mb)	Real (s)	CPU (s)
`ROOT`	529	292	12.6	12.0
`hep/rootio`	83	82	35.8	35.8

Table 2. Runtime performances of `C++/ROOT` and `hep/rootio` to read a file with 10^6 events, using no compression (up) and with default compression (down).

Table 2 shows the results obtained running the two programs with the compressed file and the uncompressed file as inputs. While the memory footprint of the `Go` program is almost an order of magnitude lower than its `C++` counterpart, it is twice as slow in the non-compressed case and almost thrice in the compressed case. The degraded performances in the compressed case have been tracked back to the decompression package from the standard library which could probably be further optimized. On the otherhand, the degraded performances in the non-compressed case come from the still young implementation of `hep/rootio` which could be also further optimized, *e.g.* preemptively loading multiple consecutive entries.

6. Conclusions

`Go` improves on `C++/Python` and addresses their deficiences with regard to code distribution, code installation, compilation, development and runtime speeds. `Go` also provides builtin facilities to tackle concurrency efficiently and easily. `Go-HEP` provides some building blocks that are already competitive with battle-tested `C++` programs, both in terms of CPU, memory usage and cores' utilization. Further improvements are still necessary in the ROOT I/O area, both in terms of performances and features, so that `Go-HEP` can be part of physics analyses pipelines.

References

[1] The `Go` programming language `https://golang.org`
[2] S Binet, E Busato, D Blyth, M Ughetto, P Waller and B Wieck, `Go-HEP`: libraries for High Energy Physics analyses in `Go`, The Journal of Open Source Software, 2017/09, `https://doi.org/10.21105/joss.00372`
[3] R Brun and F Rademakers, ROOT - An Object Oriented Data Analysis Framework, Proceedings AIHENP'96 Workshop, Lausanne, Sep. 1996, Nucl. Inst. & Meth. in Phys. Res. A 389 (1997) 81-86
[4] The DELPHES 3 collaboration, DELPHES 3: a modular framework for fast simulation of a generic collider experiment, [software] http://dx.doi.org/10.1007/JHEP02(2014)057
[5] A Buckley et al, Rivet user manual, [software] https://doi.org/10.1016/j.cpc.2013.05.021
[6] M Cacciari, G Salam, G Soyez , FastJet manual, [software] https://dx.doi.org/10.1140/epjc/s10052-012-1896-2

[3] where `T` is a `C++` builtin or a `std::string` or a `TString`

IOP Conf. Series: Journal of Physics: Conf. Series **1085** (2018) 052013 doi:10.1088/1742-6596/1085/5/052013

On the notion "significance of the difference"

Sergey Bityukov[1,2], Nikolai Krasnikov[2], Nikita Okunev[3] and Vera Taperechkina[3]

[1] National Research Center "Kurchatov Institute" Institute for High Energy Physics, Protvino, Russia
[2] Institute for Nuclear Research Russian Academy of Science, Moscow, Russia
[3] Moscow Technological University (MIREA), Moscow, Russia

E-mail: Serguei.Bitioukov@cern.ch

Abstract. The concept of the significance of a signal in presence of background in experiment is widely used in data analysis in high-energy physics. It is shown that when comparing pairs independent samples obtained from one population (Poisson flow of events), the distribution of estimates of the significance is asymptotically standard normal distribution.

1. Introduction

The concept of "the significance" of a signal in presence of background in experiment [1, 2] (or, more precisely, "the significance of the difference" between the number of signal events and zero) is widely used in data analysis in high-energy physics.

We consider the significance Type C [3]. This significance is used in works [4, 5, 6] (see. also, [7]) for analysis of experimental data.

It is shown that when comparing two independent samples obtained from one population, the distribution of estimates of "the significance of the difference" is asymptotically standard normal distribution.

Let a sample of realizations of some random variable be obtained from an infinite population within a given time. Each realization is called an event. The stream of events in experiment usually is a Poisson flow. Number of realizations, which determine by some of conditions (for example, cuts), can be either a background events, or a signal events, which are indistinguishable. Several methods exist to quantify the statistical "significance" of an signal (expected or estimated) in this sample. Following the conventions in high energy physics, the term significance usually means the "number of standard deviations" of an expected or observed signal is above expected or estimated background.

In the simplest case, this concept "significance" can be described with the help of two numbers: b - the number of background events and s - the number of signal events (signal and background events are indistinguishable) which appeared during the given time.

In a real experiment, the numbers of background and signal events are realizations of random variables. The distributions of the observed number of background events $\hat{b}$ and the observed number of signal events $\hat{s}$ in Poisson flow of events obey Poisson distributions with parameters b (expected number of background events) and s (expected number of signal events), respectively.

Note, the realization of random variable allows to estimate the parameter of Poisson distribution. It means that when we compare the estimated parameters of Poisson flows of events we compare two samples.

For example, to assess the uncertainties that arise after (or before) the measurements, the significances of $S_1 = \frac{s}{\sqrt{b}}$ or $S_2 = \frac{s}{\sqrt{s+b}}$ were often used. With a small number of events, significances S_1 and S_2 give incorrect results.

2. Classification of significances

In paper [3], a classification of significances in accordance with the scope of applicability was proposed. Let us S characterizes the significance of signal. The choice of significance to be used depends on the study. There are three types of significances.

A. If s and b are expected values then we take into account both statistical fluctuations of signal and fluctuations of background. Before observation we can calculate only an expected significance S which is a parameter of experiment. S characterizes the quality of experiment ($S_{c12} = 2(\sqrt{s+b} - \sqrt{b})$ [3] as an example).

B. If $\widehat{s+b}$ is observed value and b is expected value then we take into account only the fluctuations of background. In this case we can calculate an observed significance $\hat{S}$ which is an estimator of expected significance of experiment S. $\hat{S}$ characterizes the quality of experimental data (S_{cP} as an example). S_{cP}, probable, was proposed for using in HEP in ref. [8]. This significance corresponds a probability to observe number of events equal or greater than $s+b$ in sample with Poisson distribution with mean b which converted to equivalent number of sigmas of a Gaussian distribution.

C. If $\widehat{s+b}$ and $\hat{b}$ are observed values of signal plus background and only background with known errors of measurement then we can use the standard theory of errors to estimate the significance of signal S_d. If measured variables obey Poisson distribution or measured values have normal distributed variances the formula for S_d looks as

$$S_d = \frac{\widehat{s+b} - \hat{b}}{\sqrt{\sigma^2_{\widehat{s+b}} + \sigma^2_{\hat{b}}}}, \tag{1}$$

where $\sigma^2_{\widehat{s+b}}$ and $\sigma^2_{\hat{b}}$ are corresponding variances of errors distributions.

If samples for estimation of $\widehat{s+b}$ and $\hat{b}$ have different volumes (different integrated luminosities of experiments) then formula for significance looks as

$$S_d = \frac{\widehat{s+b} - K\hat{b}}{\sqrt{\sigma^2_{\widehat{s+b}} + K^2\sigma^2_{\hat{b}}}}, \tag{2}$$

where K is a ratio of integrated luminosities of experiments.

An important property of these significances is property that when comparing two independent samples obtained from the same population (for example, both samples are without signal events, i.e. $s = 0$) the distribution of estimates of "the significance of the difference" is close to the standard normal distribution $N(0,1)$. It is shown for several significances in paper [3] (significances S_{c12} and S_{cP}) and in paper [6] (significance Eq. 2) by Monte Carlo experiments. Fisz [9] shows that the significance (1) is asymptotically normal $N(0,1)$ in the case $s = 0$ (both samples are taken from the same population).

Let us show the presence of this property for significance in case of Eq. 2.

Let we have two independent samples which are taken from the same stationary Poisson flow of events during time t_1 and t_2, correspondingly. We study subflow with parameter λ, which satisfy the certain conditions (for example, cuts) of this flow. Let us the subflow content only background events. Let there are choosen $\hat{n}_1$ events in the first sample and $\hat{n}_2$ events in second sample. To determine the significance of the difference we use the formula

$$S_d = \frac{\hat{n}_1 - K\hat{n}_2}{\sqrt{\lambda t_1 + K^2 \lambda t_2}}, \tag{3}$$

where $K = \frac{t_1}{t_2}$. Expected numbers of events in samples λt_1 and λt_2 are variances of corresponding Poisson distributions.

Suppose that $\hat{n}_1$ and $\hat{n}_2$ are unbiased estimates of corresponding expected values n_1 and n_2 for given samples (asymptotically it is true [9]), i.e. $E(\hat{n}_i) = n_i = \lambda t_i$, $i = 1, 2$. Note, that $\lambda t_1 + K^2 \lambda t_2$ is constant. Then, due to the stationarity of the flow of events and the independence of samples, we have

$$E(S_d) = E(\frac{\hat{n}_1 - K\hat{n}_2}{\sqrt{\lambda t_1 + K^2 \lambda t_2}}) = \frac{E(\hat{n}_1) - E(K\hat{n}_2)}{\sqrt{\lambda t_1 + K^2 \lambda t_2}} = \frac{\lambda t_1 - K\lambda t_2}{\sqrt{\lambda t_1 + K^2 \lambda t_2}} = 0, \tag{4}$$

$$D(S_d) = D(\frac{\hat{n}_1 - K\hat{n}_2}{\sqrt{\lambda t_1 + K^2 \lambda t_2}}) = \frac{D(\hat{n}_1) + D(K\hat{n}_2)}{\lambda t_1 + K^2 \lambda t_2} = \frac{\lambda t_1 + K^2 \lambda t_2}{\lambda t_1 + K^2 \lambda t_2} = 1. \tag{5}$$

It means that asymptotical normality of Eq. 1 take place in the case Eq. 2.

3. Conclusion

This property allows to use significance S_d in many applications of data analysis.

Due to this property one can construct the scale for distance between, for example, histograms [10] by the use a multivariate test statistics. As a result,

- we can compare corresponding parts of two histograms,
- we can compare multidimensional histograms likewise as unidimensional histograms,
- we can compare two sets of several histograms simultaniously likewise as we compare a pair of histograms.

We are grateful to Vassili Kachanov for interest and support of this work. We thank Sergei Gleyzer, Yuri Gouz, Andrei Kataev, Nikolai Minaev and Vera Smirnova for helpful and constructive comments. The results of the work are obtained under support of Ministry of Education and Science RF (Agreement on October 17, 2014 N 14.610.21.0004, idintificator PNIER RFMEFI61014X0004).

References

[1] Linnemann J T 2003 *Proc. Conference on Staistical Problems in Particle Physics, Astrophysics and Cosmology (SLAC, Stanford)* Editors: L.Lyons, R.Mount, R.Reitmeyer, SLAC-R-703, p 35
[2] Zhu Y 2006 *High Ener. Phys. Nucl. Phys.* **30** 331 (*ePrint* arXiv:physics/0507145)
[3] Bityukov S, Krasnikov N, Nikitenko A, Smirnova V 2008 *Proc. of Science* PoS (ACAT08) 118
[4] Aubert B et al (BABAR Collaboration) 2008 *Phys.Rev.* D **78** 051102
[5] Bediaga I, Bigi I I, Gomes A, Guerrer G, Miranda J, dos Reis A C 2008 *Phys.Rev.* D **80** 096006
[6] Bityukov S, Krasnikov N, Nikitenko A, Smirnova V 2013 *Eur.Phys. J. Plus* 128:143
[7] Aaij R et al (LHCb Collaboration) 2013 *Phys. Lett.* B **726** 623
[8] Narsky I 2000 *Nucl.Instr.&Meth.*, A **450** 444
[9] Fisz M 1955 *Colloquium Mathematicum* **3** 199; also, Haight F A 1967 *Handbook of the Poisson distribution* (John Wiley & Sons, Inc.) (formula 6.4-1)
[10] Bityukov S I, Krasnikov N V, Maksimishkina A V, Smirnova V V 2016 Multidimensional test statistics and statistical comparison of histograms *Int. Journal of Economics and Statistics* **4** 98

IOP Conf. Series: Journal of Physics: Conf. Series **1085** (2018) 052014 doi:10.1088/1742-6596/1085/5/052014

Transfer Learning in Astronomy: A New Machine-Learning Paradigm

Ricardo Vilalta

Department of Computer Science, University of Houston.
Philip G. Hoffman Hall, 3551 Cullen Blvd., Houston TX, 77204-3010, USA

E-mail: rvilalta@uh.edu

Abstract. The widespread dissemination of machine learning tools in science, particularly in astronomy, has revealed the limitation of working with simple single-task scenarios in which any task in need of a predictive model is looked in isolation, and ignores the existence of other similar tasks. In contrast, a new generation of techniques is emerging where predictive models can take advantage of previous experience to leverage information from similar tasks. The new emerging area is referred to as *transfer learning*. In this paper, I briefly describe the motivation behind the use of transfer learning techniques, and explain how such techniques can be used to solve popular problems in astronomy. As an example, a prevalent problem in astronomy is to estimate the class of an object (e.g., Supernova Ia) using a generation of photometric light-curve datasets where data abounds, but class labels are scarce; such analysis can benefit from spectroscopic data where class labels are known with high confidence, but the data sample is small. Transfer learning provides a robust and practical solution to leverage information from one domain to improve the accuracy of a model built on a different domain. In the example above, transfer learning would look to overcome the difficulty in the compatibility of models between spectroscopic data and photometric data, since data properties such as size, class priors, and underlying distributions, are all expected to be significantly different.

1. Introduction

The abundance of large datasets generated for scientific research through sophisticated sensors (e.g., modern telescopes) or complex simulations, has led to a widespread interest for automated mechanisms that can analyze the data and generate models that classify events in an accurate and precise manner. A popular approach uses machine learning [8] to train a computer to recognize different events using case examples that belong to different categories. In most applications, a common assumption pervading most traditional work in machine learning is that the probability distribution from which a training sample is drawn is static; future samples must follow the same distribution for any model to remain valid. While such assumption is sensible, and has found a plethora of successful real-world applications, recent work in machine learning has shown an equally large number of applications that do not follow such assumption.

An example of such shift in distributions lies in light curve classification from star samples obtained from different galaxies [16]. An original source task may consist of identifying certain types of stars on a nearby galaxy. But if we move to galaxies lying farther away, and we try to repeat the same identification task, we will find that the distribution of stars has now changed (sometimes drastically). A major reason for such change is that at greater distances,

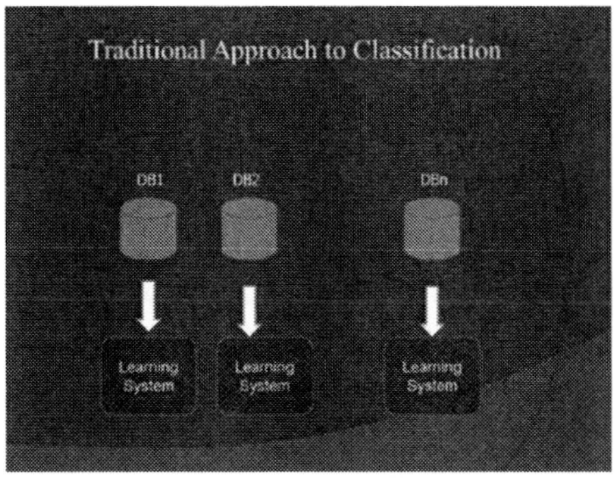

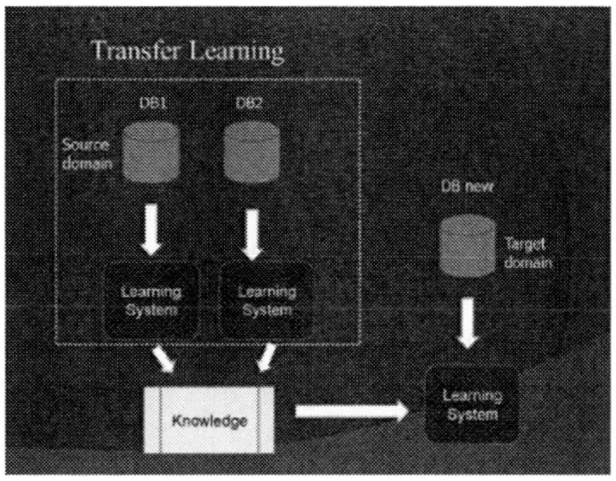

Figure 1. (left) Traditional machine learning considers each task in islation. (right) Transfer learning leverages information from previous experience.

less luminous stars fall below the detection threshold and more luminous stars are preferentially detected. The corresponding change in distribution precludes the direct utilization of one single model across galaxies; it calls for a form of model adaptation to compensate for the change in the data distribution. Many other domains exist where changes in the underlying probability distribution are primarily caused by measurements obtained under different circumstances (e.g., observing galaxies that lie at different distances), by changing the orientation or position of the same sensing device, or by utilizing a similar but more powerful device. In particle physics, for example, a model built to identify a certain particle is rendered unapplicable when we collect samples obtained using more powerful particle accelerators; this is because the range of parameter values shifts as we reach out to higher energies.

The abundance of examples exhibiting a shift in distribution as circumstances change over time calls for automated methods that adapt as a response to the dynamic nature of many learning tasks. The type of problems mentioned above have recently led to the development of a new area of study called *transfer learning*, where the mechanism leverages previous experience to increase the accuracy of predictive models. In this paper, I briefly introduce central ideas in transfer learning and describe a practical application in astronomy that captures the essence of this new paradigm.

1.1. Rationale for Transfer Learning

Before the advent of transfer learning techniques, practical solutions to the problem of shifts in probability distributions led to a continuous re-training of a predictive model over time. Each new instance of a classification or regression problem was considered in isolation, lacking the capacity to exploit information from previous experience. Figure 1(left) shows this operational mode. A training set serves as input to the process of modelling data using machine learning tools. Other datasets are treated similarly but in isolation. No information is exploited that can serve to learn experience across tasks. The traditional approach is particularly detrimental in situations where similar datasets share information that is useful across all tasks. This is clearly the case in the astronomical or particle-physics scenarios mentioned above, where a change in the sensor device or the event generator does not render any previous form of data analysis useless; instead one should be able to exploit the presence of similar patterns across tasks.

The new approach to the construction of adaptive learning models is to gather experience from previous tasks to improve on the current target task. This is also known as the transfer learning problem. Figure 1(right) depicts the new approach. Previous experience is now retained for future analysis (source domains). The repetitive use of machine learning is stored in a knowledge database that varies in nature according to the problem or task under analysis. The main difference with the previous approach becomes evident when a new target domain arrives

IOP Conf. Series: Journal of Physics: Conf. Series **1085** (2018) 052014 doi:10.1088/1742-6596/1085/5/052014

(i.e., new task). The data modelling process is now reinforced with previous experience to strengthen the predictive model, and gain leverage from patterns found in previous tasks. A brief description of these adaptive techniques is given in the following section.

2. Different Modalities in Transfer Learning

Before describing different approaches to transfer learning, I introduce some notation. It is common to assume a source dataset $T_s = \{(\mathbf{x}, y)\}$ made of training examples, where $\mathbf{x} \in \mathcal{X}$ is a feature vector, and $y \in \mathcal{Y}$ is a class or category. Dataset T_s, the source dataset, corresponds to a previous application of machine learning that produced a predictive model $f_s(\mathbf{x})$. Examples in T_s are drawn randomly from a joint distribution $P_s(\mathbf{x}, y)$; of relevance here are the marginal distribution $P_s(\mathbf{x})$ and the class-posterior distribution $P_s(y|\mathbf{x})$. We will assume model $f_s(\mathbf{x})$ needs some form of adaptation because of changes in the original data distribution. The second dataset, T_t, the target dataset, corresponds to a new application that is similar to the source task, but not identical. We assume the existence of corresponding distributions on the target task: marginal distribution $P_t(\mathbf{x})$, and class-posterior distribution $P_t(y|\mathbf{x})$. Now, rather than building $f_t(\mathbf{x})$ from scratch, transfer learning can be invoked to exploit previous experience [2, 3]. A justification for the use of transfer learning is that T_t is either small, or contains few –or none– labeled instances (e.g., due to the cost of label annotation).

Many techniques have been proposed in the area of transfer learning [10, 17]. A broad taxonomy begins by defining two approaches: homogenous and heterogenous transfer learning. In homogenous transfer learning, the feature representation for source and target is identical $\mathcal{X}_s = \mathcal{X}_t$ (i.e., both tasks share the same feature representation). In heterogenous transfer learning we dispense with such assumption.

Within homogenous transfer learning, a simple scenario assumes the difference between source and target lies on the marginal distributions $P_s(\mathbf{x}) \neq P_t(\mathbf{x})$. Other scenarios assume the marginals remain the same, while the class-posterior distributions differ $P_s(y|\mathbf{x}) \neq P_t(y|\mathbf{x})$, or that both marginals and posterior distributions differ.

A final categorization of transfer learning methods focuses on the nature of the element being transferred. One approach is to transfer instances between source and target. The strategy behind instance-based methods is to increase the weight of source instances populating regions of high density in the target domain. A piece of work along these lines is known as covariate shift [11, 15]. Under the covariance-shift assumption, the model built on the new weighted source distribution can be directly applied to the target domain. A strong assumption here requires close proximity between source and target distributions.

Another approach is to transfer features across tasks. Feature-based transfer learning projects both the source and the target datasets into a common feature space where the covariate-shift assumption holds. The new model built on the transformed space acts as the classifier on the target. Instances of this family include subspace alignment methods, where the goal is to find a common subspace that makes source and target distributions overlap.

A third approach is to transfer parameter values across tasks. This can be accomplished by simply transferring parameter values directly from source to target, or by following a Bayesian approach where model parameters on the source task are used to build a prior distribution on the target task. Additionally, multiple models can be induced from the source task, and a weighting scheme can be used to provide adequate weights to combine the learners on the target task.

The last approach is to transfer knowledge based on the nature of the relation between tasks. As an example, in relational-based transfer learning, the idea is to look for relational patterns across source and target. An example is to predict documents based on grammatical and sentence structure patterns across texts that belong to similar topics [9].

3. An Application of Transfer Learning in Astronomy

An application of transfer learning in astronomy can be found in the automated identification of Supernova Ia (SNe Ia) [7]. The task is of great importance to astronomy, because SNe Ia are considered standard candles in probing large cosmological distances; the correlation between their luminosity and distance independent quantities has led to multiple discoveries, including the accelerating expansion of the Universe [12]. The discrepancy in distribution that makes this a problem amenable to transfer learning is due to the difference between spectroscopic and photometric observations. Spectroscopy provides a high-resolution description of electromagnetic radiation and is crucial to estimate chemical composition (through spectral lines) and distances (redshift) with high precision. But spectroscopic observations are costly and time-consuming; in practice, it is more convenient to perform photometric observations that summarize radiation in a set of broad wavelength windows or filters, even at the expense of forfeiting the wealth of information otherwise available using a spectroscopic analysis.

Many modern surveys including the the Sloan Digital Sky Survey, the Dark Energy Survey, and the upcoming Large Synoptic Survey (Telescope) are designed to capture pure photometric observations for SNe Ia. A big challenge in modern astronomy is to infer spectroscopic properties from purely photometric data. In our current discussion, the source domain corresponds to spectroscopic data where Supernovae are confidently classified (as type Ia or different); the target domain corresponds to the new generation of abundant photometric light-curve datasets where class labels are scarce. Our goal is to take advantage of the source domain to attain high predictive performance on the target domain. The use of transfer learning here is crucial, since photometric observations lack a precise class label (e.g. type of SNe) to conform a reliable training set. At the same time, the naive approach of training a model $f_s(\mathbf{x})$ on the spectroscopic data T_s (source domain) and applying it directly on the photometric data T_t (target domain) is prone to failure because of the distributional shift observed between both types of observations.

Methodology

I now describe a specific study aimed at building accurate SNe Ia classifiers from photometric data, while exploiting information from spectroscopic data [7]. The first step is (almost always invariably) to reduce the dimensionality of the (source and target) data, since the original samples contain a large number of features. A recommended approach is to use kernel principal component analysis (Kernel PCA) [14]; the technique is a generalization of principal component analysis (PCA) by using non-linear components. Recently, new approaches have emerged that compute robust non-linear combinations of features, such as those found in deep-learning architectures [5, 13].

The second step is to use transfer learning to leverage experience from the source domain (spectroscopic dataset) to attain an accurate classifier on the target domain (photometric dataset). The study reported here considered two methods used: kernel mean matching KMM, and subspace alignment SA. The first method, KMM, works through a re-weighting scheme that gives more importance to those examples on the source dataset that appear closer to the target dataset [6]. Specifically, the method projects the data into a new space (reproducing Hilbert kernel space) where it minimizes the (maximum) distance between the means on each distribution (source and target). The projection helps to identify source examples that can be incorporated into the training phase while building a classifier on the target domain.

The second method, subspace alignment, is a common transfer learning method based on feature subset selection [4]. The idea is to apply PCA on source T_s and target T_t datasets separately by choosing a common space. It then attempts to align the projected source dataset with the projected target dataset in this common subspace using a subspace alignment matrix. Once source and target are aligned, a classifier is built on the transformed source dataset T_s^α, and subsequently applied to the transformed target dataset T_t^α.

Empirical Results

Experimental results are shown in Figure 2. The horizontal axis shows different classifiers including neural networks and support vector machines [8], the latter implemented with different degrees of complexity. The vertical axis shows accuracy (proportion of examples correctly classified) using 10-fold cross validation. Some interesting observations in Figure 2 are as follows. First, KMM does provide a significant advantage over a methodology that lacks any form of transfer learning. Second, the advantage is not always there; in a few cases (SVM degree 3) performance degrades; a reason for such behavior is that overly complex models capture patterns that belong exclusively to the training set and may incur in a large number of misclassifications in a validation set. Third, SA exhibits poor performance, and does not provide any advantage during transfer learning. This is a phenomenon known as *negative transfer* where the use of transfer learning leads to a loss of performance. The poor performance of SA can be explained as a result of the preliminary step to unify source and target into the same feature space; the common space does not achieve a sufficient overlap between the two distributions.

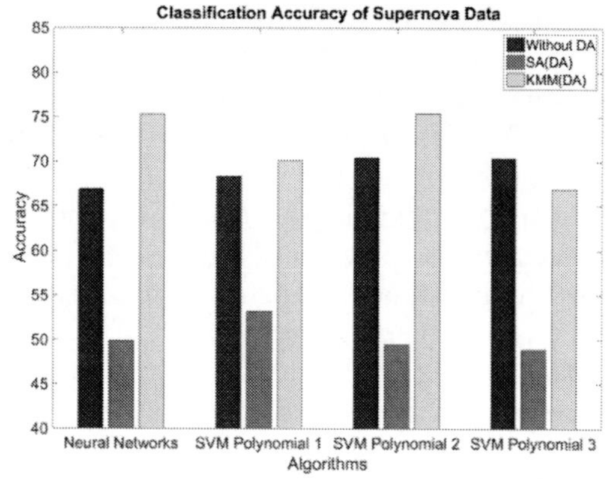

Figure 2. Accuracy on supernova target dataset with and without transfer learning (labeled as DA for Domain Adaptation).

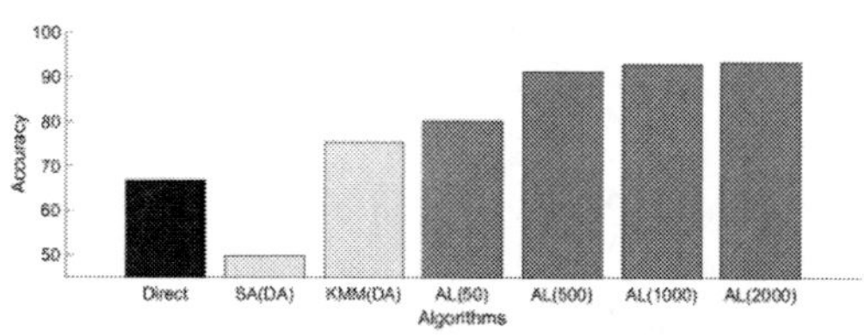

Figure 3. Accuracy on target data using transfer learning and active learning. The last 4 columns show performance when active learning is invoked (with increasing budget).

An additional observation is that further enhancement of the target model can be achieved by combining transfer learning with an area of study in machine learning known as *active-learning* [1]. Active learning points to those few instances on the target set where knowing the class label with confidence suffices to attain an accurate model. Here, the system is allowed to ask the expert for the right class label on the selected event or instance (e.g., by running a spectroscopic analysis on the selected Supernova). The process works as follows. It begins by using the model $f_t(\mathbf{x})$ created by the transfer learning algorithm as the initial model for active learning. This is followed by an iterative process that queries the next most informative instance from the target dataset, and builds a new model after the last queried instance is added to the sample. The algorithm stops when it reaches a maximum cost i.e., when it runs out of budget. Figure 3 shows results for the automated classification of Supernovae Ia when transfer learning and active learning are combined synergistically. Results show a significant advantage gained when transfer learning is combined with active learning. Even with a modest budget size, the combination yields better performance than the use of transfer learning alone. Of course, growing the budget size provides increased performance gain.

4. Summary and Conclusions

This paper shows the value behind an emerging area of study in machine learning known as *transfer learning*. The main idea is to leverage experience from previous learning tasks by transferring knowledge from a source domain to a target domain (Section 1). There are multiple approaches to transfer learning (Section 2); examples include homogenous and heterogenous transfer; transfer mechanisms based on the differences between marginal and/or class-posterior distributions; and finally, transfer learning based on the nature of the element being transferred.

The experimental study described in Section 3 shows an application of transfer learning on the identification of Supernovae Ia (SNe Ia). The study shows that while transfer learning can yield substantial gains in performance, the technology itself is not always guaranteed to succeed. In the particular study described here, kernel mean matching shows an increase in accuracy performance when identifying SNe Ia, but subspace alignment leads to a loss in accuracy; for the latter case I mentioned that some approaches to transfer learning can lead to what is known as *negative transfer*, a scenario where performance degradation is observed. The section also mentions the benefit of combining transfer learning with active learning as a means to add relevant labeled examples from the target dataset.

4.1. Acknowledgments

This work was partly supported by the Center for Advanced Computing and Data Systems (CACDS), and by the Texas Institute for Measurement, Evaluation, and Statistics (TIMES) at the University of Houston.

References

[1] Balcan, M.-F.; Beygelzimer, A.; Langford, J. 2009. Agnostic Active Learning. Journal of Computer and System Sciences, 75(1):78 – 89.
[2] Ben-David, S.; Blitzer, J.; Crammer, K.; Kulesza, A.; Pereira, F.; Vaughan, J. 2010. A Theory of Learning from Different Domains. Machine Learning, vol. 79, pp. 151175, 2010.
[3] Blitzer, J.; Crammer, K.; Kulesza, A.; Pereira, F.; Wortman, J. 2008. Learning Bounds for Domain Adaptation. Advances in Neural Information Processing Systems 20. Platt, J. C.; Koller, D.; Singer, Y.; Roweis, S. T.; Eds. Curran Associates, Inc., 2008, pp. 129136.
[4] Fernando, B.; Habrard, A.;p Sebban, M.; Tuytelaars, T. 2014. Subspace Alignment for Domain Adaptation. CoRR, vol. abs/1409.5241.
[5] Goodfellow, I; Bengio, Y; Courville, A. 2016. Deep Learning. MIT Press.
[6] Gretton, A.; Smola, A.; Huang, J.; Schmittfull, M.; Borgwardt, K.; Scholkopf, B. 2009. Covariate Shift by Kernel Mean Matching. Dataset Shift in Machine Learning, vol. 3, no. 4.
[7] Dhar Gupta, K.; Pampana, R.; Vilalta, R.; Ishida, E. E. O.; de Souza, R. S. 2016. Automated Supernova Ia Classification Using Adaptive Learning Techniques. IEEE Symposium on Computational Intelligence and Data Mining (CIDM-16), Athens, Greece.
[8] Hastie, T.; Tibshirani, R.; Friedman, J. 2009. The Elements of Statistical Learning: Data Mining, Inference and Prediction. Springer, 2nd edition.
[9] Li, F.; Pan, S.J.; Jin, O.; Yang, Q.; Zhu, X. 2012 Cross-Domain Co-Extraction of Sentiment and Topic Lexicons. Proceedings of the 50th Annual Meeting of the Association for Computational Linguistics Long Papers, vol. 1. 2012. p. 41019.
[10] Pan, S. J.; Yang, Q. 2010. A Survey on Transfer Learning. IEEE Transactions on Knowledge and Data Engineering 22(10):1345–1359.
[11] Quionero-Candela, J.; Sugiyama, M.; Schwaighofer, A.; Lawrence, N. D. 2009. Dataset Shift in Machine Learning. The MIT Press.
[12] Riess, A. G., et al. 1998. Observational Evidence from Supernovae for an Accelerating Universe and a Cosmological Constant. The Astronomical Journal, AJ, V. 116, No. 3, 1009.
[13] Sasdelli, M., et al. 2016. Exploring the Spectroscopic Diversity of Type Ia Supernovae with DRACULA: A Machine Learning Approach. Monthly Notices of the Royal Astronomical Society (MNRAS), 461, no.2, pp. 2044-2059, Oxford University Press.
[14] Scholkopf, B.; Smola, A.; Muller, K.-R. 1997. Kernel Principal Component Analysis. Berlin, Heidelberg: Springer Berlin Heidelberg, 1997, pp. 583588.
[15] Shimodaira, H. 2000. Improving Predictive Inference Under Covariate Shift by Weighting the Log-Likelihood Function. Journal of Statistical Planning and Inference, vol. 90, no. 2, pp. 227244.
[16] Vilalta, R.; Dhar Gupta K.; Macri, L. 2013. A Machine Learning Approach to Cepheid Variable Star Classification Using Data Alignment and Maximum Likelihood. Astronomy and Computing Journal, Vol. 2, pp. 46-53. Elsevier. DOI: 10.1016/j.ascom.2013.07.002.
[17] Weiss, K.; Khoshgoftar, T.; Wang, D. 2016. A Survey of Transfer Learning. Journal of Big Data 3:9 DOI 10.1186/s40537-016-0043-6.

IOP Conf. Series: Journal of Physics: Conf. Series **1085** (2018) 052015 doi:10.1088/1742-6596/1085/5/052015

GQLink: an implementation of Quantized State Systems (QSS) methods in Geant4

Lucio Santi[1,2], Federico Bergero[3], Soon Yung Jun[4], Krzysztof Genser[4], Daniel Elvira[4], Rodrigo Castro[1,2]

[1] Department of Computer Science, FCEN, University of Buenos Aires, Argentina
[2] ICC-CONICET, Argentina
[3] CIFASIS-CONICET, Argentina
[4] Fermi National Accelerator Laboratory †, PO Box 500, Batavia, IL 60510, USA

E-mail: {lsanti,rcastro}@dc.uba.ar

Abstract. Simulations in high energy physics (HEP) often require the numerical solution of ordinary differential equations (ODE) to determine the trajectories of charged particles in a magnetic field when particles move throughout detector volumes. Each crossing of a volume interrupts the underlying numerical method that solves the equations of motion, triggering iterative algorithms to estimate the intersection point within a given accuracy. The computational cost of this procedure can grow significantly depending on the application at hand. Quantized State System (QSS) is a recent family of discrete-event driven numerical methods exhibiting attractive features for this type of problems, such as native dense output (sequences of polynomial segments updated only by accuracy-driven events) and lightweight detection and handling of volume crossings. In this work we present GQLink, a proof-of-concept integration of QSS with the Geant4 simulation toolkit which stands as an interface for co-simulation that orchestrates robustly and transparently the interaction between the QSS simulation engine and aspects such as geometry definition and physics processes controlled by Geant4. We validate the accuracy and study the performance of the method in simple geometries (subject to intense volume crossing activity) and then in a realistic HEP application using a full CMS detector configuration.

1. Introduction

High energy physics (HEP) particle simulations deal intensively with the tracking of particles affected by physics processes in complex detector geometries (adjacent 3D volumes of different shapes and materials). The accuracy and efficiency of the tracking algorithms are of central interest to optimize the use of available computing resources. Every time a charged particle crosses a volume boundary, the numerical method that solves the underlying ordinary differential equations (ODEs) [1] in a magnetic field needs to be interrupted. As this situation can happen many times during the lifetime of a particle in a typical HEP setup, the underlying continuous models describing particle trajectories must cope with frequent discontinuities. These spatial discontinuities will seldom coincide with the start of a new step. Thus, custom iterative algorithms are required to approximate the timestamp of the event time of each discontinuity in an accurate way. When these events are very frequent, they can dominate the processing time of the numerical method and considerably degrade its performance.

Contemporary HEP experiments are widely based on the Geant4 [2] simulation toolkit. It uses classical numerical methods that rely on time discretization [3], in particular variations of the Runge-Kutta family of solvers [4]. The Quantized State System (QSS) methods [3, 5], on the other hand, discretize the state variables instead of discretizing the time and solve ODEs using discrete–event approximations of continuous models. A feature of QSS particularly relevant in the context of HEP simulations is that these methods handle discontinuities very efficiently [6]. In QSS, the state variables follow piecewise polynomial trajectories. The detection of a volume crossing is modeled by solving a zero-crossing function, which can be done efficiently using the aforementioned polynomials (at least for some categories of the surface boundaries). Since each new QSS step is, by definition, a discontinuity in the quantized variable –strictly speaking, a discrete event– this task is naturally supported by the numerical method.

In previous works [7, 8] we verified the potential of QSS to offer speedups in HEP simulations, in particular in scenarios with heavy volume crossing activity. Yet, our studies were limited to compare Geant4 and QSS Solver [9], a standalone simulation toolkit that provides optimized implementations of QSS methods. A salient limitation of this approach is that physics processes were turned off for comparability purposes, restricting the comparisons to simple setups conceived as baselines. In this work we enable the capability to evaluate performance in realistic HEP applications by means of a proof-of-concept integration of QSS within Geant4.

This paper is organized as follows: Section 2 briefly introduces the QSS theory and the Geant4 transportation chain and then summarizes our preliminary results on QSS suitability for HEP simulations. Section 3 provides a high-level description of GQLink. Next, Section 4 discusses the verification of the method and presents a performance comparison between pure Geant4 and GQLink in two different setups. Finally, Section 5 contains a summary, conclusions and plans for the future.

2. Background

2.1. The Quantized State System (QSS) numerical integration methods

QSS are numerical methods that solve systems of ordinary differential equations (ODEs) in the form of Eq. 1, where $\mathbf{x}(t)$ is the *state vector* and $\mathbf{u}(t)$ is the *input vector* representing independent variables for which no derivatives are present in the system.

$$\dot{\mathbf{x}}(t) = f(\mathbf{x}(t), \mathbf{u}(t)) \quad (1) \qquad \dot{\mathbf{x}}(t) = f(\mathbf{q}(t), \mathbf{u}(t)) \quad (2)$$

Traditional ODE solvers (like the Runge-Kutta family) make use of *time slicing*: given the current and past state and state derivative values, the solver estimates the next value of the state x_{k+1} one "time step" Δt into the future, i.e., at $t_{k+1} = t_k + \Delta t$. QSS solvers operate differently: they make use of *state space quantization*. Rather than discretizing the time axis, they discretize the state variable axis, evaluating the first time instant t_{k+1} in the future at which the state variable differs from its current value by one "quantum level" ΔQ, i.e., when $x_{k+1} = x_k \pm \Delta Q$. The system described by Eq. 1 is thus approximated by the quantized system shown in Eq. 2, where $\mathbf{q}(t)$ is the *quantized state vector* resulting from the quantization of the state variables $x_i(t)$. In the first-order QSS method (QSS1) each $q_i(t)$ follows a piecewise constant trajectory that is updated by a (hysteretic) *quantization function* when the difference between $q_i(t)$ and $x_i(t)$ reaches the *quantum* ΔQ_i, a value derived from the precision demanded by the user.

In QSS1, $\mathbf{q}(t)$ follow piecewise constant trajectories, which implies that $\mathbf{x}(t)$ follow piecewise linear trajectories. Higher-order QSS methods generalize this behavior: in QSSn, $\mathbf{x}(t)$ follow piecewise n-th degree polynomial trajectories and $\mathbf{q}(t)$ follow piecewise $(n-1)$-th degree polynomial trajectories [5].

In the context of HEP applications, QSS exhibit several attractive features, such as inherent asynchronicity (each state variable updates its value independently at self clocked time instants dictated by its own dynamics and the accuracy ΔQ), dense trajectory output (at any given time t_0, QSS approximates $\mathbf{x}(t_0)$ using the piecewise polynomial trajectories computed for the interval in which t_0 lies in) and efficient handling of frequent discontinuities (a discontinuity is modeled by a zero-crossing function which is in turn defined in terms of the QSS polynomials).

Regarding practical tools, *QSS Solver* [9] is an open-source standalone software that provides optimized C implementations of different QSS methods as well as other traditional algorithms (e.g., Dormand-Prince method). Equations are expressed in μ-Modelica [10], a subset of the more general Modelica language [11].

2.2. Particle transport in Geant4

Geant4 traces particle trajectories in a magnetic field by means of the Runge-Kutta family of numerical methods. Figure 1 illustrates how a charged particle q is transported along a step of length h proposed by a physics process. In this context, the step computation involves a total of 11 right-hand side ODE evaluations for the 4th-order Runge-Kutta.

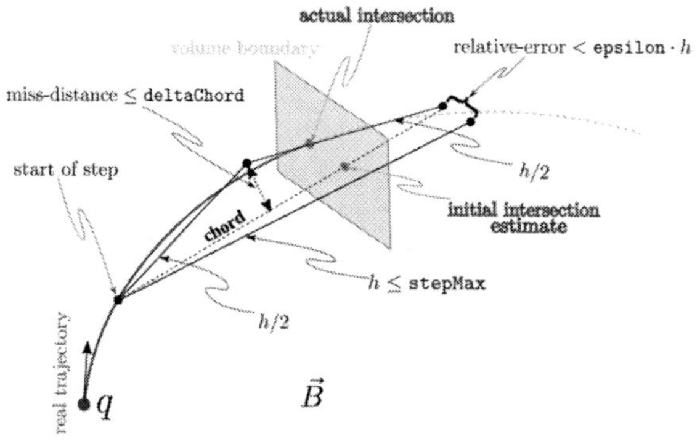

Figure 1: A charged particle trajectory in a magnetic field ($\vec{B}$) using the 4th-order Runge-Kutta method.

A comprehensive explanation of the stepping algorithm can be found in [7]. In particular, we stress that the intersection point detection algorithm can be rather expensive, as it needs to iterate back and forth until a candidate point satisfying the intersection accuracy constraints is found. This suggests that methods such as QSS, which naturally provide a lightweight handling of discontinuities, are good candidates for simulating HEP setups with frequent volume crossings. Indeed, the preliminary results summarized in the next section support this hypothesis.

2.3. Preliminary results

In [7, 8] we studied the performance of QSS Solver in the context of a simplified HEP setup consisting of a circular 2D motion of a charged particle under an uniform magnetic field. The particle crosses equidistant parallel planes as shown in Figure 2a. We found configurations with 200 planes and a track length of 100 m where QSS Solver is 6x faster than Geant4 (Figure 2b) with better accuracy, when the physics interactions are turned off.

3. Geant4 to QSS Link (GQLink): linking QSS libraries to Geant4

GQLink is a proof-of-concept mechanism to make QSS methods available within Geant4. It is currently based on the Geant4 toolkit version `10.03.p01` and the QSS Solver toolkit version 3.0. Designed as an abstract, clean, single entry point interface to the QSS Solver family of numerical integration methods, GQLink is virtually ready to support other integrators in the future. This is depicted in Figure 3a. It must be noted that GQLink was not conceived as yet another Geant4 *stepper*, although its main function is to replace the tasks typically performed by *steppers* in Geant4. Being integrated into the Geant4 building process, GQLink provides three new shared libraries: `libqss`, containing QSS core functionality, mostly derived from the QSS Solver engine, `libgqlink`, which offers the interface API between Geant4 and QSS, and `libmodel`, where the actual structure of the model to be simulated is defined (typically the Lorentz equations of motion).

IOP Conf. Series: Journal of Physics: Conf. Series **1085** (2018) 052015 doi:10.1088/1742-6596/1085/5/052015

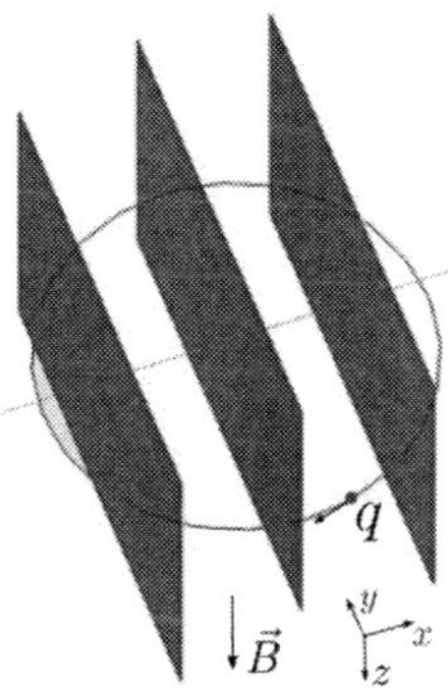

(a) Experimental setup

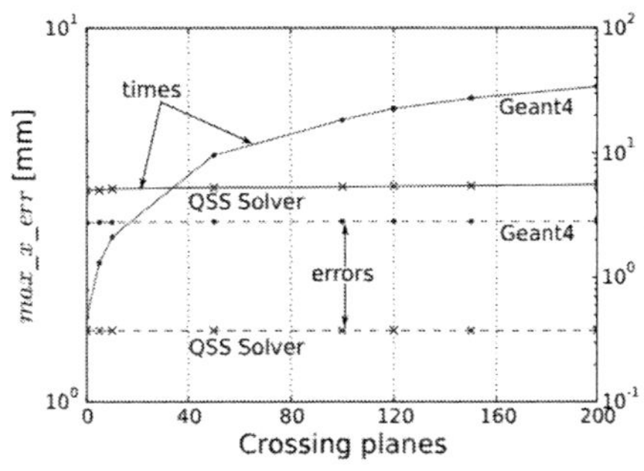

(b) Maximum error and simulation time vs. number of crossing planes

Figure 2: Preliminary comparison between Geant4 and QSS Solver

It is worth noting that currently boundary crossings are still detected through Geant4's geometry library. This is achieved by following the same call pattern as in standard Geant4 simulations: after each substep, `LocateGlobalPointWithinVolume` is called so as to notify the geometry navigator that the particle has moved to a new position inside the current volume. Then, `IntersectChord` computes an initial estimate of an intersection by means of a linear segment between the endpoints of the substep, which is later refined through an iterative procedure in `EstimateIntersectionPoint`. Still, this last function was improved by the QSS polynomial dense output information; we offer a cheaper particle transport by evaluating the polynomials instead of running the iterative, adaptive stepping algorithm implemented in `AccurateAdvance`. Figure 3b sketches how an intersection point is computed in GQLink.

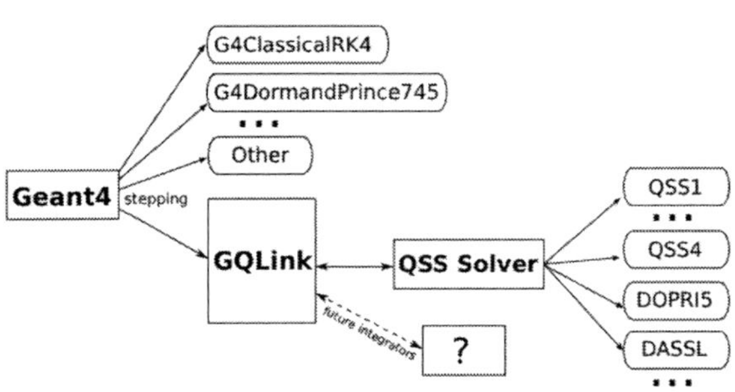

(a) High-level architecture

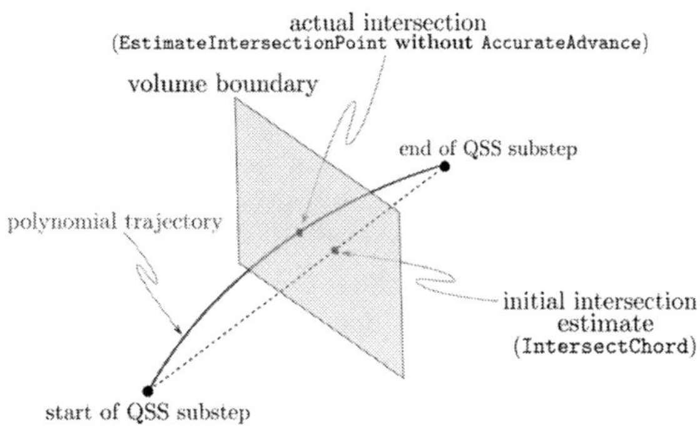

(b) Computation of intersection points

Figure 3: GQLink implementation details

We finally stress that QSS dense output is not fully exploited yet for boundary crossing detection; this is in fact the main goal driving our current work.

4. Results and discussion

We start with a statistical validation of GQLink in the context of a CMS application in order to ensure that it produces equivalent results than Geant4. After this, we compare their computing performance in this same CMS setup and also in a simplified setup with a

IOP Conf. Series: Journal of Physics: Conf. Series **1085** (2018) 052015 doi:10.1088/1742-6596/1085/5/052015

controlled, increasing number of boundary crossings. All simulations were run on a dedicated Intel Xeon E5-2620 v2 CPU (clocked at 2.10 GHz) server with 8 GB of RAM and Ubuntu 14.04.4 LTS x86_64 (`3.13.0-49-generic` kernel) OS. The compiler used was `gcc` 4.8.4 (Ubuntu 4.8.4-2ubuntu1~14.04.1).

GQLink validation was performed against a standalone Geant4 application featuring full CMS (RunI) detector geometry, volume base magnetic field excerpted from CMSSW and a particle gun shooting π^- particles (10 GeV, 10^4 events). In Figure 4, we show how the particle steps are distributed after 10^4 single π^- events, for pions (left) and secondary electrons (right). We observe that the number of steps produced are statistically consistent with those computed by Geant4. This behavior is in fact present in every other particle type as well as in the number of tracks.

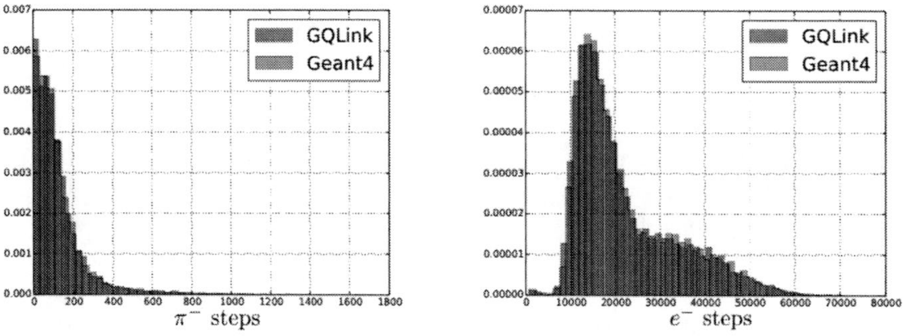

Figure 4: Number of steps for π^- (left) and secondary electrons (right) from simulation of 10^4 single π^- events

As for the computing performance, Figure 5a compares the simulation time of GQLink and Geant4 for an increasing number of events in the same CMS setup presented above. For each number of events, a total of 50 simulations were run. The plot shows their mean values along with the minimum and maximum values attained. In this case, GQLink performed about 17% slower than Geant4. We inferred that this is due to a combination of issues: a relatively small number of boundary crossings (about 8% of the total number of steps), a greater average number of substeps per integration step (up to 3x with respect to Geant4, with the set relative accuracy of 10^{-5}) and frequent reinitializations of QSS state variables after each Geant4 step (required to update the state derivatives after possible changes in the direction of the particle).

However, when the number of boundary crossings is controlled and can be forced to increase, GQLink's performance progressively improves and it eventually outperforms Geant4. This behavior is shown in Figure 5b, where we consider a simplfied experimental setup very similar to that of Figure 2 but with a single particle describing a helicoidal trajectory and with physics interactions turned off. We see that GQLink outperforms Geant4 when using at least 300 crossing planes, and can be up to ~35% faster for 700 planes. This is a consequence of the improvement derived from the QSS polynomials implemented in `EstimateIntersectionPoint`, as mentioned in Section 3.

Finally, we also analyzed the correctness and performance of the method using 50 Pythia $pp \to H \to ZZ$ events (Z to all channels) ($\sqrt{s} = 14$ TeV) in the context of the CMS application. In this case, we found that GQLink successfully simulated the 50 events in 5.86 hours, whereas Geant4 took 4.8 hours. This is an increase of approximately 22%. Note, however, that the same limitations discussed above apply in this scenario as well –namely, very few boundary crossings and greater number of QSS substeps per integration step.

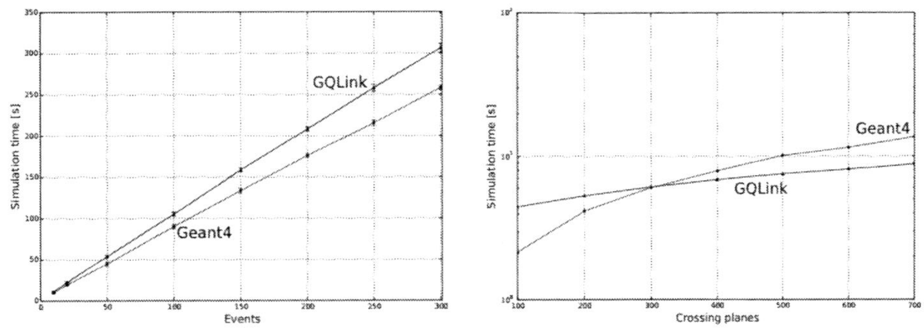

(a) CMS application (single π^- events) (b) Helix and parallel planes; physics interactions turned off

Figure 5: Performance comparison between GQLink and Geant4

5. Conclusions

We developed GQLink, a prototype for QSS methods (as implemented by the standalone simulation toolkit QSS Solver) within Geant4. It is conceived as an abstract interface that allows connecting Geant4 to arbitrary external integrators, not limited to those provided by QSS Solver. In order to validate its correctness and study its performance, we used a realistic HEP application using a full CMS detector geometry, with magnetic field and both particle gun events, shooting pions, as well as Pythia $pp \to H \to ZZ$ events (Z to all channels) ($\sqrt{s} = 14$ TeV). We found that the number of steps and tracks produced are statistically consistent with those of standard Geant4, suggesting the equivalence of the simulations. As for the performance, since these scenarios have a very light boundary crossing activity, we found that GQLink is currently about 17% slower. On the other hand, we studied a simplified scenario, with physics interactions turned off, where a single particle describing helical trajectories moves throughout an increasing, controlled number of crossing planes. In this setup, GQLink outperformed Geant4, with a 35% speedup for 700 crossing planes. This is due to an improvement in the Geant4 geometry library, where we resorted to the knowledge stored in each QSS polynomial trajectory in order to avoid some iterative stepping procedures used when locating intersection points. We are currently working on fully exploiting these QSS capabilities for efficient volume crossing detections. We aim at handling the detections completely on the QSS side, which is expected to provide considerable performance boosts.

Acknowledgments

† Operated by Fermi Research Alliance, LLC under Contract No. DE-AC02-07CH11359 with the United States Department of Energy.

References

[1] Hairer E, Nørsett S and Wanner G 1987 Solving ordinary differential equations i - nonstiff problems
[2] Allison J 2016 *Nuclear Instruments and Methods A* **835** 186–225
[3] Cellier F E and Kofman E 2006 *Continuous System Simulation* (Springer-Verlag New York, Inc.)
[4] Cockburn B and Shu C W 1998 *Journal of Computational Physics* **141** 199–224
[5] Kofman E and Junco S 2001 *Transactions of SCS* **18** 123–132
[6] Kofman E 2004 *SIAM Journal on Scientific Computing* **25** 1771–1797
[7] Santi L, Ponieman N, Jun S Y, Genser K, Elvira D and Castro R 2016 *Journal of Physics: Conference Series*
[8] Ponieman N 2015 *Aplicación de Métodos de Integración por Cuantificación al Simulador de Partículas Geant4* Master's thesis Facultad de Ciencias Exactas y Naturales. Universidad de Buenos Aires.
[9] Fernández J and Kofman E 2013 *Proc. of RPIC 2013* (Bariloche, Argentina)
[10] Bergero F, Floros X, Fernández J, Kofman E and Cellier F E 2012 *9th Int. Modelica Conference, Germany*
[11] Fritzson P 2004 *Principles of Object-Oriented Modeling and Simulation with Modelica 2.1* (Wiley-Interscience, New York)

IOP Conf. Series: Journal of Physics: Conf. Series **1085** (2018) 052016 doi:10.1088/1742-6596/1085/5/052016

Four-point function in general kinematics through geometrical splitting and reduction

Andrei I Davydychev

Institute for Nuclear Physics, Moscow State University, 119992 Moscow, Russia

E-mail: davyd@theory.sinp.msu.ru

Abstract. It is shown how the geometrical splitting of N-point Feynman diagrams can be used to simplify the parametric integrals and reduce the number of variables in the occurring functions. As an example, a calculation of the dimensionally-regulated one-loop four-point function in general kinematics is presented.

1. Introduction

In the general off-shell case, one-loop N-point diagrams (shown in figure 1) depend on $\frac{1}{2}N(N-1)$ momentum invariants $k_{jl}^2 = (p_j - p_l)^2$ and N masses of the internal particles m_i. Here and below, for the corresponding scalar integrals we follow the notation $J^{(N)}\left(n; \{\nu_i\}|\{k_{jl}^2\}, \{m_i\}\right)$ used in [1], where ν_i are the powers of the internal scalar propagators, and the space-time dimension is denoted as n, so that we can also deal with the dimensionally-regulated integrals with $n = 4 - 2\varepsilon$ [2]. Below we will mainly consider the cases when all $\nu_i = 1$.

A geometrical interpretation of kinematic invariants and other quantities related to N-point Feynman diagrams helps us to understand the analytical structure of the results for these diagrams. As an example, singularities of the general three-point function can be described pictorially through a tetrahedron constructed out of the external momenta and internal masses. Such a geometrical visualization can be used to derive Landau equations defining the positions of possible singularities [3] (see also in [4]).

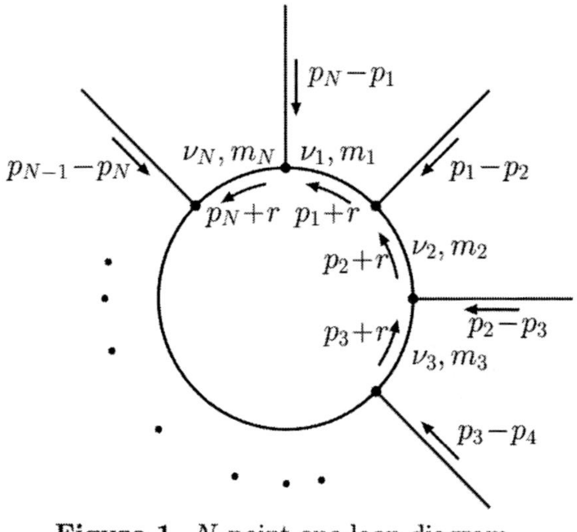

Figure 1. N-point one-loop diagram

In [5, 6, 7] it was demonstrated how such geometrical ideas could be used for an analytical calculation of one-loop N-point diagrams. For the geometrical interpretation, a "basic simplex" in N-dimensional Euclidean space is employed (a triangle for $N = 2$, a tetrahedron for $N = 3$, etc.), and the obtained results are expressed in terms of an integral over an $(N-1)$-dimensional spherical (or hyperbolic) simplex, which corresponds to the intersection

IOP Conf. Series: Journal of Physics: Conf. Series **1085** (2018) 052016 doi:10.1088/1742-6596/1085/5/052016

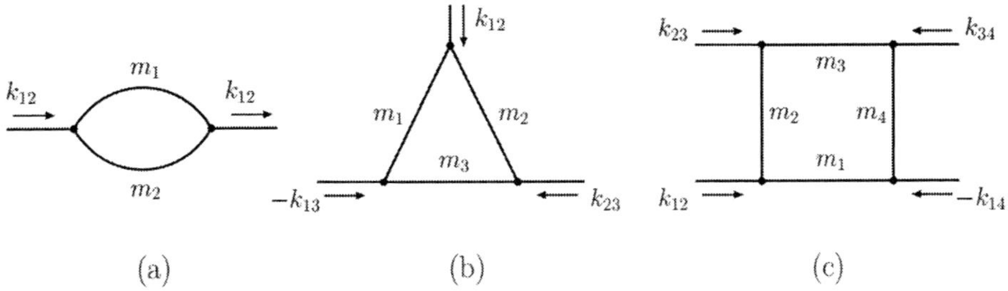

Figure 2. Momenta and masses in (a) two-point (b) three-point and (c) four-point diagrams.

of the basic simplex and the unit hypersphere (or the corresponding hyperbolic hypersurface), with a weight function depending on the angular distance θ between the integration point and the point 0, corresponding to the height of the basic simplex (see in [5]). For $n = N$ this weight function is equal to 1, and the results simplify: for the case $n = N = 3$ see in [8], and for the case $n = N = 4$ see in [9, 10]. Other interesting examples of using the geometrical approach can be found, e.g., in [11].

In this paper we will demonstrate that the natural way of splitting the basic simplex, as prescribed within the geometrical approach discussed above, leads to a reduction of the effective number of independent variables in separate contributions obtained as a result of such splitting [12]. Moreover, by considering examples with $N \leq 4$ we will show that this reduction leads to simplifications in the corresponding Feynman parametric integrals, which can be explicitly calculated in terms of the (generalized) hypergeometric functions.

2. Two-point function

For the two-point function (see figure 2a), there is only one external momentum invariant k_{12}^2, and the sides of the corresponding basic triangle are m_1, m_2 and $K_{12} \equiv \sqrt{k_{12}^2}$ (see figure 3a). The angle τ_{12} between the sides m_1 and m_2 is defined through $\cos\tau_{12} \equiv c_{12} = (m_1^2 + m_2^2 - k_{12}^2)/(2m_1m_2)$, and (in the spherical case) the integration goes over the arc τ_{12} of the unit circle, as shown in figure 3b.

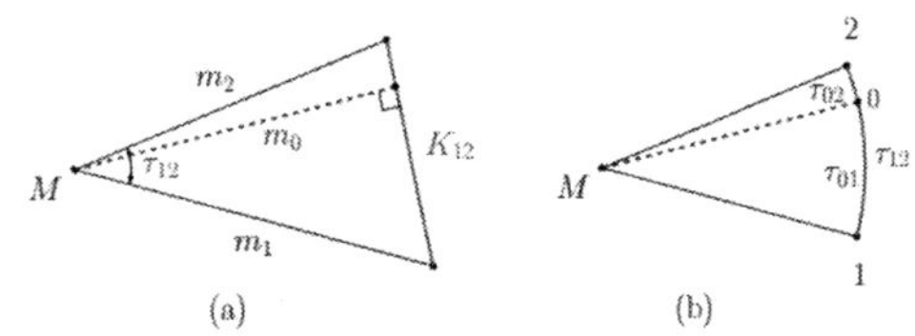

Figure 3. Two-point case: (a) the basic triangle and (b) the arc τ_{12}.

For splitting we use the height of the basic triangle, $m_0 = m_1m_2\sin\tau_{12}/\sqrt{k_{12}^2}$, and obtain two triangles with the sides $(m_1, m_0, K_{01} \equiv \sqrt{k_{01}^2})$ and $(m_2,\ m_0,\ K_{02} \equiv \sqrt{k_{02}^2})$, respectively. By construction, $K_{01} + K_{02} = K_{12}$. Here $k_{01}^2 = (k_{12}^2 + m_1^2 - m_2^2)^2/(4k_{12}^2)$ and $k_{02}^2 = (k_{12}^2 - m_1^2 + m_2^2)^2/(4k_{12}^2)$ (note that $k_{01}^2 = m_1^2 - m_0^2$ and $k_{02}^2 = m_2^2 - m_0^2$). Each of the resulting integrals can be associated with a two-point function, and we arrive at the following decomposition [12]:

$$J^{(2)}\left(n;1,1|k_{12}^2;m_1,m_2\right) = \frac{1}{2k_{12}^2}\left\{(k_{12}^2 + m_1^2 - m_2^2)J^{(2)}\left(n;1,1|k_{01}^2;m_1,m_0\right)\right.$$
$$\left. +(k_{12}^2 - m_1^2 + m_2^2)J^{(2)}\left(n;1,1|k_{02}^2;m_2,m_0\right)\right\}. \qquad (1)$$

This is an example of a functional relation between integrals with different momenta and masses, similar to those described in [13]. Moreover, as shown in [12], we can represent the right-hand side in terms of the equal-mass integrals $J^{(2)}\left(n;1,1|4k_{01}^2;m_1,m_1\right)$ and $J^{(2)}\left(n;1,1|4k_{02}^2;m_2,m_2\right)$.

Let us look at the number of variables. In the original integral $J^{(2)}\left(n;1,1|k_{12}^2;m_1,m_2\right)$ we have three independent variables: two masses and one momentum invariant (out of them we can construct two dimensionless variables). In the integral $J^{(2)}\left(n;1,1|k_{01}^2;m_1,m_0\right)$ we have one extra condition on the variables, $k_{01}^2=m_1^2-m_0^2$, so that we get two independent variables (i.e., one dimensionless variable).

This can also be seen in the integrands of Feynman parametric integrals: for the original two-point integral, the quadratic form is

$$J^{(2)}\left(n;1,1|k_{12}^2;m_1,m_2\right) \Rightarrow \left[\alpha_1\alpha_2 k_{12}^2 - \alpha_1 m_1^2 - \alpha_2 m_2^2\right], \tag{2}$$

whereas for one of the resulting integrals after splitting, remembering that $\alpha_1+\alpha_2=1$, we get

$$J^{(2)}\left(n;1,1|k_{01}^2;m_1,m_0\right) \Rightarrow \left[\alpha_1\alpha_2 k_{01}^2 - \alpha_1 m_1^2 - \alpha_2 m_0^2\right] = -\left[\alpha_1^2 k_{01}^2 + m_0^2\right]. \tag{3}$$

In this way, we obtain the following result in arbitrary dimension:

$$\begin{aligned} J^{(2)}\left(n;1,1|k_{01}^2;m_1,m_0\right) &= \mathrm{i}\pi^{n/2}\Gamma(2-n/2)\int\limits_0^1\!\!\int\limits_0^1 \frac{\mathrm{d}\alpha_1\,\mathrm{d}\alpha_2\,\delta(\alpha_1+\alpha_2-1)}{\left[\alpha_1^2 k_{01}^2+m_0^2\right]^{2-n/2}} \\ &= \mathrm{i}\pi^{n/2}\frac{\Gamma(2-n/2)}{(m_0^2)^{2-n/2}}\,{}_2F_1\left(\begin{array}{c} 1/2,\ 2-n/2 \\ 3/2 \end{array}\middle| -\frac{k_{01}^2}{m_0^2}\right), \end{aligned} \tag{4}$$

where ${}_2F_1$ is the Gauss hypergeometric function. Similar expression for the second integral, $J^{(2)}\left(n;1,1|k_{02}^2;m_0,m_2\right)$, can be obtained by permutation $1\leftrightarrow 2$. Therefore, the result for the two-point function in arbitrary dimension can be expressed in terms of a combination of two ${}_2F_1$ functions of a single dimensionless variable (see, e.g., in [5, 14]) whose ε-expansion is known to any order [15, 16].

3. Three-point function

For the three-point function (see figure 2b), there are three external momentum invariants, k_{12}^2, k_{13}^2 and k_{23}^2, and the sides of the corresponding basic tetrahedron are m_1, m_2, m_3, $K_{12}\equiv\sqrt{k_{12}^2}$, $K_{13}\equiv\sqrt{k_{13}^2}$ and $K_{23}\equiv\sqrt{k_{23}^2}$ (see figure 4a). The angles τ_{12}, τ_{13} and τ_{23} between the sides m_1, m_2 and m_3 are defined through $\cos\tau_{jl}\equiv c_{jl}=(m_j^2+m_l^2-k_{jl}^2)/(2m_jm_l)$, and (in the spherical case) the integration extends over the spherical triangle 123 of the unit sphere, see in figure 4b.

For the splitting we use the height of the basic tetrahedron, m_0, and obtain three tetrahedra, as shown in figure 5a. One of them has the sides m_1, m_2, m_0, $K_{12}\equiv\sqrt{k_{12}^2}$, $K_{01}\equiv\sqrt{k_{01}^2}$ and $K_{02}\equiv\sqrt{k_{02}^2}$, and the sides for the others can be obtained by permutation of the indices.

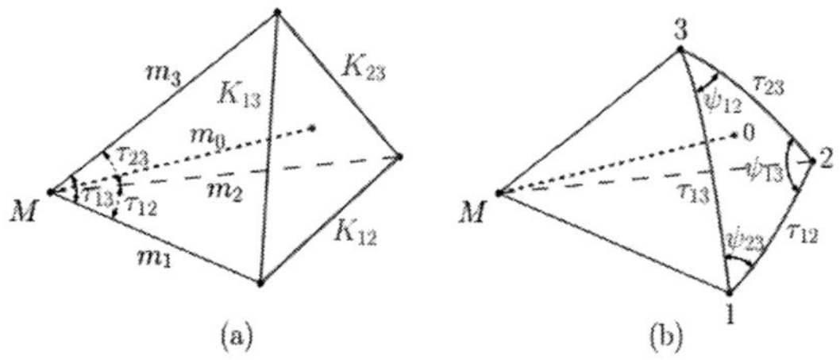

Figure 4. Three-point case: (a) the basic tetrahedron and (b) the solid angle.

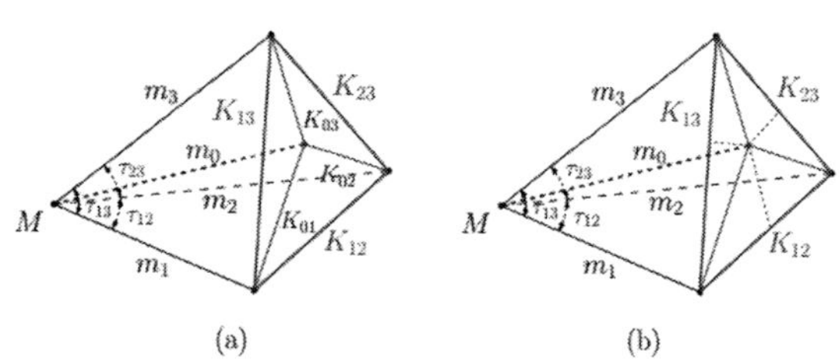

Figure 5. (a) Splitting the basic tetrahedron into three tetrahedra and (b) further splitting into six tetrahedra.

Here $k_{01}^2 = m_1^2 - m_0^2$, $k_{02}^2 = m_2^2 - m_0^2$, $k_{03}^2 = m_3^2 - m_0^2$, and $m_0 = m_1 m_2 m_3 \sqrt{D^{(3)}/\Lambda^{(3)}}$, where $\Lambda^{(3)} = \frac{1}{4}\left[2k_{12}^2 k_{13}^2 + 2k_{13}^2 k_{23}^2 + 2k_{23}^2 k_{12}^2 - (k_{12}^2)^2 - (k_{13}^2)^2 - (k_{23}^2)^2\right]$, and $D^{(3)} = \det \|c_{jl}\|$ is the Gram determinant, see in [5, 6] for more details. Each of the resulting integrals can be associated with a specific three-point function, and we arrive at the following decomposition:

$$
\begin{aligned}
J^{(3)}\left(n;1,1,1|k_{23}^2,k_{13}^2,k_{12}^2;m_1,m_2,m_3\right) = \frac{m_1^2 m_2^2 m_3^2}{\Lambda^{(3)}} & \left\{ \frac{F_1^{(3)}}{m_1^2} J^{(3)}\left(n;1,1,1|k_{23}^2,k_{03}^2,k_{02}^2;m_0,m_2,m_3\right) \right. \\
& + \frac{F_2^{(3)}}{m_2^2} J^{(3)}\left(n;1,1,1|k_{03}^2,k_{13}^2,k_{01}^2;m_1,m_0,m_3\right) \\
& \left. + \frac{F_3^{(3)}}{m_3^2} J^{(3)}\left(n;1,1,1|k_{02}^2,k_{01}^2,k_{12}^2;m_1,m_2,m_0\right) \right\}, \qquad (5)
\end{aligned}
$$

with

$$
F_3^{(3)} = \frac{1}{4m_1^2 m_2^2}\left[k_{12}^2\left(k_{13}^2+k_{23}^2-k_{12}^2+m_1^2+m_2^2-2m_3^2\right) - (m_1^2 - m_2^2)\left(k_{13}^2 - k_{23}^2\right)\right], \qquad (6)
$$

etc., so that $\sum_{i=1}^{3}(F_i^{(3)}/m_i^2) = \Lambda^{(3)}/(m_1^2 m_2^2 m_3^2)$.

By dropping perpendiculars onto the sides $K_{12} \equiv \sqrt{k_{12}^2}$, etc., each of the resulting tetrahedra can be split into two, so that in total we get six "birectangular" tetrahedra, as shown in figure 5b. In this way, we get the following relations for the integrals on the right-hand side of equation (5):

$$
\begin{aligned}
& J^{(3)}\left(n;1,1,1|k_{02}^2,k_{01}^2,k_{12}^2;m_1,m_2,m_0\right) \\
& = \frac{1}{2k_{12}^2}\left\{(k_{12}^2 + m_1^2 - m_2^2) J^{(3)}\left(n;1,1,1|k_{00'}^2,k_{01}^2,k_{10'}^2;m_1,m_{0'},m_0\right)\right. \\
& \left. +(k_{12}^2 - m_1^2 + m_2^2) J^{(3)}\left(n;1,1,1|k_{02}^2,k_{00'}^2,k_{20'}^2;m_{0'},m_2,m_0\right)\right\}, \qquad (7)
\end{aligned}
$$

etc. The notation $0'$ is explained in figure 6; in particular, $m_{0'}$ is the distance between the points M and $0'$, whereas $K_{10'} = \sqrt{k_{10'}^2}$ and $K_{20'} = \sqrt{k_{20'}^2}$ are the distances between the points $(1, 0')$ and $(2, 0')$, respectively, so that $K_{10'} + K_{20'} = K_{12}$. Note that $k_{10'}^2 = (k_{12}^2 + m_1^2 - m_2^2)^2/(4k_{12}^2)$ and $k_{20'}^2 = (k_{12}^2 - m_1^2 + m_2^2)^2/(4k_{12}^2)$, similarly to the reduction of the two-point function.

Let us analyze the number of variables in the occurring three-point integrals. In $J^{(3)}\left(n;1,1,1|k_{23}^2,k_{13}^2,k_{12}^2;m_1,m_2,m_3\right)$ we have six independent variables: three masses and three momentum invariants (out of them we can construct five dimensionless variables). In $J^{(3)}\left(n;1,1,1|k_{02}^2,k_{01}^2,k_{12}^2;m_1,m_2,m_0\right)$ we have two extra conditions on the variables, $k_{01}^2 = m_1^2 - m_0^2$ and $k_{02}^2 = m_2^2 - m_0^2$, so that we get four independent variables (i.e., three dimensionless variables). For the integral $J^{(3)}\left(n;1,1,1|k_{00'}^2,k_{01}^2,k_{10'}^2;m_1,m_{0'},m_0\right)$ we have three relations, $k_{01}^2 = m_1^2 - m_0^2$, $k_{00'}^2 = k_{01}^2 - k_{10'}^2$ and $k_{00'}^2 = m_{0'}^2 - m_0^2$. Therefore, the result for the three-point function in

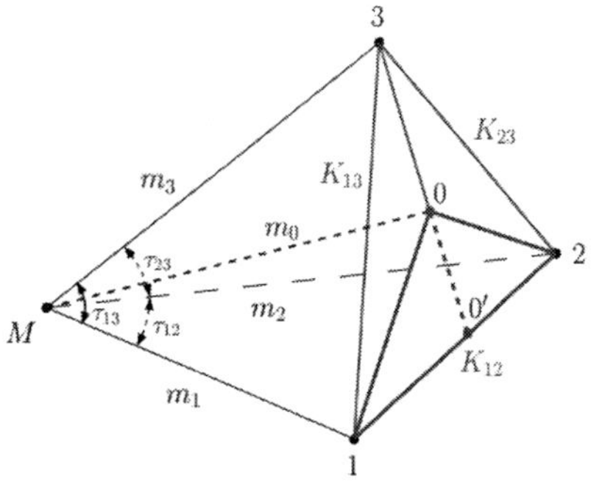

Figure 6. Last step of splitting the basic three-dimensional tetrahedron.

arbitrary dimension should be expressible in terms of a combination of functions of two dimensionless variables: indeed, we know that it can be presented in terms of the Appell hypergeometric function F_1 (see, e.g., in [6, 17, 18]).

This can also be seen in the integrands of the corresponding Feynman parametric integrals: for the original three-point integral, the quadratic form is

$$J^{(3)}\left(n;1,1,1|k_{23}^2,k_{13}^2,k_{12}^2;m_1,m_2,m_3\right) \Rightarrow \left[\alpha_1\alpha_2k_{12}^2+\alpha_1\alpha_3k_{13}^2+\alpha_2\alpha_3k_{23}^2-\alpha_1m_1^2-\alpha_2m_2^2-\alpha_3m_3^2\right], \quad (8)$$

and for one of the resulting integrals after splitting, remembering that $\alpha_1+\alpha_2+\alpha_3=1$, we get

$$\begin{aligned} J^{(3)}\left(n;1,1,1|k_{00'}^2,k_{01}^2,k_{10'}^2;m_1,m_{0'},m_0\right) &\Rightarrow \left[\alpha_1\alpha_2k_{10'}^2+\alpha_1\alpha_3k_{01}^2+\alpha_2\alpha_3k_{00'}^2-\alpha_1m_1^2-\alpha_2m_{0'}^2-\alpha_3m_0^2\right] \\ &\Rightarrow -\left[\alpha_1^2k_{10'}^2+(\alpha_1+\alpha_2)^2k_{00'}^2+m_0^2\right]. \end{aligned} \quad (9)$$

In this way, we obtain the following result in arbitrary dimension:

$$\begin{aligned} &J^{(3)}\left(n;1,1,1|k_{00'}^2,k_{01}^2,k_{10'}^2;m_1,m_{0'},m_0\right) \\ &= -\mathrm{i}\pi^{n/2}\Gamma(3-n/2)\int_0^1\!\!\int_0^1\!\!\int_0^1 \frac{\mathrm{d}\alpha_1\,\mathrm{d}\alpha_2\,\mathrm{d}\alpha_3\,\delta(\alpha_1+\alpha_2+\alpha_3-1)}{[\alpha_1^2k_{10'}^2+(\alpha_1+\alpha_2)^2k_{00'}^2+m_0^2]^{3-n/2}} \\ &= -\frac{\mathrm{i}\pi^{n/2}\Gamma(2-n/2)}{2(m_0^2)^{2-n/2}k_{00'}^2}\left\{\sqrt{\frac{k_{00'}^2}{k_{10'}^2}}\arctan\sqrt{\frac{k_{10'}^2}{k_{00'}^2}}\right. \\ &\qquad \left. -\left(\frac{m_0^2}{m_{0'}^2}\right)^{2-n/2} F_1\left(1/2,1,2-n/2;3/2\Big|-\frac{k_{10'}^2}{k_{00'}^2},-\frac{k_{10'}^2}{m_{0'}^2}\right)\right\}, \end{aligned} \quad (10)$$

where F_1 is Appell hypergeometric function of two variables,

$$F_1\left(a,b_1,b_2;c|x,y\right)=\sum_{j_1,j_2}\frac{(a)_{j_1+j_2}\,(b_1)_{j_1}\,(b_2)_{j_2}}{(c)_{j_1+j_2}}\frac{x^{j_1}\,y^{j_2}}{j_1!\,j_2!}\,.$$

Similar results for other five contributions can be obtained by permutation. Using known transformation formulae for F_1 we can see that the obtained expression (10) is equivalent to the result presented in [6].

4. Four-point function

For the four-point function (see figure 2c), there are six external momentum invariants. Out of them, k_{12}^2, k_{23}^2, k_{34}^2 and k_{14}^2 are the squared momenta of the external legs, whilst k_{13}^2 and k_{24}^2 correspond to the Mandelstam variables s and t. The sides of the corresponding basic four-dimensional simplex are m_1, m_2, m_3, m_4, and six additional sides $K_{jl}\equiv\sqrt{k_{jl}^2}$, as shown in figure 7a. The six angles τ_{jl} between the corresponding sides m_j and m_l are defined through $\cos\tau_{jl}\equiv c_{jl}=(m_j^2+m_l^2-k_{jl}^2)/(2m_jm_l)$, and (in the spherical case) the integration extends over the spherical tetrahedron 1234 of the unit hypersphere, as shown in figure 7b (for the hyperbolic case one can use analytic continuation).

For splitting we use the height of the basic simplex, m_0, and obtain four simplices, as shown in figures 8 and 8a. One of them has the sides m_1, m_2, m_3, m_0, $K_{12}\equiv\sqrt{k_{12}^2}$, $K_{13}\equiv\sqrt{k_{13}^2}$,

IOP Conf. Series: Journal of Physics: Conf. Series **1085** (2018) 052016 doi:10.1088/1742-6596/1085/5/052016

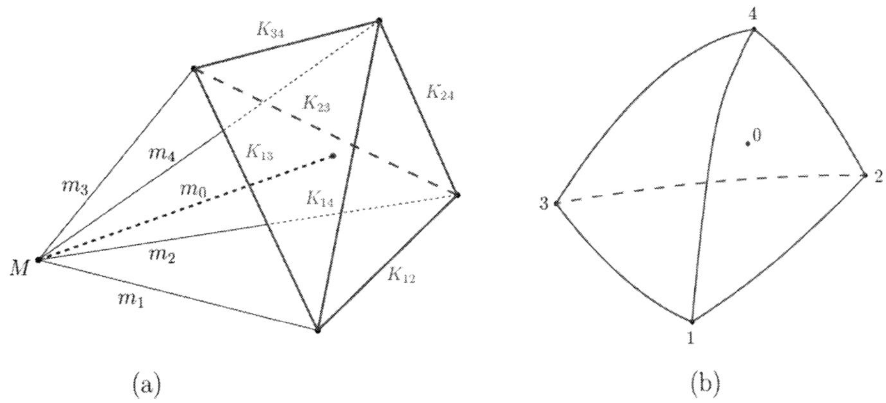

Figure 7. Four-point case: (a) the basic simplex and (b) the spherical tetrahedron.

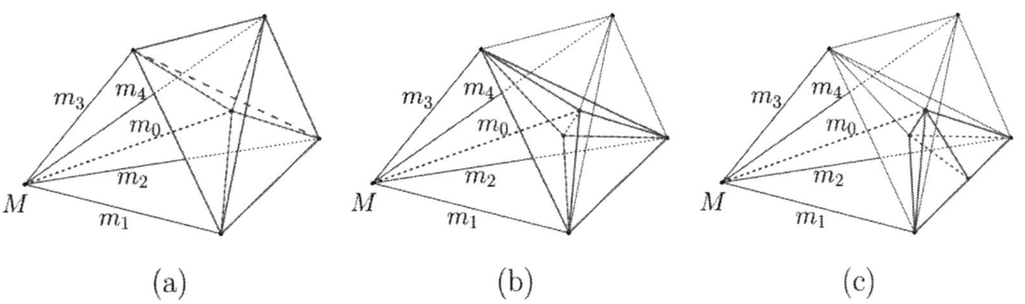

Figure 8. Four-point case: steps of splitting the basic four-dimensional simplex.

$K_{23} \equiv \sqrt{k_{23}^2}$, $K_{01} \equiv \sqrt{k_{01}^2}$, $K_{02} \equiv \sqrt{k_{02}^2}$ and $K_{03} \equiv \sqrt{k_{03}^2}$, and the sides of the others can be obtained by permutation of the indices. As before, $k_{0i}^2 = m_i^2 - m_0^2$ $(i = 1, 2, 3, 4)$, whereas $m_0 = m_1 m_2 m_3 m_4 \sqrt{D^{(4)}/\Lambda^{(4)}}$, where $D^{(4)} = \det \|c_{jl}\|$ and $\Lambda^{(4)} = \det \|(k_{j4} \cdot k_{l4})\|$, see in [5] for more details. Each of the four resulting integrals can be associated with a certain four-point function. At the next step, in each of the four tetrahedra (drawn in red) we drop the perpendiculars onto the triangle sides, as shown in figure 8b, splitting each of them into three, and then dividing each of the resulting tetrahedra into two, by dropping perpendiculars onto the $\sqrt{k_{jl}^2}$ sides, as shown in figure 8c. As a result of this splitting, we get $4 \cdot 3 \cdot 2 = 24$ simplices.

Let us look at the number of variables. In the integral $J^{(4)}\left(n; 1, 1, 1, 1 | \{k_{jl}^2\}; \{m_i\}\right)$ we have ten independent variables: four masses and six momentum invariants (out of them we can construct nine dimensionless variables). After the first step (figure 8a) we have three conditions on the variables, $k_{01}^2 = m_1^2 - m_0^2$, $k_{02}^2 = m_2^2 - m_0^2$ and $k_{03}^2 = m_3^2 - m_0^2$, so that we get seven independent variables (i.e., six dimensionless variables). After the second step (figure 8b), we get two extra conditions due to the right triangles, and after the third step (figure 8c) we get one more condition. As a result, for each of the 24 resulting four-point

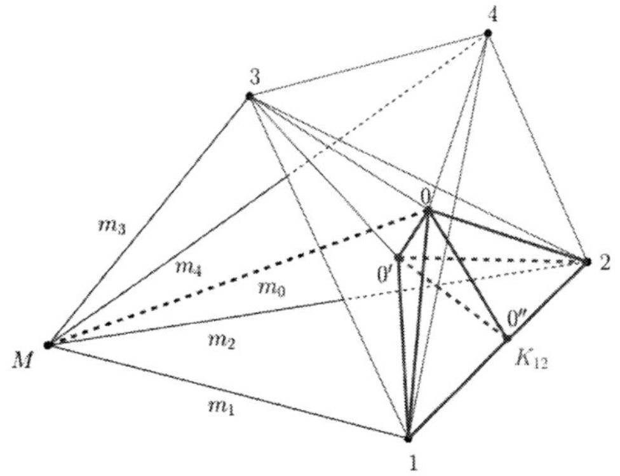

Figure 9. Last step of splitting the basic four-dimensional simplex.

IOP Conf. Series: Journal of Physics: Conf. Series **1085** (2018) 052016 doi:10.1088/1742-6596/1085/5/052016

functions we have six relations, so that we end up with four independent variables (i.e., three dimensionless variables). Therefore, the result for the four-point function in arbitrary dimension should be expressible in terms of a combination of functions of three dimensionless variables, such as, e.g., Lauricella functions and their generalizations (see, e.g., in [18, 19]).

This can also be seen in the integrands of the corresponding Feynman parametric integrals: for the original four-point integral, the quadratic form is

$$\begin{aligned} J^{(4)}&\left(n;1,1,1,1|\{k_{12}^2,k_{23}^2,k_{34}^2,k_{14}^2,k_{13}^2,k_{24}^2\};\{m_1,m_2,m_3,m_4\}\right) \\ &\Rightarrow [\alpha_1\alpha_2k_{12}^2+\alpha_1\alpha_3k_{13}^2+\alpha_1\alpha_4k_{14}^2+\alpha_2\alpha_3k_{23}^2+\alpha_2\alpha_4k_{24}^2+\alpha_3\alpha_4k_{34}^2 \\ &\quad -\alpha_1m_1^2-\alpha_2m_2^2-\alpha_3m_3^2-\alpha_4m_4^2], \end{aligned} \tag{11}$$

whereas after the last step of splitting, remembering that $\alpha_1+\alpha_2+\alpha_3+\alpha_4=1$, we get

$$\begin{aligned} J^{(4)}&\left(n;1,1,1,1|\{k_{10''}^2,k_{0'0''}^2,k_{00'}^2,k_{01}^2,k_{10'}^2,k_{00''}^2\};\{m_1,m_{0''},m_{0'},m_0\}\right) \\ &\Rightarrow [\alpha_1\alpha_2k_{10''}^2+\alpha_1\alpha_3k_{10'}^2+\alpha_1\alpha_4k_{01}^2+\alpha_2\alpha_3k_{0'0''}^2+\alpha_2\alpha_4k_{00''}^2+\alpha_3\alpha_4k_{00'}^2 \\ &\quad -\alpha_1m_1^2-\alpha_2m_{0''}^2-\alpha_3m_{0'}^2-\alpha_4m_0^2] \\ &\Rightarrow -[\alpha_1^2k_{10''}^2+(\alpha_1+\alpha_2)^2k_{0'0''}^2+(\alpha_1+\alpha_2+\alpha_3)^2k_{00'}^2+m_0^2]. \end{aligned} \tag{12}$$

The notations $0'$ and $0''$ are explained in figure 9; in particular, $m_{0'}$ is the distance between M and $0'$, $m_{0''}$ is the distance between M and $0''$, whereas $K_{00'}=\sqrt{k_{00'}^2}$, $K_{0'0''}=\sqrt{k_{0'0''}^2}$ and $K_{10''}=\sqrt{k_{10''}^2}$ are the distances between the corresponding points $(0,0')$, $(0',0'')$ and $(1,0'')$, respectively. In this way, we obtain the following result in arbitrary dimension:

$$\begin{aligned} J^{(4)}&\left(n;1,1,1,1|\{k_{10''}^2,k_{0'0''}^2,k_{00'}^2,k_{01}^2,k_{10'}^2,k_{00''}^2\};\{m_1,m_{0''},m_{0'},m_0\}\right) \\ &= \mathrm{i}\pi^{n/2}\Gamma(4-n/2)\int_0^1\!\!\int_0^1\!\!\int_0^1\!\!\int_0^1 \frac{\mathrm{d}\alpha_1\,\mathrm{d}\alpha_2\,\mathrm{d}\alpha_3\,\mathrm{d}\alpha_4\;\delta(\alpha_1+\alpha_2+\alpha_3+\alpha_4-1)}{[\alpha_1^2k_{10''}^2+(\alpha_1+\alpha_2)^2k_{0'0''}^2+(\alpha_1+\alpha_2+\alpha_3)^2k_{00'}^2+m_0^2]^{4-n/2}} \\ &= \frac{\mathrm{i}\pi^{n/2}\Gamma(3-n/2)}{2k_{0'0''}^2(m_0^2)^{3-n/2}}\left\{\sqrt{\frac{k_{0'0''}^2}{k_{10''}^2}}\arctan\sqrt{\frac{k_{10''}^2}{k_{0'0''}^2}}\;{}_2F_1\left(\begin{matrix}1/2,3-n/2\\3/2\end{matrix}\Big|-\frac{k_{00'}^2}{m_0^2}\right)\right. \end{aligned} \tag{13}$$

$$\left.-\left(\frac{m_0^2}{m_{0'}^2}\right)^{2-n/2}F_N\left(1,1,3-n/2,1/2,(n-3)/2,1/2;3/2,3/2,3/2\Big|-\frac{k_{10''}^2}{k_{0'0''}^2},-\frac{k_{00'}^2}{m_0^2},-\frac{k_{10'}^2}{m_{0'}^2}\right)\right\}, \tag{14}$$

where F_N is one of the Lauricella-Saran functions [20],

$$F_N(a_1,a_2,a_3,b_1,b_2,b_1;c_1,c_2,c_2|x,y,z)=\sum_{j_1,j_2,j_3}\frac{(a_1)_{j_1}(a_2)_{j_2}(a_3)_{j_3}(b_1)_{j_1+j_3}(b_2)_{j_2}}{(c_1)_{j_1}(c_2)_{j_2+j_3}}\frac{x^{j_1}y^{j_2}z^{j_3}}{j_1!j_2!j_3!}.$$

5. General remarks and conclusions

The geometrical approach allows us to relate the one-loop N-point Feynman diagrams to certain (hyper)volume integrals in non-Euclidean geometry. Geometrical splitting provides a straightforward way of reducing general integrals to those with lesser number of independent variables. Furthermore, in this way we can predict the set and the number of these variables in the resulting integrals. As shown in [12], for an N-point diagram (depending, in the general off-shell case, on $\frac{1}{2}(N-1)(N+2)$ dimensionless variables), after splitting in $N!$ pieces we will get a combination of $N!$ integrals, each of them depending only on $N-1$ variables. For example, in the four-point case we will get functions of three variables, rather than nine.

Geometrically, we can calculate the resulting integrals in the framework of non-Euclidean geometry or, alternatively, represent them again in terms of Feynman parameters. Since some of the variables are connected, the quadratic forms in the integrands of the resulting parametric integrals can be simplified. In this way, for $N = 2, 3, 4$ we get the following quadratic forms:

$$J^{(2)}\left(n;1,1|k_{01}^2;m_1,m_0\right) \Rightarrow -[\alpha_1^2 k_{01}^2 + m_0^2]\,,$$

$$J^{(3)}\left(n;1,1,1|k_{00'}^2,k_{01}^2,k_{10'}^2;m_1,m_{0'},m_0\right) \Rightarrow -[\alpha_1^2 k_{10'}^2 + (\alpha_1+\alpha_2)^2 k_{00'}^2 + m_0^2]\,,$$

$$J^{(4)}\left(n;1,1,1,1|\{k_{10''}^2,k_{0'0''}^2,k_{00'}^2,k_{01}^2,k_{10'}^2,k_{00''}^2\};\{m_1,m_{0''},m_{0'},m_0\}\right)$$
$$\Rightarrow -[\alpha_1^2 k_{10''}^2 + (\alpha_1+\alpha_2)^2 k_{0'0''}^2 + (\alpha_1+\alpha_2+\alpha_3)^2 k_{00'}^2 + m_0^2]\,.$$

Evaluating these integrals for an arbitrary dimension n we can get explicit expressions in terms of (generalized) hypergeometric functions: ${}_2F_1$ for $N = 2$, F_1 for $N = 3$, and F_N for $N = 4$. For $N > 4$, in the quadratic forms we should also expect sums of squares of partial sums of α's (the coefficient of m_0^2 can be understood as the square of the sum of all α's, equal to one).

Acknowledgements
I am thankful to R Delbourgo and M Yu Kalmykov with whom I started to work on this subject. I am grateful to the organizers of ACAT-2017 for their support and hospitality.

References

[1] Davydychev A I 1991 *J. Math. Phys.* **32** 1052
Davydychev A I 1992 *J. Math. Phys.* **33** 358
[2] 'tHooft G and Veltman M 1972 *Nucl. Phys.* B **44** 189
Bollini C G and Giambiagi J J 1972 *Nuovo Cim.* **12**B 20
Ashmore J F 1972 *Lett. Nuovo Cim.* **4** 289
Cicuta G M and Montaldi E 1972 *Lett. Nuovo Cim.* **4** 329
[3] Landau L D 1959 *Nucl. Phys.* **13** 181
[4] Källen G and Wightman A 1958 *Mat. Fys. Skr. Dan. Vid. Selsk.* **1**(6) 1
Mandelstam S 1959 *Phys. Rev.* **115** 1741
Cutkosky R E 1960 *J. Math. Phys.* **1** 429
Taylor J C 1960 *Phys. Rev.* **117** 261
[5] Davydychev A I and Delbourgo R 1998 *J. Math. Phys.* **39** 4299
[6] Davydychev A I 2006 *Nucl. Instr. Meth.* A **559** 293
[7] Davydychev A I 1999 *Preprint* hep-th/9908032
[8] Nickel B G 1978 *J. Math. Phys.* **19** 542
[9] Ortner N and Wagner P 1995 *Ann. Inst. Henri Poincaré (Phys. Théor.)* **63** 81
[10] Wagner P 1996 *Indag. Math.* **7** 527
[11] Davydychev A I and Delbourgo R 2004 *J. Phys.* A **37** 4871
Gorsky A and Zhiboedov A 2009 *J. Phys.* A **42** 355214
Bloch S and Kreimer D 2010 *Commun. Num. Theor. Phys.* **4** 703
Schnetz O 2010 *Preprint* arXiv:1010.5334
Mason L and Skinner D 2011 *J. Phys.* A **44** 135401
Nandan D, Paulos M F, Spradlin M and Volovich A 2013 *J. High Energy Phys.* JHEP05(2013)105
[12] Davydychev A I 2016 *J. Phys.: Conf. Series* **762** 012068
[13] Tarasov O V 2008 *Phys. Lett.* B **670** 67
Kniehl B A and Tarasov O V 2009 *Nucl. Phys.* B **820** 178
[14] Berends F A, Davydychev A I and Smirnov V A 1996 *Nucl. Phys.* B **478** 59
[15] Davydychev A I 2000 *Phys. Rev.* D **61** 087701
[16] Davydychev A I and Kalmykov M Yu 2000 *Nucl. Phys. B (Proc. Suppl.)* **89** 283
Davydychev A I and Kalmykov M Yu 2001 *Nucl. Phys.* B **605** 266
[17] Tarasov O V 2000 *Nucl. Phys. B (Proc. Suppl.)* **89** 237
[18] Fleischer J, Jegerlehner F and Tarasov O V 2003 *Nucl. Phys.* B **672** 303
[19] Bytev V V, Kalmykov M Yu and Moch S-O 2014 *Comput. Phys. Commun.* **185** 3041
[20] Saran S 1955 *Acta Math.* **93** 293

IOP Conf. Series: Journal of Physics: Conf. Series **1085** (2018) 052017 doi:10.1088/1742-6596/1085/5/052017

A minimally invasive strategy for NNLO event files

Daniel Maître

Institute for Particle Physics Phenomenology , Ogden Centre for Fundamental Physics, Department of Physics, University of Durham, Science Laboratories, South Rd, DURHAM DH1 3LE, UNITED KINGDOM

E-mail: daniel.maitre@durham.ac.uk

Abstract. In this contribution we discuss the extension of the event file format proposed in Ref [1] from NLO to NNLO. We describe a strategy that minimises the changes needed to the NNLO generator code, at the cost of increasing the CPU cost. We expect the CPU cost for the production of the event files to be amortised by reusing the event files in many different analyses.

1. Introduction

Recent advances in the calculation of one-loop amplitudes have made Next-to-Leading Order (NLO) QCD predictions for high multiplicity processes achievable (see e.g. [2] for a review). While the calculations necessary for these predictions are tractable, they come at a high CPU cost. In order to amortise the cost of the matrix element calculation and its integration over the phase-space an event file format has been proposed [1] for NLO processes. The strategy to produce and use these event files is sketched in Fig. 1. The full generator is run for a loose set of cuts and all phase-space and matrix element information is stored in a file (called nTuple file in this contribution). At a later stage the event files can be re-read and effectively replace the generator. This process is much more efficient than running the generator again and much less error prone. In this contribution we will present a strategy to obtain event files for NNLO generators. At NNLO the complexity of the generators is much higher than at NLO and it would be much harder to isolate the information needed for the event files in the code. Instead we will change strategy and try to minimise the code changes needed in the generator.

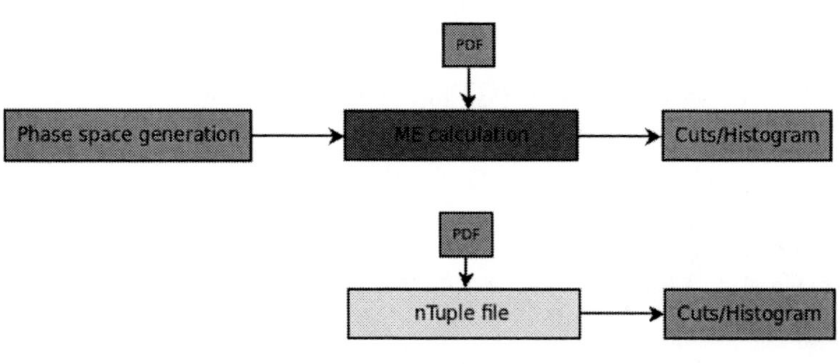

Figure 1. Overview of the production and usage of event files.

IOP Conf. Series: Journal of Physics: Conf. Series **1085** (2018) 052017 doi:10.1088/1742-6596/1085/5/052017

2. Advantages and disadvantages of event files

In a typical analysis the majority of the CPU resources are spent on the matrix element calculation. Other aspects of the analysis such as the PDF evaluation, cuts and jet algorithm tend to only take a small fraction of the time. Using event files allows to change details of the analysis after the generation, in particular the re-evaluation of the prediction with different PDF sets or scales would be extremely costly if one had to start the complete generation from the start. Beyond saving computation time this scheme allows researchers unfamiliar with the generator to use its results in a convenient fashion. In this scheme the generator expert have an opportunity to protect the users of their calculation from all the difficulties of running their program on a large scale. A significant effort is needed to validate a generator and ensure it reproduces established results after each change. Event files are a more persistent way of storing the result of a calculation and are easier to validate.

The main disadvantage of the event files is that their size is rather large. The usefulness of nTuples depends on the trade-off between storage and usage convenience. While considering the trade-offs between size and CPU time one should remember that generators have been optimised for CPU time and not optimised to produce the most efficient set of phase-space points form a storage point of view. The efficiency of the storage can generally be increased at the cost of CPU time.

3. Event files for NNLO

The event files have proven their usefulness at NLO, could they be useful NNLO too? The advantages and disadvantages are the same than at NLO, but the trade-offs are starker. While NLO event files only get uncomfortably large at large multiplicities NNLO calculations require a very large amount of phase-space points already at low multiplicities. On the other hand the CPU time for these calculations are much larger than at NLO so the potential savings are larger. To evaluate the potential of event files for NNLO generators a feasibility study [2, 3] was performed in the Les Houches workshop. For this study the authors used EERAD3 [4] to estimate the size of the event files that would be necessary to reproduce exiting results. The study's main result is shown in Fig. 2. The figure shows two storage strategies, trading some re-calculation at read-time for some storage saving, see ref. [2] for more details.

The main take away point of the feasibility study [2] is that event files could still offer a favourable size/CPU time trade off.

3.1. General structure of a NNLO generator

In order to design the least intrusive method of extracting the information we need from the NNLO generator we start form the schematic view displayed in Fig. 3. The calculation starts with the generation of the phase-space, including the parton momentum fractions in the case of hadronic initial state. This information is used in the matrix element calculation to calculate the weight for the event, this includes possible subtraction terms that are used to regulate infrared divergences. Finally the weights calculated are histogrammed according to observables calculated from the phase-space information. In the following we take the case of a hadron-hadron collider, cases with fewer hadronic initial states can be treated in a similar way.

3.2. Extracting the information

The structure shown in Fig. 3 already shows the places in the code where the phase-space information and the weights can be extracted, so that the intermediate step (the most costly part) can be replaced by reading from event files. Extracting information from the generator as it produces the NNLO cross section allows us the get the weight for a given phase-space point for the scale and PDF chosen by the generator at the time of the generation. We would like to have the flexibility to generate the weight as it would have been produced with a different scale choice

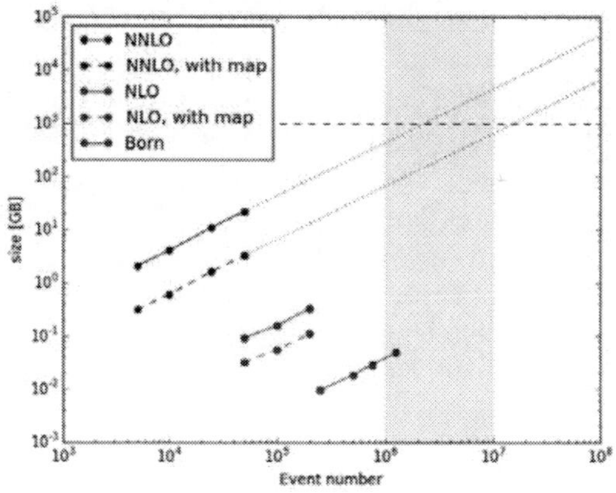

Figure 2. Size of the event files as a function of the number of events. The shaded green line represents the approximate number of events needed to match existing results. The horizontal dashed line at a size of a Terabyte is a subjective "pain threshold" where the size of the nTuples starts getting uncomfortable.

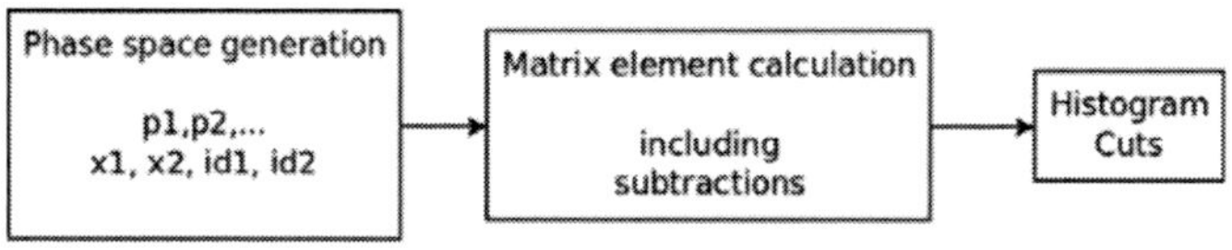

Figure 3. Schematic view of a generic generator.

or different PDF set so we need to decompose the weight ω we intercept at the histogramming stage into more specialised components. To improve the clarity of the presentation we also assume the factorisation and renormalisation scales are equal. The general form of the weight is given by

$$\omega = \alpha_S^n \, pdf(x_1, id_1) \, pdf(x_2, id_2) \times \left(c_0 + c_1 \log(\mu^2) + c_2 \log^2(\mu^2) + \ldots\right) \, . \tag{1}$$

In order to reconstruct the weight for a different PDF and scale choice we need to extract

- initial state flavours id_1, id_2
- momentum fractions x_1, x_2
- factorisation and renormalisation scales $\mu_F = \mu_R = \mu$
- The coefficients of the scale logarithms c_0, c_1, ...

The best place to extract this information is in the PDF procedure. Most generators use the LHAPDF [5] library to evaluate the PDFs, we will assume that is it the case here. If we observe the arguments passed to the LHAPDF functions we can infer the values of x_1, x_2, and μ. Because of the correlation between the PDF fit and the value of α_S used in the fit most generators also use the LHAPDF function for the evaluation of $\alpha_S(\mu)$.

Inferring the initial state flavours is more difficult, as one call to the PDF function returns the value of the PDF for all initial state flavours in a 13-element vector, we have no way to know which one the generator used (or whether multiple entries have been combined). To circumvent

IOP Conf. Series: Journal of Physics: Conf. Series **1085** (2018) 052017 doi:10.1088/1742-6596/1085/5/052017

this problem we introduce a strategy that we will reuse often for different purposes. The idea is to run several instances of the NNLO generator within slightly different environment and use the different weights obtained for each parallel instance to infer more information than is available from a single instance. To assert which of the 13 elements of the PDF vector are used in the calculation of the weight we can use 13 parallel instances of the NNLO generator, each with a modified version of the PDF function that returns zeros for all PDF values except for the one element, as sketched in Eq. 2.

$$\begin{aligned}
f_1(x) &= (*,0,0,0,0,0,0,0,0,0,0,0,0)\\
f_2(x) &= (0,*,0,0,0,0,0,0,0,0,0,0,0)\\
f_3(x) &= (0,0,*,0,0,0,0,0,0,0,0,0,0)\\
f_4(x) &= (0,0,0,*,0,0,0,0,0,0,0,0,0)\\
f_5(x) &= (0,0,0,0,*,0,0,0,0,0,0,0,0)\\
f_6(x) &= (0,0,0,0,0,*,0,0,0,0,0,0,0)\\
f_7(x) &= (0,0,0,0,0,0,*,0,0,0,0,0,0)\\
f_8(x) &= (0,0,0,0,0,0,0,*,0,0,0,0,0)\\
f_9(x) &= (0,0,0,0,0,0,0,0,*,0,0,0,0)\\
f_{10}(x) &= (0,0,0,0,0,0,0,0,0,*,0,0,0)\\
f_{11}(x) &= (0,0,0,0,0,0,0,0,0,0,*,0,0)\\
f_{12}(x) &= (0,0,0,0,0,0,0,0,0,0,0,*,0)\\
f_{13}(x) &= (0,0,0,0,0,0,0,0,0,0,0,0,*)
\end{aligned} \tag{2}$$

This method is quite expensive, in particular if one wants to identify pairs of PDF contributions as one would need to run $13 \cdot 13 = 169$ parallel versions of the generator, but it can be useful to identify potential PDF combinations in a preliminary run. In practice it is best to run two parallel instances of the generator, one with the normal PDFs and one with all the PDFs replaced by one. The ratio between the weights gives the product of the PDFs and we can compare this ratio with a table of possible combinations.

This strategy works well if the weight consists of only one PDF combination evaluated at a single pair of momentum fractions x_1, x_2. This is not always the case, especially in the subtraction method where subtraction contributions can be evaluated at different values of x_1 and x_2. In the case of multiple PDF pairs for the same weight we can use the same strategy and use instances of the generator with PDFs that are set to zero at different times. Eq. 3 illustrates this strategy. A first instance runs with each second call to the PDF set to 0, the second instance leaves the PDF vector untouched but sets the values left untouched by the first instance to zero instead. Two additional instances are run in parallel that project out the same PDF call than the first two, but replace the (non zero) values in the vector by 1. Building the ratio between the first and third instances and the second and fourth instances allows to obtain the PDF with argument x_1 or x_1' in isolation.

$$\begin{array}{rclcccc}
\omega_1 &=& c_1 & pdf(x_1,x_2,id_1,id_2,Q) & + & c_2 & 0\\
\omega_2 &=& c_1 & 0 & + & c_2 & pdf(x_1',x_2',id_1',id_2',Q)\\
\omega_1^{pdf=1} &=& c_1 & 1 & + & c_2 & 0\\
\omega_2^{pdf=1} &=& c_1 & 0 & + & c_2 & 1
\end{array} \tag{3}$$

The coefficient of the different power of scale logarithms can be obtained by running different instances of the NNLO program with different scale settings. The power of n of $\alpha_S(\mu)$ in Eq. 1

IOP Conf. Series: Journal of Physics: Conf. Series **1085** (2018) 052017 doi:10.1088/1742-6596/1085/5/052017

is usually known from the settings in the generator but if not, or if it is changing during the run we can identify it by running two parallel instances, one with α_S set to unity. Comparing the ratio of the weights with the value of α_S returned by LHAPDF we can infer n as

$$n = \frac{\log(w) - \log(w|_{\alpha_S=1})}{\log(\alpha_S)} .$$

3.3. Implementation details

To implement the strategy described above we introduce an intermediary library (called "impersonator" in Fig. 4) between the generator and the LHAPDF library that allows to set some values in the PDF vector to zero or unity. This library also reports the values of the PDF vector and α_S to be used in the comparison between generator instances. Fig. 4 shows the interactions between the generator the LHAPDF impersonator and the LHAPDF library. The only modifications needed in the generator is one line of code to report the weight where the histogramming occurs and one line of code to report the momentum configuration when the cuts are evaluated. The impersonator can be implemented in a way that is transparent to the generator, so no changes have to be made to the interface between the generator and LHAPDF. One should note here that it is important that apart from the PDFs and α_S returned to the generator everything else should be equal between instances, this means for example that no integration optimisation should be made based on the weights calculated by the instances, as this optimisation would be different for the instance using the real PDF values and those with the PDF values changed to zero or unity. In practice this means that a first warmup integration with original PDFs has to be performed and should be frozen for the event file information collection runs.

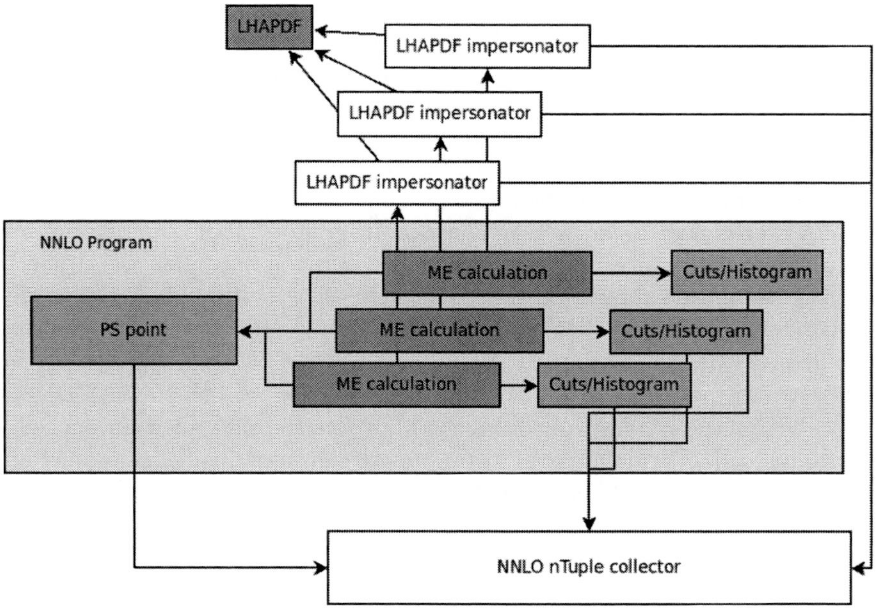

Figure 4. Schematic view of the information collection strategy: the generator is run with the same phase-space points in different environments. These different environments are implemented by the additional layer called "impersonator" in the figure. The weights, PDF and phase-space information is collected by an outside observer called "collector" in the figure.

When working with actual implementations of NNLO generators other complications arise. Most generators implement some caching of phase-space or PDF values which can interfere with our inference of the information we need to gather for the event files. It is easy to check that the information gathered is correct by running (yet another) instance with a different PDF and scale choice and checking that the information we reconstructed is able to reproduce the weights of the control instance.

4. Conclusion

We presented a strategy to collect information to build event files for processes calculated at NNLO accuracy. This method could also be applied to NLO. The strategy makes the clear choice of using some redundant CPU time in order to simplify the implementation of the information gathering. It is clear that this approach can not compete with an implementation where the generator authors extract the information directly from their code. This could be quite difficult and if attempted, the process could be aided by using the method described in this contribution as a check that the information provided in a direct way matches the empirically determined data. It should also be noted that the information extracted with the procedure above can be directly applied to construct `fastNLO/APPLgrid` [6, 7] tables for fast evaluation of histograms for PDF fits.

References

[1] Bern Z, Dixon L, Febres Cordero F, Höche S, Ita H *et al.* 2014 *Comput.Phys.Commun.* **185** 1443–1460 (*Preprint* 1310.7439)
[2] Andersen J R *et al.* 2016 *9th Les Houches Workshop on Physics at TeV Colliders (PhysTeV 2015) Les Houches, France, June 1-19, 2015* (*Preprint* 1605.04692) URL http://arxiv.org/pdf/1605.04692.pdf
[3] Maitre D, Heinrich G and Johnson M 2016 *PoS* **LL2016** 016 (*Preprint* 1607.06259)
[4] Gehrmann-De Ridder A, Gehrmann T, Glover E W N and Heinrich G 2007 *JHEP* **11** 058 (*Preprint* 0710.0346)
[5] Buckley A, Ferrando J, Lloyd S, Nordstrm K, Page B, Rfenacht M, Schnherr M and Watt G 2015 *Eur. Phys. J.* **C75** 132 (*Preprint* 1412.7420)
[6] Britzger D, Rabbertz K, Stober F and Wobisch M (fastNLO) 2012 *Proceedings, 20th International Workshop on Deep-Inelastic Scattering and Related Subjects (DIS 2012)* pp 217–221 (*Preprint* 1208.3641) URL http://inspirehep.net/record/1128033/files/arXiv:1208.3641.pdf
[7] Carli T, Clements D, Cooper-Sarkar A, Gwenlan C, Salam G P, Siegert F, Starovoitov P and Sutton M 2010 *Eur. Phys. J.* **C66** 503–524 (*Preprint* 0911.2985)

AUTHOR INDEX

AUTHOR INDEX

AUTHOR INDEX

AUTHOR INDEX

AUTHOR INDEX

AUTHOR INDEX

AUTHOR INDEX

Institute of Physics
Dirac House, Temple Back
Bristol BS1 6BE UK

ISSN: 1742-6588
ISBN 978-1-5108-7371-1

Workshop on Micro-Mini and Nano Dosimetry (MMND 2018)

Journal of Physics: Conference Series
Volume 1154

Mooloolaba, Australia
6 – 11 February 2018

IOP
Institute of Physics